Lecture Notes in Control and Information Sciences

Edited by A. V. Balakrishnan and M. Thoma

For further listing of published volumes please turn over to inside of back cover.

Lecture Notes in Control and Information Sciences
Edited by A.V. Balakrishnan and M. Thoma

IFIP

59

System Modelling and Optimization

Proceedings of the 11th IFIP Conference
Copenhagen, Denmark, July 25–29, 1983

Edited by P. Thoft-Christensen

Springer-Verlag
Berlin Heidelberg Gmb 1984

Editor
P. Thoft-Christensen
Aalborg University Centre
Sohngaardsholmsvej 57
DK-9000 Aalborg
Denmark

ISBN 978-3-540-13185-4 ISBN 978-3-540-38828-9 (eBook)
DOI 10.1007/978-3-540-38828-9

Library of Congress Cataloging in Publication Data
Main entry under title:
System modelling and optimization.
(Lecture notes in control and information sciences; 59).
Papers selected from those presented at the 11th IFIP Conference on System Modelling and Optimization, organized for Technical Committee 7 of the International Federation for Information Processing, and sponsored by the International Federation of Operational Research Societies and the International Federation of Automation Control.
Includes index.
1. System analysis--Congresses.
2. Mathematical models--Congresses.
3. Mathematical optimization--Congresses.
I. Thoft-Christensen, Palle, 1936-.
II. IFIP Conference on System Modeling and Optimization (11th : 1983 : Copenhagen, Denmark).
III. IFIP TC-7 (Organization).
IV. International Federation of Operational Research Societies.
V. International Federation of Automatic Control.
VI. Title: System modeling and optimization.
VII. Series.
QA402.S9582 1984 003 84-1282

Originally published by Springer-Verlag Berlin Heidelberg New York in 1984.

Offsetprinting: Mercedes-Druck, Berlin

2061/3020-543210

PREFACE

These proceedings contain a number of papers selected from those presented at the 11th IFIP Conference on System Modelling and Optimization, Copenhagen, Denmark, July 25 - 29, 1983.

The conference was organized for the Technical Committee 7 of the International Federation for Information Processing (IFIP) and was co-sponsored by the International Federation of Operational Research Societies (IFORS) and the International Federation of Automation Control (IFAC). It was attended by 350 participants from more than 30 countries. About 250 papers were presented.

The main work in organizing the conference was made by members of the Local Organizing Committee, namely

S. Hildebrandt	The Aarhus School of Economics and Business Administration
S. Holm	University of Odense
O. Holst	Technical University of Denmark
K. Madsen	University of Copenhagen
O. B. G. Madsen	Technical University of Denmark
P. Thoft-Christensen	Aalborg University Centre (chairman)
J. Tind	University of Aarhus
O. Tingleff	Technical University of Denmark

A preliminary screening of the proposed 450 contributions was made by the following members of the Local Organizing Committee

S. Hildebrandt	*P. Thoft-Christensen*
K. Madsen	*J. Tind*

assisted by a number of Danish scientists.

A second reviewing was performed by members of the International Program Committee

A. V. Balakrishnan, USA	*K. Madsen*, Denmark
R. F. Drenick, USA	*K. Malanowski*, Poland
E. Evtushenko, USSR	*D. H. Martin*, South Africa
M. Iri, Japan	*L. F. Pau*, Switzerland
K. Jörnsten, Sweden	*M. J. D. Powell*, United Kingdom
P. Kall, Switzerland	*H. D. Scolnik*, Argentina
R. Kluge, GDR	*J. Stoer* (chairman), FRG
J. L. Lions, France	*J. Tind*, Denmark
M. Lucertini, Italy	*G. C. Vansteenkiste*, Belgium

A final reviewing took place during the conference by the session chairman and a reviewer in the audience. A total number of 90 papers were accepted for publication.

The conference secretary was Mrs. Kirsten Aakjær, Aalborg University Centre. During the conference she was assisted by Mrs. Kirsten Bo, Technical University of Denmark, and Mrs. Vivi Læssøe, University of Copenhagen. All the above-mentioned persons are gratefully acknowledged for their assistance.

The conference was financially supported by

Danish National Science Research Council

Danish Council for Scientific and Industrial Research

Danish Social Science Research Council

Aalborg University Centre

Technical University of Denmark
DANFIP
International Business Machines A/S
European Research Office, London
Rambøll & Hannemann A/S
Knud Højgaards Fond
Thomas B. Thriges Fond

Their support is gratefully acknowledged by the conference organizer.

December 1983 P. Thoft-Christensen

TABLE OF CONTENTS

INVITED PLENARY LECTURES

MATHEMATICAL ECONOMICS

TRAFFIC AND TRANSPORTATION, REGIONAL AND URBAN PLANNING

INTEGER PROGRAMMING

NON-LINEAR PROGRAMMING, THEORY

NON-LINEAR PROGRAMMING, ALGORITHMS

NONDIFFERENTIABLE OPTIMIZATION

OPTIMAL CONTROL

OPTIMAL CONTROL (continued)

STOCHASTIC CONTROL

MODELLING OF OFFSHORE STRUCTURES

UNCERTAINTY MODELLING IN STRUCTURAL ENGINEERING

ENGINEERING APPLICATIONS

INVITED PLENARY LECTURE (late arrival)

MODELLING AND OPTIMIZATION IN SYSTEM PLANNING IN CHINA

Gui Xianyun (H.Y.Kwei)
Institute of Applied Mathematics
Academia Sinica
Beijing, China

ABSTRACT Currently, China is undergoing an economic reform. In the last few years, the Chinese government has stressed the importance of science and technology and urged the scientists in China to make contributions in the construction of China. In this paper, examples are given to show how O.R. methods are applied in system plannings in petroleum industry, regional coal base and agriculture.

INTRODUCTION

Currently, China is undergoing an economic reform. Within the next 20 years or so, we wish to develop our national economy on a rather large scale. To attain this goal there are two important factors, one is to apply advanced technology, and the other is to improve our managing ability. We, operations research workers, are mainly concerned with the latter, and wish to help our country by means of increasing economic effectiveness. In the past we have had experiences that it is difficult to convince the decision makers to apply operations research methods in economic constructions. But this situation has changed. In the last few years, our government has stressed the importance of science and technology and have encouraged scientific research, especially that which can help to improve economic results. In the past few years, methods of O.R. have been applied to industries, engineering designs, energy systems, urban plannings, transportation, pollution systems, economic plannings etc. and some of them have obtained good results. In this paper, three examples are given to show how Operations Research methods have been applied to system planning in different fields in China.

I. OPTIMAL DISTRIBUTION FOR CRUDE OIL AND PETROLEUM PRODUCTS IN CHINA

This project is a joint research work of the China Petroleum Planning Institute and the Institute of Applied Mathematics of the Academy of Sciences.

1. INTRODUCTION

A two-stage distribution problem for the crude oil and petroleum products in China is considered. The first stage is the distribution of crude oil from oil fields to various refineries with different facilities and capacities, according to their geographic positions, quality of oil, capacities and production costs of the refineries and the demands of the customers. The second stage is the distribution of petroleum products from the refineries to the customers so as to meet the customers' needs and to minimize the transportation costs. Our purpose is to maximize the total gains, that is, the total output value minus the costs of crude oil, production and transportation. It is to be noted that there is a close relation between the two stages since the production planning of the refineries will affect the transportation costs of the second stage. Therefore we propose a mathematical model in which the two stages are consdidered as a whole. The model is a large scale separable non-linear mathematical programming. It can be handled by applying Benders Decomposition Principle. The problem is decomposed into a "Master Problem" and a sequence of subproblems of the form of the classical transportation problem. An algorithm is given and its convergence proved. In the actual computation, an approximation method is used so that linear programming method is applied and the result is quite satisfactory. Comparing with the traditional method, an increase of more than 6% of net profit has been obtained.

2. MATHEMATICAL MODEL

The time period under consideration is one year. We assume that for any refinery, there are a finite number of ways for operation. For instance, there could be a project for producing more jet fuel or a project for producing more gas oil. Also for different levels of operation, the units used may be different, and thus the yields for some product may also be different. This complicates the situation since non-linearity is introduced to the problem. The way we handle this is as the following. The levels of operation are taken for discrete values as 5000KT/yr., 4500KT/yr., 4000KT/yr., 3500KT/yr., etc., the yield of each product for each of these levels is given. For any level of operation between any consecutive levels cited above, the yield can be calculated as linear combination of those for the two consecutive levels.

We introduce the following notation for our model.

j index for oil fields,

k index for refineries,

i index for petroleum products,

l index for customers,

q index for production project,

m index for levels of operation,

(k,j,q,m) the operation of refinery k, using crude oil from oil field j, operated under project q and level of operation m,

S_j supply of crude oil of oil field j,

b_m level of operation m,

r_{ikjqm} yield of product i for operation (k,j,q,m),

a_{ik} minimum allowed output of product i at refinery k,

c_{kjqm} average cost of using loKT of crude oil for operation (k,j,q,m) (the costs of crude oil and transportation are included).

P_{ik} value of unit product i (loKT) produced by refinery k,

d_{il} demand of customer l of product i,

t_{ikl} transportation cost of loKT of product i from refinery k to customer l,

x_{kjqm} a vriable denoting the amount of crude oil used in one year under operation(k,j,q,m).

y_{ikl} a variable denoting the amount of product i shipped from refinery k to customer l.

x_{kjqm}, y_{ikl} are variables we want to determine while all other quantities are known constants. Our problem can be expressed as the following mathematical programming (P):

(1) $$\text{Max } f(x) = \left\{ \sum_K \sum_j \sum_q \sum_m \left(\sum_i P_{ik} r_{ikjqm} - C_{kjqm} \right) x_{kjqm} - \sum_i \sum_k \sum_l t_{ikl} y_{ikl} \right\}$$

subject to

(2) $\sum_k \sum_q \sum_m x_{kjqm} \leq S_j$ all j

(3) $\sum_j \sum_q \sum_m r_{ikjqm} x_{kjqm} \geq d_{ik}$ for some (i,k)

(4) $\sum_j \sum_q \sum_m x_{kjqm}/b_m = 1$ all k

(5) $\sum_k \sum_j \sum_q \sum_m r_{ikjqm} x_{kjqm} \geq \sum_l d_{il} = D_i$ all i

(6) $\sum_l y_{ikl} \leq \sum_j \sum_q \sum_m r_{ikjqm} x_{kjqm} = Z_{ik}$ all (i,k)

where z_{ik} denotes the unknown total amount of product i produced at refinery k.

(7) $\sum_k Y_{ikl} = d_{il}$ all (i,l)

(8) $x_{kjqm} \geq 0$ all (k,j,q,m)

(9) $Y_{ikl} \geq 0$ all (i,k,l)

(10) for any fixed set of K, j, q, relative to m, at most two adjacent x_{kjqm} can be non-zero.

In the above model, it is understood that all summations run over the allowable combinations of the indices, since many combinations are either physically impossible (such as an ik combination which signifies product i cannot be manufactured at refinery k) or so obviously uneconomical as not to be included in the model. The meaning of (1) to (9) in the model should be clear. As we see that for a fixed policy, that is, a fixed set of values of x_{kjqm}, y_{ikl}, $r_{ikqm}\, x_{kjqm}$ is the amount of product i produced from operating (k, j,q,m). Constraints (2) are the supply constraints of the oil fields. (3) are the requirements of products at refinery k. (4) mean refinery k should be operated the whole year. (5) specify that the total amount of product i sent out from refinery k cannot exceed the amount it produces. (7) mean that the requirements of each customer must be met. (8) and (9) are the non-negativity of the variables. Constraint (10) is a stipulation that when the level of operation is between some adjacent levels mentioned above and is expressed as their linear combination, then the yield of product i will be expressed as the same linear combination of the corresponding yields of the two adjacent levels.

3. ALGORITHMS

From the model above, we notice that Benders Decomposition can be applied to our case. If the production variables x_{kjqm} of the first stage are temporarily held fixed, then the z_{ik} in constraints (6) are constants. Noting that the variables y_{ikl} are separable in (1), (6), and (7), the problem to find the optimal solutions of y_{ikl} can be solved independently for each product i from the following classical transportation problem (T_i):

$$\text{(11)} \qquad \min \sum_k \sum_l t_{ikl}\, y_{ikl}$$

subject to

$$\text{(12)} \qquad \sum_l y_{ikl} \le z_{ik} \qquad \text{for all } k$$

$$\text{(13)} \qquad \sum_k y_{ikl} = d_{il} \qquad \text{for all } l$$

$$\text{(14)} \qquad y_{ikl} \ge 0 \qquad \text{all } (k,l)$$

The fundamental idea of our method is to find a set of values for variables x_{kjqm} of the first stage satisfying constraints (2), (3), (4), (8), (10) through a so-called "master problem". From this and (6) we can compute the values of z_{ik} and thus solve the transportation problem (T_i) for each product i. It is easy to see that the set of values $\{x_{kjqm}, y_{ikl}\}$ thus obtained is a feasible solution of our original problem (P) and its corresponding value of the objective function (1) can be served as an lower bound of the optimal value of (1). After we solve a master problem and the corresponding subproblems(T_i), we say we complete an iteration. Next, we apply the informations we get from solving (T_i) to form a new master problem and get a new set of values for $\{x_{kjqm}\}$. We can prove that the value of the objective function of the master problem for each iteration can be served as a upper bound of the optimal value of (1) in (P). Thus we solve a master problem and a set of subproblems (T_i) alternatively until we learn that the value of the objective function (1) of (P) is sufficiently close to the value of the optimal solution by means of comparing the upper bound and the lower bound of (1). We have proved that the algorithm converges in a finite number of iterations.

Another alternative, an approximation method can be used to solve (P). When the computer is big enough we can apply the following method. Solve problem (P), but ignore constraint (10), by the direct method of LP. If the solution, say $\{x^*, y^*\}$, satisfies (10), then it is an optimal solution of (P). Otherwise we use $\{x^*, y^*\}$ to form another linear programming from (P) by omitting constraint (10), but allowing variables x_{kjqm} to be non-zero only those which are adjacent to non-zero variables corresponding to $\{x^*\}$.

4. COMPUTATIONAL RESULTS

We apply the approximation method on a UNIVAC 1100 computer with software FMPS. We have computed 5 different problems with different data or some modifications of the objective function (1) in (P). The solutions obtained are considered to be practical. Comparing with the traditional method, for the year 1981, an increase of more than 6% of net profit has been obtained.

II. STUDY OF A REGIONAL COAL BASE

This research is done by a group in the Institute of System Engineering, Xian Jiaotong University, China. The purpose of the research is to develop an energy planning model for a regional coal base. The model consists of five submodels regarding demand, exploitation-supply, transportation, investment and decision analysis. The submodels can be linked closely with both the national and regional economic development plans to form an overall system.

1. DECISION ANALYSIS OF THE COAL EXPLOITING SCALE FOR THE REGION IN 1990

When planning an energy base, first of all the exploiting scale should be determined. It depends on many factors such as national and regional economic strength, natural conditions and resources, transportation, economic structure, technical force etc. Naturally the problem is a complicated and multi-objective one. In this analysis, the concepts of "possibility" and "satisfiability" are introduced, and these concepts are combined to form another attributive and correlative concept of "possibility-satisfiability" to describe the degree of rationality of a certain plan at some exploiting scale so as to provide a foundation for the decision-making of a rational exploiting scale.'

In the analysis of the problem, 18 factors, 7 "possibility" and 11 "satisfiability" are considered as follows: possibilities:

capacity for coal transport P_1

capotal investment for type I and type II mines P_2, P_3

land supply P_4

water supply P_5

increase of staff and workers P_6

increase of technical personnel P_7

satisfiabilities:

capacity for coal transport q_1

capital investment for type I mines q_2, q_3

capital investment for type II mines q_4

land demand q_5

water demand (data I) q_6

water demand (data II) q_7

efficiency of staff and workers q_8

percentage of technical personnel in the total of and workers q_9

comparision of the scale with that of similar foreign coal bases q_{10}

coal supply for meeting the needs of national economy q_{11}

each p_i and q_i is defined on $(0, 1)$ · i.e. p=0 means it is not possible; p=1 means absolutely sure. Similarly for q_i. There can be different ways to combine any two factors: (Mn) means taking the minimum of the two; (.) means taking the product of the two; (+) means taking the sum of the two. With differnt combinations of the factors 12 alternatives are considered. By comparing positions of the peaks of the curves for alternatives one may conclude that capital investment, transportation, water supply and technical force are the most important factors effecting the exploiting scale.

The computational results of six of the alternatives are shown in a graph, we see that the peaks almost appear on the same exploiting scale. Thus we can be quite sure of a rational exploiting scale.

It is important that 1) the proper investment must be ensured for realizing the planned output; 2) the capacity for the coal transport should be increased as to suit the needs of coal product; 3) the water supply must be sufficient for developing the coal base; 4) the technical personnel must be increased to meet the requirements of development. If the above four factors are not being taken into full account, it would be difficult to attain the national exploiting scale or unfavorable effects on the long-term development would be experienced.

2. COAL DEMAND PREDICTION

First the demand of energy of the various regions in China is predicted by the following "state" equation

$$E(t+1)=A(t) + U(t+1) + R(t)$$

where E(t), U(t), R(t) are vectors of dimension the number of regions involved. Component $e_i(t)$ of E(t) is the amount of energy demanded by the ith region in the tth year, $u_i(t)$ in U(t) is the plan-controlled energy of the ith region in the tth year, R(t) the disturbance vector and A(t), an n×n matrix which is determined by the GNP of various region, the percentage of the output value of light and heavy industries in the GNP and the percentage of energy consumption of industry in various regions.

The coal of the base concerned mainly supplies 12 regions. The amount of coal for these regions can be derived from the E(t) predicted above using the related statistical data and the information on coal demand predicted by the ministry of Coal Industry of China. The predicted results can be used for the computation of the exploitation-supply model and transportation model.

3. DECISION ANALYSIS FOR EXPLOITATION-SUPPLY AND TRANSPORTATION PLANNING

Given the national exploiting scale and the demand of the 12 regions in 1990 of the coal base, the following exploitation-supply model and transportation model can be used to plan how to develop the coal base to reach the planned scale and how to transport the coal to suit the needs of various regions.

The problem of coal exploitation and transport is simplified into a problem with 6 production subregions in the base and 11 demanding regions, with actual transporting lines connecting the various regions. In the models 0-1 variables are used to denote the new mines and transporting lines which are likely to be built, while the continuous variables are used to represent the new increase of output of coal from various mines as well as the capacity on each transporting line. The two models are very similar with the same objective to minimize the total costs of the following three parts:

(1) costs of exploitation and production.

(2) costs of transportation.

(3) capital investment of new mines or new transportation lines.

under the constraints of meeting the demands of the 11 regions, the natural resources, conditions and capital investments etc.

However the exploitation-supply model lays particular stress on describing the coal production and exploitation of the base. It consists of 3 different types of mines with only type I of mines being considered to be newly built. The nodel assumes the possible transporting lines being fixed and decides which new mines should be built. The transportation model put more emphasis on coal transportation. Contrary to the exploitation-supply model, it assumes the production mines being all fixed and try to decide the best plan for capital investment of new railways, ports and canels. Both models are mix-integer 0-1 program. Benders decomposition principle has been applied to the models and 12 schemes for the exploitation-supply model and 8 schemes for the transport model have beencomputed on the SIMENS 7760 computer. Having analysed the computational results cases, the following results are obtained:

1) The rational proportions of outputs of mines of various types;

2) The exploiting scheduling of new mines and new transporting lines;

3) The output plan of mines of various types in 1990;

4) The rational flow direction and planned capacity on various transporting lines;

5) The analysis of investment for various projects.

These results have provided the coal base in question with a quantitative scientific foundation for its overall planning.

4. CONCLUSION

The two models mentioned above are used separately but taking information from each other to obtain a rational plan for both production and transportation. Alternatively the two models can be combined to produce a single model to get an optimal plan.

The results obtained have been proved satisfactory and they have been accepted by the government agency concerned for decision making. As a result, the method presented here provides a mean for system planning for regional energy basis.

III. FARM PLANNING OF A COUNTY

1. INTRODUCTION

The purpose of this research is to seek a farm planning of a county which fits in the overall national planning, and is constrained by the natural condition and resources, the technical condition, the economic structure, balancing the agricultural ecology, and at the same time maximizing the total profit of the county.

2. MATHEMATICAL MODEL

a) Farm Planning under Different Weather Condition

The conditions of the weather of the year are divided into four categories: normal year, drought year, waterlogging year, and both drought and waterlogging year. Under each condition a linear programming model is formed.

In the model, considering the rotating of crops, each peiod is of two years, and our objective is to maximize the total profit of two years.

According to the varieties of crops, the possible manners of rotating crops, 83 alternatives of schemes of planning are chosen. Also, according to the quality of soil, geographic features, condition of water supply, etc., the land is divided into 37 categories. We want to choose the best combination of the 83 alternatives on the 37 categories of land within the county. Thus we have 83 X 37 = 3071 variables.

The coefficients of the objective function are net profit of each farm product.

The constraints of the model are:

1) The crop recruiting plan of the nation.
2) Grain demand of the people in the county.
3) Grain required to be stored up.
4) Water supply.
5) Capacity of production.
6) Fertilizer requirements.
7) Requirements of livestock production plan.
8) Requirements of foresting.
9) Capacity of storage.
10) Capacity of transportation.

There are altogether 50 constraints.

b) Decision Analysis

For each of the four weather conditions, we have an optimal farm planning. Then according to the weather statistical data, we can get a 4 x 4 payoff matrix. Applying the matrix game method, an optimal strategy against the nature can be obtained.

3. COMPUTATIONAL RESULTS

The computation was done on a UNIVAC-1100 computer. Optimal strategies have been obtained for each weather conditions. Comparing with the actual situation of some typical years, the net profit increased by 13% to 28%.

4. SENSITIVITY ANALYSIS

The prices of farm products on the free market are not stable. Therefore diiferent prices for the coefficients have been used in the optimal solution.

5. CONCLUSION

This project is jointly done by the Agriculture Department of the city of Jinan of Shangdong Province, and the Institute of Operations Research of Qufu Teachers College. Hundreds of people have been working to obtain the proper data. Some members from the Agricultural Department of Shangdong Province have had many years of experience in agriculture, but it is through the data collecting and computation of this project that they have gained more insight of the agricultural problems in the province. They planned to use this scientific method to plan the farm production for every county in the province. Also they are considering farm management including farming, livestock production, and forestry as a whole.

UNCERTAINTY ALGEBRA. A LINEAR ALGEBRAIC SUBMODEL OF PROBABILITY THEORY

Summary of lecture by Ove Ditlevsen

Multidimensional uncertainty problems related to engineering models are most often of a complexity that prevents the practicability of most parts of conventional probability theory. Generally the tools are restricted to operations with mean values and covariances (second moment analysis). Transformation of the results of such calculations into probabilities of specified events are finally often made ad hoc by use of standard probability distribution families such as the normal distribution family or the gamma distribution family. Proper justification of this probabilistic interpretation of the second moment result is rare.

The topic of the lecture is the concept of uncertainty algebra. It is a less detailed mathematical structure than the usual Kolmogorovian probability theory. Its axioms are entirely based on elementary linear algebra.

Without using any part of traditional probability theory the uncertainty algebra is defined as a mathematical model of mean values (expectations) with attributed properties that reflect the calculational properties of sample averages. In fact, a positive linear functional which is defined on a linear space of uncertain quantities (including the constant 1) and which maps this constant 1 at the number 1 is simply called an expectation functional and this functional models sample averages.

It characterizes a deterministic model that it represents by a single number which in reality is usually an uncertain quantity. Normally one thinks of this single number as pointing to the gross location of a sample of the corresponding uncertain quantity. The sample average is indeed such a location number. Obviously, the simplest step from determinism toward describing uncertainty is to represent the uncertain quantity by one more number. This new number aims at measuring the dispersion of the sample around the location number. The usual sample standard deviation may reasonably be used as such a dispersion number. Since the square of the sample standard deviation, the sample variance, may be expressed in terms of averages, a functional that models this sample variance may be defined in a straight-forward way in terms of the expectation functional. The operational rules of the expectation functional then induce operational rules for this variance functional.

It follows immediately, when several uncertain quantities are considered and act together, that coupling numbers, covariances, between the uncertain quantities are needed. Thus an algebra of covariance operations is defined simply on the basis of the properties of the expectation functional. This algebra is powerful enough to be applicable on multidimensional linear systems. With this basis the concepts develop in a natural way. The concept of conditional expectation is introduced in a way which is motivated by an exact sample property and its operational rules follow from the operational rules of the expectation functional.

The usual algebra of events (set operations) and the assignment of probabilities to these events fall nicely within the framework of the uncertainty algebra simply as being equivalent to the algebra of a special class of uncertain quantities (zero-one uncertain quantities, that is, characteristic functions in set theory).

On basis of the properties of the class of nonnegative definite functions it is even possible to develop the concept of uncertain processes up to mean square type of operations (differentiability, integrability in m.s.) solely within the uncertainty algebra combined with elementary concepts of mathematical analysis.

Of course, a concept like convergence with probability 1 has no room within this algebraic theory. However, it is possible to supplement the axioms of uncertainty algebra by some further axioms whereby a finer mathematical structure is obtained. By proper interpretation this finer structure can be demonstrated to be equivalent to the probability space concept as defined through the axioms of Kolmogorov.

Reference

Ove Ditlevsen: Uncertainty Modeling. McGraw-Hill International Book Company, New York, 1981.

ENERGY MODELS AND ENERGY POLICY PROBLEMS

A. Voss
University of Stuttgart
Stuttgart
Federal Republic of Germany

1. INTRODUCTION

I have been involved in energy modelling for planning and policy making for more than ten years now and I am still convinced that systematic and careful modelling can contribute to better decisions in the energy policy area. I think I should make this statement right at the beginning, because my contribution will be somewhat critical. It will to some extend focus on the failures, misuses and unresolved issues in energy policy modelling rather than report about the successes, which although they are there, are still small compared with the potential benefits and prospects, that energy models can offer to the decision makers. Nevertheless I will start with a brief review of the history and methods used in energy modelling and I will describe a limited number of representative models in order to illustrate the present state of the art. The review is not intended to be exhaustive or to provide a comparative evaluation of models designed for similar purposes. Rather, the models are reviewed to illustrate the advances and the structure of recent and current efforts by energy modellers.

Thereafter I will discuss the question whether or not energy models have successfully contributed to help solving the complex problems facing the energy planner and energy policy maker. I hope to make clear, that despite the tremendous progress made in the design of complex, large-scale models, energy models were by far not as successful as they could have been in their contribution to the decision making process. And I will argue, that a new more realistic attitude, a new orientation of the preferences of the model builder is needed, that expectations must be redirected to what is needed and can be achieved, rather than to promote and construct more sophisticated or even universal models.

2. ENERGY MODEL DEVELOPMENT

The sharp increase in the price of energy in the early seventies have confronted many nations, particularly energy importers with unprecedented economic challenges they were ill-prepared for. The economies of the less affluent oil importers in the developing world were severly distorted. Even among the affluent industrialized countries, the cost of adjustment to higher energy prices in terms of higher overall price levels, unemployment, industrial restructuring, adverse distributional effects and environmental quality, have been pervasive.

Although efforts to develop energy models began in the early sixties, that is well before the first oil crisis in 1973, it was the growing awareness of the energy problem originating from this event that forced an explosion in the development of energy models. Exact figures concerning the energy models developed so far are not available, but in the reviews of energy models published by the International Institute for Applied Systems Analysis (IIASA) /1, 2, 3/ up until 1976 alone some 144 different models were characterized and classified. The individual models vary greatly in their objectives, they address a broad scope of problems for geographical areas of widely different sizes and they employ a variety of methods originating from several scientific disciplines.

The energy models developed in the sixties focused mainly upon the supply and demand of a single energy form or fuel like electricity, oil or natural gas. Faced with the complex problem of optimal allocation and routing of crude oil and oil products between different oil sources, refineries and demand centers, the petróleum companies have developed and applied particularly large allocation models, as well as models for the refining process. Another example of a successful application of models of the sectoral type are the models used for the analysis of electric utility operations and expansion plans. A large number of models have been developed and are used to evaluate the optimal expansion strategy of the power plant system required to satisfy an increased electricity demand. The models determine the optimal mix and timing of new power plants of different types so that the electricity demand over the planning horizon is satisfied at minimum discounted overall cost, including capital, fuel, as well as operating costs.

Both kind of models mentioned above focus on the supply side, that is, on the best way to satisfy an assumed energy demand. Energy is an exogenous input to these models and is often provided by econometric demand models, estimating energy or fuel demand as a function of energy prices and other determinants such as population, economic growth, etc..

A major criticism concerning sectoral, single fuel or energy form models is that they treat the development of the sector or fuel in question as isolated from the rest of the overall energy and economic system, thereby ignoring that there are many different ways to satisfy given energy service demands such as space heat, industrial process heat and transportation. A sectoral, single fuel model cannot adequately describe the interfuel substitution related to changing energy prices, technological development or environmental considerations in the different sectors of energy use.

Complying with these requirements was the main reason for the development of energy system models, describing the energy flows from different primary energy sources through various conversion and utilization processes to different end use demands. It was at the beginning of the seventies, when the work on energy system models began.

A national energy balance as shown in Fig. 1 can be viewed as a simple static model of the energy system, because it accounts at a single point in time for all energy flows from the primary energy sources, through conversion processes, to the ultimate use of various fuels and energy forms.

Most of the energy system models are based on the network representation of the energy balance approach, as it is shown in Fig. 1. Using this network of flow of resources like coal, oil, gas, nuclear or solar to various demand sectors like industry, transportation, households and the commercial sector as a simple accounting framework, the consequences of alternative ways to satisfy an estimated demand development in each of the major end-use sectors can be simulated and evaluated in terms of primary energy consumption, required conversion capacity etc.. Extensions of this type of model to analyse the impact of alternative energy supply strategies on the environment and in

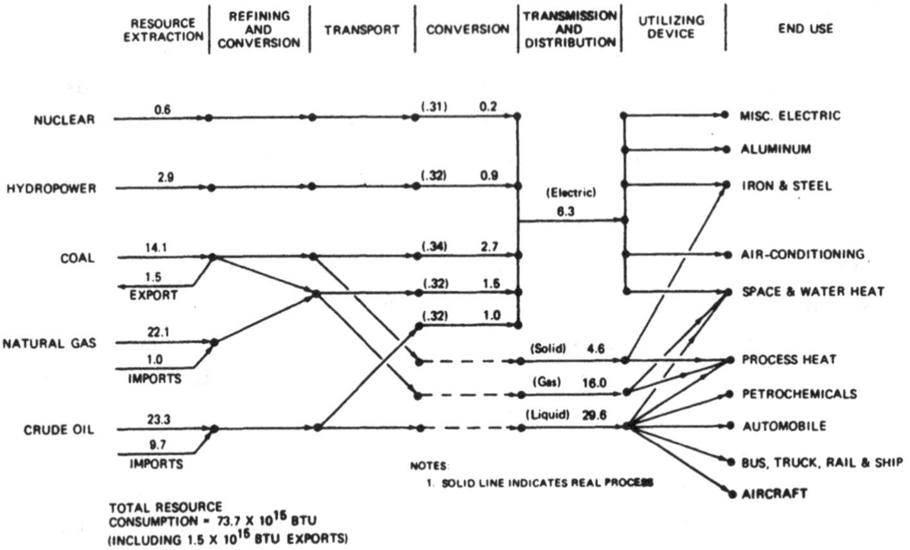

Fig. 1: National Energy Balance

terms of energy cost are easily attainable and have been used in the past. Besides these network accounting models, a series of optimizing models of whole energy systems were developed from the beginning of the seventies. These models were designed to determine the optimal allocation of energy resources and conversion technologies to end-uses using the network representation of the energy system. The models are either static with the optimization process seeking a minimization of cost for a single target year, or they are quasi dynamic and attempt to minimize the present values of costs over the whole planning horizon, subject to the demand and to a set of constraints reflecting resource availabilities and/or environmental considerations.

Accounting and optimization models of this type focus on the technical structure of the energy systems. Energy demand is usually an exogenous input to them. Therefore these models do not allow for demand adjustments due to higher energy prices or to changed GNP growth caused by rising energy cost and limited energy supplies.

Handling these issues requires models linking the energy sector with the rest of the economy. Various approaches to link economic models to models of energy demand and supply have been investigated. Generally speaking two classes of energy-economy models can be distinguished. Integrated models which explicitly describe the interrelations between the energy sector and the economy and model sets which consist of an economy and an energy system model which are linked by the transfer of data via a human interface.

This short glance back into history should show that, although the construction of energy models began only 20 years ago, there have been several important development phases as single fuel or sectoral models evolved towards models of complete energy systems and energy economy models.

This historical development pattern seems to be also a useful scheme for the classification of energy models. In the following I will distinguish between

- Single Fuel Models
- Energy System Models and
- Energy-Economy Models.

Later I will describe in some more detail typical approaches used in modelling the entire energy system and the energy-economy interactions.

But let me first comment on the methods used in energy modelling. As it was not the main goal of the energy model builders to develop new and better methods, they most often referred to the corresponding improvements and developments of other fields of science e.g. econometrics, statistics, operations research, computer science, and system science. Looking back, one can say that there are three modelling methodologies that have been applied predominantly in energy models, namely engineering process analysis, mathematical programming, and econometrics.

Econometric methods are found most often in representations of the energy demand side emphasizing the behavioral aspects of decisions on the sides of both the consumer and the supplier. Statistical techniques are used to estimate the structural parameters of the behavioral equations, e.g. macroeconomic production functions or price elasticities from observed data. Econometric models are, in general, of a higher aggregation level than process models, which often cover quite a lot of technical details of the energy supply system. This is independent of whether it is conceived as a simple accounting or as an optimization model. The linear programming technique has been used far more than other mathematical programming methods, because of its capability to solve large problems.

In addition to these methods, energy models, which make use of the input-output method, the system dynamics approach or the method of game theory were occasionally developed.

3. THE STATE-OF-THE-ART IN ENERGY MODELLING

Following the classification of energy models mentioned above, I would now like to illustrate the state of the art in energy system- and energy-economy modelling by describing typical representatives of these classes of energy models in some more detail.

	METHODOLOGY	
MODEL	SUPPLY SIDE	DEMAND SIDE
BESOM (BROOKHAVEN)	LINEAR OPTIMIZATION (STATIC)	EXOGENOUS
EFOM (GRENOBLE)	LINEAR OPTIMIZATION (QUASI DYNAMIC)	EXOGENOUS
MESSAGE (IIASA)	LINEAR OPTIMIZATION (QUASI DYNAMIC)	PARTIAL EXOGENOUS (PRICE DEPENDENT)
MARKAL (JÜLICH)	LINEAR OPTIMIZATION (QUASI DYNAMIC)	PARTIAL EXOGENOUS (PRICE DEPENDENT)

Fig. 2: Energy System Models

Fig. 2 lists several of the well-known energy system models together with the methodology used. All of these modesl use the linear programming approach. They focus on the technical, economic and environmental characteristics of the energy conversion, delivery and utilization processes that comprise the total energy system. While BESOM provides a "snapshot" of the energy system configuration, the other models are designed to analyze the evolution of the energy system ove a time period.

Let me now briefly describe the MARKAL model as a typical representative of the energy system models /4/. MARKAL was specifically designed to follow the evolution in time of the introduction of new technologies and the corresponding decline in the use of hydrocarbon resources, especially imported petroleum. Using the model, it is possible to assess the relative attractiveness of existing and new technologies and energy resources on the supply side of the system and, on the demand side, the long-range effect of conservation, of efficiency improvements in end-use devices and of inter-fuel substitution.

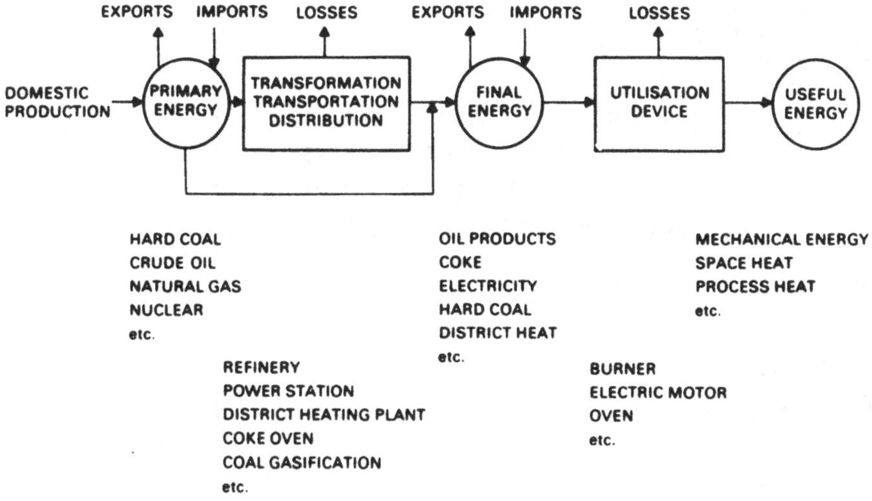

Fig. 3: The MARKAL Model

Fig. 3 shows the principal energy flows represented in MARKAL. Three types of energy are distinguished. Primary energy (e. g. domestic coal, imported crude oil) is transformed into final energy (e.g. electricity, refined oil products, district heat) through transformation and conversion, transportation and distribution processes. The final energy is then consumed in end-use devices to produce useful energy (e.g. space heat, mechanical energy) to satisfy the energy service demand, for example the demand for a warm room or the travelling from Stuttgart to Copenhagen. Useful energy or energy service demand are the exogenously specified driving variables in the MARKAL model.

MARKAL is a multiperiod linear programming model with explicit representation of some 200 technologies for energy production, conversion and end-use. The general model structure is illustrated in Fig. 4. The objective function is the sum of discounted costs of fuels, operating and maintenance, transportation and investments for adding new capacities, to satisfy the energy demand over the planning horizon. The objective function is to be minimized under a set of constraints. The constraints involve balances for individual fuels as well as limits on the installation and operation of technologies. The capacities of the

OBJECTIVE FUNCTION

MINIMIZE $\sum_{T}^{T} \beta^{T}$ [PRIMARY FUEL COSTS + OPERATING A. MAINTENANCE COSTS + TRANSPORTATION COSTS + INVESTMENT COSTS OF ADDING NEW CAPACITIES]

OVER THE TIME HORIZON T, SUBJECT TO:

- DEMAND CONSTRAINTS
- SUPPLY CONSTRAINTS
- CAPACITY CONSTRAINTS
- RESOURCE CONSTRAINTS
- IMPLEMENTATION CONSTRAINTS
- ENVIRONMENTAL CONSTRAINTS

Fig. 4: General Model Structure of MARKAL

different energy technologies depend on investments made in earlier periods and the defined lifetimes of existing technologies. Because of this representation, the model is able to describe the phasingout of existing plants and the build-up of new capacity properly. Another dynamic constraint utilized in the model limits the cumulative amount of particular resources available over the entire time horizon. The electricity and heat generating technologies have been modeled in MARKAL with explicit treatment of the load structure related to the diurnal and/or seasonal variations of the demand. Environmental considerations can also be taken into account.

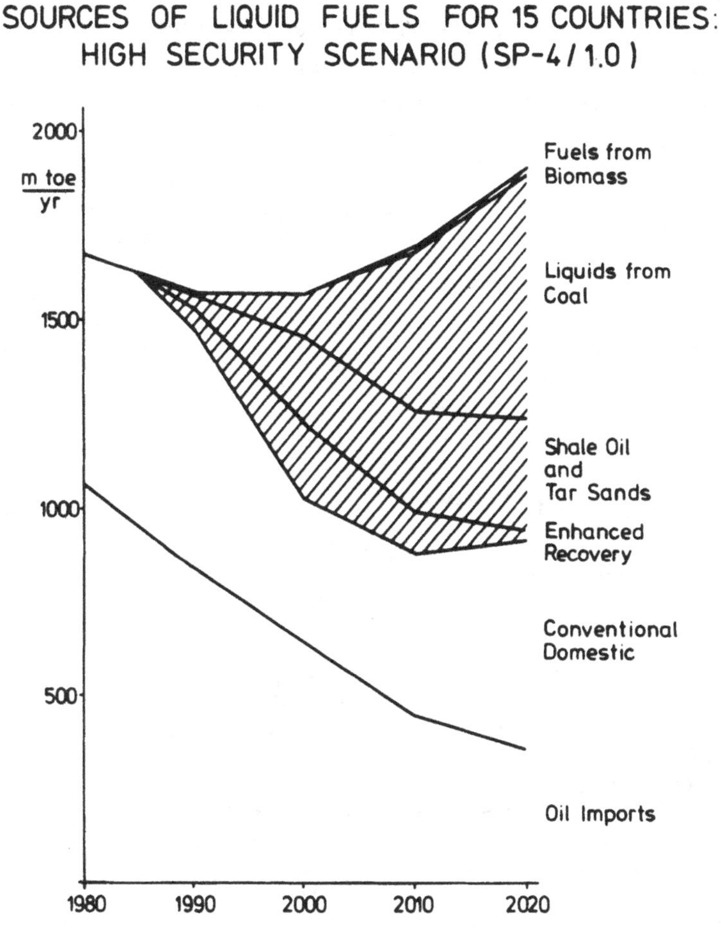

Fig. 5: Typical MARKAL output

Fig. 5 shows a typical result obtained from MARKAL indicating how the substitution of oil imports by new liquid fuels producing technologies takes place under a certain price escalation of crude oil /5/.

Another set of interesting information, which these models provide, is the trade-off between energy system costs and oil imports, as displayed in Fig. 6. The curve shows what a replacement of oil imports would cost the economy, which would have the invest in new technologies or push conservation. In the figure 6, PS-1 denotes the optimum allocation of fuels and technologies for a least cost scenario. If we move towards the lèft, the system costs increase while oil imports decline. The fact that a premium is to be paid for lower oil import energy systems is denoted by scenarios SP-1/PREM-1 and SP-1/PREM-2. Three different patterns are shown (Spain, United States, United Kingdom) illustrating differences among countries /5/.

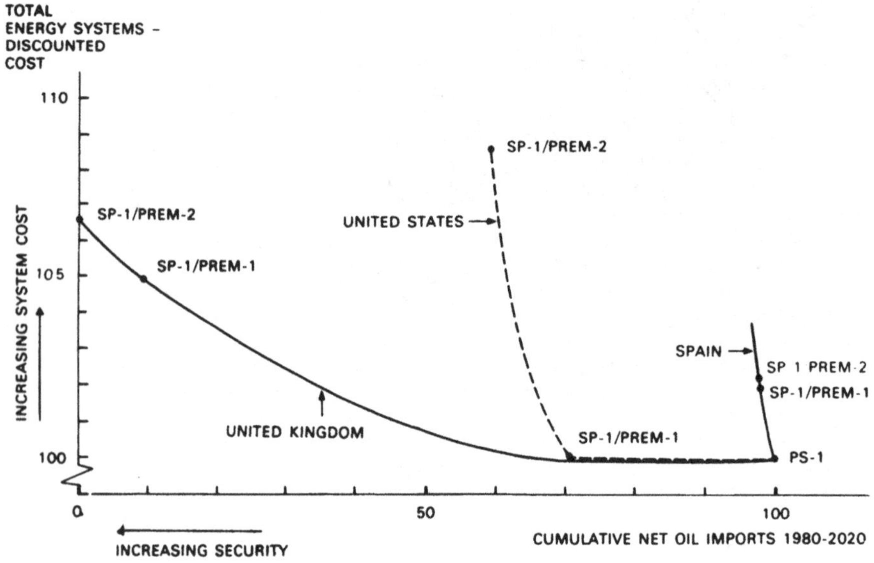

Fig. 6: Trade Off between Energy System Costs and Oil Imports

Each point of this trade-off curve represents a scenario, which itself yields a different mix of technologies and a different temporal evolution of each technology. Other trade-off, e.g. between costs and environment can be examined in a similar approach.

It should be mentioned, that this kind of linear programming models of the energy system, are able to take price demand elasticities into account. In the model the response to energy price increases is determined in three forms: investments in conservation, investments in new technologies with higher efficiencies and adjusted useful energy demand levels. This feature is typical for a model type which is often called a partial equilibrium model, where energy demand itself is a variable depending on the price of energy /6/.

The second class of models I want to discuss in some more detail are the energy-economy models. Fig. 7 lists some of the well-known models, which explicitly take into account the linkages between the energy sector and the rest of the economy.

These integrated models share some common features. They all include a macroeconomic submodel, which represents to varying degrees, the production and consumption structure in the economy. They also contain an energy supply system with depiction of energy technologies, demand and prices. Finally, there are clear linkages between the energy sector and the rest of the economy.

A distinction is made between two categories of energy-economy-models. The first category consist of models which were basically designed to study the energy-economy interactions, while the second category contains models that were desinged by linking existing energy and economy models. Fig. 7 also indicates that optimization and econometrics are the methods most often used in energy-economy models.

ETA-MACRO is an example of the first category of energy-economy models /7/. As the name suggests, it consists of two parts: ETA is a process analysis model for energy technology assessment and MACRO is a macroeconomic growth model dealing with substitution between labor, capital and energy inputs.

MODEL	METHODOLOGY
INTEGRATED MODELS	
ETA-MACRO (STANFORD UNIV.)	NON-LINEAR OPTIMIZATION ECONOMETRIC
PILOT (STANFORD UNIV.)	LINEAR OPTIMIZATION
SRI (STANFORD RES. INST.)	ECONOMETRIC OPTIMIZATION
HUDSON-JORGENSON	ECONOMETRIC
ZENCAP (ZÜRICH)	ECONOMETRIC OPTIMIZATION
MODEL SETS	
IIASA (LAXENBURG)	LINEAR OPTIMIZATION, INPUT-OUTPUT, SIMULATION
CEC (BRÜSSEL)	LINEAR OPTIMIZATION, ECONOMETRIC, ACCOUNTING
DRI-BROOKHAVEN	LINEAR OPTIMIZATION ECONOMETRIC

Fig. 7: Energy-Economy Models

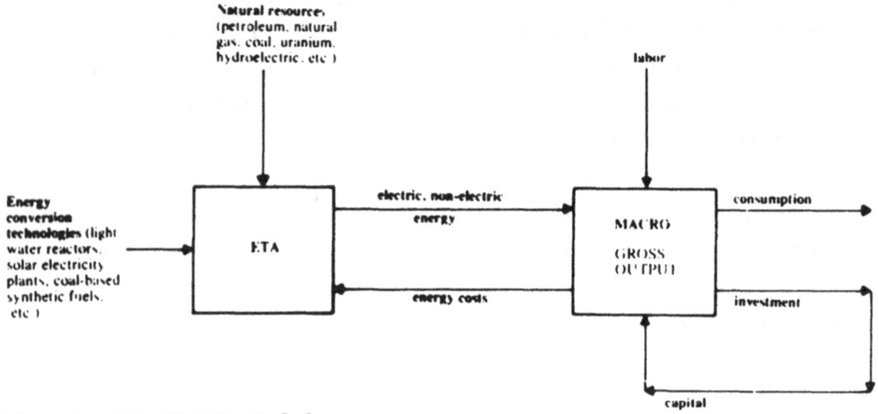

Fig. 8: ETA MACRO Model

Fig. 8 provides an overview of the principal static linkages between the energy and the macroeconomic submodels. Electric and nonelectric energy are supplied by the energy sector to the rest of the economy. Gross output depends upon the inputs of energy, labor and capital. The output is allocated between current consumption, investment in building up the stock of capital, and current payments of energy costs.

The entire model determines for each point in time an equilibrium between suply and demand, whereby substitution between labor, capital and energy inputs take place according to their availability and price. An increase in prices for energy will then affect the future level of energy demand, the fuel mix and the production structure of the economy in various ways. Price induced conservation and interfuel substitution will both have macroeconomic implications and the whole economy will adjust to the new equilibrium according to the time lags built into the model. This model is of the type which may be called a "general equilibrium model", in that it encompasses at the same time the effects, which the macroeconomy has on the energy system and vice versa the impacts of the energy system on the economy.

To be able to understand how the model works, it seems best to have a closer look to the MACRO submodel (see Fig. 9).

ALLOCATION OF ECONOMIC OUTPUT (Y)

$$Y = C + I + EC$$

LONG-RUN STATIC PRODUCTION FUNCTION

$$Y = \left[A(K^{\alpha} L^{1-\alpha})^{\rho} + B\,(E^{\beta} N^{1-\beta})^{\rho}\right]^{1/\rho}$$

WHERE $\rho = (\sigma - 1)/\sigma$ (FOR $\sigma \neq 0, 1, \infty$)

CAPITAL ACCUMULATION

$$K(T) = \lambda K(T-5) + 0.4 \cdot 5 \cdot I(T-5) + 0.6 \cdot 5 \cdot I(T)$$

$$(T = 5, \ldots, 75)$$

Fig. 9: Linkage between the Energy Sector and the Economy in the ETA-MACRO

As I mentioned already before, electric and non-electric energy are suplied by the energy sector to the rest of the economy. Like the material balance equations of an input-output model, aggregated economic output (Y) is allocated between interindustry payments for energy costs (EC) and "final demands" for current consumption (C) and investment (I) (First equation).

The production function employed assumes that the economy-wide gross output (Y) depends upon four inputs: K, L, E, N - respectively capital, labor, electric and non-electric energy. The elasticity of substitution among the input factors is separated in three fractions: substitution between capital and labor (denoted by α and $1-\alpha$), substitution between electric and non-electric energy (denoted by β and $1-\beta$), and substitution between capital/labor and electric/non-electric energy (denoted by ρ). If we were considering a static problem, the long-run production function would have the form of the second equation in Fig. 9.

In the model this production function is used in a modified form to allow for time-lags in the economy's reponse to higher energy prices. This is extremely important, because most changes concerning the adjustment to higher energy costs will be associated with new equipment and structures, and the average life-time of the capital already in place might be as high as 40 years and more as in the case of housing and urban transportation systems.

In ETA-MACRO these lags are built into the production function by appropriate growth limitations relative to previous periods. These time lags are also reflected in the equation for physical capital accumulation, which is the last on in Fig. 9. To approximate a two-year average gestation lag between investment and useable capital stocks, it is supposed that 60 % of gross investment provides an immediate increase in the capital stock, but that 40 % has a five-year delay. Capital stocks (k(t)) are expanded by gross investment (I(t)) and are reduced by the capital survival fraction.

The other submodel, ETA, is a conventional linear programming energy supply model, which for a given set of resources and technologies aims at searching an optimum energy path. The degree of detail shown here, however, is much less than in energy system models of the MARKAL type. As most of the general equilibrium models which apply aggregated functions in the economic sector and look into the energy sector with less detail, ETA-MACRO is not intended to be used as a planning tool, which produces a single set of numerical results. The merits of the model have to be seen in the fact that it enables us to check the logical consistency of competing assumptions about energy futures using a clear and straight-forward approach. In fact, the model has been found to be a useful instrument to study for instance the implications which a nuclear path would impose on the US economy, and to describe the impact of higher oil prices on economic growth.

The energy modelling approach of IIASA (the International Institute for Applied Systems Analysis) /8/ is another typical example of an energy-economy model. It is designed to analyse the energy sector as an integral part of the economy.

But unlike the integrated models (PILOT, SRI, Hudson-Jorgenson, ETA-MACRO, ZENCAP) which treat the interactions between energy and the economy within a single network of equations, IIASA has created a package containing a set of various models, applying different techniques.

IIASA's energy modelling team has adopted the philosophy that the linking of several independent and simple models has advantages over large scale model blocks involving complex functional relations. The links need not be automatic, but may involve human interference.

Fig. 10 illustrates the modelling approach adopted at IIASA. Four independent models, MEDEE-2, MESSAGE, IMPACT and MACRO are used, each applying a different methodology and having a different purpose. Every single model provides inputs to the system considered, either in the form of direct input data to other submodels or in the form of general information which is used to modify assumptions. The entire modelling approach is a highly iterative one. Initial assumptions and judgements lead to calculations and results, which provide feedback information for the alteration of the inputs until convergence is achieved.

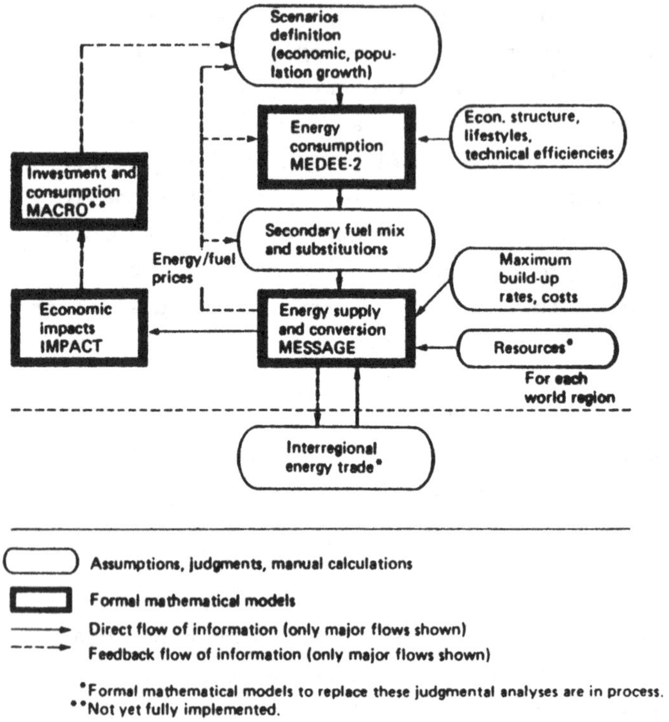

Fig. 10: The IIASA Set of Energy Model

The start of the modelling loop is determined by the definition of scenarios as indicated on top of Fig. 10. Assumptions about economic and population growth are the main parameters for the distinction of the IIASA scenarios. Information about economic and demographic developments and judgments about lifestyle changes, improvements in efficiencies of energy using devices, and the rate of penetration of new and/or improved energy-using equipment are fed into the submodel MEDEE-2. This model determines the energy demand in terms of secondary energy for major end-use categories such as space heating/cooling, water heating, cooking in the residential and commerical sector.

The technique of MEDEE-2 is simple: most of the relationships are linear combinations of variables and the model is used as a straightforward accounting framework. The resulting secondary fuel mix together with constraints on the maximum build-up rates, cost of new energy supply and conversion facilities and resource availability constraints is then inserted into the second submodel, called MESSAGE

(Model for Energy Supply System Alternatives and their General Environmental impact). MESSAGE is , like MARKAL, a time-dependent linear programming model which provides an optimum allocation of fuels to meet a given demand. It is a dynamic model and allows the explicit treatment of interfuel substitution, which takes place over time in the energy supply and conversion sector.

The third submodel, IMPACT, is a dynamic input-output based algorithm, which determines the impacts of a certain strategy on the economy in terms of:

- o Investments in energy system capacities,
- o Capacity build-up in energy related sectors of industry and corresponding capital investments,
- o Requirements for materials, equipment and services for construction and operation of the energy system and related industrial branches.

With IMPACT calculated costs, the economic feasibility of a strategy can be checked, e.g. whether or not energy will absorb unacceptably high portions of the economic products, or what amount of non-energy exports are necessary to compensate for energy imports etc.. Finally, the MACRO submodel calculates aggregated investment and consumption patterns based upon IMPACT provided cost data. This in turn leads to a revised computation of economic growth rates, which is checked with the original assumption and reentered into a new iteration loop.

It is this very broad concept of iterations within the computation routes which provide for consistent scenarios. If the full set of models are employed in iterations, we have in fact a general equilibrium approach for interactions between economic and energy sector activities.

IIASA's energy modelling set is not designed for energy planning purposes but aims at investigating the longer term perspectives for transitions to energy supply systems in a resource constrained world. It was applied in a well known study of the development of world regions between now and 2030 giving special attention to the different needs and possibilities of western industrialized countries, communist areas, developing countries and less developed countries /9/.

4. DECISION MAKING AND ENERGY MODELS

This is where the development and application of energy models stands today. I believe that the energy modelling community can look back upon a tremendously fast development over the last ten years. Great advances can be reported, such as:

- the development of models for many different issues in the energy policy and planning area

- the availability of large scale models of the entire energy system as well as of models that describe the interaction between the energy sector and the rest of the economy

- the availability of improved data bases and modelling techniques, as well as extremely powerful computers and modelling software.

But are these advances sufficient?

Is it not so,

- thatmost of the energy policy decisions and the strategic decisions in the energy industry arenot based on the outcome of an energy modelling analysis,

- that energy modellers do not have much to offer when complex real world problems require a quick answer,

- that the treatment of uncertainty, which during the last years has become the major issue in the planning process, is still unsatisfactory from the decision making point of view.

So what did the energy modellers do wrong? Nothing as yet, I believe. They developed a variety of efficient and powerful models in a reasonable short time. Methodological improvements are still possible, but as useful energy models are available yet, the attitudes of the energy modelling community must be shifted from the development of new and more detailed models to the application of the models to help to solve the problems the decision makers are confronted with.

Let me now outline some ideas how the situation can be improved.

The appreciation of energy models by the so called decision makers is characterized by up and downs. The initial phase of suspicion and skepticism that was based on ignorance was followed by a phase of overconfidence and high expectations. During that time the models, especially computer models were viewed to be able to provide answers to any question; to be not a tool for making up our minds, but the answer itself. As it turned out that the predictive power of the various energy models was not sufficient to be of empirical values in the light of events, overconfidence turned into disillusionment. Since some years we are in the phase of disillusionment. What is at stake now is

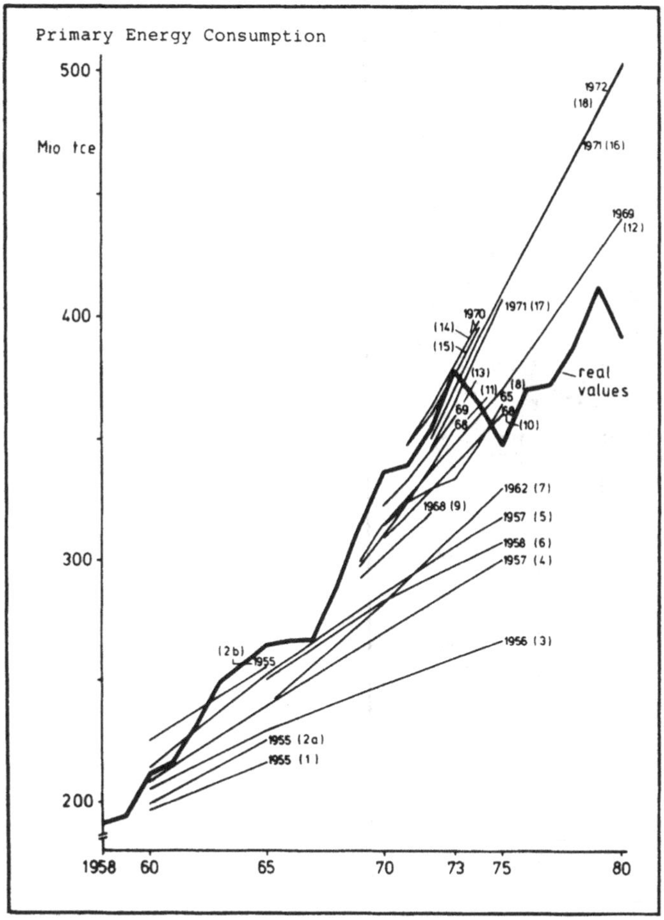

Fig. 11: Energy Forecasts (1955-1973)

to overcome the present distrust and to regain credibility. Otherwise the danger is great that energy models will never contribute to better decisions in energy policy and the energy industry.

I believe that models and modellers must adopt a more issue-oriented approach and that expectations on both sides must be reduced to what can be provided by an energy model analysis. Energy models have often been employed to provide precise numerical forecasts of the future development of the energy system. But energy forecasting is a hazardous occupation. Virtually any projection turned out to be incorrect /10/.

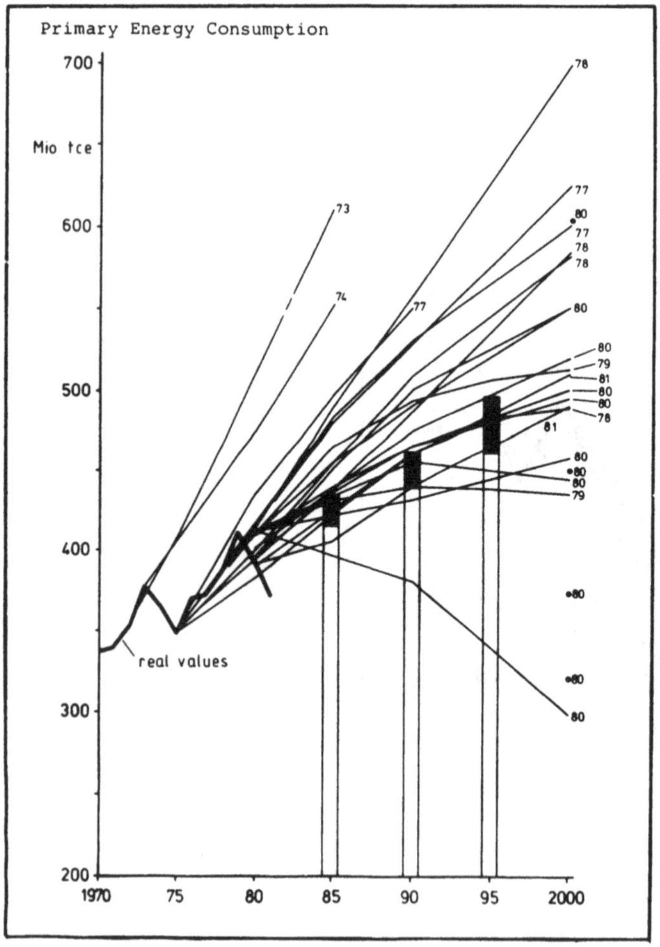

Fig. 12: Energy Forecasts (1973-1981)

Fig. 11 shows the primary energy forecast for the Federal Republic of Germany, which were published in the period from 1950 to 1972. Compared with the actual development, all forecasts turned out to be wrong. The increase of the primary energy consumption was underestimated by the forecasts of the 50's and 60's.

In Fig. 12 the primary energy forecasts published after the first oil crises in 1973 are illustrated. The figures for the primary energy consumption of the year 2000 differ by about a factor of two. Without going into further details, I think this figure demonstrates that their succes in forecasting the energy future will not be greater than that of the earlier forecasts in the 50's and 60's.

To state the point more clearly, I think that history has shown, that we can not expect any precise forecasts of the future, even if we employ very detailed and sophisticated models.

The reason for this is, that the development of the main factors determining future energy demand and supply, such as the economic growth rates or the price of crude oil, to mention only two, is to a great extent uncertain. Opinions for example about the future oil price development have changed in recent years dramatically during relatively short periods of time. The range of long term oil prices estimated published since 1973 reaches from 15 $ to 150 $ per barrel. And a recent analysis of the IIASA about the oil price estimates used in the most up-to-date long-term energy projections throughout the world showed, that the individual oil price estimates for the year 2010 differ by factor of three /11/.

Some energy modellers and energy analysisists have reacted to the increased uncertainty by generating several scenarios with different assumptions about the uncertain factors. Concerning the world oil prices uncertainty is usually reflected by assuming two or three annual growth rates, low, moderate and high. The usual recommendation to the decision maker then is: We'll give you the results under these scenarios and you make your own choice. But where does this leave the decision maker? It seem to me that this kind of analysis is not very helpful to him. If it is not possible to be more precise about the oil price development, then at least he should be provided with the information how this uncertain factors influence his near-term decisions, or with an indication of those nearterm decisions that are insensitive to these assumptions.

For the use of energy models this does mean, that rather asking what the energy demand in some future year will be, or what the contribution of different supply options in the year 2000 will be, the appropriate question is, what must an energy policy look like, if it has to be robust and flexible enough to cope with the uncertanties that lie ahead?

If energy models are to aid in decision-making, then it cannot be a meaningful aim to try to forecast the future development of the energy system. However carefully the forecast is made, the inherent uncertainty lying in the future cannot be removed. Rather the task consists in identifying with the help of the energy model and after explicit consideration of the uncertainties, what I would like to call "robust" decision steps. These are those steps relevant to the near future, that give the best possible guarantee, that the path chosen will not have been regretted at a much later point of time /12/.

I believe, that this different view of how to use energy models to provide useful information to the decision making process is a prerequisite to regain credibility and promote a more fruitful interaction between the decision makers and the model builders.

Models in general and energy models specifically should not be viewed as tools, that will predict the future more accurately. But with models we may be able to understand better the interdependances and influences of various factors - both, those that are within our control and those that are not. Making use of these potential benefits of energy models requires, that they are viewed by both the energy modellers and the decision makers as tools for developing insights rather than for forecasting numbers.

REFERENCES

/1/ Charpentier, J. P., A review of energy models No. 1, RP-74-10, IIASA, Laxenburg, Austria (1974)

/2/ Charpentier, J. P., A review of energy models No. 2, RP-75-35, IIASA, Laxenburg, Austria (1975)

/3/ Charpentier, J. P. and J. M. Beaujean, A review of energy models No. 3, RP-76-18, IIASA, Laxenburg, Austria (1976)

/4/ Rath-Nagel, S and K. Stocks, Energy Modelling for Technology Assessment: the MARKAL Approach, OMEGA, Vol. 10, No. 5 (1982)

/5/ A Group Strategy for Energy Research, Development and Demonstration, International Energy Agency, OECD, Paris (1980)

/6/ Manne, A. S., R. G. Richels and J. P. Weyant, Energy policy modelling: a survey, Operations Res. 1 (1979)

/7/ Manne, A. S., ETA-MACRO: A User's Guide, Electric Power Research Institute, EA-1724, Palo Alto, USA (1981)

/8/ Basile, P. S., The IIASA set of energy models: its design and application, IIASA-RP-80-31, Laxenburg, Austria (1980)

/9/ Häfele, W., Energy in a finite world, Ballinger Publishing Company, Cambridge, Massachusetts (1981)

/10/ Voss, A., Energieprognosen - Überflüssig oder notwendig? Brennstoff-Wärme-Kraft 35, Nr. 5 (1983)

/11/ Manne, A. S. and L. Schrattenhölzer, International Energy Workshop: A Summary of the 1983 Poll Responses, IIASA, Laxenburg, Austria (1983)

/12/ Voss, A., Nutzen und Grenzen von Energiemodellen - Einige grundsätzliche Überlegungen, Angewandte Systemanalyse, Vol. 3, No. 3 (1982)

THE ESSENTIALS OF HIERARCHICAL CONTROL

Władysław Findeisen
Institute of Automatic Control
Technical University of Warsaw
00-665 Warszawa, Poland

ABSTRACT: This paper describes the most essential features of the various decision situations that can arise whenever a system is under control of multiple decision units. Although it has been arranged in a different manner and it differs in its focal points, this paper follows a more detailed suvey of the field by the same author [Findeisen, 1982]

1. DISPERSED AUTHORITY

Let us assume that the behaviour of a system can be controlled by decisions (control actions) $c_1, \ldots, c_N$, each of the c_i belonging to the jurisdiction of a separate control unit (decision maker). We call them local decision units (LU_i) and refer to the whole as a decision structure with dispersed authority, Fig. 1..

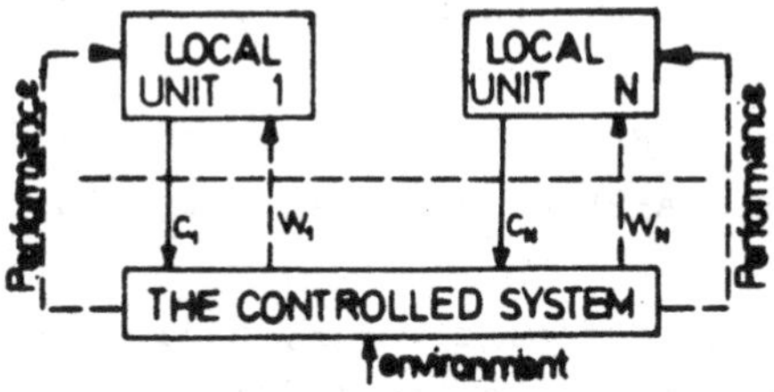

Fig.1. Organization - multiple decision units in control of a system. c_i - decision, w_i - information

We shall assume in the sequel that the local decision makers are conscious of their individual goals which means that the set of local decision problems can be put down as

$$\begin{aligned} &\underset{c_1}{\text{maximize}}\ Q_1\ (c_1, \ldots, c_N) \\ &\ldots \\ &\underset{c_N}{\text{maximize}}\ Q_N\ (c_1, \ldots, c_N) \end{aligned} \qquad (1)$$

where we had to assume that the local benefit Q_i which will be achieved in the system may depend on all local decisions $c_1, \ldots, c_N$, not only on the particular local decision c_i.

It is easy to note that the above formulation fails to be the most general: we should have started by saying that each local decision unit perceives its "local interests", which need not be maximization of something, and particular need not be maximization of a scalar benefit Q_i. Let us have this in mind and treat (1) as a simple example only of local interests.

The structure of Fig. 1, in which the goals (1) are pursued by local decision units, cannot be rationally judged upon until we say something of the overall interests that is until we find somebody who is interested in the system as a whole.

2. OVERALL INTERESTS

The arrangement of several decision units who exercise influence on a system common to them all may be referred to as an organization; the question is whether this organization has an overall goal or interests as opposed to the local goals or interests mentioned in the previous section?

There are two typical requirements to be posed by or before the organization:

(i) the system should be stable, that is the actions of individual local decision units should lead to a stable equilibrium point,

(ii) the stable equilibrium point should be satisfactory with respect to some overall goals or goal, or even optimal with respect to an overall goal.

The way in which the overall interests could be secured very much depends on two factors, namely on the structure of the controlled system and on the relation between tha individual and overall goals.

3. CONSISTENCY OR DISAGREEMENT OF INTERESTS

3.1. System and subsystems

We consider now that the controlled system known from Fig. 1 is a collection of subsystems described by input-output relations

$$y_i = F_i\,(c_i, u_i, z_i) \tag{2}$$

where y_i is output, c_i is control (decision), z_i is disturbance (environment) input and u_i is an input which may have different meanings, depending on which of the cases given below takes place.

The first of the cases is shown in Fig. 2.

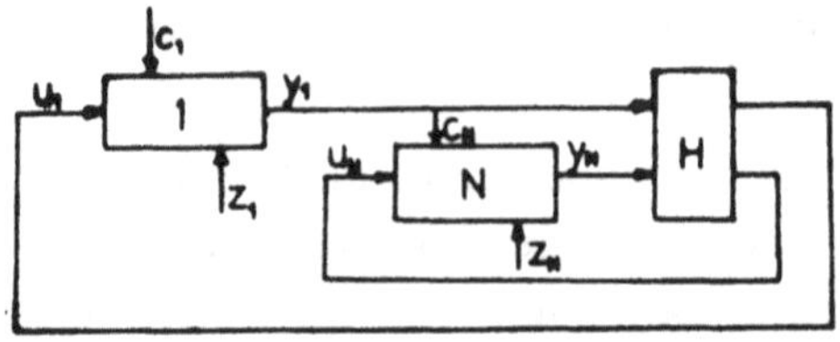

ig.2. Interconnected subsystems, $u_i = H_i y$.

ubsystems are interconnected in a way that may be described by intercon- ection matrix $H = [H_1,\ldots,H_N]^T$. The output $y = (y_1,\ldots,y_N)$ becomes function of all decisions and all disturbaces, and so does every input u_i

$$y = K(c,z), \qquad u_i = H_i K(c,z) \tag{3}$$

here K is determined by H and all F_i .

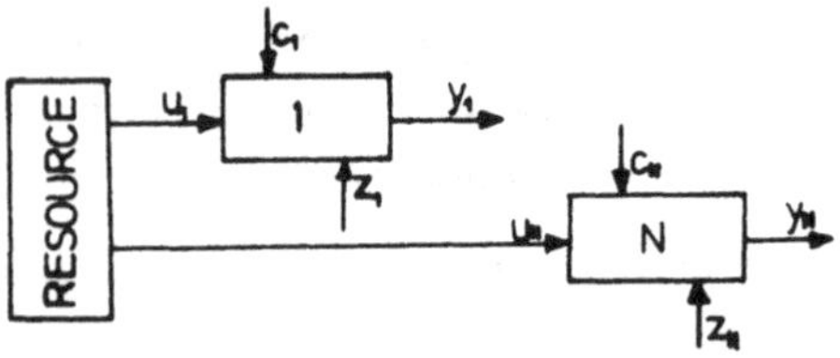

ig.3. Subsystems sharing a resource

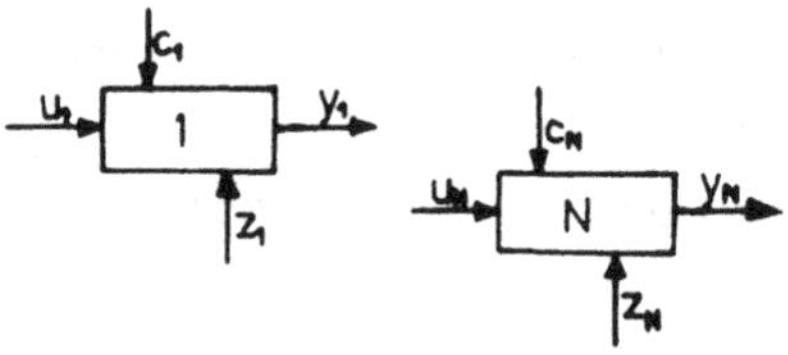

Fig.4. Independent subsystems

igure 3 shows the second case. Subsystems are not interconnected, the u_i re "resource input" and become local control decisions. The resource is mited, hence the decisions u_i are bound by a global constraint

$$u_i + \ldots\ldots + u_N \leqslant u_o \tag{4}$$

igure 4 shows the third case. Subsystems are not interconnected, resour- e inputs are limited.

ow turn to the description of the local decision problems. There will be local benefit function

$$Q_i^o\,(c_i,u_i,y_i) \tag{5}$$

s well as some local constraints, say in the form

$$(c_i,u_i) \in CU_i(z_i) \tag{6}$$

$$y_i \in Y_i \tag{7}$$

Equation (5) should be considered carefully, it says that the local benefit takes account of the cost of decision (control action c_i), cost of input u_i (note that in Figs. 3 and 4 it is a decision as well), and value of output y_i as locally accounted.

When relation (2) describing the subsystem is substituted in (5), the local benefit function becomes

$$Q_i(c_i, u_i, z_i) \tag{8}$$

This implicit form will be used most frequently in this paper.

It can be seen that in the interconnected system of Fig. 2, by virtue of (3) the local benefit function becomes a function of all controls $c_1,\ldots\ldots,c_N$ and disturbances $z_1,\ldots\ldots,z_N$

$$Q_i(c_i, H_iK(c,z), z_i) \tag{9}$$

In the Fig. 3 system the local problems are independent as long as resource constraint (4) is inactive. In general, however, the local problems are coupled in a way which may be expressed by saying that the feasible set U_i of local decision u_i depends on all other decisions u_j, $j \neq i$. One can write it as a local constraint

$$u_i \in U_i(u) \tag{10}$$

When the constraint (10) is active, the membership relation (10) becomes an equality

$$u_i = g_i(u) \tag{11}$$

and this equality substituted into (8) gives a local benefit function

$$Q_i(c_i, g_i(u), z_i) \tag{12}$$

that is once again a dependence upon decisions of all local units.

Let us turn now to the overall interests, or the interests of the organization. The resource constraint (4) was in fact an overall requirement; another example could be the requirement that some function defined on the outputs

$$J\ (y_1,\ldots\ldots,y_N) \tag{13}$$

be maximized or kept above a certain limit value. In more general terms we should state, however, that the overall benefit function would be defined on all local variables

$$J^o(c_1,\ldots,c_N,\ u_1,\ldots\ldots,u_N,\ y_1,\ldots\ldots,y_N) = J^o(c,u,y) \tag{14}$$

which gives, when the relations (2) are substituted

$$J(c_1,\ldots\ldots,c_N,\ u_1,\ldots\ldots,u_N,\ z_1,\ldots\ldots,z_N) = J(c,u,z) \tag{15}$$

Note that while the systems of Figs 2 and 3 have to be treated as "systems" by virtue of the existing interconnection and/or resource constraints, the subsystems shown in Fig. 4 are made "a system" exclusively by the existence of the overall goal (14) or (15).

We face now the following situation: the system of Fig..2, 3 or 4 is being governed by local decision units as shown in Fig. 1, we are conscious of the overall interests, what are the results?

3.2. BEHAVIOUR OF A SIMPLE SYSTEM

Let us consider the system of Fig. 4, where there are no subsystem interconnections and no limitation of resources. Assume there exist local decision units 1,...,N and compare the following three cases:

(i) the local decision makers are interested in local results only, so that the local decision problems are

$$\underset{c_i,\, u_i}{\text{maximize}}\; Q_i^o\,(c_i, u_i, y_i) \qquad i = 1,\ldots\ldots,N \tag{16}$$

(ii) the local decision makers are interested in each other´s results ("the results are common"):

$$\begin{aligned} &\underset{c_1,\, u_1}{\text{maximize}}\; Q_1^o\,(c_1, u_1, y_1,\ldots\ldots,y_N) \\ &\cdot\ \cdots\ \cdots\ \cdot \\ &\underset{c_N,\, u_N}{\text{maximize}}\; Q_N^o\,(c_N, u_N, y_1,\ldots\ldots,y_N) \end{aligned} \tag{17}$$

(iii) the goals of all local decision makers are identical:

$$\underset{c_i,\, u_i}{\text{maximize}}\; Q_o\,(c_1,\ldots\ldots,c_N;\; u_1,\ldots\ldots,u_N;\; y_1,\ldots\ldots,y_N) \qquad i = 1,\ldots\ldots, N \tag{18}$$

Case (i) corresponds to a true independence of local decision makers.

In case (ii), which is presented schematically in Fig. 5, the local decision makers are interrelated by their interests.

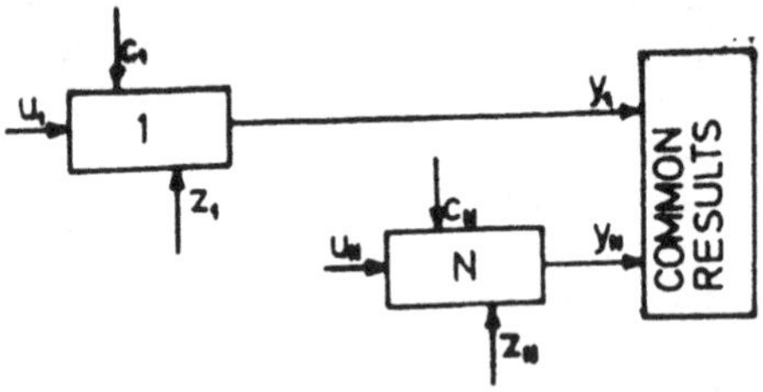

Fig. 5. The case where the results are common.

Due to the fact that the benefit function $Q_i^o(\cdot)$ of each local decision maker depends on the other outputs as well, eqn. (17), we face a typical <u>N-person game</u> situation. It can have a stable equilibrium point or not, there may be various strategies and coalitions to be advised to local decision makers on Fig. 5, and so on.

Case (iii) is a very special one: the goals of all local decision makers are identical. We can think of the present situation as of N decision makers not serving separate (although identical) goals, but one goal common to them all. The decision makers in that case will be referred to as a <u>team.</u> If we now decide to implement local decision making by means of decision rules x/ we are able to use <u>team theory,</u> originated by Marschak and Radner, [see Marschak and Radner, 1972; Ho and Chu, 1972]. The main point is that local decision makers who have a common goal have no incentive to compete and will agree to use decision rules that are team - optimal i.e. optimal from the common point of view.

Moreover, in many cases the team-optimal decision rule is also person-by-person-satisfactory that is it can be designed in a decentralized (local) way. In order to use decision rules one would have to make observations on the system or the system environment, or both. In the basic theory of teams one assumes that the local decision makers may not have the same information, although in principle they observe the same environment, see Fig. 6.

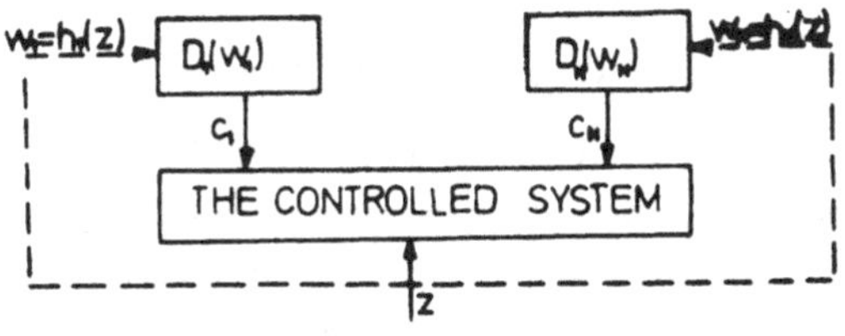

Fig.6. A team implementig decision rules.

---

x/ A decision rule is a function which assigns decision c to observation w, c=D(w). The local decision maker would be asked (or assumed) to implement decisions c_i=$D_i(w_i)$..Local observation w_i is assumed to be in a relation $w_i = h_i(z)$ to the environment input z to the whole system (Fig.6).

Decision rules are a common tool in the control theory developed in engineering applications; feedback structures, that is systems where the observation is carrying results on decisions that were taken, dominate in that framework, see [Sandell and coauthors, 1978; Śiljak, 1978; Singh, 1977] and many other places.

Up to now we did not consider any "overall goal" with respect to the single system just being discussed, except perhaps the requirement that the processes should come to an equilibrium.

Let us re-consider case (i), which was marked by a full independence of local decision makers, by adding that somebody else is interested in the common results, for example wishing to maximize an overall benefit defined on the outputs

$$\text{maximize } Q^{*} (y_1, \ldots, y_N) \tag{19}$$

If the new person had part of the decisions in his hand, he would obviously be in an N+1 - person game with all the local decision makers; he would also be in game with local decision makers of case (ii), as well as with the team of case (iii).

The new person may choose not to play the game, that is not to intervene , if the local decisions of case (i), or the equilibrium point of case (ii), or the team decision functions of case (iii) are such that the goal (19) is met in a satisfactory way.

It is obvious, for example, that if the overall goal would be identical with that of the team, case (iii), no intervention is needed. It is also easy to see that if for case (i), independent local problems, the overall goal would be

$$\text{maximize } \Psi (Q_1(c_1, u_1, y_1), \ldots, Q_N(c_N, u_N, y_N)) \tag{20}$$

where Ψ is an order - preserving function, then the locally - best decisions c_i, u_i serve at the same time best overall interests and no intervention would improve them. We say that the local and overall interests are <u>consistent</u> . In the other case they would be in <u>disagreement.</u>

In case the passive attitude is not accepeted, we must give to the new person some power. Starting at this time we will refer to him as a <u>supremal decision making unit</u>, or a <u>coordinator</u> because his aim is to coordinate decisions made by local or lower level units so as to favour the overall interests.

3.3. A SIMPLE CASE OF COORDINATION

Let us continue to consider the system of Fig. 4 and the simplest case (i) of local decision makers, eqn. (16). Assume the supremal unit pursues the goal described as (19). Let us give to him the power to prescribe (to de-

mand) the outputs y_i, $i = 1,\dots,N$. Taking it very superficially, the goal (19) can be fully optimized in that scheme, and on top of it the local decision makers would still have some freedom to act, if a required value of y_i can be obtained by various combinations of c_i, u_i. The approach will be realistic, however, only if we consider that not every value of the local outputs y_i is achieveble because of the subsystem constraints. Those constraints would have to be known to the supremal unit and taken into account when prescribing the val ues $y_1,\dots,y_N$. A rich and reliable information is required at the supremal level - a severe practical drawback of the approach.
The above would be referred to as direct method of coordination - prescribing values of some of the local variables.
The message of our example is that even in this simple case, where the subsystems are not interconnected in any way, there may exist the need to coordinate decisions and actions of local decision makers.
We are passing now to more complex situations, where the other classical method, price coordination, may as well be appropriate.

3.4. COMPETING FOR RESOURCES

Let us now assume that it is the system of Fig. 3 which is being governed by N local decision units. Assume that the decisions u_i are left to their discretion, then the set of local decision problems becomes as it was in eqn. (16)

$$\underset{c_i,u_i}{\text{maximize}}\ Q_i^o\,(c_i,\ u_i,\ y_i)$$

with the important difference that the decisions u_i are subject to the constraint of limited resources, eqn. (4).
From the point of view of a local decision maker, his decision u_i may prove to be non feasible, namely in the case when the other decisions on drawing on the resources would have gone too far.
There are two ways to think of the present situation. The first is to distinguish between the programming phase of the decisions and the execution phase. In the programming phase the local decision makers would negotiate with each other, as if being in a game. until they come to proposed decisions u_i that do not violate the constraints. Then the decisions would be executed. The other way is to set up a supremal decision unit that would allocate the scarce resource to lower level units, thus eliminating (neutralizing) the competition among them.
The supremal unit can set any values $u_1,\dots,u_N$ which satisfy the resource constraint; therefore there is an opportunity to set values which are best

from the point of view of the supremal's interests. Assume the supremal unit wants to maximize a goal defined on all system variables, so that the local and supremal problems are as follows (we drop dependence of Q_i on y_i for simplicity):

lower level (subsystems)

$$\underset{c_i \in C_i(u_i)}{\text{maximize}} \quad Q_i\ (c_i,\ u_i), \qquad i = 1,\ldots\ldots,N \tag{21}$$

supremal level (coordinator)

$$\underset{u \in U}{\text{maximize}} \quad Q\ (c_1,\ u_1,\ \ldots\ldots,\ c_N,\ u_N) \tag{22}$$

where $u \in U$ means $u_1 + \ldots\ldots + u_N \leqslant u_o$ and by $c_i \in C_i(u_i)$ we mean that decision c_i may be constrained in a way, which may depend on the allocated resource u_i.

The competition among local units is eliminated, but there is instead a game situation between the supremal unit on one side and the collection of local problems on the other.

In the situation of limited resource it will be natural to give some special right to the supremal unit - he will be the <u>leader,</u> the person who makes the first move. Thus one has an application for the theory of leader - follower games.

Consider the case in which the supremal unit knows lower level decision problems. In that case the supremal can predict the response c_i, of a local unit to decision u_i:

$$\hat{c}_i(u_i) = \arg \max_{c_i \in C_i(u_i)} Q_i(c_i,\ u_i) \tag{23}$$

and then it can find its own optimal decision by solving the problem

$$\underset{u \in U}{\text{maximize}} \quad Q\ (\hat{c}(u),\ u) \tag{24}$$

A part of theory of leader-follower games considers the case where result of (23) is a set $C_{Ri}(u_i)$ rather then a single -valued function of u_i. In such cases the supremal problem will rather be of the guaranteed result type

$$\underset{u \in U}{\text{maximize}} \quad \min_{c \in C_R(u)} Q(c,\ u) \tag{25}$$

where $C_R = C_{R1} \times \ldots\ldots \times C_{RN}$.

An example where the leader must count that the response of the follower will be somewhere within some set C_{Ri} is the case where the follower performs a satisfying approach rather than optimizing (he will then choose any c_i which gives a result $Q_i \geqslant Q_{i0}$, where Q_{i0} is his aspiration level), or the case where the follower performs multi-objective optimization, for which his choice of c_i cannot be predicted in a unique way.

The class of games with right to first move described above is called Stackelberg games [Chen and Cruz, 1972; Cruz, 1978] , or G1 games [Germeyer, 1976] .

In the case where $\hat{c}(u)$ is unique (single-valued), it is known that the result obtained by the supremal is equal or better than the Nash equilibrium solution of the same game

$$Q(\hat{c}(u), \hat{u}) \geqslant Q(c^N, u^N) \tag{26}$$

In the G1 game the "first move" of the supremal is his decision u, the followers make their decisions c_i thereafter. A further extension would be G2, G3 ..... games [Germeyer, 1976; Gorelik and Kononenko, 1982].

The G2 game differs from G1 in that the first move of the supremal unit is to reveal its decision rules $\hat{u}_i(c_i)$. Then the decisions $\hat{c}_i$ of the lower level units are taken on the basis of that information, leading ultimately to supremal decisions $\hat{u}_i(\hat{c}_i)$. Game G3 goes one step further, and so on.

Let us now consider the special case of goal consistency, that is the case where the supremal problem could be put down as

$$\underset{u \in U}{\text{maximize}} \quad \Psi\,(Q_1(c_1, u_1), \; .... \;, Q_N(c_N, u_N)) \tag{27}$$

where meaning of Ψ is the same is in (20). In that case it is known [see for example Findeisen and co-workers, 1980] that the supremal's result is fully optimal, i,e. the same as if he would maximize (27) by making the decisions c_i himself, with no intermediary of local decision units.

By allocating the resources in form of $u_1, \; .... \;, u_N$ we used the direct method of coordination; let us now consider how to use prices for the same purpose. We start again with the disagreement case, that is where the supremal goal Q differs from (27).

Assume that the supremal unit uses price η on the inputs u_i in the framework of a G1 game. Thus, the supremal calculates responses of lower level units

$$(\hat{c}_i(\eta), \hat{u}_i(\eta)) = \arg \max_{c_i, u_i} \left[Q_i(c_i, u_i) - \langle \eta, u_i \rangle \right] \tag{28}$$

and determines the optimal price $\hat{\eta}$ from

$$\underset{\eta}{\text{maximize}} \quad Q(\hat{c}(\eta), \hat{u}(\eta)) \tag{29}$$

subject to resource constraint $\hat{u}_1(\eta) + \; ..... \; + \hat{u}_N(\eta) \leqslant u_0$.

The problems (28) and (29) should be compared to (23) and (24), where direct coordination was assumed. One should note that the response $\hat{u}_i(\eta)$ is usually referred to as "demand function" and is often used in economics. A particular value $\tilde{\eta}$, such that the resource constraint is active, would be referred to as the resource-balancing price. It is known that with this price the local units are induced into decisions $(\hat{c}_i, \hat{u}_i)$ which maximize the sum $\sum_i Q_i(c_i, u_i)$. This is not, however, the goal of the supremal in the disagreement case which we are now considering. The supremal could induce the local units into somewhat different decisions, if he would use a different price η_i, for each the local units; in that case a weighted sum of Q_i's would be optimized, but still not the goal Q of the supremal.

The direct allocation of u_i can lead the local decision units into some other decisions c_i than the ones obtained with price coordination and it may happen that those decisions are more favourable for the supremal. It is, however, impossible for the supremal, in the disagreement case, to cause decisions c_i which strctly optimize the supremal benefit $Q(c, u)$.

To avoid confusion it should be stated that if the supremal benefit function does not depend on c, that is, one has $Q = Q_0(u)$, then direct coordination will strictly optimize the supremal benefit and also preserve the constraint $u_1 + \dots + u_N \leqslant u_0$. If the supremal benefit is

$$Q = Q_0(u) + \langle \eta_1, u_1 \rangle + \dots + \langle \eta_N, u_N \rangle \tag{30}$$

that is the supremal is cashing on the quantities $u_1, \dots, u_N$ allocated to the subsystems, then <u>one</u> price for all customers can optimize Q and preserve the resource constraint.

3.5. THE INTERCONNECTED SYSTEM: DIRECT COORDINATION

Let us now consider the system of Fig. 2, where the inputs of the subsystems are such as determined by outputs of other subsystems. Assume that a local decision maker has power to influence his subsystem by means of control varibles c_i and he is interested in maximizing his local goal. Even if this local goal is defined on local variables only, as $Q_i^0(c_i, u_i, y_i)$ or - equivalently - $Q_i(c_i, u_i)$, we realize that the value of local benefit which will be achieved may as well depend on all other decisions c_j, $j \neq i$, because the input u_i is dictated by other subsystems. We stated this fact at the beginning of the paper, eqn. (1).

The conclusion from (1) must be that the local decision makers are in a game with each other, with all the consequences concerning existence and stability of equilibrium point, its value, strategies to achieve it, etc. Local decision makers harm each other, are in conflict. It can be neutralized if we introduce a supremal unit that would govern over the interaction variables u_i or y_i, $i = 1, \dots, N$.

The simplest way for coordinator to neutralize the conflicts among local problems that arise because of the interconnections is to prescribe or "freeze" the interconnection variables that is the values u_i and y_i in the whole system. This would be referred to as "direct coordination", because its

instruments are directly the interconnection variables.
Prescribing the values of interconnection variables, y_d, separates the local decision problems as far as the interconnections shown in Fig. 2 are concerned. If the subsystems are also sharing a resource, then a similar separating effect will be obtained with respect to resource interdependence by prescribing, for each sybsystem, the amount r_{di} of the given resource that it is allowed to use; in what follows we drop the resource constraint for the sake of simplicity. We drop also indicating the dependecies on disturbance z_i, in order to simplify notation.
Thus a local decision problem can be put down as [cf (8), (2) and (6)], given the desired output y_d:

$$\text{maximize} \quad Q_i(c_i, u_i) \tag{31}$$

subject to

$$u_i = H_i y_d$$

$$F_i(c_i, u_i) = y_{di}$$

$$(c_i, u_i) \in CU_i$$

The solution to this problem will depend on the parameter y_d (as well as on the value z_i, which we now omit) that is one should have $\hat{c}_i(y_d)$ and $\hat{Q}_i(y_d)$.
The supremal decision problem will be - in the goal consistency case -

$$\underset{y_d}{\text{maximize}} \; Q = \Psi(\hat{Q}_1(y_d), \ldots, \hat{Q}_N(y_d)) \tag{32}$$

From the mathematical point of view the main difficulty of the direct method lies in the fact that a local problem may have no solution for some y_d because of the constraints; an output value may not be achievable or the allocated resources inadequate, or both.. Therefore, the value y_d set by the coordinator must be such that the local problems have solutions, which we express as a requirement

$$y_d \in Y \tag{33}$$

where Y is the set of feasible decisions of the coordinator.
This set cannot be easily determined because it implicitly depends on local equations, constraints and disturbances.
Let us turn now to the disagreement case, that is let us assume that the supremal goal cannot be presented as in (32) but we have to write it as

$$Q^o(c_1, u_1, y_1, \ldots, c_N, u_N, y_N) \tag{34}$$

which means that the supremal goal is defined directly on all system variables, attributing to them values which may differ from the values attributed to those variables by the local benefit functions Q_i. The supremal optimization problem becomes

$$\underset{y_d \in Y}{\text{maximize}}\ Q^0(\hat{c}_1(y_d), H_1 y_d, y_{d1}, \dots, \hat{c}_N(y_d), H_N y_d, y_{dN}) \qquad (35)$$

where $c_i(y_d)$ means solution to local problem (31) under the imposed value of output y_d and actual disturbance z_i.
The difference with respect to what was described before is that the result obtained by the supremal (the maximum value of his benefit Q^0 achieved by adjusting the coordination variable y_d) will be less than would be obtainable if the supremal could himself make the local decisions $c_1,\dots,c_N$; he fails to "preserve authority".. As opposed to it, in the goal consistency case the supremal would gain nothing by reaching himself to decisions $c_1,\dots,c_N$, the local decision makers were "following his ideas" adequately.
Let us note, that if in disagreement case the supremal goal were $Q^0(u,y)$, that is defined on the inputs-outputs only, the supremal would reach his optimum whatever the behaviour of the local units. This means, however, a case where local decisions c_i are of no cost or importance to the supremal; a case that hardly meets practical conditions.

3.6. THE INTERCONNECTED SYSTEM: PRICE COORDINATION

Let us consider the same system of Fig. 2 and the same local goals as in the preceding subsection. Assume we attempt to neutralize the competition situation among the lower level units by introducing a supremal unit, who will be given the right to set a price λ on the inputs u, where the price vector λ is composed of price vectors λ_i on every input u_i.
The main assumption of the price coordination method is that the local decision problems can be modified into

$$\text{maximize } Q_{\text{mod } i} = Q_i(c_i,u_i) - \langle \lambda_i, u_i \rangle + \langle \mu_i, y_i \rangle \qquad (36a)$$

where price λ is imposed by the supremal and the price on the output is

$$\mu_i = \sum_{j=1}^{N} H_{ji}^T \lambda_j \qquad (36b)$$

In other words, it is assumed that the local decision maker accepts to maximize the benefit function $Q_{\text{mod } i}$ rather than his "original" function Q_i.
When the local problem is formulated as (36), its solutions depend on λ .
The local decision maker is asked to determine not only $\hat{c}_i(\lambda)$ but also the desired input $\hat{u}_i(\lambda)$, subject only to his local constraint

$$(c_i,u_i) \in CU_i$$

As a consequence, the local unit determines its optimal output $\hat{y}_i(\lambda) = F_i(\hat{c}_i,\hat{u}_i)$.
The coordinator has to ensure the input-output matching (the <u>interaction balance</u>)

$$\hat{u}(\lambda) - H\hat{y}(\lambda) = 0 \qquad (37)$$

The price $\hat{\lambda}$ that satisfies (37) is called equilibrium price. The equilibrium price is known to maximize the sum of local benefits, $Q_1(\hat{c}_1,\hat{u}_1) + \dots + Q_N(\hat{c}_N, \hat{u}_N)$.

Hence, if the global benefit is a sum of local benefits (that is if Ψ in (32) is a sum), than performing the first task,i.e. neutralizing the competition by means of prices, the coordinator ensures also overall optimility (an extension to weighted sum of the Q_i's is possible, by means of differing the prices on the same thing for different local units).

It would be valuable to discuss the goal disagreement case, that is the case where the supremal is not interested in optimizing the sum $\sum_i Q_i(c_i,u_i)$ but some other goal, such as given by (34). The supremal will no longer be interested in getting equilibrium prices; however, at a non-equilibrium price the subsystems of Fig. 2 receive real inputs u_{*i} which differ from their model-based expectations u_i. Nevertheless, one can imagine operation of the local decision makers on the basis of locally measured u_{*i}, as mentioned in [Findeisen 1982] . This case seems to deserve further exploration, although one should be aware of the limitation of the interconnection scheme of Fig. 2. The subsystems in that structure are not separated by any "storages", which means that a subsystem must accept the input u_i which is being offered to him (as well as to pay for it according to price λ_i). This model may be adequate for a technological process, may fail to represent properties of a market system - in the latter case one should probably be very careful when thinking of non-equilibrium prices.

3.7. A SUMMARY OF THE CONSIDERED SITUATIONS

We can look at what has been described in this section as dealing, at first, with two questions:

(i) what is the system to be controlled (Fig.. 2, 3 or 4)?

(ii) what are the goals of local decision makers?

The first question concerns system structure (Fig. 2, 3 or 4), that is to what extent do the subsystems depend "physically" on each other. The second question concerns the individual local goals, which - as was presented, in subsection 3.2, may be:

(a) determined by local results only, eqn. (16)

(b) interfering - local decision makers interested in each other's result eqn. (17)

(c) identical, thus offering the possibility for team oper ation, eqn. (18).

The combination of system structure and local goals contributes to system behaviour under no coordination, that is the behaviour under control by local decision makers only. At this point overall interests have to come into picture: if the system is unstable, or if its equilibr ium point is not satisfactory from the overall point of view, something has to be done about it. Thus, in fact, we added two more questions:

(III) how is the system behaving under no coordination?

(IV) what are the overall interests?

and according to the answers to the latter two questions we may come to conclusion that a supremal unit with the task of coordinating the lower level decisions has to be introduced into the decision making structure - this structure will now be a hierarchical one, Fig. 7.

Having decided on introduction of the supremal unit, we ought to design its operation, above all we have to determine:

(V) what should be the goal of supremal unit?

(VI) what instruments (direct or price) will be used for coordination?

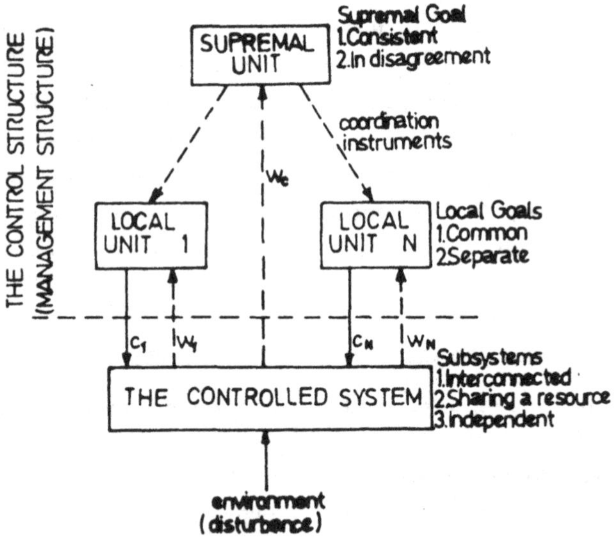

Fig. 7 The hierarchical control system.

From all the above it must follow that there is a large number of alternative situations, resulting from the various structures, local goals, overall interests, coordination methods - too many to be presented in a short paper. Moreover, several of the possible cases have hardly been investigated in the available theories, at least not to the same degree as the other cases. It must be realized, for example, that most of the available "hierarchical control theory" deals with the goal consistency case [Mesarovic, Macko and Takahara, 1970 Dirickx and Jennergren, 1979; Findeisen and co-workers, 1980], or just assumes the "common goal" situation leading to acceptance of team decision function or feedback rules [Marschak and Radner, 1972; Singh, 1977; Michel and Miller 1977; Sandell et.al., 1978; Siljak, 1978]. The class of cases where the local and supremal goals are in disagreement is much less elaborated upon; the relevant theory of "hierarchical games" seems to be quite far from completeness, cf [Germeyer, 1976; Burkov, 1977; Cruz, 1978; Gorelik and Kononenko, 1982 .

4. DECISIONS AND INFORMATION

The preceding section concentrated on the decision structure, that is on the question "who makes what decision?" If a supremal unit is used, the structure is hierarchical, as depicted in Fig. 7. Let us now eleborate on the next question: on what basis (what information) will a given decision be based, and how will the decision making in the given structure be organized?

It may be useful to start by a few remarks on decisions and the decision making nomenclature. A decision may be defined as purposeful selection of one alternative (one element) from a set of alternatives (elements). It has been indicated before that the "purposeful selection" can be expressed by a decision rule, which points out the right decision c for given observation w, or it can be performed by solving an explicit decision problem, i.e. a problem involving the benefit function or functions, the constraints etc., with the observation w as a parameter.

Which ever is the case, one should distinguish single-stage decisions and multi-stage decisions. In the latter case a decision made at time t_i takes into account that the opportunity will be given to make another decision at time t_{i+1}. As a result the decision at t_i may be less cautious or less hedged against uncertainty than would be needed if no later decision was envisaged. It seems obvious that multi-stage decisions are particularly appropriate when uncertainty is dominant, for example when the future environmetal conditions are largely unknown. It should not be inferred that a single-stage decision cannot be repeated; the point is that in multi-stage decision making one is conscious at time t_i that another decision will be made at t_{i+1} (note that time t_{i+1} need not be directly specified, but may be made dependent on a deviation in observation of the system or of the environment).

The existence of a decision rule almost automatically means an opportunity of multi-stage decision making (it can in fact be continuous in time) while in the problem - solving form of making decisions one usually has to specify whether a next decision will be made and on what incentive.

Single and multi-stage decision making is related to uncertainty in the decision problem, but not to the feature of the controlled system being static or dynamic. It also does not overlap with the distinctinction between open and closed-loop control: there may exist multi-stage open-loop control of a static or dynamic system, where control would be adjusted at times t_i, t_{i+1}, etc. to new observations of the environment only.
However, when behaviour and performance of the system plus its control is investigated, a division into open and closed-loop (feedback) structures is most relevant. Only the closed-loop structures can be unstable, and his dangerous possibility is the price to be paid for observing and using the results of past decisions in making the new ones.
There is another dimension in decision making that needs consideration, in particular in system where human decision makers are involved. A distinction of the programming phase for a decision, and of the execution (action) phase, where the prepared decision has been accepted and finally implemented, is necessary. The programming phase in a multiple-decision maker system involes an exchange of information and of "planned decisions". We mentioned it already with reference to the competition for resources, the case considered in subsection 3.4.
Let use now consider some of decision and information problems that arise in hierarchical systems, in particular the problems to be faced by the supremal unit, who decided to coordinate decisions made by lower level units.
Assume first that the direct method of coordination is used and the supremal unit is prescribing the outputs, cf. subsections 3.3 or 3.5. We stressed in the context that one should prescribr feasible outputs, that is such as can be attained in spite of all the local constraints and disturbances. This immediately produces the requirement of rather rich information on the subsystems in the hand of supremal unit. To be more specific, let us look at the decision problems formulated as (31) and (32), which reflect the two-level decision structure for the interconnected system. The process of determining a decision could be conceived as cosisting of two steps. In the first step the coordinator determines y_d, in the second step - the local decision units determine their c_i. This kind of operation does not seem very practical, since the coordinator would have to know the functional relationships $Q_i(y_d)$, the constraint set (33), and also the actual values of all disturbances. Since the constraint set (33), cannot be explicitly given except for very simple cases, all the above reduces to saying that the coordinator would have to know all local models and also the actual values of all z_i. It would be, therefore, a case of centralized information, it would probably be discarded in most applications.
The alternative is to use an iterative procedure, a process of negotitions

between the supremal and the lower level decision units. In this process, the coordinator would propose some values y_d and receive responses in form of the values of benefits Q_i. He does not have to know the local disturbances, not shape of the functions $Q_i(y_d)$. Nevertheless, he will have to keep his demands and allocations in the feasible set Y. Since he does not know neither the local models nor the disturbances, he would have to keep his decisions y_d in a "safe" region of Y, such that y_d are feasible even in the worst case of system uncertainty. This practical difficulty and disadvantage can be overcome by use of more sophisticated iterative procedure, where the local units would have the opportunity to reveal their possibilities or impossibilities to the coordinator in the consecutive steps of the iteration.
One of such approaches to coordination is the penalty function method, its idea being to use penalty type formulation in the local problems while imposing there the coordinator demands. There are also many more methods, developed in mathematical programming for effective iterative coordination of supremal and lower level problems using direct coordination instruments see Findeisen and co-workers, 1980, for references .
One can obviously refer to the coordination procedures described above as being in fact the programming phase of the decision, where the decision itself is the set of optimal values $(c_1,\ldots\ldots,c_N)$.
This programming phase consists of iterations between the supremal and infimal decision units. Their decision problems, in particular (31), may be solved on the basis of models which are formal (mathematical), or which are partly or entirely judgmental. It is not relevant for the decision structure.
In the scheme described, the decisions $(c_1,\ldots\ldots,c_N)$ were arrived at without any measurements, iterations or trials on the real system. One is, therefore, dealing with open-loop control.
It is characteristic for open-loop control that is does not have to make great distinction between static and dynamic problems; it does not have to worry, for example, whether the measurements correctly reveal the state of the system, as no feedback measurements at all are used to shape the decision.
Open-loop control would be based exclusively on observatins of the environment i.e. on measurements of the disturbance variable z. In that way, performing the programming phase of the decision as described above that is on the basis of explicity stated decision problems (31) and (32), is equivalent to implementing a decision rule: "find decision c for given z". The main differences are the following:

(I) the "decision rule" contained implicitly in decision problems (31) and (32) is much more flexible, i.e. one can change constraints, preferences, information structures, etc.;

(II) the explicit decision problems e.g. the formulation of local goals, are easier to be accepted and followed by human decision makers than the decision rules.

The decision-problem based decision making in a hierarchical structure will, most naturally, be discrete in time. In that case the open-loop decision at time t_i will always be reaching into the future, that is it will be taken with some forecast of the disturbance z in mind. As already mentioned if no particular time or event is considered in the future, at which another decision is known to be made, the decision at time t_i is single-stage, i.e. it is taken, conceptually, for the whole foreseeable future. However, most often we plan for future decision making.

It is time now to ask the question whether the lower level units, in the programming phase of the decisions, would reveal their true possibilities. Unfortunately, whenever the local decision units can recognize (distinguish) their own interests, i.e. their own benefit function, it is likely that they can be better off by misleading the coordinator.

Assume a local decision problem is

$$\begin{aligned} &\text{maximize} \quad Q_i(c_i, u_i) \\ &\text{subject to} \\ &\qquad (c_i, u_i) \in CU_i \end{aligned}$$

Assume now that u_i is given (the case of direct coordination). For this u_i, there exists an optimal c_i^x which maximizes Q_i. However, in the direct method of coordination the coordinator requires also an output

$$F_i(c_i, u_i) = y_{di}$$

If the c_i needed to produce this output differs from c_i^x, the local benefit will be lower than it could be under c_i^x. The local decision maker is, therefore, interested in preventing the coordinator from imposing y_{di}. He can do it by supplying the supremal with false information on the feasible set CU_i, namely by reporting it to be rather narrow (from the point of view of local benefit it is convenient to say that the set of all possible c_i is "concentrated" around the value c_i^x). This kind of cheating is known as "hiding the production capacities" or "hiding the activity levels" and is difficult to discover by the supremal decision unit [Green, Laffont, 1979; Ho, Luh and Olsder, 1982] . Needless to say that this or related kind of cheating can only be done if the supremal does not know the subsystem models and/or the values of local disturbances, and thus he has to rely on the data supplied by lower level units.

Cheating is not related to goal disagreement; the case described above would apply to goal consistency as well. The point is that any single local goal can be increased, at the cost of the overall goal.

The price coordination method is, in general, less demanding as far as information to be available to the supremal is concerned. Its main idea is to find equilibrium prices, cf. subsection 3.6. The supremal, who dictates the prices, need not know the subsystem constraints nor the local goals - that information can be private. All he has to know are "demand" and "supply" responses, $\hat{u}_i$ and $\hat{y}_i$, of the local decision makers. For an interconnected system the process to determine equilibrium prices - under the above mentioned privacy of information - has to be iterative. This iteration belongs to the programming phase of the actual decisions, which will be the equilibrium price $\hat{\lambda}$ and the local decisions $\hat{c}_1,\dots,\hat{c}_N$.

The mathematical conditions of existence of equilibrium prices have been extensively treated in the economic literature, or in the literature of mathematical programming as the existence conditions of Lagranian multipliers [see Findeisen and co-workers, 1980, for references] .

When the equilibrium price exists, the above mentioned iterative procedure of decision making will have to be able to find it. Here we require, that the lower level solutions $\hat{c}$, $\hat{u}$ be single-valued functions of the parameter λ, which is a vector of prices. This requirement has a simple interpretation: since the prices aim at providing a match of the outputs to the inputs of other subsystems, as well as at preservation of the resource constraint, they should have a well-defined influence on the local decisions and thus also on their respective results.

5. BEHAVIOUR UNDER UNCERTAINTY

It has been stressed, throughout this paper, that the subsystems are under the influence of some disturbance z_i. We may look at z_i as an actual external (varying) input to the subsystem, or as a parameter - which is constant but may be not exactly known. The performance of the direct and price coordination methods may differ considerably when we consider that the supremal unit decision was adjusted to wrong z_i, $i = 1,\dots\dots,N$.

Assume that at change of z_i the local units readjust their decisions, while the supremal unit keeps its coordination instruments at their previous values.

With the direct method, if it was used to allocate the resource and/or prescribe the subsystem outputs, we can expect that the local use of resources will be as it was allocated, or even below the limit, and we can expect also that local decision makers will try hard to keep to the prescribed outputs y_{di}. According to the local problem formulation, this "trying hard" may mean a local

cost which is far from what the supremal unit would otherwise like to tolerate. In summarizing: resources not overspent, outputs kept ("stable operation of the system"), but overall performance may be far from optimality.
In the price method, the danger is that the resources will be overused that is the global constraint will be violated, and also that all outputs and inputs in the system will change to values not predicted previously. The certainty of "stable operation" will disappear.. However, as there are no "strict orders" of the coordinator to be followed, there is also less danger of very uneconomic decisions in the subsystems..Thus, the direct coordination is safe but may become uneconomic, the price coordination is economic but may become unsafe.
The reader will note that even in the last discussion the assumption that the local units operate open-loop was kept to, that is they believe in their models and their estimates of z_i and do not measure the results of their decisions. The scheme is shown in Fig. 8.

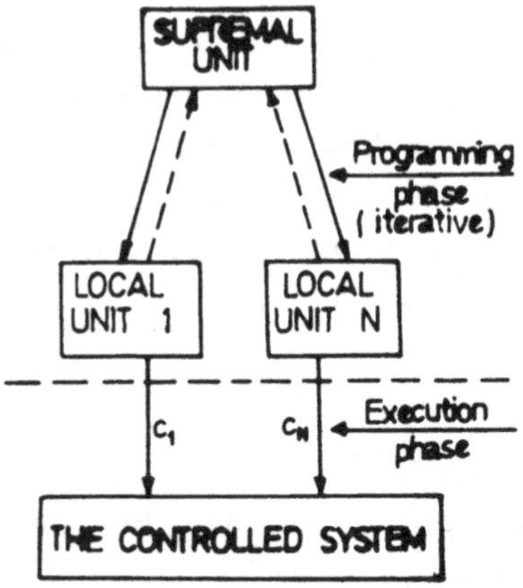

Fig. 8.. Open-loop (planning type) implementation of coordination methods.

There will be, as explained before, a programming phase in which the supremal and local units will agree on the inputs and outputs or on the prices, respectively, depending on which coordination metod is used. This phase must be iterative, unless a centralized information structure is assumed, where the supremal unit would possess all models and data available to local units. When the programming phase agreement is reached, the local decisions $c_1,\dots\dots,c_N$ can be implemented; this is execution phase.. Figure 8 can be referred to as a planning-type implementation. A mathematical model of the itera-

tive programming phase in decision making is given by decomposition methods in optimization [Dantzig, 1963; Kornai and Liptak, 1965; Lasdon, 1970; Tatjewski, 1978;, Dirickx and Jennergren, 1979] .
The open-loop control structure is of course somewhat artificial in human decision making, where the actual results of decisions are usually at least partly available. An appropriate use of feedback is also reasonable from the point of view of control theory, which suggests that, more often than not, it will improve performance of a system which is subject to uncertain inputs. There are several ways of using feedback information in the hierarchical decision making framework. Their discussuon would go beyond the intended scope of this paper. Let us note only that feedback information (in the form of measured outputs, for example) can be used by the coordinator, or by local units, or both; it may be used in connection with direct coordination instruments or with price coordination. This gives several alternative information and decision structures. Moreover, the use of feedback differs significantly for static and dynamic systems, respectively. The reader is referred to [Findeisen,1982] for generalities and to [Findeisen and co-workers, 1980] for more detailed discussion of the problem.

6. CONCLUDING REMARKS

The starting point for discussion in this paper was the multiple decision maker situation: we have assumed existence of local decision makers and assumed also that they can identify their local goals. Then, depending on those goals and on the controlled system structure (interconnections between the subsystems), various overall system behaviour may result. This behaviour has to be confronted with the overall interests: what do we demand from the system as a whole. The overall demands may lead to a necessity to introduce supremal decision making unit, which would essentially serve two purposes, neutralizing the lower level conflict and trying to get overall performance to satisfy a supremal goal, both by means of coordination instruments.
Knowing the principles of coordination, one has to consider the information aspect (centralized or decentralized information) and the disturbance (uncertainty) aspect. The relevant considerations cast additional light on the decision making problem, leading to distinction between programming phase and execution phase of a decision, as well as to the use of feedback information while controlling or managing an operating system.
As an important limitation, let us stress that the decision problems at the supremal and local levels have usually been considered to be scalar optimization; this is an important limitation of most of the theories. While the coordina-

tion mechanisms need, practically, that the lower level problems have unique solutions, it will certainly be important to consider supremal problems that would be multiobjective choice or would use the satisfacing approach.

The models and theories touched upon in this paper can be looked at from at least two points of view. First, they can be used as models to describe the functioning of control or decision systems; they display, for example, the advantages and disadvantages of direct versus price coordination in a functioning economic system, as well as some phenomena that must appear in the operation of planning mechanism..Once the models can be trusted to describe behaviour of a class of multiple decision unit systems, one can try to make the second use of the available theory to assist the design of decision making structures. This may mean specyfing the existence and role of supremal unit, choice of coordination instruments, the mode of their use, choice of control time horizon, frequencies of intervention, etc.

Little can yet be said of the practical applications of hierarchical control theory for the design of control or decision structures; there is no doubt, however, that the cognitive aspects, the improved understanding of the mechanisms, have helped people to better designs-even if not formally performed. One should be aware, in this context, of a rather severe limitation of the models and theories presented in this paper; it has been assumed throughout that a decision unit behaves rationally, that is, it is consistent in pursuing its goal or implementing the decision rule once accepted. In many human organizations this assumption may be too far from reality.

REFERENCES

Burkov, V.N.(1977). Osnovy matematicheskoy teorii aktivnych sistem (Foundations of Mathematical Theory of Active Systems), Nauka, Moskva

Chen, C.I. and J.B. Cruz (1972). Stackelberg solution for two-person games with biased information patterns. IEEE Trans. Aut.Control, AC-17, 791

Cruz, J.B. (1978). Leader-follower strategies for multilievel systems. IEEE Trans. Aut. Control. AC-23, 244

Dantzig, G. (1963). Linear Programming and Extensions. Princeton University Press, Princeton

Dirickx, Y.M.I. and L.P. Jennergren (1979). Multilevel System Analysis: Theory and Applications, John Wiley, London

Findeisen, W., F.N. Bailey, M. Brdyś, K. Malinowski, P. Tatjewski and A. Woźniak (1980), Control and Coordination in Hierarchical Systems, John Wiley, London

Findeisen, W. (1982). Decentralized and hierarchical control under consistency or disagreement of interests. Automatica, Vol . 18, 647

Germeyer, Yu.B. (1976). Igry s neprotivopolozhnymi interesami (Games with non-antagonistic interests). Nauka, Moskva

Gorelik, V.A. and A.F. Kononenko (1982). Teoretico-igrovye modeli prinyatya reshenii v ekologo-ekonomicheskich sistemackh (Game-theoretic models of decision making in ecologic-economic systems). Radio i Svyaz, Moskva

Green, R. and J.J. Laffont (1979). Incentives in Public Decision-Making, North Holland, Amsterdam

Ho, Y.C. and K.Ch. Chu (1972). Team decision theory and information structures in optimal control problems. IEEE Trans Aut. Control, AC-17, 15-22 (Part I), 22-28 (Part II)

Ho, Y.C., P.B. Luh and G.J. Olsder (1982). A control-theoretic view of incentives. Automatica, Vol. 18, 167

Kornai, J. and T. Liptak (1965).. Two-level planning. Econometrica, 33, 141

Lasdon, L.S. (1970).. Optimization Theory for Large Systems. McMillan, London

Marschak, J. and R. Radner (1972). Economic Theory of Teams. Yale University Press, New Haven

Mesarovic, M.D., D. Macko and Y. Takahara (1970). Theory of Hierarchical Multilevel Systems. Academic Press, New York

Michel, A.N. and R.K. Miller (1977). Qualitative Analysis of Large Scale Dynamic Systems. Academic Press, New York

Sandell, N.R.Jr., P. Varaiya, M. Athans and M.G. Safonov (1978). Survey of decentralized control methods for large scale systems. IEEE Trans Aut. Control, AC-23, 108

Siljak, D.D. (1978). Large-scal Dynamic Systems: Stability and Structure. North - Holland, New York

Singh, M.G. (1977). Dynamical Hierarchical Control. North-Holland, Amsterdam

Tatjewski, P. (1978). A penalty function approach to coordination in multilevel optimization problems. RAIRO, 12, 221

NEW DEVELOPMENTS IN ECONOMETRIC COMMODITY MARKET MODELING:
A MODEL OF THE WORLD COPPER MARKET

G. Wagenhals
Alfred Weber-Institut
University of Heidelberg
D 6900 Heidelberg
Federal Republic of Germany

1. INTRODUCTION

This paper deals with a new econometric model of the world copper market. It describes its main features in comparison with other copper market models and it reports about some validation experiments with the model. There are many possible applications of the model. It is used currently in a study for the Kiel Institute of World Economics to assess the effects of copper production from deep sea floor manganese nodules on land-based copper production, on the consumption and the prices of copper.

The model consists of some 60 structural equations which explain the behavior of the market participants in the world copper industry. This makes it the most highly disaggregated world copper market model currently available. It is a system of interdependent, sometimes nonlinear simultaneous difference equations, whose parameters were estimated based on annual data from 1955 to 1980. In comparison with other copper market momodels, this model has some distinctive features:

- primary supply functions are not derived from a partial adjustment approach, as in virtually all econometric copper market models, but through factor demand and restricted profit functions,
- copper mine production capacities are introduced explicitly, derived from the hypothesis that producers act to maximize their discounted net cash flow,
- the copper world market price, i.e. the London Metal Exchange price for electrolytic copper, is determined by a dynamic stock disequilibrium approach,
- a copper futures price equation is estimated for the London Metal Exchange, and it is used in
- a rational expectations approach, which is applied in modeling the inventory behavior.

Due to space restrictions, the paper deals only with the primary supply side of the model.

2. SUPPLY FUNCTIONS

Here I derive primary copper supply equations from the hypothesis that the producers maximize their profits given their short-run capacity constraints.

Institutional studies of the copper market indicate that the gestation period in mine production capacity expansions is considerable, at least a year in any case, even for short-run expansions. Thus, for any econometric model based on annual data, like the current one, it is reasonable to assume that copper producers consider their mine capacity to be given in the short run, and that they maximize their profits under this constraint.

Due to data limitations, especially - but not only - for the developing copper exporting economies, I assume that a generalized Cobb-Douglas production function sufficiently describes the technology of minerals production:

$$Q = \gamma V^{\alpha} K^{\beta},$$

where Q is the amount of copper mined, V is the amount of variable input, K is the mine production capacity, and α, β and γ denote constants.

The world copper market is essentially competitive, disregarding certain oligopolistic structures in the North American copper industry. The history of copper pricing shows that no single copper producing country has been able to influence the copper world market price by its actions effectively. We therefore assume that each producer is a competitive price-taker and that he chooses the amount of output Q and the amount of the variable factor input V which maximize his profits.

The necessary condition resulting from this sample optimization problem is $\alpha PQ = CV$,
where P is the output price and C denotes the unit cost of the variable factor. Therefore we obtain the factor demand function

$$V = (\alpha\gamma)^{\delta} P^{\delta} C^{-\delta} K^{\beta\delta},$$

where $\delta := 1/(1-\alpha)$.

This optimal factor demand gives the restricted gross profit function

$$\pi(P, C; K) = \gamma^{\delta}(\alpha^{\alpha\delta} - \alpha^{\delta}) P^{\delta} C^{-\alpha\delta} K^{\beta\delta},$$

where the costs of capacity expansions are neglected, because they are fixed in the short run and thus do not influence the producer's short-term decisions.

Assuming all prices to be positive, the supply function of a copper producer is given by the partial derivative of the gross profit function with respect to output price ("Hotelling's lemma")

$$Q = Q(P, C; K) = \frac{\partial \pi(P, C; K)}{\partial P} = \delta\gamma^{\alpha}(\alpha^{\alpha\delta} - \alpha^{\delta})(\frac{P}{C})^{\alpha\delta}K^{\beta\delta},$$

and therefore

$$\log Q = \alpha_1 + \alpha_2 \log K + \alpha_3 \log (\frac{P}{C}),$$

where

$$\alpha_1 := \log(\delta\gamma^{\delta}(\alpha^{\alpha\delta} - \alpha^{\delta})), \quad \alpha_2 := \beta\delta \text{ and } \alpha_3 := \alpha\delta.$$

Eventually, we include a stochastic error term u_t representing e.g. stochastic deviations from profit maximization and the effects of omitted variables. Therefore, the estimating equation can be written

$$\log Q_t = \alpha_1 + \alpha_2 \log K_t + \alpha_3 \log (\frac{P_t}{C_t}) + u_t.$$

The parameters of such primary supply equations were estimated for the eight most important copper mining market economies (Canada, Chile, Peru, the Philippines, South Africa, the United States, Zaire and Zambia), as well as for the rest of the Western World together. For the centrally planned economies a slightly different approach was chosen due to the lack of capacity data.

As a typical example for a primary copper supply equation, we present the estimated equation for Zaire's mine production.

Equation 1: Estimated mine production equation, Zaire

$$\log QMZI = \underset{(1.39)}{.320} + \underset{(33.4)}{.868} \log QMZIC - \underset{(-6.97)}{.149} DZI$$
$$+ \underset{(3.12)}{.0562} \log ((PCULME*REXZI)/(REXUK*COSTZI)),$$

$$\bar{R}^2 = .981, \qquad DW = 2.49,$$

where

QMZI = copper mine production, Zaire, in 1000 tons,

QMZIC = copper mine production capacity, Zaire, in 1000 tons,

PCULME = average annual price of copper, electrolytic copper, wirebars, London Metal Exchange, cash, Pound sterling per ton,

REXZI = exchange rate, Zaires per U.S. Dollar,

REXUK = exchange rate, Pound sterling per U.S. Dollar,

COSTZI = unit cost of mining index, Zaire, 1975 = 1.00,
DZI = dummy variable for Katanga war, 1 in 1963-1965, 0 otherwise.

The figures in parantheses are the t-statistics, $\bar{R}^2$ denotes the adjusted coefficient of determination and DW is the Durbin-Watson coefficient. The signs of the coefficients are as expected from economic theory; all coefficients are significantly different from zero and have magnitudes which are plausible a priori.

The complete estimation results are presented in Wagenhals [3]. Summing up, they show that not only the coefficients of Zaire's mine production equation, but all coefficients of all estimated primary supply equations have the expected signs and, with the exception of two price elasticities of supply (South Africa and "other market economies"), all coefficients are significantly different from zero (at the 5 % level).

Apart from the Katanga war, only serious labor strikes in Canada and in the United States had to be accounted for by dummy dependent variables. These, however, are the only dummy variables in the primary supply equations, contrary to most econometric copper market models with a similar degree of disaggregation.

Primary copper supply is determined mainly by available capacity. We now turn to the explanation and estimation of

3. COPPER MINE PRODUCTION CAPACITIES

The derivation of mine production capacity equations assumes that copper producers act to maximize the net worth of their firms, i.e. the present value of all future net revenues accruing to their firm over time (see Hall, Jorgenson [2], Coen [1]).

Then, for a competitive copper producer, the marginal net cash flow can be written

$$Q_K P - cd - t(Q_K P - D_i),$$

where

Q_K = marginal product of capital,
P = copper price,
c = purchase price of an additional unit of capital services,
d = rate of economic depreciation of capital goods per period,

t = corporate tax rate, and

D_i = increase in depreciation charges for tax purposes in period i.

Thus, the discounted marginal net revenue is

$$\frac{1}{r}((1-t)Q_K P - cd) + t\sum_{i=1}^{\infty}(1+r)^{-i}D_i,$$

where r is the interest rate.

We denote by B the present value of depreciation deduction on one dollar's investment

$$B = \sum_{i=1}^{\infty}(1+r)^{-i}d_i,$$

where d_i is the amount of tax depreciation permitted on an investment of one dollar i periods after the investment has been made. Then, we can show that

$$\sum_{i=1}^{\infty}(1+r)^{-i}D_i = cB(1+\frac{d}{r}).$$

We assume that the capacity is proportional to the capital stock, and that the capital stock is proportional to the flow of capital services. Then the producer increases his capacity, if the marginal discounted net revenue exceeds the purchase price of an additional unit of capital services, i.e. if

$$\frac{1}{r}((1-t)Q_K P - cd) + tcB(1+\frac{d}{r}) > c.$$

Solving for the value of the marginal product of capital, this inequality may be written equivalently

$$Q_K P > c(r+d)(1-tB)/(1-t) \quad := U,$$

where U is the implicit rental price of a unit of capital per period, i.e. the user cost of capital.

Introducing an investment tax credit and accounting for the Long Amendment gives the equation for the user cost of capital in the United States

$$U = c(r+d)(1-s-t(1-ms)B)/(1-t),$$

where s is the investment tax credit rate and m is a dummy variable, which equals 1 in 1962 and 1963, and 0 otherwise.

To calculate U, we first have to determine B, the present value of depreciation deduction resulting from a current dollar of capital expenditures. B depends on the depreciation method used.

Straight line depreciation is obligatory for most copper producing companies. In this case, the deduction is constant over the lifetime for tax purposes τ, and therefore the present value of the deduction is

$$B = \frac{1}{r\tau} (1 - e^{-r\tau}).$$

Only for the United States B is calculated according to the sum of the years' digits depreciation method. In this case, the present value of the deduction is

$$B = \frac{2}{r\tau} (1 - \frac{1 - e^{-r\tau}}{r\tau}).$$

We assume that a copper producer acts to maximize his discounted net cash flow given the Cobb-Douglas technology described above. Then, in period t, the desired capacity K_t^* is determined by the marginal productivity condition for capital, i.e. the value of the marginal product of capital equals the user cost of capital

$$K_t^* = \beta \frac{P_t Q_t}{U_t},$$

where

P_t = copper price,
Q_t = copper mine production,
U_t = user cost of capital (in period t respectively), and
β is a constant.

Generally, the desired capacity is different from the existing capacity, because copper producers cannot adjust to the optimal capacity level immediately. Therefore we assume a flexible accelerator hypothesis of stock adjustments, which gives

$$K_t = \lambda K_t^* + (1-\lambda) K_{t-1}.$$

Combining the last two equations leads to

$$K_t = \beta_0 + \beta_1 K_{t-1} + \beta_2 \frac{P_t Q_t}{U_t},$$

where $\beta_0 := 0$, $\beta_1 := 1-\lambda$ and $\beta_2 := \lambda\beta$.

Finally, a five-year sample variance of deflated copper prices σ_t^2 (to account for price risks influencing capacity decisions) and a stochastic disturbance term v_t are added. Therefore, the estimating equation reads:

$$K_t = \beta_0 + \beta_1 K_{t-1} + \beta_2 \frac{P_t Q_t}{U_t} + \beta_3 \sigma_t^2 + v_t,$$

where β_3 is an additional constant parameter.

The coefficients of copper mine production capacity equations like that were estimated for the eight most important copper producers and for the rest of the market economies together.

As a typical example, we present the mine production capacity equation for Zaire.

<u>Equation 2:</u> Estimated mine production capacity equation, Zaire

$$\begin{aligned} QMZIC = {} & \underset{(-.252)}{-5.72} + \underset{(23.8)}{.995}\, QMZIC_{-1} \\ & + \underset{(1.89)}{.00525}\, PCULME*QMZI*REXZI/(REXUK*UCZI) \\ & - \underset{(-.048)}{.0108}\, R \quad - \underset{(-1.98)}{45.5}\, D78, \end{aligned}$$

$$\bar{R}^2 = .964, \qquad DW = 1.24,$$

where

R = five year sample variance of PCULME/(REXUK*PCIF),

PCIF = international price index, unit values of manufactures (SITC 5-8), 1975 = 100,

UCZI = user cost of capital index, mining, Zaire, 1975 = 100, and

QMZI, QMZIC, PCULME, REXZI and REXUK are explained in Equation 1.

Equation 2 shows that Zaire's lagged capacity level is the main determinant of the current level. The mine capacity adjusts only very slowly to price changes. Zaire appears to be slightly risk-averse, the coefficient of the "risk variable" R is not significantly different from zero, however. The price elasticity of mine capacity is small, but significantly different from zero at the 5% level.

By and large, these results are typical for all estimated copper mine production equations. To sum up the main results:

1. With only one exception, the constants in all copper mine production capacity equations are not significantly different from zero. This supports the above a priori assumption of discounted net cash flow maximization, because it is an implication of this hypothesis as we saw above.

2. The coefficients of the lagged endogenous variables are highly significant in all equations. They always have the expected signs and magnitudes.
3. Although developing countries always adjust to price changes in the direction expected a priori, mine capacities generally adapt very slowly and only weakly. In contrast to these results, the capacity changes of the main copper producers among the industrialized market economies, namely of Canada and the United States, very highly depend on prices: the respective price elasticities are significantly different from zero at the 0.5% level. Thus, the industrialized copper producing countries react far more on price incentives than the developing economies.
4. The evidence in regard to the five-year sample price variance as a proxy for a risk variable is mixed. Generally, copper producers tend to be risk averse.

Apart from the equations described above, secondary supply equations, demand equations, price equations and an East-West trade equation were estimated. Finally, some identities close the model.

4. HISTORICAL DYNAMIC SOLUTION

To validate the model, we performed a historical dynamic simulation based on the values of the exogenous variables from 1956 to 1980 and based on the starting values of the endogenous variables in 1955. The model was solved simultaneously with the historical values of the exogenous variables and with the values predicted by the model for the lagged endogenous variables.

Calculation of many goodness of fit measures for this historical dynamic solution suggested that the model traces the main historical developments of the world copper market quite well. It captures most of the turning points in the time paths of the variables since the mid 1950s.

A series of dynamic multiplier simulation experiments, which reflected the dynamic response properties of the models, confirmed that the development of the world copper industry is described reasonably and sufficiently by our model (see Wagenhals [3], Chapter 10).

Summing up, the model is a useful tool for analytical purposes. It therefore has been used to perform a large variety of other simulation experiments and for forecasting purposes, for example, to assess the impact of copper production from manganese nodules (see Wagenhals [4]).

5. REFERENCES

Coen, R.M. [1]
"Effects of Tax Policy on Investment in Manufacturing." American Economic Review, Papers and Proceedings, 58: 1968, pp. 200-11.

Hall, R.E., Jorgenson, D.W. [2]
"Tax Policy and Investment Behavior." American Economic Review, 57: 1967, pp. 391-414.

Wagenhals, G. [3]
The World Copper Market: Structure and Econometric Model. Heidelberg: Alfred Weber-Institut, University of Heidelberg, mimeo, June 1983.

Wagenhals, G. [4]
The Impact of Copper Production from Manganese Nodules. Final Report for the Kiel Institute of World Economics, Heidelberg: Alfred Weber-Institut, University of Heidelberg, November 1983.

THE GREAT RECESSION: A CRISIS IN PARAMETERS?

U. Heilemann and H.J. Münch
Rheinisch-Westfälisches Institut für Wirtschaftsforschung
Hohenzollernstraße 1-3, D-4300 Essen 1, FRG

Beginning with the first oil crisis, the economic development in West Germany and most other industrialized countries is no longer what it used to be: growth rates sharply diminished, unemployment became a burning problem and inflation-rates did hardly respond to the slackening demand. But it came even worse: what in the mid-seventies had been called stagflation turned into a severe recession in the eighties. And even if the picture improves in the near future, the mid-term and long-term prospects are in no way very promising. As usual, it took some time until the crisis and its severity was widely recognized. But while there is some consensus as to the statistical diagnosis, there is still a wide range of opinions as to the origins (and to the therapy) of the crisis. With the risk of simplifying too much, in Germany (and elsewhere) two extreme views can be separated: one view which attributes the current problems to the growing inflexibility of the system, at least partly caused by too many direct and indirect government activities; the other view, which does not see the origin of the difficulties in changing behaviour of the economic agents which the first view tends to emphasize, but in dramatic changes of some exogenous variables, such as the various supply and demand shocks of the seventies.

The present paper examines the empirical evidence of the first view for the FRG. It tests whether there were parameter shifts during the seventies which were of such a magnitude that they can explain the bad economic performance of the last ten years. The paper is organized as follows: Section 1 presents the hypotheses to be tested, section 2 describes the data and methods used, section 3 analyses the results obtained. The paper is closed by a résumé.

1. HYPOTHESES

Generally, the assertion that something has changed during the seventies is not made with reference to specific parameters or variables. However, it seems quite natural to assume that these changes should have happened in the consumption area, the investment area and the price area. Of course, there are more areas which have obviously been hit by fundamental changes such as the area of international transactions by the transition from fixed to flexible exchange rates, or the monetary sector by the implementation of a monetary-target based policy. The limited space, however, requires to restrict our analysis to changes in consumer, investor and price behaviour, without losing the other factors out of sight.

The parameters to be tested are taken from the RWI-business cycle model [8], a quarterly econometric model which has been used for short-term forecasts and simulations since

1977. The consumption function employed in this model is a modified FRIEDMAN consumption function of the form:

$$CP70_t = \beta_1 + \beta_2 YPV70_t + \beta_3\ ZINSK_t + \beta_4\ DS1 + \beta_5\ DS2 + \beta_6\ DS3 + \beta_7\ CP70_{t-1} + u_t$$

with

CP70 : private consumption, real terms (1970 prices);

YPV70: personal disposable income, real terms;

ZINSK: short-term interest rate;

DS1-3: seasonal (dummy-) variables;

β_1-β_7: parameters;

u : stochastic term.

The investment function to be tested is of the flexible accelerator-type, making allowance for the influence of capital and wage costs:

$$IAU70_t = \beta_1 + \beta_2\ (CP70_t + IAN70_t + EX70_t) + \beta_3\ (ZINSL\text{-}PBSPJW)_{t-2} + \beta_4\ LSTK_{t-3} + \beta_5\ DS1 + \beta_6\ DS2 + \beta_7\ DS3 + u_t$$

with

IAU70 : investments in machinery and tools, real terms;

IAN70 : fixed investments, real terms;

EX70 : exports, real terms;

ZINSL : long term interest rate;

PBSPJW: growth rate of GNP-price index;

LSTK : wage costs per unit GNP (real terms).

To test the constancy of the price-behaviour we have chosen the function explaining the price index for private consumption. The function is based on a modified mark-up-theory with adaptive expectations:

$$PCP_t = \beta_1 + \beta_2\ \left(\frac{1}{4}\sum_{j=2}^{5} LDR_{t-j}\right) + \beta_3\ \left(\frac{1}{4}\sum_{j=0}^{3} KAPA_{t-j}\right) + \beta_4\ PIM_t + \beta_5\ DS1 + \beta_6\ DS2 + \beta_7\ DS3 + \beta_8\ PCP_{t-1} + u_t$$

with

PCP : price index of private consumption;

LDR : wage push (wages exceeding the increase of productivity);

KAPA: capacity utilization (deviation from a long-term trend);

PIM : price index of imports.

2. METHODS AND DATA

Having specified and estimated a regression relationship, a check of its adequacy has to be performed. This means, it has to be examined whether the equation reflects reality sufficiently and whether the requirements of the statistical tools are fulfilled. In regressions from time-series data the question of temporal stability, that is the constancy of the regression coefficients and the variance of the disturbance terms is of central importance.

Beginning with the papers of QUANDT [6] and CHOW [2], various procedures have been developed for detecting and testing a regression model's constancy over time. These methods have been classified by HACKL [5] as follows:

(1) Procedures where the regression model to be analysed is compared with an appropriate alternative model. Such a comparison generally consists in overfitting or overparameterisation of the model. The assumption of constancy has to be rejected if additional parameters contribute significantly to the explanation. Examples for this type of methods are on the one hand QUANDT's switching regression approach [6] giving information when a constant relationship changes to another, while on the other hand the hypothesis of an abruptly varying relationship can be tested by the CHOW-test [2].
(2) Procedures which are based on the analysis of the estimated residuals; as we will see later on the ordinary least squares (OLS) residuals will be transformed for statistical reasons. The analysis of residuals proves advantageous when it is uncertain which kind of non-constancy might be present.

For our purpose we decided to deal mainly with the methods suggested by BROWN, DURBIN and EVANS (hereafter BDE) [1]. Their comprehensive collection of techniques and tests contains procedures of overfitting as well as such of analysing residuals. As all the procedures are described in detail in their contribution we can restrict ourselves to the review of the most essential features.

Though including formal significance tests, BDE basicly regard their techniques as yardsticks for the understanding and interpretation of data, indicating departures from constancy mainly in a graphic way, instead of seeing the formal tests as tools for quick decisions about particular departures. Dealing with data and regression functions which in no way fulfill all the hard assumptions and restrictions underlying the proposed procedures, we agree with their opinion. Therefore, results gained by applying BDE's methods have to be interpreted carefully and confronted with each other.

As the hypotheses to be analysed are taken from a short-term econometric model, the data underlying the functions are quarterly time series (seasonally unadjusted). The period of observations comprises the 1st quarter of 1960 to the 4th quarter of 1981. Due to lags in the functions, the analysis starts with the 1st quarter of 1961.

CUSUM Test and CUSUM of Squares Test

The regression model to be analysed here can be stated as follows:

$$y_t = x_t' \beta_t + u_t, \qquad t = 1,\dots T,$$

where at time t, y_t is the observation of the dependent variable and x_t is the column vector of observations on k regressors. The disturbance terms u_t are assumed to be independent and normally distributed with means $E(u_t) = 0$ and variances σ_t^2, $t=1,\dots,T$. The vectors of parameters β_t and of variances σ_t^2 are written with the subscript t to indicate that they are allowed to vary within time. The null-hypothesis H_0 of constancy over time is:

$$\beta_1 = \beta_2 = \dots = \beta_T = \beta,$$
$$\sigma_1^2 = \sigma_2^2 = \dots = \sigma_T^2 = \sigma^2.$$

Instead of OLS-residuals z_t or their cumulative sums $Z_t = \frac{1}{\hat\sigma}\sum_{j=1}^{t} z_j$, $t=1,\dots,T$, having distributions which are hard to derive because of their complex covariance matrices, BDE use the recursive residuals

$$w_t = \frac{y_t - x_t' b_{t-1}}{\sqrt{1+x_t'(X_{t-1}' X_{t-1})^{-1} x_t}}, \qquad t = k+1,\dots,T,$$

where $X_{t-1}'=(x_1,\dots,x_{t-1})$, $Y_t'=(y_1,\dots,y_t)$ and $b_{t-1}=(X_{t-1}'X_{t-1})^{-1}X_{t-1}'Y_{t-1}$ the OLS-estimate related to the first t-1 observations. With the numerator $y_t-x_t'b_{t-1}$ as the one-step ahead forecast error for y_t, w_t can be seen as a corresponding standardized forecast error.

The CUSUM test is based on the quantities

$$W_t = \frac{1}{\hat\sigma}\sum_{j=k+1}^{t} w_j, \qquad t = k+1,\dots,T,$$

with $\hat\sigma$ being the estimated standard deviation of the overall regression. Under H_0 and from the properties of the recursive residuals, the sequence $W_{k+1},\dots,W_T$ is a sequence of approximately normal variables such that $E(W_t) = o$, $var(W_t) = t-k$ and $cov(W_s,W_t) = min(s,t) - k$. From these properties the CUSUM test is derived. The null-hypothesis of constancy is rejected if

$$W_t > a\ (T-k)^{1/2} + 2a\ (t-k)\ (T-k)^{-1/2}$$

for any $t=k+1,\dots T$, where the scalar a is chosen corresponding to the desired significance level. Graphically, this leads to pairs of straight lines through the points $(k\pm a(T-k)^{1/2})$, $(T\pm 3a(T-k)^{1/2})$, which include the path of the W_t or will be crossed by it. These lines represent proxies for the true curves of significance which are parabolic, a fact that would handicap the use of the test.

In the CUSUM of squares test the cumulative sum of squared recursive residuals

$$S_t = \sum_{j=k+1}^{t} w_j^2 \Big/ \sum_{j=k+1}^{T} w_j^2 = s_t / s_T\ , \quad t = k+1,\dots,T$$

is computed. S_t is a monotonically increasing sequence of positive numbers with $S_T = 1$. Under H_0 S_t has a Beta distribution with (t-k) and (T-k)/2 degrees of freedom and a mean value $E(S_t) = (t-k)/(T-k)$. BDE suggest the construction of an approximate confidence range

$$(t-k)/(T-k) \pm c_o \,, \qquad t = k+1,\ldots,T,$$

for S_t, where the values of c_o corresponding to the desired significance level are taken from DURBIN [4].
As DUFOUR [3] has proposed, the CUSUM of squares test can be viewed to a large extent as a test for heteroscedasticity.

Moving Regressions

Fitting the regression model on a segment of n > k observations and moving this segment along the available data, the course of the successively estimated coefficients against time provides valuable information about their supposed constancy. In order to get an idea of the magnitude of possible parameter changes, we supplemented this analysis by computing short-term elasticities and, in the case of lagged dependent variables, long-term elasticities:

$$\varepsilon_s = b_j \frac{\bar{x}_j}{\bar{y}} \quad \text{and} \quad \varepsilon_l = \frac{b_j}{1-b_{jo}} \frac{\bar{x}_j}{\bar{y}}$$

with $\bar{x}_j$ respectivily $\bar{y}$ average values of the jth regressor and the dependent variable and b_{jo} the coefficient of the lagged dependent variable. Beyond that, the estimated residual variances of the moving regressions shed light on the constancy of σ^2. As we will see in the remainder of this paper moving regressions prove to be the most important tools in investigating the constancy of the various functions.

Time-trending Regressions

A further procedure of overfitting consists in allowing the regression coefficients to become polynominals in time

$$(o)\ y_t = x_t' \beta_{(o)} + u_t$$
$$(1)\ y_t = x_t' (\beta_{(o)} + \beta_{(1)}t) + u_t$$
$$\vdots$$
$$(p)\ y_t = x_t' (\beta_{(o)} + \beta_{(1)}\, t + \ldots + \beta_{(p)}\, t^p) + u_t,$$

the β's being all vectors of length k, and p a positive integer. This approach might be preferred for detecting non-constant coefficients if the switching time is not expected to have elapsed between two observations but to be distributed over the whole observation period. The null-hypothesis of constancy H_o: p=o is to be tested against H_A: p>o. This can be done by comparing the mean-square increase in the explained variation with an estimate of the actual variance of the disturbances. The corresponding F-ratio determines whether each model gives a significantly better fit than the one before.

Quandt's Log-likelihood Ratio

If there is evidence that a regression does not follow a constant relationship but changes abruptly from one constant relationship to another, the log-likelihood ratio technique introduced by QUANDT [6,7] proves to be appropriate. For each t from t=k+1 to t=T-k-1 the quantities

$$\lambda(t) = \frac{1}{2}\, t \log \hat{\sigma}_1^2 + \frac{1}{2}\,(T-t) \log \hat{\sigma}_2^2 - \frac{1}{2}\, T \log \hat{\sigma}^2$$

have to be calculated, where $\hat{\sigma}_1^2$, $\hat{\sigma}_2^2$ and $\hat{\sigma}^2$ are the residual variances of the regressions fitted to the first t, the remaining T-t observations and the whole set of T observations, respectively.

As the exact distribution of the $\lambda(t)$ can only be derived for trivial cases where e.g. the model consists of an intercept only (cf. [5]), no tests can generally be carried out. An estimate for the unknown switching point, however, is given by the value for t where $\lambda(t)$ attains its minimum.

3. RESULTS

For reasons of space, we present and discuss the results only in graphical form, neglecting seasonal parameters. First, we summon up the findings common to all hypotheses analysed.

As to the CUSUM of squares (fig. 1-3), its path is crossed by the significance boundaries in all three cases. The CUSUM plots also show considerable departures from their zero lines indicating instabilities. Quandt's log-likelihood ratios having several local minima, mark that changes in the relationships do not occur abruptly but gradually and mainly in the first half of the 70's (in addition, the time-trending regressions improve significantly if the coefficients are allowed to change linearly with time). More generally, the results reveal some evidence of instabilities within the models. In spite of the CUSUM of squares' instabilities which might be a consequence of increasing residual variances, a fact, which is clearly indicated by the estimated residual variances of the moving regressions and which is not surprising in view of the rising level of the time series used, the deviations from constancy seem to be related to varying parameters, too. We now focus our interest on the question of the magnitude of these changes and its evaluation by a more detailed study of the functions.

When looking at the path of the root-mean-square-percentage-errors (RMSPE) for the consumption function in figure 1, they do not contradict the hypothesis of a rather constant and sufficient explaining power of the model. However, it should be noted that up to 1968 the interest-parameter, though significant in the overall regression, proves to be of no significance in the moving regressions over the 10 years periods. This means that ceteris paribus the income-parameter would have been different in most previous periods, but for the sake of simplicity we did our analysis without the reestimation of a consumption function. From figure 1, the interest- as well as the income-parameter show similar though opposite trends: till 1974 they both increase resp. decrease (in

Figure 1

EVALUATING THE CONSTANCY OF THE PRIVATE CONSUMPTION FUNCTION

1961/1 - 1981/4

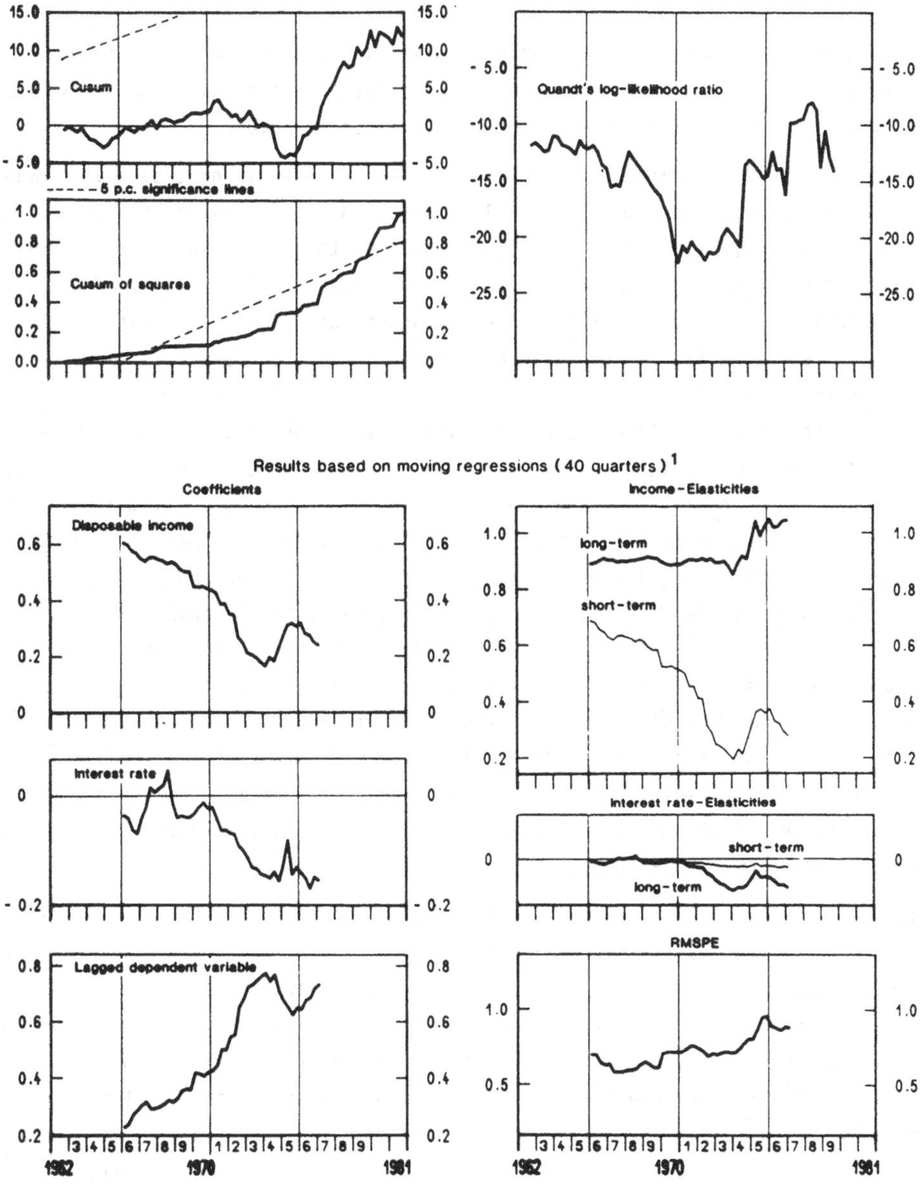

1) The results of the moving regressions are associated with the medium point of time of the corresponding sample period.

absolute values), from 1974 till the end of the sample period these developments are to a small degree reversed. The picture is different if we look at the elasticities, which are easier to evaluate. While the short-term income-elasticity reflects the same downward/upward tendency as the parameter, the interest-elasticity shows a considerable degree of constancy (however, it should be kept in mind that the explaining power of the interest variable - computed via standardized coefficients - does not exeed 3 p.c. of the total variance explained). The long-term elasticities of both variables show some constancy up to 1973, in the following period the income elasticity moves up to values of +1, while the interest elasticity moves down to values of -0.05.

How do these statistical findings correspond with our theoretical and empirical knowledge of West-Germany's private consumption? To understand the declining role of disposable income (at least up to 1973) it should be recalled, that up to the late sixties private consumption consisted to more than 55 p.c. of food, textiles and other necessities. In the following period this consumption structure changed for several reasons: private disposable income had reached a level permitting most consumers to spend more income for non-necessities like leisure or electronical goods; the tendency towards a growing part of the so called free disposable income was enlarged by the fact, that most households were well equipped with long-living consumer goods like refrigerators, washing machines and even cars [9]. These developments made household consumption decisions to an increasing degree independent of the actual income progress. The propensity to consume was not much affected by this development, since households were more guided by their long-term income-expectations. Corresponding to the changed role of actual income, the importance of the interest-parameter increased. The first oil-crisis seems to have stopped these tendencies, at least temporarely. Income-elasticities have risen again, though their level is still only half as big as in the mid-sixties. However, it is difficult to decide where these changes came from. As to the income-elasticity, the change seems mainly to be due to the augmented marginal propensity to consume (that is the income-parameter) and to a smaller degree due the lower rise of disposable income (when compared with private consumption).

Turning to the results for the investment function, it can be stated that the degree of explanation shows some constancy over the whole period (fig. 2). And all three regressors (demand, interest rate and wage costs) are of significant influence in the 10 years moving regressions. Looking at the coefficients there are some changes which clearly indicate some structural shifts, but with the exception of the demand variable these shifts only slightly exceed the standard errors of the coefficients. This picture is confirmed by the elasticities: there are only comparatively small shifts, the demand variable being an exception again. The results indicate that no "general" shift in investment behaviour can be hold responsible for the slackening investment, though it became more sensitive towards the demand-factor.

While the analysis of private consumption and investment behaviour gave no evidence of major parameter shifts (with the exception of the interest rate on private consumption)

Figure 2

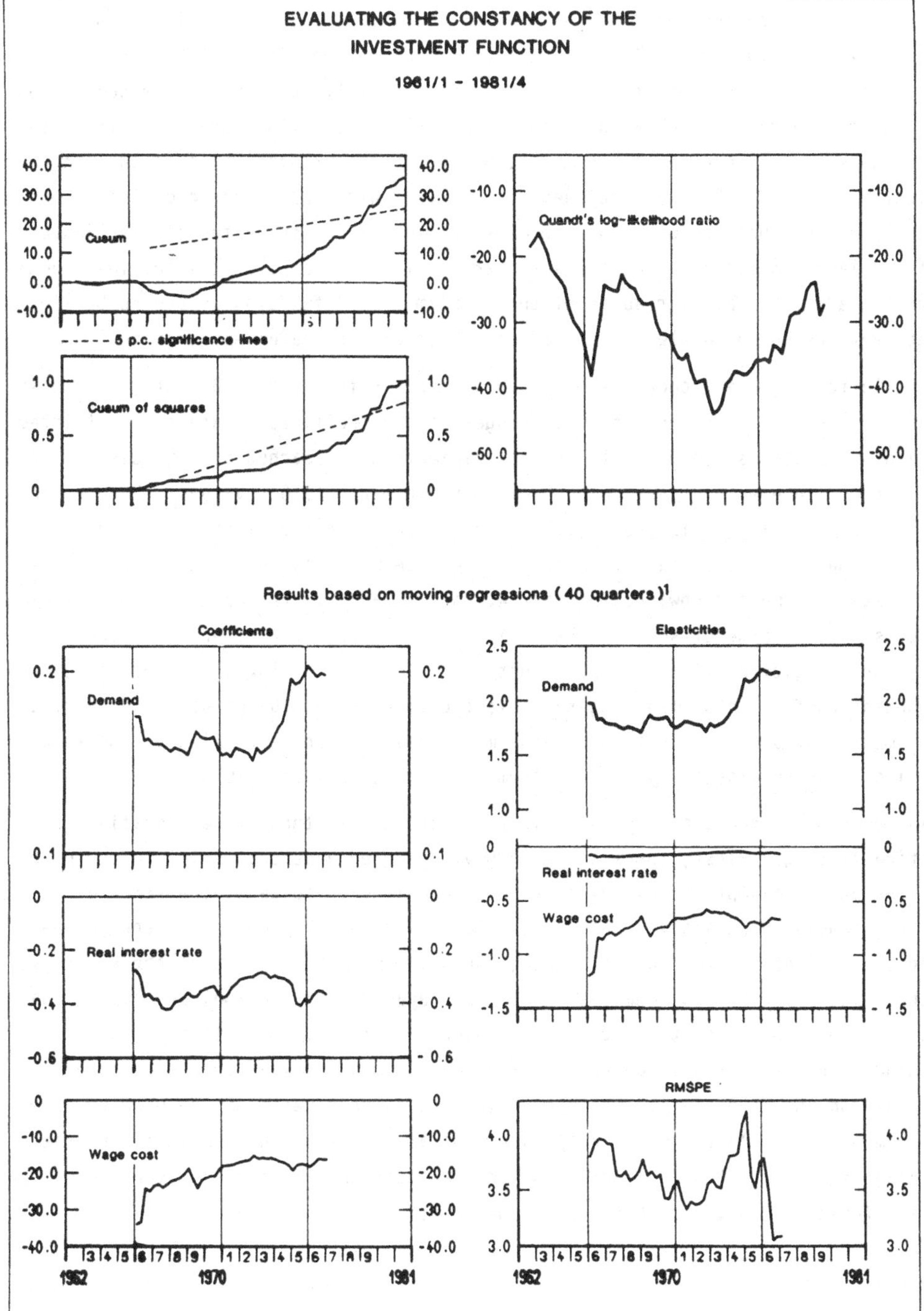

1)The results of the moving regressions are associated with the medium point of time of the corresponding sample period.

the case seems to be different for the price index of private consumption. Looking again at the moving regressions, at the mid-sixties the wage push as well as the capacity utilization did not provide a significant explanation, while at the end of this decade the price index of imports failed to be significant. Furthermore the coefficient of the lagged dependent variable shows non-acceptable fluctuations around 1. In spite of these inadequacies, especially prominent by the erratic fluctuations of the long-term elasticities (fig. 3), we continued our analysis with the introduced model, confirmed by the fact that the plots of the elasticities of the price function estimated without the lagged dependent variable were very similar to those of the original specification. In addition, all variables of the "stream lined" function proved to be significant over the whole sample period, of course showing a weaker explanation.

As to the latter, however, once again it can be derived from figure 3 that the relative explanatory power shows some constancy. Changes in the coefficients are not restricted to particular variables. In general, the coefficients have reached peak-values in 1972/73 and then declined without returning back to their early 1970's values; the coefficient of the import deflator is even still rising. Looking at the elasticities, the picture is somewhat more complicated. As to the short-term reactions the wage push elasticity is steadily moving upward. In other words, a change in the wage push/price-relationship is by and large offsetting the decline of the parameter after 1973. The elasticities for capacity utilization and import prices closely follow the movements of the corresponding coefficients. The long-term elasticities, though being affected by fluctuations mentioned above, do not show large movements with the exception of the wage push variable, whose elasticity doubles from 0.1 to 0.2 between 1970 and 1976.

Again in terms of economic theory, what is the backround of these merely statistical or descriptive findings? First, there seems to have occurred a general shift in the price index of private consumption towards a quicker reaction. In spite of the strange behaviour of the lagged dependent variable in the late sixties, the decline of the corresponding coefficient is remarkable. Second, beginning with the seventies the sensitivity of the price index towards changes in demand (capacity utilization) and supply conditions (wage push and import prices) has considerably risen. The most striking of these changes seems to be the augmented influence of import prices on the one hand and of wage push on the other. While the first might be a consequence of the growing GNP-share of imports, the latter might reflect a general change in the perception of the West German growth conditions, leading to a decrease in the willingness of employers to accept any further reductions of their income positions.

4. RÉSUMÉ

The results presented here clearly indicate that there were changes in the consumption behaviour as well as in the investment and price behaviour, particularly in the period 1972/73. But obviously the three aggregates have been affected in different ways: while in the price and consumption sector great changes occurred, the behaviour of the in-

Figure 3

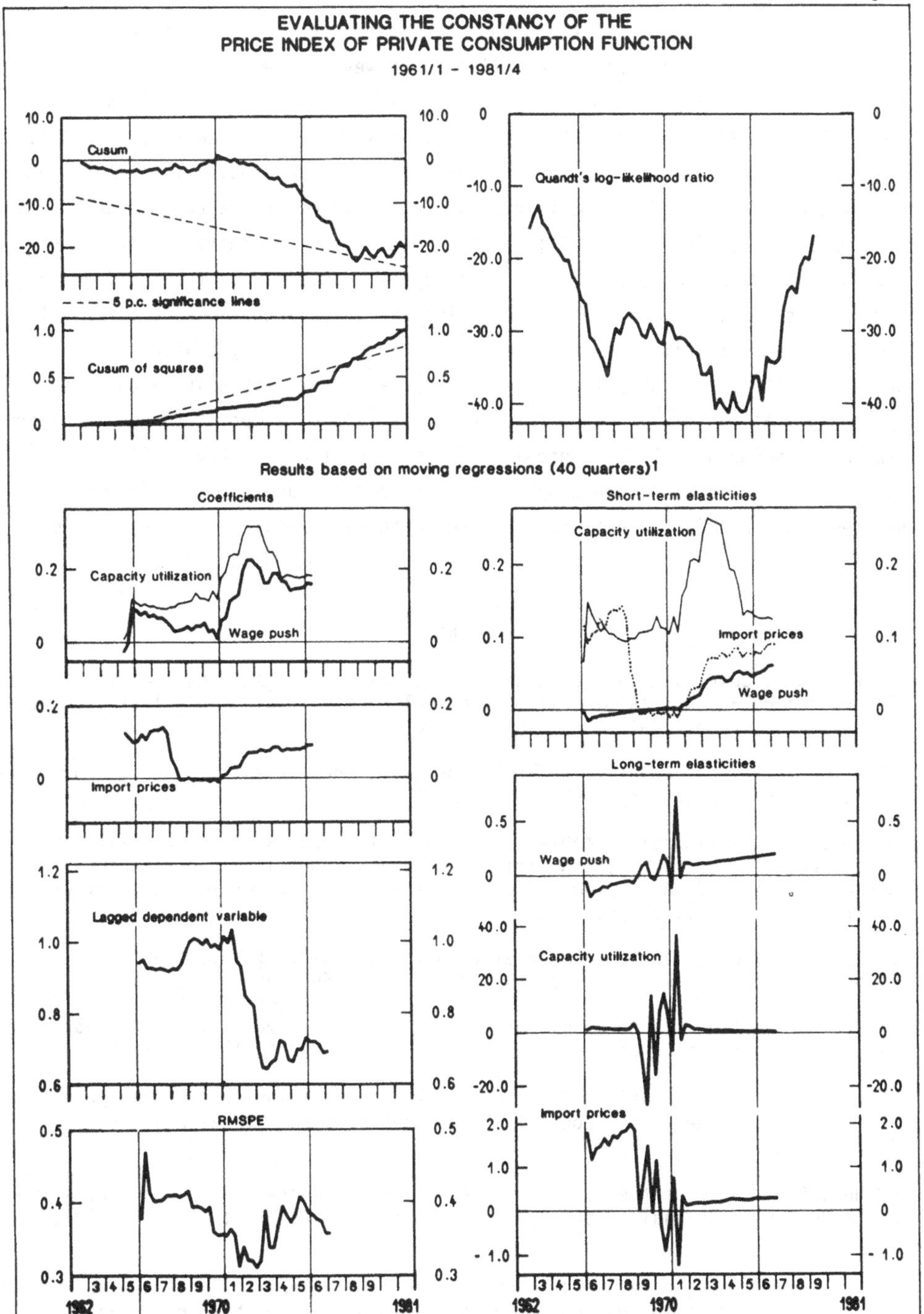

1) The results of the moving regressions are associated with the medium point of time of the corresponding sample period.

vestment sector seems to have been comparatively stable. The changes are not only related to the speed of adjustment but also to the magnitude resp. the magnitude of the elasticities towards the important determinants (however, the latter statements are still vague, a more convincing answer may probably be found in an ex-post simulation of the complete econometric system, where the functions in question were taken from). The results indicate that these parameter changes have been comparatively small (compared with the standard errors of the coefficients) in a quarter to quarter comparison. However, it must be kept in mind, that the equations, though providing rather reasonable ex-ante forecasts when employed within the complete model, may be misspecified over the earlier periods (up to the sample period ending 1977). As a consequence our results have to be judged carefully. Anyway, the results are of a merely exploratory nature and an analysis of the determining factors for the parameter shifts must be carried forward. The implications of a reduced reaction of private consumption towards an increase of disposable income as well as a growing sensitivity of prices towards alterations in supply and demand conditions are comparatively strong for any form of economic policy. Further research on this subject therefore seems to be not only promising but necessary.

REFERENCES

[1] BROWN, R.L., DURBIN, J. and J.M. EVANS: "Techniques for Testing the Constancy of Regression Relationships over Time", Journal of the Royal Statistical Society, Series B, 37 (1976), 149-163.

[2] CHOW, G.: "Tests of Equality Between Subsets of Coefficients in Two Linear Regressions", Econometrica, 28 (1960), 591-605.

[3] DUFOUR, J.-M.: "Recursive Stability Analysis of Linear Regression Relationships", in: Structural Change in Econometrics, ed. by L. Broemeling, Annals of Applied Econometrics, 1982-2, A supplement to the Journal of Econometrics, Journal of Econometrics, 19 (1982), 31-76.

[4] DURBIN, J.: "Tests for Serial Correlation in Regression Analysis Based on the Peridogram of Least Squares Residuals", Biometrika, 56 (1969), 1-15.

[5] HACKL, P.: Testing the Constancy of Regression Models over Time. Angewandte Statistik und Ökonometrie, Heft 16, Vandenhoeck & Ruprecht, Göttingen, 1980.

[6] QUANDT, R.: "The Estimation of the Parameters of a Linear Regression System Obeying Two Separate Regimes", Journal of the American Statistical Association, 53 (1958), 873-880.

[7] QUANDT, R.: "Tests of the Hypothesis that a Linear Regression System Obeys Two Separate Regimes", Journal of the American Statistical Association, 55 (1960), 324-330.

[8] RAU, R., HEILEMANN, U. and H.J. MÜNCH: "Forecasting Properties of the RWI-Model", in: Models and Decision Making in National Economies, ed. by J.M.L. Jansen, L.F. Pau and A. Straszak, Amsterdam, New York, Oxford: North-Holland, 1979, 293-300.

[9] Rheinisch-Westfälisches Institut für Wirtschaftsforschung: Analyse der strukturellen Entwicklung der deutschen Wirtschaft. Gutachten im Auftrag des Bundesministers für Wirtschaft, Band 1, Essen, 1980.

ANALYSIS AND MODELLING OF THE DEVELOPMENT ECONOMY IN THE LEAST DEVELOPED COUNTRIES

Tadashige ISHIHARA
Osaka Electro-Communication University, 18-8 Hatsumachi, Neyagawa, Osaka 572, Japan.

Michiko NISHIMURA, Hidekazu YABUUCHI, Katashi TAGUCHI
University of Osaka Prefecture, Sakai, Osaka 591, Japan.

Masaaki YONEZAWA
Kinki University, Kowakae 3-4-1, Higashi-Osaka, Osaka 577, Japan.

Abstract.
A case study to estimate national accounts containing input-output table is done for Nepal on the 1976/77 buget year.
Similar national accounts are possible for the other least developed countries, following this method. The obtained accounts are applicable and useful to many regions of informative activities such as analysis, modelling, planning of national development economy.

I. Introduction.

Planning of development plays the most important role in advancing the economy of the developing countries especially of the least less developed countries (LLDC). The economic growth of the least developed countries are being stagnate almost in this generation after the 2nd World War [1] and this is the most serious problem to be solved both on a national and international level [2]. The reasonable plannings are most required in these LLDC, where efficient uses of resources are quite necessary.

The planning of development can be done reasonablly only if it is built on the basis of the national accounts. Especially the national accounts containing I/O table manifest socio-economic structure and are very powerful tools to promote planning, feasibility studies, selection of optimal policy measures, and future forcast and decision making for all sorts of economic policies.

However in the less developed countries, minute data cannot be obtained easily and the statistics have often few reliability. The informations which are comparatively correct and easily obtainable are almost limited to various sorts of taxes and government expenditures. Not only direct and indirect taxes, it is useful also if we could find the data about gross value of production and gross domestic factor income which can

be estimated usually because the exact incomes and output value are necessary for reasonable taxation in any country. The method to make approximate input-output table for less informative LLDC, using mainly these available data, is introduced in the paper [3].
In this paper we show the actual making procedures of Nepal 1976/77 national accounts containing I/O table as the case study of the application of the method [3].

II. Making procedures of Nepal 1976/77 national accounts.
About Nepal's national accounts, Dr. R.M. Barkay, United Nation's Advisor, worked with Central Bureau of Statistics of Nepal Government and made 4 volumes of documentations [4] about Nepal's national accounts using official and all other available informations for the estimates. In this paper, we tried to estimate the interrelations of these data, i.e., input-output structures and could examine the consistency of Dr. Barkay's synthetic national accounts for the 1976/77 financial year. Following the flow chart (Fig. 1), the calculation procedures go step by step.

1). The first step is to classify the Nepal industries to practically reasonable sectors. Of course, sectors must be selected in accordance with various objects or usages of the table. We take 85 sectors classification, considering the situation of national production and economic structure and socio-economic infrastructures, international trade, main objects of development. The selection is preferable if the classes coincide with the classes or their aggregate of SITC (Commodity Indexes for the Standard International Trade Classification) and U.I.O. (Uniform Input-Output Classification for ASEAN countries prepared by I.D.E. Japan).
In our case, in addition to them, we refer Indonesia's and Thailand's classification since we think there are some resemblances of input structures between Nepal's goods and Indonesia's or Thailand's goods and we start our preliminary approximation of input structure using the input coefficient of Indonesia's or Thailand's commodities.

2) Next we fill the value added part with the data obtained from the Report of National Accounts Projects [4] and the Economic Survey [5]. We guess these data were obtained in relation to the taxation and government expenditures.

3) The third step is to estimate the final demand part using the data

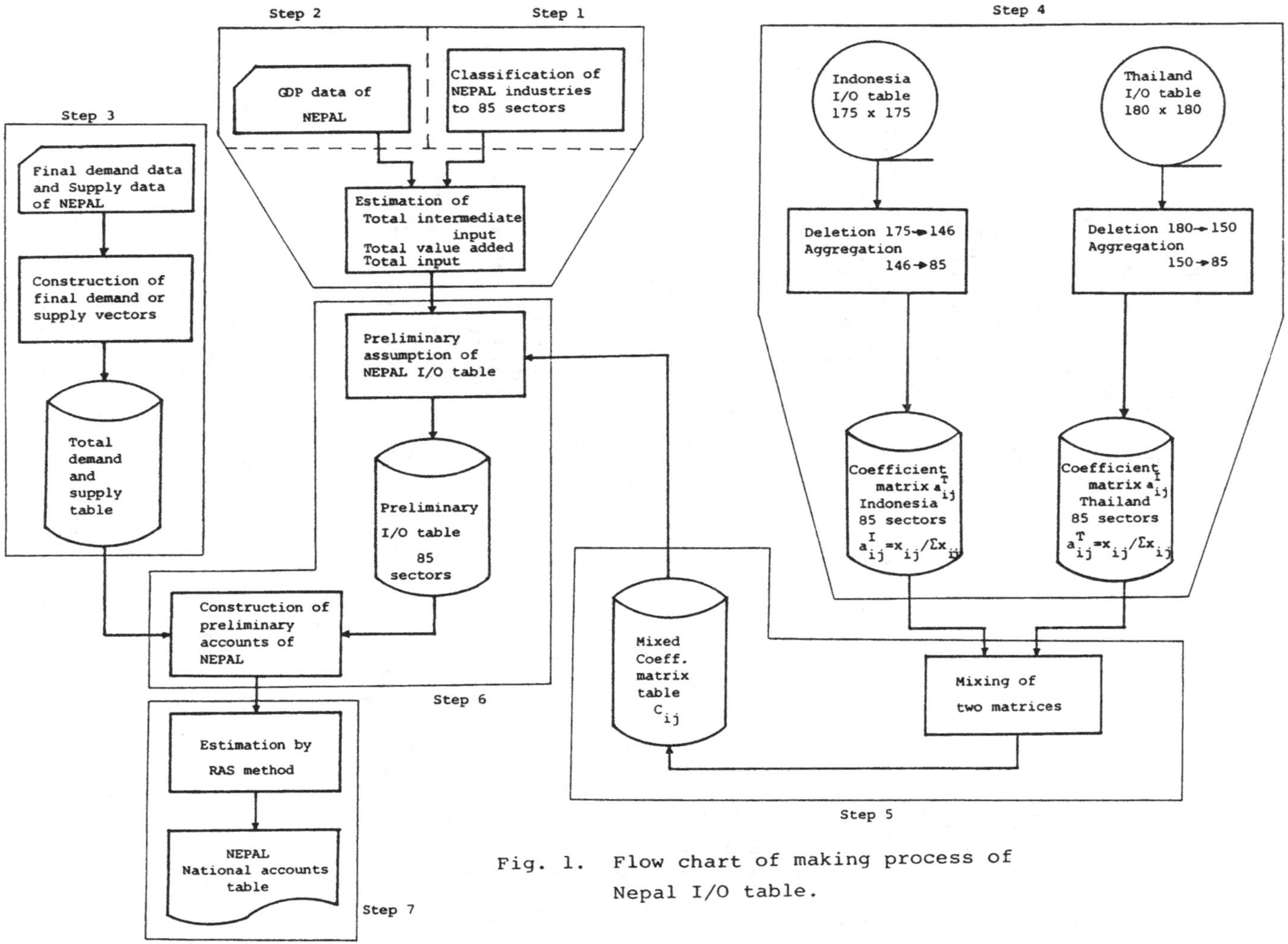

Fig. 1. Flow chart of making process of Nepal I/O table.

1	PADDY	44	MANU. OF PAPER & PAPER PRODUCT
2	MAIZE	45	MANU. OF CHEMICALS
3	BARLEY	46	OTHER MANU. OF CHEMICALS
4	WHEAT	47	MANU. OF NON-METALLIC MINERAL
5	MILLET	48	FABRICATED METAL PRODUCTS
6	POTATOES	49	OTHER FABRICATED METAL PRODUCU
7	PULSES	50	JEWELLERY
8	VEGETABLES	51	ACTIVITIES N. E. S.
9	FRUITS	52	MACHINERY
10	SUGAR CANE	53	TRANSPORT EQUIPMENT
11	JUTE	54	FOOD PROCESSING
12	OIL SEEDS	55	COTTON TEXTILES
13	TEA LEAVES	56	WOOLEN GARMENTS
14	CARDAMON	57	FOREST BASED MANUFACTURING
15	SPICES	58	METAL AND OTHER CRAFTS
16	TOBACCO	59	PETROLEUM REFINING
17	MEDICINAL HERBS	60	ELECTRICITY & GAS
18	COTTON	61	WATER
19	OTHER AGRICULTURE PRODUCTS	62	RESIDENTIAL BUILD. CONST.
20	WOOL	63	NON-RESI. BUILD. CONST.
21	HIDE AND SKIN	64	PUBLIC WORKS (AGRI. & FOREST)
22	MEAT	65	PUBLIC WORKS (NON-AGRI.)
23	MILK	66	ELEC.PLANT, WATER SUPP. CONST.
24	MILK PRODUCTS	67	OTHER CONSTRUCTION
25	POULTRY MEAT	68	TRADE
26	POULTRY EGGS	69	RESTAURANT
27	MANURE	70	HOTELS
28	BULLOCK LABOUR & LIVESTOCK	71	LAND TRANSPORT
29	TIMBER	72	WATER TRANSPORT
30	FUEL-WOOD	73	AIR TRANSPORT
31	OTHER FOREST PRODUCTS	74	TRANSPORT SERVICES
32	FISHING	75	STORAGES
33	COAL & COKE	76	COMMUNICATIONS
34	CRUDE PETROLEUM, NATURAL GAS	77	FINANCE
35	METALIC MINERAL MINING	78	INSURANCE
36	QUARRING	79	REAL ESTATE
37	OTHER NON-METALLIC MINING	80	BUSINESS SERV.& OTHER PRIV.SV.
38	FOOD MANUFACTURING	81	EDUCATION
39	OTHER MODERN FOOD MANUFACTURIN	82	SANITARY & HOSPITALS
40	BEVERAGE INDUSTRIES	83	OTHER PUBLIC ADMINI. & DEFENCE
41	TOBACCO MANUFACTURES	84	OTHER PUBLIC SERVICES
42	TEXTILE & LEATHER IND.	85	UNSPECIFIED, PROVISIONAL SECT.
43	MANU. OF WOOD & WOOD PRODUCTS		

Table 1. Sector classification of Nepal I/O table

[4] for Household Consumption Expenditure, Gross Fixed Capital Formation, Increase in Stock. For Exports and Imports, we use the Export-Import statistics [6] and the Report of the World Bank [1].

4) The fourth step is to find similar input structure in already published I/O tables of other countries for each given sector. It is possible to use many countries table at the same time, different countries corresponding to different sectors. We used Indonesia's 1971 structure [7] and Thailand's 1975 structure [8] to start preliminary assumption of the Nepal's I/O structure, because there are some simi-

larities, i.e., general economic status, weather, geography, form of cultivation and technology, etc. (these resemblances must be considered already in the first step also). Both tables of Indonesia and Thailand must be aggregated to 85 sectors corresponding to classification of Nepal's industries in the first step by deletion or aggregation.
Then we make 85 sectors coefficient matrices a^T_{ij} and a^I_{ij} of Thailand and Indonesia from them by division $a^T_{ij} = x^T_{ij}/\Sigma x^T_{ij}$ and $a^I_{ij} = x^I_{ij}/\Sigma x^I_{ij}$

5) As the fifth step, we mix both tables and make a new hypothetical I/O coefficient table as follows.
The 85 sectors are divided into two parts T and I, where for the T part sectors, Nepal input structures resembles to Thailand's and for the I part sector, Nepal's resembles to Indonesia's. For the sake of simplicity, we assume that the i-th sector belong to the T part for $1 \leq i \leq h$, and the I part for $h+1 \leq i \leq n$. Then the mixed coefficient matrix is divided to four parts as shown in Fig. 2. Here superfixes T and I of $C^{\bullet}_{ij}$ denote Thailand and Indonesia respectively.
Then element $C^{\bullet}_{ij}$ of the mixed table is calculated using following formula.

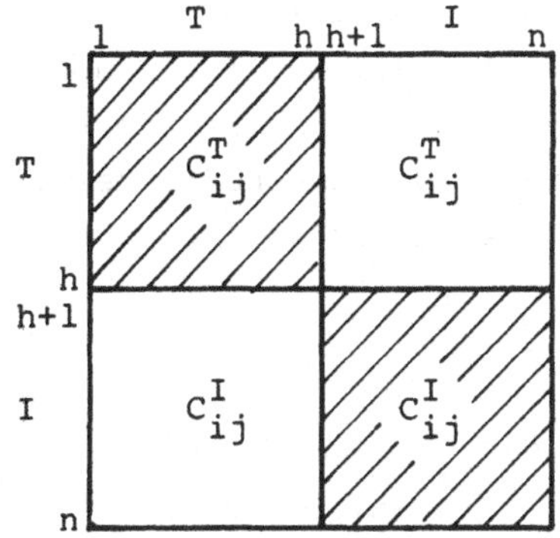

Fig. 2

$$\sum_{i=1}^{h} C^T_{ij} + \sum_{i=h+1}^{n} C^I_{ij} = 1 \quad (1 \leq j \leq n),$$

where $C^T_{ij} = a^T_{ij}$ for $\begin{pmatrix} 1 \leq i \leq h \\ 1 \leq j \leq h \end{pmatrix}$,

C^I_{ij} is proportional to a^I_{ij} for $\begin{pmatrix} h+1 \leq i \leq n \\ 1 \leq j \leq h \end{pmatrix}$ and $C^I_{ij} = a^I_{ij}$ for $\begin{pmatrix} h+1 \leq i \leq n \\ h+1 \leq j \leq n \end{pmatrix}$, C^T_{ij} is proportional to a^T_{ij} for $\begin{pmatrix} 1 \leq i \leq h \\ h+1 \leq j \leq n \end{pmatrix}$.

6) Using this coefficient matrix and real subtotal values of intermediate input which was obtained in step 2), we calculate preliminary assumed input value and fill these input parts.
Combining the intermediate part with the final demand part, we have a crude preliminary I/O table now. But the final rational I/O table must have the consistency between total supply values and total final demand values in each row compornents.

7) As the final step, to get the total consistency, we apply the RAS method.
Thus we get the consistent I/O table of Nepal and now can see her economic structure for the 1976/77 year.

III. Some problems in the making process of I/O table.
In the making process of the table, we found some problems to be solved.

1) Inaccuracy caused by the ambiguity of minute data.
In the data which we could obtain, comparatively reliable ones were slant line parts on Fig. 3.
Though they were reliable, there remained some inconsistency among data from the materials listed in the references. There were some gaps between demand and supply.

2) Inaccuracy resulted by use of Indonesia's and Thailand's structure. When we estimate the Final Demand Part (the horizontal line part of the figure above), we used principally the materials [4], but could not find minute data for each sector. For example, about Increase in Stock (304), we estimated the values for subdivision to each row components refering the Indonesia's and Thailand's structure. This caused some gaps between Demand and Supply in each row components.

IV. Application of national accounts.
Thus we could obtaine the Nepal I/O table 1976/77, though there were some problems. The follow up of this method will enable to make similar accounts for 77/78, 78/79, 79/80 of the 5th five-year plan in Nepal and they will serve for examination of plan implementation and policy measures during this period. Further, this method will be applicable for analysis of other less developed, less informative countries like Nepal and will enable to show clearly the structure of a national economy or mutual interdependencies between economic activities.
Using these accounts we will be able to make models of various socio-economic actions. Also we can estimate direct and indirect effect of some investment or government expenditures. If added by employment statistics for each sector, we can discuss the effect some expenditures of some national or international projects to the increment of employee or of GNP. Further, together with some development plans and their models, we will be able to estimate promotion behaviour of growing socio-economic system.

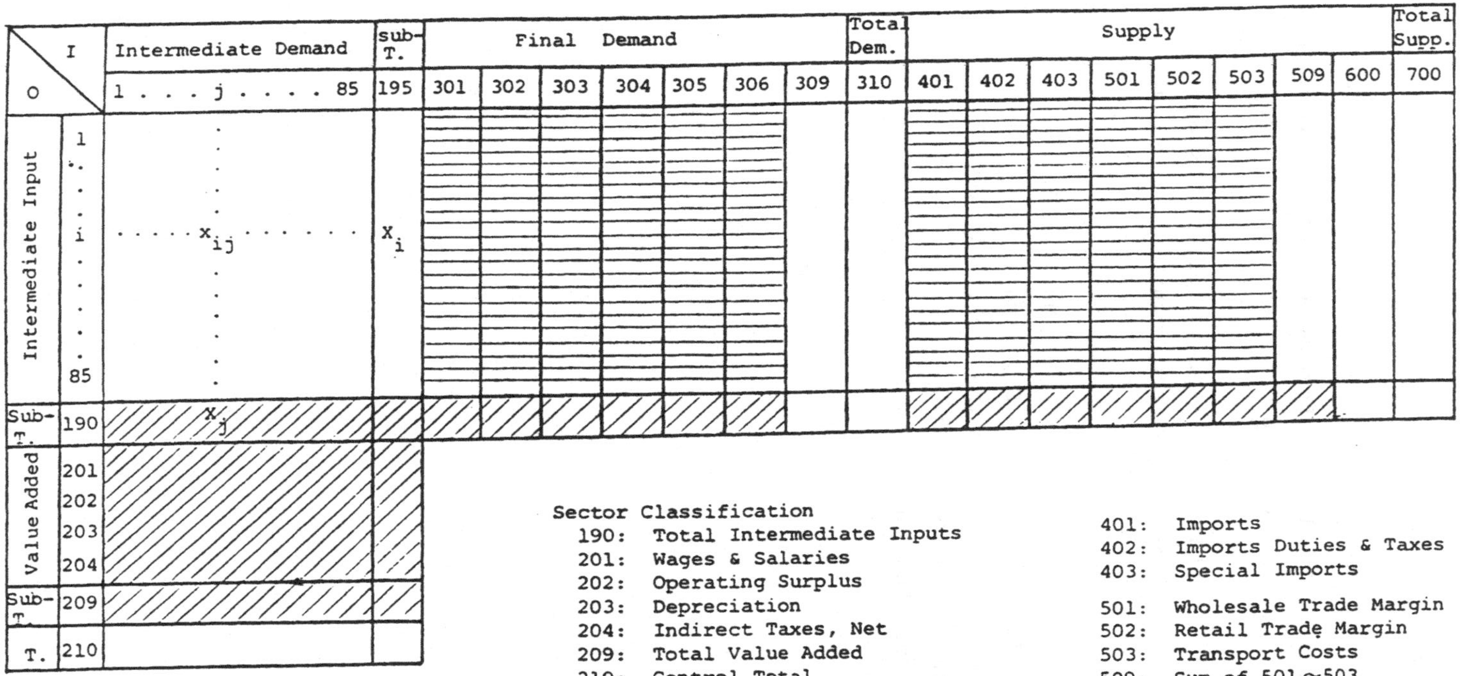

210 = 190 + 209
310 = 195 + 309
600 = 210
700 = 600 + 401 + 402 + 403 - 509

must be 700 = 310

Sector Classification
190: Total Intermediate Inputs
201: Wages & Salaries
202: Operating Surplus
203: Depreciation
204: Indirect Taxes, Net
209: Total Value Added
210: Control Total

301: Private Consumption Expenditures
302: Government "
303: Gross Fixed Capital Formation
304: Increase in Stock
305: Exports
306: Special Exports
309: Sum of 301 ~ 306
310: Total Demands

401: Imports
402: Imports Duties & Taxes
403: Special Imports

501: Wholesale Trade Margin
502: Retail Trade Margin
503: Transport Costs
509: Sum of 501 ~ 503

600: Control Total

700: Total Supply

Fig. 3. Form of national accounts (Nepal: 1976/77)

References.

[1] The World Bank, World Development Report 1979, The World Bank, Washington, D.C. (1981).

[2] T. Ishihara, M. Nishimura, H. Yabuuchi, H. Endo and T. Nishioka, Analysis and Modelling of Development Economy of the Less-Developed Countries, Proc. Asia-Pacific Conf. on Opr. Res. Soc., Singapore, pp. 59-65, 1982.

[3] T. Ishihara, H. Yabuuchi and M. Sato, A Method of Getting Approximate Input-Output Table for Less Developed Countries, Proc. of 4th International Conf. on Math. Modelling, Pergamon Press, New York, (1983).

[4] National Accounts Project, Vol. I-IV, C.B.S, Nepal, 1978.

[5] Economic Survey, 1980/81, 1981/82, C.B.S., Nepal.

[6] Quarterly Economic Bulletin, Nepal Rastra Bank, Vol. 16, 1982.

[7] Input-Output Table Indonesia 1971, I.D.E.

[8] Basic Input-Output Table Thailand 1975, I.D.E.

MACROECONOMIC EQUILIBRIUM WITH RATIONING AND VARIABLE WORKING TIME /1/

A. Battinelli
LADSEB - CNR
35100 Padova, Italy

1. INTRODUCTION

The method of fix-price equilibrium has been extensively used in the last ten years in an attempt to provide an analytical representation of the reinterpretation of the keynesian theory of unemployment put forth in the sixties by Clower and Leijonhufvud. The literature is growing and already well developed; ground-laying contributions are due to Drèze, Benassy, Barro-Grossman, Malinvaud. All these works deal with highly aggregate two-sector models and make standard convexity assumptions on consumers preferences and standard concavity assumptions on the aggregate production function.

In a recent paper (Fitoussi and Georgescu Roegen 1980) the common structure of these models is examined and some cryticisms concerning the assumed properties of the production function are put forward. The issues involved all have Georgescu Roegen's flow-fund theory of production (for which see, e.g., Georgescu Roegen 1971a and b) in the background and include, among others, the Marx-flavored distinction between labor and labor-force, the explicit analysis of the time-length of the production process, the influence of the length of the working day on the level of unemployment. In the paper, however, an alternative aggregate model is outlined but its analytical discussion is only partially carried on.

Aim of the present paper is to reformulate the assumptions underlying Fitoussi-Georgescu Roegen's two-sectoral macroeconomic model and to develop a systematic analytical discussion of its properties. The paper is organized as follows: section 2 contains the list of the assumptions characterizing the model, together with all the relevant definitions and notations; section 3 contains the discussion of the behavior of the two aggregate economic sectors, consumers and producers, in each of the possible situations occurring according to the fix-price formulation (zero, one, or two quantity constraints); section 4 contains the classification of the price-wage parameter space in the four typical regions of the fix-price literature (keynesian unemployment, repressed inflation, classical unemployment, underconsumption) together with a detailed analysis of the sign of the static employment and output multipliers and of the problem of triviality and non-existence of equilibria.

The answer to the last two problems (particularly the second: there are entire open regions in the price-wage parameter space for which no fix-price equilibria exist) is somewhat unexpected and sharpens the contrast with the existing literature.

2. ASSUMPTIONS

2.1. The period for which the market process is described is so short that: (i) the size of the population remains constant; (ii) the tastes of the population are invariable; (iii) there exists a constant number of production units operating with a given technology.

2.2. During the same period only one trading act takes place in each market.

2.3. The price vector cannot be affected by the behavior of the economic agents - consumers, producers, government. Consumers seek to maximize their utilities and firms seek to maximize their profits by adjusting the quantities they buy or sell to the given prices and to the constraints on trade in which they can possibly incur on markets where aggregate demand and supply can only be made to coincide by a rationing procedure. The government simply spends some given funds at the given prices, and is never subject to rationing.

2.4. In addition to money there are only two commodities: a homogeneous labor and a general consumer good, here after called 'commodity'. The productive process is assumed to be completely integrated, and the capital

----

/1/ The work leading to the present paper has been done when visiting the Economics Department of the European University Institute in Florence. I wish to thank prof.s M.De Cecco, J.P.Fitoussi and M.Streit for their kind hospitality. The paper is probably not much more than a rewriting of a paper by J.P.Fitoussi and N.Georgescu Roegen, and is in fact a portion of a joint effort, still waiting to see its final formulation. I am greatly indebted to prof. Fitoussi for his suggestion to work in this direction and for several discussions he had with me when the paper was being written. Moreover, I have heavily relied on the general approach outlined in V.Boehm's recent book. Finally, and according to the custom, I take complete responsability for all the shortcomings of my work.

market nonexistent. Natural resources are ignored altogether. The land on which the productive process takes place is fixed, and nothing is due for its use.

2.5. No economic agent can, or wants to, stock the commodity, either because of complete perishability or otherwise. As a consequence, the commodity is produced and consumed entirely within the period.

2.6. Money stocks at the end of the period have an indirect utility for consumers as a store of value, and constitute the primary object of concern for firms by definition of profit maximization.

2.7. In contrast with the previous two, the dimension of labor is twofold, since its contribution to the productive process as a fund element is duly emphasized /2/. On one hand the operation of any industrial plant presupposes the simultaneous utilization of some segment of the existing labor-force, measured in number of men /3/; on the other hand the description of the actual outcome of the process requires the specification of the time during which the plant is in operation within the period, measured in hours /4,5/. The distinction has its counterpart in the consumer's utility maximization problem, where the argument entering the utility function in the standard consumption - leisure trade-off is the number of working hours, but the macroeconomically relevant variable, size of the labor-force, results as the number of consumers willing to supply a positive amount of labor-time.

2.8. The aggregation problem is completely ignored: all individuals and firms are identical.

2.9. The only observable states are of the kind currently described as fix-price equilibria with rationing. In other words, there exists an invisible hand, or auctioneer, which coordinates the decisions taken by the economic agents in order to make them consistent with a given price vector and among themselves, so that the constraints on trade individually perceived by the various agents - and entering their maximization programs - coincide with the ones obtained for them on the markets where the various outcomes of those programs require the rationing procedure to be in operation.

2.10. The preferences of each consumer are represented by the following standard utility function:

$$U(x,l,m') = b \ln x + c \ln (B - l) + d \ln m' \qquad b,c,d > 0 \ , \quad b + c + d = 1 \tag{2.1}$$

where:

x is the amount of the commodity consumed within the period,
l is the working time performed within the period,
m' is the stock of money held at the end of the period,
B is the byosociological maximum number of working hours that can be performed in the period.

The budget equation, according to the assumptions above, reads as follows:

$$wl + \bar{m} = px + m' \tag{2.2}$$

where:

p is the price of the commodity,
w is the wage rate,
m is the stock of money held at the beginning of the period.

The outcome of the decisions of all consumers, identical as they are by assumption, is therefore completely

---

/2/ See Georgescu Roegen 1971a on this respect. The distinction is at least partially overlapping with the marxian distinction between labor and labor-force.

/3/ For a given plant this number may vary between a minimum, corresponding to the safety level of operation, and a maximum, corresponding to full capacity.

/4/ This coincides with the length of the working day, provided only one shift is used. The problem of the number of shifts is not studied in this paper.

/5/ The current trend, both in theoretical and empirical works, is to collapse the two dimensions in one. As understressed in Georgescu Roegen 1971b, '...the closest element revealed...from current statistical information...is the number of man-hours supplied during the year...Such data...do not even provide some indirect index of the average employment during the year. Certainly, 1600 man-hours may be the result of 400 men each working four hours as well as 200 men each working eight hours'.

specified by the triple (X,l,N'), where N' is the size of the labor-force /6/, and X is the aggregate demand of the commodity.

2.11. The aggregate production function, obtained by combination of the identical production functions of all firms, is represented by the following formula:

$$(2.3) \qquad Y = t\, f(H)$$

where:

H is the size of the labor-force employed within the period,
t is the time of operation of the plants in the period,
Y is the total amount of the commodity produced within the period,
f has the standard properties of an S-shaped function:

i) $f(H)$ is defined on $[0,+\infty)$, $f(H) \geqslant 0$, $f(H) = 0$ iff $H = 0$;

ii) $f'(H) \geqslant 0$, $f'(H) = 0$ iff $H = 0$ or $H \geqslant H^{\wedge}$;

iii) there exists H' , $0 < H' < H^{\wedge}$, such that $f''(H) > 0$ for $0 < H' < H$ and $f''(H) < 0$ for $H' < H < H^{\wedge}$.

We recall that assumptions i)-iii) imply the existence of H'' , $H' < H'' < H^{\wedge}$, such that the average product $f(H)/H$ is an increasing function of H for $0 < H < H''$, and a decreasing function of H for $H > H''$.

The aggregate profit R is therefore given by the following expression:

$$(2.4) \qquad R(t,H;p,w) = ptf(H) - twH - C = t[pf(H) - wH] - C$$

where C stays for the fixed costs.

Remark

It is convenient (for expository purposes) to observe now that profit maximization implies, given the price vector (p,w), that the decision of the productive sector relative to the variable t, time of operation in the period, is of the following type:

$t = B$ whenever there exists at least some value H, compatible with all the market constraints, such that the profit net of fixed costs per time-unit of operation $pf(H) - wH$ is greater than zero;

$t = 0$ whenever for all values of H compatible with the market constraints the unitary profit net of fixed costs is less than zero;

$t \in [0,B]$ when the only value of H compatible with the market constraints is such that the unitary profit net of fixed costs is equal to zero.

2.12. Only one shift can be used in production, either out of legal constraints or otherwise. As a consequence there can be a discrepancy between the number l of working hours supplied by the consumers and the number t of working hours demanded by the producers. When such is the case, the long side is rationed, so that the actual length of the working day is given by $\min(l,t)$. Since by inspection of the utility function one has $0 \leqslant l < B$, the remark in assumption 11 implies that the productive sector is, so to speak, quantity-taking on the market of labor-time, unless the economy is not working at all. In other words, except in the limiting case when producers decide to close down the plants, the number of working hours is completely determined by consumers. It will therefore be referred to as l from now on. Moreover, what producers really decide about is production scale $Z = f(H)$, not total production $Y = tZ$, and the outcome of their decisions will be specified by the couple (Z,H).

2.13. Rationing on the market of the labor-force is also assumed to force an adjustment of the long side on the short. Moreover, no fractioning of the working day (either of a worker among different firms, or in a plant among different workers) is allowed. In particular, when unemployment occurs, every consumer is either employed for the exact number of working hours he desires, or totally unemployed.

2.14. A similar adjustment of the long side on the short is achieved by the rationing procedure on the market

---

/6/ N' can only assume two values, depending on whether or not l is different from zero: N , the total number of consumers, or zero. By allowing the utility function, or even simply the initial money stock, to vary among the consumers, a less extreme behavior of this variable may be immediately obtained.

of the commodity. Since the model already works with an aggregate production function, nothing else needs to be specified if the productive sector is on the long side (excess supply). On the other hand, in the opposite case a distinction must be made according to the state of the market of the labor-force. If the latter is in equilibrium, or in excess demand, the excess demand on the commodity market is dealt with by uniform absolute rationing of all the consumers. If on the contrary the labor-force market is in state of excess supply, then the demand of all the unemployed consumers is fully met, and the residual quantity of the commodity produced (minus government demand which by assumption is never subject to rationing) is uniformly distributed among the employed consumers.

2.15. No economic agent can determine its behavior according to considerations of strategic interdipendence among the various agents; in particular, every agent is quantity-constraint taking as well as price taking /7/.

3. BEHAVIOR OF THE CONSUMERS AND OF THE PRODUCTIVE SECTOR IN THE VARIOUS MARKET SITUATIONS

We shall now briefly list the solutions to all the constrained maximization problems faced by consumers and producers according to the assumptions of the previous section (for details see Battinelli 1982); notice that we already work with an aggregate production function on the productive side, whereas we shall derive sectoral behavior from individual behavior on the consumption side. All the solutions listed depend upon the distributional parameters p and w, and, in the case of consumers, upon m as well. When one or two quantity constraints are present in the problem, the expressions given for the solutions are correct only for parameter values such that the constraints are effectively binding; the corresponding regions in parameter space are specified immediately afterwards, and the solutions then depend also upon the values of the constraints; (in particular, the constrained components match the respective constraints /8/, and will be omitted). We shall write $w^{\wedge}$ instead of $cm/(b+d)B$, and denote average returns $f(H)/H$ by $af(H)$; we shall write af'' instead of $af(H'')$, and denote the two labor-force levels corresponding to a given value z of marginal (resp. average) returns by $H'(z)$, $H^{\circ}(z)$ (resp. $aH'(z)$, $aH^{\circ}(z)$); finally, we shall sometimes abuse the notation $f'^{-1}(z)$ by referring it only to $H^{\circ}(z)$ /9/.

3.1 Individual consumers without quantity constraints ('walrasian consumers')

(3.1) $x(p,w,m) = b(m+Bw)/p$, $l(p,w,m) = [(b+d)Bw - cm]/w$, $m'(p,w,m) = d(m+Bw)$

(3.1a) if $w > w^{\wedge}$;

(3.2) $x(p,w,m) = bm/(b+d)p$, $l(p,w,m) = 0$, $m'(p,w,m) = dm/(b+d)$,

(3.2a) if $w \leqslant w^{\wedge}$.

3.2 Unemployed consumers

The maximization problem of the previous section is changed by the introduction of the additional constraint $l = 0$. The constraint is binding only if (3.1a) holds, in which case the solution is given by (3.2).

3.3 Individual consumers under commodity rationing

Let y denote the constraint on individual commodity demand; then we have:

(3.3) $l(p,w,m,y) = (dBw + cpy - cm)/(c+d)w$ $m(p,w,m,y) = d(Bw - py + m)/(c+d)$

/7/ This assumption has a particular form under commodity rationing, where from assumptions 11-13 employed consumers influence the total quantity of output produced by their supply of labor-time, and from assumption 14 total output produced determines the size of the constraint on commodity demand faced by the employed consumers themselves: any possibility of altering the commodity constraints by manipulation of the rationing scheme through the supply of labor-time is ruled out.

/8/ If the concept of fix-price equilibrium with rationing is in the Drèze sense (Drèze 1975); if it is in the Benassy sense (Benassy 1975), the situation is slightly different, but the distinction does not affect the present discussion.

/9/ The definition makes sense, of course, only for z positive and smaller than $f'(H')$ (resp. smaller than af''). Notice that $H'(z) < H' < H^{\circ}(z)$ and $aH'(z) < H'' < aH^{\circ}(z)$; that $H'(z) = aH'(z) = aH^{\circ}(z)$ when $z = af''$; and that $H'(z) < aH'(z) < H^{\circ}(z) < aH^{\circ}(z)$ for z smaller than af'' . Finally, at a given value w/p of the real wage, a labor-force level equal to $H'(w/p)$ always corresponds to negative net profits, whereas the opposite is true for $H^{\circ}(w/p)$, which moreover corresponds to profit maximization in the absence of quantity constraints.

(3.3a) if $w > w^{\wedge}$, $py < b(m + Bw)$, $py > m - dBw/c$

(3.4) $l(p,w,m,y) = 0$, $m(p,w,m,y) = m - py$,

(3.4a) if $w > w^{\wedge}$, $py < b(m + Bw)$, $py \leqslant m - dBw/c$

3.4 Individual consumers under commodity rationing and unemployment
This case reduces immediately to the previous two by assumptions 13 and 14; employed consumers are described by (3.3) and (3.4), and unemployed consumers by (3.2).

3.5 Production sector without quantity constraints ('walrasian producers')

(3.5) $Z(p,w) = f[f'^{-1}(w/p)]$, $H(p,w) = f'^{-1}(w/p)$,

(3.5a) if $0 < w/p \leqslant af''$;

(3.6) $Z(p,w) = 0$, $H(p,w) = 0$,

(3.6a) if $w/p > af''$;

The signs of the partial derivatives of the notional demand and supply functions of the production sector (at a positive production level) are as follows:

$$\partial H/\partial(w/p) = 1/f''[f'^{-1}(w/p)] = 1/f''(H) < 0 ;$$

$$\partial Z/\partial(w/p) = f'(H)\,[\partial H/\partial(w/p)] = f'(H)/f''(H) < 0 ;$$

(3.7)

$$\partial H/\partial p = -(w/p^2)\,[\partial H/\partial(w/p)] > 0 ; \quad \partial Z/\partial p = f'(H)\,[\partial H/\partial p] > 0 ;$$

$$\partial H/\partial w = (1/p)\,[\partial H/\partial(w/p)] < 0 ; \quad \partial Z/\partial w = f'(H)\,[\partial H/\partial w] < 0 ,$$

where for $w/p = af''$ the expressions must be taken as limits from the left.

3.6 Production sector with a constraint on the demand for labor
Let us denote by N the constraint on the size of the labor-force that can be employed; we then have:

(3.8) $Z(p,w,N) = f(N)$,

(3.8a) if $0 < w/p \leqslant af''$, $N < f'^{-1}(w/p)$, $pf(N) \geqslant wN$;

(3.9) $Z(p,w,N) = 0$,

(3.9a) if $0 < w/p \leqslant af''$, $N < f'^{-1}(w/p)$, $pf(N) < wN$.

Moreover,

(3.10) $\partial Z/\partial N = f'(N) > 0$, $\partial Z/\partial w = \partial Z/\partial p = 0$,

if (3.8a) holds.

3.7 Production sector with a constraint on the supply of the commodity
If X denotes the constraint on the output that can be sold, we have:

(3.11) $H(p,w,X) = f^{-1}(X/l)$,

(3.11a) if $0 < w/p \leqslant af''$, $X < l\,f[f'^{-1}(w/p)]$, $pX/l \geqslant wf^{-1}(X/l)$;

(3.12) $H(p,w,X) = 0$,

(3.12a) if $0 < w/p \leqslant af''$, $X < l\,f[f'^{-1}(w/p)]$, $pX/l < wf^{-1}(X/l)$.

Moreover,

(3.13) $\partial H/\partial X = 1/\{l\,f'[f^{-1}(X/l)]\} > 0$; $\partial H/\partial w = \partial H/\partial p = 0$ (if (3.11a) holds) .

3.8 Production sector with two constraints

Nothing has to be decided from producers in this case, and in general only one of the constraints is binding at a positive level of production, so that the results of one of the last two sections, whichever is relevant, still apply; in the particular case: $X = l\,f(N) < l\,f[f'^{-1}(w/p)]$, for which it is possible to say, with some abuse of language, that both constraints are binding, the results of both sections can be used indifferently.

3.9 Aggregate consumers behavior without constraints

The aggregate consumers behavior is specified by the sum of all individual demands for the commodity, by the individual supply of labor-time, and by the number of consumers wishing to sell their labor-force. We have:

$$(3.14) \qquad X(p,w,m) = Nb(m + Bw)/p \;,\quad l(p,w,m) = [(b + d)Bw - cm]/w \;,\quad N(p,w,m) = N \;,$$

$$(3.14a) \qquad \text{if} \quad w > w^{\wedge} \;;$$

$$(3.15) \qquad X(p,w,m) = (Nbm)/(b + d)p \;,\quad l(p,w,m) = 0 \;,\quad N(p,w,m) = 0 \;,$$

$$(3.15a) \qquad \text{if} \quad w \leq w^{\wedge} \;;$$

and

$$(3.16) \qquad \begin{matrix} \partial X/\partial m > 0 \;, & \partial X/\partial p < 0 \;, & \partial X/\partial w > 0 \\ \partial l/\partial m < 0 \;, & \partial l/\partial p = 0 \;, & \partial l/\partial w > 0 \end{matrix} \qquad (\text{if } w > w^{\wedge}) \;.$$

3.10 Aggregate consumers behavior in presence of unemployment

Let H denote the total amount of labor-force demanded by producers; we have:

$$(3.17) \qquad \begin{matrix} X(p,w,m,H) = H[b(m + Bw)]/p + [(N - H)bm]/[(b + d)p] = b\{mN + [(b + d)Bw - cm]H\}/(b + d)p \\ l(p,w,m,H) = l(p,w,m) = [(b + d)Bw - cm]/w \;, \end{matrix}$$

$$(3.17a) \qquad \text{if} \quad w > w^{\wedge} \;,\quad H < N \;;$$

$$(3.18) \qquad \partial X/\partial H = b[(b + d)Bw - cm]/(b + d)p = bwl(p,w,m)/(b + d)p \;,$$

if (3.17a) holds.

3.11 Aggregate consumers behavior under commodity rationing

Let Z denote the production scale (i.e., production level per unit of working time) chosen by producers; we have:

$$(3.19) \qquad py = (plZ - G)/N \;, \quad \text{and}$$

$$(3.20) \qquad l(p,w,m,Z) = l(p,w,m,y) = (dBw + cpy - cm)/(c + d)w \;,\quad N(p,w,m,Z) = N \;,$$

$$(3.20a) \qquad \text{if} \; \begin{cases} l(p,w,m)Z = Z[(b + d)Bw - cm]/w < Nb(m + Bw)/p + G/p = X(p,w,m) + G/p \;, \\ \text{and the conditions in (3.3a) are satisfied;} \end{cases}$$

$$(3.21) \qquad l(p,w,m,Z) = 0 \;,\quad N(p,w,m,Z) = 0 \;,$$

$$(3.21a) \qquad \text{if} \; \begin{cases} l(p,w,m)Z = Z[(b + d)Bw - cm]/w < Nb(m + Bw)/p + G/p = X(p,w,m) + G/p \;, \\ \text{and the conditions in (3.4a) are satisfied.} \end{cases}$$

Finally,

$$(3.22) \qquad \begin{matrix} \partial l/\partial m < 0 \;, & \partial l/\partial p > 0 \;, & \partial l/\partial w < 0 \;, \\ & \partial l/\partial y = cp/(c + d)w > 0 & \end{matrix} \qquad (\text{if (3.20a) holds}) \;.$$

3.12 Aggregate consumers behavior under commodity rationing and unemployment

We have:

$$py = [1/H]\,[plZ - G - (N - H)bm/(c + d)] \quad ; \tag{3.23}$$

$$l(p,w,m,Z,H) = l(p,w,m,y) = (dBw + cpy - cm)/(c + d)w \tag{3.24}$$

$$\text{if } \begin{cases} Z[(b + d)Bw - m]/w < b\{mN + [(b + d)Bw - m]H\}/(b + d)p + G/p = X(p,w,m,H) + G/p \\ \text{and the conditions in (3.3a) and (3.16a) are satisfied;} \end{cases} \tag{3.24a}$$

$$l(p,w,m,Z,H) = 0 \tag{3.25}$$

$$\text{if } \begin{cases} Z[(b + d)Bw - m]/w < b\{mN + [(b + d)Bw - m]H\}/(b + d)p + G/p = X(p,w,m,H) + G/p \\ \text{and the conditions in (3.4a) and (3.16a) are satisfied.} \end{cases} \tag{3.25a}$$

Finally,

$$\partial l/\partial H = (\partial l/\partial y)\,(\partial y/\partial H) \quad \text{(to be computed from (3.21) and (3.22))} \quad , \tag{3.26}$$

effect with an unambiguously negative sign, once hypothesis 14 is interpreted to mean that the analysis of the model must be restricted to values of the rationing quota for employed consumers exceeding the amount guaranteed to the unemployed ones; namely, when:

$$plZ > G + bNm/(b + d) \quad . \tag{3.27}$$

4. SPLITTING THE PRICE-WAGE PLANE IN REGIONS OF SPECIFIC TYPES OF FIX-PRICE EQUILIBRIA: RESULTS

In this section we shall state without proofs the main results concerning the existence of the walrasian equilibrium and the classification of the various fix-price equilibria associated to specific vectors of prices and wages (detailed proofs can be found in Battinelli 1982 and 1983). We recall that the fix-price classification specifies what side of the market is subject to rationing on each market; and that keynesian unemployment, repressed inflation, classical unemployment, underconsumption (or overcapitalization) respectively occur when the active constraints are on commodity supply and labor-force supply, commodity demand and labor-force demand, commodity demand and labor-force supply, commodity supply and labor-force demand.

4.1. Walrasian equilibrium

Non trivial clearing of both markets occurs if:

$$\begin{cases} w > w^{\wedge} \qquad 0 < w/p \leqslant af'' \qquad N = f'^{-1}(w/p) \\ G/p + [Nb(m + Bw)]/p = [f\{f'\ (w/p)\}]\,[(b + d)Bw - cm]/w \quad . \end{cases} \tag{4.1}$$

The condition $H'' \leqslant N < H^{\wedge}$ is necessary and sufficient for existence and uniqueness of the walrasian equilibrium.

4.2. Keynesian unemployment

The set K of price vectors (p,w) corresponding to fix-price equilibria of keynesian unemployment type, together with the corresponding levels Y,H of output and employment, is characterized by the following conditions:

$$\begin{cases} Y = l(p,w,m)f(H) = (1/w)[(b + d)Bw - m]f(H) \\ 0 = f(H) - [X(p,w,m,H) + G/p]/l(p,w,m) = f(H) - \dfrac{w}{p}\,\dfrac{b}{b + d}\left\{ H + \dfrac{mN + (b + d)G/b}{(b + d)Bw - cm} \right\} \\ H < N \qquad w > w^{\wedge} \\ 0 < w/p \leqslant af'' \qquad pf(H) \geqslant wH \qquad H < f'^{-1}(w/p) \quad , \end{cases} \tag{4.2}$$

and is always non empty. Such set is strictly contained in the angular region defined by the two conditions:

$$(4.3) \qquad w > w^{\wedge} \qquad\qquad 0 < w/p \leq af'' \quad .$$

If $N \geq H^{\wedge}$, keynesian unemployment equilibria exist for every real wage in this region; for a fixed real wage $w/p = z$, they fill a left-open segment of the corresponding line of slope z through the origin, consisting of vectors whose w-component satisfies:

$$(4.4) \qquad g(H^*(z),z) < w \leq g(aH'(z),z) \quad .$$

where the function g is defined as follows:

$$(4.5) \qquad g(H,z) = w^{\wedge} + \frac{\mathbf{m}N + (1 + d/b)G}{(b + d)B\,[(1 + d/b)zf(H) - H]} \quad ,$$

If $H'' \leq N \leq H^{\wedge}$, keynesian unemployment equilibria still exist for every real wage in the region; however, the intersection of K with each line of slope z is defined by condition (4.4) only if $z \geq f'(N)$; when $z < f'(N)$, the condition becomes:

$$(4.6) \qquad g(N,z) < w \leq g(aH'(z),z) \quad .$$

If $N \leq H''$, keynesian unemployment equilibria exist only for real wages (in the region and) strictly lower than $f(N)/N$; the intersection of K with each line of slope z is defined by condition (4.6).

In terms of comparative statics analysis, an increase in the absolute level of wages (and prices) with the real wage fixed and no accomodation in the cash holdings of both the private and the public sector is associated to a contraction of employment.

We conclude with a list of the signs of the static employment and output multipliers:

	$dH/dp\big\vert_K < 0$		$dY/dp\big\vert_K < 0$
	$dH/dw\big\vert_K$ > 0 above MM'		$dY/dw\big\vert_K$ > 0 above MM'
(4.7)	< 0 below MM'		ambiguous below MM'
	$dH/d\mathbf{m}\big\vert_K > 0$		$dY/d\mathbf{m}\big\vert_K$ ambiguous
	$dH/dG\big\vert_K > 0$		$dY/dG\big\vert_K > 0$.

where MM' is the locus of the points of vertical tangency of the constant employment curves.

4.3 Repressed inflation

The set I of price vectors (p,w) corresponding to non trivial fix-price equilibria of repressed inflation type, together with the corresponding levels Z,H,y of production scale, employment and individual constraint on commodity demand, is characterized by the following conditions:

$$(4.8) \qquad \begin{array}{lll} Z = f(N) & H = N & py = (p1Z - G)/N \\ w > w^{\wedge} & py < b(\mathbf{m} + Bw) & py > \mathbf{m} - dBw/c \\ 0 < w/p \leq af'' & N < f'^{-1}(w/p) & pf(N) \geq wN \\ \multicolumn{3}{c}{Z[(b + d)Bw - \mathbf{m}]/w < [Nb(\mathbf{m} + Bw)]/p + G/p \quad .} \end{array}$$

I is non empty only if $N < H''$ or if $H'' \leq N < H^{\wedge}$ and $af(N)/f'(N) < (1 + d/c)$; is strictly contained in the angular region defined by the two conditions:

$$(4.9) \qquad 0 < w/p \leq af'' \qquad\qquad w > c\mathbf{m}/dB \quad ,$$

and characterized further by the two conditions:

$$(4.10) \qquad N - [c/(c+d)][p/w]f(N) \geq 0$$

$$(4.11) \qquad \begin{cases} w/p \leq af(N) & (\text{if } N < H'') \\ w/p < f'(N) & (\text{if } H'' \leq N < \hat{H}) \end{cases} .$$

The intersection of R with each line through the origin, corresponding to a real wage z satisfying conditions (4.9)-(4.11), is the open segment consisting of vectors whose w-component satisfies:

$$(4.12) \qquad (c/dB)[m + z(1 + d/c)G/f(N)] < w < g(N,z) ,$$

unless z is such, say $z = z'$, that condition (4.10) is satisfied as an equality (in the last case only the vector $(w'/z',w')$ belongs to R, with w' equal to $(c/dB)(m + G/N)$).

The issue of the static employment multipliers does not arise in this region, which is characterized by full employment. The signs of the static output multipliers are the following:

$$(4.13) \qquad \begin{array}{llll} dY/dp_R > 0 & & dY/dw_R & \text{ambiguous} \\ dY/dm_R < 0 & & dY/dG_R < 0 & . \end{array}$$

4.4. Classical unemployment

The set C of price vectors (p,w) corresponding to non trivial fix-price equilibria of classical unemployment type, together with the corresponding levels Z,Y,H,y of production scale, output, employment and individual constraint on commodity demand, is characterized by the following conditions:

$$(4.14) \qquad \begin{array}{ll} H = f'^{-1}(w/p) & Z = f[f'^{-1}(w/p)] \\ Y = Z[dBw + cpy - cm]/[(c+d)w] & py = [1/H][pZ - G - (N-H)bm/(b+d)] \\ 0 < w/p \leq af'' & H < N \\ Z[(b+d)Bw - m]/w < b\{mN + [(b+d)Bw - cm]H\}/(b+d)p + G/p & \\ w > \hat{w} \qquad py < b(m + Bw) & py > m - dBw/b \qquad pZ > G + bNm/(b+d) . \end{array}$$

C is non empty only if $N \geq H''$, strictly contained in the region defined by the conditions in (4.3), and characterized further by the two conditions:

$$(4.15) \qquad H^*(w/p) - [c/(c+d)][p/w]f(H^*(w/p)) \geq 0$$

$$(4.16) \qquad w/p > f'(N) .$$

On each line through the origin, corresponding to a real wage z satisfying conditions (4.3) and (4.15)-(4.16), the vectors in C fill an open segment delimited by the curve of equation:

$$(4.17) \qquad \frac{f(H^*(z))}{z}\,\frac{cm - dBw}{c+d} + (N - H^*(z))\frac{bm}{b+d} + G = \left[\frac{f(H^*(z))}{z}\,\frac{c}{c+d} - H^*(z)\right]\frac{bmz}{(b+d)w}$$

and the curve of equation $w = g(H^*(z),z)$ (already introduced in section 4.2), unless z is such, say $z = z^*$, that condition (4.15) is satisfied as an equality (in the last case only the vector $(w^*/z^*,w^*)$ belongs to C, with $w^* = cm/dB + [c(c+d)/dBH^*(z^*)][(N - H^*(z^*))bm/(b+d) + G]$).

In this region producers are on their notional demand and supply schedules; hence the signs of the static employment multipliers have been in essence calculated in section 3.5, and will not be repeated. The signs of the static output multipliers are more involved to compute:

$$(4.18)\qquad \left.dY/dp\right|_C > 0 \qquad \left.dY/dw\right|_C \ \text{ambiguous} \qquad \left.dY/dm\right|_C < 0 \qquad \left.dY/dG\right|_C < 0 \quad .$$

4.5 Underconsumption (overcapitalization)

In this region consumers act according to their notional supply and demand schedules. It has been noticed already in section 3.8 that under the present hypotheses this can be thought to be possible only by some abuse of language. If this is done, the set U of price vectors (p,w) corresponding to fix-price equilibria of underconsumption type, together with the corresponding levels Y,H of output and employment, is characterized by the following conditions:

$$(4.19)\qquad Y = Nb(m + Bw)/p + G/p \qquad H = N \qquad w > w^{\wedge}$$
$$0 < w/p \leq af'' \qquad N < f'^{-1}(w/p) \qquad pf(N) \geq wN \quad ,$$

and coincides with the curve of equation $w = g(H^{*}(z),z)$, which separates the regions of keynesian unemployment and repressed inflation, and has been already introduced.

5. REFERENCES

BARRO,H.J. and GROSSMAN,H.I. 1971 A general disequilibrium model of income and employment. American Economic Review,vol.61,pp.82-93.

BATTINELLI,A. 1982 Fix-price equilibria with production function a la Georgescu Roegen. Preliminary version (mimeo).

BATTINELLI,A. 1983 Variable working time in macroeconomic equilibrium with rationing and fixed prices. Paper submitted at the VIII Symposium ueber Operations Research, Karlsruhe, August 22-25, 1983.

BENASSY,J.P. 1975 Neo-keynesian disequilibrium in a monetary economy. Review of Economic Studies,Vol.42,pp.503-523.

BOEHM,V. 1980 Preise, loehne und beschaeftigung. Mohr, Tubingen.

CLOWER,R.W. 1965 The keynesian counterrevolution: a theoretical appraisal. In: HAHN,F.H. and BRECHLING,F. (eds.) The theory of interest rates. Macmillan, London.

DREZE,J. 1975 Existence of an equilibrium under price rigidity and quantity rationing. International Economic Review,Vol.16,pp. 301-320.

FITOUSSI,J.P.and GEORGESCU-ROEGEN,N. 1980 An examination of the analytical foundations of disequilibrium theories. In: FITOUSSI,J.P.and MALINVAUD,E. (eds.) Unemployment in western countries, I.E.A. Conference Proceedings, pp.227-266. Macmillan, London.

GEORGESCU-ROEGEN,N.1971a The entropy law and the economic process. Harvard University Press, Cambridge, Massachussets.

GEORGESCU-ROEGEN,N.1971b Process analysis and the neoclassical theory of production. American Journal of Agricultural Economics, Vol.54,pp.279-294.

LEIJONHUFVUD,A. 1968 On keynesian economics and the economics of Keynes. Oxford University Press, Oxford.

MALINVAUD,E. 1977 The theory of unemployment reconsidered, Yrjo Jahnson Lecture at the University of Helsinki. Basil Blackwell, Oxford.

CONTINUOUS-TIME ASSET-PRICING MODELS: SELECTED RESULTS

B.A. Jensen
Institute of Finance
Copenhagen School of Economics and Business Administration
Howitzvej 60
DK-2000 Copenhagen F, Denmark

1. Introduction

The purpose of this paper is to present some selected results from the theory of continuous-time finance. It is intended for an audience familiar with stochastic calculus, but without special knowledge of the theory of finance.

The field was born around 1970 by Robert Merton, who in his two seminal papers (Merton (1971, 1973)) introduced the "expected utility approach" to continuous-time finance into the economic literature. In 1973 another seminal paper appeared by Black & Scholes (1973), who derived the first explicit formula for the pricing of European call options. Their analysis was based on an arbitrage argument. This kind of reasoning is, of course, very pervasive to all kind of economic theory, but they were apparently the first to spell out the mechanism in formulas in continuous time. We will call this the "arbitrage approach", with which we shall begin.

2. Arbitrage theory

The economic environment is supposed to be described by a certain number K of state-variables $\underline{S} = (S_1, S_2, \ldots, S_K)$ together with calendar time t. These state-variables are supposed to evolve according to a K-dimensional diffusion-process

1) $d\underline{S} = \underline{\mu}(\underline{S},t)dt + \underline{\underline{\sigma}}^T(\underline{S},t)dQ$

where Q is a Wiener-process of appropriate dimension. What these state-variables are is irrelevant for the time being; one can eventually think of them as instrumental variables. It is important, however, that a finite number K of such variables exist as sufficient statistics in order to have the environment described by a K-dimensional Markov-process.

Let $P_i(\underline{S},t)$ be the price of asset i. The stochastic differential of $P_i(\underline{S},t)$ is

2) $dP_i(\underline{S},t) = P_{it}dt + P^T_{i\underline{S}}d\underline{S} + \frac{1}{2}tr\{P_{i\underline{SS}}\,\underline{\underline{\sigma}}^T\underline{\underline{\sigma}}\}dt$

Upon substituting $d\underline{S}$ from equation 1) we have

3) $dP_i(\underline{S},t) = [P_{it} + P^T_{i\underline{S}}\underline{\mu} + \frac{1}{2}tr\{P_{i\underline{S}\underline{S}}\underline{\underline{\sigma}}^T\underline{\underline{\sigma}}\}]dt + P^T_{i\underline{S}}\underline{\underline{\sigma}}^T dQ \equiv (P_{it} + LP_i)dt + P^T_{i\underline{S}}\underline{\underline{\sigma}}^T dQ$

The basic idea in using arbitrage arguments is nothing but the <u>law of one price</u>: Identical assets, or asset portfolios that are perfect substitutes, must necessarily sell for identical prices. This in itself has no implication as to what these prices should be; it is only required that there are sufficiently many assets in the market in order to enable the construction of perfect substitutes, and thereby render an arbitrage operation a potential possibility. More precisely, it is required that one can construct a <u>riskless null-portfolio</u> or an <u>arbitrage portfolio</u>. In mathematical terms such a portfolio is any vector $\underline{X}$ with the properties

4) $\langle \underline{X}, \underline{P} \rangle = 0$ (zero net investment)

5) $\langle \underline{X}, \underline{P}_{S_j} \rangle = 0 \quad j = 1,2,\ldots,K$ (no risk)

In a competitive market, such a portfolio must have a zero expected return. Besides capital gains, this return could include continuously paid out streams of payments $\delta_i(\underline{S},t)$, so for any $\underline{X}$ satisfying 4) and 5) it must be the case that

6) $\langle \underline{X}, \frac{\partial \underline{P}}{\partial t} + L\underline{P} + \underline{\delta} \rangle = 0$

In terms of standard linear algebra this means that

$$\underline{X} \perp \text{span}\{\underline{P}, \frac{\partial \underline{P}}{\partial S_1}, \frac{\partial \underline{P}}{\partial S_2}, \ldots, \frac{\partial \underline{P}}{\partial S_K}\} \Rightarrow \underline{X} \perp (\frac{\partial \underline{P}}{\partial t} + L\underline{P} + \underline{\delta})$$

Therefore, coefficients λ and ρ_j, $j = 1,2,\ldots,K$ exist such that

7) $\frac{\partial P_i}{\partial t} + LP_i + \delta_i = \lambda P_i + \sum_{j=1}^{K} \rho_j \frac{\partial P_i}{\partial S_j}$

for any particular asset i.

To what extent can these coefficients, λ and ρ_j, be learned from the market? Consider first a riskless asset or asset portfolio, denoted as no. 0, for which $\frac{\partial P_o}{\partial S_j} = 0$, $j = 1,2,\ldots,K$. For this asset, eq. 7) is

8) $\frac{\partial P_o}{\partial t} + LP_o + \delta_o = \lambda P_o$

so <u>λ is the expected return on a riskless asset</u>. Similarly, construct portfolios $\underline{X}_j$ such that

$$X_j^T \left[\frac{\partial \underline{P}}{\partial \underline{S}}\right] = \text{j'th K-dimensional unit vector.}$$

For such portfolios eq. 7) will look like

9) $$\underline{X}_j^T \left(\frac{\partial \underline{P}}{\partial t} + L\underline{P} + \underline{\delta} \right) = \lambda(\underline{X}_j^T \underline{P}) + \rho_j$$

which means that ρ_j <u>is the payout - in excess of the riskless return - on a portfolio whose stochastic price-movements are exactly replicating the movements in the state-variable S_j</u>.

Equation 7) is a second-order p.d.e., which must be obeyed by the price P of any asset. In order to solve for P, some boundary specification is necessary. Assume that at some horizon time T it must be the case that

10) $$P(\underline{S},T) = \phi(\underline{S})$$

The simplest possible boundary specification, $\phi(\underline{S}) = 1$, is obtained for a unit discount bond maturing at time T, but for a number of contingent claims - such as options - it is equally straightforward to specify $\phi(\underline{S})$.

Leaving aside all regularity requirements, the solution to 7) can be given a stochastic representation. Define the stochastic process V as

11) $$V(u,\omega) = \exp\left(-\int_t^u (\lambda + \tfrac{1}{2}\underline{\rho}^T [\underline{\underline{\sigma}}^T \underline{\underline{\sigma}}]^{-1} \rho)\, dv - \int_t^u \underline{\rho}^T [\underline{\underline{\sigma}}^T \underline{\underline{\sigma}}]^{-1} \underline{\underline{\sigma}}^T dQ\right)$$

A standard application of Ito's lemma then shows that for any asset the stochastic process

$$P(\underline{S},u) \cdot V(u,\omega) - \int_u^T \delta(S(v),v) \cdot V(v,\omega)\, dv$$

is purely random with no drift-term, hence a martingale. Since $V(t,\omega) = 1$ it is the case that

12) $$P(\underline{S},t) = \underset{(\underline{S},t)}{E} \left\{ \phi(\underline{S}(T)) V(\underline{T},\omega) + \int_t^T \delta(\underline{S}(v),v) \cdot V(v,\omega)\, dv \right\}$$

In abstract terms, formula 12) gives the price of an asset as the expected value of the discounted future stream of payments, where the discount factors $V(v,\omega)$ constitute a stochastic process. For concrete applications of this formula, however, it is necessary to specify the process obeyed by the state-variables $\underline{S}$ and the measure induced hereby, as well as the nature of the risk-adjustment coefficients $\underline{\rho}(\underline{S},t)$.

For particular assets or classes of assets, variants of equation 7) have been derived by different authors. Its general form has been termed the "fundamental evaluation equation" by Cox, Ingersoll and Ross (1978).

3. Derivative assets

A derivative asset is an asset, whose price D is depending only upon calendar time and the price P of one other traded asset. In other words, P is the only state-variable necessary to determine the price-process D(P,t) of any asset, which is a derivative asset w.r.t. P. For this reason, P is sometimes referred to as a "valuation sufficient statistic".

Given the process

13) $dP = \mu dt + \sigma^T dQ$

we can immediately write down the p.d.e. satisfied by D, cf. equation 7):

14) $D_t + D_P\mu + \frac{1}{2}D_{PP}||\sigma||^2 + \delta = \lambda D + D_P\rho$

Since P itself is trivially a derivative asset w.r.t. P, we can determine ρ from equation 14) by inserting D(P,t) = P into eq. 14) to get

15) $\rho = \mu - \lambda P$

The famous Black- Scholes (1973) formula is easily derived from here. Consider a European call option on a stock, whose price P follows a geometric Brownian motion with constant coefficients:

16) $dP = \alpha P dt + \xi P dQ$

If the option expires at time T and has an exercise price E written upon it, the option price D(P,t) must satisfy the boundary condition

17) $D(P,T) = \max[P-E,0]$

From 15) and 16) we can determine $\mu = \alpha P$ and $\sigma = \xi P$. Assuming also that the riskless interest rate λ is constant, equation 11) takes the form

18) $V(u,\omega) = \exp[-(\lambda+\frac{1}{2}(\frac{\alpha-\lambda}{\xi})^2)(T-t) - (\frac{\alpha-\lambda}{\xi})(T-t)^{\frac{1}{2}} \cdot X(\omega)]$

where $X(\omega) \sim N(0,1)$.

Also, the solution to 16) is known to be

19) $P(T) = P(t)\exp[(\alpha - \frac{1}{2}\xi^2)(T-t) + \xi(T-t)^{\frac{1}{2}} \times X(\omega)]$

For such a European call option $\delta \equiv 0$, so applying formula 12) now tells us that its price $D(P(t),t)$ is

$$20)\quad D(P(t),t) = \int_{-\infty}^{\infty} \max\{P(t)\exp[(\alpha-\tfrac{1}{2}\xi^2)(T-t) + \xi(T-t)^{\frac{1}{2}}x]- E,0\}\times\frac{1}{\sqrt{2\pi}}\times e^{-\frac{1}{2}x^2}\times$$
$$\exp[-(\lambda+\tfrac{1}{2}(\tfrac{\alpha-\lambda}{\xi})^2)(T-t) - (\tfrac{\alpha-\lambda}{\xi})(T-t)^{\frac{1}{2}}x]dx$$

$$= \max\{P(t)\exp[(\alpha-\lambda-\tfrac{1}{2}\xi^2)(T-t) + \xi(T-t)^{\frac{1}{2}}x] - Ee^{-\lambda(T-t)},0\}\times$$
$$\frac{1}{\sqrt{2\pi}}\times\exp[-\tfrac{1}{2}(x+(\tfrac{\alpha-\lambda}{\xi})(T-t)^{\frac{1}{2}})^2]dx$$

Substituting $y=x+(\frac{\alpha-\lambda}{\xi})(T-t)^{\frac{1}{2}}$ this expression simplifies to

$$21)\quad D(P(t),t) = \max\{P(t)\exp[\xi(T-t)^{\frac{1}{2}}y - \tfrac{1}{2}\xi^2(T-t)]-Ee^{-\lambda(T-t)},0\}\frac{1}{\sqrt{2\pi}}e^{-\frac{1}{2}x^2}dx$$

The value of this integral can be calculated in a straigthforward manner in terms of the cumulative distribution function N for an N(0,1)-distribution as

$$22)\quad D(P(t),t) = P(t)\times N[\frac{\ln(P(t)/E) + (\lambda+\tfrac{1}{2}\xi^2)(T-t)}{\xi\sqrt{(T-t)}}] -$$
$$Ee^{-\lambda(T-t)}\times N[\frac{\ln(P(t)/E) + (\lambda-\tfrac{1}{2}\xi^2)(T-t)}{\xi\sqrt{(T-t)}}]$$

Equation 22) is the "Black-Scholes option pricing formula".
More complex options, like the ones treated by Geske(1979), are valued by a fundamentally similar procedure. The limitation to this procedure is that a solution to equation 13) , and thereby an exact distributional characterization of the future level of P, must be known.

Consider finally European call options on the same stock and with the same expiration date T but with variable exercise price E. For all these options the valuation equation 20) has the form

$$23)\quad D(P(t),t,E) = \int_{E}^{\infty} (x-E)pdf(x)dx$$

where pdf(x) is the density function of P(T).

By suitably combining long and short positions in options with varying exercise prices, one can create Arrow-Debreu securities w.r.t. the contingencies spanned by the level of the stock price P(T). E.g., let E be some exercise price and let h>0 be a real number. Then

$$24)\quad \frac{1}{h}[D(P,E-h)+D(P,E+h)-2D(P,E)] = \int_{E-h}^{E+h}\{1-\frac{1}{h}|x-E|\}pdf(x)dx$$

Thus pdf(x) serves as an evaluation operator for these Arrow-Debreu securities, and it can be inferred from the option prices in 23) as

$$25)\quad \frac{\partial^2 D(P(t),t,E)}{\partial E^2} = pdf(E)$$

This argument was first suggested by Breeden and Litzenberger(1979) in an intuituve fashion. A rigorous analysis is carried out in Bick(1982).

4. The term structure of interest rates

Let $P(t,T)$ be the price of a unit discount bond maturing at time T. For such a bond, $\phi(\underline{S}) \equiv 1$, so in terms of the general valuation formula its price is

$$26)\quad P(t,T) = \underset{(\underline{S},t)}{E}\,[V(T)]$$

The yield to maturity $R(t,T)$ is defined by the relation

$$27)\quad P(t,T) = \exp(-R(t,T)(T-t))$$

The term structure of interest rates is the relation between T-t and $R(t,T)$.
A number of authors have examined this relation under a variety of assumptions about valuation sufficient statistics. Assuming that the instantaneous riskless interest rate λ is such a valuation sufficient statistic, the valuation equation 7) becomes

$$28)\quad P_t + P_\lambda \alpha + \tfrac{1}{2} P_{\lambda\lambda}\xi^2 = \lambda P + \rho P_\lambda$$

where λ is assumed to follow the Ito-process

$$28)\quad d\lambda = \alpha(\lambda)dt + \xi(\lambda)dQ$$

The risk adjustment coefficient ρ cannot be determined by pure arbitrage arguments, however. Using 28) we can write the dynamics of P as

$$30)\quad dP = (\lambda P + \rho P_\lambda)dt + P_\lambda \xi dQ$$

Bond prices are subject to only one kind of risk, namely ξdQ, and the amount of risk in a bond can be measured by the magnitude P_λ. Thus, ρ is the per unit price of risk in the bond market, hence a potentially observable variable. However, it cannot be related á priori to other parameters in the model.

Vasicek(1977) examined this model, taking 29) to be the Ornstein-Uhlenbeck process with mean-reverting drift; i.e. $\alpha(\lambda) = \beta(\gamma-\lambda)$ and $\xi(\lambda)$ constant. Dothan assumed a driftless geometric random walk, i.e. $\alpha(\lambda) = 0$ and $\xi(\lambda) = \lambda\times\sigma$. Cox,Ingersoll and Ross(1978,1981) examined the socalled Bessel-process with mean-reverting drift, i.e. $\alpha(\lambda) = \beta(\gamma-\lambda)$ and $\xi(\lambda) = \sqrt{\lambda}\times\sigma$.

For reasons of space limitations, no calculations can be performed here, but as an example, Vasicek's results are given. Like other authors, he had to assume in an ad hoc manner that ξ is a constant. With this assumption, the formulas for P and R can be written as

$$31)\quad P(t,T,\lambda) = \exp[\frac{1}{\beta}(R(\infty)-\lambda)(1-e^{-\beta(T-t)})-(T-t)R(\infty)-\frac{\xi^2}{4\beta^3 T}(1-e^{-\beta(T-t)})^2]$$

$$32)\quad R(t,T,\lambda) = R(\infty) + (\lambda-R(\infty))\frac{1}{\beta T}(1-e^{-\beta T}) + \frac{\xi^2}{4\beta^3 T}(1-e^{-\beta(T-t)})^2$$

where $R(\infty) = \gamma + \frac{\xi}{\beta} - \frac{1}{2}(\frac{\xi}{\beta})^2$.

Depending upon the level of the spot rate λ, the yield curve 32) is either monotone or humped.

Richard(1978) analyzed a similar model using λ and the expected rate of commodity price inflation as choice of two valuation sufficient statistics. In a series of papers, see e.g. Brennan&Schwartz(1979,1982), the latter two authors have discussed a bond-pricing model with λ and the consol rate, i.e. the "endpoints" of the yield curve, as state variables.

5. CAPM

The fundamental valuation equation 7) must be satisfied for any asset i as well as for the price $P_{\underline{X}}$ of any portfolio of assets. I.e.

$$33)\quad \frac{\partial P_{\underline{X}}}{\partial t} + LP_{\underline{X}} + \delta_{\underline{X}} = P_{\underline{X}} + \langle \underline{\rho}, \frac{\partial P_{\underline{X}}}{\partial \underline{S}} \rangle \equiv E[dP_{\underline{X}}]$$

Defining the numbers β_i to be

$$34)\quad \beta_i = \frac{\langle \underline{\rho}, \partial P_i/\partial \underline{S} \rangle}{\langle \underline{\rho}, \partial P_{\underline{X}}/\partial \underline{S} \rangle}$$

we can combine 7) and 33) to the following expression:

$$35)\quad E[dP_i] - \lambda P_i = \beta_i \times \{E[dP_{\underline{X}}] - \lambda P_{\underline{X}}\}$$

Equation 35) is a technical implication, relating the expected payout in excess of the riskfree return on any individual asset i to the same excess expected return on the portfolio $\underline{X}$. As a defining property for the β_i's we have

$$36)\quad \langle \underline{X}, \underline{\beta} \rangle = 1$$

The Capital Asset Pricing Model - a cornerstone in modern finance - comes about when $\underline{X}$ is chosen as the market portfolio $\underline{M}$, and the risk adjustment coefficients $\underline{\rho}$ satisfy

$$37)\quad \underline{\rho} \propto [\underline{\underline{\sigma}}^T \underline{\underline{\sigma}}]\frac{\partial P_{\underline{M}}}{\partial \underline{S}}$$

Sufficient conditions to bring about 37) are examined next.

6. Expected utility maximization

The risk adjustment coefficients $\underline{\rho}$ must ultimately be linked to risk-attitudes and intertemporal consumption-preferences. Consider an investor, whose objective is to maximize the expected utility of the future stream of consumption with a possible bequest for terminal wealth:

$$38)\quad J(W,\underline{S},t) = \max_{\{\underline{X},C\}} \; E_{(\underline{S},t)} \int_t^T U(C(v),v)dv + B(W(T))$$

The decision function $\underline{X}$ is his portfolio choice concerning risky assets. The remainder of his wealth, i.e. $(1-\underline{e}^T\underline{X})\times W$, is allocated to the risk-free asset.

The investment opportunity set is given by the dynamics of asset prices P_j. To ease the notation the vector of diffusion coefficients for P_j is denoted by $\underline{\phi}_j$, and these vectors make up the matrix $\underline{\underline{\phi}}$. Similarly, we assume w.l.o.g. that $\underline{\delta} \equiv \underline{0}$. Hence the dynamics of P_j is

$$39)\quad dP_j = (P_{jt}+LP_j)dt + \phi_j^T dQ \equiv P_j(\tilde{L}P_j)dt + P_j\underline{\tilde{\phi}}_j^T dQ$$

The budget constraint to be observed by this investor is then

$$40)\quad dW = [<\underline{X},\tilde{\underline{L}}\underline{P}-\lambda\underline{e}>+\lambda]Wdt - Cdt + W\underline{X}^T\underline{\underline{\tilde{\phi}}}^T dQ$$

The Bellman optimality equation for this stochastic control problem is

$$41)\quad 0 = \max_{\{\underline{X},C\}} [U(C,v)+J_t+J_W W\times[\underline{X},\tilde{\underline{L}}\underline{P}-\lambda\underline{e}]+J_W(\lambda W-C)+J_{\underline{S}}\underline{\mu} + \tfrac{1}{2}J_{WW}W^2\underline{X}^T(\underline{\underline{\tilde{\phi}}}^T\underline{\underline{\tilde{\phi}}})\underline{X}+ \tfrac{1}{2}J_{\underline{SS}}[\underline{\underline{\sigma}}^T\underline{\underline{\sigma}}]+WJ^T_{W\underline{S}}[\underline{\underline{\sigma}}^T\underline{\underline{\tilde{\phi}}}]\underline{X}]$$

The first order conditions for this problem are

$$42)\quad U_C = J_W$$

$$43)\quad [\underline{\underline{\tilde{\phi}}}^T\underline{\underline{\tilde{\phi}}}]\underline{X} = -\frac{J_W}{J_{WW}W}[\tilde{L}\underline{P} - \lambda\underline{e}] - \frac{J^T_{W\underline{S}}}{J_{WW}W}[\underline{\underline{\sigma}}^T\underline{\underline{\tilde{\phi}}}]$$

The optimal portfolio $\underline{X}$ of risky assets can be found from 43), provided the matrix $[\underline{\underline{\tilde{\phi}}}^T\underline{\underline{\tilde{\phi}}}]$ is of full rank. We will not deal with the properties and interpretations of $\underline{X}$ in this paper; a thorough treatment hereof can be found in chapters 3 and 4 of Jensen(1982). Instead, the structure of risk premia, i.e. the expected return in excess of the riskfree return, is examined. These risk premia are found from 43) as

$$44)\quad \tilde{L}\underline{P} - \lambda\underline{e} = -\frac{J_{ww}W}{J_w}[\tilde{\underline{\underline{\phi}}}^T\tilde{\underline{\underline{\phi}}}]\underline{X}^T - \frac{J_{w\underline{S}}^T}{J_w}[\underline{\underline{\sigma}}^T\tilde{\underline{\underline{\phi}}}]$$

It turns out that everything in this setup is related to the process of the marginal utility of wealth J_w. The dynamics of this process can be calculated as

$$45)\quad \frac{dJ_w}{J_w} = \frac{1}{J_w}LJ_w dt + [\frac{J_{ww}W}{J_w}\underline{X}^T\tilde{\underline{\underline{\phi}}}^T + \frac{J_{w\underline{S}}^T}{J_w}\underline{\underline{\sigma}}^T]dQ + \frac{1}{J_w}J_{wt}dt$$

Combining 44) with 45) it follows that the risk premium on any asset is determined as minus its covariance with the rate of change in the marginal utility of wealth. This result is due to Cox, Ingersoll and Ross(1978). It can, however, also be related to the process of consumption as noted by Breeden(1979). To obtain this result we take the stochastic differential in relative terms in the first order condition 42), denoting by $(dC)_s$ the diffusion part of the process dC:

$$46)\quad \frac{1}{U_C}[U_{Ct} + LU_C]dt + \frac{1}{U_C}U_{CC}(dC)_s = \frac{1}{J_w}dJ_w$$

Comparing the diffusion parts on both sides of 46) leads to the following expression for the risk premia:

$$47)\quad \tilde{L}\underline{P} - \lambda\underline{e} = -\frac{U_{CC}}{U_C}(dC)_s^T\tilde{\underline{\underline{\phi}}}$$

In an economy where equilibrium prices are determined by the interaction of utility maximizing agents, arbitrage opportunities are obviously eliminated. This means that all relations established above using only arbitrage arguments are valid here too. From the fundamental valuation equation we have the relation

$$48)\quad \tilde{L}P_j - \lambda = \frac{1}{P_j}\times\langle\underline{\rho},\frac{\partial P_j}{\partial\underline{S}}\rangle$$

The β-coefficients in 35) can now be expressed in terms of asset price covariation with consumption or the rate of change in the marginal utility of wealth by means of 45) and 47). They are

$$49)\quad \beta_j = \frac{(dC)_s^T\phi_j}{(dC)_s^T\underline{\underline{\phi}}\underline{M}} = \frac{(dJ_w)_s^T\phi_j}{(dJ_w)_s^T\underline{\underline{\phi}}\underline{M}}$$

The standard CAPM-model, cf. equation 37), is identical to 49) only when $J_{\underline{wS}} = \underline{0}$. For a more thorough discussion of the kind of investor preferences and/or stochastic environment that will bring about this, the reader is referred to chapter 3 of Jensen(1982). However, the point is that the portfolio choices $\underline{X}$ of different individuals necessarily become proportional and consequently proportional to the market portfolio in equilibrium. Besides, the standard CAPM-model assumes stationarity of the involved distributions, which leads to the following fundamental problem.

7. Rosenberg-Ohlson's dilemma

Let there be a number of investors indexed by i, and let investor i have a fraction ω_i of aggregate wealth. Then as an accounting identity we have

$$50)\quad \underline{P} = \underline{M}(\sum_i W_i)$$

Taking the stochastic differential on both sides of 50) and focusing on the diffusion part hereof we have

$$51)\quad \underline{\tilde{\Phi}}_j^T = \left(\frac{dM_j}{M_j}\right)_s^T + (\sum_i \omega_i \underline{X}_i^T)\underline{\tilde{\Phi}}_j^T$$

This can also be stated as

$$52)\quad [I - (\sum_i \omega_i \underline{X}_i^T)']\underline{\tilde{\Phi}}_j^T = \left(\frac{dM_j}{M_j}\right)_s^T$$

and a little reflection shows that the rank of the covariance matrix of market shares must be the same as the rank of $[\underline{\tilde{\Phi}}^T\underline{\tilde{\Phi}}]$. In particular, stationarity of distributions leading to constant portfolio choices $\underline{X}_i$ and thereby constant market shares M_j leads to a degeneracy of the covariance matrix of asset price movements. This rank must necessarily be one in case of stationary distributions and portfolio choices.

8. References

Bick, Avi(1982): Comments on the Valuation of Derivative Assets. Journal of Financial Economics, vol.10 no.3, pp.331-345.

Black, F.& Scholes, M. (1973): The Pricing of Options and Corporate Liabilities. Journal of Political Economy, vol.81, pp.673-654.

Breeden, D.T.(1979): An Intertemporal Asset Pricing Model with Stochastic Consumption and Investment Opportunities. Journal of Financial Economics, vol 7, pp.265-296.

Breeden, D.T.& Litzenberger, R.H. (1978): Prices of state-contingent claims implicit in option prices. Journal of Business, vol.51, pp. 621-651.

Brennan, M.& Schwartz, E. (1979): A Continuous Time Approach to the Pricing of Bonds. Journal of Banking and Finance, vol.3, pp.133-175.

Brennan,M.& Schwartz,E. (1982): An Equilibrium Model of Bond Pricing and a Test of Market Efficiency. Journal of Financial and Quantitative Analysis,vol.17,no.3,pp.301-330.

Cox,J.C.,Ingersoll,J.E.& Ross,S.A. (1978): A Theory of the Term Structure of Interest Rates. Research Paper no. 468, GSB, Stanford University.

Cox,J.C.,Ingersoll,J.E.& Ross,S.A. (1981): A Re-examination of Traditional Hypotheses about the Term Structure of Interest Rates. Journal of Finance,vol.36,no.4,pp.769-799.

Dothan,L.U.(1978): On the Term Structure of Interest Rates. Journal of Financial Economics,vol.6,pp. 59-69,

Geske,Robert (1979): The Valuation of Compound Options. Journal of Financial Economics,vol.7,pp.63-81.

Jensen,B.A. (1982): Contributions to the Theory of Portfolio Selection and General Equilibrium in Exchange Economies. Unpublished thesis,Aarhus and Copenhagen.

Richard,S.F. (1978): An Arbitrage Model of the Term Structure of Interest Rates. Journal of Financial Economics,vol.6,pp 3-57.

Rosenberg,B.& Ohlson,J.A. (1976): The Stationary Distribution of Return and Portfolio Separation in Capital Markets: A Fundamental Contradiction. Journal of Financial and Quantitative Analysis,vol.11,pp. 393-402.

Vasicek,O.A. (1977): An Equilibrium Characterization of the Term Structure of Interest Rates. Journal of Financial Economics, vol.5,pp.177-188.

THE NATIONAL INVESTMENT MODEL - "N.I.M."

J. CANETTI and D. KOHN
ETUDES ECONOMIQUES GENERALES
ELECTRICITE DE FRANCE
2, rue Louis Murat
75008 - PARIS

1 - INTRODUCTION

The basic aim of the NIM is to help in the choice of thermal (conventional and nuclear) generating facilities investments and to draw up a picture of the possible trend of the national park of generating facilities in the future. Apart from weekly pumping stations, it is not concerned with the choice of hydropower stations.

The main output of this model is the optimum investment programme i.e. the schedule of the capacities to be commissioned (or decommissioned) each year for the various types of equipments.

The criterion of choice corresponds to the objective set for a quasimonopolistic establishment, as EDF is, of the official sector : to meet demand at least cost.

The purpose is to determine the capital flow which minimizes the global discounted cost of all expenses related to electricity supply :

- investment costs,
- operating costs,
- fuel costs,
- shortage costs,

on a long period of time.

This optimization problem is solved by optimal control theory through a "steepest descent" algorithm (and a process to speed convergence).

Thanks to the large number of units of each type of the French electric system, we may aggregate the units into homogeneous groups of plants which allows us to do a continuous optimization.

The model uses an iterative process requiring the criterion calculus at each step, for all the concerned years (this is done by simulation of the matching of production and consumption).

In this regard, the NIM starts from an initial investment programme and modifies it until the optimum is reached.

The optimization process consists of a back and forth shuttling beween two main sub-fonctions :

- one is in charge of calculating the total cost (which is the economic criterion) for a given park,

- the other has to distort the configuration of the park, so that the cost is reduced ; the mathematical procedure is an algorithm of gradient (steepest descent algorithm).

The new park thus calculated is returned to the initial function up to the moment when a convergence test shows that the proposed programme has reached virtually optimum conditions (as no further worthwhile reduction in costs can be achieved).

The approach is schematised in the diagram below :

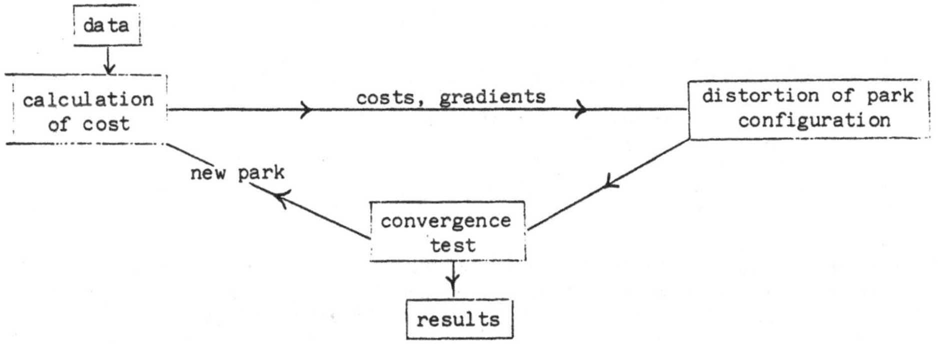

The main sub-functions and the general problems encountered together with the data needed and the desired output are briefly described in the following paragraphs.

The mathematical formulation of the problem is given in the annexe.

2 - ARCHITECTURE OF THE MODEL

2.1 - Calculation of the criterion : total cost

This function calculates the aggregate present discounted value and the corresponding gradients of a given park, i.e. the derivatives of this value with respect to installed power and yearly capital costs.

Total cost has two components : the first is fixed costs grouping those that are independent of the management of facilities ; the second component is the management cost associated with their use.

2.11 - Fixed costs

These include :

- capital costs which must be granted to build the various units of equipments ;
- fixed operating costs, which are maintenance costs of all in service equipments, whether they are in use or not (mainly maintenance and staff costs).

2.12 - Operating costs

They are the fuel costs corresponding to the effective use of the various units of equipments and possibly a shortage cost.

This management cost depends on the demand forecast for each year and each week, as well as on the availability of equipments. These two parameters are subject to annual, hourly and seasonal variations, as well as to a certain number of random phenomena.

The expected management cost is calculated by simulation of the matching of production and consumption. The operating of the park is simulated for each week in the event of various occasional occurences in different hourly brackets.

2.2 - The distortions of the park : the optimization algorithm

This function, given the computed aggregate discounted cost and its gradients, determines a new investment programme.

2.21 - Preliminary calculations

Given each additionnal unit of equipment available at the date t, the sum of management and shortage savings is calculated by adding the gradients backwards over time ; this quantity ψ_i^t is the marginal utilisation value of the unit of equipment i at the date t.

ψ_i^t represents the reduction of the present discounted value of future management and shortage costs per kW produced by an additional unit of equipment i.

This reduction must be balanced against the discounted cost of investment j_i^t in this unit of equipment to assess the corresponding advantage or disadvantage (i.e. the positive or negative yield $\psi_i^t - j_i^t$).

The selection criterion between units of equipement and commissioning dates can be expressed as follows : without external constraints, the marginal yield is nul at optimum.

By contrast, it may be necessary to accept negative yields by reason of constraints on the minimum investment flow or positive yields because of constraints relating to maximum flow.

2.22 - The principle of the algorithm

The principle of the algorithm is to do away with negative yields and to capture positive yields to the extent that the constraints allow : investment is changed proportionally to yields.

2.23 - The regulating parameter

A regulating parameter modulates the magnitude of the change. The choice of this parameter must be made with care as it conditions the convergence speed of the algorithm.

2.24 - Decommissioning of certain types of equipment

It is rather more complex to calculate the changes to be made to the park of units of equipment going out of commission, because each unit in a technical level that is being phased out should have disappeared at the end of its physical life. These contraints induce a coupling of all the years over which these technical levels and units are present. An algorithm allows the value of the duals attached to these constraints to be assessed, and consequently, determination of corrections to be made to the yields of these levels.

The schedule and types of equipments to be decommissionned every year is given as well as an output of the model.

2.3 - The convergence test

At optimum conditions, the marginal yields are nul for all the optimized technical levels in the absence of constraints.

The stop test verifies whether this condition is fulfilled, but, as the optimum cannot be reached exactly, it is mereley made sure that the "error" -defined as the highest relative yield- is below a certain value.

When this test is positive, the proposed programme is optimum to within the limit value. Otherwise, a further iteration must be performed : the proposed programme is modified and the new one is returned to the "cost calculation" function.

2.4 - The "steps" of the NIM

2.41 - Because of the great number of simulation needed before the convergence is declared, the NIM has been divided into three "steps".

2.42 - These are not internal sub-functions of the model, but on the contrary, increasingly finer versions of the NIM (for the simulation).

This system has been designed to reconcile the precision sought with a reasonable computational burden.

2.43 - Costs and yields must be recalculated for each new investment policy specified by the iteration of the gradient algorithm.

We have seen that the expected costs related to management and shortages are calculated by simulating the operation of the park on a weekly basis, allowing for various occasional occurences. This calculation is very cumbersome given the number of occasional occurences to be studied to obtain a realistic simulation.

In addition, expected costs must be calculated as accurately as possible. Accordingly, the occasional conditions influencing the management of the electric system must be finely described.

2.44 - A system for progressive scaling down has been devised, and has been implemented in several versions of the NIM corresponding to increasingly fine models of internal management.

Each step provides a solution which is taken as starting point for the following one.

The final step allows the optimum to be reached with the desired degree of accuracy, and the investment programme determined by earlier steps is precise enough for the definitive optimum conditions to be calculated rapidly.

3 - GENERAL PROBLEMS

3.1 - Problems linked to the definition of the criterion

3.11 - The nature of the criterion

The criterion is economic in nature as it comprises capital, fixed and fuel costs. It is not an accounting or financial criterion.

The reasoning is conducted in constant prices to avoid the difficulty of inflation.

To take the distribution over time of expenditure and receipts into account the rate of discount determined by the Public Authorities, 9 % per year in constant currency, is used, and not the actual cost of money on the financial market.

3.12 - The time horizon of the model

The NIM covers a very long period of time (up to 30 years).

3.13 - Modelling demand

Demand is a parameter which is set a priori, and is not a variable in the model. The effect -via the level of marginal costs, tariffs and trading policy-of the investment programme on consumption is not taken into account.

3.2 - The choice of variables

3.21 - The vairables associated with fixed costs

The investment variables should comprise all units of equipment which might be developed. At present in France, this involves taking into account conventional and nuclear thermal installations and pumping stations.

Investment and decommissionings (considered as negative investment) are optimized simultaneously.

Units of equipment are grouped into homogeneous technical levels. It is assumed that the installed power within each level can vary continuously, whereas in reality, it varies discretely each time a new unit is commissioned.

3.22 - Management variables

As noted the criterion of the NIM includes the management and shortage costs. The management of the electric system should, therefore, be described within the model. Management basically relates to the level of use of the various facilities available.

- The future use to be made of thermal facilities bears a direct influence on the desirable volume of investment. The level of use of the various types of generating facilities are thus indispensable management variables.

- The maintenance of thermal facilities and the optimum investment level are interdependent.

- The management of pumping facilities is very sensitive to the changes in the structure of the thermal park and affects the equipments rentability. This aspect of management is described in the NIM using rules as to optimum placement.

- Gravity-fed hydro-production has for practical purposes reached its ceiling in France and its mode of management is less influenced by the structure of the thermal park than in the case of pumping facilities. Its management is thus located upstream of the NIM, by the "system P" (1).

- Flexibility in the case of equipment is not represented in the NIM because it is too complex to take into account the various "kinetic" needs (teleregulation, tertiary reserve).

In addition, power transfert stresses have an effect on the management of production facilities. They have no doubt a limited impact on the choice of investment, but it is not negligible. This very complex effect and the kinetic aspects are represented by an "unsuitability coefficient".

3.3 - The uncertainties of the future and random factors

3.31 - Uncertainties

The NIM operates on the basis of a set of well determined hypotheses. However, some parameters are uncertain : costs, long-term demand, commissioning time, etc.

Different approaches (minimax regret, minimax expectation, etc...) allow this type of problem to be dealt with using the results of the NIM for the different scenarios envisaged.

3.32 - Random factors

A certain number of random factors affect the electricity system : random climatic conditions which have an effect on demand, random hydraulic conditions, and random conditions of equipment availability. A "medium-term occasional factor" can also be added to this list, albeit this factor pertains more to uncertainty.

Random conditions disturb the operation of the electric system. The use of the thermal park and pumping facilities depends on their occurence.

By contrast, the investment variables can react only to the whole range of cases which can be envisaged because of expectations covering all possible cases.

---

(1) the "system P" is a model which optimizes the management of hydropower plants every year, once given an idea of the thermal park.

3.4 - Shortage

Meeting demand is a necessity for a monopolistic public establishment. But as we have seen, random factors affect the system and demand can consequently not be met in every conceivable situation. The choice of a tolerable failure threshold is therefore inevitable.

This choice of a policy can relate to the expectation of the energy foregone every year or the duration of a shortage. It is also possible to define the critical situations for which a reaction is indispensable, or the critical periods during which the probability of failure should remain limited ; these criteria which boil down to identifying a threshold not to be exceeded are quite difficult to handle in the optimization models as they lack flexibility : they lead to the limitation of risk to the desired level, whatever the cost involved in terms of the volume of equipement to be constructed be. In addition, the scope of these criteria depends on the characteristics of the system studied and there is no reason for these to remain stable over time.

This is why shortage is no longer determined by a threshold, but by a "shortage cost" which leads to its limitation. This "cost" has an economic justification : shortage is initially a condition in which the electricity producer is obliged to resort to exceptional procedures involving a certain overcost in comparison with normal situations (which corresponds to the internal approach). If the phenomenon persists, the exceptional procedures suffice no longer and the customer is directly affected by the effects of the shortage : this entails a certain cost for the community. This latter cost cannot be estimated whit accurracy ; all we can say is that it increases with the magnitude of the shortage ; a parabolic portrayal has appeared to be satisfactory.

As this explicit approach is unprecise, the cost of shortage has been determined by studying past situations in such a way that the economic balancing of shortage and equipment costs will produce the same result as the earlier definition of a certain threshold. Shortage cost thus basically remains a matter of policy decision.

4 - THE OUTPUT OF THE MODEL

4.1 - The optimum investment and decommissioning programme

The objective of the model is to define this programme. The programme is given for each of the units of equipment under study, in terms of annual capital and decommissioning flows, together with the corresponding figures for installed capacity guaranteed available.

The best policy of limited availability is also determined for the concerned thermal levels.

The elements of the criterion at optimum conditions are presented, especially annual capital and fuel costs.

4.2 - Additional information

The NIM includes a MANAGEMENT sub-function. It is thus able to provide information on the future management of the electric system. These outputs are a by-product of the model

It is particularly possible to secure knowledge :

(a) - of marginal costs (possibly aggregated) and the utilisation costs of the different units of equipment ;

(b) - of the expected annual costs of shortage and average durations of shortage ;

(c) - of the marginal duration of calls on capacity in each thermal level and the annual energy generated by each.

These informations are used for tariff studies : since M. BOITEUX's conference about principles for selling at marginal cost, the tariff studies are based on the marginal production costs values. Thus the NIM is a necessary tool for tariffication.

Energy and power valorisations are now calculated from marginal costs that are outputs of the NIM ; they are particularly used for the dimensionning of the differents projects (especially the hydraulic ones) which are marginal projects.

ANNEX

The generation system, in year t, can be characterized by a state vector with n components (i = 1 to n) :

X_i^t = capacity of type i equipment existing in year t.

The addition of capacity in year t will be represented by a control vector with n components :

U_i^t = addition of capacity available in year t for type i equipment.

Consequently the state equations describing the evolution of the power system take a particularly simple form :

$$\left| \begin{array}{ll} X_i^{t+1} & = X_i^t + U_i^t \\ X_i^1 & = X_i \end{array} \right. \qquad (i = 1 \text{ to } n)$$

with constraints : $U^t \geqslant 0$.

The economic criterion can now be specified. The discounted investment cost of year t is simple to write :

$$J^t = \sum_{i=1}^{n} J_i^t U_i^t$$

where J_i^t is the discounted unit cost for an investment in type i equipment in year t, including the discounted cost of the identical replacement of this equipment at the end of its life.

The expected discounted operating and shortage costs corresponding to the available capacity X^t+U^t during year t will be respectively denoted $G^t(X^t+U^t)$ and $D^t(X^t+U^t)$. Functions G^t and D^t result from the suboptimization of the equipement operation.

The cost function will be completed by a terminal criterion $S(X^T)$ that will evaluate the residual discounted value of capacity X^T existing at time T.

It now becomes possible to state the simplified optimum planning problem of the power system :

$$\text{MIN} \sum_{t=1}^{T-1} \left[\sum_{i=1}^{n} J_i^t U_i^t + G^t(X^t + U^t) + D^t(X^t + U^t) \right] + S(X^T)$$

$$\left| \begin{array}{ll} X_i^{t+1} & = X^t + U^t \\ X_i^o & = \text{fixed} \\ U_i^t & \geqslant 0 \end{array} \right\} \quad i = 1,n$$

This is a optimum control problem. The main difficulty of this problem arises from its large scale ; one must not forget that the operating costs G^t and the shortage costs D^t result from the optimization of the operating model with fixed equipments (X^t+U^t).

1.2 - Optimality conditions

We use the Pontryagin Maximum Principle.

One can decompose the dynamic problem into several static problems. The Hamiltonian is :

$$\mathcal{H}^t = - \left[\sum_{i=1}^{n} J_i^t U_i^t + G^t(X^t + U^t) + D^t(X^t + U^t) \right] + \sum_{i=1}^{n} \psi_i^{t+1} U_i^t$$

The adjoint system is :

$$(1) \quad \left| \begin{array}{l} \psi_i^{t+1} = \psi_i^t + \dfrac{\partial G^t}{\partial X_i^t} + \dfrac{\partial D^t}{\partial X_i^t} \\ \\ \psi_i^T = 0 \end{array} \right.$$

It can be solved as follows :

$$\psi_i^t = - \sum_{\tau=t}^{T-1} \left[\frac{\partial G^\tau}{\partial X_i^\tau} + \frac{\partial D^\tau}{\partial X_i^\tau} \right] \qquad (i = 1 \text{ to } n)$$

The optimum control is given by the maximization of the Hamiltonian

$$\text{Max} \left[\sum_{i=1}^{n} (\psi_i^{t+1} - J_i^t) U_i^t - G^t(X^t + U^t) - D^t(X^t + U^t) \right]$$

Using the Kühn and Tücker theorem

$$(2) \quad \left| \begin{array}{l} \psi_i^{t+1} - J_i^t - \dfrac{\partial G^t}{\partial U_i^t} - \dfrac{\partial D^t}{\partial U_i^t} = 0 \text{ if } U_i^t > 0 \\ \\ U_i^t = 0 \text{ if } \psi_i^{t+1} - J_i^t - \dfrac{\partial G^t}{\partial U_i^t} - \dfrac{\partial D^t}{\partial U_i^t} < 0 \ . \end{array} \right. \qquad (i = 1 \text{ to } n)$$

1.3 - Economic interpretation

Considering the following formulation

$$\psi_i^t = - \sum_{\tau=t}^{T-1} \left[\frac{\partial G^\tau}{\partial X_i^\tau} + \frac{\partial D^\tau}{\partial X_i^\tau} \right] \qquad (i = 1 \text{ to } n)$$

the component ψ_i^t of the co-state vector ψ^t appears as the sum of futur savings of operating cost and shortage cost provided by the additional unit of equipment i at time t. This is the definition of the use value of equipment i

$$\boxed{\psi_i^t = \text{use value of equipment } i \text{ at time } t}$$

Since economic depreciation is precisely the loss of use value during year t, we can write :

$\psi_i^{t+1} - \psi_i^t$ = economic depreciation of the equipment i at time t

Net marginal gains

We have stated the Kühn and Tücker conditions related to the maximization of the Hamiltonian with respect to U^t. Since G and D are functions of (X^t+U^t) we have

$$\left|\begin{array}{l} \dfrac{\partial G^t}{\partial X^t} = \dfrac{\partial G^t}{\partial U^t} \\[2ex] \dfrac{\partial D^t}{\partial X^t} = \dfrac{\partial G^t}{\partial U^t} \end{array}\right. \qquad i = 1 \text{ to } n$$

Thus the Kühn and Tücker condition (2) becomes :

$$\left|\begin{array}{l} \psi^{t+1} - J_i^t - \dfrac{\partial G^t}{\partial X_i^t} - \dfrac{\partial D^t}{\partial X_i^t} = 0 \quad \text{if} \quad U_i^t = 0 \\[3ex] U_i^t = 0 \quad \text{if} \quad \psi_i^{t+1} - J_i^t - \dfrac{\partial G^t}{\partial X_i^t} - \dfrac{\partial D}{\partial X_i^t} < 0 \end{array}\right. \qquad (i = 1 \text{ to } n)$$

Taking the dual system (1) into account

$$\psi_i^{t+1} = \psi_i^t + \frac{\partial G^t}{\partial X_i^t} + \frac{\partial D^t}{\partial X_i^t}$$

we obtain

$$(3) \qquad \boxed{\begin{array}{ll} \psi_i^t - J_i^t = 0 & \text{if} \quad U_i^t > 0 \\[2ex] U_i^t = 0 & \text{if} \quad \psi_i^t - J_i^t < 0 \end{array}}$$

At the optimum the investment cost and the use value are equal, except when the constraint $U_i^t \geqslant 0$ is active.

4 - THE OPTIMIZATION ALGORITHM

The steepest descent algorithm which is used, consists in minimizing the first order approximation of the criterion, with an initial given point at each iteration.

Practically the command has to be modified proportionnally to yields.

The principle of the algorithm is to do away with negative yields and to capture positive yields to the extent thats the constraints allow : investment is changed proportionnally to yields.

A NONLINEAR ECONOMETRIC MODEL WITH BOUNDED CONTROLS AND AN ENTROPY OBJECTIVE

K.O. Jörnsten, Linköping Institute of Technology, Linköping, Sweden.
C.L. Sandblom, Concordia University, Montreal, Canada.

1. INTRODUCTION

A number of studies have been done concerning optimal stabilization and control of economic models. The optimal policies in such studies have the undesirable tendency of fluctuating with an amplitude which makes them unacceptable from an economic viewpoint. A way to avoid this is to require the controls to move only within prespecified upper and lower bounds. The bounds, which may be either constant or time varying, are then selected so that the resulting optimal policies will be realistic, and may be regarded as restrictions self-imposed by the decision maker for political (i.e. external) reasons.

The presence of policy bounds makes the optimal control problem more difficult to solve; this is of course one reason why few results have been presented in this area. One way to deal with the issue of bounded controls is to regard the discrete time optimal control problem as an ordinary mathematical programming problem with equality as well as inequality constraints and in its standard form with a convex (quadratic) objective function.

This mathematical programming problem can be tackled in a number of different ways. We can either use special purpose codes developed for large scale dynamic nonlinear optimization of the reduced gradient type; or we may try to apply some decomposition scheme and thereby make use of the inherent structure of the econometric model.

In this paper we present a completely different approach based on the entropy concept. Thus our econometric model has an entropy objective instead of the commonly used quadratic objective function. It is shown how this objective can be used to determinate an optimal policy and also how one may use the entropy concept to avoid the undesirable property of widely fluctuating optimal policies. Furthemore, a discussion is held on how to take computational advantage of the dynamic structure of the econometric model.

This work was supported, in part, by the Natural Sciences and Engineering Research Council of Canada under grant number A8523 which is gratefully acknowledged.

An outline of the paper is as follows. In section 2 we present the linear discrete time model. Section 3 discusses the choice of objective and the criteria used to guide this choice with emphasis on applications in economic policy modelling. Section 4 is focussed on models with control bounds, the need for such bounds and their impact on the complexity of the model. In section 5 we come back to the objective function choice and present the entropy function as a possible objective in economic policy modelling. Sections 6 and 7 discuss econometric models with entropy objectives, with as well as without control bounds. Section 8 is devoted to a solution method discussion and in the concluding section we discuss the benefits of the entropy approach and present some ongoing research in this field.

2. THE LINEAR DISCRETE TIME MODEL

Consider the following reduced form of a linear econometric model

$$\begin{cases} x_t = Ax_{t-1} + Bu_t + b_t + e_t & t = 1, \ldots, N \\ x_0 \text{ given} & \end{cases}$$

The state vector x_t (an n-dimensional column vector) refers to endogenous variables, and the control vector u_t (an m-dimensional column vector) to policy variables. The coefficient matrices A (n by n) and B (n by m) are considered to be constant and known. The n-vector b_t accounts for all exogenous variables (not subject to control) of the system as well as for any constants, so that the error term e_t can be assumed to have zero mean.

Although our system is written in first order form, this formulation covers general higher order systems as well (see e.g. Chow, 1975, ch. 2). We shall also put the error term e_t identically equal to zero, and consider only the purely deterministic case.

This is not as serious a simplification as it may first appear. Since our model is linear the so-called certainty equivalence principle holds (Simon, 1956; see also Chow, 1975, ch. 7) with the solution to the deterministic model in feedback form. Therefore, the stochastic case is solved by inserting the expected values of the random variables in the deterministic formulas.

The general solution to systems of the type

$$\begin{cases} x_t = Ax_{t-1} + Bu_t + b_t & t = 1, \ldots, N \\ x_o \text{ given} & \end{cases} \tag{1}$$

can be written

$$x_t = A^t x_0 + \sum_{j=1}^{t} A^{t-j} Bu_j + \sum_{j=1}^{t} A^{t-j} b_j , \qquad t = 1, \ldots, N \tag{2}$$

which can be verified by simple substitution. This is only a slight generalization of the fact that the "no control" trajectory

$(\bar{x}_t)_{t=1}^{N}$, i.e. the solution to the system

$$\begin{cases} \bar{x}_t = Ax_{t-1} + b_t & t = 1, \ldots, N \\ x_0 \quad \text{given} & \end{cases} \tag{3}$$

can be expressed as

$$\bar{x}_t = A^t x_0 + \sum_{j=1}^{t} A^{t-j} b_j \qquad t = 1, \ldots, N \tag{4}$$

3. THE OBJECTIVE FUNCTION

The selection of a proper objective function is an important question. First of all it is necessary to consider which variables to include in the objective. Should the objective be chosen such that it consists of both state variable as well as control variable components?

Before we discuss these questions in detail we give some results that can be obtained by choosing a classical objective function, i.e. a quadratic objective, which includes both control and state variables.

Consider the following welfare cost function J (to be minimized):

$$J = \sum_{t=1}^{N} (x_t - \tilde{x}_t)^T Q(x_t - \tilde{x}_t) + (u_t - \tilde{u}_t)^T R(u_t - \tilde{u}_t) \tag{5}$$

where $(\tilde{x}_t)_{t=1}^{N}$, $(\tilde{u}_t)_{t=1}^{N}$ are given "target" or "nominal" trajectories, and Q and R are given symmetric matrices.

The econometric model with the objective function (5) and the system equations (1) can be interpreted as a model in which the aim is to steer x_t and u_t as close as possible towards the "target" trajectories.

If Q is assumed to be positive semidefinite and R positive definite, one can show that there is a unique trajectory

$(u_t)_{t=1}^{N}$

that will minimize J subject to (1).

Since one can include $A\,\tilde{x}_{t-1} + B\,\tilde{u}_t$ in the b_t - vector we can, without loss of generality and for notational ease, restrict ourselves to welfare cost functions J of the type

$$J = \sum_{t=1}^{N} x_t^T Q x_t + u_t^T R u_t \tag{6}$$

In a similar fashion it is easy to show that linear terms in the welfare cost function can be included in the formulation (6); see Jörnsten and Sandblom, 1982.

Thus, without loss of generality, we shall study the classical discrete time linear-quadratic problem.

P1. $$\min_{u_1, \ldots, u_N} J(x, u) = \sum_{t=1}^{N} x_t^T Q x_t + u_t^T R u_t$$

subject to $$\begin{cases} x_t = Ax_{t-1} + Bu_t + b_t & t = 1, \ldots, N \\ x_0 \text{ given.} \end{cases}$$

With Q positive semidefinite and R positive definite one can show (see e.g. Sandblom, 1977) that a unique solution $(x_t^*, u_t^*)_{t=1}^{N}$ exists to the system given above and that this solution is determined by:

$$u_t^* = G_t x^*_{t-1} + g_t \qquad t = 1, \ldots, N \tag{7}$$

$$G_t = -(R + B^T P_t B)^{-1} B^T P_t A \qquad t = 1, \ldots, N \tag{8}$$

$$g_t = -(R + B^T P_t B)^{-1} B^T (P_t b_t + p_t) \qquad t = 1, \ldots, N \tag{9}$$

$$\begin{cases} P_{t-1} = Q + A^T P_t (A + BG_t) & t = 1, \ldots, N \\ P_N = Q \end{cases} \tag{10}$$

$$\begin{cases} p_{t-1} = A^T P_t (Bg_t + b_t) + A^T p_t & t = 1, \ldots, N \\ p_N = 0 \end{cases} \tag{11}$$

By inserting the expressions (8) and (9) in (10) and (11) we obtain the matrix Riccati difference equations:

$$\begin{cases} P_{t-1} = Q + A^T P_t A - A^T P_t B (R + B^T P_t B)^{-1} B^T P_t A, & t = 1, \ldots, N \\ P_N = 0 \end{cases} \tag{12}$$

and

$$\begin{cases} p_{t-1} = A^T (P_t b_t + p_t) - A^T P_t B (R + B^T P_t B)^{-1} B^T (P_t b_t + p_t), & t = 1, \ldots, N \\ p_N = 0 \end{cases} \tag{13}$$

An optimal solution to the optimal control problem P1 can now be obtained as follows. First the Riccati equation (12) is solved recursively backwards in time for $t = N, N-1, \ldots, 2, 1$. The equation (13) is then also solved recursively, backwards in time. With $x_0^* = x_0$ given we then obtain, successively, $u_1^*, x_1^*, u_2^*, x_2^*, \ldots, u_N^*, x_N^*$ by alternatively using

$$u_t^* = -(R + B^T P_t B)^{-1} B^T (P_t A x_{t-1}^* + P_t b_t + p_t), \qquad t = 1, \ldots, N,$$

and the system equations (1).

In a number of studies optimal controls of macroeconometric country models have been investigated using classical linear-quadratic control models. Although the econometric models employed may have been fairly realistic, the optimal policies derived from such models often tend to be intolerable from an economic policy point of view since the optimal policies have the undesirable tendency of fluctuating wildly with a large amplitude.

In what follows we will discuss some alternative ways out of this dilemma.

4. MODELS WITH CONTROL BOUNDS

Sandblom, Banasik and Parlar (1981) presented an econometric control model, in which upper and lower bounds (constraints) are imposed on the control variables. These bounds can be interpreted as bounds within which the decision maker will operate. The reason to introduce these bounds is to ensure that the resulting optimal control policies are realistic and politically acceptable.

The new model is thus

$$\text{P2.} \qquad \min_{u_1, \ldots, u_N} J(x, u) = \sum_{t=1}^{N} x_t^T Q x_t + u_t^T R u_t$$

$$\text{subject to} \quad \begin{cases} x_t = Ax_{t-1} + b_t & t = 1, \ldots, N \\ x_0 \text{ given} \end{cases}$$

$$\underline{M}_t \leq u_t \leq \overline{M}_t \qquad t = 1, \ldots, N$$

With the presence of control bounds (in control engineering referred to as nonlinearities) the problem as it stands cannot be solved by the Riccati equation approach.

However, the model can be solved (although not as efficiently as the classical unconstrained model) by the use of mathematical programming methods based on sparse matrix techniques and reduced gradient methods such as used in the well known CONOPT code (Drud and Meeraus, 1980). Alternatively one can use the model structure by the use of decomposition techniques, see for instance Jörnsten and Sandblom, 1982.

Although the control bounds introduced in the model P2 make the solution more realistic, the model still has a number of properties which are questionable from an economic point of view.

In econometric models the control variables will usually be fiscal or monetary policy instruments, such as government nonwage spending or interest rates. In an economist's thinking there is no real welfare cost incurred by using a control variable of this type in order to approach the goals of full employment and stable prices.

Including control costs (i.e. control variables) in the objective function (welfare cost function) would rather "contaminate" the criterion function by adding to the economically meaningful part of it a component expressing purely control theoretic computational aspects. This is so since the results presented above requires the matrix R to be positive definite.

Thus a model which is preferred from an economic point of view would be the model

P3. $$\min_{u_1, \ldots, u_N} \quad J = \sum_{t=1}^{N} x_t^T Q x_t$$

subject to
$$x_t = A x_{t-1} + Bu_t + b_t \qquad t = 1, \ldots, N$$
$$x_0 \text{ given}$$
$$\underline{M}_t \leqq u_t \leqq \overline{M}_t \qquad t = 1, \ldots, N$$

This is the model presented by Sandblom, Banasik and Parlar (1981) with the objective function modified so as to include piece-wise quadratic functions of the type described in figure 1. Here we have illustrated the function

$$f(z) = \begin{cases} 1.5\ (z-3)^2, & z > 3 \\ 0, & 1 \leqq z \leqq 3 \\ 0.5\ (z-1)^2, & z < 1 \end{cases} \qquad :$$

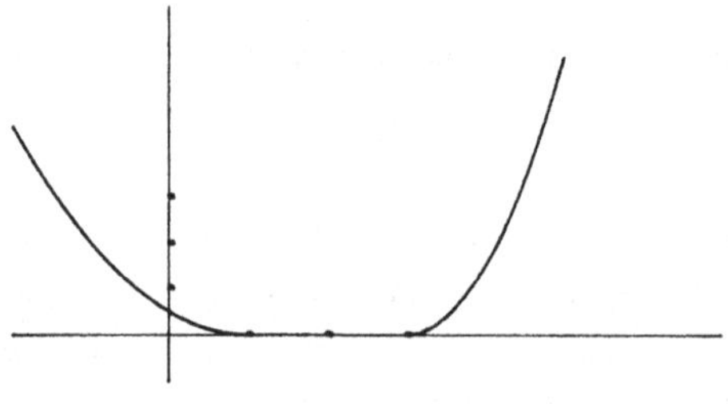

Figure 1. Piece-wise quadratic objective function

This model (P3) is such that the criticism against the linear quadratic formulation is met (Livesey 1973, Bock von Wülfingen and Pauly, 1978).

In Sandblom, Banasik and Parlar (1981), a minimum principle for the problem P3 is developed giving necessary and sufficient optimality conditions. This minimum principle holds equally well for problem P3 without the control bounds. However, the problem P3 is rather difficult to solve and it appears that no particular recursive solution procedure is readily available. To solve the problem P3 we have to use mathematical programming based solution methods for convex problems with linear constraints or use some kind of decomposition technique making it possible to take the structure of the model into account.

5. AN ALTERNATIVE OBJECTIVE FUNCTION

The objective function used in the above model is a quadratic function developed from the classical optimal control objective (5)

$$J = \sum_{t=1}^{n} (x_t - \tilde{x}_t)^T Q (x_t - \tilde{x}_t) + (u_t - \tilde{u}_t)^T R (u_t - \tilde{u}_t)$$

where $(\tilde{x}_t)_{t=1}^{N}$, $(\tilde{u}_t)_{t=1}^{N}$

are given "target" or "nominal" trajectories. Models with this type of objective will aim to get the control and state trajectories as "close" as possible to the target trajectories.

Instead of measuring closeness by the use of the quadratic objective we can choose other functions to measure this proximity.

One such function is the entropy function $f(z) = z_0 + z \log \frac{z}{z_0 e}$ where $0 \log 0$ is defined as 0. $f(z)$ has its unconstrained minimum at z_0.

Here we have illustrated the function

$$f(z) = 2 + z \log \frac{z}{2e} , \; z > 0 :$$

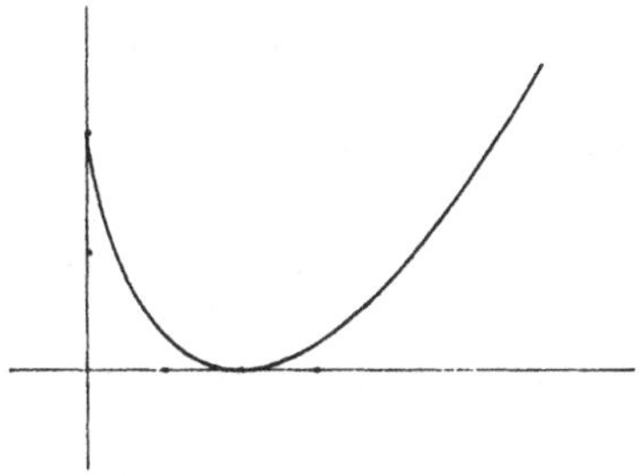

Figure 2. Entropy objective function

Thus the entropy function f(z) can be used as an objective instead of the quadratic function (5) with the same interpretation. The term z_0 can be left out of f(z), as from a mathematical point of view this is only a simple translation.

That is, the function

$$H = \sum_{t=1}^{N} \sum_{i=1}^{n} x_{it} \log \frac{x_{it}}{\tilde{x}_{it} \cdot e} + \sum_{t=1}^{N} \sum_{i=1}^{m} u_{it} \log \frac{u_{it}}{\tilde{u}_{it} \cdot e} \tag{14}$$

can be used as objective instead of the quadratic objective function.

The interpretation of this model with the objective (14) is that we try to generate a solution, that is, trajectories

$\{x_t{*}\}_{t=1}^{N}$ and $\{u_t{*}\}_{t=1}^{N}$ as close as possible to the "target" trajectories

$\{\tilde{x}_t\}_{t=1}^{N}$, $\{\tilde{u}_t\}_{t=1}^{N}$. We must obviously require that the target trajectory components are all positive, because of the logarithmic terms. This gives us the "linear-entropy" dynamic model

P4. $$\min_{u_1, \ldots, u_N} \quad H = \sum_{t=1}^{N} \sum_{i=1}^{n} x_{it} \log \frac{x_{it}}{\tilde{x}_{it} \cdot e} + \sum_{t=1}^{N} \sum_{i=1}^{m} u_{it} \log \frac{u_{it}}{\tilde{u}_{it} \cdot e}$$

subject to $x_t = Ax_{t-1} + Bu_t + b_t$, $\quad t = 1, \ldots, N$

x_0 given

$x_t, u_t \geqq 0 \quad t = 1, \ldots, N$

It can be shown, using elementary Kuhn-Tucker theory, that the optimal solution to the model P4 is unique (due to the objective function strict convexity) and that the solution can be expressed as explicit functions of the dual variables corresponding to the system equations. Let the dual variables to the system equations be the n-dimensional row vectors π_t.

The optimal solution (optimal trajectories) can be written

$$x_{it}^{*} = \tilde{x}_{it} \exp\left[-\pi_{it}^{*} + \sum_{j=1}^{n} a_{ji}\, \pi_{j,\, t+1}^{*}\right], \qquad t = 1, \ldots, N$$

where a_{ji} is the ji-th element of the matrix A; and

$$u_{it}^{*} = \tilde{u}_{it} \exp\left[\sum_{j=1}^{n} b_{ji}\, \pi_{jt}^{*}\right], \qquad t = 1, \ldots, N$$

where b_{ji} is the ji-th element of the matrix B. Also, $\pi_0 = \pi_{N+1} = 0$.

Is this model P4 easy to solve?

The answer to this question is "yes". Entropy programs are easy to solve; especially equality constrained entropy programs. (See for instance Erlander (1981), Eriksson (1980).) Why is this so? The reason is that entropy programming problems belong to a class of problems which has dual programs that can be expressed in dual variables only. Thus instead of solving the primal problem P4 we solve the following Lagrangian dual problem D4

D4.
$$\max_{\pi} \quad L(\pi) = -\sum_{t=1}^{N}\sum_{i=1}^{n}(\tilde{x}_{it}\exp\left[-\pi_{it}+\sum_{j=1}^{n}a_{ji}\pi_{j,\,t+1}\right] - \\ -\sum_{t=1}^{N}\sum_{i=1}^{n}\tilde{u}_{it}\exp\left[\sum_{j=1}^{n}b_{ji}\pi_{jt}\right]-\sum_{t=1}^{N}\pi_t b_t \;;$$

which is an unconstrained concave nonlinear program with dimension given by the number of system equations (for details, see Eiselt, Pederzoli and Sandblom, section IV.3.3).

6. AN ENTROPY ECONOMETRIC MODEL WITH CONTROL BOUNDS

By the introduction of control bounds $\underline{M}_t \leq u_t \leq \overline{M}_t$ $\qquad t = 1, \ldots, N$

the following model P5 is created:

P5.
$$\min_{u_1,\ldots,u_N} \quad H = \sum_{t=1}^{N}\sum_{i=1}^{n} x_{it}\log\frac{x_{it}}{\tilde{x}_{it}\cdot e} + \sum_{t=1}^{N}\sum_{i=1}^{m} u_{it}\log\frac{u_{it}}{\tilde{u}_{it}\cdot e}$$

subject to
$$x_t = Ax_{t-1} + Bu_t + b_t \qquad t = 1, \ldots, N$$
$$x_0 \text{ given}$$
$$\underline{M}_t \leq u_t \leq \overline{M}_t \qquad t = 1, \ldots, N$$

Using the optimality conditions we find that the optimal solution can be written:

$$x^*_{it} = \tilde{x}_{it}\exp\left[-\pi^*_{it}+\sum_{j=1}^{n}a_{ji}\pi^*_{j,\,t+1}\right], \qquad t = 1, \ldots, N$$

$$u^*_{it} = \tilde{u}_{it}\exp\left[\sum_{j=1}^{n}b_{ji}\pi^*_{jt}+\underline{\gamma}^*_{it}-\overline{\gamma}^*_{it}\right], \qquad t = 1, \ldots, N$$

where $\underline{\gamma}_{it}$, $\overline{\gamma}_{it}$ are dual variables corresponding to the lower and upper control bounds.

Thus the model with control bounds is also such that the optimal solutions can be written as explicit functions of the dual variables. As with the model P4 we can solve P5 by solving its Lagrangian dual D5.

D5.
$$\max_{\substack{\pi \\ \underline{\gamma},\,\overline{\gamma}\,\geq\,0}} \quad L'(\pi,\underline{\gamma},\overline{\gamma}) = -\sum_{t=1}^{N}\sum_{i=1}^{n}\tilde{x}_{it}\exp\left[-\pi_{it}+\sum_{j=1}^{n}a_{ji}\pi_{j,\,t+1}\right] - \\ -\sum_{t=1}^{N}\sum_{i=1}^{m}\tilde{u}_{it}\exp\left[\sum_{j=1}^{n}b_{ji}\pi_{jt}+\underline{\gamma}_{it}-\overline{\gamma}_{it}\right] - \\ -\sum_{t=1}^{N}\left(\pi_t b_t + \underline{\gamma}_t\underline{M}_t - \overline{\gamma}_t\overline{M}_t\right)$$

This is a nonlinear program with simple nonnegativity bounds on the variables $\underline{\gamma}$, $\overline{\gamma}$.

To create an efficient solution method to D4 and D5 one needs to take the model structure (i.e. the primal problems dynamic structure) into account and not just the sparsity. This is the subject of current research.

7. AN ENTROPY ECONOMETRIC MODEL WITH AN OBJECTIVE FUNCTION THAT INCLUDES ONLY STATE VARIABLES

As discussed in Sandblom, Banasik and Parlar (1981), one obtains a more realistic econometric model if the objective includes only the state variables. The reason for this is that the control variables usually are fiscal and monetary policy instruments and it is difficult to put a welfare cost on such variables.

The model P5 is thus relaxed slightly by discarding the control variable part of the objective.

The model P6 can thus be written

P6.
$$\min_{u_1, \ldots, u_N} H' = \sum_{t=1}^{N} \sum_{i=1}^{n} x_{it} \log \frac{x_{it}}{\tilde{x}_{it} \cdot e}$$

$$\text{subject to} \quad x_t = Ax_{t-1} + Bu_t + b_t \qquad t = 1, \ldots, N$$

$$x_0 \text{ given}$$

$$\underline{M}_t \leqq u_t \leqq \overline{M}_t \qquad t = 1, \ldots, N$$

The Kuhn-Tucker optimality conditions for the model P6 are

$$\log \frac{x_{it}}{\tilde{x}_{it} \cdot e} + 1 + \pi_{it} - \sum_{j=1}^{n} a_{ji} \pi_{j,t+1} = 0 \qquad i = 1, \ldots, n \quad t = 1, \ldots, N$$

$$\sum_{j=1}^{n} b_{ji} \pi_{jt} + \underline{\gamma}_{it} - \overline{\gamma}_{it} = 0 \qquad i = 1, \ldots, m \quad t = 1, \ldots, N$$

$$x_t = Ax_{t-1} + Bu_t + b_t \qquad t = 1, \ldots, N$$

$$x_0 \text{ given}$$

$$\underline{M}_t - u_t \leqq 0 \qquad t = 1, \ldots, N$$

$$\underline{\gamma}_t (\underline{M}_t - u_t) = 0 \qquad t = 1, \ldots, N$$

$$u_t - \overline{M}_t \leqq 0 \qquad t = 1, \ldots, N$$

$$\overline{\gamma}_t (u_t - \overline{M}_t) = 0 \qquad t = 1, \ldots, N$$

Using the first Kuhn-Tucker condition we get

$$x^*_{it} = \tilde{x}_{it} \exp \left[- \pi_{it} + \sum_{j=1}^{n} a_{ji} \pi_{j,t+1} \right]$$

from which we can see that the optimal state trajectory can be written as an explicit function in the dual variables only. This explicit functional form is not as easy to derive for the control of this model.

Since the model P6 is a convex programming problem with linear constraints it can be solved using standard techniques for nonlinear programming. Of course one should in that case use special purpose algorithms developed for nonlinear econometric models, such as the CONOPT code by Drud and Meeraus.

An alternative solution method is to use an entropy perturbation technique, thus creating a model with the objective function

$$\sum_{j=1}^{N} \sum_{i=1}^{n} x_{it} \log \frac{x_{it}}{\tilde{x}_{it} \cdot e} + \mu \sum_{t=1}^{N} \sum_{i=1}^{m} u_{it} \log \frac{u_{it}}{\tilde{u}_{it} \cdot e}$$

where $\mu > 0$ is a given small number.

By doing this we create subproblems which have explicit easily solvable duals like problem D5. This entropy function approach has been very successful in solving multiple objective programming problems; see for instance Hallefjord, Jörnsten (1983), Eriksson, Hallefjord and Jörnsten (1982).

Currently we are investigating other more direct solution methods to the problem P6 based on the explicit functional form for the optimal state trajectories x^*_{it}. Such methods more or less directly solve the system of equations

$$\sum_{j=1}^{m} b_{ij} u_{jt} = \tilde{x}_{it} \exp\left[- \pi_{it} + \sum_{j=1}^{n} a_{ji} \pi_{j,\, t+1} \right] - \sum_{j=1}^{n} a_{ij} \tilde{x}_{j,\, t-1} \cdot$$

$$\cdot \exp\left[- \pi_{j,\, t-1} + \sum_{k=1}^{n} a_{kj} \pi_{kt} \right] - (b_t)_i$$

$$\pi_t B + \underline{\gamma}_t - \overline{\gamma}_t = 0 \qquad t = 1, \ldots, N$$

$$\underline{\gamma}_t (\underline{M}_t - u_t) = 0 \qquad t = 1, \ldots, N$$

$$\overline{\gamma}_t (u_t - \overline{M}_t) = 0 \qquad t = 1, \ldots, N$$

in the variables u_t and π_t, $\underline{\gamma}_t$, $\overline{\gamma}_t$.

CONCLUSIONS

In this paper we have discussed nonlinear econometric models with entropy objective functions. It has been shown that the selection of an entropy objective has two major advantages compared with the commonly used quadratic objective function. First of all, this objective function choice makes it possible to explicitly solve all linearly constrained models, i.e. with linear equality as well as inequality constraints. Secondly, the entropy objective derived from the entropy concept of information theory has a very nice property of only allowing changes in the variables that are of moderate size; thus the objective has one of the most valuable properties that we look for when modelling economic policy situations. In addition the objective function choice also offers attractive algorithmic features, i.e. the possibility to use fast and easily implemented solution methods with only small storage requirements.

REFERENCES

Bock v. Wülfingen, G. and Pauly, P. (1978): An Optimization Approach in Multiple Target Problems using Inequality Constraints: The Case Against Weighted Criterion Functions, Annals of Economic and Social Measurement, Vol. 6, pp. 613-630.

Chow, G.C. (1975): Analysis and Control of Dynamic Economic Systems. John Wiley & Sons, New York.

Drud, A. and Meeraus, A. (1980): CONOPT - A System for Large Scale Dynamic Nonlinear Optimization - User's Manual, Version 0.105. Development Research Center, World Bank, Washington, D.C.

Eiselt, H.A., Pederzoli, G. and Sandblom, C.L. (1984): Continuous Optimization Models, Walter de Gruyter & Co., Berlin-New York.

Eriksson, O., Hallefjord, A. and Jörnsten, K. (1982): A Long Range Planning Problem with Multiple Objectives, Report LiTH-MAT-R-82-33, Linköping Institute of Technology, Linköping, Sweden.

Eriksson, J. (1980): A Note on Solution of Large Sparse Maximum Entropy Problems with Linear Equality Constraints, Mathematical Programming, Vol. 18, pp. 146-154.

Erlander, S. (1981): Entropy in Linear Programs, Mathematical Programming, Vol. 21, pp. 137-151.

Hallefjord, A. and Jörnsten, K. (1982): An Entropy Approach to Multiobjective Programming, Report LiTH-MAT-R-82-13, Linköping Institute of Technology, Linköping, Sweden.

Jörnsten, K. and Sandblom, C.L. (1982): Optimization of Economic Systems using Nonlinear Decomposition, Report LiTH-MAT-R-82-28, Linköping Institute of Technology, Linköping, Sweden, and in Journal of Information and Optimization Sciences, 1984.

Livesey, D.A. (1973): Can Macro-Economic Planning Problems Ever be Treated as a Quadratic Regulator Problem?. Proceedings of the IFAC/IFORS International Conference on Dynamic Modelling and Control of National Economies, pp. 1-14.

Sandblom, C.L. (1977): Optimization of Economic Policy using Lagged Controls, Journal of Cybernetics, Vol. 7, pp. 257-267.

Sandblom, C.L., Banasik, J.L. and Parlar, M. (1981): Economic Policy with Bounded Controls, Department of Quantitative Methods, Concordia University, Montreal.

Simon, H.A. (1956): Dynamic Programming under Uncertainty with a Quadratic Criterion Function, Econometrica, Vol. 24, pp. 74-81.

A MODEL OF COAL TRANSPORT MANAGEMENT IN A RAIL NETWORK

M. Bielli, G. Calicchio, M. Cini and L. Giuliani
National Research Council & National Railways
Rome, Italy

1. INTRODUCTION

Coal transportation by the Italian railway system will constitute one of the main traffic in the next years, due to well known energy problems relative to oil.

In Italy coal transportation will be characterized by the following peculiarities:

- coal must be almost entirely imported so that supply will mainly concern seaports;
- most users, both public (production of electric power and heating) and private (industries) will be located in the inland territory;
- coal is to be carried essentially by unit (blocked) trains on the railway network.

In a first phase rail traffic of coal will overlap the existing traffic of passengers and goods, without affecting it, and this will be possible owing to the actual limited coal demand.

In fact Italian network is not - homogeneous with high technological main lines, almost saturated by the actual traffic, and secondary lines, single track or not - electrified, with low flows and then available for coal transportation.

Therefore in this phase the problem is that of using the present rail network in a balanced and rationale way by an efficient management of coal transport, with the aim of minimizing transportation costs without interferring with other kinds of flows.

In a second phase, foreseeable increase in coal demand due to plants conversion and new installation will make necessary to plan appropriate network expansions.

In this case will be convenient to determine a new global allocation of different traffics, in order to assure a better performance of the system.

In this paper we restrict our analysis to the first phase and present a linear optimization model on multicommodity flow network for the management of coal transportation, which is also suitable for tackling problems relative to the second phase.

2. MODEL OF COAL TRANSPORTATION BY RAIL

Coal transportation has been widely considered in the literature [1,2] and in particular models of rail network management have been developed [3,4,5,6], which follow network programming approaches.

Nevertheless, no one of such models is completely suitable to deal with the peculiarities of coal management on the Italian rail network, so that a specific model has been developed.

The main peculiarity of the model is that only the coal traffic is considered, as a decision variable. As a consequence, in order of not affecting the allocation of already existing traffic of passengers and goods, line-capacities available for coal transportation are defined, expressed in trains per day.

Another difference concerns the definition of commodity flows. In fact it is not necessary to consider a different commodity for each origin-destination pair, because any demand point can be served by any supply point.

On the contrary it is necessary to introduce commodities in order to take into account of distinct kinds of unit trains to be considered, with respect to the tons carried and to engine employed.

In fact the railway lines have different levels of performance depending on their technological and geographical characteristics, that essentially limit the maximum load, which can be carried by a single train.

As a consequence the commodities introduced are related to the different class of lines, which are fixed in a limited number (about a tenth).

Since each line allows transit only for those trains with useful load smaller than the maximum permissible, each commodity flow will affect only a subnetwork of the entire railway system.

Finally, a fixed transit time is associated to each line, depending on the kind of unit train and independent of the traffic on the line.

This hypothesis is acceptable for ranges of traffic levels fixed in advance and in absence of congestion phenomena.

A further peculiarity of the model is the particular handling of demands and supplies, by the introduction of a logistic network [7], which allows to define an optimality of flow on the network, also in absence of matching between demand and supply.

This feature is particularly convenient in view of the goals to be achieved by the model, which are not only to determine optimal routing and optimal make up (blocking) of trains, but also to evaluate, in general, the impact of coal demand on railway performance and to analyse possible saturation of the network.

The main differences between this model and the other ones proposed in the literature are summarized in Table 1.

3. MATHEMATICAL FORMULATION

In this section it is shown how the model previously introduced can be formulated as a multicommodity flow optimization problem. As far as the network is concerned dummy nodes and arcs are added to physical ones in order to characterize the demands and supplies, with respect to the kinds of multicommodity flows considered.

In particular demand and supply nodes are splitted and connected to suitable supersource and supersink nodes relative to the commodities introduced (Figure 1).

Flow variable x_{ij}^{k}, relative to commodity k (blocked train of type k), represents the number of trains on the generic arc (i,j) of the network.

Global flow T^k represents the number of trains of type k passing through the network.

The goal of the model is to maximize the amount of tons of coal delivered on the network and simultaneously to determine the optimal traffic assignment, with respect to the global transit time.

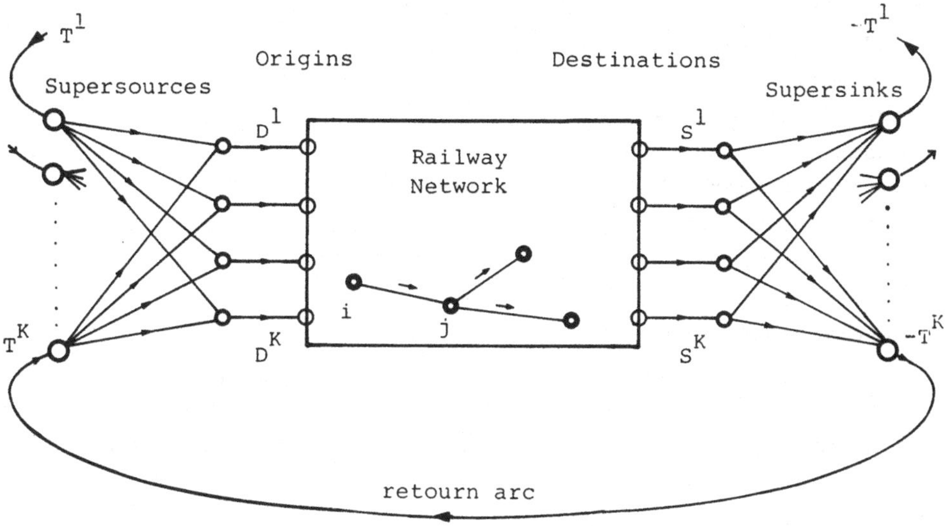

Figure 1: Model of rail transportation and logistic networks

The mathematical formulation of the model is expressed as:

$$\max \sum_k (c^k T^k - \sum_{(i,j)\in A} t^k_{ij} x^k_{ij}) \tag{1}$$

$$E^k x^k = \begin{cases} T^k & \text{at supersource node} \\ -T^k & \text{at sypersink node} \\ 0 & \text{otherwise} \end{cases} \qquad \forall\, k \in K \tag{2}$$

$$\sum_k x^k_{ij} \leq n_{ij} \qquad \forall\, (i,j) \in A \tag{3}$$

$$\sum_k a^k x^k_q \leq S_q \qquad \forall\, q \in Q \tag{4}$$

$$\sum_k a^k x^k_r \leq D_r \qquad \forall\, r \in R \tag{5}$$

$$x^k \geq 0 \text{ and integer } \forall\, k \in K \tag{6}$$

$$T^k \geq 0 \text{ and integer } \forall\, k \in K \tag{7}$$

with the following notations:

A = set of rail network arcs
a^k = tons carried by a unit train of k-commodity
c^k = wheigth coefficient of the k-commodity global flow
D_r = generic coal demand
E^k = node-arc incidence matrix relative to k-subnetwork
i,j = indices of rail network nodes
(i,j) = generic arc of set A
K = set of commodities
k = index of commodity
n_{ij} = maximum number of unit trains admitted on line (i,j)
Q = set of supply arcs
R = set of demand arcs
r,q = indices of logistic network nodes
S_q = generic coal supply
t^k_{ij} = transit time of line (i,j) by unit trains of k-type
T^k = global k-commodity flow
x^k_{ij} = k-commodity flow on arc (i,j), in unit (blocked) trains per day
x^k_r = k-commodity flow on demand arc r
x^k_q = k-commodity flow on supply arc q
x^k = vector of k-commodity flows, relative to all arcs of the k-sub network

It must be remarked that problem (1) - (7) corresponds to the clas sical formulation of linear multicommodity models, but has the peculiarity that global follows T^k on retourn arcs are control variables as well and not fixed data.

Coefficients c^k in (1), in general depending on a^k, must be big with respect to other ones, in order to insure the maximal flow on the network.

Constraints (2) represent flow equilibrium equations at nodes for each commodity.

Constraints (3) express limitations on the number of trains on the lines according to the residual capacity for coal transportation.

Constraints (4) impose that the amount of tons delivered from any supply point is less than the avaliable quantity.

Constraints (5), in connection with the first term of the objective function, insure the best matching between single demands and relative flows.

It can be remarked that constraints (3) are the usual capacity constraints on the physical network, while constraints (4) and (5), which implicitly limit global flows T^k, can be interpreted as capacity constraints on logistic arcs. This formulation insures the feasibility of the problem, since external flows are not imposed, and at the same time tends to satisfy demands, consistently with capacity constraints on the network.

4. SOLUTION METHOD AND COMPUTATION ALGORITHM

Several solution methods for linear multicommodity flow problems have been developed and related algorithm compared [8,9]. In particular the resource directive decomposition method results very efficient [10] and presents some advantages for our specific problem.

Following this approach the problem (1) - (7) can be reformulated in this two level decomposed form:

$$\min \sum_k g^k(y^k) \tag{8}$$

$$\sum_k y_{ij}^k = n_{ij} \qquad \forall\ (i,j) \in A \tag{9}$$

master problem

$$\sum_k a^k y_r^k = D_r \qquad \forall\ r \in R \tag{10}$$

$$\sum_k a^k y_q^k = S_q \qquad \forall\ q \in Q \tag{11}$$

$$y^k \geq 0 \quad \text{and integer} \tag{12}$$

where

$$g^k(y^k) = \max(c^k T^k - \sum_{(i,j)\in A} t_{ij}^k x_{ij}^k) \tag{13}$$

$$E^k x^k = \begin{cases} T^k & \text{at supersource node} \\ -T^k & \text{at supersink node} \\ 0 & \text{otherwise} \end{cases} \tag{14}$$

single commodity problems

$$0 \leq x^k \leq y^k \quad \text{and integer} \tag{15}$$

$$T^k \geq 0 \quad \text{and integer} \tag{16}$$

and where y^k is the vector of arc-resources allocations to k-commodity flows.

In this method, a master program solves upper level problem (8) - (12) and gives the vector of capacity allocations to lower level problems (13) - (16), that for fixed values of y^k are not-interacting single commodity problems.

From solutions determined at lower level it can be derived the subgradient component of objective function (8). The process iterates until an optimal solution, or at least a good approximation, is reached.

In particular allocation y_r^k and y_q^k is constraints (10) - (11) represent variables more directly related to the blocking of trains at demand an supply nodes.

Nevertheless if an optimal solution does exist the convergence is assured only for continuous values of flow variables, while in our case all flow variables must be integer.

However, since problem (1) - (7), or its equivalent (8) - (16), physically always admits integer solutions, the method can be efficiently utilized as well.

In fact it is straightforward to obtain suboptimal integer solution starting from continuous one, by iterating with balanced approximation to integers of variables y^k and by solving with such values the local single commodity problems [11].

5. CONCLUSIONS

The proposed method to deal with management of coal trasportation, is actually experimentally applied to a regional rail Italian network.

An idea of possible further development of the model for dealing with capacity expansions, relative to the second phase of the project, is given in [12].

TABLE 1

MAIN DIFFERENCES BETWEEN MULTICOMMODITY NETWORK MODELS OF COAL TRANSPORTATION

	OUR MODEL	OTHERS
kinds of traffics	rail coal only	- combined rail freight - multimodal transport
kinds of commodities	different kinds of coal unit trains	- different kinds of rail transport (passengers and freight) - origin/destination pairs
demand and supply	elastic (control variables)	fixed patterns from origins to destinations
arc-capacity constraints	residual capacity of lines for coal transportation	real global capacity of lines
value of flow variables	integer (number of trains)	continuous (tons carried)

REFERENCES

[1] C.J. CHANG, R.D. MILES & K.C. SINHA "A regional railroad network optimization model for coal transportation", Transpn. Res. B, vol. 15B, n. 4, pp. 227-238, 1981.

[2] R.D. WOLLMER "Investment in Stochastic Minimum Cost Generalized Multicommodity Networks with Application to Coal Transport", Networks, vol. 10, pp. 351-362, 1980.

[3] A.A. ASSAD "Models for rail transportation", Transpn. Res. A, Vol. 14A, pp. 205-220, 1980.

[4] A.A. ASSAD "Modelling of rail networks: toward a routing/makeup model", Transpn. Res. B, vol. 14B, pp. 101-114, 1980

[5] L.D. BODIN, B.L. GOLDEN & A.D. SCHUSTER "A model for the blocking of trains", Transpn. Res. B, vol. 14B, pp. 115-120, 1980.

[6] Z.F. LANSDOWNE "Rail freight traffic assignment", Transpn. Res. A, vol. 15A, pp. 183-190, 1981.

[7] M.H. WAGNER "Supply-Demand Decomposition of the National Coal Model", Operations Research, vol. 29, n. 6, pp. 1137-1153, Nov-Dec. 1981.

[8] A.A. ASSAD "Multicommodity network flows: Computational experience", Working Paper OR-058-76, MIT Operations Research Center, Oct. 1976.

[9] J.R. EVANS "A network decomposition/aggregation procedure for a class of multicommodity transportation problems", Networks vol. 13, pp. 197-205, 1983.

[10] J.L. KENNINGTON & R.V. HELGASON "Algorithms for network programming", John Wiley & Sons, N.Y., 1980.

[11] M. BIELLI & M. CINI "Multicommodity network flow models: applications and computational experience", NETFLOW 83, International Workshop on network flow optimization theory and practice, Pisa, March 1983.

[12] M. BIELLI, G. CALICCHIO, M. CINI & S. IOZZIA "Optimal investment policies in regional transport infrastructure", XXI European Congress of Regional Science Association, Barcelona, August 1981.

DECOMPOSITION OF OPTIMAL CONTROL IN ENERGY MINIMISATION IN RAILWAY TRAFFIC

Branko Lušičić, dipl. ing.
Mihailo Pupin Institute
Volgina 15
11000 Belgrade, Yugoslavia

1. INTRODUCTION

Costs of railway power consumption account for about a third of direct traction expenditures in railway traffic. Economy can be achieved by selecting the proper mix of tractive stock and of fixed railway installations and through the operational organization of railway traffic. This paper will address the operational problem of optimal scheduling of a train in railway traffic, formulated as follows:

For given trains and a given traffic route, schedule a timetable, i.e., determine running times and station stopping times such that traction power costs be minimal.

The stated problem is overly complex and practically insolvable for several reasons: the task of forming a timetable is in itself sufficiently difficult and time-consuming, algorithmic complexity involved in solving such a problem cannot be handled by algorithms and computer capabilities available today. For these reasons, the problem has to be decomposed into partial optimisation problems:

1. Determination of optimum running times of single trains,
2. Determination of optimum station stopping times for all the trains participating in traffic.

A solution to the second problem is proposed in this paper (the first was considered in previous papers). In solving this problem, one starts from the fixed timetable defined in accordance with nonenergy criteria, into which new trains (usually freight trains which do not have to follow a strict timetable) should be inserted. For these trains, it is necessary to find such departure times from source stations and station stopping times on route that will ensure minimum energy costs. The methodology to be described includes all traction types. For electric traction seasonal fares and extra charge for peak loads are taken into account and attention is given to achieving uniform energy consumption. Application of the developed dynamic programming algorithms is illustrated on an actual example of the daily traffic cycle on the main Yugoslav route (Beograd-Zagreb line).

2. OPTIMAL ALLOCATION OF STATION STOPPING TIMES

The problem of economizing electric railway traffic cannot be said to have been completely solved, if only optimum running times on subsections and optimum trajectories have been determined. Energy consumption is computed from summary watt-hour meters that take into consideration the time of day, season, a charge for active and reactive power, and a peak power charge. Therefore, to actieve minimum power cost, it is necessary to determine the optimum departure times from source stations as well as the optimum stations stopping times where trains halt for all the timetabled electric-locomotive hauled trains under consideration. At the same time it is to be noted that the power cost cannot be viewed as the criterion, but only as one of the criteria for timetable scheduling. As far as passenger and direct freight transportation are concerned, traffic speed and fitting into international timetables are of primary importance. So, out of the total traffic, the only trains left for optimisation are freight ("free") trains which are, in a majority of cases, subject to a single constraint - fitting into the existing timetable. The task to be solved is how to insert newly formed (or "free") trains into the existing timetable meeting, in addition to traffic conditions, the condition of minimum electric power cost.

3. LOAD CALCULATION

A program based on Rich's approximative formula has been developed for calculating the loads of transformer substations (TS). Step by step calculation of the instantaneous configuration of the overhead contact system, due to the moving trains, and the determination of voltage profiles for each configuration forms the basis of this program. The instantaneous layout is solved iteratively starting from the rated voltage value in the TS. Assuming the power on a section to be supplied to trains from one end it is possible to replace all the trains on this section by a single equivalent train. The distance between this train and TS is determined applying the center of gravity method, whereby

$$d_v = \frac{\sum_{i=1}^{I} S_i d_i}{S_v} , \qquad (1)$$

where:

- I - the total number of trains on the section under consideration
- S_i - complex power of the i-th train [kW]
- S_v - complex power of the equivalent train [kW]

The active, reactive and complex powers of the equivalent train will be, respectively:

$$P_v = \sum_{i=1}^{I} P_i , \qquad Q_v = \sum_{i=1}^{I} Q_i , \qquad S_v = \sqrt{P_v^2 + Q_v^2} , \qquad (2)$$

where P_i and Q_i stand for the active and reactive power at the pantograph of the i-th train. Grid impedance from the TS to the equivalent train is:

$$Z = \sum_{j=1}^{J} r_j d_j + j \sum_{j=1}^{J} x_j d_j = R + jX , \qquad (3)$$

with

$$\sum_{j=1}^{J} d_j = d_v , \qquad (4)$$

where:

- r_j - per-unit-length resistance for the j-th subsection [m/km]
- x_j - per-unit-length reactance for the j-th subsection [m/km]
- d_j - length of the subsection (with constant per-unit-length impedance value)[km]

In a two corridor line, the per-unit-length impedance varies with the variable current flowing through the overhead power carrier from one corridor to the other due to mutual impedance. The value of the per-unit-length impedance is corrected in this case.

Voltage at the pantograph of the equivalent train is determined by Rich's approximate formula

$$U_v = \frac{1}{2} \left[\sqrt{U_{EVP}^2 + 2S_v(Z-W)} + \sqrt{U_{EVP}^2 - 2S_v(Z+W)}\right] \qquad (5)$$

where U_{EVP} is TS output voltage, and W is given by

$$W = R \cos\rho_e + X \sin\rho_e . \qquad (6)$$

Active and reactive power losses at the TS output for each section are determined by expressions

$$P_{EVP}^{(k)} = \frac{|U_{EVP} - U_v^{(k)}|^2}{|Z^{(k)}|^2} R^{(k)} + P_v^{(k)} , \qquad (7)$$

$$Q_{EVP}^{(k)} = \frac{|U_{EVP} - U_v^{(k)}|^2}{|Z^{(k)}|^2} X^{(k)} + Q_v^{(k)} . \qquad (8)$$

The total active and reactive power losses at the TS will then be

$$P_{EVP} = \sum_{k=1}^{K} P_{EVP}^{(k)} , \qquad Q_{EVP} = \sum_{k=1}^{K} Q_{EVP}^{(k)} , \qquad (9)$$

where K denotes the total number of supply sections.

4. ELECTRICITY RATE STRUCTURE

Electric power consumption is billed according to the following two basic elements: quantities billed and rate items.

The quantities billed include:

1. basic charge for rated power
2. consumed active power
3. excessive reactive power consumed

The rated values depend on the voltage supplied at the billing site, on the season and the time of day.

There are two voltage levels of the transmission network at billing sites: 110 kV and 35 kV. The 110 kV voltage is in use in single-phase traction, and 35 kV voltage in electric d.c. railway traction systems.

There are two seasonal rates, the higher-rate and lower-rate season, for all three billed quantities. As for the time of day, there exist the higher and lower rate periods for billing the consumed active and excessive reactive power. It is to be noted that reactive power is billed only if excess consumption occurs, i.e., if the power factor is lower than 0.9. The basic accounting period is one month.

The basic charge for rated power is applied to the maximum 15-minute peak load achieved over the monthly accounting period. This load is usually achieved during the higher daily rate and it will be assumed here that it is billed with higher rate.

The power cost for one monthly accounting period and for the portion of railway network supplied from the set of transforming substations for which a summary power consumption account is made may now be presented by the following expression

$$C = \sum_{j=1}^{J} \sum_{i=1}^{I} a_i [E_{ji}^A + r_i E_{ji}^R] + d P_{15max} , \qquad (10)$$

where:

E_{ji}^A - active power consumed at the accounting site in the i-th time unit in the j-th day of the montly accounting period [kWh]

E_{ji}^R - excess reactive power consumed at the accounting site in the i-th time unit in the j-th day of the monthly accounting period [kVArh]

P_{15max} - maximum 15-minute peak load [kW]

J - number of days in the optimization interval

I - number of discrete time intervals within one day

a_i - rate coefficient for active power [Para*/kWh]

r_i - rate coefficient for reactive power [Para/kWArh]

d - rate coefficient for rated power consumption [Din*/kW]

5. OPTIMAL SCHEDULING OF A SINGLE TRAIN

A part of electrified track defined by segment $[x_p, x_k]$ on which M trains are timetabled to run will be considered. Of this total number of trains, M_1 are hauled by electric locomotives, and M_2 by diesel-electric locomotives. A new train is to be inserted into the existing timetable. Apart from the traffic conditions to be met, another condition imposed is that the power cost resulting from the extended timetable be minimal.

The main accounting period is one month. Assuming the timetable cycle (T) of 24 hours is repeated daily with no deviations, it is possible to introduce a reduced accounting period corresponding to one cycle, T. In this case, the minimal cost incurred in one cycle provides a guarantee that minimal cost will also be incurred over the monthly accounting period. If a different freight train is to be inserted daily, the procedure is then repeated daily for each train and again guarantees that the monthly costs are minimal. Optimality criterion will be established taking into account requirements for both minimal power cost and uniform power consumption. The following relation is then obtained for the optimality criterion

$$J = \sum_{i=1}^{I} a_i E_i^A + r_i E_i^R + d_i (P_N - P_i)^2 + c_i \tag{11}$$

where:

P_i - active power at the accounting site in the i-th time unit [kW]

P_N - rated power [kW]

d_i - rate coefficient to account for nonuniformities in power consumption [Din/kW^2]

c_i - a constant

The remaining parameters are as stated in Section 3.

Assuming no overconsumption of reactive power to occur under the current traffic volume $\cos\varphi > 0.9$, relation (11) becomes

$$J = \sum_{i=1}^{I} a_i E_i^A + d_i (P_N - P_i)^2 + c_i \ . \tag{12}$$

Active power consumption over the time interval Δt, in the i-th time unit, will be

$$E_i = \int_{(i-1)\Delta t}^{i\Delta t} P(t)\, dt \ , \tag{13}$$

where P(t) is the instantaneous value of active power at the accounting site, and is given by the expression

***Din - the basic Yugoslav monetary unit*

Para - 1/100 part of the basic Yugoslav monetary unit Dinar

$$P(t) = \sum_{m=1}^{M_1+1} P_m(t) + g[P_1(t),\dots,P_{M_1+1}(t),\ \ell_1(t),\dots,\ell_{M_1+1}(t)] \tag{14}$$

where:

$P_m(t)$ - the instantaneous active power at the pantograph of the m-th train locomotive [kW]

g.. - active power losses in the overhead contact system [kW]

$\ell_m(t)$ - distances at time t between trains and TS [km]

M_1+1 - index corresponding to the train being scheduled

Since the sum of the instantaneous active power of trains from the existing timetable is considerably larger than the instantaneous active power value at the pantograph of the new-train locomotive, the losses in overhead contact system may be regarded as resulting mainly from trains from the existing timetable. Expression (14) then becomes

$$P(t) = \sum_{m=1}^{M_1} P_m(t) + g[P_1(t),\dots,P_{M_1}(t),\ \ell_1(t),\dots,\ell_{M_1}(t)] + P^V(t) \tag{15}$$

i.e.

$$P(t) = P^F(t) + P^V(t)\ , \tag{16}$$

where:

$P^F(t)$ - instantaneous active power at the accounting site (due to trains from the existing timetable) [kW],

$P^V(t)$ - instantaneous active power at the pantograph due to the newly scheduled train [kW].

Substituting equation (16) into equation (13), one obtains

$$E_i = \int_{(i-1)\Delta t}^{i\Delta t} P^F(t)\,dt + \int_{(i-1)\Delta t}^{i\Delta t} P^V(t)\,dt \tag{17}$$

i.e.

$$E_i = \Delta t\ (\bar{P}_i^F + \bar{P}_i^V). \tag{18}$$

Substituting equation (18) into (12) the final form of optimality criterion is obtained after arranging, as

$$J = \sum_{i=1}^{I} d_i\,(\bar{P}_i^V)^2 + b_i\ \bar{P}_i^V\ , \tag{19}$$

where:

$$c_i = 2\ d_i\ P_N\ \bar{P}_i^F - d_i(\bar{P}_i^F)^2 - a_i\ \Delta t\ \bar{P}_i^F - d_i\ P_N^2 \tag{20}$$

$$b_i = 2\ d_i\ \bar{P}_i^F + a_i\ \Delta t - 2\ d_i\ P_N\ . \tag{21}$$

The instantaneous active power at the pantograph of the new-train, and the instantaneous distances of this train from the TS, are thus computed in discrete intervals for the complete portion of railway network for which summary power consumptions accounts are made, i.e., for which optimization is performed. These functions represent the solutions of the equations of motion of the inserted train and depend on various parameters. Only one type of parameters is of interest for the problem treated - departure times from stations at which the train halts. These times are subject to severe constraints defined by the traffic schedule on the route under consideration. In addition to the departure times and running times of trains from the existing timetable, these elements also include the so-called intervals determining the shortest time intervals between two trains on the same corridor. Two types of intervals are distinguishable here - in-station intervals and intervals between train departures. For each station in which the new train halts, it is possible to determine a set of time instants Ω, for which these conditions are satisfied with respect to all the timetabled trains. Accordingly, one may write for the set of allowable departure times of the new train from station n

$$\{t_p : t_p \in \Omega_n\} . \qquad (22)$$

The following relation must hold between two successive departure times t_{pn} and t_{pn+1}

$$t_{pn+1} = t_{pn} + t_{vn} + t_{bn} , \qquad (23)$$

where:

t_{vn} - running time on the subsection between stations S_n and S_{n+1}

t_{bn} - stopping time at station S_{n+1}

If T_u is the maximum allowable running time of the new train on the whole route, the following boundary condition must also be met

$$\sum_{n=1}^{N} t_{vn} + t_{bn} \leq T_u \qquad (24)$$

The total running time must not be longer than one timetable cycle, i.e. 24 hours.

Assuming changes in the voltage of the overhead grid due to the motion of the new train, do not cause significant changes in the speeds and powers on the pantograph of trains participating in traffic, one may give the final statement of the optimization problem considered.

Find the optimal values t_{pi} of departure times t_{pi} of the new train, for which the minimal value of functional 19 is obtained, provided the constraints defined by traction model and electric power system model are satisfied as well as the constraints given by (22) and (23), and the boundary condition (24).

The problem of optimal scheduling of several new trains into the existing timetable subject to the constraint of minimal power cost may be viewed as a K-dimension, N-stage decision process. The problem may be solved by the method of successive approximations. The method consists essentially in decomposing a K-dimension problem into K one-dimension subproblems, each of the subproblems having fewer state coordinates than the basic problem.

6. RECURRENT FORMULA

Optimality criterion (19), constraints (22) to (24) and the mathematical model of traction and electric power system represent a set of extremely complex relations. The number of stations N in the case considered is comparatively large. That is why the solution of the problem cannot be sought by analytical procedures. Dynamic programming method will therefore be applied.

To reduce the optimization process to an N-stage decision process, the traffic route should be discretized. If the stations in which the new train halts are accepted to be decision making points, the route under consideration will be divided into N subsections. Active power diagram at the pantograph of the new-train locomotive, defined on the whole route, divides then into N functions each of which corresponds to a single subsection. On the basis of the afore-stated assumption that changes in the voltage of overhead contact system due to the motion of the new train cause no significant changes in the speeds and powers at the pantograph of the remaining trains participating in traffic, these functions may be regarded as depending only on the departure times from the source station at the subsection. Optimality criterion (19) assumes then the following form

$$J = J(t_{p1},\dots,t_{pn}) = \sum_{n=1}^{N} G_n(t_{pn}) , \tag{25}$$

where

$$G_n(t_{pn}) = \sum_{i=t_{pn}}^{t_{pn}+t_{vn}} [d_i(\bar{P}^V_{i-t_{pn}})^2 + b_i\ \bar{P}^V_{i-t_{pn}}] . \tag{26}$$

It is then possible applying the dynamic programming philosophy to write the recurrent relation for generating the embedded minima functions

$$F_n(t_{pn}) = \min_{t_{bn}} \{G_n(t_{pn}) + F_{n+1}(t_{pn} + t_{vn} + t_{bn})\} \tag{27}$$

$$n = N-1,\ N-2,\dots,1$$

with

$$F_N(t_{pN}) = G_N(t_{pN}) .$$

The solution is obtained in the form of a series of embedded optimal station stopping times. Departure time from the source station on the route is determined using, the minimal value of optimallty criterion in the first stage

$$\hat{t}_{p1} = \arg \{\min_{\underline{t}_{p1}} F_1(\underline{t}_{p1})\} . \tag{28}$$

Optimal departure times for the remaining subsections are obtained from the recurrent relation

$$\hat{t}_{pn+1} = \hat{t}_{pn} + t_{vn} + \hat{t}_{bn}(\hat{t}_{pn}) \ ; \quad n = 1,2,\dots,N-1 . \tag{29}$$

7. PROGRAM ORGANIZATION

Optimal allocation of station stopping times represents a complex problem in view of both the mathematical model and the large number of input data. Having this in mind, a programmed system has been developed within which certain processing stages have been separated and made independent. The programmed system consists of 9 independent computer programs which, altogether define five processing stages:

1. data preparation
2. railway traction simulation
3. load calculation of the TS
4. optimal allocation of station stopping times
5. preparation of output documents

Brief descriptions of these five processing stages will be given in the sequel. A general block diagram of the procedure for solving the optimal allocation of station stopping times is shown in Figure 1.

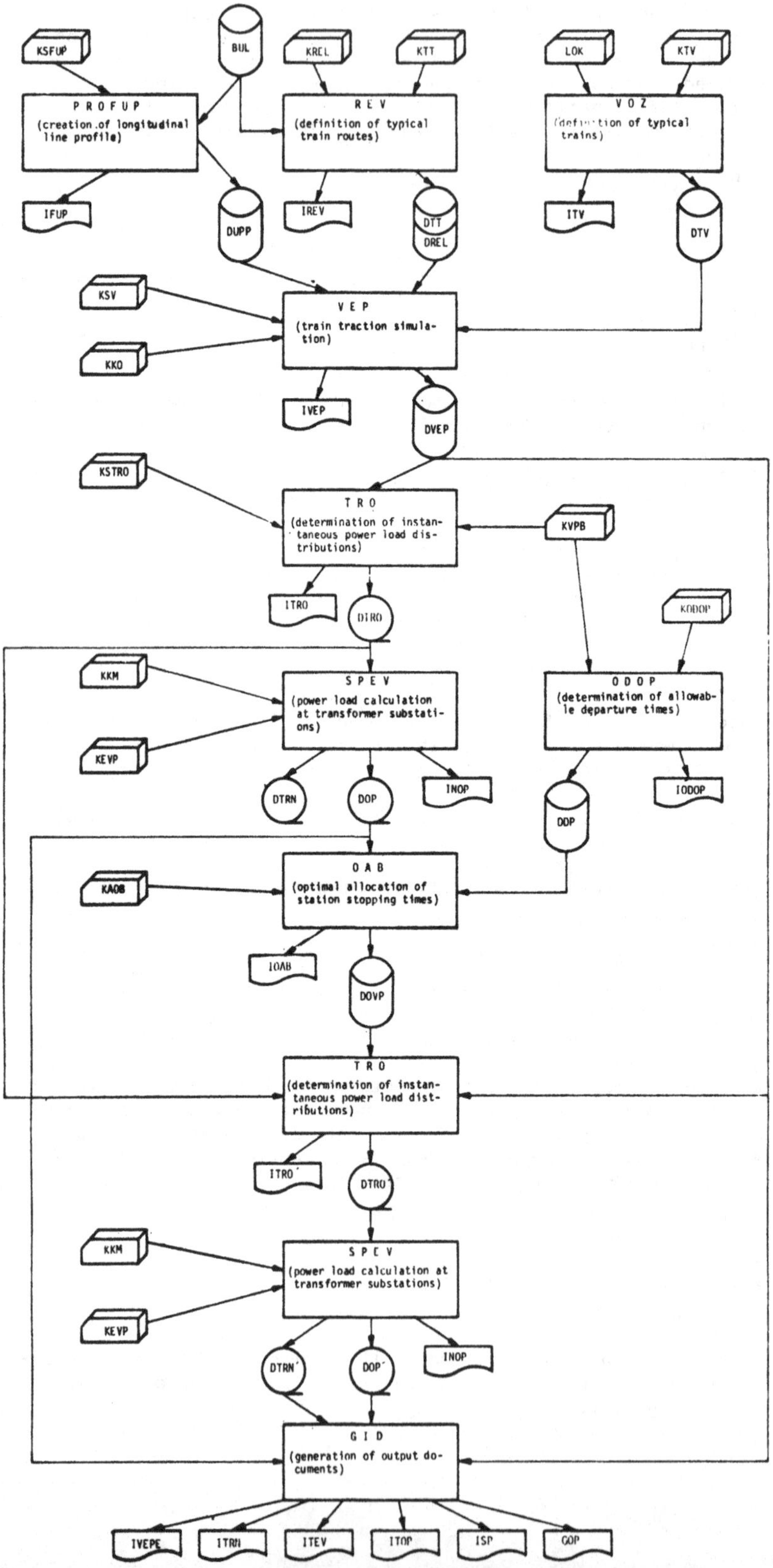

Figure 1
General block diagram of the procedure for solving the optimal allocation of station stopping times

Data preparation stage includes the creation of the longitudinal track profile, timetable prescription and train definition. The longitudinal track profile (DUPP file) is formed by the program PROFUP using the data obtained from longitudinal track profile data bank (BUL). Timetable data are prescribed by program REV. Train routes are formed in the file DREL and typical train subsections are created in the file DTT. Train routes file contains a list of all stations on the route considered. The file of typical train subsections contains data on train subsections. These data include inter-station running times. The notion of a typical train subsection has been introduced to avoid the repetition of subsections having identical running times. The program VOZ is used to create the file DTV in which data on typical trains, processed and prepared for use in traction simulation program, are stored. The notion of typical trains has been introduced for the same reason as the notion of typical subsections.

In the second processing stage, program VEP performs traction simulation, successively for all the trains participating in traffic on the route under consideration. The new train is also covered by traction simulation. Specification is done by associating typical train subsections to typical trains. In addition, the lists of stations in which trains stop and are switched are also prescribed. Trains do not halt in stations during simulation, and the appropriate stopping times are taken into account in the next calculation stage. All running times are relative, i.e., all trains start at 00.00. Simulation results are stored on a disk in the file DVEP.

Time displacement of the results of traction simulation is done by program TRO by adding the actual departure times and station stopping times. Program results, (the instantaneous load distribution), are stored in the file DTRO. Computation time savings are thus achieved, since no repeated simulations are performed for the trains identical in all respects except for departure times and station stopping times. These modified simulation results (DTRO file) are employed for load calculation of the TS by means of program SPEV. Data on transformer stations and overhead grid system are defined by cards. Calculation results for the existing timetable are stored on magnetic tapes, in the files DOP and DTRN. The first file stores data on loads and other relevant data for all TS, while the second file stores voltage values at the pantograph of all locomotives. Results are obtained for each time instant in which calculation is made. The length of time iteration is adjustable (its minimal value is 12 sec). This stage also includes program ODOP whose task is to determine allowable departure times for the new train. They are determined on the basis of departure times of trains participating in traffic on the route considered according to the existing timetable, station stopping times and shortest times betwen trains on a corridor. Program results, (allowable departure times) are stored in file DDP.

The central stage in the application of the programming system consists of solving the problem of optimal allocation of station stopping times using program OAB. Program inputs include the loads of transformer substations for the existing timetable (file DOP), power at the pantograph of the new-train obtained as the result of traction simulation (file DVEP) and allowable departure times (file DDP). The program provides the optimal departure times from the stations in which the new train halts.

The process of optimal allocation of station stopping times is completed by determining the optimal departure times. The last processing stage is intended to provide for suitable presentation of the obtained results. To this end it is necessary to repeat the load calculation of trainsformer substations, but now for the extended timetable. Program TRO is used to create a new file, DTRO´, which contains also the results of traction simulation for the new train, modified in time according to the optimal departure times. New files DOP´ and DTRN´ are created by program SPEV. Presentation of results is performed by program GID. The program provides 6 types of reports allowing a complete analysis of the performed calculations:

1. catalog of the file DVEP
2. table of per-train power consumptions
3. table of loads of the TS and
4. summary results by the TS
5. graphical representation of load diagrams of the TS on a 15-minute basis
6. voltage table

8. NUMERICAL EXAMPLE

Application of the dynamic programming algorithm and programing system is illustrated using the example of Beograd-Zagreb line (to Tovarnik). This is a 132-km long subsection, and power is supplied to the overhead contact system from 3 transforming substations. Traffic includes 150 trains, 137 of which are hauled by single-phase electric locomotives and 13 by diesel-electric locomotives. Optimization of a 1400 Mp freight train hauled by single-phase electric locomotive (JŽ-441 series) within the existing timetable will be considered. The train starts from Topčider Freight station and is scheduled to halt in the following stations: Novi Beograd, Zemun, Zemun Polje, Batajnica, Nova Pazova, Stara Pazova, Golubinci, Putinci, Ruma, Voganj, Sremska Mitrovica, Martinci, Kuk, Erdevik and Šid.

Data required for solving the problem of optimal allocation of station stopping times may be classified into: line data, train data, timetable data, data on transforming substations and overhead contact system.

The new train was inserted into the existing timetable with the optimal allocation of its stopping times. T_u = 24 hours was accepted as the maximum allowable running time of the new train on the whole route, and the optimum departure time from the source station and optimum stopping times in the stations in which the train halts were obtained. These time instants together with inter-station running times define the optimum schedule of the new train. Insertion of the new train into the existing timetable caused changes in the loads of transforming substations. A summary diagram of active power for the new, optimally extended, timetable is shown in Figure 2. The portion of load due to the existing ("old") timetable is marked by "." in the diagram, while the increase in load resulting from the insertion of the new train is marked by "*". The optimal schedule of the new train is drawn in the same Figure, below the diagram.

Using the calculated loads of transforming substations for the existing timetable and for the extended timetable, the consumptions of active and reactive power during the higher and lower daily rate item for the whole hour cycle were calculated and the value of maximum 15-minute peak load recorded during the higher daily rate was determined. Assuming this timetable to repeat daily with no deviations, power consumption over one monthly accounting period was calculated and power costs were accounted. Since the condition $\cos \rho > 0.9$ (no over-consumed reactive power) was met in all three cases, the consumed reactive power was not taken into account. Savings resulting from optimal insertion of the new train into the existing timetable as compared with nonoptimal insertion were determined on the basis of calculated costs. The value of savings, given in Table 1, was determined for both seasonal rate items. These savings were achieved under current power consumption billing regulations. If a charge for deviation from the mean power were accepted as the quantity billed instead of the charge for maximum 15-minute load, the achieved savings would be considerably higher. In addition, this would also result in a more uniform power consumption. More uniform consumption may be of great importance as regards both consumption quality and an improved voltage state in the overhead contact system. Too large load variations result in high voltage drops in the contact system and these may, in turn, endanger the execution of the timetable. In addition, high voltage drops affect the operation of auxiliary equipment in trains.

Evaluation of power consumption uniformity, i.e., the deviation of 15-minute peak loads from the mean value may be expressed by the following relation

$$\Delta P_{15} = \sqrt{\sum_{i=1}^{I} (P_i - P_S)^2 / (I-1)} \qquad (30)$$

P_i - active power value at the accounting site in the i-th time unit (kW)
P_S - mean power value (kW)
I - the number of time units in one timetable cycle

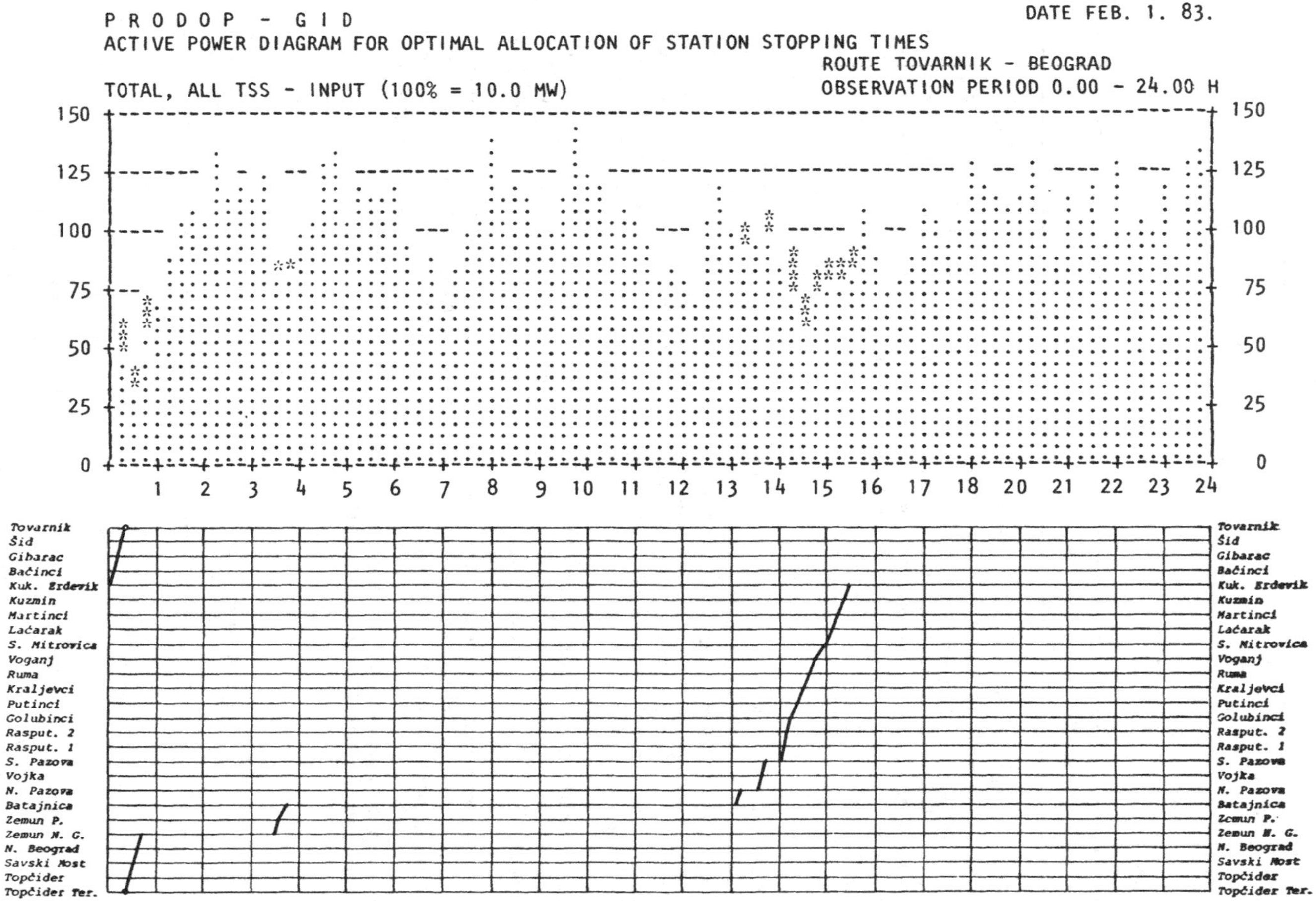

Figure 2 Active power diagram for optimal allocation of station stopping times

Table 1

	Reduction in electric energy cost for the new train [%]		Mean deviation of 15-minute peak loads from the mean value [kW]
	Lower seasonal rate item	Higher seasonal rate item	
Nonoptimized timetable	0	0	22,39
Optimized timetable	50,09	31,83	18,28

If relation (30) is applied to cases of optimal and nonoptimal insertion of the new train into the existing timetable, values that are also given in Table 1 will be obtained. As may be seen from this example, optimal insertion of a single train, which accounts for about 1.5 percent of the total active power consumption, into the existing timetable reduces the mean deviation of 15-minute peak loads from the mean power by about 4 kW, or 18 percent, as compared with nonoptimal insertion.

CONCLUSION

The problem of power cost minimization in railway traction was considered. Since direct solution of this problem is a too extensive task, the problem was decomposed into two problems: *1. Determination of optimum running times of single trains, 2. Determination of station stopping times for all the trains participating in traffic.* As the first problem was considered in previous papers, only solution to the second problem was presented in this paper. The methodology presented can be employed for all traction types: steam, diesel-electric, electric d.c. and single-phase electric.

The presented example shows that slight timetable corrections may result in considerable savings in both traction power and electric energy consumption costs as well as significantly more uniform distribution of the loads of transforming substations.

The developed programs and the supporting data base are implementable in any computing center. Highly-automated algorithms require minimum engagement of a user and operator during program running. In view of all the stated points, one may conclude that energy criteria should also be taken into account in railway traffic organization in addition to purely traffic-based criteria. The methodology developed and presented in this paper makes this entirely attainable.

ACKNOWLEDGEMENTS

Author expresses his sincere appreciation to Dr M. Vušković of Mihailo Pupin Institute for initialization and constant support of this work. A special acknowledgement is due to I. Tomašević of Mihailo Pupin Institute for assistance in system model development.

REFERENCES

1. R.Bellman and S.Drayfus, "Applied Dynamic Programming", Princeton Univ.Press, Princeton, New Jersey, 1961.
2. M.Vušković i B.Lušičić, "Determination of Optimum Running Times of Single Trains", (in Serbian), Proc.of the 15th annual ETAN Conf. Split, June 1971.
3. B.Lušičić i M.Vušković, "Determination of Optimal Interstation Running Time due to Minimal Energy Consumption",(in Serbian),Proc.of the 16th annual ETAN Conf. Velenje, June 1972.
4. M.Vušković i B.Lušičić, "Methodology and Programming Package for Railway Traction Simulation", Second IFAC/IFIP/IFORS Symposium on Traffic Control and Transportation System, Cote D'Azur, 1974.
5. M.Vušković i B.Lušičić, "Simulation Package for Railway Traction Simulation", 4the ORE Colloquium on Technical Computer Programs, München, May 1974.

OPTIMAL URBAN BUS ROUTING WITH SCHEDULING FLEXIBILITIES

F. Soumis, J. Desrosiers, M. Desrochers
Ecole des Hautes Etudes Commerciales de Montréal
Montréal, Canada H3T 1V6

1. INTRODUCTION

Bus fleet route planning is often carried out in the following two sequential stages:

1) Based on the demand, determine the trips to be carried out.
2) Assign buses to the trips so as to minimize total costs.

Step 1 fixes the starting time of each trip taking into account the demand, but without considering bus assignment. Step 2 optimizes bus assignment without modifying the trip starting times established in step 1. The resulting operating plan is suboptimal: costs can be reduced without significantly affecting the quality of service by slightly modifying the schedules of certain trips a posteriori to reduce the number of vehicles required and the total travelling time.

We propose to fix the departure times during cost minimization in step 2. Step 1 involves only the determination of an interval during which each trip must begin to ensure an adequate quality of service.

The bus assignment problem with fixed departure times can be formulated as a minimum cost flow problem whose optimal solution is easily obtained [8]. With flexible departure times, the problem becomes more difficult. Good solutions have been obtained within acceptable computation times using heuristic methods. One method which is particularly suitable for this type of problem is that developed by Bokinge and Hasselstrom [3] and integrated into the Volvo Traffic Planning Package. The optimal method proposed here is capable of solving problems encountered in practice. Results for problems with 128 and 158 departures are presented in section 10. The problems come from two Swedish cities and were suggested to us by the authors of [3].

We now introduce the terminology used. A <u>trip</u> is an itinerary which must be carried out by the same bus. Trip i is characterized by an origin, a destination, a duration, a cost and a time interval $[a_i, b_i]$ during which the trip must begin. An <u>intertrip</u> arc is an empty run which may be carried out by a vehicle. Intertrip arc (i,j) goes from the destination of trip i to the origin of trip j. Its duration t_{ij} and its cost c_{ij} include respectively the duration and cost of trip i. <u>A route</u> is a sequence of trips and intertrip arcs carried out by a vehicle between two visits to the depot. A <u>route block</u> is a sequence of routes carried out by the same vehicle during the day.

2. MATHEMATICAL FORMULATION

The buses flow through a network made up of a set of nodes representing the trips $i = 1,\ldots,n$ and the depot (node 0) and joined by a set of arcs which include the intertrip arcs (i,j), $i,j = 1,\ldots,n$, arcs $(0,j)$ $j = 1,\ldots,n$ joining the depot to the origin of each trip, and arcs $(i,0)$, $i = 1,\ldots,n$ joining the destination of eacl trip to the depot. The structure of the network varies depending on whether there is only one or several vehicle depots, and whether vehicles visit the depot only once or several times in a day. The network with multiple exits from a single depot is presented in section 4.

The mathematical formulation includes two types of variables: flow variables x_{ij} taking the value 1 when arc (i,j) is used by a vehicle, and continuous time variables t_i, associated with the departure time of each trip. The variable travel costs are associated with the flow variables corresponding to the intertrip arcs, and a fixed cost per vehicle is associated with the flow variables corresponding to the depot exit arcs.

The optimal routes respecting the scheduling constraints are the solution of the following problem:

$$\text{Min} \sum_{i=0}^{n} \sum_{j=0}^{n} c_{ij} x_{ij} + W \sum_{i=0}^{n} \sum_{j=0}^{n} x_{ij}(t_j - t_{ij} - t_i) \qquad (1)$$

$$\sum_{j=0}^{n} x_{ij} = \sum_{i=0}^{n} x_{ji} = 1 \qquad i=1,\ldots,n \qquad (2)$$

$$x_{ij} \geq 0 \qquad i,j=0,1,\ldots,n \qquad (3)$$

$$x_{ij} > 0 \Rightarrow t_i + t_{ij} \leq t_j \qquad i,j=1,\ldots,n \qquad (4)$$

$$a_i \leq t_i \leq b_i \qquad i=1,\ldots,n \qquad (5)$$

$$x_{ij} \text{ binary} \qquad i,j=0,1,\ldots,n \qquad (6)$$

The first term in the objective function includes the fixed costs of the vehicles and the travel costs while the second term evaluates the cost of waiting between trips. (W is the cost of a one minute wait). Note that the waiting cost term is non-linear but becomes linear if the schedule is fixed.

If the waiting costs are dropped, relations (1), (2) and (3) form a routing problem without scheduling constraints. This is a minimum cost flow problem which is easily solved, and whose solution is integer. Relations (4) describe the compatibility requirements between the routes and the schedule. Constraints (5) establish the time intervals within which trips must begin.

3. LITERATURE REVIEW

One possible heuristic solution method involves the discretization of the time intervals and the replacement of each variable t_i by a set of binary variables associated with each decision to begin trip i or not at a discrete point in time. Levin [10] used this approach for an air transportation problem and Swersey and Ballard [11] used it for school transportation; they found that an optimal integer solution is often obtained when using the simplex algorithm on the problem without integrality constraints. Note, however, that these authors restricted the objective function to deal with the number of vehicles only, after discovering that it was more difficult to obtain integrality when travel costs were also included. Other methods suggested by Bodin et al [2] include travel costs but additional approximations have to be made.

Bokinge and Hasselstrom [3] have developed a heuristic network algorithm for the problem which involves the solution of several fixed schedule problems using the minimum cost flow algorithm. An initial fixed schedule problem considers the "nucleus" of each trip in order to identify peak periods and to obtain a lower bound for the number of vehicles. The trip "nucleus" is defined by fixing the departure time for the trip at the end of the starting time interval, and by fixing the ending time as if the trip began at the beginning of the time interval. The authors then continue with the original trips fixing the schedules so as to increase the chance of obtaining good routes. The fixed schedule problem is handled using the minimum cost flow algorithm. The schedule is then modified and the solution process is repeated until a satisfactory solution is obtained.

Branch-and-bound approaches are the most commonly used methods for the optimal solution of this type of problem. A relaxed problem is solved and branching is carried out to eliminate infeasible solutions. Several relaxations of the problem are possible; each represents a different compromise between solution cost and quality of approximation.

A very good approximation was obtained with the simplex algorithm by relaxing the integrality of the covering problem whose columns are routes satisfying the scheduling constraints [6]. To avoid enumerating all possible routes, a column generation procedure is used in which the subproblems to be solved are shortest path problems with scheduling constraints [5]. Excellent results were obtained with this approach in the single depot case with only one exit from the depot per vehicle and where vehicle waiting time was not considered. On the other hand, this approach required 200 000_8 words of central memory to handle a problem with 151 trips.

An approximation which is easier to solve is obtained by relaxing constraints (4) and (5) and by suppressing the waiting time term; an integer assignment problem is obtained and this is easily solved without excessive memory requirements. This relaxation has been successfully used by Carpaneto and Toth [4] for the travelling salesman problem - a special case of the problem studied here. The present authors [7] have also used this approach to handle this problem in the case of a single exit per vehicle from the depot, and no waiting costs. In this article, we present an adaptation of this method to handle the more general case of multiple exits from the depot and an objective function which includes waiting costs as well as empty running costs and fixed vehicle costs.

4. THE NETWORK

For the problem with multiple exits from a single depot, intertrip arcs (i,j) are of two types: direct, or via the depot. A direct intertrip arc exists if

$$a_i + t_{ij} \leq b_j \ . \tag{7}$$

Its cost includes the travel cost c_{ij} and the waiting cost $W(t_j - t_{ij} - t_i)$. An intertrip arc via the depot is possible if

$$a_i + t_{i0} + t_{0j} \leq b_j \ . \tag{8}$$

Its cost is $c_{i0} + c_{0j}$ as there is no waiting cost at the depot. Note that in general, distances are Euclidean ($t_{i0} + t_{0j} \geq t_{ij}$) and that a direct intertrip journey is always possible if one via the depot is possible. When the two types of movement are possible only the one with the least cost is retained as an intertrip arc (i,j). In urban transit problems where the workday considered is much longer than individual trip length, the number of arcs increases rapidly as the number of trips increases. A morning trip can be followed by any afternoon trip, so the graph is almost complete and the number of intertrip arcs is of the order of $\frac{1}{2}n^2$. The networks obtained with the 128 trip and 158 trip problems had 6812 and 8073 arcs respectively. A reduced network (Figure 1) is obtained by replacing the arcs passing through the depot by a new set of arcs and nodes as defined below:

1- For each node i which has an intertrip arc leading from it via the depot, define a depot node with an associated time of $a_i + t_{i0}$, and an arc leading from i to the new node.

2- For each node j which has an intertrip arc via the depot arriving at it, define a depot node at time $b_j - t_{0j}$ and an arc from the new node to j.

3- Construct the sequence $\{N_k\}$ $k = 1,\ldots,K$ by classifying all the depot nodes in increasing time order ($K \leq 2n$).

4- Replace the old depot node by N_0 at the beginning of the period and by N_{K+1} at the end. Drop the 2n incident arcs.

5- Define the arcs (N_k , N_{k+1}) for $k = 0,\dots,K$.

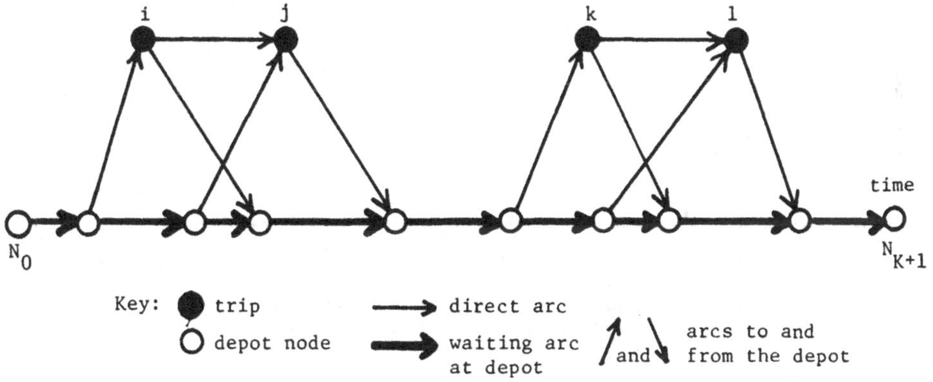

Figure 1: The network with multiple depot exits

This procedure replaces the 2n arcs incident to the depot and all the intertrip arcs via the depot (of the order of $\frac{1}{2}n^2$) by a maximum of 2n nodes and 3n arcs. For the 128 and 158 trip problems the reduced networks have 979 and 1320 arcs respectively. Note also that this reduced network retains all the essential information: there exists a path from i to j following the direction of the arcs in the reduced network if and only if there is such a path in the initial network.

5. SOLUTION OF THE RELAXED PROBLEM

Constraints (4) and (5) are relaxed so as to retain only the network constraints (2) and (3). Furthermore the waiting cost which depends on both the time variables and the flow variables is replaced by a lower bound which is a linear function of the flow variables only. This bound is $Wb_{ij}\, x_{ij}$ where $b_{ij} = \max\{0, a_j - b_i - b_{ij}\}$. Figure 2 illustrates the relationship between b_{ij} and the exact waiting time.

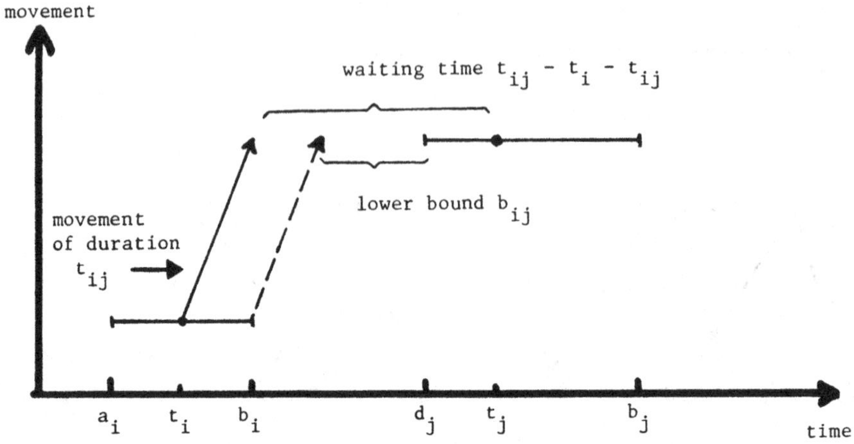

Figure 2: Lower bound on waiting time used in the network relaxation

A network problem is thus obtained and constraint (6) can be dropped as the problem will have an integer solution. This problem was solved using the RNET software [9] which is a version of the simplex algorithm designed especially for networks.

Note that the flows on the arcs incident to the trips are binary but that the flows on 2 consecutive depot nodes are integer. These flows represent the number of buses waiting at the depot at each time. To minimize the fixed cost associated with fleet size as well as travel and waiting costs, it is sufficient to add the fixed cost per vehicle to the arc (N_0 , N_1) with the variable x_{N_0,N_1} representing the number of buses used.

The solution of this problem gives a lower bound on the fixed costs, travel costs and waiting costs for the original problem.

6. TRIP SCHEDULES

In order to obtain a solution to the original problem from the preceding network solution, a schedule must be defined.

Note that in the network solution, all the trips are on unitary flow routes beginning at a depot node N_r, using an exit arc, a sequence of direct intertrip arcs and a depot entrance arc leading to a depot node N_s. The duration of each route is first determined, along with the time interval within which it must begin to be feasible with minimum waiting time. For a route of v trips we define iteratively for each subroute made up of a depot node N_r and trips $1,2,\dots,k (k \le v)$:

- $[\alpha^k, \beta^k]$ an interval within which it must begin at node N_r.
- d^k its duration up to the beginning of trip k.

In order for the schedule to be feasible with minimum waiting time

$$\begin{cases} [\alpha^1, \beta^1] = [a_1 - t_{01}, b_1 - t_{01}] \\ d^1 = t_{01} \end{cases} \tag{9}$$

For $k \geq 1$, if $\beta^k + t_{k,k+1} + d^k \leq a_{k+1}$ there is no waiting between k and k+1, and

$$\begin{cases} [\alpha^{k+1}, \beta^{k+1}] = [\alpha^k, \beta^k] \cap [a_{k+1} - d^k - t_{k,k+1}, b_{k+1} - d^k - t_{k,k+1}] \\ d^{k+1} = d^k + t_{k,k+1} . \end{cases} \tag{10}$$

Otherwise the minimum wait subroute is obtained with

$$\begin{cases} [\alpha^{k+1}, \beta^{k+1}] = [\beta^k, \beta^k] \\ d^{k+1} = d^k + (a_{k+1} - b_k) . \end{cases} \tag{11}$$

Note that if $[\alpha^k, \beta^k] = \emptyset$ there is no feasible schedule for the subroute $N_r \to 1 \to \dots \to k$. On the other hand for each time $t \in [\alpha^k, \beta^k]$, a feasible schedule with minimum waiting time can be defined for the subroute $N_r \to 1 \to \dots \to k$ by taking for $i = 1, \dots, k$, $t_i = t + d^i$. In addition, if the route is complete and includes v trips, arrival at the depot N_s can occur at time:

$$\begin{cases} \alpha^* = \alpha^v + d^v + t_{v0} \text{ at the earliest and} \\ \beta^* = \beta^v + d^v + t_{v0} \text{ at the latest.} \end{cases} \tag{12}$$

(t_{v0} includes the duration of trip v and the return to the depot)

7. BRANCHING TO SATISFY SCHEDULING CONSTRAINTS ON THE DIRECT ARCS

Branching is carried out when a route has no feasible schedule. For example, consider a route $N_r \to 1 \to 2 \to 3 \to 4 \to 5 \to N_s$ which is infeasible for $k \geq 4$. Belmore and Malone's [1] branching method is used. Figure 3 shows how the different ways of fixing the network variables x_{12} , x_{23} and x_{34} at 0 or 1 can be grouped into 3 branches.

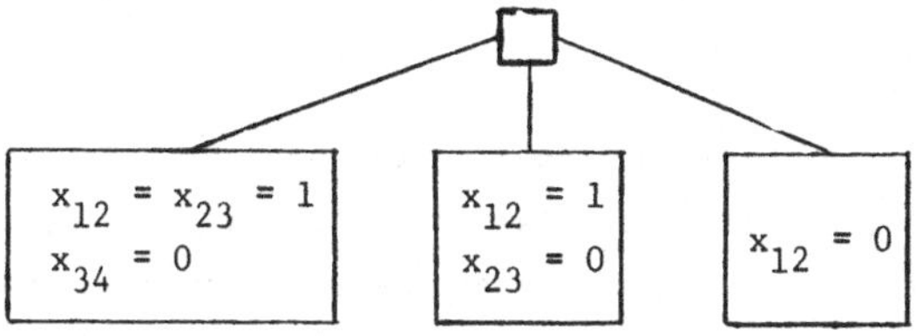

Figure 3: Branching on the arcs of the network : $1 \to 2 \to 3 \to 4$ infeasible

The problems to be handled at each branch are of the same type as the original problem but the network is modified. For example, in the first branch, arc (3,4) is dropped and nodes 1, 2 and 3 are combined while all arcs entering nodes 2 and 3 and leaving nodes 1 and 2 are also dropped.

When, for each node in the tree, each route has a non-empty starting time interval, a feasible solution to the original problem is obtained for each choice of a set of route starting times within the corresponding intervals $[\alpha, \beta]$. The number of vehicles used depends on the choice of these times and is greater than or equal to the number used by the network solution at this node.

8. DEPOT EXIT AND ENTRANCE SCHEDULE MINIMIZING THE NUMBER OF VEHICLES

To determine the number of vehicles used, the set of routes to be covered by the same vehicle must be identified. This problem is of the same type as the original problem except that the trips are replaced by routes with their corresponding time intervals $[\alpha, \beta]$. This problem is however much easier than the original because of its data structure, so the same algorithm is not used. The routes begin and end at times of the day where vehicles have time to go to the depot i.e. outside peak periods. Thus the choice of times within the intervals $[\alpha, \beta]$, and the sequencing of routes to form route blocks influences only slightly the maximum number of vehicles required at peak periods. The following method produces a solution with the minimum number of vehicles.

A lower bound $L(t)$ on the number of vehicles outside the depot at each time t is calculated. This is obtained by supposing that vehicles leave the depot as late as possible and return as early as possible i.e. they leave at time β^v and arrive at time α^*. $L(t)$ is piecewise constant and its maximum L is a lower bound on the number of vehicles necessary to cover all the routes constructed at this node of the branch-and-bound tree.

An upper bound $U(t)$ on the number of vehicles outside the depot at each time t is then estimated. This is obtained by supposing that all vehicles leave the depot as early as possible and return as late as possible for each route, i.e. they leave at time α^v and return at time β^*. The maximum U of this function is an upper bound on the number of vehicles necessary.

If $L = U$, this gives the number of vehicles required to cover all the routes independantly of their starting times. In particular, beginning all routes at starting times α^v with FIFO vehicle allocation to the routes would constitute a feasible solution with the minimum number of vehicles at this node of the branch-and-bound tree. Otherwise, a better upper bound U^1 on the minimum number of vehicles required to cover the routes can be defined by constructing the following feasible solution. The departure times for routes leaving the depot before the peak period are set at α^v and the departure times for routes leaving after the peak period are set at β^v. Boundary cases are set arbitrarily. This bound can be improved by obtaining U^*, the minimum number of vehicles required amongst all combinations of route departure times whose intervals $[\alpha^v, \beta^v]$ or $[\alpha^*, \beta^*]$ touch the time interval for which L(t) is maximum.

In the examples tested, we always obtained L = U or L = U' with only one feasible solution to be constructed. In general, U^* can easily be calculated as the number of routes whose beginning or ending time intervals touch the rush hour period $\{t | L(t) = L\}$ is small. Note that this method is unsuitable for solving the original problem as several trips may begin or end in the period where L(t) is maximum, thus making combinatorial exploration very laborious.

In addition to finding the minimum number of vehicles required at this node of the branch-and-bound tree, this method identifies a feasible solution having this number of vehicles, and the same travel costs as the network solution at this node.

9. BRANCHING TO REDUCE WAITING COSTS

The exact waiting cost of this solution can be calculated as the schedule is now known. If the waiting time coincides with the lower bound used in the network relaxation we have an optimal solution for this node in the branch-and-bound tree. This was the case for all branch-and-bound nodes explored for our test problems. When the waiting time does not coincide, and the node cannot be eliminated as bounded, branching is continued on a route whose waiting time is greater than the approximation used. The same type of branching is used as that shown in Figure 3 except that an initial branch is added for which all flow variables making up the route are set at 1. For each branch, the reduced network is constructed and the upper bound on the waiting time b_{ij} is recalculated for all arcs touching the node where several trips are aggregated. The bounds are generally increased for these arcs as the aggregated node has a shorter time interval than the individual nodes composing it.

10. TESTS

The two problems of 128 and 158 trips proposed by Bokinge and Hasselstrom have in general very narrow time intervals of 0, 2 or 4 minutes with a few trips having 10-12 minute intervals. Numerical results are presented in table 1.

PROBLEM	128 TRIPS $t_i=a_i$ (3)	128 TRIPS TIME WINDOWS		158 TRIPS $t_i=a_i$ (3)	158 TRIPS TIME WINDOWS	158 TRIPS (4) TIME WINDOWS × 2	158 TRIPS (4) TIME WINDOWS × 3
Algorithm	Network	B. et H. [3]	Branch-and-bound	Network	Branch-and-bound	Branch-and-bound	Branch-and-bound
OPTIMAL SOLUTION							
- Nb. buses	34	32	32	41	41	41	40
- Empty travel time	2337	2424	2307	4695	4671	4660	4638
- Waiting time	345	152	327	447	331	292	227
- Cost (1)	5019	5000	4941	9837	9726	9612	9503 (5)
- Time (CDC173) (seconds)	10	--	5	19	20	22	24
- Nodes in branch-and-bound	--	--	5	--	1	5	23
NETWORK RELAXATION(2)							
- Nb. buses	32	--	32	41	41	41	40
- Cost	5019	--	4928	9837	9726	9582	9490

Table 1: Numerical Results

(1) Cost was defined as suggested by Bokinge and Hasselstrom:

Cost = waiting time + 2 * empty running time

(2) Network solution at first node of branch-and-bound

(3) Problems with fixed schedules at beginning of interval $[a_i, b_i]$

(4) More difficult problems constructed by multiplying the width of the time intervals by 2 or 3.

(5) This solution has not been proved optimal as the waiting time is 6 minutes over the network bound. To obtain an optimal solution, the branching described in Section 9 should be carried out.

10. CONCLUSION

These results show that urban transit routing problems with flexible schedules can be solved optimally. Problems with wider time intervals could be solved, and the network aggregation used allows systems of up to 500 trips to be handled without memory problems.

Savings of the order of 5% in vehicle numbers, and of 1-3% in travel and waiting times could represent appreciable economies on a large network. The savings in travel and waiting times are much higher with our exact method than with the heuristic method. The comparison is made on only one example but this would seem to be generally indicative as the example given was selected by the authors of the heuristic method.

11. REFERENCES

[1] Bellmore, Malone: Pathology of travelling salesman subtour elimination algorithm. Operations Research 19 (1971) 278-307.

[2] Bodin, Golden, Assad, Ball: Routing and Scheduling of Vehicles and Crews: The State of the Art. Computer and Operations Research 10 (1983), 69-211.

[3] Bokinge, Hasselstrom: Improved Vehicle Scheduling in Public Transport Throught Systematic Changes in the Time Table. EJOR 5 (1980) 388-395.

[4] Carpaneto, Toth: Some new branching and bounding criteria for the asymmetric travelling salesman problem. Management Science 26 (1980), 736-743.

[5] Desrosiers, Pelletier, Soumis: Plus court chemin avec contraintes d'horaires. RAIRO, Rech. Opér. 17 (1983), 1-21.

[6] Desrosiers, Soumis, Desrochers: Routing with Time Windows by Column Generation. Ecole des Hautes Etudes Commerciales de Montréal, no G83-15, presented at EURO VI, Vienna (1983).

[7] Desrosiers, Soumis: Routes sur un réseau espace-temps. Administrative Sciences Association of Canada, Man. Sci./Rech.op., vol. 3 no 2 (1982), 24-32.

[8] Desrosiers, Soumis, Desrochers, Sauvé: Routing and Scheduling with Time Windows Solved by Network Relaxation and Branch-and-bound on Time Variables. To appear in Computer Scheduling of Public Transport, vol. II, North Holland, J.-M. Rousseau Ed.

[9] Grigoriadis, Tau Hsu: RNET - The Rutgers Minimum Cost Network Flow Subroutines. Rutgers University, New Jersey, 1979.

[10] Levin: Scheduling and fleet routing models for transportation systems, Transportation Science 5, 232-255.

[11] Swersey, Ballard: Scheduling School Buses. Published at Yale School of Organization and Management and presented at TIMS/ORSA Conference, San Diego (1982).

DEVELOPMENT OF DEMAND-RESPONSIVE STRATEGIES FOR URBAN TRAFFIC CONTROL

Nathan H. Gartner
University of Lowell
Lowell, Massachusetts, U.S.A.

1. INTRODUCTION

Intersections of urban arterial streets are the critical elements in most urban street systems. The safe and efficient movement of traffic through these points is largely a function of traffic signalling equipment and traffic control strategies. Many advances have occurred in the last decade which has seen the introduction of computer-based traffic control systems in ever-increasing numbers. Several hundred such systems have already been installed and many more are under development throughout the world.

Strategies are commonly calculated off-line by arterial or network optimization techniques and are then stored in the computer's memory for implementation by various on-line criteria. A number of attempts have been made to develop strategies that are calculated on-line in response to the prevailing traffic conditions. The goal has been to relieve the traffic engineer from the constant burden of data collection and strategy revision. These attempts have met with mixed success. (See Tarnoff (1)and Gartner (2)).

One of the major experiments with computer-based traffic control systems was the Urban Traffic Control System (UTCS) research project which was conducted by the U.S. Department of Transportation (DOT) in Washington, D.C. (3). The project was directed toward the development and testing of a variety of network control concepts and strategies, divided into three generations of control. The different generations can be briefly characterized as follows:

First-Generation Control (1-GC) - This mode of control uses prestored signal timing plans which are calculated off-line based on historical traffic data. The plan controlling the traffic system can be selected on the basis of time-of-day (TOD), by direct operator selection, or by matching from the existing library a plan best suited to recently measured traffic conditions (TRSP). The matching criterion is based on a network threshold value composed of volumes and occupancies. Frequency of update is 15 minutes. Plans can be calculated by any off-line signal optimization method; TRANSYT-generated plans were selected for testing in UTCS.

Second-Generation Control (2-GC) - This is an on-line strategy that computes and implements in real-time signal timing plans based on surveillance data and predicted volumes. The optimization process (an on-line version of SIGOP) is repeated at 5-minute intervals.

Third-Generation Control (3-GC) - This strategy was conceived to implement and evaluate a fully responsive, on-line traffic control system. Similar to 2-GC it computes control plans to minimize a network-wide objective using for input predicted traffic conditions. The differences are that the period after which timing plans are revised is shorter (3-5 minutes), and that cycle length is required to vary in time and space.

The different UTCS control strategies were designed to provide an increasing degree of traffic responsiveness through a reduction of the update interval, with a view to improving urban street network performance. However, results of field testing showed that the expectations were not entirely fulfilled (see Henry et al (4)). 1-GC, in its various modes of operation, performed overall best and demonstrated that it can provide measurable reductions in total travel time over that which could be attained with a well-timed three-dial system. The traffic-responsive mode of 1-GC plan selection is generally more effective than the time-of-day mode. 2-GC had a mixed bag, but was overall inferior compared to 1-GC. These results are generally consistent with those experienced in other places (e.g., Glasgow (5) and Toronto (6)). 3-GC, in the form tested in the UTCS system, degraded traffic flow under almost all the conditions for which it was evaluated.

From the results of the studies cited above it became clear that an effective demand-responsive traffic control system requires the development of new concepts and not merely the extension of existing concepts toward shorter time frames and using predicted values that are less and less reliable. Ways must be devised to use the available detector information to provide good control for future traffic.

This paper describes the development and testing of strategies toward this end. Three different computer programs were developed; they are briefly described in the following sections. A detailed description is available is a U.S. Department of Transportation research report by Gartner (7)).

2. A DYNAMIC PROGRAMMING APPROACH

The first approach for calculating demand-responsive traffic signal control strategies is based on Dynamic Programming (8,9). Consider a single intersection with signal phases that consist of effective green times and effective red times only. All traffic arrivals on the approaches to the intersection are assumed to be known for a finite horizon length. The optimization process is decomposed into N stages, where each stage represents a discrete time interval (such as, 5-second long). A typical stage i is illustrated below:

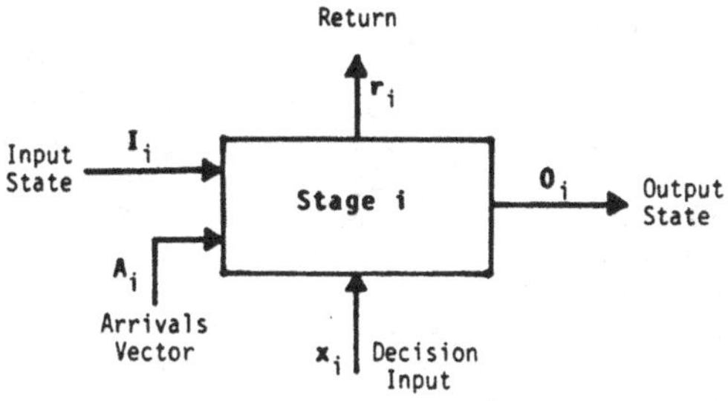

At stage i we have an input state vector I_i, an arrivals vector A_i, output state vector O_i, input decision variable x_i, economic return (cost) output r_i, and a set of transformations:

$$O_i = T_i(I_i, A_i, x_i)$$

$$r_i = R_i(I_i, A_i, x_i)$$

The state of the intersection is characterized by the state of the signal (green or red) and by the queue-length on each of the approaches. Assuming a two-phase signal, the input decision variable indicates whether the signal is to be switched at this stage (x = 1) or remain in its present state (x = 0). The return cost output is the intersection's index of performance (the total delay time), which has to be minimized. The functional relationship between the input and output variables is based on the queueing-discharge processes at the intersection, i.e., the inflow and outflow relative to the signal settings.

Dynamic programming optimization is carried out backwards, i.e., starting from the last time interval and back-tracking to the first, at which time an optimal switching policy for the entire time horizon can be determined. The switching policy consists of the sequence of phase switch-ons and switch-offs throughout the horizon.

The recursive optimization functional is given by the following equation:

$$f_i^*(I_i) = \min_{x_i} \{ R_i(I_i,A_i,x_i) + f_{i+1}^* (I_i,A_i,x_i) \}$$

The return at state i is the queueing delay incurred at this stage and is measured in vehicle-interval units. Thus, when the optimization is complete at stage i = 1 we have $f_1^*(I_1)$ which is the minimized total delay over the horizon period for a given input state I_1. The optimal policy is retraced by taking a forward pass through the stored arrays of $x_i^*(I_i)$. The policy consists of the optimal sequence of switching decisions (x_i^*; i = 1, . . . ,N) at all stages of the optimization process.

An example of the demand-responsive control strategy calculated by this approach is shown in Figure 1, for a 5-minute horizon length. The signal is two-phase and only two approaches are considered, A and B. The figure shows the arrivals on the approaches, the optimal switching policies and the resulting queue-length histories. The signal timings appear as hatched (red) and blank (green) areas, including an all-red overlapping red interval at each switching point. The total Performance Index (PI) is 196 vehicle-intervals.

3. PSEUDO DYNAMIC PROGRAMMING APPROACH

The Dynamic Programming (DP) method for calculating demand-responsive control policies requires advance knowledge of arrival data for the entire horizon period. This is usually beyond what can be obtained from available surveillance systems. Moreover, DP optimization requires an extensive computational effort and, since it is carried out backwards in time, precludes the opportunity for modification of forthcoming control decisions in light of updated traffic data. Thus the DP approach, while assuring global optimality of the calculated control strategies, is unsuitable for on-line use. Also, it is noted that this approach produces a good deal of information that is not used. Optimal policies are obtained for all possible initial conditions, yet only one of these policies applies in practice.

Consequently we set out to develop a simplified optimization procedure that would be amenable to on-line implementation, yet would provide results of comparable quality to those obtained via Dynamic Programming. The procedure, a Pseudo Dynamic Programming (PDP) approach, has the following basic features:

1. The optimization process is divided into sequential stages of T-seconds. The stage length is in the range of 50-100 seconds (i.e. similar to a cycle length for a fixed-time traffic signal) and

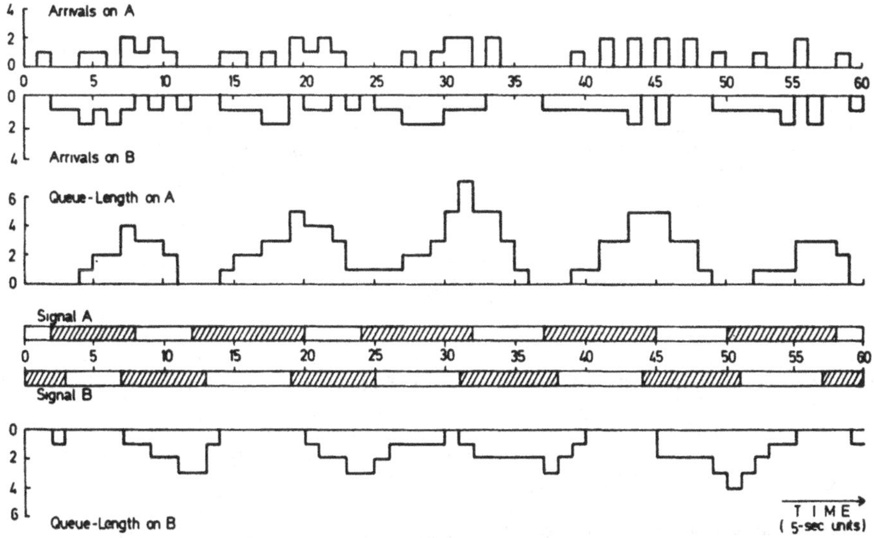

Figure 1: Arrivals and queues for demand-responsive control strategy at a two-phase signal.

consists of an integral number of the basic time intervals.

2. During each stage we require at least one signal change (switchover) and allow up to three switchovers. This is designed to provide sufficient flexibility for deriving an optimal demand-reponsive policy.

3. For any given switching sequence at stage n we define a performance function on each approach that calculates the total delay during the stage (in vehicle-intervals):

$$\Phi_n(t_1,t_2,t_3) = \Sigma_i (Q_o + A_i - D_i)$$

where Q_o = initial queue; A_i = arrivals during interval i; D_i = departures during interval i; and (t_1,t_2,t_3) are the possible switching times during this stage.

4. The optimization procedure consists of a sequential constrained search (see Rao (10)). The objective function (total delay) is evaluated sequentially for all feasible switching sequences. At each iteration, the current performance index (objective value) is compared with the previously stored value and, if lower, replaces it. The corresponding switching point times and final queue-lengths are also stored. At the end of the search, the values in storage are the optimal solution.

The optimal switching policies are calculated independently for each stage, in a forward sequential manner for the entire process (i.e., one stage after another). Therefore, this approach is amenable for use in an on-line system (unlike the DP approach). The information and decision flow at a typical stage n is illustrated at the top of the next page.

A comparison of computational results indicates that the PDP approach provides results that are very close to the optimum obtained by the Dynamic Programming approach. In most cases the difference in the Performance Index is under 10%. This is very encouraging since the

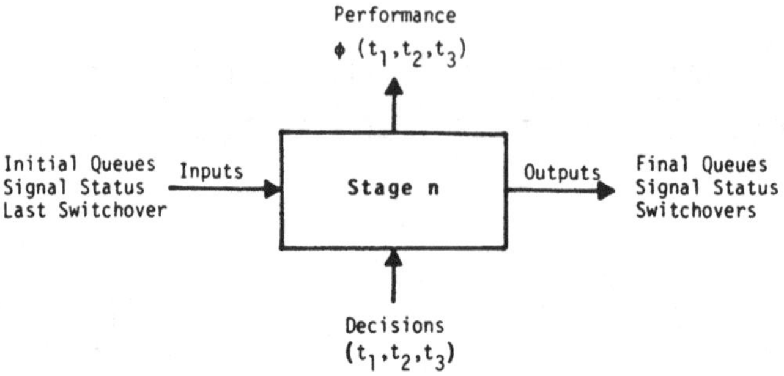

computational requirements (and the traffic data that are needed) are much reduced.

4. THE ROLLING HORIZON APPROACH

The previous section identified a basic building block for demand-responsive decentralized control. The technique that is used requires future arrival information for the entire stage which is difficult to obtain. To reduce these requirements in such a way that we can utilize only available flow data we introduce the rolling horizon concept. This concept is used by operations research analysts in production-inventory control (see Wagner (11)). We apply the same concept to the traffic control problem. The stage length consists of k intervals, which is the Projection Horizon, i.e., the period for which we need traffic flow information. From upstream detectors we can obtain arrival data for a near term period of r intervals at the "head" of the state. For the next (k-r) intervals, the "tail" of the stage, we supply flow data from a model. We calculate an optimal policy for the entire stage, but implement it only for the head section. We then shift (roll) the Projection Horizon r-units ahead, obtain new flow data for the stage (head and tail) and repeat the process, as shown in the figure below:

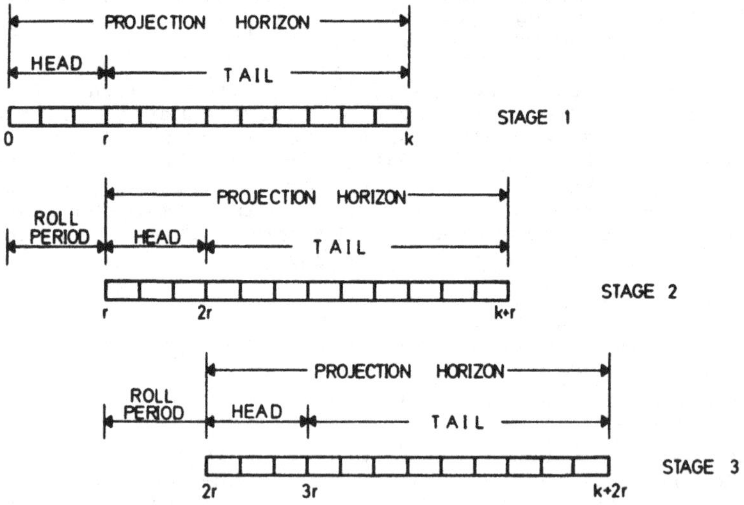

The basic steps in the process are as follows:

Step

0. Determine stage length k and roll period r.
1. Obtain flow data for first r intervals (head) from detectors and calculate flow data for next k-r intervals (tail) from model and detectors.
2. Calculate optimal switching policy for entire stage by PDP.
3. Implement switching policy for roll period (head) only.
4. Shift Projections Horizon by r units to obtain new stage. Repeat steps 1-4.

The computer program which implements this process is named OPAC: Optimization Policies for Adaptive Control and is described by Gartner (12).

The OPAC strategy was tested using actual arrivals for the head of the stage and two types of models for the tail:

1. Variable-Tail (V-T) - projected actual arrivals are taken for tail
2. Fixed-Tail (F-T) - the tail consists of a fixed flow, equal to the average flow rate during the period.

The first model was only used to test the rolling horizon concept and compare the results with previous experimentations. The second model is of primary interest as it represents a practical approach to implementing OPAC. We can use measurements from upstream detectors for head data and smoothed average flows for tail data, both of which are readily available. The head data are continuously updated in the rolling process. As one would expect, the variable-tail OPAC produces policies that are better than those produced by the simplified approacl and in most cases, replicate the standards obtained by the Dynamic Programming approach. Fixed-tail OPAC, although using smoothed data, comes very close to the optimal and represents a feasible and promisin approach to real-time control. As shown in Figure 2, OPAC offers rather substantial savings when compared with a fixed-time strategy such as Webster's (13) and comes very close to the possible optimum.

5. CONCLUSIONS

On-line traffic control strategies should be capable of providing results that are better than those produced by the off-line methods. The studies reported in this paper indicate that substantial benefits can be achieved with truly responsive strategies. The reason that previous experiments have failed is not because their rationale was wrong (that traffic-responsive control should provide benefits over fixed-time control), but because of a failure of the models and procedures which were implemented to deliver the desired results. The SCOOT strategy (14), which was recently implemented in the U.K., has made significant strides in this direction, but still seems to restrict the range of possible control options.

As indicated above, OPAC offers a feasible and promising approach to real-time control. The strategy is designed to make use of readily available data, produces control policies that are almost as effective as those that would be obtained uner ideal conditions, and has very reasonable computational requirements. What is, perhaps, even of greater significance is the OPAC flow model. It considers the entire projection horizon in the optimization process and, therefore, should be amenable for application in a demand-responsive decentralized flexibly-coordinated system. In such a system one would use the analysis capabilities of OPAC to structure the flows in the traffic network so that coordination can be preserved on the one hand, while taking advantage of the ever present variations in flows on the other. Thus, the system would require both local analysis capabilities and communication with adjacent controllers. A sketch of the envisioned information flow is illustrated in Figure 3. The result would be a hierarchical system of the general type described by Findeisen (15). The development of such a system is the goal of the next phase of this research.

ACKNOWLEDGEMENT

This paper is based, in part, on a research project conducted under sponsorship of the Office of University Research, U.S. Department of Transportation. The opinions expressed in the paper, however, are those of the author and not necessarily those of the sponsoring agency.

REFERENCES

1. P.J. Tarnoff, "Concepts and Strategies - Urban Street Systems", Proc. Intern. Symposium on Traffic Control Systems, Berkeley, California, pp. 1-12, August 1979.

2. N.H. Gartner, "Urban Traffic Control Strategies: The Generation Gap", Proc. 2nd Intern. ATEC Congress, Paris, April 1980.

3. J. MacGowan and I.J. Fullerton, "Development and Testing of Advanced Control Strategies in the Urban Traffic Control System" (3 articles), Public Roads, Vol. 43 (nos. 2,3,4), 1979-1980.

4. R.D. Henry, R.A. Ferlis and J.L. Kay, "Evaluation of UTCS Control Strategies - Executive Summary", Report No. FHWA-RD-76-149, FHWA, Washington, D.C., August 1976.

5. J. Holroyd and D.I. Robertson, "Strategies for Area Traffic Control Systems Present and Future", TRRL Report LR569, 1973.

6. Corporation of Metropolitan Toronto, "Improved Operation of Urban Transportation Systems", Vol. 1 (March 1974), Vol. 2 (Nov. 1975), Vol. 3 (Nov. 1976), Toronto, Canada.

7. N.H. Gartner, "Demand-Responsive Decentralized Urban Traffic Control, Part I--Single Intersection Policies", Office of University Research, U.S. Dept. of Transportation, Rept. DOT-RSPA-DPB-50-81-24, February 1982.

8. R.B. Grafton and G.F. Newell, "Optimal Policies for the Control of an Undersaturated Intersection," Proc. Third Intern. Symp. on Theory of Traffic Flow (L.C. Edie et al., Eds.), American Elsevier, New York, 1967.

9. D.I. Robertson and R.D. Bretherton, "Optimum Control of an Intersection for any Known Sequence of Vehicle Arrivals", 2nd IFAC/IFIP/IFORS Symp. on Traffic Control and Transportation Systems, North-Holland, Amsterdam, pp. 3-17, 1974.

10. S.S. Rao, Optimization Theory and Applications, Wiley Eastern, New Delhi, 1978.

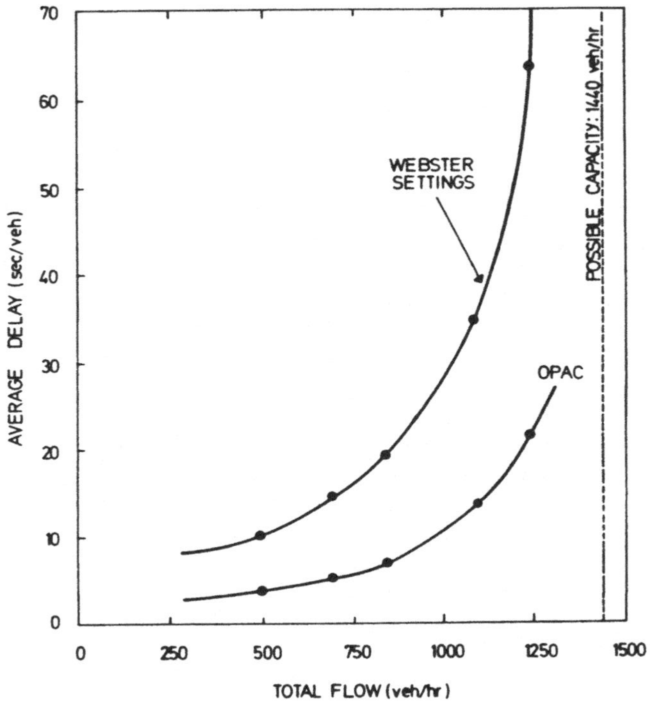

Figure 2: Comparison of control strategies at a single intersection.

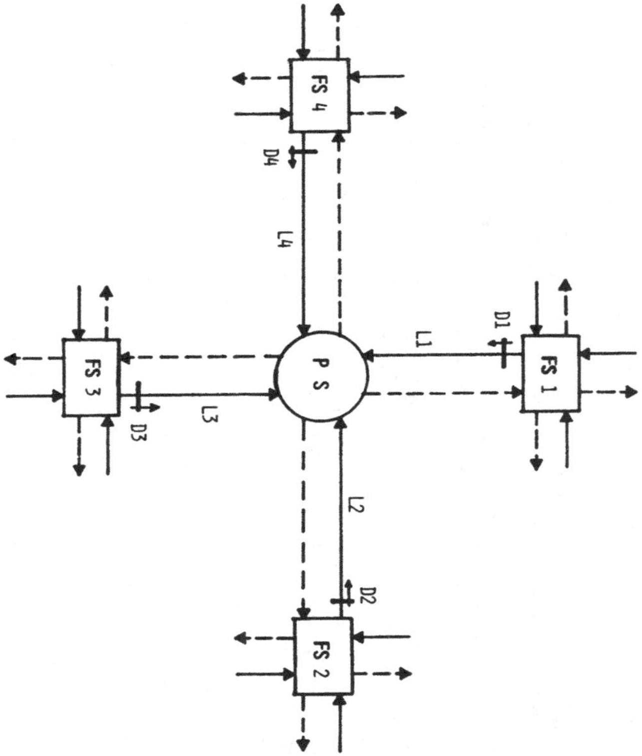

Figure 3: Information flow for decentralized traffic control system. (PS-primary signal, ES-feeder signal, D-detector, L-link).

11. H.M. Wagner, Principles of Operations Research, 2nd Edition, Prentice Hall, 1977.

12. N.H. Gartner, "OPAC: A Demand-Responsive Strategy for Traffic-Signal Control", Transportation Research Record 906, TRB, 1983.

13. F.V. Webster, "Traffic Signal Settings", Road Research Technical Paper No. 39, H.M. Stationery Office, London, 1958.

14. P.B. Hunt et al, "SCOOT - A Traffic Responsive Method of Coordinating Signals", TRRL Laboratory Report 1014, 1981.

15. W. Findeisen, "The Essentials of Hierarchical Control", 11th IFIP Conf. on System Modelling and Optimization, Copenhagen, July 1983 (in these Proceedings).

AN ALGORITHM FOR MULTIPLE CHOICE KNAPSACK PROBLEM

K. Dudziński and S. Walukiewicz

Systems Research Institute
Polish Academy of Sciences
Newelska 6, 01-447 Warsaw, Poland

1. INTRODUCTION

The Multiple Choice Knapsack Problem (MCK) is given by

$$\text{maximize} \sum_{k=1}^{m} \sum_{j \in N_k} c_j x_j, \tag{1}$$

subject to

$$\sum_{k=1}^{m} \sum_{j \in N_k} a_j x_j \leq b, \tag{2}$$

$$\sum_{j \in N_k} x_j = 1 \quad \text{for} \quad k=1,\ldots,m, \tag{3}$$

$$x_j = 0 \text{ or } 1 \text{ for } j \in N = \bigcup_{k=1}^{m} N_k, \tag{4}$$

where all a_j, c_j and b are positive and the choice classes N_k are mutually disjoint.

Such a problem is a natural generalization of a single 0-1 knapsack problem given by (1), (2) and (4), if we have to determine only one item from a given class, to fill the knapsack. The MCK problem has many practical applications as mentioned eg. in Armstrong at al. [2] and Sinha and Zoltners [6].

There are a lot of different approaches to solve MCK and one of them is the branch and bound method using bounds obtained from solving linear relaxations LMCK of MCK, in which in place of (4) we require

$$0 \leq x_j \leq 1 \quad \text{for} \quad j \in N. \tag{5}$$

In this method it is very important to have an efficient algorithm for solving LMCK problem, as it is applied many times during solving MCK.

In Section 2 some properties, of optimal solutions of LMCK and the rules of variable reduction are given. In Section 3 a fast algorithm for

LMCK is presented and in Section 4 the branch and bound method for MCK is described. The computational results are reported in Section 5.

2. PROPERTIES OF OPTIMAL SOLUTIONS TO LMCK AND MCK.

We say that x_j is P dominated, if there is an optimal solution x^* to a problem P such that $x_j^*=0$, i.e., the dominated variables may be apriori removed from a problem P.

There are two rules of variable dominance (see |6| for details) for the MCK problem.

Integer dominance: if for $p,q \in N_k$ $a_p \leq a_q$ and $c_p \geq c_q$ then x_q is MCK and LMCK dominated.

Linear dominance: if for i,j,l N_k $a_i < a_j < a_l$, $c_i \leq c_j \leq c_l$ and $(c_j-c_i)/(a_j-a_i) \leq (c_l - c_j)/(a_l-a_j)$, then x_j is LMCK dominated.

The reduction of variables may be done by sorting variables in each class and thus its complexity is of the order

$$0(\sum_{k=1}^{M} |N_k|\log|N_k|) = 0(|N|\log|N|) \tag{6}$$

And now without loss of generality we may assume that all variables in each class N_k are MCK undominated and ordered such that for $i \in N_k$

$$a_i < a_{i+1} \quad \text{and} \quad c_i < c_{i+1} \tag{7}$$

Let $M_k \subseteq N_k$ be the sets of LMCK undominated variables. Without loss of generality we may assume that M_k is reindexed such that for each $i \in M_k$ (7) is satisfied and

$$d_i > d_{i+1}, \quad \text{where} \quad d_i = (c_{i+1}-c_i)/(a_{i+1}-a_i) \tag{8}$$

We may assume that $|M_k| \geq 2$, otherwise class M_k may be removed and b decreased. Due to (7) and (8), the line connecting the points (a_i,c_i) $i \in M_k$ for each $k=1,\ldots,m$ is a concave piecewise linear function where m refer now to the number of reduced classes.

The reduced LMCK problem has three important properties (see [6]).

(a) there are at most two fractional variables in a basic feasible solution (having at most m+1 nonzero variables),

(b) in an optimal solution the fractional variables must be adjacent in some M_p,

(c) if i_k for $k=1,\dots,m$ and i_p+1 are variable indices such that $i_k \in M_k$, $i_p+1 \in M_p$ (i_p, i_p+1 are indices of fractional variables) then they are indices of nonzero variables of optimal solution, if

$$\sum_{k=1}^{m} a_{i_k} \leq b < \sum_{k=1}^{m} a_{i_k} + a_{i_p+1} - a_{i_p} \tag{9}$$

and for $j \in M_k$ respectively

$$\text{if } j < i_k \ (j > i_k) \text{ then } (c_j - c_{i_k})/(a_j - a_{i_k}) \geq d_{i_p} \ (\leq d_{i_p}). \tag{10}$$

3. SUBLINEAR ALGORITHM FOR LMCK

It is obvious that any algorithm for solving LMCK is of the complexity at least $0(|N|)$. However, if we reduce the problem with the computational effort of (6) (which should be done only once during the branch and bound method) then, if n denotes the number of LMCK undominated variables then the reduced problem may be solved with

$$0\,(m \log^2(n/m)) \tag{11}$$

running time, what is better then $0(n)$.

We have a following corollary of (a), (b), (c):

Corollary. If $d=d_{i_p}$ for some $i_p \in M_p$ then for $k \neq p$, if

$$i_k = \begin{cases} \max\{j \in M_k : d_j \geq d\}, & \text{if there exists } j \in M_k : d_j \geq d \\ \min\{j \in M_k\} & \text{, otherwise,} \end{cases}$$

then (10) is satisfied.

Thus it is enough to find such a value of d, say d^*, that i_k defined as in Corollary satisfy (9), (and therefore together with i_p, i_p+1 are indices of nonzero variables of the optimal solution to LMCK), by the following dual method.

Algorith LMCK.

Step 0. $L:=\{1,\dots,m\}$; $B:=b$;

Step 1. $d:=d_{i_p}$ for some $i_p \in M_p$ and $p \in L$;

for $k \in L \setminus \{p\}$ determine i_k as in Corollary;

$E:=\{k \in L : d_{i_k}=d\}$;

$S:= \sum_{k \in L} a_{i_k}$; $T:= \sum_{k \in L \setminus E} a_{i_k} + \sum_{k \in E} a_{i_k+1}$;

Step 2. if $S \leq B \leq T$ then go to Step 5;

Step 3. for $k \in L$ do
begin if $S > B$ then $M_k:=\{j \in M_k : j \leq i_k\}$
else begin if $k \in E$ then $i_k:=i_k+1$; $M_k:=\{j \in M_k : j \geq i_k\}$
end;
if $|M_k|=1$ then begin $B:=B-a_{i_k}$, $L:=L-\{k\}$ end;
end;

Step 4. if $L \neq \emptyset$ and $B > 0$ then go to Step 1 else stop;

Step 5. for $k \in E$ do
begin $S:=S+a_{i_k+1}-a_{i_k}$;
if $S < B$ then $i_k:=i_k+1$
else begin $p:=k$; stop end;
end.

We look for d^* in the main iteration formed by Steps 1 to 4. In Step 1 we determine d and i_k, $k \in L$. If conditions of Step 2 are satisfied then $d^*=d$ and we have to find the optimal solution in Step 5, such to satisfy (9). Otherwise we decrease each set M_k in Step 3. If $S > B$ then choosen d is smaller than d^* and we have to look for a greater value of d, which may be only in the left part of M_k i.e. for $j < i_k$. Similarly if $T < B$ then d is too large and we look for a smaller one. In Step 4 we check, if we should stop, since there is no feasible solution to the problem.

The computational complexity of Algorithm LMCK strongly depends of the determination of d in Step 1. If we determine d as the greatest one of all d_i, then indeed we get the algorithm of Sinha and Zoltners |6|. And with the modification of Glover and Klingman |5|, in which set L is implemented as a heap with the greatest d_i on the top, then the computational complexity is of the order $O(n \log m)$.

The best way of choosing d is to determine the median of all d_i. In such a way at each iteration at least half of variables is eliminated. If we apply the linear time selection algorithm presented e.g. in Aho at. al. [1] then we get the computational complexity of the order $O(n)$

and indeed we get the algorithm of Zemel [7]. However, as we look for the median in ordered sets, due to (8), we can apply the $O(m\log(r/m))$ algorithm of Fredrickson and Johnson [4] for selecting the r-th largest element in m ordered sets, heve $r=\lceil n/2\rceil$. The indices i_k may be determined during searching for the median. It is important that algorithm of Frederickson and Johnson [4] is optimal with respect to a constant value, and if $n=O(m)$ then its complexity is $O(n)$. Thus the computational complexity of Algorithm LMCK can be bounded by (11).

The presented algorithm is general in such a way that algorithms of Sinha and Zoltners [6] and Zemel [7] are indeed its special cases and with a good strategy of selecting d is even sublinear with respect to n.

The desribed method is a two phase algorithm for the LMCK problem, first the reduction is performed then the reduced problem is solved. Recently Dyer [3] developed an one phase algorithm having the complexity $O(|N|)$, so is optimal with respect to a constant value. However it is hard to develope an efficient branch and bound method for MCK exploiting Dyer's algorithm, as his algorithm does not perform so strong reduction.

4. THE BRANCH AND BOUND METHOD FOR MCK

First we should reduce the problem by applying rules of integer dominance to satisfy (7) for $i\in N_k$, with $O(|N|\log|N|)$ running time.

Let N refer now to the set of all MCK undominated variables. The reduction to LMCK undominated variables (selecting M_k) can be done in $O(|N|)$ (see e.g. Glover and Klingman [5]) to satisfy for $i\in M_k$ both (7) and (8).

Let i_k for $k=1,\dots,m$ and $i'_p > i_p$ be the indices of the optimal solution of LMCK such that $i_k\in N_k$ and $i'_p\in N_p$ (notice that indices in M_k are different than in N_k) where i_p, i'_p corresponds to indices of fractional variables.

The branching scheme (see [6]) is to separate the fractional variables i.e. to obtain two MCK subproblems having respectively N_p^1 and N_p^2 in place of N_p, such that $i_p\in N_p^1$, $i'_p\in N_p^2$ and $N_p=N_p^1\cup N_p^2$, and for $k\neq p$ sets N_k are unchnged. This can be done in several ways e.g. such that N_p^1 and N_p^2 are possibility of the same cardinality. So let

$N_p^1 = \{j \in N_p : j \leq r\}$ and $N_p^2 = \{j \in N_p : j > r\}$, where $i_p \leq r < i_p'$.

Two LMCK subproblems should be solved. Both of then have the same M_k (as in orginal LMCK) for $k \neq p$, but M_p^1 and M_p^2 should be computed. From the concavety of piecewise linear function describing M_k, we have to reduce only sets $S_1 \subseteq N_p^1$ and $S_2 \subseteq N_p^2$, such that $S_1 = \{j \in N_p : i_p \leq j \leq r\}$ and $S_2 = \{j \in N_p : r < j \leq i_p'\}$, to sets T_1 and T_2 respectively satisfying (7) and (8).
Then we have

$$M_p^1 = \{j \in M_p : j \leq i_p\} \cup T_1$$

$$M_p^2 = T_2 \cup \{j \in M_p : j \geq i_p'\}$$

The presented branching scheme gives us the possibility to obtain the reduced LMCK subproblems with a very small, computational effort bounded by $0(|S_1|)+0(|S_2|)$, and both of them may be efficiently solved by Algorithm LMCK.

5. COMPUTATIONAL RESULTS

As the main effort of the algorithm for the MCK problem is carrying out by the algorithm for LMCK we compare different methods of solving LMCK. The algorithm of selecting the median is usually expensive in practice, even it has a low bound of its computational complexity. In the Algorithm LMCK instead of selecting the median we apply selecting as d the middle element of some three elements and i_k are computed by binary search method presented e.g. in Aho at. al. [1] and we obtain the expected computational complexity of the order (11).

We compare the algorithm of Sinha and Zoltners [6] (SZ), the algorithm of Dyer [3] (the median is not selected but defined as described above) with Algorithm LMCK for uniformly distributed randomly generated integers a_j, c_j ordered to satisfy (7) and (8) i.e. all variables are LMCK undominated (the ordering time is not included in results presented in a table) since indeed only such problems are solved durring solving the MCK problem. We determine

$$b = 0.5 \sum_{k=1}^{m} \left(\min_{j \in N_k} a_j + \max_{j \in N_k} a_j \right).$$

The algorithm were coded in FORTRAN IV for the IBM 370/145. The reported times given in CPU seconds are mean times for 10 testing problems for each m and n. As it can be seen in the table, if the ratio n/m is large then the algorithm SZ and LMCK are much better than one of Dyer.

Table
Comparison of different algorithms for LMCK

Number of classes m	Number of variables per class m/n	Number of all variables n	Dyer	SZ	LMCK
5	5	25	0.04	0.01	0.01
	10	50	0.15	0.02	0.02
	20	100	0.47	0.03	0.02
	50	250	2.87	0.08	0.03
	100	500	11.35	0.16	0.03
10	5	50	0.10	0.02	0.02
	10	100	0.27	0.04	0.03
	20	200	1.48	0.09	0.04
	50	500	5.85	0.18	0.04
20	10	200	0.62	0.09	0.04
	20	400	2.16	0.17	0.07
50	5	250	0.53	0.13	0.08
	10	500	1.46	0.26	0.10
100	5	500	1.00	0.30	0.16

REFERENCES

[1] A.V.Aho, J.E.Hopcroft and J.D.Ullman, "The Design and Analysis of Computer Algorithms", Addison-Vesley, Reading, Mass. 1974.

[2] R.D.Armstrong, D.S.Kung, P.Sinha and A.A.Zoltners, "A Computational Study of a Multiple Choice Knapsack Algorithm", ACM Trans. on Math. Software 2 (1983), 184-198.

[3] M.E.Dyer, "An 0(n) Algorithm for the Multiple Choice Knapsack Linear Program", Research Report, Tesside Polytechnic (1982).

[4] G.N.Frederickson and D.B.Johnson, "The Complexity of Selection and Ranking in X+Y and Matrices with Sorted Columns", J. Comput. and System. Sci. 24, 197-208 (1982).

[5] F.Glover and D.Klingman, "An 0(n log n) Algorithm for LP Knapsack with GUB Constraints", Math. Programmnig 17, 345-361, (1979).

[6] P.Sinha and A.A.Zoltners, "The Multiple Choice Knaspsack Problem",

Opus. Res. 27, 503-515 (1979).

[7] E.Zemel, "The Linear Multiple Choice Knapsack Problem", Opns. Res. 28, 1412-1423 (1980).

AGGREGATION OF EQUALITIES IN INTEGER PROGRAMMING

G. PLATEAU and M.T. GUERCH
Université de Lille 1 - IEEA Informatique
Bât M3 - 59655 Villeneuve d'Ascq Cedex
France

1. INTRODUCTION

Given a set of m integer-valued functions f_i i=1,...,m defined on a set X of $\mathbb{R}^n$, the methods which consist in replacing the m diophantine equations :

$$f_i(x) = 0 \qquad i=1,\dots,m \; ; \; x \in X \tag{1}$$

by a single one

$$\sum_{i=1}^{m} \lambda_i \, f_i(x) = 0 \; ; \; x \in X \tag{2}$$

whose each solution is a solution of (1) - the weights $\lambda_i \in \mathbb{Z}_*$ i=1,...,m will be called *feasible multipliers* - may be classified as follows :

(i) the first class includes the so-called *2-aggregration* methods [1,2,4,5,9,10,11,16]: the single equation (2) equivalent to the system (1) is constructed by a cascade of two by two linear combinations of the m equations of (1).

All of these methods have been presented by their authors in an algebraic framework. A geometrical interpretation of them allows a theoretical comparison and an improvement of Bradley's method (section 2).

(ii) the second class includes the so-called *G-aggregation* methods [7,12,14] : the integer coefficients λ_i i=1,...,m are now globally obtained by the construction of an integer matrix M of size m × (m-1) which satisfies the two conditions :

- its columns generate a basis of the space $\{z \in \mathbb{Z}^m \mid \sum_{i=1}^{m} \lambda_i z_i = 0\}$
- the set $\{(k,x) \in \mathbb{Z}_*^{m-1} \times X \mid f_i(x) = Mk \;\; i=1,\dots,m\}$ is empty.

New types of matrix M are proposed. For some classes of functions f_i they lead to better results than those obtained by the two known matrices (section 3).

The reader is requested to see [6] for more details, and [8] for a time complexity analysis of aggregation and related problems. Computational results complete this study ; they are devoted to the determination of feasible multipliers (by implementing an improvement of Bradley's method), the solving of 0-1 linear knapsack problems with an equality constraint (by an adaptation of the code presented in [3]), the solving of 0-1 linear problems with m (≥2) equality constraints (by using a so-called relaxation-aggregation method which allows to get an optimal solution by actually aggregating a subset of constraints (see [13]).

2. THE 2-AGGREGATION METHODS

For this first class of methods, the resulting single equation is constructed by a cascade of two by two linear combinations of the m equations.

For an algebraîc point of view, at each iteration, two equations

$$g_1(x) = 0 \; ; \; g_2(x) = 0 \; ; \; x \in X \qquad (3)$$

(g_1 and g_2 may be two functions among the f_i $i=1,\dots,m$ or, more generally, two combinations of a part of these m functions) are aggregated to a single one

$$\lambda_1 \, g_1(x) + \lambda_2 \, g_2(x) = 0 \; ; \; x \in X \qquad (4)$$

equivalent to (3) when the integer weights λ_1 and λ_2 satisfy the hypotheses of the following result :

THEOREM 1

The relatively prime integers λ_1 and λ_2 are feasible multipliers if and only if the set

$$E = \{(k,x) \in \mathbb{Z}_* \times X \mid g_1(x) = k\,\lambda_2 \; ; \; g_2(x) = -k\,\lambda_1\}$$

is empty.

Up to now, in most cases, authors compare their own sufficient conditions for the emptiness of the set E with the previous ones by considering only few numerical examples. Instead of constructing a difficult global algebraîc study of the known results, we propose here a simple geometrical interpretation (section 2.1) which allows a theoretical comparison (section 2.2) and an improvement of Bradley's results which we claim to be the best ones (section 2.3).

2.1 Geometrical interpretation

By introducing, as in [16], the so-called *spectra* associated with the integer-valued functions g_1 and g_2 :

$$S_i = \{z \in \mathbb{Z} \mid \exists \; x \in X : g_i(x) = z\} \qquad i=1,2$$

and

$$S_{12} = \{(z_1,z_2) \in \mathbb{Z}^2 \mid \exists \; x \in X : g_i(x) = z_i \qquad i=1,2\}$$

the geometrical aspect of the aggregation of (3) to (4) consists in finding in $\mathbb{R}^2$ a straight line which passes through the origin without including any non-zero point of the *spectrum* S_{12} (see figure 1).

Two types of approaches may be distinguished :

(i) Weinberg [16] has proposed that we call an *interior method* because the spectrum S_{12} is actually constructed in a first time in order to find a line of the type

$\lambda_1 g_1 + \lambda_2 g_2 = 0$. This method works when the cardinality of S_{12} is not too large, but chiefly when the cardinality of $S_1 \times S_2$ is also not too large.

(ii) All of the other existing methods may be called *exterior methods* because they consist in finding a point z^* of $\mathbb{Z}^2$ whose components are relatively prime integers and such that z^* and $-z^*$ are outside a set-denoted by C - which contains all of the points of the spectrum S_{12} ; of course, the straight line which passes through the origin and z^* surely avoids any point of this spectrum. Thus $\lambda_1 = z_2^*$ and $\lambda_2 = -z_1^*$ are feasible multipliers for the aggregation of (3) to (4).

An improvement of this scheme consists in applying a *hybrid method* (see figure 2) which, at first, enumerates a part P of the spectrum S_{12} and then finds a point z^* of $\mathbb{Z}^2$ satisfying three conditions :

a) z_1^* and z_2^* are relatively prime integers

b) z^* and $-z^*$ do not coincide with the elements of P

c) z^* and $-z^*$ are outside a set C which contains the points of $S_{12} \backslash P$.

2.2 Theoretical comparison

Weinberg's method, unique element of the class of the *interior methods*, cannot be compared with the *exterior methods* whose class may be divided into two subclasses by distinguishing methods for which the set C includes the product of spectra $S_1 \times S_2$ (and thus S_{12}) and those for which the set C refers only to S_{12}. Some ideas of the comparison are given ; for a lot of details and proofs see [6].

2.2.1 The $S_1 \times S_2$ - aggregation methods

This subclass includes three methods proposed by Anthonisse [1], Glover-Woolsey [5] and Kendall-Zionts [9] ; for each of them, the initial functions f_i $i=1,\ldots,m$ are supposed to be affine.

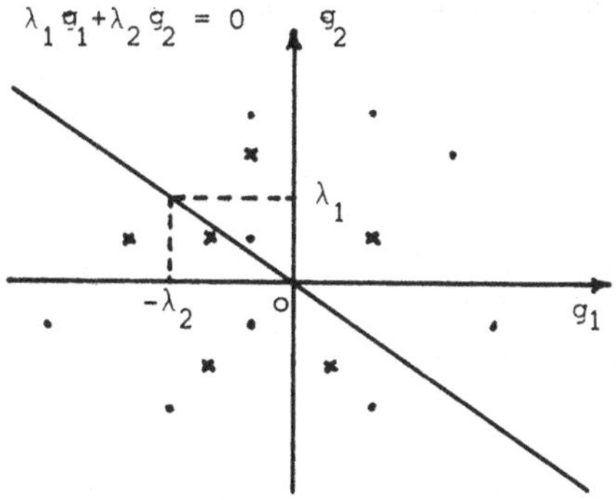

figure 1

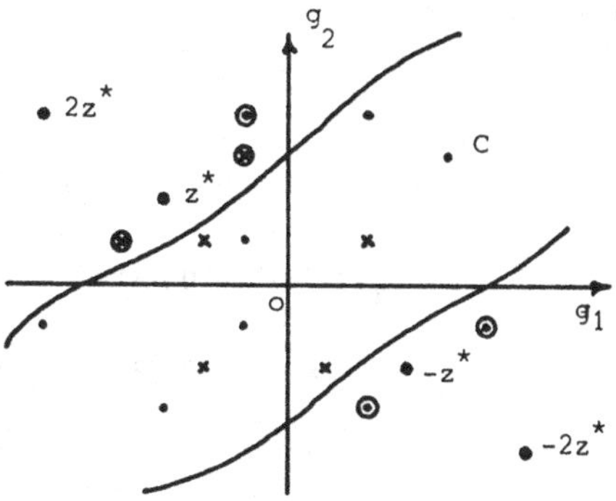

figure 2

• elements of $S_1 \times S_2$; × elements of S_{12} ; ⊛ and ⊙ elements of P

By letting $\ell_i = \min_{x \in X} g_i(x)$ and $u_i = \max_{x \in X} g_i(x)$ $i=1,2$

the set C considered by Anthonisse is nothing else that the rectangle $[\ell_1,u_1]\times[\ell_2,u_2]$.

For the hybrid method of Glover-Woolsey and the generalization of Kendall-Zionts, it is possible to go inside the product of spectra $S_1 \times S_2$ and, as a consequence, to improve the previous result : for each function g_i $(i=1,2)$, a set of greatest values over X are enumerated between u_i and a lower bound greater than $u_i/2$ in order to select a point $z^* \in \mathbb{Z}^2$ whose components avoid the enumerated values. These hybrid methods may be summarized by this result :
Let $B(g_i,k_i)$ $i=1,2$ be the set of the $k_i \in \mathbb{N}_*$ greatest values of $g_i(x)$, $x \in X$:

THEOREM 2 [9]

The relatively prime integers λ_1 *and* λ_2 *are feasible multipliers if the following conditions are satisfied :*

(i) $\lambda_1 \notin B(g_2,k_2)$ *and* $\lambda_2 \notin B(g_1,k_1)$

(ii) $\lambda_1 > \max\left(\min_{y \in B(g_2,k_2)} y, u_2/2\right)$ *and* $\lambda_2 > \max\left(\min_{y \in B(g_1,k_1)} y, u_1/2\right)$

2.2.2 The S_{12} - aggregation methods

This subclass includes five methods proposed by Anthonisse [1], Bradley [2], Glover [4], Mathews [10] and Meyer [11] ; for each of them (except for Bradley's method), the construction of the set C consists in finding two functions γ_1 and γ_2 such that

$$S_{12} \subset \{(z_1,z_2) \in \mathbb{Z}^2 \mid \gamma_1(z_1) \le z_2 \le \gamma_2(z_1)\}.$$

For Glover's method which generalizes Mathews' results, the functions γ_1 and γ_2 are affine. For Anthonisse's method, γ_1 and γ_2 are piece-wise linear functions, one convex and the other concave, obtained by the solvings of a lot of continuous knapsack problems. For Meyer's method, which takes into account non linear integer-valued functions, γ_1 and γ_2 are non-decreasing functions and the result is as follows :

THEOREM 3

Given γ_1 *and* γ_2 *two non-decreasing functions such that*

$$\gamma_1(g_1(x)) \le g_2(x) \le \gamma_2(g_1(x)) \qquad \forall x \in X$$

the relatively prime integers λ_1 *and* λ_2 *are feasible multipliers if*

$$\lambda_1 > \max(\gamma_2(-\lambda_2), -\gamma_1(\lambda_2))$$

Finally, Bradley's method allows to construct two orthants symetrical as refer to the origin by solving only two knapsack problems (either in the integer version or in the continuous version) in order to choose z^* in one of these orthants which exclude the elements of S_{12} ; for example, by denoting $v(\cdot)$ the value of a problem $(\cdot)$, a result of Bradley may be stated as follows :

THEOREM 4 [2]

> *Given* $\lambda_2 \in \mathbb{Z}$ *and*
>
> (K_1) $\max g_2(x)$ s.t. $g_1(x) \le -\lambda_2$; $x \in X$
>
> (K_2) $\min g_2(x)$ s.t. $g_1(x) \ge \lambda_2$; $x \in X$
>
> *the relatively prime integers* λ_1 *and* λ_2 *are feasible multipliers if*
>
> $$\lambda_1 > \max (v(K_1), -v(K_2))$$

Bradley's results are theoretically the best ones (as concern the magnitude of multipliers) in the sense that for any given integer λ_2, it is always possible to obtain orthants which get in the sets C proposed by the other authors.

2.3 Improvement of Bradley's results

A straightforward improvement of Bradley's results is achieved by applying a hybrid method which consists in searching several best solutions of the two knapsack problems ((K_1) and (K_2) for example) in order to get new orthants which contain the previous ones (they are closer to the origin) ; thus, it is possible to get in the spectrum S_{12} and to choose a better prime point z^* :

By denoting $v^j(K_i)$ $i=1,2$; $j=1,\ldots,k_i \in \mathbb{N}$ the best consecutive objective values associated with the k_i best solutions of problem (K_i) - $v(K_i) = v^1(K_i)$ and $v^j(K_i) > v^{j+1}(K_i)$ for $j=1,\ldots,k_i-1$ and $i \in \{1,2\}$ - the new result may be stated :

THEOREM 5

> *The relatively prime integers* λ_1 *and* λ_2 *are feasible multipliers if the following conditions are satisfied :*
>
> (i) $\lambda_1 > v^{k_1}(K_1)$ *and* λ_1 *does not divide* $v^j(K_1)$ $j=1,\ldots,k_1-1$
>
> (ii) $\lambda_1 > -v^{k_2}(K_2)$ *and* λ_1 *does not divide* $-v^j(K_2)$ $j=1,\ldots,k_2-1$

3. THE G-AGGREGATION METHODS

The methods of this class generate globally the components of the feasible multiplier $\lambda \in \mathbb{Z}_*^m$ by constructing a $m\times(m-1)$ integer matrix M which satisfies the two following properties :

(F) M is a λ-*fundamental matrix,* that means

$$\{z \in \mathbb{Z}^m \mid \lambda z = 0\} = \{z \in \mathbb{Z}^m \mid z = Mk,\ k \in \mathbb{Z}^{m-1}\}$$

(A) M is an *aggregating matrix,* that means

$$\{(k,x) \in \mathbb{Z}_*^{m-1} \times X \mid f_i(x) = M_i k \quad i=1,\dots,m\} = \emptyset.$$

Historically, Padberg [12] has been the first to exploit this idea by using a theoretical result due to Smith [15] who proposed a type of matrix - denoted by $\overset{1}{M}$ - λ-fundamental. The works of Padberg consist in choosing a sufficiently simple multiplier λ in order to simplify the general form of $\overset{1}{M}$ with the aim to prove that this simplified form satisfies property (A) :

THEOREM 6 [*12*]

The $m\times(m-1)$ matrix $\overset{1}{M} = \begin{bmatrix} -1 & \dots & -1 \\ q & & \\ & \ddots & (q-1) \\ 0 & & \ddots \\ & & q \end{bmatrix}$ where $q \in \mathbb{N}_*$

is aggregating if $q > \max\limits_{i=2,\dots,m} \max\limits_{x\in X} |f_i(x)|$.

Corallary

Under the assumptions of theorem 6, $\lambda_i = q^{m-i}$ $i=1,2,\dots,m$ are feasible multipliers for the aggregation of (1) to (2).

Kaliszewski-Libura [7] have taken up the reverse order to verify properties (F) and (A) ; they construct in a first time a structure of matrix such that property (A) is easily satisfied and then find a single feasible multiplier λ for which this matrix satisfies property (F) whose characterization may be formulated as follows :

THEOREM 7

Given a $m\times(m-1)$ *integer matrix* M *with a maximal rank, there exists* $\lambda \in \mathbb{Z}_*^m$ *with relatively prime integer components such that* M *is a* λ*-fundamental matrix, if and only if*

(i) $\lambda M = 0$

(ii) $\exists\, p \in \mathbb{Z}^m : \lambda p = 1$ *and* (p,M) *is unimodular.*

The associated single feasible multiplier (generated by condition (i)) is defined as :

$$\lambda_i = (-1)^{i+1} \det M_{[i]} \qquad i=1,\dots,m$$

where $M_{[i]}$ denotes matrix M without its ith row.
By noting that the unimodularity of (p,M) may be characterized by

$$\gcd_{i=1,\dots,m} (\det M_{[i]}) = 1 \qquad (*)$$

(see [14] for another type of characterization), consequently Kaliszewski-Libura and us propose matrices M such that det $M_{[i]}$ i=1,...,m are easily to compute and condition (*) is simple to satisfy. This leads to the following summarized results :

Kaliszewski-Libura propose this type of matrix :

$$\overset{2}{M} = \begin{bmatrix} \alpha_1 & & 0 \\ 0 & \alpha_2 & \\ & & \ddots \ \alpha_{m-1} \\ -\alpha_m & -\alpha_m \cdots & -\alpha_m \end{bmatrix}$$

with $\alpha_i \in \mathbb{N}_*$ i=1,...,m and the following multipliers :

$$\overset{2}{\lambda}_i = (-1)^{m-1} \prod_{k=1}^{m} \alpha_k/\alpha_i \qquad \forall\, i$$

THEOREM 8 [7]

(i) *$\overset{2}{M}$ is a λ-fundamental matrix if*

$$\gcd(\alpha_i,\alpha_j) = 1 \qquad \forall\, i,\, j \in \{1,\dots,m\} \qquad i\neq j$$

(ii) *$\overset{2}{M}$ is an aggregating matrix if*

$$\alpha_i > \max_{x\in X} f_i(x) \quad \forall\, i \in \{1,\dots,m\}$$

This last result may be improved as follows :

THEOREM 9

Given $k_i \in \mathbb{N}_*$ i=1,...,m, *$\overset{2}{M}$ is an aggregating matrix if*

$$\forall\, i \qquad \alpha_i > \min_{y\in B(f_i,k_i)} y$$

and α_i *does not divide* y $\quad \forall\, y \in B(f_i,k_i)$.

Now, we propose two new types of matrices

$$\overset{3}{M} = \begin{bmatrix} \alpha_1 & & & \\ \beta_2 & \alpha_2 & & 0 \\ \vdots & \vdots & \ddots & \\ \beta_{m-1} & \beta_{m-1} & \cdots & \alpha_{m-1} \\ -\alpha_m & -\alpha_m & \cdots & -\alpha_m \end{bmatrix} \quad \text{and} \quad \overset{4}{M} = \begin{bmatrix} \alpha_1 & & & & 0 \\ -\alpha_2 & 1 & & & \\ & -\alpha_3 & \ddots & & \\ & & \ddots & 1 & \\ 0 & & & -\alpha_{m-1} & 1 \\ & & & & -\alpha_m \end{bmatrix}$$

where $\alpha_i \in \mathbb{N}_*$ $i=1,\dots,m$ and $\beta_i \in \mathbb{N}$ $i=2,\dots,m-1$. The associated sufficient conditions for satisfying properties (F) and (A) are the followings :

<u>*THEOREM 10*</u>

(i) *$\overset{3}{M}$ is a λ-fundamental matrix if*

$$\gcd(\alpha_i,\alpha_m) = 1 \quad i=1,\dots,m-1 \; ; \; \gcd(\alpha_i,\alpha_j-\beta_j) = 1 \quad \forall\, i < j \in \{2,\dots,m-1\}$$

(ii) *by letting $\beta_1 = \beta_m = 0$, $\overset{3}{M}$ is an aggregating matrix if*

$$\alpha_i - \beta_i > - \min_{x\in X} f_i(x) \qquad i=1,\dots,m$$

The associated feasible multiplier is such that

$$\overset{3}{\lambda}_i = (-1)^{m-1} \alpha_1 \alpha_2 \dots \alpha_{i-1} (\alpha_{i+1}-\beta_{i+1}) \dots (\alpha_{m-1}-\beta_{m-1}) \alpha_m \qquad \forall\, i$$

An improved condition such that $\overset{3}{M}$ is aggregating is :

$$(**) \quad \begin{cases} \alpha_{i+1} - \beta_{i+1} > \\ \max\{\max\{f_{i+1}(x) \mid f_i(x) \le \alpha_i \; ; \; x \in X\}, \\ \qquad - \min\{f_{i+1}(x) \mid f_i(x) \ge -\alpha_i \; ; \; x \in X\} \\ i=1,\dots,m-1 \qquad (\beta_1 = \beta_m = 0) \end{cases}$$

<u>*THEOREM 11*</u>

(i) *$\overset{4}{M}$ is a λ-fundamental matrix if*

$$\gcd(\alpha_i,\alpha_1) = 1 \qquad i=2,\dots,m$$

(ii) *$\overset{4}{M}$ is an aggregating matrix under condition (**) with $\beta_i = 0$ $\forall i$.*

The associated feasible multiplier is such that

$$\overset{4}{\lambda}_1 = (-1)^{m-1} \prod_{j=2}^{m} \alpha_j \; ; \; \lambda_i = (-1)^{m-1} \alpha_1 \prod_{j=i+1}^{m} \alpha_j \qquad i=2,\dots,m.$$

<u>Numerical example</u> : [14]

$$f_1(x) = x_1 + 2\,x_2 + 100\,x_3 + 10\,x_4 - 103$$

$$f_2(x) = 2x_1 + x_2 + 100\,x_3 + 5\,x_4 - 103$$

$$f_3(x) = 2x_1 + x_2 + x_3 + x_4 - 4$$

$$X = \{x \in \mathbb{R}^4 \mid x_j \in \{0,1\} \quad j=1,\dots,4\}$$

$$\overset{2}{M} = \begin{bmatrix} 11 & 0 \\ 0 & 7 \\ -2 & -2 \end{bmatrix} \quad \overset{2}{\lambda} = (14,22,77) \;\text{(theorem 8)} \;; \quad \overset{2}{M}{}' = \begin{bmatrix} 6 & 0 \\ 0 & 7 \\ -5 & -5 \end{bmatrix} \quad \overset{2}{\lambda}{}' = (35,30,42) \;\text{(theorem 9)}$$

$$\overset{3}{M} = \begin{bmatrix} 1 & 0 \\ 0 & 3 \\ -4 & -4 \end{bmatrix} \quad \overset{3}{\lambda} = (12,4,3) \;\text{(theorem 10 (i) and condition (**))} \;; \quad \overset{4}{M} = \begin{bmatrix} 1 & 0 \\ 3 & 1 \\ 0 & 4 \end{bmatrix} \quad \overset{4}{\lambda} = (12,4,1) \;\text{(theorem 11)}$$

Obviously $\overset{4}{\lambda} < \overset{3}{\lambda} < \overset{2}{\lambda}{}'$ and $\sum_{i=1}^{3} \overset{2}{\lambda}{}'_i < \sum_{i=1}^{3} \overset{2}{\lambda}_i$.

REFERENCES

[1] Anthonisse J.M. "A note on equivalent systems of linear diophantine equations", Operations Research 17 (1973) 167-177.

[2] Bradley G.H. "Transformation of integer programs to knapsack problems", Discrete Mathematics 1 (1971) 29-45.

[3] Fayard D., Plateau G. "An algorithm for the solution of the 0-1 knapsack problem", Computing 28 (1982) 269-287.

[4] Glover F. "New results on equivalent integer programming formulations", Mathematical Programming 8 (1975) 84-90.

[5] Glover F., Woolsey R.E. "Aggregating diophantine equations", Zeitschrift für Operations Research (1972).

[6] Guerch M.T. "La contraction d'équations diophantiennes", Doctorat de 3° cycle (1983).

[7] Kaliszewski I., Libura M. "Constraints aggregation in integer programming" Report MPD 5-77, Systems Research Institute, Polish Academy of Sciences, Warszawa (Poland) (1977).

[8] Kannan R."Polynomial-time aggregation of integer programming problems", Journal of the Association for Computing Machinery 30 (1983) 133-145.

[9] Kendall K.E.,Zionts S."Solving integer programming problems by aggregating constraints", Operations Research 25 (1977) 346-351.

[10] Mathews G.B. "On the partition of numbers", Proceedings of the London Mathematical Society 28 (1896) 486-490.

[11] Meyer R.R. "Equivalent constraints for discrete sets", Discrete Applied Mathematics 1 (1979) 31-50.

[12] Padberg M.W. "Equivalent knapsack-type formulations of bounded integer linear programs : an alternative approach", Naval Research Logistics Quarterly 19 (1972) 699-708.

[13] Plateau G., Guerch M.T. "Aggregation of equalities in integer programming : a computational study", Publication ANO Lille (1983).

[14] Rosenberg I.G. "Aggregation of equations in integer programming", Discrete Mathematics 10 (1974) 325-341.

[15] Smith H.J.S. "On systems of linear indeterminate equations and congruences", Philosophical Transactions CLI (1861) 293-326.

[16] Weinberg F. "A necessary and sufficient condition for aggregation of linear diophantine equations", IFOR - Studienberichte 4 (1976).

ON JOB-SHOP SHEDULING WITH RESOURCES CONSTRAINTS

J. Grabowski and A. Janiak
Technical University of Wrocław
Wrocław, Poland

1. INTRODUCTION

This paper deals with a class of job-shop problems with allocation of continuously-divisible constrained nonrenewable resources. This problem appears in many branches of industry in which the production process is characterized by a flow of elements in the technological sequence. These elements are processed on succesive machines. The processing times of operations depend on an amount of a resource (energy, catalyst, fuel) alloted to these operations. Therefore the problem arises to determining such a sequence of operations on each machine and such an allocation of constrained resources that total time of performing all operations is minimal.

The classic job-shop problem (without resource constraints) was formulated by Conway at al. in [3]. This problem was represented by using a disjunctive graph by Roy and Sussmann [8] and solved by Balas [1] and Florian [4] by using the branch-and-bound theory. The best result has been obtained by Florian's algorithm in which active schedules was applied. Next in 1982 Bouma showed in [2] that, the best algorithm for solving job-shop problem was suggested by Grabowski in [5]. This algorithm is also based on brach-and-bound technique and disjunctive graphs theory but its theory is based on the critical path concept using the block system approach. The block was such a sequence of operations, that a better solution cannot be obtained by an interchange of processing order of operations inside the block. Now, the question arises: "is it possible to use the block approach for solving job-shop problem with resources constraints?" It is not directly possible, since in the classic job-shop problem there was only one critical path and in our problem there are many critical paths. Therefore, we must introduce segment approach defined in section 3. It has been appeared that the formulae evaluating all descendants of current solution, defined in [5], are not useffull enough in our problem, and therefore, we shall apply the lower bound evaluations of descendants. We shall also modify branching rule.

2. PROBLEM FORMULATION

The paper is devoted to the general job-shop problem with allocation of continuously-divisible constrained nonrenewable resources, indicated by $n|m|G,\ Res \geqslant 0|C_{max}$. The problem can be formulated as follows. There are n jobs $J_1,\dots,J_n$ that have to be processed on m machines $M_1,\dots,M_m$. A job J_i $(i = 1,2,\dots,n)$ consists of a sequence of n_i operations O_j; these operations are indexed by $j = N_{i-1}+1,\dots,N_i$, where $N_i = \sum_{l=1}^{i} n_l$. Machine M_v $(v \in M \triangleq \{1,2,\dots,m\}$ can handle only one job at a time; the set of operations to be performed on M_v is denoted by N^v. An operations O_j $(j = 1,2,\dots,N_n)$ corresponds to the processing of the job λ_j on the machine μ_j during an uninterrupted processing time p_j. We shall assume that the processing time p_j for $j \in N^{v'}$, $v' \in M^1 \subset M$ is constant and for $j \in N^v$, $v \in M^2 \subset M$, $M^1 \cup M^2 = M$, $M^1 \cap M^2 = \emptyset$, $p_j \triangleq p_j(u_j) \triangleq a_j u_j + b_j$, where $a_j < 0, b_j > 0$ are known and u_j is the amount of continuous resource alloted to O_j. We assume, moreover the following set of feasible allocations of a resources:

$$U \triangleq \Big\{ \bar{u} \in R^{\sum_{v \in M^2} \bar{\bar{N}}^v} : \bar{u} = [\bar{u}_{v_1}, \bar{u}_{v_2}, \dots, \bar{u}_{v_i}, \dots, \bar{u}_{v_{\bar{\bar{M}}^2}}]' \wedge$$

$$\wedge\ v_i \in M^2 \wedge \forall (v_i \in M^2)(\bar{u}_{v_i} = [u_{v_{i_1}}, u_{v_{i_2}}, \dots, u_{v_{i_j}}, \dots, u_{v_{i_{\bar{\bar{N}}^{v_i}}}}]' \wedge$$

$$\wedge\ v_{i_j} \in N^{v_i}) \wedge \sum_{i \in N^v} u_i \leqslant \hat{U}_v \wedge v \in M^2 \wedge \alpha_i \leqslant u_i \leqslant \beta_i \wedge i \in N^v \Big\}$$

where $\hat{U}_v$ is the global amount of the continuously divisible resource alloted for performing operation from the set N^v on machine M_v, $v \in M^2$ and $0 \leqslant \alpha_i \leqslant \beta_i \leqslant \infty$ are known. We want to find such a processing order on each machine and such an allocation of resource amount $\hat{U}_v$ among the operations O_i, $i \in N^v$, for $v \in M^2$ that the maximum completion time C_{max} is minimized. It is obvious that for an arbitrarily chosen resource allocation our problem reduces to the classic job--shop problem - $n|m|G|C_{max}$ which was considered in preceding section.

The job-shop problem with resources constraints is NP-complete since with the assumption that $p_j(u_j) = const$ for $j \in N^v$, $v \in M^2$, it can be reduced to the classic job-shop problem which is NP-complete. This complexity result serves as a formal justification to use enumerative methods to solv the problem.

The problem of allocation of resources for the settled schedule of operations (i.e. without sequencing the operations) have been considered in details in many papers. For example, the problem in which for every operation the function relating the performing speed to the alloted amount of continuous resource is known and the state which has to be reached in order to complete the operation is also given was considered e.g. in [6]. Problem of sequencing of operations on identical parallel machines for the case of renewable resource, splittable operations, for operations models given in the form of performing speed-resource amount functions was considered by Węglarz [10]. Problems in which operation resource requirements are discrete, i.e. concern resource amount belonging to given finite sets, with renowable, nonrenewable and doubly - constrained resources for splittable and non-splittable operations under multicriteria approach were considered by Słowiński [9].

3. MATHEMATICAL MODEL

An $n|m|G$, $Res \geq 0|C_{max}$ problem can be conveniently represented by a disjunctive graph $D = \langle A, V^o \cup V \rangle$ where:

- A is the set of vertices representing the operations including fictitious initial and final operations O_0 and O_*: $A = \{0,1,\ldots ,N_n,*\}$;
- V^o is the set of directed conjunctive arcs representing the given machine orders of the jobs:

$$V^o = \{\langle j,j+1\rangle \mid l_j = l_{j+1}\} \cup \{\langle 0,N_{i-1}+1\rangle, \langle N_i,*\rangle \mid i = 1,\ldots,n\};$$

- V is the set of directed disjunctive arcs, representing the possible processing orders an consecutive machines: $V = \{\langle j,j'\rangle \mid \mu_j = \mu_{j'},\ j \neq j'\}$. There is attached a weigth p_j to each vertex j with $p_0 = p_* = 0$.

The disjunctive graph for the example $5|2|G$, $Res\geq 0|C_{max}$ from section 5 is drawn in Fig. 1.

Let $S_r \subset V$ contain exactly one disjunctive arc from each disjunctive pair $\{\langle j,j'\rangle , \langle j',j\rangle\}$ and the graph $D_r(S_r) = \langle A,V^o \cup S_r\rangle$ has no circuits. (By chosing $\langle j,j'\rangle$ we mean that the operation O_j precedes $O_{j'}$ on the machine M_v).

Let $R_s = \{S_1,S_2,\ldots,S_r,\ldots,S_a\}$ be the family of all such subsets and $R_D \triangleq \{D_r = \langle A,V^o \cup S_r\rangle\}$ be the family of all graphs associated with these subsets.

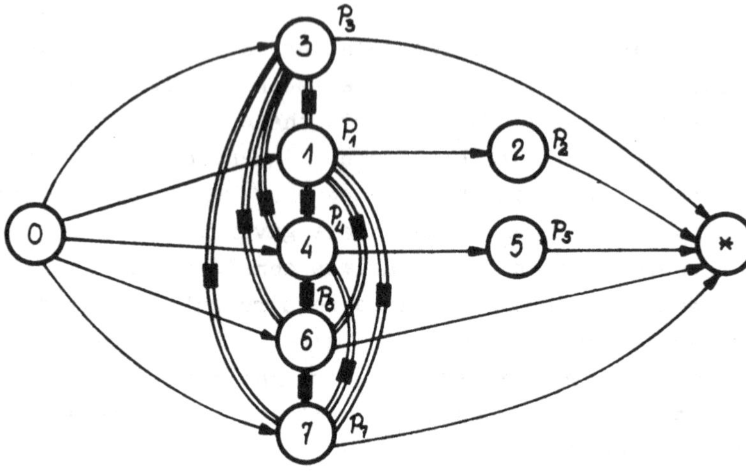

Fig. 1. Disjunctive graph

Hence, R_D is the family of graphs corresponding to all feasible schedules of operations in an $n|m|G$, $Res \geqslant 0|C_{max}$ problem. This problem is equivalent to the problem of finding the minimaximal path

$$L^* = L_{S_r^*}(\bar{u}^*) = \min_{D_r \in R_D, \bar{u}_r \in U} L_r(\bar{u}_r)$$

with the optimal selection S_r^* of disjunctive arcs and optimal allocation of resources $\bar{u}_r^* \in U$ (L_r $(\bar{u}_r)$ is the length of the critical path in D_r under the allocation of resources $\bar{u}_r^* \in U$).

4. SOME PROPERTIES OF THE PROBLEM

The following denotations will be used.

Let C_r^l denote the set of arcs of the l-th critical path in $D_r \in R_D$.

A sequence of operations $\langle j_1, j_2, \ldots, j \rangle$ containing the maximal number of operations is called a segment in $D_r \in R_D$ if there exists in $D_r \in R_D$ a path: $B = \langle\langle j_1, j_2\rangle, \langle j_2, j_3\rangle, \ldots, \langle j_{q-1}, j_q\rangle\rangle$ such that each arc of this path belongs to the same critical paths, i.e.

$$\langle j_i, j_{i+1}\rangle \in S_r \cap C_r^l, \quad i = 1,2,\ldots,g-1;$$

$l = 1,2,\ldots,l_s$ where l_s denotes the number of critical paths passing through B.

Let P_k be the set of indexes of operations of the k-th segment and k_r be the number of segments in D_r. A subsequence $\langle j_2, j_3, \ldots, j_{g-1} \rangle$ of the operations of the segment P_k which contains neither the first operation j_g nor the last operation j of the segment P_k is called an internal segment.

Theorem 1

For each $D_r \in R_D$, if the graph $D_s \in R_D$ has been obtained from D_r by an interchange of a performance order of operations and if $L_s(\bar{u}_s^*) < L_r(\bar{u}_r^*)$, then in D_s at least one operations of a k-th segment of D_r is moved before (after) its first (last) operation for k = 1 or 2... or k_r.

The above theorem can be proved taking in view that:

- any interchange of operations performance order in internal segments of graph D_r, without changing the resource allocation, cannot decrease the length of the critical path in D_r;
- for $\bar{u}_r^* \in U$, if there exists at least one operation j, which does not belong to any segment in D_r and such that $u_{jr}^* > \alpha_j$, then u_{ir} is equal to β_i for every $i \in P_k$, $k = 1, 2, \ldots, k_r$ and vice versa;
- if the models of operations are linear then the global amount of the resource assigned to the operations of each internal segment is at the same time allocated optimally with respect to the length of the path of the internal segment.

This theorem implies the enumeration scheme and branching rule which will be used while solving the $(n|m|G, Res \geq 0|C_{max})$ problem by a branch and bound procedure with mixed strategy of bound. Starting with any initial graph $D_1 = \langle A, V^o \cup S_1 \rangle \in R_D$ we generate a sequence of acyclic graphs $D_r = \langle A, V^o \cup S_r \rangle \in R_D$. For each D_r from the sequence we compute the optimal allocation of resource $\bar{u}_r^* \in U$ and the critical paths and we identify all the segments P_k enumerating them $k = 1, 2, \ldots$ $\ldots, k_r$. Each new graph $D_s \in R_D$ is obtained from a preceding graph D_r of the sequence by moving one operation of some segment in D_r. Each operation j of the k-th segment is moved before the first (or after the last) operation of this segment. Moving of the operation in D_r is equivalent to the replacement of some disjunctive arcs of a selection S_r into the corresponding reverse disjunctive arcs from $V \setminus S_r$.

For each $D_r \in R_D$ generated under the algorithm the set $F_r \subset S_r$ some disjunctive arcs (precedence constraints) is fixed. Now, we will find a lower bound $LB_k^b(j)$, $(LB_k^a(j))$ for the graph $D_s \in R_D$ which is generated from the graph D_r by moving the operation j before the first (after the last) operation of the k-th segment in this graph. Since the set F_s is fixed in any descendants of D_s, then the lower bound on the descendants of D_s can be taken as the length of the critical path $L(F_s, \bar{u}^*_{F_s})$ in the graph $D_s(F_s) = \langle A, V^0 \cup F_s \rangle$, where $\bar{u}^*_{Fs}$ is the optimal allocation of resources in this graph. Anallogically to Lageweg, Lenstra and Rinnooy Kan [7] lower bound can be also obtained by the relaxation of the capacity constraints on all the machines except for the chosen one.

Let us denote by $E^b_k(E^a_k)$ the set of operations which are to be moved before the first (after the last) operation of the k-th segment under fixed precedence constraints - F_r. We want to choose on operation generating a descendant D_s with the smallest possible $L_s(\bar{u}^*_s)$ in order to obtain quickly a good upper bound. Hence, the operations from the sets E^b_k and E^a_k will be chosen in order of the nondecreasing lower bounds.

5. ALGORITHM, EXAMPLE

Algorithm

Step 1. Compute $\bar{u}^*_r \in U$ and $L_r(\bar{u}^*_r)$ for D_r. If $L_r(\bar{u}^*_r) < L_*(*)$ ($L_*(*)$ is the current upper bound), then set $L_*(*) := L_r(\bar{u}^*_r)$. Identify the E^a_k and E^b_k sets of candidates in D_r. If $E^a_k = \emptyset$ and $E^b_k = \emptyset$, $k = 1,2,\ldots$ $\ldots, k_r$, then go to Step 3. Otherwise, for each operation j of these sets compute $LB^a_k(j)$ and $LB^b_k(j)$ and go to Step 2.

Step 2. Modify the sets $E^a_k \neq \emptyset$ and $E^b_k \neq \emptyset$ in the following way:

$$E^a_k := \left\{ j \in E^a_k \mid LB^a_k(j) < L_*(*) \right\},$$

$$E^b_k := \left\{ j \in E^b_k \mid LB^b_k(j) < L_*(*) \right\}.$$

If $E^a_k = \emptyset$ and $E^b_k = \emptyset$, $k = 1,2,\ldots,k_r$, then go to Step 3. Otherwise, among the candidates in D_r, choose an operation j with the smallest value of $LB^\lambda_k(j)$, $\lambda \in \{a,b\}$. If $LB^b_k(j)$ (or $LB^a_k(j)$) is chosen, then generate a new graph D_s by moving this operation before the first (or after the last) operation in the k-th segment in D_r, and fixing the precedence constraints F_s. Then set $D_r := D_s$ and go to Step 1.

Step 3. Backtrack to the predecessor D_p of D_r. If D_r has no predecessor i.e., if we are instructed to backtrack from the graph D_1, then the algorithm terminates: the graph D_r associated with the current

$L_{*}(*)$ is optimal. Otherwise, eliminate the graph and all its date, set $D_r := D_p$ and go to Step 2.

Example

Consider the $5|2|G$, $Res \geq 0|C_{max}$ problem specified by the following dates:

$$J_1 = \{O_1, O_2\},\ J_2 = \{O_3\},\ J_3 = \{O_4, O_5\},\ J_4 = \{O_6\},\ J_5 = \{O_7\}.$$

The models of the operations are following: $p_1 = 9-4u_1$, $0 \leq u_1 \leq 2$; $p_2 = 9-2u_2$, $0 \leq u_2 \leq 2$; $p_3 = 11-3u_3$, $0 \leq u_3 \leq 1$; $p_4 = 5-3u_4$, $0 \leq u_4 \leq 1$; $p_5 = 2-u_5$, $0 \leq u_5 \leq 1$; $p_6 = 4-2u_6$, $0 \leq u_6 \leq 1$; $p_7 = 3-2u_7$, $0 \leq u_7 \leq 1$;

$$\hat{U} = U_1 + U_2 \doteq 6.\ N^1 = \{1,3,4,6,7\},\ N^2 = \{2,5\}.$$

The disjunctive graph for the example is drawn in Fig. 1. We start with a graph D_1. The optimal graph D_x with the optimal performance times of operations is drawn in Fig. 2. It was found in the secound iteration but we had to compute 14 lower bounds for this example. The critical path is represented by means of thick lines and its length $L_{*}(*) = C_{max} = 14$.

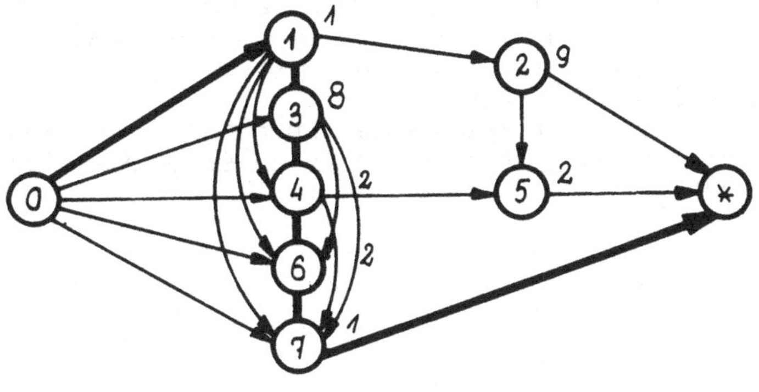

Fig. 2. The optimal graph $D_2 = \langle A, V^0 \cup S_2 \rangle$

6. CONCLUSION

Some of the concepts introduced can be applied to sequencing problems in which the critical path method can be used, i.e. flow-shop, parallel shop with resource constraints under criteria: maximum completion time, maximum lateness, maximum tardines and maximum penalty cost.

REFERENCES

[1] Balas, E. (1967). Discrete programming by the filter method. Opns. Res., 15, pp 915-967.

[2] Bouma, R.W. (1982). Job-shop scheduling: A comparison of three enumeration schemes in a branch-and-bound approach. Master´s thesis, Erasmus University Rotterdam, Faculty of Economics, Department of Econometrics/Operations Research.

[3] Conway, R.W., W.L. Maxwell, L.W. Miller (1967). Theory of scheduling, Addison Wesley, Reading, Mass.

[4] Florian, M., P. Trepant, G.B. McMahon (1971). An implicit enumeration algorithm for the machine scheduling problem. Management Science, 17, pp 782-792.

[5] Grabowski, J. (1980). On two-machine scheduling with release and due dates to minimize maximum lateness. Opsearch, 17, pp 133-154.

[6] Janiak, A., A. Stankiewicz (1983). On the equivalence of local and global time-optimal control of a complex of operations. Int. J. Control (to appear).

[7] Lageweg, B.J., J.K. Lenstra and A.H.G. Rinnooy Kan (1977). Job shop scheduling by implicit enumeration. Management Sci, 24, pp 441-450.

[8] Roy, B., B.Sussmann (1964). Les Problemes d´Ordonnancement avae Contraintes Disjonctives. Note DS No 9 bis, SEMA Montrouge.

[9] Słowiński, R. (1981). Multiobjective network scheduling with efficient use of renewable and nonrenewable resources. Europ. J. Opl. Res., 7, pp 265-273.

[10] Węglarz, J. (1979). Project scheduling with discrete and continuous resources. IEEE Trans. Syst., Man and Cybern., SMC-9, pp 644-650.

SPACE COVERING TECHNIQUE FOR MULTICRITERION OPTIMIZATION

Y.Evtushenko and M.Potapov
Computing Centre, USSR Academy of Sciences
Vavilova 40
Moscow, USSR

The following form of multicriterion optimization problem is considered:

minimize $f^1(x),\dots,f^m(x)$ subject to $x \in X$,

where $x = (x^1,\dots,x^n)$, x^i are real numbers. X is a bounded constraint set in n-dimensional Euclidean space E^n.

The set of all Pareto - optimal solutions is denoted by X_*. Let Y denote the image of X in the space E^m under the mapping $f(x) = \left[f^1(x),\dots,f^m(x)\right]$; that is $Y = f(X)$. Analogously $Y_* = f(X_*)$ is the image of X_*. We suppose that the vector function f(x) is Lipschitzian with known constant L on X, i.e.

$$\|f(x_1) - f(x_2)\| \leq L \|x_1 - x_2\| .$$

The set Y_* is the set of all Pareto-optimal solution of the following elementary multicriterion optimization problem:

$$\text{minimize } y \text{ subject to } y \in Y ,$$

with Y being the constraint set.

Determine the finite approximate ε-set in E^m:

$$A_p = \left[y_1,\dots,y_p\right] , \text{ where } y_i = f(x_i).$$

This set satisfies the following conditions:

1) for any $y \in Y_*$ there exists such $y_i \in A_p$ that $\| y - y_i \| < \varepsilon$;
2) for any $y_i \in A_p$, there exists no other $y_q \in A_p$ such that $y_i \geq y_q$ and $y_i \neq y_q$.

We describe algorithm for finding ε-approximation set. Suppose that the current ε-approximation set is

$$A_s = \left[y_1,\dots,y_s\right] , \text{ where all } y_i \in Y.$$

Suppose that we calculate some new vector $\bar{y} = f(\bar{x})$ in a point $\bar{x} \in X$. Introduce index set

$$J(\bar{y}) = \left\{ i \subset [1:s] \; : \; \bar{y} > y_i \right\} .$$

If the set $J(\bar{y})$ is empty and $\|\bar{y} - y_i\| > \varepsilon$ for all $i \in [1:s]$ then we include $\bar{y}$ in the approximation set and obtain $A_{s+1} = \{y_1, \ldots, y_s, y_{s+1}\}$ $y_{s+1} = \bar{y}$.

If the set $J(\bar{y})$ is not empty then $\bar{y}$ can be omitted. Moreover some neighbourhood of $\bar{y}$ can be omitted. Lets determine the set $M(\bar{y})$ in objective space

$$M(\bar{y}) = \left\{ y : \max_{i \in I(\bar{y})} \min_{j \in [1:m]} (y^j - y_i^j) \leq 0 \right\}$$

and denote g the distance between y and the set $M(\bar{y})$:

$$g = \min_{y \in M(\bar{y})} \| \bar{y} - y \| \quad , \; y = f(x)$$

All the points $y \in E^m$, which satisfy the condition

$$\| \bar{y} - y \| \leq g + \varepsilon \tag{1}$$

are not important, because they do not improve current approximation of Pareto set. We can rewrite the inequality (1) in the following form

$$\| \bar{y} - f(x) \| \leq g + \varepsilon \quad .$$

From Lipschitzian condition we have

$$\bar{y}^j + L \| \bar{x} - x \| \geqslant f^j(x) \geqslant \bar{y}^j - L \| \bar{x} - x \| \; .$$

The inequality (1) holds if x satisfies

$$\| \bar{x} - x \| \leq \frac{g + \varepsilon}{L} \tag{2}$$

Numerical algorithm consists of a covering set X with cubes inscribed in the spheres which are defined similarly to (2). This procedure is close to one which is described in detail in [1]. It may easily be seen that the sequence of cubes thus constructed will cover the restricted set X after a finite number of steps.

REFERENCES

1. Evtushenko Y.G.: Zh. Vychisl.Mat.mat.Fiz., 11,6,pp.1390-1403 (1971).

LEXICOGRAPHICAL ORDER, INEQUALITY SYSTEMS AND OPTIMIZATION

J.E. Martínez-Legaz
University of Barcelona
Barcelona, Spain

1. INTRODUCTION

In [5] we have proved a separation theorem by a general type of halfspaces defined by means of a lexicographical inequality on $\overline{R}^n$ instead of the usual scalar relation valid for any convex set, to construct a conjugacy theory for quasiconvex functions which is exact in two senses: the second conjugate of any function coincides with its quasiconvex hull and the generalized subdifferential of a quasiconvex function is not empty at any point. The essential of this theorem can be obtained from Th.1 in [3] and Th.6 in [4], although the proofs are by a different method. But this notion of separation seems have not been exploited in inequality systems and optimization theories. To do this, a somewhat more detailed study of the lexicographical order is needed.

In Section 2 we study the lexicographical order on $\overline{R}^n$ and extend it to matrices columnwise. The main result on this section is Prop. 2, giving a characterization of lexicographical nonnegativity in terms of termwise nonnegativity. We also obtain a necessary and sufficient condition for the lexicographical nongegativity of a continuous vector function over a compact set.

In Section 3 we obtain a Farkas type theorem for lexicographical consequences of linear inequality systems.

In Section 4 we apply our Separation Theorem to the theory of convex inequality systems, giving a necessary and sufficient condition for inconsistency.

In Section 5 we apply the results obtained in the preceding section to optimization theory. For an optimal solution of a convex problem we can always find a matricial Lagrange multiplier of a certain vector Lagrangian. We also get a necessary and sufficient condition for a point to be Chebyshev solution of an inconsistent optimization problem. Finally, we show that the exact quasiconvex dual (see [5]) of a mathematical programming problem may be restricted to the set of lexicographically nonnegative variables.

2. LEXICOGRAPHICAL ORDER

In this paper the elements of $\overline{R}^n$, $\overline{R} = R \cup \{+\infty, -\infty\}$ will be considered column vectors, and the superscript ^{T} will mean transpose. First, we introduce the lexicographical order on $\overline{R}^n$.

Definition: Let $x = (x_1, \ldots, x_n)^T$, $y = (y_1, \ldots, y_n)^T \in R^n$, $x \neq y$.

$$x >_L y \text{ if } x_k > y_k, \text{ with } k = \min i/x_i \; y_i .$$

We put $x \geq_L y$ if $x >_L y$ or $x=y$.

The next proposition gives an useful characterization of lexicographical nonnegativity. The vector inequality $\geq$ means componentwise.

<u>Proposition 1</u>

Let $x \in R^n$. $x \geq_L 0$ if and only if $Lx \geq 0$ for some unitary lower-triangular matrix L.

<u>Proof</u>: For $x=0$ the statement is trivial. Let $x=(x_1,\dots,x_n)^T \in R^n \setminus \{0\}$, $L=(l_{ij})$ with $l_{ii}=1$, $l_{ij}=0$ for $i<j$, $Lx \geq 0$, $k=\min\{i/x_i\ 0\}$. Since we have $x_k = \sum_{j=1}^{k-1} l_{kj}x_j + x_k \geq 0$, the ralation $x \geq_L 0$ must hold.

To prove the converse we will use induction on n. The statement is obvious for n=1. Let us suppose that it is true for vectors of length n-1, n>1. Let $x=(x_1,\dots,x_{n-1},x_n)^T \geq_L 0$. Since $\bar{x}=(x_1,\dots,x_{n-1}) \geq_L 0$, there exists a $(n-1)\times(n-1)$ unitary lower-triangular matrix $\bar{L}$ such that $\bar{L}\bar{x} \geq 0$. If the matrix $L = \begin{pmatrix} L & 0 \\ 0 & 1 \end{pmatrix}$ does not satisfy the required condition we can consider $L = \begin{pmatrix} \bar{L} & 0 \\ 0 \dots l_{nk} \dots & 1 \end{pmatrix}$, with $k=\min\{i/x_i \neq 0\}$, $l_{nk} \geq -x_n/x_k$

<u>Corolario 1.1</u>

$x \geq_L 0$ if and only if $x=\tilde{L}p$ for some unitary lower-triangular matrix $\tilde{L}$ and some $p \geq 0$.

<u>Proof</u>: We can put $p=Lx$, $\tilde{L}=L^{-1}$, with L given by Prop.1.

Next we define the (columnwise) lexicographical order for matrices, $\geq_L$. The relation $\geq$ will mean termwise.

<u>Definition</u>: Let $A=(a_{ij})=(c_1,\dots,c_n)=(r_1,\dots,r_m)^T$, B be m n matrices.

$A \geq_L 0$ if and only if $c_j \geq_L 0$ for all $j=1,\dots,n$.

$A \geq_L B$ if and only if $A-B \geq_L 0$.

<u>Lemma 1</u>

Let $A \geq_L 0$. If $m>1$ there exist $\alpha_1,\dots,\alpha_{m-1} \in R$ such that $\sum_{j=1}^{m-1} \alpha_j r_j + r_m \geq 0$.

<u>Proof</u>: By induction on m. For m=2, we can take $\alpha_1 \geq \max\{-a_{2j}/a_{1j}, a_{1j}>0\}$ if $r_1 \neq 0$, and arbitrary α_1 otherwise.

Now we consider $m>2$ and suppose that the statement is true for any $(m-1)\times n$ matrix. Let $J = \{j/m-1=\min\{i/a_{ij} \neq 0\}\}$. Put $\alpha_{m-1}=1$ if $J=\emptyset$ and $\alpha_{m-1} \geq \max\{-a_{mj}/a_{m-1,j}, j \in J\}$ if $J \neq \emptyset$. Since $A=(r_1,\dots,r_{m-2},\alpha_{m-1}r_{m-1}+r_m) \geq_L 0$, the proof is obtained by applying the induction hypothesis to A.

<u>Proposition 2</u>

$A \geq_L 0$ if and only if $LA \geq 0$ for some unitary lower-triangular matrix L.

<u>Proof</u>: The "if" part is an inmediate consequence of Prop.1. The converse will be proved by induction on m. For m=1 the statement is just the same that in Prop.1. Let m>1 and the assertion be true for any $(m-1)\times n$ matrix. Since $\bar{A}=(r_1,\dots,r_{m-1})^T \geq_L 0$ an unitary lower-triangular matrix L must exist such that $\bar{L}\bar{A} \geq 0$. The matrix $L = \begin{pmatrix} L & 0 \\ \alpha & 1 \end{pmatrix}$, $\alpha=(\alpha_1,\dots,\alpha_{m-1})$, with the α_i's given in the previous lemma satisfies the required condition.

Corollary 2.1

$A \geq_L 0$ if and only if $A=\tilde{L}P$ for some unitary lower-triangular matrix $\tilde{L}$ and some $P \geq 0$.

Proof: We can take $\tilde{L}=L^{-1}$, $P=LA$, with L given in Prop. 2.

Corollary 2.2

a) If $A \geq_L 0$ and $B \geq 0$, the $AB \geq_L 0$ provided that this product exists.

b) If $A \geq_L 0$ and L is an unitary lower-triangular matrise then $LA \geq_L 0$ provided that this product exists.

Proof: By Prop.2 and the fact that the sets of termwise nonnegative matrices and unitary lower-triangular matrices are closed under multiplication.

The next corollary gives another easy characterization of lexicographical nonnegativity of matrices ($\min_L$ means minimum in a lexicographical sense).

Corollary 2.3

$A \geq_L 0$ if and only if $\min_L\{As/s \geq 0\}=0$.

Proof: If $A \geq_L 0$ and $s \geq 0$, by corollary 2.2a) we have $As \geq_L 0=A0$. To prove the converse, taking e_j the jth unitary vector we obtain that the jth column of A, $c_j = Ae_j \geq_L \min_L\{As/s \geq 0\} = 0$; hence $A \geq_L 0$.

Some of these results for matrices have an analogous counterpart for vector functions. The next lemma is the corresponding to lemma 1.

Lemma 2

Let C be a compact subset of R^n, $f_i: C \to R$ continuous for $i = 1,\dots,m$, $f(x) = (f_1(x),\dots,f_m(x))^T$ for $x \in C$. If $f(x) >_L 0$ for every $x \in C$, there exist real numbers $\alpha_1,\dots,\alpha_{m-1}$ such that

$$\sum_{i=1}^{m-1} \alpha_i f_i(x)+f_m(x) > 0 \text{ for all } x \in C.$$

Proof: By induction on m. For m=1 the statement is clearly true. Let $m>1$, and suppose that the lemma holds for m-1. We consider the set $X=\{x \in C/f_i(x)=0, i=1,\dots,m-2, f_m(x) < 0\}$. For any $x \in X$ we have $f_{m-1}(x) > 0$, hence $k=\inf\{f_{m-1}(x)/x \in X\} \geq 0$. If k=0, there is a sequence $x_n \in X$ such that $f_{m-1}(x_n) \to 0$. We can choose a subsequence x_{n_j} converging to some $\bar{x} \in C$. We obtain:

$$f_i(\bar{x}) = \lim f_i(x_{n_j}) = 0 \quad \text{for} \quad i=1,\dots,m-2,$$

$$f_{m-1}(x) = \lim f_{m-1}(x_{n_j}) = \lim f_{m-1}(x_n) = 0,$$

thus $f_m(x) > 0$. But by the other hand, $f_m(x) = \lim f_m(x_n) \leq 0$, which yields a contradiction. Therefore $k>0$. Let $k': \in (0,k)$, $1< \inf\{f_m(x)/x \in C\}$, $\alpha_{m-1} \geq \max\{0,-1/k'\}$. For any $x \in X$,

$$\alpha_{m-1}f_{m-1}(x)+f_m(x) > \alpha_{m-1}k'+1 > 0.$$

If $x \in C$, $f_i(x)=0$, $i=1,\dots,m-2$, but $x \notin X$, obviously

$$\alpha_{m-1}f_{m-1}(x) + f_m(x) > 0.$$

Thus we have proved that

$$(f_1(x),\dots,f_{m-1}(x),\alpha_{m-1}f_{m-1}(x)+f_m(x))^T >_L 0 \text{ for all } x \in C.$$

Now we can apply the induction hypothesis.

Proposition 3

Let f,C as in lemma 2. f is lexicographically nonnegative over C if and only if for any $\varepsilon > 0$ there is an unitary lower-triangular matrix L_ε such that $L_\varepsilon f(x) + \varepsilon(1,\dots,1)^T > 0$ for all $x \in C$.

Proof: Let us suppose that the existence of matrices L_ε holds, and let $x_o \in C$, $i \in \{1,\dots,m\}$. Let $\varepsilon > 0$, L_ε the corresponding unitary lower-triangular matrix, and α^ε_{ij}, $j=1,\dots,i-1$, the first elements of the ith row of L . We have

$$f_i(x_o) + \varepsilon = \sum_{j=1}^{i-1} \alpha^\varepsilon_{ij} f_j(x_o) + f_i(x_o) + \varepsilon > 0,$$

hence $f_i(x_o) \geq 0$.

Conversely, if f is lexicographically nonnegative over C and $\varepsilon > 0$, for each $i=1,\dots$ $\dots,m$ the relation $(f_1(x),\dots,f_i(x)+\varepsilon)^T >_L 0$ holds; thus by lemma 2 there are real numbers α_{ij}, $j=1,\dots,i-1$, such that

$$\sum_{j=1}^{i-1} \alpha^\varepsilon_{ij} f_j(x) + f_i(x) + \varepsilon > 0 \text{ for all } x \in C.$$

The matrix

$$L = \begin{pmatrix} 1 & 0 & \dots & 0 \\ \alpha^\varepsilon_{21} & 1 & \dots & 0 \\ \dots & \dots & \dots & \dots \\ \alpha^\varepsilon_{m1} & \alpha^\varepsilon_{m2} & \dots & 1 \end{pmatrix}$$

satisfies the desired conditions.

Remark 1: The "if" part in Prop. 3 does not require the compactness and continuity assumptions.

Remark 2: From this proposition we get that a sufficient condition for the lexicographical nonnegativity of f over C is the existence of an unitary lower-triangular matrix L such that $Lf(x) \geq 0$ for any $x \in C$. But this is not a necessary condition, as shows the following example: n=1, m=2, $C=[-1,0]$, $f_1(x)=x^2$, $f_2(x)=x$. If $L=\begin{pmatrix} 1 & 0 \\ \alpha & 1 \end{pmatrix}$ is such that $Lf(x) \geq 0$ for all $x \in [-1,0]$, i.e., $\alpha x^2 + x \geq 0$ for all $x \in [-1,0]$, by taking $x=-1$ we obtain $\alpha \geq 1$; hence $-\frac{1}{2}\alpha \in C$, but for this point the second component of $Lf(x)$ is $-1/4\alpha < 0$, which is a contradiction.

The next corollary may be regarded as the continuous version of corollary 2.2a)

Corollary 3.1

Let C,D be compact subsets of R^n and R^p, respectively, $f_i: C \to R$, $i=1,\dots,m$, $g: C \times D \to R$ continuous functions, $g(x,y) \geq 0$ for $(x,y) \in C \times D$, $h_i: D \to R$ defined by $h_i(y) = \int_C f_i(x) g(x,y) dx$, $h(y) = (h_1(y),\dots,h_m(y))^T$. If f is lexicographically nonnegative over C, then h is lexicographically nonnegative over D.

Proof: If g vanishes identically over $C \times D$ or $v(C) = \int_C dx = 0$, then $h(y)=0$ for all $y \in C$. Otherwise let $y \in D$, $\varepsilon > 0$, K an upper bound of g over $C \times D$, $\delta = \varepsilon / Kv(C)$, L_δ an unitary lower-triangular matrix such that $L_\delta f(x) + \delta(1,\dots,1)^T > 0$ for all $x \in C$, $\alpha_{i1}, \dots$

$\ldots,\alpha_{i,i-1}$ the first terms of the ith row of L. We have

$$\sum_{j=1}^{i-1}\alpha_{ij}h_j(y)+h_i(y)+\varepsilon = \sum_{j=1}^{i-1}\alpha_{ij}\int_C f_j(x)g(x,y)dx + \int_C f_i(x)g(x,y)dx + \delta Kv(C) \geq \int_C (\sum_{j=1}^{i-1}\alpha_{ij}f_j(x)+f_i(x)+\delta)g(x,y)dx > 0.$$

Thus we have proved that $L_\delta h(y)+\varepsilon(1,\ldots,1)^T > 0$ for all y D and we can apply Prop.3.

3. LINEAR INEQUALITY SYSTEMS

The following theorem extends the Farkas theorem by considering lexicographical consequences of a linear inequality system. The standard Farkas theorem is recovered by letting p=1.

Proposition 4 (Generalized Farkas Theorem)

Let A,C be m×n and p×n matrices respectively. Then $Cx \leq_L 0$ is a consequence of $Ax \leq 0$ if and only if WA=C for some $W \geq_L 0$.

Proof: First we consider the case where WA=C and $W \geq_L 0$. If x is such that $Ax \leq 0$ by corollary 2.2a) we have $WAx \leq_L 0$, i.e., $Cx \leq_L 0$.

Suppose now that $Cx \leq_L 0$ is a consequence of $Ax \leq 0$. Let $r_1,\ldots,r_p$ be the rows of C. Clearly $r_1^Tx \leq 0$ is a consequence of $Ax \leq 0$. By the Farkas Theorem there is $w_1 \geq 0$ such that $w_1^TA=r_1^T$. Let $k\in\{2,\ldots,p\}$ and suppose we have got $w_1,\ldots,w_{k-1}$ such that $W_{k-1} = (w_1,\ldots,w_{k-1})^T \geq_L 0$, $W_{k-1}A=(r_1,\ldots,r_{k-1})^T$. By Prop. 2 $L_{k-1}W_{k-1} \geq 0$ for some unitary lower-triangular matrix L_{k-1}. Since $r_k^Tx \leq 0$ is a consequence of the linear inequality system $Ax \leq 0$, $r_i^Tx=0$, $i=1,\ldots,k-1$, again by Farkas Theorem $v^TA+\sum_{i=1}^{k-1}\alpha_i r_i = r_k^T$ holds for some $v \geq 0$, $\alpha_i \in R$, $i=1,\ldots,k-1$; hence $(v+\sum_{i=1}^{k-1}\alpha_i w_i)^TA=r_k^T$. Let $w_k = v+\sum_{i=1}^{k-1}\alpha_i w_i$. The matrices $L_k=\begin{pmatrix}L_{k-1} & 0\\ \alpha & 1\end{pmatrix}$, with $\alpha=(-\alpha_1,\ldots,-\alpha_{k-1})$, $W_k=(w_1,\ldots,w_{k-1},w_k)^T$ verify $L_kW_k \geq 0$, whence $W_k \geq_L 0$ by Prop. 2, and $W_kA=(r_1,\ldots,r_k)^T$. Thus this process can continue until k=p.

Corollary 4.1

Let A, $C=(r_1,\ldots,r_p)^T$ be m×n and p×n matrices, respectively, $p>1$, d R^m, $c=(c_1,\ldots,c_p)^T \in R^p$, and suppose that the system $Ax \leq d$ is consistent. Then $Cx \leq_L c$ is a consequence of $Ax \leq d$ if and only if $WA=(r_1,\ldots,r_k)^T$, $Wd <_L (\text{or} \leq_L) (c_1,\ldots,c_k)^T$ for some $k <$ (respectively =) p, $W \geq_L 0$.

Proof: First, suppose that for some $k <$ (or =) p there is $W \geq_L 0$ with $WA=(r_1,\ldots,r_k)^T$ and $Wd <_L$ (respectively $\leq_L$) $(c_1,\ldots,c_k)^T$. Let x be such that $Ax \leq d$. We have $(r_1,\ldots,r_k)^Tx = WAx \leq_L Wd <_L$ (respectively $\leq_L$) $(c_1,\ldots,c_k)^T$. Hence $Cx \leq_L c$.

Conversely, define $I=\{i/Ax \leq d \Rightarrow (r_1,\ldots,r_i)^Tx <_L (c_1,\ldots,c_i)^T\}$, k the minimum of $I\cup\{p\}$. Let x,t be a solution of the system $Ax-td \leq 0$, $-t \leq 0$. If $t>0$ we obtain $Ax/t \leq d$, whence $(r_1,\ldots,r_k)^Tx/t \leq_L (c_1,\ldots,c_k)^T$, that is, $(r_1,\ldots,r_k)^Tx-t(c_1,\ldots,c_k)^T \leq_L 0$. If t=0 we have $Ax \leq 0$. Take x_o such that $Ax_o \leq d$, $(r_1,\ldots,r_{k-1})^Tx_o = (c_1,\ldots,c_{k-1})^T$. For any $\lambda>0$, $A(x_o+\lambda x) \leq d$ must hold; therefore $(r_1,\ldots,r_k)^Tx_o + \lambda(r_1,\ldots,r_k)^Tx \leq_L (c_1,\ldots,c_k)^T$ and thus $(r_1,\ldots,r_k)^Tx \leq_L 0$, i.e.,

$(r_1,\dots,r_k)^T x - t(c_1,\dots\ \dots,c_k)^T \leq_L 0$. We have proved that the homogeneous relation $(r_1,\dots,r_k)^T x - t(c_1,\dots\ \dots,c_k)^T \leq_L 0$ is a consequence of the homogeneous system $Ax-td\leq 0$, $-t\leq 0$. By the Generalized Farkas Theorem there is $W\geq_L 0$, $w\in R^k$, $w\geq_L 0$ such that $(W,w)\begin{pmatrix}A & -d\\ 0 & 1\end{pmatrix} = \begin{pmatrix} r_1 \dots r_k\\ -c_1\dots -c_k\end{pmatrix}^T$, i.e., $WA=(r_1,\dots,r_k)^T$, $-Wd-w = (-c_1,\dots,-c_k)^T$. The last equality implies $Wd \leq_L (c_1,\dots,c_k)^T$. If $k<p$, this inequality can be strengthened to $<_L$, because we could replace in this reasoning c_k by $c_k = \max\{r^T x/Ax\leq d, (r_1,\dots,r_{k-1})^T x= =(c_1,\dots,c_{k-1})^T\}$.

4. CONVEX INEQUALITY SYSTEMS

In [5] we have proved the following Separation Theorem (see also [3],[4] and the comments in the introduction):

Separation Theorem

Let C be a convex subset of R^n. For any $x_o\in R^n\setminus C$ there is a n×n matrix A, $t\in \bar{R}^n$, such that $Ax <_L t \leq_L Ax_o$ for all $x\in C$.

Next we are going to use this theorem to obtain some new results about the consistency of convex inequality systems. In [2] properties of this kind are given, but assuming lower semicontinuity of the corresponding functions.

Proposition 5

Let C be a convex subset of R^n, f_i: $C\to R$ convex functions for $i=1,\dots,m$. Let $f(x)= =(f_1(x),\dots,f_m(x))^T$ for $x\in C$. The inequality system $f(x)\leq 0$ is inconsistent if and only if there is a p×m matrix $A\geq_L 0$, $1\leq p\leq m$, such that $Af(x) >_L 0$ for all x C.

Proof: The "if" part is an easy consequence of corollary 2.2a) and does not require the convexity assumptions. Let us suppose that the system $f(x)\leq 0$ is inconsistent, and consider the set $W=\{w\in R^m/f(x)\leq w$ for some $x\in R^n\}$. W is convex and $0\notin W$. By the Separation Theorem there is a square matrix $\tilde{A}=(a_1,\dots,a_m)^T$ such that $\tilde{A}w>_L 0$ (we can reverse the sense of inequalities in its statement). In particular $\tilde{A}f(x)>_L 0$ for all $x\in C$. We also obtain $a_1^T f(x)+a_1^T s\geq 0$ for $x\in C$, $s\geq 0$. This implies $a_1\geq 0$, or $A_1\geq_L 0$ calling $A_k=(a_1,\dots,a_k)$ for $k=1,\dots,m$. Let $p=\max\{k/A_k\geq_L 0\}$. By Prop.2 $L_pA_p\geq 0$ for some unitary lower-triangular matrix L_p. If $p=m$ we can take $A=\tilde{A}$ and the proposition is proved. If $p<m$, since $A_pf(x)\geq_L 0$ for all $x\in C$, it is enough to prove that this inequality is always strict. Otherwise take $x\in C$ such that $A_pf(x)=0$, $s\geq 0$ with $A_ps= =0$. Since $f(x)+s\in W$, the relation $a^T_{p+1}f(x)+a^T_{p+1}s\geq 0$ must hold; therefore $a^T_{p+1}s\geq 0$ (otherwise we could multiply s by some large enough $\lambda>0$ to violate the inequality). By the Farkas theorem there are real numbers $\alpha_1,\dots,\alpha_p$ such that $\sum_{i=1}^p \alpha_i a_i \leq a_{p+1}$. The matrix $L=\begin{pmatrix}L_p & 0\\ \alpha & 1\end{pmatrix}$, with $\alpha=(-\alpha_1,\dots,-\alpha_p)$ verifies $LA_{p+1}\geq 0$, i.e., $A_{p+1}\geq_L 0$, thus contradicting the definition of p. Hence we can take $A=A_p$ and the proposition is proved.

Corollary 5.1

Let C, f be as in Prop.5, f_o:$C\to R$ convex, and the system $f(x)\leq 0$ be consistent over C. The system $f_o(x)\leq 0$, $f(x)\leq 0$ is inconsistent if and only if there is a p×(m+1) matrix $A \geq_L 0$, $1 \leq p \leq m+1$, with its first column being nonzero and having 1 as its first nonzero element, such that

$$A(f_o(x), f_1(x), \dots, f_m(x))^T >_L 0 \quad \text{for all} \quad x \in C.$$

Proof: If the first column of A would vanish the system $f(x) \leq 0$ would be inconsistent. Hence we can divide A by the first nonzero element of its first column.

In the succesive we will denote the vector $(0,\dots,0,a)^T \in R^k$ by $<a>_k$.

Corollary 5.2

Let C, f, f_o be as in corollary 5.1. The system $f(x) \leq 0$, $f_o(x) < 0$ is inconsistent if and only if there is a $k \times m$ matrix $B \geq_L 0$, $1 \leq k \leq m+1$, such that $<f_o(x)>_k$ + + $B(f_1(x),\dots,f_m(x))^T \geq_L 0$ for all $x \in C$.

Proof: If the existence of B holds we have $<f_o(x)+e^t>_k + B(f_1(x),\dots,f_m(x))^T >_L 0$ for all $x \in C$, $t \in R$. By corollary 5.1 this means that the system $f(x) \leq 0$, $f_o(x)+e^t \leq 0$ is inconsistemt and hece we get that the system $f(x) \leq 0$, $f_o(x) < 0$ is also inconsistent.

Conversely, if the system $f(x) \leq 0$, $f_o(x) \leq 0$ is inconsistent so is the system $f(x) \leq 0$, $f_o(x)+e^t \leq 0$. By corollary 5.1 there is a $p \times (m+1)$ matrix $A \geq_L 0$, $1 \leq p \leq m+1$, with 1 being the first nonzero element of the first column of A (in place k), such that $A(f_o(x)+e^t, f_1(x),\dots,f_m(x))^T >_L 0$ for all $x \in C$, $t \in R$. Let us call B the matrix consisting of the intersection of the first k rows and the last m columns of A. We have

$$<f_o(x)+e^t>_k + B(f_1(x),\dots,f_m(x))^T \geq_L 0 \text{ for any } x,t.$$

If the first k-1 components of $<f_o(x)>_k + B(f_1(x),\dots,f_m(x))^T$ are 0 for some $x \in C$, the last one must be greater or equal than $-e^t$ for all $t \in R$; hence it is nonnegative. This completes the proof.

Remark: We can always consider k=m+1 by adding m+1-k zero rows to B (as its first rows).

5. OPTIMIZATION

a) Convex optimization

We begin considering the following optimization problem: (P) minimize $f_o(x)$ subject to $f(x) \leq 0$, where $C \subset R^n$, $f_i: C \to R$, $i=0,1,\dots,m$, $f(x)=(f_1(x),\dots,f_m(x))^T$ for $x \in C$. We define the lexicographical Lagrangian of this problem, L, by

$$L(x,B) = <f_o(x)>_{m+1} + Bf(x), \quad x \in C, \; B \geq_L 0,$$

and (x,B) is a lexicographical saddlepoint of L if

$$L(x,B) \leq_L L(x,B) \leq_L L(x,B) \text{ for all } x,B.$$

In the next proposition no constraint qualification is required.

Proposition 6

If $(\bar{x},\bar{B})$ is a lexicographical saddlepoint of L, $\bar{B}=(\bar{b}_1,\dots,\bar{b}_m)$, then $\bar{x}$ is an optimal solution for (P), and $f_i(\bar{x})\bar{b}_i=0$ for $i=1,\dots,m$. Conversely, if $\bar{x}$ is an optimal solution of (P) and (P) is a convex problem (that is, C and f_i, $i=0,1,\dots,m$, are convex), there exists $\bar{B} \geq_L 0$ such that $(\bar{x},\bar{B})$ is a lexicographical saddlepoint of L.

Proof: If $(\bar{x},\bar{B})$ is a lexicographical saddlepoint of L, by taking B=0, $B=2\bar{B}$ in the definition of lexicographical saddlepoint we obtain $L(\bar{x},\bar{B})=\langle f(\bar{x})\rangle_{m+1}$. By corollary 2.2 we also obtain $Bf(\bar{x}) \leq_L 0$ for any $B \geq_L 0$, which clearly implies $f(x) \leq 0$. Let x verify $f(x) \leq 0$. By corollary 2.2a), $\bar{B}f(x) \leq_L 0$. Hence $\langle f_o(\bar{x})\rangle_{m+1} = L(\bar{x},\bar{B}) \leq_L L(x,\bar{B}) = \langle f_o(x)\rangle_{m+1}+\bar{B}f(x) \leq_L \langle f_o(x)\rangle_{m+1}$ and thus $f_o(\bar{x}) \leq f_o(x)$, i.e., $\bar{x}$ is an optimal solution for (P). The complementary slackness condition is a consequence of the fact that all the terms in the right hand side of $0=\bar{B}f(\bar{x})=f_i(\bar{x})b_i$ are lexicographically nonpositive and hence they must vanish.

To prove the converse we apply corollary 5.2 to the system $f(x) \leq 0$, $f_o(x)-f_o(\bar{x}) < 0$. There is a matrix $\bar{B} \geq_L 0$ such that $\langle f_o(x)-f_o(\bar{x})\rangle_{m+1}+\bar{B}f(x) \geq_L 0$ for all $x \in C$. The condition of lexicographical saddlepoint for (x,B) is easily verified.

Remark: From Prop.6 we can obtain the usual saddlepoint theorem for the ordinary Lagrangian of a convex program under the Slater constraint qualification, because in this case the m first rows of B must be zero. Otherwise the second inequality in the definition of a lexicographical saddlepoint would not hold for a point x_o satisfying $f(x_o) < 0$.

Corollary 6.1

Let $\bar{x}$ be a feasible point of (P) (that is, $f(\bar{x}) \leq 0$). If there is a matrix $P \geq 0$ such that $Pf(\bar{x})=0$ and for any $\varepsilon > 0$ there is an unitary lower-triangular matrix L satisfying

$$\langle f_o(x)\rangle_{m+1}+L_\varepsilon Pf(x)+\varepsilon(1,\dots,1)^T > \langle f_o(\bar{x})\rangle_{m+1}$$

for any $x \in C$, then $\bar{x}$ is an optimal solution for (P). If (P) is a convex problem, C is compact and the f_i's are continuous, the converse also holds.

Proof: We can rewrite the inequality in the statement as

$$L_\varepsilon(\langle 1\rangle_{m+1}P)\begin{pmatrix} f_o(x) & - & f_o(\bar{x}) \\ f(x) & - & f(\bar{x}) \end{pmatrix} + \varepsilon\,(1,\dots,1)^T > 0$$

By Prop.3 (see the remark 1 following it) we obtain

$$(\langle 1\rangle_{m+1}P)\begin{pmatrix} f_o(x) & - & f_o(\bar{x}) \\ f(x) & - & f(\bar{x}) \end{pmatrix} \geq_L 0 \quad \text{for all} \quad x \in C$$

that is, $\langle f_o(x)\rangle_{m+1}+Pf(x) \geq_L \langle f_o(\bar{x})\rangle_{m+1}+Pf(\bar{x}) = \langle f_o(\bar{x})\rangle_{m+1}$. If $f(x) \leq 0$, then $Pf(x) \leq 0$ and by the above inequality necessarily the m first rows of Pf(x) vanish, whence $f_o(x) \geq f_o(\bar{x})$.

To prove the converse let us consider $\bar{B}$ the matrix given by Prop.6. By corollary

2.1, $\bar{B}=LP$ for some unitary lower-triangular matrix L and some $P \geq 0$. From $\bar{B}f(\bar{x})=0$ and the nonsingularity of L we deduce that $Pf(x)=0$. Furthermore the saddlepoint condition for $(\bar{x},\bar{B})$ means that the expression

$$L(<1>_{m+1}P) \begin{pmatrix} f_o(x) - f_o(\bar{x}) \\ f(x) - f(\bar{x}) \end{pmatrix}$$

is lexicographically nonnegative over C. Now we can apply Prop.3, the fact that the set of unitary lower-triangular matrices is closed under multiplication and the assertion made at the beginning of the proof.

b) Inconsistent problems

In [1] a Chebyshev solution of the inconsistent convex finite-dimensional problem

$$(P) \text{ minimize } f_o(x) \text{ subject to } f_i(x) \leq 0, \ i=1,\ldots,m, \quad x \in C$$

has been defined as an optimal solution of the problem

$$(P') \text{ minimize } f_o(x) \text{ subject to } f_i(x) \leq a, \ i=1,\ldots,m, \ x \in C,$$

with $a=\min_x \max_i f_i(x) > 0$. As observed there, the problem (P') does not verify the Slater's constraint qualification; hence a lexicographical approach will be interesting. By Prop.6 we have that $\bar{x} \in C$ is a Chebyshev solution of (P) if and only if there is a $(m+1)\times m$ matrix $B=(b_1,\ldots,b_m) \geq_L 0$ such that $(\bar{x},\bar{B})$ is a lexicographical saddlepoint of L given by

$$L(x,B) = <f(x)>_{m+1}+B(f_1(x)-a,\ldots,f_k(x)-a)^T \text{ for } x \in C, \ B \geq_L 0.$$

Moreover the complementary slackness condition gives $b_i=0$ if $f_i(\bar{x}) < a$. By especifying the definition of lexicographical saddlepoint to this Lagrangian we get:

Proposition 7

$\bar{x}$ C is a Chebyshev solution of (P) if and only if it verifies $f_i(\bar{x}) \leq a$ for i=1,... ...,m and is a lexicographical minimum of

$$< f(x) >_{m+1}+\bar{B}(f_{i_1}(x),\ldots,f_{i_p}(x))^T \text{ for some } \bar{B} \geq_L 0,$$

$i_1<\ldots<i_p$ being the indexes satisfying $f_{i_j}(\bar{x})=a$.

From corollary 6.1 we also obtain:

Corollary 7.1

Let $\bar{x} \in C$, $i_1,\ldots,i_p$ as in Prop.9. If there is a matrix $P \geq 0$ such that $P(f_{i_1}(\bar{x}),\ldots$ $\ldots,f_{i_n}(\bar{x}))^T=0$, and for any $\varepsilon > 0$ there is an unitary lower-triangular matrix L_ε satisfying

$$<f(x)>_{m+1}+L_\varepsilon P(f_{i_1}(x),\ldots,f_{i_p}(x))^T+\varepsilon(1,\ldots,1)^T > <f(\bar{x})>_{m+1} \quad \forall x \in C,$$

then $\bar{x}$ is a Chebyshev solution of (P). The converse also holds if C is compact and the f_i's are continuous.

c) Quasiconvex optimization

In [5] we have introduced the exact quasiconvex dual of the problem

$$\text{(P) minimize } f(x) \text{ subject to } g(x) \geq 0,$$

with $f: R^n \to R$, $g: R^n \to R^m$, as

$$\text{(D) maximize } d(A) = \inf\{f(x)/Ag(x) \geq_L As \text{ for some } s \geq 0\}$$
$$\text{over the set of all } m\times m \text{ matrices } A.$$

We can equivalently write $d(A) = \inf\{f(x)/Ag(x) \geq_L \min_L\{As/s \geq 0\}$, where $\min_L$ means lexicographical minimum.

Proposition 8

The dual problem can be stated as

$$\text{(D) maximize } d(A) = \inf\{f(x)/Ag(x) \geq_L 0\}$$
$$\text{subject to } A \geq_L 0$$

Proof: Let $A=(a_1,\dots,a_m)^T$, $(v_1,\dots,v_m)^T = \min_L\{As/s \geq 0\}$, $k=\max\{l/v_i=0\ \forall i \leq l\}$. If $k=m$, by corollary 2.3 $A \geq_L 0$; otherwise the condition $Ag(x) \geq_L (v_1,\dots,v_m)^T$ is equivalent to $\tilde{A}g(x) \geq_L 0 = \min_L\{As/s \geq 0\}$ with $\tilde{A}=(a_1,\dots,a_k,0,\dots,0)^T$.

By an argument similar to the one in the last proof we can see that the exact quasiconvex Lagrangian L of this problem (see [5] verifies:

$$L(x,A) = \begin{cases} f(x) & \text{if } a^Tg(x) > 0 \text{ or } (a_1^Tg(x)=0, a_2^Tg(x) > 0) \text{ or } \dots \text{ or} \\ & (a_1^Tg(x)=\dots=a_{k-1}^Tg(x)=0,\ a^Tg(x) \geq 0) \\ +\infty & \text{otherwise} \end{cases}$$

where $A=(a_1,\dots,a_m)^T$, $k=\max\{i/(a_1,\dots,a_i)^T \geq_L 0\}$. We can consider that $k=m$ if we restrict A to be lexicographically nonnegative.

REFERENCES

1. A.Ben-Israel, A. Ben-Tal, S. Zlobec, Optimality in Nonlinear Programming: A feasible Directions Approach (Wiley-Interscience, New York, 1981).

2. C. Berge, A. Ghouila-Houri, Programmes, jeux et réseaux de transport (Dunod, Paris, 1969).

3. P.C.Hammer, "Maximal convex sets", Duke Math. J. 22(1955) 103-106

4. V.L.Klee, Jr., "The structure of semispaces", Math. Scand. 4(1956) 54-64.

5. J.E. Martínez-Legaz, "Exact quasiconvex conjugation", paper presented at the XI International Symposium on Mathematical Programming (Bonn, 1982).

STABILITY OF GENERALIZED EQUATIONS AND KUHN-TUCKER POINTS OF PERTURBED CONVEX PROGRAMS

B. Kummer
Humboldt-Universität zu Berlin, PF 1297
1086 Berlin, GDR

1. Introduction

The paper deals with the set K(f) of the Kuhn-Tucker points of a convex program

(f) $\min \{ f_o(x): \quad f_i(x) \leq 0 , \quad x \in C \quad (i = 1,\ldots,m) \}$

where $f_o, f_1,\ldots,f_m$ are real-valued convex functions on R^n and C is a closed convex subset of R^n. As it is well-known K(f) coincides with the solution set of the generalized equation

(1.1) $\quad 0 \in \partial f_o(x) + \sum_{i=1}^{m} y_i \partial f_i(x) + N_C(x)$

(1.2) $\quad 0 \in -(f_1(x), \ldots ,f_m(x)) + N_{R^m_+}(y)$

$\qquad x \in C , \; y \in R^m_+$

which we briefly denote by

(1) $\quad 0 \in \Gamma_f(t) + N_T(t) \quad , \quad t \in T$

setting $t = (x , y)$, $T = C \times R^m_+ \subset R^{n+m}$ and defining $N_T(t)$ to be the outward normal cone of T at the point t.

In order to study the stability behavior of the Kuhn-Tucker mapping K as a function of f results of the implicit-function-type for generalized equation have been appeared to be very useful (see [1] ,[2],[6],[7],[8],[9],[10]). In all of these models the allowed variation of the right-hand-side of (1) is a continuously translation by a single-valued function. When one considers non-differentiable convex functions these variations, however, will be too restrictiv .

In the present paper we will point out what types of stability of generalized equations would be of particular interest in this (non-differentiable) case and will give some basic results. In order to get a brief and simple representation our suppositions will be not of greatest generality as the reader easily confirmes.

2. The stability problem and motivations

Induced by the properties of the mappings Γ_f and N_T in (1) let T be a non-empty closed subset of R^p and M_1 and M_2 the following families of closed multifunctions doing from T into the set of the non-empty, closed and convex subsets of R^p:

$M_1 = \{ \Gamma_1: \; \Gamma_1$ is upper- semicontinuous and compact-valued on T $\}$

$M_2 = \{ \Gamma_2: \; \Gamma_2(t)$ is a cone with the vertex 0 for all t in T $\}$.

In the generalized equation

(2) $\quad 0 \in \Gamma_1(t) + \Gamma_2(t) \; , \quad t \in T$

we consider $\Gamma_2 \in M_2$ as fixed and study the set of solutions $S(\Gamma_1)$ as a function of $\Gamma_1 \in M_1$.

Let $\{N_k^\delta : \delta > 0\}$ be a family of neighbourhoods of Γ_1 to be specified later, let $\varepsilon > 0$ and set

$$S_\varepsilon(\Gamma_1) = \{z \in R^p : \ \mathrm{dist}(z, S(\Gamma_1)) \leq \varepsilon\}.$$

Then, we will say that N_k^δ solves the stability problem defined by Γ_1 and ε if

(3) $\quad \emptyset \neq S(G_1) \subset S_\varepsilon(\Gamma_1) \quad$ for all $G_1 \in N_k^\delta$.

The following families of neighbourhoods of Γ_1 seem to be of interest having in mind the generalized equation (1) and the variation of the convex program (f):

N_1^δ formed by continuous translations

$$N_1^\delta = \{G_1 \in M_1 : \ G_1(t) = h(t) + \Gamma_1(t), \ h: T \rightarrow R^p \text{ continuous}, \|h(t)\| \leq \delta\},$$

N_2^δ formed by variations of the ranges

$$N_2^\delta = \{G_1 \in M_1 : \ d_H(G_1(t), \Gamma_1(t)) \leq \delta\}$$

where d_H denotes the Hausdorff - distance of compact sets,

N_3^δ formed by variations of the graph of the mapping $\Gamma_1 + \Gamma_2$

$$N_3^\delta = \{G_1 \in M_1 : \ d_H(\mathrm{graph}(G_1 + \Gamma_2), \mathrm{graph}(\Gamma_1 + \Gamma_2)) \leq \delta\}.$$

Obviously, $N_1^\delta \subset N_2^\delta \subset N_3^\delta$. The reason why even these neighbourhoods are introduced lies in the behavior of the subdifferential-mapping of convex functions.

<u>Lemma 1</u>: Let $F, G: R^n \rightarrow R$ be convex functions and $C \subset R^n$ be a non-empty, closed and convex subset. Then, there is $q > 0$ depending on the considered norm only such that

(4) $\quad d_H(\mathrm{graph}(\partial F + N_C), \ \mathrm{graph}(\partial G + N_C)) \leq q \cdot \sqrt{\varepsilon}$

whenever

$|F(x) - G(x)| \leq \varepsilon \quad$ for all $x \in C$.

If, in addition, F is differentiable and C is compact, for each $\varepsilon > 0$ there is $\delta = \delta(F, C, \varepsilon) > 0$ such that

(5) $\quad d_H(\partial F(x), \partial G(x)) \leq \varepsilon \quad$ for all $x \in C$

whenever G satisfies

$|G(x) - F(x)| \leq \delta \quad$ for all x, $\mathrm{dist}(x, C) \leq 1$.

A proof of the first part is given by R.Schultz [11], the second one is shown in [3], prop.(8), where also the relation between ε and δ is pointed out. It should be noted that the Lemma may be verified very easily by applying Ekeland's variational principle and that $q = \sqrt{2}$ if the Hausdorff-distance in (4) bases on the Euclidean norm.

We return now to the program (f) and suppose the set $K(f)$ to be non-empty and bounded. Denoting by $K_\varepsilon(f)$ the closed ε-neighbourhood of $K(f)$ we are interested in positive r and β such that the relations

(6) $\quad \emptyset \neq K(g) \subset K_\varepsilon(f)$

hold for all convex programs (g) satisfying

(7) $\quad \| g(x) - f(x) \| \leq \beta$ for all x, $\|x\| \leq r$.

Since K(g) is always convex standard arguments show that (6) holds if and only if the solution set Q(g) of the generalized equation

(8) $\quad 0 \in \Gamma_g(t) + N_{T_o}(t)$, $\quad t \in T_o := T \cap K_{2\varepsilon}(f)$,

whose domain is compact, fulfils

(9) $\quad \emptyset \neq Q(g) \subset K_{\varepsilon}(f)$.

Regarding now Γ_f and Γ_g as mappings on the domain T_o and r as sufficiently large (We need $\|z\| \leq r+1$ for all $z \in K_{2\varepsilon}(f)$) one finds that (7) implies

(10.1) $\quad \Gamma_g \in N_1^{O_1(\beta)}(\Gamma_f)$ $\quad$ if f and g are differentiable

(10.2) $\quad \Gamma_g \in N_2^{O_2(\beta)}(\Gamma_f)$ $\quad$ if f is differentiable

(10.3) $\quad \Gamma_g \in N_3^{O_3(\beta)}(\Gamma_f)$.

Here the values $O_k(\beta)$ are tending to zero as $\beta \rightarrow 0$ and may be determined by using Lemma 1.

3. Reduction of the stability problems

First at all we note that the stability problem defined by Γ_1 and ε (and concerned with (2)) will become more or less difficult depending on the type of neighbourhoods we have to consider. For the first type N_1^{δ} the reader will find many important results in [1] , [2] ,[6] ,...,[10] . As we will see it they may be applied to the second type N_2^{δ} without any differences. For the third type N_3^{δ} the stability problem is not yet solved completely. Nevertheless, the behavior of the Kuhn-Tucker points of a perturbed non-differentiable convex program can be described (in the sense of (6) and (7)) if one knows the behavior in relation to small linear perturbations of the objective function and the right-hand-sides of the restrictions only. This will follow from (10), Lemma 1 and Theorem 2.

Lemma 2: Let T be a non-empty, closed subset of R^p, $\Gamma_1 \in M_1$, $\Gamma_2 \in M_2$ and let the generalized equation

(2) $\quad 0 \in \Gamma_1(t) + \Gamma_2(t)$, $\quad t \in T$

have no solution. Then, there exists a continuous function $u: T \longrightarrow R^p$ such that

$$\inf \langle u(t) , \Gamma_1(t) + \Gamma_2(t) \rangle := \inf \{ \langle u(t), z_1 + z_2 \rangle : z_j \in \Gamma_j(t) \} > 0$$

for all $t \in T$.

Proof: We put $\Gamma(t) = \Gamma_1(t) + \Gamma_2(t)$ and $A(t) = \{ v: \inf \langle v, \Gamma(t) \rangle > 0 \}$. Since $\Gamma(t)$ is closed and convex, A(t) is convex and non-empty. In order to show that A is lower semicontinuous, let $v \in A(t)$, $t_\nu \rightarrow t$, $t_\nu \in T$ (ν = 1,2 ...). We have $\text{dist}(v , A(t_\nu)) \longrightarrow 0$ to verify. If this would be false for some $r > 0$ and (without loss of generality) all ν we had

$(v + rB) \cap A(t_\vartheta) = \emptyset$ where B denotes the closed unit ball in R^p.

Hence

(11) $$\sup_{w \in v+rB} \inf_{z \in \Gamma(t_\vartheta)} \langle w , z \rangle \leq 0.$$

Since $\Gamma_1 \in M_1$, $v \in A(t)$ and $0 \in \Gamma_2(t)$, for some $\lambda \in (0 , r)$ and all $\vartheta > \vartheta(\lambda)$, the inequality

$$\inf \{ \langle w , \Gamma_1(t_\vartheta) \rangle : w \in v + \lambda B \} > 0$$

holds. Therefore, (11) gives

$$\sup_{w \in v + \lambda B} \inf_{z_2 \in \Gamma_2(t_\vartheta)} \langle w , z_2 \rangle < 0$$

and, since $\Gamma_2(t_\vartheta)$ is a cone with the vertex 0, we observe

$$\sup_{w \in v + \lambda B} \inf_{z_2 \in \Gamma_2(t_\vartheta) \cap B} \langle w , z_2 \rangle < 0 .$$

Using the minimax-theorem and well-known continuity arguments one finds that there is $\bar{z}_2 \in \Gamma_2(t) \cap \mathrm{bd}\, B$ with

$$\sup_{w \in v + \lambda B} \langle \bar{z}_2 , w \rangle \leq 0.$$

The letter implies $\langle v , \bar{z}_2 \rangle < 0$ and contadicts $v \in A(t)$. That means A is in fact lower semicontinuous and possesses, by Michael's selection theorem [5] , a continuous selection what completes the proof.

The previous Lemma is the key for the following

<u>Theorem 1</u> : If any neighbourhood N_1^δ solves the stability problem defined by Γ_1 and ε (related to the generalized equation (2)) then N_2^δ solves it too.

Proof: Let $G_1 \in N_2^\delta$ and put $G(t) = G_1(t) + \Gamma_2(t)$. If $0 \in G(s)$, i.e. $s \in S(G_1)$, we have for some z in $\delta \cdot B$: $z \in \Gamma(s) := \Gamma_1(s) + \Gamma_2(s)$.

Since the mapping $t \longrightarrow -z + \Gamma_1(t)$ belongs to N_1^δ we obtain $s \in S_\varepsilon(\Gamma_1)$.

In order to show $S(G_1) \neq \emptyset$ we assume conversely $S(G_1) = \emptyset$. By Lemma 2 there is a continuous function u satisfying

$\| u(t) \| = 1$ and $\inf \langle u(t) , G(t) \rangle > 0$ for all $t \in T$.

Because of $d_H (G_1(t) , \Gamma_1(t)) \leq \delta$ one easily confirmes

(12) $$\inf \langle u(t) , \Gamma(t) \rangle > - \delta$$

and therefore $- \delta \cdot u(t) \notin \Gamma(t)$.

Since the mapping $t \longrightarrow \delta \cdot u(t) + \Gamma_1(t)$ belongs to N_1^δ we get obviously a contradiction.

<u>Remark 1</u>: If N_1^δ solves the stability problem defined by Γ_1 and ε and if $G_1 \in N_3^\delta$ then $S(G_1) \subset S_{\varepsilon + \delta}(\Gamma_1)$.

Indeed, if $0 \in G(s)$ and d_H (graph G , graph Γ) $\leq \delta$ there are $z \in R^p$ and $s' \in T$ such that $\| (o , s) - (z , s') \| \leq \delta$ and $z \in \Gamma(s')$. Since the mapping $t \rightarrow -z + \Gamma_1(t)$ lies in N_1^δ we may conclude $s' \in S_\varepsilon(\Gamma_1)$ and $s \in S_{\varepsilon + \delta}(\Gamma_1)$.

For studying the behavior of Kuhn-Tucker points the next theorem is of particular interest.

<u>Theorem 2</u>: Let T be a non-empty, convex and compact subset of R^p, $\Gamma_1 \in M_1$, $\Gamma_2 \in M_2$ and suppose the existence of some $\delta > 0$ such that, for z in $\delta \cdot B$, the sets $\Gamma^-(z) := \{t \in T : z \in \Gamma_1(t) + \Gamma_2(t)\}$ are non-empty and convex. Then, $S(G_1) \neq \emptyset$ whenever $G_1 \in N_2^{\delta}$.

If, in addition, $\Gamma^-(z) \subset S_{\varepsilon}(\Gamma_1)$ for all z in $\delta \cdot B$ so N_2^{δ} solves the stability problem defined by Γ_1 and ε.

Proof: The second statement follows from the first one as it is seen in the proof of theorem 1. We follow the same proof to verify $S(G_1) \neq \emptyset$ and consider the function u we have constructed there. Since $\|u(t)\| = 1$ the mapping H as $H(t) = \Gamma^-(-\delta \cdot u(t))$ possesses non-empty and convex ranges in T. Because of the continuity of u the mapping H is closed. By Kakutani's fixed point theorem then there is some $s \in T$ satisfying $s \in H(s)$. This means $-\delta \cdot u(s) \in \Gamma(s)$ and contradicts (12). Hence, $S(G_1) \neq \emptyset$, the theorem is true.

<u>Remark 2</u>: Even if all the suppositions of theorem 2 are fulfilled and additionally $\Gamma_2(t) \equiv \{0\}$ holds the stability problem with respect to neighbourhoods of the third type may be unsolveable.

<u>Example1</u>: Let T be the unit ball in R^2, $\Gamma_2(t) \equiv \{0\}$ and let $\Gamma_1(t)$ be the convex hull $\operatorname{conv}\{0, t\}$ if $t \in \operatorname{bd} T$, and $\{0\}$ otherwise. Obviously, the suppositions os theorem2 are satisfied with $\varepsilon = \delta = 1$. We fix now any z in bd T and put

$$G_{1\delta}(t) = \begin{cases} \{-\delta \cdot z\} & \text{if } t \in \operatorname{int} T \text{ or } t \in \operatorname{bd} T,\ \|t - z\| < \delta \\ \operatorname{conv}\{t, -\delta z\} & \text{otherwise}. \end{cases}$$

Then : $d_H(\text{graph } G_{1\delta}, \text{graph }_1) \rightarrow 0 \quad (\delta \rightarrow 0)$ and $0 \notin G_{1\delta}(t)$ for all $t \in T$.

The next example shows that , in theorem 2, convexity of the sets $\Gamma^-(z)$ cannot be removed.

<u>Example 2</u>: Let T be the closed interval $[0, 2\pi]$ and $x(t) = (\cos t, \sin t)$. Define $\Gamma_1(t) = \operatorname{conv}\{0, x(t)\}$ $(\subset R^2)$, $\Gamma_2(t) = \{0\}$, $\varepsilon = 1$.

Anew all suppositions of theorem2, without convexity of the sets $\Gamma^-(\gamma, 0)$, $0 < \gamma \leq 1$ are fulfilled. For $G_{1\delta}(t) = \delta \cdot x(t) + \Gamma_1(t)$ we obtain $G_{1\delta} \in N_1^{\delta}$ and $0 \notin G_{1\delta}(t)$.

Let us apply now theorem 2 to the stability of Kuhn-Tucker points where we assume K(f) to be non-empty and bounded and $\varepsilon > 0$ to be fixed. We introduce the problem of finding some $\alpha > 0$ such that for all $z = (c, b) \in R^{n+m}$ with $\|z\| \leq \alpha$ the set $K^z(f)$ of the Kuhn-Tucker points for

$\min\{f_o(x) - \langle c, x\rangle : f_i(x) \leq b_i, x \in C \ (i = 1,\dots,m)\}$

is non-empty and contained in $K_{\varepsilon}(f)$.

Setting $T_o = (C \times R^m_+) \cap K_{2\varepsilon}(f)$ and , in theorem 2, $T = T_o$, $\Gamma_1 = \Gamma_f$, $\Gamma_2 = N_{T_o}(t)$ we see that the conditions

$$\emptyset \neq K^z(f) \subset K_\varepsilon(f) \qquad \text{for all } z,\ \|z\| \leq \alpha$$

and

$$\emptyset \neq \Gamma^-(z) \subset S_\varepsilon(\Gamma_1) \qquad \text{for all } z,\ \|z\| \leq \alpha$$

are the same. The convexity of $\Gamma^-(z)$ is trivial.

Therefore, if such α would be known we could conclude

$$\emptyset \neq S(G_1) \subset S_\varepsilon(\Gamma_1)$$

if G_1, defined on T_0, belongs to N_2^α.

Having in mind the equivalence of (6) and (9) we may then summarize:

If f is differentiable the relations (6) hold for all convex problems

(g) $\min\{g_0(x) : g_i(x) \leq 0,\ x \in C\ (i = 1,\dots,m)\}$

which satisfy (7) with such $\beta > 0$ that $O_2(\beta)$ in (10.2) is not greater than α.

If f is not differentiable we have to choose $\beta > 0$ such that $O_3(\beta)$ from (10.3) is not greater than α. In view of the remark 1 then the inclusion

$$K(g) \subset K_{\varepsilon+\alpha}(f)$$

holds whenever g satisfies (7). More detailed conditions which imply additionally $K(g) \neq \emptyset$ are to be found in [4].

References

[1] J.-P. AUBIN, "Lipschitz behavior of solutions to convex minimization problems", Working Paper WP-81-76 (1981), IIASA Laxenburg

[2] HOANG TUY, "Stability property of a system of inequalities", Math. Operationsforschung und Statistik, Ser. Optimization 8 (1977) 27-39

[3] B. KUMMER, "Generalized equations:solvability and regularity", Preprint Nr.30, Humboldt-Univ. Berlin, Sekt. Math. (1982)

[4] B. KUMMER and R.SCHULTZ,"Kuhn-Tucker points of parametric convex programs as solutions of perturbed generalized equations" (in russian) to appear

[5] E. MICHAEL, "Continuous selections I", Annals Math. 63 (1956) 361-382

[6] S.M. ROBINSON, "Generalized equations and their solutions,Part I: Basic theory", Mathematical Programming Study 10 (1979) 128-141

[7] S.M. ROBINSON, "Generalized equations and their solutions,Part II: Applications to nonlinear programming", Mathematical Programming Study 19 (1982) 200-221

[8] S.M. ROBINSON, "Stability theory for systems of inequalities, Part I: Linear systems", SIAM Journ. Num.Anal. 12 (1975) 754-769

[9] S.M. ROBINSON,"Stability theory for systems of inequalities, Part II: Differentiable nonlinear systems", SIAM Journ. Num.Anal.13 (1976) 497-513

[10] S.M. ROBINSON,"Strongly regular generalized equations", Mathematics of Operations Research 5 (1980) 43-62.

[11] R.SCHULTZ, "An approach to stability in convex programming using the topological degree of set-valued mappings",Proceedings Opt.Conf.Sellin 1983

DUALITY AND STABILITY THEOREMS FOR CONVEX MULTIFUNCTIONAL PROGRAMS

Ewa Bednarczuk
Systems Research Institute of PAS
01-447 Warsaw, Newelska 6

1. INTRODUCTION

In this paper we consider a minimization problem of the form

$$P \qquad \inf\{f(x):\ x\in\Gamma y_0\}$$

where $f:X\to R$ is a real-valued convex function defined on a space X and $\Gamma:Y\to X$ is a convex multifunction defined on a space Y, i.e. for every $y\in Y$, $\Gamma y\subset 2^X$ and $\lambda\Gamma y_1+(1-\lambda)\Gamma y_2\subset\Gamma\lambda y_1+(1-\lambda)y_2)$ for every $y_1,y_2\in Y$ and $\lambda\in<0,1>$. By multifunction Γ the problem P may be embedded into a family of perturbed problems

$$P_y \qquad \inf\{f(x):\ x\in\Gamma y\}\ .$$

In the present paper we formulate the dual problem for P with y as a dualization parameter. The proposed duality generalizes the existing scheme of Tind and Wolsey /10/ and the earlier one of Gould /5/ by assuming more general form of description of constraints set and by admitting dualization parameters other than right-hand-side vectors. This last fact is of particular importance in stability problems.

2. CONVEX MULTIFUNCTIONS

Let X and Y be any real linear spaces and $\Gamma:Y\to X$ be a convex multifunction. The set $\mathrm{dom}\Gamma=\{y\in Y:\ \Gamma y\neq\emptyset\}$ is called the domain of Γ; the image of an arbitrary $y\in Y$ under Γ is denoted by Γy or $\Gamma(y)$.

Convexity of Γ is equivalent to convexity of its graph $G(\Gamma)$, $G(\Gamma)\subset Y\times X$ and $G(\Gamma)=\{(y,x)\in Y\times X:\ x\in\Gamma y\}$. Multifunctions of this kind were considered in context of optimization and stability problems e.g. by Borwein /1,2/ and Robinson /6,7/ .

We say that Γ recedes in the direction $z\neq 0$, if and only if, $\Gamma y\subset\Gamma(y+\lambda z)$ for every $\lambda>0$ and $y\in\mathrm{dom}\Gamma$. The set of all vectors $z\in Y$, including $z=0$, satisfying the latter condition forms the recession cone of Γ. The recession cone of Γ will be denoted by $\Gamma 0^+$.

Proposition 1

Let X and Y be any real linear spaces. If $\Gamma: Y \to X$ is a convex multifunction then the following conditions hold:

(i) $\Gamma 0^+ = \{z \in Y: \Gamma y \subset \Gamma(y+z) \text{ for } y \in \operatorname{dom}\Gamma\}$

(ii) $\Gamma 0^+ = 0^+ G(\Gamma) \cap \{(y,x) \in Y \times X: x=0\}$

(iii) $\Gamma 0^+$ is convex .

Examples

1. Let K be a closed convex cone contained in a topological vector space Y and let $g: X \to Y$ be a closed convex function defined on a topological vector space X . A multifunction $\Gamma b = \{x \in X: b-g(x) \in K\}$ recedes in every direction belonging to K and $\Gamma 0^+ = K$.

2. Let $\Gamma: R^m \to R^n$, $\Gamma y = \{x \in R^n: g_i(x,y) < 0 \quad i=1,\dots,r\}$, all g_i are closed and convex in x and y . Then $\Gamma 0^+ = \{z \in R^m: (0,z) \in g_i 0^+$ for $i=1,\dots,r\}$ where $g_i 0^+$ denotes the recession cone of g_i .

We say that the multifunction $0^+\Gamma: Y \to X$ is the <u>recession multifunction of Γ</u> if and only if $G(0^+\Gamma) = 0^+G(\Gamma)$.

Proposition 2

Let X and Y be finite dimensional spaces. Then $0^+\Gamma(0) = \{0\}$ if and only if Γy is bounded for every $y \in Y$.

Proposition 3

Let X and Y be any pair of linear spaces. Then $z \in \Gamma 0^+$ if and only if $0 \in 0^+\Gamma(z)$.

An extended-real-valued convex function $f\Gamma: Y \to \bar{R} = R \cup \{\pm\infty\}$ defined as $f\Gamma(y) = \inf\{f(x): x \in \Gamma y\}$ with the convention $f\Gamma(y) = +\infty$ whenever $\Gamma y = \emptyset$ will be called the <u>perturbation function</u> of the family of problems $\{P_y\}$.

The following fact relates a feasible set multifunction to the perturbation function recession cone.

Proposition 4

Let X and Y be any real linear spaces. If $\Gamma: Y \to X$ is a feasible set multifunction of the family of problems $\{P_y\}$ then $\Gamma 0^+ \subset 0^+ f\Gamma$ where $0^+ f\Gamma$ is the recession cone of the perturbation function .

Proof. If $z \in 0^+\Gamma$ then $\Gamma y \subset \Gamma(y+\lambda z)$ for every $y \in \mathrm{dom}\Gamma$ and $\lambda > 0$. It implies that $\inf\{ f(x) : x \in \Gamma y\} \geq \inf\{ f(x) : x \in \Gamma(y+\lambda z)\}$ for every $y \in \mathrm{dom}\Gamma$ and $\lambda > 0$.

Let us assume now that Y and U are two locally convex Hausdorff linear topological spaces in duality with respect to the bilinear form $\langle y,u \rangle$ $y \in Y$, $u \in U$ with topologies compatible with this duality.

We use standard notations of domf, for the effective domain of arbitrary convex function $f:Y \to R$ and f for its conjugate; $f : U \to R$, $f^*(u) = \sup_{y \in Y}\{\langle y,u \rangle - f(y)\}$; by Q^O we denote the polar cone of arbitrary cone $Q \subset Y$, $Q^O = \{u: \langle u,z \rangle\ 0$ for every $z \in Q\}$.

By Proposition 4 $\Gamma 0^+$ is contained in the recession cone of the perturbation function. The following proposition gives the dual relation.

Proposition 5

Let Y,U be a dual pair of spaces. Then $\mathrm{dom} f\Gamma^* \subset (\Gamma 0^+)^O$.

Proof. Let us consider $\sup_{y \in Y}\{\langle y,u \rangle - f\Gamma(y)\}$ for arbitrary $u \in U$ and let $z=t+k$ for arbitrary but fixed $t \in Y$ and $k \in \Gamma 0^+$. Then $\langle z,u \rangle - f\Gamma(z) = \langle t,u \rangle + \langle k,u \rangle - f\Gamma(t+k) \geq \langle t,u \rangle + \langle k,u \rangle - f\Gamma(t)$. Thus, $\sup_{y \in Y}\{\langle y,u \rangle - f\Gamma(y)\} \geq \sup_{k \in \Gamma 0^+}\{\langle k,u \rangle + \langle t,u \rangle - f\Gamma(t)\}$ If there exists $k \in \Gamma 0^+$ such that $\langle k,u \rangle > 0$ then $\sup_{k \in \Gamma 0^+} \langle k,u \rangle = +\infty$. So, if $u \notin (\Gamma 0^+)^O$ then $u \notin \mathrm{dom} f$.

3. DUAL PROBLEM

Denoting the class of all affine functionals defined on Y by Aff(Y) we introduce the set $\mathcal{F} \subset \mathrm{Aff}(Y)$, $\mathcal{F} = \{F \in \mathrm{Aff}(Y): Fy = \langle y,u \rangle + u_o,\ u \in (\Gamma 0^+)^O,\ u_o \in R\}$. Now we can define the dual to the problem P as

$$\text{D} \qquad \begin{array}{l} \text{supremize}_F Fy_o \\ \text{subject to:} \\ \qquad f(x) \geq \sup_{\{y:x \in \Gamma y\}} Fy \quad \text{for every } x \in X, \\ \qquad F \in \mathcal{F}. \end{array}$$

Example

Let us consider the problem

$$\text{P} \qquad \inf\{f(x):\ b_o - g(x) \in D\}$$

where $f:X \to R$, $g:X \to Y$ are closed mappings, $b \in Y$, $D \subset Y$ is a closed convex cone. If we define multifunction $\Gamma:Y \to X$, $\Gamma b = \{x:\ b - g(x) \in D\}$ then $\Gamma 0^+ = D$ and $\sup_{\{b:x \in \Gamma b\}} Fb = Fg(x)$ for every F belonging to $\mathcal{F}$ and the dual D takes the form of the dual proposed by Tind and Wolsey /10/.

By simple calculations we obtain

Proposition 6

The conjugate function of the perturbation function f can be expressed by the following formula for arbitrary $u\varepsilon U$

$$-f\Gamma^*(u)=\inf_{x\varepsilon X}\{f(x)-\sup_{\{y:x\varepsilon\Gamma y\}}<y,u>\}.$$

The above proposition allows to formulate the dual problem D in the equivalent form

$$D' \qquad \begin{array}{ll} \text{supremize}_{u,u_o} & <y_o,u>+u_o \\ \text{subject to:} & -f\Gamma^*(u) \geqslant u_o \\ & u\varepsilon(\Gamma 0^+)^o,\ u_o\varepsilon R \end{array}$$

The formulation D' of the dual problem admits a usual geometric interpretation. Namely, the problem D might be viewed as a problem of finding among all supporting hyperplanes of $f\Gamma$ such one which has the maximal value at y_o. In fact, the condition $-f\Gamma^*(u) \geqslant u_o$ can be rewritten as $f\Gamma(y)\geqslant<y,u>+u_o$ for every $y\varepsilon Y$.

For the sake of completeness we should notice that for a primal minimization problem of the form

$$\inf\{f(x):\ x\varepsilon\Gamma y_o,\ x\varepsilon C\}$$

where C is a convex subset of X the dual D takes the form

$$\begin{array}{ll} \text{supremize}_F\ Fy_o & \\ \text{subject to:} & \\ & f(x) \geqslant\sup_{\{y:x\varepsilon\Gamma y\}} Fy \quad \text{for every} \quad x\varepsilon C \\ & F\varepsilon\mathcal{F} \end{array}$$

Denoting by $\gamma(0)$ the optimal value of the dual problem we may formulate

Theorem 3.1 (weak duality)

If $\underline{x}$ is feasible for P and $\underline{F}$ is feasible for D then $f\Gamma(y_o)\geqslant\gamma(0)$.

The following properties follows immediately from the formulation of the dual problem and the conditions assuring the existence of nonvertical support for convex functions

Property 1. D is feasible if and only if $f\Gamma(y)>-\infty$ for every $y\varepsilon Y$.

Property 2. If y_o belongs to the interior of $\operatorname{dom}\Gamma$ and $f\Gamma(y_o)$ is finite then the problem D has a solution.

The question of equality between optimal values of P and D can be reduced to the question whether the equality $f\Gamma(y_o)=f\Gamma^{**}(y_o)$ holds since

$\sup\{\langle y_o,u\rangle+u_o : \ -f\Gamma^*(u)\geqslant u_o,\ u\in(\Gamma 0^+)^o,\ u_o\in R\} =$
$\sup\{\langle y_o,u\rangle-f\Gamma^*(u) : u\in(\Gamma 0^+)^o\}= \sup_{u\in U}\{\langle y_o,u\rangle-f\Gamma^*(u)\}= f\Gamma^{**}(y_o)$.

Corresponding theorems are contained in Rockafellar /13/, Dolecki /3/, Joly and Laurent /11/.

3a. Symmetry of the dual.

Let X and V be in duality with respect to the bilinear form $\langle x,v\rangle$ where $x\in X$ and $v\in V$. A multifunction $\Gamma^*:V\to \mathrm{Aff}(Y)$ defined as

$$\Gamma^* v = \{F\in \mathcal{F}:\ f(x)+\langle x,v\rangle\geqslant\sup_{\{y:x\in\Gamma y\}} Fy \ \text{ for every } x\in X\}$$

will be called the dual multifunction of the problem P .

Now we may introduce the perturbed dual problem D_v for $v\in V$ as

D_v $\quad \sup\{Fy_o : F\in\Gamma^*v\}$, $/D_o$ corresponds to D / .

The dual perturbation function $\gamma:V\to R$ defined as $\gamma(v) = \sup\{Fy_o : F\in\Gamma^*v\}$ is concave with $\gamma(0)$ equal to the optimal value of D .

Basing on the similar arguments as in the proposition 5 we obtain that $\mathrm{dom}\gamma\subset(\Gamma^*0^+)^*$ where $(\Gamma^*0^+)^*$ is the dual cone of Γ^*0^+ , i.e. $(\Gamma^*0^+)^*=\{x\in X : \langle x,t\rangle\geqslant 0$ for every $t\in\Gamma\ 0^+\}$.

If we introduce $\mathfrak{X}\subset\mathrm{Aff}(V),\mathfrak{X}=\{\mathfrak{x}\in\mathrm{Aff}(V): \mathfrak{x}(v)=\langle x,v\rangle+x_o, x\in(\Gamma^*0^+)^*, x_o\in R\}$ then the dual to the problem D may be formulated as

DD $\quad$ infimize$_{\mathfrak{x}}$ $\mathfrak{x}(0)$

subject to:

$$Fy_o\leqslant\inf_{\{v:F\in\Gamma\ v\}}\mathfrak{x}(v) \quad \text{for every } F\in\mathcal{F}$$
$$\mathfrak{x}\in\mathfrak{X} \ .$$

Similarily to the proposition 6, for the dual perturbation function, we have the expression that $-\gamma^*(x)= \sup_{F\in\mathcal{F}}\{Fy_o-\inf_{\{v:F\in\Gamma^*v\}}\langle x,v\rangle\}$.

Thus the problem DD may be restated as

DD, $\quad$ infimize x_o

subject to: $x_o\geqslant-\gamma^*(x)$

$x\in(\Gamma^*0^+)^*$, $x_o\in R$.

Theorem 3.2

The duality induced by D is symmetric in the sense that the optimal values of DD and P are equal under the assumption that the equality between optimal values of P and D holds.

3b. Solutions of the dual.

A convex function $f:Y\to R$ is said to be subdifferentiable at $y_o \in Y$ if there exists $u \in U$ such that $f(y) \geq f(y_o) + \langle y-y_o, u\rangle$ for every $y \in Y$, i.e. there exists a continuous affine minorant of f which takes the value $f(y_o)$ at y_o. Such an element $u \in U$ is called a <u>subgradient of f at y_o</u>.

Theorem 3.3

If $\hat{u} \in U$ is a subgradient of $f\Gamma$ at y_o then $(\hat{u}, -f\Gamma^*(\hat{u}))$ is a solution of D.

Proof. Let us observe that the pair $(\hat{u}, -f\Gamma^*(\hat{u}))$ is feasible and $f\Gamma^*(\hat{u}) + f\Gamma(y_o) = \langle y_o, \hat{u}\rangle$. So $f\Gamma^*(\hat{u})$ must be finite, what implies that $\hat{u} \in (\Gamma 0^+)^o$ and obviously $u_o = -f\Gamma^*(\hat{u}) \leq -f\Gamma^*(\hat{u})$. The optimality follows from the relation $\langle \hat{u}, y_o\rangle - f\Gamma^*(\hat{u}) = f\Gamma(y_o) + f\Gamma^*(\hat{u}) - f\Gamma^*(\hat{u}) = f\Gamma(y_o)$ what completes the proof.

Theorem 3.4

If the equality of the optimal values of P and D holds and $(\hat{u}, \hat{u}_o)$ is a solution of the dual problem D then $\hat{u}$ is a subgradient of $f\Gamma$ at y_o and $\hat{u}_o = -f\Gamma^*(\hat{u})$.

4. RELATIONS TO OTHER DUALS

4a. Lagrangean duals

For the problem P we consider the Lagrangean $L: X\times U \to R$, $L(x,u) = f(x) - \sup_{\{y: x\in\Gamma y\}} \langle y,u\rangle + \langle y_o, u\rangle = L(x,u,y_o)$. The Lagrangean of this form was introduced by Kurcyusz /12/ and investigated by Kurcyusz and Dolecki /4/, Dolecki /3/. The dual pair of problems connected with $L(x,u)$ can be written in the usual form as

LP $\quad \inf_{x\in X} \sup_{u\in U} L(x,u) = LP_{opt}$

LD $\quad \sup_{u\in U} \inf_{x\in X} L(x,u) = LD_{opt}$

Theorem 4.1

The problem LD is equivalent to D.

Proof. The problem LD may be rewritten as $\sup_{u\in U}\{\langle y_o,u\rangle + \inf_{x\in X}\{f(x) + -\sup_{\{y:x\in\Gamma y\}}\langle y,u\rangle\}\}$. If $\inf_{x\in X}\{f(x) - \sup_{\{y:x\in\Gamma y\}}\langle y,u\rangle\} = -\infty$ for all $u\in U$ then $f\Gamma^*(u) = +\infty$ and D is infeasible, so, $\gamma(0) = -\infty$. Otherwise, there exists $u \in U$ such that $\inf_{x\in X}\{f(x) - \sup_{\{y:x\in\Gamma y\}}\langle y,u\rangle\} > -\infty$. According to the proposition 5 such u satisfies also the condition that $u \in (\Gamma 0^+)^o$ so D is feasible and $\gamma(0) = \sup\{\langle y_o,u\rangle - f\Gamma^*(u) : u\in(\Gamma 0^+)^o\} =$
$= \sup_{u\in U}\{\langle y_o,u\rangle + \inf_{x\in X}\{ f(x) - \sup_{\{y:x\in\Gamma y\}}\langle y,x\rangle\}\} = LD_{opt}$.

Abstract minimization problems with perturbations were investigated also by Rockafellar /13/ in the form

$$\inf_{x\varepsilon X} F(x,y)$$

where $F:X\times Y\to R$ is an extended-real-valued function with $F(x,0)=f(x)$. The minimization over all the space X is obtained by redefining minimized function f of the problem P so that $f(x)=+\infty$ for $x\notin\Gamma y_o$, $y_o=0$. The Lagrangean function $K:X\times U\to R$ connected with such family of problems is defined in /13/ as

$$K(x,u)= \inf_{y\varepsilon Y}\{F(x,y)+\langle y,u\rangle\} .$$

Perturbation of the original problem P by the function $F(x,y)$ admits various forms of perturbations in constraints set and minimized function.

If we consider the problem P with $y_o=0$ then the function $F(x,y)$ connected with the family of perturbed problems P_y may expressed as $F(x,y)=f(x)+\delta(y:\ \Gamma^{-1}x)$ where $\delta(y:\ \Gamma^{-1}x)=0$ if $y\varepsilon\Gamma^{-1}x$ and $+\infty$ otherwise, $\Gamma^{-1}x=\{y:\ x\varepsilon\Gamma y\}$. For such function $F(x,y)$ the Lagrangean $K(x,u)$ takes the form

$$K(x,u)= \inf_{y\varepsilon Y}\{f(x)+\delta(y:\ \Gamma^{-1}x)+\langle y,u\rangle\}=f(x)+\inf_{y\varepsilon Y}\{\delta(y:\Gamma^{-1}x)+\langle y,u\rangle\}=$$
$$=f(x)-\delta^*\{-u:\ \Gamma^{-1}x\}= L(x,-u,0).$$

4b. Surrogate dual

For the problem P a surrogate problem may be formulated as

SP $\qquad \inf\{f(x):\ Fy_o\leqslant\sup_{\{y:x\varepsilon\Gamma y\}} Fy\}\overset{..}{=} SP_{opt}$

where $F\varepsilon\mathcal{F}$. It is an immediate observation that $Fy_o\leqslant\sup_{\{y:x\varepsilon\Gamma y\}}Fy$ for $x\varepsilon\Gamma y_o$ and consequently $f\Gamma(y_o)\geqslant SP_{opt}$. If we denote

$$v(F)= \inf\{f(x):\ Fy_o\leqslant\sup_{\{y:x\varepsilon\Gamma y\}} Fy\}$$

then the surrogate dual problem may defined as

SD $\qquad \sup\{v(F):\ F\varepsilon\mathcal{F}\}$.

Theorem 4.2

If the equality $f\Gamma(y_o)= LD_{opt}$ holds then the surrogate dual problem SD is equivalent to D.

5. FINAL REMARKS

Throughout the paper we considered convex problems P assuming convexity of a minimized function f and a feasible multifunction Γ. The most important consequence of these assumptions is convexity of the perturbation function $f\Gamma$.

Let us observe firstly that the dual problem remains well defined if we remove the convexity assumptions. However, we cannot expect the existance of the nonvertical support closing the duality gap in nonconvex case. One of the usually treated way to avoid this difficulty is to consider larger classes of dual price functions $\mathcal{F}$. This approach, suggested by Gould /5/, was developed by Dolecki and Kurcyusz /4/ and Dolecki /3/. The analysis of different price functions classes $\mathcal{F}$ and the resulting duals was also given by Tind and Wolsey /10/ for problems with inequality constraints. These ideas may used also in the context of duality proposed in the present paper.

There exists several ways of introducing dual multifunction Γ to a given multifunction Γ. In finite dimensional spaces dual multifunction Γ^* may be considered as having the graph $G(\Gamma^*)$ which is the polar set to the graph of Γ. The properties of such dual multifuntion and corresponding pair of linear dual problems were analysed recently by Ruys and Weddepohl /14/. Most recently this idea was considered by Dolecki /15/ in general spaces.

The notion of the recession cone of multifunction, introduced in section 2, and appearing in the formulation of the dual problem D gives rise to the more detailed description of the set of price functions $\mathcal{F}$. It gives additional information about the structure of the set $\mathcal{F}$.

BIBLIOGRAPHY

1. J.M.Borwein, Convex relations in analysis and optimization, in: Generalized Concavity in Optimization and Economics, ed. S.Schaible, W.T.Ziemba, Academic Press 1981, pp.335-377
2. J.M.Borwein, Multivalued convexity and optimization: a unified approach to inequality and equality constraints, Math.Progr. vol.13 /1977/ pp.183-199
3. S.Dolecki, Abstract study of optimality conditions, J.Math.Anal.Appl. vol.73 /1980/ pp.24-48
4. S.Dolecki, S.Kurcyusz, On Φ-convexity in extremal problems, SIAM J.Cont.Opt. vol.16 /1978/ pp.277-300
5. F.J.Gould, Nonlinear duality theorems, Cahier du Centre dEtudes de Rech.Oper. vol.14 /1972/ pp.196-212
6. S.M.Robinson, Regularity and stabilita for convex multivalued functions, Math.OR vol.1 /1976/ pp.205-210
7. S.M.Robinson, Normed convex processes, Trans.Amer.Math.Soc. vol.174 /1972/ pp.127-140

8. R.T.Rockafellar, Convex Analysis, Princeton University Press, Princeton 1969

9. J.Sikorski, On surrogate constraint duality, Working Paper IBS,ZPM1980

10. J.Tind, L.A.Wolsey, An elementary survey of general duality theory in mathematical programming, Math.Progr. vol.21 /1981/ pp.241-261

11. J.L.Joly, P.J.Laurent, Stability and duality in convex minimization problems, R.I.R.O. vol.15 /1971/ pp.3-42

12. S.Kurcyusz, Some remarks on generalized Lagrangians, Proc.7th IFIP Conf. Nice, Sept.1975 Part II, Springer 1976

13 R.T.Rockafellar, Conjugate duality and optimization, Reg.Conf.Ser. Appl.Math 1974

14. P.H.M.Ruys, H.N.Weddepohl, Economic theory and duality, in:Convex Analysis and Mathematical Economics, ed. J.Krein, Springer 1979

15 S.Dolecki, Duality theory, manuscript /61 pages/

PARAMETRIZING THE VALUE FUNCTIONS IN DYNAMIC PROGRAMMING

P.O. Lindberg
Dept. of Mathematics
Royal Institute of Technology
S-100 44 Stockholm, SWEDEN

1. Introduction

A wellknown phenomenon in Dynamic Programming is the so called "curse of dimensionality". Essentially it says that the computational effort in Dynamic Programming grows exponentially with the dimension of the state space.

In some instances it is possible to circumvent this curse. One such case is the quadratic one, i.e. when the involved cost functions are quadratic. Then the value functions happen to be quadratic and the recursion may be done in terms of the parameters of the quadratic functions.

In this paper we will survey some results on when such a finite dimensional parametrization of the value functions is possible. Due to space limitations we cannot include all proofs and partial results. We will rather aim at presenting the problems and motivating some results. For full details the reader is referred to Lindberg (1981).

Initially we will restrict ourselves to discrete stage Dynamic Programming with functional equations of the form

$$V_{n+1}(x) = \inf_u \{c_n(u) + V_n(x-u)\} \qquad (1)$$

where the cost functions c_n are convex.

For this situation, when the c_n have identic recession functions a finite dimensional parametrization of the V_n turns out to be possible if and only if there are a finite number of convex function $f_o,\ldots,f_k$ such that all c_n^*, the convex conjugates of c_n, (see e.g. Rockafellar (1970)), are of the form

$$c_n^* = \sum_{i=1}^{d} \lambda_i(f_i - f_o) + \sum_{i=d+1}^{k} n_i f_i$$

for some real λ_i and nonnegative integer n_i. These results are used to study the case of stochastic Dynamic Programming of the form leading to the recursive equation

$$V_{n+1}(x) = \inf_u \{c_n(u) + E_{\xi_n} V_n(x - u - \xi)\} \tag{2}$$

For this situation, one can show that when the c_n are defined on the real line, and the ξ_n may be any random variables with sufficiently small support, then all c_n must be of the form

$$c_n(x) = p_1 x + \delta(x \mid [p_2, \infty)) + p_3$$

or all of the form

$$c_n(x) = p_1 x^2 + p_2 x + p_3$$

or all of the form

$$c_n(x) = p_1 e^{p_2 x} + cx + p_3 \quad \text{(all with the same c)}$$

for some parameters $(p_1, p_2, p_3) \in R^3$.
In the last two cases the c_n must be defined on the whole real line.

2. Conditions on the parametrization

We want to find a finite dimensional parameter set, say P in R^d, such that the functions V_n arising in the recursion (1) are of the form

$$V_n = V(p)$$

for some $p \in P$.
What may be demanded of such a parametrization?

It is wellknown that the unit interval I in R may be mapped one-to-one onto the unit cube I^k in R^k, (since these point sets have the same cardinality).
If the dimension of the parameter space shall have any meaning, such mappings must be excluded.

The above mapping of I onto I^k is not continuous though. But, for the parameter mapping $p \rightarrow V(p)$ to have any use at all it must of course be continuous in some sense.

However, demanding $V: p \rightarrow V(p)$ to be continuous does not solve the case, since the equally wellknown space filling curves map I continuously onto I^k (see e.g. Hocking & Young (1961), §3.3).

The solution seems to be to demand the mapping $p \rightarrow V(p)$ to be one-to-one and continuous, i.e. a homeomorphism.
Hence we must introduce a topology for the functions V.

Let us assume that the functions c_n, which we assume convex, are defined on a finite dimensional space R^k. We will allow the functions to take the value $+\infty$, so that they are essentially only defined on their effective domain, dom c_n. (We will follow Rockafellar (1970) in notation and terminology for convex analysis. For any undefined terms, please consult this reference.)

Before introducing a suitable topology let us note that (1) may be written

$$V_{n+1} = c_n \square V_n \tag{3}$$

where $\square$ denotes infimal convolution. Taking convex conjugates (3) becomes

$$V^*_{n+1} = c^*_n + V^*_n \tag{4}$$

where * denotes conjugation.

Since addition is easier to work with than conjugation, it is natural to look for a topology for convex functions that behaves well under conjugation.

The natural topology to use for the convex functions, at least in an optimization setting, seems to be the one introduced by Wijsman (1966). Let us shortly review the definition and main properties of this topology, since it does not seem to be quite wellknown.

Let B be the open unit ball in R^k. For a sequence $\{A_i\}$ of sets in R^k we define (see Kuratowski (1966), § 29) the *lower limit*

$$\operatorname*{Li}_{i\to\infty} A_i \overset{\Delta}{=} \{x \in R^k \mid \text{for all } \varepsilon > 0, x \in A_i + \varepsilon B \text{ eventually}\}$$

and the *upper limit*

$$\operatorname*{Ls}_{i\to\infty} A_i \overset{\Delta}{=} \{x \in R^k \mid \text{for all } \varepsilon > 0, x \in A_i + \varepsilon B \text{ infinitely often}\}.$$

Note that $\mathrm{Li}A_i$ and $\mathrm{Ls}A_i$ are closed sets and that $\mathrm{Li}A_i$ is convex if the A_i are convex.

Further we say that A_i *converges* to A, denoted $A_i \to A$, if

$$\mathrm{Li}A_i = \mathrm{Ls}A_i = A.$$

For elementary properties of these limits consult Kuratowski (1966, § 29).

For convex and concave functions we say that f_i converges to f, denoted $f_i \to f$, if

$$\text{epi } f_i \to \text{epi } f.$$

The following theorem by Wijsman (1966) shows that this convergence concept fulfills our needs.

<u>Theorem 1</u>. If $\{f_i\}$ is a sequence of convex functions on R^k such that $f_i \to f$, then

$$f_i^* \to f^*.$$

(Here f^* is the convex conjugate of f.)

Wijsman further notes that these limits correspond to a metrizable topology on the set of closed convex functions (see below).

If we assume the functions involved to be closed, then we may use V_n^* as a representation for V_n, since the mapping $V_n \to V_n^*$ is one to one in this case (Rockafellar(1970), thm 12.2). Due to (4) this simplifies the analysis.

If a convex function does not take the value $-\infty$, then it is closed if and only if it is lower semicontinuous. This is a very natural condition in an optimization setting. In the lack of it, it would be hard to guarantee that infima are attained even over compact sets.

Moreover, (Rockafellar thm 7.4), the closure of a convex function agrees with the function itself, except possibly on the relative boundary of the domain. Hence, one might perceive the closure of a convex function as a well behaved representative of the equivalence class of convex functions agreeing with the given function except possibly on the relative boundary of the domain.

Thus we will generally restrict ourselves to proper closed convex functions, i.e. closed convex functions, not identically $+\infty$ or $-\infty$. Let $Cx(R^k)$ denote the set of such functions on R^k.

To make full use of (4) however, our topology must behave well under algebraic operations such as addition and convex combinations.

This is not easily achieved though. Since convex functions take the value $+\infty$ outside their effective domain not even the law of cancellation is valid for addition. I.e.

$$f + g = f + h$$

does not necessarily imply $g = h$.

For convex combinations we in fact have the following result

Theorem 2 (Lindberg (1981)). For $f,g \subset Cx(R^k)$ the mapping $\lambda \to \lambda f + (1-\lambda)g$ is continuous on $[0,1]$ if and only if $\mathrm{ri\,dom}\,f = \mathrm{ri\,dom}\,g$. □
(Here $\mathrm{ri\,dom}\,f$ is the relative interior of the effective domain of f.)

Due to this result, we will try to achieve common relative interiors of domains for the functions studied. Then we in fact get continuity of addition at the same time.

3. Finite dimensional sets of convex functions closed under addition

Now let us state the parametrization problem in more mathematical terms. We have a set F of convex functions, F consisting of the functions V_n. We want F to be finitely parametrized. With this we will understand that F is *locally euclidean*, i.e. each point in F has a neighbourhood homeomorphic to an open set in some R^d. We will also assume that the dimension d is the same over all F.
Taking conjugates F is mapped onto some F^*, which we want to be closed under addition by (4). Thus we have the following

Problem 1. Suppose F^* is a d-dimensional locally euclidean subset of $Cx(R^k)$. Under what conditions is F^* closed under addition? A partial answer to this is the following

Theorem 3 (Lindberg (1981)). Let F^* be a d-dimensional locally euclidean set in $Cx(R^k)$. Further suppose F^* closed under addition.
Let $g_o \in F^*$ and set
$F_o^* = \{g \in F^* \mid \mathrm{ri\,dom}\,g = \mathrm{ri\,dom}\,g_o\}$.
Then locally around g_o, F_o^* is "flat", i.e. there are $g_1,\dots,g_m$ such that each g in F_o^* sufficiently close to g_o is of the form $g = \sum_{j=0}^{m} \gamma_j g_j$.

Proof sketch: Take g_1 close to g_o in F_o^* (see figure below).
Since F_o^* is closed under addition, $2g_o$, $2g_1$ and $g_o + g_1$ belong to F_o^*.
Now taking g_1 sufficiently close to g_o, one can assure $\frac{1}{2}(g_o + g_1)$ to belong to F_o^*.
Repeating the argument, all "binary" points in the "interval" $[g_o,g_1]$ will belong to F_o^* and by continuity the whole interval itself. Iterating with a g_2 outside the line through g_o and g_1, and so on, we get a simplex $\Delta_m = \mathrm{conv}\{g_o,\dots,g_m\} \subset F_o^*$. Since the dimension is bounded by d, the process must stop. □

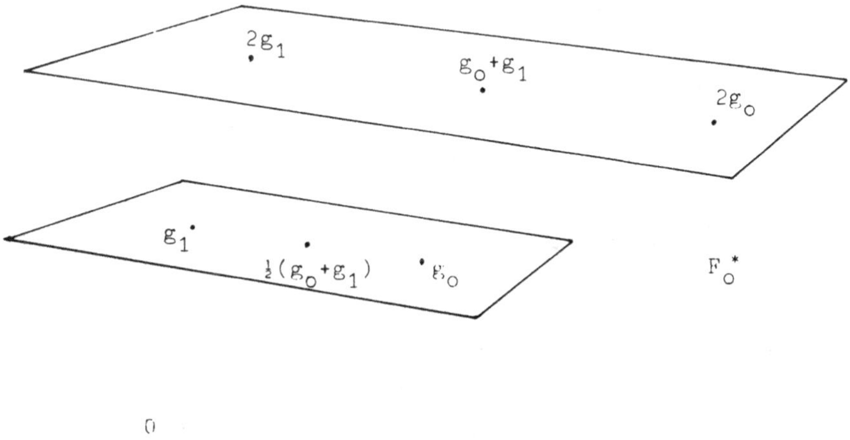

The result mentioned in the introduction is a simple corollary of the theorem.

4. Closedness under means

In stochastic dynamic programming we have the recursive equation (2):

$$V_{n+1}(x) = \inf_u \{c_n(u) + E_{\xi_n} V_n(x - u - \xi_n)\} \qquad (2)$$

Depending on what classes we allow for ξ_n we get different parametrization problems. We could of course restrict ourselves to some finitely parametrized class of distributions. We will not do this, however, but rather put restrictions on the support of the distributions, in order to have the mean value in (2) be defined when the effective domain of the functions involved is not all R^k.

Let F denote our finitely parametrized set in $Cx(R^k)$ from which our functions c_n are taken, and in which V_n lie.
Then taking means, we get new functions $\overline{V}_n^{\xi}$, defined by

$$\overline{V}_n^{\xi}(x) = E_{\xi} V_n(x-\xi).$$

Thus taking all possible means of functions in F, we get a set $\overline{F}$.
If, however, we choose $\xi \equiv 0$, then

$$\overline{V}_n^{\xi} = V_n, \text{ whence } F \subset \overline{F}.$$

Our finite parametrization now demands

$$F \square \bar{F} \subset F.$$

It can be shown (see Lindberg (1981)) that this is not possible unless in fact $\bar{F} \subset F$ and hence $\bar{F} = F$.

Let $f \in F$ be given and define f_ξ by

$$f_\xi(x) = E_\xi f(x-\xi).$$

It is clear that if the effective domain of f has no interior, then $f_\xi = \infty$ unless the support of ξ is parallel to dom f. To avoid technicalities in the presentation we will therefore assume dom f to have nonempty interior for all $f \in F$. In this case $f_\xi \neq \infty$ if only ξ has sufficiently small support. Thus we have the following

Problem 2. Let F be a locally euclidean d-dimensional subset of $Cx(R^k)$. Under what conditions do we have $f_\xi \in F$ for all $f \in F$ and all random variables ξ with sufficiently small support?

The solution to this problem we will give with partly intuitive arguments in the case of Cx(R), i.e. convex functions on the real line.

Let $f \in F$ be given and let $X = \{\xi | \xi$ random variables with sufficiently small support$\}$. Further let $f_X = \{f_\xi | \xi \in X\}$. Then it is easily seen that f_X is convex. Since moreover $f_X \subset \bar{F} = F$, f_X is finite dimensional and it has a representation of the form

$$f_X = \{g = \sum_{i=1}^{m} \gamma_i g_i | \Sigma \gamma_i = 1, \gamma \in \Gamma\}$$

for some functions $g_i \in F$ and some set $\Gamma \subseteq R^m$.

Now consider $t \in R$ as a constant random variable. Then $(g_i)_t \in f_X$ for sufficiently small t (and if g_i are chosen sufficiently close to f). Thus

$$(g_i)_t = \sum_{j=1}^{m} \gamma_{ij}(t) g_j.$$

In particular

$$(g_i)_{s+t} = ((g_i)_t)_s = (\sum_{j=1}^{m} \gamma_{ij}(t) g_j)_s = \sum_j \gamma_{ij}(t)(g_j)_s =$$

$$= \sum_{j,k} \gamma_{ij}(t)\gamma_{jk}(s) g_k.$$

But we also have

$$(g_i)_{s+t} = \sum_{k=1}^{m} \gamma_{ik}(s+t)g_k.$$

Thus, given that we have chosen the g_i affinely independent,

$$\gamma_{ik}(t+s) = \sum_j \gamma_{ij}(t)\gamma_{jk}(s),$$

or letting $G(t)$ be the matrix with components $\gamma_{ij}(t)$

$$G(t+s) = G(t)G(s).$$

Further G is continuous and $G(0) = I$. Hence we can define $L(s)$ such that (at least for small s) $\exp(L(s)) = G(s)$. Then L is continuous and satisfies

$$L(s+t) = L(s) + L(t)$$

which is the Cauchy equation whose only continuous solutions are linear. Therefore $L(s) = Cs$ for some constant matrix C. Performing a coordinate transformation if necessary (in the space spanned by the g_i) we may assume that C is on Jordan canonical form. But then $G(s) = \exp(Cs)$ has components of the form $s^n\exp(\lambda s)/n!$

Now take a $g \in f_X$, say $g = \sum_i \gamma_i g_i$. Then $g(x) = g(x-(x_o-x)) = g_{x_o-x}(x_o) = $
$= \sum_i \gamma_i (g_i)_{x_o-x}(x_o) = \sum_{i,j} \gamma_i\gamma_{ij}(x_o-x)g_j(x_o)$. We have "proved"

Theorem 4. For F to solve problem 2 in the case $R^k = R$ the functions in F must be finite linear combinations of terms of the form $x^n\exp(\lambda x)$.

5. Closedness under infimal convolutions and means

Now we are in position to combine the results of sections 4 and 5.
For F to solve our problem of finite parametrization in the recursion (2) we have the following condition

(i) F^* closed under addition

(ii) F closed under means.

For the case of $F \subseteq Cx(R)$, (ii) by theorem 4 gives a concrete representation of the possible functions in F. On the conjugate side (i) by theorem 3 gives the structure of F^*.

It is now possible though very technical to go through the different possibilities for F, checking growth rates on both sides, enforcing convexity etc. We then arrive at the result stated in the introduction:

Theorem 5. Let F be a d-dimensional locally euclidean subset of $Cx(R)$, closed under infimal convolution and under means with respect to random variables with sufficiently small support.

Then the f in F are parametrized by $p = (p_1,p_2,p_3) \in R^3$ in that they are all of the form

$$f(x) = p_1x + \delta(x|[p_2,\infty)) + p_3$$

or all of the form

$$f(x) = p_1x^2 + p_2x + p_3$$

or all of the form

$$f(x) = p_1\exp(p_2x) + cx + p_3 \quad \text{(for some constant c).}$$

References

J.G. Hocking & G.S. Young (1961): "Topology", Addison-Wesley, Reading,Mass.

K. Kuratowski (1966): "Topology", Academic Press, New York.

P.O. Lindberg (1981): "Parametrizing the Value Functions in Dynamic Programming", Technical Report TRITA-MAT-1981-13, Dept. of Mathematics, Royal Institute of Technology, S-100 44 Stockholm, Sweden.

R.T. Rockafellar (1970): "Convex Analysis", Princeton University Press, Princeton.

R.A. Wijsman (1966): "Convergence of Convex Sets, Cones and Functions II", Trans. Amer. Math. Soc. 123, 32-45.

A SMOOTH SEQUENTIAL PENALTY FUNCTION METHOD FOR SOLVING NONLINEAR PROGRAMMING PROBLEMS

C.G. Broyden and N.F. Attia
Department of Computer Science
University of Essex
Wivenhoe Park
Colchester
Essex
England

1. Introduction

We consider the problem

$$\text{minimise} \quad F(x), \quad x \in R^n \tag{1.1a}$$

$$\text{subject to} \quad c_i(x) = 0, \quad i=1, \ldots, m \tag{1.1b}$$

where F and all c_i are twice continuously differentiable nonlinear functions of x.

The algorithm solves the problem by minimising the composite function $\Phi(x,r)$, where

$$\Phi = F(x) + \psi(c,r), \tag{1.2}$$

where ψ is a standard penalty or barrier term, e.g.

$$\psi = r^{-1} \sum_{i=1}^{m} c_i^2$$

or

$$\psi = - r \sum_{i=1}^{m} \ln(c_i)$$

and where r is a sufficiently small positive parameter. It is known that if x* is the solution of problem (1.1) and x*(r) is the unconstrained minimum of (1.2) then, under mild conditions [3]

$$\lim_{r \to 0} x^*(r) = x^*$$

The algorithm minimises Φ for decreasing positive values of r using a Quasi-Newton method combined with an orthogonal transformation based on the Jacobian of the constraints. If f and g denote respectively the gradients of Φ and F, and J denotes the Jacobian of $c = [c_i]$ (assumed to have full row rank), the necessary condition for the existence of a minimum implies that

$$f = g + J^T \nabla_c \psi = 0 \qquad (1.3)$$

at the solution. This equation may in principle be solved by Newton's method where the correction Δx to an approximate solution is computed as the solution of

$$[H + J^T D J] \Delta x = -f, \qquad (1.4)$$

where H = Hessian of $F + \sum_i \left(\frac{\partial \psi}{\partial c_i} \cdot \text{Hessian of } c_i \right)$ and where $D = \nabla^2_{cc} \psi$.

H is generally well-behaved but for most common penalty and barrier functions $J^T D J$ is of rank m and $\|J^T D J\| \to \infty$ as $r \to 0$. The condition number of $H + J^T D J$ thus increases indefinitely as $r \to 0$, with singularity occurring in the limit [5]. The difficulty of solving (1.3) using a standard unconstrained minimization algorithm is due to this ill conditioning and is a major reason for this approach's lack of preferment.

This paper describes an effective algorithm for solving equation (1.3) which avoids the weaknesses of the methods which use straightforward unconstrained optimisation algorithms. It generalizes in an obvious way when inequality constraints are present.

2. The algorithm

The solution of

$$f = g + J^T \nabla_c \psi = 0, \quad r \to 0 \qquad (2.1)$$

is based on Newton's method, where

$$[H + J^TDJ]\Delta x = -f \tag{2.2}$$

Let now $M = D^{-1}(JJ^T)^{-1}J$. Multiplying (2.2) by M gives

$$[MH + J]\Delta x = -Mf \tag{2.3a}$$

where (2.3a) consists of m equations. To obtain the remaining (n-m) equations, let P be an (n-m)xm matrix whose columns form a basis for the null-space of J, so that JP = 0.

Multiplying (2.2) by P^T gives

$$P^TH\Delta x = -P^Tf \tag{2.3b}$$

and this, with equation (2.3a), becomes

$$\begin{bmatrix} MH + J \\ P^TH \end{bmatrix} \Delta x = - \begin{bmatrix} M \\ P^T \end{bmatrix} f \tag{2.4}$$

Since as r -> 0, ||M|| -> 0 for all common penalty and barrier functions, we have approximately

$$\begin{bmatrix} J \\ P^TH \end{bmatrix} \Delta x = - \begin{bmatrix} M \\ P^T \end{bmatrix} f \tag{2.5}$$

This approximation is good enough (if r -> 0) for Newton's method to converge. The coefficient matrix is essentially independent of r and is ill-conditioned only if the original problem is in some way pathological.

To solve (2.5), note that there exists an orthogonal matrix Q such that

$$QJ^T = \begin{bmatrix} U \\ 0 \end{bmatrix} \tag{2.6}$$

where U is nonsingular upper triangular. From (2.5)

$$J\Delta x = JQ^TQ\Delta x = -Mf \tag{2.7}$$

If $$Q\Delta x = \Delta z = \begin{bmatrix} \Delta z_1 \\ \Delta z_2 \end{bmatrix}$$

where $\Delta z_1 \epsilon R^m$ and $\Delta z_2 \epsilon R^{n-m}$, equation (2.7) becomes

$$U^T \Delta z_1 = -Mf \tag{2.8}$$

Now since $JP = 0$, $JQ^T QP = 0$ and hence

$$[U^T \ \ 0] \begin{bmatrix} K_1 \\ K_2 \end{bmatrix} = 0 \tag{2.9}$$

where $$QP = K = \begin{bmatrix} K_1 \\ K_2 \end{bmatrix} \tag{2.10}$$

Thus, from (2.9) (since U is nonsingular), $K_1 = 0$. Now since K_2 may be chosen arbitrarily, set it equal to I so that $K = \begin{bmatrix} 0 \\ I \end{bmatrix}$

From (2.10) $P = Q^T K$ and equation (2.3b) becomes

$$K^T QHQ^T Q\Delta x = -P^T f \tag{2.11}$$

If we now denote QHQ^T by G, where

$$G = \begin{bmatrix} G_{11} & G_{12} \\ G_{21} & G_{22} \end{bmatrix}$$

then equation (2.11) may be written

$$G_{22}\Delta z_2 = -(P^T f + G_{21}\Delta z_1) \tag{2.12}$$

Thus if we compute Q, U and G we may easily compute Δz_1 and Δz_2 and hence Δx, since $\Delta x = Q^T \Delta z$.

Equations (2.8) and (2.12) can be simplified further.

If $$Qg = \begin{bmatrix} h_1 \\ h_2 \end{bmatrix}$$ then it can easily be shown that

$$U^T \Delta z_1 = - D^{-1}(U^{-1} h_1 + \nabla_c \psi) \tag{2.13}$$

and $$G_{22}\Delta z_2 = -(h_2 + G_{21}\Delta z_1) \tag{2.14}$$

Since the matrix D is diagonal for all common penalty and barrier functions and the matrix U is upper triangular, it is very easy to compute Δz_1 and hence Δz_2 if G_{22} is nonsingular.

The basic algorithm may thus be stated:

1) Choose an initial value of x

2) Compute f, J, and H. Compute Q and U from J using elementary orthogonal transformations

3a) Compute $G = QHQ^T$

3b) Compute $h = Qg$

4) Solve equation (2.13) for Δz_1

5) Solve equation (2.14) for Δz_2

6) Compute $\Delta x = Q^T \Delta z$.

7) Compute $x := x + \Delta x\theta$, where θ is chosen to reduce Φ by a line search, and repeat from step 2.

3. Estimating H and G

We notice from the above description that we have to compute the matrix H at every iteration of the algorithm. To do this explicitly would require an excessive amount of calculation but a possible alternative is to use an estimate B to the matrix H using gradient information gained during each iteration. If we were to update B using the BFGS formula [1]

$$B_1 = B - \frac{Bss^TB}{s^TBs} + \frac{yy^T}{s^Ty}$$

(the subscript 1 denotes new values)

where $y = f_1 - f$

$s = x_1 - x$

and $f = g + J^T \nabla_c \psi$

then B_1 would, along s, approximate the ill-conditioned Hessian $H + J^T DJ$ of the composite function. To approximate H, all that is necessary is to keep $\nabla_c \psi$ constant when computing y (thereby ensuring that D is null) so that

$$y = g_1 - g + (J_1^T - J)(\nabla_c \psi)_1$$

This is essentially the device described by Powell [6].

We note further that step (3a) is also time consuming and we look at ways of speeding it up. We show that we can use the orthogonally transformed vectors Qy and Qs to update the matrix $G = QHQ^T$ directly. Assume that J, and hence Q, is constant. Since the BFGS formula is

$$B_1 = B - \frac{Bss^T B}{s^T Bs} + \frac{yy^T}{s^T y}$$

multiplying by Q and Q^T gives

$$QB_1Q^T = QBQ^T - \frac{QBss^T BQ^T}{s^T Bs} + \frac{Qyy^T Q^T}{s^T y}$$

Let now u = Qy and v = Qs and denote QBQ^T by G. Then

$$G_1 = G - \frac{Gvv^T G}{v^T Gv} + \frac{uu^T}{v^T u}$$

and this update gives the matrix G_1 directly, eliminating step (3a). Since in general Q is not constant we use in practice the most recent value when computing u and v. This seems to work reasonably well.

4. Experimental Results

As stated in Section 1 (above), the algorithm may be extended to handle inequality constraints and we outline its performance on one such problem. This is the alkylation process (Problem 3) quoted by Dembo [2]. It involves 7 variables and 28 constraints, 14 of which are simple bounds on the variables.

The extended algorithm successfully solves this problem, selecting the correct set of active constraints at the solution and using an active set strategy to handle the inequality constraints.

The initial point chosen

$$x_0 = [1745.0,\ 110.0,\ 3048.0,\ 89.0,\ 93.0,\ 8.0,\ 145.5]^T$$

differs from that of Dembo, in which the fifth and final components were 92.0 and 145.0 respectively, in order that no constraints were violated, nor even active, at the initial solution.

The algorithm then yielded the correct solution

$$x = [1698.184,\ 53.665,\ 3031.298,\ 90.109,\ 95.000,\ 10.449,\ 153.535]$$

after 63 iterations. Iterations per r-value were

r	No of iterations
10^{-7}	55
10^{-13}	5
10^{-17}	2
10^{-21}	1

The high number of iterations for the first value of r is due to the fact that during this phase eight constraints became active, with two subsequently becoming inactive. The six remaining active constraints stayed unchanged throughout the three final phases of the algorithm. Their corresponding Lagrange multipliers at the solution were

[-13, -221, -10764, -1287, -1773, -175, -20182].

5. Conclusions

i) The experimental results show that the algorithm converges well and avoids problems due to ill-conditioning of the Hessian of Φ as $r \rightarrow 0$. Values of $r < 10^{-30}$ have been successfully used.

ii) It is possible to update G_{22}^{-1} as well as B so that the algorithm does not have to solve any systems of linear equations. Experiment shows that this substantially reduces the cpu time.

iii) The Lagrange multipliers are calculated at the solution and decisions as to which constraints to drop (in the inequality constraints case) are then based on the signs and magnitudes of these computed multipliers. This is perhaps the least satisfactory part of the algorithm as the calculation becomes sensitive to rounding errors as the constraints decrease in magnitude.

Note: The above is only an outline description of the algorithm; many significant details have been omitted in the interests of brevity.

Subsequent to the presentation of this algorithm at the IFIP Conference at Copenhagen in July 1983, the authors' attention was drawn to a paper by Gerencser [4]. This gives an alternative derivation of equations (2.13) and (2.14), but includes no discussion of the problems of evaluating H and G, nor any experimental results.

References

1. Broyden, C.G. (1970). The Convergence of a Class of Double-rank Minimization Algorithms, 2, The New Algorithm, J. Inst. Maths. Applics, Vol 6, pp. 222 - 231.

2. Dembo, R.S. (1976). A Set of Geometric Programming Test Problems and their Solutions, Mathematical Programming 10, pp. 192 - 213.

3. Fiacco, A.V. and McCormick, G.P. (1968). Nonlinear Programming: Sequential Unconstrained Minimization Techniques, John Wiley and Sons, New York and Toronto.

4. Gerencser, L. (1974). A Second-order Technique for the Solution of Nonlinear Optimization Problems, Colloquia Mathematica Societatis Janos Bolyai, 12, Progress in Operations Research, EGER (Hungary).

5. Gill, P.E., Murray, W. and Wright, M.H. (1981). Practical Optimization, Academic Press, London and New York.

6. Powell, M.J.D. (1978), "A Fast Algorithm for Nonlinearly Constrained Optimization Calculations", in Numerical Analysis, Dundee 1977 (Ed. G.A Watson), Lecture notes in Mathematics 630, Springer-Verlag, Berlin.

A CLASS OF CONTINUOUSLY DIFFERENTIABLE EXACT PENALTY FUNCTION ALGORITHMS FOR NONLINEAR PROGRAMMING PROBLEMS

G. Di Pillo*, L. Grippo**

* Dipartimento di Informatica e Sistemistica
Università di Roma "La Sapienza"
Via Eudossiana, 18
00184 Roma - Italy

** Istituto di Analisi dei Sistemi ed Informatica del C.N.R.
Viale Manzoni, 30
00185 Roma - Italy

ABSTRACT

In this paper we describe Newton-type and Quasi-Newton algorithms for the solution of NLP problems with inequality constraints, which are based on the use of a continuously differentiable exact penalty function.

1. INTRODUCTION

Continuously differentiable exact penalty functions for the solution of nonlinear programming problems were introduced for the first time in [1], with reference to the equality constrained case.

More recently, it was shown in [2] that a continuously differentiable exact penalty function can be constructed also for problems with inequality constraints. More specifically, it was proved that, under mild regularity assumptions, the augmented Lagrangian obtained by employing the multiplier function introduced by Glad and Polak in [3], turns out to be an exact penalty function.

In the present paper we define unconstrained minimization algorithms for the solution of the inequality constrained problem based on consistent approximations of Newton's direction for the exact penalty function. In particular we describe Newton-type algorithms, which require second order derivatives of the problem functions, and a structured Quasi-Newton algorithm, employing only first order derivatives. Finally some preliminary numerical results are reported.

2. PROBLEM FORMULATION

The problem under consideration is the nonlinear programming problem:

$$\text{minimize } f(x), \text{ subject to } g(x) \leq 0 \tag{1}$$

where the function $f: R^n \to R^1$ and $g: R^n \to R^m$ are three times continuously differentiable on R^n.

Given any $x \in R^n$ we define the index sets $I_o(x) \triangleq \{i: g_i(x) = 0\}$, $I_\pi(x) \triangleq \{i: g_i(x) \geq 0\}$, $I_\nu(x) \triangleq \{i: g_i(x) < 0\}$.

We assume that the following hypothesis is satisfied:

ASSUMPTION A. *For any* $x \in R^n$, *the gradients* $\nabla g_i(x)$, $i \in I_o(x)$ *are linearly independent.*

Moreover, we shall make use, when needed, of the following assumption, where X is a given subset of R^n:

ASSUMPTION B. *At any point* $x \in X$ *where* $\sum_{i\in I_\pi(x)} \nabla g_i(x) g_i(x) = 0$ *it results* $g_i(x) = 0$ *for all* $i \in I_\pi(x)$.

The algorithms proposed here for the solution of problem (1) are based on the unconstrained minimization of the exact penalty function:

$$U(x;\varepsilon) = f(x) + \lambda(x)'(g(x) + Y(x;\varepsilon)y(x;\varepsilon)) + \frac{1}{\varepsilon}\|g(x) + Y(x;\varepsilon)y(x;\varepsilon)\|^2, \quad \varepsilon > 0 \tag{2}$$

where $\lambda(x)$ is the multiplier function introduced in [3] and defined by:

$$\lambda(x) = -M^{-1}(x)\frac{\partial g(x)}{\partial x}\nabla f(x)$$

with:

$$M(x) \triangleq \frac{\partial g(x)}{\partial x}\frac{\partial g(x)'}{\partial x} + \gamma^2 G^2(x), \qquad G(x) \triangleq \mathrm{diag}(g_i(x)), \qquad \gamma > 0;$$

the vector function $y(x;\varepsilon)$ is given, componentwise, by

$$y_i(x;\varepsilon) \triangleq \left[-\min\left(0, \frac{2g_i(x) + \varepsilon\lambda_i(x)}{2}\right)\right]^{\frac{1}{2}}, \qquad i = 1,\dots,m;$$

and $Y(x;\varepsilon) = \mathrm{diag}[y_i(x;\varepsilon)]$.

Let us denote by $L(x,\lambda) \triangleq f(x) + \lambda' g(x)$ the Lagrangian function for problem (1) and define:

$$\nabla_x L(x,\lambda(x)) \triangleq [\nabla_x L(x,\lambda)]_{\lambda=\lambda(x)}, \qquad \nabla_x^2 L(x,\lambda(x)) \triangleq [\nabla_x^2 L(x,\lambda)]_{\lambda=\lambda(x)}.$$

It can be easily verified that function (2) is continuously differentiable with respect to x and that its gradient is given by:

$$\begin{aligned}\nabla U(x;\varepsilon) = {} & \nabla f(x) + \frac{\partial g(x)'}{\partial x}\lambda(x) + \frac{\partial \lambda(x)'}{\partial x}(g(x) + Y(x;\varepsilon)y(x;\varepsilon)) \\ & + \frac{2}{\varepsilon}\frac{\partial g(x)'}{\partial x}(g(x) + Y(x;\varepsilon)y(x;\varepsilon))\end{aligned} \tag{3}$$

where:

$$\frac{\partial \lambda(x)}{\partial x} = -M^{-1}(x)\left[\frac{\partial g(x)}{\partial x}\nabla_x^2 L(x,\lambda(x)) + \sum_{j=1}^m e_j \nabla_x L(x,\lambda(x))'\nabla^2 g_j(x) + 2\gamma^2 \Lambda(x) G(x)\frac{\partial g(x)}{\partial x}\right], \tag{4}$$

being e_j the j-th column of the $m \times m$ identity matrix, and $\Lambda(x) \triangleq \mathrm{diag}(\lambda_i(x))$.

We will refer to any pair $(\bar{x},\bar{\lambda}) \in R^n \times R^m$ satisfying the Kuhn-Tucker necessary conditions: $g(\bar{x}) \leq 0$; $\bar{\lambda}'g(\bar{x}) = 0$; $\bar{\lambda} \geq 0$; $\nabla_x L(\bar{x},\bar{\lambda}) = 0$ as a K - T pair for problem (1).

We introduce the index sets:

$$I_+(x) \triangleq \{i: 2g_i(x) + \varepsilon\lambda_i(x) > 0\}, \qquad I_-(x) \triangleq \{i: 2g_i(x) + \varepsilon\lambda_i(x) \leq 0\};$$

accordinlgy, we define the subvectors:

$$g_+(x) \triangleq [g_i(x)]_{i\in I_+(x)} \qquad g_-(x) \triangleq [g_i(x)]_{i\in I_-(x)},$$

$$\lambda_+(x) \triangleq [\lambda_i(x)]_{i\in I_+(x)} \qquad \lambda_-(x) \triangleq [\lambda_i(x)]_{i\in I_-(x)}$$

whose components appear in the same order as in g and λ, and we refer to the vectors g and λ reordered in the form:

$$g(x) = [g_+(x)'g_-(x)']' \qquad \lambda(x) = [\lambda_+(x)'\lambda_-(x)']'.$$

A similar convention is adopted for the reordering of the vectors g and λ, induced by the index sets $I_o(x)$ and $I_\nu(x)$.

3. PROPERTIES OF THE EXACT PENALTY FUNCTION U

Some of the main results concerning the relationships between local solutions of problem (1) and local unconstrained minima of U are summarized in this section.

PROPOSITION 1. *Let* $(\bar{x},\bar{\lambda})$ *be a* K - T *pair for problem* (1). *Then, for any* $\varepsilon > 0$, $\bar{x}$ *is a stationary point of* U, $\lambda(\bar{x}) = \bar{\lambda}$ *and* $U(\bar{x};\varepsilon) = f(\bar{x})$.

PROPOSITION 2. *Let* X *be a compact subset of* R^n *and suppose that Assumption* B *holds on* X. *Then, there exists an* $\bar{\varepsilon} > 0$ *such that for all* $\varepsilon \in (0,\bar{\varepsilon}]$, *if* $\bar{x} \in X$ *is a stationary point of* $U(x;\varepsilon)$, *the pair* $(\bar{x},\lambda(\bar{x}))$ *satisfies also the* K - T *conditions for problem* (1).

PROPOSITION 3. *Let* $(\bar{x},\bar{\lambda})$ *be a* K - T *pair for problem* (1) *and assume that strict complementary holds at* $(\bar{x},\bar{\lambda})$.

Then, for any $\varepsilon > 0$, *the function* $U(x;\varepsilon)$ *is twice continuously differentiable in a neighbourhood of* $\bar{x}$, *and the Hessian matrix of* $U(x;\varepsilon)$ *evaluated at* $\bar{x}$ *is given by:*

$$\nabla^2 U(\bar{x};\varepsilon) = \nabla_x^2 L(\bar{x},\lambda(\bar{x})) + \frac{\partial\lambda_o(\bar{x})'}{\partial x}\frac{\partial g_o(\bar{x})}{\partial x} + \frac{\partial g_o(\bar{x})'}{\partial x}\frac{\partial\lambda_o(\bar{x})}{\partial x} + \frac{2}{\varepsilon}\frac{\partial g_o(\bar{x})'}{\partial x}\frac{\partial g_o(\bar{x})}{\partial x} - \frac{\varepsilon}{2}\frac{\partial\lambda_\nu(\bar{x})'}{\partial x}\frac{\partial\lambda_\nu(\bar{x})}{\partial x}.$$

PROPOSITION 4. *Let* $(\bar{x},\bar{\lambda})$ *be a* K-T *pair for problem* (1) *and assume that:* (i) *strict complementarity holds at* $(\bar{x},\bar{\lambda})$; (ii) $\bar{x}$ *is an isolated local minimum point for problem*

(1) *satisfying the second order sufficiency condition:*

$$x'\nabla_x^2 L(\bar{x},\bar{\lambda})x > 0 \quad \textit{for all} \quad x: \frac{\partial g_o(\bar{x})}{\partial x} x = 0, \quad x \neq 0 .$$

Then, there exists an $\bar{\varepsilon} > 0$ *such that for all* $\varepsilon \in (0,\bar{\varepsilon}]$, $\bar{x}$ *is an isolated local minimum point for* $U(x;\varepsilon)$, *and the Hessian matrix* $\nabla^2 U(\bar{x};\varepsilon)$ *is positive definite.*

PROPOSITION 5. *Let* X *be a compact subset of* R^n*; suppose that Assumption* B *holds on* X *and that strict complementary holds at any* K - T *pair* $(\bar{x},\bar{\lambda})$ *with* $\bar{x} \in X$.

Then, there exists an $\bar{\varepsilon} > 0$ *such that, for all* $\varepsilon \in (0,\bar{\varepsilon}]$, *if* $\bar{x} \in X$ *is a local unconstrained minimum point of* $U(x;\varepsilon)$ *with positive definite Hessian* $\nabla^2 U(\bar{x};\varepsilon)$, $\bar{x}$ *is an isolated local minimum point of problem* (1), *satisfying the second order sufficiency conditions.*

The proofs of the preceding propositions, as well as additional results can be found in [2].

4. NEWTON-TYPE ALGORITHMS

The minimization of $U(x;\varepsilon)$ by Newton's method would require the evaluation of third order derivatives of the problem functions. This can be avoided by employing consistent approximations of the Newton's direction which require only the evaluation of second order derivatives.

A first possibility, which was already considered in [3], is that of replacing $\nabla^2 U(x;\varepsilon)$ with the approximating matrix $H(x;\varepsilon)$ defined by:

$$H(x;\varepsilon) \triangleq \nabla_x^2 L(x,\lambda(x)) + \frac{\partial \lambda_+(x)'}{\partial x}\frac{\partial g_+(x)}{\partial x} + \frac{\partial g_+(x)'}{\partial x}\frac{\partial \lambda_+(x)}{\partial x} + \frac{2}{\varepsilon}\frac{\partial g_+(x)'}{\partial x}\frac{\partial g_+(x)}{\partial x} - \frac{\varepsilon}{2}\frac{\partial \lambda_-(x)'}{\partial x}\frac{\partial \lambda_-(x)}{\partial x}. \tag{5}$$

The properties of $H(x;\varepsilon)$ are established in the following proposition.

PROPOSITION 6. *Let* $(\bar{x},\bar{\lambda})$ *be a* K - T *pair for problem* (1). *Then, under the assumptions of proposition* 4, *it results* $H(\bar{x};\varepsilon) = \nabla^2 U(\bar{x};\varepsilon)$. *Moreover, there exists an* $\bar{\varepsilon} > 0$ *such that for all* $\varepsilon \in (0,\bar{\varepsilon}]$, *the matrix* $H(x;\varepsilon)$ *is positive definite in a neighbourhood of* $\bar{x}$.

PROOF. By Proposition 1, we have $\lambda(\bar{x}) = \bar{\lambda}$; then, by the strict complementary assumption, it results $I_+(\bar{x}) = I_o(\bar{x})$ and $I_-(\bar{x}) = I_\nu(\bar{x})$, so that, recalling Proposition 3, we have $H(\bar{x};\varepsilon) = \nabla^2 U(\bar{x};\varepsilon)$. Then, the positive definiteness of $H(x;\varepsilon)$ in a neighbourhood of $\bar{x}$ for sufficiently small values of ε, follows from the positive definiteness of $\nabla^2 U(\bar{x};\varepsilon)$ stated in Proposition 4. ◀

By employing the approximation H of $\nabla^2 U$ we can define the following Newton-type algorithm:

ALGORITHM 1.

$\hat{x} = x + \alpha d$

$H(x;\varepsilon)d = -\nabla U(x;\varepsilon)$

where x, $\hat{x}$ are respectively the present and the next iterate, d is the search direction and α is the stepsize.

On the basis of Proposition 6, direction d is a consistent approximation of Newton's direction for $U(x;\varepsilon)$.

We note that in Algorithm 1 second order derivatives of the problem functions appear in both members of the system which yields the search direction.

This prevents the possibility of deriving from Algorithm 1 a Quasi-Newton algorithm employing only first order derivatives.

It can be shown, however, that a consistent approximation of the Newton's direction can be obtained by solving a system which does not contain second order derivatives in the r.h.m.

In fact, consider the following system:

$$\begin{bmatrix} \nabla_x^2 L(x,\lambda(x)) & \frac{\partial g_+(x)'}{\partial x} & \frac{\partial g_-(x)'}{\partial x} \\ \frac{\partial g_+(x)}{\partial x} & 0 & 0 \\ 0 & 0 & I \end{bmatrix} \begin{bmatrix} d_x \\ d_+ \\ d_- \end{bmatrix} = - \begin{bmatrix} \nabla_x L(x,\lambda(x)) \\ g_+(x) \\ \lambda_-(x) \end{bmatrix} \tag{6}$$

and suppose that it admits a solution $d \triangleq [d_x' \; d_+' \; d_-']'$.

We shall prove that the component d_x of d is a consistent approximation of Newton's direction for U.

To this aim we preliminarly state the following lemma.

LEMMA 1. *Let* $d \triangleq (d_x' \; d_+' \; d_-')'$ *be a solution of system* (6); *then it results:*

$$P(x)\frac{\partial \lambda(x)}{\partial x}d_x = P(x)\begin{bmatrix} d_+ \\ d_- \end{bmatrix} + (Q(x) - R(x))d_x \tag{7}$$

where:

$$P(x) \triangleq \frac{\partial g(x)}{\partial x}\frac{\partial g(x)'}{\partial x} + \gamma^2 \begin{bmatrix} 0 & 0 \\ 0 & G_-^2(x) \end{bmatrix},$$

$$Q(x) \triangleq \gamma^2 \begin{bmatrix} G_+(x)\Lambda_+(x)\frac{\partial g_+(x)}{\partial x} \\ 0 \end{bmatrix},$$

$$R(x) \triangleq \sum_{j=1}^{m} e_j \nabla_x L(x,\lambda(x))'\nabla^2 g_j(x) + 2\gamma^2\Lambda(x)G(x)\frac{\partial g(x)}{\partial x} + \gamma\begin{bmatrix} G_+^2(x) & 0 \\ 0 & 0 \end{bmatrix}\frac{\partial \lambda(x)}{\partial x},$$

with:

$G_+(x) \triangleq \mathrm{diag}(g_i(x))_{i \in I_+(x)}$, $\quad G_-(x) \triangleq \mathrm{diag}(g_i(x))_{i \in I_-(x)}$,

$\Lambda_+(x) \triangleq \mathrm{diag}(\lambda_i(x))_{i \in I_+(x)}$.

PROOF. The first equation of system (6), premultiplied by $\frac{\partial g}{\partial x}$, gives:

$$\frac{\partial g(x)}{\partial x}\nabla_x^2 L(x,\lambda(x))d_x + \frac{\partial g(x)}{\partial x}\frac{\partial g(x)'}{\partial x}\begin{bmatrix} d_+ \\ d_- \end{bmatrix} + \frac{\partial g(x)}{\partial x}\nabla_x L(x,\lambda(x)) = 0$$

Then, adding and subtracting the term $\gamma^2 G^2\lambda$ and recalling the definition of $\lambda(x)$, we have:

$$\frac{\partial g(x)}{\partial x}\nabla_x^2 L(x,\lambda(x))d_x + \frac{\partial g(x)}{\partial x}\frac{\partial g(x)'}{\partial x}\begin{bmatrix} d_+ \\ d_- \end{bmatrix} - \gamma^2\begin{bmatrix} G_+(x)\Lambda_+(x)g_+(x) \\ G_-^2(x)\lambda_-(x) \end{bmatrix}$$

from which, taking into account the second and the third equation of system (6) we obtain, after simple calculations:

$$\frac{\partial g(x)}{\partial x}\nabla_x^2 L(x,\lambda(x))d_x = -P(x)\begin{bmatrix} d_+ \\ d_- \end{bmatrix} - Q(x)d_x. \tag{8}$$

On the other hand, recalling (4), it can be written:

$$\frac{\partial g(x)}{\partial x}\nabla_x^2 L(x,\lambda(x)) = -M(x)\frac{\partial\lambda(x)}{\partial x} - \sum_{j=1}^{m} e_j\nabla_x L(x,\lambda(x))'\nabla^2 g_j(x) - 2\gamma^2\Lambda(x)G(x)\frac{\partial g(x)}{\partial x}. \tag{9}$$

Then, by substituting the r.h.m. of (9) into (8), and by writing

$$M(x) = P(x) + \gamma^2\begin{bmatrix} G_+^2(x) & 0 \\ 0 & 0 \end{bmatrix}$$

we obtain (7). ◀

Then we can state the following proposition.

PROPOSITION 7. *Let* $(\bar{x},\bar{\lambda})$ *be a* K - T *pair for problem* (1) *and assume that strict complementarity holds at* $(\bar{x},\bar{\lambda})$. *Then, there exists a neighbourhood* Ω *of* $\bar{x}$ *such that, for all* $x \in \Omega$, *if* $d \triangleq (d_x' \; d_+' \; d_-')'$ *is a solution of system* (6), *it results:*

$$\tilde{H}(x;\varepsilon)d_x = -\nabla U(x;\varepsilon),$$

where $\tilde{H}(x;\varepsilon)$ *is a continuous matrix defined on* Ω *and satisfying* $\tilde{H}(\bar{x};\varepsilon) = \nabla^2 U(\bar{x};\varepsilon)$.

PROOF. Let $(\bar{x},\bar{\lambda})$ be a K - T pair for problem (1). Then, by the strict complementary assumption, there exists a neighbourhood Ω_1 of $\bar{x}$, where $I_+(x) = I_0(\bar{x})$ and $I_-(x) = I_\nu(\bar{x})$.

Recalling the definition of the matrix $P(x)$ introduced in (7), we have that $P(\bar{x}) = M(\bar{x})$ and that, by continuity, $P(x)$ is nonsingular in some neighbourhood $\Omega \subseteq \Omega_1$.

Therefore, by the preceding Lemma, we have, for all $x \in \Omega$:

$$\frac{\partial\lambda(x)}{\partial x}d_x - \begin{bmatrix} d_+ \\ d_- \end{bmatrix} = P^{-1}(x)[Q(x) - R(x)]d_x. \tag{10}$$

Moreover, it can be easily verified that on the neighbourhood Ω, the gradient

$\nabla U(x;\varepsilon)$ reduces to:

$$\nabla U(x;\varepsilon) = \nabla f(x) + \frac{\partial g_+(x)'}{\partial x}\lambda_+(x) + \frac{\partial \lambda_+(x)'}{\partial x} g_+(x) - \frac{\varepsilon}{2}\frac{\partial \lambda_-(x)'}{\partial x}\lambda_-(x) + \frac{2}{\varepsilon}\frac{\partial g_+(x)'}{\partial x} g_+(x). \quad (11)$$

Let now $H(x;\varepsilon)$ be the matrix defined in (5); then, taking (11) into account and recalling that $(d_x' \; d_+' \; d_-')'$ solves system (6), we have:

$$H(x;\varepsilon)d_x + \nabla U(x;\varepsilon) = \begin{bmatrix} \frac{\partial g_+(x)'}{\partial x} & -\frac{\varepsilon}{2}\frac{\partial \lambda_-(x)'}{\partial x} \end{bmatrix} \left\{ \frac{\partial \lambda(x)}{\partial x} d_x - \begin{bmatrix} d_+ \\ d_- \end{bmatrix} \right\}. \quad (12)$$

Therefore, by (10) and (12), letting:

$$\tilde{H}(x;\varepsilon) \triangleq H(x;\varepsilon) - \begin{bmatrix} \frac{\partial g_+(x)'}{\partial x} & -\frac{\varepsilon}{2}\frac{\partial \lambda_-(x)'}{\partial x} \end{bmatrix} P^{-1}(x)[Q(x) - R(x)]$$

it results:

$$\tilde{H}(x;\varepsilon)d_x = -\nabla U(x;\varepsilon) \quad \text{for all} \quad x \in \Omega.$$

Finally, since $Q(\bar{x}) = 0$ and $R(\bar{x}) = 0$ it follows from Proposition 6 that $\tilde{H}(\bar{x};\varepsilon) = \nabla^2 U(\bar{x};\varepsilon)$. ◀

Then we can define the following algorithm:

ALGORITHM 2.

$$\hat{x} = x + \alpha d_x$$

$$\begin{bmatrix} \nabla_x^2 L(x,\lambda(x)) & \frac{\partial g_+(x)'}{\partial x} \\ \frac{\partial g_+(x)}{\partial x} & 0 \end{bmatrix} \begin{bmatrix} d_x \\ d_+ \end{bmatrix} = - \begin{bmatrix} \nabla_x f(x) + \frac{\partial g_+(x)'}{\partial x}\lambda_+(x) \\ g_+(x) \end{bmatrix}$$

On the basis of the preceding results both Algorithm 1 and 2 employ a consistent approximation of Newton's direction; thus superlinear convergence rate can be established provided that the stepsize α goes to 1.

On the other hand, by taking into account Propositions 6 and 7 it can be shown ([4], p. 36, Prop. 1.15) that, assuming convergence, a unit stepsize eventually satisfies some standard stopping rule for the line search, such as Armijo's rule.

5. A QUASI-NEWTON ALGORITHM

A Quasi-Newton algorithm for the minimization of U can be derived from Algorithm 2 by approximating only the matrix $\nabla_x^2 L(x,\lambda(x))$.

More specifically we define the following algorithm.

ALGORITHM 3.

$$\hat{x} = x + \alpha d_x$$

$$\begin{bmatrix} D & \frac{\partial g_+(x)'}{\partial x} \\ \frac{\partial g_+(x)}{\partial x} & 0 \end{bmatrix} \begin{bmatrix} d_x \\ d_+ \end{bmatrix} = - \begin{bmatrix} \nabla f(x) + \frac{\partial g_+(x)'}{\partial x}\lambda_+(x) \\ g_+(x) \end{bmatrix} \tag{13}$$

where the matrix D is defined by means of an updating process of the form $\hat{D} = D + \Delta D$, such that

$$\hat{D}(\hat{x} - x) = \nabla_x L(\hat{x},\lambda(\hat{x})) - \nabla_x L(x,\lambda(\hat{x})).$$

A suitable updating formula for D could be one which ensures that D remains positive definite (see, for instance [5]).

In this case it can be shown, under suitable compactness assumptions, that the search direction d_x satisfies an angle condition with respect to the line search function U.

PROPOSITION 8. *Let $\mathcal{D}$ be a compact set of symmetric positive definite* $n \times n$ *matrices and let* X *be a compact subset of* R^n. *Suppose that the matrix* P(x) *appearing in* (7) *is non singular on* X.

Then, there exist numbers $\rho > 0$ *and* $\bar{\varepsilon} > 0$ *such that, for all* $\varepsilon \in (0,\bar{\varepsilon}]$, *if system* (13) *admits a solution for* $(D,x) \in \mathcal{D} \times X$, *the component* d_x *satisfies:*

$$d_x'\nabla U(x;\varepsilon) \leq - \rho\|d_x\|\|\nabla U(x;\varepsilon)\|.$$

PROOF. We proceed by contradiction. If the proposition is false, then, for each integer $k > 0$ there exist numbers $\varepsilon_k \leq \frac{1}{k}$, $\rho_k \leq \varepsilon_k^2$ and a point $(D_k,x_k) \in \mathcal{D} \times X$ such that system (13) has a solution $(d_{xk}' \; d_{+k}')'$, with $d_{xk} \neq 0$, and:

$$d_{xk}'\nabla U(x_k,\varepsilon_k) + \rho_k\|d_{xk}\|\|\nabla U(x_k;\varepsilon_k)\| > 0. \tag{14}$$

Now, since the number of different index sets in the sequence $\{I_+(x_k),I_-(x_k)\}$ is finite, we can extract a subsequence, which we relabel again $\{(D_k,x_k)\}$ such that $\{I_+(x_k),I_-(x_k)\}$ remain unchanged for all k.

Then, recalling (3), inequality (14) can be put into the form:

$$d_{xk}'[\nabla f(x_k) + \frac{\partial g_+(x_k)'}{\partial x}\lambda_+(x_k) + \frac{\partial \lambda_+(x_k)'}{\partial x}g_+(x_k) - \frac{\varepsilon_k}{2}\frac{\partial \lambda_-(x_k)'}{\partial x}\lambda_-(x_k) + \frac{2}{\varepsilon_k}\frac{\partial g_+(x_k)'}{\partial x}g_+(x_k)]$$

$$+ \rho_k\|d_{xk}\|\|\nabla U(x_k;\varepsilon_k)\| > 0,$$

whence, by (13):

$$-d'_{xk}[D_k d_{xk} + \frac{\partial g_+(x_k)'}{\partial x} d_{+k} + \frac{\partial \lambda_+(x_k)'}{\partial x}\frac{\partial g_+(x_k)}{\partial x} d_{xk} + \frac{\varepsilon_k}{2}\frac{\partial \lambda_-(x_k)'}{\partial x}\lambda_-(x_k)$$

$$+ \frac{2}{\varepsilon_k}\frac{\partial g_+(x_k)'}{\partial x}\frac{\partial g_+(x_k)}{\partial x} d_{xk}] + \rho_k \|d_{xk}\| \|\nabla U(x_k;\varepsilon_k)\| > 0. \tag{15}$$

Now, proceding in a way similar to that followed in the proof of Lemma 1 for establishing (8), it can be verified that the solution of system (13) satisfies:

$$\begin{bmatrix} d_{+k} \\ -\lambda_-(x_k) \end{bmatrix} = -\bar{P}^1(x_k)\frac{\partial g(x_k)}{\partial x} D_k d_{xk} - P^{-1}(x_k)Q(x_k)d_{xk} \tag{16}$$

where P and Q are the matrices appearing in (7) and P is nonsingular by assumption.

Then, by substituting (16) into (15) and by letting $\delta_k = \frac{d_{xk}}{\|d_{xk}\|}$ we can write:

$$-\delta'_k D_k \delta_k - \delta'_k \frac{\partial \lambda_+(x_k)'}{\partial x}\frac{\partial g_+(x_k)}{\partial x}\delta_k - \frac{2}{\varepsilon_k}\delta'_k \frac{\partial g_+(x_k)'}{\partial x}\frac{\partial g_+(x_k)}{\partial x}\delta_k$$

$$+\delta'_k \left[\frac{\partial g_+(x_k)'}{\partial x} - \frac{\varepsilon_k}{2}\frac{\partial \lambda_-(x_k)}{\partial x}\right]\left[\bar{P}^1(x_k)\frac{\partial g(x_k)}{\partial x} D_k + P^{-1}(x_k)Q(x_k)\right]\delta_k$$

$$+ \rho_k \| D_k\delta_k + \frac{\partial \lambda_+(x_k)'}{\partial x}\frac{\partial g_+(x_k)}{\partial x}\delta_k - \frac{2}{\varepsilon_k}\frac{\partial g_+(x_k)'}{\partial x}\frac{\partial g_+(x_k)}{\partial x}\delta_k$$

$$+ \left[\frac{\partial g_+(x_k)'}{\partial x} - \frac{\varepsilon_k}{2}\frac{\partial \lambda_-(x_k)'}{\partial x}\right]\left[\bar{P}^1(x_k)\frac{\partial g(x_k)}{\partial x} D_k + P^{-1}(x_k)Q(x_k)\right]\delta_k \| > 0. \tag{17}$$

Now, by compactness of $\mathcal{D} \times X$ and by the fact that $\|\delta_k\| = 1$ we can extract a subsequence D_k, x_k, δ_k converging to a point $(\hat{D},\hat{x},\hat{\delta})$, with $\hat{D} \in \mathcal{D}$, $\hat{x} \in X$, $\|\hat{\delta}\| = 1$.

Letting $\varepsilon_k \to 0$ and recalling that $\rho_k \leq \varepsilon_k^2$ we have that the last term in l.h.s. of (17) goes to zero; then it must result $\frac{\partial g_+(\hat{x})}{\partial x}\hat{\delta} = 0$ which implies, in turn $-\hat{\delta}'\hat{D}\hat{\delta} \geq 0$. Thus, since $\|\hat{\delta}\| = 1$, we get a contradiction with the positive definiteness of $\hat{D}$. ◀

6. NUMERICAL RESULTS

The performance of the algorithms described before has been evaluated by solving two standard test problems (TP) for different values of the penalty parameter ε and for $\gamma=1$.

TEST PROBLEM 1. (Rosen and Suzuki)

Minimize $f(x) = -5(x_1 + x_2) + 7(x_4 - 3x_3) + x_1^2 + x_2^2 + 2x_3^2 + x_4^2$ subject to:

$$(\sum_{i=1}^{4} x_i^2) + x_1 - x_2 + x_3 - x_4 - 8 \leq 0$$

$$x_1^2 + 2x_2^2 + x_3^2 + 2x_4^2 - x_1 - x_4 - 10 \leq 0$$

$$2x_1^2 + x_2^2 + x_3^2 + 2x_1 - x_2 - x_4 - 5 \leq 0.$$

Solution: $\bar{x} = (0,1,2,-1)'$ with $f(\bar{x}) = -44$.

Starting point: $x^\circ = 0$.

TEST PROBLEM 2. (Wong)

Minimize $f(x) = (x_1 - 10)^2 + 5(x_2 - 12)^2 + x_3^4 + 3(x_4 - 11)^2 + 10x_5^6 + 7x_6^2 + x_7^4 - 4x_6x_7 - 10x_6 - 8x_7$

subject to:

$$2x_1^2 + 3x_2^4 + x_3 + 4x_4^2 + 5x_5 - 127 \leq 0$$

$$7x_1 + 3x_2 + 10x_3^2 + x_4 - x_5 - 282 \leq 0$$

$$23x_1 + x_2^2 + 6x_6^2 - 8x_7 - 196 \leq 0$$

$$4x_1^2 + x_2^2 - 3x_1x_2 + 2x_3^2 + 5x_6 - 11x_7 \leq 0.$$

Solution: $\bar{x} = (2.33050, 1.95137, -0.47754, 4.36573, -0.62448, 1.03813, 1.594)'$ with $f(\bar{x}) = 680.630$.

Starting point: $x^\circ = (1,2,0,4,0,1,1)'$.

The numerical experiments were performed by employing Algorithms 1, 2 and 3 with the same line search procedure. Algorithm 3 was implemented by using the BFGS updating formula, modified as proposed in [5].

The results obtained are reported in Table 1. For each case we report the number LS of line searches and the number NU of function evaluations needed to attain the solution with an accuracy on the objective function of the order 10^{-6}. On the basis of a limited computational experience the algorithms considered here appear to be competitive with the most effective techniques presently available, at least for small dimensional problems with highly nonlinear constraints. In fact, the main difficulty in the minimization of U lies in the matrix inversion required at each function evaluation. For problems with a large number of constraints the exact augmented

Table 1

	ε	ALG 1		ALG 2		ALG 3	
		LS	NU	LS	NU	LS	NU
TP1	1.	8	22	6	9	15	32
	0.1	10	26	9	24	16	32
	0.01	19	47	16	42	40	79
TP2	1.	11	25	11	31	47	84
	0.1	12	25	12	35	52	100
	0.01	19	59	18	32	84	165

Lagrangian approach proposed in [6] could be more advantageous.

As regards the selection of the penalty coefficient we observe that in the test problems worked out all algorithms were successful for relatively large values of ε. In any case, it is possible to employ the procedure for the automatic selection of the penalty coefficient described in [3].

Finally, we remark that the numerical ill conditioning which may arise in the matrix inversion required for the computation of the multiplier function, can be avoided by a proper selection of the parameter γ.

REFERENCES

[1] R. FLETCHER, *A Class of Methods for Nonlinear Programming with Termination and Convergence Properties*, in: Integer and Nonlinear Programming, J. Abadie ed., North Holland, 1970.

[2] G. DI PILLO and L. GRIPPO, *A Continuously Differentiable Exact Penalty Function for NLP Problems with Inequality Constraints*, Techn. Rep. IASI-CNR, n. 48, Dec. 1982.

[3] T. GLAD and E. POLAK, *A Multiplier Method with Automatic Limitation of the Penalty Growth*, Math. Programming, vol. 17, pp. 140 - 155, 1979.

[4] D.P. BERTSEKAS, *Constrained Optimization and Lagrange Multiplier Methods*, Academic Press, 1982.

[5] M.J.D. POWELL, *The Convergence of Variable Metric Methods for Nonlinearly Constrained Optimization Calculations*, in: Nonlinear Programming 3, O.L. Mangasarian, R.R. Meyer, S.M. Robinson eds., Academic Press, 1978.

[6] G. DI PILLO and L. GRIPPO, *A New Augmented Lagrangian Function for Inequality Constraints in Nonlinear Programming*, J. of Optimization Theory and Applications, vol. 36, pp. 495 - 519, 1982.

ON THE EFFECTIVENESS OF THE BAYESIAN NONPARAMETRIC APPROACH TO GLOBAL OPTIMIZATION

B. Betro'
CNR-IAMI
via Cicognara 7
I-20129 Milano, Italy

1. INTRODUCTION

Algorithms based on random sampling are now currently accepted as effective tools for solving optimization problems in which the objective function cannot be assumed to be unimodal (for a recent survey, see Archetti and Schoen (1983)). The basic scheme for such algorithms consists of the following steps

(i) a certain number of points, say n, are uniformly drawn in the search domain;
(ii) a number of "promising points" are selected from the n and a local search is started from each of them, leading to a set of local optima which hopefully includes the global ones;
(iii) the best obtained value of the objective function is tested to be a satisfactory approximation to the global optimum.

Step (ii) can be effectively performed by means of techiques aimed at identifying clusters of points which presumably coincide with the so called regions of attraction of the optima (Boender et al., 1982). The crucial point in the design of global optimization algorithms is (iii) as, due to the lack of manageable analytical criteria for the global optima, an "exacxt" test cannot be built up.

In order to provide a statistical framework for decision about the achievement of a satisfactory approximation, the Bayesian nonparametric approach has been developed in Betro' (1981), Betro' (1983), Betro' and Rotondi (1983), Betro' and Vercellis (1983). The underlying idea is to model the distribution of the sampled values of the objective function by a suitable family of random distribution functions, and to infer about this distribution, according to a Bayesian scheme, conditioning upon the sample. Then the accuracy of a global optimum estimate can be defined in terms of a quantile of suitable order of the unknown distribution and decision about the achievement of a prescribed accuracy can be handled in the framework of decision theory.

In this paper attention is focussed on two aspects closely influencing the overall effectiveness of the approach. First, as the Bayesian scheme requires the specification in some form of prior information about the optimization problem to be solved, the problem is considered of requiring prior information in a form which is more naturally available to the "optimizer" and hence after all better tailored to the problem itself. Moreover, considering the fact that scheme (i)-(iii) is in practical implementations repeated until the test in (iii) yields a positive answer, the test itself is given a sequential formulation wich takes explicitly into account the computational cost of function evaluations.

Some numerical experiences are finally exhibited illustrating the results obtained on standard test functions.

2. THE BAYESIAN NONPARAMETRIC APPROACH

Let f be a continuous function on a compact set $K\subset R^{\ell}$ and f^* be its maximum over K. The global optimization problem can be formulated as the problem of finding a point $\hat{x}_\varepsilon$ such that $\hat{f}_\varepsilon=f(\hat{x}_\varepsilon)$ is an approximation to f^* in some sense, within a prefixed accuracy $\varepsilon>0$.

In Betro' (1981) it was observed that, considering the function F(t)

$$F(t)=\text{meas}(\{x\in K:\ f(x)\leq t\})/\text{meas}(K) \tag{1}$$

then an approximation to f^* with accuracy $\varepsilon>0$ can be defined in terms of F as the value t_ε such that

$$t_\varepsilon=\min\{t:\ F(t)\geq 1-\varepsilon\}. \tag{2}$$

For small ε, the set of points in K corresponding to values of f exceeding t_ε is, by (1), of negligible measure, so that we can define as approximation to f^* within the accuracy ε any function value $\hat{f}$ such that

$$\hat{f}\geq t_\varepsilon \text{ or, equivalently, } F(\hat{f})\geq 1-\varepsilon. \tag{3}$$

There is no practical possibility of evaluating analytically F and hence t_ε, but inference about t_ε can be provided from step (i), observing that F is the distribution function of the random variable $Y=f(X)$, where X is a uniform random variable in K. By definition (2) itself, t_ε is then the quantile of order $1-\varepsilon$ of the distribution function F.

In the Bayesian nonparametric approach to inference (Ferguson, 1973; Doksum, 1974) a model for an unknown distribution function is introduced by means of a stochastic process whose trajectories are distribution functions. A relevant example is given by the process

$$F(t)=1-\exp(-Y(t)) \tag{4}$$

when Y(t) is a nondecreasing right continuous independent increments process such that Y(t) is Gamma distributed with moment generating function

$$M_{Y(t)}(v)=E(\exp(-vY(t)))=\left(\frac{\lambda}{\lambda+v}\right)^{\gamma(t)},\quad v\geq 0; \tag{5}$$

$\gamma(t)$ is a right continuous nondecreasing function such that $\gamma(-\infty)=0$ and $\gamma(+\infty)=+\infty$. Y(t) is called Gamma process (Doksum, 1974). When $y_1,\ldots,y_n$ is a sample from F given by (4), then the posterior moment generating function

$$M_{Y(t)}(v|y_1,\ldots,y_n)$$

has a manageable analytical expression (Betro', 1981), which enables to compute probabilities of the type

$$Pr\{F(t)\geq q|y_1,\ldots,y_n\}$$

by standard inversion formulas.

In order to build up a test for hypothesis (3), it must be observed that, according to step (ii), the estimate $\hat{f}$ which is actually tested cannot be seen as a constant or a function of the sampled values $y_1,\ldots,y_n$, being the best observed function value obtained after the local searches. A model for such an $\hat{f}$ in the form of a random variable $\hat{F}$ has been introduced in Betro' (1983), after which it is possible to obtain an expression for

$$M_{Y(t)}(v|y_1,\ldots,y_n,\hat{F}=\hat{f})$$

(Betro' and Rotondi, 1983) and hence compute probabilities of the type

$$Pr\{F(\hat{f})\geq q|y_1,\ldots,y_n,\hat{F}=\hat{f}\} \quad (6)$$

which are the basis for any test on hypothesis (3).

3. ASSESSMENT OF PRIOR PROBABILITIES

The implementation of the Bayesian nonparametric approach requires, once F(t) is given through (4) by a Gamma process, the specification of the parameters λ and $\gamma(t)$ in (5), which should reflect prior information about the problem to be solved.

In previous implementations, this was done specifying the form of $\beta_0(t)=E(F(t))$ and adjusting two unspecified parameters according to a preliminary sample of 100 points. Then the parameter λ was given a value asking the "optimizer" to provide a guess of the optimum value f^*. This procedure is motivated by the fact that the specification of a "prior guess" of F(t), i. e. $\beta_0(t)$, seems outside the possibilities and the interests of the optimizer; indeed it would require the assessment of a prior distribution for Y=f(X), that is for the result of a single evaluation of f at a random point.

Being the object of the inferential process the unknown quantile t_ε, viewed as the lowest acceptable approximation to f^*, it seems more natural to incorporate prior available information into a prior distribution for t_ε, which can be thought as the prior distribution of the optimum, or better of the approximation to the optimum which can be reasonably achieved through the random sampling.

Let then $\beta_0(t)$ be now this distribution; then, by (4) and (5),

$$\beta_0(t)=Pr\{t_\varepsilon\leq t\}=Pr\{1-\varepsilon\leq F(t)\}$$

$$=Pr\{Y(t)\geq -\log\varepsilon\}$$

$$=1-\Gamma(-\lambda\log\varepsilon;\gamma(t)), \quad (7)$$

where $\Gamma(x;a)$ is the standard Gamma distribution with shape parameter a. Equation (7) does not yield $\gamma(t)$ in a closed form, but it can be solved numerically for any required value of $\gamma(t)$, given $\beta_0(t)$ an actual expression and λ a value, whose determination is to be left to numerical experience.

An acceptable form for $\beta_0(t)$ appears to be for example the Gumbel distribution,

$$\beta_0(t)=\exp(-\exp(-(t-a)/b)), \tag{8}$$

where a is a location parameter and b is a scale parameter. Observe that the distribution is not symmetric, and that values larger than a are more "probable" than their symmetric around a.

4. A SEQUENTIAL DECISION RULE

A simple test of hypothesis (3) consists in testing whether $\Pr\{F(\hat{f})\geq 1-\varepsilon|y_1,\ldots,y_n,\hat{F}=\hat{f}\}$ is larger than some prefixed treshold, say p. Although good results have been obtained with this decision rule when p=0.5 (Betro' and Rotondi, 1983), in view of the sequential use of scheme (i)-(iii) with the enlargement of the sample size n until the test yields an affirmative answer, some difficulties arise because of the fact that there is no attempt to control the computational cost which increases with the amount of f. e. collected. On the other hand, the developement of an optimal sequential decision rule, taking precisely into account the expected losses of deciding that the prescribed accuracy has been achieved ($\hat{f}\geq t_\varepsilon$) or that new f. e. have to be collected, seems very hard, in consideration also of the fact that the hypothesis to be tested (3) may change when the sample is increased because of the possibility of updating $\hat{f}$.

It' s author's opinion that a decision rule with no claim of optimality but better tailored to the sequential nature of optimization procedures would be adequate for practical purpose. One of such rules can be developed as follows.

Let' s assume that N is the number of function evaluations (f. e.) spent after the execution of steps (i)-(ii) with a sample of size n. N is given by n plus the number (possibly null) of f. e. spent during the local searches in (ii). Let $\hat{f}_N$ be the current estimate of f* (best evaluated function value out of the N); then in a sequential view the possible actions are to retain $\hat{f}_N$ as the final estimate to f^* or to get more f. e., update $\hat{f}_N$ and take a new action. If w_0 is the loss for accepting hypotesis (3) when it is false, expressed in the same units of the computational cost of obtaining a single function evaluation, the expected loss for the action "accept" is given by

$$L_n=N+w_0\Pr\{\hat{f}_N<t_\varepsilon|n\},$$

where n briefly indicates the n sampled values and the event $\hat{F}_N=\hat{f}_N$ as in (6).

Let' s assume that decision is to collect more f. e., executing again steps (i),(ii) with an enlarged sample of size n'. Let N' the updated total number of function evaluations after step (ii) and $\hat{f}_{N'}$ the updated optimum estimate. Then the new expected loss is

$$L_{n'}=N'+w_0\Pr\{\hat{f}_{N'}<t_\varepsilon|n'\}.$$

Thus a sequential decision rule can be obtained deciding to accept $\hat{f}_{N'}$ and to stop if the condition

$$L_{n'}>L_n \tag{9}$$

is satisfied, that is when the effect of adding new f. e. has been that of increasing the expected loss. At the beginning, if there is no value of the objective function available for giving the estimate $\hat{f}^*$, we may assume $f_0=-\infty$, so that $\Pr\{\hat{f}_0<t_\varepsilon\}=1$ and $L_0=w_0$.

(9) can be rewritten as

$$\Pr\{F(\hat{f}_{N'})\geq 1-\varepsilon|n'\}-\Pr\{F(\hat{f}_N)\geq 1-\varepsilon\ |n\}<(N'-N)/w_0,$$

which shows how the stopping rule depends on the relative cost of accepting a bad estimate to the cost of a batch of f. e..

We remark that if $1-\Pr\{F(\hat{f}_N)\geq 1-\varepsilon|n\}<(n'-n)/w_0$, then it is not necessary to proceed further, as after new evaluations the expected loss will be surely increased; indeed

$$\Pr\{F(\hat{f}_{N'})\geq 1-\varepsilon|n'\}-\Pr\{F(\hat{f}_N)\geq 1-\varepsilon|n\}$$

$$\leq 1-\Pr\{F(\hat{f}_N)\geq 1-\varepsilon|n\}<(N'-N)/w_0.$$

5. NUMERICAL EXAMPLES

In order to illustrate the effect of the stopping rule introduced in Section 4 as well as the feasibility of assessing prior probabilities as proposed in Section 3, two examples will be considered drawn from the standard set of global optimization test functions in Dixon and Szego (1978), Branin's function and Hartman's six variables function.

According to the fact that the scale of tne test functions, changed in sign for dealing with maximization, is such that the absolute value of f^* is few units large, a reasonable choice of the parameters a and b in (6) is to give the median m=a+0.36651b the value zero and to set b=4, so that, roughly speaking, a positive value of the maximum is considered as probable as a negative one, belonging with probability about 0.99 to the interval (-8,20).

The same sampling strategy of Betro' and Rotondi (1983) was used for step (i), starting with a sample of n=100 points and subsequently incrementing n, if needed, by 50. Also the clustering procedure was the same, resulting in at most one local search for each enlargement of the sample.

The required accuracy ε was given the value 0.0005 and the loss w_0 was assumed to be equivalent to 2000 f. e..

For Branin's function, after the initial sample, the local search started from the best sampled value converged to one of the three global maxima, corresponding to the value $f^*=-.397887$, requiring 33 f. e. (each gradient evaluation was considered as 4 f. e.), so that $\hat{f}_{133}=-.397887$. Now, as $\Pr\{\hat{f}_{133}\geq t_\varepsilon|100\}$ turned out to be

.441>133/2000=0.0665, the sample was enlarged to 150 points, and a second local search was started obtaining another global maximum after 32 f. e.. Now $Pr\{\hat{f}_{215} \geq t_{\varepsilon} | 150\} = 0.444$, so that the difference with the previous probability is less than 82/2000=0.041 and (8) is satisfied. Observe that stopping occurs with a probability less than 0.5, so that the previously adopted decion rule would have let sampling continue.

For the Hartman's six variables function, after the initial sample and 234 f. e. spent by the local search, a local maximum with $\hat{f}_{334} = -3.2032$ was found. Conditional probability of $\hat{f}_{334} \geq t_{\varepsilon}$ was computed as .391 and hence new 50 random points were drawn. No local search was performed at this point and conditional probability increased to .456, with an increment of .065>50/2000=0.025. After the subsequent enlargement of the sample, a new local search reached the global maximum $f^* = 3.3224$ after 173 f. e.; now $Pr\{\hat{f}_{607} \geq t_{\varepsilon} | 200\} = 0.512$, so that the difference with the previous probability is less than 223/2000=0.112 and (8) is satisfied. The previously adopted decision rule would have stopped sampling exactly at the same point, so that the new rule is not performing worse than the old.

REFERENCES

ARCHETTI F. and SCHOEN F., 1983, A survey on the global optimization problem: general theory and computational approaches, to appear in Annals of Operations Research, 1.

BETRO' B., 1983, A Bayesian nonparametric approach to global optimization, in Methods of Operations Research, Stahly ed., 45,47-59, Athenaum.

BETRO' B. and ROTONDI R., 1983, A Bayesian algorithm for global optimization, to appear in Annals of Operations Research, 1.

BETRO' B. and VERCELLIS C., 1983, Bayesian nonparametric inference and Monte Carlo optimization, presented at II International Meeting on Bayesian Statistics, 5-10 Sep. 83, Valencia, Spain.

BOENDER C. G. E., RINNOOY KAN A. H. G., TIMMER G. T., 1982, A stochastic method for global optimization, Mathematical Programming, 22, 125-140.

DIXON L. C. W. and SZEGO G. P. eds., 1978, Towards Global Optimization 2, North Holland.

DOKSUM K., 1974, Tailfree and neutral random probabilities and their posterior distributions, Annals of Probability, 2, 163-201.

FERGUSON T. S., 1973, A Bayesian analysis of some nonparametric problems, Annals of Statistics, 1, 209-230.

CONVERGENT CUTTING PLANES FOR LINEAR PROGRAMS WITH ADDITIONAL REVERSE CONVEX CONSTRAINTS

by

M.C. Böhringer
Bell Telephone Labs
and
S.E. Jacobsen
Department of System Science
University of California, Los Angeles

1. INTRODUCTION

Consider the mathematical program, denoted by (P),

$$\begin{aligned} \min\ & f(x) \\ & Ax \geq b \\ & g_i(x) \geq 0, \qquad i=1,\dots,r \\ & g_i(x) \geq 0, \qquad i=r+1,\dots,K \end{aligned}$$

where

1. A is mxn, m>n
2. f and each g_i map R^n into R^1
3. each g_i, i=1,...r, is quasi-convex
4. each g_i, i=r+1,...,K is quasi-concave

Additionally, define

$$F_o = \{x \in R^n \mid Ax \geq b\}$$

$$G_i = \{x \in R^n \mid g_i(x) \geq 0\}$$

$$G_{RC} = \bigcap_{i=1}^{r} G_i$$

$$G_c = \bigcap_{i=r+1}^{K} G_i$$

the feasible region can then be written as: $F = F_o \cap G_{RC} \cap G_c$

(Note: since it is the presence of the nonlinear quasi-convex constraints, $g_i(x) \geq 0$, i=1,...,r, which causes F to be generally nonconvex, we shall assume that such constraints are always present in problems of "the type P" (i.e., r≥1). Also, K = r means that all the nonlinear constraints are quasi-convex and, in this case, we define $G_c = R^n$.)

Problems of type P are extremely difficult to solve since the feasible region F is usually disconnected into several pieces and some of the latter may be nonconvex. As a result, any feasible direction algorithm will, at best, find a local minimum for just one of the "connected pieces" of the feasible region F. Moreover, the number and location of the other "connected pieces" are unknown.

Problems of type (P) have been called "reverse convex programs" (e.g., see Mangasarian [11], Meyer [12], Avriel [1], and Hillestad and Jacobsen [9]) since each of the first r nonlinear constraints has the property that $G_i^c = \{x \in R^n | g_i(x)<0\}$ is convex (assume each g_i is continuous). Therefore, F is the intersection of a convex set (i.e., $F_o \cap G_c$) with the intersection of the complements of convex sets (i.e., with $G_{RC} = \bigcap_{i=1}^{r} G_i$). Such problems have also been called "complementary convex programs" (e.g., see Avriel and Williams [3], and Avriel [2]).

Problems of the type P were first studied by Rosen [13] in a control theoretic setting and, subsequently, by Avriel and Williams [3] in an engineering design (i.e., signomial or complementary geometric programs) setting. These authors developed a linearization procedure which, as shown also by Meyer [12], converges to a Kuhn-Tucker point. However, it is relatively easy to construct examples where this latter procedure does not produce an optimal point; in fact, the procedure often produces points which are not even contained in a connected piece which also contains an optimal solution.

Bansal and Jacobsen [4] studied the problem of maximizing network flow capacity when economies-of-scale are present. In particular, each arc's cost function for increasing its capacity is concave. Such a problem is of type P where f is linear and K=r=1 (i.e., there is one nonlinear constraint and it is of the quasi-convex type. In particular, let there be N arcs, let $c_i(x_i)$ be the cost function for increasing the capacity of arc i by x_i units, and let B>0 be a given budget level. The budget constraint can then be written as $g_1(x)=B - \sum_{i=1}^{N} c_i(x_i) \geq 0$). Bansal and Jacobsen produced a finite pivot method for finding a "basic local minimum" and then showed how to find a better one. Also, there is a finite number of such basic local minima.

Hillestad [8] produced a branch and bound edge search procedure for solving problems of type P when f is linear and there is one nonlinear constraint and it is quasi-convex.

Hillestad and Jacobsen [9] studied problems of type P where K=r (i.e., all nonlinear constraints are quasi-convex). They defined "quasi-local vertex" and "basic solutions" and showed the two notions are equivalent. When F_o is bounded Hillestad and Jacobsen showed that conv F is a convex polytope. Therefore, when f is quasi-concave an optimal solution can be found among the vertices (finite in number) of convF and, hence, among the basic solutions of F. These types of results show that reverse convex programs are natural generalizations of linear programs. In [9] the method of Ueing [15] for problems of type P where

K=r and f is concave, is generalized, discussed, and shown to be theoretically interesting.

Hillestad and Jacobsen [10] continued the study of problem P where f is linear and K=r=1 (i.e., a linear program with one additional nonlinear reverse convex constraint). The main theoretical result of that paper can be described as follows. Let $V(F_o)$ denote the set of vertices of F_o, $f(x) = c^Tx$, and let $\bar{x}$ solve $\min \{c^Tx | x \varepsilon V(F_o) \cap G_{RC}\}$. The hyperplane $c^Tx = c^T\bar{x}$ intersects several edges of F_o and at least one such edge contains an optimal solution (also, it is not generally true that an optimal solution is contained on an edge which contains $\bar{x}$). This result then provides [10] a finite pivot method for optimizing a linear program with one reverse convex constraint.

2. THE BASIC IDEA OF THE ALGORITHM

In what follows we assume that f(x)=cx and $r=^K$. That is, P is a problem of minimizing a linear function subject to a set of linear constraints and an additional set of quasi-convex constraints, $g_j(x) \geq 0$, $j=1,\ldots r$. Also, for a convex polyhedral set P_k and for a vertex x^k of P_k so that $g_j(x^k)<0$ we let $H(x^k,j)$ denote a half space with the property that $x^k \notin H(x^k,j)$ and $\text{conv}F \subset H(x^k,j)$. We discuss later how to compute $H(x^k,j)$.

Algorithm 1

Step 0. Set $k=0$, $P_k=F_0$

Step 1. Let $x^K \varepsilon V(P_k)$ solve $\min\{c^Tx | x\varepsilon P_k\}$, where $V(P_k)$ denotes the set of vertices of P_k. If x^k F, stop --x^k is optimal for problem P.

Step 2. Let $L = \{j | g_j(x^k)<0\}$. Choose a subset $C \subset L$ and let

$$P_{k+1} = P_k \cap \left[\bigcap_{j \varepsilon C} H(x^k,j) \right].$$

Set k to be k+1 and return to Step 1.

Figures 1,2, and 3 show, respectively, that algorithm 1 may converge finitely to an optimal solution, may converge infinitely to an optimal solution, or may converge to an infeasible solution.

More specifically, the situation is as follows. For this type of cutting plane procedure we have $F \subset \text{conv}F \subset P_{k+1} \subset P_k$ for all iterations k. Also, $P_k \to \cap P_\nu = P^*$ (in the sense of the Hausdorff metric) and, therefore, $P^* \supset \text{conv}F$. Let $x^k \to x^*$, subsequentially if necessary. Then, since $x^k \varepsilon V(P_k)$ solves $\min \{c^Tx | x\varepsilon P_k\}$, we have that $x^* \varepsilon P^*$ and x^* solves $\min \{c^Tx | x\varepsilon P^*\}$. Therefore, <u>if</u> $P^* = \text{conv}F$ then x^* is optimal for $\min \{c^Tx | x\varepsilon \text{conv}F\}$ which, in turn, is equal to $\min\{c^Tx | x\varepsilon F\}$. Therefore, since $V(\mathrm{conv}F) \subset F$, if $x^* \varepsilon V(\mathrm{conv}F)$ then $x^* \varepsilon F$ and x^* solves $\min\{c^T x \mid x \varepsilon F\}$. If $x^* \notin V(\mathrm{conv}F)$ then $x^* \notin F$ is possible. However, it is generally not the case that $P^* = \mathrm{conv}F$. For instance, Figure 3 clearly shows that $P^* \neq \mathrm{conv}F$. Figure 2 shows that the cutting plane method does, for that example, have the property that $P^* = \mathrm{conv}F$; however, examples can be constructed where the method converges to an optimal point and $P^* \neq \mathrm{conv}F$. In fact, Figure 1 provides such an example.

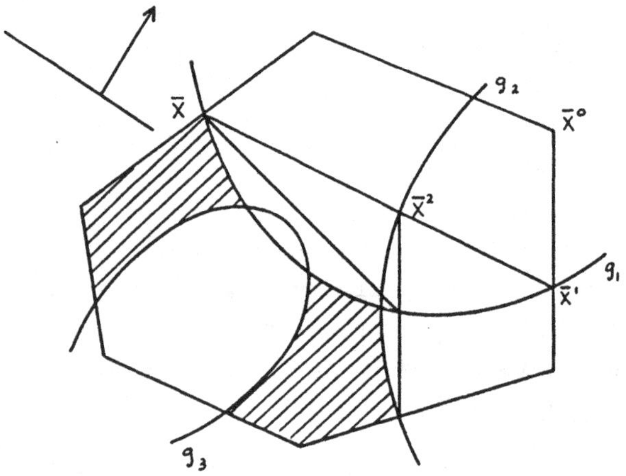

Figure 1: Algorithm 1 converges finitely

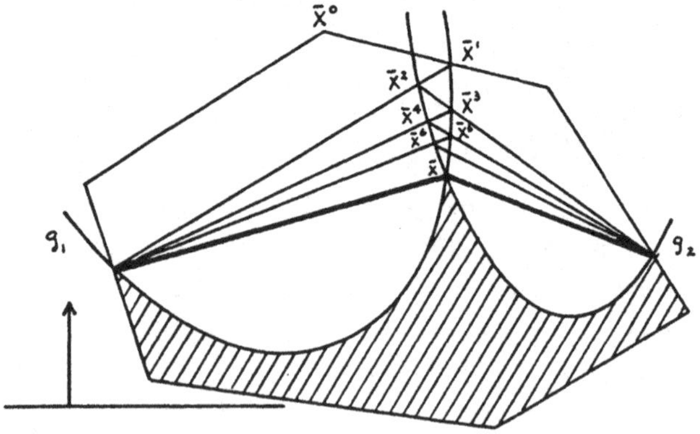

Figure 2: Algorithm 1 converges infinitely

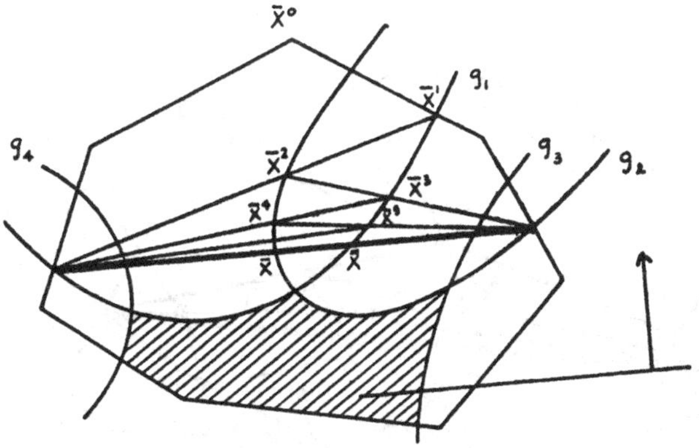

Figure 3: Algorithm 1 converges to an infeasible point

3. MODIFICATION OF THE ALGORITHM

Hillestad and Jacobsen [9] use algorithm 1 where each $H(x^k,j)$ is defined by the Tuy cut [14] with respect to $x^k \varepsilon V(P_k)$ (note that the cuts of Figures 1,2, and 3 are not Tuy cuts). They also presented an example of convergence to an infeasible point. Their example and the example of Figure 3 suggest that as the successive cutting planes are becoming parallel to the objective it pays to include cuts with respect to vertices of P_k which are nearly optimal.

More specifically, convergence to an optimal point will occur for both examples if we modify the procedure somewhat. In particular, let x^k be defined as in Step 1 and let $x^{k\ell}$, $\ell \varepsilon I$, be a set of vertices of P_k which are "almost as good" as x^k (i.e., the $x^{k\ell}$, $\ell \varepsilon I$, are ε-optimal for for the kth linear program, where $\varepsilon > 0$ is preassigned). Now, construct cuts with respect to a subset of the $x^{k\ell}$, $\ell \varepsilon I$. In fact, consider Figure 4 which is the example of Figure 3 and for which we incorporate the above modification. Consider the linear program for which $\bar{x}^5$ is optimal. Let $\varepsilon > 0$ be such that the point y^* is ε-optimal and note that $g_3(y^*)<0$. If we perform the same type of cut as previously, the cut denoted by $C(y^*)$ is generated (i.e., $C(y^*)$ is the bounding hyperplane of $H(y^*,3)$) and therefore the point u^* is ε-optimal at the next iteration. We see that $g_2(u^*)<0$ and the cut denoted by $C(u^*)$ is generated which, in turn, leads to v^* as a vertex for the linear program at the next iteration. Since v^* is below the heavy line we see that, as this process continues, convergence to an optimal point will occur.

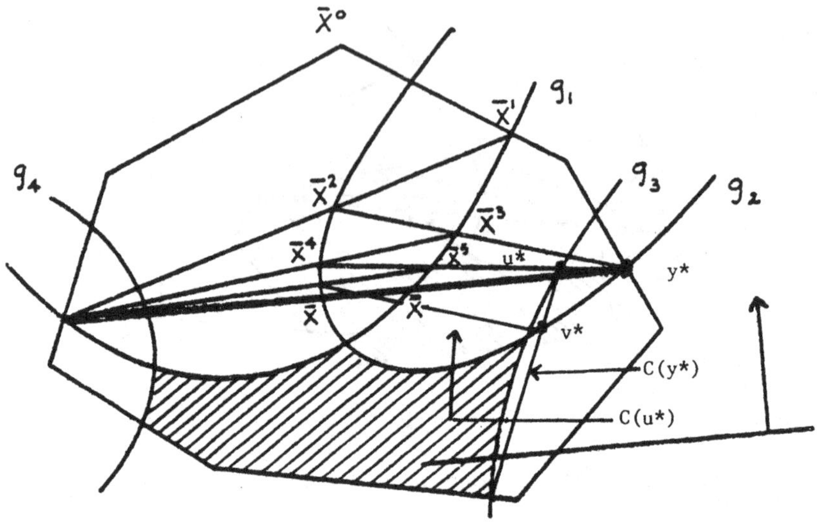

Figure 4: Convergent Behavior of Modified Algorithm

The above example motivates the idea that, as the method proceeds, ε-optimal vertices of the P_k should be sought in order to generate deep cuts and to guarantee convergence to a feasible point and, hence, to an optimal point. To be specific, let $\mathcal{O}(x^k,\varepsilon)=\{x\varepsilon V(P_k)\mid c^Tx\le c^Tx^k+\varepsilon\}$, let $\mathcal{O}\subset\mathcal{O}(x^k,\varepsilon)$, for $x\ \varepsilon\ \mathcal{O}$ let $L(x)=\{j\mid g_j(x)<0\}$, and let $C(x)\subset L(x)$. Note that if $x\ \varepsilon\ O(x^k,\varepsilon)$, $x\neq x^k$, is such that $x\ \varepsilon\ F$ then x is an ε-optimal solution for P. The fundamental idea is that step 2 of the cutting plane method should be changed.

Algorithm 2

Step 2´ If for any $x\ \varepsilon\ \mathcal{O}(x^k,\ \varepsilon)$, we have $x\varepsilon F$, stop -- any such x is ε-optimal for P. Otherwise let $\mathcal{O}\subset\mathcal{O}(x^k,\varepsilon)$ be such that $x\ \varepsilon\ \mathcal{O}$ implies $x\notin F$.

Set $P_{k+1} = P_k\cap\left(\bigcap_{x\varepsilon\mathcal{O}}\ \bigcap_{j\varepsilon C(x)} H(x,j)\right)$

Set k to be k+1 and return to Step 1.

4. CUT GENERATION

From a computational point-of-view, about the easiest cut to generate is the Tuy-cut. To be specific, let $x^k\ \varepsilon\ V(P_k)$ solve the kth linear program and assume $g_j(x^k)<0$. Assume also that P_k is n-dimensional and nondegenerate and, therefore, let x^{ki}, $i=1,\ldots,n$, denote the n neighboring vertices, in P_k, of x^k. For $i=1,\ldots,n$, let α_i solve

$$\min\{\alpha\ge 0\mid g_j(x^k+\alpha d^{ki})\ge 0\}$$

where $d^{ki}=x^{ki}-x^k$ and where we assume each of the above n line search problems is feasible. Let $z^{ki}=x^k+\alpha_i d^{ki}$, $i=1,\ldots,n$, and let the nxn matrix D be defined by $D=[z^{k}i-x^k,\ldots,z^{kn}-x^k]$ and let e denote a row vector of n "ones" (i.e., $e=(1,1,\ldots 1)$). The Tuy cut is given by $eD^{-1}(x-x^k)\geq 1$ and we take $H(x^k,j)=\{x|eD^{-1}(x-x^k)\geq 1\}$.

Enumerative Methods in Cut Generation

The Tuy cut separates $\{x^k\}$ from $conv\{z^{k1},\ldots,z^{kn}\}$ and is clearly the deepest such cut. However, the z^{ki} lie on rays emanating from x^k and therefore deeper cuts, with respect to P_k, can be generated at an additional computational burden.

Definition: Let $x^k \varepsilon V(P_k)$ and assume $x^k \notin G_j$. We say z is a first point of x^k with respect to P_k and G_j if z is on an edge of P_k and there is a sequence of adjacent vertices $v^o,v^1,\ldots,v^s,u,v,v\varepsilon G_j$, so that $x^k=v^o$, $x\varepsilon[v^{i-1},v^i]$ implies $x\notin G_j$, $i=1,\ldots,s$, $z\varepsilon(u,v]\cap G_j$, and $x\varepsilon[u,z)$, implies $x\notin G_j$. Note that $z=u+\bar{\lambda}(v-u)$ where $\bar{\lambda}$solves

$$\min\ \{\lambda\geq 0|g_j(u+\lambda(v-u))\geq 0\}$$

Let $F_j(x^k)$ denote the set of all such first points. Note that $F_j(x^k)$ may be found by considering all vertices on paths leaving x^k until reaching a vertex $v\ \varepsilon\ V(P_k)$, $g_j(v)\geq o$, and then solving the above line search problem. The method of Dyer and Proll [6], because of the method used to find vertices, is ideally suited for finding $F_j(x^k)$. Also note that $conv(P_k\cap G_j)$ is a convex polytope whose vertices are among the edges of P_k [10].

The idea is to now generate a hyperplane which separates $\{x^k\}$ and $convF_j(x^k)$. By duality we know that the minimum distance from x^k to $convF_j(x^k)$ is also the maximum distance from x^k to all separating hyperplanes which support $convF_j(x^k)$. More precisely, we can compute that supporting and separating hyperplane which is farthest from x^k by solving

$$\begin{aligned}&\max_{\lambda,\alpha}\ \alpha\\ &\qquad\alpha\leq\lambda^Tz^{ki}-\lambda^Tx^k\quad,\quad i=1,\ldots,f\\ &||\lambda||^*\quad\leq\ 1\end{aligned}$$

where $F_j(x^k)=\{z^{k1},\ldots,z^{kf}\}$ and $||\cdot||^*$ is the dual norm. Let $(\lambda,\alpha)=(y,\alpha^*)$ be optimal. Then the hyperplane $y^Tx=\bar{\alpha}$, where $\bar{\alpha}=\alpha^*+y^Tx^k$, separates x^k from $F_j(x^k)$ and we take $H(x^k,j)=\{x|y^Tx\geq\alpha\}$

While the above hyperplane is the farthest from x^k it does not necessarily provide the deepest cut. We can generally improve the depth of the cut by replacing the constraint $||\lambda||^*\leq 1$ with the constraint $\sum_{j=1}^{f}\lambda^T(z^{kj}-x^k)\leq 1$. Intuitively, if f=n and the z^{ki} are affinely independent, then the solution of the cut finding problem is a Tuy cut.

However, generally $f \neq n$ and the cut finding problem can be thought of as a generalization of the Tuy cut (e.g., see Carvajal-Moreno [5]). In fact it can be shown that, under rather general conditions, the hyperplane $y^T x = \alpha^* + y^T x^k$ contains at least n of the elements of $F_j(x^k)$. Also note that each of the cuts of Figures 1-4 is of this type.

4. CONVERGENCE

This section outlines the proof of convergence for Algorithm 2. Actually, an additional technical modification to Step 2′ is required to prove convergence; however, this modification can be ignored for most iterations and be used only every so often during the course of the algorithm.

Let $d_j(x^k)$ denote the distance, measured by $||\cdot||$, of the point x^k to conv$F_j(x^k)$. Given the set $F_j(x^k)$, $d_j(x^k)$ is the optimal value of the cut finding problem. Let

$$J(x^k) = \{j \mid g_j(x^k) \le g_\nu(x^k), \text{ all } \nu\}$$

and let

$$d(x^k) = \max\{d_j(x^k) \mid j \in J(x^k)\} .$$

At Step 2′, if there is an $x \in O(x^k,\varepsilon) \cap G_{RC_k}$ then an ε-optimal solution for P has been found. Otherwise, choose $\bar{x}^k \in O(x^k,\varepsilon)$ so that $d(\bar{x}^k) \ge d(x)$ for all $x \in O(x^k,\varepsilon)$. Now let $J_k \subset \{1,2,\ldots,r\}$ be any index set of the reverse convex constraints so that $d_i(\bar{x}^k) = d(\bar{x}^k)$ for some $i \in J_k$ and $g_j(\bar{x}^k) \le 0$ for all $j \in J_k$. Now, for all $j \in J_k$, compute $H(\bar{x}^k,j)$ and set

$$P_{k+1} = P_k \cap \left[\bigcap_{j \in J_k} H(\bar{x}^k,j) \right] .$$

Define the metric space

$$\mathcal{F}(F_0) = \{P \mid P \subset F_0, \text{ P nonempty, closed, and convex}\},$$

where the Hausdorff metric [7] is employed. Let $X_{kj} = P_k \cap \text{bd}G_j$ and note that $F_j(x^k) \subset \text{bd}G_j$. The following can easily be shown.

Lemma: Let $P_k \in \mathcal{F}(F_0)$ be a sequence of sets so that $P_{k+1} \subset P_k$, and let $x^k \in P_k$ be such that $g_j(x^k) < 0$ and $x^k \to \bar{x}$. Then $P_k \to \bar{P} = \bigcap_\nu P_\nu$, $\bar{x} \in \bar{P}$, and $X_{kj} \to \bar{P} \cap \text{bd}G_j$ and, moreover, $d_j(x^k) \to d_j(\bar{x})$.

Convergence can be shown by employing the algorithmic map approach developed by Zangwill [16]. The algorithmic point-to-set mapping A: $\mathcal{F}(F_0) \to \mathcal{F}(F_0)$ is defined by $P' \in A(P)$ if P' may be obtained from P by one iteration of the above described modification of Algorithm 2. We define the solution set S to be

$$S = \{P \in \mathcal{F}(F_0) \mid \exists \bar{x} \in V(P) \cap G_{RC} \text{ s.t. } \bar{x} \text{ is optimal}\}$$

In order to be able to define a descent function let

$$\tilde{A}_j = \{A \subset R^n | A \text{ is an affine set of dimension } j\}$$

Define $f_j(P)$, $P \in \mathcal{F}(F_0)$, as $f_j(P) = \sup_{A \in \tilde{A}_j} \iint\limits_{P \cap A} \cdots \int dv$, the j-fold integral giving the volume of $A \cap P$ (i.e., $f_j(P)$ gives the "biggest area" of any j-dimensional slice through P). It can be shown that for fixed A the above integral is a continuous function in P. From the continuity of the j-fold integral in P follows immediately the continuity of $f_j(P)$). Also, clearly $f_j(P) \leq f_j(P')$ if $P \subset P'$. Now define $f(P) = (f_n(P), \ldots, f_1(P)$. We shall say that $f(P) \leq f(P')$ if for some $k \in \{1, \ldots, n\}$ $f_k(P) \leq f_k(P')$ and $f_j(P) = f_j(P')$ for $j = k+1, \ldots, n$. Thus $f(P)$ is continuous and $f(P') \leq f(P)$ if $P' \in A(P)$. If $P \notin S$ then P' is strictly contained in P implying $f(P') < f(P)$. Therefore $f(P)$ is a descent function.

It can also be shown that the algorithmic map is closed at points $P \notin S$ and, therefore, convergence of the modified Algorithm 2 follows. We have left out the detailed proofs of the above comments because of space limitations and because we chose to emphasize the basic idea behind generating appropriate cuts.

Naturally, a good deal of computation is involved in the above procedure and while the method has worked well for some problems it is clear that such cutting plane methods in conjunction with other enumerative methods need to be developed.

REFERENCES

1. Avriel, M., "Methods for Solving Signomial and Reverse Nonconvex Programming Problems," in Optimization and Design (eds. M. Avriel, M.J. Rijckaert, D.J. Wilde), Prentice-Hall, Englewood Cliffs, N.J. (1973).

2. Avriel, M., Nonlinear Programming, Prentice-Hall (1976).

3. Avriel, M. and A.C. Williams, "Complementary Geometric Programming," SIAM J. Appl. Math., 19, pp. 125-141 (1970).

4. Bansal, P.P., and S.E. Jacobsen, "Characterization of Local Solutions for a Class of Nonconvex Programs," J. Opt. Theory and Appl., Vol. 15, No. 5, pp. 549-564, (May 1975).

5. Carvajal-Moreno, R., "Minimization of Concave Functions Subject to Linear Constraints," Operations Research Center, ORC 72-3, University of California, Berkeley (Feb. 1972).

6. Dyer, M.E. and L.G. Proll, "An Algorithm for Determining All Extreme Points of a Convex Polytope," Mathematical Programming, 12, pp. 81-96 (1977).

7. Hausdorff, F., Set Theory, 2nd ed., Chelsea, New York (1962).

8. Hillestad, R.J., "On Solving Optimization Problems Subject to a Budget Constraint," J. ORSA, Vol. 23, No. 6, pp.1091-1098 (Nov.-Dec. 1975).

9. Hillestad, R.J., and S.E. Jacobsen, "Reverse Convex Programming," J. Applied Mathematics and Optimization, 6, pp.63-78 (1980).

10. Hillestad, R.J. and S.E. Jacobsen, "Linear Programs with an Additional Reverse Convex Constraint," J. Applied Mathematics and Optimization, 6, pp. 257-268 (1980).

11. Mangasarian, O.L., Nonlinear Programming, McGraw-Hill (1969).

12. Meyer, R., "The Validity of a Family of Optimization Methods," SIAM J. Control, Vol. 8, pp. 41-54 (1970).

13. Rosen, J.B., "Iterative Solution of Nonlinear Optimal Control Problems," SIAM J. Control, Vol. 4, pp. 223-244 (1966).

14. Tuy, H., "Concave Programming under Linear Constraints," Soviet Math. 5, pp. 1437-1440 (1964).

15. Ueing, U., "A Combinatorial Method to Compute a Global Solution of Certain Non-Convex Optimization Problems," in Numerical Methods for Nonlinear Optimization, F.A. Lootsma (ed.), Academic Press (1972).

16. Zangwill, W.I., Nonlinear Programming: A Unified Approach, Prentice-Hall (1969).

A FAST VORONOI-DIAGRAM ALGORITHM WITH APPLICATIONS TO GEOGRAPHICAL OPTIMIZATION PROBLEMS

Masao Iri
University of Tokyo, Tokyo, Japan
Kazuo Murota
University of Tsukuba, Ibaraki, Japan
Takao Ohya
Central Research Institute of Electric Power Industry, Tokyo, Japan

Abstract There are many kinds of facility location problems, or "geographical optimization problems", which are appropriately formulated in terms of the Voronoi diagram. Except few quite special problems, we can get a solution to a problem of that kind only by means of numerical approach which involves a large number of function evaluations, where each evaluation requires constructing the Voronoi diagram for a tentative distribution of facilities and computing integrals with reference to that diagram. In such a case, the practical feasibility of the numerical solution of the problem depends largely upon the efficiency of the algorithm to be used for constructing the Voronoi diagram. In this paper, we shall formulate a class of location problems and show that, if we use the Voronoi-diagram algorithm recently proposed by the authors, we can numerically solve considerably large problems within a practicable time.

1. Introduction

The location problem has many difficult "faces", among which the "combinatorial face" would be the most essential. However, when we deal with the problem in the continuum, i.e., in the two-dimensional Euclidean plane, we may encounter another, no less difficult, face of it. In fact, the objective function to be minimized or maximized of a problem is often defined in terms of the Voronoi diagram for the given location of facilities (or the like), i.e. the partition of the entire plane into the territories of the facilities. Naturally, we have to resort to some numerical approach to the solution, except for very few particular cases, and the numerical solution will ordinarily consist of iterative evaluation of the objective function as well as of its partial derivatives.

If, as is widely believed, the construction of the Voronoi diagram for given n points in the plane were itself a big computational problem for n large, an algorithm which calls a subroutine for the Voronoi-diagram construction many times would be far from being practical. However, recent progress in computational geometry has produced a number of fast algorithms for the Voronoi diagram, which gives us a bright prospect for the possibility of bringing the numerical solution of the location problem into the practically tractable family.

In the following, we shall formulate a fundamental class of location problems, or "geographical optimization problems", and show that they can be solved numerically within a practicable time for as many as over one hundred points or facilities, if we use a Voronoi-diagram algorithm which is fast enough. The algorithm which the authors recently proposed and which they call the "quaternary incremental method" [14] was found to be qualified as such one.

2. Mathematical and Computational Background for the Voronoi Diagram

For n distinct points $\mathbf{x}_i=[x_i^\kappa]$ $(i=1,\ldots,n)$ given in the N-dimensional Euclidean space $\mathbf{R}^N$ $(\kappa=1,\ldots,N)$, we define a partition of $\mathbf{R}^N$ into n convex polyhedral regions V_i $(i=1,\ldots,n)$ in such a way that every point $\mathbf{x}=[x^\kappa]$ in V_i is closer to $\mathbf{x}_i$ than to any other $\mathbf{x}_j$ $(j\neq i)$:

$$V_i = \bigcap_{j:j\neq i} \{\mathbf{x} \mid \|\mathbf{x}-\mathbf{x}_i\| < \|\mathbf{x}-\mathbf{x}_j\|\}. \tag{2.1}$$

Obviously we have

$$\overline{\bigcup_i V_i} = \mathbf{R}^N. \tag{2.2}$$

V_i is a kind of "territory" of point $\mathbf{x}_i$ and is called the <u>Voronoi region</u> belonging to $\mathbf{x}_i$. The partition determines in an obvious manner a polyhedral complex, which is called the <u>Voronoi diagram</u> for the given n points $\mathbf{x}_i$'s. We shall denote the common boundary of V_i and V_j by

$$W_{ij} = \partial V_i \cap \partial V_j. \tag{2.3}$$

In the two-dimensional case, i.e. in $\mathbf{R}^2$, the one-dimensional section of a Voronoi diagram is a planar graph, of which the vertices we shall call the <u>Voronoi points</u> and the edges the <u>Voronoi edges</u>. The points $\mathbf{x}_i$'s will be called <u>generators</u>. A Voronoi edge is part of the perpendicular bisector of some couple of generators, and a Voronoi point is the excenter of some triple of generators. Since the degree of a Voronoi point, as a vertex of the graph, is three except for the degenerate case, we may consider the dual graph whose vertices are generators and which has an edge between two generators iff the corresponding Voronoi regions have a boundary edge in common. The dual graph is called the Delaunay triangulation of $\mathbb{R}^2$.

The Voronoi diagram, as well as the Delaunay triangulation, has now been recognized as a concept of fundamental importance in many kinds of problems in geography, urban planning, environmental control, physics, biology, ecology, etc. [2], [3], [9], [16]. Many computational algorithms for constructing the Voronoi diagram have also been proposed, but it has been known that the Voronoi diagrams in a space whose dimensionality is higher than two are rather hard to construct practically, as we are usually concerned with the diagrams with very many points, say more than hundreds or even thousands in number. (See, e.g., [1].) However, if we confine ourselves to the case of two dimensions, a rather primitive incremental algorithm such as the one proposed in [5] sometimes works fairly well, and several "theoretically optimal" algorithms of the divide-and-conquer type with the worst-case time complexity $O(n \log n)$ (and with the linear space complexity $O(n)$) have been proposed, improved and generalized [7], [10], [11], since the appearance of the epoch-making paper [15] in computational geometry.

It is interesting in this context to see that, for almost all problems that we might encounter in practical situations, the more primitive algorithm of the incremental type can be improved to run substantially faster than the theoretically optimal algorithms of the divide-and-conquer type. In fact, the algorithm of the incremental type which makes use of a quaternary tree (which was proposed by the authors [13], [14] and will be outlined in Appendix) has been proved to run in linear time $O(n)$ on the average——more specifically, in about 2.5 ms per point on a typical large digital computer——and to be quite robust against nonuniformity of the distribution of points [14].

3. Problem Formulation

To be specific, let us consider n "facilities" placed, respectively, at points $\mathbf{x}_1,\ldots,\mathbf{x}_n$ of the N-dimensional Euclidean space $\mathbb{R}^N$ ($\mathbf{x}_i=[x_i^\kappa]$, $\kappa=1,\ldots,N$). The territories of those facilities are the Voronoi regions V_i defined by (2.1). We consider furthermore a distribution of "inhabitants" represented by a certain measure $d^N\mu(\mathbf{x})$ on $\mathbb{R}^N$, to be called the measure of population density.

It will not be very unrealistic to assume that an individual inhabitant living in region V_i will enjoy a service of the i-th facility and that the cost for him to gain access to a facility is a function (ordinarily, monotone increasing, and even convex) of the Euclidean distance of $\mathbf{x}$ and the location $\mathbf{x}_i$ of the nearest facility. Then the total cost connected with the serviceability of the facilities is written qua function of $\mathbf{x}_1,\ldots,\mathbf{x}_n$ as

$$F(\mathbf{x}_1,\ldots,\mathbf{x}_n) = \frac{1}{2}\int_{\mathbb{R}^N} f(\min_j \|\mathbf{x}-\mathbf{x}_j\|^2)\, d^N\mu(\mathbf{x})$$

$$= \frac{1}{2}\sum_{i=1}^{n}\int_{V_i} f(\|\mathbf{x}-\mathbf{x}_i\|^2)\, d^N\mu(\mathbf{x}), \qquad (3.1)$$

where the factor 1/2 has been introduced for convenience' sake without loss of generality. Thus we have been led to an optimization problem by means of which we may determine the optimum locations of a given number of facilities for a given population distribution, i.e., the problem of minimizing the function F in (3.1). Since each $\mathbf{x}_i$ has N coordinates x_i^κ ($\kappa=1,\ldots,N$), the problem is essentially the minimization with respect to as many as Nn variables.

4. Solution Algorithm

Before entering into the discussion of a solution algorithm, we should make some observation on the properties of the objective function F of (3.1). Firstly, even if f is convex, F is in general nonconvex, as is evident from the obvious fact that F has several local minima and/or stationary points. Therefore, unless we dare to tackle the combinatorial difficulty connected with the existence of many local minima of a nonconvex function, we have to be contented with "a local minimum". Secondly, F is in general nondifferentiable, although the value of F will depend smoothly on $\mathbf{x}_i$'s for smooth f so long as the locations $\mathbf{x}_i$ of the generators (facilities) are varying in such a way that the topology of the Voronoi diagram may remain unchanged. (This will be seen more concretely in terms of the explicit formula for the partial derivatives of F which we shall calculate in the following.) We can expect at best the existence of its subgradient. Since no general optimization algorithm is available at present which works for such functions better than the most primitive class of descent methods, we have to resort to a variant of primitive descent algorithm.

Thus, we shall investigate the algorithms of the following type, where we denote by $\mathbf{X}$ the Nn-dimensional unknown vector whose components are the components of n vectors $\mathbf{x}_1,\ldots,\mathbf{x}_n$:

$$\mathbf{X}^T = [\mathbf{x}_1^T, \mathbf{x}_2^T,\ldots, \mathbf{x}_n^T]\,. \qquad (4.1)$$

<0> Initial guess:——Start from an arbitrarily given initial guess $\mathbf{x}^{(0)}$, and repeat <1>-<3> for $\nu=0,1,2,\ldots$ until some stopping criterion is satisfied.

<1> Search direction:——Compute the gradient $\nabla_{\mathbf{x}}F(\mathbf{x}^{(\nu)})$, or the partial derivatives, of F at the ν-th approximate solution $\mathbf{x}^{(\nu)}$. Then determine the search direction $\mathbf{d}^{(\nu)}$ from $\nabla_{\mathbf{x}}F(\mathbf{x}^{(\nu)})$ (and from some auxiliary quantities if we want).

<2> Line search:——Determine $\hat{\alpha}$ (up to a certain degree of approximation) such that

$$F(\mathbf{x}^{(\nu)} + \hat{\alpha}\,\mathbf{d}^{(\nu)}) = \min_{\alpha} F(\mathbf{x}^{(\nu)} + \alpha\,\mathbf{d}^{(\nu)}). \tag{4.2}$$

<3> New approximation:——Set

$$\mathbf{x}^{(\nu+1)} = \mathbf{x}^{(\nu)} + \hat{\alpha}\,\mathbf{d}^{(\nu)}. \tag{4.3}$$

There are a number of variants of the algorithm of the above type corresponding to the choice of the search direction in <1>, the manner of performing the line search in <2> and the rule for stopping the iteration. We have tested several variants which will be described in section 6.

5. Calculation of the Partial Derivatives

Since the objective function in (3.1) is not familiar in form it may not be useless to write down its partial derivatives (the gradient and the Hessian). In so doing, we adopt the tensor notation in $\mathbf{R}^N$ in order to maintain the geometrical meanings of the relevant expressions as clear as possible. Thus, we denote by $g_{\lambda\kappa}$ the metric tensor in $\mathbf{R}^N$ (which may be regarded simply as the N×N identity matrix if one would not like to be involved in tensor notation) and use Einstein's summation convention (i.e., if one and the same index appears at two places of a term, one as a contravariant (upper) index and the other as a covariant (lower) index, then the summation with respect to that index over the range $\{1,2,\ldots,N\}$ is understood). For example, we write, for two vectors $\mathbf{x}$ and $\mathbf{y}$ in $\mathbf{R}^N$,

$$\|\mathbf{x} - \mathbf{y}\|^2 = g_{\lambda\kappa}x^{\kappa}y^{\lambda} \quad (= \sum_{\lambda=1}^{N}\sum_{\kappa=1}^{N} g_{\lambda\kappa}x^{\kappa}y^{\lambda}) \ . \tag{5.1}$$

We need some more auxiliary symbols to make the expressions handsome. The distance between two generators $\mathbf{x}_i$ and $\mathbf{x}_j$ will be denoted by

$$\alpha_{ij} \equiv \|\mathbf{x}_i - \mathbf{x}_j\| \ , \tag{5.2}$$

and the (N-1)-dimensional measure induced by $d^N\mu(\mathbf{x})$ on a hypersurface in $\mathbf{R}^N$ will be denoted by $d^{N-1}\mu(\mathbf{x})$.

The partial derivative of the F of (3.1) with respect to an x_i^{λ} will consist of two components, one coming from the derivative of the integrand function and the other from the variation of the region V_i itself as well as of the regions V_j's adjacent to V_i due to the variation of $\mathbf{x}_i$. The first component takes the form:

$$\frac{1}{2}\int_{V_i}\frac{\partial}{\partial x_i^\lambda}f(\|\mathbf{x}-\mathbf{x}_i\|^2)d^N\mu(\mathbf{x}) = \int_{V_i}g_{\lambda\kappa}(x_i^\kappa-x^\kappa)f'(\|\mathbf{x}-\mathbf{x}_i\|^2)d^N\mu(\mathbf{x}) \quad (5.3)$$

and the second component is the sum, taken over all the boundary hyperplanes W_{ij}'s of V_i, of the terms of the form due to the variation of the V_i:

$$\int_{W_{ij}} f(\|\mathbf{x}-\mathbf{x}_i\|^2)\frac{1}{\alpha_{ij}}g_{\lambda\kappa}(x_i^\kappa-x^\kappa)d^{N-1}\mu(\mathbf{x}) \quad (5.4)$$

and the terms due to the variation of an adjacent region V_j:

$$-\int_{W_{ij}} f(\|\mathbf{x}-\mathbf{x}_j\|^2)\frac{1}{\alpha_{ij}}g_{\lambda\kappa}(x_i^\kappa-x^\kappa)d^{N-1}\mu(\mathbf{x}). \quad (5.5)$$

However, since $\|\mathbf{x}-\mathbf{x}_j\|=\|\mathbf{x}-\mathbf{x}_i\|$ on W_{ij}, the terms constituting the second component cancel themselves, so that only the first component will remain, i.e., we have

$$\frac{\partial F}{\partial x_i^\lambda} = \int_{V_i}g_{\lambda\kappa}(x_i^\kappa-x^\kappa)f'(\|\mathbf{x}_i-\mathbf{x}\|^2)d^N\mu(\mathbf{x}). \quad (5.6)$$

The second derivatives of F may be calculated in a similar but rather complicated way to give

$$\frac{\partial^2 F}{\partial x_i^\kappa \partial x_j^\lambda} = \begin{cases} H_{\lambda\kappa}^i + G_{\lambda\kappa}^i & \text{for } j=i, \\ G_{\lambda\kappa}^{ji} & \text{for } j\neq i \text{ with } W_{ij}\neq\emptyset, \\ 0 & \text{for } j\neq i \text{ with } W_{ij}=\emptyset, \end{cases} \quad (5.7)$$

where

$$H_{\lambda\kappa}^i=\int_{V_i}[g_{\lambda\kappa}f'(\|\mathbf{x}-\mathbf{x}_i\|^2)+2g_{\lambda\nu}g_{\kappa\mu}(x_i^\mu-x^\mu)(x_i^\nu-x^\nu)f''(\|\mathbf{x}-\mathbf{x}_i\|^2)]d^N\mu(\mathbf{x}),$$

$$G_{\lambda\kappa}^{ji} = K_{\lambda\kappa}^{jji}, \qquad G_{\lambda\kappa}^i = -\sum_{\substack{j:j\neq i \\ W_{ij}\neq\emptyset}} K_{\lambda\kappa}^{iji},$$

$$K_{\lambda\kappa}^{kji}=\int_{W_{ij}}\frac{1}{\alpha_{ij}}g_{\lambda\nu}g_{\kappa\mu}(x_i^\mu-x^\mu)(x_k^\nu-x^\nu)f'(\|\mathbf{x}-\mathbf{x}_i\|^2)d^{N-1}\mu(\mathbf{x}). \quad (5.8)$$

The special case where f(t)=t is worth noting, because, in that case, the partial derivatives of F are written in a form which admits a direct physical interpretation as follows. For the components of the gradient of F, we have

$$\frac{\partial F}{\partial x_i^\lambda} = \int_{V_i} g_{\lambda\kappa}(x_i^\kappa - x^\kappa)\, d^N\mu(\mathbf{x}) = \mu(V_i)\cdot g_{\lambda\kappa}(x_i^\kappa - \bar{x}_i^\kappa), \tag{5.9}$$

where

$$\mu(V_i) = \int_{V_i} d^N\mu(\mathbf{x}) \tag{5.10}$$

is the total measure of the region V_i, and

$$\bar{x}_i^\kappa = \int_{V_i} x^\kappa d^N\mu(\mathbf{x})/\mu(V_i) \tag{5.11}$$

is the "centroid" of V_i with respect to the measure $d^N\mu(\mathbf{x})$. For the components of the Hessian, we have

$$H^i_{\lambda\kappa} = \mu(V_i)\cdot g_{\lambda\kappa}, \tag{5.12.1}$$

$$K^{kji}_{\lambda\kappa} = \frac{\mu_{ij}(W_{ij})}{\alpha_{ij}} g_{\lambda\nu} g_{\kappa\mu} [x_i^\mu x_k^\nu - x_i^\mu \bar{x}_{ij}^\nu - \bar{x}_{ij}^\mu x_k^\nu + \overline{xx}_{ij}^{\mu\nu}], \tag{5.12.2}$$

where

$$\mu_{ij}(W_{ij}) = \int_{W_{ij}} d^{N-1}\mu(\mathbf{x}) \tag{5.13}$$

is the total measure of W_{ij} with respect to $d^{N-1}\mu(\mathbf{x})$,

$$\bar{x}_{ij}^\mu = \int_{W_{ij}} x^\mu d^{N-1}\mu(\mathbf{x})/\mu_{ij}(W_{ij}) \tag{5.14}$$

is the centroid of W_{ij} with respect to the measure $d^{N-1}\mu(\mathbf{x})$, and

$$\overline{xx}_{ij}^{\mu\nu} = \int_{W_{ij}} x^\mu x^\nu d^{N-1}\mu(\mathbf{x})/\mu_{ij}(W_{ij}) \tag{5.15}$$

is the mean second moment of W_{ij} with respect to $d^{N-1}\mu(\mathbf{x})$.

6. Numerical Examples

Choosing the simplest case of $f(t)=t$, we performed a number of experimental computations for two-dimensional problems ($N=2$) having generators $n=16$ to 256 in number, with respect to three different population distributions, by means of various algorithms with two different search directions and two different line search methods. We considered the following population distributions confined in the unit square $\mathbf{S}=[-\frac{1}{2}, \frac{1}{2}]\times[-\frac{1}{2}, \frac{1}{2}]$:

1° $d^2\mu(\mathbf{x})$ is uniform in $\mathbf{S}$ and vanishes outside;

2° $d^2\mu(\mathbf{x})$ is proportional to $\exp(-25\|\mathbf{x}\|^2)$ in $\mathbf{S}$ and vanishes outside (Tanner-Sherratt type distribution of the urban population [12]);

3° $d^2\mu(\mathbf{x})$ is proportional to $\exp(-\|\mathbf{x}\|\cdot(25\|\mathbf{x}\|-10))$ in $\mathbf{S}$ and vanishes outside (Newling type population distribution [12]).

(See Fig. 1 for these distributions.)

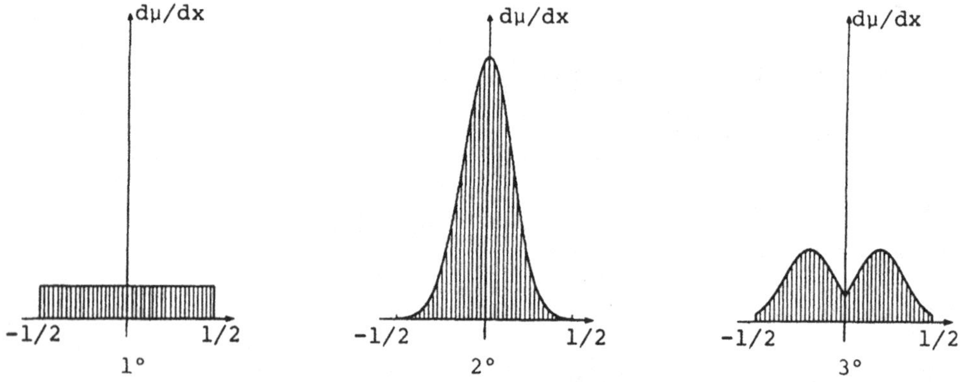

Fig.1. Population distributions used in the experiments (one-dimensional section)

In Fig. 2, examples for n=128 are shown. Starting from a random initial distribution of n generators such as shown in Fig. 2(a) together with the corresponding Voronoi diagram, we had, after a sufficiently large number of iterations, the configurations shown in Figs. 2(b)-(d), respectively, for the population distributions of types 1°-3°. (The number of iterations depends on the variant of the algorithm used; see Fig. 3.)

As we mentioned in Introduction, we used a fast Voronoi diagram algorithm we recently proposed [14] for the construction of the Voronoi diagram. As the criterion for terminating the iteration the condition:

$$\max_{i,\kappa}|x_i^{(\nu+1)\kappa} - x_i^{(\nu)\kappa}| \leqq 10^{-5} \tag{6.1}$$

was adopted in the experimental computation.

For the line search, we tested two variants of the so-called "Goldstein method" [4], which chooses $\hat{\alpha}$ so as to satisfy the inequalities

$$\mu_2\hat{\alpha}\mathbf{d}^{(\nu)}\cdot\nabla_{\mathbf{X}}F(\mathbf{X}^{(\nu)}) \leqq F(\mathbf{X}^{(\nu)}+\hat{\alpha}\mathbf{d}^{(\nu)}) - F(\mathbf{X}^{(\nu)}) \leqq \mu_1\hat{\alpha}\mathbf{d}^{(\nu)}\cdot\nabla_{\mathbf{X}}F(\mathbf{X}^{(\nu)}) \tag{6.2}$$

with appropriately prescribed parameters μ_1 and μ_2 $(0<\mu_1<\mu_2<1)$. (We adopted the values $\mu_1=0.2$ and $\mu_2=0.9$ throughout our experiment.) In searching for such an $\hat{\alpha}$, we begin with an initial guess $\alpha^{(0)}$, increase

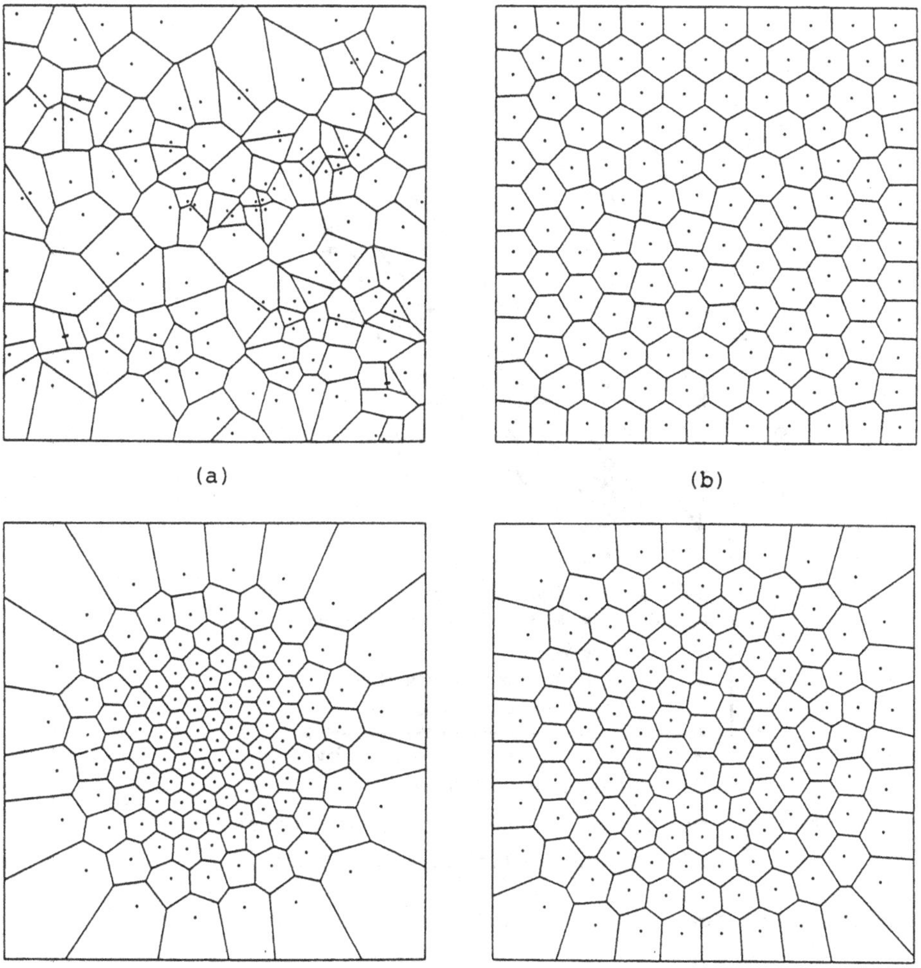

(a) (b) (c) (d)

Fig.2. (a) Initial distribution

(b) Near optimum distribution for the population density of type 1°

(c) Near optimum distribution for the population density of type 2°

(d) Near optimum distribution for the population density of type 3°

the value step by step as $\alpha^{(1)}$, $\alpha^{(2)}$,... until the left inequality in (6.2) holds, say at $\alpha^{(m)}$. If $\alpha^{(m)}$ satisfies the right inequality as well, then we stop, setting $\hat{\alpha}=\alpha^{(m)}$. If $\alpha^{(m)}$ does not satisfy the right inequality, then we search for an $\hat{\alpha}$ by means of the binary search in the interval $(\alpha^{(m-1)},\alpha^{(m)})$. Actually we tested the following two kinds of sequences $\alpha^{(0)}$, $\alpha^{(1)}$,...; one is in an "arithmetic progression"

$$A: \quad \alpha^{(i)} = i\ \alpha^{(0)}, \tag{6.3}$$

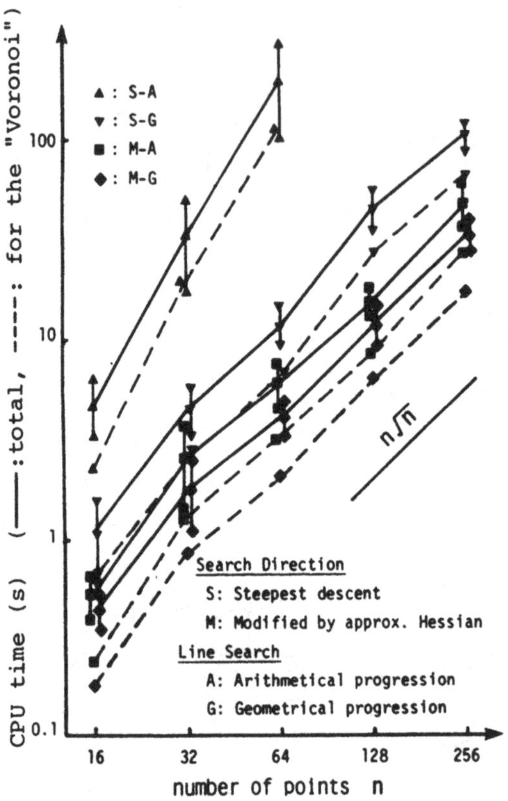

Fig.3. Comparison of the computation times by different solution algorithms

and the other in an "geometric progression"

$$G: \quad \alpha^{(i)} = 2^{i}\alpha^{(0)}. \tag{6.4}$$

(Incidentally, we set $\alpha^{(0)}=1$ rather arbitrarily, and assumed $\alpha^{(-1)}=0$.) In the following, we shall abbreviate the line search based on (6.3) as "A", and that based on (6.4) as "G".

For the simplest search direction $\mathbf{d}^{(\nu)}$, we took the direction of "steepest descent" (to be abbreviated as "S" in the following"), or more specifically (see (5.9)),

$$S: \mathbf{d}^{(\nu)} = -\nabla_{\mathbf{x}}F(\mathbf{x}^{(\nu)}),$$

$$\mathbf{x}_i^{(\nu+1)} = \mathbf{x}_i^{(\nu)} + \hat{\alpha}(\bar{\mathbf{x}}_i^{(\nu)} - \mathbf{x}_i^{(\nu)})\cdot\mu(V_i). \tag{6.5}$$

We also investigated a more sophisticated direction $\mathbf{d} = -H^{-1}\cdot\nabla_{\mathbf{x}}F$, which is obtained by modifying the steepest-descent direction with a certain approximation H of the Hessian (2n×2n matrix) of F. Since the exact Hessian (see (5.7) and (5.12)) is too complicated to incorporate in the iteration process, we adopted the part $H^i_{\lambda\kappa}$ (see (5.12.1)) for the "approximation". (It is numerically not a good approximation, but it is qualitatively qualified as such because it is symmetric and positive definite and has an adequate physical dimension.) Specifically, we have from (5.7) and (5.12.1)

$$\text{M:}\quad \mathbf{d}^{(\nu)} = -H^{-1}\cdot\nabla_{\mathbf{x}}F(\mathbf{x}^{(\nu)}),$$

$$\mathbf{x}_i^{(\nu+1)} = \mathbf{x}_i^{(\nu)} + \hat{\alpha}(\bar{\mathbf{x}}_i^{(\nu)} - \mathbf{x}_i^{(\nu)}). \tag{6.6}$$

We shall abbreviate this choice of search direction as "M" (standing for "modified"). It may be interesting to see that the only difference between the steepest-descent direction and the modified direction in this case consists in whether or not there is the factor $\mu(V_i)$ in the modification term of (6.5) and (6.6).

The results of computational experiments for the population density of type 1° are summarized in Fig. 3. (The results for the other population densities looked quite similar.) From these results, we can get the following observations.

(a) From the viewpoint of computational time, the algorithm (M-G) using the approximate Hessian with the geometric-progression line search performs best. (The difference in performance of various algorithms is rather conspicuous.)

(b) Except for the algorithm (S-A) using the steepest-descent direction with the arithmetic-progression line search, the computation time seems to increase with the number of generators in the order of about $n^{3/2}$.

(c) The larger part of computation time is spent for the construction of the Voronoi diagrams.

Furthermore, we counted the number of function evaluations in each line search, and observed that

(d) The average number of function evaluations needed in one line search is $O(n)$ by the algorithm S-A, $O(\log n)$ by the algorithm S-G, and a constant independent of n by the algorithms M-A and M-G. (The "constant" in the last case is about 2, and about seven line searches out of ten required indeed less than 3 function evaluations.)

Thus, we may conclude that we can solve the geographical optimization problems (of the kind discussed in the present paper) with as many as hundreds of facilities within a practicable time if we make use of a sufficiently fast Voronoi-diagram algorithm.

Acknowledgement

The authors are indebted to many colleagues and friends for their substantial collaboration in computational experiments and valuable advices from various standpoints. Among them, the authors would like to mention, with cordial gratitude, the names of Prof. D. Avis of McGill University, Prof. T. Koshizuka of the University of Tsukuba, and Prof. A. Okabe, Dr. T. Asano and Mr. H. Imai of the University of Tokyo.

The research was supported in part by the Grant in Aid for Scientific Research of the Ministry of Education, Science and Culture of Japan.

References

[1] Avis, D., and Bhattacharya, B.K.: Algorithms for computing d-dimensional Voronoi diagrams and their duals. Technical Report SOCS 82.5, McGill University, Montreal, Canada, 1982.

[2] Boots, B.N.: Some observations on the structure of socio-economic cellular networks. Canadian Geographer, Vol.19, 1975, pp.107-120.

[3] Finney, J.L.: Random packings and the structure of simple liquids —— I. The geometry of random close packing. Proceedings of the Royal Society of London, Vol.A-319, 1970, pp.479-493.

[4] Goldstein, A.A.: Constructive Real Analysis. Harper & Row, 1968.

[5] Green, P.J., and Sibson, R.: Computing Dirichlet tessellation in the plane. The Computer Journal, Vol.21, 1978, pp.168-173.

[6] Horspool, R.N.: Constructing the Voronoi diagram in the plane. Technical Report SOCS 79.12, McGill University, Montreal, Canada, 1979.

[7] Imai, H., Iri, M., and Murota, K.: Voronoi diagram in the Laguerre geometry and its applications. SIAM Journal on Computing (to appear).

[8] Iri, M., Murota, K., and Ohya, T.: Geographical optimization problems and their practical solutions (in Japanese). Proceedings of the 1983 Spring Conference of the OR Society of Japan, C-2, pp.92-93.

[9] Iri, M., et al.: Fundamental Algorithms for Geographical Data Processing (in Japanese). Technical Report T-83-1, Operations Research Society of Japan, 1983.

[10] Lee, D.T.: Two-dimensional Voronoi diagrams in the L_p-metric. Journal of the ACM, Vol.27, 1980, pp.604-618.

[11] Lee, D.T., and Drysdale, R.L., III.: Generalization of Voronoi diagrams in the plane. SIAM Journal on Computing, Vol.10, 1981, pp.73-87.

[12] Newling, B.E.: The spatial variation of urban population densities. Geographical Reviews, 1969, 4.

[13] Ohya, T., Iri, M., and Murota, K.: A fast incremental Voronoi-diagram algorithm with quaternary tree. (Submitted to Information Processing Letters)

[14] Ohya, T., Iri, M., and Murota, K.: Improvements of the incremental method for the Voronoi diagram with computational comparison of various algorithms. (Submitted to the Journal of the OR Society of Japan)

[15] Shamos, M.I., and Hoey, D.: Closest-point problems. Proceedings of the 16th Annual IEEE Symposium on the Foundations of Computer Scinece, Berkeley, 1975, pp.151-162.

[16] M. Tanemura and M. Hasegawa: Geometrical models of territory I —— Models for synchronous and asynchronous settlement of territories. Journal of Theoretical Biology, Vol.82, 1980, pp.477-496.

Appendix: Quaternary Incremental Algorithm for the Voronoi Diagram

We shall give a brief description of the method of constructing the Voronoi diagram for n points (or, generators) given in a unit square of the Euclidean plane, according to [14]. (See also [13].)

The method is basically of the incremental type [5]; i.e., it starts from the Voronoi diagram $\mathbf{V}_3$ and augments it to $\mathbf{V}_4, \mathbf{V}_5, \ldots$ until the complete diagram $\mathbf{V}_n$ is obtained. In augmenting the diagram $\mathbf{V}_{m-1}$ to $\mathbf{V}_m$ by adding a new generator $\mathbf{x}_m$, we first determine the Voronoi region $V_{N[m]}$ to which $\mathbf{x}_m$ belongs in $\mathbf{V}_{m-1}$ (Phase 1:" Nearest Neighbor Search"). Then we construct the new region V_m in $\mathbf{V}_m$ which is the territory of the new $\mathbf{x}_m$ as follows (Phase 2:"Modification of the Diagram"). (See Fig. A1, and also [5].)

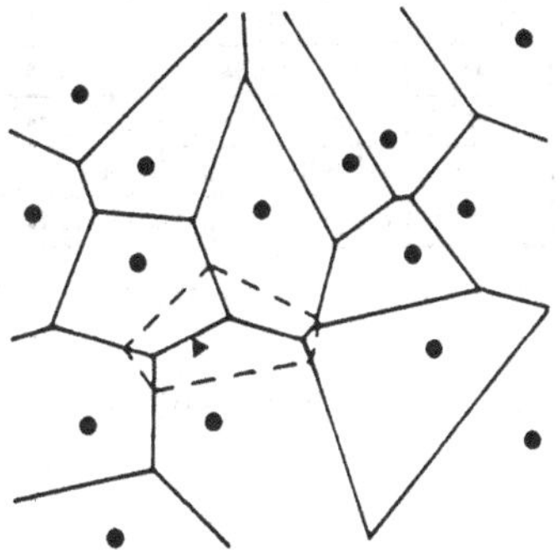

Fig.A1. Construction of the new region for a newly added point

Solid lines: Old diagram
●,●,... : Old generators
▲ : New generator
Broken lines: The boundary of the region for the new generator

Let ℓ_1 be the perpendicular bisector of $\mathbf{x}_m$ and $\mathbf{x}_{N[m]}$. The ℓ_1 intersects two of the boundary edges of $V_{N[m]}$. Let the region adjacent to $V_{N[m]}$ at one of these two edges be $V_{N_1[m]}$ (the territory of generator $\mathbf{x}_{N_1[m]}$), and the perpendicular bisector of $\mathbf{x}_m$ and $\mathbf{x}_{N_1[m]}$ be ℓ_2. The ℓ_2 intersects one of the boundary edges of $V_{N_1[m]}$, other than that which ℓ_1 intersects. Let the region adjacent to $V_{N_1[m]}$ at that edge be $V_{N_2[m]}$ (the territory of $\mathbf{x}_{N_2[m]}$), and the perpendicular bisector of $\mathbf{x}_m$ and $\mathbf{x}_{N_2[m]}$ be ℓ_3. Repeating in this manner, we shall return to the initial region $V_{N[m]}$ to complete the new region of $\mathbf{x}_m$.

What is crucial for the efficiency of the method is (i) in what order to add generators, and (ii) how to find the region $V_{N[m]}$ in the old diagram $\mathbf{V}_{m-1}$ to which the newly added generator $\mathbf{x}_m$ belongs. It may be intuitively not very difficult to admit that, if we augment the diagram keeping it as uniform and similar as possible at every stage, then we have an algorithm which runs in linear time O(n) on the average. In order to realize this uniformity and similarity, we make use of a quaternary structure in the following way.

<1> Divide each side of the unit square into 2^M equal parts to get 4^M small square "buckets", where M is chosen such that $4^M \fallingdotseq n$.

```
c:=⌊2^M/3⌋;
I(0):=c; J(0):=c;
u:=(-1)^(M+1); m:=1; r:=0; s:=2c+u;
for t:=1 to M do
  for p:=1 to m do
    r:=r+1;
    I(r):=s-I(r-m);
    J(r):=J(r-m);
  m:=2m;
  for p:=1 to m do
    r:=r+1;
    I(r):=I(r-m);
    J(r):=s-J(r-m);
  m:=2m;
  u:=-2u;
  s:=s+u;
```

Fig.A2. Bucket-numbering algorithm

J \ I	0	1	2	3	4	5	6	7	8	9	10	11	12	13	14	15
15	191	190	186	187	171	170	174	175	239	238	234	235	251	250	254	255
14	189	188	184	185	169	168	172	173	237	236	232	233	249	248	252	253
13	181	180	176	177	161	160	164	165	229	228	224	225	241	240	244	245
12	183	182	178	179	163	162	166	167	231	230	226	227	243	242	246	247
11	151	150	146	147	131	130	134	135	199	198	194	195	211	210	214	215
10	149	148	144	145	129	128	132	133	197	196	192	193	209	208	212	213
9	157	156	152	153	137	136	140	141	205	204	200	201	217	216	220	221
8	159	158	154	155	139	138	142	143	207	206	202	203	219	218	222	223
7	31	30	26	27	11	10	14	15	79	78	74	75	91	90	94	95
6	29	28	24	25	9	8	12	13	77	76	72	73	89	88	92	93
5	21	20	16	17	1	0	4	5	69	68	64	65	81	80	84	85
4	23	22	18	19	3	2	6	7	71	70	66	67	83	82	86	87
3	55	54	50	51	35	34	38	39	103	102	98	99	115	114	118	119
2	53	52	48	49	33	32	36	37	101	100	96	97	113	112	116	117
1	61	60	56	57	41	40	44	45	109	108	104	105	121	120	124	125
0	63	62	58	59	43	42	46	47	111	110	106	107	123	122	126	127

Fig.A3. Example of numbering buckets (M=4)

<2> Number the 4^M buckets (from 0 to 4^M-1) by the "bucket-numbering algorithm" shown in Fig. A2 (see also an example for M=4 in Fig. A3), where it should be noted that this algorithm runs in O(n) time.

<3> Make a quaternary tree, with leaves (ordered from <u>left to right</u>) corresponding to the 4^M buckets (ordered according to the bucket numbers). Associate each generator to the bucket (=leaf) to which it belongs. (This can be done by multiplying the coordinates by 2^M and then truncating the fractional parts off.)

Scanning the leaves from left to right, if a leaf contains a generator, put it in all the ancestors of the quaternary tree in which no generator has yet been put. (See Fig. A4 for <3>.)

<4> Scanning the nodes of the tree starting from the top (root) in the breadth-first manner (from <u>right to left</u> on the same level), as soon as we find a generator i that has not been added to the diagram, add it to the diagram and construct its territory (the new Voronoi region) as has been described in the above (Phase 2).

In Phase 1, adopt the Voronoi region V_j of the generator $\mathbf{x}_j$ in the

J/I	0	1	2	3	4	5	6	7
7	[47] 24	[46] 59	[42] 60	[43] 25	[59]	[58]	[62] 61 14	[63] 37
6	[45]	[44] 64 16	[40] 57	[41] 52	[57]	[56]	[60] 15 5	[61]
5	[37] 23	[36]	[32] 17	[33] 18 7	[49] 33 8	[48] 29 9 1	[52] 36 32	[53] 53 51
4	[39]	[38] 22 13	[34]	[35] 19	[51] 31	[50] 54	[54]	[55] 50 49
3	[7] 45 43 3	[6]	[2] 30 27	[3] 2	[19] 63	[18] 62	[22] 20 4	[23]
2	[5] 46	[4]	[0] 11	[1] 10	[17] 55	[16] 58 28	[20]	[21] 48
1	[13] 6	[12] 26	[8] 44 38	[9] 21 12	[25]	[24] 39	[28] 35	[29] 65
0	[15] 47	[14]	[10] 34	[11] 66	[27] 42 41	[26]	[30]	[31] 56 40

Fig.A4. Example of bucketing 66 generators distributed in the unit square in constructing the corresponding quaternary tree. (Bucket-numbers are indicated in brackets, whereas generator-numbers are indicated without brackets.)

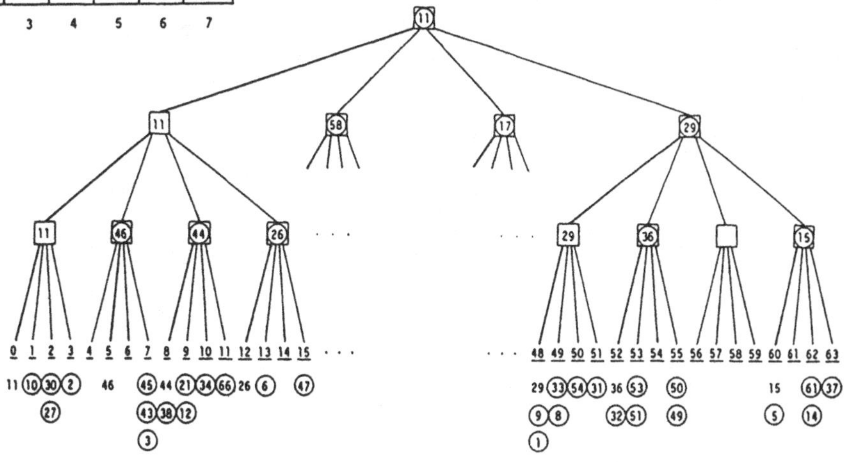

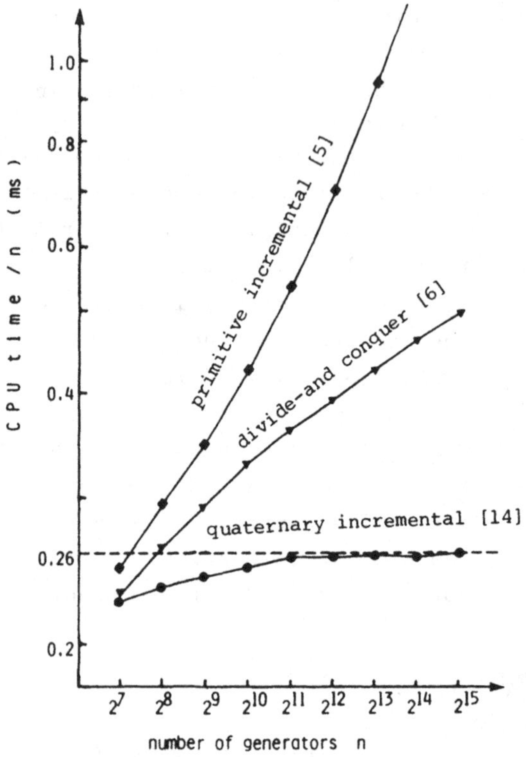

Fig.A5. Comparison of the performances of three different Voronci algorithms

father node as the initial guess for the $V_{N[m]}$ to which the new generator $\mathbf{x}_m$ belongs. (The $V_{N[m]}$ is determined by starting from the V_j and moving from a region to another in such a way that the distance of generator $\mathbf{x}_m$ to the generator of the region may decrease monotone.)

NONLINEAR OPTIMIZATION BY A CURVILINEAR PATH STRATEGY

L. GRANDINETTI
Dipartimento di Sistemi
Università della Calabria
87036 Arcavacata (Cosenza), Italy

1. INTRODUCTION

To find a local solution of the unconstrained minimization problem

$$\min F(x), x \in R^n$$

it is not a straightforward calculation in many difficult situations. These situations, associated -for instance- to the solution of nonlinear programs via penalty methods, may be pictorially described as those in which the objective function behaves like a long curved valley with very steep walls.

Particularly in these situations it may be suitable to find a curved trajectory in R^n which passes through the minimizer and, then, to follow this trajectory by curvilinear searches with the aim to reach the solution in few *long* steps. This strategy is intended to overcome the typical behaviour of classical descent methods which, in these situations, necessarily perform many *short* steps along linear monodimensional manifolds. In fact, most of them are based on the iterative model

$$x_{k+1} = x_k + \alpha_k d_k \qquad (1)$$

where k denotes the iteration index, x_k is the current iterate, $d_k \in R^n$ defines the direction of a linear trajectory, and $\alpha_k \in R_+$ is a steplength along d_k suitably chosen in such a way that $f(x_{k+1}) < f(x_k)$.

In this paper an original way to construct a trajectory is pointed out; it is based on the concept of parallel tangent hyperplanes and possesses some interesting theoretical properties.

In principle this trajectory may be derived as solution of a system of differential equations; however it is shown how a suitable approximation of it can be obtained without any explicit solution of the differential equations.

Lastly, a prototype implementation of this curvilinear path strategy is devised; the relevant numerical experiments, although limited, seem to indicate that the method may be capable of very high efficiency, at least for certain classes of objective functions.

2. GENERALLY DESIRABLE TRAJECTORY FEATURES

A trajectory $x(\alpha)$, i.e. a nonlinear monodimensional manifold parametrized by the scalar $\alpha \in R_+$, may be associated in many ways to $f(x)$ for the purpose of its minimization; therefore it is sensible to devise a framework of suitable general properties to be possessed by trajectories of practical interest.

Here the following properties are taken into consideration, having assumed $f:R^n \to R$ globally differentiable.

(i) Regularity

A basic natural requirement for the trajectory is that

$x(\alpha):R_+ \to R^n$

be a one-to-one continuous mapping.
This excludes the possibility for discontinuities, bifurcation points and loops to occur; the benefits of avoiding such situations can be remarkable whenever a numerical treatment of the trajectory has to be done. And in fact this is the case for trajectories of practical interest.

The additional stronger requirement that $x(\alpha)$ be globally differentiable, can be also considered generally desirable; this means, in other words, that for each value of α is defined the tangent linear variety to the trajectory at $x(\alpha)$.

(ii) Suitability

Straightforward requirements for a trajectory $x(\alpha)$, $\alpha \in R_+$, to be suitable for the purpose of local minimization of $f(x)$, can be considered the following:

(a) the convergence to the local minimizer x^* from an initial estimate $x(o)=x_0$;

(b) the possibility to determine α^* (a finite or a limit value) such that $x(\alpha^*)$ be the local optimizer.

Let us consider, for example, the following objective function:

$$f(x) = \tfrac{1}{2}x^T Ax+b^T x \qquad (2)$$

$A \in R^{n\times n}$, $A > 0$ and $b \in R^n$

and associate to it the trajectory implicitly defined by the system of ordinary differential equations (ODE):

$$x'(\alpha) = -A^{-1}g_0$$

with

$$x(o) = x_0$$

and

$$g_0 \equiv \nabla f(x_0) \quad .$$

It is immediate to see that the trajectory is regular and suitable; i.e. it passes through the minimizer and, in addition, for $\alpha = 1$, we get:

$x(1) = x_0 - A^{-1}g_0 \equiv x^*$.

It is worth while to observe that f(x) is always decreasing on $x(\alpha)$, for $\alpha \in [o,\alpha^*]$.

Remark 2.1

The attribute (b) is an ideal one; generally neither $x(\alpha)$ is a known function of α nor it is possible to determine analytically $x(\alpha^*)$ like in previous simple example. Therefore the weaker requirement (more realistic than that stated in (b)) that f(x) decreases as α increases to α^* can be, more conveniently, assumed.
This descent property of the trajectory may result particularly useful in its practical numerical processing.

(iii) Characterizability

It is crucial that $x(\alpha)$ be characterized in terms of properties of f(x).
An useful way for defining the trajectory may be that of exploiting differential properties of f(x) (e.g. $x(\alpha)$ expressed as solution of a system of ODE). A particular characterization of this type is discussed in the sequel of this paper.

(iv) Linear invariance

It is sensible that the trajectory described into the domain of the variable x be invariant under a linear transformation of variables defined by:

$$x = J\xi$$

$$\xi \in R^n \ ; \quad J \in R^{nxn} \ , \quad \det J \neq 0.$$

This precisely means that trajectories constructed in terms of x and ξ with the same procedure are related in such a way that:

$$x(\alpha) = J\ \xi(\alpha) \ , \quad \forall\alpha \ .$$

3. A TRAJECTORY DERIVED VIA PARALLEL TANGENT HYPERPLANES

A way to construct a trajectory based on the concept of parallel tangent hyperplanes is pointed out. Several desirable properties and features are possessed by this trajectory for classes of functions of practical interest.

The basic idea on which the method is founded can be usefully introduced by means of a geometric sketch in R^2, illustrated in Fig.1.

Let us consider, in correspondence to a starting point x_0, where $g_0 \neq o$, the tangent plane to the level curve defined by $f(x)=f(x_0)$.

Furthermore consider the parallel planes

$$g_0^T(x-x_0) = -\alpha$$

parametrized by the scalar $\alpha \geq 0$.

Finally consider those points where the planes defined before are tangent to the level curves.

The locus of such points (parametrized by the nonnegative scalar α) defines the trajectory $x(\alpha)$.

If, for example, we consider again the strictly convex function given by (2), then it is easy to recognize that the trajectory associated to it via parallel tangent planes is:

(i) regular (precisely, it is linear);

(ii) characterizable;

(iii) suitable $:\alpha^* \ni x(\alpha^*)=x^*$, easily determined;

(iv) linearly invariant.

As an immediate consequence, we have that the trajectory derived on the basis of the previous concept possesses excellent behaviour on convex quadratic functions (precisely the same as the Newton method) and thus, typically, near the solution of a general smooth function.

However, it is worthwhile to observe that, in addition, a useful behaviour is possessed for nonconvex quadratic functions (e.g. concave, saddle), differently of the simple Newton method. In fact, in these cases a descent trajectory, linearly invariant, is still provided.

Since all desirable properties are generally guaranteed at least near the solution of a smooth function, it seems natural to extend this procedure to general cases.

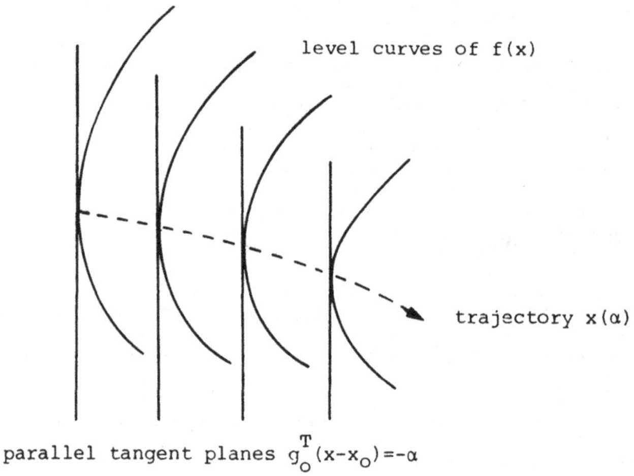

Fig.1. Parallel tangent planes and related trajectory.

The generalization of the method to the n-dimensional case and its proper formulation lead to formalize the definition of the trajectory as the locus of points satisfying the following canonical conditions:

$$g_o^T (x-x_o) = -\alpha \tag{3a}$$

$$g(x(\alpha)) = \lambda(\alpha) g_o \tag{3b}$$

where:

$\alpha \geq o$; $x(o) = x_o$; $g_o \equiv g(x_o) \neq o$

and $\lambda(\alpha)$ is a scalar function such that $\lambda(o)=1$ and $\lambda(\alpha)=o$ whenever $g(x(\alpha))=o$ (i.e. at any stationary point on the trajectory).
It is worth noting that $\lambda(\alpha)$ can be interpreted as the Lagrange multiplier associated to the subproblem of minimizing $f(x)$ constrained by the generic tangent hyperplane.

In addition to (3), the continuity of $x(\alpha)$ is assumed as a canonical side-condition; in fact it guarantees (at least for broad classes of problems) local unicity of the trajectory.

Finally, algebraic manipulations lead to the following characterization:

$$x'(\alpha) = -\frac{u(\alpha)}{g_o^T u(\alpha)} \tag{4a}$$

$$x(o) = x_o \tag{4b}$$

Here the n-dimensional vector $u(\alpha)$ satisfies:

$Gu = g_o$

where $G \equiv \nabla^2 f(x(\alpha))$.

In principle the characterization is not always possible (i.e. for any value of α) and may seem quite restrictive. However, in practice, when the trajectory is treated by some suitable numerical technique, there are no severe consequencies at least for broad classes of functions (provided that the trajectory exists and is well defined).

The *properties* of the trajectory formally defined here, generally agree, at least for some large classes of functions, with those desirable features described in section 2.
In fact $x(\alpha)$ is *regular*, if $f(x)$ is sufficiently well-behaved. In addition, it is *linearly invariant* since the following result holds

Fact 3.1.

Let us assume

$x = J\,\xi$

$\xi \in R^n$, $J \ni \det J \neq o$.

Then, on the basis of the defining conditions (3), the trajectories in the domain of x and ξ are related by:

$x(\alpha) = J\,\xi(\alpha)$,

showing linear invariance. □

Moreover a *descent property* relies on the following result

Fact 3.2.

If $x(\alpha)$ is globally differentiable, then it is a straightforward consequence of (3) that

$g^T x'(\alpha) = -\lambda(\alpha)$. □

It is worth noting that, since $\lambda(o)=1$, initially the trajectory's slope is negative; then, whenever $\lambda(\alpha) > o$, the trajectory is a descent one. This happens, at least, on the arc between $\alpha=o$ and $\bar{\alpha}$ (for which $\lambda(\bar{\alpha})=o$) where a stationary point is encountered.

Lastly, the trajectory is generally *characterizable*. Except "pathological" situations, this happens almost everywhere and, if necessary, the trajectory can be described by continuation; as a typical example, the following objective function can be considered:

$f(x) = 1 - \exp(-x^T x)$.

It is worth noting that most of the above mentioned properties can be precisely established for some important classes of functions.

For strictly convex and globally differentiable functions, it can be proved, in particular, that $\lambda(\alpha)$ is monotonically decreasing; in fact, it stands the following:

Lemma 3.1.

Given $f(x)$ strictly convex and globally differentiable, then for any (α_1, α_2) with $\alpha_1 < \alpha_2$ we have

$\lambda_2 < \lambda_1$. □

With essentially the same assumptions on $f(x)$, then the stronger results that the trajectory is regular, suitable, linearly invariant and characterizable can be proved.

For the class of pseudoconvex functions in the sense of Mangasarian [1] , which are interesting for their applications in several contexts, the trajectory still possesses the desirable properties although in weaker sense ($\lambda(\alpha)$ non monotonically decreasing; characterizability almost everywhere).

On the basis of the many sound results, it seems possible to draw the conclusion that the method generally produces trajectories well-behaved, is descent and essentially "becomes" the Newton method near the solution of smooth functions, thus showing good potentialities for effective numerical implementations.

4. QUADRATIC MODEL OF THE TRAJECTORY

It is not realistic, in general situations, to derive the trajectory $x(\alpha)$ as solution of a system of nonlinear ODE of the type (4).

To overcome the problem of explicitly solving differential equations, here the following quadratic model of $x(\alpha)$ is taken into consideration:

$$x(\alpha) \simeq x(o) + \alpha s + \tfrac{1}{2}\alpha^2 t \qquad (5)$$

where

$s \triangleq x'(o)$; $t \triangleq x''(o)$.

If f(x) is such that the trajectory is locally characterizable at x_0, then s can be explicitly obtained;indeed,in this case, it is an obvious consequence of (4) the following

Fact 4.1.

At the initial point x_0 the trajectory's first derivative is given by:

$$s = - \frac{G_0^{-1} g_0}{g_0^T G_0^{-1} g_0} \quad . \tag{6}$$

□

If, in addition, f(x) possesses continuous third derivatives, then the following important result holds

Proposition 4.1.

If $f(x) \in C^3$ and, in addition, the trajectory is locally characterizable at x_0, then the vector t can be derived by solving:

$$G_0 t = - \overset{s}{G}_0 s \tag{7}$$

where $\overset{s}{G}_0$ is a tensor defined by:

$$\overset{s}{G}_0 = \left[G_1 s \;\vdots\; G_2 s \;\vdots\; \dots \;\vdots\; G_n s \right]^T$$

with

$$G_i \triangleq \nabla^2 g_i(x_0) \;;\; g_i \triangleq \frac{\partial f}{\partial x_i} \;,\; i=1,\dots,n.$$

Proof.

Let us assume the following second order approximation of the gradient g along the trajectory:

$$g(x(\alpha)) = g(x(o)) + G_0 \left[x(\alpha)-x(o) \right] + \tfrac{1}{2}\overset{x}{G}_0[x(\alpha)-x(o)]$$

where

$$x \equiv [x(\alpha) - x(o)] = \alpha s + \tfrac{1}{2}\alpha^2 t$$

$$\overset{x}{G}_0 \equiv \left[G_1 x \;\vdots\; \dots \;\vdots\; G_n x \right]^T .$$

It follows that

$$g(x(\alpha)) = g(x(o)) + \alpha G_0 s + \tfrac{1}{2}\alpha^2 G_0 t + \tfrac{1}{2}(\alpha \overset{s}{G}_0 + \tfrac{1}{2}\alpha^2 \overset{t}{G}_0)(\alpha s + \tfrac{1}{2}\alpha^2 t)$$

and, by neglecting terms of high degree in α, the following gradient approximation is obtained:

$$g(x(\alpha)) = g(x(o)) + \alpha G_0 s + \tfrac{1}{2}\alpha^2 G_0 t + \tfrac{1}{2}\alpha^2 \overset{s}{G}_0 s \quad . \tag{8}$$

But, on the other hand, from conditions (3) it can be easily derived the following approximation for $\lambda(\alpha)$:

$$\lambda(\alpha) = 1 - \frac{\alpha}{g_0^T G_0^{-1} g_0}$$

which, utilized in the canonical condition

$$g(x(\alpha)) = \lambda(\alpha)\, g_o ,$$

permits to obtain:

$$g(x(\alpha)) = g_o - \frac{\alpha\, g_o}{g_o^T G_o^{-1} g_o} . \qquad (9)$$

By equating the expressions given by (8) and (9), we finally get:

$$G_o t + \overset{s}{G_o} s = o . \qquad \square$$

It derives from the above results that, under relatively mild assumptions on $f(x)$, the quadratic model of the trajectory is completely specified.

5. COMPUTATIONAL ASPECTS

The calculation of third derivatives, needed in general situations to obtain the quadratic model of the trajectory, seems to be a serious computational complication. In principle these calculations could be avoided by approximating third derivative matrices in a suitable sense (e.g. by quasi-Newton techniques). However explicit computation of third derivatives is challenging, at least for problems of moderate size, especially if the expected benefits are noticeable. Under this respect, modern advanced computational techniques, like parallel processing and symbolic manipulation, may result very helpful.

Apparently a second source of difficulties may be the occurrence of singularities of the Hessian matrix of $f(x)$; in practice this is not a serious one, at least for objective functions sufficiently regular, provided that an appropriate initialization of the trajectory is done and a suitable curvilinear search along the quadratic model is performed.

When the approximating quadratic trajectory is used in an iterative scheme, it becomes crucial to devise effective steplength algorithms for the computation of a suitable value of α, at any iteration. The present quadratic model can benefit in many situations of the Goldfarb [2] curvilinear path steplength algorithms; these generalize the Armijo and Goldstein conditions to the case

$$x(\alpha) = x(o) + \alpha s + \alpha^2 t$$

providing, under mild assumptions, convergence to a stationary point.

Additional aids to a practical numerical implementation of the curvilinear search are:

(i) a third order approximation of $f(x)$, restricted to the quadratic trajectory, can be computed without any extra computational effort;

(ii) a good initial guess of the steplength can be provided, at least near the solution, by the value of $\alpha \ni \lambda(\alpha) = o$.

6. PROTOTYPE IMPLEMENTATION

An implementation has been devised, following previous ideas, and tested on functions which seem to possess features suitable for a meaningful validation of the method.

A very effective behaviour has been achieved, by using accurate curvilinear search for the following test functions:

Test 1. Rosenbrock standard

$$f(x) = 100\,(x_1^2-x_2)^2 + (1-x_1)^2 \quad ,$$

starting point $x_0 = (-1.2\ ,\ 1.0)^T$.

Test 2. Powell singular

$$f(x) = (x_1+10x_2)^2 + 5(x_3-x_4)^2 + (x_2-2x_3)^4 + 10(x_1-x_4)^4 \quad ,$$

starting point $x_0 = (3,\ -1\ ,\ 0\ ,\ 1)^T$.

The numerical results for the Rosenbrock function, which can be considered a good model of long curved valley with steep walls, are reported in Table 1.

Table 1. Results for the test function *Rosenbrock standard*

ITERATION	x		f(x)
Initial Conditions	-1.20	1.00	24.2
1st	0.568	0.322	1.8×10^{-1}
3rd	0.996	0.991	9.5×10^{-5}
5th	1.000	1.000	1.4×10^{-11}

These can be fully appreciated if compared with analogous results of very well established classical algorithms. A good quasi-Newton algorithm takes 28 iterations for decreasing f(x) to the value of 10^{-5}; similar results are provided by a good implementation of the newest collinear scaling algorithm [3] . Moreover, a trust region algorithm, of the Hebden-Moré type [4] , takes 21 iterations to minimize f(x), and a good implementation of the modified Newton method (of the Gill and Murray type [4]) takes 24 iterations.

Computational results relative to the Powell function, whose Hessian is singular at the minimizer, are reported in Table 2. The numerical performance of the curvilinear path strategy can be still considered very good; in fact a good quasi-Newton algorithm takes 15 iterations for decreasing f(x) to the value of 10^{-5} and 40 iterations to minimize it.

Table 2. Results for the test function *Powell singular*

ITERATION	x				f(x)
Initial Conditions	3.0	-1.0	0.0	1.0	$2.1x10^{2}$
3rd	$1.6x10^{-1}$	$-7.5x10^{-3}$	$3.2x10^{-2}$	$1.9x10^{-2}$	$1.3x10^{-2}$
5th	$3.9x10^{-2}$	$-4.3x10^{-3}$	$6.1x10^{-3}$	$6.6x10^{-3}$	$2.6x10^{-5}$
9th	$1.1x10^{-3}$	$-1.1x10^{-4}$	$1.8x10^{-4}$	$1.8x10^{-4}$	$3.1x10^{-11}$

7. CONCLUDING REMARKS

The first computational experience with the new curvilinear path strategy presented here seems encouraging; the testing, although very limited, has been carried out on functions which can be considered representative of those for which this method has been designed and may result particularly appropriate (e.g., very ill-conditioned, like the Powell function near the solution).

At present, the effort in computing third derivatives can be a limiting factor, at least for large size problems. However, it is worth noting that, among various possibilities to avoid this calculations, the following tensor approximation has been taken into consideration:

$$\alpha \overset{s}{G}_0 = G(x_0 + \alpha s) - G_0 \ .$$

Computations carried out with this approximation (which, therefore, produces a second derivative method), provided numerical results very close to those reported in the Tables 1 and 2.

Although the results presented here are essentially based on an *abstract algorithm* and further research is needed on it, nevertheless it seems possible to draw the conclusion that the underlying theory is sound, indicating the potentiality of the method in achieving high performances.

REFERENCES

[1] MANGASARIAN,O.L., *Nonlinear Programming*, McGraw-Hill, New York, 1969.

[2] GOLDFARB,D., *Curvilinear path steplength algorithms for minimization which use directions of negative curvature*, Mathematical Programming, Vol.18, pp.31-40, 1980.

[3] AL-DHAHIR,J., *A descent method for unconstrained optimization based on a conic model*, University of Dundee, MS Thesis, 1982.

[4] FLETCHER,R., *Practical Methods of Optimization, Vol.1*, Wiley, Chirchester, 1980.

A UNIFIED NONLINEAR PROGRAMMING THEORY FOR PENALTY, MULTIPLIER, SQP AND GRG METHODS

K. Schittkowski
Institut für Informatik
Universität Stuttgart
Azenbergstraße 12
D-7000 Stuttgart 1

Abstract: The four most important approaches for solving the constrained nonlinear programming problem, are the penalty, multiplier, sequential quadratic programming, and generalized reduced gradient method. A general algorithmic frame will be presented, which realizes any of these methods only by specifying a search direction for the variables, a multiplier estimate, and some penalty parameters in each iteration. This approach allows to illustrate common mathematical features and, on the other hand, serves to explain the different numerical performance results we observe in practice. It will be shown that the algorithm is well-defined and approximates a Kuhn-Tucker point of the underlying problem.

1. INTRODUCTION

In the following, we consider solution methods for the nonlinear programming problem

$$x \in \mathbb{R}^n: \begin{array}{l} \min \ f(x) \\ g(x) = 0 \end{array} \qquad (1)$$

with equality constraints. The functions $f : \mathbb{R}^n \to \mathbb{R}$ and $g : \mathbb{R}^n \to \mathbb{R}^m$ are assumed to be continuously differentiable. Although the subsequent analysis is applicable to inequality constraints as well, we restrict ourselves to the above problem type. In both cases, the same underlying mathematical ideas are used, but equality constraints lead to a formally simpler presentation and to an easier understanding of the results.

The basic tool in nonlinear programming theory is the Lagrange function

$$L(x,v) := f(x) - v^T g(x) \qquad (2)$$

defined for all $x \in \mathbb{R}^n$ and $v \in \mathbb{R}^m$. In particular, the function $L(x,v)$ serves to define a so-called augmented Lagrange function

$$\psi_r(x,v) := L(x,v) + \frac{1}{2} r\, g(x)^T g(x) \ , \qquad (3)$$

which depends on an additional penalty parameter $r > 0$. This augmented Lagrange function will be used furtheron to derive a unified nonlinear programming theory. The intention is to develop one algorithm for solving (1), which realizes either a penalty, a multiplier, a sequential quadratic programming, or a generalized reduced gradient method only by specifying a search direction, a multiplier guess, and two penalty parameters in each iteration step. The algorithms mentioned above, represent the most important approaches to solve the nonlinear programming problem (1) both from the theoretical and practical point of view.

The unified approach allows to find out common features of the algorithms, in particular their attempts to approximate the Kuhn-Tucker conditions for the nonlinear programming problem. On the other hand, it is also possible to see the fundamental differences between the four algorithms which allow now to explain the divergent numerical performance behaviour we observe in practice, cf. Schittkowski [9] and Hock, Schittkowski [5]. Furthermore we obtain a unified global convergence result in the sense that starting from arbitrary initial values, a Kuhn-Tucker point for (1) will be approached.

Note however, that the previously mentioned nonlinear programming methods define only four classes of related algorithms. In each case, there exist various modifications to realize the corresponding method, e.g. by defining different multiplier estimates, penalty parameters, line search procedures, approximations of Hessian matrices, etc. Instead of presenting a general theory covering all variants proposed in the literature, we pick out one specific realization in each case, which seems to be representatives for the method, and which has been or could be implemented directly in form of a computer program.

2. THE UNIFIED NONLINEAR PROGRAMMING ALGORITHM

We present the general model of a unified nonlinear programming algorithm. Only a search direction for the variables, a multiplier guess, and some penalty parameters have to be determined, to specify one of the methods mentioned in Section 1, i.e. a penalty (PE), multiplier (MU), sequential quadratic programming (SQP), or a generalized reduced gradient (GRG) algorithm. Subsequently all these algorithms proceed in the same way: A steplength is computed, the new iterates are formulated and a certain approximation formula of a Hessian matrix is updated.

To realize the algorithm, we need a special line search to determine a steplength, i.e. a procedure that minimize a given one-dimensional function. Here one could implement any algorithm known from unconstrained nonlinear programming theory which guarantees convergence. The following method will be proposed because of its simple structure.

(2.1) Algorithm (Line search): Let $\phi : \mathbb{R} \to \mathbb{R}$ be a continuously differentiable function with $\phi'(0) < 0$. Furthermore let μ, β, and $\overline{\beta}$ be some constant real numbers with $0 < \mu < \frac{1}{2}$ and $0 < \beta \leq \overline{\beta} < 1$.
Search for the first index i_o, i.e. the first iterate α_{i_o} from a monotone decreasing sequence $\{\alpha_i\}$ of positive numbers with $\beta^i \leq \alpha_i \leq \overline{\beta}^i$, such that

$$\phi(\alpha_i) \leq \phi(0) + \mu\alpha_i\phi'(0). \tag{4}$$

As a further module of the nonlinear programming algorithm, we need a procedure to update a given approximation of a Hessian matrix. This could be any well-known quasi-Newton method, e.g. the BFGS-formula which is repeated here for the matter of completeness.

(2.2) Algorithm (BFGS-update): Let $B \in \mathbb{R}^{n\times n}$ be a symmetric matrix and $d,q \in \mathbb{R}^n$ be given data. Then compute a new update matrix by

$$\overline{B} := \Pi(B,q,d) := \begin{cases} B & , \text{ if } d^Tq = 0 \text{ or } d^TBd = 0, \\ B + \dfrac{qq^T}{d^Tq} - \dfrac{Bdd^TB}{d^TBd} & , \text{ otherwise}. \end{cases} \tag{5}$$

The matrix $\overline{B}$ remains symmetric and positive definite, whenever B is symmetric and positive definite and $d^Tq > 0$. This latter condition could be enforced by a simple modification of the BFGS-formula (5), cf. Powell [7].

Now we are able to present the general nonlinear programming algorithm.

(2.3) Algorithm (Unified nonlinear programming method): Choose arbitrary initial vectors $x_o \in \mathbb{R}^n$, $v_o \in \mathbb{R}^m$, some non-negative penalty parameters r_o^{ls}, r_o^{qN}, and a positive semi-definite matrix $B_o \in \mathbb{R}^{n \times n}$.
For $k = 0,1,2,\ldots$ compute x_{k+1}, v_{k+1}, r_{k+1}^{ls}, r_{k+1}^{qN}, and B_{k+1} as follows:

1. According to the method chosen, determine a search direction $d_k \in \mathbb{R}^n$, an approximation of the multiplier $u_k \in \mathbb{R}^m$, and new penalty parameters r_{k+1}^{ls}, r_{k+1}^{qN}.

2. Apply Algorithm (2.1) to the merit function

$$\phi_k(\alpha) := \psi_{r_{k+1}^{ls}}\left(\begin{pmatrix} x_k \\ v_k \end{pmatrix} + \alpha \begin{pmatrix} d_k \\ u_k - v_k \end{pmatrix}\right), \tag{6}$$

to obtain a steplength α_k.

3. Define new iterates

$$x_{k+1} := x_k + \alpha_k d_k \quad ,$$

$$v_{k+1} := v_k + \alpha_k (u_k - v_k) \quad .$$

4. If allowed by the method chosen in Step 1, compute new multiplier estimates u_{k+1} and let $u_k' := u_{k+1}$. Otherwise set $u_k' := u_k$.

5. Compute a new matrix B_{k+1} by the BFGS-formula (5) of Algorithm (2.2), i.e. let

$$B_{k+1} := \Pi(B_k, \nabla_x \psi_{r_{k+1}^{qN}}(x_{k+1}, u_k') - \nabla_x \psi_{r_{k+1}^{qN}}(x_k, u_k), x_{k+1} - x_k) \ .$$

Note that the algorithm iterates the variables x_k and the multiplier estimates v_k simultaneously to allow a more straightforward convergence analysis. The line search will be performed only with respect to the variables, if $u_k = v_k$.

3. ADAPTION OF PENALTY, MULTIPLIER, SEQUENTIAL QUADRATIC PROGRAMMING AND GENERALIZED REDUCED GRADIENT METHODS

In the proceeding section, the general model of a nonlinear programming method has been outlined. There remains the question how to choose a search direction d_k, multiplier estimates u_k, and penalty parameters r_k^{ls}, r_k^{qN}, to realize one of the optimization methods under consideration, i.e. a penalty, multiplier, sequential quadratic programming, or generalized reduced gradient algorithm. Any of the resulting four sub-algorithms can then be implemented in Step 1 of Algorithm (2.3).

First the classical penalty method will be considered, which is obtained by realizing the following calculations for d_k, u_k, r_{k+1}^{ls}, and r_{k+1}^{qN} in Step 1 of Algorithm (2.3). To simplify the analysis, we introduce the abbreviations

$$g_k := g(x_k) \quad , \quad A_k := A(x_k)$$

for any given x_k, where $A(x) \in \mathbb{R}^{mxn}$ is the derivative matrix of $g(x)$.

<u>(3.1) Algorithm (Penalty method):</u> Let ϵ_o, ϵ, and ρ be positive numbers with $\epsilon < 1$ and $\rho > 1$. Assume that Algorithm (2.3) provides the data x_k, v_k, r_k^{ls}, and B_k.

(i) $\| \nabla_x \psi_{r_k^{ls}}(x_k, v_k) \| > \epsilon_k$: Let $r_{k+1}^{ls} := r_{k+1}^{qN} := r_k^{ls}$ and

$$\epsilon_{k+1} := \epsilon_k \ .$$

(ii) $\| \nabla_x \psi_{r_k^{ls}}(x_k, v_k) \| \leq \epsilon_k$: Let $r_{k+1}^{ls} := r_{k+1}^{qN} := \rho r_k^{ls}$ and

$$\epsilon_{k+1} := \epsilon \epsilon_k \ .$$

(iii) Determine a solution d_k of the linear system

$$B_k d + \nabla f(x_k) = -r_{k+1}^{ls} A_k^T g_k$$

and define $u_k := v_k$.

When starting with $v_o = 0$, we get $u_k = v_k = 0$ for all k. If B_k is non-singular, the search direction d_k can be written in the form

$$d_k = -B_k^{-1} \nabla_x \psi_{r_{k+1}}(x_k, 0)$$

where the index 'ls' is omitted, since both penalty parameters are identical. Obviously we obtain a quasi-Newton algorithm to minimize this penalty function for a fixed penalty parameter, until the optimality condition (ii) is satisfied. Then the penalty parameter is enlarged, and the whole unconstrained optimization process is repeated.

Therefore Algorithm (3.1) realizes a typical penalty method as for example described in Fiacco and McCormick [2].

Multiplier methods have been developed with the intention to improve the multiplier estimates and to obtain at least moderate penalty parameters, cf. Fletcher [3] or Rockafellar [8].

<u>(3.2) Algorithm (Multiplier method):</u> Let ϵ_o and ϵ be positive numbers with $\epsilon < 1$. Assume that Algorithm (2.3) provides the data x_k, v_k, r_k^{ls}, and B_k, so that B_k is nonsingular.

(i) $\| \nabla_x \psi_{r_k^{ls}}(x_k, v_k) \| > \epsilon_k$: Let $r_{k+1}^{ls} := r_{k+1}^{qN} := r_k^{ls}$,

$\epsilon_{k+1} := \epsilon_k$, and $u_k := v_k$.

(ii) $\| \nabla_x \psi_{r_k^{ls}}(x_k, v_k) \| \le \epsilon_k$: Define

$$r_{k+1}^{ls} := \max\left(\frac{2\,|g_k^T g_k - g_k^T A_k B_k^{-1} \nabla_x L(x_k, v_k)|}{g_k^T A_k B_k^{-1} A_k^T g_k},\; r_k^{ls} \right), \tag{7}$$

and

$$u_k := v_k - r_{k+1}^{ls}\, g_k \; . \tag{8}$$

Let $r_{k+1}^{qN} := r_{k+1}^{ls}$ and $\epsilon_{k+1} := \epsilon\epsilon_k$.

(iii) Determine a solution d_k of the linear system

$$B_k d + \nabla f(x_k) = A_k^T (v_k - r_{k+1}^{ls}) \; .$$

Again the penalty parameter for the line search and the quasi-Newton procedure are identical. From

$$d_k = -B_k^{-1} \nabla_x \psi_{r_{k+1}}(x_k, v_k)$$

we conclude that Algorithm (3.2) define a quasi-Newton algorithm for solving the unconstrained optimization problem

$$\min \{ \psi_{r_{k^*}}(x, v_{k^*}) : x \in \mathbb{R}^n \} \quad ,$$

where k* denotes a previous iteration index. As soon as the stopping condition (ii) is satisfied, the penalty parameters and the multipliers are updated and the whole unconstrained minimization algorithm is restarted. Although various rules have been proposed in the past to iterate multipliers, most practical implementations of multiplier methods realize (8) because of its numerical simplicity.

Some precautions are necessary to guarantee the satisfaction of the constraints at a limit point and to obtain a descent direction for the merit function. Therefore the penalty parameters are updated by (7) and they are chosen to satisfy the condition

$$r_{k+1}^{ls} \|g_k\|^2 \leq \frac{1}{2} d_k{}^T B_k d_k \quad .$$

Penalty and multiplier methods have both the disadvantage that an internal loop is constructed in the algorithm leading to a lot of additional iterations. To avoid this inner loop, to obtain better multiplier estimates, and to avoid furthermore ill-conditioned quasi-Newton matrices B_k depending on a possibly too large penalty parameter, so-called sequential quadratic programming methods have been developed in the last years, cf. Han [4] or Powell [7]. They possess a quite simple structure and fit into the general frame (2.3).

<u>(3.3) Algorithm (Sequential quadratic programming method):</u> Assume that Algorithm (2.3) provides the data x_k, v_k, r_k^{ls}, and B_k.

(i) Determine a solution d_k, u_k of the linear system

$$B_k d + \nabla f(x_k) = A_k{}^T u \quad ,$$
$$A_k d + g_k = 0 \quad .$$

(ii) Define

$$r_{k+1}^{ls} := \max\left(\frac{2\|u_k - v_k\|^2}{d_k{}^T B_k d_k} \;,\; r_k^{ls} \right) \quad ,$$

$$r_{k+1}^{qN} := 0 \quad .$$

Equivalently the search direction d_k can be considered as the solution of a quadratic subproblem of the form

$$d \in \mathbb{R}^n: \quad \begin{array}{l} \min \frac{1}{2} d^T B_k d + \nabla f(x_k)^T d \\[1em] A_k d + g_k = 0 \quad , \end{array}$$

and u_k is the corresponding multiplier vector. This explains the notation 'sequential quadratic programming method', since a sequence of successive quadratic programming subproblems must be solved. The quasi-Newton formula is applied now to the Lagrange function indicating that B_k is an approximation of the Hessian of the Lagrange function.

The formal description of a generalized reduced gradient method is more complicated, cf. Abadie [1] or Lasdon, Waren [6]. We proceed from a classification of the variables into basic and non-basic variables, i.e. we let

$$x = (x^b, x^{nb}) \ ,$$

where $x^b \in \mathbb{R}^m$, $x^{nb} \in \mathbb{R}^{n-m}$. The decision is made under the assumption that the first m columns of A(x) are linearly independent. To facilitate the description of the algorithm, we suppose furtheron that this condition is valid for all iterates x_k, i.e. that for all k

$$A_k = (A_k^b : A_k^{nb})$$

with a nonsingular matrix $A_k^b \in \mathbb{R}^{m x m}$ and a matrix $A_k^{nb} \in \mathbb{R}^{m x (n-m)}$. As in linear programming, the intention is now to iterate the non-basic or independent variables x_k^{nb} with the aim to minimize the objective function, where the basic or dependent variables x_k^b are adjusted to satisfy the constraints. In the framework of Algorithm (2.3), the search directions and penalty parameters will be calculated as follows.

(3.4) Algorithm (Generalized reduced gradient method): Let ϵ be a positive number. Assume that the iterates $x_k = (x_k^b, x_k^{nb})$ with $x_k^b \in \mathbb{R}^m$, $x_k^{nb} \in \mathbb{R}^m$, $x_k^{nb} \in \mathbb{R}^{n-m}$, v_k, r_k^{ls}, and

$$B_k = \begin{pmatrix} 0 & : & 0 \\ 0 & : & B_k^{nb} \end{pmatrix}$$

are given, so that A_k^b is nonsingular, and $B_k^{nb} \in \mathbb{R}^{(n-m)x(n-m)}$.

(i) Determine multiplier estimates u_k by solving the linear system

$$A_k^{b^T} u = \nabla_{x^b} f(x_k) \ .$$

(ii) $\|g_k\| > \epsilon$: Let $d_k^{nb} := 0$.

(iii) $\|g_k\| \leq \epsilon$: Let d_k^{nb} be the solution of the linear system

$$B_k^{nb} d^{nb} + \nabla_{x^{nb}} f(x_k) = A_k^{nb^T} u_k \ .$$

(iv) The search direction d_k^b for the basic variables is obtained by solving

$$A_k^b d^b + A_k^{nb} d_k^{nb} + g_k = 0 .$$

Define new penalty parameters

$$r_{k+1}^{ls} := \max\left(\frac{2\|u_k - v_k\|^2}{{d_k^{nb}}^T B^{nb} d_k^{nb} + g_k{}^T g_k} + 1 , r_k^{ls} \right) ,$$

$$r_{k+1}^{qN} := 0, \text{ and set } d_k := (d_k^b, d_k^{nb}) .$$

In Step (ii) of Algorithm (3.4), a Newton method is defined by

$$d_k^b = -{A_k^b}^{-1} g_k ,$$

to solve the system of m nonlinear equations

$$g(x^b, x_{k^*}^{nb}) = 0 ,$$

where k^* denotes a previous iteration index. This inner loop is repeated until the constraints are satisfied with respect to a tolerance ϵ. Then a search direction d_k^{nb} with respect to the non-basic variables x^{nb} is computed by

$$d_k^{nb} = -{B_k^{nb}}^{-1} (\nabla_{x^{nb}} f(x_k) - {A_k^{nb}}^T {A_k^b}^{-T} \nabla_{x^b} f(x_k)) .$$

The vector in the parantheses is the so-called reduced gradient.

Since the computation of u_k is independent from B_{k+1}, we are able to determine u_{k+1} in Step 4 of Algorithm (2.3) and to set $u_k' = u_{k+1}$. Algorithm (3.4) retains the special structure of B_k, i.e. the matrix B_{k+1} is decomposed in the same way as B_k, whenever we start with

$$B_0 := \begin{pmatrix} 0 & : & 0 \\ 0 & : & B_0^{nb} \end{pmatrix} , \quad B_0^{nb} \in \mathbb{R}^{(n-m)\times(n-m)} .$$

4. THE GLOBAL CONVERGENCE THEOREM

An additional advantage of the unified presentation of the four approaches to solve the nonlinear programming problem is the possibility to derive a common convergence result. It will be shown in this section that, starting from a arbitrary initial point, all methods will approximate a Kuhn-Tucker point of (1). First the assumptions are gathered which are partially mentioned in the introduction and the comments of Algorithm (2.3).

<u>(4.1) Assumption:</u> The following statements are assumed to be valid:

(i) The problem functions f and g are continuously differentiable on $\mathbb{R}^n$.

(ii) For all iterates x_k of Algorithm (2.3) and all possible accumulation points of $\{x_k\}$, the matrices A(x) are of full rank. In the GRG-case, the first m columns of A(x) are linearly independent.

(iii) The quasi-Newton matrices B_k and B_k^{nb}, respectively, are positive definite.

(iv) Algorithm (2.1) consists of infinitely many.iteration steps.

The assumption that the first m columns of A(x) are linearly independent when using a GRG-method, is stated only to simplify the analysis. Otherwise one has to determine the linearly independent columns by any standard decomposition technique. As noted in Section 2, a simple modification of the BFGS-formula (5) guarantees positive definite matrices B_k and B_k^{nb}, respectively. The specific choice of B_k is not required to prove the convergence theorem, so that Assumption (4.1) (iii) is not crucial for the subsequent analysis. The last condition can always be assumed without loss of generality, since otherwise, we suppose that a Kuhn-Tucker point has been reached.

The first question is of fundamental importance for Algorithm (2.3). We have to know whether the search directions computed in Step 1 will lead to a descent direction for the merit function (6), so that the algorithm is well-defined. First we introduce the notation

$$z_k := \begin{pmatrix} x_k \\ v_k \end{pmatrix} \qquad p_k := \begin{pmatrix} d_k \\ u_k - v_k \end{pmatrix}$$

k = 0,1,2,..., where x_k, v_k are iterates of Algorithm (2.3) and d_k, u_k are determined by the special subalgorithm implemented in Step 1. It is then possible to show, cf. Schittkowski [10], that

$$\nabla\psi_{r_{k+1}^{ls}}(z_k)^T p_k \le \begin{cases} -\frac{1}{2} d_k{}^T B_k d_k & \text{for PE, MU, SQP.} \\ \\ \frac{1}{2}(d_k{}^T B_k d_k + g_k{}^T g_k) & \text{for GRG.} \end{cases} \tag{9}$$

In particular we conclude from (9) that $\phi_k'(0) < 0$, cf. (6). The GRG-algorithm (3.4) must be handled separately, since the only intention behind Step (ii) is to satisfy the constraints and no attempt is made to approximate the Kuhn-Tucker conditions. In that case, (9) is equivalent to

$$\nabla\psi_{r_{k+1}^{ls}}(z_k)^T p_k \le -\frac{1}{2} g_k{}^T g_k \quad .$$

The right-hand side of (9) will serve as a basic tool to prove the convergence of Algorithm (2.3). But first we need a lemma which is well-known from unconstrained optimization and which is to be used for the subsequent convergence theorem.

<u>(4.2) Lemma:</u> Under Assumptions (4.1) let z_k, p_k, and r_{k+1}^{ls} be generated by Algorithm (2.3) for $k = 0,1,2,\ldots$, so that the following conditions are satisfied:

(i) There are positive and bounded numbers ν_k with

$$\nabla\psi_{r_{k+1}^{ls}}(z_k)^T p_k \le -\nu_k$$

for all k.

(ii) The sequences $\{z_k\}$, $\{p_k\}$, and $\{r_{k+1}^{ls}\}$ are bounded.

Then

$$\lim_{k\to\infty} \nu_k = 0 \ .$$

Lemma (4.2) can be applied to prove a convergence theorem for Algorithm (2.3) by specifying ν_k and by interpreting (9) in an appropriate way, cf. Schittkowski [10].

<u>(4.3) Theorem:</u> Let z_k, p_k, and r_{k+1}^{ls} be computed by Algorithm (2.3), and B_k be the corresponding quasi-Newton matrix for $k = 0,1,2,\ldots$. Assume furthermore that Assumption (4.1) and the following conditions are satisfied:

(i) There is a $\gamma > 0$ with

$$d_k^T B_k d_k \geq \gamma \, d_k^T B_o d_k$$

for all k.

(ii) The sequences $\{z_k\}$, $\{d_k\}$, and $\{B_k\}$ are bounded.

Then there exists an accumulation point x^* of $\{x_k\}$ and a corresponding multiplier vector $w^* \in \mathbb{R}^m$ with

$$\nabla_x L(x^*,w^*) = 0 \ ,$$

$$g(x^*) = 0 \ .$$

REFERENCES

[1] J. Abadie, The GRG method for non-linear programming, in: Design and implementation of optimization software, H.J. Greenberg ed., Sijthoff and Noordhoff, Alphen aan den Rijn, The Netherlands, 1978.

[2] A.V. Fiacco, G.P. McCormick, Nonlinear sequential unconstrained minimization techniques, J. Wiley & Sons, New York, 1968.

[3] R. Fletcher, An ideal penalty function for constrained optimization, in: Nonlinear programming 2, O.L. Mangasarian, R.R. Meyer, S.M. Robinson eds., Academic Press, New York, 1975.

[4] S.P. Han, A globally convergent method for nonlinear programming, J. Optimization Theory and Applications, Vol. 22 (1977), 297-309.

[5] W. Hock, K. Schittkowski, A comparative performance evaluation of 27 nonlinear programming codes, to appear: Computing.

[6] L.S. Lasdon, A.D. Waren, Generalized reduced gradient software for linearly and nonlinearly constrained problems, in: Design and implementation of optimization software, H.J. Greenberg ed., Sijthoff and Noordhoff, Alphen aan den Rijn, The Netherlands, 1978.

[7] M.J.D. Powell, A fast algorithm for nonlinearly constrained optimization calculations, in: Proceedings of the 1977 Dundee Conference on Numerical Analysis, Lecture Notes in Mathematics, Springer-Verlag, Berlin, Heidelberg, New York, 1978.

[8] R.T. Rockafellar, Augmented Lagrange multiplier functions and duality in non-convex programming, SIAM Journal on Control, Vol. 12 (1974), 268-285.

[9] K. Schittkowski, Nonlinear programming codes, Lecture Notes in Economics and Mathematical Systems, No. 183, Springer-Verlag, Berlin, Heidelberg, New York, 1980.

[10] K. Schittkowski, On a unified algorithmic description of some nonlinear programming algorithms, submitted for publication.

A LINEARIZATION ALGORITHM FOR CONSTRAINED NONSMOOTH MINIMIZATION

K.C.Kiwiel

Systems Research Institute, Polish Academy of Sciences
Newelska 6, 01-447 Warsaw, Poland

1. INTRODUCTION

We consider the problem of minimizing f on $S=\{x\in R^N : F(x)\leq 0\}$, where f and F are real-valued functions defined on R^N that are locally Lipschitz continuous, but not necessarily continuously differentiable or convex. We assume that one can compute $f(x)$ and a certain subgradient of f $g_f(x)\in\partial f(x)$ at each $x\in S$, and $F(x)$ and a certain subgradient of F $g_F(x)\in\partial F(x)$ at each $x\notin S$, where ∂f and ∂F denote the subdifferentials (generalized gradients) of f and F [1]. This problem abounds with applications and has been treated in several papers; see, e.g. [5,6,7].

We present a readily implementable feasible point method of descent for minimizing f on S. To deal with nondifferentiability of f and F the method accumulates subgradients of f and F calculated at many trial points. At each iteration a search direction is found by solving a quadratic programming problem with linear inequalities generated by the past subgradients. Then a line search procedure finds the next approximation to a solution and the next trial point. The two-point line search is employed to detect discontinuities in the gradients of f and F.

The method is an extension of Kiwiel's aggregate subgradient method [3] to the nonconvex and constrained case. It differs significantly from Mifflin's extension [6] of an algorithm of Lemarechal [4]. More specifically, instead of using all the past subgradients for each search direction finding, we either select N+1 of them, or use only two so-called aggregate subgradients of f and F, respectively. The aggregate subgradients, which are certain convex combinations of the past subgradients of f and F, are recursively updated as the algorithm proceeds. Also our techniques for dealing with nonconvexity differ from those in [6], and can be regarded as improvements on those in [5,7].

The algorithm is globally convergent in the sense that all its accumulation points are stationary for f on S. Moreover, if f and F are convex and the Slater constraint qualification holds (the interior of S is nonempty) then the algorithm generates a minimizing sequence for f on S, which converges to a solution whenever f attains its infimum on S. No such strong convergence results are known for other comparable methods.

In Section 2 we derive the method with subgradient selection, which is described in full detail in Section 3. In Section 4 we present the method with subgradient aggregation, which requires less storage and work per iteration than the version with subgradient selection.

Due to lack of space, we omit the proofs of the results of this paper [2], which will appear elsewhere.

2. DERIVATION OF THE METHOD

We start by reviewing necessary optimality conditions for our problem, see [1].

For any fixed $x \in S$, define the improvement function

$$H(y;x)=\max\{f(y)-f(x),F(y)\} \quad \text{for all } y.$$

Note that if $H(y;x) < H(x;x)=0$ then $f(y) < f(x)$ and $F(y) < 0$, and hence $y \in S$ is better than x. It follows that if $\bar{x} \in S$ minimizes f locally on S then $H(\cdot;\bar{x})$ attains its local minimum at $\bar{x}$. In this case $0 \in \partial H(\bar{x};\bar{x})$ (the subdifferential of $H(\cdot;\bar{x})$ at $\bar{x}$). Since $\partial H(x;x) \subset M(x)$ for

$$M(x)=\begin{cases} \partial f(x) & \text{if } F(x) < 0, \\ \operatorname{conv}\{\partial f(x) \cup \partial F(x)\} & \text{if } F(x) = 0, \\ \partial F(x) & \text{if } F(x) > 0, \end{cases}$$

the necessary condition for optimality of $\bar{x}$ is $0 \in M(\bar{x})$. For this reason, a point $\bar{x} \in S$ is called stationary for f on S if $0 \in M(\bar{x})$. Moreover, if f and F are convex and $F(\tilde{x}) < 0$ for some $\tilde{x}$, then stationarity is equivalent to optimality, and $\bar{x}$ minimizes f on S if and only if

$$\inf\{H(y;\bar{x}) : y \in R^N\} = H(\bar{x};\bar{x}) = 0.$$

In view of the above remarks, one may test whether any point $x \in S$ is stationary by trying to find a point y such that $H(y;x) < H(x;x)$, i.e. a direction $d=y-x$ of descent for $H(\cdot;x)$ at x. Finding a descent direction for $H(\cdot;x)$ is difficult, since $H(\cdot;x)$ is nondifferentiable and we only have the following approximation to its subdifferential

$$g(x) = \begin{cases} g_f(x) & \text{if} \quad x \in S, \\ g_F(y) & \text{if} \quad x \notin S, \end{cases}$$

$g(x) \in M(x)$, instead of $M(x)$. However, $M(x)$ may be approximated by calculating $g(y)$ at many points y close to x. This idea of bundling the subgradients of f and F will be used in our method.

The algorithm will generate a sequence of points $x^k \in S$, $k=1,2,\ldots$, search directions $d^1, d^2, \ldots$ in R^N and nonnegative stepsizes t_L^1, t_L^2, ..., related by $x^{k+1}=x^k+t_L^k d^k$ for $k=1,2,\ldots$, where $x^1 \in S$ is a given starting point. The sequence $\{x^k\}$ is intended to converge to the required solution. The method will also calculate a sequence of trial points $y^{k+1}=x^k+t_R^k d^k$ for $k=1,2,\ldots$, and subgradients $g^k=g(y^k)$ for all $k \geq 1$, where $y^1=x^1$ and the trial stepsizes t_R^k satisfy $t_L^k \leq t_R^k$.

For simplicity of exposition, we shall temporarily assume that at each trial point y^j the algorithm computes $f(y^j)$ and $F(y^j)$, and the subgradients $g_f^j=g_f(y^j)$ and $g_F^j=g_F(y^j)$. Thus each point y^j defines the linearizations

$$f_j(x)=f(y^j)+\langle g_f^j, x-y^j \rangle \quad \text{and} \quad F_j(x)=F(y^j)+\langle g_F^j, x-y^j \rangle \tag{1}$$

for all x. At the k-th iteration the subgradient information collected at the j-th iteration $(j<k)$ is characterized by the linearization values $f_j^k=f_j(x^k)$ and $F_j^k=F_j(x^k)$, and the distance measure

$$s_j^k = |y^j-x^j| + \sum_{i=j}^{k-1} |x^{i+1}-x^i|, \tag{2}$$

since

$$f_j(x) = f_j^k + \langle g_f^j, x-x^k \rangle \quad \text{and} \quad F_j(x) = F_j^k + \langle g_F^j, x-x^k \rangle, \tag{3}$$

$$|y^j-x^k| \leq s_j^k. \tag{4}$$

The use of the easily updated variables f_j^k, F_j^k and s_j^k enables us not to store y^j.

Suppose that at the k-th iteration we have several past subgradients (g_f^j, f_j^k), $j \in J_f^k$, and (g_F^j, F_j^k), $j \in J_F^k$, where $J_f^k \subset \{j : y^j \in S\}$ and $J_F^k \subset \{j : y^j \in S\}$. Since we want to find a descent direction for the function

$$H^k(x) = \max\{f(x)-f(x^k), F(x)\}$$

at x^k, we may construct the following piecewise linear (polyhedral) approximations

$$\hat{f}^k(x) = \max\{f_j(x) : j \in J_f^k\} \quad \text{and} \quad \hat{F}^k(x) = \max\{F_j(x) : j \in J_F^k\},$$

$$\hat{H}^k(x) = \max\{\hat{f}^k(x)-f(x^k), \hat{F}(x)\}$$

to f, F and H^k, respectively. Note that in terms of the available data $\hat{H}^k$ can be expressed as

$$\hat{H}^k(x) = \max\{-\alpha_{f,j}^k + \langle g_f^j, x-x^k \rangle : j \in J_f^k \; ; \; -\alpha_{F,j}^k + \langle g_F^j, x-x^k \rangle : j \in J_F^k\}, \tag{5}$$

where $\alpha_{f,j}^k = f(x^k)-f_j^k$ and $\alpha_{F,j}^k = -F_j^k$ are the linearization errors. If f and F are convex, then $f(x) \geq f_j(x)$ and $F(x) \geq F_j(x)$ for all x, and $f(y^j) = f_j(y^j)$ and $F(y^j) = F_j(y^j)$ for all j, hence $\hat{H}^k$ is a global lower approximation to H^k. Hence one could try to find a descent direction for H^k at x^k by solving the problem

$$\text{minimize } \hat{H}^k(x^k+d) \quad \text{over all } d \in R^N. \tag{6}$$

This is the cutting plane approach [4]. However, subproblem (6) may have no solution, and $\hat{H}^k(x^k+d)$ is a doubtful approximation to $H^k(x^k+d)$ if $|d|$ is large. Therefore we shall consider the following regularized version of (6)

$$\text{minimize } \hat{H}^k(x^k+d) + \tfrac{1}{2}|d|^2 \quad \text{over all } d \in R^N, \tag{7}$$

where the penalty term $\frac{1}{2}|d|^2$ serves to keep x^k+d^k, with d^k being the solution of (7), in the region where $\hat{H}^k$ agrees with H^k. In practice, d^k may be found by solving the following quadratic programming problem for $(d^k, v^k) \in R^N \times R$

$$\begin{aligned} &\underset{(d,v) \in R^{N+1}}{\text{minimize}} && \tfrac{1}{2}|d|^2 + v, \\ &\text{subject to} && -\alpha_{f,j}^k + \langle g_f^j, d \rangle \leq v, \quad j \in J_f^k, \\ &&& -\alpha_{F,j}^k + \langle g_F^j, d \rangle \leq v, \quad j \in J_F^k. \end{aligned} \tag{8}$$

Moreover,

$$v^k = \hat{H}^k(x^k+d^k) = \hat{H}^k(x^k+d^k) - H^k(x^k)$$

may be regarded as an approximate directional derivative of H^k at x^k, and can be used at line searches.

Observe that $v^k \le 0$ if all the linearization errors are nonnegative, since then $\hat{H}^k(x^k) \le 0$ by (5), while $\hat{H}^k(x^k+d^k)+\frac{1}{2}|d^k|^2 \le \hat{H}^k(x^k)$, so that $v^k \le -\frac{1}{2}|d^k| \le 0$. This is the case if f and F are convex, since then $\alpha^k_{f,j}=f(x^k)-f^k_j \ge 0$ and $\alpha^k_{F,j}=-F^k_j \ge -F(x^k) \ge 0$. To retain this property in the nonconvex case, we shall use the following definitions of linearization errors in (5) and (8)

$$\alpha^k_{f,j} = |f(x^k)-f^k_j| \quad \text{and} \quad \alpha^k_{F,j} = |F^k_j|, \tag{9}$$

which automatically reduce to the previous definitions in the convex case. Thus we shall always have $v^k \le 0$.

When the next trial point y^{k+1} is found, we have to select the next subgradient index sets of the form

$$\begin{aligned} J^{k+1}_f &= \hat{J}^k_f \cup \{k+1\} \quad \text{and} \quad J^{k+1}_F = \hat{J}^k_F \quad \text{if} \quad y^{k+1} \in S, \\ J^{k+1}_f &= \hat{J}^k_f \quad \text{and} \quad J^{k+1}_F=\hat{J}^k_F \cup \{k+1\} \quad \text{if} \quad y^{k+1} \notin S, \end{aligned} \tag{10}$$

where $\hat{J}_f \subset J^k_f$ and $\hat{J}^k_F \subset J^k_F$. The obvious choice $\hat{J}_f=J^k_f$ and $\hat{J}^k_F=J^k_F$, resulting in total subgradient accumulation $(J^k_f \cup J^k_F=\{1,\dots,k\}$ for all k) as in [6], would lead to difficulties with storage and computation after a large number of iterations. Therefore we need a rule for dropping irrelevant past subgradients without impairing convergence. To this end, let $\lambda^k_{f,j}$, $j \in J^k_f$, and $\lambda^k_{F,j}$, $j \in J^k_F$, denote (possibly nonunique) Lagrange multipliers of (8) such that the sets

$$\hat{J}^k_f = \{j \in J^k_f : \lambda^k_{f,j} \ne 0\} \quad \text{and} \quad \hat{J}^k_F = \{j \in J^k_F : \lambda^k_{F,j} \ne 0\} \tag{11a}$$

satisfy

$$|\hat{J}^k_f \cup \hat{J}^k_F| \le N+1 \tag{11b}$$

(recall that $(d^k,v^k) \in R^{N+1}$). The required Lagrange multipliers can be found, for instance, by solving the following linear programming problem by the simplex method

$$\underset{\lambda}{\text{minimize}} \quad \sum_{j \in J^k_f} \lambda_{f,j}\alpha^k_{f,j} + \sum_{j \in J^k_F} \lambda_{F,j}\alpha^k_{F,j},$$

$$\text{subject to} \quad \sum_{j \in J_f^k} \lambda_{f,j}(1, g_f^j) + \sum_{j \in J_F^k} (1, g_F^j) = (1, -d^k), \tag{12}$$

$$\lambda_{j,f} \geq 0,\ j \in J_f^k,\ \lambda_{F,j} \geq 0,\ j \in J_F^k.$$

The subgradients indexed by $\hat{J}_f^k$ and $\hat{J}_F^k$ embody all the information that determined (d^k, v^k), since (d^k, v^k) also solves subproblem (8) with J_f^k and J_F^k replaced by $\hat{J}_f^k$ and $\hat{J}_F^k$.

In the nonconvex case, $\hat{H}^k$ is a good local approximation to H^k around x^k only if its defining points y^j, $j \in J_f^k \cup J_F^k$, are close to x^k, i.e. if the value of the locality radius

$$a^k = \max\{s_j^k : j \in J_f^k \cup J_F^k\} \tag{13}$$

(see (4)) is sufficiently small. Therefore, together with (10) we shall use subgradient deletion rules for reducing $J_f^k \cup J_F^k$ at some iterations, and additional line search requiremens to ensure that $s_{k+1}^{k+1} = |y^{k+1} - x^{k+1}|$ is sufficiently small.

3. THE METHOD WITH SUBGRADIENT SELECTION

We now state the method. Its line search procedure is given below.

Algorithm 3.1.

Step 0 (Initialization). Select the starting point $x^1 \in S$ and a final accuracy tolerance $\varepsilon_s \geq 0$. Choose fixed positive line search parameters $m_L < m_R < 1$, $\bar{a}, \bar{t}$ and $\bar{\theta} < 1$, a predicted shift in x at the first iteration $s^1 > 0$, and a locality tolerance $m_a > 0$. Set $y^1 = x^1$, $s_1^1 = 0$, $J_f^1 = \{1\}$, $g_f^1 = g_f(y^1)$, $f_1^1 = f(y^1)$ and $J_F^1 = \emptyset$. Set $\theta^1 = \bar{\theta}$ and $k=1$.

Step 1 (Direction finding). Solve the k-th subproblem (8), finding (d^k, v^k) together with Lagrange multipliers $(\lambda_f^k, \lambda_F^k)$ and sets $\hat{J}_f^k$ and $\hat{J}_F^k$ satisfying (11). Compute a^k by (13).

Step 2 (Stopping criterion). If $\max\{|d^k|, m_a a^k\} \leq \varepsilon_s$ terminate; otherwise, proceed.

Step 3 (Delection test). If $|d^k| \leq m_a a^k$ go to Step 4, otherwise go to Step 5.

Step 4 (Subgradient deletion). Delete from J_f^k and J_F^k all j smaller then k-N. Delete the smallest numbers from J_f^k and J_F^k. If $J_f^k \cup J_F^k = \emptyset$ set $J_f^k = \{k\}$, $g_f^k = g_f(x^k)$, $f_k^k = f(x^k)$ and $s_k^k = 0$. Go to Step 1.

Step 5 (Line search). By a line search procedure as given below, find two stepsizes t_L^k and t_R^k such that $0 \le t_L^k \le t_R^k$ and such that the two points $x^{k+1} = x^k + t_L^k d^k$ and $y^{k+1} = x^k + t_R^k d^k$ satisfy

$$f(x^{k+1}) \le f(x^k) + m_L t_L^k v^k,$$

$$F(x^{k+1}) \le 0,$$

$$t_R^k = t_L^k \quad \text{if} \quad t_L^k \ge \bar{t},$$

$$-\alpha(x^{k+1}, y^{k+1}) + \langle g(y^{k+1}), d^k \rangle \ge m_R v^k \quad \text{if} \quad t_L^k < \bar{t}, \tag{14}$$

$$|y^{k+1} - x^{k+1}| \le \bar{a},$$

$$|y^{k+1} - x^{k+1}| \le \theta^k s^k \quad \text{if} \quad t_L^k = 0,$$

$$|y^{k+1} - x^{k+1}| \le \bar{\theta} |x^{k+1} - x^k| \quad \text{if} \quad t_L^k > 0,$$

where

$$\alpha(x,y) = \begin{cases} |f(x) - f(y) - \langle g_f(y), x-y \rangle| & \text{if} \quad y \in S, \\ |-F(y) - \langle g_F(y), x-y \rangle| & \text{if} \quad y \notin S. \end{cases}$$

Step 6. If $t_L^k = 0$ (null step) set $s^{k+1} = s^k$ and $\theta^{k+1} = \bar{\theta}\theta^k$. Otherwise, i.e. if $t_L^k > 0$ (serious step), set $s^{k+1} = |x^{k+1} - x^k|$ and $\theta^{k+1} = \bar{\theta}$.

Step 7 (Subgradient updating). Find J_f^{k+1} and J_F^{k+1} by (10). Set $g_f^{k+1} = g_f(y^{k+1})$ if $y^{k+1} \in S$, $g_F^{k+1} = g_F(y^{k+1})$ if $y^{k+1} \notin S$. Compute $f_j^{k+1} = f_j(x^{k+1})$, $j \in J_f^{k+1}$, $F_j(x^{k+1})$, $j \in J_F^{k+1}$, and s_j^{k+1}, $j \in J_f^{k+1} \cup J_F^{k+1}$, by (1)-(3).

Step 8. Increase k by 1 and go to Step 1.

Line Search Procedure 3.2.

(i) Set $t_L = 0$ and $t = t_U = \min\{1, \bar{a}/ d^k\}$.

(ii) If $f(x^k+td^k) \le f(x^k)+m_L tv^k$ and $F(x^k+td^k) \le 0$ set $t_L=t$; otherwise set $t_U=t$.

(iii) If $t_L \ge \bar{t}$ set $t_L^k=t_R^k=t_L$ and return.

(iv) If $-\alpha(x^k+t_L d^k, x^k+td^k)+\langle g(x^k+td^k),d^k\rangle \ge m_R v^k$ and either $t_L=0$ and $t|d^k| \le \theta^k s^k$ or $t-t_L \le \bar{\theta} t_L$, then set $t_L^k=t_L$, $t_R^k=t$ and return.

(v) Set $t=t_L+\bar{\theta}(t_U-t_L)$ and go to (ii).

The above procedure terminates in a finite number of iterations if f satisfies the following "semismoothess" hypothesis: for any $x \in R^N$, $d \in R^N$ and sequences $\{\bar{g}_f^i\} \subset R^N$ and $\{t^i\} \subset R_+$ satisfying $\bar{g}_f^i \in \partial f(x+t^i d)$ and $t^i \downarrow 0$, one has

$$\limsup_{i\to\infty} \langle \bar{g}_f^i, d\rangle \ge \liminf_{i\to\infty} [f(x+t^i d)-f(x)]/t^i,$$

and a similar condition holds for F; see [5,6].

Our stopping criterion is based on the fact that at Step 2 we have

$$-d^k \in \mathrm{conv}\{g(y^j) : j \in J_f^k \cup J_F^k\} \subset \{M(y) : |y-x^k| \le a^k\},$$

so that x^k is approximately stationary if the values of $|d^k|$ and a^k are small. Of course, the asymptotic convergence results mentioned in Section 1 assume that $\varepsilon_s=0$.

4. THE METHOD WITH SUBGRADIENT AGGREGATION

We shall now describe a modification of Algorithm 3.1 in which two aggregate subgradients

$$(p_f^{k-1}, f_p^k) \in \mathrm{conv}\{(g_f^j, f_j^k)\}_{j=1}^{k-1} \quad \text{and} \quad (p_F^{k-1}, F_p^k) \in \mathrm{conv}\{(g_F^j, F_j^k)\}_{j=1}^{k-1}$$

replace the past subgradients at the k-th iteration.

In Step 0 we set two deletion indicators $r_f^1=r_F^1=1$. In Step 1 find (d^k, v^k) to

$$\text{minimize } \tfrac{1}{2} d^{\,2}+v \quad \text{over all } (d,v) \in R^N \times R \text{ satisfying}$$

$$-\alpha_{f,j}^k+\langle g_f^j, d\rangle \le v, j \in J_f^k, \; -\alpha_{f,p}^k+\langle p_f^{k-1}, d\rangle \le v \quad \text{if } r_f^k=0, \tag{15}$$

$$-\alpha^k_{F,j} + \langle g^j_F, d\rangle \le v,\ j \in J^k_F, \quad -\alpha^k_{F,p} + \langle p^{k-1}_F, d\rangle \le v \quad \text{if} \quad r^k_F = 0,$$

where $\alpha^k_{f,p} = |f(x^k) - f^k_p|$ and $\alpha^k_{F,p} = |F^k_p|$. Find any Lagrange multipliers $\lambda^k_{f,j}$, $\lambda^k_{f,p}$, $\lambda^k_{F,j}$, $\lambda^k_{F,p}$ of (15), setting $\lambda^k_{f,p}=0$ if $r^k_f=1$, and $\lambda^k_{F,p}=0$ if $r^k_F=1$. Set

$$\nu^k_f = \sum_{j \in J^k_f} \lambda^k_{f,j} + \lambda^k_{f,p} \quad \text{and} \quad \nu^k_F = \sum_{j \in J^k_F} \lambda^k_{F,j} + \lambda^k_{F,p},$$

$$\tilde\lambda^k_{f,j} = \lambda^k_{f,j}/\nu^k_f,\ j \in J^k_f,\ \tilde\lambda^k_{f,p} = \lambda^k_{f,p}/\nu^k_f,\ \tilde\lambda^k_{F,j} = \lambda^k_{F,j}/\nu^k_F,\ j \in J^k_F,\ \tilde\lambda^k_{F,p} = \lambda^k_{F,p}/\nu^k_F$$

$$(p^k_f, \tilde f^k_p) = \sum_{j \in J^k_f} \tilde\lambda^k_{f,j}(g^j_f, f^k_j) + \tilde\lambda^k_{f,p}(p^{k-1}_f, f^k_p),$$

$$(p^k_F, \tilde F^k_p) = \sum_{j \in J^k_F} \tilde\lambda^k_{F,j}(g^j_F, F^k_j) + \tilde\lambda^k_{F,p}(p^{k-1}_F, F^k_p).$$

If $r^k_f = r^k_F = 1$, calculate a^k by (13). In Step 4 set $r^k_f = r^k_F = 1$. In Step 5 and Line Search Procedure 3.2 replace v^k by

$$\tilde v^k = -\{|d^k|^2 + \nu^k_f|f(x^k) - f^k_p| + \nu^k_F|\tilde F^k_p|\}.$$

In Step 7 set $r^{k+1}_f = 1$ if $r^k_f = 1$ and $\nu^k_f = 0$, $r^{k+1}_f = 0$ otherwise, $r^{k+1}_F = 1$ if $r^k_F = 1$ and $\nu^k_F = 0$, $r^{k+1}_F = 0$ otherwise. Set $\tilde f^{k+1}_p = \tilde f^k(x^{k+1})$ and $\tilde F^{k+1}_p = \tilde F^k(x^{k+1})$, where

$$\tilde f^k(x) = \tilde f^k_p + \langle p^k_f, x - x^k\rangle \quad \text{and} \quad \tilde F^k(x) = \tilde F^k_p + \langle p^k_F, x - x^k\rangle,$$

and $a^{k+1} = \max\{a^k + |x^{k+1} - x^k|,\ |y^{k+1} - x^{k+1}|\}$.

In the above method with subgradient aggregation one may choose any sets $\hat J^k_f$ and $\hat J^k_F$ in (10). This allows the user to control storage and work per iteration. For instance, if $\hat J^k_f = \hat J^k_F = \emptyset$ for all k, then only four subgradients are used for each search direction finding. Of course, using more subgradients leads to faster convergence.

REFERENCES

[1] Clarke, F. (1976). A new approach to Lagrange multipliers. Math. Oper. Res., 1, 165-174.

[2] Kiwiel, K.C. (1982). Efficient algorithms for nonsmooth optimization and their applications. Ph.D.Thesis, Department of Electro-

nics, Technical University of Warsaw.

[3] Kiwiel, K.C. (1983). A. aggregate subgradient method for nonsmooth convex minimization. To appear in Math. Programming.

[4] Lemarechal, C. (1978). Nonsmooth optimization and descent methods. RR 78-4, International Institute for Applied Systems Analysis, Laxenburg, Austria.

[5] Mifflin, R. (1977). An algorithm for constrained optimization with semismooth functions. Math. Oper. Res., 2, 191-207.

[6] Mifflin, R. (1982). A modification and an extension of Lemarechal's algorithm for nonsmooth minimization. In: D.Sorensen and R.Wets, eds., Nondifferential and variational techniques in optimization, Mathematical Programming Study 17, 77-90.

[7] Polak, R., D.Q. Mayne and Y. Wardi (1983). On the extension of constrained optimization algorithms from differentiable to nondifferentiable problems. SIAM J. Control Optim., 21, 179-203.

BETTER THAN LINEAR CONVERGENCE AND SAFEGUARDING IN NONSMOOTH MINIMIZATION

Robert Mifflin
Washington State University, Department of Pure and Applied Mathematics
Pullman, WA 99164-2930 USA

1. INTRODUCTION

We consider the problem of minimizing f on R^n where f is locally Lipschitz and one subgradient [1] $g(x) \in \partial f(x) \subset R^n$ is known at each $x \in R^n$. The problem is nonsmooth if an optimal point x^* (necessarily satisfying the stationarity condition $0 \in \partial f(x^*)$) is a point of discontinuity of g. For example if

$$f(x) = \max [f_1(x), f_2(x), \ldots, f_m(x)] \tag{1}$$

where each f_j is C^1 then a suitable choice for g is given by

$$g(x) = \nabla f_i(x) \quad \text{for some } i \text{ such that } f_i(x) = f(x).$$

Such a g is usually not continuous at points x where $f(x) = f_j(x)$ for more than one index j. If this max function is minimized at such a point then for an algorithm to converge it must find the corresponding ∇f_j's and convex multipliers to form the n-dimensional zero vector. To be rapidly convergent an algorithm must also generate and use some second order information on the active f_j's.

We are interested in algorithms that do not explicitly exploit any underlying structure of f but do employ safeguarded piecewise quadratic approximation. At the kth iteration such an algorithm generates a step vector d_k by solving a quadratic programming subproblem. The subproblem depends on previously generated points $y_1, y_2, \ldots, y_k$ via a subset of $n+1$ or less elements of $\{1,2,\ldots, k\}$ denoted by I_k and via $f(y_i)$ and $g(y_i)$ for $i \in I_k$ and some second order information. A particular y_i (usually one with the smallest f-value) is denoted also by x_k. To have convergence from any starting point for $n > 1$ it may be necessary to have a line search from x_k along d_k to find y_{k+1}. For infinitely many k it is possible to have $x_{k+1} = x_k$ when $y_{k+1} \neq x_{k+1}$, so we cannot show Q-superlinear convergence as in smooth minimization.

---

*Research sponsored by the Air Force Office of Scientific Research, Air Force System Command, USAF, under Grant Number AFOSR-83-0210. The U.S. Government is authorized to reproduce and distribute reprints for Governmental purposes notwithstanding any copyright notation thereon.

If f is piecewise $-C^2$ in a certain sense then we give evidence in this paper for the possibility of obtaining better than linear convergence to a stationary point x^* in the sense that

$$|x_k + d_k - x^*| \leq \sum_{i \in I_k} a_{ik} \, |y_i - x^*| \tag{2}$$

where for each $i \in I_k$ $\{a_{ik}\} \to 0$ if $\{y_j\} \to x^*$ for all $j \in I_k$ as $k \to \infty$. For the case when $n > 1$ a subsequent paper will discuss problem assumptions and line search stopping criteria to insure convergence to a stationary point and to show that for all k sufficiently large y_{k+1} may be defined as $x_k + d_k$. Part of the safeguarding useful for obtaining stationarity in the limit is introduced here in section 3. For single variable problems where $n = 1$ the desired convergence and rate of convergence results are obtained in [2] for the convex case, in [5] for the constrained convex case and in [6] for the constrained nonconvex case.

2. SUBPROBLEM MOTIVATION

To motivate many of our ideas it suffices to consider first the example function given by (1) with $m = 2$ and f_1 and f_2 quadratic functions, i.e. f_1 and f_2 have constant second partial derivative matrices denoted by $\nabla^2 f_1$ and $\nabla^2 f_2$, respectively.

Suppose we know points x and y corresponding to different functions, i.e. $f(x) = f_1(x)$, $g(x) = \nabla f_1(x)$, $f(y) = f_2(y)$ and $g(y) = \nabla f_2(y)$. For $i = 1,2$ let H_i be an estimate of $\nabla^2 f_i$. Also, suppose that $f(x) \leq f(y)$ and we are interested in a move away from the better point x of the form $x + d$. Expanding about x where f_1 is known gives

$$f_1(x+d) \simeq f(x) + g(x)^T d + \frac{1}{2} d^T H_1 d$$

and expanding about y where f_2 is known gives

$$\begin{aligned} f_2(x+d) &\simeq f(y) + g(y)^T (x+d-y) + \frac{1}{2} (x+d-y)^T H_2 (x+d-y) = \\ &= f(y) + [g(y) + \tfrac{1}{2} H_2 (x-y)]^T (x-y) + [g(y) + H_2 (x-y)]^T d + \tfrac{1}{2} d^T H_2 d. \end{aligned}$$

In order to exhibit the special role of x we let

$$g_1 = g(x),\ g_2 = g(y) + H_2(x-y),\ p_1 = 0 \ \text{ and}$$

$$p_2 = f(x) - f(y) - [g(y) + \frac{1}{2} H_2 (x-y)]^T (x-y) .$$

Then for $i = 1,2$

$$f_i(x+d) \simeq f(x) - p_i + g_i^T d + \frac{1}{2} d^T H_i d .$$

So,

$$f(x+d) = \max[f_1(x+d),\ f_2(x+d)] \simeq f(x) + \max_{i=1,2} [-p_i + g_i^T d + \tfrac{1}{2} d^T H_i d].$$

In order for $x+d$ to approximately minimize f suppose we choose d to minimize the rightmost max function. The optimality conditions for such a d are the following:

There exists a multiplier $\lambda \in [0,1]$ such that

$$(1-\lambda)(g_1 + H_1 d) + \lambda(g_2 + H_2 d) = 0 \tag{3a}$$

and

$$-p_1 + g_1^T d + \tfrac{1}{2} d^T H_1 d = -p_2 + g_2^T d + \tfrac{1}{2} d^T H_2 d \quad \text{if } 0 < \lambda < 1. \tag{3b}$$

If we let G estimate $(1-\lambda)H_1 + \lambda H_2$ in (3a) and let zero estimate the second order d terms in (3b), then we are led to the conditions

$$(1-\lambda) g_1 + \lambda g_2 = -Gd$$

and

$$p_2 - p_1 = (g_2 - g_1)^T d \qquad \text{if } 0 < \lambda < 1.$$

These are the optimality conditions for the subproblem

$$\underset{d \in R_n}{\text{minimize}} \{\max_{i=1,2} [-p_i + g_i^T d] + \tfrac{1}{2} d^T G d\}.$$

What we have in mind for the choice of G is

$$G = (1-\lambda^-) H_1^- + \lambda^- H_2^-$$

where H_1^- and H_2^- are previous Hessian estimates and λ^- is the multiplier from the corresponding subproblem solution. There may have to be a modification to make G positive definite, unless we impose additional bounds on d and have a subroutine that solves indefinite quadratic programming problems.

3. THE GENERAL CASE WITH SAFEGUARDING

For a general locally Lipschitz function f the above analysis leads us to consider algorithms having a subproblem at the kth iteration of the form

$$\underset{d \in R^n}{\text{minimize}} \ \max_{i \in I_k} [-p_{ik} + g_{ik}^T d] + \tfrac{1}{2} d^T G_k d \tag{4}$$

where I_k is a subset of $\{1,2,\ldots,k\}$,

$$g_{ik} = g(y_i) + H_{ik}(x_k - y_i), \tag{5a}$$

$$p_{ik} = f(x_k) - f(y_i) - g(y_i)^T (x_k - y_i) - \tfrac{1}{2} (x_k - y_i)^T H_{ik}^s (x_k - y_i), \tag{5b}$$

y_i and $g(y_i)$ are points and corresponding subgradients generated at previous iterations, x_k is usually a y_i having the smallest f-value and H_{ik}, H^S_{ik} and G_k are symmetric $n \times n$ matrices with G_k usually positive definite. Let d_k be the optimal solution to (4) and w_k be the corresponding optimal value, i.e.,

$$w_k = \max_{i \in I_k} [-p_{ik} + g_{ik}^T d_k] + \frac{1}{2} d_k^T G_k d_k .$$

From related convergence analysis in [4] and [6] and rate of convergence analysis in [6] it appears that the <u>safeguard matrix</u> H^S_{ik} in (5b) should be chosen as a modification of the Hessian estimate H_{ik} in (5a) such that

$$p_{ik} \geqq \max[\alpha|x_k - y_i|^2, \ f(x_k) - f(y_i) - g(y_i)^T (x_k - y_i)] \tag{6}$$

for some parameter $\alpha > 0$. However, if f is convex H^S_{ik} and α may be chosen to be zero. Choosing $p_{ik} \geqq 0$ guarantees that the subproblem optimal objective value w_k is nonpositive, since $p_{jk} = 0$ for the index j such that $y_j = x_k$ and, thus, $d = 0$ gives a subproblem objective value of zero. From the subproblem optimality conditions given in the next section (6) guarantees that if $w_k = 0$ then x_k is stationary. Furthermore, $\alpha > 0$ implies that if $\{p_{ik}\} \to 0$ then $\{x_k - y_i\} \to 0$. These safeguard implications are useful results for showing convergence when the new points y_{k+1} and x_{k+1} are determined by executing a line search from x_k along d_k with line search stopping criteria depending on w_k, including the following:

Define x_{k+1} by $x_{k+1} = x_k + td_k$ where t is a nonnegative stepsize such that

$$f(x_k + d_k) \leqq f(x_k) + tmw_k \tag{7}$$

and $m \in (0,1)$ is a parameter. A second condition to guarantee that w_{k+1} is sufficiently closer to zero than w_k if t is too small will be the subject of a subsequent paper.

The order 2 nature of the safeguard is useful for the rate of convergence analysis considered next.

4. BETTER THAN LINEAR CONVERGENCE

For ease of reading we temporarily drop the iteration index k and consider the quadratic programming subproblem

$$\underset{(d,v)\in R^{n+1}}{\text{minimize}} \quad v + \frac{1}{2} d^T G d$$
$$\text{subject to} \quad v \geq g_i^T d - p_i \qquad \text{for } i \in I$$

which is equivalent to subproblem (4) with $G = G_k$, $g_i = g_{ik}$, $p_i = p_{ik}$ and $I = I_k$. Assume that G is positive definite.

Let v and d be optimal for this subproblem. Then there exist non-negative multipliers λ_i for $i \in I$ that sum to one and satisfy

$$\lambda_i [v - (g_i^T d - p_i)] = 0 \tag{8a}$$

and

$$-Gd = \sum_{i \in I} \lambda_i g_i \,. \tag{8b}$$

Let $J \subseteq I$ be the set of active subproblem constraint indices, i.e.,

$$v = g_i^T d - p_i \quad \text{if and only if } i \in J \,.$$

Then (8a) implies $\lambda_i = 0$ for $i \notin J$.

J is nonempty, because if J is empty then v could be reduced with d unchanged to obtain a feasible solution with a lower objective value which contradicts the optimality of v and d.

Suppose $\ell \in J$. Then

$$g_i^T d - p_i = g_\ell^T d - p_\ell \qquad \text{for all } i \in J \,,$$

or

$$(g_i - g_\ell)^T d = p_i - p_\ell \qquad \text{for all } i \in J \,. \tag{9}$$

Choose the largest number of indices i from J such that the n-vectors $g_i - g_\ell$ are linearly independent and let A be the full column rank matrix whose columns are these vectors. Let $M \subset J$ be the corresponding index set and note that $\ell \notin M$. Let m be the number of elements in M with $m = 0$ if M is empty. Let q be the corresponding m-vector whose components are $p_i - p_\ell$ for $i \in M$, i.e., such that from (9)

$$A^T d = q \,. \tag{10}$$

Let Z be an n by $n-m$ matrix with $n-m$ linearly independent columns orthogonal to those of A. Then

$$A^T Z = 0 \quad \text{and} \quad Z^T A = 0 \tag{11}$$

and there exist vectors $d_A \in R^m$ and $d_Z \in R^{n-m}$ such that d may be decomposed as

$$d = Ad_A + Zd_Z \ . \tag{12}$$

If $m = n$ then Z and d_Z are vacuous and $d = Ad_A$. If $m = 0$ then A, d_A and q are vacuous, and Z may be taken as the identity matrix, in which case $d = d_Z$.

The following lemma gives expressions for d_A and d_Z in terms of the subproblem data and some arbitrary multipliers.

Lemma 1:

$$d_A = (A^T A)^{-1} q \tag{13a}$$

and

$$d_Z = -(Z^T G Z)^{-1}(Z^T G A d_A + \sum_{i \in J} \bar{\lambda}_i Z^T g_i) \tag{13b}$$

where $\bar{\lambda}_i$ for $i \in J$ are any numbers satisfying

$$\sum_{i \in J} \bar{\lambda}_i = 1 \ . \tag{14}$$

Proof. Multiplying (12) on the left by A^T and using orthogonality (11) gives

$$A^T d = A^T A d_A + A^T Z d_Z = A^T A d_A . \tag{15}$$

Since A has full column rank, (13a) follows from (10) and (15).

The subproblem optimality conditions (8) imply that

$$-Gd = \sum_{i \in J} \lambda_i g_i ,$$

so multiplying on the left by Z^T gives

$$-Z^T G d = \sum_{i \in J} \lambda_i Z^T g_i \ . \tag{16}$$

Orthogonality of the columns of Z and A implies that

$$Z^T (g_i - g_\ell) = 0 \qquad \text{for all } i \in M.$$

This result actually holds for all $i \in J$, because the vectors $g_i - g_\ell$ that are not columns of A can be expressed as linear combinations of the columns of A. Thus

$$Z^T g_i = Z^T g_\ell \qquad \text{for all } i \in J. \tag{17}$$

Therefore (16), (17) and the fact that the mulipliers sum to one give

$$-Z^T G d = \sum_{i \in J} \lambda_i Z^T g_i = \sum_{i \in J} \lambda_i Z^T g_\ell = Z^T g_\ell \ . \tag{18}$$

Suppose $\bar{\lambda}_i$ for $i \in J$ are any numbers satisfying (14). Then (17) and (18) imply

$$-Z^T G d = \sum_{i \in J} \bar{\lambda}_i Z^T g_i \ . \tag{19}$$

Since Z has full column rank and G is positive definite, (13b) follows from (12) and (19). ∎

To use the previous lemma to show how much $x+d$ deviates from a stationary point x^* requires the following preliminary result that assumes some underlying piecewise-C^2 structure for f:

Lemma 2: Suppose there is a C^2-function f_{j_i} defined on a convex set containing y_i and x^* such that

$$f(x^*) = f_{j_i}(x^*), \quad f(y_i) = f_{j_i}(y_i) \quad \text{and} \quad g(y_i) = \nabla f_{j_i}(y_i).$$

Let

$$e_i = f(x^*) - f(x) + p_i - g_i^T (x^*-x).$$

Then

$$e_i = \tfrac{1}{2}(x^*-x)^T[2H_i - H_i^S](x^*-x) + (x^*-y_i)^T[H_i^S - H_i](x^*-x) + \\ + \tfrac{1}{2}(x^*-y_i)^T[\hat{H}_{j_i} - H_i^S](x^*-y_i) \tag{20a}$$

where

$$\hat{H}_{j_i} = \nabla^2 f_{j_i}(y_i + \hat{t}_i(x^*-y_i)) \quad \text{and} \quad \hat{t}_i \in (0,1). \tag{20b}$$

Furthermore

$$g_i = \nabla f_{j_i}(x^*) + [H_i - \tilde{H}_{j_i}](x^*-y_i) - H_i(x^*-x) \tag{21a}$$

where

$$\tilde{H}_{j_i} = \int_0^1 \nabla^2 f_{j_i}(x^* + t(y_i - x^*))dt \ . \tag{21b}$$

Proof: By the assumptions on f_{j_i} and (20b)

$$f(x^*) = f(y_i) + g(y_i)^T(x^*-y_i) + \tfrac{1}{2}(x^*-y_i)^T \hat{H}_{j_i}(x^*-y_i). \tag{22}$$

The definition of e_i, (22) and (5) with k deleted imply that

$$e_i = f(y_i) + g(y_i)^T(x^*-y_i) + \tfrac{1}{2}(x^*-y_i)^T \hat{H}_{j_i}(x^*-y_i) - f(x) + \\ + f(x) - f(y_i) - [g(y_i) + \tfrac{1}{2} H_i^S(x-y_i)]^T(x-y_i) - [g(y_i) + H_i(x-y_i)]^T(x^*-x).$$

The terms involving f sum to zero, so, by adding and subtracting x^* in appropriate places

$$e_i = g(y_i)^T(x^*-y_i) + \tfrac{1}{2}(x^*-y_i)^T \hat{H}_{j_i}(x^*-y_i) - \\ - [g(y_i) + \tfrac{1}{2} H_i^S(x-x^* + x^*-y_i)]^T(x-x^* + x^*-y_i) - \\ - [g(y_i) + H_i(x-x^* + x^*-y_i)]^T(x^*-x) \ .$$

Finally, the terms involving $g(y_i)$ sum to zero and the remainder is equivalent to the desired result (20a).

The second result (21a) follows from the definition of g_i, (21b) and the fact that

$$g(y_i) - \nabla f_{j_i}(x^*) = \int_0^1 \nabla^2 f_{j_i}(x^* + t(y_i - x^*))(y_i - x^*)dt. \quad \blacksquare$$

Now we may give the principal result of this paper, an explicit representation for $x + d - x^*$ in terms of three types of errors.

<u>Theorem</u>: Suppose the assumption of Lemma 2 holds for each $i \in J$ and that x^* and $\lambda^*_{j_i}$ for $i \in J$ satisfy the following stationarity conditions involving the corresponding functions f_{j_i}:

$$\sum_{i \in J} \lambda^*_{j_i} = 1 \tag{23a}$$

$$\sum_{i \in J} \lambda^*_{j_i} \nabla f_{j_i}(x^*) = 0 . \tag{23b}$$

Then

$$x + d - x^* = A(A^TA)^{-1}e^1 - Z(Z^TGZ)^{-1}(Z^TGA(A^TA)^{-1}e^1 + e^2 + e^3)$$

where

e^1 is an m-vector whose components are $e_i - e_\ell$ for $i \in M$,

$$e^2 = \sum_{i \in J} \lambda^*_{j_i} [Z^TH_i - Z^T \tilde{H}_{j_i}](x^* - y_i) , \tag{24}$$

$$e^3 = [Z^TG - Z^T\overline{G}](x^* - x), \tag{25}$$

and

$$\overline{G} = \sum_{i \in J} \lambda^*_{j_i} H_i . \tag{26}$$

<u>Proof</u>: From the definition of e_i for $i \in J$

$$e_i - e_\ell = p_i - p_\ell - (g_i - g_\ell)^T(x^* - x) \qquad \text{for } i \in M,$$

so from the definitions of e^1, q and A

$$e^1 = q - A^T(x^* - x).$$

Then, from (13a),

$$d_A = (A^TA)^{-1}(A^T(x^* - x) + e^1) .$$

Decomposing $(x^* - x)$ as

$$x^* - x = A(x^* - x)_A + Z(x^* - x)_Z \tag{27}$$

gives

$$d_A = (A^TA)^{-1}(A^TA(x^* - x_A) + A^TZ(x^* - x)_Z) + (A^TA)^{-1} e^1$$

or, since $A^TZ = 0$,

$$d_A = (x^*-x)_A + (A^TA)^{-1} e^1 . \tag{28}$$

From (21a)

$$\sum_{i\in J} \lambda^*_{j_i} Z^Tg_i = Z^T \sum_{i\in J} \lambda^*_{j_i} \nabla f_{j_i}(x^*) + \sum_{i\in J} \lambda^*_{j_i} [Z^TH_i - Z^T\tilde{H}_{j_i}](x^*-y_i) - Z^T \sum_{i\in J} \lambda^*_{j_i} H_i(x^*-x).$$

By the stationarity condition (23b) and definitions (24) and (26)

$$\sum_{i\in J} \lambda^*_{j_i} Z^Tg_i = e^2 - Z^T\bar{G}(x^*-x) = e^2 + [Z^TG - Z^T\bar{G}](x^*-x) - Z^TG(x^*-x) .$$

By definition (25) and decomposition (27)

$$\sum_{i\in J} \lambda^*_{j_i} Z^Tg_i = e^2 + e^3 - Z^TG\,A(x^*-x)_A - Z^TGZ(x^*-x)_Z . \tag{29}$$

Combining (13b) with $\bar{\lambda}_i = \lambda^*_{j_i}$ for $i\in J$, (28) and (29) gives

$$d_Z = -(Z^TGZ)^{-1}(Z^TGA(x^*-x)_A + Z^TGA(A^TA)^{-1}e^1 + e^2 + e^3 - Z^TGA(x^*-x)_A - Z^TGZ(x^*-x)_Z) ,$$

so

$$d_Z = -(Z^TGZ)^{-1}(Z^TGA(A^TA)^{-1} e^1 + e^2 + e^3) + (x^*-x)_Z . \tag{30}$$

Since

$$x + d - x^* = Ad_A + Zd_Z - A(x^*-x)_A - Z(x^*-x)_Z$$

the desired result follows from (28) and (30). ∎

Returning the iteration index k to appropriate places we see from lemma 2 and the theorem that better than linear convergence of type(2) can be attained if, for example, as $k \to \infty$

$\{(A_k^TA_k)^{-1}\}$, $\{(Z_k^T G_k Z_k)^{-1}\}$, $\{H_{ik}\}_{i\in J_k}$ and $\{H^S_{ik}\}_{i\in J_k}$ are bounded,

$$\{x_k - x^*\} \to 0, \quad \{y_i - x^*\}_{i\in J_k} \to 0 ,$$

$$\{Z_k^T H_{ik} - Z_k^T \nabla^2 f_{j_i}(x^*)\}_{i\in J_k} \to 0$$

and

$$\{Z_k^T G_k - Z_k^T \bar{G}_k\} \to 0 .$$

Ongoing research deals with determining choices for I_k, H_{ik}, H^S_{ik} and G_k and problem assumptions to obtain the above conditions for better than linear convergence.

5. THE SINGLE VARIABLE CASE

For the single variable case when $n = 1$ we do not want to have a line search, so in [6] there are alternative safeguards to insure convergence to a stationary point. Also in [6] there are, what now can be viewed as, slight modifications of the above ideas to take advantage of the two-sided nature of the univariate problem. More specifically, the scalar step d_k is not always exactly the subproblem solution defined above. The set corresponding to I_k has two elements with the corresponding y_i's denoted by x_k and y_k and the new points x_{k+1} and y_{k+1} are determined in a simple manner as two out of the three points $x_k + d_k$, x_k and y_k. The better than linear convergence result (2) is implied by the proofs in [6] which rely heavily on the two-sided nature of the one variable problem (and, hence, are not suitable for the multivariable case). In fact, we can define the two a_{ik}'s in (2) so that $a_{ik} \leq 0$ for the one i corresponding to the y_i (either x_k or y_k) that is on the opposite side of x^* from $x_k + d_k$. The rate of convergence proof assumes that the rather weak hypotheses that to the right (resp. left) of the algorithm's stationary limit point either f is convex or f is C^2 with a limiting right (resp. left) second derivative existing and if the limiting right (resp. left) first derivative is zero then f is strongly convex and C^2 on the right (resp. left).

A BASIC implementation of the single variable algorithm appears in [3].

6. REFERENCES

[1] F. H. Clarke, Generalized gradients and applications, Transactions of the American Mathematical Society 205 (1975) 247-262.

[2] C. Lemarechal and R. Mifflin, Global and superlinear convergence of an algorithm for one-dimensional minimization of convex functions, Mathematical Programming 24 (1982) 241-256.

[3] I. Mifflin and R. Mifflin, A BASIC program for solving univariate constrained minimization problems, Dept. of Pure and Applied Mathematics, Washington State University (Pullman, WA 1983).

[4] R. Mifflin, A modification and extension of Lemarechal's algorithm for nonsmooth minimization, in Nondifferential and Variational Techniques in Optimization, R. Wets and D. Sorensen, eds. Mathematical Programming Study 16 (1982) 77-90.

[5] R. Mifflin, A superlinearly convergent algorithm for one-dimensional constrained minimization problems with convex functions, Mathematics of Operations Research 8 (1983) 185-195.

[6] R. Mifflin, Stationarity and superlinear convergence of an algorithm for univariate locally Lipschitz constrained minimization, Department of Pure and Applied Mathematics, Washington State University (Pullman, 1982), to appear in Mathematical Programming.

ON THREE APPROACHES TO THE CONSTRUCTION OF NONDIFFERENTIABLE OPTIMIZATION ALGORITHMS

E. Polak and D. Q. Mayne*
Department of Electrical Engineering and Computer Sciences
and the Electronics Research Laboratory
University of California, Berkeley, CA 94720

ABSTRACT
We present three approaches to the construction of nondifferentiable optimization algorithms. The first consists of extending differentiable optimization algorithms, the second consists of replacing the nondifferentiable functions with smoothed out, differentiable ε-approximations, and the final approach consists in transforming a constrained nondifferentiable optimization problem into a minmax problem tractable by outer approximation techniques.

1. INTRODUCTION

Nondifferentiable optimization problems occur frequently in engineering design [P1, P6]. In the process of devising algorithms for solving these nondifferentiable engineering design problems we have developed three distinct approaches. The first, and best tried approach, is based on a systematic extension of differentiable optimization algorithms. Referring to [P2,P3,G2,P8] we see that it is particularly successful when the constraints have the form $\max_{\alpha \in A(x)} \phi(x,\alpha) \leq 0$, where $x \in \mathbb{R}^n$ is the design $\alpha \in A(x)$ vector, $\phi : \mathbb{R}^n \times \mathbb{R}^m \to \mathbb{R}$ is locally Lipschitz continuous and A is an upper semi-continuous set valued map. Typically, the elements of $A(x)$ are frequencies, times, temperatures, production tolerances, etc.

The second approach is based on approximating a nondifferentiable function $f : \mathbb{R}^n \to \mathbb{R}$ by a differentiable one obtained by integrating $f(\cdot)$ over an ε-hypercube, with $\varepsilon > 0$, driven to zero according to an appropriate law [M1]. It is an approach that should work well in the case where $f(x)$ is reasonably easy to compute, and when its generalized gradient [C1] does not have a convenient description.

The last approach is an approach of last resort and consists of approximating a function $f : \mathbb{R}^n \to \mathbb{R}$ by a piecewise "conical" function [M2]. This approach is based on the theory of outer approximations [P5,G2,Z1].

2. EXTENSION OF DIFFERENTIABLE OPTIMIZATION ALGORITHMS

The approach to be described in this section evolved from the Armijo gradient method [A1] and the methods of feasible directions described in [P4,P7]. It can be explained adequately by considering only unconstrained problems of the form

$$\min\{\psi(x) \mid x \in \mathbb{R}^n\} \tag{2.1}$$

with $\psi : \mathbb{R}^n \to \mathbb{R}$ locally Lipschitz continuous. The development of algorithms for constrained nondifferentiable optimization problems is quite similar to the one for unconstrained optimization, as the reader will find in [P10]. We recall that when $\psi(\cdot)$ is differentiable, its directional derivative at x, in the direction h is given by

$$d\psi(x;h) \triangleq \lim_{t \to 0} \frac{\psi(x+th)-\psi(x)}{t}$$

$$= \langle \nabla\psi(x), h \rangle \tag{2.2}$$

*Department of Electrical Engineering, Imperial College, London SW7 2BT, England.

Clearly, if $\psi(\cdot)$ is continuously differentiable, then $d\psi(\cdot;\cdot)$ is continuous. In this case we have the following

Armijo Gradient Method [A1]

Parameters: $\alpha,\beta \in (0,1)$.

Data: $x_0 \in \mathbb{R}^n$.

Step 0: Set i = 0.

Step 1: Compute the search direction at x_i:

$$h_i = \arg\min_h\{\tfrac{1}{2}\|h\|^2+d\psi(x;h)\} = -\nabla\psi(x_i) \tag{2.3}$$

Step 2: Compute the step length at x_i:

$$\lambda_i = \max_{k\in\mathbb{N}_+}\{\beta^k|\psi(x_i+\beta^k h_i)-\psi(x_i) \le -\beta^k\alpha\|h_i\|^2\} \tag{2.3}$$

Step 3: Update: $x_{i+1} = x_i + \lambda_i h_i$, i = i+1 and go to Step 4. □

The convergence properties of the Armijo method can be stated as follows.

Theorem 2.1: Suppose that $\{x_i\}_{i=0}^{\infty}$ is any sequence constructed by the Armijo gradient method. If for any $K \subset \{0,1,2,\ldots,\}$ $x_i \overset{K}{\to} \hat{x}$, as $i \to \infty$, then $\nabla\psi(\hat{x}) = 0$. □

Note: An examination of the proof of this theorem (see [P4]) shows that it depends crucially on the continuity of $d\psi(\cdot;\cdot)$. □

Now, when $\psi(\cdot)$ is only locally Lipschitz continuous, $d\psi(\cdot;\cdot)$ need not exist. However, the Clarke generalized directional derivative [C1] does exist. It is defined by

$$d_0\psi(x;h) \triangleq \overline{\lim_{\substack{t\to 0\\ x'\to x}}} \frac{\psi(x'+th)-\psi(x')}{t} \tag{2.4}$$

Referring to [C1] we find that

$$d_0\psi(x;h) = \max_{\eta\in\partial\psi(x)} \langle \eta,h \rangle \tag{2.5}$$

where $\partial\psi(x)$ is the Clarke generalized gradient of ψ at x. Unlike in the continuously differentiable case, see [C1], we can now assert only that $\partial\psi(\cdot)$ and $d_0\psi(\cdot;\cdot)$ are upper semi-continuous (u.s.c.) inthe sense of Berge [B2]. Furthermore, the differentiable case optimality condition $\nabla\psi(\hat{x}) = 0$ becomes replaced with $0 \in \partial\psi(\hat{x})$ [C1].

A naive generalization of the Armijo method consists of replacing the search direction computation (2.3) with

$$h_i = \arg\min_h\{\tfrac{1}{2}\|h\|^2+d_0\psi(x_i;h)\}$$

$$= -\arg\min\{\|h\|^2|h\in\partial\psi(x_i)\} \tag{2.6}$$

and keeping the original step length and update formulas.

Unfortunately, when this is done, because $d_0\psi(\cdot,\cdot)$ is only u.s.c., one can no longer assert the natural generalization of Theorem 2.1, and hence $x_i \overset{K}{\to} \hat{x}$ does not imply that $0 \in \partial\psi(\hat{x})$.

A closer examination reveals that the naive generalization does work under the assumption that $\partial\psi(\cdot)$ is locally uniformly u.s.c. Since this assumption is basically equivalent to requiring that $\psi(\cdot)$ be continuously differentiable, it is of little intrinsic interest. However, it motivates us to postulate the construction of a family of convergent direction finding maps $\{G_\varepsilon\psi(\cdot)\}_{\varepsilon\geq 0}$, which are locally uniformly u.s.c. with respect to $\partial\psi(\cdot)$ [P10], to be used as replacements for $\partial\psi(\cdot)$ in (2.6). In addition, to make the process well defined, we need to introduce an ε selection rule.

Definition 2.1 [P10]: For every $\varepsilon \geq 0$, let $G_\varepsilon : \mathbb{R}^n \to 2^{\mathbb{R}^n}$ be a set valued map. Then $\{G_\varepsilon\psi(\cdot)\}_{\varepsilon\geq 0}$ is said to be a family of convergent direction finding maps for $\psi(\cdot)$ if

(i) For all $x \in \mathbb{R}^n$, $\partial\psi(x) = G_0\psi(x)$;

(ii) For all $x \in \mathbb{R}^n$, $0 \leq \varepsilon < \varepsilon'$, $G_\varepsilon\psi(x) \subset G_{\varepsilon'}\psi(x)$ holds;

(iii) For all $\varepsilon \geq 0$, $G_\varepsilon\psi(x)$ is convex and bounded on bounded sets.

(iv) For any $\hat{x} \in \mathbb{R}^n$, $G_\varepsilon\psi(x)$, as a function of (ε,x), is upper-semi-continuous at $(0,\hat{x})$ in the sense of Berge [B2].

(v) For every $x \in \mathbb{R}^n$, $\varepsilon > 0$, $\delta > 0$, there exists a $\rho > 0$ such that for any $x', x'' \in B(x,\rho) \triangleq \{\hat{x} \in \mathbb{R}^n \mid \|x-\hat{x}\| \leq \rho\}$ $\quad \partial\psi(x') \subset \{G_\varepsilon\psi(x'')+B(0,\delta)\}$. □

For example, when $\psi(x) = \max_{\alpha\in A} \phi(x,\alpha)$ with $\phi : \mathbb{R}^n \times \mathbb{R} \to \mathbb{R}$ continuously differentiable, and $A = [\alpha_0,\alpha_1]$, we can take

$$G_\varepsilon\psi(x) \triangleq \underset{\alpha\in A_\varepsilon(x)}{\mathrm{co}} \{\nabla_x d(x,\alpha)\} \tag{2.7a}$$

where

$$A_\varepsilon(x) = \{\alpha\in A \mid \psi(x) \leq \phi(x,\alpha)+\varepsilon,\ \alpha \text{ local maximizer of } \phi(x,\cdot) \text{ in } A\} \tag{2.7b}$$

For other examples, including max eigenvalue problems, see [P8,P10].

In choosing a search direction by means of a direction finding map $G_\varepsilon\psi(\cdot)$, one must obviously decide on what value of ε to use. The following rule is suggested by the methods of feasible directions described in [P7]. For any $\varepsilon > 0$, $x \in \mathbb{R}^n$ let $h_\varepsilon(x)$ be defined by

$$h_\varepsilon(x) = -\operatorname{argmin}\{\|h\|^2 \mid h \in G_\varepsilon\psi(x)\} \tag{2.8}$$

Let $\gamma, \varepsilon_0 > 0$, $\nu \in (0,1)$ be given. We define $\varepsilon : \mathbb{R}^n \to \mathbb{R}$ by

$$\varepsilon(x) = \max\{\varepsilon \mid \varepsilon=\varepsilon_0\nu^k, k \in \mathbb{N}_+, \|h_\varepsilon(x)\| \geq \gamma\varepsilon\} \tag{2.9}$$

we can then define

$$h(x) = h_{\varepsilon(x)}(x) \tag{2.10}$$

Setting $h_i = h(x_i)$ in (2.3) results in a convergent algorithm, i.e., we get

Theorem 2.2: Consider problem (2.1) with $\psi : \mathbb{R}^n \to \mathbb{R}$ locally Lipschitz continuous. Suppose that $\{x_i\}_{i=0}^\infty$ is any sequence constructed by the Armijo Method with the substitution $h_i = h(x_i)$, as given by (2.10). If for any $K \subset \mathbb{N}_+$, $x_i \overset{K}{\to} \hat{x}$ as $i \to \infty$, then $0 \in \partial f(\hat{x})$. □

For extensions of these ideas to constrained optimization see [P10]. These extensions are quite straightforward.

3. NONDIFFERENTIABLE OPTIMIZATION VIA ADAPTIVE SMOOTHING

Next we describe a smoothing approach [M1].

Let $\psi : \mathbb{R}^n \to \mathbb{R}$ be locally Lipschitz continuous and for any $x \in \mathbb{R}^n$ and $\varepsilon \geq 0$, let

$$N_\varepsilon(x) \triangleq \{x' \in \mathbb{R}^n \mid \|x'-x\|_\infty \leq \varepsilon\} \tag{3.1}$$

Next, for any $\varepsilon \geq 0$, we define the <u>smoothing function</u> $\psi_\varepsilon : \mathbb{R}^n \to \mathbb{R}$, by

$$\psi_\varepsilon(x) = a(\varepsilon) \int_{N_\varepsilon(x)} \psi(x')dx' \tag{3.2a}$$

where

$$a(\varepsilon) = [\int_{N_\varepsilon(x)} 1 \cdot dx']^{-1} = \frac{1}{(2\varepsilon)^n} \tag{3.2b}$$

Clearly, $\psi_\varepsilon(x)$ is continuously differentiable for every $\varepsilon > 0$. When $\psi : \mathbb{R} \to \mathbb{R}$, we see that

$$\psi_\varepsilon(x) = \frac{1}{2\varepsilon} \int_{x-\varepsilon}^{x+\varepsilon} \psi(x')dx' \tag{3.3a}$$

and

$$\frac{d}{dx} \psi_\varepsilon(x) = \frac{1}{2\varepsilon} [\psi(x+\varepsilon)-\psi(x-\varepsilon)] \tag{3.3b}$$

i.e., the derivative is given by a finite difference. It is not difficult to show that the following holds true.

<u>Theorem 3.1</u>:

(i) For every $\varepsilon > 0$, $\nabla\psi_\varepsilon(\cdot)$ is well defined and continuous.

(ii) For every $\varepsilon > 0$,

$$\nabla_\varepsilon \psi(x) \in \partial_{2\varepsilon}\psi(x) \triangleq \underset{x' \in N_{2\varepsilon}(x)}{\text{co}} \partial f(x') \tag{3.4}$$

(iii) $\nabla_\varepsilon \psi(x) = 0$ for all $\varepsilon > 0$ implies that $0 \in \partial\psi(x)$. □

Theorem 3.1 suggests the following algorithms. First, consider the unconstrained problem

$$P_u : \min\{\psi(x) \mid x \in \mathbb{R}^n\} \tag{3.5}$$

with $\psi : \mathbb{R}^n \to \mathbb{R}$ locally Lipschitz continuous.

Let $\{\varepsilon_i\}_{i=0}^\infty$ $\{\gamma_i\}_{i=0}^\infty$ be any two sequences such that $\varepsilon_i > 0$, $\gamma_i > 0$ for all i, $\varepsilon_i \downarrow 0$, $\gamma_i \downarrow 0$ as $i \to \infty$. Construct a sequence $\{x_i\}_{i=0}^\infty$, by means of a descent algorithm (e.g., the Armijo Gradient method) such that

$$\|\nabla\psi_{\varepsilon_i}(x_i)\| \leq \gamma_i \quad i = 0, 1, 2, \ldots \tag{3.6}$$

The following is then true.

<u>Theorem 3.2</u>: Consider $\{x_i\}_{i=0}^\infty$ constructed as above. Suppose that $x_i \overset{K}{\to} \hat{x}$ as $i \to \infty$

for some index set $K \subset \{0,1,2,\ldots\}$. Then $0 \in \partial f(\hat{x})$ (i.e., $\hat{x}$ is a stationary point for (3.5)). □

Next consider the constrained problem

$$P_c : \min\{f(x) \mid g^j(x) \leq 0, j=1,2,\ldots,m\} \quad (3.7a)$$

where f, $g^j : \mathbb{R}^n \to \mathbb{R}$ are locally Lipschitz continuous functions. We replace P_c by the family of differentiable problems, with $\varepsilon > 0$,

$$P_{c\varepsilon} : \min\{f_\varepsilon(x) \mid g^j_\varepsilon(x) \leq 0, j=1,2,\ldots,m\} \quad (3.7b)$$

Again choose two sequences $\varepsilon_i \downarrow 0$, $\varepsilon_i \downarrow 0$ and use an algorithm such as a phase I-phase II method [P7], to construct a sequence of points $\{x_i\}_{i=0}^{\infty}$ together with multipliers $\mu_i^0, \mu_i^1, \ldots, \mu_i^m$ such that

$$\text{(i)} \quad g^i_{\varepsilon_i}(x_i) \leq \gamma_i, \; j = 1, 2, \ldots, m \quad (3.8a)$$

$$\text{(ii)} \quad \|\mu_i^0 \nabla f_{\varepsilon_i}(x_i) + \sum_{j=1}^{m} \mu_i^j \nabla g^i_{\varepsilon_i}(x_i)\| \leq \gamma_i \quad (3.8b)$$

$$\text{(iii)} \quad \mu_i^j \geq 0, \; \sum_{j=0}^{m} \mu_i^j = 1 \quad (3.8c)$$

$$\text{(iv)} \quad |\mu_i^j g^j_{\varepsilon_i}(x_i)| \leq \gamma_i \quad (3.8d)$$

<u>Theorem 3.3</u>: Consider the sequence $\{x_i\}_{i=0}^{\infty}$ constructed as above. Suppose that $x_i \xrightarrow{K} \hat{x}$ as $i \to \infty$ for some $K \subset \{0,1,2,3,\ldots\}$. Then

$$g^i(\hat{x}) \leq 0 \text{ for } j = 1,2, \ldots, m \quad (3.9a)$$

$$0 \in \underset{j \in I(\hat{x})}{\text{co}} \{\partial f(\hat{x}) \cup \partial g^j(\hat{x})\} \quad (3.9b)$$

where $I(\hat{x}) = \{j \in \underline{m} \mid g^j(\hat{x}) = 0\}$ and $\underline{m} \overset{\Delta}{=} \{1,2,\ldots,m\}$ (i.e., x satisfies the "standard" necessary optimality condition in [M1]). □

We thus see that a nondifferentiable optimization problem can be solved by solving a sequence of possibly progressively more ill conditioned differentiable optimization problems.

The main difficulty with this approach lies in the need to evaluate the integrals defining $\psi_\varepsilon(x)$, $\nabla\psi_\varepsilon(x)$. It appears that these can be approximated by making use of Monte Carlo techniques [K1].

4. AN OUTER APPROXIMATIONS APPROACH TO NONDIFFERENTIABLE OPTIMIZATION

This approach [M2] makes the most sense for a problem of the form

$$P : \min\{f(x) \mid x \in X\} \quad (4.1)$$

where $f : \mathbb{R}^n \to \mathbb{R}$ is locally Lipschitz continuous, X is a compact set in $\mathbb{R}^n$, and when we have a Lipschitz constant L for $f(\cdot)$ an X, i.e.,

$$|f(x)-f(y)| \leq \|x-y\| \quad \forall\, x, y \in X \quad (4.2)$$

We now define a probing function $g : \mathbb{R}^n \times \mathbb{R}^n \to \mathbb{R}$ by

$$g(x,y) \stackrel{\Delta}{=} f(y) - L\|x-y\|_\infty \qquad (4.3)$$

Note the similarity of these functions to Balder's needle function [B1]. The following result is easy to obtain [M2]:

Theorem 4.1: Let $g : \mathbb{R}^n \times \mathbb{R}^n \to \mathbb{R}$ be defined by (4.3) then

(i) $f(x) = \max\{g(x,y) \mid y \in X\}$ (4.4a)

(ii) $x = \operatorname{argmax}\{g(x,y) \mid y \in X\}$ (4.4b)

(iii) $\hat{x}$ solves P (4.1) if and only if $\hat{x}$ solves

$$\min_{x\in X} \max_{y\in X} g(x,y) \qquad (4.4c)$$

□

The advantage of this observation lies in the fact that methods such as those in [G1,P5] of outer approximations now permit the decomposition of (4.4c) into a sequence of more manageable problems

$$P_i : \min_{x\in X} \max_{y\in X_i} g(x,y_i) \qquad (4.5)$$

where the X_i contain only a finite number of points. The simplest scheme for constructing X_i is defined recursively as follows: Let X_0 be a set containing a finite number of points. Then we define, for i = 0, 1, 2, ...,

$$x_i \in \arg \min_{x\in X} \max_{J\in X_i} g(x,y_i) \qquad (4.6a)$$

$$y_i \in \arg \max_{y\in X} g(x_i,y) \qquad (4.6b)$$

$$X_{i+1} = \{y_i\} \cup X_i \qquad (4.6c)$$

As defined, the cardinality of X_i increases monotonically with i. In [P5,G1,E1] we find schemes for dropping points from X_i thus reducing its cardinality.

5. CONCLUSION

We have demonstrated that there are at least three distinct approaches to the construction of nondifferentiable optimization algorithms.

Acknowledgement

Research sponsored by National Science Foundation Grants ECS-7913148 and CEE-8105790, the Air Force Office of Scientific Research (AFSC) United States Air Force Contract F49620-79-C-0178, Semiconductor Research Consortium, and the Science and Engineering Research Council.

REFERENCES

[A1] Armijo, L., "Minimization of functions having Lipschitz continuous first partial derivatives," Pacific Journal of Mathematics, Vol. 16, pp. 1-3, 1966.

[B1] Balder, E. J., "An extension of duality-stability relations to nonconvex optimization problems," SIAM J. Contr. and Opt., Vol. 15, No. 2, pp. 329-343, 1977.

[B2] Berge, C., Topological Spaces, Macmillan Co., New York, 1963.

[C1] Clarke, F. H., Nondifferentiable Analysis and Optimization, J. Wiley and Sons, New York, 1983.

[E1] Eaves, B. C., and W. I. Zangwill, "Generalized cutting plane algorithms," SIAM J. Contr. and Opt., Vol. 9, pp. 529-542, 1971.

[G1] Gonzaga, C., and E. Polak, "On constraint dropping schemes and optimality functions for a class of outer approximations algorithms," SIAM J. Contr. and Opt., Vol. 17, No. 4, pp. 477-493, 1979.

[G2] Gonzaga, C., E. Polak, and R. Trahan, "An improved algorithm for optimization problems with functional inequality constraints," IEEE Trans. on Automat. Contr., Vol. AC-25, No. 1, pp. 49-54, 1979.

[K1] Kushner, H., and D. S. Clark, Stochastic Approximation Methods for Constrained and Unconstrained Systems, Springer Verlag, New York, 1978.

[M1] D. Q. Mayne and E. Polak, "Non-differentiable optimization via adaptive smothing," University of California, Electronics Research Laboratory Memo UCB/ERL M82/82, Nov. 12, 1982, to appear in JOTA.

[M2] Mayne, D. Q., and E. Polak, "Outer approximations algorithms for nondifferentiable optimization problems," University of California, Electronics Research Laboratory Memo UCB/ERL M83/40, July 12, 1983, to appear in JOTA.

[P1] Polak, E., "Semi-infinite optimization in engineering design," International Symposium on Semi-Infinite programming, Univ. of Texas, Austin, Texas, Sept. 8-10, 1981.

[P2] Polak, E., "Algorithms for optimal design," in: Optimization of Distributed Parameter Structures: Vol. 1, E. J. Haug and J. Cea, eds., Sijthoff & Noordhoff, 1981, pp. 586-602.

[P3] Polak, E., and D. Q. Mayne, "An algorithm for optimization problems with functional inequality constraints," IEEE Trans. on Automat. Contr., Vol. AC-21, No. 2, 1976.

[P4] Polak, E., Computational Methods in Optimization: A Unified Approach, Academic Press, New York, 1971.

[P5] Polak, E., D. Q. Mayne and R. Trahan, "An outer approximations algorithm for computer aided design problems," JOTA, Vol. 23, No. 3, 1979, pp. 331-352.

[P6] Polak, E., and A. Sangiovanni Vincentelli, "Theoretical and computational aspects of the optimal design centering, tolerancing and tuning problem," IEEE Trans. on Circuits and Systems, Vol. CAS-26, No. 9, pp. 795-813, 1979.

[P7] Polak, E., R. Trahan and D. Q. Mayne, "Combined phase I - phase II methods of feasible directions," Mathematical Programming, Vol. 17, No. 1, pp. 32-61, 1979.

[P8] Polak, E., and Y. Wardi, "A nondifferentiable optimization algorithm for the design of control systems subject to singular value inequalities over a frequency range," Automatica, Vol. 18, No. 3, pp. 267-283, 1982.

[P9] Polak, E., D. Q. Mayne and Y. Wardi, "On the extension of constrained optimization algorithms from differentiable to nondifferentiable problems," SIAM J. Contr. and Opt., Vol. 21, No. 2, pp. 179-204, 1983.

[P10] Polak, E., and D. Q. Mayne, "Algorithm models for nondifferentiable optimization," University of California, Electronics Research Laboratory Memo UCB/ERL M82/34, May 10, 1982.

AN ALGORITHM FOR MINIMIZING NONDIFFERENTIABLE CONVEX FUNCTIONS UNDER LINEAR CONSTRAINTS

J.-J. Strodiot and V.H. Nguyen
Facultés Universitaires N.-D. de la Paix
Namur, Belgium

1. INTRODUCTION

Non-smooth optimization is concerned with mathematical programming problems whose objective function and/or whose constraints are non-differentiable. In this paper, we shall study a class of minimization problems of the following type :

$$(P) \quad \begin{cases} \text{Minimize} \quad f(x) \\ \text{subject to} \quad A\,x = b\ , \\ \qquad\qquad\quad\ \ x \geqslant 0\ , \end{cases}$$

where $x \in \mathbb{R}^n$ (n - dimensional Euclidean space), $f : \mathbb{R}^n \to \mathbb{R}$ is a given convex *but not necessarily differentiable* function, A is a given $n \times n$ matrix or rank m and b is a given vector in $\mathbb{R}^m$.

Our aim is to develop a minimization algorithm which begins at any arbitrary feasible point, i.e., satisfying $A\,x = b$ and $x \geqslant 0$ and which generates an infinite sequence of feasible points converging to a solution of (P) .

When f is differentiable, a very well-known method for solving (P) is the reduced gradient method (see, for example, [6]). It consists to choose a basis among the columns of A and then by using the equality constraints $A\,x = b$, to eliminate m of the n variables. A reduced problem with only bound constraints is obtained and the gradient of the reduced problem, called the reduced gradient, is used to get a feasible descent direction for problem (P) . An exact or inexact line search is then performed.

When the objective function f is not necessarily differentiable, Nguyen and Strodiot have recently proposed an iterative method for solving a problem of the same type but with only linear *inequality* constraints [9, 12] . The method is of bundle type and is a generalization to the linearly constrained problem of a bundle method due to Wolfe [13] and Lemarechal [4] . The method is based on the fact that the only information available at any point x is the value and a single subgradient of the objective function. The calculated subgradients are stored by means of a bundle. Our aim in this paper is to adapt the philosophy of the reduced gradient method to the nondifferentiable problem (P) by using the bundle method for solving the reduced problem.

In the next section, we develop a procedure which, given certain positive tolerances, terminates in a point which approximates the solution to a degree measured by the tolerances. Thereafter we discuss the convergence of the proposed algorithm and give some numerical results. The details of the proofs are omitted in this short paper. They can be found in a forthcoming paper [1].

2. THE ALGORITHM

Following the philosophy of the reduced gradient method, we partition the matrix A in two submatrices B and N such that B is a nonsingular matrix of order $m \times m$ and N a matrix of order $m \times (n-m)$. B is called the basis and, without loss of generality, we can suppose that it is constituted by the first m columns of A. To this partition of A corresponds a partition of each vector x denoted by (x_B, x_N). The components of x_B are called the basic variables and those of x_N the nonbasic variables. Moreover $A x = b$ can be written under the form $x_B = B^{-1}(b - N x_N)$ and then f becomes $\hat{f}$ defined by

$$\hat{f}(x_N) = f(B^{-1} b - B^{-1} N x_N, x_N) .$$

It is obvious that $\hat{f}$ remains convex.

The reduced problem associated to (P) and to the basis B is then the problem of minimizing $\hat{f}(x_N)$ subject to $x_N \geqslant 0$ and $B^{-1} b - B^{-1} N x_N \geqslant 0$. It is obvious that, if $\bar{x}_N$ is a solution of this problem, then $\bar{x} = (B^{-1} b - B^{-1} N x_N, x_N)$ is a solution of (P). Now let $\bar{x}$ be a feasible point for (P) and assume that the corresponding basic variables $\bar{x}_B$ are all positive. Then $B^{-1} b - B^{-1} N x_N > 0$ and the search of a feasible direction for the reduced problem at $\bar{x}_N$ only involves to deal with the bound $x_N \geqslant 0$. Thus we are interested in solving the following problem

$$(\hat{P}) \quad \begin{cases} \text{Minimize} \quad \hat{f}(x_N) \\ \\ \text{subject to} \quad x_N \geqslant 0 . \end{cases}$$

For numerical reasons, it is easier to seek "approximate" solutions of $(\hat{P})$ than exact ones. Let $\varepsilon > 0$ be fixed. An ε - solution $\bar{x}_N$ is a feasible solution for $(\hat{P})$ which satisfies the inequality $\hat{f}(\bar{x}_N) \leqslant \inf\{\hat{f}(x_N) \mid x_N \geqslant 0\} + \varepsilon$. It is obvious that $\bar{x}_N$ is an ε - solution for $(\hat{P})$ if and only if $\bar{x}_N$ is an ε - solution to the unconstrained problem of minimizing $\hat{f}(x_N) + \psi_S(x_N)$ over all $x_N \in \mathbf{R}^{n-m}$ where $S = \{x_N \in \mathbf{R}^{n-m} \mid x_N \geqslant 0\}$ and ψ_S denotes the indicator function of S. Moreover, as $\hat{f} + \psi_S$ is convex, $\bar{x}_N$ is an ε - solution for $(\hat{P})$ if and only if $0 \in \partial_\varepsilon(\hat{f}+\psi_S)(\bar{x}_N)$, the ε - subdifferential of $\hat{f} + \psi_S$ at $\bar{x}_N$ [10].

We can now use the techniques developed for the unconstrained problem. If $\bar{x}_N$ is not an ε - solution, then a descent direction can be obtained by projecting the origin onto $-\partial_\varepsilon(\hat{f}+\psi_S)(\bar{x}_N)$. Then a line search is performed along this direction. This method supposes that the whole ε - subdifferential of $\hat{f}+\psi_S$ is known at $\bar{x}_N$. Very often it is not the case.

However, as is usual in non-smooth optimization, it will be supposed that one subgradient of f can be computed at each point. Numerically this means that at each iteration a subset of $\partial_\varepsilon(\hat{f}+\psi_S)(\bar{x}_N)$ has to be built which should approximate it as much as possible. But first we express $\partial_\varepsilon(\hat{f}+\psi_S)(\bar{x}_N)$ in terms of $\partial_\varepsilon f(\bar{x})$.

Proposition 1.

$$\partial_\varepsilon(\hat{f}+\psi_S)(\bar{x}_N) = \bigcup_{0\leqslant\varepsilon_0\leqslant\varepsilon} \{\partial_{\varepsilon_0}\hat{f}(\bar{x}_N) - \{\nu\in\mathbf{R}^{n-m} \mid \nu \geqslant 0 \,,\ \nu^T\bar{x}_N \leqslant \varepsilon-\varepsilon_0\}\}$$

and

$$\partial_\varepsilon\hat{f}(\bar{x}_N) = \{\hat{g} = g_N - N^T(B^{-1})^T g_B \mid g = (g_B\,, g_N) \in \partial_\varepsilon f(\bar{x})\}$$

where $\bar{x}$ denotes the point $(B^{-1}b - B^{-1}N\bar{x}_N\,,\ \bar{x}_N)$.

The vectors $\hat{g}\in\partial_\varepsilon\hat{f}(\bar{x}_N)$ will be called the *reduced* ε *- subgradients*. Now approximating $\partial_\varepsilon(\hat{f}+\psi_S)(\bar{x}_N)$ amounts to approximate $\partial_{\varepsilon_0}f(\bar{x})$ for each ε_0 , $0\leqslant\varepsilon_0\leqslant\varepsilon$. This can be done exactly as in the unconstrained case. Let ε_0 be fixed and suppose we are at iteration k . The points $x_1,\dots,x_k$ are known and also $g_i\in\partial f(x_i)$, $i=1,\dots,k$. Then $\partial_{\varepsilon_0}f(\bar{x})$ is approximated by the set

$$\{\sum_{i=1}^k \lambda_i g_i \mid \lambda_i \geqslant 0\,,\ i=1,\dots,k\,,\ \sum_{i=1}^k \lambda_i = 1\,,\ \sum_{i=1}^k \lambda_i\,\alpha(\bar{x},x_i) \leqslant \varepsilon_0\}$$

where

$$\alpha(\bar{x},x_i) = f(\bar{x}) - f(x_i) - \langle g_i\,, x_i - \bar{x}\rangle .$$

Hence $\partial_{\varepsilon_0}\hat{f}(\bar{x}_N)$ is approximated by the set

$$\{\sum_{i=1}^k \lambda_i \hat{g}_i \mid \lambda_i \geqslant 0\,,\ i=1,\dots,k\,,\ \sum_{i=1}^k \lambda_i = 1\,,\ \sum_{i=1}^k \lambda_i\,\alpha(\bar{x},x_i) \leqslant \varepsilon_0\}$$

and finally $\partial_\varepsilon(\hat{f}+\psi_S)(\bar{x}_N)$ is approximated by the set

$$\hat{G}_\varepsilon(\bar{x}_N) = \left\{ \sum_{i=1}^k \lambda_i \hat{g}_i - \nu \;\middle|\; \begin{array}{l} \lambda_i \geqslant 0\,,\ i=1,\dots,k\,,\ \Sigma\lambda_i = 1\,,\ \nu\geqslant 0\,, \\ \Sigma\lambda_i\,\alpha(\bar{x},x_i) + \nu^T\bar{x}_N \leqslant \varepsilon \end{array} \right\}.$$

To summarize : projecting 0 onto $\partial_\varepsilon(\hat{f}+\psi_S)(\bar{x}_N)$ is approximated by projecting 0 onto $\hat{G}_\varepsilon(\bar{x}_N)$. This can be done operationally by solving the following quadratic problem :

$$(Q)\quad \begin{cases} \text{Minimize} & \frac{1}{2}\| \sum_{i=1}^{k} \lambda_i \hat{g}_i - \nu \|^2 \\ \text{s.t.} & \sum_{i=1}^{k} \lambda_i = 1 \ , \quad \lambda_i \geq 0 \ \text{ for } \ i = 1, \dots, k \ , \quad \nu \geq 0 \ , \\ & \Sigma \, \lambda_i \, \alpha(\bar{x}, x_i) + \nu^T \, \bar{x}_N \leq \varepsilon \ . \end{cases}$$

If we denote by λ_i^* , $i = 1, \dots, k$, and ν^* the optimal solution of (Q) , then $d_N = \nu^* - \sum_{i=1}^{k} \lambda_i^* \hat{g}_i$ is the opposite of the projection of 0 onto $\hat{G}_\varepsilon(\bar{x}_N)$. It can be proven that d_N is a feasible direction for the reduced problem $(\hat{P})$.

Once the direction d_N has been obtained, it remains to set $d_B = -B^{-1} N \, d_N$ to get a feasible direction $d = (d_B, d_N)$ for problem (P) . A line search is then performed along this direction in order to get, if possible, an improving point. Recall that the line search problem is to find a step $t > 0$ and a subgradient $g \in \partial f(x_k + t\,d)$ such that

$$\langle d, g \rangle \geq m_1 \, a_k \ , \tag{1}$$

$$f(x_k + t\,d) \leq f(x_k) + t \, m_2 \, a_k \ , \tag{2}$$

$$x_k + t\,d \ \text{ feasible} \ , \tag{3}$$

where m_1 and m_2 satisfy $0 < m_2 < m_1 < 1$ and $a_k = -\|d\|^2 - u\,\varepsilon$ (u is the Lagrange multiplier associated with the last constraint in (Q)).

A step satisfying (1) and (2) is called a *serious step* [13] . However it is not always possible to get a serious step. In this case we set $x_{k+1} = x_k$ and we improve the approximation $\hat{G}_\varepsilon(\bar{x}_N)$ by adding a new subgradient to $\{g_1, \dots, g_k\}$. Such a step is called a *null step* [13] . Now, if we want the point $x_k + t\,d$ to remain feasible, the steplength $t > 0$ cannot take values greater than t^{max} defined by

$$t^{max} = \begin{cases} \min \{ -\frac{\bar{x}_i}{d_i} \mid d_i < 0 \} & \text{if } \ d \ngeq 0 \ , \\ +\infty & \text{if } \ d \geq 0 \ . \end{cases}$$

Of course, if t^{max} is too small, it can happen that (1) , (2) and (3) cannot all be satisfied for some $t \leq t^{max}$. In this case, only equation (3) will be considered with $t = t^{max}$. The resulting step will be called the maximum feasible step.

Once the line search has been performed, a new feasible point x^+ is obtained and it can happen that one of its basic variables is equal to zero. In that case, in order to start again the process, we have to perform a basic change, i.e., to find a new basis B^+ such that the new basic variables are all positive. This can be done if we impose the classical non-degeneracy assumption :

At all feasible point x *of* (P) *, there exists a basis* B *such such that the corresponding basic variables are all positive.*

However this assumption is not sufficient to guarantee the convergence of the algorithm (exactly as in the differentiable case). Indeed, if B is a non-degenerated basis at each point of a sequence of feasible points, then it can happen that a basic variable tends to 0^+ . But then, t_k^{max} can approach zero and the algorithm could converge to a non-optimal solution (jamming phenomenon [6]). Thus we may accept a degenerated basis at the limit point if the sequence of stepsizes (t_k) satisfies $t_k < t_k^{max}$ for k large enough. In order to get this property, we impose the following condition due to Huard [3] :

For every convergent subsequence of feasible points generated by the algorithm, we have that, if an index s *is a candidate to enter into the basis infinitely many times, then the limit value of the corresponding component* x_s *is positive.*

This condition is very weak because it allows the following rule for the basis change :

1. *A basis change is performed only when a basic variable hits its bound.*

2. *The zero basic variables are successively replaced by the non-basic variables which have the largest values.*

To summarize one iteration : We start by solving problem (Q) in order to get a reduced direction d_N . If the norm of d_N is less than a tolerance η , then we stop. Otherwise we build $d = (d_B , d_N)$ and we perform a line search along d . We update, if necessary, the current point, the bundle, the weights and the basis and the process is started again.

We have the following convergence theorem.

Theorem 1. If f is bounded from below on the constraint set, if the non-degeneracy assumption and the condition on the basis change are satisfied, then the algorithm stops after a finite number of iterations. The stopping point x_k is an ε - solution up to η for (P) , i.e.

$$f(x_k) \leq f(x) + \varepsilon + \eta \| x - x_k \|$$

for each feasible point x .

3. COMPUTATIONAL RESULTS

The proposed algorithm has been implemented and tested on four linearly constrained problems. The parameters have been chosen as follows : $\varepsilon = 10^{-5}$, $\eta = 10^{-5}$, $m_1 = 0.2$ and $m_2 = 0.1$. Problem (Q) was solved using Mifflin's method [8] and the results are summarized in Table 1, where the c.p.u. times are exclusive of input and output. The code was written by A. Bihain and the runs were performed on a DEC-2060 computer.

Example 1. Consider the following five functions

$$f_j(x) = 2 \sum_{i=1}^{5} c_{ij}\, x_{10+i} + 3 \cdot d_j\, x_{10+j}^2 + e_j - \sum_{i=1}^{10} a_{ij}\, x_i \,, \quad j = 1, \ldots, 5 .$$

The objective function is $f(x) = \max_{1 \le j \le 5} f_j(x)$ and the linear constraints are : $x_i + x_{i+1} \le 1$ for $i = 1, \ldots, 4$, $0 \le x_i \le 10$ for $i \ne 12$ and $0 \le x_{12} \le 70$. (for more details, see [11, problem 3]).
Our starting point is $x_i = 10^{-3}$ ($i \ne 12$) , $x_{12} = 60$.

Example 2. The problem is to minimize a maximum of 326 functions under 7 linear constraints where we have chosen $s = 0.4$; for more details, see [7, example 6] .

Example 3. It is a minisum problem with 6 variables and one linear constraint. For more details, see [2, problem 4] .
Our starting point is $x_i = 0$, $1 \le i \le 6$.

Example 4. The objective function is a maximum of 5 quadratic functions (see [5, test problem 1]) and the linear constraints are : $\sum_{i=1}^{10} x_i = 1$, $x_i \ge 0$ for $i = 1, \ldots, 10$.
The starting point is $x_i = 0.1$ ($1 \le i \le 10$).

Table 1. Summary of results

Test	1	2	3	4
Number of functions evaluations	77	40	68	49
Computed value	-53.2307	0.08541	68.8295	0.2610
Number of basis changes	9	1	2	3
c.p.u. time(s)	10	19	3	3

REFERENCES

[1] A. Bihain, V.H. Nguyen and J.-J. Strodiot, "A reduced subgradient algorithm", submitted for publication.
[2] J. Chatelon, D. Hearn and T. Lowe, "A subgradient algorithm for certain minimax and minisum problems", *SIAM Journal on Control and Optimization* 20 (1982) 455-469.
[3] P. Huard, "Un algorithme général de gradient réduit", *Bulletin de la Direction des Etudes et Recherches*, E.D.F., Série C, n° 2 (1982) 91-109.
[4] C. Lemarechal, "An extension of Davidon methods to non-differentiable problems", *Mathematical Programming Study* 3 (1975) 95-109.
[5] C. Lemarechal and R. Mifflin, eds., *Nonsmooth Optimization* (Pergamon Press, New York, 1977).

[6] D. Luenberger, *Introduction to Linear and Nonlinear Programming* (Academic Press, New York, 1973).
[7] K. Madsen and H. Schjaer-Jacobsen, "Linearly constrained minimax optimization", *Mathematical Programming* 14 (1978) 208-223.
[8] R. Mifflin, "A stable method for solving certain constrained least squares problems", *Mathematical Programming* 16 (1979) 141-158.
[9] V.H. Nguyen and J.-J. Strodiot, "A linearly constrained algorithm not requiring derivative continuity", *Engineering Structures* (in press).
[10] R. Rockafellar, *Convex Analysis* (Princeton Unviersity Press, Princeton NJ, 1970).
[11] R. Saigal, "The fixed point approach to nonlinear programming", *Mathematical Programming Study* 10 (1979) 142-157.
[12] J.-J. Strodiot, V.H. Nguyen and N. Heukemes, "ε-optimal solutions in nondifferentiable convex programming and some related questions", *Mathematical Programming* 25 (1983) 307- 328.
[13] Ph. Wolfe, "A method of conjugate subgradients for minimizing nondifferentiable functions", *Mathematical Programming Study* 3 (1975) 145-173.

ON SINGULAR AND BANG-BANG PROCESSES IN OPTIMAL CONTROL

G. Warnecke
GMD Schloß Birlinghoven
Postfach 1240
D-5205 St. Augustin 1
FRG

Abstract. It is characteristic of process P that it, sometimes, becomes singular. In practice, however, it is very difficult to prove in advance that process P is singular or bang-bang. Sufficient conditions for various sets of assumptions are derived.

1. INTRODUCTION

1.1. Setting of the problem

Consider the optimal control process P the evolution of which is governed by the bilinear system

$$\frac{dx}{dt} = A(t)x + B(t)xu + C(t)u \qquad (1.1)$$

and the cost of which is given by

$$I(u)=\int_{t_0}^{t_f} (a(x(t),t) + b(x(t),t)u(t))dt \qquad (1.2)$$

where the data is as follows: Real n×n matrices A(t), B(t) and real n×1 matrix C(t), continuously differentiable with respect to time $t\in\mathbb{R}$ (the real line), initial point $x(t_0)=x_0\neq 0$ from Euclidean n-space $\mathbb{R}^n$ and target set $\mathbb{R}^n$; set U of admissible controllers consists of all measurable functions $u(t)\in\mathbb{R}$ on the finite closed interval $I=[t_0,t_f]$, $t_0<t_f$, such that for each u(t) the corresponding (absolutely continuous) response $x(u)=x(t,u(t))=x(t)\in\mathbb{R}^n$ by (1.1) on I is steering $x(t_0)$ to $x(t_f)\in\mathbb{R}^n$; absolute value $|u(t)|$ of u(t) on I is ≤ 1; real functions a(x,t), b(x,t), continuously differentiable with respect to $(x,t)\in\mathbb{R}^{n+1}$. Here, a controller $u^*(t)$ in the admissible set U is called optimal, with respect to the cost functional I(u), in case $I(u^*)\leq I(u)$ for all u(t) in U.
Denote by $U^*\subset U$ the subset of all optimal controllers u^* of process P.

Remark 1.1. By [6], P.262, $U^*\neq\emptyset$ (empty set).

In what follows * denotes quantities determined by an optimal controller u^*. In general matrix notation will be used, matrices of one column or one row will be called vectors.
The adjoint response p(t)=p(t,u(t))=p(u) on I corresponding to the res-

ponse $x(t)=x(u)$, $u \in U$, is given as (the unique absolutely continuous) solution of

$$\frac{dp}{dt} = a_x(x(t),t) + b_x(x(t),t)u(t) - p(A(t)+B(t)u(t)) \qquad (1.4)$$

$$p(t_f) = 0 \qquad (1.5)$$

with gradient $a_x(x,t)$ and $b_x(x,t)$ of $a(x,t)$ and $b(x,t)$ with respect to $x \in \mathbb{R}^n$, respectively.
The coefficient of the control variable in the Hamiltonian function of process P at $x=x(u)$ $(=x(t)$, $p=p(u)$ $(=p(t))$, $u \in U$, will be denoted by $S(t)$ (=switching function) and it writes

$$S(t)=-b(x(t),t)+p(t)B(t)x(t)+p(t)C(t), t \in I. \qquad (1.6)$$

Remark 1.2. For an optimal controller u^* $x(u^*)=x^*(t)$, $p(u^*)=p^*(t)$ and $S^*(t)$ instead of $S(t)$ will be written while numbering is $(1.4)^*$, $(1.5)^*$, $(1.6)^*$ instead of (1.4), (1.5), (1.6) in this case.
By the maximal principle of Pontryagin ([6], p.319, remark 3 of theorem 2) problem $(1.4)^*$, $(1.5)^*$ has a non trivial solution $p^*(t)$ on I while

$$u^*(t) = \text{sgn}\, S^*(t) \text{ if } S^*(t) \neq 0. \qquad (1.7)$$

This leads to the following definitions. For motivation see 1.2.
The (optimal control) process P is said to be *bang-bang* with respect to an optimal controller u^*, and u^* is called a *bang-bang optimal controller* in case, the corresponding switching function

$$S^*(t) \neq 0 \quad \text{on } I. \qquad (1.8)$$

Correspondingly, process P is called *partially singular* with respect to u^*, u^* is said to be a *partially singular optimal controller* in case, the corresponding switching function

$$S^*(t) \equiv 0, \text{ on } \omega, \qquad (1.9)$$

where $\omega \subset I$ is a set of positive measure, mes $\omega > 0$.

Remark 1.3. By subsection 2.3., it is seen that, in general, a partially singular optimal controller consists of subsets of singular controlling combined with subsets of bang-bang controlling.
As a result of that,

$$\omega = I \qquad (1.10)$$

in (1.9), is called special attention to by speaking of *totally singular* instead of partially singular, in this case.
Denote by U_p^* $(U_t^*, U_b^*) \subset U^*$ the set of partially singular (totally singular, bang-bang) optimal controllers of process P. By (1.8), (1.9), (1.10), these sets U_λ^*, $\lambda \in \square = \{p,t,b\}$, have no elements in common, $\bigcap_\square U_\lambda^* = \emptyset$, $U^* = \bigcup_\square U_\lambda^*$;

and so the *problem* is to *prove* how the optimal controllers $u^* \in U^*$ are distributed to the sets U_λ^*, $\lambda \in \square$.

1.2. Motivation

The relation between sales and advertising expenditures found empirically by Vidale and Wolfe [3] can be conceived of as a special case of the Cauchy problem (1.1), $x(t_0)=x_0$. The Vidale and Wolf relation has given rise [3] to establishing optimal control processes in management of the typ under consideration, but with the bilinear term omitted for, inspite of all that is known about the nature of optimal strategies for various sets of assumptions, the solution of particular problems remains an art rather than a science [3], p.33, a fact which necessitates and may motivate further mathematical research in this field including the study of singular optimal control processes. In modern optimal control theory it soon became apparent that the mathematical theory available was not sufficient for certain special problems in which Pontryagin's Principle yielded no additional information on the stationary control. These processes were described as singular processes. See e.g. [1] p.484 where singular processes are formulated which, in the event of target $\mathbb{R}^n$, include process P in case, (1.1) and the integrand in (1.2) are autonomous. Except that singular processes have arisen in various areas of actual interest [2], they are also observed to exhibit many interesting and deep theoretical niceties. Because of this it is important that singular processes be amenable to mathematical analysis. However, since Pontryagin's Principle does not yield any information directly on singular controls the task for researchers was to discover new necessary, necessary and sufficient, and sufficient conditions for optimality in the singular case [2],[5]. In case of process P, by $U^* \neq \emptyset$, optimality of certain $u \in \mathcal{U}$ is assured which implies to prove how these u's are distributed to the sets U_λ^*, $\lambda \in \square$. Sufficient conditions for various sets of assumptions will be given.

2. SUFFICIENT CONDITIONS

2.1. Preparation

Let matrix norm, [6], be $|D| = \sum_{i,j=1}^{n} |d_{ij}|$, real matrix $D=(d_{ij})_{1\leq i,j\leq n}$,

$\| D \| = \max_{I} |D(t)|$ for continuous $D(t)$ on I.

CONCLUSION 2.1. For process P is

$$|x(t)-x_0|<|x_0|e^{2(\|A\|+\|B\|+\|C\|)|t_f-t_0|},\quad \forall t\in I,\ \forall u\in U. \tag{2.1}$$

This says process P is uniformly bounded. This result follows by standard analysis from the integral representation of the Cauchy problem (1.1), $x(t_0)=x_0$.

Let

$$K=\{x\in \mathbb{R}^n \mid |x-x_0|\leq|x_0|e^{2(\|A\|+\|B\|+\|C\|)|t_f-t_0|}\} \tag{2.2}$$

with data from process P, and let

$$B_\varepsilon(x_0)=\{x\in \mathbb{R}^n \mid |x-x_0|<\varepsilon,\ \varepsilon>0\}. \tag{2.3}$$

For $\varphi(x,t)$ continuous in $\mathbb{R}^{n+1}$ define for fixed $t\in I$ $\overset{o}{\operatorname{supp}}\,\varphi(.,t)=\{x\in\mathbb{R}^n \mid \varphi(x,t)\neq\emptyset\}$ the complement (with respect to $\mathbb{R}^n$) of which, denoted by $C\overset{o}{\operatorname{supp}}\,b(.,t)$, being closed (in $\mathbb{R}^n$) with interior not necessarily nonempty. For φ independent of t $\overset{o}{\operatorname{supp}}\,\varphi$ will be written. Let M_n=set of real n×n matrices, and let ω a nonempty subset of $\mathbb{R}$.

LEMMA 2.1. Let row vector $b(s)\in\mathbb{R}^n$, $s\in\omega$, and assume $B(t)$, $R(s,t)\in M_n$ satisfy for $s,t\in\omega$

$$\operatorname{rk} B(s)=r, \tag{2.4}$$

r,n independent of $s,t\in\omega$.

$$\operatorname{rk} R(s,t)=n, \tag{2.5}$$

Here rk=*rank*.

Then $b(s)R(s,t)B(t)=0\quad \forall(s,t)\in\omega\times\omega$ iff $b(s)B(s)=0\quad \forall s\in\omega$.

By (2.4), (2.5) $\operatorname{rk}R(s,t)B(t)=r$, and by linear algebra, for fixed parameters, both equations have the solution space $\mathbb{R}^d$, $d=n-r$.

$L_n(t_0,t_f)$=set of vectors $z=(z_1,\dots,z_n)$ of functions $z_i\in L(t_0,t_f)$=set of (real) integrable functions on I for the measure dt.

LEMMA 2.2. Let $A(t)$, $B(t)\in M_n$ continuous on I, suppose $B(t)$ to satisfy (2.4) on I, rank r independent of t, and let $\varphi\in L_n(t_0,t_f)$. For $u\in U$ assume $p(t)=(p_1(t),\dots,p_n(t))$ to be the unique solution on I of the Cauchy problem

$$\frac{dp}{dt}=\varphi(t)-p(A(t)+B(t)u), \tag{2.6}$$

$$p(t_f)=0. \tag{2.7}$$

If

$$\varphi(s)B(s)=0 \quad \text{for a.a.(}\textit{almost all}\text{)}\ s\in I \tag{2.8}$$

then

$$p(t)B(t)\equiv 0 \quad \text{on } I. \tag{2.9}$$

To show this lemma is true, remaind [8], p.795 that for the final condition t_f, $p(t_f)=0$ the unique absolutely continuous function $p(t)$ of (2.6), (2.7) is given on I by the variation of parameters formula

$$p(t)=\int_t^{t_f}\varphi(s)R(s,t)ds$$

which implies

$$p(t)B(t)=\int_t^{t_f}\varphi(s)R(s,t)B(t)ds, \qquad (2.10)$$

where $R(s,t)\in M_n$ is the absolutely continuous resolvant on $I\times I$ of the associated homogeneous equation. We have $\mathrm{rk}R(s,t)=n$, independent of s,t. Let (2.8) hold. There is $\omega\subset I$, $\mathrm{mes}(I\setminus\omega)=0$ such that $\varphi(s)B(s)=0$ on ω. By lemma 2.2.

$$\varphi(s)R(s,t)B(t)\equiv 0 \quad \text{on } \omega\times\omega. \qquad (2.11)$$

Integration of (2.11) with respect to $s\in I$ provides (2.9) by (2.10).

Remark 2.1. Consider process P. Herein, let $B(t)$ satisfy (2.4) on I. Then (2.9) follows if $a_x(x,.)$, $b_x(x,.)$ are solutions of the linear equation (in the commutative ring $K(I)$ of continuous functions on I) $yB=0$.

In $K=K(I)$ let ring operations of addition and multiplication be defined pointwise. As customary, [4], define for K the finitely generated (e.g. by n elements of a canonical basis) free) K-modul K^n (sometimes written $K^n(I)$). Let $K_d(I)$ a K-submodul of K^n generated by d, $1\leq d\leq n$, elements of the basis of K^n.

LEMMA 2.3. Let continuous $B(t)\in M_n$ on $\mathbb{R}$ satisfy (2.4) on I for fixed rank r independent of t, $1\leq r<n$. Suppose $B(t)$ is the coefficient matrix of the linear equation (in the ring $K(I)$) $yB=0$.
Then there exists a system of d linearily independent elements $y_i\in K_d(I)$ satisfying $y_i(t)B(t)=0$ on I, $1\leq i\leq d=n-r$.
Equation $yB=0$ in $K(I)$ has general solution y generated by the linear combinations of $y_1,\ldots,y_d$ in $K_d(I)$, that is

$$y=\sum_{i=1}^{n}\lambda_i y_i,\quad \lambda_i\in K(I),\quad 1\leq i\leq d.$$

Remark 2.2. If r=n then the determinant of B is a unit, [4], in $K(I)$, and therefore, $yB=0$ has unique solution $y=0$ in K.
For proof of lemma 2.3 see [9].
In what follows it is convenient to introduce notation $\mathbb{K}_d(B)=K(I)$-modul $K_d(I)$ the basis of which is given by the system y_i from lemma 2.3. Furtheron, let $\mathbb{K}_d^k(B)=K(I)$-submodul of $\mathbb{K}_d(B)$ with k elements of system y_i as a basis where fixed k may take values 1,2,...d. By remark 2.2 define $\mathbb{K}_0(B)=\{0\}$. Let set $A\subset\mathbb{R}^n$ and suppose real function $f(x,t)$ is defined

on an appropriate subset of $\mathbb{R}^{n+1}$. Then $f(x,t)\in \mathbb{K}_d(B)$ for each $x\in A$ means: For each such x there is a system $\lambda_i\in K(I)$ such that

$$f(x,t)=\sum_{i=1}^{n}\lambda_i(t)y_i(t) \text{ on } I.$$

Now, consider process P. Vector C(t) in (1.1) is said to depend linearily on the columns b_i $(\in K^n(I))$ of B(t) in (1.1) in case, there is a system $\lambda_i\in K(I)$ such that C is satisfying a linear relation

$$C = \sum_{i=1}^{n}\lambda_i y_i \text{ in } K^n(I).$$

2.2. Sufficient condition for $U^*=U_t^*$

THEOREM 2.1. Consider process P. Let the following assumptions hold:

H1. $\mathrm{rk}B(t)=r$ on I for fixed r independent of t, $1\leq r\leq n$, $d=n-r$.

For each $x\in K$ (2.12)

is

$$a_x(x,.)\in \mathbb{K}_d(B) \quad (2.13)$$

$$b_x(x,.)\in \mathbb{K}_d^k(B) \text{ for fixed } k,\ 1\leq k\leq d, \text{ if } d>0, \quad (2.14)$$

$$b_x(x,.)\in \mathbb{K}_0(B), \text{ if } d=0. \quad (2.15)$$

C(t) depends linearily on the columns of B(t). (2.16)

H2. There exists $\varepsilon>0$ such that the relations

$$B_\varepsilon(x_0)\subset \bigcap_{a.a.\, t\in I} C\,\overset{0}{\mathrm{supp}}\, b(.,t) \quad (2.17)$$

$$\|A\|+\|B\|+\|C\|\leq |t_f-t_0|^{-1}\ln\sqrt{\frac{\varepsilon}{|x_0|}} \quad (2.18)$$

hold.

Then $U^*=U_t^*$. (2.19)

Proof. Take $u^*\in U^*$ arbitrary. By (2.1), (2.18)

$$x^*(t)\in B_\varepsilon(x_0)\subset K,\ \forall t\in I. \quad (2.20)$$

Consequently, by (2.17), this gives, $\forall s\in I$, $x^*(s)\in B_\varepsilon(x_0)\subset C\,\overset{0}{\mathrm{supp}}\, b(.,t)$ almost everywhere on I and therefore $b(x^*(t),t)=0$ for a.a. $t\in I$. By continuity of x^* and b

$$b(x^*(t),t)\equiv 0 \text{ on } I. \quad (2.21)$$

Put $\varphi(s)=a_x(x^*(s),s)+b_x(x^*(s),s)u^*(s)$ which is a measurable function on I. By (2.20), (2.12), (2.13), (2.14) or (2.15) one has $\varphi(s)B(s)=0$ for a.a. $s\in I$. For this choice of φ in lemma 2.2, problem (2.6), (2.7) becomes problem $(1.4)^*$, $(1.5)^*$ for p^*. Consequently, this lemma yields

$$p^*(t)B(t)\equiv 0 \text{ on } I. \quad (2.22)$$

By continuity of x^*, finally,

$$p^*(t)B(t)x^*(t)\equiv 0 \quad \text{on } I. \tag{2.23}$$

Let $B(t)$ have columns $b_i(t)$. Then (2.22) says $p^*(t)b_i(t)\equiv 0$ on I, $1\leq i\leq n$. By (2.16)

$$p^*(t)C(t)\equiv 0 \text{ on } I. \tag{2.24}$$

which together with (2.21) and (2.23) implies the statement $S^*(t)\equiv 0$ on I.

2.3. Sufficient condition for $U^*=U_p^*$

In the preceding proof take

$$b(x,t)=\beta(x)\gamma(t), \tag{2.25}$$

β, γ being continuously differentiable with respect to their arguments in their respective domains $\mathbb{R}^n$, $\mathbb{R}$ of definitions. Let $\varepsilon>0$ exist such that the relations

$$B_\varepsilon(x_0)\subset \overset{0}{\operatorname{supp}}\,\beta \tag{2.26}$$

and (2.18) hold. Let

$$\begin{aligned} &\gamma(t)\equiv 0 \quad \text{on } \omega\in B \\ &\gamma(t)\neq 0 \quad \text{on } I\setminus \textstyle\bigcup_B \omega \end{aligned} \tag{2.27}$$

where $B=\{\omega\}$ is a family of closed subsets of I, $\text{mes}\,\omega>0$, $\bigcup_B \omega\subset I$, and with any two elements of B being seperated from each other by, at least, one subset of positive measure, not contained in B. It is $\text{mes}(I\setminus\bigcup_B\omega)>0$. Let H1 hold.The proof of theorem 2.1 gives

$$\beta(x^*(t))\neq 0 \quad \forall t\in I, \tag{2.28}$$

because $x^*(t)\in B_\varepsilon(x_0)\subset K$, $\forall t\in I$. Consequently,by (2.26) $x^*(t)\in \overset{0}{\operatorname{supp}}\,\beta$, $\forall t\in I$, which is (2.28). By (2.23), (2.24) $S^*(t)=-b(x^*(t),t)$, $\forall t\in I$, which by (2.25), (2.28), (2.27) gives

THEOREM 2.2. Consider process P. Let H1 hold, and let $b(x,t)$ be given by (2.25) with β, γ defined as above. Let H2 hold with (2.17) replaced by (2.26).
Then $U^*=U_p^*$.

See also remark 1.3.

2.4. Sufficient condition for $U^*=U_b^*$

Replace (2.17) by

$$B_\varepsilon(x_0) \subset \bigcap_{a.a.\,t\in I} \overset{0}{\text{supp}}\, b(.,t)\,, \tag{2.29}$$

and let the other assumptions of theorem 2.1 be unchanged. Then the proof of this theorem changes in one part, and this part reads as follows: By (2.18) and (2.1) $x^*(t)\in B_\varepsilon(x_0)\subset K$, $\forall t\in I$. Consequently, by (2.29), $\forall s\in I$, $x^*(s)\in B_\varepsilon(x_0)\subset \overset{0}{\text{supp}}\, b(.,t)$ for a.a. $t\in I$ which implies $b(x^*(s),t)\neq 0$ for all $s\in I$ and for a.a. $t\in I$. Thus $b(x^*(t),t)\neq 0$ on I which by (2.23), (2.24) gives the statement $S^*(t)=-b(x^*(t),t)\neq 0$ on I of the following

THEOREM 2.3. Consider process P. Let H1 hold. Assume H2 with (2.17) replaced by (2.29).
Then $U^*=U_b^*$.

3. NUMBER OF SWITCHES OF $u^*\in U_b^*$

In practice, bang-bang optimal controller with a finite set of switches are of great significance. If the statement of [7],p.152, were true this would imply each $u^*\in U_b^*$ should have a finite set of switches in I. Actually, this statement is wrong. The correct statement is, that process P does not assure that the switches of the bang-bang optimal controller should be finite or a particularly simple subset of I. To illustrate this, turn over to theorem 2.2. Herein define $\gamma(t)$ anew by

Real function $\gamma(t)$ is continuously differentiable on $\mathbb{R}$.
For $\Omega\subset I$, $\text{mes}\,\Omega=0$, define $\gamma(\Omega)=0$, and $\gamma(I\setminus\Omega)\neq 0$ (3.1)

Remark 3.1. $\gamma(t)=(t-\tau)^3\sin(t-\tau)^{-1}$, $t_0<\tau<t_f$, is such a function.

Replacing of (2.27) by (3.1) in the proof of theorem 2.2 yields the

COROLLARY 3.1. In the assumptions of theorem 2.2 replace (2.27) with (3.1).
Then $u^*\in U^*$ $(=U_b^*)$ have switching set Ω.

Consequently, the Switching-Time-Variation Method,[7],is inapplicable, in general, to the computation of $u^*\in U_b^*$ for it requires the switching set to be finite and not too large,[7], p.74. For computation of singular optimal controls and for open problems in this respect see [2], chap. 5 and 6.

4. REFERENCES

M. ATHANS, P.L. FALB [1] Optimal control. N.Y. 1966

D.J. BELL, D.H. JOHNSON [2] Singular optimal control problems. London, 1975.

A. BENSOUSSAN, E.G. HURST Jr., B. NÄSLUND [3] Management applications of modern control theory. Amsterdam, 1974.

R. GODEMENT [4] Algebra. Paris, 1968.

H.W. KNOBLOCH [5] Higher order necessary conditions in optimal control theory. Lecture Notes in Control and Information Sciences, 34. Berlin, 1981.

L. MARKUS, E.B. LEE [6] Foundations of optimal control theory. N.Y., 1967

S.F. MOON [7] Optimal control of bilinear systems and systems linear in control. Diss. Univ. New Mexico, August 1969.

L. SCHWARTZ [8] Analyse mathématique. VolI. Paris, 1967.

SHAPE CONTROLABILITY FOR FREE BOUNDARIES

J.P. Zolésio
Département de Mathématiques
Parc Valrose 06034 Nice Cedex France

We study the evolution of the free boundary Σ arising in the classical obstacle problem. When the boundary Γ of the domain is perturbed by a vector field V we explicit the vector field W which built the deformation of Σ.

In the classical obstacle problem for the membrane the free boundary appears as being the zero level set of the function $Y=y-\psi$ (where y is the displacement of the membrane and ψ the obstacle, the constraint being $y \geq \psi$).

This situation is general in many problems then we first study the evolution of a level curve when some parameters, may be the geometry, have given variations. As an introduction we recall some previous results then, given a speed deformation V of the shape Ω of the membrane, we explicit the speed deformation W of the free boundary Σ.

I. Deformation of shape

Let Ω be a smooth bounded open set in $R^n, n\geq 2$, and V a given vector field in a neigbourhood of $o \times \Omega$. We will consider V(t) the mapping $x \mapsto V(t,x)$ defined on a neigbourhood of $\bar{\Omega}$. For any X in a neigbourhood of $\bar{\Omega}$ we consider the ordinary differen tial equation $(d/dt)x(t)=V(t,x(t))$ with the initial condition $x(o)=X$. The mapping $T_t(V):\ X \mapsto x(t)$ is a smooth one to one transformation. If V belongs to the space

$C^o([o,t_1[\,, C^k(R^n, R^n))$, with $k\geq 1$, then $T_t(V)$ and its inverse mapping $T_t(V)^{-1}$ belong to $C^k(R^n, R^n)$. For mor precisions we refer to J.P.Zolésio [5], [6], [7]

If Γ is the boundary of Ω we consider $\Omega_t=T_t(V)(\Omega)$ and $\Gamma_t=T_t(V)(\Gamma)$ the perturbated domain and boundary, built by the field V at time t. (of course $\Omega_o=\Omega$ and $\Gamma_o=\Gamma$). Now for any $s>o$ the domain Ω_{t+s} and its boundary Γ_{t+s} are built by the field V_t given by $V_t(s) = V(t+s)$: $\Omega_{T+s} = T_{t+s}(V_t)(\Omega_t)$ and the same for the boudaries.

I.I The vector field W which built the level curves of a given function u

Suppose that:

(1) u is a smooth function, $u \in C^2(\bar{\Omega})$ and u=o on Γ, reaches its maximum at a single point x_u in Ω with $M=\max u=u(x_u)$ o

(2) $|\nabla u(x)| > o \quad \forall x \in \Omega - x_u$

Theorem 1 (J.P.Zolésio[7])

The level curves $\Sigma_t=u^{-1}(t)=\{x \in \bar{\Omega} / u(x)=t\}$, fot $o\leq t<M$ are C^1 manifolds which are built by the autonomeous vector field $W = |\nabla u|^{-2}\ \nabla u$ in the sens that :

$$\Sigma_t = T_t(W)(\Gamma) \quad \text{for} \quad o\leq t<M .$$

Any other field builting these level curves can be written W+S where S(x) is tangent to Σ_t, with t=u(x).

Remark 1. $T_t(W)$ is the exponential e^{tW} for W is autonoeous.

Remark 2. The basic assumption(2) is obviously necessary because of the expression of W. It can be replaced, in any dimension $n\geq 2$ by an easy criteria:

Proposition 1. Assume that:

(3) $\Delta u(x) \geq o$ or $\Delta u(x) \leq o$ in Ω

then (2) holds for any dimension $n\geq 2$

proof: given any $t, o<t<M$, consider t_1 and t_2 such that $o\leq t_1<t<t_2<M$ and $\Omega_1=\{x\in\Omega \,/\, t_1<u(x)<t\}$, $\Omega_2=\{x\in\Omega \,/\, t<u(x)<t_2\}$. If $\Delta u\geq o$ then u reaches its maximum on Ω_1 at any point P of $\Sigma_t=u^{-1}(t)$; if $\Delta u\leq o$ $-u$ reaches its maximum at any point P of Σ_t. In both cases, by the Maximum Principle (seeH.H Protter and H.F.Weimberger [1])we know that at any point P of Σ_t we have $|(\partial/\partial n)u(P)|>o$. Now on Σ_t we have $|\nabla u| = |(\partial/\partial n)u|$, then (2) holds.

Application 1

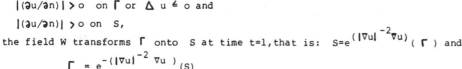

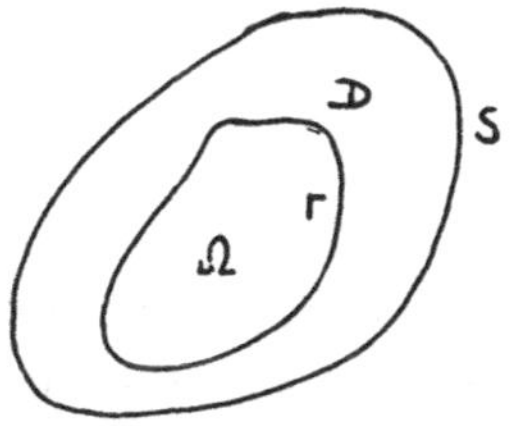

If Γ and S are two smooth manifolds in $R^n, n\geq 2$, such that Γ (Rep.S) is the boundary of a regular open set Ω (Rep.D) $\bar{\Omega}\subset D$, then for any function $u\in C^2(A)$ with $A = D\setminus\bar{\Omega}$, such that $\Delta u\geq o$ and $|(\partial u/\partial n)|>o$ on Γ or $\Delta u\leq o$ and $|(\partial u/\partial n)|>o$ on S,

the field W transforms Γ onto S at time t=1, that is: $S=e^{(|\nabla u|^{-2}\nabla u)}(\Gamma)$ and

$$\Gamma = e^{-(|\nabla u|^{-2}\nabla u)}(S)$$

III. The field W which builts a given level curve of a one parameter family functions

s is a fixed given constant, say $s>o$.

Ω_t is a domain depending on the parameter t, for example $\Omega_t=T_t(V)(\Omega)$, $\Omega=\Omega_o$, and y_t belongs to $H^m(\Omega_t)$, $y=y_o$, may be the solution of some boundary value problem posed in the geometry Ω_t. Assume that for t little enough,

(4) $|t|<\varepsilon$, $\Sigma_t=y_t^{-1}(s)=\{x\in\Omega_t \,/\, y_t(x)=s\}$ (and $\Sigma_o=\Sigma$) is non empty and is a C^1 manifold in Ω_t.

The problem we look now is to find the field W(t,w) which builts Σ_t that is such that $\Sigma_t = T_t(W)(\Sigma)$

To handle this problem we need to introduce the derivative of y_t on the parameter t.

If the domain Ω_t is built by the field V(t,x), the MATERIAL derivative (see J.P.Zolésio[7]) is given by :

(5) $$\dot{y}(V) = \lim_{t\to o} (y_t\circ T_t(V) - y)/t \quad \text{(limit in the } H^m(\Omega) \text{ norm)}$$

and the SHAPE derivative of y in the direction of the vector field V is

(6) $$y'(V) = \dot{y}(V) - \nabla y.V(o) \quad \text{element of } H^{m-1}(\Omega)$$

It is shown (see J.P.Zolésio [7])that y'(V) actually does not depends on the field V but just on $v = V(o).n$, the normal component of $V(o)$ on the boundary Γ of Ω.

Following the notations of the end of the first section we will write $\dot{y}_t(V_t)$ and $y'_t(V_t)$ the material and shape derivatives of y_t at Ω_t in the direction of the field V_t. When no confusion is possible we simply write $\dot{y}$, y', $\dot{y}_t$ and y'_t.

We now formulate some hypothesis on y_t :

(7) y_t belongs to $C^2(\bar{\Omega}_t)$

(8) $|\nabla y_t| > o$ on $\Sigma_t = y_t^{-1}(s) = \{x \in \Omega_t \;/\; y_t(x) = s\}$

(9) the material derivative $\dot{y}_t(V_t)$ exists in $H^m(\Omega_t)$, $m \geq 1$

(1o) the shape derivative $y'_t(V_t)$ belongs to $C^1(\Omega_t)$

Theorem 2 . Assuming (7) to (1o), the level curve $\Sigma_t = y_t^{-1}(s)$ are C^1 manifolds in Ω_t which are built by the (non autonomeous) vector field

(11) $W(t,x) = - y'_t(V_t)(x) \; |\nabla y_t(x)|^{-2} \; \nabla y_t(x)$ in the sens that:

$\Sigma_t = T_t(W)(\Sigma)$, $t \geq o$, with $\Sigma = \Sigma_o$; where $y'_t(V_t)$ is the shape derivative at Ω_t in the direction V_t.

proof: for any g given in $C^\infty(R^n)$ consider

$$f(t) = \int_{T_t(Z)(\Sigma)} g \circ T_t(Z)^{-1} \; (\det DT_t(Z))^{-1} \circ T_t(Z)^{-1} \quad y_t \; da_t$$

where da_t is the superficial measure, DT is the jacobian matrix of the transformation T and $Z = Z(t,x)$ is any smooth vector field such that :

$Z(t) = V(t)$ in a neigbourhood of Γ_t ,

$Z(t) = W(t)$ in a neigbourhood of Σ_t (W given by (11))

By change of variable we have:

$$f(t+r) = \int_{\Sigma_t} g \; y_{t+r} \circ T_{t+r}(Z_t) \; da , \quad \text{and}$$

$f'(t) = \int_{\Sigma_t} g \; \dot{y}_t(Z_t) \; da$, as $m \geq 1$ this material derivative is correct in $H^{1/2}(\Sigma_t)$;

But $\dot{y}_t(Z_t) = o$ on Σ_t; effectively we have

$\dot{y}_t(Z_t) = y'_t(Z_t) + \nabla y_t \,.\, Z(t)$; $\quad y'(Z_t)$ just depends on the value of Z_t on the boundary Γ_t then:

$y'_t(Z_t) = y'_t(V_t)$ and using (11) and $Z(t) = W(t)$ on Σ_t we get $\dot{y}_t(Z_t) = o$.

Then $f(t) = f(o)$ is a constant function

$f(o) = \int_{\Sigma} g\, y\, da = s \int_{\Sigma} g\, da$ (for $y=s$ on $\Sigma = y^{-1}(s)$)

and again by change of variable $f(o) = \int_{T_t(Z)(\Sigma)} s\, g \circ T_t(Z)^{-1} \det(D(T_t(Z)^{-1}))\, da_t$,

but $\det(DT_t(Z)^{-1})) = (\det DT_t(Z))^{-1} \circ T_t(Z)^{-1}$,

and identifying with the initial expression of $f(t)$ we have $y_t = s$ on $T_t(Z)(\Sigma)$ that is $T_t(Z)(\Sigma) = \Sigma_t = y_t^{-1}(s)$.

Remark 3 .It seems that (8) is not enough for the validity of the expression (11) for the field W.We don't want to go here in more details but just say that in fact the expression (11) of W(t) is just necessary in a neigbourhood of Σ_t and we can change it by any smooth prolongation out this neigbourhood so that the expression (8) is enough.

IV. The Free Boundary Problem (for the membrane)

We consider an obstacle problem,the model problem being the following variational inequality (see J.L.Lions,R.Glowinski,H.Tremolière [3]):

f and g being smooth functions on R^n, $n \geq 2$ (the forcing term and the obstacle) Ω a bounded smooth open set (the shape of the membrane),V(t,x) a given vector field (the speed of deformation of the shape of the membrane), $\Omega_t = T_t(V)(\Omega)$ the shape of the memb ane at time t and y_t the displacement of the Ω_t-shaped membrane under the loading f.The convex set of admissible displacements is:

$$K(\Omega_t) = \{ h_t \in H_o^1(\Omega_t) \;/\; h_t \geq g \}$$

we assume that:

(12) $f < o$, $\max_{\Omega} g > o$ and $g < o$ in a neigbourhood of Γ (then for t little enough these conditions hold on Ω_t and Γ_t).

Y_t is the solution of

(13) $y_t \in K(\Omega_t)$, $\int_{\Omega_t} \nabla y_t . \nabla(h_t - y_t) dx \geq \int_{\Omega_t} f(h_t - y_t)\, dx$, $\forall h_t \in K(\Omega_t)$

From H.Brezzis,G.Stampacchia [8] we know that $y_t \in W^{2,\infty}(\Omega_t)$. We consider the closed subset of Ω_t (the contact or coincidence set):

$A_t^c = \{ x \in \Omega_t \;/\; y_t(x) = g(x) \}$ and Σ_t its (free) boundary is a part of the boundary of following opensubset of Ω_t

$$\overset{\circ}{\Omega}_t = \{ x \in \Omega_t \;/\; y_t(x) > g(x) \}$$

$\partial\overset{\circ}{\Omega}_t = \Gamma_t \cup \Sigma_t$, Σ_t is the free boundary of the problem;if Σ_t is a C^1 manifold then y_t is solution of the problem:

(14) $-\Delta y_t = f$ in $\overset{\circ}{\Omega}_t$, $y_t = o$ on Γ_t , $y_t = g$ and $(\partial/\partial n) y_t = (\partial/\partial n) g$ on Σ_t

The material derivative $\dot{y}_t$
exists (derivative in the $H^1_o(\Omega_t)$ norm) and the shape derivative y'_t is the solution of the following variational inequality: (see [4])

$$y'_t \in S_{V_t}(\Omega_t)=\Big\{ h_t \in H^1(\Omega_t) \,/\, h_t = -(\partial/\partial n) y_t \; V(t).n_t \text{ on } \Gamma_t \,,$$

(15) $$h_t \geq o \text{ quasi-evry where on } A^c_t \,,$$

$$\int_{A^c_t} f h_t \, dx = o \Big\}$$

$$\int_{\Omega_t} \nabla y'_t . \nabla (h_t - y'_t)\, dx \geq o \quad , \forall\, h_t \in S_{V_t}(\Omega_t)$$

Now, the expression of the convex set S_{V_t} can be very simple:

Proposition 2.

If $f > 0$ or $f < o$ on R^n then

(15') $$S_{V_t}(\Omega_t) = \Big\{ h_t \in H^1_o(\Omega_t) \,/\, h_t = -(\partial/\partial n) y_t \; V(t).n_t \text{ on } \Gamma_t \text{ and } h_t = o \text{ a.e. on } A^c_t \Big\}$$

proof: We have $h_t = o$ quasi-evry where on A^c_t that is: one can find $A_n \subset A_{n-1} \subset \ldots \subset A^c_t$ (sequence of mesurable sets) such that $h_t = o$ on $A^c_t \setminus A_n, \forall n$, with capacity $ca(A_n) \rightarrow o$ that implies measure$(A_n) \rightarrow o$;

$(h_t)^- = \sup(o, -h_t)$ belongs to $H^1(\Omega_t)$ and $\int_{A^c_t} h_t^- \, dx = \int_{A_n} h_t^- \, dx, \forall n$

then $\int_{A^c_t} h_t^- \, dx = o$; but $h_t^- \geq o$ a.e. then $h_t^- = o$ a.e. on A^c_t, then $h_t \geq o$ a.e. on A^c_t .

Corollary 1

y'_t is then the solution of the classical Dirchlet problem:

(16) $$\Delta y'_t = o \text{ in } \mathring{\Omega}_t \,,\quad y'_t = -(\partial/\partial n) y_t \; V(t).n_t \text{ on } \Gamma_t, y'_t = o \text{ on } \Sigma_t \;.$$

We now look to the field W which built the free boundary Σ_t .The cons traint being $y_t \geq g$ on Ω_t we introduce

(17) $$Y_t = y_t - g \geq o \quad \text{on } \Omega_t$$

The free boundary is the zero level curve: $\Sigma_t = Y_t^{-1}(o) = \{ x \in \mathring{\Omega}_t \,/\, Y_t(x) = o \}$

Taking s=o in the second section we know that the field which built such a level curve is $W(t,x) = - Y'_t \, |\nabla Y_t|^{-2} \, \nabla Y_t$

Now, the shape derivative $Y'_t = y'_t$ is caracterized by (16) for g'=o (as the obstacle function g does not depend on the domain but is given over all R^n).

The main difficulty comes fromthe fact that the condition $|\nabla Y_t| > o$ on Σ_t vanishes for the free boundary condition is precisely $\nabla y_t = o$ on Σ_t .

The tool is to consider, for $r > o$ little, the neigbourhing level curve of Σ_t :

(18) $\Sigma_t^r = Y_t^{-1}(r) = \left\{ x \in \mathring{\Omega}_t \ / \ Y_t(x) = r \right\}$

Proposition 3 .
If $\Delta Y_t = -(f + \Delta g) \geq o$ in a neigbourhood (in $\mathring{\Omega}_t$) of Σ_t then $|\nabla Y_t| > o$ on Σ_t^r .

proof:
The maximum of Y_t on the compact set $\left\{ x \in \mathring{\Omega}_t \ / \ o \leq Y_t(x) \leq r \right\}$ is reached at each point P of Σ_t^r and by the maximum principle (see again [1]) we have $|\nabla Y_t(P)| > o$ and this result holds for any dimension $n \geq 2$.

Then the field $W(t) = - y_t' \ |\nabla Y_t|^{-2} \ \nabla Y_t$ is well defined at the neig bourhood of Σ_t^r and then by the theorem 2 we have

$$\Sigma_t^r = T_t(W)(\Sigma_o^r)$$

To reach the situation r=o ,that is the free boundary Σ_t, we make a Taylor expension. Our basic assumption is the regularity of both the free boundary Σ_t and the displacement y_t (or equivalentely Y_t).

From Kindherlerer-Stampacchia [2] we know that assuming

(19) y_t belongs to $C^2(\overline{\Omega}_t)$

then Σ_t is a smooth manifold (the smoothness depending in our case on the smoothness of f and g, the loading term and the obstacle).

Considering r little, each point x of Σ_t^r can be written as $x = z + s n_t(z) + o(s)$ for some unique z on Σ_t .And $r \to o$ as $s \to o$; $n_t(z)$ is the normal to Σ_t at z.

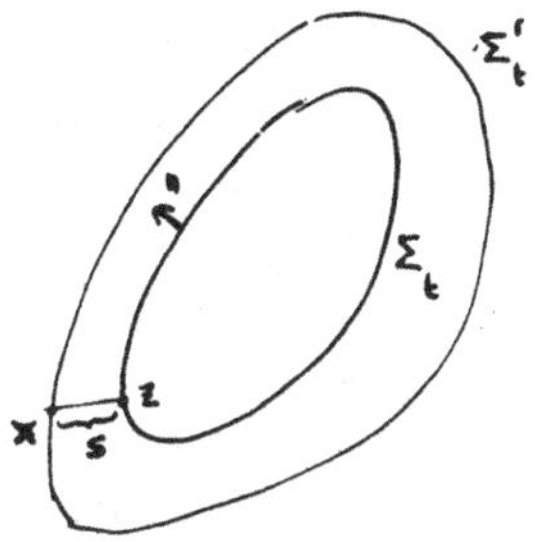

We write the Taylor expensions :

$$y_t'(x) = y_t'(z) + s \ \nabla y_t'(z) \ . \ n(z) + o(s)$$

$$\nabla Y_t'(x) = \nabla Y_t(z) + s \ D^2 Y_t(z).n(z) + o(s)$$

Now on the free boundary Σ_t we have $Y_t = o$, $\nabla Y_t = o$, $Y_t' = y_t' = o$, then

$$\nabla Y_t(x) = s \ D^2 Y_t(z).n_t(z) + o(s) \quad , \quad y_t'(x) = s \ \nabla y_t'(z).n_t(z) + o(s)$$

and for x belonging to Σ_t^r we have

$$W(t,x) = - y_t'(x) \ |\nabla Y_t(x)|^{-2} \ \nabla Y_t(x) =$$

$$\left((\partial/\partial n) y_t'(z) + o(1) \right) \ \left\| D^2 Y_t(z).n_t(z) + o(1) \right\|^{-2} D^2 Y_t(z).n_t(z) + o(1)$$

When $r \to o$ then $x \to z$ andthe field W can be defined by continuity at $z \in \Sigma_t$ by:

$$W(t,x) \longrightarrow W(t,z) = (\partial/\partial n) y'_t(z) \; \| D^2 Y_t(z).n_t(z) \|^{-2} \; D^2 Y_t(z).n_t(z)$$

The problem is now to identify this last term for z belonging to Σ_t .
But on Σ_t we have also :

$$\nabla(\partial_i Y_t) = (\partial/\partial n)(\partial_i Y_t) n_t \quad , \quad \partial_{ij} Y_t = (\partial/\partial n)(\partial_i Y_t) n_j$$

$$(D^2 Y_t . n_t)_j = \partial_{ij} Y_t (n_t)_i = (\partial/\partial n)(\partial_i Y_t) \; n_i n_j$$

that is :

$$D^2 Y_t . n_t = \langle D^2 Y_t . n_t \, , \, n_t \rangle \, n_t \tag{20}$$

But we know also (see J.P.Zolésio [7]) that on the level curve Σ_t we have:

$$\Delta Y_t = \Delta_{\Sigma_t} Y_t + H_t \; (\partial/\partial n) Y_t + \langle D^2 Y_t . n_t \, , n_t \rangle \tag{21}$$

Where Δ_{Σ_t} is the Laplace Beltrami operator on the **manifold** Σ_t

($\Delta_{\Sigma_t} h = \mathrm{div}_{\Sigma_t} (\nabla_t \; h)$ where div_{Σ_t} is the tangential divergence and ∇_t h the tangential gradient of h on the manifold , see [7])

H_t is the mean curvature of Σ_t

In the three terms of the decomposition (12) the first one is equal to zero for Y_t is constant on Σ_t, the second one does not contribute for $(\partial/\partial n) Y_t = o$ on Σ_t; then from (20) and (21) we get:

$$D^2 Y_t(z) . n_t(z) = (\; f(z) + \Delta g(z) \;) \; n_t(z) \quad \text{on} \quad \Sigma_t , \tag{22}$$

and we have got the

Thorem 3 .

Assuming that the data f and g are smooth enough with $f \geq o$ or $f \leq o$, $f + \Delta g < o$ and (19) holds for any t, the free boundary Σ_t is built by the vector field

$$W(t,x) = (f(x) + \Delta g(x))^{-1} \qquad \nabla y'_t(x) \tag{23}$$

where y'_t is the solution of the Dirichlet problem (16) posed on $\mathring{\Omega}_t$.

Application 2 . A very simple example is to use a gradient method to minimize (or maximize) the volume of the contact set A^c. Following [7] we introduce the shape functional $J(\Omega_t) = $ =measure(A^c_t) . The boundary of A^c_t is built by the vec tor field W given at (23), then the Eulerian derivative is

$$dJ(\Omega_t \, ; V_t) = \frac{d}{dt} \text{measure}(\; A^c_t \;) = \int_t W(t).n_t \; da_t$$

Introducing the adjoint equation as

(24) $-\Delta P_t = o$ in $\mathring{\Omega}_t$, $P_t = o$ on Γ_t , $P_t = (F + \Delta g)^{-1}$ on Σ_t

we get:

$$dJ(\Omega_t, V_t) = \int_{\Sigma_t} (\partial/\partial n) y'_t \, P_t \, da_t$$

$$= - \int_{\Gamma_t} (\partial/\partial n) y_t \, (\partial/\partial n) P_t \, V(t).n_t \, da_t$$

Then we see that the functional J is shape-differentiable (following the terminology introduced in [7] [5] for the mapping $V \longmapsto dJ(\Omega, V)$ is linear then the structure theorem for this situation is verified: $dJ(\Omega, V) = \langle G, V(o) \rangle$ where G is a vectorial distribution -an element of $\mathcal{D}'(R^n, R^n)$ -which can be written as $G = {}^t\gamma_\Gamma (g\ n)$ where n is the normal on Γ and g a (scalar) distribution on the manifold Γ ,here $g = (\partial/\partial n) y \ (\partial/\partial n) P$.

Remark 4 If the basic assumption $f \leq$ or $f \geq o$ fails then the shape derivative y' remains the solution of the variational inequality (15) which **cannot** be reduced to the dirichlet problem (16) and then the linearity of $V \longmapsto dJ(\Omega; V)$ can fail (that is there is no gradient G and of course no density gradient g on Γ) Such non differentiable situations are now investigated

References

1 **M.H.Protter and H.F. Weimberger.** Maximum principle in differential equations Englewood Cliffs, N.J., Prentice-Hall, 1967

2 **D.Kinderlehrer, G.Stampacchia.** An introduction to variational inequalities... Academic Press, 1980

3 **J.L.Lions, R.Glowinski, Trémolière.** Analyse numerique des inequations variationnelles, tome 2, Paris, Dunod, 1976

4 **J.Sokolowsi, J.P.Zolésio.** Dérivation par rapport au domaine dans les problèmes unilatéraux. Rapport INRIA n°132, Paris 1982

5 **J.P.Zolésio** Identification de domaine par déformation. Thèse de do doctorat d'état, Nice, 1979.

6 ------------ Un résultat d'éxistence de vitesse convergente; C.R.Acad.Sci. Série A, t.283, p.855, 1976

7 ------------- The Material Derivative(or speed method) for shape optimization in "Optimization of distributed parameter structure(vol.2) J.Céa, Ed.Haug Eds., Sijthoff and Noordhoff 1981, Alphen aan den Rijn

8 **H.Brézzis, G.Stampacchia** Bull. Soc. Math. de France , t. 86 , 1968

APPROXIMATION OF BOUNDARY CONTROL PROBLEMS FOR EVOLUTION VARIATIONAL INEQUALITIES ARISING FROM FREE BOUNDARY PROBLEMS

Irena Pawlow
Systems Research Institute
Polish Academy of Sciences
Newelska 6, 01-447 Warsaw
Poland

1. INTRODUCTION

The paper is concerned with a semi-discrete approximation of boundary control problems for systems governed by a class of evolution variational inequalities of the second kind arising in particular from multi-phase Stefan problems.

We present a convergence analysis, so far not studied enough in the literature, for the approximating optimal control problems. The proposed approximation procedure is composed of two stages: (i) construction of a family of auxiliary optimal control problems, involving regularization of the variational inequalities, (ii) construction of semi-discrete Galerkin approximations in space to the regularized problems. Both optimal controls for the approximating problems and the corresponding minimal values of the cost functionals are shown to be convergent to those for the original problem. The results concern equally iterative and collective convergences with respect to regularization and discretization parameters.

The paper develops approximation ideas presented in [9]. The results obtained can be compared with those due to Tiba [16,17] who has considered Neumann boundary control problem for two-phase Stefan problem, using its enthalpy formulation. That approach is advantageous compared to ours as it employs $L^2(\Sigma)$ control space whereas we take $H^1(0,T;L^2(\Gamma))$, conditioned by the arguments ensuring existence of a weak solution to the variational inequality. On account of this we get however more regularity in time of that weak solution, permitting a similar treatment of control problems with distributed, as well as boundary and terminal observations in L^2-spaces. Moreover, we are able to prove some properties of the approximating control problems that do not seem to be known in the case of controls in $L^2(\Sigma)$. Among them of particular interest are error estimates for the approximations (see (11),(14)) and the result on the collective convergence given in Thm. 2.

There are results on approximation of distributed control problems for systems governed by evolution variational inequalities of the first kind due to Arnǎutu [1]. Various aspects of control problems for systems of Stefan type, including existence questions, optimality conditions and approximation methods were studied in [2,3,8,9,11,15]. The survey [6] offers a state-of-art overview of control of parabolic free boundary problems.

2. BOUNDARY CONTROL PROBLEM FORMULATION

Let Ω be a bounded domain in R^n, $n \geq 1$ with a sufficiently regular boundary Γ. For a fixed $T > 0$ let $Q = \Omega\times(0,T)$ and $\Sigma = \Gamma\times(0,T)$.

The systems under consideration are assumed to be governed by a pseudoparabolic boundary value problem of the form

$$\begin{cases} \gamma(y') - \Delta y \ni f & \text{in } Q & (1.a)\\ \frac{\partial y}{\partial \nu} + g_o\, y = \int_0^t u(\tau)\, d\tau & \text{on } \Sigma & (1.b)\\ y(0) = 0 & \text{in } \Omega & (1.c)\end{cases}$$

where γ is a maximal monotone graph in $R\times R$, in general multivalued, $\vec{\nu}$ denotes the unit outward vector normal to Γ, $y' = \partial y/\partial t$, $f = f(x,t)$, $g_o = g_o(x)$, $u = u(x,t)$ are given functions, with u representing the control variable.

Problems of this type arise, in particular, as models of various nonlinear heat conduction and diffusion processes associated with phase transitions. The most classical models of phase transition processes are referred to as one- or multiphase Stefan problems. In that context system (1) may be considered as a generalized formulation of two-phase Stefan problem. Then the graph γ, physically representing enthalpy of the system, takes the form

$$\gamma(r) = \tilde{\gamma}(r) + L\,H(r) , \qquad r \in R . \tag{2.a}$$

Here $L \geq 0$ is a given constant, $H(r)$ is the Heaviside's function (extended to the whole vertical segment $[0,1]$ for $r = 0$),

$$\tilde{\gamma}(r) = \int_0^r \rho(\xi)\, d\xi , \tag{2.b}$$

where the coefficient ρ may exhibit a finite jump discontinuity at $r = 0$ corresponding to the phase transition, and is assumed to be bounded

$$0 < \bar{\rho} \leq \rho(r) \leq \bar{\bar{\rho}} < +\infty , \qquad r \in R . \tag{3.a}$$

More precisely, we assume that ρ admits representation

$$\rho(r) = \tilde{\rho}(r) + \tilde{\tilde{\rho}}\, H(r) , \qquad r \in R \tag{3.b}$$

where $\tilde{\rho} \in C^1(R)$, $|D\tilde{\rho}(r)| \leq c_\rho < +\infty$, $\tilde{\tilde{\rho}}$ is a constant.

Physically, L corresponds to the latent heat of phase transition, $\rho = c/k$ with c, k representing specific heat and heat conductivity, respectively.

In (1), the variable y may be interpreted as freezing index, related to temperature θ by the relationship

$$y(x,t) = \int_0^t \theta(x,\tau)\, d\tau , \qquad (x,t) \in Q . \tag{4}$$

Clearly, $y' = \theta$ in Q . The mixed type condition on Σ corresponds then to performing the control action by means of the boundary heat flux u , with g_o measuring permeability to heat at the boundary Γ . The right-hand side term f admits the representation

$$f(x,t) = \int_0^t \lambda(x,\tau)\, d\tau + w_o(x) \tag{5}$$

where λ corresponds to distributed heat sources and w_o is the initial enthalpy, compatible with the initial temperature θ_o so that $w_o \in \gamma(\theta_o)$ in Ω . For more detailed exposition of the links of (1) with the classical formulation of two-phase Stefan problem we refer to [10] .

In order to assure the existence of a solution to problem (1) (see assumptions to Thm. 1) , we shall take $\mathcal{U} = H^1(0,T;L^2(\Gamma))$ equipped with the standard norm $\|\cdot\|_{\mathcal{U}}$ as the control space.

The following control problem with distributed observation will be considered:

$$\begin{cases} \text{Minimize} \quad J(y',u) = \frac{1}{2} \| y' - \theta_d \|^2_{L^2(Q)} + \frac{\alpha}{2} \| u \|^2_{\mathcal{U}} \\ \text{over} \quad u \in \mathcal{U}, \quad \text{subject to dynamics (1)} ; \quad \theta_d \in L^2(Q) , \ \alpha > 0 \quad \text{are given.} \end{cases} \tag{6}$$

Referring again to the phase transition processes, the above performance index may correspond to an attempt of ensuring a desired temperature regime θ_d , directly, or to pursuing a prescribed pattern of the phase change interface movement, indirectly.

To specify the concept of solution to problem (1) , let us introduce the notations:

$H = L^2(\Omega)$, $V = H^1(\Omega)$; $(\cdot,\cdot)$, $(\cdot,\cdot)_\Gamma$ the scalar products in $L^2(\Omega)$ and $L^2(\Gamma)$, respectively;

$$a(y,z) = (\nabla y , \nabla z) + (g_o\, y , z)_\Gamma ;$$

$$\Psi(z) = L \int_\Omega \psi_o(z(x))\, dx , \quad \psi_o(z) = z^+ .$$

The solution y of (1) is then understood as satisfying:

$$\text{(VI)} \quad \begin{cases} y , y' \in L^\infty(0,T;V) , \\ (\tilde{\gamma}(y'(t)) - f(t) , z - y'(t)) + a(y(t) , z - y'(t)) - \\ \qquad - (\int_0^t u(\tau)\, d\tau , z - y'(t))_\Gamma + \Psi(z) - \Psi(y'(t)) \geqq 0 \\ \qquad\qquad \forall\, z \in V , \quad \text{a.a. } t \in [0,T] , \\ y(0) = 0 \quad \text{in } \Omega . \end{cases}$$

This formulation is usually classified as evolution variational inequality of the second kind [4] .

The solution y of (VI) can be proved to exist and be unique.

Theorem 1 [10]. Assume that

(A1) γ is defined by (2),(3) ;

(A2) $u \in \mathcal{U}$;

(A3) f is defined by (5) , $\lambda \in L^2(Q)$;

(A4) $w_o \in H$, $\theta_o \in V \cap L^\infty(\Omega)$, $w_o = \gamma^o(\theta_o)$ where γ^o denotes the minimal norm section of the graph γ ;

(A5) $g_o \in L^\infty(\Gamma)$, $g_o \geqq 0$, the set $\{ x \in \Gamma \mid g_o(x) > 0 \}$ is of Lebesgue measure positive with respect to Γ .

Then there exists the unique solution y of (VI) such that

$$y' \in L^\infty(0,T;V) \cap H^1(0,T;L^2(\Omega)) \quad , \quad y'(0) = \theta_o \quad \text{in } \Omega \ , \tag{7.a}$$

satisfying the estimates

$$\| y \|_{L^\infty(0,T;V)} + \| y' \|_{L^\infty(0,T;V)} + \| y'' \|_{L^2(Q)} \leqq C \tag{7.b}$$

with a finite constant C dependent only upon bounds on the data.

Let $\Xi : \mathcal{U} \to L^2(Q)$ denote the state observation operator defined by $\Xi(u) = y'$ where y is the solution of (VI) corresponding to control u . The control problem (6) can be expressed then in the form

$$\text{(P)} \quad \underset{u \in \mathcal{U}}{\text{Min}} \ \{ I(u) = J(\Xi(u),u) = \tfrac{1}{2} \| \Xi(u) - \theta_d \|^2_{L^2(Q)} + \tfrac{\alpha}{2} \| u \|^2_{\mathcal{U}} \} .$$

Remark 1. Due to (7) , we can similarly consider also the control problems with boundary observation in $L^2(\Sigma)$ or terminal observation in $L^2(\Omega)$.

3. PROPERTIES OF THE STATE OPERATOR. EXISTENCE RESULTS

We recall now some properties of the operator Ξ , resulting from a study of (VI) .

Proposition 1 [11]. (i) Ξ is compact from $\mathcal{U}$ into $L^2(Q)$,
(ii) Ξ is Lipschitz continuous from $L^2(\Sigma)$ into $L^2(Q)$.

Remark 2. The only result available relative to more regularity of Ξ guarantees its Gateaux differentiability on a dense subset of $L^2(\Sigma)$, as a consequence of the above assertion (ii) .

The existence of optimal solutions for problem (P) follows immediately by standard arguments in view of the compactness of Ξ and weak lower semicontinuity of the norms.

Corollary 1. Problem (P) admits at least one optimal solution $\hat{u}$.

Remark 3. In general, no uniqueness of optimal controls for problem (P) can be assured. Nevertheless, after a suitable choice of the parameter α one can get **strict convexity of the functional** I and therefore uniqueness of the optimal control for (P) . We refer to Kluge [7] for a study of regularization of control problems, in particular for estimates on α permitting its choice so that to provide this convexity.

4. REGULARIZED CONTROL PROBLEMS

The lack of differentiability of Ξ makes numerical solution of the control problem more time consuming as descent or nondifferentiable minimization techniques have to be applied. An alternative approach consists in using regularization techniques to the problem so that to fall into a differentiable framework. The approximation procedure we propose exploits regularization of the state operator Ξ prior to discretization of the problem.

For $\varepsilon > 0$ let us introduce the operator $\Xi_\varepsilon : \mathcal{U} \to L^2(Q)$ defined by $\Xi_\varepsilon(u) = y_\varepsilon'$ where y_ε is the solution of problem $(VI)_\varepsilon$ being a "smoothed" counterpart of (VI). It has the same structure as (VI) with Ψ, $\tilde{\gamma}$ and w_o replaced respectively by

$$\Psi_\varepsilon(z) = L \int_\Omega \psi_{o\varepsilon}(z(x))\, dx \quad , \quad \tilde{\gamma}_\varepsilon(z) = \int_0^z \rho_\varepsilon(\xi)\, d\xi \quad , \tag{8.a}$$

$$w_{o\varepsilon} = \tilde{\gamma}_\varepsilon(\theta_o) + L\, D\psi_{o\varepsilon}(\theta_o) \tag{8.b}$$

where $\psi_{o\varepsilon}$ is a smooth approximation of ψ_o, defined in particular as

$$\psi_{o\varepsilon}(r) = \begin{cases} 0 \quad , & r \leqq 0 \\ \dfrac{1}{\varepsilon^5} r^6 - \dfrac{3}{\varepsilon^4} r^5 + \dfrac{5}{2\varepsilon^3} r^4 \quad , & 0 < r \leqq \varepsilon \\ r - \dfrac{\varepsilon}{2} \quad , & r > \varepsilon \end{cases} \tag{8.c}$$

and $\rho_\varepsilon(r) = \tilde{\rho}(r) + \tilde{\tilde{\rho}}\, D\psi_{o\varepsilon}(r) \quad , \quad r \in R\,.$ (8.d)

Apparently, $(VI)_\varepsilon$ corresponds to problem (1) with a compatible, smooth, single-valued approximation γ_ε of the graph γ :

$$\gamma_\varepsilon(r) = \tilde{\gamma}_\varepsilon(r) + L\, D\psi_{o\varepsilon}(r) \quad , \quad r \in R\,. \tag{9}$$

We shall recall now some basic properties of the regularized problem.

<u>Proposition 2 [12]</u>. Let (A1)-(A5) be satisfied. Then there exists the unique solution y_ε of $(VI)_\varepsilon$, such that $y_\varepsilon \in L^\infty(0,T;V)$, $y_\varepsilon' \in L^\infty(0,T;V) \cap H^1(0,T;L^2(\Omega)) \cap L^2(0,T;H^2(\Omega))$, $y_\varepsilon(0) = 0$, $y_\varepsilon'(0) = \theta_o$ in Ω, satisfying estimate (7.b) and

$$\| y_\varepsilon' \|_{L^2(0,T;H^2(\Omega))} \leqq \frac{C}{\varepsilon^{1/2}} \tag{10}$$

with a finite constant C independent of ε.

<u>Proposition 3 [11]</u>. (i) Ξ_ε is compact from $\mathcal{U}$ into $L^2(Q)$;

(ii) Ξ_ε is Lipschitz continuous from $L^2(\Sigma)$ into $L^2(Q)$ with the Lipschitz constant independent of ε ;

(iii) Ξ_ε is Gateaux differentiable in $L^2(\Sigma)$.

The regularization influence upon the solution may be estimated in the following way.

Proposition 4 [11,14]. Assume, in addition to (A1)-(A5), that

(A6) $\text{mes}\ \{ x \in \Omega \mid 0 < \theta_0(x) < \varepsilon \} \leqq C\varepsilon$.

Then for any $u \in \mathcal{U}$

$$\| \Xi(u) - \Xi_\varepsilon(u) \|_{L^2(Q)} \leqq C\varepsilon^{1/2} \tag{11}$$

where C is a constant independent of ε .

The control problem related to $(VI)_\varepsilon$ has the form

$$(P)_\varepsilon \qquad \underset{u\in\mathcal{U}}{\text{Min}} \{ I_\varepsilon(u) = \frac{1}{2} \| \Xi_\varepsilon(u) - \theta_d \|^2_{L^2(Q)} + \frac{\alpha}{2} \| u \|^2_{\mathcal{U}} \} .$$

Evidently, $(P)_\varepsilon$ has at least one optimal solution $\hat{u}_\varepsilon$.

Remark 4. An edge of the applied regularization approach is inherent in the resulting differentiability of the state operator. This in turn implies the differentiability of the cost functional and therefore gives rise to optimality conditions in an explicit constructive form .

5. SEMI-DISCRETE CONTROL PROBLEM

Having constructed the regularized problem, we shall approach it by semi-discrete Galerkin approximations exploiting piecewise linear finite elements.

From now on we shall assume, for simplicity, that

(A7) Ω is a convex polygonal domain in R^2 .

Let us introduce a regular triangulation of $\Omega = \bigcup_{T\in\mathcal{T}_h} T$ with the parameter h denoting length of the largest triangle edge. We shall approximate $(P)_\varepsilon$ assuming that $y_\varepsilon(t)$ and $u(t)$ for $t \in [0,T]$ are in finite dimensional subspaces of V and $L^2(\Gamma)$, respectively. The approximation of V is taken in the form

$$V_h = \{ z_h \in V \cap C^0(\bar{\Omega}) \mid z_h \text{ is a polynomial of order} \leqq 1 \text{ for every } T \in \mathcal{T}_h \} .$$

Let U_h be the space generated by the traces of elements of V_h on Γ endowed with the $L^2(\Gamma)$-norm. We introduce also the corresponding approximation $\mathcal{U}_h$ of the control space $\mathcal{U}$:

$$\mathcal{U}_h = \{ u_h \in \mathcal{U} \mid u_h, u'_h \in L^2(0,T;U_h) \} .$$

With these approximations we arrive at the semi-discrete problem corresponding to $(VI)_\varepsilon$ taking the form of a system of ordinary differential equations

Determine $y_{\varepsilon h} : [0,T] \to V_h$, satisfying

$$(VI)_{\varepsilon,h} \begin{cases} (\gamma_\varepsilon(y'_{\varepsilon h}(t)) - f_{\varepsilon h}(t), z_h) + a(y_{\varepsilon h}(t) , z_h) - \\ - (\int_0^t u_h(\tau)\, d\tau , z_h)_\Gamma = 0 \quad \forall\ z_h \in V_h , \quad \text{a.a. } t \in [0,T] , \\ y_{\varepsilon h}(0) = 0 \quad \text{in } \Omega \end{cases}$$

where $f_{\varepsilon h}(x,t) = \int_0^t \lambda(x,\tau)\, d\tau + w_{o\varepsilon h}(x)$, $w_{o\varepsilon h} = \gamma_\varepsilon(\theta_{oh})$,

$\theta_{oh} = I_h\, \theta_o$, $I_h : C^0(\bar{\Omega}) \to V_h$ is an interpolation operator.

There exists the unique solution $y_{\varepsilon h}$ of $(VI)_{\varepsilon,h}$, satisfying estimates (7.b) with a constant C independent of ε and h, moreover $y'_{\varepsilon h}(0) = \theta_{oh}$ in Ω (see [10]).

Let $\Xi_{\varepsilon h} : \mathcal{U}_h \to L^2(0,T;V_h)$ denote the discrete state observation operator, defined by $\Xi_{\varepsilon h}(u_h) = y'_{\varepsilon h}$.

Proposition 5 [14] . (i) $\Xi_{\varepsilon h}$ is compact from $\mathcal{U}_h$ into $L^2(Q)$;
(ii) $\Xi_{\varepsilon h}$ is Lipschitz continuous from $L^2(\Sigma)$ into $L^2(Q)$ with the Lipschitz constant independent of ε , h ;
(iii) $\Xi_{\varepsilon h}$ is Gateaux differentiable in $L^2(\Sigma)$ and its Gateaux differential $D\Xi_{\varepsilon h}(u_h)v_h$ at the point u_h in the direction v_h ($u_h , v_h \in \mathcal{U}_h$) satisfies

$$D\Xi_{\varepsilon h}(u_h)\, v_h = \xi'_h \tag{12}$$

where $\xi_h \in L^\infty(0,T;V_h)$ is the solution of the problem

$$\begin{cases} (D\gamma_\varepsilon([\Xi_{\varepsilon h}(u_h)](t))\, \xi'_h(t) , z_h) + a(\xi_h(t) , z_h) = \\ \quad = (\int_0^t v_h(\tau)\, d\tau , z_h)_\Gamma , \quad \forall\ z_h \in V_h , \quad \text{a.a. } t \in [0,T] , \\ \xi_h(0) = 0 \quad \text{in } \Omega . \end{cases} \tag{13}$$

Proposition 6 [13] . Let (A1)-(A7) be satisfied. Then for any $u \in \mathcal{U}$ there exists a constant C independent of ε , h , such that

$$\| \Xi_\varepsilon(u) - \Xi_{\varepsilon h}(u) \|_{L^2(Q)} \leqq C \frac{h}{\varepsilon^{1/2}} . \tag{14}$$

Let us formulate now the semi-discrete analogue of the control problem $(P)_\varepsilon$:

$$(P)_{\varepsilon,h} \qquad \min_{u_h \in \mathcal{U}_h} \{ I_{\varepsilon h}(u_h) = \frac{1}{2} \| \Xi_{\varepsilon h}(u_h) - \theta_d \|^2_{L^2(Q)} + \frac{\alpha}{2} \| u_h \|^2_{\mathcal{U}} \} .$$

Again, due to the compactness of $\Xi_{\varepsilon h}$ the existence of at least one optimal solution $\hat{u}_{\varepsilon h}$ of $(P)_{\varepsilon,h}$ is ensured.

We introduce the following adjoint problem corresponding to $(P)_{\varepsilon h}$:

$$(AP)_{\varepsilon,h} \begin{cases} \text{Determine } p_{\varepsilon h} : [0,T] \to V_h , \text{ satisfying} \\ (D\gamma_\varepsilon(y'_{\varepsilon h}(t))\, p'_{\varepsilon h}(t) , z_h) - a(p_{\varepsilon h}(t) , z_h) = \\ \quad = (y'_{\varepsilon h}(t) - \theta_d(t) , z_h) \quad \forall\ z_h \in V_h , \text{ a.a. } t \in [0,T] , \qquad (15) \\ p_{\varepsilon h}(T) = 0 \quad \text{in } \Omega \end{cases}$$

where $y'_{\varepsilon h} = \Xi_{\varepsilon h}(u_h)$.

There exists the unique solution $p_{\varepsilon h}$ of $(AP)_{\varepsilon,h}$ such that $p_{\varepsilon h} \in L^\infty(0,T;V_h) \cap \cap H^1(0,T;L^2(\Omega))$.

By (12),(15) , simple calculations yield the following characterization of the gradient of $I_{\varepsilon h}$:

$$\begin{aligned} DI_{\varepsilon h}(u_h)\, v_h &= (\Xi_{\varepsilon h}(u_h) - \theta_d , D\Xi_{\varepsilon h}(u_h)\, v_h)_{L^2(Q)} + \alpha\, (u_h, v_h)_{L^2(\Sigma)} + \\ &+ \alpha\, (u'_h, v'_h)_{L^2(\Sigma)} = \int_0^T (-p_{\varepsilon h}(t) + \alpha\, u_h(t) , v_h(t))_\Gamma\, dt + \\ &+ \alpha \int_0^T (u'_h(t) , v'_h(t))_\Gamma\, dt = \int_0^T (\int_0^t [p_{\varepsilon h}(\tau) - \alpha\, u_h(\tau)]\, d\tau + \\ &+ \alpha u'_h(t) , v'_h(t))_\Gamma dt + (\int_0^T [-p_{\varepsilon h}(t) + \alpha\, u_h(t)]\, dt , v_h(T))_\Gamma \\ &\forall\ u_h , v_h \in \mathcal{U}_h . \qquad (16) \end{aligned}$$

Note that the knowledge of $u'_h \in L^2(\Sigma)$ and $u_h(T) \in L^2(\Gamma)$ uniquely determines $u_h \in \mathcal{U}_h$. In this connection (16) suggests directly the following gradient type algorithm for solving problem $(P)_{\varepsilon,h}$ (see also [1] for an algorithm of the same type).

Let n denote the iteration number, and $\mathfrak{u}_h = \{ u'_h , u_h(T) \} \in L^2(0,T;U_h) \times U_h$.

<u>Step 0</u> : Set $n = 0$, fix an arbitrary $\mathfrak{u}_h^{(0)}$.

<u>Step 1</u> : Compute $y'^{(n)}_{\varepsilon h} = \Xi_{\varepsilon h}(u_h^{(n)})$ by solving $(VI)_{\varepsilon,h}$.

<u>Step 2</u> : Compute $I_{\varepsilon h}(u_h^{(n)})$ and verify if $u_h^{(n)}$ is satisfactory as solution. If YES then STOP , otherwise go to Step 3.

<u>Step 3</u> : Compute $p^{(n)}_{\varepsilon h}$ by solving problem $(AP)_{\varepsilon,h}$.

<u>Step 4</u> : Compute $\mathfrak{u}_h^{(n+1)}$ from the equalities

$$u'^{(n+1)}_h(t) = u'^{(n)}_h(t) - \rho_n \{ \int_0^t [p^{(n)}_{\varepsilon h}(\tau) - \alpha\, u^{(n)}_h(\tau)]\, d\tau + \alpha\, u'^{(n)}_h(t) \}, \quad t \in (0,T) ,$$

$$u^{(n+1)}_h(T) = u^{(n)}_h(T) - \rho_n \int_0^T [-p^{(n)}_{\varepsilon h}(t) + \alpha\, u^{(n)}_h(t)]\, dt$$

where ρ_n is an appropriately chosen real parameter, e.g., according to the line search method.

Step 5 : Set $n := n+1$ and go to Step 1 .

Problems $(VI)_{\varepsilon,h}$ and $(AP)_{\varepsilon,h}$ may be solved numerically, e.g., by the method described in [12] .

6. CONVERGENCE RESULTS

By the referred properties of Ξ_ε and $\Xi_{\varepsilon h}$ it is possible to prove the following iterative and conditionally collective convergence results for problem $(P)_{\varepsilon,h}$.

Theorem 2 [14] .

I. Iterative convergence :

(i) For any fixed $\varepsilon > 0$ there exists a subsequence $\{\hat{u}_{\varepsilon h}\}$ (denoted again by the same indices) such that for $h \to 0$

$$\hat{u}_{\varepsilon h} \to \hat{u}_\varepsilon \quad \text{strongly in } \mathcal{U} ,$$

$$\Xi_{\varepsilon h}(\hat{u}_{\varepsilon h}) \to \Xi_\varepsilon(\hat{u}_\varepsilon) \quad \text{strongly in } L^2(Q) ,$$

$$I_{\varepsilon h}(\hat{u}_{\varepsilon h}) \to \hat{I}_\varepsilon \; , \quad I_\varepsilon(\hat{u}_{\varepsilon h}) \to \hat{I}_\varepsilon \quad \text{with the rate of convergence } O(\frac{h}{\varepsilon^{1/2}})$$

where $\hat{u}_\varepsilon$ is an optimal control for $(P)_\varepsilon$ and $\hat{I}_\varepsilon$ denotes the minimal value of the functional I_ε .

(ii) There exists a subsequence $\{\hat{u}_\varepsilon\}$ such that for $\varepsilon \to 0$

$$\hat{u}_\varepsilon \to \hat{u} \quad \text{strongly in } \mathcal{U} ,$$

$$\Xi_\varepsilon(\hat{u}_\varepsilon) \to \Xi(\hat{u}) \quad \text{strongly in } L^2(Q) ,$$

$$I_\varepsilon(\hat{u}_\varepsilon) \to \hat{I} \; , \quad I(\hat{u}_\varepsilon) \to \hat{I} \quad \text{with the rate of convergence } O(\varepsilon^{1/2})$$

where $\hat{u}$ is an optimal control for problem (P) and $\hat{I}$ the corresponding minimal value of the functional I .

II. Collective convergence :

Provided $h \leqq h_o \varepsilon^{\mu+1/2}$, $h_o , \mu > 0$, there exists a subsequence $\{\hat{u}_{\varepsilon h}\}$ such that for $h , \varepsilon \to 0$

$$\hat{u}_{\varepsilon h} \to \hat{u} \quad \text{strongly in } \mathcal{U} ,$$

$$\Xi_{\varepsilon h}(\hat{u}_{\varepsilon h}) \to \Xi(\hat{u}) \quad \text{strongly in } L^2(Q) ,$$

$$I_{\varepsilon h}(\hat{u}_{\varepsilon h}) \to \hat{I} \; , \quad I(\hat{u}_{\varepsilon h}) \to \hat{I} \quad \text{with the rate of convergence } O(\varepsilon^{1/2}+h^\mu)$$

where $\hat{u}$ is an optimal control for problem (P) .

Clearly, the optimal convergence rate in (II) is achieved at $h = h_o \varepsilon$.

The outline of the proof, similar for all the statements of the theorem, is the following. First, by the coercitivity of the cost functional the sequence of optimal controls $\hat{u}_{\varepsilon h}$ (resp., $\hat{u}_{\varepsilon}$) is shown to be bounded in $\mathcal{U}$ independently of the parameters ε and h (resp., ε) . This implies, after going to a subsequence, for $h \to 0$ (resp., $\varepsilon \to 0$) weak convergence of the corresponding optimal controls in $\mathcal{U}$ with some limit elements $\bar{u}$. Appropriate compactness results for $\Xi_{\varepsilon h}$ (resp., Ξ_{ε}) and weak lower semi-continuity of the functional $J : L^2(Q) \times \mathcal{U} \to R$ permit to claim these limits $\bar{u}$ to be in fact optimal controls $\hat{u}_{\varepsilon}$ (resp., $\hat{u}$) .

Essential for the proof of the strong convergence of controls in $\mathcal{U}$ are the estimates

$$| I(\hat{u}_{\varepsilon}) - I_{\varepsilon}(\hat{u}_{\varepsilon}) | \leqq C \varepsilon^{1/2} ,$$

$$| I_{\varepsilon}(\hat{u}_{\varepsilon h}) - I_{\varepsilon h}(\hat{u}_{\varepsilon h}) | \leqq C \frac{h}{\varepsilon^{1/2}} ,$$

resulting from (11) and (14) .

Eventually, the statement concerning the conditionally collective convergence is again a consequence of (11),(14) .

<u>Remark 5</u> . The presented approximation method combined with time discretization can form foundations for complete numerical solution of the problem under consideration.

REFERENCES

[1] V. Arnǎutu , Approximation of optimal distributed control problems governed by variational inequalities , Numerische Math. , <u>38</u> (1982) , 393-416 .

[2] V. Barbu , Necessary conditions for distributed control problems governed by parabolic variational inequalities , SIAM J. Control Optim. , <u>19</u> (1981) , 64-86 .

[3] V. Barbu , Boundary control problems with nonlinear state equation , SIAM J. Control Optim. , <u>20</u> (1982) , 125-143 .

[4] G. Duvaut , J.-L. Lions , Les Inéquations en Mécanique et en Physique , Dunod, Paris , 1972 .

[5] C.M. Elliott , J.R. Ockendon , Weak and Variational Methods for Moving Boundary Problems , Res. Notes in Math. , Vol. 59 , Pitman , Boston , 1982 .

[6] K.-H. Hoffmann , M. Niezgódka , Control of parabolic systems involving free boundaries , Free Boundary Problems : Theory and Applications , A. Fasano , M. Primicerio , Eds. , Research Notes in Math. , Vol. 79 , Pitman , Boston , 1983 , 431-462 .

[7] R. Kluge , paper to be published in Mathematische Nachrichten (personal communication) .

[8] Z. Meike , D. Tiba , Optimal control for a Stefan problem , Analysis and Optimization of Systems , A. Bensoussan , J.-L. Lions , Eds. , Lecture Notes in Control and Information Sci. , Vol. 44 , Springer-Verlag , Berlin , 1982 , 776-787 .

[9] M. Niezgódka , I. Pawlow , Optimal control for parabolic systems with free boundaries - Existence of optimal solutions, Approximation results , Optimization Techniques , K. Iracki , K. Malanowski , S. Walukiewicz , Eds. , Lecture Notes in Control and Information Sci. , Vol. 22 , Springer-Verlag , Berlin , 1980 , 412-420 .

[10] I. Pawlow , A variational inequality approach to generalized two-phase Stefan problem in several space variables , Annali Matem. Pura Applicata , 131 (1982) , 333-373 .

[11] I. Pawlow , Variational inequality formulation and optimal control of nonlinear evolution systems governed by free boundary problems , Applied Nonlinear Functional Analysis , R. Gorenflo , K.-H. Hoffmann , Eds. , Methoden und Verfahren der Mathematischen Physik , Vol. 25 , Verlag Peter Lang , Frankfurt am Main , 1983 , 213-250 .

[12] I. Pawlow , Approximation of an evolution variational inequality arising from free boundary problems , Optimal Control of Partial Differential Equations , K.-H. Hoffmann , W. Krabs , Eds. , ISNM , Birkhäuser-Verlag , Basel , in print .

[13] I. Pawlow , Discrete approximation of a pseudo-parabolic variational equation , to be published .

[14] I. Pawlow , Approximation of boundary control problems for two-phase Stefan type processes , to be published .

[15] Ch. Saguez , Contrôle optimal de systèmes à frontière libre , Thèse d'Etat , Université Technologique de Compiegne , 1980 .

[16] D. Tiba , Boundary control for a Stefan problem , Optimal Control of Partial Differential Equations , K.-H. Hoffmann , W. Krabs , Eds. , ISNM , Birkhäser-Verlag , Basel , in print .

[17] D. Tiba , A finite element discretization of boundary control of a two-phases Stefan problem , Preprint Series in Math. , No. 34 , Institutul de Matematica , Bucureşti , 1983 .

OPTIMAL CONTROL OF GENERALIZED FLOW NETWORKS

Markos Papageorgiou
DORSCH CONSULT IGmbH
P.O.Box 210243
8000 München 21, FR Germany

1. INTRODUCTION

In the last decade much research work has been devoted independently to the development of control algorithms for several processes having the following common characteristics:

- a network structure with
- some kind of flow along the network's links.
- The flowing medium passes through storage units and
- is directed from specific origin locations to specific destination locations.
- The flow is subject to several capacity constraints.
- There are some time varying load or demand variables at origin or destination locations which must be taken into account.

The following processes could be considered as belonging to that class:
Multireservoir water resources systems and sewer networks: The control problem in these systems consists in minimizing overflows through maximum utilization of the existing storage capacity (see Jamshidi and Mohseni[1], Trotta et.al. [2], Papageorgiou [3]).

Water distribution and gas transportation networks. The task of the control system is to deliver water or natural gas from a set of sources to a set of demand points over complex networks of compressors and pipelines that often extend over several hundred kilometers minimizing operation cost (Joalland and Cohen [4], Osiadacz and Bell [5]).

Freeway corridors and road traffic networks. The problem consists in setting traffic lights at intersections of a street network so as to minimize delay caused by waiting queues or congestions (Capelle [6], Lin et.al. [7], Papageorgiou [8]). An important difference between these two processes and the previous ones is due to the fact that flow consists of destination oriented subflows and cannot be arbitrarily manipulated.

A general control problem applying to all mentioned cases could be stated as follows:

Specify some control inputs influencing the flow process in the network so as to minimize a performance criterion subject to the capacity constraints and the time varying demands.

In view of the common characteristics of the above processes and the adjoined control problems, it seems reasonable to ask the question: Is it possible, and if yes, up to which extent, to apply a *unified* approach to control system design ? In this paper a first attempt is made towards an answer of this question.

2. A MULTILAYER CONTROL STRUCTURE

Above problem could be principally formulated as a mathematical optimization problem for each of the mentioned processes but the high dimension and complexity do not permit a practical solution, which should be obtained on-line and with moderate computational cost. Hence, there are two alternative ways of handling the problem:

- either consider a *decentralized* problem formulation in which the overall process is split into independent subprocesses
- or consider the *overall* system in a *simplified* form.

Decentralization and control of independent subsystems doesn't seem reasonable for the most of the mentioned processes, because optimal decisions in subnetworks depend on the flows arriving from adjacent subnetworks and hence the overall optimum cannot be approximated accurately enough by the sum of the suboptima. On the other hand, results obtained by reasonable simplifications might be near-optimal if the *main* features of the process are included in the simplified process description and this is the way followed along this paper.

A reasonable way of simplifying a network process with storage capabilities is to subdivide it into a number of subnetworks with or without storage capacities. Each subnetwork is considered as a single reservoir for the purpose of overall optimization. Links connecting the subnetworks are considered as pure delay elements according to the results of the kinematic waves theory, Lighthill and Whitham [9]. The optimization problem resulting through these simplifications is described in the following section. All flows leaving a subnetwork are considered as control variables whereas volumes of subnetworks are taken as state variables. The optimal trajectories for volumes and flows obtained in this way are then used as reference trajectories for decentralized subnetwork control. The aim of this *direct control layer* is to translate the optimization decisions into real control actions (see Findeisen and Lefkowitz [10], Papageorgiou [11]). This corresponds to making subnetworks' behavior as has been assumed in the optimization procedure. Predicted load trajectories as well as further model parameters needed for the optimization are provided by an adaptation layer according to figure 1.

The operation of the multilayer control system is summarized in the following:

(i) The adaptation layer specifies the predicted load trajectories and further parameter values and initiates an optimization run.

(ii) Solution of the simplified optimization problem provides optimal state and control trajectories for the simplified network.

(iii) Local regulators, one for each subnetwork, operate in a closed-loop mode so as to make volumes and flows of the subnetworks follow their reference trajectories.

(iv) If a significant deviation of the predicted load from the real load occurs, or if local regulators are not capable of following the reference trajectories, the whole procedure is repeated with new predictions or new parameter values.

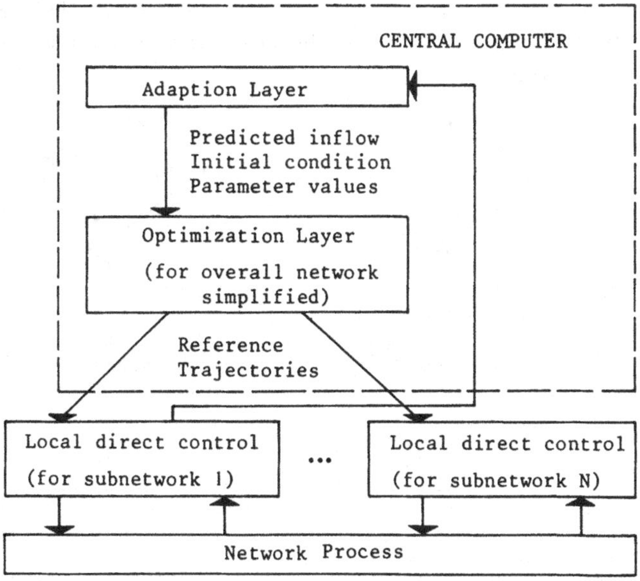

Figure 1. A multilayer control structure for network processes

The main benefits of the multilayer control structure are:

- Near-optimal results are achieved for the overall network.
- Because of the simplification introduced, the computational effort for the solution of the optimization problem is kept low.
- Because of the decentralized structure of the direct control layer, low implementation cost and high reliability will be achieved. Note that interconnections between the subnetworks are considered in the optimization procedure and this is the reason why they can be ignored in the direct control layer.
- Since the optimization problem can be solved on-line repeatedly, sensitivity of the results with respect to inaccurate load predictions is kept low.
- Design of each control layer can be performed independently.

- The overall control system can be implemented in a distributed computer system. A minicomputer might be necessary for the central tasks of adaptation and optimization. Microcomputers connected with the central computer in a star system might perform the decentral direct control tasks.

3. THE OPTIMIZATION LAYER

We are interested in developing a control system which can be applied to several network systems. With respect to the optimization layer this means that we must formulate and solve a fairly general optimization problem. In order to do that, we will define some generalized elements shown in fig. 2. The simplified overall network should be built up as an arbitrary combination of these elements.

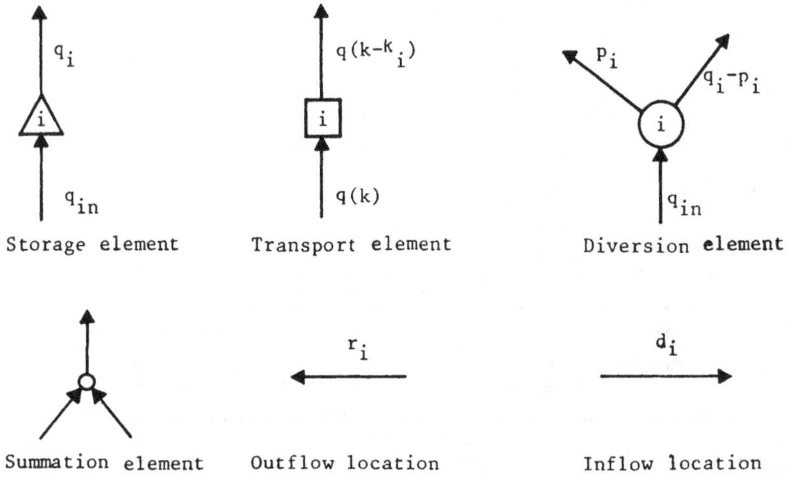

Figure 2. Elements of the simplified network

3.1 Elements of the generalized flow network

Storage element. It represents a small subnetwork with storage capacity. The dynamic equation governing the element's behavior is a simple conservation equation

$$V_i(k+1) = V_i(k) + T\,[q_{in}(k) - q_i(k)],\ i = 1, \ldots, n_v, \tag{1}$$

with T the discrete-time period. Outflow q_i is considered as a control variable which should be selected from an admissible control region

$$0 \leq q_i(k) \leq q_{i,max}. \tag{2}$$

The state variables V_i should not exceed the subnetwork's capacity

$$0 \leq V_i(k) \leq V_{i,max}.$$

Transport element. It represents a connecting link of the simplified network and is a delay element with delay κ_i, $i = 1, \ldots, n_T$, as indicated in figure 1.

Diversion element. It represents a small subnetwork without storage capacity. The inflow q_{in} to the diversion element is divided into two outflows p_i and $q_{in} - p_i$. The admissible control region for the control variable p_i is given by

$$0 \leq p_i(k) \leq \min \{p_{i,max}, q_{in}(k)\}, \; i = 1, \ldots, n_p \tag{4}$$

Summation element. Several inflows are summed up to a unique outflow.

Outflow and Inflow element. They represent a load or delivery location depending on the specific process under consideration. In the following we will assume that inflow trajectories $d_i(k)$ are predicted and outflow trajectories $r_i(k)$ are control variables.

3.2 Formulation of the optimization problem

We define the following vectors:

$$\underline{V} = [V_1 \ldots V_{n_v}]^T; \; \underline{q} = [q_1 \ldots q_{n_v}]^T; \; \underline{r} = [r_1 \ldots r_{n_r}]^T; \underline{d} = [d_1 \ldots d_{n_d}]^T$$

where $\underline{V}$ is the state vector, $\underline{q},\underline{r}$ are control vectors and $\underline{d}$ is a known input vector. The state equation of the simplified network can now be given in a general form

$$\underline{V}(k+1) = \underline{V}(k) + \sum_{\nu=o}^{m} [B_\nu \underline{q}(k-\nu) + D_\nu \underline{r}(k-\nu) + F_\nu \underline{d}(k-\nu)] \tag{5}$$

where B_ν, D_ν, F_ν, $\nu = 0, \ldots, m$ are time invariant matrices. The control objective depending on the particular process requirements is supposed to be

$$J = \theta[\underline{V}(K)] + \sum_{k=o}^{K-1} g[\underline{V}(k), q(k), \underline{r}(k), d(k)]. \tag{6}$$

The mathematical optimization problem for a generalized network can now be formulated:

> For given $\underline{d}(k)$, $k=-m, \ldots, K-1$; $\underline{r}(k)$, $\underline{q}(k)$, $k=-m, \ldots -1$;
> initial condition $\underline{V}(0) = \underline{V}_o$; optimization horizon K;
> Find the optimal control trajectories $\underline{r}(k)$, $\underline{q}(k)$, $k=0, \ldots, K-1$;
> Minimizing the control objective (6) subject to eqs. (2)-(5).

3.3 Solution of the optimization problem

Above optimization problem can be principally solved by application of dynamic programming (DP) algorithms. However, the computer storage space needed for DP algorithms becomes enormous, if the optimization problem includes more than 3-4 stage variables and more than 10-15 sample time periods. For this reason, the discrete maximum principle has been applied in order to reduce computational effort. The main drawbacks of the discrete maximum principle are:

- State variable constraints, like the ones of eq. (3) lead to extreme difficulties.
- The optimization problem is solved by deriving the necessary conditions for optimality and formulating a Two-Point-Boundary-Value-Problem (TPBVP), Pearson and Sridhar [12]. Solution of the TPBVP fulfills only the necessary conditions and might correspond to a local minimum instead of the global one.
- The TPBVP is solved by use of numerical algorithms, convergence of which must be investigated.

We will now briefly discuss each of these difficulties in connection to our optimization problem.

State Variables Constraints. In order to overcome difficulties connected with state variables constraints, a penalty function method is used (see Russell [13], Okamura [14]). Instead of considering eq. (13) in the optimization problem, we extend the control objective as follows

$$J' = J + \frac{1}{2} \sum_{k=0}^{K-1} \{ ||\underline{\Psi}[\underline{V}(k)]||_Q^2 + ||\underline{\Psi}[\underline{V}_{max} - \underline{V}(k)]||_Q^2 \} \tag{7}$$

with $Q > 0$, diagonal and

$$\Psi_i(\underline{\eta}) = \begin{cases} 0 & \text{if } \eta_i \geq 0 \\ \eta_i & \text{if } \eta_i < 0 \end{cases} \tag{8}$$

Thus, exceeding of the constraints of eq. (3) is penalized by the two additional terms in the new control objective. If V_i lies in the admissible state region, Ψ_i becomes zero and has no influence on the performance index J' . It should be noted that the implicit consideration of stage variable constraints by use of this penalty method does not prohibit exceeding of constraints in a strict way but can reduce the constraints violation down to an amount which is of little practical significance. This is achieved by increasing the weighting matrix Q up to a sufficient level. In the following the prime notation will be dropped from the modified performance functional (7).

Necessary Conditions. If the performance index J is selected to be *convex* and since all constraints of the optimization problem are linear, necessary conditions for optimality are sufficient conditions as well (Lasdon [15]) and the stationary point will be a global minimum. In order to derive the optimality conditions, we define the Hamiltonian

$$H = g[\underline{V}(k),\underline{q}(k), \underline{r}(k), \underline{d}(k)]+||\underline{\Psi}[\underline{V}(k)]||_Q^2 + ||\Psi[\underline{V}_{max} - \underline{V}(k)]||_Q^2 + \underline{\lambda}(k+1)^T \{\underline{V}(k)+ \sum_{\nu=o}^{m} [B_\nu \underline{q}(k-\nu)+D_\nu \underline{r}(k-\nu)+F_\nu \underline{d}(k-\nu)]\} \quad (9)$$

with $\underline{\lambda}$ the costate vector. The necessary and sufficient conditions for optimality are given by

$$\underline{V}(k+1)= \partial H/\partial\underline{\lambda}(k+1)= \underline{V}(k)+ \sum_{\nu=o}^{m} [B_\nu \underline{q}(k-\nu)+ D_\nu \underline{r}(k-\nu)+F_\nu \underline{d}(k-\nu)] \quad (10)$$

$$\underline{\lambda}(k) = \partial H/\partial\underline{V}(k)= \partial g/\partial\underline{V}(k)+\underline{\lambda}(k+1)+Q\{\underline{\psi}[\underline{V}(k)]- \underline{\Psi}[\underline{V}_{max}-\underline{V}(k)]\} \quad (11)$$

$$H[\underline{V}^*(k), \underline{u}^*(k),\underline{\lambda}^*(k)]\geq H[\underline{V}^*(k), \underline{u}(k), \lambda^*(k)] \;\forall u\epsilon\Omega \quad (12)$$

with Ω the admissible control region and $\underline{u}^T = [\underline{q}^T\underline{r}^T]$. The boundary conditions are given by

$$\underline{V}(0)= \underline{V}_o \quad (13)$$

$$\underline{\lambda}(K) = \partial\theta/\partial\underline{V}(K) \quad (14)$$

$$\underline{\lambda}(k) = \underline{0} \; , \; k > K \; . \quad (15)$$

Eqs. (10) - (12) constitute a TPBVP, solution of which is equivalent to the solution of the optimization problem.

Solution of the TPBVP. The following iterative algorithm is used:

Step 1: Guess some control variable trajectories $\underline{r}^1$, $\underline{q}^1$. Set the Iteration index L = 1.

Step 2: Integrate eq. (10) forwards in time starting from eq. (13) and eq. (11) backwards in time starting from eq. (14) in order to specify $\underline{x}^L$, $\underline{\lambda}^L$.

Step 3: Improve the control variable trajectories by a gradient method.

Step 4: If optimality conditions are fulfilled with desired accuracy, stop and record $\underline{u}^L$ as the optimal solution. Else, set L : = L + 1 and go to step 2.

A conjugate gradient method (Lasdon et.al.[16]) has been used in step 3 in order to achieve high convergence rate for a wide range of Q-values (Pagurek and Woodside [17], Luenberger [18]). Calculations in step 3 include

$$\underline{u}^{L+1}(k) = \text{sat}\,\{\underline{u}^L(k) - \tau^L \underline{\phi}^L\} \tag{16}$$

with τ the gradient step, ϕ the conjugate gradient and

$$\text{sat}_i\{\underline{\eta}\} = \begin{cases} \eta_{i,max} & \text{if } \eta_i \geq \eta_{i,max} \\ \eta_i & \text{if } \eta_{i,min} < \eta_i < \eta_{i,max} \\ \eta_{i,\,min} & \text{if } \eta_{i,} \leq \eta_{i,min} \end{cases} \tag{17}$$

The conjugate gradient is calculated iteratively by use of

$$\underline{\phi}^L = \underline{\Delta}^L + \beta^{L-1}\underline{\phi}^{L-1} \tag{18}$$

$$\beta^{L-1} = \sum_{k=o}^{K-1} (\underline{\Delta}^L)^T \underline{\Delta}^L / \sum_{k=o}^{K-1} (\underline{\Delta}^{L-1})^T \underline{\Delta}^{L-1} \tag{19}$$

where $\underline{\Delta} = \partial H/\partial \underline{u}$ is the common gradient.

Selection of gradient step τ is very important for convergence of the iterative algorithm. Choosing τ too big might lead to divergence, whilst a small τ leads to slow convergence. In absence of inequality constraints, above iterative procedure would represent a linear dynamic discrete time system and a *unique* choise of gradient step τ guaranteeing quick convergence would be possible. However, since inequality constraints are considered, the structure of the discrete time system and consequently the stability region of the iterative procedure may change after each iteration step. For this reason, it is reasonable to adapt the gradient step value τ to changing stability conditions so as to speed up convergence and avoid divergence of the algorithm. For this purpose, the following rule for the specification of τ^L has been applied:

If $J^L \leq J^{L-1}$: set $\tau^{L+1} = a_1 \tau_L^L$; $a_1 > 1$
If $J^L > J^{L-1}$: set $\tau^{L+1} = a_2 \tau^L$; $a_2 < 1$;
go back to step 3 of the previous iteration; do not increase τ for the next 30 iteration steps.

This means that τ is being increased as long as the performance functional J decreases from iteration step to iteration step. If an increasing J is obtained, achievement of the stability bounds is assumed and for this reason the gradient step is reduced and is kept constant for the next 30 iteration steps.

The algorithm was applied to the flow optimization of a network with 8 storage elements, 8 outflow locations, 8 inflow locations, 10 transport elements for an optimization horizon of K = 60. It was found that the conjugate gradient algorithm was five times quicker than common gradient algorithms. Solution was achieved in 90 s in a DEC 2060 digital computer. Results are reported by Papageorgiou [3], [19].

4 CONCLUSIONS

The multilayer control structure presented in section 2 seems to be a suitable way of handling system's complexity in network processes. The adaptation layer provides load predictions which are used in the optimization procedure for the overall network. Direct control layer translates the optimal control decisions into real control actions. Results obtained for the optimization layer can be utilized for the network processes mentioned in section 1. In the contrary, design of the adaptation layer and the direct control layer must be performed on the basis of the particular features of the process under consideration. Application of these ideas to sewer network control is reported by Papageorgiou [3], [19].

REFERENCES

[1] Jamshidi, M., and M. Mohseni: On the optimization of water resources systems with statistical inputs. In "System simulation in Water Resources", Vansteenkiste, G. C., Ed., North-Holland Publ. Co., 1976, pp. 393 - 408.

[2] Trotta, P.D., Labadie, J.W., and N.S. Grigg: Automatic control strategies for urban stormwater. Proc. ASCE J. of the Hydraulics Div. 103 (1977), pp. 1443 - 1459.

[3] Papageorgiou, M: Optimal control of large scale combined sewer systems. 3rd IFAC/IFORS Symp. on Large Scale Systems, Warsaw, Poland, 1983, pp. 387 - 393.

[4] Joalland, G., and G. Cohen: Optimal control of a water distribution network by two multilevel methods. Automatica 16 (1980), pp. 83 - 88.

[5] Osiadacz, A., and D. J. Bell: A simplified algorithm for large scale gas networks. 8th IFAC Congress, Kyoto, Japan, August 24 - 28, 1981, pp. x-59-65.

[6] Capelle, D.G.: Freeway Corridor Systems. Proc. Intern. Symp. on Traffic Control Systems, Berkeley, California, Aug. 6 - 9, 1979, pp. 33 - 53.

[7] Lin, D., Ulrich, F., Kinney, L.L., and K.S.P. Kumar: Hierarchical Techniques in Traffic Control. Proc. 3rd IFAC/IFIP/IFORS Symp. on Control in Transportation Systems, Columbus, Ohio, 1976, pp. 163 - 171.

[8] Papageorgiou, M.: Applications of Automatic Control Concepts to Traffic Flow Modeling and Control. Springer-Verlag, Berlin-Heidelberg-New York-Tokyo, 1983.

[9] Lighthill, M.J., and G.B. Whitham: On kinematic waves I. Flood movement on long rivers. Proc. Royal Society of London, Series A 229 (1955), pp.281 - 316.

[10] Findeisen, W., and I. Lefkowitz: Design and application of multilayer control. 4th IFAC World Congress, Warsaw, Poland, 1969, pp. 3 - 22.

[11] Papageorgiou, M.: Multilayer control of freeway traffic. 21st IEEE Conf. on Decision and Control, Orlando, Florida, 1982, pp. 604 - 610.

[12] Pearson J.B., Jr. and R. Sridhar: A discrete optimal control problem. IEEE Trans. on Automatic Control AC-11 (1966), pp. 171 - 174.

[13] D.L. Russell: Penalty functions and bounded phase coordinate control. J. SIAM Control Ser. A 2 (1965), pp. 409 - 422.

[14] Okamura, K.: Some mathematical theory of the penalty method for solving optimum problems. J. SIAM Control Ser. A 2 (1965), pp. 317 - 331.

[15] Lasdon, L. S.: Duality and decomposition in matematical programming. IEEE Trans. on System Science and Cybernetics SSC-4 (1968), pp. 86 - 100.

[16] Lasdon, L.S., Mitter, S. K., and A.P. Waren: The conjugate gradient method for optimal control problems. IEEE Trans. on Automatic Control AC-12 (1966), pp. 132 - 138.

[17] Pagurek, B., and C.M. Woodside: The conjugate gradient method for optimal control problems with bounded control variables. Automatica 4 (1968), pp. 337 - 349.

[18] Luenburger, D.G.: Convergence Rate of a Penalty Function Scheme. J. Optimization Theory and Applications 7 (1971) pp. 39 - 51.

[19] Papageorgiou, M.: Automatic control strategies for combined sewer systems. Proc. ASCE J. of the Environmental Engng. Div. 109 (1983), to appear.

HÖLDER CONDITION FOR THE MINIMUM TIME FUNCTION OF LINEAR SYSTEMS

E. Gyurkovics
Computing Center of Eötvös Lorand University
H-1093. Budapest, Dimitrov tér 8, Hungary

1. Introduction

Let us consider the linear time-invariant system, which is described by the differential equation

$$\dot{x} = Ax + u \quad , \tag{1}$$

where x is the n-dimensional state vector, u is the n-dimensional control vector and A is an nxn constant matrix. Let Ω be a given compact subset of the n-dimensional Euclidean space E^n. A bounded and measurable function u, defined on the interval $[0,T]$ is said to be an admissible control if it satisfies the constraint $u(t)\in\Omega$ almost everywhere on $[0,T]$. We shall refer to this controlled system as system (A,Ω).

The problem is here that of reaching the origin in least time. We will study the <u>minimum time function</u> $\tau: R\to E^1$, defined by

$$\tau(x) = \inf\{t : -x\in R(t)\},$$

where $R(t)$ is the reachable set of the (A,Ω) at time $t\geq 0$,

$$R(t) = \int_0^t e^{-sA}\Omega \, ds \;, \qquad \text{and} \quad R = \bigcup_{t\geq 0} R(t) \quad .$$

Thus τ is taken finite-valued.

The properties of the function τ, such as its continuity, differentiability as well as the Lipschitz or Hölder condition for it, have been studied extensively. The knowledge of these properties is important, among others, for the construction of computational algorithms. One of the most important results concerning the minimum time function is due to N. N. Petrov [8] and it states that the necessary and sufficient condition for the continuity of the function τ is the local N-controllability (or, in other words, arbitrary-interval null-controllability [3]) of the system (A,Ω). We remember that the system (A,Ω) is said to be locally N-controllable if for each T>0 there exists an open set $V(T)$ in E^n which contains the origin and for which any $x_0\in V(T)$ can be controlled to the origin at time T by an admissible control. The investigations of [2] have shown that for a wide class of locally N-controllable systems the function τ is not differentiable, moreover, it isn't Lipschitz-continuous. Conditions by which τ satisfies the local Hölder condition are known only for special systems such as systems in E^2 with finite constraint set [6], [7], or linear systems with finite symmetric constraint set [5].

The aim of this paper is to give a more general condition by which τ is a locally Hölder function.

Definition. A function $f: W \to E^1$, $W \subset E^n$ is said to be a locally Hölder function if for any $x_0 \in W$ there exist such positive numbers $M(x_0)$, α and $\delta(x_0)$ that for any $x_1, x_2 \in S_{\delta(x_0)}(x_0)$

$$|f(x_1) - f(x_2)| \le M(x_0) \| x_1 - x_2 \|^{\alpha} \quad .$$

Here and in the following $S_\delta(x) = \{y \in E^n : \|y-x\| \le \delta\}$ and $S = \{x \in E^n : \|x\| = 1\}$.

Throughout this paper we assume that the following condition holds for the system (A, Ω).

Condition A. There exist an n-dimensional vector χ_0 and an nxn matrix B with the properties:

(i) $B\chi_0 = 0$;

(ii) $\chi_0 + (C+B)\, S_1(0) \subset \operatorname{convhull}(\Omega)$,

where

$$C = \begin{cases} 0 & , \text{ if } \chi_0 = 0 , \\ \dfrac{\chi_0 \chi_0'}{\|\chi_0\|} & , \text{ if } \chi_0 \neq 0 ; \end{cases}$$

(iii) $\operatorname{rank}(B, AB, \ldots, A^{n-1}B) = n$.

2. Auxiliary lemmata

Lemma 1. If the Condition A holds, then for any $t>0$

$$S_{q(t)}(0) \subset R(t),$$

where

$$q(t) := \min_{\chi \in S} \int_0^t \{(\chi_0, e^{-sA'}\chi) + [(\chi_0, e^{-sA'}\chi)^2 + \|B' e^{-sA'}\chi\|^2]^{\frac{1}{2}}\}\, ds \,, \tag{2}$$

and $q(t)>0$ for any $t>0$.

The proof of this lemma can be found in [1].

Remark 1. From this lemma it follows immediately that if the Condition A holds then the system (A, Ω) is locally N-controllable and the minimum time function is continuous.

Let m denote the smallest number for which

$$\operatorname{rank}(B, AB, \ldots, A^m B) = n . \qquad (3)$$

Lemma 2. Suppose that the Condition A holds with $X_o = 0$ and m is given in (3). Then there exist such positive numbers t_o, γ_1, γ_2 that for all $t \in (0, t_o]$

$$\gamma_1 t^{m+1} < q(t) < \gamma_2 t^{m+1} . \qquad (4)$$

Proof. If the Condition A holds with $X_o = 0$ then

$$q(t) = \min_{X \in S} \int_0^t \| B' e^{-sA'} X \| \, ds .$$

For any $t > 0$ and $X \in S$ the inequality holds

$$\frac{q(t)}{t^{m+1}} \leq \frac{1}{t^{m+1}} \int_0^t \| B' e^{-sA'} X \| \, ds .$$

From the definition of the number m it follows that the system of algebraic equations

$$\|X\| = 1, \; B'X = 0, \; B'A'X = 0, \; \ldots, \; B'A'^{m-1} X = 0$$

has at least one solution $\bar{X}$. (If m=0, then we have only the equation $\|X\|=1$ and we can take an arbitrary $\bar{X} \in S$.) For this $\bar{X}$ one obtains the relations

$$\int_0^t \| B' e^{-sA'} \bar{X} \| ds = \int_0^t \| \sum_{i=m}^{\infty} B'(-A')^i \, \bar{X} \frac{s^i}{i!} \| ds \leq$$

$$\leq \| B'(-A')^m \| \frac{t^{m+1}}{(m+1)!} + \| B'(-A')^{m+1} \| e^{\|A'\| t} \frac{t^{m+2}}{(m+2)!} .$$

In consequence of this, one can easily show the existence of a $t_{o1} > 0$ and a $\gamma_2 > 0$ such that the right inequality in (4) holds.

Now we shall prove the left inequality in (4). For any $t > 0$ and $X \in E^n$ we have

$$\frac{1}{t^{m+1}} \int_0^t \|B' e^{-sA'} \chi\| \, ds \geq$$

$$\geq \frac{1}{t^{m+1}} \int_0^t \|\sum_{i=0}^{m} B'(-A')^i \chi \frac{s^i}{i!}\| ds - \frac{1}{t^{m+1}} \int_0^t \|\sum_{i=m+1}^{\infty} B'(-A')^i \chi \frac{s^i}{i!}\| ds .$$

Since the second term on the right hand side of this inequality tends to zero as $t \to 0$, it is enough to estimate the first term. For simplicity let us introduce the notation

$$b_{i,\chi} = \frac{1}{i!} B'(-A')^i \chi .$$

Substituting $s = tz$ and using the basic property of the Euclidean norm we get

$$\frac{1}{t^{m+1}} \int_0^t \|\sum_{i=0}^{m} B'(-A')^i \chi \frac{s^i}{i!}\| ds \geq \frac{1}{t^m} \int_0^1 \left|\sum_{i=0}^{m} b_{i,\chi}^{(j)} t^i z^i\right| dz , \tag{5}$$

where $b_{i,\chi}^{(j)}$ denotes the jth component of the vector $b_{i,\chi}$. From the condition (3) it follows that

$$\eta := \min_{\chi \in S} \sum_{i=0}^{m} \|b_{i,\chi}\| > 0 .$$

Let $S^{(m)}$ be that subset of S that for any $\chi \in S^{(m)}$ the vector $b_{m,\chi}$ has a component $b_{m,\chi}^{(j)}$ that

$$\left|b_{m,\chi}^{(j)}\right| \geq \frac{\eta}{m \cdot n} . \tag{6}$$

For an arbitrary $\chi \in S^{(m)}$ let us take in the estimation (5) that component of the vector $\sum_{i=0}^{m} b_{i,\chi} t^i z^i$ for which (6) is fulfilled. Since the $L_1(0,1)$ norm of polynomials of degree not greater than m, with the main coefficient equal to 1 has a positive lower bound (see, e.g. [4], p. 36), there exists a constant $c_0 > 0$ such that

$$\frac{1}{t^m} \int_0^1 \left|\sum_{i=0}^{m} b_{i,\chi}^{(j)} t^i z^i\right| dz = \left|b_{m,\chi}^{(j)}\right| \int_0^1 \left|\sum_{i=0}^{m} \frac{b_{i,\chi}^{(j)}}{|b_{m,\chi}^{(j)}|} t^{i-m} z^i\right| dz \geq c_0 \frac{\eta}{m \cdot n} \tag{7}$$

for any $\chi \in S^{(m)}$. (We remark that $\left|b_{m,\chi}^{(j)}\right| < \frac{\eta}{m \cdot n}$ for all $j=1,\dots,n$ and $\chi \in S \setminus S^{(m)}$.)

Suppose that the sets $S^{(m)}, \dots, S^{(m-k+1)}$ have already been defined for some k, $1 \le k \le m$, and an estimation of type (7) has been proven on these sets. Then let

$$S^{(m-k)} \subset S \setminus \left\{ \bigcup_{i=0}^{k-1} S^{(m-i)} \right\}$$

be such that for any $\chi \in S^{(m-k)}$ the vector $b_{m-k,\chi}$ has a component $b^{(j)}_{m-k,\chi}$ for which

$$\left| b^{(j)}_{m-k,\chi} \right| \ge \frac{\eta}{m \cdot n} \; . \tag{8}$$

(If $S^{(m-k)} = \emptyset$ for some k, then we have nothing to prove and we can continue with k+1.) Now for an arbitrary $\chi \in S^{(m-k)}$ we take in (5) the j-th component for which (8) is true and then we get the following inequalities:

$$\frac{1}{t^m} \int_0^1 \left| \sum_{i=0}^{m} b^{(j)}_{i,\chi} \, t^i z^i \right| dz \ge \frac{1}{t^m} \int_0^1 \left| \sum_{i=0}^{m-k} b^{(j)}_{i,\chi} \, t^i z^i \right| dz \; -$$

$$- \left(\int_0^1 \left| b^{(j)}_{m,\chi} \right| z^m \, dz + \frac{1}{t} \int_0^1 \left| b^{(j)}_{m-1,\chi} \right| z^{m-1} \, dz + \dots + \frac{1}{t^{k-1}} \int_0^1 \left| b^{(j)}_{m-k+1,\chi} \right| z^{m-k+1} \, dz \right) \ge$$

$$\ge \frac{\eta}{m \cdot n} \left[\frac{1}{t^k} \int_0^1 \left| \sum_{i=0}^{m-k} \frac{b^{(j)}_{i,\chi}}{b^{(j)}_{m-k,\chi}} \, t^{i+k-m} z^i \right| dz - \sum_{i=0}^{k-1} \frac{1}{t^i (m+1-i)} \right] \; .$$

Since the integral on the right hand side of the last inequality has a positive lower bound c_k for any $\chi \in S^{(m-k)}$, we get

$$\frac{1}{t^m} \int_0^1 \left| \sum_{i=0}^{m} b^{(j)}_{i,\chi} \, t^i z^i \right| dz \ge c_0 \frac{\eta}{m \cdot n} \tag{9}$$

for any $\chi \in S^{(m-k)}$ and $t \in (0, t_k]$, where t_k is defined by the condition

$$\frac{1}{t^k} \left(c_k - \sum_{i=0}^{k-1} \frac{t^{k-i}}{(m+1-i)} \right) \ge c_0 \quad , \qquad t \in (0, t_k] \; .$$

Obviously, such a positive t_k always exists. Continuing this procedure until k=m, we can get the inequality (9) for all $\chi \in \bigcup_{i=0}^{m} S^{(i)}$ and $t \in (0, t_{0,2}]$, where $t_{0,2} = \min\{t_k : 1 \le k \le m\}$. To complete the proof we have only to show that

$$S = \bigcup_{i=0}^{m} S^{(i)} .$$

However by the definition of the number η this relation is evident.

Lemma 3. If the Condition A holds with $X_0 \neq 0$ and m is given in (3) then there exist such positive numbers γ and t_0 that for all $t \in (0,t_0]$

$$\gamma\, t^{2m+1} < q(t) .$$

Proof. Let $\chi \in S$ and

$$q(t,\chi) := \int_0^t \{(\chi_0,\, e^{-sA'}\chi) + [(\chi_0,\, e^{-sA'}\chi)^2 + ||B'e^{-sA'}\chi||^2]^{\frac{1}{2}}\}\, ds ,$$

then

$$q(t,\chi) \geq \int_0^t ||B'e^{-sA'}\chi||^2 \left\{[||\chi_0|| + (||\chi_0||^2 + ||B'||^2)^{\frac{1}{2}}]\, e^{s||A||}\right\}^{-1} ds .$$

Therefore there exist a positive constant c that for all $\chi \in S$, $t \in (0,1]$

$$\frac{q(t,\chi)}{t^{2m+1}} \geq ct \int_0^t \left(\frac{||B'e^{-sA'}\chi||}{t^{m+1}}\right)^2 ds \geq c\left(\int_0^t \frac{||B'e^{-sA'}\chi||}{t^{m+1}}\, ds\right)^2 .$$

We have already seen in the proof of lemma 2 that there exist such numbers $\gamma_1 > 0$ and $t_0 > 0$ that

$$\int_0^t \frac{||B'e^{-sA'}\chi||}{t^{m+1}}\, ds > \gamma_1$$

for all $\chi \in S$, $t \in (0,t_0]$, from which the assertion of the lemma immediately follows.

3. Basic results

Theorem 1. If the Condition A holds and m is given in (3) then the minimum time function of the system (A,Ω) is a locally Hölder function with the exponent

$$\alpha = \frac{1}{2m+1} .$$

Proof. From lemma 1 and lemma 3 it follows that the reachable set $R(t)$ contains a ball with the radius γt^{2m+1} for all $t \in (0,t_0]$. This means that for all $x \in R(t)$, $t \in (0,t_0]$

$$||x|| \geq \gamma\tau(x)^{2m+1}$$

i.e.

$$\tau(x) \leq M||x||^{\alpha} \quad , \tag{10}$$

where $\alpha = (2m+1)^{-1}$, $M = \gamma^{-\alpha}$ and $x \in S_\delta(x)$, $\delta = \gamma t_0{}^{2m+1}$. Then the assertion of the theorem follows from the lemma 1 of [7]. This lemma states that from (10) it follows that τ is a locally Hölder function.

Remark 2. In general, for systems satisfying the Condition A, the exponent α in theorem 1 cannot be increased. The following example shows that there exist systems for which the exponent $\alpha = (2m+1)^{-1}$ is the exact one.

Example 1. Let

$$A = \begin{pmatrix} 0 & -1 \\ 1 & 0 \end{pmatrix} \quad , \qquad \Omega = \{u \in E^2 : u_1^2 + (u_2 - 1)^2 \leq 1\} \quad .$$

For this system the Condition holds with

$$B = \begin{pmatrix} 1 & 0 \\ 0 & 0 \end{pmatrix} \quad , \qquad \chi_0 = \begin{pmatrix} 0 \\ 1 \end{pmatrix}$$

and m=1. A simple calculation shows that for the support function of the reachable set of this system the following relation is true

$$\min_{\chi \in S} c(R(t), \chi) = q(t) = t - 2\sin\frac{t}{2} \quad .$$

If $t_0 > 0$ is small enough then for any $t \in (0, t_0]$ we have

$$\frac{1}{48}t^3 < q(t) < \frac{3}{48}t^3 \quad .$$

Thus for any $t \in (0, t_0]$ there exists such a point $\tilde{x}$ that

$$2\sqrt[3]{2}\,||\tilde{x}||^{\frac{1}{3}} < \tau(\tilde{x}) = t \quad .$$

Theorem 2. If the Condition A holds with the vector $\chi_0 = 0$ and m is given in (3) then the minimum time function of the system (A, Ω) satisfies the local Hölder condition with the exponent

$$\alpha = \frac{1}{m+1} \quad .$$

Proof. Using lemma 2 instead of lemma 3, the proof is the same as for the theorem 1.

Remark 3. Suppose that for the system (A,Ω) the Condition A holds with $\chi_0=0$ and there exists an nxn matrix $\bar{B}$ for which

$$\Omega \subset \bar{B}\, S_1(0)$$

and for which the relation (3) is fulfilled with the same m as for the matrix B. Then $\alpha = (m+1)^{-1}$ is the exact exponent in the Hölder condition for the minimum time function of the system (A,Ω).

Indeed, let $\bar{q}(t)$ be defined by the formula (2) with the matrix $\bar{B}$. Then for any $t>0$ there exists such an element $\tilde{x}$ on the boundary of $R(t)$ that

$$||\tilde{x}|| \le \bar{q}(t) \quad .$$

Moreover, from lemma 2 it follows that there exist such positive numbers t_0 and γ_2 that for any $t\in(0,t_0]$

$$\bar{q}(t) < \gamma_2 t^{m+1} \quad .$$

From this it can be concluded that for any $t\in(0,t_0]$ there exists an $\tilde{x}\in\partial R(t)$ for which

$$\gamma_2^{-\frac{1}{m+1}} ||\tilde{x}||^{\frac{1}{m+1}} \le \tau(\tilde{x}) = t \ .$$

4. A locally N-controllable system for which the minimum time function doesn't satisfy the local Hölder condition

In Ref. [5] the conjecture is stated that for any locally N-controllable system with a finite constraint set Ω, the minimum time function satisfies the local Hölder condition. Here it will be shown that a similar conjecture is not true without the assumption that Ω is finitely generated.

Example 2. Let

$$A = \begin{pmatrix} 0 & -1 \\ 1 & 0 \end{pmatrix} \quad , \quad \Omega = \{u\in E^2 \ : \ -1\le u_1\le 0, \ |u_2|\le y(|u_1|)\} \ ,$$

where the function y is defined on the interval [0,1] by

$$y(x) = k(\ell)\, x^{(1-\frac{1}{\ell})} \quad , \qquad \text{if} \quad x \in [0.5^{\ell-1},\ 0.5^{\ell-2}]$$

and

$$k(\ell) = 2^{\sum_{i=2}^{\ell} \frac{1}{i} - (1-\frac{1}{\ell})} \quad , \qquad \ell = 2, 3, \ldots \ .$$

Since $y(x)/x\to\infty$ as $x\to 0$, we have

$$C\ell\ (\text{Conichull}\ (\Omega)) = E_-^2 \ ,$$

where

$$E^2_- = \{u \in E^2 : u_1 \le 0\} \quad .$$

Therefore the system (A,Ω) is locally N-controllable (see, e.g. [3] p. 179). We remark that for this system the Condition A doesn't hold.

Let us define the function y_ℓ for an arbitrary integer $\ell>2$ by

$$y_\ell(x) = \begin{cases} y(x) & , \text{ if } \ x \ge 0.5^{\ell-1} \ , \\ k(\ell)x^{(1-\frac{1}{\ell})} & , \text{ if } \ 0 \le x \le 0.5^{\ell-1} \ , \end{cases}$$

and let

$$\Omega_\ell = \{u \in E^2 : -1 \le u_1 \le 0, \ |u_2| \le y_\ell(|x_1|)\} \quad .$$

One can easily verify that $y_\ell(x) \ge y(x)$ for all $x \in [0,1]$, and therefore $\Omega_\ell \supset \Omega$. Denoting the support function of a nonempty compact set F by $c(F, .)$, we get the inequalities

$$c(\Omega_\ell, \chi) \ge c(\Omega, \chi)$$

and

$$\int_0^t c(\Omega, e^{-sA'}\chi)ds \le \int_0^t c(\Omega_\ell, e^{-sA'}\chi)ds \tag{11}$$

for all $t>0$ and $\chi \in S$. Let us take $\chi = (\cos\beta, \sin\beta)$ with a $\beta \in [-\Pi,\Pi]$. Then

$$\min_{\chi \in S} \int_0^t c(\Omega_\ell, e^{-sA'}\chi)ds \doteq \min_{\beta \in [-\Pi,\Pi]} \int_\beta^{\beta+t} c\left(\Omega_\ell, \begin{pmatrix}\cos s\\ \sin s\end{pmatrix}\right)ds \quad .$$

By a closer investigation of the function $c(\Omega_\ell,\chi)$ one can prove that if t is small enough then

$$\min_{\beta \in [-\Pi,\Pi]} \int_\beta^{\beta+t} c\left(\Omega_\ell, \begin{pmatrix}\cos s\\ \sin s\end{pmatrix}\right)ds = \frac{2}{\ell}\, k(\ell)^\ell \left(1-\frac{1}{\ell}\right)^{\ell-1} \int_0^t \mathrm{tg}^{\ell-1} s \ \sin s \ ds =$$

$$= \frac{2}{\ell}\, k(\ell)^\ell \left(1-\frac{1}{\ell}\right)^{\ell-1} \int_0^t s^\ell \left(\frac{\mathrm{tg}\ s}{s}\right)^{\ell-1} \frac{\sin s}{s}\, ds \quad . \tag{12}$$

Since $\mathrm{tg}\ s/s \to 1$ and $\sin s/s \to 1$ as $s \to 0$, from (11) and (12) one can conclude that there exist such positive numbers $t_{0,1}$ and $K(\ell)$ that

$$\min_{\chi \in S} \int_0^t c(\Omega, e^{-sA'}\chi)\, ds \le K(\ell)\, t^{\ell+1} \quad , \text{ for any } \quad t \in (0,t_{0,1}] \ .$$

Let M and ε be arbitrarily given positive numbers. We shall show that in any neighbourhood of the origin there exists such a point $\bar{x}$, that for the system (A,Ω)

$$M\|\bar{x}\|^\varepsilon < \tau(\bar{x}) \quad .$$

Let ℓ_1 be a fixed integer, for which $1/\ell_1 \leq \varepsilon$. Then there exists a positive number $t_{0,2}$ that for any $0 \leq t \leq t_{0,2}$

$$K(\ell_1)t < M^{-\ell_1}.$$

Since

$$c(R(t),\ \chi) = \int_0^t c(\Omega,\ e^{-sA'}\chi)\ ds \quad ,$$

from the previous considerations it follows that

$$\min_{\chi \in S} c(R(t),\ \chi) < M^{-\ell_1}\ t^{\ell_1}$$

for all $t \in (0, t_0]$, where $t_0 = \min\{t_{0,1},\ t_{0,2}\}$. This means that for any $t \in (0, t_0]$ there exists a point $\bar{x}$, that $\tau(\bar{x}) = t$ and

$$||\bar{x}|| < M^{-\ell_1}\ t^{\ell_1}\ ,$$

i.e.

$$M||\bar{x}||^{\varepsilon} < \tau(\bar{x}) \quad .$$

This shows that for the locally N-controllable system (A,Ω) the minimum time function doesn't satisfy the Hölder condition.

References

1. Gyurkovics, E., in "Problems of Computational Mathematics", ed. Dmitriev V. I. and Olah Gy., (Russian), Comp. Center of Eötvös L. University, Budapest, 1982., 33-62.
2. Hajek, O., Funkcial. Ekvac. 20 (1976), 97-114.
3. Jacobson, D. H., Extensions of Linear-Quadratic Control, Optimization and Matrix Theory, Acad. Press, London, 1977.
4. Krylov, V. I., Approximate calculation of integrals (Russian), Nauka, Moscow, 1967.
5. Liverovskii, A. A., Differencial'nye Uravnenija (Russian), 16 (1980) 3, 414-423.
6. Liverovskii, A. A., Differencial'nye Uravnenija (Russian), 17 (1981) 4, 604-613.
7. Petrenko, T. Yu., Differencial'nye Uravnenija (Russian), 9 (1973) 7, 1244-1255.
8. Petrov, N. N., Prikl. Mat. Meh. (Russian), 34 (1970) 5, 620-626.

The Quadratic Cost Problem for $L_2[0,T;\ L_2(\Gamma)]$ Boundary Input Hyperbolic Equations

I. Lasiecka and R. Triggiani
Mathematics Department
University of Florida
Gainesville, Florida 32611

1. Introduction

The aim of the present paper is a study of the quadratic optimal control problem on a finite, fixed interval $[0,T]x$, $T < \infty$, for second order linear hyperbolic partial differential equations, where L_2 - boundary controls act either in the Dirichlet or in the Neumann Boundary Conditions (B.C.).

Let Ω be an open, bounded domain in R^n with boundary Γ. Let $A(\xi,\partial)$ be a partial differential operator at order two in Ω, uniformly strongly elliptic, with smooth real coefficients. We consider the mixed hyperbolic problems

$$\frac{\partial^2 y}{\partial +2}(t,\xi) = -A(\xi,\partial)y(t,\xi), \text{ in } (0,T] \times \Omega = Q$$

$$y(0,\xi) = y_0(\xi);\ \frac{\partial y}{\partial t}(0,\xi) = y_1(\xi)\quad \xi \in \Omega \qquad [1.1]$$

and either the Dirichlet B.C.

$$y(t,\sigma) = u(t,\sigma) \text{ in } (0,T] \times \Gamma \equiv \Sigma \qquad [1.1D]$$

or else the Nuemann B.C.

$$\frac{\partial y}{\partial \eta}(t,\sigma) = u(t,\sigma) \text{ in } (0,T) \times \Gamma \equiv \Sigma \qquad [1.1N]$$

where the control function $u(t,\sigma)$ is assumed to belong to $L_2(0,T;\ L_2(\Gamma)) \equiv L_2(\Sigma)$. Three optimal Control Problems are investigated.

1) Optimal Control Problem for Dirichlet boundary controls with distrubuted or interior observation: (O.C.P.$_I$)
Here the cost functional is

$$J_I(u,y) = \int_0^T [|y(t)|_\Omega^2 + |u(t)|_\Gamma^2]dt + |y(T)|_\Omega^2$$

and the corresponding problem is:
mimimize $J_I(u,y)$ over all $u \in L_2(\Sigma)$, where $y(u)$ is the solution of (1.1) with Dirichlet boundary conditions (1.1D).

2) Optimal Control Problem for Neumann boundary controls, with distributed or interior observation (O.C.P.$_{II}$).
Here, with the same cost functional J_I as above, the corresponding problem is:
minimize $J_I(u,y)$ over all $u \in L_2(\Sigma)$, where $y(u)$ is the solution of (1.1) with Neumann boundary condition (1.1N).

3) Optimal Control Problem for Neumann boundary controls with boundary observation (O.C.P.$_{III}$).
Here, the cost functional is

$$J_{III}(u,y) = \int_0^T [|y(t)|_\Gamma^2 + |u(t)|_\Gamma^2]dt + |y(T)|_\Omega^2$$

which contains the trace of the solution y over Σ. The corresponding problem is:
minimize $J_{III}(u,y)$ over all $u \in L_2(\Sigma)$, where $y(u)$ is the solution of (1.1) with Neumann boundary condition (1.1N).
Existence and uniqueness of the optimal pair (u^0,y^0) for each optimal control problem can be ascertained by standard arguments in optimization theory. Our interest here, however, lies in investigating the "Riccati feedback synthesis" of each optimal problem, i.e. the possibility of realizing or synthesizing the optimal control $u^0(t)$ as a pointwise (perhaps only a.e. in t) feedback of the optimal solution $(y^0(t), \frac{dy^0}{dt})$ of the type:

$$u^0(t) = C\ \mathcal{P}(t) \begin{vmatrix} y^0(t) \\ \frac{dy^0(t)}{dt} \end{vmatrix} \qquad [1.2]$$

where C is a time independent operator, known in terms of the original hyperbolic equation (1.1) while the operator $\mathcal{P}(t)$ is expected (by analogy) with other known dynamics) to satisfy in a suitable sense a

Riccati Differential (or perhaps only Integral) Equation. This synthesis will, indeed hold true, as explained in more details in the next section. Before turning to the statement of our results, we review existing literature on the quadratic optimal problems for hyperbolic partial differential equations with boundary controls. It should be noted that a basic preliminary difficulty encountered in the study of the O.C.P. is the question of regularity of the solutions to the mixed problem (1.1); in particular, whether the costs J(u,y) are well-set. Thus, the crux of the cases that we study here is that, in the Dirichlet case, we penalize both the Dirichlet boundary control and the corresponding solution in the $L_2(0T; L_2(\Gamma))$ and $L_2(0T; L_2(\Omega))$ respectively, while, in the Neumann case, we penalize the trace of the solution in $L_2(0,T; L_2(\Gamma))$. This is the distinguishing feature, which differentiates our present results from those already existing in the literature. The only relevant references are [C-P.1] and [L.1]. To be more specific, the problem O.C.P.$_I$ was considered, however with smoother boundary controls, e.g. $u \in H_0^2(\Sigma)$ [L.1] or $u \in L_2(0,T; H^{1/2}(\Gamma)$ as in [C-P.1]. However, Lions [L.1] studies neither the Riccati synthesis, in particular the associated Riccati equations for the problems P_I and P_{II} nor the regularity of the optimal pair (u^0,y^0) for the problem O.C.P.$_{II}$ which are precisely the object of our analysis. Similarily, the abstract evolution approach of [C-P.1] neither covers the L_2 - theory for boundary observation studied here, nor studies regularity properties of the O.C.P.$_{II}$ problem.

2. Main Results

Let $A_D: L_2(\Omega) \supset \mathscr{D}(A_D) \to L_2(\Omega)$ be defined as follows:

$$A_D x = A(x,\partial)x \text{ for all } x \in \mathscr{D}(A_D)$$

$$\mathscr{D}(A_D) = \{x \in L_2(\Omega);\ A_D x \in L_2(\Omega);\ x|_\Gamma = 0\}$$

Similarily, we define A_N by

$$\mathscr{D}(A_N) = \{x \in L_2(\Omega);\ A_N x \in L_2(\Omega);\ \frac{\partial x}{\partial \eta}\Big|_\Gamma = 0\}$$

For simplicity of notation only we assume AD and AN to be selfadjoint. It is well known that both A_D and A_N generate (analytic semigroups on $L_2(\Omega)$ and moreover) cosine operators $C_D(t)$ and $C_N(t)$ on $L_2(\Omega)$. The explicit solution to problem (1.1) with B.C. (1.1D) is:

$$y_D(t) = C_D(t)y_0 + S_D(t)y_1 + A_D \int_0^t S_D(t-\tau)Du(\tau)d\tau \qquad [2.1]$$

and to problem (1.1) with B.C. is

$$y_N(t) = C_N(t)y_0 + S_N(t)y_1 + A_N \int_0^t S_N(t-\tau)Nu(\tau)d\tau \qquad [2.2]$$

where S_D and S_N stand for the corresponding sine operators, while D and N are the Dirichlet and Neumann maps, defined, respectively, by

$$\begin{cases} Dg \equiv v \text{ iff} \\ -A(x,\partial)v = 0 \text{ and } v|\Gamma = g \end{cases} \qquad [2.3]$$

and

$$\begin{cases} Ng \equiv v \text{ iff} \\ A(x,\partial)v = 0 \text{ and } \frac{\partial v}{\partial \eta}\Big|_\Gamma = 0 \end{cases} \qquad [2.4]$$

The following regularity results will be crucially used in the sequel: The map

$$(L_Du)(t) \equiv A_D \int_0^t S_D(t-\tau)Du(\tau)d\tau$$

is a linear continuous from

$L_2(\Sigma)$ into $C([0,T]; L_2(\Omega)]$; see [L-T.1] [2.5]

and the map

$$(L_Nu)(t)|_\Gamma \equiv [A_N \int_0^t S_N(t-\tau)Nu(\tau)d\tau]|_\Gamma$$

is a linear continuous from

$L_2(\Sigma)$ into $L_2(\Sigma)$ see [L-M.I], and also [L-T.2]
($y|_\Gamma$ = Dirichlet trace of y).

These regularity results, (2.5) and (2.6), imply that the cost functionals for all control problems are well set. In order to write the Riccati equation, we need to introduce the operator

$$\mathscr{A}_D \equiv \begin{vmatrix} 0 & I \\ -A_D & 0 \end{vmatrix}$$

which is well known to generate a C_0 - semigroup on

$$E \equiv L_2(\Omega) \otimes [\mathscr{D}(A_D^{1/2})]' \equiv L_2(\Omega) \otimes H^{-1}(\Omega)$$

We similarily defined A_N, by replacing A_D with A_N. We also introduce the operator

$$\mathscr{B}_D^*: E \supset \mathscr{D}(\mathscr{B}_D^*) \; L_2(\Gamma) \text{ defined by } \mathscr{B}_D^* v = D^* v_2, \; v = [v_1, v_2]$$

The operator $\mathscr{B}_D^*$ is clearly unbounded. Now, we are in a position to formulate our results.

Problem O.C.P.$_I$ Theorem 1 The unique control u^0 of the optimal control problem (O.C.P.$_I$) can be expressed in feedback form as:

$$\text{(i)} \quad u^0(t) = -\mathscr{B}_D^* \mathscr{P}(t) \begin{vmatrix} y^0(t) \\ y_t^0(t) \end{vmatrix}$$

where $\mathscr{P}(t)$ is a selfadjoint positive definite operator on E and satisfies the following Riccati Differential Equation:

$$\begin{aligned} \frac{d}{dt}(\mathscr{P}(t)x,y)_E = -(x_1,y_1)_\Omega - (\mathscr{P}(t)x, \mathscr{A}_D y)_E \\ - (\mathscr{P}(t)\mathscr{A}_D x,y)_E - (\mathscr{B}_D^*\mathscr{P}(t)x, \mathscr{B}_D^*\mathscr{P}(t)y)_\Gamma \end{aligned} \qquad [2.7]$$

for all $x,y \in \mathscr{D}(\mathscr{A}_D)$ and all $t \in [0,T]$ with terminal condition $\mathscr{P}(T) = I$;

(ii) the solution to the above Riccati equation (2.7) is unique within the class of selfadjoint operators $\mathscr{P}(t): E \to E$ and such that

$$\mathscr{B}_D^* \mathscr{P}(t)x \in L_\infty(0,T; L_2(\Gamma) \text{ for } x \in E;$$

(iii) the operator $\mathscr{P}(t)$ satisfies the regularity condition:
$\mathscr{B}_D^*\mathscr{P}(t)$ continuous: $E \to C([0,T]; L_2(\Gamma))$ □

As a corollary of the above Theorem 1, we obtain immediately

Corollary to Theorem 1 For the O.C.P.$_I$ - problem the Optimal control u^0 is continuous; i.e.: $u^0 \in C([0,T]; L_2(\Gamma))$ □

This follows easily from (i), (iii) and the regularity results (2.5).

Problem O.C.P.$_{II}$ For the Neumann B.C. "control problems", we introduce the space

$$E \equiv \mathcal{D}(A_N^{1/4}) \otimes [\mathcal{D}(A_N^{1/4})]' \equiv H^{1/2}(\Omega) \otimes [H^{+1/2}(\Omega)]'$$

and the operator $\mathcal{B}_N^* : E \subset \mathcal{D}(\mathcal{B}_N^*) \to L_2(\Gamma)$ defined by

$$\mathcal{B}_N^* v = N^* A_N^{*1/2} v_2, \quad v = [v_1, v_2] \in \mathcal{D}(\mathcal{B}^*)$$

Theorem 2(a) (regularity for O.C.P.$_{II}$) The optimal solution for the problem (O.C.P.$_{II}$) satisfies the following regularity properties: with $y_0 \in H^{\frac{3}{2}-2\varepsilon}(\Omega)$ and $y_1 \in H^{1/2-\varepsilon}(\Omega)$ we have:

(i) $u^0 \in C([0,T]; H^{2-2\varepsilon}(\Gamma))$; $u_t^0 \in C([0,T]; H^{1-2\varepsilon}(\Gamma))$

(ii) $y_t^0 \in L_1(0,T; H^{\frac{3}{2}}(\Omega)) \cap C([0,T]; H^{\frac{3}{2}-2\varepsilon}(\Omega))$

$y_t^0 \in C([0,T]; H^{1/2-2\varepsilon}(\Omega))$

with $\varepsilon > 0$ arbitraily small:

Theorem 2(b) (Riccati synthesis for O.C.P.$_{II}$) The optimal control $u^0(t)$ to the O.C.P.$_{II}$ - problem can be expressed in a feedback form as

$$\text{(i)} \quad u^0(t) = -\mathcal{B}_N^* \mathcal{P}(t) \begin{vmatrix} y^0(t) \\ y_t^0(t) \end{vmatrix}$$

where $\mathcal{P}(t)$ satisfies the Riccati Differential equation

$$\frac{d}{dt}(\mathcal{P}(t)x,y)_E = -(x_1,y_1)_\Omega - (\mathcal{P}(t)\mathcal{A}_N x,y)_E$$
$$- (\mathcal{P}(t)x, \mathcal{A}_N y)_E + (\mathcal{B}_N^* \mathcal{P}(t)x, \mathcal{B}_N^* \mathcal{P}(t)y)_\Gamma \qquad [2.8]$$

for all $x,y \in E$; $0 < t < T$ and the terminal condition $\mathcal{P}(T) = I$

(iv) The solution to the above Riccati equation (2.8) is unique within the class of all self-adjoint operators $\mathcal{P}(t)$ on E, which satisfy the property

$$\mathcal{B}^*\mathcal{P}(t)x \in L_\infty(0,T; L_2(\Gamma)) \text{ for } x \in E$$

(iii) the solution $\mathcal{P}(t)$ to (2.8) satisfies $\mathcal{B}^*\mathcal{P}(t)$ continuous: $E \to C([0,T]; L_2(\Gamma))$ □

Problem O.C.P.$_{II}$ Mathematiclly speaking, the most challenging problem is, of course, the problem with boundary observation. In this case, we have likewise

Theorem 3 The unique control $u^0 \in L_2(\Sigma)$ of problem O.C.P.$_{II}$ can be expressed in a feedback form as:

$$\text{(i)} \quad u^0(t) = -\mathcal{B}_N^*\mathcal{P}(t)\begin{vmatrix} y^0(t) \\ y_t^0(t) \end{vmatrix}$$

where $\mathcal{P}(t): E \to E$ satisfies the Riccati Equation

$$\frac{d}{dt}(\mathcal{P}(t)x,y)_E = -(N^*A^*x_1, N^*A^*y_1)_\Gamma - (\mathcal{P}(t)\mathcal{A}_N x,y)_E -$$

$$(\mathcal{P}(t)x, \mathcal{A}_N y)_E + (\mathcal{B}^*\mathcal{P}(t)x, \mathcal{B}^*\mathcal{P}(t)y)_\Gamma$$

for all $x,y \in \mathcal{D}(\mathcal{A}_N)$ and a.e. in $t \in [0,T]$ with terminal condition $\mathcal{P}(T) = 0$;

(i) the operator $\mathcal{P}(t)$ satisfies the following regularity condition

$$\mathcal{B}^*\mathcal{P}(t) \text{ continuous: } E \to L_2(\Sigma) \square$$

Remarks

1. Notice that in the case of problem O.C.P.$_{II}$ we do not obtain the uniqueness of the corresponding Riccati equation. This fact is due to low regularity of $\mathcal{B}^*\mathcal{P}(t)x$ with $x \in E$, which is only in $L_2(\Sigma)$, as opposed to being in $C([0,T]; L_2(\Gamma))$ in the case of problems O.C.P.$_I$ and O.C.P.$_{II}$. Nevertheless, $L_2(\Sigma)$ of $\mathcal{B}^*\mathcal{P}(t)x$ is sufficient to give a meaning to the quadratic term in the Riccati Equation.

2. In the case of problem O.C.P.$_{III}$, no extra regularity of the optimal control is available - in contrast with the two other problems.

3. Comments on the proofs of the Theorems

It should be stressed that our approach here to the Riccati synthesis of the optimal control is both "explicit" and "constructive", in the sense that an operator is first defined by an explicit formula in terms of the given dynamics, and only subsequently proved to be a solution of a Riccati Differential Equation. In fact, for problem $O.C.P._I$, we define $\mathscr{P}(t)$ to be

$$\mathscr{P}(t)x = \int_t^T \begin{vmatrix} C_D(\tau-t) & \Phi_1(\tau,t)x \\ A_D S_D(\tau-t) & \Phi_1(\tau,t)x \end{vmatrix} d\tau \tag{3.1}$$

where $\Phi_1(\tau,t)$ is the first coordinate of the corresponding evolution operator given by

$$\Phi_1(\tau,t)x = [I_t + L_t L_t^*]^{-1} [C_D(\cdot-t)x_1 + S_D(-t)x_2] \qquad [3.2]$$

where L_s stands for L_D starting at time s; i.e.

$$(L_s u)(t) \equiv A_D \int_s^t S_D(t-\tau)\, Du(\tau)d\tau$$

For problem $O.C.P._{III}$, we define instead

$$\mathscr{P}(t)x = \int_t^T \begin{vmatrix} A_N^{1/2} C_N(\tau-t)N\, N^* A_N^* & \Phi_1(\tau,t)x \\ A_N^{\frac{3}{2}} S_N(\tau-t)N\, N^* A_N^* & \Phi_1(\tau,t)x \end{vmatrix} d\tau \qquad [3.3]$$

with $\Phi_1(\tau,t)$ defined similarily as before, by replacing the Dirichlet map by the Neumann map. As mentioned before, the major difficulty in deriving the Riccati equation is related to low regularity of the optimal solution (particularily for problems $O.C.P._I$ and $O.C.P._{III}$, and in particular to the low regularity at the corresponding Riccati operator. In order to give a meaning to the Riccati equation, one must show - as a necesary step - that $\mathscr{B}^* \mathscr{P}(t)$ is well defined in some sense. This fact is equivalent to saying that the trace of $\mathscr{P}(t)$ is well defined in $L_2(\Gamma)$. On the other hand, regularity of the trace can not stem from the standard trace theory, as the solution $\mathscr{P}(t)$ does not have enounqh interior regularity. Therefore, the fact that $\mathscr{B}^* \mathscr{P}(t)x$ is well defined on the boundary (see Thm. 1(iii), Thm. 2 iii and Thm. 3 (iii)) is not obvious by any means: it is established in the paper by developing an adequate trace theory for the solutions of homogeneous hyperbolic problems. To be more precise, we first prove that.

Lemma 3.1 (see [L-T.3]) the operators $D^*A_D^* S_D^*(t)$ and $D^*A_D^{*\,1/2} C_D^*(t)$ are linear bounded from $L_2(\Omega)$ into $L_2(\Sigma)$ □

Remark: Notice that the above result" translated" into P.D.E. says that normal derivative on the boundary of the solution to:

$$u_{tt} = A(x,\partial)u$$

$$u|_\Gamma = 0$$

$$u(0) = u_0 \in H_0^1(\Omega)$$

$$u_t(0) = u_1 \in L_2(\Omega)$$

is well defined in $L_2(\Sigma)$. This fact, again does not follow from the standard trace theory as $u(t) \in H^1(\Omega)$ and $\frac{\partial\mu}{\partial\eta}$ in general may be not well defined. As for problem O.C.P.$_{II}$, we have the following counterpart of Lemma 3.1.

Lemma 3.2 [see [L-T.2]) The operators

$$N^*A_N^{1+1/4} S_N^*(t) \text{ and } N^*A_N^{\frac{3}{4}} C_N^*(t)$$

are linear bounded from $L_2(\Omega)$ into $L_2(\Sigma)$ □

The results of Lemmas 3.1 and 3.2 are crucial in establishing that

$$\mathscr{B}_D^* \mathscr{P}(t) \text{ continuous: } E \to C([0,T]; L_2(\Gamma)) \qquad [3.4]$$

and

$$\mathscr{B}_N^* \mathscr{P}(t) \text{ continuous: } E \to L_2(\Sigma) \text{ for O.C.P.}_{III} \text{ - problem} \qquad [3.5]$$

which is the necessary first step in establishing the Riccati equations for $\mathscr{P}(t)$. For the remaining of the proof we refer to [L-T.2] and [L-T.3] □

4. References

[C-P.1] R. Curtain-A. Pritchard, "An abstract theory for unbounded control action for distributed parameter systems" SIAM J. Control Opt. 15 (1977) 566-611.

[L.1] J. L Lions, Optimal Control of Systems Goverened by Partial Differential Equations. Springer-Verlag, 1971.

[L-M] J. L. Lions and E. Magenes, "Nonhomogeneous Boundary Value Problems and Applications" Vols. I, II Springer-Verlag Berlin-Heidelberg-New York, 1972.

[L-T.1] I. Lasiecka-R. Triggiani, "Regularity of hyperbolic equations under $L_2[0,T; L_2(\Gamma)]$ - Dirichlet boundary terms" Appl. Math. Optim. (1983) 275-286.

[L-T.2] I. Lasiecka-R. Triggiani, "Riccati Equations for Hyperbolic partial differential equations with $L_2(0,T; L_2(\Gamma))$ - Dirichlet boundary terms" submitted to SIAM J. on Control and Optimiz.

[L-T.3] I. Lasiecka-R. Triggiani, "Hyperbolic equations with nonhomogeneous boundary Neumann terms: Part I: Regularity; Part II: Riccati Equations for interior and boundary observations" to be submitted.

Modelling and Control of Water Quality in a River Section

A. Bogobowicz
Institute of Geophysics
Polish Academy of Sciences
00-973 Warszawa, ul.Pasteura 3
Poland

J. Sokołowski
Systems Research Institute
Polish Academy of Sciences
01-447 Warszawa, ul.Newelska 6
Poland

Notation

A area of cross section $|m^2|$
B width of channel $|m|$
C_S concentration of dissolved oxygen in saturation $|mgl^{-1}|$
H depth of water in channel $|m|$
H_o steady state depth of water in channel $|m|$
h increment of depth of water $|m|$
K_1 biodegradation and sedimentation coefficient $|h^{-1}|$
K_2 atmospheric reaeration coefficient $|h^{-1}|$
K_{21} BOD removal coefficient $|h^{-1}|$
S_o, S_1, S_2, S_3 lateral sources of pollution $|mg\,l^{-1}\,h^{-1}|$
t time $|h|$
u control variable $|m|$
V river flow velocity $|mh^{-1}|$
V_o steady state river flow velocity $|mh^{-1}|$
v increment of river flow velocity $|mh^{-1}|$
w concentration of chlorides $|mg\,l^{-1}|$
x longitudinal river dimension $|m|$
y concentration of BOD $|mg\,l^{-1}|$
z concentration of dissolved oxygen $|mg\,l^{-1}|$
η_1, η_2 control variables $|mg\,l^{-1}|$
ξ control variable
$\zeta_1, \zeta_2, \zeta_3, \zeta_4$ constants
$\phi_1, \phi_2, \phi_3, \phi_4, \phi_5$ adjoint state variables

Function spaces:

Sobolev space $H^1(Q)$:

$$H^1(Q) = \{\phi \in L^2(Q) \mid \frac{\partial\phi}{\partial t}, \frac{\partial\phi}{\partial x} \in L^2(Q)\}$$

where $\phi = \phi(x,t)$, $(x,t) \in Q = (0,L) x (0,T)$

Banach space $W(Q) \subset H^1(Q)$ with the norm:

$$||\phi||_{W(Q)} = ||\phi||_{L^2(Q)} + ||\phi||_{C(0,T;L^2(0,L))} + ||\phi||_{C(0,L;L^2(0,T))}$$

Linear spaces $\Phi_o(Q)$, $\Phi_1(Q)$:

$$\Phi_o(Q) = \{\phi \in H^1(Q) | \phi |_{x=o}=0,\ \phi|_{t=o}=0\}$$

$$\Phi_1(Q) = \{\phi \in H^1(Q) | \phi |_{x=L}=0,\ \phi|_{t=T}=0\}$$

with the norms:

$$||\phi||_{\Phi_o(Q)} = (||\phi||^2_{L^2(Q)} + ||\phi(\cdot,T)||^2_{L^2(0,L)} + ||\phi(L,\cdot)||^2_{L^2(0,T)})^{1/2}$$

$$||\phi||_{\Phi_1(Q)} = (||\phi||^2_{L^2(Q)} + ||\phi(\cdot,0)||^2_{L^2(0,L)} + ||\phi(0,\cdot)||^2_{L^2(0,T)})^{1/2}$$

Spaces $W(Q_1)$, $\Phi_o(Q_1)$, $\Phi_1(Q_1)$, $Q_1=(x_o,L)x(0,T)$, $x_o \in (0,L)$, are defined in exactly the same way.

Introduction

The severe deterioration of the water quality in many basins is caused by the increasing amount of wastes from the recent expansion in socio-economic developments. This paper considers this problem with respect to a section of the Upper Vistula located in the Upper Silesia industrial region. A mathematical model of the water quality in this section is given by a system of partial differential equations. A performance index and the constraints in formulating an optimal control policy are strictly limited to the water quality in the river. The control problem is formulated within the framework of a hierarchical water management system in the Upper Silesia industrial region developed by Malinowski et al. (1979) (see Fig.1).

The outline of the paper is the following. Section 2,3 describe Water Quality Model and Open Channel Flow Model respectively. Section 4 contains results obtained for an optimal control problem. In section 5 an example of computations is presented.

Water quality model

The model includes transport and sedimentation but neglect other phenomena such as photosynthesis. It is assumed that the water quality of

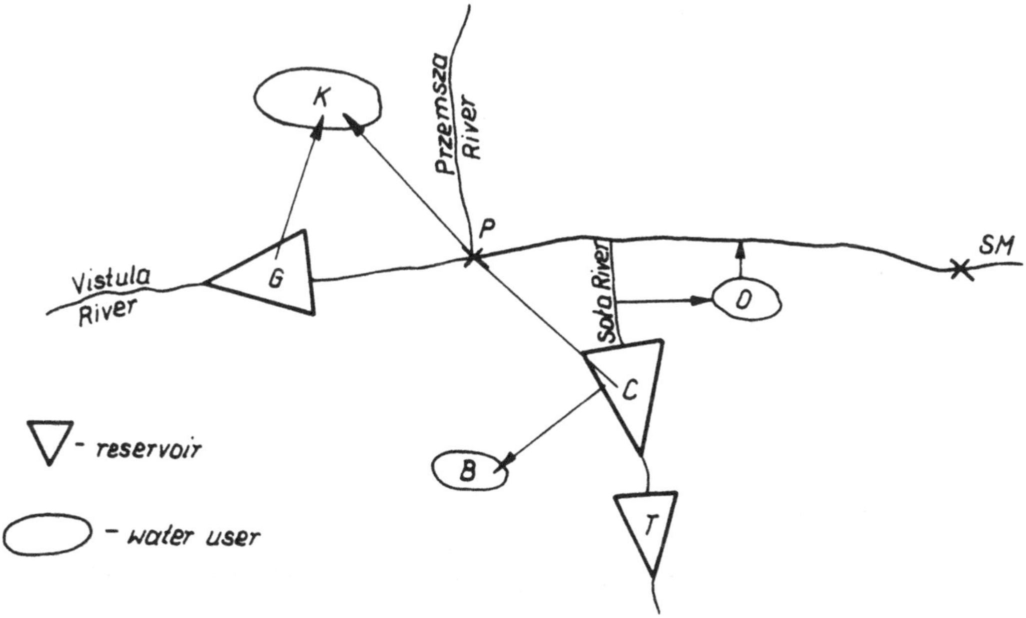

FIGURE 1. Operation of the multireservoir system on the Upper Vistula.

the river depends mostly on the biochemical oxygen demand (BOD), concentrations of dissolved oxygen (DO) and chlorides. This is justified by the presence of chemical factories and neighbouring coal mines in the Upper Silesia industrial region. Concentrations, denoted by the symbols y,z,w are assumed to be functions of one space variable $x \in (0,L)$ and a time variable $t \in (0,T)$. It should be noted that because of the high flow velocity of water in the Upper Vistula the diffusion terms in the equations of the water quality model i.e. components containing second order derivatives with respect to space variable x can be neglected. Functions $y(\cdot,\cdot)$, $z(\cdot,\cdot)$ can be determined by solving the following systems of hyperbolic equations:

BOD

$$(1) \quad L_{11}(V)y \overset{\text{def}}{=} \frac{\partial y}{\partial t} + \frac{\partial (Vy)}{\partial x} + K_1 y = S_1 A, \text{ in } Q$$

DO

$$(2) \quad L_{22}(V)z + K_{21}y \overset{\text{def}}{=} \frac{\partial z}{\partial t} + \frac{\partial (Vz)}{\partial x} + K_2 z + K_{21} = K_2\, C_S\, A + S_2 A, \text{ in } Q$$

chlorides

.3) $L_{33}(V)w \overset{def}{=} \frac{\partial w}{\partial t} + \frac{\partial (Vw)}{\partial x} = S_3A$, in Q

with the boundary conditions:

(4) $y(o,t) = \eta_1(t)$

(5) $z(o,t) = z_1(t)$

(6) $w(o,t) = \eta_2(t)$

and the initial conditions:

(7) $y(x,0) = y_o(x)$

(8) $z(x,0) = z_o(x)$

(9) $w(x,0) = w_o(x)$

where $\eta_1(.), \eta_2(.), z_1(.), y_o(.), z_o(.), w_o(.)$, are given functions.

We assume also that the jump of function $y(.,.)$ occurs at a given point $x_1 \in (0,L)$:

(10) $y(x_1^+,t) = y(x_1^-,t) + AS_o(1-\xi(t))$

Condition (10) describes the presence of a source of pollution at the point $x=x_1$.

To apply the model in practice, the model parameters have to be determined. Some can be calculated empirically as proposed by Hullet (1974) or estimated on the basis of meteorogical and hydrological forecasts.

It is assumed that the river flow velocity V can be calculated from an Open Channel Flow Model.

Open Channel Flow Model

The model is used for description of the flow regulation by means of multi-purpose reservoir located at point $x=x_o$. Releases from the reservoir are denoted by $u(t)$, $t \in (0,T)$.

It is assumed that river flow velocity V and river depth H take on the form:

(11) $$V(x,t) = \begin{cases} V_o, & (x,t) \in (0,x_o)x(0,T) \\ V_o+v(x,t), & (x,t) \in (x_o,L)x(0,T) \end{cases}$$

$$(12)\qquad H(x,t) = \begin{cases} H_o & (x,t)\in(0,x_o)x(0,T) \\ H_o + h(x,t), & (x,t)\in[x_o,L)x(0,T) \end{cases}$$

where V_o, H_o denotes steady state velocity and water depth of the open channel respectively. Functions v,h can be determined by solving the system of linear hyperbolic equations:

$$(13)\qquad L_{44}h + L_{45}v \overset{def}{=} \frac{\partial h}{\partial t} + V_o\frac{\partial h}{\partial x} + H_o\frac{\partial v}{\partial x} = 0$$

$$(14)\qquad L_{54}h + L_{55}v \overset{def}{=} g\frac{\partial h}{\partial x} + C_2 h + \frac{\partial v}{\partial t} + V_o\frac{\partial v}{\partial x} + C_1 v = 0$$

$$\text{in}\quad Q_1 = (x_o,L)x(0,T)$$

with appropriate boundary and homogeneous initial conditions. Equations (13), (14) are obtained |2| by linearization of Saint - Venant equations.

In Eq. (14) g denotes gravity constant, $c_1=2V_o/KH_o^N$, $c_2=-V_o|V_o|Ng/(KH_o^{N+1})$ where constants K,N are selected on the basis of Manning formula for the friction slope.

In order to characterize weak solutions to the system (13), (14) with the boundary conditions:

$$(15)\qquad h(x_o,t) = u(t),\quad v(x_o,t) = 0,\ t\in(0,T)$$

we introduce the following notation:

Denote by $E_{ij}(\cdot,\cdot):L^2(Q_1)xH^1(Q_1)\longrightarrow R$

linear form:

$$E_{ij}(r,\phi) = \int_{x_o}^{L}\int_{o}^{T} r(x,t)(L_{ij}^{*}\phi)(x,t)dxdt$$

where L_{ij}^{*}, i,j=4,5, denote adjoint operators e.g. $L_{45}^{*}\phi = -H_o\frac{\partial\phi}{\partial x}$

Weak solution $(h,v)\in[W(Q_1)]^2$ satisfies the integral identity:

$$(16)\qquad \begin{cases} E_{44}(h,\phi)+E_{45}(v,\phi) = \int_o^T u(t)\phi(x_o,t)dt & \forall\phi\in\Phi_1(Q_1) \\ E_{54}(h,\psi)+E_{55}(v,\psi) = 0, & \forall\psi\in\Phi_1(Q_1) \end{cases}$$

Lemma 1

Assume that $V_o > 0$, $H_o > 0$, $V_o^2 + gH_o \neq 0$. Then there exists a unique weak solution to (16) such that:

(17) $$\|h\|_{W(Q_1)} + \|v\|_{W(Q_1)} \leq C\,\|u\|_{L^2(0,T)}$$

Furthermore, if $u \in H^2(0,T)$, $u(0)=u'(0)=0$ then

(18) $$\|v\|_{C(\bar{Q}_1)} + \left\|\frac{\partial v}{\partial x}\right\|_{C(\bar{Q}_1)} \leq C\,\|u\|_{H^2(0,T)}$$

The proof of (17) is classical |1|. Estimation (18) can be obtained by standard argument.

Let us now characterize weak solutions to Water Quality Model (1)-(9). It can be shown by application of projection theorem |8|, taking into account (17), (18), that there exists a unique weak solution $(y,z,w) \in [W(Q)]^3$ to (1)-(9), which satisfies the following integral identities:

(19) $$\begin{cases} E_1(u;y,\phi)=F_1(u,\xi,\eta_1;\phi), & \forall\phi \in \Phi_1(Q) \\ E_2(u;z,\phi)+E_{21}(y,\phi)=F_2(u;\phi), & \forall\phi \in \Phi_1(Q) \\ E_3(u;w,\phi)=F_3(u,\eta_2;\phi), & \forall\phi \in \Phi_1(Q) \end{cases}$$

where

(20) $$E_i(u;r,\phi)= \int_o^L \int_o^T r(x,t)(L_{ii}^*(V)\phi)(x,t)dxdt \qquad i=1,2,3$$

(21) $$E_{21}(y,\phi)= \int_o^L \int_o^T K_{21}y(x,t)\phi(x,t)dxdt$$

(22) $$F_1(u,\xi,\eta_1;\phi)= \int_o^T A(x,t)S_o(1-\xi(t))\phi(x_1,t)dxdt + \int_o^T v_o\eta_1(t)\phi(o,t)dt+ + \int_o^L y_o(x)\phi(o,x)dx$$

(23) $$F_2(u;\phi)= \int_o^L \int_o^T A(x,t)(K_2C_S-S_2)\phi(x,t)dxdt + \int_o^L z_o(x)\phi(x,0)dx + + \int_o^T v_oz_1(t)\phi(0,t)dt$$

(24) $$F_3(u,\eta_2;\phi)= \int_o^L \int_o^T S_3A(x,t)\phi(x,t)dxdt + \int_o^L w_o(x)\phi(x,0)dx + + \int_o^T v_o\eta_o(t)\phi(0,t)dt$$

$$(25)\qquad A(x,t) = \begin{cases} A_o & (x,t) \in [0,x_o) \times [0,T] \\ A_o + Bh(x,t), & (x,t) \in [x_o,L] \times [0,T] \end{cases}$$

Optimal Control Problem

The optimal control of the water quality in the studied section of the Vistula can be formulated by finding the admissible functions $\hat{u}(t)$, $\hat{\eta}_1(t)$, $\hat{\eta}_2(t)$, $\hat{\xi}(t)$ such that the corresponding value of the performance index $J(u,\eta_1,\eta_2,\xi)$ over time T along the segment L(L=23km) is minimized.

Denote by $J(u,\eta_1,\eta_2,\xi)$ the following functional:

$$(26)\qquad J(u,\eta_1,\eta_2,\xi) = \frac{1}{2}\int_o^T\int_o^L \{\zeta_1(C_S - z(x,t))_+^2 + \zeta_2\ (w(x,t)-w_d)_+^2\}dxdt + $$
$$+ \frac{1}{2}\int_o^T \{\zeta_3(u\ (t)-u_d(t))^2 + \zeta_4(\xi(t))^2\}dt$$

where w_d is the desired concentration of chlorides, u_d is the desired water depth obtained from the upper layer of the control structure policy; ζ_i, i=1,..,4 denote weighting coefficients.

Control inputs of the system can be described as follows:

- $\eta_1(t)$, $\eta_2(t)$ define BOD and chlorides pollution load from the pollution source located at x=0, for $t \in (0,T)$.
- $\xi(t)$ define BOD pollutant load from the pollution source located at $x=x_1$, for $t \in (0,T)$.
- u(t) denotes increment of the water depth obtained by release from the multi-purpose reservoir located at point $x=x_o$, for $t \in (0,T)$.

Denote by U_{ad} the set of admissible functions $u(\cdot)$, $\eta_1(\cdot)$, $\eta_2(\cdot)$, $\xi(\cdot)$ of the form:

$$(27)\qquad U_{ad} = \{(u(.),\eta_1(.),\eta_2(.),\xi(.)) \in H^2(0,T) \times [L^2(0,T)]^3 :$$

$$0 \le u(t) \le u_{max},\ 0 \le \eta_1(t) \le \overline{y},$$

$$0 \le \eta_2(t) \le \overline{w},\ 0 \le \xi(t) \le \xi_{max}$$

for a.e. $t \in (0,T)$,

$$\int_o^T u(t)dt \le u_1,\ u(0) = u'(0) = 0,$$

$$\int_0^T (u''(t))^2 dt \le C\}$$

where u_{max}, $\bar{y}$, $\bar{w}$, ξ_{max}, u_1, C are given constants.

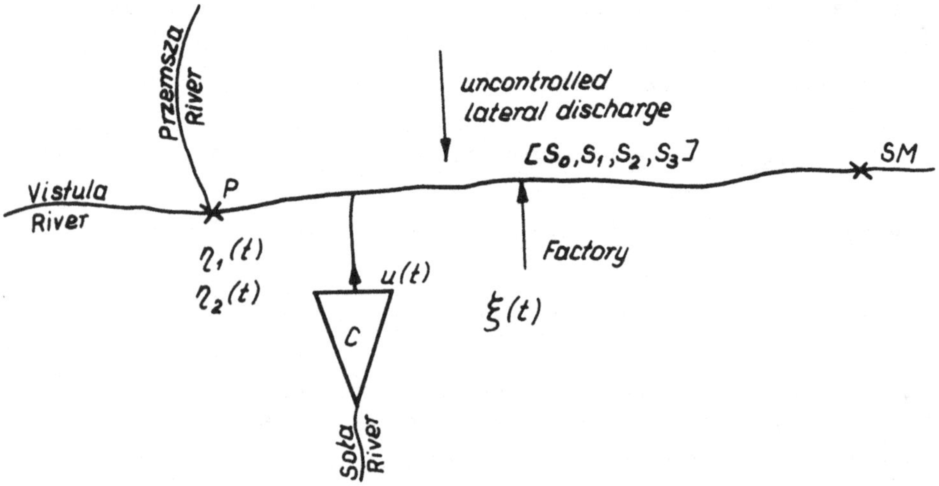

FIGURE 2. The section of the Upper Vistula studied here.

Theorem 1

There exists an optimal control $(\hat{u},\hat{\eta}_1,\hat{\eta}_2,\hat{\xi}) \in U_{ad}$ minimizing the performance index (26) over the set of admissible controls U_{ad}. The optimal control satisfies the following optimality system:

State equations:

$$\begin{cases} \hat{v},\hat{h} \in W(Q_1) \\ E_{44}(\hat{h},\phi)+E_{45}(\hat{v},\phi)=\int_0^T \hat{u}(t)\phi(x_o,t)dt \ , \quad \forall\phi \in \Phi_1(Q_1) \\ E_{54}(\hat{h},\psi)+E_{55}(\hat{v},\psi) = 0 \ , \quad \forall\psi \in \Phi_1(Q_1) \end{cases}$$

$$(28)\quad \begin{cases} \hat{y},\ \hat{w},\ \hat{z} \in W(Q) \\ E_1(\hat{u};\hat{y},\phi)=F_1(\hat{u},\hat{\xi},\hat{\eta}_1;\phi), \quad \forall\phi \in \Phi_1(Q) \\ E_2(\hat{u};\hat{z},\phi)+E_{21}(\hat{y},\phi)=F_2(\hat{u};\phi), \quad \forall\phi \in \Phi_1(Q) \\ E_3(\hat{u};\hat{w},\phi)=F_3(\hat{u},\hat{\eta}_2;\phi), \quad \forall\phi \in \Phi_1(Q) \end{cases}$$

Adjoint state equation:

$$(29)\quad \begin{cases} p_1,p_2 \in W(Q_1) \\ E^*_{44}(p_1,\phi)+E^*_{45}(p_2,\phi)= \int_0^T BS_o\,(1-\hat{\xi}(t))p_3(x_1,t)\phi(x_1,t)dt+ \\ + \int_{x_o}^L \int_0^T B\,((K_2C_S-S_2)p_4(x,t)-S_3p_5(x,t))\phi(x,t)dxdt\ , \quad \forall\phi \in \Phi_o(Q_1) \\ E^*_{54}(p_1,\phi)+E^*_{55}(p_2,\phi)=\int_{x_o}^L \int_0^T\{\hat{y}(x,t)\ \frac{\partial p_3}{\partial x}(x,t)+\hat{z}\,(x,t)\ \frac{\partial p_4}{\partial x}(x,t)\ + \\ + \hat{w}(x,t)\ \frac{\partial p_5}{\partial x}(x,t)\}\phi(x,t)dxdt\ , \quad \forall\phi \in \Phi_o(Q_1) \end{cases}$$

$$(30)\quad \begin{cases} p_3,p_4,p_5 \in W(\,Q\,) \\ E^*_1(\hat{u};p_3,\phi)+E_{21}(p_4,\phi)=0, \quad \forall\phi \in \Phi_o(Q) \\ E^*_2(\hat{u};p_4,\phi)=\int_o^L \int_o^T \zeta_2(C_S-\hat{z}(x,t))_+\phi(x,t)dxdt\ , \quad \forall\phi \in \Phi_o(Q) \\ E^*_3(\hat{u};p_5,\phi)=-\int_o^L \int_o^T\zeta_3(\hat{w}(x,t)-w_d(x,t))_+\phi(x,t)dxdt\ , \quad \forall\phi \in \Phi_o(Q) \end{cases}$$

<u>Optimality Condition</u>:

$$(31)\quad \int_o^T\{\zeta_4(\hat{u}(t)-u_d(t))-gp_2(x_o,t)-V_op_1(x_o,t)\}(u(t)-\hat{u}(t))dt\ +$$

$$\int_o^T\{A(x_1,t)S_op_3(x_1,t)-\zeta_5\hat{\xi}(t)\}(\xi(t)-\hat{\xi}(t))dt$$

$$\int_o^T V_o\,\{p_3(0,t)(\eta_1(t)-\hat{\eta}_1(t))+p_5(0,t)(\eta_2(t)-\hat{\eta}_2(t))\}dt \geq 0$$

$$\forall(u(\cdot),\eta_1(\cdot),\eta_2(\cdot),\xi(\cdot)) \in U_{ad}$$

The proof of existence of an optimal solution follows by application of the abstract result given in |12| (Lemma 1, p.286). Necessary optimality conditions can be obtained by standard argument and application of the abstract result given in |13| (Lemma 2, p.292).

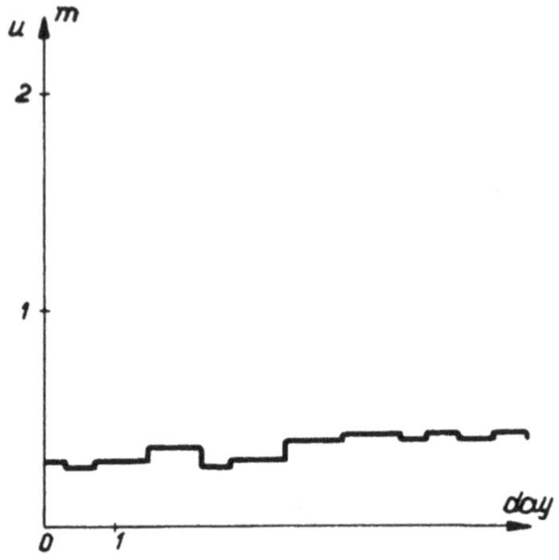

Fig. 3 Optimal values of increment of flow at the point $x = x_o$ (u)

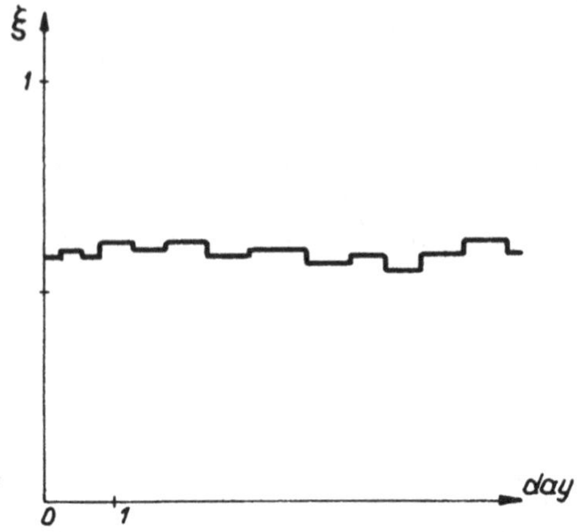

Fig. 4 Optimal values of discharges of BOD at the point $x = x_1$ (ξ)

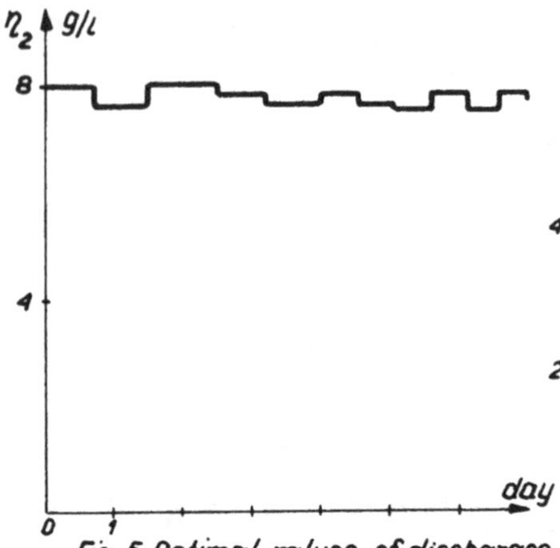

Fig. 5 Optimal values of discharges of BOD at the point $x = 0$

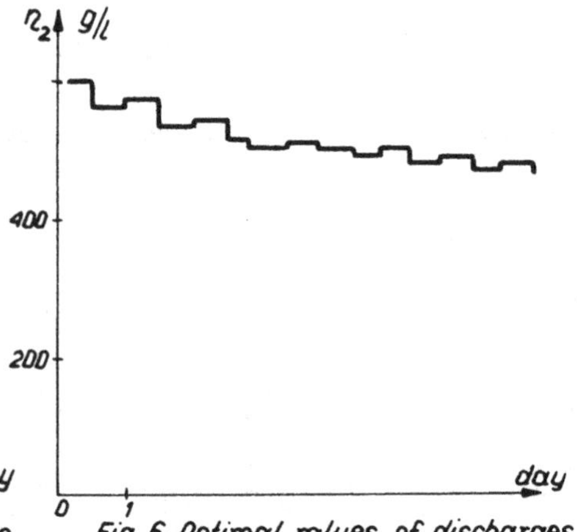

Fig. 6 Optimal values of discharges of chlorides at the point $x = 0$

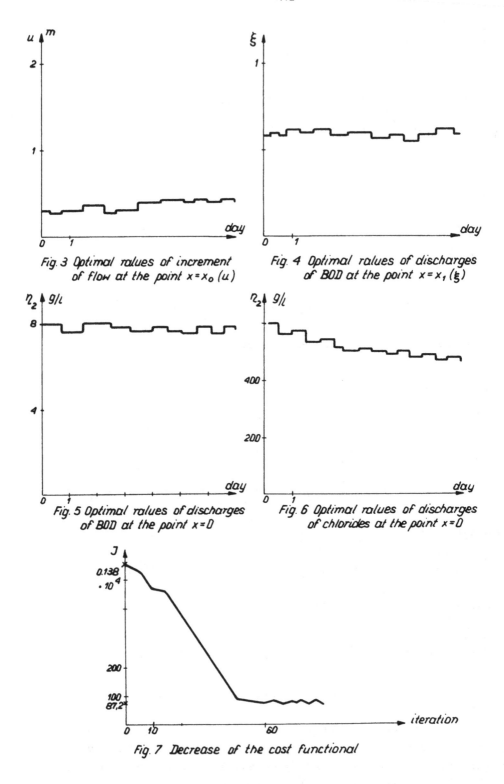

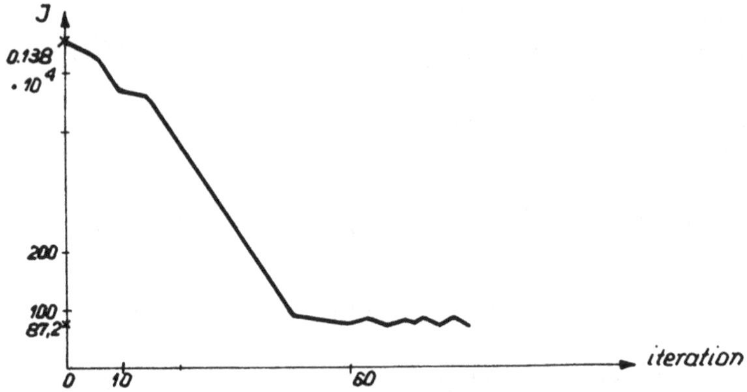

Fig. 7 Decrease of the cost functional

Numerical results

The problem of water quality control is solved numerically after formulation of the discrete form of equations (1)-(9), (13)-(15) and objective functional (26). For the five hyperbolic equations describing the system the finite difference method (implicite scheme) and the double sweep")(Cunge 1964) are used. The minimization of the given performance index subject to constraints (27) is achieved by applying the penalty shifted method combined with the conjugate gradient method.

The water quality control problem is solved numerically with respect to a section of the Upper Vistula, shown in Fig.2.

$L = 23$ km	$V_o = 0.85$ m/s	$\bar{y} = 12$ mg/l	$x_o = 1{,}8$ km
$T = 7$ days	$H_o = 2.5$ m	$\bar{w} = 400$ mg/l	$x_1 = 10{,}7$ km

The coefficients for the equations modelling the system are the following:

$$K_1 = 0.19\exp(0.029 \cdot S)$$

where S - denotes the initial concentration of BOD at point $x=0$.

$$K_{21} = K_1$$

$$K_2 = 1.72 \cdot V_o^{0.5}/H_o^{1.5}$$

Results of computations for the example are shown in Fig.3-7. The starting point was selected as follows:

$$u=0, \xi=0, \eta_1 = 8\text{ mg/l}, \eta_2 = 600\text{ mg/l}$$

References

|1| N.U.Ahmed and K.L.Teo, Optimal Control of Distributed Parameter Systems, Elsevier North Holland, New York (1981).

|2| M.Benahmed, Identification de non-linearities ou de parameters reparties dans deux equations aux derivees partielles evalutive non lineaires, These d'Etat, Universite Paris 9 (1978).

|3| A.Bogobowicz and J.Sokołowski, Optimal water quality control by treating the effluent at the pollution source and by flow regulation, IAHS-AISH Publ. No.129, pp.139-143, (1980).

|4| J.A.Cunge and Wegner, Integration numerique des equations d'ecoulement de Barre Saint - Venant par un scheme implicite de diffe-

rence finies, La Mouille Blanche No.1 (1967).

|5| J.G.I.Dooge, Linear theory of hydrologic systems, Techn. Bull. No.1468, Agricultural Research Service, USDA Washington (1973).

|6| W.Hullet, Optimal estuary aeration: an application of distributed parameter systems control theory, Appl. Math. Optim., No.1, pp.20-63 1974 .

|7| Z.Kundzewicz, Parameters approximation of hydrologic models (Thesis), Institute of Geophisics, Polish Academy of Sciences, Warsaw (1979).

|8| J.L.Lions, Equations differentielles operationelles et problems aux limites, Springer Verlag, Berlin (1961).

|9| J.L.Lions, Controle optimal de systems gouvernes par des equations aux derivees partielles, Dunod, Paris (1968).

|10| K.Malinowski, K.Salewicz and T.Terlikowski, Two-layer hierarchical optimizing control structure for multireservoir water systems, Tech. Report of the Institute of Automatic Control, Technical University of Warsaw, Poland, pp.115-155 (1979).

|11| S.Rinaldi, R.Soncini-Sessa, H.Stehfest and H.Tamura, Modelling and Control of River Quality, McGraw-Hill, New York (1978).

|12| J.Sokołowski, Optimal Control in Coefficients for Weak Variational Problems in Hilbert Space, Appl. Math. Optim. 7, pp. 283-293, (1981).

|13| J.Sokołowski, Control in Coefficients of PDE, Abhandlungen der Akademie der Wissenschaften der DDR, Jahrgang 1981, No.2N, pp. 287-295.

|14| R.V.Thomann, Systems Analysis and Water Quality Management, McGraw Hill, New York (1971).

A GALERKIN APPROXIMATION FOR THE ZAKAI EQUATION

A. Germani* and M. Piccioni**

* Istituto di Analisi dei Sistemi ed Informatica del CNR
Viale Manzoni 30, 00185 Roma, Italy

** Istituto di Matematica, Informatica e Sistemistica, Università di Udine
Via Mantica 3, 33100 Udine, Italy

1. INTRODUCTION

In the last years a large amount of work in filtering of diffusion processes has been devoted to the study of the evolution of the so-called "unnormalized conditional density". Such evolution is described by the well known Zakai equation [1,2,3]. This is a bilinear PDE forced by the observation process. The problem of constructing finite-dimensional approximations of such equation appears to be the major issue for practical applications.

The usual approach in the literature [4,5,6] consists in approximating the diffusion model by a finite-state Markov chain and writing down the Zakai equation for it, which turns out to be a stochastic ODE. In this paper a different approach is proposed, namely the Zakai equation is directly approximated by means of the classical Galerkin method for solving deterministic PDE's [7,8]. In particular, under rather usual assumptions, the convergence in mean square will be proved. To prove such a result, some general properties of stochastic differential equations in Hilbert spaces will be largely exploited [9].

2. THE FILTERING PROBLEM

The following filtering problem will be considered. Let $\{X_t\}$ be the unknown d-dimensional state and $\{Y_t\}$ the measured output (supposed scalar for sake of simplicity) of the following stochastic differential system in $[0,T]$

$$dX_t = b(X_t)dt + \sigma(X_t)dB_t \ , \quad dY_t = h(X_t)dt + dV_t \ , \tag{1}$$

where $\{B_t \ , V_t\}$ is a standard (d+1)-dimensional Wiener process defined on some $(\Omega,F,P,\{F_t\})$, X_o is an F_o-measurable random variable with density $p_o \in L^2(R^d)$, and $Y_o = 0$.

Let us assume that:

i) the entries of b, σ and h belongs to $L^\infty(R^d)$;

ii) the entries of b and σ are also continuous with bounded derivatives;

iii) once defined $a(x) = \sigma(x)\sigma^*(x)$, there exists $\alpha > 0$ such that

$$z^* a(x) z \geq \alpha z^* z \qquad \forall x, z \in R^d \tag{2}$$

Such hypotheses firstly guarantee that the system (1) has a unique strong solution, moreover the infinitesimal generator of the process $\{X_t\}$

$$L = \frac{1}{2} \sum_{i,j=1}^{d} a_{ij}(x) \frac{\partial^2}{\partial x_i \partial x_j} + \sum_{i=1}^{d} b_i(x) \frac{\partial}{\partial x_i}$$

defines a bounded linear operator from $H^1(R^d)$ to $H^{-1}(R^d)$[10]. Finally multiplication by h defines a bounded self-adjoint operator on $L^2(R^d)$.

The derivation of the Zakai equation needs the introduction of a new probability measure $\tilde{P}$ on (Ω, F), under which $\{Y_t\}$ is a standard Wiener process independent of $\{X_t\}$, which is defined by $d\tilde{P}/dP = Z_T^{-1}$, where:

$$Z_t = \exp\{\int_0^t h(X_s) dY_s - \frac{1}{2} \int_0^t h^2(X_s) ds\}, \ t \in [0,T] \tag{3}$$

Next let F_t^y be the σ-algebra generated by the observation process up to time $t \in [0,T]$, and recall the Kallianpur-Striebel formula [11]

$$E(v(X_t)|F_t^y) = \tilde{E}(v(X_t)Z_t|F_t^y)/\tilde{E}(Z_t|F_t^y) \quad , \ v \in L^\infty(R^d). \tag{4}$$

The Zakai equation describes the evolution of the numerator of the r.h.s. of (4).

For any Hilbert space H (with norm $|\cdot|$ and scalar product $(\cdot,\cdot)$) denote by $Z(0,T;H)$ the space of all H-valued F_t^y-adapted processes continuous in the mean square norm. Let now H be $L^2(R^d)$ and denote also by $\langle \cdot,\cdot \rangle$ the duality pairing between $H^1(R^d)$ and $H^{-1}(R^d)$. The Zakai equation

$$d(q_t, v) = \langle q_t, Lv \rangle dt + (q_t, hv) dY_t, \ v \in H^1(R^d), \ q_0 = p_0 \in L^2(R^d) \tag{5}$$

has a unique solution in $Z(0,T;L^2(R^d))$, which satisfies [3]

$$(q_t, v) = \tilde{E}(v(X_t)Z_t|F_t^y), \ v \in L^\infty(R^d) \tag{6}$$

so that q_t has the meaning of an "unnormalized conditional density".

3. THE MILD FORM OF THE ZAKAI EQUATION

In this section the variational framework is replaced by a purely Hilbert space one in order to utilize explicitly C_0-semigroups of operators on $L^2(R^d)$. First of all note that the hypotheses i),ii),iii) implies that there exists $\gamma > 0$ and λ real such that

$$\lambda|v|^2 - \langle v, Lv \rangle \geq \gamma\|v\|^2, \quad v \in H^1(R^d) \tag{7}$$

where $\|\cdot\|$ is the norm in $H^1(R^d)$ [10]. Under such condition it is well known that the restriction of L to $\{u \in H^1(R^d): |\langle v,u\rangle| \leq K_u|v|\}$ is the infinitesimal generator L of a C_o-semigroup S of bounded operators on $L^2(R^n)$ [7], such that

$$|S(t)| \leq e^{\lambda t}. \tag{8}$$

THEOREM 1. The process $\{q_t\}$, unique solution of the equation (5) in $Z(0,T;L^2(R^d))$ is such that for $t \in [0,T]$

$$(q_t,v) = (S^*(t)q_o,v) + \int_o^t (q_\tau, hS(t-\tau)v)dY_\tau \quad , \quad v \in L^2(R^d) \tag{9}$$

PROOF. The main step of the proof is in proving the following Itô formula: for any $f \in C^1([0,T]; D(L))$, where D(L) is endowed with the Hilbert structure of the graph norm [10], it results

$$d(q_t,f(t)) = (q_t,Lf(t))dt + (q_t,hf(t))dY_t + (q_t,\dot{f}(t))dt \tag{10}$$

Such a formula is obtained by expanding f(t) along an orthonormal basis for D(L), applying the Itô formula to each component and going to the limit in mean square.

Applying (10) to the function $f(\tau) = S(t-\tau)v$, $v \in D(L^2)$ and integrating in $\tau \in [0,t]$ it results

$$(q_t,v) = (S^*(t)q_o,v) + \int_o^t (q_\tau, hS(t-\tau)v)dY_\tau$$

and being $D(L^2)$ dense in $L^2(R^n)$, equation (9) is obtained by a mean square limit. □

Equation (9) can be expressed in a mild form as soon as a suitable definition of stochastic integral is given.

Let H be a real separable Hilbert space and, as in the previous section, let Z(0,T;H) be the space of H-valued adapted process on some $(\bar{\Omega}, G, Q, \{G_t\})$, which is Banach with respect to the norm

$$|||\, z \,||| = \sup_{t\in[0,T]} (E_Q|z_t|^2)^{1/2} \tag{11}$$

Now let $\{W_t, G_t\}$ be a standard scalar Wiener process on $\bar{\Omega}$. Let $\{\phi_i\}$ be any orthonormal basis in H. Then it is easily seen that for any z in Z(0,T;H) the sequence $\{I_t^{(n)}\} \subset Z(0,T;H)$ defined by

$$I_t^{(n)} = \sum_{i=1}^n \left[\int_o^t (z_s,\phi_i)dW_s\right]\phi_i \quad , \quad t \in [0,T] \tag{12}$$

converges to a limit in Z(0,T;H) which is independent of the chosen basis, which is denoted by $\int_0^t z_s dW_s$. Moreover

$$(\int_0^t z_s dW_s, \phi) = \int_0^t (z_s, \phi) dW_s , \quad t \in [0,T], \ \phi \in H. \tag{13}$$

For more about such kind of integrals see [12].

With such a definition the Zakai equation (9) can be rewritten as a mild equation in $Z(0,T;L^2(R^d))$

$$q_t = S^*(t) q_o + \int_0^t S^*(t-\tau) h q_\tau dY_\tau \tag{14}$$

Equations of this kind has been largely studied in the literature: in particular the uniqueness of the solution in $Z(0,T;L^2(R^d))$ is well known [13].

4. A CONTINUITY THEOREM

In order to study Galerkin approximations for equations of the Zakai type the main tool is a continuity theorem, whose lenghty proof is contained in [9]. The result is stated for a general Hilbert space H.

Let S_λ be the space of all C_o-semigroups of linear bounded operators on H such that the estimate (8) holds, and let B be an arbitrary linear bounded operator on H. Then the map $\chi : S_\lambda \to Z(0,T;H)$, which associates to each G in S_λ the unique solution in $Z(0,T;H)$ of the mild equation

$$\chi_t = G(t)\chi_o + \int_0^t G(t-\tau) B \chi_\tau dW_\tau , \quad t \in [0,T] \tag{15}$$

for a fixed $\chi_o \in H$, is well defined.

THEOREM 2. The map χ is continuous when S_λ is endowed with the topology of uniform strong convergence on $[0,T]$. □

By consequence, if $G_n \in S_\lambda$ for each integer n, then for n going to the infinity

$$\sup_{t\in[0,T]} |(G(t)-G_n(t))x| \to 0, \ \forall x \in H \Rightarrow |||\chi(G_n)-\chi(G)||| \to 0.$$

REMARK. The main consequence of theorem 2 for our purposes is that, once an approximation scheme is considered for the stochastic bilinear equation in mild form (15), its mean square convergence is guaranteed by the convergence for the linear deterministic equation obtained for B = 0 with any initial condition. Conversely this condition is also necessary, as it results by taking expectations in equation (15).

The problem of convergence in S_λ is completely characterized by the following result, known as the Trotter-Kato theorem [14].

THEOREM 3. Let $\{G_n\}$ be a sequence of C_o-semigroups in S_λ with their associated infinitesimal generators $\{A_n\}$. Then $\{G_n\}$ converges in S_λ to G, whose generator is A, if and only if there exists $\mu > \lambda$ such that

$$\lim_{n\to\infty} (\mu I-A_n)^{-1}x = (\mu I-A)^{-1}x \ , \ \forall x \in H \tag{16}$$

□

5. GALERKIN APPROXIMATIONS

For the particular case of the Zakai equation (14) Galerkin approximations are considered and using the theorems of the previous section their convergence is established.

Let $\{v_n\}$ be a sequence in $D(L^*)$ and define $V_n=\text{span}\{v_1,\dots,v_n\}$. Let Π_n and P_n be the orthogonal projections of $H^1(R^d)$ and $L^2(R^d)$ on V_n respectively. Suppose that $\{\Pi_n\}$ converges strongly to the identity in $H^1(R^d)$. By consequence, the same is true for $\{P_n\}$ in $L^2(R^d)$. For, let $h \neq 0$ in $L^2(R^d)$ be orthogonal to V_n for each n; then denoting by $((\cdot,\cdot))$ the scalar product in $H^1(R^d)$ and by Δ the Laplacian operator, one has: $(((I-\Delta)^{-1}h,v_n))=(h,v_n)=0$, $n=1,2,\dots$, from which $h = 0$, because $(I-\Delta)^{-1}$ is one-to-one.

Let us consider the following equation in the space V_n, where $\bar{x} \in L_2(R^d)$

$$\frac{d}{dt}(u^{(n)}(t),v)=(L^*u^{(n)}(t),v), \ v\in V_n, \ u^{(n)}(0)=P_n\bar{x}, \ t \in [0,T] \tag{17}$$

which has the unique solution $u^{(n)}(t) = \exp(P_nL^*P_nt)\bar{x}$, where by (7)

$$|e^{P_nL^*P_nt}| \le e^{\bar{\lambda}t} \ , \ \bar{\lambda} = \max(0,\lambda). \tag{18}$$

Equation (17) represents the Galerkin method applied to the following linear evolution equation known as the Fokker-Planck equation

$$\frac{d}{dt}(u(t),v) = \langle u(t), Lv \rangle , \ u(0) = \bar{x}, \ v \in H^1(R^d), \ t \in [0,T] \tag{19}$$

Such a method is easily extended to the case of the equation (5), yielding the equation in $Z(0,T;V_n) \subset Z(0,T;L^2(R^d))$

$$d(q_t^{(n)},v)=(L^*q_t^{(n)},v)dt+(hq_t^{(n)},v)dY_t, \ v\in V_n \ , \ q_o^{(n)}= P_nq_o \tag{20}$$

which, by theorem 1, is equivalent to

$$q_t^{(n)} = e^{P_nL^*P_nt} q_o + \int_0^t e^{P_nL^*P_n(t-\tau)} hq_\tau^{(n)} dY_\tau . \tag{21}$$

At this point the main result can be obtained.

THEOREM 4. The process $\{q_t^{(n)}\}$ converges to the solution q_t of the Zakai equation in the space $Z(0,T;L^2(R^d))$.

PROOF. By theorem 2, because of the bound (18), it suffices to prove that, for any $\bar{x} \in L^2(R^d)$, the solution $u^{(n)}$ of (17) converges to the solution u of (19). By Trotter-Kato theorem it is enough to prove that for some $\mu > \bar{\lambda}$ the solution y_n of the following elliptic equations

$$\mu y_n - P_nL^*P_ny_n = x \quad , \quad x \in L^2(R^d) \tag{22}$$

Converges for $n \to \infty$ to the solution y of

$$\mu y - L^*y = x \ . \tag{23}$$

In view of (8) the solution y_n of (22) is unique and decomposing it in the form

$$y_n = y_n' + y_n'' \ , \ y_n' \in V_n \ , \ y_n'' \in V_n^\perp$$

then the equation (22) is equivalent to the following system

$$\mu y_n' - P_nL^*P_ny_n' = P_nx \ , \quad y_n'' = \mu^{-1}(I-P_n)x \tag{24}$$

which immediately shows that y_n'' goes to zero in $L^2(R^d)$. By (23) and (24) it is obtained

$$(\mu(y_n' - y) - L^*(y_n' - y), v) = 0 \ , \ \forall v \in V_n$$

from which

$$\alpha\|y_n'-y\|^2 \leq \mu|y_n'-y|^2 - (y_n'-y, L(y_n'-y)) = ((\mu I-L^*)(y_n'-y), \Pi_ny-y) +$$

$$+((\mu I-L^*)(y_n'-y), y_n'-\Pi_ny) = \mu(y_n'-y, \Pi_ny-y) - \langle y_n'-y, \mathcal{L}(\Pi_ny-y_n')\rangle \leq$$

$$\leq (\mu + \|\mathcal{L}\|)\|y_n' - y\|\|\Pi_ny - y\|$$

which shows that

$$|y_n - y| \leq |y_n' - y| + |y_n''| \leq \|y_n' - y\| + |y_n''| \to 0$$

□

The previous result is easily generalizable to the case in which V_n is no longer contained in $D(L^*)$ but the generators cannot be expressed in the form P_nAP_n. This is more understandable if the equation (20) is expressed coordinate wise.

If the Gram-Schmidt procedure is applied on $\{v_n\}$ an orthonormal basis $\{w_n\}$ is obtained for $L^2(R^d)$. The solution of equation (20) can be written as

$$q_t^{(n)} = \sum_{i=1}^{n} \xi_t^{(n,i)} w_i$$

where the coefficients $\xi_t^{(n,i)}$ are obtained by solving in $[0,T]$ the following stochastic ODE in R^n

$$d\xi_t^{(n,i)} = \sum_{j=1}^{n} a_{ij}\xi_t^{(n,j)} + \sum_{j=1}^{n} b_{ij}\xi_t^{(n,j)} dY_t \qquad (25)$$

$$\xi_o^{(n,i)} = (p_o, w_i),\ a_{ij} = (L^* w_j, w_i),\ b_{ij} = (w_j, hw_i),\ i,j=1,\dots,n.$$

An interesting feature which is retained from the deterministic case is that, when the dimension of the approximating system (25) is increased, the preexisting coefficients are not reset. Equation (25) is really a finite-dimensional approximate filter for the system (1), as it will be pointed out in the following section.

6. FINAL REMARKS

Up to now the convergence of Galerkin approximations $\{q_t^{(n)}\}$ to the "unnormalized conditional density" q_t has been proved in the space $Z(0,T;L^2(R^d))$ on the "artificial" probability space $(\Omega, F, \tilde{P})$. It turns out that only epsilon is lost with reference to the original measure P. In fact, the process $\{Z_t\}$ defined by (4) satisfies the stochastic differential equation on $(\Omega, F, \tilde{P})$

$$dZ_t = Z_t h(X_t) dY_t\ ,\quad Z_o = 1,\quad t \in [0,T],$$

so that all its moments are finite [15].

This implies that for any arbitrarily small $\varepsilon > 0$

$$E(|q_t - q_t^{(n)}|^{2-\varepsilon}) = \tilde{E}(|q_t - q_t^{(n)}|^{2-\varepsilon}\ Z_T) \le (E|Z_T|^{\frac{2}{\varepsilon}})^{\frac{\varepsilon}{2}} |\!|\!| q - q^{(n)} |\!|\!|^{2-\varepsilon}$$

which shows the uniform convergence in $L^{2-\varepsilon}(L^2(R^d))$ of the sequence of Galerkin approximations.

It is well known that the solution of the Zakai equation (5) has a version which is continuous with respect to the paths of Y_t in the uniform topology [4].

This is no longer true for equation (25) when m-valued observations are considered: ($m \geq 2$), because equation (25) becomes

$$d\xi_t^{(n,i)} = \sum_{j=1}^{n} a_{ij}\xi_t^{(n,i)} + \sum_{k=1}^{m} \sum_{j=1}^{n} b_{ij}^k \xi_t^{(n,i)} dY_t^k \tag{26}$$

$$\xi_o^{(n,i)} = (p_o, w_i), \quad a_{ij} = \langle w_j, Lw_i \rangle \quad b_{ij}^k = (h_k w_j, w_i). \tag{27}$$

Also in this case the convergence in $Z(0,T;L^2(R^d))$ of $q_t^{(n)} = \sum_{i=1}^{n} \xi_t^{(n,i)} w_i$ to the solution q_t of the general Zakai equation

$$d(q_t, v) = \langle q_t, Lv \rangle dt + \sum_{k=1}^{m} (q_t, h_k v) dY_t^k, \quad v \in H^1(R^d), \; q_o = p_o$$

can be proved with slight modifications of the previous proofs, even without the hypothesis that $w_i \in D(L^*)$ (see the definition of a_{ij}'s in (27)). The pathwise unsolvability of (26) depends on the fact that the m matrices $\{b_{ij}^k \; i,j = 1,\dots,n\}$, $k = 1,\dots,m$, in general do not commute for any finite integer n [16], whereas the multiplication operators by h_k on $L^2(R^d)$ clearly do.

This is not a serions accident, in that the proposed approximations hold in mean square. For a practical point of view, here the main issue concerns the implementability of the equation (26). But this can be achieved by standard approximation schemes for stochastic ODE's which converge again uniformly in the mean square to the solution of (26), once this has been rewritten in Stratonovich form [17].

ACKNOWLEDGEMENTS

The authors wish to thank Prof. G. Da Prato for having suggested a shorter proof for Theorem 1.

REFERENCES

[1] M. ZAKAI: *On the Optimal Filtering of Diffusion Processes*, Z. Wahrschein. verw. Geb., 11, 1969, pp. 230-243.

[2] N.V. KRYLOV, B.L. ROZOVSKII: *On the Cauchy Problem for Linear Stochastic Partial Differential Equations*, Math. USSR Izvestija, 11, 1977, pp. 1267-1284.

[3] E. PARDOUX: *Stochastic Partial Differential Equations and Filtering of Diffusion Processes*, Stochastics, 3, 1979, pp. 127-167.

[4] J.M.C. CLARK: *The Design of Robust Approximations to the Stochastic Differential Equations of Nonlinear Filtering*, in J.K. Skwirzynski ed.,Communication Systems and Random Process theory, Sijthoff and Noordhoff, 1978, pp. 721-734.

[5] H.J. KUSHNER: *A Robust Discrete-State Approximation to the Optimal Nonlinear Filter for a Diffusion*, Stochastics, 3, 1979, pp. 75-83.

[6] G.B. DI MASI, W.J.RUNGGALDIER : *Continuous-time Approximations for the Nonlinear Filtering Problem*, Appl. Math. and Opt., 7, 1981, pp. 233-245.

[7] R.E. SHOWALTER: *Hilbert Space Methods for Partial Differential Equations*, Pitman, San Francisco, 1979.

[8] J.T. ODEN, J.N. REDDY: *An Introduction to the Mathematical Theory of Finite Elements*, Wiley-Interscience, New York, 1976.

[9] A. GERMANI, M. PICCIONI: *Finite Dimensional Approximation for Stochastic Bilinear Differential Equations on Hilbert Spaces*, Report R. 61, IASI-CNR, 1983.

[10] J.L. LIONS: *Equations Differentielles-Operationnelles*, Springer Verlag, Berlin, 1961.

[11] G. KALLIANPUR: *Stochastic Filtering Theory*, Springer Verlag, New York, 1980.

[12] M. METIVIER, J. PELLAUMAIL: *Stochastic Integration*, Academic Press, New York, 1980.

[13] G. DA PRATO, M. IANNELLI, L. TUBARO: *Linear Stochastic Differential Equations in Hilbert Spaces*, Rend. Acc. Naz. Lincei, LXIV, 1978, pp. 22-29.

[14] E.B. DAVIES: *One-parameter Semigroups*, Academic Press, London, 1980.

[15] A. FRIEDMAN: *Stochastic Differential Equations and Applications*, Vol. 1, Academic Press, New York, 1975.

[16] H. SUSSMANN: *On the Gap Between Deterministic and Stochastic Ordinary Differential Equations*, Ann. of Prob., 1978, pp. 19-41.

[17] N. IKEDA, S. WATANABE: *Stochastic Differential Equations and Diffusion Processes*, North-Holland, Amsterdam, 1981.

SOME SINGULAR CONTROL PROBLEM WITH LONG TERM AVERAGE CRITERION

J.L. MENALDI
Departments of Mathematics
Wayne State University
Detroit, Michigan 48202
U.S.A.

M. ROBIN
INRIA - Domaine de Voluceau
Rocquencourt - B.P. 105
78153 Le Chesnay Cedex
France

Abstract : in this paper we consider an optimal control problem for a diffusion process $y(t) = y_x^v$ solution of $dy(t) = g(y)dt + \sigma(y)dw_t + dv_t$, where v_t is an increasing positive adapted process, with the long term average cost

$$J(v) = \liminf_{T\uparrow\infty} \frac{1}{T} E\int_0^T f(y_x^v(t))dt.$$

The paper gives, for one dimensional processes, existence result for an optimal control in relation with a reflected diffusion process as in Karatzas [16], and characterization of the optimal cost. Moreover, the asymptotic analysis of the discounted cost problem is carried out.

I. INTRODUCTION

This paper consider the following stochastic control problem : v_t being an increasing adapted process, the state process $y_x(t)$ is the solution of a stochastic differential equation

$$dy_x(t) = g(y_x)dt + \sigma(y_x)dw_t + dv_t \quad , \quad y_x(o) = x \;, \qquad (1.1)$$

and we want to minimize the long term average cost, (for a given f)

$$J(v) = \liminf_{T\uparrow\infty} \frac{1}{T} E\int_0^T f(y_x(t))dt \;. \qquad (1.2)$$

That kind of control problems is motivated by several applications, for example :

- the asymptotic problem of impulse control or continuous control, when the cost of control vanishes (cfr. Menaldi et al.[18]) ;
- "monotone follower" problems like in Benes et al [5], Karatzas [16];
- impulse control problem with linear control cost (see [19];
- multiechelon inventory problems (cfr. Menaldi-Rofman [20]).

Such situations have been studied mainly in the context of finite horizon ([5], [8], [9], [10], [11], [5], [16]) of infinite horizon discounted criteria. On the other hand, long term average criterion for continuous time Markov processes have been studied in many references (cfr. for example Doshi [12], Lasry [17] , Robin [22] and also [23] for a survey).

The monotone follower problem with long term average criterion has been studied by Karatzas [16] for the Wiener process (cfr. also [2], [3], [13]).

2. ONE DIMENSIONAL DIFFUSION PROCESS

2.1. Statement of the problem

Let us consider the one dimensional diffusion process,

$$dy_x(t) = g(y_x(t))dt + \sigma(y_x(t))dw_t + dv_t \quad , \quad y_x(o) = x \in \mathbb{R} \quad (2.1)$$

we will assume that

g, σ are Lipschitz continuous functions from $\mathbb{R}$ into itself, (2.2)

$$g(x) \le -\gamma_1 - \gamma_2 x^+ \quad , \quad \gamma_1, \gamma_2 > 0 \quad (2.3)$$

$$\sigma^2(x) \ge \beta > 0 \quad , \quad \sigma \text{ bounded,} \quad (2.4)$$

$v_t \in \mathcal{V}$ set of positive, adapted process, having locally bounded variation . (2.5)

On the other hand, we are given a function f such that :

$$f \in C^1(\mathbb{R}), \quad 0 \le f \le k(1 + x^2), \qquad f(x) \to +\infty \text{ as } |x| \to \infty \quad (2.6)$$

and $\exists$ R > 0 such that $\forall x \ge R$ then $f'(x) \ge \delta > 0$. We also define $\mu = (1 + x^2)^{-1}$, and $L_\mu^\infty\{h | \mu h \in L^\infty(\mathbb{R})\}$. We define the α- discounted problem as follows : to minimize, over $v \in \mathcal{V}$

$$J_x^\alpha(v) = E\int_0^\infty e^{-\alpha t} f(y_x(t))dt \ . \quad (2.7)$$

Denoting $u_\alpha(x) = \inf_{v \in \mathcal{V}} J_x^\alpha(v)$, we recall a result of Menaldi-Robin [19]for $\alpha > 0$.

<u>THEOREM 2.1</u> : <u>under the assumptions</u> (2.1), (2.2), (2.4), (2.5), (2.6), <u>then</u> u_α <u>is the unique solution of</u>

$$-\frac{1}{2}\sigma^2 u_\alpha'' - g u_\alpha' + \alpha u_\alpha \le f \quad , \quad u_\alpha' \ge 0, \qquad (Lu_\alpha + \alpha u_\alpha - f)u_\alpha' = 0 \quad , \quad u_\alpha \in W_\mu^{2,\infty} \quad (2.8)$$

<u>where</u> $Lu_\alpha = -\frac{1}{2}\sigma^2 u_\alpha'' - g u_\alpha'$, <u>and</u> $W_\mu^{2,\infty} = \{h | h, h', h'' \in L_\mu^\infty(\mathbb{R})\}$. <u>Moreover this solution</u> $u_\alpha \in C^2(\mathbb{R})$. □

The limit problem when $\alpha \to 0$, if we assume that $\lim_{\alpha\to 0} \alpha u_\alpha(x) = \lambda$ a constant, and $\lim_{\alpha\to 0} u_\alpha(x) - u_\alpha(0) = w(x)$, should be

$$Lw + \lambda \le f \quad , \quad w' \ge 0 \quad , \quad (Lw + \lambda - f)w' = 0 \tag{2.9}$$

we are going first to use a direct approach for (2.9) and next to study the convergence of (2.8) when $\alpha \to 0$.

2.2 Direct study of the undiscounted inequality

Defining $h = w'$, (2.9) can be written as

$$\begin{aligned} &Ah = -\frac{\sigma^2}{2} h' - gh \le f - \lambda \quad , \quad h \ge 0, \\ &(Ah + \lambda - f)h = 0 \ . \end{aligned} \tag{2.10}$$

Let us also define

$$\mathcal{V}_1 = \{v \in \mathcal{V} \mid E(y_x^v(t))^2 \le \text{constant}, \ \forall t \ge 0 \} \tag{2.11}$$

THEOREM 2.2 : under the assumptions (2.1) to (2.6) and if f is convex then,

(i) there exists a unique solution $(\lambda, w) \in \mathbb{R}^+ \times C^2$ of (2.9) such that

$$0 \le w(x) \le k(1 + x^2) \tag{2.12}$$

(ii) $\lambda = \inf_{v \in \mathcal{V}_1} J(v)$, where $J(v) = \liminf_{T \uparrow \infty} \frac{1}{T} E \int_0^T f(y_x^v(t))dt$.

(iii) there exists $\hat{v} \in \mathcal{V}_1$ such that $\lambda = J(\hat{v})$

(iv) there exists $\bar{x} \in \mathbb{R}$ such that

$$Lw + \lambda = f \text{ for } x > \bar{x} \ , \ w(x) = w(\bar{x}) \text{ elsewhere.} \tag{2.13}$$

The proof will consist of several lemmas.

Lemma 2.1 : under the assumptions of Theorem 2.2 (i) there exists a solution $(\lambda, h) \in \mathbb{R}^+ \times C^1(\mathbb{R})$ of (2.10), such that $h \le k(1 + x^+)$ for some constant k, (ii) there exists $\bar{x}$ such that $Ah + \lambda = f$ for $x \ge \bar{x}$, and $h(x) = 0$ otherwise.

Proof : let R be a fixed number and let us consider the problem : to find $(\lambda_R, q_R) \in \mathbb{R}^+ \times C^1_\mu$ (*) such that

$$-\frac{\sigma^2}{2} q' - gq = f - \lambda \quad , \quad x \in]R,\infty[\quad , \quad q(R) = 0 \ . \tag{2.14}$$

One can see that, for fixed λ, the only C^1_μ solution of $Aq = f - \lambda$, $x \in \mathbb{R}$ is, (with $\tilde{g} = -2g/\sigma^2$), $q(x) = \int_0^\infty \exp(-\int_0^s \tilde{g}(x+r)dr)\frac{2}{\sigma^2}(f-\lambda)(x+s)ds$.

In order to have $q_R(x)$ such that $q_R(R) = 0$, we must take

$$\lambda_R = \frac{\int_0^\infty \exp(-\int_0^s \tilde{g}(R+r)dr) \ \frac{2}{\sigma^2} f(R+s)ds}{\int_0^\infty \exp(-\int_0^s \tilde{g}(R+r)dr) \ \frac{2}{\sigma^2} \ (R+s)ds} \tag{2.15}$$

(*) $C^1_\mu = \{h \mid \mu h, \mu h' \in C^o\}$

and then

$$q_R(x) = \int_0^\infty \exp(-\int_0^s \tilde{g}(x+r)dr)\, \frac{2}{\sigma^2}(f-\lambda_R)(x+s)ds. \qquad (2.16)$$

Then one can see that λ_R given by (2.15) is continuous w.r.t R and that,

$$\lambda_R > k'.\int_0^\infty e^{-ks} f(R+s)ds.$$

Therefore, there exists $\bar{R}$ such that $\lambda_{\bar{R}} = \inf_R \lambda_R$.

We can also check that

$$\frac{d}{dR}\lambda_R = \frac{2}{\sigma^2(R)F(R)}(\lambda_R - f(R)) \qquad (2.17)$$

(where R is the denominator in (2.15)) and therefore

$$\lambda_{\bar{R}} = f(\bar{R}), \qquad (2.18)$$

and since f is convex, we have $\lambda_{\bar{R}} = f(R)$ at most in two points $R_1 \le R_2$. Then from (2.17), we deduce that $\bar{R} = R_1$. It is not difficult to check that

$$q_{\bar{R}}(x) \ge 0 \quad , \quad \forall x \in [\bar{R},\infty[. \qquad (2.19)$$

Now collecting (2.14), (2.18), (2.19) and defining

$$h(x) = q_{\bar{R}}(x) \quad , \quad \text{for } x \in [\bar{R},\infty[\quad , \quad h(x) = 0 \text{ elsewhere}, \qquad (2.20)$$

then h satisfies the equation on $[\bar{R},\infty[$, $h \ge 0$ everywhere, and $h = 0$ on $]-\infty,\bar{R}[$, moreover, from the equation, thanks to (2.18),

$$\lim_{x\to R+} h'(x) = 0$$

therefore $h \in C^1_\mu(\mathbb{R})$, $h'(\bar{R}) = 0$ and since $f(x)-\lambda \ge 0$ for $x \le \bar{R}$, we have $Ah \le f - \lambda$, $\forall x$ and therefore (λ,h) is solution of (2.10).
Let us now prove that $h \le k(1+x^+)$. From (2.16), since σ is bounded, and $\lambda \ge 0$, we clearly have, for <u>$x \ge 0$</u> :

$$h(x) \le k_1\int_0^\infty \exp(-\int_0^s(\delta_1+\delta_2(x+r))dr)(1+(x+s)^2)ds$$

for some $k_1,\delta_1, \delta_2 > 0$. Then elementary computations give $h(x) \le k_2(1+x)$ for some $k_2 > 0$, $\forall x \ge 0$, which yields $h(x) \le k(1+x^+)$ for some k large enough. □

<u>REMARK 2.1</u>. In order to obtain the existence of (λ,h), $g(x) \le -\gamma_1$, is sufficient. But (2.3) is needed to get $h(x) \le k(1+|x|)$. □

<u>Lemma 2.2.</u> λ <u>being fixed</u>, (2.10) <u>has at most one solution</u> h.

Proof : let $H = -h$, for fixed λ, H is solution of

$$H' - \tilde{g} H \leq F_\lambda,\ H \leq 0\ ,\ (H' - \tilde{g} H - F_\lambda)H = 0\ , \tag{2.21}$$

where $F_\lambda = \frac{2}{\sigma^2}(f-\lambda)$. H can be interpreted as the optimal cost of a simple deterministic stopping problem, namely :

$$H(x) = \inf_T \int_0^T \exp\left(\int_0^r \tilde{g}(x-s)ds\right)F_\lambda(x-r)dr. \tag{2.22}$$

And then, the uniqueness follow from the usual theory of optimal stopping. □

REMARK 2.2. For $\lambda = \lambda_{\bar{R}}$ as in Lemma 2.1, one could check that $H(x)= 0$ $\forall x \leq \bar{R}$ and $H(x) < \bar{R}$. □

Lemma 2.3. : let $w(x) = \int_{\bar{x}}^x h(r)dr$, then $(\lambda,w) \in \mathbb{R}^+ \times C^2$ is the solution of (2.9) satisfying (2.12). Moreover $\lambda = \inf(J(v),\ v \in \mathcal{V}_1)$ and there exists an optimal control.

Proof : clearly, (λ,w) satisfies the required properties. Applying Ito's formula to $w(y_x^v(t))$, for any v, we have

$$E_w(y_x^v(t) = w(x) + E\int_0^t\left(\frac{\sigma^2}{2}w'' + gw'\right)(y_x^v(s))ds + E\int_0^t w'(y_x^v(s^-))dv_s^c$$
$$+ E\sum_{s\leq t} (w(y_w^v(s)) - w(y_x^v(s^-)))$$

(cfr [21]), where v^c is the continuous part of v and where the last term involves only the jumps of v. Since $w'\geq 0$, and v_t is increasing the two last terms are non negative, and using (2.9) we obtain

$$\lambda \leq \frac{1}{T} E\int_0^T f(y_x^v(t))dt - \frac{w(x)}{T} + \frac{1}{T} Ew(y_x^v(T)). \tag{2.23}$$

If $v \in \mathcal{V}_1$, then the last term in (2.23) goes to zero and therefore

$$\lambda \leq J(v)\quad ,\quad \forall v \in \mathcal{V}_1\ . \tag{2.24}$$

Now let $\hat{v}$ the control defined as follows :

$$\hat{v}_t = \xi_t \text{ if } x \geq \bar{x}\ , \tag{2.25}$$

where ξ_t is the increasing process associated with the reflected diffusion with coefficient (g,σ), on the set $[\bar{x},\infty[$ (see [7], [24]): and

$$d\hat{v}_t = (-x + \bar{x})\delta_o(t)dt + d\xi_t \text{ if } x < \bar{x}\ , \tag{2.26}$$

meaning that there is an immediate jump at $\bar{R}$ and then $\hat{v}_t$ is such that $y_x(t)$ is reflected on $[\bar{R},\infty[$, then we show that $\hat{v} \in \mathcal{V}_1$.

Indeed applying Ito's formula to $z^2(t) = (\hat{y}_x(t) - \bar{x})^2$, with $\hat{y}_x(t) = y_x^{\hat{v}}(t)$,

$$E\, z^2(t) = z^2(0) + E\int_0^t[\sigma^2 + 2g(\hat{y}_x(s))]\, z(s)ds.$$

But since $\hat{y}_x(t) \geq \bar{x}$, one has, a.s.,

$$g(\hat{y}_x(t)) \leq -\gamma_1 - \gamma_2(\hat{y}_x(t))^+ = -\gamma_1 - \gamma_2(\hat{y}_x(t) - \bar{x}) + \gamma_2\bar{x} \ .$$

Then, taking ε such that $\beta_1 = 2\gamma_2 - \varepsilon > 0$, we obtain

$$E\ z^2(t) + \beta_1 \int_0^t z^2(s)ds \leq z^2(0) + \beta_2 t,$$

from which we get $E\ z^2(t) \leq K \quad \forall t \geq 0$, which gives $v \in \mathcal{V}_1$.

Then, since for $\hat{v}$, we have the equality in (2.23), we obtain :

$$\lambda = J(\hat{v}) \quad \square \tag{2.27}$$

Collecting the previous Lemmas, the Theorem is proved.

REMARK 2.3. The relationship between problem similar to (2.8) and the reflected diffusion process was shown in several works for the Wiener process in [5], [15] , and for an undiscounted problem for the Wiener process in [16] . □

2.3 Asymptotic study of the discounted problem

The main result of this section is the following

THEOREM 2.3 : *under the assumptions* (2.1) *to* (2.6) *then*

$$\lim_{\alpha\to 0} \alpha u_\alpha(x) = \lambda \tag{2.28}$$

$$\lim_{\alpha\to 0} u_\alpha(x) - u_\alpha(0) = w(x) \tag{2.29}$$

where (λ,w) *is the solution of* (2.9) *as shown in Theorem* 2.2.

Proof : we first show that

$$\alpha u_\alpha(x) \leq k \text{ for some } k > 0 \ . \tag{2.30}$$

Let v_o the control defined as $v_o(t) = \xi_t$ if the initial state $x \in [0,+\infty[$, and $dv(t) = (-x)\delta_o(t)dt + d\xi_t$ if $x < 0$. This means as previously that $y_x^v(t) = \bar{y}_{x_{vo}}(t)$, if $\bar{y}_x(t)$ is the reflected diffusion process on $[0,\infty[$, $(x \in \mathbb{R}^+)$. We obviously have

$$\alpha u_\alpha(x) \leq \alpha J_x^\alpha(v_o) \leq \alpha\ E \int_0^\infty e^{-\alpha t}\, k(1 + \bar{y}^2_{x_{vo}}(t))dt. \tag{2.31}$$

And applying Ito's formula to $z^2(t) = \bar{y}^2_{x_{vo}}(t)$, we obtain $Ez^2(t) \leq K$, which gives (2.30).

Now, as shown in Menaldi-Robin [19], $\forall\alpha > 0$, there exists $x_\alpha \in \mathbb{R}$ such that

$$Lu_\alpha + \alpha u_\alpha = f \text{ , on }]x_\alpha, +\infty[\text{ , } u'_\alpha(x) = 0 \text{ elsewhere.} \tag{2.32}$$

But the equation $Lu_\alpha + \alpha u_\alpha = f$ is also satisfied at x_α (u_α is C^2) and

therefore we have $\alpha u_\alpha(x_\alpha) = f(x_\alpha)$. Moreover $\alpha u_\alpha(x) \le f(x)$, $\forall x \le x_\alpha$ giving $f(x) \ge f(x_\alpha)$ on $]-\infty, x_\alpha]$, and so $f'(x_\alpha) \le 0$, implying, since f is convex $x_\alpha \le \bar{y} = \arg\min f$, and $f(x_\alpha) = \alpha u_\alpha(x_\alpha) \le \alpha u_\alpha(\bar{y}) \le C(1+\bar{y}^2)$.

Therefore, we deduce that x_α stays in a compact subset of $\mathbb{R}$ for any $\alpha \le \alpha_o$.

Moreover, from (2.32), we have defining $h_\alpha = u'_\alpha$

$$\begin{aligned} &-h'_\alpha + \tilde{g}\, h_\alpha = \frac{2}{\sigma^2}(f - \alpha u_\alpha) = F_\alpha \text{ ,} \\ &h_\alpha(x) = \int_0^\infty \exp(-\int_0^\infty \exp(-\int_0^y \tilde{g}(x+s)ds)F_\alpha(x+s)ds \end{aligned} \tag{2.33}$$

and from (2.30) and some computations like in Lemma 2.2., we have,

$$0 \le h_\alpha(x) \le k(1+x^+) \text{, uniformly w.r.t. } \alpha \text{, for some } k > 0. \tag{2.34}$$

Therefore, using (2.33) on $[x_\alpha, +\infty]$, and $h'_\alpha = 0$ on $]-\infty, x_\alpha[$, we obtain

$$|h'_\alpha| \le k'(1 + x^2) \text{ unformly w.r.t. } \alpha. \tag{2.35}$$

Then defining $w_\alpha(x) = u_\alpha(x) - \min u_\alpha$, we deduce from (2.30) and (2.34), taking in account that $x_\alpha = \arg\min u_\alpha$ and that x_α stays in a compact, $\lim_{n\to\infty} \alpha_n u_{\alpha_n}(x) = \bar{\lambda}$ (a constant) for a subsequence $\alpha_n \to 0$.

Moreover, collecting (2.30), (2.34), (2.35), w_α is uniformly bounded in $W^{2,\infty}_\mu$ and therefore, one can take the limit in

$$Lw_\alpha + \alpha u_\alpha \le f,\ w'_\alpha \ge 0 \text{ , } (Lw_\alpha + \alpha u_\alpha - f)\, w'_\alpha = 0 \text{ ,}$$

(like in Tarres [25] for instance), in order to obtain that $\bar{w}(x) = \lim w_\alpha(x)$ satisfies (2.9).

Then we conclude that $(\bar{\lambda}, \bar{w}) = (\lambda, w)$ from the uniqueness result of § 2.2.

□

REFERNCES

[1] E.N. BARRON and R. JENSEN, Optimal Control Problems with No Turining Back, J. Diff. Equations, 36 (1980), pp. 223-248.

[2] J.A. BATHER, A Diffusion Model for the Control of a Dam, J. Appl. Prob., 5 (1968), pp. 55-71.

[3] J.A. BATHER and H. CHERNOFF, Sequential Decisions in the Control of a Spaceship, Proc. 5th Berkeley Symp. on Mathematical Statistics and Probability, Berkeley, University of California Press, 1967, Vol.3, pp. 181-207.

[4] J.A. BATHER and H. CHERNOFF, Sequential Decisions in the Control of a Spaceship(Finite Fuel), J.Appl. Prob., 4 (1967), pp. 584-604.

[5] V.E. BENES, L. A. SHEPP and H.S. WITSENHAUSEN, Some Solvable Stochastic Control Problems, Stochastics, 4 (1980), pp. 39-83.

[6] A. BENSOUSSAN and J.L. LIONS, Applications des inéquations variationnelles en contrôle stochastique, Dunod, Paris, 1978.

[7] A. BENSOUSSAN and J.L. LIONS, Contrôle Impulsionnel et Inéquations quasi-variationnelles, Dunod, Paris, 1982.

[8] M.I. BORODOWSKI, A.S. BRATUS and F.L. CHERNOUSKO, Optimal Impulse Correction Under Random Perturbations, Appl. Math. Mech. (PMM), 39 (1975), pp. 797-805.

[9] A.S. BRATUS, Solution of Certain Optimal Correction Problems with error of Execution of the Control Action, Appl. Math. Mech. (PMM), 38 (1974), pp. 433-440.

[10] F.L. CHERNOUSKO, Optimum Correction Under Active Disturbances, Appl. Math. Mech. (PMM), 32 (1968), pp. 203-208.

[11] F.L. CHERNOUSKO, Self-Similar Solutions of the Bellman Equation for Optimal Correction of Random Disturbances, Appl. Math. Mech. (PMM), 35 (1971), pp. 333-342.

[12] T.B. DOSHI, Continuous Time of Markov Processes on an Arbitrary State Space : Average Return Criterion. Stoch. Proc. 4 (1976) pp. 55-77.

[13] M.J. FADDY, Optimal Control of Finite Dams : Continuous Output Procedure, Adv. Appl. Prob., 6 (1974), pp. 689-710.

[14] V.K. GORBUNOV, Minimax Impulsive Correction of Perturbations of a Linear Damped Oscillator, Appl. Math. Mech. (PMM), 40 (1976) pp. 252-259.

[15] I. KARATZAS, The Monotone Follower Problem in Stochastic Decision Theory, Appl. Math. Optim., 7 (1981) pp. 175-189.

[16] I. KARATZAS, A class of Singular Stochastic Control Problems, Adv. Appl. Prob.,

[17] J.M. LASRY, Contrôle stochastique ergodique. Thèse, Université de Paris IX, 1974.

[18] J.L. MENALDI, J.P. QUADRAT and E. ROFMAN, On the Role of the Impulse Fixed Cost in Stochastic Optimal Control : An Application to the Management of Energy Production, Lecture Note in Cont. and Inf. Sci.,.38 (1982), Springer-Verlag, New York, pp. 671-679.

[19] J.L. MENALDI and M. ROBIN, On some Cheap Control Problems for Diffusion Processes, Trans. Am. Math. Soc., to appear. See also C.R. Acad. Sc. Paris, Série I, 294 (1982), pp. 541-544.

[20] J.L. MENALDI and E. ROFMAN, A Continuous Multi-Echelon Inventory Problem, Proc. 4th IFAC-IFIP Symp. on Information Control Problems in Manufacturing Technology, Gaithersburg, Maryland, USA, October 1982, pp. 41-49.

[21] P.A. MEYER, Cours sur les intégrales stochastiques, Lectures notes in Mathematics, 511 (1976), Springer-Verlag, Berlin, pp. 245-400.

[22] M. ROBIN, On some Impulse Control Problem with Long Run Average Cost. SIAM J. Control and Opt. 19 §1981) n° 3, pp. 333-358.

[23] M. ROBIN, Long Term Average Cost Control Problems for Continuous Time Markov Processes : a Survey, Acta Appl. Math. , to appear.

[24] D.W. STROOCK - S. WARADHAN, Diffusion Processes with Boundary Conditions - Comm. Prure Appl. Math., 24 (1971) pp. 147-225.

[25] R. TARRES, Comportement asymptotique d'un problème de contrôle stochastique. TR N° 8215, Unviersité de Paris IX, 1982.

ON ERGODIC CONTROL PROBLEMS ASSOCIATED WITH OPTIMAL MAINTENANCE AND INSPECTION

Ł. Stettner
Institute of Mathematics Polish Academy of Sciences
Warsaw, Poland

1. INTRODUCTION

Recently there has been a great deal of activity in optimal replacement theory. Below we will consider the optimal replacement model with costly observation and long run average cost criterion. This problem was investigated first for Poisson and Wiener processes in [8] and [2] respectively. The case with discounted functional was considered in [1] and [7].

Let $X=(\Omega, F_t, F, x_t, P_x)$ be a right continuous Feller Markov process with with values in locally compact separable space $(E, \mathcal{E})$, where $\Omega = D(R^+, E)$ is canonical space, F_t, F are universally completed σ fields of $F_t^o = \sigma\{x_s\ s \leq t\}$ and $F^o = \sigma\{x_s,\ s \geq 0\}$ respectively, and Feller property means that semigroup $\Phi(t)$, $\Phi(t) f(x) \equiv E_x\ f(x_t)$, transforms C_o (the space of continuous, vanishing at infinity functions) into C_o.

Suppose we know the true state of the process only at the special times called inspections. These times can be chosen and the observation is costly and if process is in state x is equal to $g(x) \geq k > 0$, $g \in C$. Knowing the real position of the process we can decide to continue with the motion or to innovate the process from any point in E. The shift from the state x to state y costs $c(x) + d(y)$, $0 \leq c, d \in C$.

To describe the evolution of our system we have to introduce the new canonical space $\tilde{\Omega} = \Omega^N$, $F. = F.^N$. Then our strategy V is the sequence of stopping times τ_i and random variables ξ_i , where

$$\tau_1 = \sigma = \text{const.}, \ldots, \quad \tau_{n+1} = \tau_n + \sigma_n(x^n_{\tau_n}), \ldots, \quad \xi_i = \zeta_i(x^i_{\tau_i})$$

$\sigma_1, \ldots, \sigma_n, \ldots, \zeta_1, \ldots, \zeta_n, \ldots,$ are $\mathcal{E}$ measurable, real or E valued respectively functions and for $\omega = (\omega_1, \ldots, \omega_n, \ldots)$, $x^i_t(\omega) = \omega_i(t)$.

Let $\eta_1 = \inf\{n: \ \xi_n \neq x^n_{\tau_n}\}$ and $\eta_n = \inf\{m \geq \eta_{n-1}: \xi_m \neq x^n_{\tau_m}\}$.

The strategy $V = (\nu_1, \nu_2, \ldots)$ where $\nu_i = (\tau_{\eta_{i-1}+1}, \tau_{\eta_{i-1}+2}, \ldots, \tau_{\eta_i}, \xi_{\eta_i})$

is the evolution of the system in i^{th} cycle, i.e. we inspect at times $\tau_{\eta_{i-1}+1}, \ldots, \tau_{\eta_i - 1}$ and after the inspection at time τ_{η_i} we shift the process to random point ξ_{η_i} and the evolution starts afresh.

Similarly as in paper [5] we construct probability measure P^V on $\tilde{\Omega}$. Let $G_{\tau_n} = \sigma\{F^{n+1}_{\tau_n-}, F^n_{\tau_n} \otimes \{\emptyset,\Omega\}\}$. The projections P^n of the measure P^V on the spaces $\Omega^n = \overset{n+1}{\underset{1}{\times}} \Omega$ have the following properties

$$P^0_x = P_x, \ldots, P^n_x = P^{n-1}_x \otimes \varepsilon_{\varphi_y} \quad \text{on } G_{\tau_n}$$

$$P^n_x(\Theta^{-1}_{n,\tau_n} B \mid G_{\tau_n}) = \varepsilon_{\varphi_{\omega_1(\tau_1)}}(\omega_1) \otimes \ldots \otimes \varepsilon_{\varphi_{\omega_{n-1}(\tau_{n-1})}}(\omega_{n-1}) \otimes P_{\xi_n(\omega_1,\ldots,\omega_n)}(B)$$

where $\varphi_y(t) = y$, $t \geq 0$ is the constant trajectory and $B \in F^{n+1}_\infty$.

The trajectory (y_t) of the controlled process X is of the form

$$y_t(\omega) = x^n_t(\omega_{n+1}) \text{ for } t \in [\tau_n, \tau_{n+1}[, \; \tau_0 = 0, \; y_{\tau_{n+1}}(\omega) = \xi_{n+1}(\omega_{n+1})$$

With each strategy V is associated the long run average cost functional

$$(1.1) \quad J_x(V) = \liminf_{t\uparrow\infty} t^{-1} E^V_x\{\int_0^t f(y_s)\,ds + \sum_{i=1}^{\infty} \chi_{\tau_i \leq t}\, g(x^{i-1}_{\tau_i}) + \sum_{i=1}^{\infty} \chi_{\tau_{\eta_i} \leq t} [c(x^{\eta_i-1}_{\tau_{\eta_i}}) + d(\xi_{\eta_i})]\}$$

2. GENERAL THEOREM

Let

$$(2.1) \quad \lambda = \inf_x \inf_V (E_x \tau_\eta)^{-1} E_x\{\int_0^{\tau_\eta} f(x_s)\,ds + \sum_{i=1}^{\eta} g(x_{\tau_i}) + c(x_{\tau_\eta}) + d(x)\}$$

Then using the technics of paper [9] we can prove the following general theorem

<u>Theorem 2.1.</u> Suppose there exists a continuous solution w of the equation

$$(2.2) \quad w(x) = \inf_{t \geq 0} E_x\{\int_0^t (f(x_s) - \lambda)\,ds + Mw(x_t)\}$$

where $Mw(x) = g(x) + \min\{w(x), c(x)\}$, with the boundary condition

$$(2.3) \quad \inf_x [w(x) + d(x)] = 0$$

Then $\inf_V J_x(V) = \lambda$ and the optimal inspections are of the form

$$(2.4) \quad \tau^*_1 = t^*(x) = \inf\{t \geq 0: w(x) \geq E_x\{\int_0^t (f(x_s) - \lambda)\,ds + Mw(x_t)\}\}, \ldots,$$
$$\tau^*_{n+1} = \tau^*_n + t^*(x_{\tau_n})$$

If x^ε is such that $w(x^\varepsilon) + d(x^\varepsilon) \leq \varepsilon$, then the strategy $V = (\tau^*_i, \xi^\varepsilon_i)$ where

$$(2.5) \quad \xi^\varepsilon_i = \begin{cases} x^\varepsilon & \text{if } x^i_{\tau_i} \in \Pi_w \\ x^i_{\tau_i} & \text{otherwise} \end{cases} \qquad \Pi_w = \{x: w(x) > c(x)\}$$

is $\varepsilon(t^*(x^\varepsilon))^{-1}$ optimal.

The Theorem 2.1 does not cover all possible situations. The following Proposition and Corollary also characterize optimal control in special cases.

Proposition 2.1. If there exists a continuous solution of the equation (2.2) and

(a) $\exists_{y\in E}\quad E_y\tau^*_\eta=\infty$, then the strategy $V=(\tau^*_1,\dots,\tau^*_{\eta_1},y,\tau^*_{\eta_1+1},\dots,\tau^*_{\eta_2},y,\dots)$ is optimal one,

(b) $\sup_y E_y\tau^*_\eta=\infty$, then the strategy $V=(\tau^*_1,\dots,\tau^*_{\eta_1},y,\tau^*_{\eta_1+1},\dots,\tau^*_{\eta_2},y,\dots)$ is $\frac{w(y)+d(y)}{E_y t^*(y)}$ -optimal.

Corollary 2.1. If $z(x)=\liminf_{t\uparrow\infty} t^{-1}E_x\int_0^t f(x_s)\,ds$, and $\inf_{x\in E} z(x)=\lambda$, then the strategy to jump in x^ε, where $z(x^\varepsilon)\le\lambda+\varepsilon$, and to do nothing is ε-optimal.

To prove the existence of the continuous solution of the equation (2.2) we need extra assumptions. In the next two sections we will impose additional assumptions on the time of replacement. Later on we will return to the general case and we solve the replacement problem for processes satisfying the exponential stability property.

3. FINITE TIME MODEL

In this section we will assume that from the technical reasons we have to replace in a finite constant time T, and the strategy depends on the time from the last innovation z.

Theorem 3.1. There exists (x,z) continuous function w_T satisfying the equation

$$(3.1)\qquad w_T(x,z)=\inf_{0\le t\le T-z} E_x\{\int_0^t (f(x_s)-\lambda_T)\,ds+Mw_T(x_t,z+t)\}$$

$$Mw_T(x,z)=\min\{w_T(x,z)+g(x),\,c(x)+g(x)\}$$

with the boundary condition

$$(3.2)\qquad \inf_x\{w_T(x,0)+d(x)\}=0\ ,\ \text{where}$$

$$(3.3)\qquad \lambda_T=\inf_x\inf_{\nu_T}(E_x\tau_\eta)^{-1}E_x\{\int_0^{\tau_\eta} f(x_s)\,ds+\sum_{i=1}^{\eta} g(x_{\tau_i})+c(x_{\tau_\eta})+d(x)\}$$

$$\nu_T=(\tau_1,\tau_2,\dots,\tau_\eta,x)$$

$$(3.4)\qquad \tau_1=\sigma=\text{const.},\dots,\tau_{n+1}=\tau_n+\sigma_n(x^n_{\tau_n},\tau_n),\dots$$

$$\zeta_i=\zeta_i(x^i_{\tau_i},\tau_i)$$

$\sigma_1,\dots,\sigma_n,\dots,\zeta_1,\dots,\zeta_n,\dots$ are $\mathcal{E}\otimes\mathcal{R}$ measurable real or **E** valued functions.

The optimal inspections are of the form

$$(3.5)\qquad \tau^*_1=t^*_T(x,0),\dots,\ \tau^*_{n+1}=\tau^*_n+t^*_T(x_{\tau_n},\tau_n),\dots,\text{where}$$

$$(3.6)\quad t^*_T(x,z) = \inf\{T-z \geq t \geq 0:\ w_T(x,z) \geq E_x\{\int_0^t (f(x_s)-\lambda)\,ds + Mw_T(x_t,t+z)\}$$

Moreover $\inf_V J_x(V)=\lambda$, and if x^ε is such that $w_T(x^\varepsilon,0)+d(x^\varepsilon)\leq \varepsilon$, then the strategy $V=(\tau^*_i, \xi^\varepsilon_i)$ where

$$\xi^\varepsilon_i = \begin{cases} x^\varepsilon & \text{if } x^i_{\tau^*_i} \in \Pi_{w_T} \\ x^i_{\tau^*_i} & \text{otherwise} \end{cases} \qquad \Pi_{w_T} = \{x:\ w_T(x,0) > c(x)\}$$

is $\dfrac{\varepsilon}{t^*(x^\varepsilon,0)}$ optimal.

Proof. We will show only the existence of the continuous solution of (3.1), since the second part of the theorem follows (similarly as Theorem 2.1) from the paper [9].

Proposition 3.1. If $m(x,z)$ is (x,z) continuous, then

$$h(x,z) = \inf_{0\leq t\leq T-z} E_x\{\int_0^t (f(x_s)-\lambda)\,ds + m(x_t, z+t)\}$$

is also (x,z) continuous. Moreover there exists optimal time $t^*(x,z)$

$$t^*(x,z) = \inf\{0\leq t\leq T-z:\ h(x,z)\geq E_x\{\int_0^t (f(x_s)-\lambda)\,ds + m(x_t, z+t)\}$$

Proof of Proposition 3.1. Let $x_n \to x$, $z_n \to z$. Then

$$|h(x_n,z_n)-h(x,z)| \leq |h(x_n,z_n)-h(x,z_n)| + |h(x,z_n)-h(x,z)| \leq$$

$$\leq \sup_{0\leq t\leq T-z_n} \int_0^t |\Phi(u)f(x_n)-\Phi(u)f(x)|\,du + \sup_{0\leq t\leq T-z} |E_{x_n} m(x_t,z+t) - E_x m(x_t,z+t)|$$

$$+ \int_{z\wedge z_n}^{z\vee z_n} |\Phi(u)f(x)-\lambda|\,du + \sup_{t\geq 0} E_x|m(x_t,(z+t)\wedge T) - m(x_t,(z_n+t)\wedge T)|$$

To finish the proof we need the following two lemmas

Lemma 3.1. If $(E,\mathcal{E})$ is locally compact separable space, X is right continuous Feller Markov process, then for $v\in C$, $\Phi(t)v(x)$ is (t,x) continuous function.

Proof. Suppose $x_n\to x$, $t_n\to t$. If $v\in C_0$ we have

$$|\Phi(t)v(x)-\Phi(t_n)v(x_n)| \leq |\Phi(t)v(x)-\Phi(t)v(x_n)| + |\Phi(t)v(x_n)-\Phi(t_n)v(x_n)|$$

$$\leq |\Phi(t)v(x)-\Phi(t)v(x_n)| - \|\Phi(|t-t_n|)v-v\| \longrightarrow 0$$

because the semigroup $\Phi(t)$ is strongly continuous on C_0. If $v\in C$ only it is enough to show that $|\Phi(t)v(x_n) - \Phi(t_n)v(x_n)| \longrightarrow 0$ as $n\to\infty$.

Let $K=\{x_n,\ n\in N,\ x\}$. Since K is compact we know from [10]

$$\sup_{x\in K} P_x\{\sup_{0\leq s\leq T} d(x_s,x)\geq R\} \longrightarrow 0 \text{ as } R\to\infty$$ where d is a metric on E, compatible with the topology such that every closed ball is compact. Thus there exists sufficiently large compact set L_ε, $K\subset L_\varepsilon$ that

$$\sup_{x\in K} P_x\{\exists_{s\in[0,T]}\ x_s\in L^c_\varepsilon\} \leq (4\|v\|)^{-1}\cdot\varepsilon$$

Let $\overline{v}_\varepsilon(x) = v(x)$ on L_ε, and $\overline{v}_\varepsilon(x) = d(x,\{y:\ d(y,L_\varepsilon)\leq 1\}^c)\cdot v(x)$ on L^c_ε

So $\bar{v}_\varepsilon \in C_0$ and $|\Phi(t)v(x_n) - \Phi(t_n)v(x_n)| \le |\Phi(t)v(x_n) - \Phi(t)\bar{v}_\varepsilon(x_n)| + |\Phi(t)\bar{v}_\varepsilon(x_n) - \Phi(t_n)\bar{v}_\varepsilon(x_n)| + |\Phi(t_n)\bar{v}_\varepsilon(x_n) - \Phi(t_n)v(x_n)| \le E_{x_n}|v(x_t) - \bar{v}_\varepsilon(x_t)| + |\Phi(t)\bar{v}_\varepsilon(x_n) - \Phi(t_n)\bar{v}_\varepsilon(x_n)| + E_{x_n}|v(x_{t_n}) - \bar{v}_\varepsilon(x_{t_n})| \le 4\|v\| \sup_{x\in K} P_x\{\exists_{s\in[0,T]}\ x_s \in L_\varepsilon^c\} + |\Phi(t)\bar{v}_\varepsilon(x_n) - \Phi(t_n)\bar{v}_\varepsilon(x_n)| \le \varepsilon + |\Phi(t)\bar{v}_\varepsilon(x_n) - \Phi(t_n)\bar{v}_\varepsilon(x_n)| \to \varepsilon$ as $n \to \infty$

Since ε can be chosen arbitrarily small the proof is complete.
Now one can easy notice that the process (x_t, t) is Feller, Markov on $E\times[0,\infty)$ with the semigroup $\bar{\Phi}(t)v(x,z) = E_{x,z}[v(x_t, z+t)]$. Therefore

$$|E_{x_n} m(x_t, z+t) - E_x m(x_t, z+t)| = |\bar{\Phi}(t)m(x_n,z) - \bar{\Phi}(t)m(x,z)|$$

$$|E_x m(x_t, z+t) - E_x m(x_t, z_n+t)| = |\bar{\Phi}(t)m(x,z) - \bar{\Phi}(t)m(x,z_n)|$$

and the following lemma finishes the proof of Proposition 3.1

<u>Lemma 3.2.</u> If $v(x,t)$ is (x,t) continuous and $x_n \to x$ as $n\to\infty$, then

$$\sup_{0\le t\le T} |v(x_n,t) - v(x,t)| \to 0 \text{ as } n\to\infty$$

The proof is left for the reader.

For a given strategy $\nu = (\tau_1, \tau_2, \dots, \tau_\eta, \xi_\eta)$ define the strategy $\nu_{|n} = (\tau_1, \tau_2, \dots, \tau_{\eta\wedge n}, \xi_{\eta\wedge n})$. Let

$$I_x(\nu) = E_x\{\int_0^{\tau_\eta}(f(x_s) - \lambda)ds + \sum_{i=1}^{\eta} g(x_{\tau_i}) + c(x_{\tau_\eta})\}$$

$$w_T^1(x,z) = \inf_{0\le t\le T-z} E_x\{\int_0^t (f(x_s)-\lambda)\,ds + c(x_t) + g(x_t)\},\dots,$$

$$w_T^{n+1}(x,z) = \inf_{0\le t\le T-z} E_x\{\int_0^t (f(x_s)-\lambda)\,ds + Mw_T^n(x_t, z+t)\},\dots$$

Then

<u>Proposition 3.2.</u> For each n w_T^n is (x,z) continuous and

$$w_T^n(x,z) = \inf_{\substack{\nu_{|n} \\ \tau_{n\wedge\eta}\le T-z}} I_x(\nu)$$

The proof is standard (see [7]) and follows from the Proposition 3.1.
Now we are in position to complete the proof of Theorem 3.1.
Since $\|c\| \ge \inf_\nu I_x(\nu) \ge -T(\|f\|+\lambda) + k\,E_x\eta$, so $E_x\eta \le (\|c\| + T(\|f\|+\lambda))k^{-1} = K$, and we can restrict ourselves to such strategies that $E_x\eta \le K$.
For a strategy ν such that $\tau_\eta \le T-z$ we have

$$I_x(\nu_n) - I_x(\nu) = E_x\{-\int_{\tau_{n\wedge\eta}}^{\tau_\eta}(f(x_s)-\lambda)\,ds + \sum_{i=1}^{\eta} g(x_{\tau_i}) - c(x_{\tau_\eta}) + c(x_{\tau_{n\wedge\eta}})\}$$

$$\lambda E_x(\tau_\eta - \tau_{\eta\wedge n}) + E_x\{\chi_{n<\eta}\,c(x_{\tau_n}) - c(x_{\tau_\eta})\} \le \lambda T\, E_x\{\chi_{n<\eta}\} + 2\|c\|E_x\{\chi_{n<\eta}\}$$

But $nE_x\{\chi_{n<\eta}\} \le E_x\eta \le K$, so $I_x(\nu_{|n}) - I_x(\nu) \le Kn^{-1}(\lambda T + 2\|c\|)$ and

$$0 \le w_T^n(x,z) - \inf_{\nu,\tau_\eta\le T-z} I_x(\nu) \le Kn^{-1}(\lambda T + 2\|c\|)$$

Thus $w_T^n(x,z) \downarrow w_T(x,z) = \inf_{\nu,\tau_\eta\le T-z} I_x(\nu)$ uniformly as $n\to\infty$.

Therefore w_T is the continuous solution of (3.1) and by the definition of λ_T the boundary condition (3.2) is also satisfied.

4. FAILURE MODEL

Let us suppose that Markov process X is killed with an intensity $\alpha + u(x)$, $0<\alpha$, $0 \leq u \in C$. This means if X denotes production of a machine, we have the situation that our system is subject to random failure. Upon failure the system is replaced by a new, similar one, starting from arbitrary chosen random point and a fixed failure cost h and replacement cost d are incurred. If we replace the machine before failure the replacement cost $c(x) + d(y)$ is incurred. To describe our model we begin with some definitions.

Let $\hat{\Omega} = \Omega \times [0,\infty]$, $\hat{F} = F \times \mathcal{R}$, for a fixed ω, α_ω be the measure on $([0,\infty], \mathcal{R})$ such that $\alpha_\omega(\gamma, \infty] = \exp(-\int_0^\gamma (\alpha + u(x_s))ds) \equiv M_\gamma(\omega)$, and for $\Gamma \in F \times \mathcal{R}$, $\Gamma^\omega = \{\gamma, (\omega,\gamma) \in \Gamma\}$

(4.1) $\quad \hat{P}_x(\Gamma) = E_x[\alpha_\omega(\Gamma^\omega)]$

Define $\hat{X}_t(\omega,\gamma) = X_t(\omega)$ if $t<\gamma$ and $\hat{X}_t(\omega,\gamma) = \Delta$ if $t \geq \gamma$, where Δ is a failure state, and for $\hat{\omega} = (\omega,\gamma)$ put $\hat{\Theta}_t \hat{\omega} = (\Theta_t \omega, (\gamma - t) \vee 0)$. This way the process $\hat{X} = (\hat{\Omega}, \hat{F}_t, \hat{F}, \hat{x}_t, \hat{\Theta}_t, \hat{P}_x)$ is obtained by killing X_t at a rate $\alpha + u(x_t)$, and is called subprocess of X associated with the multiplicative functional M. The following proposition basing on Theorem III 3.3 [3] and Lemma 5.4 [4] characterizes the properties of the process $\hat{X}$

<u>Proposition 4.1.</u> X is Feller Markov process and

(4.2) $\quad \hat{E}_x f(\hat{x}_t) = E_x\{f(x_t) M_t(\omega)\}$ for bounded, measurable function f, if we put $f(\Delta)=0$.

Similarly as in [5] with each strategy V we associate the new probability space $\tilde{\Omega} = \hat{\Omega}^N$, $\tilde{F} = \hat{F}^N$, $\hat{P}^V$. The functional is of the form

$$(4.3) \quad J_x(V) = \liminf_{t\uparrow\infty} t^{-1} \tilde{E}_x^V \{ \int_0^t f(y_s)ds + \sum_{i=1}^{\infty} \chi_{\tau_i \leq t}\, g(x_{\tau_i}^{i-1}) + \sum_{i=1}^{\infty} \chi_{\tau_{\eta i} \leq t} [\chi_{\tau_{\eta i} \leq \tau_{\eta i-1} + T_i}\, c(x_{\tau_{\eta i}}) + \chi_{\tau_{\eta i-1}+T_i < \tau_{\eta i}}\, h + d(\xi_{\eta_i})]\}$$

where $\xi_i = \zeta(\hat{x}_{\tau_i})$, $\zeta : E \cup \{\Delta\} \longrightarrow E$ is measurable function and T_i is the lifetime of the process in i-th cycle. The optimal stationary cost per one cycle is

$$(4.4) \quad \lambda = \inf_x \inf_V \{\hat{E}_x(\tau_\eta \wedge T)\}^{-1} \hat{E}_x \{ \int_0^{\tau_\eta \wedge T} f(x_s)ds + \sum_{i=1}^{\eta} g(x_{\tau_i}) \chi_{\tau_i < T} + c(x_{\tau_\eta})\chi_{\tau_\eta < T} + h\, \chi_{T \leq \tau_\eta} + d(x) \}$$

Let us define another space $\bar{\Omega}$ and subproces $\bar{X}$ with the multiplicative functional $\bar{M}_t = \exp(-\int_0^t u(x_r)dr)$. Then

$$(4.5)\qquad \lambda = \inf_{x} \inf_{\nu} \left(\bar{E}_x \int_0^{\tau_\eta} e^{-\alpha s}\, ds\right)^{-1} \bar{E}_x \Big\{ \int_0^{\tau_\eta} F(\bar{x}_s)\, e^{-\alpha s}\, ds + \sum_{i=1}^{\eta} g(\bar{x}_{\tau_i}) e^{-\alpha \tau_i} + c(\bar{x}_{\tau_\eta}) e^{-\alpha \tau_\eta} \Big\}$$

where $F(x) = (\alpha + u(x))h + f(x)$ and $c(\Delta) = 0$

Denote by z the elapsed time from the last decision time. Define

$$w^1(x,z) = \inf_{t \geq 0} \Big\{ \int_0^t (F(x,z+s) - \lambda)\, e^{-\alpha s} ds + e^{-\alpha t} [g(x,z+t) + c(x,z+t)] \Big\}$$

$$F(x,z) = \bar{E}_x F(\bar{x}_z), \quad g(x,z) = \bar{E}_x g(\bar{x}_z), \quad c(x,z) = \bar{E}_x c(\bar{x}_z)$$

$$(4.6)\qquad w^{n+1}(x,z) = \inf_{t \geq 0} \Big\{ \int_0^t (F(x,z+s) - \lambda) e^{-\alpha s} ds + e^{-\alpha t} M w^n(x,z+t) \Big\}$$

where $Mw(x,z) = \bar{E}_x \{ g(\bar{x}_z) + w(\bar{x}_z, 0) \wedge c(\bar{x}_z) \}$

Using technics from the previous section ,[9] and [6] we can prove the following theorem

Theorem 4.1. The function w^n is continuous and $w^n \downarrow w$ uniformly as $n \to \infty$, where w is the solution of the equation

$$(4.7)\qquad w(x,z) = \inf_{t \geq 0} \Big\{ \int_0^t (F(x,z+s) - \lambda)\, e^{-\alpha s}\, ds + e^{-\alpha t} Mw(x,z+t) \Big\}$$

with the condition $\inf \{ w(x,0) + d(x),\ x \in E \} = 0$.

If at the last inspection the process was in the state x, then the inspection after time $t^*(x,0)$

$$(4.8)\qquad t^*(x,z) = \inf\{ s \geq 0 : w(x,z+s) = Mw(x,z+s) \}$$ is optimal one.

The innovations from the points x^ε, for which $w(x^\varepsilon,0) + d(x^\varepsilon) \leq \varepsilon$ are $\varepsilon \cdot t^*(x^\varepsilon,0)^{-1}$ optimal and the optimal value of the functional (4.3) is equal to λ.

5. REMARKS ABOUT GENERAL CASE

In this section we will summarize the remarks relating the general model, which can be observed after the consideration of the special cases 3,4. Namely we can introduce the decreasing sequence of functions

$$(5.1)\qquad \begin{aligned} w^1(x) &= \inf_{t \geq 0} E_x \Big\{ \int_0^t (f(x_s) - \lambda) ds + c(x_t) + g(x_t) \Big\}, \ldots, \\ w^{n+1}(x) &= \inf_{t \geq 0} E_x \Big\{ \int_0^t (f(x_s) - \lambda)\, ds + Mw^n(x_t) \Big\} \end{aligned}$$

$$w^n \downarrow w$$

$$(5.2)\qquad w(x) = \inf_{t \geq 0} E_x \Big\{ \int_0^t (f(x_s) - \lambda)\, ds + Mw(x_t) \Big\}$$

and write down the following observations

1. w^n are uppersemicontinuous, so w is also uppersemicontinuous,
2. for w^1 there exists optimal time $t_1^*(x) = \inf\{ t \geq 0 : w^1(x) = E_x \{ \int_0^t f(x_s) - \lambda)\, ds + c(x_t) + g(x_t) \}$, if $t_1(x)$ is bounded then $w^1(x)$ is continuous,

   let $t_n^*(x) = \inf\{ t \geq 0 : w^n(x) = E_x \{ \int_0^t (f(x_s) - \lambda)\, ds + Mw^{n-1}(x_t) \}$

   if $w^{n-1} \in C$ and $t_n(x)$ is bounded, then $w^n \in C$,

3. if there exist $i \in N$, $\bar{x} \in E$ such that $w^i(\bar{x}) + d(\bar{x}) = 0$, then it is enough to use only $\inf\{j: w^j(\bar{x}) + d(\bar{x}) = 0\}$ inspections,
4. if we may consider only the strategies ν satisfying the property

(5.3) $\sup_x E_x \tau_\eta \leq l$, l constant independent of the strategy,

then we can adapt Menaldi idea [5] to prove the continuity of w, i.e. similarly as in the proof of Theorem 3.1, using (5.3) we obtain $E_x \eta \leq (\|g\| + \|c\| + \lambda l)\ k^{-1}$, so $E_x\{\chi_{n<\eta}\} \leq n^{-1} E_x \eta$, and thus $0 \leq w^n(x) - w(x) \leq \lambda E_x\{\tau_\eta - \tau_{\eta \wedge n}\} + 2\|c\| E_x\{\chi_{n<\eta}\} \leq E_x\{\chi_{n<\eta}\}(\lambda l + 2\|c\|) \longrightarrow 0$ uniformly with respect to $x \in E$,

5. under the assumption that the class of suboptimal strategies satisfies the property (5.3), the boundary condition $\inf_x [w(x) + d(x)] = 0$ is also satisfied - thus we can apply Theorem 2.1.

6. ERGODIC CASE

To show an example of Markov processes for which suboptimal strategies satisfy (5.3), we consider compact state space E and impose the following ergodic assumption

(6.1) there exists invariant probability measure μ and constants $B, \gamma > 0$, such that for $\Gamma \in \mathcal{E}$, $|P_x(x_t \in \Gamma) - \mu(\Gamma)| \leq B\ e^{-\gamma t}$

The last assumption is very restrictive. Nevertheless is satisfied for a class of diffusions with reflection in a bounded regular domain or jump Markov processes for details see [6] . As a consequence of (6.1) we have

Lemma 6.1.[6]. The equation $-Av = f - \bar{f}$, where $\bar{f} = \int_E f(x)\mu(dx)$ and A is infinitesimal operator of X, has the continuous, bounded, unique up to constants solution.

Therefore since $w(x) = \inf_\nu I_x(\nu) \leq \|c\| + \|g\|$ and

$$E_x\Big\{\int_0^{\tau_\eta} (f(x_s) - \lambda)\, ds + \sum_{i=1}^{\eta} g(x_{\tau_i}) + c(x_{\tau_\eta})\Big\} = v(x) + E_x\Big\{(\bar{f} - \lambda)\,\tau_\eta + \sum_{i=1}^{\eta} g(x_{\tau_i}) + c(x_{\tau_\eta}) - v(x_{\tau_\eta})\Big\}$$

for $\bar{f} > \lambda$ we may restrict ourselves to strategies satisfying (5.3). Thus using Remarks 5.4 and 5.5 we obtain

Theorem 6.1. Under the above assumptions $\inf_V J_x(V) = \lambda$. If $\lambda < \bar{f}$ the optimal strategy is characterized by Theorem 2.1. Otherwise the strategy to do nothing is optimal.

The last part of the paper is devoted the study of the asymptotic behaviour of $w_\alpha(x) = \inf_V J_x^\alpha(V)$, when the discount factor α goes to zero

(6.2) $$J_x^\alpha(V) = E_x\Big\{\int_0^\infty e^{-\alpha s} f(y_s)\, ds + \sum_{i=1}^{\infty} g(x_{\tau_i}^{i-1})\, e^{-\alpha\tau_i} + \sum_{j=1}^{\infty} c(x_{\tau_{\eta_j}}^{\eta_j - 1})\, e^{-\alpha\tau_{\eta_j}}\Big\}$$

First let us notice that w_α satisfies the equation

$$w_\alpha(x) = \inf_{\nu} E_x\{\int_0^{\tau_\eta} e^{-\alpha s} f(x_s)\,ds + \sum_{i=1}^{\eta} g(x_{\tau_i})\, e^{-\alpha\tau_i} + e^{-\alpha\tau_\eta}[\min w_\alpha(x) + c(x_{\tau_\eta})]\} \tag{6.3}$$

So for $v_\alpha(x) = w_\alpha(x) - \min w_\alpha(x)$

$$v_\alpha(x) = \inf_{\nu} E_x\{\int_0^{\tau_\eta} e^{-\alpha s}(f(x_s) - \alpha \min w_\alpha)\,ds + \sum_{i=1}^{\eta} g(x_{\tau_i})\, e^{-\alpha\tau_i} + e^{-\alpha\tau_\eta} c(x_{\tau_\eta}) \quad \text{and} \quad 0 \le v_\alpha(x) \le \|g\| + \|c\| \tag{6.4}$$

The following theorem holds

Theorem 6.2. $\lim_{\alpha\downarrow 0} w_\alpha(x) = \lambda$. Moreover, if $\lambda < \bar{f}$ then v_α converges in C as $\alpha \to 0$ to w the solution of the equation (2.2)

$$w(x) = \inf_{\nu} E_x\{\int_0^{\tau_\eta}(f(x_s) - \lambda)\,ds + \sum_{i=1}^{\eta} g(x_{\tau_i}) + c(x_{\tau_\eta})\} \tag{6.5}$$

Proof. Since $w_\alpha \le R_\alpha f$, $\alpha R_\alpha f \to \bar{f}$ in C as $\alpha \to 0$, $\limsup_{\alpha\uparrow 0} \alpha w_\alpha \le \bar{f}$.
If for a sequence $\alpha_n \to 0$, $\lim \alpha_n w_{\alpha_n} = \bar{\lambda} < \bar{f}$, then we can show identically as M. Robin in Theorem 3.2 [6] that $v_{\alpha_n} \to w$ defined in (6.5) with λ replaced by $\bar{\lambda}$, and therefore $\bar{\lambda} = \lambda$.
If $\alpha_n w_{\alpha_n} \to \bar{f}$, but $\lambda < \bar{f}$, then for the optimal cycle strategy ν^* of the undiscounted problem (which there exists by Theorem 6.1) we have from (6.4)

$$E_x\{\int_0^{\tau_\eta^*} e^{-\alpha_n s}\alpha_n \min w_{\alpha_n}\,ds\} \le E_x\{\int_0^{\tau_\eta^*} f(x_s)\, e^{-\alpha_n s}\,ds + \sum_{i=1}^{\eta} e^{-\alpha_n\tau_i^*} g(x_{\tau_i^*}) + e^{-\alpha_n\tau_\eta^*} c(x_{\tau_\eta^*})\}$$

Since $E_x \tau_\eta < \infty$, letting $n \to \infty$ we obtain

$$\bar{f}\, E_x \tau_\eta^* \le I_x(\nu^*) + \lambda\, E_x \tau_\eta^*$$

but for $\bar{x} = \arg\min w$, $E_{\bar{x}} \tau_\eta^* > 0$ and $w(\bar{x}) = 0$ so $\bar{f} \le \lambda$. The reverse inequality is also satisfied $\lambda \le \bar{f}$. Thus $\lambda = \bar{f}$.

REFERENCES

1. Anderson, R. F., Friedman, A.: Optimal inspections in a stochastic control problem with costly observations. Math. Oper. Res. 2, 155-190 (1977)
2. Antelman, G. R., Savage, I. R.: Surveillance problems: Wiener processes. Naval Res. Log. Quat. 12, 35-56 (1965)
3. Blumenthal R. M., Getoor, R. K.: Markov Processes and Potential Theory. Academic Press 1968
4. Gihman, I. I., Skorohod, A. V.: The Theory of Stochastic Processes. Vol. II. Berlin, Springer-Verlag 1975
5. Robin, M.: Controle impulsionnel des processus de Markov. Thesis.

University of Paris IX (1978)
6. Robin, M.: On some impulsive control problems with long run average cost. SIAM on Control 19, 333-358 (1981)
7. Robin, M.: Optimal maintenance and inspection: an impulsive control approach. Proc. 8 IFIP Symp. Optimiz. Lect. Notes in Contr. Inf. Sc. 6
8. Savage, I. R., Richard, I.: Surveillance problems. Naval Res. Log. Quat. 9, 187-209 (1962)
9. Stettner, Ł.: On impulsive control with long run average cost criterion. To appear in Studia Mathematica (1983)
10. Stettner, Ł.: A note on optimal stopping for Feller processes. Preprint (1983)

CONVERGENCE OF A STOCHASTIC VARIABLE METRIC METHOD WITH APPLICATION IN ADAPTIVE PREDICTION

L. Gerencsér
Computer and Automation Institute of the Hungarian Academy of Sciences
H 1502, Budapest XI. Kende utca 13-17, Hungary

1. INTRODUCTION

Since the basic convergence theorem of Ljung |4| for the analysis of recursive estimators had been established there has been a growing interest in exploiting this theorem for the design of new algorithms. The aim of this paper is to show how the quasi-Newton method of Broyden |1| can be converted into a stochastic approximation procedure. The full development of this idea is fairly technical, hence we restrict ourselves to the exposition of procedures, theorems and basic ideas. The research is motivitated by problems in adaptive prediction. Adaptive prediction is a problem which fits into the scheme of Ljung, as has been shown by e.g. Wittenmark |6| and Holst |3|. In section 2 we define a continuous-time version of Broyden's method. In section 3 we describe a general scheme and a stability theorem connected with it. It is applied to the continuous time Broyden-method in Section 4. Section 5 contains a stochastic version of the sheme given in Section 3. The main theorem is Theorem 5.2. We apply these results to the design and analysis of a certain-type of stochastic Broyden-method in Section 6, while in Section 7 we summarize some of our critical remarks.

2. DERIVATION OF A CONTINUOUS-TIME BROYDEN METHOD

Let $f=(f_1,\dots,f_n)^T$ be a vector-valued linear function $f(\Theta)=G^*\Theta+b^*$ of the n dimensional vector variable $\Theta=(\Theta_1,\dots,\Theta_n)^T$ and let Θ^* be a solution of the linear algebraic equation

$$f(\Theta) = 0. \tag{2.1}$$

We shall estimate the Jacobian matrix G^* on the basis of observations $f(\eta(t))$, where $\eta(t)$ is a sequence converging to Θ^*. We have

$$f(\eta(t))=G^*\eta(t)+b^* . \tag{2.2}$$

Let us select a single line in (2.2) say the i-th line. It will be written in the form

$$f_i(\eta(t))=g_i^{*T}\eta(t)+b_i^* \tag{2.3}$$

where g_i^T denotes the i-th row of G . Introducing the compact notations

$$k_i^* = \begin{pmatrix} g_i^* \\ b_i^* \end{pmatrix} \qquad \Psi(t) = \begin{pmatrix} \eta(t) \\ 1 \end{pmatrix} \tag{2.4}$$

we have

$$f_i(\eta(t)) = \Psi^T(t) k_i^* . \tag{2.5}$$

The estimate of k_i^* after t observations will be denoted by $k_i(t)$. It will be computed recursively by projecting $k_i(t-1)$ on the hyperplane $f_i(\eta(t)) - \Psi^T(t)k = 0$:

$$k_i(t) = k_i(t-1) + (1/\alpha(t))\Psi(t)(f_i(\eta(t)) - \Psi^T(t)k_i(t-1)) \tag{2.6}$$

where $\alpha(t) = \Psi^T(t)\Psi(t)$.

Returning to our original problem, let G(t), b(t) denote the current estimates of G^*, b^*. We have get from (2.6) the following algorithm:

$$G(t) = G(t-1) + (1/\alpha(\eta(t))(f(\eta(t)) - G(t-1)\eta(t) - b(t-1))\eta^T(t) \tag{2.7}$$

$$b(t) = b(t-1) + (1/\alpha(\eta(t))(f(\eta(t)) - G(t-1)\eta(t) - b(t-1)) \tag{2.8}$$

where $\alpha(\eta) = 1 + \eta^T\eta$.

This updating formula is a modification of BROYDEN's updating formula |1|. The difference is that in the new method the constant term b^* is also estimated.

The measurement points $\eta(t)$ will be generated recursively by quasi-Newton steps. Thus we get the following algorithm:

$$\Theta(t) = \Theta(t-1) - G^{-1}(t)f(\Theta(t-1)) \tag{2.9}$$

$$G(t) = G(t-1) + (1/\alpha(\Theta(t)))(f(\Theta(t)) - G(t-1)\Theta(t) - b(t-1))\Theta^T(t) \tag{2.10}$$

$$b(t) = b(t-1) + (1/\alpha(\Theta(t)))(f(\Theta(t)) - G(t-1)\Theta(t) - b(t-1)) \tag{2.11}$$

where $\alpha(\Theta) = 1 + \Theta^T\Theta$.

The advantage of this modification of BROYDEN's method is that we can easily define a continuous-time analogue of the discrete time algorithm as follows.
Continuous time Broyden-method:

$$\dot{\Theta}(t) = -G^{-1}(t)f(\Theta(t)) \tag{2.12}$$

$$\dot{G}(t) = (1/\alpha(\Theta(t)))(f(\Theta(t)) - G(t)\Theta(t) - b(t))\Theta^T(t) \tag{2.13}$$

$$\dot{b}(t) = (1/\alpha(\Theta(t)))(f(\Theta(t)) - G(t)\Theta(t) - b(t)). \tag{2.14}$$

Here $\dot{}$ denotes differentiation with respect to the continuous time t.
This algorithm extends to nonlinear f.

Let us itroduce the new variables $H(t)=G^{-1}(t)$ and $\eta(t)=-H(t)b(t)$. Taking into account the equality $\dot{H}(t)=-H(t)\dot{G}(t)H(t)$ we get the following alternative algorithm:

$$\dot{\Theta}(t)=-H(t)f(\Theta(t)) \tag{2.15}$$

$$\dot{H}(t)=-(H(t)f(\Theta(t))-\Theta(t)+\eta(t))\Theta^T(t)H(t)/\alpha(\Theta(t)) \tag{2.16}$$

$$\dot{\eta}(t)=-(H(t)f(\Theta(t))-\Theta(t)+\eta(t))/\alpha(\Theta(t)). \tag{2.17}$$

In the next section we consider a general continuous-time algorithmic scheme which covers both forms of the continuous-time Broyden method.

3. STABILITY THEOREMS FOR ORDINARY DIFFERENTIAL EQUATIONS

Let a system of differential equations

$$\dot{x}(t)=\varepsilon p(x(t),y(t)) \tag{3.1}$$

$$\dot{y}(t)= q(x(t),y(t)) \tag{3.2}$$

be given where $x,p(x,y)\in R^n$ and $y,q(x,y)\in R^m$. Let us assume that the following conditions are satisfied.

Condition 3.1. The functions p is once, q is twice continously differentiable on an open set $\Omega_{xy}\in R^{n+m}$.

Condition 3.2. The nonlinear equation for y

$$q(x,y)=0 \tag{3.3}$$

has a continuously differentiable solution $y=Y(x)$ such that $(x,Y(x))\in\Omega_{xy}$, where $x\in\Omega_x\subset R^n$ where Ω_x is some open set and $q_y(x,Y(x))$ is nonsingular along this hypersurface.

Condition 3.3. $q(x,y)=0$ implies $p(x,y)=0$ more exactly

$$p(x,Y(x))=0 \tag{3.4}$$

for $x\in\Omega_x$.

We shall use the notation

$$\Pi=\{(x,Y(x)):x\in\Omega_x\}. \tag{3.5}$$

Thus Π constitutes a hypersurface of equilibrium points for the system of d.e. (3.1) (3.2).

Condition 3.4. The closure of Π lies in Ω_{xy}.
Let Ω_{xy} be an open set and let $a(x,y)$, $b(x,y)$ be vector valued functions defined on Ω_{xy}. We shall use the abbreviated notations

$$a=O(b) \quad \text{or} \quad a(x,y)=O(b(x,y)) \quad \text{on} \quad \Omega_{xy} \quad \text{if} \tag{3.6}$$

an estimation $\|a(x,y)\| < c\| b(x,y)\|$ holds in Ω_{xy}.

Definition 3.1.
The system of differential equation (3.1), (3.2) is exponentially stable on Π if there exists a continuously differentiable function W defined on a neighbourhood $\Omega^{o}_{xy} \subset \Omega_{xy}$ of Π such that

(i) $W(x,Y(x))=0$ for $x\in\Omega_x$ and $W(x,y)>0$ if $y\neq Y(x)$.
(ii) $y-Y(x)=O(W^{1/2}(x,y))$ in Ω^{o}_{xy}.
(iii) If $(x(t), y(t))$ is a piece of a solution trajectory of (3.1), (3.2), then

$$\frac{d}{dt} W(x(t),y(t))<-cW(x(t),y(t)) \qquad c>0, \quad \text{for } (x,y)\in\Omega^{o}_{xy} \tag{3.7}$$

Remark 3.1. Ω^{o}_{xy} can be assumed to have the following form:

$$\Omega^{o}_{xy} =\{(x,y):x\in\Omega_x, \quad W(x,y)<\delta \quad \delta>0\}.$$

The significance of the above definition is that if exponential stability can be established, then convergence of the solution trajectories $(x(t),y(t))$ to Π can be proved. More exactly, we have the following:

Theorem 3.1. If a system of differential equation is exponentially stable, then for any $x(0)\in\Omega_x$ there existis a neighbourhood S of $Y(x)$, such that if $y(0)\in S$, then the solution of (3.1), (3.2) is defined for all positive t-s and $x(t)$, $y(t)$ converges to some $(x^*,y^*)\in\Pi$ as $t\to\infty$. Moreover $W(x(t),y(t))$ decreases exponentially.

To verify exponential stability the following condition will be useful:

Condition 3.5. The differential equation

$$\dot{y}(t)=q(x,y(t)) \tag{3.8}$$

is asymptotically stable in $Y(x)$ for all $x\in\Omega^{o}_{x}$, moreover the Jacobian $q_y(x,Y(x))$ is stable for all $x\in\Omega_x$. We have the following

Theorem 3.2. If Conditions (3.1)-(3.5) are satisfied, then the system of differential equation (3.1), (3.2) is exponentially stable for sufficiently small ε.

Remark 3.2. We can choose $\varepsilon=1$ if W_x is itself small. This is the case if $Y(x)\equiv$ const.

Theorem 3.1 and 3.2 will be proved in Gerencsér | 2 |

4. CONVERGENCE OF THE CONTINUOUS-TIME BROYDEN METHOD

We shall apply the results of the previous section to show that the differential equation defining the second form of the continuous time Broyden method is exponentially stable on some region. We use the following correspondence:

$$x=H \qquad y=\binom{\eta}{\Theta}. \tag{4.1}$$

Let $\Gamma(H)=-HG^*$ $\quad \Omega_H=\{H:\mathrm{Re}\lambda(\Gamma(H))<-\delta<0\}$ (4.2)

and let $\Omega_{H\Theta\eta} = \Omega_H \times \Omega_\Theta \times \Omega_\Theta$.

For any fixed $H\in\Omega_H$ the differential equation

$$\dot{\Theta}(t)=-Hf(\Theta(t)) \tag{4.3}$$

$$\dot{\eta}(t)=-(Hf(\Theta(t))-\Theta(t)+\eta(t)) \tag{4.4}$$

is asymptotically stable in $\Theta=\Theta^*$, $\eta=\Theta^*$ because of the block trinagular structure. The equilibrium surface Π is defined by

$$\Pi=\{(H,\Theta^*,\Theta^*):H\in\Omega_H^O\}. \tag{4.5}$$

where Ω_H^O is some open bounded subset of Ω_H such that $\Omega_H^O\subset\Omega_H$.

According to Remark 3.2., Theorem 3.1 can be applied with $\varepsilon=1$. Thus we get the following

Theorem 4.1. If Conditions 2.1, 2.2. are satisfied then the system of differential equation (2.15)-(2.17) describing the algorithm is exponentially stable on Π. As a consequence, for any initial value $H(0)$ there exists a neighbourhood S of Θ^*, such that if $\Theta(0)$, $\eta(0)\in S$, then $\Theta(t)$, $\eta(t)\to\Theta^*$ and $H(t)\to H^{**}$ as $t\to\infty$, where H^{**} is some matrix in Ω_H.

5. A STABILITY THEOREM FOR STOCHASTIC DIFFERENTIAL EQUATIONS

Now we shall extend Theorem 4.1 for stochastic systems. Let a system of stochastic differential equations

$$dx(t)=\frac{\varepsilon}{t}(p(x(t),y(t))\,dt+\sigma_x(x(t),y(t))dw(t)) \quad \varepsilon>0,\ t\geq 1 \tag{5.1}$$

$$dy(t)=\frac{1}{t}(q(x(t),y(t))dt+\sigma_y(x(t),y(t)dw(t)) \tag{5.2}$$

be given with a constant (deterministic) initial condition $x(1),y(1)$. Here x,

$p(x,y)\in R^n$ $y,q(x,y)\in R^m$ and $w(t)$ is an n+m-dimensional Wiener process defined on the probability space $(\Omega,\mathcal{F}_t,P)$. The σ-algebra generated by $w(s)$ $1\le s\le t$ is denoted by t.

Condition 5.1. The function p,q,σ_x,σ_y are continously differentiable in R^{n+m}, moreover q is twice differentiable.

Let W be the Lyapunov function defined in Definition (3.1). To apply Ito calculus we need

Condition 5.2. W is twice continously differentiable.

Let L_t denote the infinetisemal generator of the process (5.1), (5.2). Then Ito-calculus gives after elementary calculations the estimation:

$$L_tW(z)\le-(c_1/t)W(z)+d/t^2 \tag{5.3}$$

where c_1, $d>0$, for $z\in\Omega^o_{xy}$

Let τ be the stopping time defined as the first moment of hitting the boundary of Ω^o_{xy}. Furthermore let

$$\tilde{W}(t,z)=W(z)+\alpha_o t^{-\beta} \tag{5.4}$$

where $0<\alpha_o$, $0<\beta<1$ are appropriate positive constants, and let

$$U(t)=\tilde{W}(t\wedge\tau,z/t\wedge\tau)) \tag{5.5}$$

where $t\wedge\tau=\min(t,\tau)$. Finally let $A(t)=\{\omega:t\le\tau\}$, where ω stands for the elementary events of the probability space. Then we have the following.

Theorem 5.1. The process $\{U(t),\mathcal{F}_t\}$ is a supermatingale and for any $0<\rho\le1$ we have

$$E_{A(t)}U^\rho(t)\leqq ct^{-\gamma} \tag{5.6}$$

with some $c,\gamma>0$.

Remark. E_A denotes integration an the subset A of the probability space with respect to the probability measure.

The theorem is a slight modification of a theorem given in Nevelson and Khasminskii |5|. The convergence speed given in (5.6) is sufficient to prove a convergence with probability 1 theorem. We have:

Theorem 5.2. If $(x(t),y(t))(\omega)$ is a separable solution of (5.1), (5.2) for $0<\varepsilon<\varepsilon_o$ then the solution trajectory $(x(t\wedge\tau),y(t\wedge\tau))(\omega)$. converges with probability 1 and for almost all $\omega\in A$ $(x(t),y(t))(\omega)$ converges to some point of the equilibrium surface. The probability of the event A can be made arbitrarily close to 1 by starting the process from a large $t=t_o$ instead of $t=1$ and choosing $(x(t_o),y(t_o))$ close enough to the equilibrium surface.

The main idea of the proof is that using an integral representation of (5.1), (5.2) and the moment's estimation in (5.6) we can show that for any unbounded subdivision $t_o<t_1<t_2 \ldots$ of the semiinfinite interval $|t_o,\infty)$ we have

$$\sum_{i=0}^{\infty} E\,\|z(t_{i+1})-z(t_i)\| \leq c\, t_o^{-\beta/2}$$

with some $c>0$.

6. STOCHASTIC BROYDEN METHOD

The results of the previous section can be used to formulate a stochastic Broyden method. We get the following algorithm and theorem:

$$d\Theta(t)=-\frac{1}{t}\, G^{-1}(t)(f(\Theta(t))dt+\sigma(\Theta(t))dw(t)) \tag{6.1}$$

$$dG(t)=\frac{1}{t}\, d\varepsilon(t)\Theta^T(t)/\alpha(\Theta(t)) \tag{6.2}$$

$$db(t)=\frac{1}{t}\, d\varepsilon(t)/\alpha(\Theta(t)) \tag{6.3}$$

where $d\varepsilon(t)=f(\Theta(t))dt+\sigma(\Theta(t))dw(t)-G(t)\Theta(t)dt-b(t)dt$.

Theorem 6.1. Assume that the conditions of Theorem 4.1 and Condition 5.1. are satisfied. Then for any initial approximation $G^o\in\Omega_G$ of G^* and any $\rho>0$ there exists a t_o and a neighborhood S of Θ^* such that if $\Theta(t_o)\in S$ and for $b(t_o)$, $G(t_o)=G^o$ $G(t_o)\Theta(t_o)+b(t_o)$ is sufficiently small then the solution trajectory $(\Theta(t), G(t), b(t))$ starting hom $(\Theta(t_o), G(t_o), b(t_o)$ will not leave $\Omega_{\Theta G,b}$ with probabilty not less than $1-\rho$. Whenever $(x(t), G(t), b(t))$ does not leave $\Omega_{\Theta Gb}$ it converges to some point of the equilibrium surface except for a set of zero measure.

A completely similar theorem holds for the stochastic version of the inverse form. We shall state only the algorithm.

$$d\Theta(t)=-\frac{1}{t}\, H(t)(f(\Theta(t))dt+\sigma(\Theta(t))dw(t) \tag{6.4}$$

$$dH(t)=-\frac{1}{t}\, d\varepsilon(t)\Theta^T(t)H(t)/\alpha(\Theta(t)) \tag{6.5}$$

$$d\eta(t)=-\frac{1}{t}\, d\varepsilon(t)/\alpha(\Theta(t)), \tag{6.6}$$

where $d\varepsilon(t)=H(t)(f(\Theta(t))dt+\sigma(\Theta(t))dw(t))-\Theta(t)dt+\eta(t)dt$

Remark: The inverse form given here was obtained by adding noise terms to the deterministic inverse. We get a slightly different form if we invert the stochastic algorithm (6.1), (6.2), (6.3) itself.

7. DISCUSSION

The results of Section 2. 3. 4. 5. are of independent interest. But the final result in Section 6. is unsatisfactory. The expectation for deterministic quasi-Newton methods is that they should have superlinear speed of convergence. The philosophy for a stochastic quasi-Newton method should be concerned with the statistical efficieny of the method. Efficieny depends on $\lim_t G(t)$. Thus we have to ask whether the proposed algorithm is one which improves the efficiency that would have been obtained for the initial G(0). This we can not answer.

The following research is in progress: we can directly try to find efficient methods that is for which $\lim_t G(t)=G^*$. Aqain for deterministic methods it was not necessary and perhaps not possible to estimate G^* exactly. But for a stochastic method the sequence $\Theta(t)$ is sufficiently rich, so that observations at $\Theta(t)$ are sufficient to find a consistent estimator of G^*. The technical implementation of these ideas is slightly involved. The basic tool is a theorem of Ljung |4|. Our result is probably the first non-standard application of Ljung's theorem. For details we refer to Gerencsér | 2 |.

ACKNOWLEDGEMENT

The financial support of IFIP Organizing Committee to participate at the conference is gratefully acknowledged. I am indebted to I. Gyöngy and G.Zs. Vágó for helpful discussion.

REFERENCES

1. Broyden, C.G.: A class of methods for solvin nonlinear equations. Maths. Comput., 19, 577-593 (1965).

2. Gerencsér, L.: Continuous-time and stochastic quasi-Newton methods. Submitted to SIAM J. Control and Optimization. Under revision.

3. Holst, I.: Adaptive prediction and recursive estimation. Report TFRT-1013 Dept. Aut. Cont., Lund Inst. Techn., Lund, Sweden. (1977).

4. Ljung, L.: Analysis of recursive stochastic algorithms. IEEE Trans. Aut. Cont. AC-22 551-575 (1977).

5. Nevelson, M.B.; Khasminskii, R.Z.: Stochastic approximation and recurrent estimation. (in Russian), Navka, Moscow (1972).

6. Wittenmark, B.: A self-tuning predictor. IEEE Trans. Aut. Cont., AC-19, 848-851 (1974).

MODELIZATION AND FILTERING OF DISCRETE SYSTEMS AND DISCRETE APPROXIMATION OF CONTINUOUS SYSTEMS

H. KOREZLIOGLU

E.N.S.T.
46, Rue Barrault
75 634 - PARIS - FRANCE

G. MAZZIOTTO

C.N.E.T.
38-40, Rue du G. Leclerc
92 131 - ISSY - FRANCE

INTRODUCTION

This paper gives a description of time-discrete stochastic systems and discusses filtering problems concerning them. A discrete system is a pair of discrete-time random processes X and Y described by a state equation : $X_{n+1} = f_n(X_n, B_n)$ and an observation equation $Y_n = h_n(X_n, W_n)$ where the pair (B,W) is a sequence of mutually independent random variables. It is shown that the weak markovian representation problem reduces to the construction of a discrete system. The representation of the state and the observation processes in terms of noise processes B and W, as in the classical Kalman model ([5]), renders the realization and filtering problems of discrete stochastic system clearer.

The second part of the paper deals with the approximation by periodic sampling of the classical continuous system consisting of a p-dimensional state process X, strong solution of the equation : $dX_t = a(X_t)\,dt + b(X_t)\,dB_t$ and a q-dimensional process Y defined by : $dY_t = c(X_t)\,dt + dW_t$, where B and W are independent Brownian motions; a, b and c are time-independent, bounded Lipschitzian functions. The following approximation steps are considered.

1. Y is sampled with a sampling period of length h.
2. Moreover, $c(X_t)$ is replaced by a process $\tilde{C}$ taking the constant value $c(X_{(n-1)h})$ in the interval $[(n-1)h, nh[$.
3. After all these, X is replaced by a process $\bar{X}^h$ which is a discrete-time Markov chain approximating it at the sampling points and takes the constant value $\bar{X}^h_{(n-1)h}$ in the interval $[(n-1)h, nh[$.

For each step the corresponding recursive filtering algorithm is derived and the order of the approximation of the continuous filter is given here without its proof. Further details will appear in a forthcoming paper.

1. DISCRETE SYSTEMS

$(\Omega, A, \mathbb{P})$ is a probability space on which random variables and processes of this paragraph are defined. $(E,\underline{\underline{E}})$, $(F,\underline{\underline{F}})$, $(E',\underline{\underline{E}}')$, $(F',\underline{\underline{F}}')$ are measurable spaces. Whenever a cartesian product of these spaces is considered it is understood that the corresponding σ-algebra is the product σ-algebra of the factors. If a space E is a Polish space, $\underline{\underline{E}}$ is the σ-algebra of Borel sets of E. If E is a finite space, then its topology is the discrete topology and $\underline{\underline{E}}$ is then the set of all subsets of E.

DEFINITION 1.1: A discrete stochastic system is a pair of random processes $X = (X_n\,;\, n \in \mathbb{N})$ and $Y = (Y_n\,;\, n \in \mathbb{N})$ with values in E and F, respectively, and satisfying the following state and observation equations

(1.1) $X_{n+1} = f_n(X_n, B_n) \qquad n \geq 0$

(1.2) $Y_n = h_n(X_n, W_n) \qquad n \geq 0$,

where $(B,W) = ((B_n, W_n)\,;\, n \in \mathbb{N})$ is a sequence of $E' \times F'$-valued independent random variables, independent of X_0, f_n and h_n are measurable non-random mappings of $E \times E'$ into E and $E \times F'$ into F, respectively.

The system (X,Y) is said to be finite if E, E', F, F' are finite spaces.

The above definition has an obvious stationary version.

In the applications we shall deal with here, spaces E, E', F and F' will be either finite spaces or $\mathbb{R}^n$-spaces. Before considering these applications we give the following elementary lemma.

LEMMA 1.2: Let U be a random variable with values in a Polish space $(R,\underline{\underline{R}})$ and V a random variable with values in a space $(S,\underline{\underline{S}})$. There is then a random variable W, independent of V, with values in some space $(T,\underline{\underline{T}})$, and there is a measurable mapping f of $S \times T$ into R such that U and $U' = f(V,W)$ have almost surely the same conditional probability distribution given V.

Proof: Let P_v be the conditional probability distribution of U given V. P_v is defined for all $v \in S$ except on a negligible set S_0. For $v \in S_0$, we can choose the conditional probability distribution equal to an arbitrary distribution.
Let us define $(T,\underline{\underline{T}},Q)$ by $(R^S, \underline{\underline{R}}^{\otimes S}, \otimes_{v\in S} P_v)$, W by the generic point of T, and f by : $\forall\, v \in S$, $f(v,.) : T \to R$ is the projection on the vth component of T. If $U' = f(U,W)$, then U' is a random variable on the product probability space $(\Omega \times T, \underline{\underline{A}} \otimes \underline{\underline{T}}, P \otimes Q)$. It is easily seen that U and U' have a.s. the same conditional probability distribution given V.

REMARK 1.3: The above proof shows that the space T need not to be larger than R^S. The choice of a smaller space depends of course on the given conditional distribution. In the case where R and S are finite spaces, as far as we know, the construction of the minimal space T is an open problem. There is an interesting case for which the minimal space T is easy to construct. Let the conditional probability of U given V be defined by the transition matrix $(P_v(u))_{(v,u)\in S\times R}$ and suppose that all the rows $(P_v\,;\,v\in S)$ are permutations of each other. Then T can be chosen in such a way that the number of its points is equal to the number of nonzero elements of the set $(P_v(u)\,;\,u\in R)$ for fixed v. In fact, let v_0 be fixed in S and let W be the generic point of the support $T\subset R$ of $P_{v_0}(.)$ with the distribution $(P_{v_0}(u)\,;\,u\in R)$. f can be chosen in such a way that for fixed v the mapping f(v,.) of T into R is one-to-one, so that the image of P_{v_0} under f(v,.) is the permutation of P_{v_0} into P_v when restricted to their nonzero elements.

The application of Lemma 1.2 to transition probabilities of a discrete-time Markov process X shows that X can be weakly realized by an automaton.

PROPOSITION 1.4: Let $\tilde{X}=(\tilde{X}_n\,;\,n\in\mathbb{N})$ be a Markov process with values in a Polish space E. Then there is a sequence $B=(B_n\,;\,n\in\mathbb{N})$ of independent random variables, independent of X_0, and measurable mappings $f_n : E\times E \to E$ such that the E-valued Markov chain X defined by

$$(1.3)\qquad X_{n+1}=f_n(X_n,B_n)\quad\text{with}\quad X_0=\tilde{X}_0$$

has the same probability distribution as $\tilde{X}$.

Finite stationary Markov chains transmitted through memoryless stationary communication channels have been successfully used for speech modelization and identification (cf. for instance ([7])). Applying again Lemma 1.2 we see that such situations can be modelled by a state equation and an observation equation as follows:

$$X_n=f(X_{n-1},B_n)\quad,\qquad Y_n=h(X_n,W_n)\qquad n\in\mathbb{Z}\quad,$$

where $((B_n,W_n)\,;\,n\in\mathbb{N})$ is a sequence of independent random variables with values in a finite space $E'\times F'$.

This brings us to the finite Markovian realization problem of finite-space valued stationary processes. We refer to ([11]) for a comprehensive introduction to the subject.

DEFINITION 1.5: Let F be a finite space and let $\tilde{Y}=(\tilde{Y}_n\,;\,n\in\mathbb{Z})$ be an F-valued stationary process. $\tilde{Y}$ is said to have a weak finite Markovian realization if there exists a Markov chain X with values in a finite space E and a process Y with values in F, such that (X,Y) is stationary, the σ-algebra generated by $(X_k, Y_k\,;\,k>n)$, the future, and the σ-algebra generated by $(X_k, Y_k\,;\,k\le n)$, the past, are conditionally independent given X_n, $\forall\, n\in\mathbb{Z}$, and that $\tilde{Y}$ and Y have the same distribution.

The following proposition shows that the two processes X and Y in the above definition are equivalent to a finite stochastic system.

PROPOSITION 1.6: Let $\tilde{Y}$ be as in Definition 1.5. Then it has a finite weak Markovian realization iff there is a stationary stochastic finite system

(1.4) $X'_n = f(X'_{n-1}, V_n)$, $Y'_n = h(X'_{n-1}, V_n)$, $n \in \mathbf{Z}$

with an F-valued Y' and independent identically distributed (i.i.d.) V_n's such that Y' and $\tilde{Y}$ have the same distribution.

Proof: It is proved in ([11]) that the distribution of the pair (X,Y) in Definition 1.5 is entirely determinated by the transition probability matrix $(P(X_{n+1} = j\ ,\ Y_{n+1} = y\ /\ X_n = i))$. According to Lemma 1.2 it is possible to find i.i.d. random variables V_n's and mappings f and h such that the finite system (X',Y') of (1.4) has the same distribution as the pair (X,Y) if X_0 and X'_0 have the same (stationary) distribution. Conversely, it is obvious that if there is a finite system (X',Y') satisfying the conditions of the proposition, then it has the same properties as the pair (X,Y) of Definition 1.5. Therefore, $\tilde{Y}$ has a weak finite Markovian realization.

We would like to remark that a finite stochastic system was defined in ([11]) as the pair (X,Y) of Definition 1.5. The above proposition justifies our Definition 1.1.

The filtering algorithm for a finite discrete system (1.1) and (1.2) can be carried out as in ([1]). We denote by $P(x_{s_1},..,x_{s_m}, y_{t_1},..,y_{t_n} / x_{u_1},..,x_{u_p}, y_{v_1},..,y_{v_q})$ the conditional probability of $\{X_{s_1} = x_{s_1},..,X_{s_m} = x_{s_m}, Y_{t_1} = y_{t_1},...,Y_{t_m} = y_{t_m}\}$ given $\{X_{u_1} = x_{u_1},..,X_{u_p} = x_{u_p}, Y_{v_1} = y_{v_1},..,Y_{v_q} = y_{v_q}\}$ and by y^n the sequence $(y_0,..,y_n)$. The algorithm is then the following.

For $n \geq 1$ determine $\sigma_{n+1/n}(x_{n+1}/y^n)$ and $\sigma_{n/n}(x_n/y^n)$ by

$$(1.5)\quad \begin{cases} \sigma_{n+1/n}(x_{n+1}/y^n) = \sum_{x_n \in E} P(x_{n+1}, y_n/x_n)\ \sigma_{n/n-1}(x_n/y^{n-1}) \\ \sigma_{n/n}(x_n/y^n) = P(y_n/x_n)\ \sigma_{n/n-1}(x_n/y^{n-1}) \\ \text{with}\quad \sigma_{0/-1}(x_0/y^{-1}) = P(x_0) \end{cases}$$

and then compute

$$(1.6)\qquad P(x_n/y^n) = \sigma_{n/n}(x_n/y^n)\ /\ \sum_{x_n \in E} \sigma_{n/n}(x_n/y^n)\ .$$

In the case where B and W are independent processes, the algorithm for $\sigma_{n/n}$ becomes

$$(1.7)\qquad \sigma_{n/n}(x_n/y^n) = P(y_n/x_n) \sum_{x_{n-1} \in E} P(x_n/x_{n-1})\ \sigma_{n-1/n-1}(x_{n-1}/y^{n-1})$$

$$\text{with}\quad \sigma_{0/0}(x_0/y^0) = P(y_0/x_0)\ P(x_0)\quad .$$

If $E = \mathbb{R}^p$ and $F = \mathbb{R}^q$ and if all the probability distributions are defined by densities, then the filtering algorithms are the same as those above where conditional probabilities are to be replaced by conditional densities and sums by integrals.

2. APPROXIMATIONS OF CONTINUOUS SYSTEMS

All the processes considered in this section will be indexed on the bounded interval $[0,T] \subset \mathbb{R}_+$.

Let $B = (B_t ; t \in [0,T])$ and $Y = (Y_t ; t \in [0,T])$ be two independent Brownian motions with values in $\mathbb{R}^m$ and $\mathbb{R}^q$, and defined on their canonical probability spaces $(\Omega^B, \underline{\underline{B}}_T, \mathbb{P}_0^B)$ and $(\Omega^Y, \underline{\underline{Y}}_T, \mathbb{P}_0^Y)$. $\underline{\underline{F}}^B = (\underline{\underline{B}}_t ; t \in [0,T])$ and $\underline{\underline{F}}^Y = (\underline{\underline{Y}}_t ; t \in [0,T])$ will denote their natural filtrations. The reference probability space on which the "state and observation" model is constructed is $(\Omega, \underline{\underline{F}}_T, \mathbb{P}_0)$ where $\Omega = \Omega^B \times \Omega^Y$, $\underline{\underline{F}}_T$ is the completion of $\underline{\underline{B}}_T \otimes \underline{\underline{Y}}_T$ with respect to $\mathbb{P}_0^B \otimes \mathbb{P}_0^Y$, and $\mathbb{P}_0$ is the extension of $\mathbb{P}_0^B \otimes \mathbb{P}_0^Y$ to $\underline{\underline{F}}_T$. $(\Omega, \underline{\underline{F}}_T, \mathbb{P}_0)$ is given the filtration $\underline{\underline{F}} = (\underline{\underline{F}}_t ; t \in [0,T])$ where $\underline{\underline{F}}_t$ is the σ-algebra generated by $\underline{\underline{B}}_t \otimes \underline{\underline{Y}}_t$ and the negligible sets of $\underline{\underline{F}}_T$. We denote by E_0, E_0^B and E_0^Y the expectations on $(\Omega, \underline{\underline{F}}_T, \mathbb{P}_0)$, $(\Omega^B, \underline{\underline{F}}_T^B, \mathbb{P}_0^B)$ and $(\Omega^Y, \underline{\underline{F}}_T^Y, \mathbb{P}_0^Y)$, respectively.

Let $X = (X_t ; t \in [0,T])$ be a Markov diffusion with values in $\mathbb{R}^p$, defined on the probability space $(\Omega^B, \underline{\underline{F}}_T^B, \mathbb{P}_0)$ (or equivalently on $(\Omega, \underline{\underline{F}}_T, \mathbb{P}_0)$) as the unique strong solution of the stochastic differential equation

$$(2.1) \qquad X_t = x_0 + \int_0^t a(X_s)\, ds + \int_0^t b(X_s)\, dB_s \quad , \quad t \in [0,T]$$

where a and b are bounded and Lipschitzian functions on $\mathbb{R}^p$ with values in $\mathbb{R}^p$ and the space of $p \times m$-matrices, respectively, and $x_0 \in \mathbb{R}^p$. We denote by $(P_t ; t \in [0,T])$ the transition semigroup of X.

Let $L = (L_t ; t \in [0,T])$ be defined on $(\Omega, \underline{\underline{F}}_T, \mathbb{P}_0)$ by

$$(2.2) \qquad L_t = \exp\left\{ \int_0^t c(X_s).dY_s - \frac{1}{2} \int_0^t |c(X_s)|^2\, ds \right\}$$

where c is a Lipschitzian and $\mathbb{R}^q$-valued bounded function on $\mathbb{R}^p$. ($|.|$ denotes the Euclidian norm on $\mathbb{R}^n$ and a.b the scalar product of two elements a, b $\in \mathbb{R}^n$).

It is known ([8]) that L is positive $(\underline{\underline{F}}, \mathbb{P}_0)$-martingale such that $E_0(L_T) = 1$, $L_T > 0$ a.s. and $L_T \in \mathbb{L}^p(\Omega, \underline{\underline{F}}_T, \mathbb{P}_0)$ for $p \in [1, \infty[$. Then $\mathbb{P} = L_T \cdot \mathbb{P}_0$ is a probability measure on $(\Omega, \underline{\underline{F}}_T)$ equivalent to $\mathbb{P}_0$. B is also $(\underline{\underline{F}}, \mathbb{P})$-Brownian motion and Y has the following decomposition

$$(2.3) \qquad Y_t = \int_0^t c(X_s)\, ds + W_t$$

where W is another $(\underline{\underline{F}}, \mathbb{P})$-Brownian motion independent of B.

The state and observation processes are then described on $(\Omega, \underline{\underline{F}}_T, \mathbb{P})$ by Equations (2.1) and (2.3).

The expectation on $(\Omega, \underline{\underline{F}}_T, \mathbb{P})$ will be denoted by E. $b(\mathbb{R}^n)$ will denote the space of all bounded real Borel functions on $\mathbb{R}^n$ with the sup-norm written by $\|.\|$.

The unnormalized filtering process is the process $\sigma = (\sigma_t \,;\, t \in [0,T])$ with values in the space of positive measures on $\mathbb{R}^p$, defined by

$$(2.4) \qquad \sigma_t(f)(\omega^Y) = \int_{\Omega^B} L_t(\omega^B,\omega^Y)\, f(X_t(\omega^B))\, \mathbb{P}_0^B(d\omega^B) \qquad t \in [0,T]$$

for all $f \in b(\mathbb{R}^p)$. The filtering process is then the process $\Pi = (\Pi_t, t \in [0,T])$ with values in the space of probability measures on $\mathbb{R}^p$ given by the Kallianpur-Striebel formula:

$$(2.5) \qquad \Pi_t(f) = \sigma_t(f) \,/\, \sigma_t(1) \quad , \quad t \in [0,T]$$

for all $f \in b(\mathbb{R}^p)$, (4).

Suppose that the observation process Y is sampled with a constant period of length h and that the information about Y is given by the samples $Y_0 = 0, Y_h, \ldots, Y_{nh}, \ldots$ with $n \leq [T/h]$. ($[a/b]$ denotes the integer part of a/b). We denote by $\underline{\underline{Y}}^h_t$ the σ-algebra generated by $(Y_{nh} \,;\, n \leq [t/h])$ and we note that it is also generated by the set $(\Delta Y^h_n \,;\, n \leq [t/h])$ where $\Delta Y^h_n = Y_{nh} - Y_{(n-1)h}$ and $\Delta Y^h_0 = Y_0 = 0$.
Under $\mathbb{P}_0$ these random variables are mutually independent Gaussian variables with mean 0 and covariance matrix hI_q.

The filtering process Π^h with respect to sampled observations $(Y_{nh} \,;\, n \leq [T/h])$ is defined by

$$\Pi^h_t(f) = E(f(X_t) \,/\, \underline{\underline{Y}}^h_t) \quad \text{a.s.} \quad \text{for all} \quad f \in b(\mathbb{R}^p) .$$

As Π in (2.5), Π^h is given by

$$(2.6) \qquad \Pi^h_t(f) = \sigma^h_t(f) \,/\, \sigma^h_t(1) \quad , \quad \text{for all} \quad f \in b(\mathbb{R}^p) ,$$

where

$$(2.7) \qquad \sigma^h_t(f) = E_0(L_t\, f(X_t) \,/\, \underline{\underline{Y}}^h_t) = E_0(\sigma_t(f) \,/\, \underline{\underline{Y}}^h_t) \quad \text{a.s.} \quad .$$

σ^h is the unnormalized filtering process corresponding to sampled observation Y^h.

<u>THEOREM 2.1</u>: Let the kernel H_n be defined by: $\forall\, f \in b(\mathbb{R}^p)$,

$$(2.8) \qquad H_n(X_{(n-1)h}, f) = $$
$$= E^B_0(\exp\{(\tfrac{1}{h}\int_{(n-1)h}^{nh} c(X_s)\, ds).\Delta Y^h_n - \tfrac{1}{2h}\left|\int_{(n-1)h}^{nh} c(X_s)\, ds\right|^2\}\, f(X_{nh}) \,/\, X_{(n-1)h})$$

Then for $t = nh$, $n = 1,2,\ldots,[T/h]$, the unnormalized filter σ^h_t is obtained by the following recurrence formula

$$(2.9) \qquad \sigma^h_{nh}(dx) = \int_{\mathbb{R}^p} H_n(u,dx)\, \sigma^h_{(n-1)h}(du) \quad , \text{ and}$$

$$(2.10) \qquad \sigma^h_t(dx) = \int_{\mathbb{R}^p} P_{t-(n-1)h}(u,dx)\, \sigma^h_{(n-1)h}(du) \qquad \text{for} \quad (n-1)h < t < nh \quad .$$

Proof: In order to shorten the notations, we put

$$Z^{nh}_{(n-1)h} = \exp\{ \int_{(n-1)h}^{nh} c(X_s).dY_s - \frac{1}{2} \int_{(n-1)h}^{nh} |c(X_s)|^2 \, ds\}$$

We first remark the following

$$(2.11) \qquad \sigma^h_t(f)(\omega^Y) = \int_{\Omega^B} E^Y_0(L_t(\omega^B,.) / \underline{\underline{Y}}^h_t)(\omega^Y) \; f(X_t(\omega^B)) \; \mathbb{P}^B_0(d\omega^B)$$

On the other hand, we have

$$(2.12) \qquad E^Y_0(L_t / \underline{\underline{Y}}^h_t) = \prod_{k=1}^{[t/h]} E^Y_0(Z^{kh}_{(k-1)h} / \Delta Y^h_k) .$$

Then it is proved that

$$(2.13) \qquad E^Y_0(Z^{kh}_{(k-1)h} / \Delta Y^h_k) = \exp\{c^h_k.\Delta Y^h_k - \frac{h}{2}|c^h_k|^2\} \quad \text{where} \quad c^h_k = \frac{1}{h} \int_{(k-1)h}^{kh} c(X_s) \, ds.$$

Now, let us put

$$\hat{Z}^{kh}_{(k-1)h} = \exp\{c^h_k.\Delta Y^h_k - \frac{h}{2}|c^h_k|^2\} \quad , \quad \hat{L}_{nh} = \prod_{k=1}^{n} \hat{Z}^{kh}_{(k-1)h}$$

According to (2.11), (2.12), (2.13), for $f \in b(\mathbb{R}^p)$ we have

$$\sigma^h_{nh}(f) = E^B_0(\hat{L}_{(n-1)h} \; \hat{Z}^{nh}_{(n-1)h} \; f(X_{nh})) = E^B_0(\hat{L}_{(n-1)h} \; E^B_0(\hat{Z}^{nh}_{(n-1)h} \; f(X_{nh}) / \underline{\underline{B}}_{(n-1)h}))$$

$$= E^B_0(\hat{L}_{(n-1)h} \; E^B_0(\hat{Z}^{nh}_{(n-1)h} \; f(X_{nh}) / X_{(n-1)h})) .$$

From this, Formula (2.9) is deduced. Formula (2.10) is derived in the same way.

The above theorem improves the results obtained in ([9]) and in ([14]).

We give here without proof the degree of the approximation when the filter Π is replaced by the filter Π^h corresponding to the sampled observation Y^h.

THEOREM 2.2: For fixed $f \in b(\mathbb{R}^p)$ and $t \in [0,[T/h]h]$ the following assertions hold.

i) The net $(\sigma^h_t(f) \, ; h > 0)$ (resp. $(\Pi^h_t(f) \, ; h > 0)$) converges to $\sigma_t(f)$ (resp. $\Pi_t(f)$) in the space $\mathbb{L}^n(\Omega^Y, \underline{\underline{Y}}_T, \mathbb{P}^Y_0)$ (resp. $\mathbb{L}^n(\Omega^Y, \underline{\underline{Y}}_T, \mathbb{P})$) for $n \in [1,\infty[$ as h decreases to 0. The convergence is a.s. on a set of totally ordered partitions $(0 < h_i < 2h_i < \ldots < [T/h_i]h_i \; ; \; i \in \mathbb{N})$.

ii) $\quad E^Y_0(|\sigma_t(f) - \sigma^h_t(f)|^2) = O(h)$ and $E^Y(|\Pi_t(f) - \Pi^h_t(f)|) = O(\sqrt{h})$.

Within an error margin of order $\sqrt{h}$, the recursive filtering formula (2.9) replaces the well-known Zakai equation for the non-linear filtering, ([15]). For a complete computation of σ^h the main difficulty is then the computation of the kernel H_n of (2.8), which is computed by means of the transition semigroup of X. But in most of the cases this semigroup is not easy to compute. To ease matters, one naturally suggests the approximation of the Markov diffusion X by means of a discrete time Markov process $(\bar{X}^h_{nh}\ ; n=0,1,..,[T/h])$ with $\bar{X}^h_0 = x_0$, such that

$$(2.14) \qquad E(|X_{nh} - \bar{X}^h_{nh}|^2) = O(h)$$

and we may approximate X by the process $\bar{X}^h$ defined by

$$(2.15) \qquad \bar{X}^h_t = \bar{X}^h_{(n-1)h} \quad \text{for} \quad (n-1)h \leq t < nh \ .$$

It is easily computed that $E(|X_t - \bar{X}^h_t|^2) = O(h)$ for all t, because $E(|X_t - X_{(n-1)h}|^2) = O(h)$ for $t \in [(n-1)h,nh]$, ([3]).

An example of such a Markov chain is given by the classical Euler scheme, scheme, ([12]), ([13]):

$$(2.16) \qquad \bar{X}^h_{nh} = \bar{X}^h_{(n-1)h} + a(\bar{X}^h_{(n-1)h})\ h + b(\bar{X}^h_{(n-1)h}).\Delta B^h_n$$

where $\Delta B^h_n = B_{nh} - B_{(n-1)h}$, for $n \geq 1$, with $\bar{X}^h_0 = x_0$.

We put

$$(2.17) \qquad \bar{L}^h_t = \exp\{\int_0^t c(\bar{X}^h_s).dY_s - \frac{1}{2}\int_0^t |c(\bar{X}^h_s)|^2\ ds\}$$

where $\bar{X}^h$ is derived as in (2.15) from a Markov chain satisfying (2.14). We denote by $\bar{\sigma}^h$ and $\bar{\Pi}^h$ the corresponding unnormalized and normalized filters, defined by

$$\bar{\sigma}^h_t(f) = E_0(\bar{L}_t\ f(\bar{X}^h_t)\ /\ \underline{\underline{Y}}^h_t) \quad \text{and} \quad \bar{\Pi}^h_t(f) = \bar{\sigma}^h_t(f)\ /\ \bar{\sigma}^h_t(1) \qquad \text{for } f \in b(\mathbb{R}^p).$$

We then have the following results.

<u>THEOREM 2.3</u>: i) Let $\bar{P}^h$ denote the one-step transition probability of the chain $(\bar{X}^h_{nh}\ ; n=0,1,..,[T/h])$. Then for $t = nh$; $n=1,..,[T/h]$, $\bar{\sigma}^h$ is given by

$$(2.18) \qquad \bar{\sigma}^h_{nh}(dx) = \int_{\mathbb{R}^p} L(u,\Delta Y^h_n)\ \bar{P}^h(u,dx)\ \bar{\sigma}^h_{(n-1)h}(du)$$

with $\bar{\sigma}^h_0(dx) = \delta_{x_0}(dx)$, where

$$L(u,v) = \exp\{c(u).v - \frac{h}{2}|c(u)|^2\} \qquad u \in \mathbb{R}^p,\ v \in \mathbb{R}^q \ .$$

For $(n-1)h < t < nh$, we have $\bar{\sigma}^h_t(dx) = \bar{\sigma}^h_{(n-1)h}(dx)$

ii) For a Lipschitzian $f \in b(\mathbb{R}^p)$ the following holds.

$$(2.19) \qquad E_0(|\sigma_t(f) - \bar{\sigma}^h_t(f)|^2) = O(h)$$

$$(2.20) \qquad E(|\Pi_t(f) - \bar{\Pi}^h_t(f)|) = O(\sqrt{h}) \ .$$

REMARK 2.4: The approximation of Step 2 mentioned in the introduction, corresponds to the case where $\overline{X}^h_t = X_t$ for all t, with $c(\overline{X}^h)$ in (2.17) replaced by the process $\tilde{C}$ described there. In this case, $\overline{P}^h$ in (2.18) coincides with P_h and for $(n-1)h < t < nh$ we have

$$\overline{\sigma}^h_t(dx) = \int_{\mathbb{R}^p} P_{t-(n-1)h}(u,dx)\, \overline{\sigma}^h_{(n-1)h}(du)$$

For the computation of the convergence rates of Part ii) in this case, f does not need to be Lipschitzian. In a recent work ([10]), J. Picard obtained a bound of order h^2 for (2.19) and of order h for (2.20), when B, X and Y are real processes under the supplementary conditions that f should be Lipschitzian and c should be a C^2-function with bounded first and second derivatives.

In this work we have not discussed the quantization problem for Y^h and $\overline{X}^h$. Some approximation results on the subject can be found in ([6]) and ([2]). We think that the quantization of X may be considered as a source coding problem and deserves to be looked deeper into.

REFERENCES

(1) K.J. ASTRÖM : "Optimal control of Markov processes with incomplete state information". J. Math. Anal. and Appl. 10, 174-205 (1965).

(2) G.B. DI MASI and W.J. RUNGGALDIER : "Approximations and bounds for discrete-time filtering". Preprint (1982).

(3) I.I. GIKHMAN and A.V. SKOROKHOD : "Stochastic differential equations". Springer Verlag, Berlin (1972).

(4) G. KALLIANPUR and C. STRIEBEL : "Estimation of stochastic systems". Ann. Math. Stat. 39, 785-801 (1968).

(5) R.E. KALMAN : "A new approach to linear filtering and prediction problems". Trans. ASME, J. Basic Eng. 82 D, 34-45 (1960).

(6) H.J. KUSHNER : "Probability methods for approximations in stochastic control and for elliptic equations". Academic Press, New York (1977).

(7) S.E. LEVINSON , L.A. RABINER and M.M. SONDHI : "An introduction to the application of the theory of probabilistic functions of a Markov process to automatic speech recognition". The Bell Syst. Tech. J. 62-4, 1035-1074 (1983).

(8) R.S. LIPTSER and A.N. SHIRYAYEV : "Statistics of random processes 1: general theory". Springer Verlag, Berlin (1977).

(9) J.T. LO : "Optimal nonlinear estimation 2: discrete observation". Information Sc. 7, 1-10 (1974).

(10) J. PICARD : "Approximation of nonlinear filtering problems and order of convergence". To be published in "Filtering and control of random processes" Proceedings of the ENST-CNET Colloquium, Paris February 1983.

(11) G. PICCI : "On the internal structure of finite state stochastic processes". Lect. N. in Economics and Math. Syst. 162, Springer Verlag, Berlin (1979).

(12) E. PLATEN : "Approximation of Ito integral equations". Lect. N. in Control and Inf. Sc. 28, 172-176, Springer Verlag, Berlin (1980).

(13) E. PLATEN : "Approximation method for a class of Ito processes". Lietuvos Mat. Rinkinys XXI-1, 121-133 (1981).

(14) Y. TAKEUCHI and H. AKASHI : "Nonlinear filtering formulas for discrete time observations". SIAM J. Control and Opt. 12-2, 224-261 (1981).

(15) M. ZAKAI : "On the optimal filtering of diffusion processes". Z. Wahr. V. Geb. 11, 230-249 (1969).

EXTREMALS IN STOCHASTIC CONTROL THEORY*

U. G. Haussmann
Department of Mathematics
University of British Columbia
Vancouver, Canada, V6T 1Y4

1. INTRODUCTION

The problem of optimally controlling a stochastic system has been studied for well over twenty years. Some success has been achieved in applying the theory to management science problems (e.g. inventory theory), where usually the model is discrete in the time variable and the solution is obtained via dynamic programming. For models which are continuous in the time variable (i.e. for diffusions), the method of dynamic programming reduces to solving a quasilinear parabolic partial differential equation - a very difficult proposition. Hence it is not surprising that success has been very sparse. However there is another approach which has been quite successful in the deterministic theory: apply the necessary conditions, i.e. the Pontryagin maximum principle, to eliminate most non-optimal controls leaving only the "extremal" controls, and then establish optimality by possibly appealing to existence and uniqueness. Our aim is to show that a similar technique can be applied to the problem of optimally controlling a diffusion.

In section two we state the problem and give the necessary conditions. In sections three and four we consider two examples.

2. THE PROBLEM AND NECESSARY CONDITIONS

Let us recall a simple deterministic problem which will serve as a model for the stochastic case.

Find a function $u:[0,T] \to U$ so as to minimize

$$J_0(u) = c_0(x(T)) + \int_0^T \ell_0(t, x(t), u(t))\, dt$$

subject to

*This work was supported by NSERC under grant A 8051.

$$J_1(u) = c_1(x(T)) + \int_0^T \ell_1(t, x(t), u(t))\, dt \leq 0 \qquad (1)$$

$$J_2(u) = c_2(x(T)) + \int_0^T \ell_2(t, x(t), u(t))\, dt = 0 \qquad (2)$$

$$\frac{dx}{dt} = f(t, x, u(t)) \, , \; x(0) = x_0 . \qquad (3)$$

Here T is fixed, finite, U is a closed subset of d-dimensional Euclidean space $\mathbb{R}_d$, x and f are $\mathbb{R}_n$-valued and c_i, ℓ_i, i = 0, 1, 2 are scalar (so c and ℓ are $\mathbb{R}_3$-valued).

A control function u is _optimal_ if it solves the above problem.

A control function u is _feasible_ if there exists $x:[0,T] \to \mathbb{R}_n$ such that (1), (2), (3) hold.

Let ' denote transpose. The _Hamiltonian_ of the problem is

$$H(t, x, u, p, \lambda) = p'f(t,x,u) + \lambda'\ell(t,x,u)$$

for p in $\mathbb{R}_n$, λ in $\mathbb{R}_3$.

The celebrated maximum principle of Pontryagin et al. states

_If_ $\hat{u}$ _is optimal with_ $\hat{x}$ _the corresponding solution_ of (3), _then_

(i) _there exists_ $\lambda = (\lambda_0, \lambda_1, \lambda_2)'$ in $\mathbb{R}_3$, $\lambda \neq 0$, $\lambda_0 \leq 0$, $\lambda_1 \leq 0$, $\lambda_1 J_1(\hat{u}) = 0$,

(ii) _there exists an_ $\mathbb{R}_n$_-valued function_ p _such that_

$$\frac{dp'}{dt} = - \frac{\partial H}{\partial x}(t, \hat{x}(t), \hat{u}(t), p(t), \lambda),$$

$$p(T)' = \lambda' \frac{\partial c}{\partial x}(\hat{x}(T)),$$

(iii) _for almost all_ t

$$H(t,\hat{x}(t), \hat{u}(t), p(t), \lambda) = \sup_{u \in U} H(t,\hat{x}(t),u,p(t),\lambda).$$

We say that $\hat{u}$ is _extremal_ if it is feasible and (i), (ii), (iii) hold. Now the theorem can be restated.

Theorem 1. _If_ $\hat{u}$ _is optimal, then it is extremal_.

Let us now turn to the stochastic problem. Suppose that the state equation (3) is perturbed by noise, i.e.

$$\frac{dx}{dt} = f(t,x,u(t)) + \text{noise}.$$

Now the functions J_i are random variables so we replace them by their expectations. More precisely the problem, which we call (P), is:

<u>Find a control law</u> u <u>assuming values in the closed set</u> U <u>so as to minimize</u> $J_0(u)$ <u>subject to</u>

$$J_1(u) \leq 0, \quad J_2(u) = 0 \tag{4}$$

$$x(t) = x_0 + \int_0^t f(s,x(s),\ u(s,x(\cdot)))\ ds + \int_0^t \sigma(s)\ dw(s) \tag{5}$$

<u>where</u>

$$J_i(u) = E\{c_i(x(T)) + \int_0^T \ell_i(t,x(t),u(t,x(\cdot)))\ dt\},\ i = 0,\ 1,2.$$

Here w is a standard n-dimensional Brownian motion and $\int \sigma(s)dw(s)$ is an Itô integral with $\sigma(t)$ an $n \times n$ matrix. We could let σ depend on x but this would only complicate and obscure the presentation. The control function u is to be a feedback law on the past state; a convenient way to satisfy this requirement is to assume that

$$u : [0,T] \times \underline{C}([0,T]; \mathbb{R}_n) \to U$$

is Borel measurable and adapted to the canonical Borel filtration on $\underline{C}([0,T]; \mathbb{R}_n)$, the space of all $\mathbb{R}_n$-valued continuous functions defined on [0,T] with the topology of uniform convergence. E stands for expectation. The equation (5) is also written as

$$dx = f(t,x,u(t,x(\cdot)))dt + \sigma(t)dw(t),\ x(0) = x_0.$$

Observe that if $\sigma = 0$ then (P) is just the deterministic problem considered above. We carry on the analogy with just one modification.

The control law u is <u>optimal</u> if it solves (P).

The control law u is <u>feasible</u> if there exists a stochastic process $\{x(t): 0 \leq t \leq T\}$ such that (4) and (5) hold.

The <u>Hamiltonian</u> of (P) is

$$H(t,x,u,p,\lambda) = p'\ f(t,x,u) + \lambda'\ \ell\ (t,x,u).$$

The control law $\hat{u}$ is <u>extremal</u> if it is feasible (with corresponding $\hat{x}$) and

(i) there exists $\lambda = (\lambda_0,\lambda_1,\lambda_2)'$ in $\mathbb{R}_3$, $\lambda \neq 0, \lambda_0 \leq 0, \lambda_1 \leq 0, \lambda_1 J_1(\hat{u}) = 0$,

(ii) there exists an $\mathbb{R}_n$-valued stochastic process $\overline{p}$ such that

$$\frac{d\overline{p}'}{dt} = \frac{\partial H}{dx}(t,\ \hat{x}(t),\ \hat{u}(t,\ \hat{x}(\cdot)),\ p(t),\lambda),$$

$$\overline{p}(T)' = \lambda' \frac{\partial c}{\partial x}(\hat{x}(T)), \quad \text{w. p. 1},$$

(iii) if $p(t) = E\{\overline{p}(t)|\hat{x}(s),\ 0 \leq s \leq t\}$ then for almost all t

$$H(t,\hat{x}(t),\hat{u}(t,\hat{x}(\cdot)),\lambda) = \sup_{u \in U} H(t,\hat{x}(t),\ u,\ p(t),\ \lambda) \quad \text{w. p. 1.}$$

Observe that in (iii) we have replaced $\overline{p}$ by p, its conditional expectation given the past. Since $\overline{p}$ depends on the future, then the maximization of the Hamiltonian containing $\overline{p}$ would be impossible to carry out. We remark that if either or both of the constraints J_1 and J_2 are absent then the corresponding λ is just in $\mathbb{R}_1$ or $\mathbb{R}_2$.

We list the required assumptions, giving them the label (A):

$f,\ \ell$ are continuous in u for each (t,x) (A1)

$f,\ \ell,\ c$ are continuously differentiable in x for each (t,u) (A2)

$\frac{\partial f}{\partial x},\ \sigma,\ \sigma^{-1}$ are bounded (A3)

$|f(t,x,u)| \leq k_1(1 + |x| + |u|)$ (A4)

$|\ell(t,x,u)|+|c(x)| \leq k_2(1+|x|^q+|u|^q), \quad 1 \leq q < \infty,$ (A5)

$|\frac{\partial \ell}{\partial x}(t,x,u)|+|\frac{\partial c}{\partial x}(x)| \leq k_3(1+|x|^{q-1}+|u|^q)$ (A6)

$|u(t,x(\cdot))| \leq k(u)(1 + \sup_{0 \leq s \leq t}|x(s)|)$ where k(u) is a constant depending on u (A7)

Theorem 2. <u>Assume</u> (A). <u>If</u> $\hat{u}$ <u>is optimal for</u> (P) <u>then it it extremal for</u> (P).

The proof can be found in [5], or under slightly stronger hypotheses in [1], [2], [3].

3. THE LINEAR REGULATOR

The problem is

$$\min E\{x(T)'Dx(T) + \int_0^T x(t)'M(t)x(t) + u(t)'N(t)u(t)\ dt\}$$

subject to

$$dx = [A(t)x + B(t)u]\, dt + \sigma(t)\, dw(t), \quad x(0) = x_0.$$

We assume that $M(t)\geq 0$, $D\geq 0$, $N(t)>0$, $\sigma(t)\sigma(t)'>0$, and the control set U is all of $\mathbb{R}_d$. Our aim is to find an extremal control; this we do by looking at (i), (ii), (iii), making an educated guess at u and then verifying that this u is indeed extremal. We have

$$H(t,x,u,p,\lambda) = p'[A(t)x + B(t)u] + \lambda_0[x'M(t)x + u'N(t)u].$$

Now (i) holds if and only if $\lambda = \lambda_0 \neq 0$ (since there are no constraints (4)), so we can rescale λ and $\overline{p}$ to obtain $\lambda = \lambda_0 = -1$. Next (ii) holds if for some $\overline{p}$

$$\frac{d\overline{p}'}{dt} = - \overline{p}'\, A(t) - 2\lambda_0\hat{x}(t)'M(t),$$

$$\overline{p}(T)' = 2\lambda_0\hat{x}(T)'D,$$

$$\text{i.e. } \overline{p}(t)' = 2\lambda_0\hat{x}(T)'D\, \Phi\, (T,t) + \int_t^T 2\lambda_0\hat{x}(s)'M(s)\, \Phi\, (s,t)\, ds \qquad (6)$$

where $\Phi(s,t)$ is the fundamental matrix solution of

$\frac{dx}{ds} = A(s)x$ with $\Phi(t,t) = I$. If we now set

$$\overline{x}_t(s) = E\{\hat{x}(s) \mid \hat{x}(r),\ 0 \leq r \leq t\}$$

then from the definition of p we obtain

$$p(t)' = 2\lambda_0\{\overline{x}_t(T)'D\, \Phi(T,t) + \int_t^T \overline{x}_t(s)'M(s)\, \Phi\, (s,t)\, ds\}. \qquad (7)$$

Finally (iii) holds if and only if $\frac{\partial H(t,\hat{x}(t),u,p(t),\lambda)}{\partial u} = 0$ at $\hat{u} = u(t,\hat{x}(\cdot))$, i.e. if

$$\hat{u}(t,x(\cdot)) = \frac{1}{2}\, N(t)^{-1}\, B(t)'\, p(t). \qquad (8)$$

Observe that $\hat{u}$ is linear in p, c.f.(8), that p is linear in $\overline{x}_t(\cdot)$, c.f. (7), and that $\overline{x}_t$ satisfies

$$\frac{d\overline{x}_t(s)}{ds} = A(s)\overline{x}_t(s) + B(s)E\{\hat{u}(s,\hat{x}(\cdot))|\hat{x}(r),\ 0<r<t\},$$

$$\overline{x}_t(t) = \hat{x}(t),$$

so that $\overline{x}_t(s)$ is linear in $\hat{x}(t)$ provided $\hat{u}(s,x(\cdot)) = K(s)x(s)$. Since this linear structure is so tempting, let us guess that indeed $\hat{u}(t,x(\cdot)) = K(t)x(t)$, and proceed to show that this $\hat{u}$ is extremal for a suitable matrix K(t).

Certainly if K(•) is bounded and measurable then (5)

i.e. $dx = [A(t) + B(t)K(t)]xdt + \sigma(t)dw(t)$,

has a solution, so $\hat{u}$ is feasible. We take $\lambda = \lambda_0 = 1$ and define $\overline{p}$ by (6). Then (i) and (ii) hold. If $\Phi_k(s,t)$ is the fundamental matrix solution of

$$\frac{dx}{ds} = [A(s) + B(s)K(s)]x, \text{ then } \overline{x}_t(s) = \Phi_k(s,t)\hat{x}(t) \text{ and}$$

$$p(t)' = -2\, x(t)'\{\Phi_k(T,t)'D\ \Phi(T,t) + \int_t^T \Phi_k(s,t)'M(s)\ \Phi(s,t)\, ds\}$$

$$= -2\, x(t)'P(t)'$$

if we define P(t)' as the expression in braces above. Now (iii) holds if, c.f. (8),

$$K(t)\hat{x}(t) = -\frac{1}{2} N(t)^{-1}B(t)'\ 2P(t)\hat{x}(t)$$

i.e. $\quad K(t) = -N(t)^{-1}\ B(t)^{-1}\ P(t).$ (9)

From (9) and the definition of P we can characterize it as the solution of

$$\frac{dP}{dt} + A(t)'P + PA(t) - P\ B(t)\ N(t)^{-1}B(t)'P + M(t) = 0 \tag{10}$$

$$P(T) = D.$$

Since K is bounded, we conclude that $\hat{u}(t,x(\cdot)) = N(t)^{-1}B(t)'P(t)x(t)$ is extremal.

Suppose we wish to add a constraint to the problem, which now becomes

$$\min E\{x(T)'Dx(T) + \int_0^T x(s)'M(s)x(s) + u(s)'N(s)u(s)\, ds\}$$

subject to

$$E|x(T)|^2 \leq \beta$$

$$dx = [A(t)x + B(t)u(t)]\ dt + \sigma(t)\ dw$$

The assumptions on the system parameters remain as before. The Hamiltonian does not change but (7) is replaced by

$$p(t)' = 2\lambda_0\{\bar{x}_t(T)'D\ \Phi(T,t) + \int_t^T \bar{x}_t(s)'M(s)\Phi(s,t)\ ds\} + 2\lambda_1\ \bar{x}_t(T)'\Phi(T,t)$$

$$= 2\lambda_0\{\bar{x}_t(T)'(D + \frac{\lambda_1}{\lambda_0} I)\Phi(T,t) + \int_t^T \bar{x}_t(s)'\ M(s)\Phi(s,t)\ ds\}.$$

Conceivably we could have $\lambda_0 = 0$, $\lambda = (0,\lambda_1)' \neq 0$, but we shall only look for <u>normal</u> extremals, i.e. those for which $\lambda_0 \neq 0$, and hence by rescaling for which $\lambda = (-1, -\rho)'$ with $\rho \geq 0$. But now ρ is exactly the same as for the unconstrained problem with terminal cost $x'(D + \rho I)x$. Recall also the condition $\lambda_1 J_1(\hat{u}) = 0$, so either $\rho = 0$ or $J_1(\hat{u}) = 0$. Hence we have three cases:

(a) The extremal control for the unconstrained problem happens to satisfy $E|\hat{x}(T)|^2 \leq \beta$. Then this control is also extremal for the constrained problem with $\lambda = (-1,0)'$ and with $\bar{p}$ as in (6).

If however for the unconstrained problem we have $E|x(T)|^2 > \beta$, then write P_ρ for the solution of (10) with terminal condition $P_\rho(T) = D + \rho I$, and let $K_\rho = N^{-1}B'P_\rho$, $u_\rho(t,x(\cdot)) = K_\rho(t)x(t)$. Using u_ρ solve (5) for x_ρ, i.e.

$$dx = (A + B\ K_\rho)x\ dt + \sigma\ dw.$$

(b) If it is possible to choose $\rho = \hat{\rho} > 0$ such that

$$E|x_{\hat{\rho}}(T)|^2 = \beta,$$

then $\hat{u} \equiv u_{\hat{\rho}}$ is extremal with $\lambda = (-1,-\hat{\rho})'$ and $\bar{p}$ is given by (6) with D replaced by $D + \rho I$.

(c) If it is impossible to choose such $\hat{\rho} > 0$, then no normal extremals exist.

To see this observe that if we write

$$E|x_\rho(T)|^2 \equiv R(\rho)$$
$$= x_0' \, \Phi_\rho(T,0)'\Phi_\rho(T,0)x_0$$
$$+ \int_0^T \text{trace}[\Phi_\rho(T,s)'\sigma(s)\sigma(s)'\Phi_\rho(T,s)] \, ds$$

where Φ_ρ is Φ_{K_ρ}. Then (a) holds if $R(0) < \beta$, (b) holds if $R(0) > \beta > R(\infty) \equiv \lim_{\rho\to\infty} R(\rho)$. Note that R is monotone decreasing. Now (c) holds if $R(\infty) > \beta$, i.e. no matter how large a penalty $(\rho|x(T)|^2)$ we add into the cost of the unconstrained problem, the second moment at time T is greater than β. Hence it is impossible to satisfy the constraint.

There remains the question of whether the extremal controls are in fact optimal. Here is a useful result whose proof can be found in [4]. Let $H^*(t,x,p,\lambda) = \sup_{u\in U} H(t,x,u,p,\lambda)$.

Theorem 3. Assume (A) and

(I) $\hat{u}$ is a normal extremal

(II) c_0, c_1 are convex, c_2 is affine,

(III) $H^*(t,\cdot,p(t),\lambda)$ is concave for each t, w.p.1.

Then $\hat{u}$ is optimal.

In our example $c_0(x) = x'Dx$, $c_1(x) = x'x - \beta$, $c_2 = 0$, so that (II) is satisfied. Moreover $(\lambda_0 = -1)$

$$H^*(t,x,p,\lambda) = p'A(t)x - x'M(t)x + \frac{1}{4}\, p'B(t)N(t)^{-1}B(t)'p$$

so H^* is concave in x for each (t,p). Hence in the cases (a), (b) the extremal controls which we found are in fact optimal. This is also true in the unconstrained case.

4. A NON-LINEAR EXAMPLE

Let us now consider a scalar but non-linear problem:

$$\min E|x(T)|^2$$

subject to

$$dx = [F(t,x) + G(t,x)u]dt + \sigma(t)dw, \quad x(0) = x_0.$$

We assume

$|F(t,x)|+|G(t,x)| \le k_1(1 + |x|)$,

$|\frac{\partial F}{\partial x}(t,x)|, |\frac{\partial G}{\partial x}(t,x)|, |\sigma(t)|$ bounded,

$|\sigma(t)| > 0$, $F(t,\cdot)$ is odd, $G(t,\cdot)$ is even or odd, and $U = [-1,1]$.

Since we wish to drive $x(T)$ to 0 it is probable that at each t we should push toward 0 as much as possible, i.e. we guess

$$\hat{u}(t,x(\cdot)) = -\operatorname{sgn}[x(t)G(t,x(t))]$$

where $\operatorname{sgn} a = a/|a|$ if $a \neq 0$, and $\operatorname{sgn} 0 = 0$. Then $\hat{x}$ satisfies

$$dx = [F(t,x) - |G(t,x)|\operatorname{sgn} x]\, dt + \sigma(t)\, dw;$$

since $|\sigma(t)| > 0$ then a solution exists so $\hat{u}$ is feasible. We have $H(t,x,u,p,\lambda) = p[F(t,x) + G(t,x)u]$, and we take $\lambda = \lambda_0 = -1$, $\bar{p}(t) = -2\hat{x}(T)\Phi(T,t)$ where

$$\Phi(S,t) = \exp \int_t^S \frac{\partial F}{\partial x}(r,\hat{x}(r)) - \frac{\partial G}{\partial x}(r,\hat{x}(r)) \operatorname{sgn}[\hat{x}(r)G(r,\hat{x}(r))]\, dr.$$

Then (i) and (ii) are satisfied. Moreover (iii) is satisfied if and only if $\hat{u}(t,\hat{x}(\cdot)) = \operatorname{sgn}[p(t)G(t,\hat{x}(t))]$, i.e. if and only if

$$-\operatorname{sgn} \hat{x}(t) = \operatorname{sgn} p(t) \text{ where}$$

$$p(t) = E\{\bar{p}(t)|\hat{x}(s),\ 0 \le s \le t\}$$

$$= E\{\bar{p}(t)|\hat{x}(t)\}.$$

The fact that $\operatorname{sgn} \hat{x}(t) = \operatorname{sgn} E\{\hat{x}(T)\Phi(T,t)|\hat{x}(t)\}$ is established in [3], so it follows that $\hat{u}$ is extremal.

Unfortunately theorem 3 does not apply so we must proceed differently to prove optimality. Here are two ways to do this. Since p is unique (it is the negative of the gradient of the value function c.f. [2]) and since (iii) gives u uniquely in terms of p, then the extremal control is unique. Since an optimal feedback control exists (follows from dynamic programming), then the unique extremal must be optimal.

One could also appeal to another result, [5], which states that if $\hat{u}$ is extremal with an adjoint process, i.e. p, defined not through (ii) but rather in a more abstract manner, then $\hat{u}$ is optimal. With some work this result allows us to conclude again that $\hat{u} = -\operatorname{sgn}(xG)$ is optimal.

REFERENCES

[1] U. G. Haussmann, "General necessary conditions for optimal control of stochastic systems", in Stochastic Systems: Modelling, Identification and Optimization II (R.J.B. Wets ed.), Mathematical Programming Study 6, (1976), pp. 30-48.

[2] ———— , "On the adjoint process for optimal control of diffusion processes", SIAM J. Control Option. 19, (1981), pp. 221-243.

[3] ———— , "Some examples of optimal stochastic controls", SIAM Review 23 (1981), pp. 292-303.

[4] ———— , "Extremal controls for completely observable diffusions," in Advances in filtering and optimal stochastic control (W. H. Fleming and L. G. Gorostiza ed.), Lecture Notes in Control and Information Sciences 42, (1982), pp. 149-160.

[5] ———— , Optimal control of diffusions, in preparation.

DESIGN WAVE DETERMINATION BY FAST INTEGRATION TECHNIQUE

Henrik O. Madsen and Ove Bach-Gansmo
Section for Structural Reliability
Research Division, Det norske Veritas
Høvik, Oslo, Norway

1. INTRODUCTION

The largest sea surface elevation or wave height above mean water level in a time period is an important design parameter for offshore structures. The surface elevation above mean level is a result of tidal effects and waves. Only waves are considered in this paper. The largest wave height is of interest both in a short term period e.g. during installation and in a long term period during production.

A standard procedure for calculation of the extreme wave height distribution is based on a description of the sea as a sequence of stationary sea states. Within each sea state the surface elevation is modeled as a narrow band Gaussian process with a spectral density function determined from characteristic sea state parameters, traditionally the significant wave height, H_S, and the mean zero upcrossing period, T_Z. Individual maxima, M, within a sea state are assigned a Rayleigh distribution according to the assumption of a narrow band process. Alternatively, a Rice distribution is sometimes used. The cumulative distribution function for M conditional upon the sea state parameters H_S and T_Z is denoted by

$$P(M \leq m \mid h_S, t_Z) = F_M(m \mid h_S, t_Z) \tag{1}$$

Individual maxima within a sea state are assumed independent. The distribution for the largest maximum, M_D, within a sea state is therefore

$$F_{M_D}(m \mid h_S, t_Z) = F_M(m \mid h_S, t_Z)^{\frac{D}{t_Z}} \tag{2}$$

D is the duration of the sea state and D/t_Z is the (mean) number of individual waves within the sea state for the narrow band process. The distribution function for M_D within an arbitrary sea state is obtained by integrating (2) with the probability density function of (H_S, T_Z) as weight function.

$$F_{M_D}(m) = \int_0^\infty \int_0^\infty F_M(m \mid h_S, t_Z)^{\frac{D}{t_Z}} f_{H_S,T_Z}(h_S, t_Z)\, dh_S\, dt_Z \tag{3}$$

The largest maxima from sea state to sea state are assumed independent. Provided all sea states are of the same duration, the maximum wave height M_T within an anticipated period T has the distribution function

$$F_{M_T}(m) = F_{M_D}(m)^{\frac{T}{D}} \tag{4}$$

$$= \left[\int_0^\infty \int_0^\infty F_M(m \mid h_S, t_Z)^{\frac{D}{t_Z}} f_{H_S,T_Z}(h_S, t_Z)\, dh_S\, dt_Z \right]^{\frac{T}{D}}$$

The result depends on the somewhat arbitrary value of D.

The design wave is generally defined as a wave height which is exceeded with a small probability.

Very often the sea scatter diagram represented by the joint probability density function $f_{H_S,T_Z}(\ ,\ )$ is not known with certainty due to a short measurement period or measurement uncertainty. Such statistical or measurement uncertainties can also be included in the analysis. Assume as an example that the probability density function $f_{H_S,T_Z}(\ ,\ )$ contains two unknown parameters μ_1 and μ_2 with probability density functions $f_{\mu_1}(\mu_1)$ and $f_{\mu_2}(\mu_2)$. For independent parameters (4) is then extended as

$$F_{M_T}(m) = \int_{-\infty}^{\infty} \int_{-\infty}^{\infty} F_{M_T}(m \mid \mu_1, \mu_2)\, f_{\mu_1}(\mu_1)\, f_{\mu_2}(\mu_2)\, d\mu_1 d\mu_2 \tag{5}$$

Three points can be criticized in the analysis procedure:

i) Dependencies between wave heights within a sea state are ignored

ii) A double numerical integration is necessary to evaluate the extreme value distribution

iii) Inclusion of statistical and measurement uncertainty in the sea scatter involves additional numerical integration.

This paper presents an analysis method in which the above mentioned points are dealt with in an efficient way. The method was a fast integration technique based on a first order reliability method and results for the extreme wave distribution are obtained with very good accuracy at a very low cost of computation. In addition, information of the most likely sea state in which the extreme wave occurs is obtained.

2. SEA ELEVATION MODEL

The sea surface elevation $X(t)$ above mean level is modeled as a sequence of mean zero stationary Gaussian processes each described by two parameters, the significant wave height H_S and the mean zero crossing period T_Z. The duration, D, of one stationary period is first taken as 8 hours and the distribution for the extreme wave height, M_D, in this 8 hour period is determined.

The one sided power spectral density $S_X(\omega)$ for $X(t)$ is taken as the Pierson and Moskowitz spectrum

$$S_X(\omega) = \frac{1}{4\pi}\left(\frac{2\pi}{T_Z}\right)^4 H_S^2\, \omega^{-5} \exp\left(-\frac{1}{\pi}\left(\frac{2\pi}{T_Z}\right)^4 \omega^{-4}\right), \quad \omega > 0 \tag{6}$$

The procedure works for any choice of spectrum and not solely for the one selected.

From the theory of extremes of stationary Gaussian processes, the cumulative distribution function for M_D can then be approximated as, [6]

$$F_{M_D}(m \mid h_S, t_Z) \approx F_R(m \mid h_S, t_Z) \exp(-\nu_{R,q}(m \mid h_S, t_Z) D) \tag{7}$$

$R(t)$ is the envelope process defined in [1]

$$R(t)^2 = X(t)^2 + \hat{X}(t)^2 \tag{8}$$

where $\hat{X}(t)$ is the Hilbert transform of $X(t)$. $\nu_{R,q}(\ )$ is the mean rate of qualified envelope upcrossings which is obtained as an interpolation between the mean rate of process upcrossings $\nu_X(\ )$ for high levels and the mean rate of envelope upcrossings $\nu_R(\ )$ for low levels as, [6]

$$\nu_{R,q}(m) = \nu_X(m) \left[1 - \exp\left(-\frac{\nu_R(m)}{\nu_X(m)}\right)\right] \tag{9}$$

For a stationary Gaussian process $R(t)$ has a Rayleigh distribution and (7) becomes

$$F_{M_D}(m \mid h_S, t_Z) = \left(1 - \exp\left(-\frac{m^2}{2\lambda_0}\right)\right) \exp\left(-\frac{1}{2\pi}\frac{\sqrt{\lambda_2}}{\sqrt{\lambda_0}} \exp\left(-\frac{m^2}{2\lambda_0}\right) D \left(1 - \exp\left(-\sqrt{2\pi}\, \frac{m}{\sqrt{\lambda_0}} \sqrt{1 - \lambda_1^2/\lambda_0\lambda_2}\right)\right)\right) \tag{10}$$

The spectral moments λ_i are defined by

$$\lambda_i = \int_0^\infty \omega^i S_X(\omega) d\omega, \quad i = 0,1,2,... \tag{11}$$

The three lowest order spectral moments are computed upon inserting $S_X(\omega)$ from (6) in (11) and integrating

$$\lambda_0 = \frac{1}{16} H_S^2 \tag{12}$$

$$\lambda_1 = \frac{1}{16} H_S^2 \frac{2\pi}{T_Z} \frac{1}{\pi^{1/4}} \Gamma\left(\frac{3}{4}\right) \tag{13}$$

$$\lambda_2 = \frac{1}{16} H_S^2 \left(\frac{2\pi}{T_Z}\right)^2 \tag{14}$$

$\Gamma(\ )$ denotes the gamma function.

As an example, the distribution of (H_S, T_Z) is here taken as a joint lognormal distribution with

$$E[H_S] = 3.00\ m \qquad E[T_Z] = 7.00\ s \tag{15}$$

$$Var[H_S] = 3.60\ m^2 \qquad Var[T_Z] = 1.80\ s^2 \tag{16}$$

$$Cov[H_S, T_Z] = 2.00\ ms \tag{17}$$

The distribution of $(\ln H_S, \ln T_Z)$ is joint normal with

$$E[\ln H_S] = \mu_1 = 0.9304 \qquad E[\ln T_Z] = \mu_2 = 1.9279 \tag{18}$$

$$Var[\ln H_S] = \sigma_1^2 = 0.5801^2 \qquad Var[\ln T_Z] = \sigma_2^2 = 0.1899^2 \tag{19}$$

$$Cov[\ln H_S, \ln T_Z] = \rho\sigma_1\sigma_2 = 0.8258 \cdot 0.5801 \cdot 0.1899 \tag{20}$$

The numerical values have been determined from the following relations valid for joint lognormal distributed random variables

$$E[H_S] = \exp(\mu_1 + \frac{1}{2}\sigma_1^2) \qquad E[T_Z] = \exp(\mu_2 + \frac{1}{2}\sigma_2^2) \tag{21}$$

$$Var[H_S] = E[H_S]^2(\exp(\sigma_1^2) - 1) \qquad Var[T_Z] = E[T_Z]^2(\exp(\sigma_2^2) - 1) \tag{22}$$

$$Cov[H_S, T_Z] = E[H_S]E[T_Z](\exp(\rho\sigma_1\sigma_2) - 1) \tag{23}$$

The probability density function for (H_S, T_Z) is

$$f_{H_S T_Z}(h_S, t_Z) = \frac{1}{h_S t_Z \sigma_1 \sigma_2} \varphi(\frac{\ln h_S - \mu_1}{\sigma_1}, \frac{\ln t_Z - \mu_2}{\sigma_2}; \rho) \tag{24}$$

where $\varphi(\ , \ ;\rho)$ is the standardized joint normal probability density function with correlation coefficient ρ.

$$\varphi(x,y;\rho) = \frac{1}{2\pi\sqrt{1-\rho^2}} \exp(-\frac{x^2 + y^2 - 2\rho xy}{2(1-\rho^2)}) \tag{25}$$

Again, the procedure works for any continuous joint distribution function and not solely for the one selected.

The unconditional distribution function for M_D is

$$F_{M_D}(m) = \int_0^\infty \int_0^\infty F_{M_D}(m \mid h_S, t_Z)\, f_{H_S T_Z}(h_S, t_Z) dh_S\, dt_Z \tag{26}$$

involving a double integration for each value of m. The integration must be done numerically.

3. APPLICATION OF FIRST ORDER RELIABILITY METHOD

A considerable amount of computational work can be saved by the use of a first order reliability method (FORM), [3]. By this method the complementary cumulative distribution function

$$G_{M_D}(m) = 1 - F_{M_D}(m) = P(M_D > m) \tag{27}$$

is determined for any value of m. Using the FORM terminology the limit state function is

$$g(m_D, h_S, t_Z) = m - m_D \tag{28}$$

and the probability $P(g(m_D, h_S, t_Z) < 0)$ that the limit state function takes a negative value is needed. Instead of operating in the (M_D, H_S, T_Z)-space a transformation into a space of independent and standardized normal variables is carried out. The so-called Rosenblatt transformation, [4], is used as suggested in [3]. The variables are numbered with H_S as variable no. 1, T_Z as variable no. 2 and M_D as variable no. 3. The conditional distribution functions used in the transformation are

$$F_{H_S}(h_S) = \Phi(\frac{\ln h_S - \mu_1}{\sigma_1}) \tag{29}$$

$$F_{T_Z}(t_Z \mid h_S) = \Phi(\frac{\ln t_Z - (\mu_2 + \rho\frac{\sigma_2}{\sigma_1}(\ln h_S - \mu_1))}{\sigma_2\sqrt{1-\rho^2}}) \tag{30}$$

$$F_{M_D}(m_D \mid h_S, t_Z) = (1 - \exp(-\frac{1}{2}(\frac{4m_D}{h_S})^2)) \exp[-\frac{1}{t_Z}\ D \exp(-\frac{1}{2}\ (\frac{4m_D}{h_S})^2) \tag{31}$$

$$(1 - \exp(-\sqrt{2\pi}\ \frac{4m_D}{h_S} \sqrt{1 - \frac{1}{\sqrt{\pi}}\ \Gamma(\frac{3}{4})^2}))]$$

The transformation into a set of independent and standardized normal variables U is

$$\Psi: \quad U_1 = \Phi^{-1}(F_{H_S}(H_S)) \tag{32}$$

$$U_2 = \Phi^{-1}(F_{T_Z}(T_Z \mid H_S)) \tag{33}$$

$$U_3 = \Phi^{-1}(F_{M_D}(M_D \mid H_S, T_Z)) \tag{34}$$

In u-space the limit state surface has the equation

$$g_u(\mathbf{u}) = g(\Psi^{-1}(\mathbf{u})) = 0 \tag{35}$$

The inverse transformation Ψ^{-1} is simple for H_S and T_Z while it must be carried out numerically for M_D.

The limit state surface in u-space is illustrated in Fig.1. The probability in (27) is equal to the probability content in the hatched area of Fig.1. In a first order reliability method the limit state surface is replaced by the tangent hyperplane at the point closest to the origin. This point is called the 'design point', $\mathbf{u}^*$, and has the coordinates

$$\mathbf{u}^* = \beta\boldsymbol{\alpha}^* \tag{36}$$

$\boldsymbol{\alpha}^*$ is the unit directional vector to $\mathbf{u}^*$ and β is the reliability index. The probability content outside the tangent hyperplane is $\Phi(-\beta)$ and the first order approximation is therefore

$$G_{M_D}(m) \approx \Phi(-\beta(m)) \tag{37}$$

The approximation is generally very good, in particular for large arguments m, i.e. for small probabilities.

Improved approximations are obtained by approximating the limit state surface by a hyperparaboloid at the design point in such a way that the limit state surface and the approximating paraboloid have the same main curvatures at the design point, [5]. The corresponding approximation is then called a second order approximation. Here the first order approximations are considered sufficient.

The design point can be found by geometrical considerations. In the numerical calculations presented in this paper the method described in [3] has been used. This method avoids the exact inversion of the transformation Ψ.

Fig.2 shows a comparison between the FORM results and results from a numerical integration procedure. For the numerical integration (27) is rewritten as

$$G_{M_D}(m) = \int_{-\infty}^{\infty}\int_{-\infty}^{\infty} \frac{1}{\pi}[1-(1-\exp(-\frac{1}{2}(\frac{4m}{h_S})^2))\exp[-\frac{D}{t_Z}\exp(-\frac{1}{2}(\frac{4m}{h_S})^2) \tag{38}$$

$$(1-\exp(-3.9192\frac{m}{h_S}))]] \; \exp(-u^2)\exp(-v^2)du dv$$

where

$$h_S = \exp(0.9304 + 0.8204\,u) \tag{39}$$

$$t_Z = \exp(1.9279 + 0.2217\,u + 0.1515\,v) \tag{40}$$

A 45 times 45 Gauss-Hermite numerical integration is used. The two curves in Fig.2 are almost coinciding for small arguments m. For larger arguments the numerical integration results start to oscillate slightly illustrating the insufficiency of the numerical procedure. Naturally, a better numerical integration procedure can be selected but the improvement of the results compared to the FORM results is very marginal while the computational efforts are significantly increased. The good agreement indicates that the first order approximation is sufficient.

From the definition of $\beta(\mathbf{u})$ as the length of the vector $\mathbf{u}$ follows that

$$\frac{\partial\beta(\mathbf{u})}{\partial u_i} = \alpha_i \tag{41}$$

The numerical value α_i^* is thus a measure of the sensitivity of the reliability index to inaccuracies in the value of u_i at the design point. Fig.3 shows the value of α_i^* as a function of m. It is observed that α_2^* is very small indicating that a small error is introduced by considering u_2 as fixed equal to 0, i.e. by having a one to one correspondence between H_S and T_Z. The relative error by fixing $u_2=0$ is less than 1% on $G_{M_D}(m)$ and is not visible in Fig.2.

The design point gives the most probable value of $\mathbf{u}$ if the wave height m is exceeded. The corresponding set of (H_S,T_Z) therefore in a sense gives the most probable sea state prevailing when the wave height is exceeded.

4. INCLUSION OF STATISTICAL UNCERTAINTY

The analysis can be extended to include statistical or measurement uncertainty in some of the distributional parameters in the joint distribution for (H_S, T_Z). Let e.g. the expected values $E[H_S]$ and $E[T_Z]$ be modeled as independent random variables μ_H and μ_T with distribution functions $F_{\mu_H}(\;)$ and $F_{\mu_T}(\;)$, respectively. The distribution parameters μ_1 and μ_2 in the lognormal distribution are then also random variables. The total number of random variables is thus 5 and the computation of the unconditional distribution function for M_D involves a quadruple integration.

When the first order reliability technique is applied the computational work is only slightly increased. As an example, μ_H and μ_T are taken as independent and normally distributed with expected values as in (15) and with a coefficient of variation of 10%. The five distribution functions needed in the transformation Ψ are given as

$$F_{\mu_1}(\mu_1) = F_{\mu_H}(\exp(\mu_1 + \frac{1}{2}\sigma_1^2)) = \Phi(\frac{\exp(\mu_1 + \frac{1}{2}\sigma_1^2) - 3}{0.3}) \tag{42}$$

$$F_{\mu_2}(\mu_2) = F_{\mu_T}(\exp(\mu_2 + \frac{1}{2}\sigma_2^2)) = \Phi(\frac{\exp(\mu_2 + \frac{1}{2}\sigma_2^2) - 7}{0.7}) \tag{43}$$

$$F_{H_S}(h_S \mid \mu_1, \mu_2) = \Phi(\frac{\ln h_S - \mu_1}{\sigma_1}) \tag{44}$$

$$F_{T_Z}(t_Z \mid \mu_1, \mu_2, h_S) = \Phi(\frac{\ln t_Z - (\mu_2 + \rho\frac{\sigma_2}{\sigma_1}(\ln t_S - \mu_1))}{\sigma_2\sqrt{1-\rho^2}}) \tag{45}$$

$$F_{M_D}(m_D \mid \mu_1, \mu_2, h_S, t_Z) = (1 - \exp(-\frac{1}{2}(\frac{4m_D}{h_S})^2)) \tag{46}$$

$$\exp[-\frac{1}{t_Z}\, 28800 \exp(-\frac{1}{2}\,(\frac{4m_D}{h_S})^2)\,(1 - \exp(-\sqrt{2\pi}\,\frac{4m_D}{h_S}\,\sqrt{1 - \frac{1}{\sqrt{\pi}}\,\Gamma(\frac{3}{4})^2}))]$$

Based on these five distribution functions the FORM applies directly. Result are shown in Fig.4, where also the corresponding curve from Fig.2 is given for comparison. The sensitivity factors are shown in Fig.5. The sensitivity factors for the statistical uncertainty variables are not very large for this short term sea state.

5. MAXIMUM DISTRIBUTION IN ANTICIPATED LIFE TIME

The distribution of the maximum M_T in the anticipated life time T is easy to compute if there is no statistical or measurement uncertainty in the sea scatter diagram. Assuming independent maxima from sea state to sea state, the result is

$$F_{M_T}(m) = F_{M_D}(m)^{\frac{T}{D}} \tag{47}$$

where $F_{M_D}(m)$ is approximated by the FORM result.

If the sea scatter diagram is not exactly known, a FORM can still be applied and numerical integration be avoided. As in the previous section, let the expected values of H_S and T_Z be random values μ_H and μ_T. The distribution function $F_{M_T}(m)$ is then given in (5). Use of a FORM can be carried out with the limit state function as

$$g(m_T, \mu_1, \mu_2) = m - m_T \tag{48}$$

and the conditional distribution functions

$$F_{\mu_1}(\mu_1) = F_{\mu_H}(\exp(\mu_1 + \frac{1}{2}\sigma_1^2)) \tag{49}$$

$$F_{\mu_2}(\mu_2 \mid \mu_1) = F_{\mu_T}(\exp(\mu_2 + \frac{1}{2}\sigma_2^2)) \tag{50}$$

$$F_{M_T}(m_T \mid \mu_1, \mu_2) = F_{M_D}(m_T \mid \mu_1, \mu_2)^{\frac{T}{D}} \tag{51}$$

The third distribution function (51) is not known exactly but is computed by a FORM as described in the previous sections.

Fig.6 shows a comparison of (47) with the results obtained by including statistical uncertainty in the sea scatter diagram. T is taken as 50 years. Fig.7 shows the sensitivity factors in the case with statistical uncertainty in the sea scatter diagram. It is observed that the relative importance of statistical uncertainty is much larger for this long term extreme than for the short term extreme in Fig.5.

6. INFLUENCE OF DURATION OF SHORT TERM SEA STATE

The results presented so far depend on the somewhat arbitrary assumption of a sea state duration of 8 hours. To investigate the effect of this assumption Fig.8 shows result similar to those in Fig.6 for $D=4,8$ and 16 hours. The observed differences are due to the assumption of independent sea state parameters from sea state to sea state.

The distribution of M_D can also be determined for the case in which D is taken as a random variable. The results are again obtained by a FORM. An additional random variable is included compared to the previous section. The limit state functions and conditional distribution functions are

Without uncertainty in the sea scatter diagram:

$$g(m_T,d) = m - m_T \tag{52}$$

$$F_D(d) = \Phi(\frac{\ln d - 10.238}{0.25}) \tag{53}$$

$$F_{M_T}(m_T \mid d) = F_{M_D}(m)^{\frac{T}{d}} \tag{54}$$

With uncertainty in the sea scatter diagram:

$$g(m_T,\mu_1,\mu_2,d) = m - m_T \tag{55}$$

$$F_D(d) = \Phi(\frac{\ln d - 10.238}{0.25}) \tag{56}$$

$$F_{\mu_1}(\mu_1 \mid d) = F_{\mu_H}(\exp(\mu_1 + \frac{1}{2}\sigma_1^2)) \tag{57}$$

$$F_{\mu_2}(\mu_2 \mid \mu_1,d) = F_{\mu_T}(\exp(\mu_2 + \frac{1}{2}\sigma_2^2)) \tag{58}$$

$$F_{M_T}(m_T \mid \mu_1,\mu_2,d) = F_{M_D}(m_T \mid \mu_1,\mu_2)^{\frac{T}{d}} \tag{59}$$

7. CONCLUSIONS

The paper presents an application of first order reliability theory in the evaluation of design waves for structural analysis of offshore structures. Uncertainties in sea scatter diagram, sea state duration and the inherent uncertainty in the description of the sea as a random process are included in the analysis.

Time consuming numerical integrations are completely avoided and the computations are extremely efficient. This will have an even greater importance in a complete probabilistic structural analysis where the number of random variables is much higher.

The relative importance of the various sources of uncertainty are determined both for a short term period and for a long term period.

8. ACKNOWLEDGEMENT

The authors wish to thank Det norske Veritas for allowing them to prepare and present this paper. The opinions stated in the paper are those of the authors and do not necessarily reflect those of Det norske Veritas.

9. REFERENCES

[1] Cramer,H. and Leadbetter,M.R. : *Stationary and Related Stochastic Processes,* John Wiley and Sons, New York, 1967.

[2] Ditlevsen,O. :'System Reliability Bounding by Conditioning', *Journal of Engineering Mechanics Division,* ASCE, Vol.108, No.EM5, Oct.1982, pp.708-718.

[3] Hohenbichler,M. and Rackwitz,R. :'Non-Normal Dependent Variables in Structural Reliability', *Journal of the Engineering Mechanics Division,* ASCE, Vol.107, No.EM6, Dec.1981.

[4] Rosenblatt,M. :'Remarks on a Multivariate Transformation', *Annals of Mathematical Statistics,* Vol.23, 1954, pp.470-472.

[5] Tvedt,L. :'Two Second-Order Approximations to the Failure Probability', submitted to *Structural Safety,* July, 1983.

[6] Vanmarcke,E.H. :'On the Distribution of the First-Passage Time for Normal Stationary Random Processes', *Journal of Applied Mechanics,* ASME, No.42, 1975, pp.215-220.

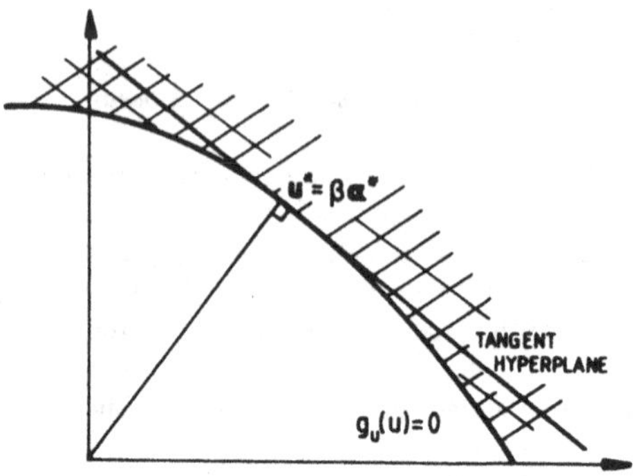

Fig.1 Limit state surface in u-space.

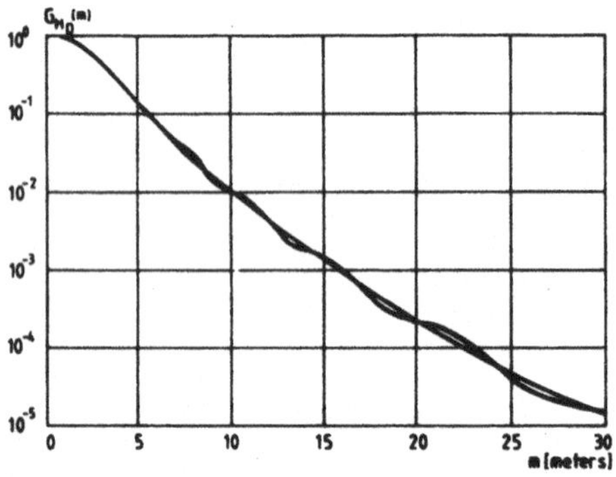

Fig.2 Comparison of FORM result and numerical integration result for sea state maximum distribution.

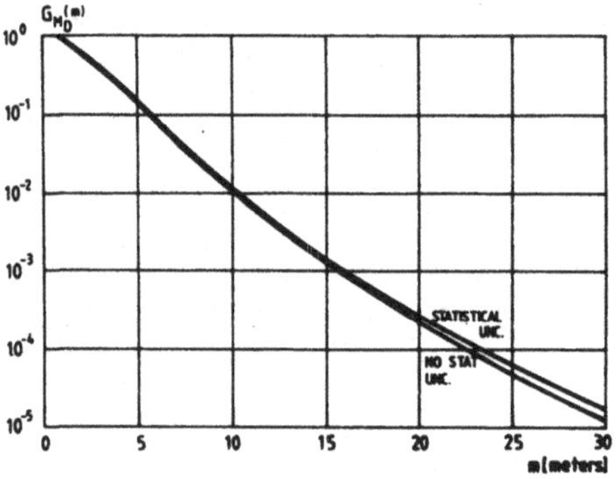

Fig.4 Comparison of FORM results for sea state maximum distribution with and without statistical uncertainty in the sea scatter diagram.

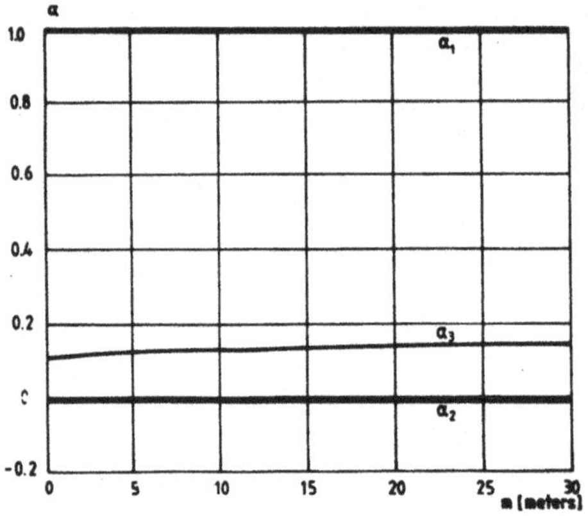

Fig.3 Sensitivity factors in FORM analysis of sea state maximum distribution.

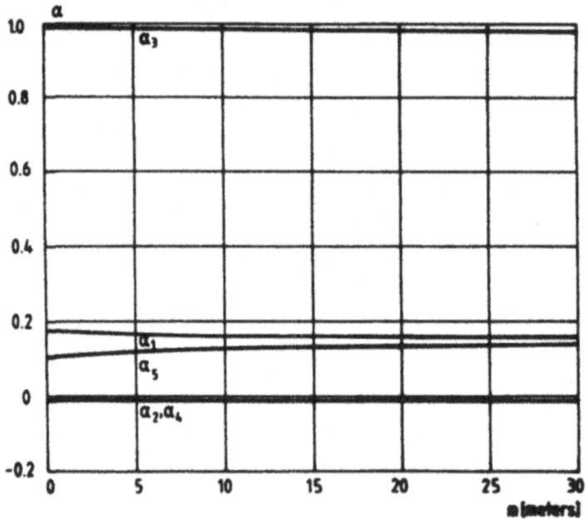

Fig.5 Sensitivity factors in FORM analysis for sea state maximum distribution including statistical uncertainty in sea scatter diagram.

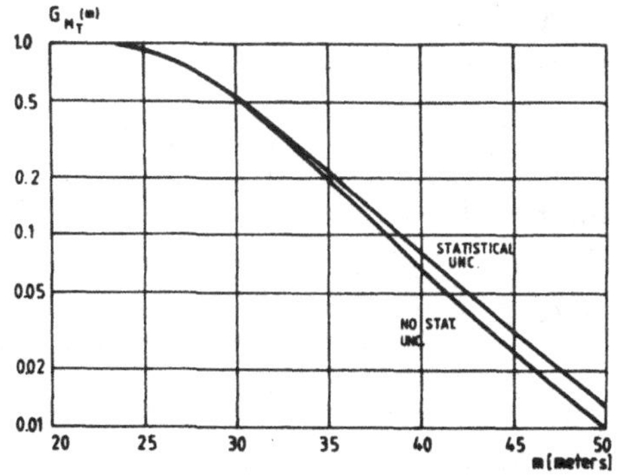

Fig.6 Comparison of FORM results for life time maximum distribution with and without statistical uncertainty included in the sea scatter diagram.

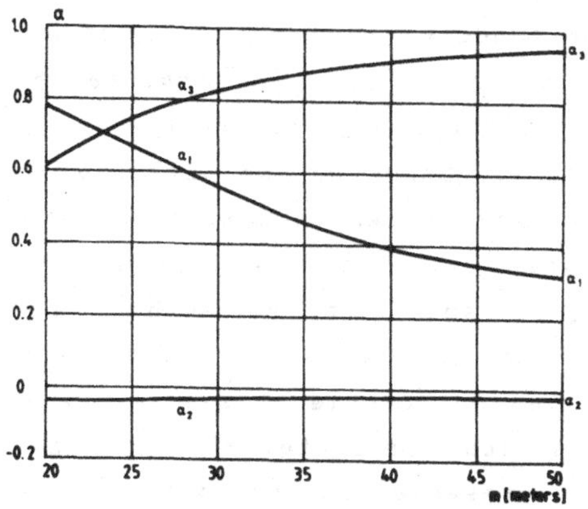

Fig.7 Sensitivity factors in FORM analysis for life time maximum with statistical uncertainty included in sea scatter diagram.

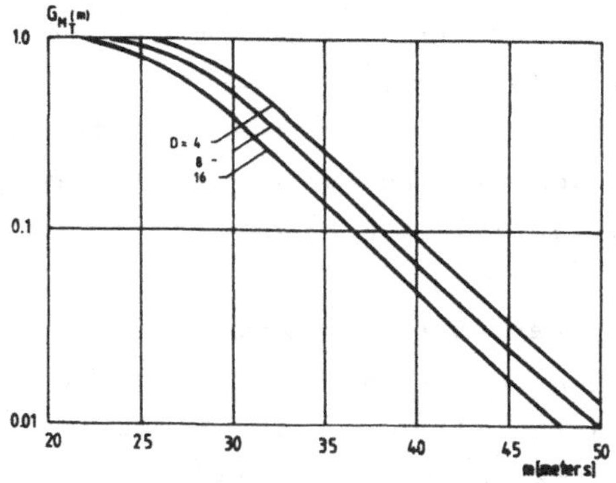

Fig.8 Illustration of effect of sea state duration on life time maximum distribution.

A METHOD TO EVALUATE THE CONSEQUENCES OF MEMBER FAILURE IN JACKET-TYPE OFFSHORE PLATFORM STRUCTURES

James K. Nelson, Jr.
Civil Engineering Department
Texas A&M University
College Station, TX 77843

William J. Graff
Department of Civil Engineering
University of Houston
Houston, TX 77004

SUMMARY

Determining the importance of a member to a structure is a necessary part of reliability based design. The importance of a member can be determined by computing the change in the behavior of the structure if the member were to fail. Once the importance of the member is known, the member should be categorized in a rational manner so that similar members can be designed with the same system consequence factor. To determine the significance of a member to a structure, a computationally efficient structural reanalysis algorithm is needed so the cost of design does not become excessive. A reanalysis algorithm and a proposed classification system for the members in jacket-type offshore platforms is presented in this paper.

INTRODUCTION

An evaluation of the consequences in a structure caused by the failure of a component within the structure is a necessary part of reliability based design. Presented herein is a method to determine the consequences of member failure in jacket-type offshore platforms. In this paper, failure is defined as the complete loss of a member's resistance to load. Such failures occur as the result of fatigue, poor construction, excessive load, and mechanical damage.

To evaluate the consequences of a member failing, the behavior of the remaining structure was determined. This was accomplished by removing a single member from the structure and computing the changes in the displacements and forces in the remaining structure. This member then was replaced and another member was removed; the process was repeated for all members in the structure. Each member can be classified according to the change in the forces in the other members in the platform and the serviceability of the platform.

Jacket-type offshore platforms are composed of many members. Graff (1981) presents typical platform configurations. If a complete analysis of the platform is required each time a member is removed, the cost of design will become exorbitant. A computationally efficient procedure to remove members from the structure has been developed and is presented in the section entitled "Reanalysis Algorithm." This algorithm is based upon the initial strain concept of structural reanalysis. Advantage was taken of the fact that the degrees-of-freedom (DOF) to be modified during reanalysis for the removal of a single member are well defined; they are the DOF associated with the member being removed. The computer time required for reanalysis is less than would be required if new analyses of the platform were performed. For the jacket structures studied, the time required for reanalysis was approximately one-third of the time required for the original analysis.

Presented in the section entitled "Member Categorization" is a proposed classification system for the members in jacket-type offshore platforms. The classification system is quantitative and considers both the force changes in the components and the serviceability of the platform. Each time a member is removed, the forces in the remaining members are computed and are compared to the forces in the members in the original structure. The original structure was assumed to be designed in the elastic range of the material. This assumption is consistent with the American Petroleum Institute (API) design specification for fixed offshore platforms (RP2A). The serviceability of the platform is measured by the inclination of a reference plane. Four member categories have been proposed: Non-Redundant Members, Primary Structural Members I, Primary Structural Members II, and Redundant Members. The significance of each category is discussed in this paper.

AN OVERVIEW OF STRUCTURAL RELIABILITY

All structures have an inherent risk of failure caused by uncertainty in the design process. The objective of reliability based design is to quantify this uncertainty so the risk can be evaluated. Risk is defined as the probability, P_f, that the load on the structure exceeds the resistance of the structure. Risk is expressed as:

$$P_f = P(R < L) \qquad (1)$$

The load, L, and the resistance of the structure, R, are assumed to be independent random variables which have been modelled successfully with Gaussian, Weibull, and lognormal distribution functions. Much of the work reported in the literature assumes the distribution of R and L to be lognormal. A random variable with a lognormal distribution is one in which the logarithms of the variate are normally distributed. The reliability of a structure is the probability that the load on the system does not exceed the resistance of the system. Reliability, P_s, which is the complement of risk, is:

$$P_s = P(R > L) = 1 - P_f \qquad (2)$$

Associated with each structure is a safety index, SI. The safety index is a function of the random variables L and R. The safety index is assumed to have a standard normal distribution with a mean value of zero and a standard deviation of one. The safety index is computed by:

$$SI = (u_R - u_L)/(s_R{}^2 + s_L{}^2)^{0.5} \qquad (3)$$

In Equation 3, u_R and u_L are the natural logarithms of the 50th percentile of the resistance and load, respectively. The standard deviation of the logarithms of the resistance and load, s_R and s_L, respectively, are computed by:

$$s_x = 0.39(\ln x_{90} - \ln x_{10}). \qquad (4)$$

In Equation 4, x is the variate R or L. Reliability is the cumulative probability of the safety index being less than some value X. This concept is expressed as:

$$P_s = P(SI < X) \qquad (5)$$

Cumulative probability is equal to the area under the probability distribution curve between negative infinity and X. The probability of failure is equal to the area under the probability distribution curve between X and positive infinity. This area is approximated, for typical values of the safety index, by (Bea, 1980):

$$P_f = 0.475\exp(-SI^{1.6}). \qquad (6)$$

The probability of success, then, is approximated by:

$$P_s = 1 - 0.475\exp(-SI^{1.6}). \qquad (7)$$

The calculated risk is a very small number generated using a statistical model. The safety index for typical structures ranges from 2.5 to 4.0 (Moses, 1980) which correspond to probabilities of failure of 0.62 percent and 0.003 percent, respectively. These results are affected by the ability to model the system and by limitations in the statistical data base. The calculated risk should be viewed as speculative and should not be compared with actuarial tables. The calculated risk, though, is suitable as a comparison between different platforms in different operating environments.

LOAD AND RESISTANCE FACTOR DESIGN

The reliability based design concept discussed above provides an estimate of the reliability of a platform, but is not useful for everyday design. The reliability based design procedure has been cast in a load and resistance factor design format. Such a design format removes from the engineer the responsibility of determining the mean values of load and resistance, the standard deviation of each of these variables, and the appropriate safety index. These items are accounted for in coefficients that modify the nominal load and resistance of the system. Equations for computing nominal resistance and load and the coefficients can be specified in design codes.

The load and resistance factor design format that has been recommended for incorporation by the API into specification RP2A is (Moses, 1980):

$$r_i r_s R = l_i l_a L \qquad (8)$$

The left side of Equation 8 represents the resistance effects. The term r_i is the resistance intensity factor. This factor is intended to account for variations in the nominal resistance of the system. The term r_s is the system consequence factor and is intended to cover the nature of the component failure. The system consequence factor and the component resistance factor decrease from unity as the importance and the uncertainty of the member increases.

The load factor also is divided into two parts. The load intensity factor, l_i, accounts for variations in the nominal load and experience in specifying the nominal load. The load analysis factor, l_a, is intended to cover the variation in load effects due to design and theoretical assumptions, and also due to variations in measurement. This factor reflects the limitations of the current analysis methods. The load factors increase from unity as the confidence in the load decreases.

The purpose of this paper is to present a method for classifying the members in jacket-type platforms so the system consequence factor can be assigned in a rational, quantitative, and consistent manner. Determining the magnitude of the system consequence factor associated with each member category is beyond the scope of this work.

REANALYSIS ALGORITHM

The reanalysis algorithm that has been developed for both the member removal process and the nonlinear analysis after member removal is based upon the initial strain concept of structural reanalysis. When using the initial strain concept, the structure is made to act as a modified structure by the application of an additional load. When a member is removed from a structure or when the stiffness of the system changes because of material nonlinearity, the structural stiffness matrix is modified. When the loads applied to the system remain the same, the displacements of the structure change. This is shown by the equation following:

$$[K - dK]\{X_o + dX\} = \{F\} \tag{9}$$

The term [K] in Equation 9 is the original structural stiffness matrix and [dK] is a modification to the stiffness matrix. [K] and [dK] are matrices of the same size. The displacements in the original structure are $\{X_o\}$ and the changes in these displacements when the stiffness matrix changes are {dX}. The loads applied to the structure are {F}. The modified displacements of the system, {X}, are equal to the quantity $\{X_o + dX\}$. When using the initial strain concept, these displacements are computed from:

$$[K]\{X\} = \{F + dF\} \tag{10}$$

The vector {dF} is selected such that the displacements computed using Equation 10 are the same as those computed using Equation 9.

<u>Modification for Member Removal</u>. Structural modification can involve any of the DOF in the system, but modification for the removal of a single member involves only the DOF associated with that member.

Advantage was taken of this fact when the algorithm for member removal was developed.

The strain energy stored in the original structure under the influence of the modified load {F + dF} is greater than the strain energy stored in the modified structure, i.e., the original structure with a member removed, under the influence of the actual loads {F} by an amount U_j. The quantity U_j is the strain energy in the member that is being removed mathematically in the original structure under the influence of the modified loads. Recalling that the strain energy stored in a structure is equal to the work done on the structure by the external loads, this relationship can be expressed as:

$$\{F + dF\}^t\{X\}/2 - U_j = \{F\}^t\{X\}/2 \tag{11}$$

From Equation 11, the strain energy in member j can be shown to be:

$$U_j = \{dF\}^t\{X\}/2 \tag{12}$$

The strain energy in member j computed using the elemental forces and displacements is:

$$U_j = \{P_j\}^t\{x_j\}/2 \tag{13}$$

In Equation 13, $\{P_j\}$ are the elemental forces and $\{x_j\}$ are the elemental displacements associated with member j. When Equations 12 and 13 are set equal to each other, and recalling that:

$$\{x_j\} = [B_j]\{X\} \tag{14}$$

the modification load vector {dF} can be shown to be:

$$\{dF\} = [B_j]^t\{P_j\} \tag{15}$$

Recall that the elemental forces are computed by:

$$\{P_j\} = [k_j]\{x_j\} \tag{16}$$

When Equation 10 is solved for {X} and is substituted along with Equation 14 into Equation 16, $\{P_j\}$ is found to be:

$$\{P_j\} = [k_j][B_j][K^{-1}]\{F\} + [k_j][B_j][K^{-1}]\{dF\}$$

The first term to the right of the equal sign in Equation 17 is the force in member j, $\{P_{jo}\}$, in the original structure under the influence of the actual loads. If this substitution is made, Equation 17 can be reduced to:

$$\{P_j\} = \{P_{jo}\} + [k_j][B_j][K^{-1}][B_j]^t\{P_j\} \tag{18}$$

After $\{P_j\}$ is computed using Equation 18, {dF} can be computed using Equation 15. {dF} then can be combined with the actual loads and the modified displacement vector can be computed.

Modification for Nonlinear Behavior. If the stresses in any of the remaining members in the platform become inelastic, the stiffness matrix must be modified. The initial strain concept was used for the nonlinear analysis procedure, also. However, more DOF need to be modified than did for member removal and the DOF to be modified are not necessarily known apriori. As such, the solution procedure needs to deal with the entire structure. A solution cannot be obtained in terms of the member forces and displacements as was done in the modification procedure for member removal.

If the displacements that are computed using Equation 9 are set equal to the displacements that are computed using Equation 10, the following expression for the modification vector {dF} can be obtained:

$$\{dF\} = [dK]\{X_o\} + [dKK^{-1}]\{dF\} \tag{19}$$

Equation 19 is inconvenient to use when computing {dF}. It can be solved using the method of successive approximations. This approach reduces to an infinite series that involves the powers of a matrix.

Through substitution of the following expression:

$$\{dX\} = [K^{-1}]\{dF\} \tag{20}$$

into Equation 19, the following equation can be obtained:

$$\{dF\} = [dK]\{X_o\} + [dK]\{dX\} \tag{21}$$

{dF} now can be computed through iterative improvement using Equations 20 and 21. An initial estimate of {dX} is made; this estimate is commonly taken to be zero. This estimate is substituted into Equation 21 and an estimate of {dF} is computed. This estimate of {dF} is substituted into Equation 20 and an improved estimate of {dX} is computed. This improved estimate of {dX} is substituted then into Equation 21 and an improved estimate of {dF} is computed. This process is repeated until the difference between successive calculations of {dF} is acceptably small.

MEMBER CATEGORIZATION

A quantitative member classification system has been developed to typify the consequences of member failure in jacket-type platforms. The classification system involves four member categories, namely:

Non-Redundant Member. The failure of a member in this category necessitates a shutdown of the platform due to a loss of serviceability or due to a collapse of the structural system.

Primary Structural Member I. The failure of a member in this category causes stress beyond the elastic limit to occur in other members of the platform but the platform does not collapse. The nonlinear behavior of the system is considered when evaluating collapse.

Primary Structural Member II. The failure of a member in this category causes a significant change in the magnitude of the forces in the other members in the platform but the stresses in all of the other members remain elastic.

Redundant Member. The failure of a member in this category has a small influence on the magnitude of the forces in the other members in the platform and the stresses in the remaining members remain elastic.

When a platform is subjected to more than one set of loads, it is possible that a particular member can be placed in more than one category. The consequences of failure for a member can be different for each of the load conditions. The member is placed in the most severe category of the categories to which it can belong. A Non-Redundant Member is the most severe member category and a Redundant Member is the least severe member category.

The serviceability of the platform is measured by the inclination of a reference plane. The reference plane is selected by the analyst and should be indicative of the serviceability of the structure. Although other measures of serviceability can be used, the inclination of a reference plane was selected because many of the operations that occur on platforms are dependent upon inclination of a surface. The limits for allowable inclination have not been established because these are dependent upon the operation being performed and upon the operating philosophy of the drilling contractor and platform owner.

First yield is used as an indication of the transition from an elastic to an inelastic stress state. This transition is the boundary between the second and third categories and is determined using the interaction equations presented by API with the explicit safety factors removed. After first yield, the members that have yielded are assumed to behave in an elastic-perfectly plastic manner; the forces in the member remain constant. If the level of strain in a member exceeds twenty times the yield strain of the material, the structure is assumed to be entering into a plastic collapse mode. Although these assumptions may not be valid for the analysis of a platform, they are valid for the purpose of classifying the members in a platform.

APPLICATION OF MEMBER CLASSIFICATION SYSTEM

The application of the member classification system that has been presented herein can be demonstrated by considering a longitudinal frame from a platform. This frame, which is typical of longitudinal frames that can be found on platforms, is shown in Figure 1a. The platform this frame came from was subjected to two loading conditions. The first loading condition was a quartering wave approaching the platform from one direction and the second loading condition was a quartering wave approaching the platform from an orthogonal direction. A quartering wave is one which approaches the platform at a 45 degree angle to the principal axis of the structure. Superimposed on each of these loading conditions was a vertical load that was typical of operating loads that occur on the structure. The members of the frame shown in Figure 1a were classified for each of the loading conditions. These results are presented in Figures 1b and 1c. Figure 1b is for the wave approaching from one direction and Figure 1c is for the wave approaching from the other direction.

As can be seen in Figures 1b and 1c, some of the members are classified differently for each of the two loading conditions. As stated previously, the final classification of a member is the most severe of the member categories to which it can belong. The final classification of the members in this frame is presented in Figure 5c.

REFERENCES

American Petroleum Institute, Specification for the Design, Planning, and Construction of Fixed Offshore Platforms, January, 1982.

Bea, R. G., "Reliability in Offshore Platform Criteria," Journal of the Structural Division, ASCE, Vol. 106, No. ST9, September, 980, pp. 1835-1853.

Graff, W. J., Introduction to Offshore Structures: Design Fabrication Installation, Gulf Publishing Company, Houston, 1981.

Moses, F., and Russell, L., Applicability of Reliability Analysis in Offshore Design Practice, API-PRAC Project 79-22, A Final Report Submitted to the American Petroleum Institute, Dallas, June, 1980.

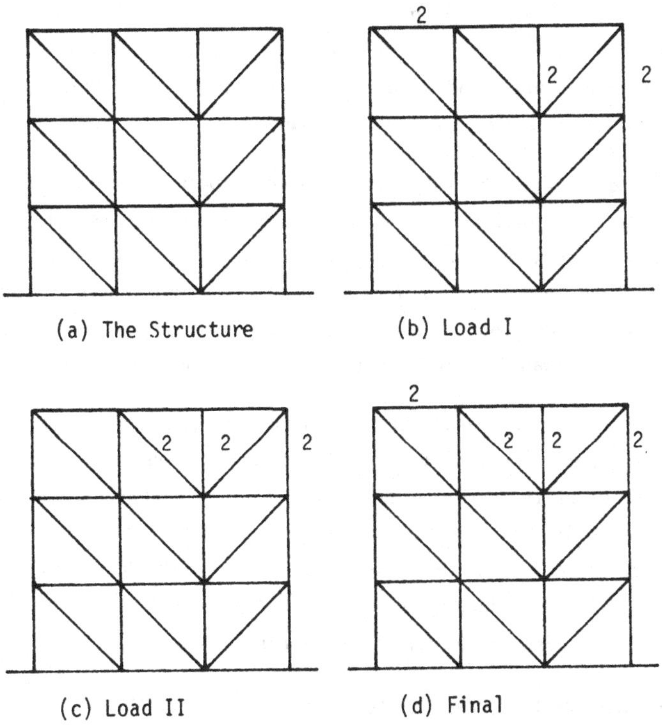

Figure 1 Classification of Members in a Frame
(Members Not Labeled Are Cat. 3)

ON SELECTING A TARGET RELIABILITY FOR DEEP WATER TENSION LEG PLATFORMS

D. Faulkner
Department of Naval Architecture and Ocean Engineering
The University of Glasgow
Glasgow, Scotland

1. INTRODUCTION

Most engineering structures have evolved from previous "successful" past practice, ships and bridges being typical examples. What constitutes success is clearly discernible so far as fatigue cracking is concerned. But when it comes to ultimate collapse limit states, which often govern the main disposition of material, and hence also in many cases cost, success is much harder to define. For an acceptable level of safety (non-failure) the best design is surely that which meets the service or performance requirements within the specified constraints and with the minimum economy. Economy usually relates to initial material and construction costs, although total life costs should not be ignored. These would include maintenance and repair and, for moving structures, fuel and other costs as they relate to structural arrangements and weight.

Success defined in these rational terms is, for most steel structures, seldom if ever modelled or quantified. Nor is structural safety measurable, except of course for failures. So, the usual evolutionary process of design based on so-called successful past practice may be perfectly fine for the drag, sea-worthiness and other performance requirements for ships, for example, where bad designs are immediately obvious, but the author believes the process is more illusory than real for ships [1] and many other steel structures.

This is mentioned because ship classification societies have, by and large, become the arbiters for certifying the structural worthiness of mobile and semi-buoyant offshore structures. Some caution is therefore necessary in any attempt to rationally choose reliability levels for such structures since values derived from current practice have only limited value and may in fact be quite misleading. Fortunately, both Det norske Veritas [2,3] and American Bureau of Shipping [4,5] are steadily moving toward ultimate load limit state reliability based design rules for offshore structures.

The world's first Tension Leg Platform (see figure 1) described by Mercier [6] was, however, certified using working stress design rules (Lloyd's Register [7]) whose safety factors are virtually indistinguishable from those used in ships and which are generally quite inapprop-riate and unduly conservative when applied to thin stiffened cylindrical shells. This, coupled with the perhaps inevitable prima donna effect, whereby conservatism usually creeps in at every stage in a new concept, led to a rapid escalation in weight. Figure 2 illustrates that when compared with large semi-submersibles with which the TLP could reasonably be compared the structure was about 40% heavier than might be expected. This is unfortunate, not only for costs, but also because first designs would normally become the trend setter for the future.

Conoco recognised the scope for improved structural design. Coupled with the prospect of production platforms being needed in much deeper water where structural weight becomes much more critical, they decided as a matter of some urgency to promote buckling tests to fill in the gaps in existing knowledge. At the same time ABS also recognised the need and were invited to establish a TLP Rule Case Committee (RCC) whose mandate was to propose a Model Code using extreme load limit state methods and Level I partial safety factors derived using Level II first order - second moment reliability methods to ensure that the relative uncertainties in loading and strength are better balanced and represent the best state-of-the-art.

The author was invited to chair the RCC which met thirteen times in fifteen months ending with a draft Model Code [5] for final technical committee comment and hopefully approval. An early task of the committee was to recommend a code format and safety index (a reliability index, beta) which was progressively refined. The overall activity has been described by Faulkner et al [8].

This paper is concerned with the studies leading to the initial selection and final recommendation for the target safety index. Against the background just described it was felt that three things became immediately important:

a) Because of the novel nature of the TLP concept which is compliant (partially free to move) and of thin shell construction (D/t ratios 500 to 1000) a return to fundamental considerations of loading, strength and reliability was important. These have been described by Chen et al [9] and Faulkner et al [10].

b) The activity provided a rare opportunity to take stock of improved technical knowledge over the years so that where ignorance has been eliminated, or at least significantly improved, safety factors may reasonably be reduced resulting in more efficient designs.

c) The normal process of "calibration" directly based upon previous design rules and experience was impossible; therefore a more indirect approach was adopted, as will now be described.

A fourth important factor which could not be pursued satisfactorily in the time available was a rational economic basis for selecting **safety** levels, for example, based on Marshall [11]. All that can be said is that the reliability levels finally recommended are thought to be somewhat conservative compared with that which would be recommended if a total economic appraisal could be made based on today's market, salvage and clean up prices.

2. BASIC CONCEPTS

2.1 Safety Index (beta)

In the classical approach to structural reliability, as discussed for example by Leporati [12], load effects or actions Q and resistances R are described by distinct random variables. The safety margin is then defined by:

$$Z = R - Q \qquad (1)$$

The failure surface is defined by $Z = 0$ so that $Z < 0$ defines the unsafe domain. The safety or reliability index of the design is defined as the ratio:

$$\beta = m_z/s_z = v_z^{-1} \qquad (2)$$

where m_z and s_z are the mean and standard deviation of Z and v_z the coefficient of variation of Z. In simple terms it is equal to the number of standard deviations that the mean value of the safety margin Z falls in the safe region. In most cases β can be determined by a simple iterative procedure [12]. Given the safety index the notional probability of failure for the structure can be found from:

$$p_f = \Phi(-\beta) \qquad (3)$$

where Φ is the standardised normal distribution function. From a practical point of view the safety or reliability index is generally accepted as being the most adequate parameter for assessing the safety of a structure.

2.2 Code Format

On the basis of the Level II reliability approach it is possible to derive partial safety factors (PSFs) for use in safety check equations. This provides the basis of the Level I method for use in design which makes little if any reference to statistical properties beyond those necessary for defining the more important variables such as nominal yield stress, etc. In its simplest form the safety check equation can be written [3,10]:

$$\sum_i \gamma_{fi} Q_{ki} \leq R_k/\gamma_m \gamma_c \qquad (4)$$

which essentially relates nominal or characteristic values of extreme design load effects Q_k and ultimate strength R_k of the structural component. The gammas are the partial safety factors which reflect the uncertainties in loads (γ_{fi}), in as-built strength (γ_m) and also the nature of the structure (see redundancy later) and the seriousness of the consequencies of failure (γ_c) - this latter of course being linked to socio-economic factors.

The actual code format adopted as an interim measure [4,8] is naturally somewhat more complex than equation (4) and checking for component safety is generally done with a three term interaction equation:

$$\sum_{i=1,2,3} \left\{\frac{\gamma_s Q_{si} + \gamma_q Q_{qi} + \gamma_d B Q_{di}}{R_{ki}/\gamma_m \gamma_i (\phi_s)}\right\}^{j_i} \leq 1 \qquad (5)$$

where subscripts s, q and d refer to static, quasi-static and dynamic load effect components and B is a systematic modelling or bias factor for the dynamic component; subscripts i = 1, 2 and 3 refer to the equivalent or resolved axial, shear and pressure load and resistance effects; the γs are PSFs all greater than unity, γ_m accounts for uncertainties in the material properties, the γ_is are modelling uncertainties for the three strength components; and j_i is an interaction exponent for each of these three strength components. ϕ_s is the redundancy or systems factor normally less than unity but which is taken as unity for the present. γ_c for socio-economic consequences of failure has been omitted because the more rational approach of selecting the target β from a minimisation of total costs is preferred, and this in turn would lead to an adjustment of all the PSFs. Both of these factors require further study.

2.3 Ultimate "Failure"

The developed code is based upon the capability of the main structural components, such as columns, pontoons and braces (see figure 1), to withstand the most damaging combinations of extreme forces estimated to occur in the lifetime of the structure. However, as in most codes, component capability is then defined in local failure terms. The combination of average in-plane forces and lateral loads and pressures acting over four to six longitudinal stiffeners and one transverse frame space are assessed from linear elastic response of the whole structure. These local forces or stress fields are then checked to ensure that the proposed structure satisfies the safety check equation (5).

2.4 Redundancy Considerations

The structural components designed, as just described, will in general have a reserve strength against total component failure beyond these local failure conditions by virtue of moment distribution type redundancy from two possible effects:

a) cross-section or internal redundancy analogous to the plastic shape factor in beam bending

b) the need generally for a mechanism of hinges to develop along the component before final collapse can occur

There may also be further reserve strength arising from simplifications often made even with comprehensive finite element modelling of the structure.

In addition, and especially if the TLP were to include diagonal bracing, there would be further reserve of strength before total collapse of the whole structural system could occur. This arises from the more widely understood space frame type redundancy associated with the concepts of system stability and determinacy. This has been studied for fixed jacket structures in a safety context by Marshall and Bea [13], Moses [14],

Lloyd and Clawson [15] and others. An understanding of the nature of the residual strength of components following their own maximum load is then important and research continues in this subject.

de Oliveira and Zimmer [16] have discussed all these aspects and make a plea for more attention to continuous structures. They also reminded us that the structural redundances of a complex structure largely depends on the quality of the connections which enable the several components to act together. The principle of ensuring that the connections should be able to transfer the full axial, shear and moment strengths of the weakest member, as put forward for ship stiffener connections by Faulkner [17], would seem to be appropriate for offshore structures also. Much work has been done in this respect for connections between unstiffened tubulars, but little if any work has been carried out to ensure there is smooth and efficient load transfer between the larger stiffened nodes, or for connections between these larger members and unstiffened braces, for example.

Until such studies are complete the full advantages of redundancy cannot be realised. A closely related and clearly important concept is that of damage tolerance. Mistree [18] has shown that a damage tolerant structure must have both reserve strength and residual strength in adequate measure to reduce the consequences of component failure or loss. Gorman and Moses have also studied this topic formally and proposed methods for establishing appropriate PSFs [19].

Preliminary studies by the author show that for practical TLPs the reserve strengths beyond element failure are unlikely to be less than 50% and in most cases will be appreciably more, well in excess of 100% at times. With well designed connections it seems quite reasonable to expect a 2:1 ratio, although clearly more work is required.

2.5 Redundancy and Reliability

Some attention has been given to the question of redundancy in order to highlight the evidently quite conservative definition of ultimate "failure" used. We should then bear this in mind when finally choosing a target reliability based on such localised limit states. These limit states may nevertheless in themselves be important characteristics of failure in many cases, as loads beyond these levels may start rendering the structure unserviceable - possibly even leading to progressive collapse associated with cracking, etc.

Such reasoning leads one to expect that, even in the fullness of time when our knowledge concerning redundancy in practical structures has improved, the element or component basis for design may still prove attractive at some higher acceptable level of probability. Thus, although a more exact determination of the probability of total or system collapse should follow from studies such as those of Moses [14], Thoft-Christensen and Baker [20] and many others, it seems that a move toward component safety checking which incorporates approximate systems PSF ϕ_s has appreciable merit. It should be examined alongside the more logical (in principle) use of a systems β_s to derive all the PSFs as the best way of including safety and redundancy considerations in the design process.

If we accept the classical approach to reliability as in 2.1 and assume the idealised distributions for lifetime load effect Q and large population component strength R are normal and independent then it follows that:

$$\beta = \frac{\theta - 1}{\sqrt{(\theta v_R)^2 + v_Q^2}} \qquad (6)$$

where $\theta = m_R/m_Q$, the central safety factor

Then, assuming the mean values are characteristic values, it follows that an acceptable multiplicative resistive partial safety factor (less than unity)is ϕ_R defined by:

$$\phi_R = R^*/m_R = 1 - \beta\alpha_R v_R \qquad (7)$$

where $\alpha_R = \theta v_R/\sqrt{(\theta v_R)^2 + v_Q^2}$, the resistive sensitivity parameter

and R^* is the "design point" value for maximum p_f

If we now define the load for systems collapse as being n times that for component collapse then (using subscript s for system values) it follows that $\theta_s = n\theta$ and the new safety index β_s can be evaluated from (6). An approximation to the system factor ϕ_s to use with design based on component failure can then be estimated from:

$$\phi_s = \frac{\phi_{Rs}}{\phi_R} = \frac{1 - \beta_s\alpha_{Rs}v_{Rs}}{1 - \beta\alpha_R v_R} \qquad (8)$$

Evaluating ϕ and β for realistic TLP ranges of n = 1.5 to 2.0 and β = 3.0 to 4.0 for typical assumed values of v_Q = 0.2 and v_R = 0.15 = v_{Rs} yields:

component	n = 1.5		n = 2.0	
β	ϕ_s	β_s	ϕ_s	β_s
3.0	0.67	3.2	0.49	4.9
4.0	0.64	4.4	0.47	5.4

Not surprisingly, perhaps, the system reduction factor is approximately inversely proportional to reserve strength factor n, with rather lower values for the safer structures. The scope for structural weight and cost saving is clearly large, although the earlier cautionary remarks could be relevant. Certainly the large increase in the lifetime safety index (of about 1.0 on average) supports any argument for erring toward a low value as a target for element or component based design having a moderate but nevertheless still acceptable risk of happening.

3. INDIRECT CALIBRATION

In choosing a lifetime target safety index the normal process of "calibration" based directly on previous design levels and service experience for the structural type was impossible. Therefore, an indirect approach was adopted in which closely related structures and codes were examined so that a "feel" for a sensible choice of safety index could be achieved bearing in mind:

- the improved loading and strength modelling recommended [9,10]
- arguments such as redundancy just covered and other neglected factors

Current recommendations for other related structures were also reviewed.

3.1 North Sea TLP

The North Sea TLP was designed using the familiar 100-year design wave and a range of acceptable stress and strength requirements. The DnV rules [3] were the only ones sufficiently comprehensive in an ultimate strength sense to enable worthwhile reliability studies to be undertaken. The main difficulty was what statistical properties to give to the 100-year design wave, and so three definitions were examined to bracket the reliability levels implicit in the design to DnV rules. The following statistical descriptions were assumed:

Variable	Bias	COV %	Distribution
Yield stress	1.16	7	Log normal
Young's modulus	1.0	4	Log normal
Dynamic loads	1.0	30	Log normal
Other loads	1.0	10	Normal
Radius	1.0	5	Normal
All other dimensions	1.0	4	Normal

In addition, "modelling"uncertainties in terms of bias and covs were judged from available buckling test data for axial compression and external pressure loads to incorporate with the strength formulations for reliability purposes. These subjective covs varied from 8 to 15% depending upon load, structural type and slenderness parameters.

The lifetime safety indices implicit in the design of the ring-stiffened corner columns (see figure 1) were calculated for the top of the column where static and quasi-static axial loads dominate, and also for the bottom of the column just above the pontoon where pressure forces also become important. The results varied widely with β values from about 3 to 8 depending upon the loading assumptions. Likely average values would be $\beta = 5.2$ for the top of the column and $\beta = 5.4$ for the bottom as reported by Das [21].

The exercise was repeated by Chen using the more realistic interim loading models [9] proposed by the RCC which suggest that an average value of $\beta \simeq 5.0$ for the corner column.

More efficient ring and stringer stiffened columns would be considered for future similar designs where axial compression loads are high. An indication of the range of β values which could arise is provided by an interesting analysis made by Das et al [22]. In this the UK stringer stiffened buckling models were analysed using the DnV-ECCS strength formulations and axial compression loads statistically defined to be representative of extreme lifetime conditions. The safety indices ranged from $\beta = 3.0$ to 6.8; but when a modelling error x_m was included the spread was much smaller with $\beta = 2.6$ to 3.8. The strength modelling has recently been improved yet further [10] even with the larger population of test data from the Conoco buckling tests. Flat structure is now included.

3.2 An Existing Semi-submersible

A reliability analysis using similar statistical assumptions and interim loading model as just described was completed for an eight column twin hull unit with a displacement of about 25,000 T at the operating draft. For the ring and stringer stiffened corner column immediately above the lower hull the derived lifetime safety indices varied between $\beta = 4.2$ and 4.6. For an intermediate column the average safety index for the ring and stringer stiffened shell just above the lower hull was $\beta = 5.0$, whereas for the ring stiffened shell further up the same column $\beta = 3.9$.

The strength formulations used were those developed by the RCC [10] which probably accounts for the somewhat higher value for the stringer stiffened structure as this allows advantage to be taken of greater strength arising from curvature effects. This is a good example of where our improved knowledge now allows us to remove some of the existing conservatism. It is also interesting to note that the average value of the safety index is about 4.4 compared with 5.3 say for the North Sea TLP. This is a very substantial difference which by itself would quite easily account for excess weight in the North Sea TLP of at least 40%.

3.3 Columns and Beam-Columns

Unstiffened tubulars may be used for braces in TLP structures and are well covered in offshore design codes. It was therefore felt useful to apply reliability methods to estimate safety indices which would arise using a range of 100-year wave load definitions.

DnV rules [3] and API recommended practice [23] were used for the column analyses, although the DnV slenderness dependent safety factor (kappa) was ignored for inconsistency reasons [22]. With statistical assumptions similar to those in section 3.1, safety indices were estimated using the more conservative mean-value approach (first order first moment) for 54 columns designed with slenderness values 0.2, 1.0 and 2.0 and wave/dead load ratios of 1, 5 and 20. The results are shown in figure 3 and are summarised in the following table but note that if advanced Level II methods were used the safety indices would be expected to be raised by about 0.5 on average above those shown.

Code	Mean β	cov%	Max β	Min β
API RP2A (1981)	3.30	23	4.6	1.9
DnV Rules (1977)	3.62	13	4.5	2.8

A much more limited range of calculations was also completed for beam-columns for the API formulations and for a range of end:lateral loads. The resulting β values were on average 6% lower than for the columns. The DnV rules were considered to be too cumbersome and difficult to apply. Some calculations using the latest Load Resistance Factor Design (LRFD) formulations of the AISC specifications gave β values in the range 3.7 to 4.7.

3.4 Fixed Platform Studies

Anzai and Bryant [24] have compared a recent Danish (DOR) limit state code [25] with a traditional API working stress design approach for a North Sea fixed jacket platform calibrated by Moses [26]. From Table 3 of reference [24] the results may be summarised:

- API range of β = 1.6 to 3.3
- DOR range of β = 2.1 to 2.6

API is less conservative overall and their recommendations seem to imply a wider range of safety which varies appreciably between different components and locations in the structure. As the wave/dead load ratio increases β decreases by more than 1.0 with API, and much less so than for DOR. The Danish code is altogether more consistent, no doubt because of its ultimate strength approach.

It is interesting to note that the wave induced load effect assumed a bias value of 0.8 and a cov = 37%. This will be discussed again later, but the large cov no doubt accounts for the somewhat low β values. However, the absolute values of β may not be directly comparable with earlier values as the load definitions are unstated.

API have been extending this initial work for calibrating API RP2A for reliability based design and for introducing LRFD for fixed platforms. Moses [26] has evaluated safety indices for tubular members having L/D ratios about 20 for two such platforms in the Gulf of Mexico. The wave/gravity load ratios varied between 2 and 40 and three different modes of failure were considered. The following table gives LRFD β values (with working stress values in parenthesis) evaluated for covs for Q_g = 10%, Q_w = 40% and R = 20%.

Failure mode	Mean β	Max β	Min β
Yield	2.15 (1.95)	2.3 (2.1)	2.1 (1.8)
Bending	2.7 (2.9)	2.9 (3.3)	2.5 (2.6)
Column	2.4 (2.1)	2.5 (2.3)	2.2 (1.9)

In both cases β decreases as the ratio wave/gravity loads increase - as in the earlier study - but with less variation. The relatively low values will in some cases be due to the large cov for wave loading as many braces experience high wave induced loads. Wirsching [27] has been studying for the API probability based fatigue design criteria for offshore structures. For fixed jacket structures this suggests an average lifetime safety index of about 2.7, with of course the expected large spread. But this result seems to include a safety factor of two on life, and the definition of local failure (even if it were through cracking) may not be compatible in terms of seriousness with ultimate buckling collapse of a major component. So some caution is necessary when comparing the two cases.

3.5 UK Offshore Thinking

When Flint and Baker finished their rationalisation work for structural safety [28] the UK Department of Energy commissioned a supplement to this work for offshore installations [29]. This includes a fairly arbitrary suggestion that target reliability levels be set at a notional annual probability of failure $p_f = 2.5 \times 10^{-6}$. For a 25 year life, for example, this would give $p_f = 6.25 \times 10^{-5}$ which is equivalent to $\beta = 3.83$. This is an appreciably lower lifetime reliability than the same authors recommended for the new code for steel bridges [30,31] which is based on an overall structural average $p_f = 0.63 \times 10^{-6}$ ($\beta = 4.85$) derived from the old bridge code BS 153.

The rationale for the lower target safety suggested for offshore structures appears to be that, even so, the risk to life implied by the annual $p_f = 2.5 \times 10^{-6}$ would be only 1/400 of that due to other hazards associated with working on offshore structures which is about 10^{-3} per annum. This rather arbitrary argument reminds one of a socio-economic criterion advanced by the Building Research Station in the UK which gives a target lifetime probability of failure as eq. (43) of reference [28]:

$$p_{ft} = 10^{-4} K_s n_d / n_r \tag{9}$$

where K_s is a social criterion

n_d is the design lifetime in years

n_r is the number of people at risk

The lowest value of K_s suggested is 0.005 for dams and the highest value is 5 which might be relevant for marine structures whose safety does not impinge on the general public. Taking this value and $n_r = 200$ people leads to the annual target sugested above of $p_f = 2.5 \times 10^{-6}$. This corresponds to $\beta = 3.7$ for a lifetime n_d of 50 years.

3.6 Ship Comparisons

Some caution has already been expressed about any value to be derived from so-called successful ship comparisons, but with care something useful can be said. Using modern methods Mansour derived lifetime safety indices for deck collapse (all longitudinally stiffened) in 16 merchant ships with a worldwide mission profile and 2 for North Atlantic operations [32]. Lengths varied from 160 m to 330 m and β values varied from 4.2 to

6.4. Faulkner extended this work [33] and included a brief look at implied safety levels for bottom structure. The lowest value for the outer bottom collapse of a 160 m cargo ship was β = 4.2.

However, the position for transversely framed bottom structures is much worse. For the extensively studied MARINER class the likely average values for the outer bottom range from β = 1.7 to 2.1 although large variabilities are present. Using scantlings derived from current Classification Society rules for ships of length 120 m and less Faulkner designed a double bottom structure similar to that being considered by the RN for a seabed operations vessel. The lifetime safety index for the inner bottom structure extreme hogging condition was between β = -0.5 and 1.8, depending upon the compression strength analysis used. With a large still water hogging stress it follows that the IB would undoubtedly have been in a state of near collapse even before the ship left harbour. The navy wisely resorted to longitudinal stiffening, but the anomaly in the ship rules still remains.

Leaving aside these inconsistencies for longitudinally stiffened merchant ships, Meek's statement that if they ever were strength critical they are not so now [34] must surely be generally agreed. Merchant ships have become more vibration and fatigue critical, but that is another story. Faulkner [35] recommended values of β = 3.0 to 4.5 for use in merchant ships design. On further reflection lifetime values at the lower end of this range would appear to be perfectly acceptable for ultimate bending collapse - but may not be possible unless good detail design is incorporated to reduce fatigue trouble. In the absence of this, or indeed of a rational reliability approach yet for ship fatigue, a target value of β = 4.0 should perhaps be used. Fortunately, in TLP main structural components fatigue does not seem to have as strong an influence on scantlings and hence on structural weight as in ships.

The same level of over-safety does not, however, apply in most naval designs. Faulkner and Sadden [35] derived for five RN frigates lifetime safety indices for ultimate strength of the upper decks which varied from β = 0.9 to 3.4. It is perhaps no great surprise that after many years of service the weaker decks are showing distress and are having to be stiffened. β = 3.0 would seem to be a suitably cautious target value for naval design again to minimise fatigue trouble even though the structural details are notably better than in merchant ships.

3.7 Steel Bridge Codes

The new UK code for steel bridges [30,31] was referred to earlier. The weighted average lifetime target reliability level was based upon that for the old BS 153 code which is $p_f = 0.63 \times 10^{-6}$ equivalent to β = 4.85. Baker in discussing reference [30] showed that components designed to the old code but now analysed using the modern recommended methods would have notional probabilities of failure ranging from:

- p_f = 3×10^{-3} from the lower end for plate panels, to
- p_f = 10^{-17} from the top end for struts

This corresponds to a lowest safety index $\beta = 2.75$ and an upper value which is quite meaningless.

As those components with the lowest notional safety have not failed in past designs, it would seem that when considering the target reliability for the new British steel bridge code there was appreciable scope to have chosen a lifetime target nearer the lower envelope of safe past designs, thereby benefiting from improved knowledge. For example, $\beta = 3.5$ say would seem very reasonable rather than the 4.85 selected, especially as the variation in β due to the limited range of the PSFs is quite small ($\approx \pm$ 0.5) due to the improved modelling of loads and strength over recent years. It is understood that when considering the new German bridge design codes, safety indices in the range $\beta = 3.5$ to 4.0 were felt to be approrpriate. This means the resulting designs would be rather more than two-orders notionally less safe than British ones (but still adequately safe); they would also be about 15 to 20% lighter and appreciably cheaper.

3.8 North American Building Codes

The design of steel buildings in North America is largely governed by specifications of the American Institute of Steel Construction (AISC). Load Resistance Factor Design has been actively studied and debated for some while. Reference [36] contains eight papers on the subject prepared by members of the ASCE committee on LRFD to stimulate professional debate and review. The first paper by Ravindra and Galambos presented some data for implied safety indices using latest loading and strength modelling which varied upwards from $\beta = 2.6$. They

a) recommended a target value of $\beta = 3$ be generally adopted, and

b) suggested that beta may be varied to account for the importance of the structure with $\beta = 3.0$ for ordinary buildings
$\beta = 4.5$ for very important buildings
$\beta = 2.5$ for temporary structure

Recommendation (a) was based upon the approximate reliability of presently designed simple beams and columns which constitute most building members. This recommendation was accepted by the AISC but suggestion (b) was not, a global value of $\beta = 3.0$ being preferred throughout. This has so far not been formally implemented so far as the author is aware.

Recommended loading combinations are proposed by Ellingwood et al [37] in the 1980 revision of the American National Standard A58 which is meant to complement LRFD. The target β values recommended are 2.5 and 3.0 if earthquakes are excluded in the design and $\beta = 1.75$ if earthquakes are included. Generally speaking these values are claimed to be more conservative than indicated by current practice when the variable loads are dominant and less conservative when permanent loads dominate the design.

3.9 Summary

For convenience the approximate average safety indices derived above are listed, but the text should be consulted concerning details and relevance:

1)	North Sea TLP	5.3
2)	Existing semi-submersible	4.4
3)	Columns and beam-columns	3.3
4)	Fixed jacket platforms	2.3
5)	UK offshore thinking	3.7
6)	Merchantship deck collapse	5.3
7)	RN frigate deck collapse	2.2
8)	UK steel bridges	4.8
9)	North American buildings	3.0

It is not certain whether the last one has been implemented as a target. Some of these are indicated on figure 4 which shows the safety index plotted against the notional probability of failure.

4. OTHER STUDIES

4.1 Environmental Modelling

The neglect of Redundancy was discussed in section 2. Similar largely unknown errors with significant systematic components on the conservative side also arise from the present attempts to model reality in the process of deriving load effects from the prescribed environmental descriptions. Uncertainties exist for example in:

- the assumptions made in the environmental descriptions, e.g. statistical wave models, neglect of directionality and spreading co-linear wave, wind and current assumptions, etc.
- the methods used for computing loads on the structure, e.g. Stokes 5th order waves, Morison's equation vs diffraction theory, neglect of interference effects, consideration of marine growth
- idealising the environmental loads which vary in space and time by equivalent design loads, e.g. uniformly distributed or concentrated
- modelling the complicated three-dimensional rigid body response including wave-structure interaction, damping, lumped mass assumptions, etc.
- the structural analysis, e.g. modelling the complex structure by a set of simplified members having rigid joints, uniform properties, neglect of secondary members, etc.
- the transformation of internal nominal loads in a component to load effects or field stresses on structural elements

- the distributional assumptions for load effects
- the selection of the most critical of these simultaneously acting load effects for designing the shell, stiffening and connections.

The Rule Case Committee collected information regarding these largely subjective uncertainties by means of a questionnaire sent to experts having significant related experience in floating structures. In spite of the relatively small sample the results [5] are considered to be of good quality. The customary description of modelling errors is with a bias B defined as:

$$B = \frac{\text{the real, or actual load effect}}{\text{the predicted load effect by the model used}} \qquad (10)$$

The results of the questionnaire were analysed statistically assuming a multiplicative model for overall bias and independent random errors. The results for the quasi-static and dynamic load effects for use in the safety equation (5) are:

Load effect	Bias B	cov %
quasi-static Q_q	0.75	26
dynamic Q_d	0.71	25

The RCC were divided in their own opinions, for example, as to the correctness of a multiplicative model which was felt by some to be too non-conservative. They were under no doubt, however, that for probabilistic design all sources of uncertainty which can be identified should be quantified and that significant uncertainties must be quantified. All three load effect random errors were therefore quanitified but only the bias for the dynamic (wave-induced) load effect was quantified at anything other than unity. For the present a rather cautious value of $B_d = 0.9$ is assumed as being a token move in the right direction to cover the neglect of directional spreading, etc.

4.2 Parametric Studies

The target reliability initially selected by the Rule Case Committee was $\beta \simeq 3.7$ which corresponds to $p_f \simeq 10^{-4}$ and which approximately represents influential UK thinking. Using this target and the established Rackwitz and Fiessler transformation for non-normal distributions [20] partial safety factors were evaluated corresponding to those shown in the safety checking equation (5). The parameters varied were:

- two TLP platform types for the North Sea and Gulf of Mexico
- cylindrical columns with a practical range of stiffening
- unstiffened tubulars and flat structure
- a range of realistic load combinations
- differing statistical distributions for load effect and resistance

The effect of differing reliability targets was also examined. A very wide range of PSFs resulted. For example, for the chosen initial target the dynamic load effect PSF γ_d varied from about 1.5 to 2.3 and the modelling PSF γ_N for axial compression loads varied from 1.0 to 1.5. It was thought to be impractical to specify PSFs for all possible combinations.

5. EXPERIENCED RISKS

5.1 Statistics

It is of interest to know what the risks experienced by structures in service are, although acceptable levels for design should not be based on experienced levels for many reasons. For marine structures some values are given in the following table:

Marine Structure	Annual $p/10^{-3}$	Source
Fixed Gulf of Mexico Platforms:		
. waves exceeding 100-year design level in one hurricane (HILDA 1964)	1.3	Marshall [11]
. structural loss (over 16 years)	1.3	Marine Board [38]
. collision losses (one operator)	5	Marshall [11]
. loss - all causes	3	Historical
Mobile Drilling Units, worldwide:		
. jack-ups - loss	12	DnV 12 yr review [39]
- severe structural damage	13	"
. drillships and barges:		
- loss	4.9	"
- severe structural damage	13.5	"
. semi-submersibles:		
- loss	2.1	"
- severe structural damage	6.2	"
. total for MDUs:		
- loss	8.0	"
- severe structural damage	11.2	"
Merchant Ships, worldwide:		
. loss - all causes	5.9	"
. severe structural damage	$\simeq 1$	Gran, DnV [40]
. loss - due to structure	$\simeq 0.3$	Author's judgement

For offshore structures these experienced values are anything from 0 to 4 orders of magnitude greater than the notional probabilities of failure which correspond to the typical safety indices discussed in section 3. For merchant ships the discrepancy is 3 to 8 orders of magnitude greater. This is much greater than for offshore structures but for most ships the notional p_f is in any case quite meaningless.

5.2 Gross Errors

There are many reasons for the discrepancies just referred to and these include the necessarily notional nature of various models used and assumptions made in evaluating safety, in particular the quite large uncertainties in the tails of the distributions, the generally small samples of losses which can truly be traced to structural weakness as distinct from other causes, and so on. But overriding all this is the wide recognition that the great majority of structural failures occur for unexpected reasons, often in ways not previously encountered. Such events are classified as gross errors or blunders, often of human origin, which radically alter the probability of failure in an essentially unpredictable way. Thoft-Christensen and Baker [20] have classified such errors and nicely summarise their link with quality assurance and reliability theory. The analysis of many structural failures shows that the majority could not have been prevented by minor increases in safety margins or PSFs.

6. DISCUSSION

There is virtually no evidence that any significant loss of life in marine structures arises from inadequate safety factors. Indeed, statistics show that the number of deaths arising from structural failure (most of which arise from gross errors and blunders anyway) are many orders of magnitude lower than the voluntary risks people accept in their daily lives. That being so, the level of reliability should rationally be based on economic optimisation considerations alone. Accepting that at present the data needed for this does not appear to be available then it is certainly proper to be guided by past practice.

Most codes can be seen to be evolutionary in nature. The philosophy usually advocated [20,30] is to aim for a target reliability level based on the average of the previous code. This British preference for "the average" is, the writer submits, misguided and weak. It is uncompetitive and does not allow full benefits to be gained from our improved knowledge. As ignorance diminishes so should the safety factors.

An efficient structure is one which does not fail, has adequate but not excessive safety and which minimises cost. Inevitably cost is very closely linked to safety factors, especially in structures whose scantlings are governed by ultimate strength considerations. So aiming at the lower safety envelope of accepted past designs has considerable merit - not the average.

In some ways it can be fortunate when a new concept like a TLP comes along for which there is no past experience, although what might be called the "prima donna" danger has been mentioned. The RCC was forced to return to fundamentals. The paper has examined a wide variety of factors which should properly influence any choice of target reliability level.

The initial choice of $\beta = 3.7$ ($p_f \simeq 10^{-4}$) for the lifetime of the structure would, on the basis of all these studies, appear to be a good one offering as it does useful scope for weight and cost saving with do doubts concerning inadequate safety. It would appear to be a little less conservative than for an existing semi-submersible, though apparently still more conservative than for fixed platforms, and certainly more conservative than for naval frigates which also in general have a more severe fatigue problem.

Indeed, on these bases it would appear there is scope for further reduction in the target reliability level. Further arguments in favour of this are:

a) the strength model is still quite pessimistic based as it is on very localised collapse and totally ignoring reserve strength and redundancy

b) the load effect modelling still neglects wave directionality and spreading, non simultaneous occurrence of maxima, and assumes the rather conservative log-normal distribution

c) although a bias of 0.9 has been adopted for the wave-induced loads, this is still appreciably higher than the values suggested by the analysis of the questionnaire, which were 0.75 for quasi-static and 0.71 for dynamic load effects

d) most structural failures (80-90% is typically quoted) arise from gross errors or blunders.

However, further examination would be required in all these model uncertainties before any substantial reduction in beta could be contemplated, below 3.0 say. But a reduction to 3.25 has been suggested in view of the conservative assumptions still adopted.

The most cogent argument for an immediate lowering of the target beta arises if a direct derivation of PSFs was mandatory for primary structure using the well established Level-II algorithms. For a fixed $\beta = 3.7$ we saw a wide spread in PSFs to cover the range of designs and load combinations. Conversely, if it were possible to agree on a set of fixed PSFs the range of β values throughout the structural components of various designs would itself be large, probably covering about two orders of magnitude, say from 3.1 to 4.3. Such a direct approach would eliminate any scatter and a target value of $\beta = 3.0$ would then seem to be very reasonable and no less safe.

7. CLOSURE

7.1 Final Recommendations

These were:

1) that a direct derivation of the PSFs be adopted for all primary structures with a sugested target safety index of 3.0 and using approved Level-II algorithms

2) for secondary structure the use of tabulated PSFs would be acceptable based on a target safety index of 3.7.

Primary structure consists of the outer surfaces of the hull where structural integrity is vital to the whole structure. This includes stiffening members, braces and also deck stiffening and plating. Secondary structure is largely internal which can be easily inspected.

7.2 Acknowledgements

The permission of ABS and Conoco to publish this paper is gratefully acknowledged, as is the help and enthusiasm freely given by staff of both firms who colaborated in this endeavour to make it all possible.

REFERENCES

1. FAULKNER, D.: "Safety Factors?", Steel Plated Structures, Crosby Lockwood Staples, 1977.
2. FJELD, S.: "Reliability of Offshore Structures", J.Pet.Tech., October 1978.
3. Rules for the Design, Construction and Inspection of Offshore Structures, DnV, Oslo, 1977.
4. STIANSEN, S.G.: "Development of Reliability-Based Structural Design Criteria for Tension Leg Platforms", Marine Safety Conference, University of Glasgow, September 1983.
5. Model Code for Structural Design of Tension Leg Platforms (Draft), Conoco/ABS Rule Case Committee, ABS, New York, July 1983.
6. MERCIER, J.A. et al: "Design of the Hutton TLP", Offshore SE Asia 82 Conf., Singapore, February 1982.
7. Preliminary Guidance Rules Tension Leg Installations, Lloyd's Register of Shipping, London, May 1980.
8. FAULKNER, D., BIRRELL, N.D. and STIANSEN, S.G.: "Development of a Reliability Based Code for the Structure of Tension Leg Platforms", Paper OTC 4648, OTC Houston, 1983.
9. CHEN, Y.N., LIU, D. and SHIN, Y.S.: "Probabilistic Analysis of Environmental Loading and Motion of a Tension Leg Platform for Reliability-Based Design, Marine Safety Conference, University of Glasgow, September 1983.
10. FAULKNER, D., CHEN, Y.N. and de OLIVEIRA, J.G.: "Limit State Design Criteria for Stiffened Cylinders of Offshore Structures", ASME 4th National Congress of Pressure Vessels and Piping Technology, Portland, Or, June 1983.
11. MARSHALL, P.W.: "Risk Evaluation for Offshore Structures", J.Str.Div., ASCE, No. ST12, December 1969.
12. LEPORATI, E.: The Assessment of Structural Safety, Research Studies Press, Letchworth, 1979.
13. MARSHALL, P.W. and BEA, R.G.: "Failure Modes of Offshore Platforms", Proc. BOSS'76 Conf., NIT, Trondheim, 1976.
14. MOSES, F.: "System Reliability Developments in Structural Engineering", Structural Safety, vol. 1, no. 1, September 1982.
15. LLOYD, J.R. and CLAWSON, W.C.: "Reserve and Residual Strength of Pile-Founded Offshore Platforms", Design-Inspection-Redundancy Symposium, Ship Structure Committee, Williamsburg, Va, November 1983.

16. de OLIVEIRA, J.G. and ZIMMER, R.A.: "Redundancy Considerations in the Structural Design of Floating Offshore Platforms", Design-Inspection-Redundancy Symposium, Ship Structure Committee, Williamsburg, Va, November 1983.

17. FAULKNER, D.: "Welded Connections used in Warship Structures", Trans. RINA, vol. 106, 1964.

18. MISTREE, F.: "Design of Damage Tolerant Structural Systems", Euromech Colloq. 164, University of Siegen, W.Germany, 1982.

19. GORMAN, M.R. and MOSES, F.: "Partial Factors for Structural Damage", ASCE Proceedings on Probabilistic Methods in Structural Engineering, St. Louis, ASCE, 1981.

20. THOFT-CHRISTENSEN, P. and BAKER, M.J.: "Structural Reliability Theory and its Applications", Springer-Verlag, Berlin, 1982.

21. DAS, P.K. and FAULKNER, D.: "Safety Factor Evaluation for Cylindrical Components of Floating Platforms in Extreme Loads", Marine Safety Conference, University of Glasgow, September 1983.

22. DAS, P.K., FRIEZE, P.A. and FAULKNER, D.: "Reliability of Stiffened Cylinders to resist Extreme Loads", Proc. BOSS'82 Conf., MIT, Cambridge, 1982.

23. Recommended Practice for Planning, Designing, and Constructing Fixed Offshore Platforms, API RP 2A, Twelfth ed., API, Dallas, Tx., January 1981.

24. ANZAI, T. and BRYANT, L.M.: "Comparison of a Limit State Design Code with API RP 2A", Paper OTC 4191, OTC Houston, 1982.

25. Regulations for Permanent Offshore Steel Construction, Danish Engineering Association, Copenhagen, May 1980.

26. MOSES, F.: "Guidelines for Calibrating API RP 2A for Reliability-Based Design", Final Report, API-PRAC Project 80-82, October 1981.

27. WIRSCHING, P.W.: "Probability Based Fatigue Design Criteria for Offshore Strucures", Final Report, API-PRAC Project 81-15, API, Dallas, Tx., January 1983.

28. Rationalisation of Safety and Serviceability Factors in Structural Codes, CIRIA Report 63, London, October 1976.

29. UK Department of Energy Supplement to Reference 2 : Supplementary Report on Offshore Installations, by A.R. Flint and M.J. Baker, October 1976.

30. FLINT, A.R. and BAKER, M.J.: "The Derivation of Safety Factors for Design of Highway Bridges", Proc. Conf. on the New Code for the Design of Steel Bridges, Cardiff, March 1980.

31. New Code for the Design of Steel Bridges, Part 3, BS 5400, London, 1982.

32. MANSOUR, A.E.: "Approximate Probabilistic Method of Calculating Ship Longitudinal Strength", J.Ship Res., vol. 18, no. 3, September 1974.

33. FAULKNER, D.: "Semi-Probabilistic Approach to the Design of Marine Structures", SNAME Extreme Loads Response Symposium, Arlington, Va, October 1981.

34. MEEK, M.: "Respective Approaches to Safety Design and Construction - Ships", RINA and Roy.Aero.Soc. joint one-day Symposium, London, April 1983.

35. FAULKNER, D. and SADDEN, J.A.: "Toward a Unified Approach to Ship Structural Safety", Trans RINA, vol. 121, 1979.

36. RAVINDRA, M.V. and GALAMBOS, T.V.: "Load and Resistance Factor Design for Steel", J.Str.Div., ASCE, vol. 104, no. ST9, September 1978.

37. ELLINGWOOD, B. et al: "Development of a Probability Based Load Criterion for American National Standard A 48", National Bureau of Standards, NBS Special Publication 577, June 1980.

38. "Safety in Offshore Oil", National Research Council, Marine Board Report, Washington, DC, 1981.

39. RÖREN, E.M.Q. et al, Eighth ISSC Joint Session II, ISSC, Paris, September 1982.

40. GRAN, S.: "Reliability of Ship Hull Structures", DnV Report No. 78-216, Oslo, 1978.

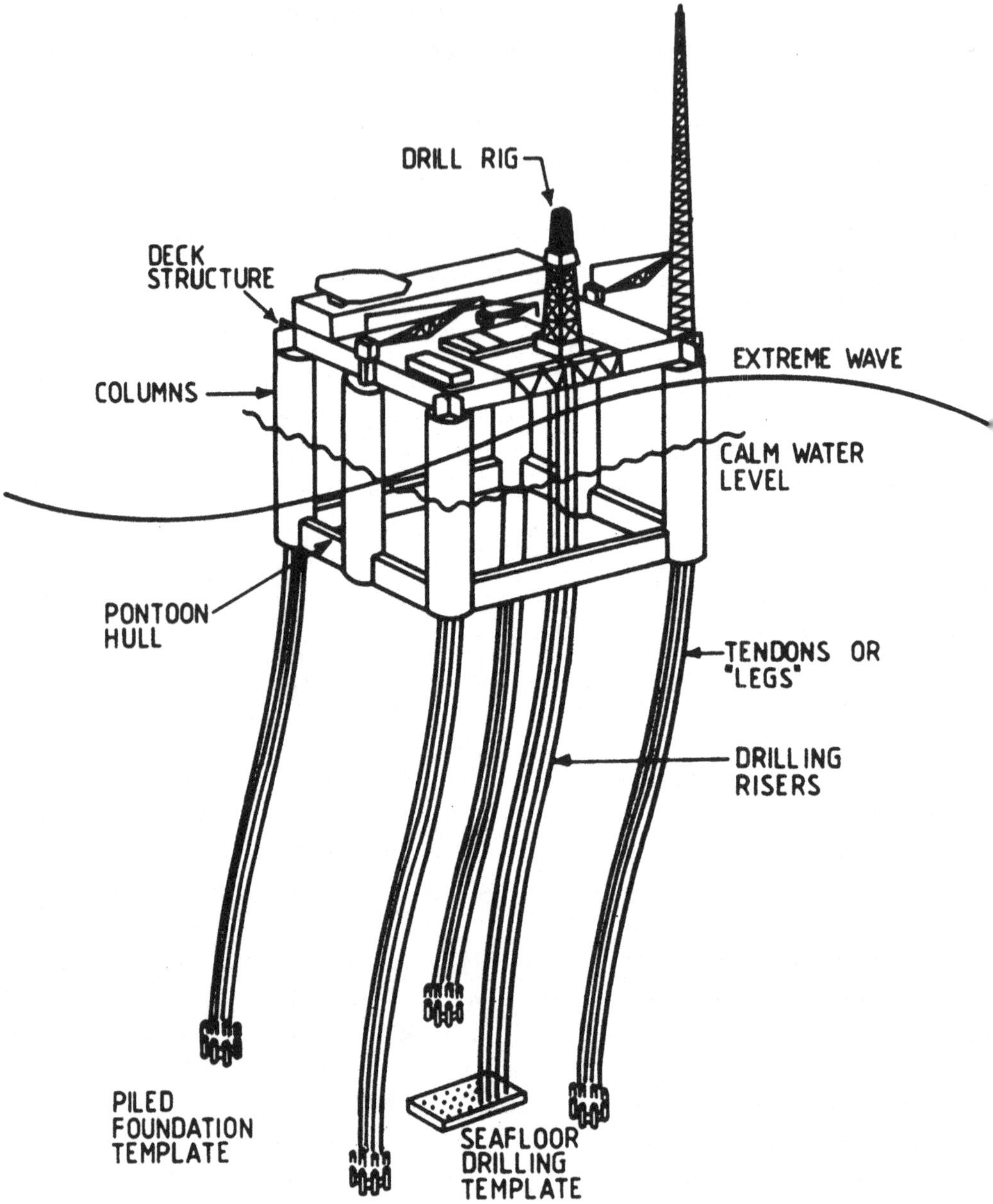

Fig. 1. NORTH SEA TLP UNDER EXTREME WAVE ACTION

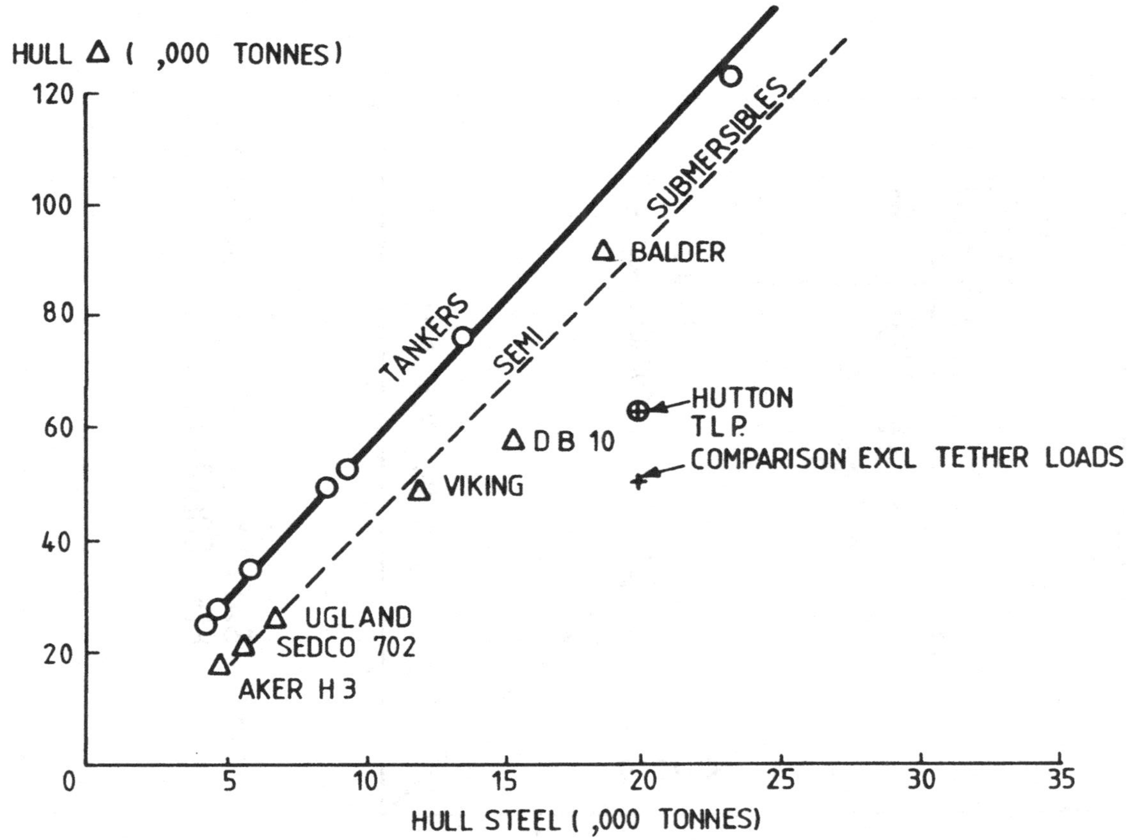

Fig. 2. STEEL WEIGHT vs DISPLACEMENT OF SEMI-SUBMERSIBLES ETC.

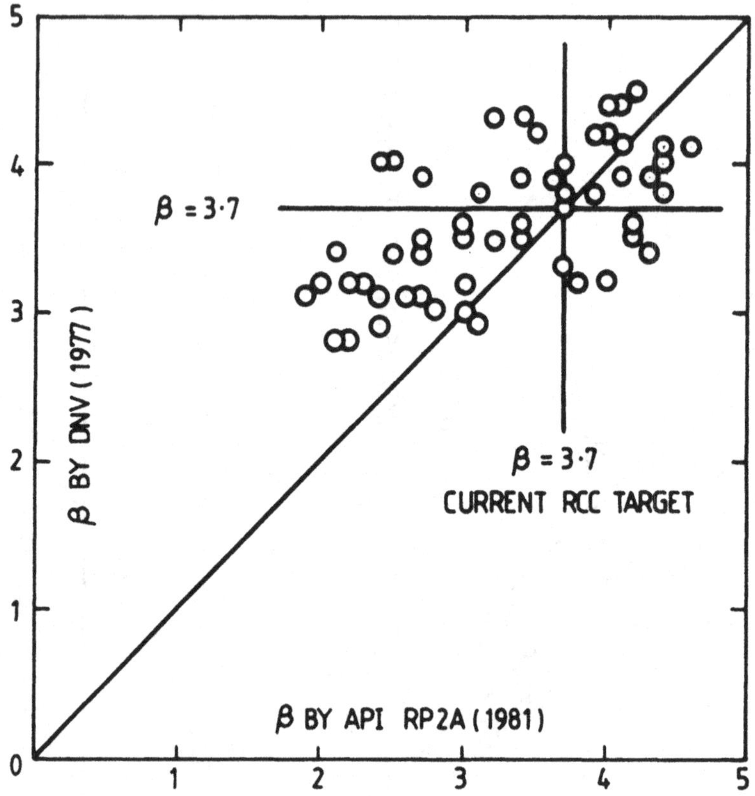

Fig. 3. SAFETY INDEX COMPARISONS FOR COLUMNS

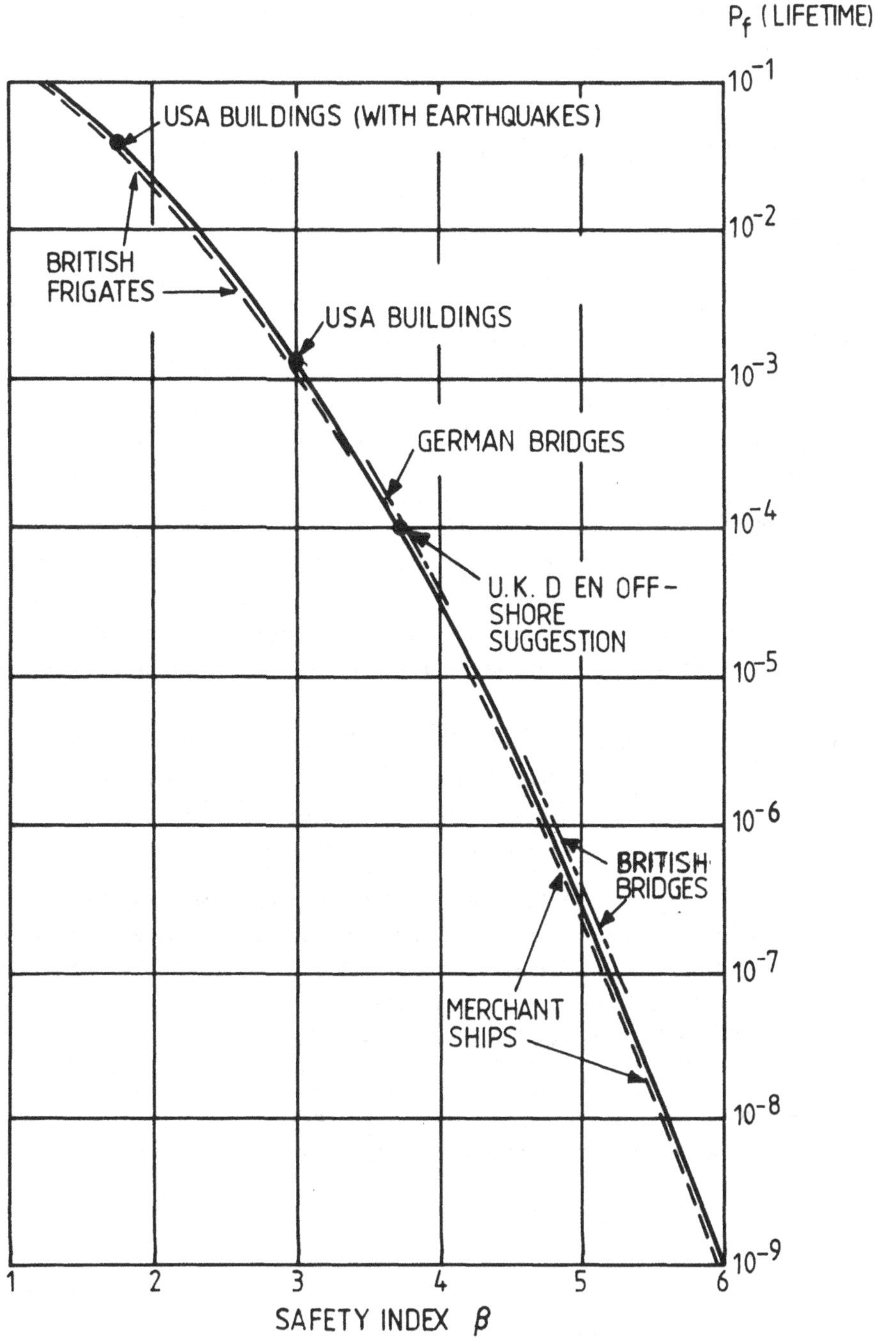

Fig. 4. SAFETY INDEX VS PROBABILITY OF FAILURE

FREQUENCY VERSUS TIME DOMAIN IDENTIFICATION OF COMPLEX STRUCTURES MODAL SHAPES UNDER NATURAL EXCITATION

M. Prevosto*, B. Barnouin* and C. Hoen**
*CNEXO/COB
B.P. 337
29273 BREST CEDEX (France)
**SINTEF
TRONDHEIM (Norway)

1. INTRODUCTION

Vibration monitoring is a basic tool for detecting damages or partial failures in a vibrating structure and can be used to study modelisation of such structures.

Most of the techniques used in vibration monitoring try to identify the modal characteristics of a part or the whole structure through its dynamic responses.

These techniques can be classified into two parts :

. Artificial excitation : swept sine testing, impulse techniques are often used for the study of ground and air vehicles. At the scale of an offshore structure, they are heavy and expensive methods.

. Natural excitation : in the case of offshore structures, the natural excitation (swell, wind and current) is "free" and powerful. Unfortunately the transfer from the excitations to the sollicitations is not well known. So we shall have to make some hypotheses about these sollicitations, the accuracy of the signal processing methods depending on the accuracy of those hypotheses.

In the sequel of this paper, we shall be interested in dynamic characteristics identification of a complex structure with natural excitation.

2. HYPOTHESES

i) Hypothesis on the behaviour of the structure :

We suppose that the dynamic behaviour of the structure can be described by an infinite order dynamical system.

ii) Hypothesis on the input force :

In the frequency range we are interested in, the input force of the dynamical equation can be considered like a non-stationary white noise (i.e. with time varying covariance matrix).

Thus behaviour of the structure can be broken up into an infinite number of unimodal vibrations.

By the combination of the interesting frequency range, the geometry of the input force, the layout and finite number of the observation points, only few of these unimodal vibrations are sufficiently powerful to be detected.

3. PROBLEMS

A lot of studies has been devoted to the vibration frequency identification methods and comparison of such methods (Fourier method, Maximum Eutropy method, Maximum Likelihood, ...) [1] [2] [3]. In this paper we are interested in two of their multichannels equivalents (Fourier method, ARMA modelling method). We compare the efficiency of these two algorithms only concerning the modal shape identification. (The modal frequencies are supposed to be known or previously identified). It is mainly an experimental comparison.

The very difficult problems we encounter on real signal and on which the comparison will be proceeded are :

- . closed vibration modes
- . slightly excited vibration modes
- . non stationarity of the excitation

4. PHYSICAL MODEL

Under the hypotheses i) and ii) we can describe the behaviour of our structure by the system :

$$\text{(1)} \qquad \begin{cases} (Mp^2 + Cp + K)Z_s = E_s \\ Y_s = LZ_s \end{cases}$$

with p the derivative operator and s the continuous time

M, C, K mass, damping and stiffness matrices

Y_s the observation vector

E_s is a non-stationary white noise with time-varying covariance matrix

The corresponding state space model is :

$$\text{(2)} \qquad \begin{cases} pX_s = AX_s + B_s \\ Y_s = HX_s \end{cases}$$

where

$$\text{(3)} \qquad X_s = \begin{pmatrix} Z_s \\ pZ_s \end{pmatrix} \qquad A = \begin{pmatrix} 0 & I \\ -M^{-1}K & -M^{-1}C \end{pmatrix} \qquad B_s = \begin{pmatrix} 0 \\ M^{-1}E_s \end{pmatrix}$$

$$H = \begin{pmatrix} L & 0 \end{pmatrix}$$

The matrix A is supposed to be asymptotically stable and has the following spectral decomposition : $A = UDU^{-1}$

$$\text{(4)} \qquad U = \begin{pmatrix} \phi & \bar{\phi} \\ \phi\Delta & \bar{\phi}\bar{\Delta} \end{pmatrix} \qquad D = \begin{pmatrix} \Delta & 0 \\ 0 & \bar{\Delta} \end{pmatrix}$$

$$\phi = \begin{pmatrix} \varphi_1 \varphi_2 \dots \varphi_n \end{pmatrix} \qquad \Delta = \begin{pmatrix} \sigma_1 + j\omega_1 & \cdots & 0 \\ 0 & \cdots & \sigma_2 + j\omega_2 \end{pmatrix}$$

where ϕ is the matrix whose columns are the mode shapes and Δ contains the poles of the system.

The problem we are interested in is to identify Δ and $L\phi$ the observed part of the mode shapes.

4.1. Discrete time model

Since we are interested in discrete time identification, let us introduce the discrete time system obtained by sampling (2) at a sampling rate $1/\Delta t$ and denote by t the corresponding discrete time :

$$\text{(5)} \qquad \begin{cases} X_{t+1} = FX_t + V_{t+1} \\ Y_t = HX_t \end{cases}$$

with

$$F = e^{A\Delta t} \qquad \mathrm{Cov}(V_t) = Q_V(t) = \int_t^{t+1} e^{As} Q_B(s) e^{A^T s} ds$$

Now, our previous problem can be translated in this discrete time context in an obvious way, since Δ has to be replaced by $\exp(\Delta.\Delta t)$ whereas ϕ remains unchanged.

5. IDENTIFICATION ALGORITHMS

5.1. Frequency domain method

5.1.1. Diagonalization of spectral matrices [4]

In the stationary case the spectrum of the observation vector Y_Δ can be written:

$$S_{YY}(p) = L\phi(pI-\Delta)^{-1}(pI-\bar{\Delta})^{-1}Q_E(pI+\Delta)^{-1}(pI+\bar{\Delta})^{-1}\phi^* L^T$$

or its discrete equivalent

$$S_{YY}(z) = L\phi(I-z^{-1}e^{\Delta})^{-1}(I-z^{-1}e^{\bar{\Delta}})^{-1}Q(z)(I-ze^{\Delta})^{-1}(I-ze^{\bar{\Delta}})^{-1}\phi^* L^T$$

If the spectrum of Y is computed for $z = e^{j\omega_1}$, where $p_1 = \sigma_1 + j\omega_1$, corresponds to an isolated mode, with low damping and well excited, we obtain :

$$(7) \quad S_{YY}(z) = L\phi\begin{pmatrix}(1-\bar{z}z_1)^{-1}(1-\bar{z}\bar{z}_1)^{-1} & 0 \\ 0 & \ddots \text{ Very Small Values} \ddots\end{pmatrix} Q(e^{j\omega_1})\begin{pmatrix}(1-zz_1)^{-1}(1-z\bar{z}_1)^{-1} & 0 \\ 0 & \ddots \text{ Very small values} \ddots\end{pmatrix}\phi^* L^T$$

if $Q(e^{j\omega_1})$ does not destroy the sparsity of the diagonal matrices (i.e. p_1 well excited), then we can approximate spectrum by

$$S_{YY}(z) \simeq \lambda L\varphi_1\varphi_1^* L^T$$

So the diagonalization of the spectrum at $z = e^{j\omega_1}$ gives us the modal vector $L\varphi_1$ (We will see further what happens when the pole has not the proprieties previously mentioned).

5.1.2. Identification algorithm

In a practical point of view, the modal parameters identification algorithm can be divided into three steps.

. By an FFT method, the computation of the spectrum S_{YY} which supplies a set of SDM (Spectral Density Matrices)

. Detection of frequencies where the spectrum reaches a maximum of energy, for example by observation of the largest eigenvalue

. At these detected frequencies, the computation of the eigenvector corresponding to the largest eigenvalue.

5.2. Time domain method

5.2.1. Stochastic realization [5]

The approach here is different. Starting from the state system (5), let be n the smallest index such that there exist matrices $A_1, \ldots, A_n$ such that

$$LF^n = \sum_{i=1}^{n} A_i LF^{n-i} \tag{8}$$

we can show that the pair $(\hat{H}, \hat{F})$, where

$$\hat{L} = (I \; 0 \ldots 0) \qquad \hat{F} = \begin{pmatrix} 0 & I & & 0 \\ & & \ddots & \\ & 0 & & I \\ A_1 & \cdots & \cdots & A_n \end{pmatrix} \tag{9}$$

is a (non-minimal) solution of our problem, in the sense that the pairs $(\lambda, L\varphi_\lambda)$, where λ are the eigenvalues of F and φ_λ the corresponding eigenvectors belong to the set of the pairs $(\hat{\lambda}, \hat{L}\hat{\varphi}_{\hat{\lambda}})$ obtained by diagonalization of $\hat{F}$.

Orthogonality conditions

Then assuming the model (5) to be exact, we get for every t and $i \geq 0$

$$y_{t+i} = LF^i . X_t + \sum_{j=1}^{i} LF^{i-j} V_{t+j} \tag{11}$$

Hence using (8) and the fact that V_t is a white noise, we get the following orthogonality conditions :

for every t

$$y_{t+n} - \sum_{i=1}^{n} A_i \, y_{t+n-i} \perp \operatorname{span}\{y_t, y_{t-1}, \ldots\}$$

where $x \perp y$ means $\mathbb{E}(xy^T) = 0$ ($\mathbb{E}$ denoting expectation)

By using the minimum number of orthogonality constraints we get what we called Instrumental Variable Method (IVM) which corresponds to the identification of the AR part of an ARMA model

IVM solve for $\hat{A}_1, \ldots, \hat{A}_n$ the orthogonality conditions.

$$y_{t+n} - \sum_{i=1}^{n} \hat{A}_i \, y_{t+n-i} \perp y_t, \ldots, y_{t-n+1}$$

5.2.2. Identification algorithm

Two steps can be defined :

. Identification of the AR part of an ARMA model (solution of a non symetric Toeplitz system)

. Extraction of the modal parameters from the polynomial $A(z^{-1})$ by solving a kernel problem :

- $\det\left(A(z_i^{-1})\right)=0$ which gives us $z_i = e^{(\sigma_i + j\omega_i)\Delta t}$

- $A(z_i^{-1}).V = 0$ which gives us $V = H\varphi_i$

(This problem is solved using a reduction to Hermite form) [6]

6. NON ISOLATED MODE

We have seen in 5.1.1. what results from the diagonalization of the SDM in the case of an isolated mode, with low damping and well excited. We may ask the question of what occurs when it is not the case, that is to say a mode which is perturbed by another mode, either very close to it or very powerful compared to it.
At the vicinity of this mode $z = e^{j\omega_1}$, the spectrum has the form :

$$S_{yy}(z) = L\phi \begin{pmatrix} (1-z^{-1}z_1)^{-1}(1-z^{-1}\bar{z}_1)^{-1} & & 0 \\ & (1-z^{-1}z_2)^{-1}(1-z^{-1}\bar{z}_2)^{-1} & \\ 0 & & \text{...very small values...} \end{pmatrix} Q(z) \begin{pmatrix} \cdots \end{pmatrix} \phi^{*} L^{T}$$

for two neighbouring modes
the same form that (7) for two modes with one powerful.

In the two cases the MDS at this frequency can be approximated by :

$$S_{yy}(z) \simeq \begin{pmatrix} H\varphi_1 & H\varphi_2 \end{pmatrix} R \begin{pmatrix} \varphi_1^{*} H^{T} \\ \varphi_2^{*} H^{T} \end{pmatrix}$$

So the diagonalization of the MDS gives us two orthogonal vectors which are linear combination of the two corresponding modal vectors

$$\varphi_{1_{ident}} = \alpha.\varphi_1 + \beta.\varphi_2$$

the coefficient of the linear combination depends on :

. The vibration parameters of the two modes

. The geometry of the excitation

So, in the stationary case, even if an exact estimation of the Fourier spectrum is provided,

. The MDS method gives biased estimate of modal shapes as previously described

. The ARMA modelling identifies the exact modal shapes.

7. NON STATIONARITY

For the kind of signal we are studying (excitation generated by the swell), one can consider two types of non stationarity,

. a long term one, which corresponds to a change in the geometry of the excitation due for example to a change in the swell direction,

. a short term one, which is an evolution of the excitation power. It is the case when we have a more powerful group of waves.

Some results have been obtained [7] which proved the robustness of the IVM in presence of this two types of non stationarities. Such a robustness does not exist for MDS method (case of first non stationarity type) due to the problem studied in §6.

8. EXPERIMENTAL COMPARISON

For more clearness, we have chosen to proceed the experimental comparison on a very simple mechanical system. Here it is a spring-mass system with proportional damping.

We have adjust the coefficients of structural matrices such that we were in front of difficult cases : very closed, weakly excited modes.

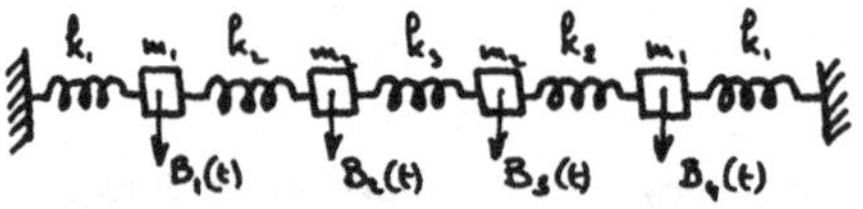

For each test, to compute mean and variance of the estimators, we have run the identification on 10 samples of 6144 points (response of the mechanical system to a white noise $B(t)$).
Fig. 1 represents the logarithmic spectrum of the first two masses.

8.1. Stationary case

- Identification of the first mode

	Spectrum Method		ARMA Method	
Theoretical Mode Shapes	Mean	SD*	Mean	SD*
.134	.144	.004	.134	.002
.694	.693	$< 10^{-3}$	.694	$< 10^{-3}$
.694	.693	$< 10^{-3}$	.694	$< 10^{-3}$
.134	.137	.002	.134	$< 10^{-3}$

*SD = Standard Deviation

The 1st mode is well identified by the two methods but one can see a light bias in the MDS method case.

FIGURE 1

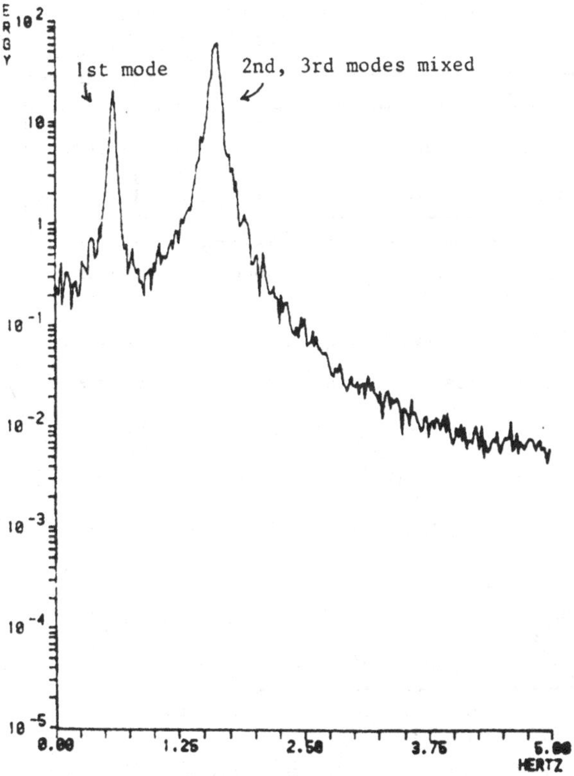

POWER SPECTRUM - CHANNEL 1

MASS 1

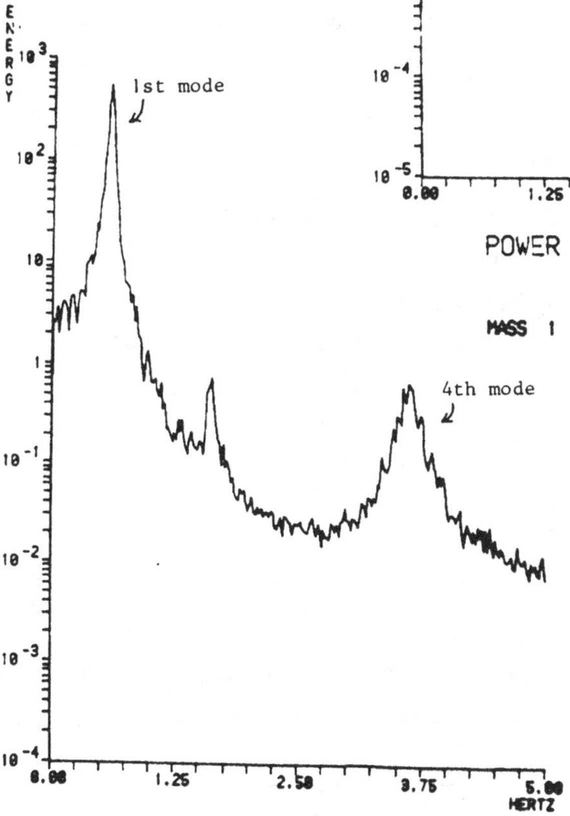

POWER SPECTRUM - CHANNEL 2

MASS 2

- Identification of the fourth mode

Theoretical Mode Shape	Spectrum Method		ARMA Method	
	Mean	SD	Mean	SD
.028	.06	.02	.03	.008
-.707	-.719	.017	-.708	.003
.707	.691	.017	.705	.003
-.028	-.03	.003	-.028	$< 10^{-3}$

We observe yet a bias in the MDS method case and for the two methods an increasing of the variance due to the power of the mode.

- For the second mode, we have plotted the coefficients α and β identified. The ideal case would be obviously $\alpha = 1$, $\beta = 0$

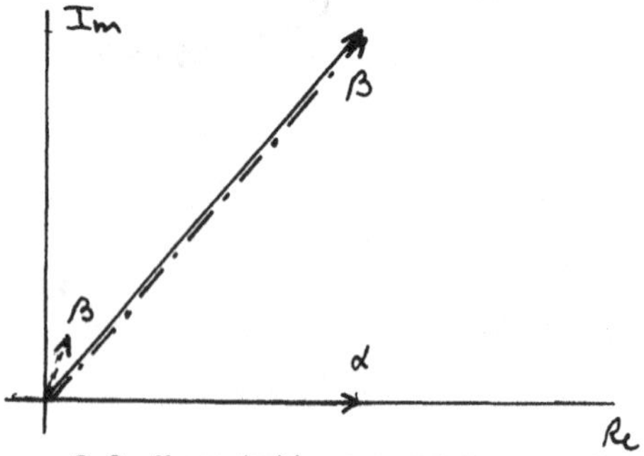

—— Theoritical Spectrum Method result

— - — Spectrum method

----- ARMA method

Fig. 2.

8.2. Non stationary case

The covariance matrix of the excitation, constant in the previous case, is now varying with time. This variation function is plotted in fig. 3.

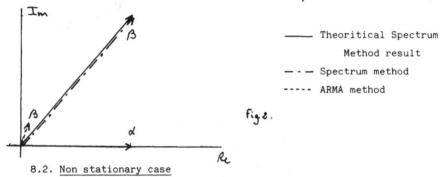

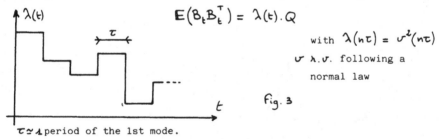

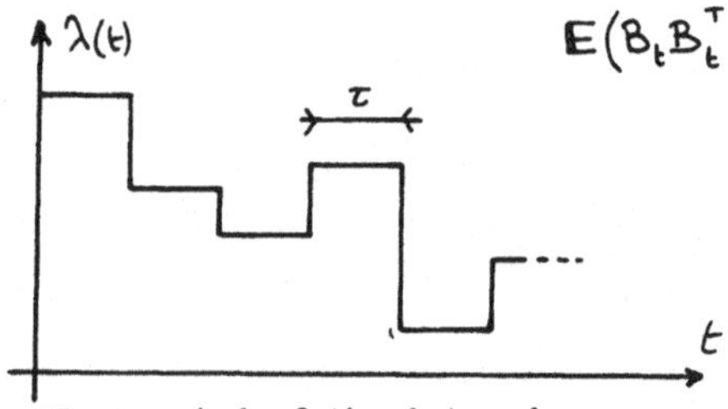

$$E(B_t B_t^T) = \lambda(t).Q$$

with $\lambda(n\tau) = \sigma^2(n\tau)$

σ r.v. following a normal law

Fig. 3

$\tau \simeq 1$ period of the 1st mode.

The choice of the σ law and of the time τ has not been guided by any physical reality.

We have observed no change in the identification of the last three modes. Change in variance occurs only on the first mode and for the MDS method.

Spectrum Method (1st mode)	
Stationary case	Non-Stationary case
SD	SD
.004	.019
$< 10^{-3}$	.003
$< 10^{-3}$	.003
.002	.008

9. CONCLUSION

The accuracy of the modal shape estimation has a great importance for the modelization and the monitoring of structures. These modal shapes more than giving complementary information to the modal frequencies, permit to reach a better sensitivity to some modification in the structure.
We have compared two algorithms, MDS and IVM methods, on their capabilities to identify such modal shapes. The results presented in this paper have been confirmed by some more tests on large system synthetic signal and real ones. So we can conclude by the fact that on all the problems uncountered in the case of offshore structure excited by swell the time domain approach had a better behaviour than those of the frequency domain approach.

ACKNOWLEDGMENT : Results presented herein are part of a project entrusted jointly to the OTTER Group (Norway) and to Centre National pour l'Exploitation des Océans (France) by Elf Aquitaine Norge A/S whose financial and technical support is herewith acknowledged.

10. REFERENCES

[1] Damping and natural frequency estimation using the least Pth optimization technique.
X. Lu, J.K. Vandiver,
OTC 4283 (1982).

[2] Maximum Likelihood estimation of structural parameters from random vibration data.
W. Gersh, N.N. Nielsen, H. Akaike,
Journal of Sound and Vibration (1973), 31(3), p. 295-308.

[3] The estimation of natural frequencies and damping ratios of offshore structures.
R.B. Campbell, J.K. Vandiver,
OTC 3861 (1980).

[4] Estimation of modal shapes by diagonalization of interspectral densities matrices.
B. Barnouin,
Rapport Technique CNEXO (1983).

[5] Modélisation et identification des caractéristiques d'une structure vibratoire: Un problème de réalisation stochastique d'un grand système non stationnaire (In English text).
M. Prevosto, A. Benveniste, B. Barnouin,
Rapport de recherche INRIA n° 130 (1982).

[6] Solution of the Kernel problem for polynomial matrices using the Hermite reduction.
M. Olagnon,
Rapport Technique CNEXO (1983)

[7] Non stationary stochastic realization, and single sample identification.
A. Benveniste, J.J. Fuchs,
IRISA (INRIA) Report (1982).

FATIGUE OF OFFSHORE PLATFORMS: A METHOD OF ANALYSIS (1)

by W. J. Graff (2) and T. M. Koudelka (3)

INTRODUCTION

Metals subjected to fluctuating stress variations or reversals may develop small cracks at discontinuities in the material, leading ultimately to fatigue failure. Fatigue, different from static overload, is the term used to describe material behavior under repeated cycles of stress or strain. This deterioration results from stable crack growth which gradually destroys the load resisting properties of the cross section. Fatigue failures can occur at stresses well within the elastic range, as measured in the uniaxial static tension test.

Fatigue is a localized phenomena and, for steel jacket offshore structures, it is principally a consideration at the intersections of the tubular members, i.e., at the tubular joints; see Figure 1. Stresses in the joints are continuously fluctuating under the influence of the continuously acting wave loads. Joints for primary consideration are those that are highly stressed or have high stress concentration factors which amplify the nominal stresses. Regions of high stress are called hot spots and can experience three-dimensional crack initiation.

Following crack initiation, which can go undetected until a through-thickness crack is developed, crack propagation occurs. The crack can achieve significant length before it develops into a through-thickness crack. The rate of crack propagation depends on the stress gradient, the restraint surrounding the cracked region, the future pattern of fluctuating loads, corrosiveness of environment, etc. While a tubular joint has large reserve strength enabling it to continue to function during crack propagation, this ability diminishes as the crack length increases. The crack spreads progressively until the stressed section becomes so small that this remaining section cannot sustain the load and large plastic deformation occurs. Failure of a joint is often said to have occurred when a crack has propagated to such an extent that an incoming brace is unable to carry or transmit load, thus exhausting the joint's fatigue life.

The difficulties in treating fatigue behavior of tubular joints in offshore structures arise as one attempts to describe the applied forces and the specifics of the structural response. The five major elements of the analysis are the wave spectra, the nominal stress history of the joint in

---

(1) A paper presented at the Eleventh International Federation For Information Processing (IFIP) Conference on System Modeling and Optimization, Copenhagen, Denmark, July 25-29, 1983.

(2) Exxon Company, U.S.A., Consultant; Professor of Civil Engineering at the University of Houston, Houston, Texas.

(3) Senior Project Engineer, Platform Design Group, Exxon Company, U.S.A., Houston, Texas.

question, the stress concentrations that produce the hot spot stresses, the capability of the material to endure the stress cycles, and a criterion for determining cumulative damage to the joint. Each of the elements is important to the analysis, and there are aspects of each element which remain in a developmental state analytically. Each of these elements will be reviewed briefly in the sections that follow.

WAVE SPECTRA

The dynamic loading on offshore structures comes from wave action. In order to establish the maximum structural displacement, or the structural response as a function of time, it is necessary to determine the characteristics of the waves that routinely strike the structure.

Waves are generated by the wind blowing over the water surface; the state of disturbance is called a sea. In the open ocean a large number of waves are almost always superimposed on each other giving the water surface its chaotic appearance. Energy of a wave system is usually plotted versus frequency of the waves. Small waves of short period die out and give their energy to waves of longer period, which in turn grow in amplitude. A graph of the energy density in the waves versus wave frequency is called a wave spectrum, and represents the energy envelope of many waves of different periods, directions and heights superimposed on one another.

Ocean wave profiles (i.e., water surface elevations above and below still water level) are considered to have normal (Gaussian) statistical distribution with as many points below the mean as above it.[(1)*] Waves vary continuously with time and are considered to represent statistically stationary processes over intervals of a few hours. According to Vughts and Kinra[(2)], three hours is generally accepted as the maximum time interval for the sea to remain stationary. To be able to treat a random phenomenon mathematically the process must, additionally, be ergodic. Pan et al.[(3)] emphasize that due to the passage of storms the average value of the wave heights can change from time to time, and the wave profiles are non-ergodic random processes. However, this difficulty may be overcome by assuming that the wave environment may be represented by a series of sea states, each a statistically stationary process having constant mean wave height; see Figure 2. This procedure requires that for the particular ocean location there must be sufficent knowledge of the oceanographic conditions to permit determination of significant wave height, peak wave period, spectral width, directionality, and percent occurrence for each of the sea states.

The parameters that describe a unidirectional sea, significant wave height, peak wave period, and spectral width are related to the moments of the wave spectrum about the origin of the spectrum plot. Spectral width is a measure of the irregularity of the sea and ranges from zero to unity. A spectral width of zero corresponds to a wave spectrum having an infinitely narrow frequency range;

---

* Numbers refer to references.

a spectral width of unity corresponds to an infinitely broad wave spectrum. If the component frequencies making up the spectrum are concentrated over a narrow range, the spectral width approaches zero, and if the process is Gaussian, then the wave amplitude will be characterized by a Rayleigh probability density distribution. For a narrow frequency range wave spectrum, an individual wave will have a wave height very nearly twice the wave amplitude, therefore it is assumed that wave heights follow a Rayleigh density distribution also.

In reality, the unidirectional wave spectrum cannot be considered a narrow-band frequency distribution. Thus, a random sea is represented by an infinite sum of sinusoidal components and the wave height versus frequency curve may follow an empirical relationship such as the Pierson-Moskowitz spectrum with the spectral width normally considered to be large. This wave spectrum applies to fully developed seas.

When all of the distributions representing the various sea states are summed with the respective occurrences applied, one gets the overall wave height distribution or spectral density function for the ocean wave and the mean zero-crossing wave period. The percent occurrence of each sea state, defined by a significant wave height and a mean or peak period, is determined from a wave scatter diagram. The percent occurrence is simply 100 times the fraction formed by dividing the number of occurrences of a given sea state by the total number of sea states.

NOMINAL STRESS HISTORY OF A PARTICULAR JOINT

The underlying principles of determining dynamic response to random excitation are the concepts of Fourier Series, Fourier Integral, and Fourier Transform, all expressed in complex form.[(4)] For a single-input-single-output system, an equation can be written for spectral density of nominal member stress, $S_s(f)$, where the right hand side shows excitation by the spectral density, $S\eta(f)$, of a single sea state and $|H(f,u)|^2$ is the magnitude of the wave (frequency and direction dependent) stress transfer function,

$$S_s(f) = |H(f,u)|^2 \, S\eta(f)$$

The stress transfer function is determined in the following manner. For a wave component of given direction (u) incident onto the jacket structure and of given amplitude and period, the amplitudes of the nominal stress response at all points of interest within the jacket must be determined in order to obtain the ratio of the nominal stress amplitude to the wave amplitude at each point. For the same direction (u), this process is repeated for a sufficient number of wave periods to define the ratio throughout the range of wave frequencies for each point of interest.

Any large structural analysis computer program may be used to determine the stress transfer function. The total number of load cases to be solved depends on the number of wave directions, wave periods, and wave positions relative to the structure. Solving for the ratios of nominal stress amplitude to wave amplitude is an entirely deterministic calculation under the action of periodic waves. For a given wave approach, a plot of the ratios versus frequency constitutes the required $|H(f,u)|^2$ for the chosen direction at the chosen point in the structure.

Figure 3 shows the spectral density of the waves, $S\eta(f)$, the transfer function, and the spectral density of the response, i.e., the nominal member stress. The mean square of the response is obtained by determining the area under the response spectral density curve by numerical integration. The square root of the mean square member stress, σrms, represents the nominal stress history for the particular degree of freedom at the joint in question.

Since the transfer process is linear (or linearized), the variation in stress for a particular degree of freedom at a particular joint is assumed to be Gaussian inasmuch as the excitation is essentially Gaussian. It is generally assumed that on a sea-state by sea-state basis the peak of the structural response will follow the Rayleigh probability density distribution. The loads that may be considered static for the purpose of a fatigue analysis include: gravity forces, buoyancy forces, operational loads, and wind and current forces. The only significant dynamic load is caused by wave and the forces categorized as static may be ignored in the analysis.

Presently used force calculation methods were developed initially for use with the design wave or maximum storm wave. The force exerted by waves on a cylindrical member, according to Morison's equation, consists of a drag component, C_d, related to the kinetic energy of the water, and an inertial force component, C_m, related to the acceleration of the water particles. The total horizontal force on the structure is represented by the sum of the forces acting on the individual members for a given position of the wave relative to the structure.

For fatique calculations, the ordinary waves, developed in low to moderate seas, are generally of interest and are represented with sufficient accuracy by the linear (Airy) wave theory. The hydrodynamic equations for both water particle velocity and acceleration are directly related to wave height which together with C_d are adjusted to give the proper "linearized" total force; C_m is usually considered known.

HOT SPOT STRESS HISTORY FOR A PARTICULAR TYPE OF STRESS

Stress history refers to the stress ranges and the corresponding numbers of stress cycles that a given location is expected to experience. It reflects the overall response of the structure, as well as the local joint geometry.

Tubular joints are so proportioned that if failure occurs, it will normally happen in the chord wall.[5] Those locations on the chord most susceptible to fatigue failure are shown in Figure 1. The branch fastens to the chord wall with a curved or saddle-like shape; consequently, the nominal stresses in both the branch and the chord are nonuniformly distributed. Due to the curvature of the chord surface, there is high local bending; it is this bending that produces the stress concentrations that cause the high localized stresses at the hot spots. The wall of the chord frequently experiences hot spot stress several times larger than the nominal brace stress. Hot spot stresses are caused by nominal axial forces and bending moments in the brace and chord (minor), as well as local stiffness variations and abrupt geometric changes in the joint.

The rms nominal stress values are multiplied by predetermined stress concentration factors to produce the rms hot spot stress. A stress concentration factor, or SCF, must be determined for each type of applied load of interest on the chord at each joint. SCFs, defined as the ratio of the hot spot stress in the chord to the nominal stress in the branch member that caused it, are influenced by joint geometry, the nature of the loading, and the local weld toe effects. Several important geometric factors affecting stress concentration factors are:

- the thickness-to-diameter ratio of the chord member, T/D
- the branch-to-chord diameter ratio, d/D
- the branch-to-chord thickness ratio, t/T
- the angle of branch inclination, θ, and
- the separation distance or gap between branches measured along the surface of the chord.

Figure 1 illustrates the joint parameters mentioned above. Formulae that estimate SCFs for simple T-, Y-, K-, and TK-joints commonly used in offshore platforms have been developed.(6) More complex joints are either broken down into simple configurations for which the semi-empirical SCF equations are applicable, or they are studied in more elaborate detail.

MATERIAL CAPABILITY TO ENDURE STRESS CYCLES

Experience over the last 100 years and many laboratory tests have proven that a metal may crack at a relatively low stress if that stress is applied a great number of times. Fatigue cracks initiate and grow under the repeated application of nominal tensile stresses that are usually lower than the material yield strength.

As random wave loading on offshore platforms came into consideration as a fatigue problem, three important differences were noted from earlier fatigue situations. Firstly, the rates of loading are vastly slower. Ocean waves pass the offshore platform at the rate of about 10 cycles per minute or 600 per hour. Secondly, offshore structural joints contain high residual stresses as a result of their having been welded together. These residual stresses often reach or exceed the yield stress in regions of the chord wall where hot spots occur. Thus, when a tubular joint is put into service, there may be a local yielding of the material, and a redistribution of stresses (applied and residual together) may occur. In studying fatigue of welded specimens it has been found that because of the high residual stresses, plotting stress range rather than maximum stress, better describes the behavior of the welded connection. And thirdly, because of the corrosiveness of sea water, S-N curves for steels in the ocean subjected to repeated or random loading do not appear to have fatigue limits as specimens tested in air. The American Petroleum Institute's Recommended Practice[7] considers the X-curve, which makes allowance for the high cycle corrosion aspects, as the lower bound S-N curve applicable to most conventional tubular joints.

CUMULATIVE DAMAGE CRITERIA

Over the span of its design life, an offshore structure is subjected to a continuous spectrum of cyclic wave loading which require that consideration be given to cumulative fatigue damage. It is the low-stress, high-cycle load situation that is the significant contributor to cumulative fatigue damage. The measure of cumulative damage most often used is the Palmgren-Miner rule:

$$DR = \sum_{i=1}^{k} \frac{n(S_{ri})}{N(S_{ri})} = \frac{1.0}{SF}$$

where: DR is the cumulative damage ratio equal to the damage sum divided by the design lifetime of the structure,

k is the number of stress range intervals considered,

$n(S_{ri})$ is the number of cycles of applied stress having a stress range of S_{ri},

$N(S_{ri})$ is the number of cycles that cause failure using a stress range S_{ri},

SF is a safety factor, commonly equal to 2.0, applied to the structural design lifetime.

The principal virtue of the rule is simplicity; its most significant shortcoming is that no recognition is given to the sequence of application of the various stress levels, and damage is assumed to accumulate at the same rate at a given stress level, regardless of the order of application. Neither is there recognition of the difference between crack initiation and crack propagation in the rule.

The hot spot stress history for a particular type of stress in a particular joint is broken into a number of intervals of stress range; the number of applications of each stress range is determined. These stress range intervals all together describe the number of stress cycles which the particular type of stress in the joint will experience during the lifetime of the joint. The API-X S-N design curve indicates the number of cycles of various stress ranges a joint may experience before failure occurs. Thus, the Palmgren-Miner rule affords a means of comparing the actual number of stress cycles with the allowable number of stress cycles, Figure 4.

CLOSING REMARKS

Offshore platforms which are located in deep water exhibit significant dynamic response and require a fatigue analysis; shallow water platforms may require a fatigue analysis depending on their wave environment. The Palmgren-Miner approach to predict fatigue damage is the most widely used procedure. This approach requires the hot spot stress history of every joint in the platform.

Typically, stress transfer functions are obtained and multiplied by the wave spectra to yield stress response spectra. These spectra provide the nominal stress ranges which are modified by the SCFs to produce hot spot stress ranges. For each stress range the number of cycles to failure is obtained from the appropriate S-N curves.

Given the stress history and S-N curve input, the Palmgren-Miner rule may be applied to evaluate the damage in each joint due to the cyclic loading affecting the platform. Once an initial fatigue analysis has been conducted and results indicate that certain joints may have a fatigue problem, several options are available that may increase the fatigue lives of these joints. Increasing the chord thickness in the joint area is generally the first option. Yet, this increase is limited to practical dimensions. Increasing branch diameters and/or the branch thicknesses, stiffening the joint, and grinding the welds are steps that may also be taken. If no serious fatigue problem exists, utilizing these options will generally make the design more acceptable.

Avoiding fatigue damage completely is realistically impossible. Cracks will occur in connections, but their growth to critical stages is designed to occur at a time much greater than the structural design life. Some cracking, for example, due to an unforeseen wave environment or fabrication flaw, may occur. In these cases, built-in structural redundancy means that cracking should not jeopardize the structure's integrity. Certain cracks can be avoided or controlled by close inspection during and after fabrication of the structure. Repairs can be made if damages are not severe; but repairs are costly, particularly after the platform has been installed. To date, offshore platforms have had few problems due to fatigue damage. By continuing with good design, fabrication, and inspection procedures, it is certain that future platforms will provide the dependable service required during their lifetimes.

REFERENCES

1. Sarpkaya, T., and Isaacson, M., "Mechanics of Wave Forces on Offshore Structures," Chapter 7, Von Nostrand Reinhold Company, New York, 1981.
2. Vughts, J. H., and Kinra, R. K., "Probabilistic Fatigue Analysis of Fixed Offshore Structures," 1976 Offshore Technology Conference, Paper 2608.
3. Pan, R. B., Maddox, N. R., and Plummer, F. B., "Fatigue Analysis of Offshore Structures," Ocean Engineering, November 15, 1975.
4. Hallam, M. G., Heaf, N. J., and Wootton, L. R., "Dynamics of Marine Structures," Report UR-8, Second Edition, CIRIA Underwater Engineering Group, London, October 1978.
5. Graff, W. J., "Introduction to Offshore Structures: Design, Fabrication, Installation," Chapter 10, Gulf Publishing Company, Houston, Texas, 1981.
6. Kuang, J. G., Potvin, A. B., and Leick, R. D., "Stress Concentration in Tubular Joints," 1975 Offshore Technology Conference, Paper 2205.
7. "Recommended Practice For Planning, Designing, and Constructing Fixed Offshore Platforms," Thirteenth Edition, American Petroleum Institute, Washington, D. C., January, 1982.

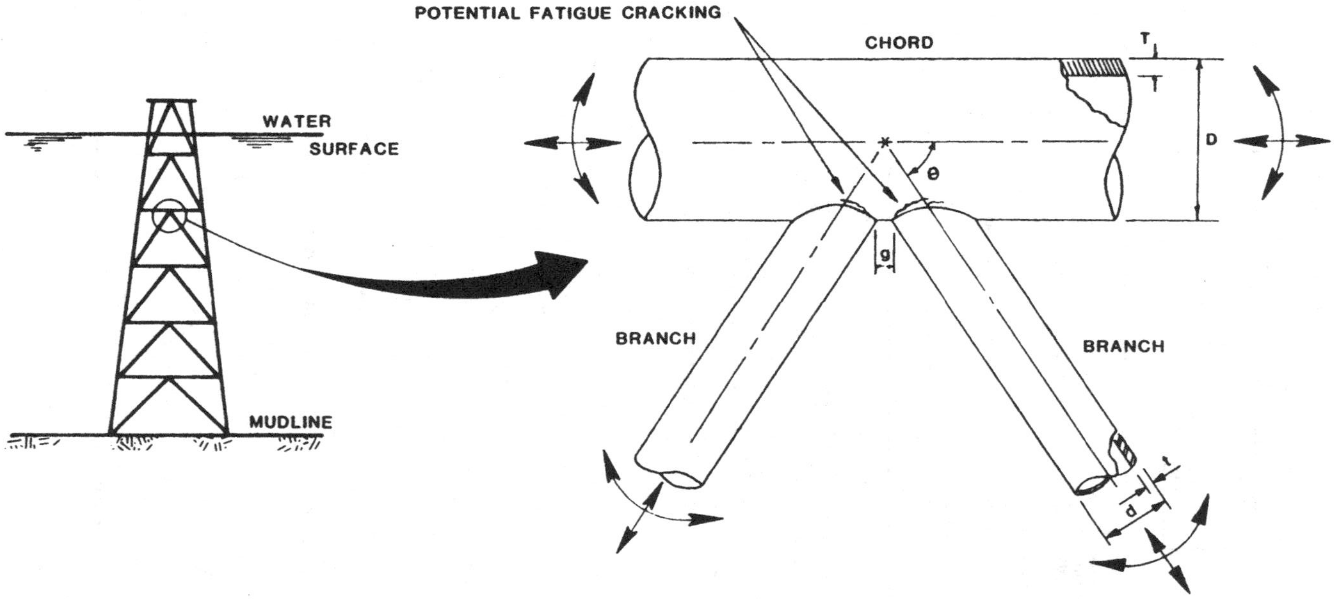

FIGURE 1 - TUBULAR CONNECTION: PARAMETERS & POTENTIAL CRACKING LOCATIONS

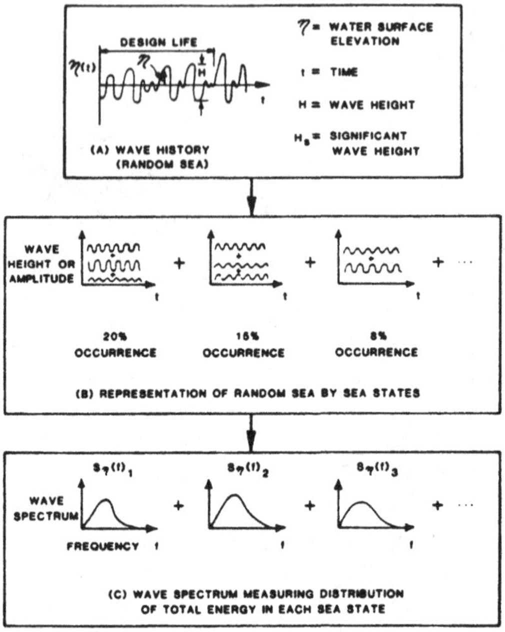

FIGURE 2 - GENERATION OF WAVE SPECTRA

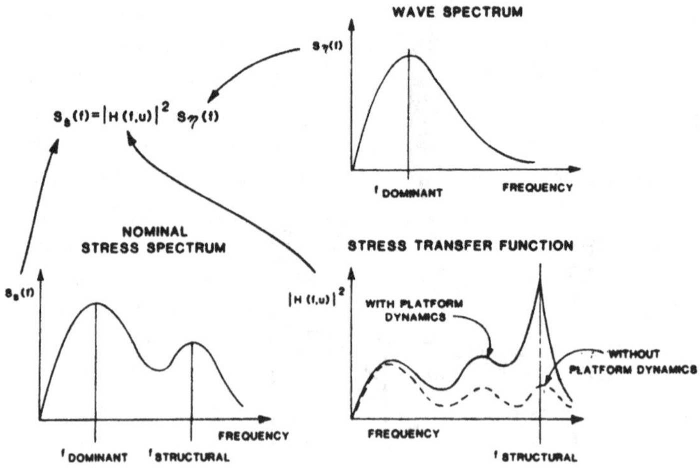

FIGURE 3 - FREQUENCY DOMAIN ANALYSIS

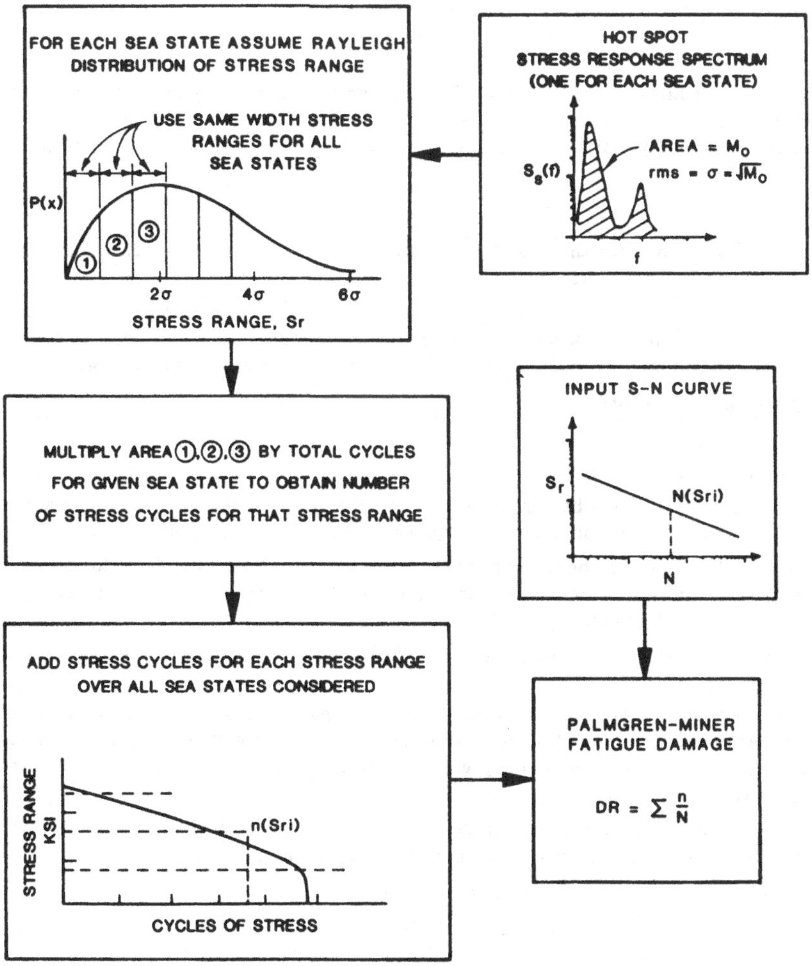

FIGURE 4 - FATIGUE DAMAGE COMPUTATION

STOCHASTIC DESIGN OF RUBBLE MOUND BREAKWATERS

S. R. K. Nielsen, Civil Engineer, Ph.D., Rambøll & Hannemann, Consulting Engineers and Planners, Nørresundby, Denmark, and
H. F. Burcharth, Professor of Marine Civil Engineering, Aalborg University, Aalborg, Denmark.

ABSTRACT

A level III reliability method for the determination of the optimum mass of a specific concrete armour unit for rubble mound breakwaters is presented.

Two failure modes are considered. Damage can occur due to displacement of the units leading to exposure of the filter layers and the core or due to breakage of the units by heavy rocking or accumulated fatigue damages. Emphasis is put on the probabilistic method rather than on the physical processes.

A numerical example is presented.

1. INTRODUCTION

A rubble mound breakwater may be considered as a structural system made up of a finite number of structural components such as the armour layer, wave screen, front toe, back slope, etc.

The reliability of the system in the interval $[0,T]$, T being the intended lifetime of the structure, is determined by the vector process of basic variables $\{\underset{\sim}{X}(t),\ t \in [0,T]\}$, $X_i(t) : \Omega \curvearrowright R$, such as significant wave height $H_s(t)$, peak period $T_p(t)$, etc.

A failure mode of the system is defined as a prescribed combination of joint failure events of structural components.

The analysis of componental reliability is somewhat troublesome partly because damages are caused by accumulated effects from prior samples of the basic variable processes, partly because the damage rates in component i depends on the actual failure mode j, e.g. the rate of armour unit displacements is increased if the front toe is damaged at the same time.

The state of component i in case of failure mode j at time t is determined by the failure indicator $I_{ij}(t)$: $\Omega \curvearrowright \{0,1\}$ with functional dependency on the set of stochastic variables $\{\underset{\sim}{X}(\tau), \tau \in [0,t]\}$

$$I_{ij}(t) = \mathcal{F}_{ij}(\{X(\tau), \tau \in [0,t]\}) = \begin{cases} 0 & \text{if failure at time t} \\ 1 & \text{if not failure at time t} \end{cases} \tag{1}$$

The probability of failure of the system can then be written

$$P_f(T) = P(\exists\, t \in [0,T] : \prod_{j=1}^{J} \sum_{i=1}^{I} I_{ij}(t) = 0) \tag{2}$$

I is the number of components, and J is the number of failure modes.

Unfortunately the failure functionals of all relevant components in all relevant combinations are not at present known. The necessary model testing and data collection still need to be done. The present paper describes how the safety margin (1) for the armour layer can be modelled synthesising available test data, and assuming no interaction with other structural components.

The time dependent basic variables indicated above includes parameters affecting both the hydraulic and the mechanical stability of the armour layer.

The hydraulic stability is influenced by

I Geometry of the structure.
Armour unit geometry and relative density.
Armour layer thickness and packing density.
Filter layer permeability and thickness.
Core permeability.
Cross-section profile (slope angle, berms, wave walls, etc.).

II Bottom bathymetry and character of the seabed.

III Deep and shallow water wave climate.
Joint probability distributions of wave height, -period, and -direction, wave grouping, long and short term statistics and persistence of waves, shoaling effects.

IV Water level variations (storm surge, tides, etc.).

Since the influence of each separate parameter on the stability is not yet well established a semiempirical design approach is used in practice. Traditionally a number of important parameters are combined into overall parameters, identified by physical reasonings and dimensional analysis.

The hydraulic stability is the resistance to displacement of armour units. Usually the hydraulic stability is studied in physical models. Abrasion might cause gradual reduction in the hydraulic stability. This can be modelled by a deterministic function or a stochastic process, which increases the damage rates with time.

The mechanical stability is the resistance to breakage of the units. Impact due to heavy rocking and rolling might cause instant fracture. Moderate rocking during less severe wave actions may give rise to accumulated fatigue damage of the armour unit material.
The two failure modes of the armour layer dealt with in the present model correspond to the lack of mechanical and hydraulic stability. The damage rates in both failure modes are determined by model tests. The scatter of these results due to uncontrolled parameters is considered in the final variance analysis of the damage rates.

2. BASIC ASSUMPTIONS AND RESULTS

The armour layer is considered as a system of M basic unit areas, each with separate strength and loading. Damage rates to specific wave loadings are determined for some or all of these basic areas. It is taken as failure of the breakwater armour layer when at least one basic unit area fails, i.e. the system can be identified as a series system.

Two failure modes of a basic unit area are considered. The first failure mode, $k = 1$ of the armour layer specifies the structural failure of the armour units due to impacts from heavy rocking and rolling and accumulated fatigue damages from more moderate rocking in smaller storms. This failure mode is relevant especially for the slender complex types of units such as Dolosse and Tetrapods. The second failure mode, $k = 2$ indicates displacement of armour units at least of the magnitude of one characteristic diameter, resulting in exposure of the underlayers to the waves.

The accumulated damage percentage processes $\{D_{k,i}(t), t \in [0, T]\}$ signifies the relative number of armour blocks in the i'th basic unit area, failing in the mode k.

These quantities are assumed on the form

$$D_{k,i}(t) = \sum_{n=1}^{N_{k,i}(t)} \Delta D_{k,i,n} \cdot s_{k,i,n} \tag{3}$$

Damages at basic unit area i in failure mode k take place at random times $0 \leqslant \tau_{k,i,1} < \ldots . < \tau_{k,i,N_{k,i}(t)} \leqslant t$. $\{N_{k,i}(t)\}$ are homogeneous Poisson counting processes specifying the random number of such damage increments in the interval $[0,t]$, ref. [4]. The intensities $\lambda_{k,i}$ of the counting processes indicate the expected numbers of storms per unit of time capable of giving rise to any damage contributions in mode k at basic unit area i.

The damage increments $\Delta D_{k,i,n}$, $n \in 1, 2, \ldots$, are stochastic variables, assumed to be mutually independent and identically distributed as the stochastic variable $\Delta D_{k,i}$. These quantities will be further explained in a succeeding section. The samples of $\Delta D_{k,i}$ are positive. Hence the realizations of (3) are non-decreasing functions with probability 1.

$s_{k,i} : [0,t] \cap R_+$, $k \in 1, 2$ are deterministic, monotonic, increasing functions of time, specifying long term tendencies of increased damage rates, e.g. due to abrasion. These functions are normalized as follows

$$s_{k,i}(0) = 1 \tag{4}$$

A typical realization of the accumulated damage percentage processes is shown on figure 1.

The distribution function of the accumulated damage percentage $D_{k,i}(t)$ at time t can now be determined. Actually this quantity has the characteristic function

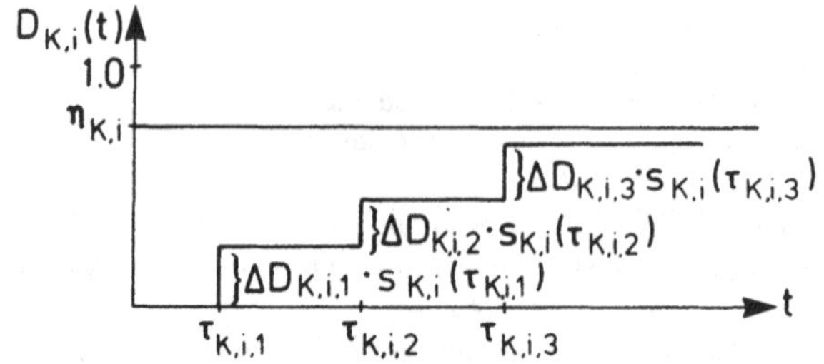

Figure 1: Realization of accumulated damage percentage process.

$$M_{D_{k,i}(t)}(\theta) = \exp\left(\lambda_{k,i} \int_0^t (M_{\Delta D_{k,i}}(s_{k,i}(\tau)\,\theta) - 1)\, d\tau\right) \tag{5}$$

where $M_{\Delta D_{k,i}} : R \cap C$ is the characteristic function of $\Delta D_{k,i}$.
The expectation and standarddeviation of $D_{k,i}(t)$ is termed $\mu_{D_{k,i}(t)}$ and $\sigma_{D_{k,i}(t)}$.

The distribution of $D_{k,i}(t)$ is of the mixed type with a concentrated, although diminishing probability mass $\epsilon_{k,i}(t)$ at $D_{k,i}(t) = 0$:

$$\epsilon_{k,i}(t) = P(D_{k,i}(t) = 0) = P(N_{k,i}(t) = 0) = \exp(-\lambda_{k,i} \cdot t) \tag{6}$$

The remaining probability mass $(1 - \epsilon_{k,i}(t))$ is continuously distributed as specified by the frequency function $f^{\circ}_{D_{k,i}(t)}(x)$, with expectation $\mu^{\circ}_{D_{k,i}}(t)$ and standard deviation $\sigma^{\circ}_{D_{k,i}}(t)$, see figure 2

$$\mu^{\circ}_{D_{k,i}}(t) = \frac{\mu_{D_{k,i}}(t)}{1 - \epsilon_{k,i}(t)} \tag{7}$$

$$\sigma^{\circ}_{D_{k,i}}(t) = \left(\frac{\sigma^2_{D_{k,i}}(t)}{1 - \epsilon_{k,i}(t)} - \frac{\epsilon_{k,i}(t)}{(1 - \epsilon_{k,i}(t))^2}\, \mu^2_{D_{k,i}}(t)\right)^{\frac{1}{2}} \tag{8}$$

The moments (7) and (8) provide that $D_{k,i}(t)$ will have the moments $\mu_{D_{k,i}}$ and $\sigma_{D_{k,i}}$.

The distribution function of $D_{k,i}(t)$ approaches a normal distribution, as $t \to \infty$, due to the central limit theorem. Instead of the exact result (7), it then seems reasonable to apply the following approximation, which will be asymptotically correct

$$F_{D_{k,i}(t)}(x) \cong (1 - \epsilon_{k,i}(t))\, \Phi\left(\frac{x - \mu^{\circ}_{D_{k,i}}(t)}{\sigma^{\circ}_{D_{k,i}}(t)}\right) + \epsilon_{k,i}(t) \tag{9}$$

Φ indicates the distribution function of a standardized normal variable.

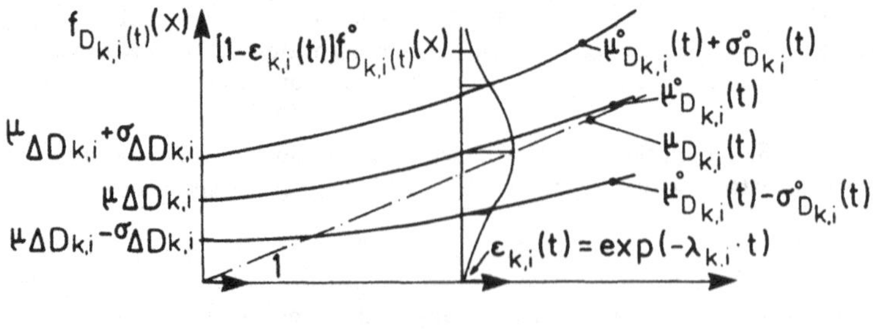

Figure 2: First order probability density function of accumulated damage percentage process.

Failure of basic unit area i occurs, when either of the accumulated damage percentages $\{D_{k,i}(t)\}$ for the first time exceed the allowable limits $\eta_{k,i}$ which can be specified separately for any basic unit area depending on its position and importance. The reliability problem of the armour layer system then becomes an ordinary first passage probability problem.

The distribution function of first passage time T of the system and the first passage times $T_{i,k}$ of the basic unit area related as follows

$$F_T(t) = P(T \leqslant t) = 1 - P\left(\bigcap_{\substack{i=1,M \\ k=1,2}} \{T_{i,k} > t\}\right) \tag{10}$$

Because the damage percentage processes have realizations which are non-decreasing with probability one, the events $\{T_{i,k} > t\}$ and $\{D_{k,i}(t) \leqslant \eta_{k,i}\}$ are identical. Hence

$$F_T(t) = 1 - P\left(\bigcap_{\substack{i=1,M \\ k=1,2}} \{D_{k,i}(t) \leqslant \eta_{k,i}\}\right) \tag{11}$$

The damage accumulations in failure modes k = 1 and k = 2 take place at different storm levels, and may reasonably be considered independent stochastic variables, at least at long term. Further, if the correlation length of local weakenings of the armour layer is small compared to the dimensions of the basic units, all stochastic variables in (11) can be considered mutually independent. Hence

$$F_T(t) \cong 1 - \prod_{\substack{i=1,M \\ k=1,2}} \left[(1 - \epsilon_{k,i}(t))\, \Theta\left(\frac{\eta_{k,i} - \mu_{\overset{\circ}{D}_{k,i}}(t)}{\sigma_{\overset{\circ}{D}_{k,i}}(t)}\right) + \epsilon_{k,i}(t)\right] \tag{12}$$

If any basic area is fully repaired after a failure, and the intervals between succeeding failures are assumed to be mutually independent, the failure rate ν_i is related to the frequency function f_{T_i} of the first passage time within basic area i through the inhomogeneous Volterra integral equation of 2. kind

$$\nu_i(t) = f_{T_i}(t) + \int_0^t f_{T_i}(t-\tau)\, \nu_i(\tau)\, d\tau \tag{13}$$

$$f_{T_i}(t) = -\frac{\partial}{\partial t}\left(\prod_{k=1}^{2} F_{D_{k,i}(t)}(\eta_{k,i})\right) \tag{14}$$

The first term i (13) represents crossings at time t, which are first passages. $f_{T_i}(t-\tau)\, \nu_i(\tau)$ represents the joint probability density of a crossing at time τ and a first passage after repair at time τ at the elapsed time $t - \tau$. Hence the second term i (14) represents the number of crossings at time t, which have had their first passages prior to time t.

Under sinilar conditions, the failure rate ν of the entire armour layer equally fulfils eq. (13), with f_T given by (10. Obviously, as may be deduced from (10) and (13)

$$\nu(t) \leqslant \sum_{i=1}^{M} \nu_i(t) \tag{15}$$

The assumption leading to (12) reduces the reliability problem to componental level. The parameters $\nu_i : [0,T] \cap R_+$ defined by (13) turn out to be essential in the design strategy outlined below.

3. MODELLING OF DAMAGE INCREMENT

In the load model significant storm arrivals are specified by a homogeneous Poisson counting process with intensity λ_o. The wave condition in the basic unit area i during any of these storms is specified by the parameters $(H_{s,i}, T_{p,i})$, where $H_{s,i}$ is the significant wave height in front of structure exclusive reflected waves, and $T_{p,i}$ is the peak period (most probable wave period). A storm is characterized by the growth and the succeeding decrease of wave heights, cf. figure 3. The maximum wave height $H_{s,i,max}$ is assumed to be Weibull distributed, i.e.

$$F_{H_{s,i,max}}(h) = 1 - \exp\left(-\left(\frac{h - h_1}{h_2}\right)^{h_3}\right), \quad h \in [h_1, \infty[\tag{16}$$

The significant wave heights within a certain storm, on condition of the maximum wave height $H_{s,i,max} = x$, is assumed to be uniformly distributed in the interval $[h_1, x]$, i.e.

$$F_{H_{s,i} \mid H_{s,i,max}}(h \mid x) = \begin{cases} 0 & , \quad h \in [\,0, h_1\,] \\ \dfrac{h - h_1}{x - h_1} & , \quad h \in\,]\,h_1, x\,] \\ 1 & , \quad h \in\,]\,x, \infty\,[\end{cases} \tag{17}$$

The uniform distribution is tantamount to the triangular growth and decrease curve shown in figure 3.

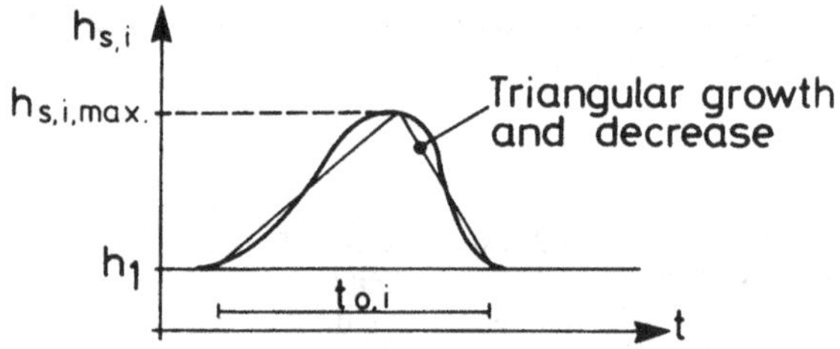

Figure 3: Variation of significant wave height during a single storm.

Rocking induces impact stresses in the armour units. This is considered the main reason of fatigue damage. Contributions from wave loadings incapable of rocking armour units are ignored. It is assumed that any unit rocking in a storm will also be rocking in a later and severer storm. At a certain limiting storm level $\eta^{\circ}_{1,i}$ percentage of the armour units will be rocking. Obviously, the accumulated fatigue damage will be smallest in the last activated armour unit. Hence, when this unit is failing due to fatigue exactly $\eta^{\circ}_{1,i}$ percentage of the armour units have failed.

$\Delta D_{1,i}$ signifies the fatigue damage increment in the last activated armour unit among all $\eta^{\circ}_{1,i}$ percentage of units, rocking at the limiting storm level. For a storm with $H_{s,i,max} = x$, this quantity is assumed on the form

$$\Delta D_{1,i} = \int_{h_1}^{x} \frac{dt}{T_{0,i}(\Delta\bar{\sigma})} \tag{18}$$

dt is the time interval with significant wave heights in the interval $]h, h + dh]$. In accordance with (17), dh and dt will be linearly dependent

$$dt = t_{0,i} \frac{dh}{x - h_1} \tag{19}$$

$t_{0,i}$ is the duration of a storm with $H_{s,i} = h_1$, see figure 4. (17) and (19) have been based on the data samples in figure 4. As seen $t_{0,i}$ turns out to be independent of x. The data represent the largest storms in a 20 years period for a certain location. The smaller storms represent sea states where movements are negligible in respect to fatigue damage.

It is assumed that the stress ranges from a single wave within a certain armour unit can be identified by one parameter. The average value of this quantity at a time where the storm is specified by $(H_{s,i}, T_{p,i})$ is termed $\Delta\bar{\sigma}_i$. The dependency of $\Delta\bar{\sigma}_i$ on $(H_{s,i}, T_{p,i})$ can generally be established only by model- or full scale tests. The following explicit relationship has been assumed in what follows

$$\Delta\bar{\sigma}_i = \Delta\bar{\sigma}_{i,1} \frac{H_{s,i} - h_{s,i,0}}{h_{s,i,1} - h_{s,i,0}} \quad , \quad H_{s,i} \in [h_{s,i,0}, \infty[\tag{20}$$

$\Delta\bar{\sigma}_i$,1 is the average stress range at the wave height $h_{s,i,1}$ and depends to some extent on the magnitude of the armour units. $h_{s,i,0}$ is the wave height, at which $\eta^{\circ}_{1,i}$ percentage of the armour units is rocking.

$T_{0,i}(\Delta\bar{\sigma}_i)$ is the fatigue life of the armour units in a wave condition where the average stress range is $\Delta\bar{\sigma}_i$. Consequently (18) in combination with (3) is a formulation of the conventional Palmgren-Miner accumulated damage theory.

$T_{0,i}$ as a function of $\Delta\bar{\sigma}_i$ is a material property, which has been assumed on the form

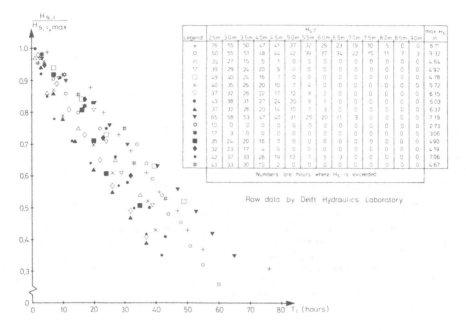

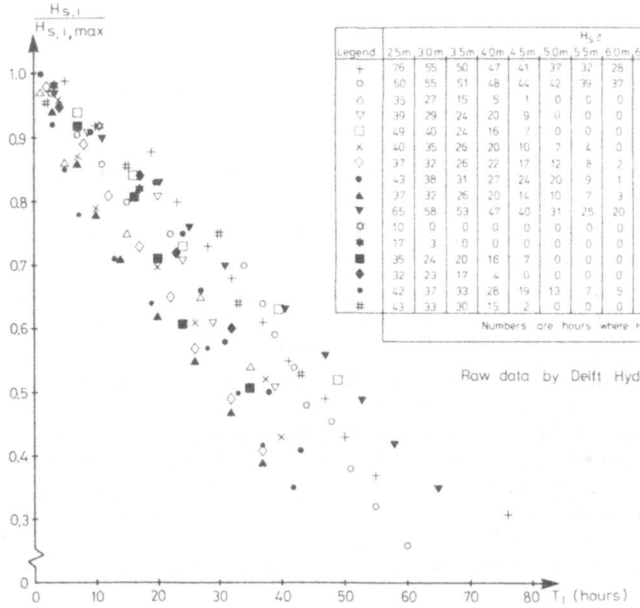

Legend	2.5m	3.0m	3.5m	4.0m	4.5m	5.0m	5.5m	6.0m	6.5m	7.0m	7.5m	8.0m	8.5m	9.0m	max H_s m
+	76	55	50	47	41	37	32	28	23	19	10	5	0	0	8.11
○	60	55	51	48	44	42	39	37	34	22	15	11	7	3	9.32
△	35	27	15	5	1	0	0	0	0	0	0	0	0	0	4.64
▽	39	29	24	20	9	0	0	0	0	0	0	0	0	0	4.92
□	49	40	24	16	7	0	0	0	0	0	0	0	0	0	4.78
×	40	35	26	20	10	7	4	0	0	0	0	0	0	0	5.72
◇	37	32	26	22	17	12	8	2	0	0	0	0	0	0	6.15
●	43	38	31	27	24	20	9	1	0	0	0	0	0	0	5.03
▲	37	32	26	20	14	10	7	3	0	0	0	0	0	0	6.37
▼	65	58	53	47	40	31	25	20	11	3	0	0	0	0	7.19
✡	10	0	0	0	0	0	0	0	0	0	0	0	0	0	2.73
✱	17	3	0	0	0	0	0	0	0	0	0	0	0	0	3.06
■	35	24	20	16	7	0	0	0	0	0	0	0	0	0	4.90
◆	32	23	17	4	0	0	0	0	0	0	0	0	0	0	4.19
•	42	37	33	28	19	13	7	5	3	0	0	0	0	0	7.06
#	43	33	30	15	2	0	0	0	0	0	0	0	0	0	4.67

Numbers are hours where H_s is exceeded

Raw data by Delft Hydraulics Laboratory

Figure 4: Duration of storms T_i, with significant wave height exceeding level $H_{s,i}$.

$$T_{0,i}(\Delta\bar{\sigma}_i) = T_{z,i} \cdot \begin{cases} \dfrac{1}{\left(\dfrac{\Delta\bar{\sigma}_i}{\bar{\sigma}_T}\right)^{m_i}} & , \quad \Delta\sigma_{i,min} \leqslant \Delta\bar{\sigma}_i < \infty \\ \infty & , \quad 0 \leqslant \Delta\bar{\sigma}_i < \Delta\sigma_{i,min} \end{cases} \tag{21}$$

$T_{z,i}$ is the expected zero crossing period, $\bar{\sigma}_T$ is the expected dynamic tensile strength, $\Delta\sigma_{i,min}$ is the minimum stress range capable of inducing any fatigue damage in the material, and m_i is a material parameter. $\Delta\sigma_{i,min}$ is obtained at a certain limiting wave height $h_{i,min}$. As seen from (20), $h_{i,min} > h_{s,i,0}$.

$T_{p,i}$ and $T_{z,i}$ are assumed to be dependent on the significant wave height $H_{s,i}$ as given by the following explicit relationship

$$T_{p,i} = t_{p,i,1}\left(\frac{H_{s,i}}{h_{s,i,1}}\right)^{h_4} \quad , \quad T_{z,i} = t_{z,i,1}\left(\frac{H_{s,i}}{h_{s,i,1}}\right)^{h_4} \tag{22}$$

$t_{p,i,1}$ and $t_{z,i,1}$ are the wave period and crossing periods at the referential wave height $h_{s,i,1}$.

On condition of the sample $H_{s,i,max} = x$, the displacement damage increment $\Delta D_{2,i}$ is suggested in the form, in which (10) has been applied

$$\Delta D_{2,i} = \int_{h_1}^{x} \frac{dt}{t_{2,i}} \cdot \Delta D^{\circ}_{2,i}(h, t(h)) = \frac{t_{0,i}}{t_{2,i}} \cdot \frac{1}{x - h_1} \int_{h_1}^{x} \Delta D^{\circ}_{2,i}(h, t(h))\, dh \tag{23}$$

$\Delta D^{\circ}_{2,i}(H_{s,i}, T_{p,i})$ is the relative number of blocks within the basic unit area i, displaced at least one characteristic diameter relative to their initial position in a storm $(H_{s,i}, T_{p,i})$ during the interval $t_{2,i}$. Hence $\Delta D^{\circ}_{2,i}/t_{2,i}$ indicates the damage rate.

The quantity $\Delta D^{\circ}_{2,i}$ can be determined by model tests. As suggested in [3] the results may, due to lack of more precise formulae, be presented by a stability number S_i and the surf similarity parameter ξ_i [5], defined as follows

$$S_i = \frac{H_{s,i}}{\left(\frac{\overline{W}_i}{\gamma_{B,i}}\right)^{1/3}\left(\frac{\gamma_{B,i}}{\gamma_w} - 1\right)} \quad , \quad \xi_i = tg\alpha_i \cdot T_{p,i}\left(\frac{g}{2\pi H_{s,i}}\right)^{1/2} \tag{24}$$

where $\overline{W}_i$ = average weight of armour units in basic area i, $\gamma_{B,i}$ = specific weight of armour units in basic area i, γ_w = specific weight of water, α_i = angle of slope at basic area i ($\alpha_i <$ natural angle of repose), and g = acceleration of gravity.

For fixed armour unit weight, (24) represents simply a one-to-one mapping of the basic wave load parameters ($H_{s,i}$, $T_{p,i}$) into a non-dimensional representation.

Similar to displacement stability curves also curves for the rocking percentage R_i in the basic unit area i can be established, see figure 6. Note that the rocking percentage is assumed independent of the time exposure, whereas the displacement percentage increases linearly with this quantity.

$T_{0,i}$ and $\Delta D^{\circ}_{z,i}$ in (18) and (19) have been defined as combined stochastic variables depending on ($H_{s,i}$, $T_{p,i}$). Actually these quantities are regression models with random noise, because some parameters affecting stability have not been controlled. As seen from the approximation (9) only second order moments of $\Delta D_{k,i}$ need to be known. Consequently this inherent model uncertainty needs only to be known until second order moments. The regressions $\Delta \bar{D}^{\circ}_{2,i}(h)$ and $\bar{T}_{0,i}(h)$ of $\Delta D^{\circ}_{2,i}$ and $T_{0,i}$ on $H_{s,i}$, and corresponding variational coefficients $\kappa_i(h)$ and $\zeta_i(h)$ are estimated from a sample of repeated test results. Further assuming the stochastic variables $\Delta D_{2,i}(h_1)$ and $\Delta D_{2,i}(h_2)$ respectively $T_{0,i}(h_1)$ and $T_{0,i}(h_2)$ to be fully correlated for every h_1, h_2 $\in [0,x]$ corresponding to unchanged relative material response to different sea states, the moments searched for can be calculated, applying conventional second moment algebra in the stochastic integrals (18) and (23).

The unconditioned expectations $E[\Delta D_{k,i}]$ and $E[\Delta D^2_{k,i}]$ are finally obtained by taking the expectation with respect to the distribution of $H_{s,i,max}$.

The intensities $\lambda_{k,i}$ of the counting processes are obtained from the following expression

$$\lambda_{k,i} = \lambda_0 \int_{\Omega_{k,i}} f_{H_{s,i,max}, T_{p,i,max}}(h,t)\, dh\, dt \tag{25}$$

$f_{H_{s,i,max}, T_{p,i,max}}$ is the joint frequency function of the maximum significant wave heigt $H_{s,i,max}$ and associated peak period $T_{p,i,max}$ during a storm. The domain of integration $\Omega_{k,i}$ indicates the subset of the sampling space of ($H_{s,i,max}$, $T_{p,i,max}$) for storms, which gives rise to any damage in the k'th failure mode in the basic unit area i. $\Omega_{1,i}$ is given by the following expression, cf. (16), (20), (21)

$$\Omega_{1,i} = \{(h,t) \mid h > h_o\} \tag{26}$$

$$h_0 = \max(h_1 , h_{i,min}) \tag{27}$$

4. THE APPLICATION TO OPTIMUM DESIGN

The repair expenses P_1 during the stipulated lifetime T_0 of the structure, discounted to the time of construction with the inflation-regulated rate of interest r can be written as follows

$$P_1 = \sum_{i=1}^{M} \sum_{\ell=1}^{L_i(T_0)} C_{i,\ell} \cdot (1 + r)^{-\tau_{i,\ell}} \tag{28}$$

$\{L_i(t), t \in [0,T_0]\}$ are inhomogeneous Poisson counting processes, specifying the random number of failures at times $0 \leq \tau_{i,1} < \tau_{i,2} < \ldots < \tau_{i,L_i(T_0)} \leq T_0$ in the basic unit area i.

The intensities $\nu_i : [0,T_0] \curvearrowright R_+$ of the counting processes, i.e. the expected number of failures per unit of time, are determined from (13).

$C_{i,\ell}$ is the total inflation-regulated cost of the ℓth failure within basic unit area i, made up of costs of site establishment, down time, social expenses, and repair. These quantities are assumed to be mutually independent stochastic variables, identical distributed as the stochastic variable C_i. C_i depends on the magnitude and duration of the storm at the instant of failure. Consequently C_i may be considered as a combined stochastic variable depending on ($H_{s,i,max}$, $T_{p,i,max}$), i.e.

$$C_i = C_i(H_{s,i,max} , T_{p,i,max}) \tag{29}$$

The characteristic function of the stochastic variable P_1 becomes, cf. (5)

$$M_{P_1}(\theta) = \exp\left(\sum_{i=1}^{M} \int_0^{T_0} \left(M_{C_i}\left((1+r)^{-\tau} \cdot \theta\right) - 1 \right) \cdot \nu_i(\tau)\, d\tau \right) \tag{30}$$

$M_{C_i} : R \curvearrowright C$ is the characteristic function of the stochastic variable C_i.
Hence the expectation p_1 becomes

$$p_1 = E[P_1] = \sum_{i=1}^{M} E[C_i] \int_0^{T_0} (1+r)^{-\tau} \cdot \nu_i(\tau)\, d\tau \tag{31}$$

The moment $E[C_i]$ can be calculated from the regression model (29).
Let $M \subset R^n$ be the set of feasible designs and $p_o : M \curvearrowright R$ be the construction price. The optimal design $\underset{\sim}{x}_o$ is then selected according to the design strategy

$$p_o(\underset{\sim}{x}_o) + p_1(\underset{\sim}{x}_o) = \min_{\underset{\sim}{x} \in M} (p_o(\underset{\sim}{x}) + p_1(\underset{\sim}{x})) \tag{32}$$

5. NUMERICAL EXAMPLE

The outlined theory will be demonstrated for a breakwater specified by the succeeding parameters

Slope of angle	: $\tan\alpha = 1/1.5$
Type of armour	: Complex, slender un-reinforced concrete units $\gamma_B = 23.8$ kN/m³ Dynamic tensile strength $\bar{\sigma}_T = 6000$ kPa
Reference armour unit	: $\bar{W}_0 = 150$ kN Number of units = 0.17/m²
Unit area	: 32 m x 32 m = 1024 m²
Number of unit areas	: M = 20
Length of breakwaters	: $\ell = 20 \times 32 = 640$ m
Stipulated lifetime	: $T_0 = 100$ years
Specific weight of water	: $\gamma_w = 10.0$ kN/m³
Long term distribution of maximum wave heights, eq. (16)	: $h_1 = 0.1$ m $h_2 = 6.0$ m $h_3 = 2.8$
Duration of storms, eq. (19)	: $t_{0,i} = 70$ hours
Stress range relation, eq. (20)	: $h_{s,i,1} = 12.9$ m $\Delta\bar{\sigma}_{i,1} = 6000$ kPa
Fatigue parameters, eq. (21)	: $m_i = 6.36$ $\Delta\bar{\sigma}_{i,min} = 1250$ kPa

The corresponding Wöhler curve, obtained from test series with 5 repeated tests is shown in figure 5.

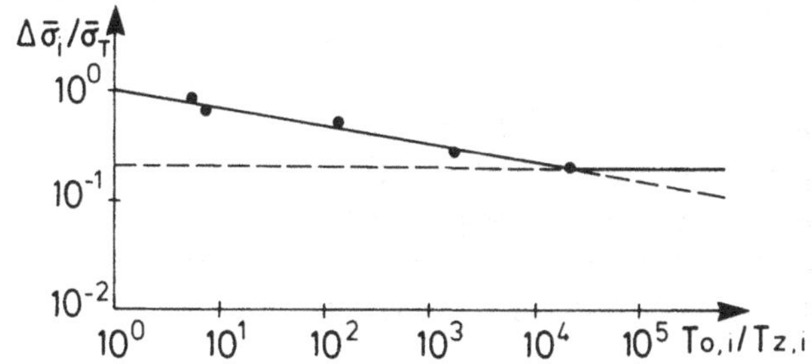

Figure 5: Wöhler curve for impact loaded armour units (flexural stress). Number of stress ranges to first sign of crack. Each point represents average from 5 tests (Burcharth 1983).

The variational coefficient ζ_i of $T_{0,i}/T_{z,i}$, estimated from the above test series, has been indicated in table 1 as a function of $\Delta\bar{\sigma}/\Delta\bar{\sigma}_T$.

Table 1: The variational coefficient of fatigue life duration as a function of average stress range.

$\Delta\bar{\sigma}\,/\,\bar{\sigma}_T$	0.208	0.302	0.500	0.698	0.854
$\zeta_i(\Delta\bar{\sigma})$	2.103	1.550	1.171	0.146	0.404

The functional relationship has been represented by the following linear fit:

$$\zeta_i\,(h) = 2.3 - 2.1\,\frac{\Delta\bar{\sigma}_i}{\bar{\sigma}_T}$$

The uncertainty contributions from the determination of the limiting wave height $h_{s,i,0}$ have been ignored in the present study.

Peak period relation, eq. (22) :	$t_{p,i,1} = 16.7$ s	Zero crossing period relation, eq. (22) :	$t_{z,i,1} = 11.5$ s
Power indication, eq. (22):	$h_4 = 0.45$	Intensity of significant storms :	$\lambda_0 = 0.85$/year

The average rocking percentage $\bar{R}$ and the average damage increment $\Delta\bar{D}^{\circ}_{2,i}$ and the associated variational coefficient κ_i have been determined by model tests, see figure 6. Notice that the damage was not equally distributed over the quadratic model test area. The damage within a zone of 25% amounted to approximately the double of the indicated values.

The slope of the stability curves has been taken from [3]. The stability curve corresponding to p percentage of damage can be formulated analytically as follows

$$S_i = s_{p,i} - r_{0,i}\xi_i \quad , \quad r_{0,i} = 0.325 \tag{32}$$

where $s_{p,i}$ has been tabulated below.

From a given wave height $H_{s,i}$ and armour unit weight $\bar{W}$, S_i and ξ_i are calculated from (24). $s_{p,i}$ is then obtained from (32), and the rocking percentage $\bar{R} = p$, respectively damage percentage $\Delta\bar{D}^{\circ}_{2,i} = p$ and variational coefficient κ_i, are obtained by linear interpolation or extrapolation from table 2.

Legend: Each data set obtained from 20 tests

			Rocking R_i in %		Displacement $\Delta D^{\circ}_{2,i}$ in %	
	S_i	ξ_i	μ	σ	μ	σ
△	1.67	4.30	1.46	0.90	0.11	0.32
▲	2.26	3.68	3.28	1.74	0.54	0.79
○	2.58	3.46	5.03	2.57	1.73	2.26
●	3.19	3.10	7.21	2.82	3.45	2.74
×	3.47	2.98	10.15	3.16	5.98	3.24

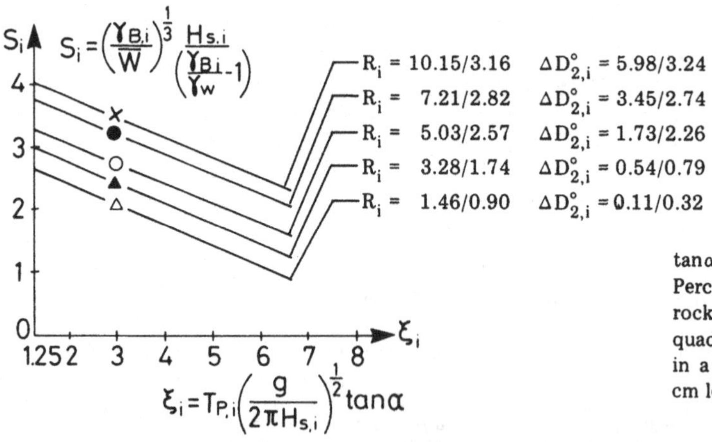

$\tan\alpha = 1/1.5$ in all tests. Percentages are no. of units rocking or displaced within a quadratic area of 0.4 x 0.4 m in a model where $H_s \cong 13.0$ cm leads to rapid destruction.

Figure 6: Hydraulic rocking and displacement stability curves. Damage after $t_{2,i} \cong 3$ hours in prototype, corresponding to 1200 waves with expected zero crossing period $T_{z,i} \sim 9$ s. (Burcharth and Brejnegaard 1982, 1983).

Table 2: Hydraulic stability data.

	Rocking		Displacement	
$s_{p,i}$	p (%)	κ_i	p (%)	κ_i
3.07	1.46	0.616	0.11	2.91
3.46	3.28	0.530	0.54	1.46
3.70	5.03	0.511	1.73	1.306
4.20	7.21	0.391	3.45	0.794
4.48	10.15	0.311	5.98	0.542

$\eta^{\circ}_{1,i} = 0.1$. Hence the 10% rocking stability curve is determined by $s_{p,i} = 4.466$.

The allowable damage limits $\eta_{k,i}$, $\eta^{\circ}_{1,i}$, and the functions $s_{k,i}$ in (4) have been selected as follows

$$\eta_{1,i} = 1 \quad , \quad \eta^{\circ}_{1,i} = 0.1$$
$$\eta_{2,i} = 0.1$$
$$s_{k,i}(t) \equiv 1 \quad , \quad k \in 1, 2$$

The first passage density function of a basic unit area can now be calculated. The result has been shown in figure 7 as a function of the armour unit weight. Notice the shift of the probability mass towards higher failure times, indicating increased reliability, as the average armour unit weight is increased.

As an example the construction price of the entire breakwater and the expected inflation regulated repair price of a basic unit area with length 32 m are taken as dependent on the average armour unit weight as follows

$$E[C_i] = 32\,\mathrm{m} \cdot c \left(\frac{\bar{W}}{\bar{W}_0}\right)^{\frac{1}{3}} \quad , \quad c = 0.2 \cdot 10^6\ \mathrm{Dkr./m} \quad , \quad p_0 = (a + b\left(\frac{\bar{W}}{\bar{W}_0}\right)^{\frac{1}{3}})\ell$$

$$a = 0.35 \cdot 10^6\ \mathrm{Dkr./m}, \quad b = 0.15 \cdot 10^6\ \mathrm{Dkr./m}$$

The authors are fully aware of the simplicity of these cost estimates.

The expected total expenses made up of construction price and expected repair prices can now be calculated for a specific rate of interest r and an average armour unit weight $\bar{W}$. The relationship is shown in figure 8. The minimum of the curves specifies the optimum armour unit weight W_{opt}, according to the applied design criteria. The dependency of optimum W_{opt} on the rate of interest r is seen.

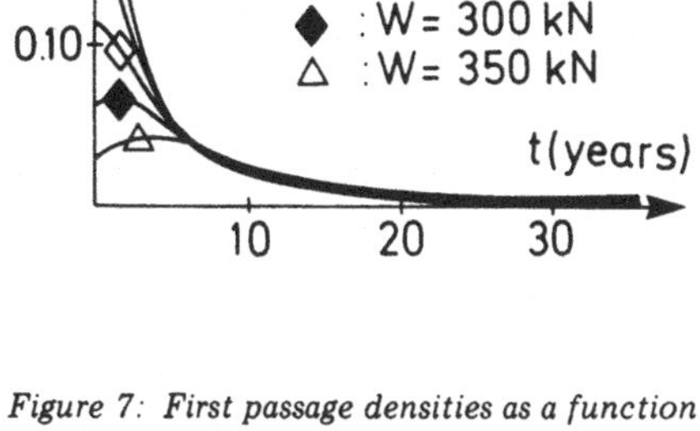

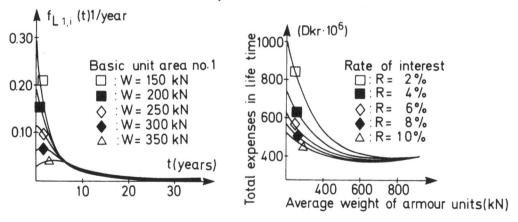

Figure 7: First passage densities as a function of the average weight of armour units in the basic unit area i = 1.

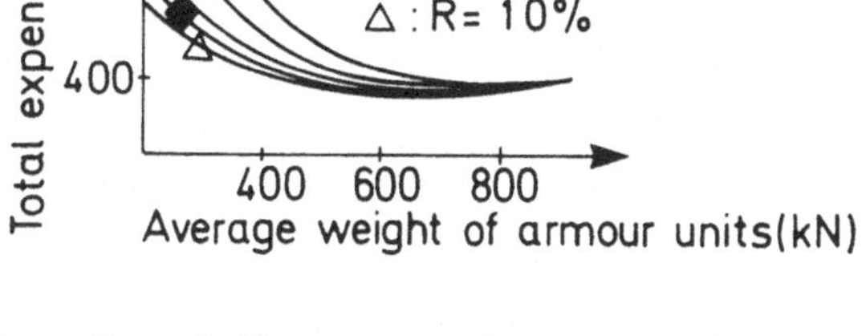

Figure 8: The variation of expected total expenses of breakwater armour layer during stipulated lifetime T_0 = 100 year, depending on armour unit weight $\bar{W}$ and inflation regulated rate of interest r.

6. CONCLUDING REMARKS

A level III reliability method has been developed from which the armour layer of a rubble mound breakwater can be designed, so that the total costs made up of construction price and expected maintenance expenses are minimized. The theory is demonstrated by a numerical example, where the armour unit weight is the only design parameter. The cost estimates for construction and repair used in the numerical example are primitive and used as an illustrative example only. More evaluated cost functions can be implemented and extension to more complex problems where several parameters (slope of angle, material strength parameters etc., etc.) are introduced in the cost function, is straightforward. A computer programme has been developed, from which all relevant quantities in the theory can be determined.

7. REFERENCES

[1] GRAVESEN, H., and SØRENSEN, T.: »Stability of Rubble Mound Breakwaters«. Proc. 24th Int. Nav. Congress, PIANC, Leningrad 1977, p. 65 - 74.

[2] GRAVESEN, H., JENSEN, O. J., and SØRENSEN, T.: »Stability of Rubble Mound Breakwaters II«, Danish Hydraulic Institute (DHI), 1980.

[3] BURCHARTH, H. F.: »The Effect of Wave Grouping on On-Shore Structures«, Coastal Engineering, 2 (1979), p. 189 - 199.

[4] PARZEN, E.: »Stochastic Processes«, Holden - Day, San Fransisco 1962.

[5] BRUUN, P., and GÜNBAK, A. R.: »New Design Principles for Rubble Mound Structures«, Proc. 15th Int. Coastal Eng. Conf., 1976.

PROBABILISTICALLY OPTIMUM DESIGN OF FRAME STRUCTURE

Y. Murotsu*, M. Kishi*, H. Okada*, M. Yonezawa**, K. Taguchi*
* University of Osaka Prefecture, College of Engineering, Sakai, Osaka 591, JAPAN
** Kinki University, Faculty of Science and Technology, Higashi-Osaka, Osaka 577, JAPAN

ABSTRACT

An optimum design problem is treated to determine a frame structure minimizing the structural cost or weight under the constraints on the allowable failure probabilities of critical sections specified in the structure. The design variables are the geometrical dimensions of the structure while its configuration and the loading conditions are specified together with their probabilistic nature. The optimization problem is reduced to a nonlinear programming and it is solved by using the sequential linear programming.

1. INTRODUCTION

There are many studies on optimum design of frame structures based on reliability concept[1)~4)]. These are mainly concerned with minimization of structural weight or cost under the constraint on structural reliability or the expected total cost composed of the structural cost and the cost caused by failure of the structure. However, they have a common limitation that the failure criteria, that is, failure modes and their mode equations, should be specified *a priori*. In order to perform the optimum design of large-scale structures which have too many modes of failure to identify all of them beforehand, a new design procedure is to be developed which enables us to optimize the design variables by automatically generating the failure criteria of the structure.

This paper considers an optimum design problem to minimize the structual weight under the constraints on the failure probabilities of the critical sections in the frame structure. The configuration of the structure and loading conditions together with their probabilistic nature are assumed to be given. The design variables are the geometrical dimensions of the elements. The failure criterion of the critical section is expressed as a function of the strength of the element and the internal force, which is generated by a matrix method. The optimum design problem is reduced to a nonlinear programming problem. The solution to the optimization problem is obtained by applying the sequential linear programming, and numerical examples are given to illustrate the applicability of the proposed method.

2. OPTIMUM DESIGN PROBLEM

2.1. Design Formulation

Consider a structural system which consists of n elements with l applied loads. The configuration, the materials to be used and the loading conditions together with their probabilistic nature are assumed to be given. When the shape of the cross section of each element is specified, there are one-to-one correspondences among its cross sectional area, geometrical moment of inertia and plastic section modulus. This means that by taking one geometrical dimension of the each element, e.g., thickness, as the design variable, the structural cost or weight W is uniquely determined as a function of the design variables:

$$W = W(\mathbf{X}) \tag{1}$$

where $\mathbf{X}$ denotes an n-dimensional vector whose elements are the design variables X_j ($j = 1,2,...,n$). Since the statistical properties of the materials and load are given, the failure probability P_f of the structure is also determined by specifying $\mathbf{X}$, i.e. ,

$$P_f = P_f(\mathbf{X}) \tag{2}$$

The minimum-weight design problem is in general formulated as follows:

Find $\mathbf{X}$

such that

$$W(\mathbf{X}) = \sum_{i=1}^{n} \rho_i l_i A(X_i) \rightarrow \text{minimize} \tag{3}$$

subject to

$$P_f(\mathbf{X}) \leq P_{fa} \tag{4}$$

where ρ_i : density of the i-th element

l_i : length of the i-th element

$A(X_i)$: cross sectional area of the i-th element

P_{fa} : allowable failure probability of the structure.

In the case of large-scale structures, there are too many modes of failure to identify all of them beforehand, and thus it is impossible to calculate the structural failure probability exactly. Therefore, an upper bound of the structural failure probability is often evaluated by using the stochastically dominent failure modes[5), 6)], which also needs much computation time for a large structure. Moreover, the evaluation of the failure probability is repeated many times in the optimum design processes, and the total computation time will become too enormous to be practical.

Introduced in this study are the constraints on the failure probabilites of the critical sections whose failure may trigger structural failure, instead of the constraint on structural failure probability. Then, a new optimum design problem is formulated as follows :

Find $\mathbf{X}$ such that

$$W(\mathbf{X}) = \sum_{i=1}^{n} \rho_i l_i A(X_i) \rightarrow \text{minimize} \tag{5}$$

subject to

$$\begin{aligned} P_{fj}(\mathbf{X}) &\le P_{faj} \qquad (\ j = 1, 2, \ldots, m\ ) \\ g_k(\mathbf{X}) &= 0 \qquad (\ k = 1, 2, \ldots, p\ ) \end{aligned} \tag{6}$$

where $P_{fj}(\mathbf{X})$: failure probability of the j-th critical section

P_{faj} : allowable failure probability of the j-th critical section

m : number of the critical sections

p : number of the relations among the design variables.

2.2. Failure Criteria of Critical Sections

Failure of critical section [7)] is assumed to occur if

$$\frac{M_j}{M'_{pj}} + \frac{N_j}{N'_{pj}} \ge 1 \qquad (\ j = 1, 2, \cdots, m\ ) \tag{7}$$

where M_j : bending moment of the j-th critical section

N_j : axial force of the j-th critical section

$M'_{pj} = \text{sign}(M_j)\sigma_{Yi} AZ_{pj}$

$N'_{pj} = \text{sign}(N_j)\sigma_{Yi} A_{pj}$

σ_{Yi} : yield stress of the i-th element

$AZ_{pj} = AZ_p(X_i)$: plastic section modulus of the j-th critical section

$A_{pj} = A_p(X_i)$: cross sectional area of the j-th critical section.

AZ_{pj} and A_{pj} are the function of the design variable X_i, where subscript i denotes the element which contains the j-th critical sections. Buckling failure under compression is assumed to occur if

$$\frac{M_j}{M'_{pj}} + \frac{N_j}{N'_{cj}} \geq 1 \qquad (j = 1, 2,\cdots, m)$$

$$N'_{cj} = \mathrm{sign}(N_j)\sigma_{ci}A_{pj} \tag{8}$$

σ_{ci} : buckling stress of the i-th element (see Appendix 1).

The bending moment and axial force of the critical section are calculated by using a matrix method and written in the form :

$$M_j = \sum_{k=1}^{l} b_{jk}^{(m)}(\mathbf{X}) L_k$$

$$N_j = \sum_{k=1}^{l} b_{jk}^{(a)}(\mathbf{X}) L_k \qquad (j = 1, 2,\cdots, m) \tag{9}$$

where L_k are the applied loads, and $b_{jk}(\mathbf{X})$ are the load effect coefficients. From Eqs. (7) and (9), the limit state function (Z_j) of the j-th critical section is expressed in the form :

$$Z_j = \sigma_{Yi}^2 AZ_p(X_i)A_p(X_i) - \sum_{k=1}^{l}\{[\mathrm{sign}(M_j)b_{jk}^{(m)}(\mathbf{X})A_p(X_i) + \mathrm{sign}(N_j)b_{jk}^{(a)}(\mathbf{X})AZ_p(X_i)]\sigma_{Yi}L_k\}$$

$$(j = 1, 2,\cdots, m) \tag{10}$$

Yielding failure of the critical section occurs if Z_j is not positive, i.e. , $Z_j \leq 0$. In the case of a compression element, the following limit state function (Z_{cj}) corresponding to buckling failure follows from Eqs. (8) and (9):

$$Z_{cj} = \sigma_{Yi}\sigma_{ci}AZ_p(X_i)A_p(X_i) - \sum_{k=1}^{l}\{[\mathrm{sign}(M_j)b_{jk}^{(m)}(\mathbf{X})\sigma_{ci}A_p(X_i) + \mathrm{sign}(N_j)b_{jk}^{(a)}(\mathbf{X})\sigma_{Yi}AZ_p(X_i)]L_k\}$$

$$(j = 1, 2,\ldots, m) \tag{11}$$

Hence, buckling failure of the critical section occurs when $Z_{cj} \leq 0$.

2.3. Failure Probabilities of Critical Sections

Limit state functions of the critical section are given by Eqs. (10) and (11), where σ_{Yi}, σ_{ci}, $\mathrm{sign}(M_j)$, $\mathrm{sign}(N_j)$ and L_k are random variables. Therefore, Z_j and Z_{cj} are also random variables. Although $\mathrm{sign}(M_j)$ and $\mathrm{sign}(N_j)$ are binary indicator random variables, they are treated for simplicity as deterministic quantities and given the values corresponding to the mean values of the loads. σ_{Yi} is a positive random valiable, and thus the failure criterion $Z_j \leq 0$ is simplified as follows :

$$Z'_j = \sigma_{Yi} AZ_p(X_i) A_p(X_i) - \sum_{k=1}^{l} \{[\mathrm{sign}(M_j) b_{jk}^{(m)}(\mathbf{X}) A_p(X_i) + \mathrm{sign}(N_j) b_{jk}^{(a)}(\mathbf{X}) AZ_p(X_i)] L_k\} \le 0 \quad (j = 1, 2, \ldots, m) \tag{12}$$

Consequently, Z'_j is expressed as a linear combination of the random variables.

The failure probability $P_{fj}(\mathbf{X})$ of the critical section is calculated by

$$P_{fj}(\mathbf{X}) = \mathrm{Prob}[(Z'_j \le 0) \cup (Z_{cj} \le 0)] \quad (j = 1, 2, \ldots, m) \tag{13}$$

Since Z'_j and Z_{cj} are not independent random variables, the joint probability must be calculated to evaluate Eq.(13) exactly. However, no efficient methods are available for calculating it except the case of Gaussian distribution. Eventually, an upper bound of P_{fj} is estimated by

$$P_{fj}(\mathbf{X}) \le \mathrm{Prob}[Z'_j \le 0] + \mathrm{Prob}[Z_{cj} \le 0] \quad (j = 1, 2, \ldots, m) \tag{14}$$

Data for describing the random variables σ_{Yi}, σ_{ci}, and L_k are usually not so amply provided that their distributions can be exactly specified. In practice, information about them may be limited to their first- and second-order moments. Even when the distribution function of these random variables are exactly specified, it is very difficult to know the distribution of Z'_j and especially that of Z_{cj} which is a nonlinear function of the random variables. By taking account of these facts, a first-order second-moment approximation method is applied to the reliability analysis and optimum design of the structural system.

3. SOLUTION TO PROBLEM

The optimum design problem formulated in the previous section is a nonlinear programming problem, and it may be effectively solved by using SLP (Sequential Linear Programming)[8),9)]. SLP uses a LP (Linear Programming) algorithm sequentially in such a way that in the limit the successive solutions of the linear programming problems converge to those of nonlinear programming problems. Let $\mathbf{X}^{(k)}$ be the design variables in the k-th stage. By using Taylor's expansion, it follows that

$$W(\mathbf{X}) \approx W(\mathbf{X}^{(k)}) + \nabla W(\mathbf{X}^{(k)})\Delta\mathbf{X} \tag{15}$$

$$P_{fj}(\mathbf{X}) \approx P_{fj}(\mathbf{X}^{(k)}) + \nabla P_{fj}(\mathbf{X}^{(k)})\Delta\mathbf{X} \qquad (j = 1, 2, \ldots, m) \tag{16}$$

Hence, the original nonlinear programming problem is reduced to the following linearized problem :

Find $\Delta\mathbf{X}$

such that

$$\nabla W(\mathbf{X}^{(k)})\Delta\mathbf{X} \rightarrow \text{minimize} \tag{17}$$

subject to

$$\nabla P_{fj}(\mathbf{X}^{(k)})\Delta\mathbf{X} \leq P_{faj} - P_{fj}(\mathbf{X}^{(k)}) \qquad (j = 1, 2, \ldots, m) \tag{18}$$

The optimum solution ($\Delta\mathbf{X}$) to this problem is easily obtained by using LP, and the design variable $\mathbf{X}^{(k+1)}$ in the next stage is given by

$$\mathbf{X}^{(k+1)} = \mathbf{X}^{(k)} + \Delta\mathbf{X} \tag{19}$$

Adaptive move limits which limit the step size of $\Delta\mathbf{X}$ are used to secure the validity of the linear approximation and termination conditions on the successive changes in the design variables, and the weighting factors are also introduced to exclude the oscillation phenomena.

4. NUMERICAL EXAMPLES

Numerical examples for two types of frame structures are presented to demonstrate the validity of the proposed method. It is assumed that the applied loads and the strengths of the sections are statistically independent while the strengths in the same elements are completely correlated. All elements have tubular cross sections with constant ratio of outside radius to wall thickness. The design variables are the outside radiuses of the elements.

4.1. Portal Frame

An optimum design problem is solved for a portal frame with horizontal and vertical loads applied as shown in Fig. 1. The outside radiuses of the columns and the beam are taken as the two independent design variables, considering a symmetric structure. The results are listed in Table 1 for various combinations of coefficients of variation

($CV_{\sigma Yi}$, CV_{Lk}). Table 2 shows the effects of the correlations between the loads on the optimum solutions. It is seen that the structural weight becomes large as the coefficients of variation and correlation coefficients are large. For the calculations presented so far, no consideration has been given to the initial deflections of the elements. Table 3 shows the effects of the initial deflections on the optimum design. Further, the effects of the allowable failure probabilities of the critical sections are investigated. The results are given in Table 4. In the optimization problem, there are two design variables, and the optimal solution lies on the extreme point of intersection of the two constraints. In case 1, the active two constraints are the critical sections 6 and 8. In case 3, however, they are the sections 4 (or 5) and 8.

4.2. Jacket Structure

Another optimum design problem is solved for a jacket structure with 15 elements shown in Fig. 2, where the eight independent design variables are taken, corresponding to the elements numbered in the parenthesises. Data concerned are specified as shown in Table 5. Piles inside the jacket legs are ignored and complete fixing is assumed at the base. The optimum solution is given in Table 6.

5. CONCLUDING REMARKS

This paper is concerned with an optimum design of structural systems to minimize the structural weight subject to the allowable failure probabilities of the critical sections of the elements. SLP (Sequential Linear programming) is effectively applied to determining the optimum values of the design variables. Through numerical examples, the validity of the proposed method is demonstrated.

In the proposed method, the failure constraints are imposed on the critical sections; therefore, the method is easily applicable to the optimum design of a large-scale structure where the objective function is a total cost composed of the structural cost and those of inspections and repairs. Further, the proposed method will be extended to solve the optimum design problem under a constraint on the structural system failure probability, by sequentially changing the allowable failure probabilities of the critical sections so as to confine the system failure probability within the allowable value.

REFERENCES

1. Vanmarcke, J., *et al.*, Reliability-Based Optimum Design of Simple Plastic Frames, R72-46, Dept.of Civil Eng., Massachusetts Institute of Technology, Cambridge, Mass., (1972).
2. Gallagher, R.H., and Zienkiewicz, O.C.(Ed.), Optimum Structural Design, John Wiley, New York, (1973).
3. Murotsu, Y., *et al.*, Optimum Design of Structures Using Second-Moment Approximation of Reliability, Theoretical and Applied Mechanics, Vol. 26, (1976).
4. Murotsu, Y., *et al.*, Optimum Structural Design under Constraint on Failure Probability, ASME paper 79-DET-114, (1979).
5. Murotsu, Y., Reliability Analysis of Frame Structure Through Automatic Generation of Failure Modes, in:Thoft-Christensen, P. (Ed.), Reliability Theory and its Application in Strucural and Soil Mechanics, Martinus Nijhof Publishers (1983).
6. Murotsu, Y., *et al.*, Identification of Stochastically Dominant Failure Modes in Frame Structure, Proc. of 4th International Conf. on Application of Statistics and Probability in Soil and Structural Engineering , Florence (1983).
7. Okada, H., *et al.*, Safety Margins for Reliability Analysis of Frame Strucrures, Bull. Univ. Osaka Pref., Ser. A, 32, 2, (1983), (to appear).
8. Himmelblau, D.H., Applied Nonliear Programming, McGraw-Hill, New York, (1972).
9. Murotsu, Y., *et al.*, Optimum Structural Design based on Extended Reliability Theory, Proc. 11th Congr. of International Council of the Aeronautional Sciences, Vol. 1, (1978).
10. Timoshenko, S.T., and Gere, J.M., Theory of Elastic Stability, McGraw-Hill, New York, (1961).

Appendix 1. Buckling Stress[10)]

Buckling stress of the member with initial deflection is given as follows :

$$\sigma_C = \frac{1}{2}(\sigma_Y+\sigma_E+\frac{w_0}{s}\sigma_E)\{1-\sqrt{1-(4\sigma_E\sigma_Y)/(\sigma_Y+\sigma_E+\frac{w_0}{s}\sigma_E)^2}\} \tag{A1}$$

where $\sigma_E = \pi^2 E/(l/s)^2$: Euler's buckling stress

E : modulas of elasticity

$s = \sqrt{I/A}$: radius of gyration

w_0 : initial deflection

I : geometrical moment of inertia

A : cross sectional area

l : length.

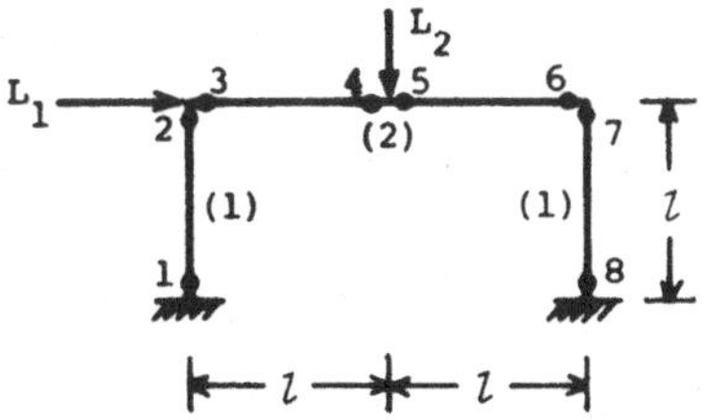

$\bar{L}_1$ = 50 kN, $\bar{L}_2$ = 40 kN, l = 5 m

Mean value of yield stress $\bar{\sigma}_{Yi}$ = 274 MPa

Young's modulus E_i = 210 GPa

Density ρ_i = 7.85 × 10^{-3} kg/cm^3

(wall thickness)/(outside radius) = 0.05

Fig. 1. Portal frame.

Table 1. Effects of the coefficients of variation (portal frame).

CV_{L_k}	coefficient of variation of yield stress $CV_{\sigma_{Yi}}$					
	0.05			0.1		
	outside radius (m)		structural weight	outside radius (m)		structural weight
	column x_1^*	beam x_2^*	W* (kg)	column x_1^*	beam x_2^*	W* (kg)
0.15	0.155	0.146	1097	0.164	0.155	1224
0.3	0.169	0.157	1279	0.175	0.163	1383
0.6	0.192	0.175	1621	0.196	0.180	1710

allowable failure probability P_{faj} = 0.001
correlation coefficient $\rho_{L_1L_2}$ = 0.0

Table 2. Effects of the correlation coefficients between the loads (portal frame).

CV_{L_k}	correlation coefficient between the loads $\rho_{L_1L_2}$								
	0.0			0.5			1.0		
	outside radius (m)		structural weight	outside radius (m)		structural weight	outside radius (m)		structural weight
	column x_1^*	beam x_2^*	W* (kg)	column x_1^*	beam x_2^*	W* (kg)	column x_1^*	beam x_2^*	W* (kg)
0.15	0.164	0.155	1224	0.165	0.157	1246	0.166	0.158	1267
0.3	0.176	0.163	1383	0.178	0.168	1435	0.180	0.171	1480
0.6	0.196	0.180	1710	0.200	0.188	1811	0.204	0.193	1894

$CV_{\sigma_{Yi}}$ = 0.1, allowable failure probability P_{faj} = 0.001

Table 3. Effects of the initial deflections of the elements (portal frame).

CV_{L_k}	mean of initial deflection $\bar{\omega}_0/s$ (s : radius of gyration)								
	0.0			0.05			0.1		
	outside radius (m)		structural weight	outside radius (m)		structural weight	outside radius (m)		structural weight
	column x_1^*	beam x_2^*	W* (kg)	column x_1^*	beam x_2^*	W* (kg)	column x_1^*	beam x_2^*	W* (kg)
0.15	0.164	0.155	1224	0.166	0.157	1252	0.168	0.159	1288
0.3	0.176	0.163	1383	0.177	0.165	1405	0.178	0.166	1432
0.6	0.196	0.180	1710	0.197	0.181	1728	0.198	0.182	1747

$CV_{\sigma_{Yi}}$ = 0.1, $\rho_{L_1L_2}$ = 0.0, CV_{ω_0} = 0.1, P_{faj} = 0.001

Table 4. Effects of the allowable failure probabilities of the critical sections (portal frame).

	outside radius (m)		
	case 1	case 2	case 3
column X_1^*	0.196	0.197	0.195
beam X_2^*	0.180	0.180	0.188
structural weight W* (kg)	1710	1711	1759

($CV_{\sigma_{Yi}}/CV_{L_k}$) = (0.1/0.6), $\rho_{L_1L_2}$ = 0.0

case 1. $P_{faj} = 0.001$ ($j = 1, 2, \ldots, 8$)

case 2. $P_{fa4} = P_{fa5} = 0.0001$
$P_{faj} = 0.001$ ($j \notin \{4, 5\}$)

case 3. $P_{fa4} = P_{fa5} = 0.00001$
$P_{faj} = 0.001$ ($j \notin \{4, 5\}$)

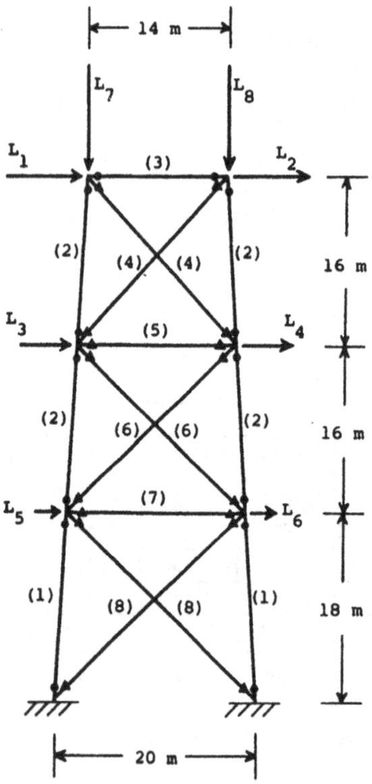

Fig. 2. Jacket structure.

Table 5. Data of jacket structure.

(1) Data of materials

Yield stress $\bar{\sigma}_{Yi}$ (MPa)	$CV_{\sigma_{Yi}}$	Young's modulus E_i (GPa)	Density ρ_i (kg/cm^3)
274	0.1	210	7.85×10^{-3}

(2) Data of loads

	L_k (kN)	CV_{L_k}
L_1 , L_2	240	0.3
L_3 , L_4	120	0.3
L_5 , L_6	40	0.3
L_7 , L_8	2940	0.1

$\rho_{L_iL_j} = 1.0$ $(i,j \in \{1,\ldots,6\})$

$\rho_{L_7L_8} = 1.0$

$\rho_{L_iL_j} = 0.0$ $\begin{pmatrix} i \in \{1,\ldots,6\} \\ j \in \{7,8\} \end{pmatrix}$

Table 6. Optimum solution to jacket structure.

design variable	outside radius (m)
X_1^*	0.363
X_2^*	0.321
X_3^*	0.101
X_4^*	0.163
X_5^*	0.093
X_6^*	0.172
X_7^*	0.116
X_8^*	0.132
W* (kg)	3.69×10^5

$P_{fa(\bullet)} = 0.001$
$P_{fa(\blacktriangle)} = 0.005$
(For the critical sections, see Fig. 2.)

(wall thickness)/(outside radius) = 0.05

RELIABILITY ANALYSIS OF ELASTO-PLASTIC STRUCTURES

P. Thoft-Christensen and J. D. Sørensen
Aalborg University Centre
Aalborg, Denmark

1. INTRODUCTION

This paper only deals with framed and trussed structures which can be modelled as systems with ductile elements. The elements are all assumed to be linear-elastic perfectly plastic. The loading is assumed to be concentrated and time-independent. The strength of the elements and the loads are modelled by normally distributed stochastic variables. This last assumption is not essential, since non-normally distributed variables can be approximated by equivalent normally distributed variables by well-known methods. All geometrical dimensions and stiffness quantities are assumed to be deterministic.

Failure of this type of system is defined either as formation of a mechanism or by failure of a prescribed number of elements. In the first case failure is independent of the order in which the elements fail, but this is not so by the second definition.

The reliability analysis consists of two parts. In the first part significant failure modes are determined. Non-significant failure modes are those that only contribute negligibly to the failure probability of the structure. Significant failure modes are determined by the β-unzipping method by Thoft-Christensen [1]. Two different formulations of this method are described and the two definitions of failure can be used by the first formulation, but only the failure definition based on formation of a mechanism by the second formulation.

The second part of the reliability analysis is an estimate of the failure probability for the structure on the basis of the significant failure modes. The significant failure modes are as usual modelled as elements in a series system (see e.g. Thoft-Christensen & Baker [2]). Several methods to perform this estimate are presented including upper- and lower-bound estimates.

Upper bounds for the failure probability estimate are obtained if the failure mechanisms are used. Lower bounds can be calculated on the basis of series systems where the elements are the non-failed elements in a non-failed structure (see Augusti & Baratta [3]).

2. IDENTIFICATION OF SIGNIFICANT FAILURE MODES

A simple, but also computer-time consuming method to determine significant failure modes is *simulation*, where realizations of relevant stochastic variables are simulated. Corresponding to each set of realizations failure modes are determined by analysis of the structure (see Ferregut-Avila [4]).

Heuristic search has been suggested by several authors. An ideal heuristic search method will disclose the significant failure modes in a sequence with decreasing failure probabilities. Moses [5] has suggested an incremental method where one significant failure mode is determined. A drawback by this method is that all loads are fully correlated. Further failure modes can be determined by changing the strength of the elements (increasing the variance) or by simulating realizations of the strength of the elements (see Gorman [6]). Identification of significant failure modes can also be formulated as an *optimization problem.* For linear structures the safety margins of the mechanisms are linear in load and strength variables. The individual mechanisms are identified by the coefficients to the stochastic variables and the reliability index β for the mechanisms is a non-linear function of these coefficients. Ma & Ang [7] consider identification of significant failure modes as the problem of finding minimum reliability index β. The variables in this optimization problem are the coefficients to the load and strength variables. A local minimum of the reliability index β corresponds to a significant failure mechanism. All mechanisms are linear combinations of a set of fundamental mechanisms. This observation has been used by Ma & Ang [7] in formulating the problem of determining significant failure modes as a non-linear optimization problem. Klingmüller [8] has used a linear programming method to determine significant failure modes. *Failure tree* methods have also been used for this purpose. Each node (branching point) in the failure tree corresponds to a failure element. Murotsu et al. [9] calculates the failure probabilities for all elements and selects those with the highest failure probabilities. These elements are supposed to fail one by one and additional fictitious loads are applied to the structure corresponding to the yield capacity of the failed elements. By this method a number of failure modes are determined. A failure tree method has also been used by Kappler [10].

The *β-unzipping method* (Thoft-Christensen [1], Thoft-Christensen & Sørensen [11]) is a failure tree method to identify significant failure modes. Each branch in the failure tree is chosen on the basis of reliability indices for the elements. In this paper two formulations of the β-unzipping method are presented.

It is assumed that the structure can be modelled by n so-called failure elements with safety margins

$$M_i = \min\{R_i^+ - S_i,\ R_i^- + S_i\} \tag{1}$$

where S_i is a stochastic variable describing the loading of the failure element i and where R_i^+ and R_i^- are stochastic variables describing the yield capacity in »tension» and »compression». Often $R_i^+ = R_i^-$. By the linear elastic analysis

$$S_i = \sum_{j=1}^{k} a_{ij} P_j \tag{2}$$

where P_j, j = 1, 2, . . . , k are the loads on the structure and a_{ij} are the coefficients of influence. The failure criterion for failure element i

$$M_i \leq 0 \tag{3}$$

and the corresponding reliability index β_i can be calculated.

In the β-unzipping method failure elements are assumed to fail one at a time until a mechanism has been formed. After failure of an element the structure is modified by putting the corresponding strength equal to zero and adding a fictitious loading P_{k+i} corresponding to the yield capacity of the ith element in the failure sequence $\{j_1, j_2, \ldots, \}$. Formation of a mechanism can be unveiled by the fact that the corresponding stiffness matrix is singular. In the computer programme used in this investigation two alternatives can be used for the fictitious loads $P_{k+i} = f(R_{j_i})$, $i = 1, 2, \ldots; j_i \in \{1, 2, \ldots, n\}$

$$(1)\quad f(R_{j_i}) = R_{j_i} \tag{4}$$

i.e. equal to the yield capacity of the element.

$$(2)\quad f(R_{j_i}) = \gamma_{j_i} R_{j_i} = \frac{r_{j_i}}{\mu_{R_{j_i}}} R_{j_i} \tag{5}$$

where r_{j_i} is the j_i-coordinate of the design point and where $\mu_{R_{j_i}}$ is the expected value at R_{j_i}. Let element i be the last element to fail before the mechanism is formed. Then the safety margin M_i is equal to the safety margin for the mechanism provided all the elements have failed (during the unzipping) in the same manner as the actual mechanism indicates. Experience shows that this is not always the case. Therefore, instead of (1) the following safety margin is used

$$M_i = \min\{(R_i^+ + \sum_{j=k+1}^{k+m} |a_{ij}|\ P_j - \sum_{j=1}^{k} a_{ij}P_j),\ (R_i^- + \sum_{j=k+1}^{k+m} |a_{ij}|\ P_j + \sum_{j=1}^{k} a_{ij}P_j)\} \tag{6}$$

where m is the number of failed elements. By using $|a_{ij}|$ and not a_{ij} in (6) the correct safety margin is obtained because then all coefficients to yield forces and yield moments are positive.

In figure 1 part of a typical failure tree determined by the β-unzipping method is shown. In each circle the upper number is equal to the number of the failure element and the lower number to the corresponding β-index. Each branch ending in a box indicates a failure mechanism or failure mode. At each branching point the structure is re-analyzed as described above, and failure elements with reliability indices within a prescribed distance from the lowest index for non-failed elements define the new branches of the failure tree. More detailed description of the automatic generation of the failure tree is given in Thoft-Christensen [1] and Thoft-Christensen & Sørensen [11].

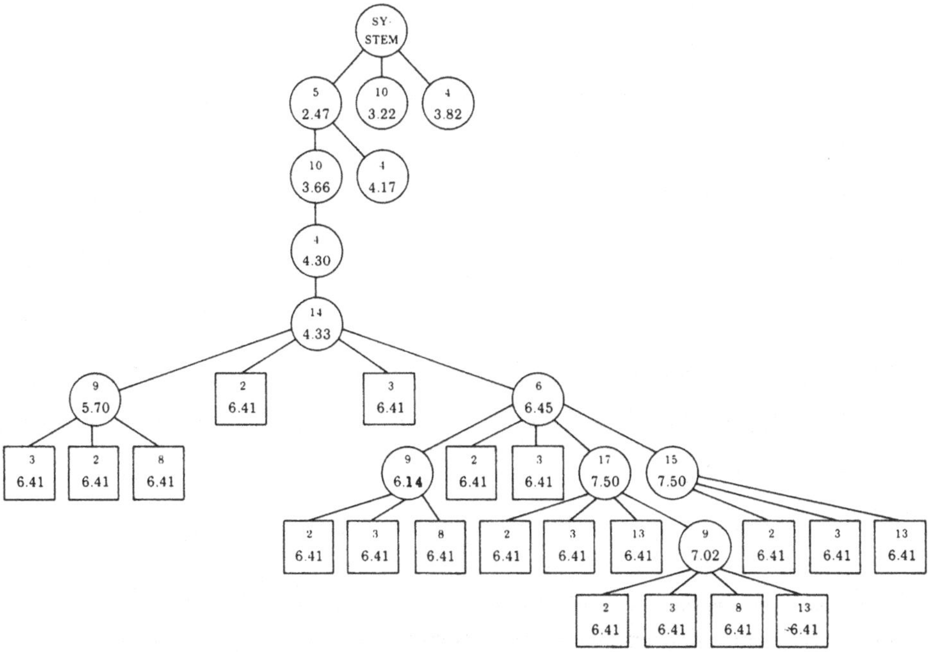

Figure 1. Failure tree for the structure in figure 5. ○ indicates non-failure of the structure and □ failure of the structure.

With regard to the safety margin the first parenthesis in (6) is chosen if

$$\sum_{j=1}^{k} a_{ij}\mu_{P_j} \geqslant 0$$

and the second parenthesis if

$$\sum_{j=1}^{k} a_{ij}\mu_{P_j} \leqslant 0$$

It is of interest to note that the general β-unzipping method can also be used when failure of a structure is defined as failure of a number of elements and also when the behaviour of some elements is modelled by brittle elements ($f(R_{j_i}) = 0$).

In the alternative formulation based on fundamental mechanisms only ductile elements can be taken into account and failure is defined as formation of a mechanism. Fundamental mechanisms can be automatically generated by a method suggested by Watwood [12]. The number of fundamental mechanisms is $m = n - r$, where n is the number of potential failure elements (yield hinges) and r the degree of redundancy. Let the number of real mechanisms be $m_e > 0$, then the number of joint mechanisms is $m_k = m - m_e$. The safety margin for the fundamental mechanism i is

$$M_i = \sum_{j=1}^{n} |a_{ij}|\, R_j - \sum_{j=1}^{k} b_{ij} P_j \tag{7}$$

where $\underset{\sim}{a}$ and $\underset{\sim}{b}$ are influence matrices. Let $\beta_1 \leqslant \beta_2 \leqslant \ldots \leqslant \beta_{m_e}$ be an ordered set of reliability indices for the m_e real mechanisms. Further mechanisms can be constructed by linear combinations of fundamental mechanisms. The corresponding failure margins will be like (7). The problem is now to combine the fundamental mechanisms so that the significant mechanisms can be obtained in an efficient way. Here a failure tree formulation can be used. First the real fundamental mechanism 1 is chosen. Then it is combined with all the other fundamental mechanisms and the corresponding reliability indices are calculated. The smallest

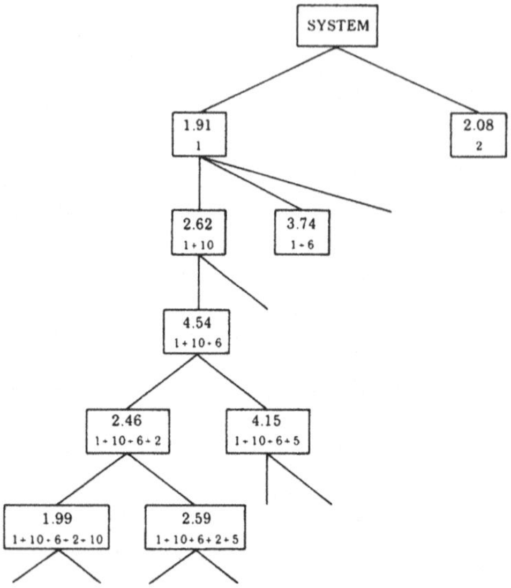

Figure 2. Part of the failure tree for the structure considered in example 1. [2.62 / 1+10] indicates that the mechanism obtained by combining the fundamental mechanisms 1 and 10 has a reliability index β = 2.62.

index β_{min} among those is determined and mechanisms with indices in the interval $[\beta_{min}, \beta_{min} + \epsilon_1]$, where $\epsilon_1 \geq 0$ is a chosen constant, define the branches in the failure tree based on the mechanism in question. The combined mechanisms are now starting points for new branches where each branch symbolizes that the branching point at the end of the branch is constructed by combination of a combined mechanism and a fundamental mechanism. Corresponding to each branching point a new β_{min} value is determined. In general, β_{min} will decrease when more and more mechanisms are combined. The branching is terminated when the number of levels is greater than a number N_{max} or when $\beta_{min} > \beta_u + \epsilon_2$, where β_u is the reliability index for the mechanism and where $\epsilon_2 > 0$. In this way each branching point symbolizes a mechanism. This is illustrated in figure 2, where part of the failure tree for the structure in figure 3 is shown. There are in this case 5 real mechanisms and 5 joint mechanisms.

Significant failure mechanisms are determined on the basis of the failure tree. Let the smallest realiability index for a mechanism be β_0. Then the significant mechanisms are defined as those with reliability indices in the interval $[\beta_0, \beta_0 + \epsilon_3]$ where $\epsilon_3 > 0$.

3. EVALUATION OF THE PROBABILITY OF FAILURE

A measure of the reliability of the structure can be obtained on the basis of the identified significant failure mechanisms by modelling failure of the structure as a series system where the elements are the significant failure mechanisms (see Thoft-Christensen & Baker [2]). The probability of failure then is

$$P_s = P(F_1 \cup F_2 \cup \ldots \cup F_n) \tag{8}$$

where $P(\cdot)$ is the probability measure and F_i, $i = 1, 2, \ldots, n$ is the event that failure occurs by failure of mechanism i. In general F_i and F_j, $i \neq j$, will be correlated due to correlation between the failure elements and due to common failure elements in the failure mechanisms. Therefore, estimate of P_s will involve time-consuming calculation of multi-integrals. However, a number of approximate methods have been suggested (see e.g. Thoft-Christensen & Sørensen [11] or Grimmelt & Schuëller [13]).

Upper and lower bounds for P_s have been suggested by Ditlevsen [14] and Kounias [15]. Calculation of these bounds (here called Ditlevsen bounds) requires estimate of the probability of the intersection of F_i and F_j, i.e. $P(F_i \cap F_j)$. In this paper $P(F_i \cap F_j)$ is calculated by numerical integration. Ang & Ma [16] have suggested an approximate method called the PNET-method, to evaluate P_s. This method is based on a grouping of all elements according to mutual correlation.

In the following, safety margins are assumed normally distributed and linear. Then from (8)

$$P_s = 1 - \int_{-\infty}^{\beta_1} \int_{-\infty}^{\beta_2} \cdots \int_{-\infty}^{\beta_n} \varphi_n(\underset{\sim}{x}\,;\underset{\sim}{\rho})dx_1 dx_2 \ldots dx_n = 1 - \Phi_n(\underset{\sim}{\beta}\,;\underset{\sim}{\rho}) \tag{9}$$

where $\varphi_n(\underset{\sim}{x}\,;\underset{\sim}{\rho})$ is the n-dimensional density function for n standardized stochastic variables and $\underset{\sim}{\rho}$ the matrix of correlation coefficients. When all correlation coefficients are equal, $\rho_{ij} = \rho \geqslant 0$, then (see Dunnet & Sobel [18])

$$P_s = 1 - \int_{-\infty}^{\infty} \varphi(u) \prod_{i=1}^{n} \Phi\left(\frac{\beta_i - \sqrt{\rho}\,u}{\sqrt{1-\rho}}\right) du \tag{10}$$

Plackett [17] has shown that

$$P_s = 1 - \Phi_n(\underset{\sim}{\beta}, \underset{\sim}{\kappa}) - \sum_{i<j} \int_{\kappa_{ij}}^{\rho_{ij}} \frac{\partial \Phi_n(\underset{\sim}{\beta}; t\underset{\sim}{\rho} + (1-t)\underset{\sim}{\kappa})}{\partial \lambda_{ij}} d\lambda_{ij} \tag{11}$$

where $\underset{\sim}{\rho}$ and $\underset{\sim}{\kappa}$ are regular matrices of correlation and

$$\lambda_{ij} = t\rho_{ij} + (1-t)\kappa_{ij} \quad , \quad 0 \leqslant t \leqslant 1 \tag{12}$$

and where e.g.

$$\frac{\partial \Phi_n(\underset{\sim}{\beta}; \underset{\sim}{\lambda})}{\partial \lambda_{12}} = \varphi_2(\beta_1, \beta_2; \lambda_{12}) \Phi_{n-2}(\beta_3 - \mu_{21}, \beta_4 - \mu_{22}, \ldots, \beta_n - \mu_{2\,n-2}; \underset{\sim}{m}_{22}) \tag{13}$$

$\underset{\sim}{\mu}_2$ and $\underset{\sim}{m}_{22}$ are the conditional expected value vector and covariance matrix for $X_3, X_4, \ldots, X_n$ given $X_1 = \beta_1$ and $X_2 = \beta_2$.

A simple approximate value for P_s can be obtained from (10) even if the correlation coefficients are unequal, namely by putting $\rho = \bar{\rho}$, where $\bar{\rho}$ is the average correlation coefficient defined by

$$\bar{\rho} = \frac{1}{n(n-1)} \sum_{i \neq j} \rho_{ij} \tag{14}$$

This corresponds to neglecting the last term in (11) and putting $\kappa_{ij} = \bar{\rho}$

$$P_s \cong 1 - \Phi_n(\underset{\sim}{\beta}\,;\{\bar{\rho}\}) \tag{15}$$

where $\{\bar{\rho}\}$ is a correlation matrix with $\rho_{ij} = \bar{\rho}$, $i \neq j$.

From equation (13) it is seen that

$$0 \leqslant \frac{\partial \Phi_n(\underset{\sim}{\beta}\,;\underset{\sim}{\rho})}{\partial \rho_{ij}} \leqslant \varphi_2(\beta_i, \beta_j; \rho_{ij}) \tag{16}$$

Let $\rho_{min} = \min_{i<j} \{\rho_{ij}\}$ and $\rho_{max} = \max_{i<j} \{\rho_{ij}\}$ then from (11), (13), and (16) the following bounds for Φ_n can be derived (Φ_n is an increasing function of ρ_{ij}):

$$\Phi_n(\underset{\sim}{\beta}; \underset{\sim}{\rho}) \leqslant \min\{\Phi_n(\underset{\sim}{\beta}; \{\rho_{max}\}), [\Phi_n(\underset{\sim}{\beta}; \{\rho_{min}\}) + \sum_{i<j} (\Phi_2(\beta_i, \beta_j; \rho_{ij}) - \Phi_2(\beta_i, \beta_j; \rho_{min}))]\} \tag{17}$$

$$\Phi_n(\underset{\sim}{\beta}; \underset{\sim}{\rho}) \geqslant \max\{\Phi_n(\underset{\sim}{\beta}; \{\rho_{min}\}), [\Phi_n(\underset{\sim}{\beta}; \{\rho_{max}\}) + \sum_{i<j} (\Phi_2(\beta_i, \beta_j; \rho_{ij}) - \Phi_2(\beta_i, \beta_j; \rho_{max}))]\} \tag{18}$$

It follows from (12) - (16) that P_s can also be approximated by

$$P_s \cong 1 - \Phi_n(\underset{\sim}{\beta}\,;\{\bar{\rho}\}) - \sum_{i<j} \{\Phi_2(\beta_i, \beta_j; \rho_{ij}) - \Phi_2(\beta_i, \beta_j; \bar{\rho})\} \tag{19}$$

The approximation (15) is suggested by Thoft-Christensen & Sørensen [19] (see also Ditlevsen [20]). An approximation based on the so-called equivalent correlation coefficient $\bar{\bar{\rho}}$ indirectly defined by

$$P_s \cong 1 - \Phi_n(\underline{\beta}; \{\bar{\rho}\}) - \Phi_2(\beta_i, \beta_j; \rho_{max}) + \Phi_2(\beta_i, \beta_j; \bar{\rho}) = 1 - \Phi_n(\underline{\beta}; \{\bar{\bar{\rho}}\}) \qquad (20)$$

has also been suggested by Thoft-Christensen & Sørensen [19]. This approximation can be derived from (19) by neglecting all terms in the summation except the one with the maximum correlation coefficient.

Let n_s and n_r be the number of kinematically admissible mechanisms and the number of identified significant failure mechanisms, respectively. Then it follows from the upper-bound theorem of plasticity (see Augusti & Baratta [3]) that P_u^{ℓ} given by

$$P_s^{\ell} = P(\bigcup_{i=1}^{n_r} F_i) \qquad (21)$$

is a lower bound of P_s. This can easily be seen from (8) when n is substituted by n_s and n_r. An upper bound of P_s can be derived from the lower-bound theorem of plasticity (see [3])

$$P_s^u = P(\bigcap_{i=1}^{n_m} \bigcup_{j=1}^{n_{u,i}} F'_{ij}) \qquad (22)$$

where n_m is a number of statically admissible stress distributions, $n_{u,i}$ is the number of non-failed failure elements in the structure i corresponding to a statically admissible stress distribution, and F'_{ij} is the event that failure element j in such a structure fails. If the numerical signs in (6) are removed, then F'_{ij} is equal to the event that $M_i \leq 0$ and (22) can be written

$$P_s^u = P(\bigcap_{i=1}^{n_m} \bigcup_{j=1}^{n_{u,i}} (F_{ij}^C \cup F_{ij}^T)) = P(\bigcap_{i=1}^{n_m} F_i^n) \qquad (23)$$

where F_{ij}^C and F_{ij}^T correspond to the linear safety margins in the first and second parenthesis in (6). Note that (23) corresponds to a parallel system with elements which are series systems with $2n_{u,i}$ elements. It will often be useful to use the following upper bound of P_s^u

$$P_s^u \leq \min_i P(F_i^n) \qquad (24)$$

Finally

$$P(\bigcup_{i=1}^{n_r} F_i) \leq P_s \leq \min_i P(\bigcup_{j=1}^{n_{u,i}} (F_{ij}^C \cup F_{ij}^T)) \qquad (25)$$

The bounds in (25) can be estimated by the approximations and bounds for series systems shown earlier in this chapter.

The safety margins corresponding to the right hand side of (25) can only be determined if the first formulation of the β-unzipping method is used.

In the next chapters some of the methods shown above will be used in two examples. The results will be presented by the so-called generalized β-index, β_G, defined by (see Ditlevsen [21])

$$\beta_G = -\Phi^{-1}(P_s) \qquad (26)$$

4. EXAMPLE 1

Consider the frame structure in figure 3 with corresponding expected values and coefficients of variation for the stochastic variables in table 1. Yield moments in the same line are considered fully correlated and yield moments in different lines are mutually independent.

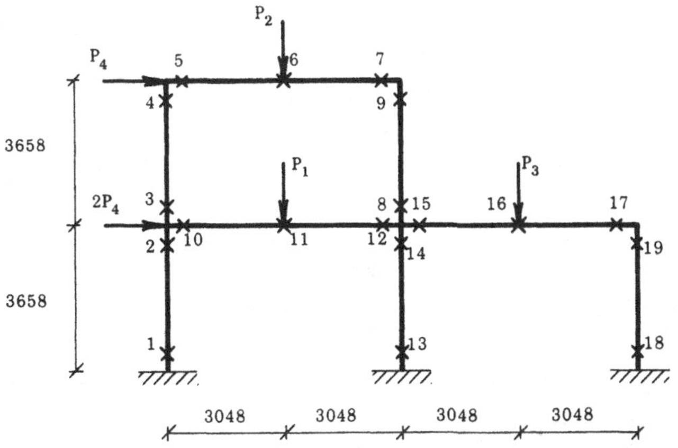

Figure 3. Geometry, loading and potential yield hinges (×) for the structure analysed in example 1.

Table 1.

Variables	Expected values	Coefficients of variation
P_1	169×10^3	0.15
P_2	89×10^3	0.25
P_3	116×10^3	0.25
P_4	31×10^3	0.25
$R_1, R_2, R_{13}, R_{14}, R_{18}, R_{19}$	95×10^6	0.15
R_3, R_4, R_8, R_9	95×10^6	0.15
R_5, R_6, R_7	122×10^6	0.15
R_{10}, R_{11}, R_{12}	204×10^6	0.15
R_{15}, R_{16}, R_{17}	163×10^6	0.15

Table 2. The 12 most significant failure mechanisms for the structure in figure 3.

No.	Failure elements										β
1	1	6	8	9	11	12	13	14	18	19	1.88
2	1	2	13	14	18	19					1.91
3	1	6	9	11	12	13	16	18	19		1.94
4	1	3	11	12	13	14	18	19			1.99
5	4	6	9								1.99
6	15	16	19								1.99
7	1	6	7	8	11	12	13	14	18	19	2.08
8	10	11	12								2.08
9	1	6	7	11	12	13	16	18	19		2.10
10	1	4	9	11	12	13	16	18	19		2.17
11	1	4	8	9	11	12	13	14	18	19	2.18
12	5	6	7								2.18

The 12 most dominant failure mechanisms are shown in table 2 (from Ma & Ang [7]). By the first formulation of the β-unzipping method and by (4) the mechanisms shown in table 3 are determined. The calculations are performed on a CDC Cyber 170-720 computer. 10 of the most dominant failure mechanisms are determined after 2000 sec. computer time, but the remaining mechanisms 10 and 11 are closely correlated with mechanisms 3 and 1 and will therefore have a negligible influence on the reliability of the structure.

Table 3. Failure mechanisms by the β-unzipping method.

Run No.	Significant failure mechanisms	Computer time (sec.)
1	8 12	125
2	7 8 12	150
3	7 8 9 12	300
4	2 4 7 8 9 12	750
5	2 4 5 6 7 8 9 12	1000
6	1 2 3 4 5 6 7 8 9 12	2000

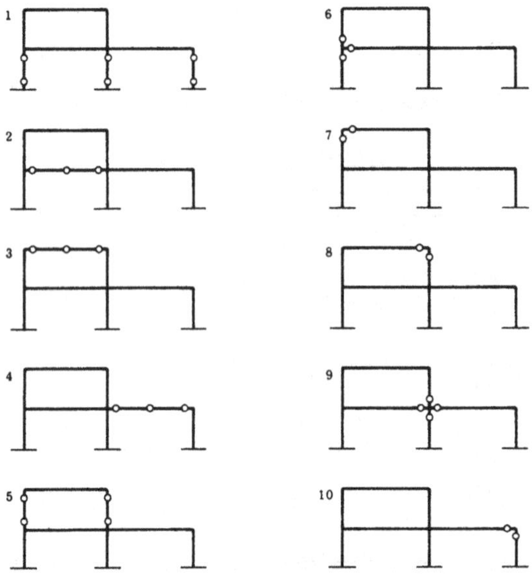

Figure 4. Set of fundamental mechanisms used in example 1.

Table 4. Generalized reliability index β_G for the structure in figure 3.

Run No.	Ditlevsen bounds	$\beta_G(\bar{\rho})$	$\beta_G(\bar{\bar{\rho}})$	PNET
1	1.83 - 1.83	1.79	1.83	1.83
2	1.69 - 1.69	1.67	1.69	1.62
3	1.62 - 1.68	1.58	1.66	1.62
4	1.39 - 1.52	1.35	1.42	1.50
5	1.17 - 1.30	1.14	1.20	1.29
6	1.15 - 1.36	1.08	1.16	1.15

By the second formulation of the β-unzipping method based on the set of fundamental mechanisms shown in fig. 4 all 12 significant failure mechanisms were identified in only 17 sec. $N_{max} = 20$, $\epsilon_1 = 0.5$, $\epsilon_2 = 2$ and $\epsilon_3 = 0.3$ were used. Although the choice of fundamental mechanisms and of N_{max}, ϵ_1, ϵ_2, ϵ_3 values may be of importance for this result it seems to indicate that the formulation based on fundamental mechanisms is superior to the more direct first formulation. The main reason for the reduced computer time is of course that only one analysis of the structure is necessary by the second formulation, whereas the first formulation requires such an analysis at each branching point.

The generalized reliability index β_G will in this example be estimated on the basis of some of the methods presented in chapter 3. The mechanisms in table 3 are used as elements in a series system and the results are shown in table 4.

Ma & Ang [7] have calculated the Ditlevsen bounds on the basis of all different real mechanisms with the result $0.53 \leq \beta_G \leq 1.47$ and by Monte-Carlo simulation determined the estimate $\beta_G = 1.20$.

On the basis of the safety margins determined by the first formulation of the β-unzipping method one gets $-1.17 \leq \beta_G \leq 1.30$. These bounds are determined as Ditlevsen bounds for the series systems in the right and left hand side of (25). The lower bound is for a structure with failure in elements 1, 6, 11, 12, 15, 18, and 19. Computer time is 2000 seconds.

The upper bound 1.30 for β_G calculated by (25) is close to the simulated estimate 1.20, but the lower bound is useless. It follows from table 4 that the approximate estimate $\beta_G(\bar{\bar{\rho}}) = 1.16$ and β_G(PNET) = 1.15 only differ by 3% from the simulated estimate.

5. EXAMPLE 2

Consider the structure in figure 5 and the data in table 5. All stochastic variables are considered independent. This structure has been investigated by Grimmelt & Schuëller [13]. By the first formulation of the β-unzipping method and (6) all significant failure mechanisms (see table 6) were identified in 154 seconds. The number of fundamental mechanisms is $n - r = 17 - 9 = 8$. By the second formulation of the β-unzipping method based on the set of fundamental mechanisms in figure 6 and the same data as in example 1 all significant failure mechanisms were identified in 8 seconds. This result confirms the conclusion in example 1.

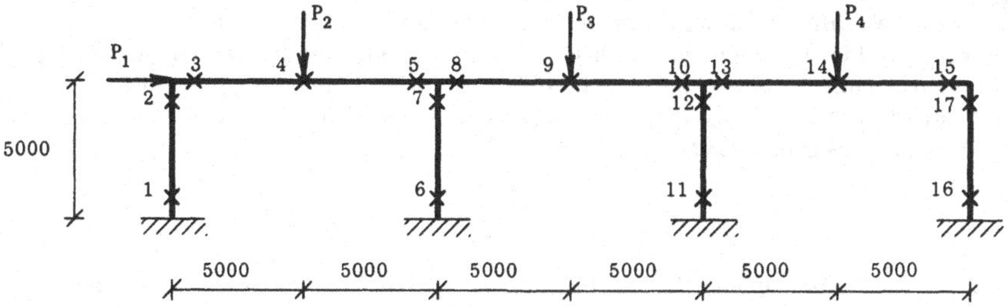

Figure 5. Geometry, loading and potential yield hinges for the structure analysed in example 2.

Table 5.

Variables	Expected values	Coefficient of variation
P_1	$50 \cdot 10^3$	0.1
P_2, P_3, P_4	$40 \cdot 10^3$	0.1
$R_1, R_2, \ldots, R_{17}$	$101 \cdot 10^6$	0.1

Table 6. Significant failure mechanisms

No.	Failure elements			β
1	2	4	5	6.41
2	3	4	5	6.41
3	8	9	10	6.41
4	13	14	15	6.41
5	13	14	17	6.41

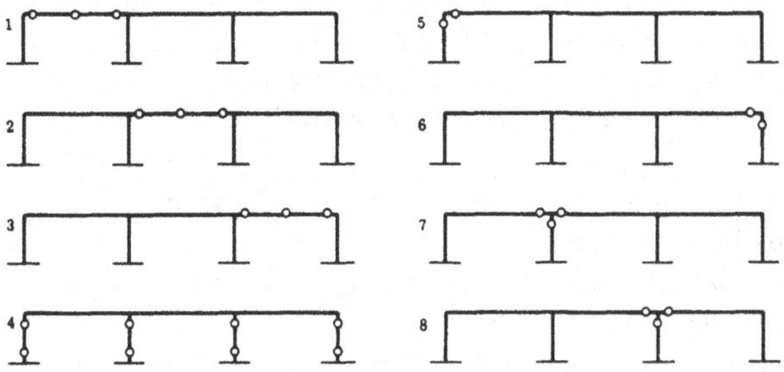

Figure 6. Set of fundamental mechanisms used in example 2.

On the basis of the significant mechanisms in table 6 the Ditlevsen bounds are $6.17 \leq \beta_G \leq 6.17$. Further, $\beta_G(\bar{\bar{\rho}}) = 6.16$ and $\beta_G(\text{PNET}) = 6.24$. These values are close to the results by Grimmelt & Schuëller [13] calculated on the basis of all possible mechanisms. It can therefore be concluded that the 5 mechanisms in table 6 really are the most dominant.

By the first formulation of the β-unzipping method the following bounds corresponding to the right hand side of (25) can be calculated: $3.94 \leq \beta_G \leq 3.94$. They correspond to a structure with failure in elements 6, 9, 18, and 19 or 6, 9, 19, and 20. Then by (25) one gets $3.94 \leq \beta_G \leq 6.17$. Again only the upper bound is close to the correct value.

6. CONCLUSIONS

In the first part of this paper two formulations of the so-called β-unzipping method to identify significant failure modes were presented. Both methods are failure tree methods. The efficiency of the two methods is evaluated on the basis of two examples. The conclusion is that the second formulation seems to be superior when it can be used. It is based on fundamental mechanisms and is therefore only applicable to failure definitions based on mechanisms. The first formulation is more general. It can be used for ductile and brittle elements and failure of the structure need not be defined as formation of a mechanism.

In the second part of the paper the estimate of the failure probability is investigated. A lower bound for the probability of failure is calculated on the basis of the identified significant failure mechanisms. These failure mechanisms are elements in a series system. A number of different methods to estimate the failure probability of series systems are discussed, namely Ditlevsen bounds, the PNET-method, and approximate methods based on the average and the equivalent correlation coefficient. New bounds and a new approximation are suggested. The reliability of two different structures are estimated by some of these methods and it can be concluded that the upper bound for the failure probability P_s (lower bound for the generalized index β_G) is a bad estimate for the correct value. However, the lower bound for P_s (upper bound for β_G) is close to the correct value. For both structures approximate estimates of β_G based on the equivalent correlation coefficient $\bar{\bar{\rho}}$ and the PNET-method are good estimates.

7. REFERENCES

[1] Thoft-Christensen, P.: *The β-unzipping Method.* Institute of Building Technology and Structural Engineering, Aalborg University Centre, Report 8207, Aalborg 1982.

[2] Thoft-Christensen, P. & Baker, M. J.: *Structural Reliability and Its Applications.* Springer Verlag, Berlin - Heidelberg - New York, 1982.

[3] Augusti, G. & Baratta, A.: *Limit Analysis of Structures with Stochastic Strength Variations.* Journal of Structural Mechanics, Vol. 1, No. 1, 1972, pp. 43-62.

[4] Ferregut-Avila, C. M.: *Reliability Analysis of Elasto-Plastic Structures.* NATO Advanced Study Institute, P. Thoft-Christensen (ed.), Martinus Nijhoff, The Netherlands, 1983, pp. 445-451.

[5] Moses, F.: *Structural System Reliability and Optimization.* Computers & Structures, Vol. 7, 1977, pp. 283-290.

[6] Gorman, M. R.: *Reliability of Structural Systems.* Report No. 79-2, Case Western Reserve University, Ohio, Ph.D. Report, 1979.

[7] Ma, H.-F. & Ang, A. H.-S.: *Reliability Analysis of Ductile Structural Systems.* Civil Eng. Studies, Structural Research Series No. 494, University of Illinois, Urbana, August 1981.

[8] Klingmüller, O.: *Anwendung der Traglastberechnung für die Beurteilung der Sicherheit von Konstruktionen.* Forschungsberichte aus dem Fachbereich Bauwesen. Gesamthochschule Essen, Heft 9, Sept. 1979.

[9] Murotsu, Y., Okada, H., Yonezawa, M. & Taguchi, K.: *Reliability Assessment of Redundant Structure.* ICOSSAR '81, Trondheim, Norway, 1981, pp. 315-329.

[10] Kappler, H.: *Beitrag zur Zuverlässigkeitstheorie von Tragwerken unter Berücksichtigung nichtlineare Verhaltens.* Dissertation, Technische Universität München, 1980.

[11] Thoft-Christensen, P. & Sørensen, J. D.: *Calculation of Failure Probabilities of Ductile Structures by the β-unzipping Method.* Institute of Building Technology and Structural Engineering, Aalborg University Centre, Report 8208, Aalborg 1982.

[12] Watwood, V. B.: *Mechanism Generation for Limit Analysis of Frames.* ASCE, Journal of the Structural Division, Vol. 105, NO. ST1, Jan. 1979, pp. 1-15.

[13] Grimmelt, M. & Schuëller, G. I.: *Benchmark Study on Methods to Determine Collapse Failure Probabilities of Redundant Structures.* SFB 96, Heft 51, Technische Universität München, 1981.

[14] Ditlevsen, O.: *Narrow Reliability Bounds for Structural Systems.* Journal of Structural Mechanics, Vol. 7, No. 4, 1979, pp. 453-472.

[15] Kounias, E. G.: *Bounds for the Probability of a Union, with Applications.* The Annals of Mathematical Statistics, Vol. 39, 1968, pp. 2154-2158.

[16] Ang, A. H.-S. & Ma, H.-F.: *On the Reliability Analysis of Framed Structures.* Proc. ASCE Specialty Conference on Probabilistic Mechanics and Structural Reliability, Tucson 1979, pp. 106-111.

[17] Plackett, R. K.: *A Reduction Formula for Normal Multivariate Integrals.* Biometrika, Vol. 41, 1954, pp. 351-360.

[18] Dunnet, C. W. & Sobel, M.: *Approximations to the Probability Integral and Certain Percentage Points of a Multivariate Analogue of Student's t-Distribution.* Biometrika, Vol. 42, 1955, pp. 258-260.

[19] Thoft-Christensen, P. & Sørensen, J. D.: *Reliability of Structural Systems with Correlated Elements.* Applied Mathematical Modelling, Vol. 6, 1982, pp. 171-178.

[20] Ditlevsen, O.: *Taylor Expansion of Series System Reliability.* DCAMM Report No. 235, Lyngby 1982.

[21] Ditlevsen, O.: *Generalized Second Moment Reliability Index.* Journal of Structural Mechanics, Vol. 7, 1979, pp. 435-451.

THRESHOLD CROSSINGS IN NONLINEAR SYSTEMS AND SHIP CAPSIZE PREVENTION

J.F. Dunne* and J.H. Wright
Department of Engineering Mathematics
University of Bristol
Bristol BS8 1TR
United Kingdom

Summary

Ships capsize when the roll response in a random seaway exceeds a critical threshold. The ability to predict the frequency of occurrence of crossings of high roll thresholds should enable design criteria for capsize prevention to be formulated on a more scientific basis than has been possible before.

We present a method for making such predictions. A conventional model of roll motion is adopted: a second-order system with nonlinear damping and a restoring function with zeros at finite positive and negative displacements, corresponding to capsize points. The roll excitation is assumed to be a stationary Gaussian process with known spectral density function. The mean time between upcrossings of a high threshold by the response process is predicted by an iterative scheme. The scheme assumes that the crossing properties of the response to the correlated excitation can be approximated by replacing the excitation with a white noise process of suitable intensity. Extensive simulations show that this assumption can give good results, even when the excitation is a rather narrow-band process. The prediction method reveals a remarkable sensitivity of risk of capsize to changes in damping and restoring functions.

1. INTRODUCTION

Following the common practice in research into ship roll motion, we adopt an equation of the form

$$\ddot{\theta} + D(\dot{\theta}) + R(\theta) = n(t) \qquad (1)$$

where θ denotes the angle of roll, $D(\dot{\theta})$ is a nonlinear damping function, and $R(\theta)$ is a nonlinear restoring function with zeros at $\pm\theta_v$, the angle

* Now at: Koninklijke/Shell Exploratie en Produktie Laboratorium, Postbus 60, 2280 AB, Rijswijk, Netherlands.

of vanishing stability. The excitation n(t) is realistically assumed to be stationary and Gaussian with spectrum $S(\omega)$. For simplicity, let the time scale be such that the undamped natural roll frequency is unity. The restoring function can be written as an odd-power polynomial

$$R(\theta) = \theta + k_3\theta^3 + k_5\theta^5 + \dots \tag{2}$$

and the damping function as a cubic polynomial [1]

$$D(\dot{\theta}) = 2\xi\dot{\theta} + c_3\dot{\theta}^3 \tag{3}$$

Capsize is a serious problem resulting in many losses of small ships such as trawlers and coasting vessels. Existing design criteria [2] specify lower bounds on various aspects of the restoring function, namely areas up to certain angles, the ordinate at a specific angle, the angle of maximum restoring moment, and the slope at the origin. These stipulations ensure that normally when a wave rolls a ship over it has sufficient static stability to right itself. For continuing dynamic rolling in severe seas, however, these criteria have proved to be inadequate. For instance they make no reference to the damping function, and as we shall show the frequency of occurrence of dangerous roll angles is sensitively dependent upon roll damping.

The major difficulty with the analysis of system (1) - (3) is that the response $\theta(t)$ is a nonlinear tranformation of a non-white process [3]. The method of equivalent linearisation [4, 5] allows response statistics to be obtained for non-white excitation, but the study of capsize presupposes large-amplitude responses for which the system nonlinearities are dominant. The Fokker-Planck method on the other hand [6, 7] requires either a white noise excitation or a simple transformation thereof, and this does not fit the wave excitation experienced by ships. In the present approach these two methods are combined into an iterative scheme [8] which allows useful results to be obtained for the nonlinear, non-white case.

2. THRESHOLD-CROSSING PREDICTION METHOD

Let $f(\theta, \dot{\theta})$ represent the stationary joint probability density function of θ and $\dot{\theta}$. The rate at which upcrossings occur for the $\theta(t)$ - process at threshold $a > 0$ is [9]

$$\lambda(a) = \int_0^\infty \dot{\theta} f(a, \dot{\theta}) d\dot{\theta} \tag{4}$$

and the mean time between upcrossings is then

$$\mu(a) = [\lambda(a)]^{-1} \tag{5}$$

By symmetry this is also the mean downcrossing time for the threshold -a. For linear damping and white noise ($S(\omega) \equiv J$) the Fokker-Planck can be solved [9]:

$$f(\theta, \dot{\theta}) = A\exp\left\{-\frac{4\xi}{J}\left[\int_0^{\theta} R(u)du + \tfrac{1}{2}\dot{\theta}^2\right]\right\} \tag{6}$$

where A is a normalising constant. Using this in (4), (5) gives

$$\mu(a) = \frac{\displaystyle\int_{-\infty}^{\infty} \exp\left\{-\frac{4\xi}{J}\int_0^{v} R(u)du\right\} dv}{\displaystyle\tfrac{1}{2}\sqrt{\frac{J}{2\pi\xi}}\exp\left\{-\frac{4\xi}{J}\int_0^{a} R(u)du\right\}} \tag{7}$$

The structure of the prediction method is shown in Fig.1. The roll equation (1) is first linearised [4, 5] to give

$$\ddot{\theta} + 2\xi_e\dot{\theta} + k_e\theta = n(t) \tag{8}$$

where $\xi_e = \xi + \tfrac{1}{2}c_3E(\dot{\theta}^4)/E(\dot{\theta}^2)$

and $k_e = E[\theta R(\theta)]/E(\theta^2)$

These are approximate expressions from which the moments $E(\theta\dot{\theta})$, $E(\dot{\theta}R(\theta))$ and $E(\theta\dot{\theta}^3)$ have been eliminated. For the special case represented by (6) these are all zero, and simulations confirm that their values are small for more realistic situations also. Computation of mean upcrossing times for the linearised system with correlated and white excitation is straightforward [10]:

$$\mu_{S(\omega)}(a) = 2\pi\frac{\sigma_{S1}}{\sigma_{S2}}\exp\left(\frac{a^2}{2\sigma_{S1}^2}\right), \quad \mu_J(a) = 2\pi\frac{\sigma_{J1}}{\sigma_{J2}}\exp\left(\frac{a^2}{2\sigma_{J1}^2}\right)$$

where $\sigma_{S1}^2 = \int_{-\infty}^{\infty}[H(\omega)]^2S(\omega)d\omega$, $\sigma_{S2}^2 = \int_{-\infty}^{\infty}\omega^2[H(\omega)]^2S(\omega)d\omega$

$$\sigma_{J1}^2 = J\int_{-\infty}^{\infty}[H(\omega)]^2d\omega, \qquad \sigma_{J2}^2 = J\int_{-\infty}^{\infty}\omega^2[H(\omega)]^2d\omega$$

$$H(\omega) = \left[(k_e - \omega^2)^2 + 4\xi_e{}^2\omega^2\right]^{-\frac{1}{2}}$$

The value of J is chosen to minimise the square-error criterion

$$\int_0^{a_{max}} [\mu_{S(\omega)}(a) - \mu_J(a)]^2 da$$

for a suitable high threshold a_{max}. This intensity of white noise is now applied to the system with nonlinear restoring function:

$$\ddot{\theta} + 2\xi_e\dot{\theta} + R(\theta) = \sqrt{J}Z(t)$$

where in order to avoid the Ito informulation, Z(t) is regarded as the generalised derivative of a Wiener process [11]. For this system the joint density function (6) applies, from which the moments for equivalent linearisation (8) can be obtained. The iteration continues until J converges, and the predicted mean crossing times for the nonlinear system (1) with excitation $S(\omega)$ are then given by (7) with $\xi = \xi_e$.

3. COMPARISON WITH SIMULATED RESULTS

The simulations consist of direct integration of the roll equation by conventional methods, the excitation being derived from a linear congruential random number generator with Box-Müller transformation [12] and filtering [13] to provide a correlated normal sequence. Only crossings of high thresholds are of interest, because these represent potentially dangerous roll conditions, but such crossings are rare. Excitation intensities substantially higher than those which would be encountered in actual seas are needed if statistically useful samples of data are to be obtained in a reasonable time. In addition, numerical instabilities and problems of convergence arise which will be discussed elsewhere. The exact result (7) for white noise is a valuable check on the accuracy of the simulation procedure.

For all the results shown here a cubic resoring function (2) is used with $k_3 = -0.5$. Only weak dependence of the accuracy of predictions on strength of nonlinearity in restoration has been found.

Fig.2 shows mean crossing time against threshold for white noise ($J = 0.07$) with $\xi = 0.1$ For linear damping ($c_3 = 0$) the good agreement with the exact result (7) shows that the simulation procedure is sound. As a further check, the response of the linear system ($c_3 = k_3 = 0$) is also shown. Agreement remains good as nonlinear damping is introduced. The restoring function $R(\theta)$ is plotted on the same horizontal scale to show that the simulations are concentrated within the nonlinear region of interest. A 70% confidence interval for the sample mean is superimposed upon one of the simulation points, indicating the sparsity of data for the most extreme threshold.

The following test spectrum (Fig.3a) allows the excitation to be varied from white noise to a rather narrow-band process by changing a single parameter:

$$S(\omega) = \begin{cases} \dfrac{pJ}{1 + (p-1)|\omega - \omega_0|} & \omega < \omega_c \\ 0 & \omega > \omega_c \end{cases}$$

where J, ω_0 and p are scale, location and shape parameters respectively. The value of $\omega_0 = 0.75$ was chosen to give significant variation in spectral amplitude within the system passband. Fig.3b shows results with varying input bandwidth, for $\xi = 0.1$, $c_3 = 0.6$. The agreement is reasonably good, even for $p = 2$. As a further test, the results in Fig.4b were obtained using the more realistic spectrum [14]

$$S(\omega) = \frac{r}{\omega^8} e^{-0.2/\omega^4}$$

This is shown in Fig.4a for $r = 0.006$, together with the actual spectrum of the excitation generated as described above. For such a narrow spectrum this method is not ideal, and for an accurate comparison the approximating spectrum was also adopted for the predictions. This time $\xi = 0.035$, $c_3 = 0.4$. Figs.3b and 4b indicate that the accuracy of the prediction method is only weakly dependent upon input bandwidth.

4. SENSITIVITY OF CROSSING STATISTICS

The crossing rate at a high threshold is sensitively dependent upon intensity of excitation. This is indicated in Fig.4b, in which a 17% reduction in intensity causes approximately a four-fold increase in mean crossing times.

Sensitivity to aspects of the $R(\theta)$ curve is illustrated in Fig.5 in which the inset shows three curves with varying initial slope and position of peak, and the same total area, height of peak, and vanishing angle. The excitation has a more realistic level although still severe, so the vertical scale is in hours rather than seconds and simulation is now impractical. A particularly notorious loss, that of the Gaul in 1974 [15], occurred in circumstances corresponding closely to curve (2) in Fig.5. This predicts a mean crossing time to a 60^0 roll angle of approximately 1½ hours, a clear indication of danger.

Roll damping exhibits the most extreme sensitivity. For the three damping curves in Fig.6, the mean crossing times at the vanishing angle are 28 minutes, 83 days, and 1600 years. Quantitatively these figures do not mean a great deal, because the roll equation will not accurately model such extreme behaviour. Qualitatively, however, the increase of 7½ orders of magnitude brought about by a 10-fold increase in damping demonstrates the importance of including this parameter in future design criteria for capsize prevention.

5. CONCLUSIONS

The prediction method assumes that for the purpose of predicting threshold-crossings, the correlated excitation may be replaced by a white-noise process. The intensity is optimised not at the input to the system, where the excitation may in fact be narrow-band, but at the output, in terms of the response statistic of interest. The main objective is reliable prediction of crossings of high thresholds, between the peak of the restoring function R(θ) and the vanishing angle. Because such crossings are rare in practice, the simulations used for testing the method involved unrealistically severe excitation which reduced the time-scale (in real time) from days to seconds. Despite this, the prediction method was of order 1000 times faster than simulation. Agreement is generally good, but further development of the method is required to reduce errors when the damping is strongly non-linear.

The speed and flexibility of the method should enable the risk of capsize to be assessed while a ship is being designed, with limited recourse to simulations and model experiments for testing purposes. The method provides a direct link between the phenomenon which is to be controlled, and the parameters through which control is to be exercised. The existing design criteria consist of a set of *ad hoc* rules, which can be evaluated and enhanced now this link is established. This presupposes further development of the roll equation to incorporate parametric excitation and other modes of motion, further research into the relationship between wave height and roll excitation, and the extension of the prediction method to accommodate these advances. It also presupposes some general agreement on an acceptable level of risk.

References

1. J.F. DALZELL: "A note on the form of ship roll damping", Journal of Ship Research 22(1978) 178-185.
2. "Recommendations on Intact Stability of Fishing Vessels", A168, IMCO 1968.
3. J.B. ROBERTS: "A stochasic theory of nonlinear ship rolling in irregular seas", Journal of Ship Research 26(1982) 229-245.
4. T.K. CAUGHEY: "Equivalent linearisation techniques", Journal of the Acoustical Society of America 35(1963) 1706-1711.
5. L.VASSILOPOULOS: "Ship rolling at zero speed in random beam seas with nonlinear damping and restoration", Journal of Ship Research 15 (1971) 289-294.
6. M.R. HADDARA: "A modified approach for the application of the Fokker-Planck equation to nonlinear ship motion in random waves", International Shipbuilding Progress 21 (1974) 283-287.

7. B.J.DE JONG: "Some aspects of ship motions in irregular beam and longitudinal waves", Ph.D. Dissertation, Delft, 1970.

8. J.F. DUNNE: "Ship roll response and capsize prediction in random beam seas", Ph.D. Dissertation, Bristol, 1982.

9. T.T. SOONG: *Random differential equations in science and engineerin* Academic Press, 1973.

10. H. CRAMER and M.R. LEADBETTER: *Stationary and related stochastic processes,* John Wiley, 1967.

11. L. ARNOLD: *Stochastic differential equations: theory and applications,* Wiley Interscience, 1973.

12. W.J. KENNEDY and J.E. GENTLE: *Statistical computing,* Marcel Dekker, 1980.

13. E.M. SCHEUER and D.S. STOLLER: "On the generation of normal random vectors", Technometrics 4 (1962) 278-281.

14. M.K. OCHI: "Wave statistics for the design of ships and ocean structures", Trans. SNAME 86 (1978) 47-76.

15. A. MORRALL: "The GAUL disaster: an investigation into the loss of a large stern trawler", Trans RINA 123 (1981) 391-440.

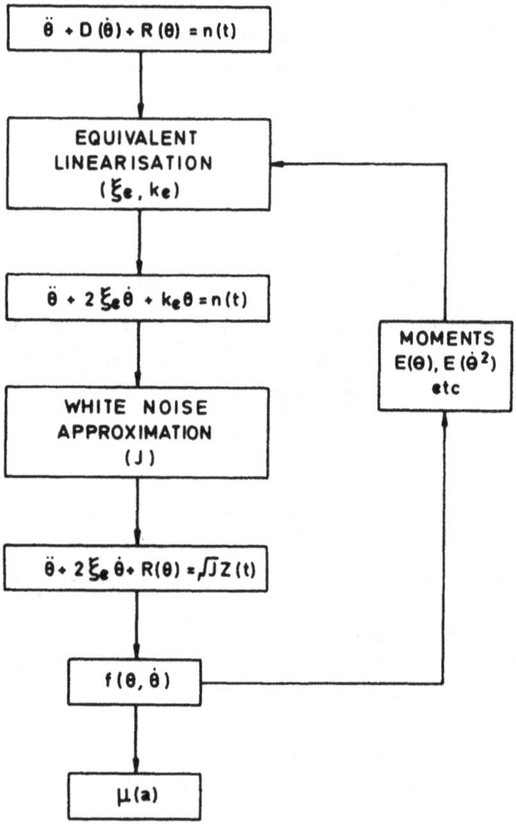

Fig.1 Structure of the prediction method

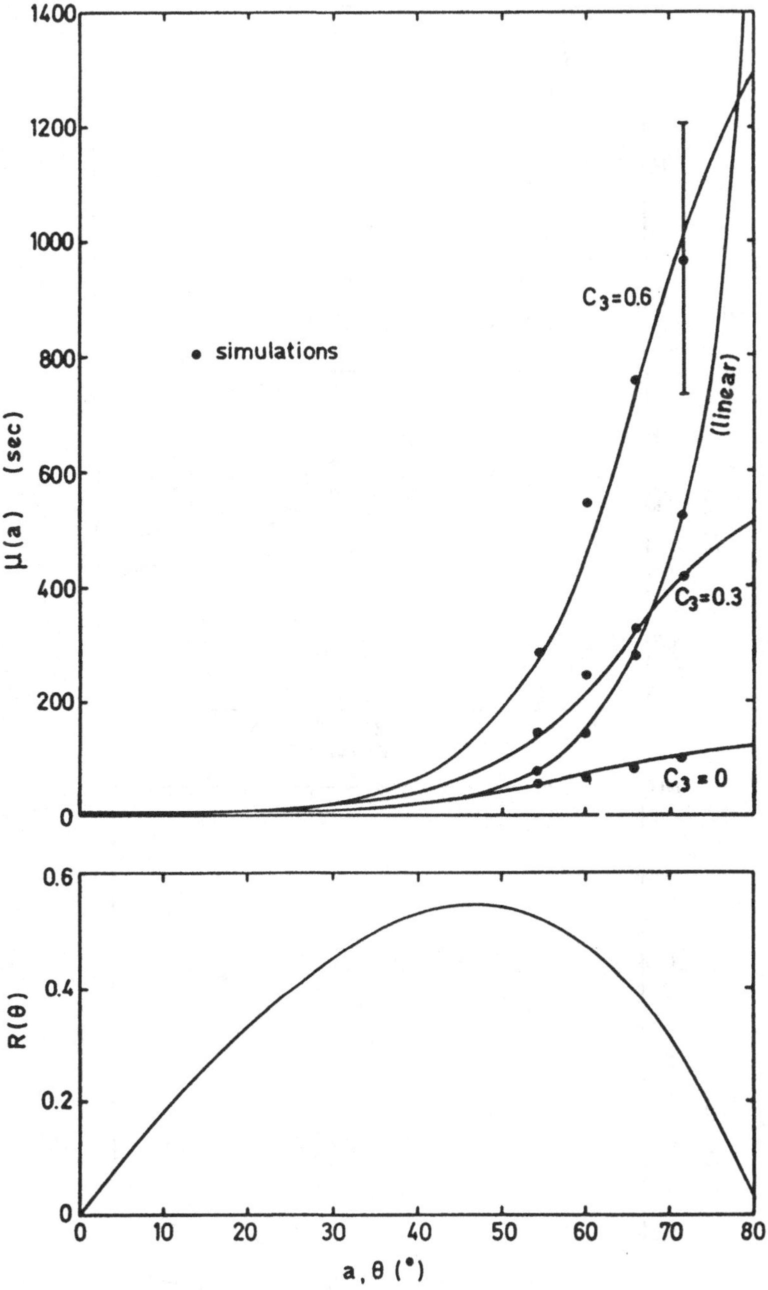

Fig.2 Results for white noise excitation

(a)

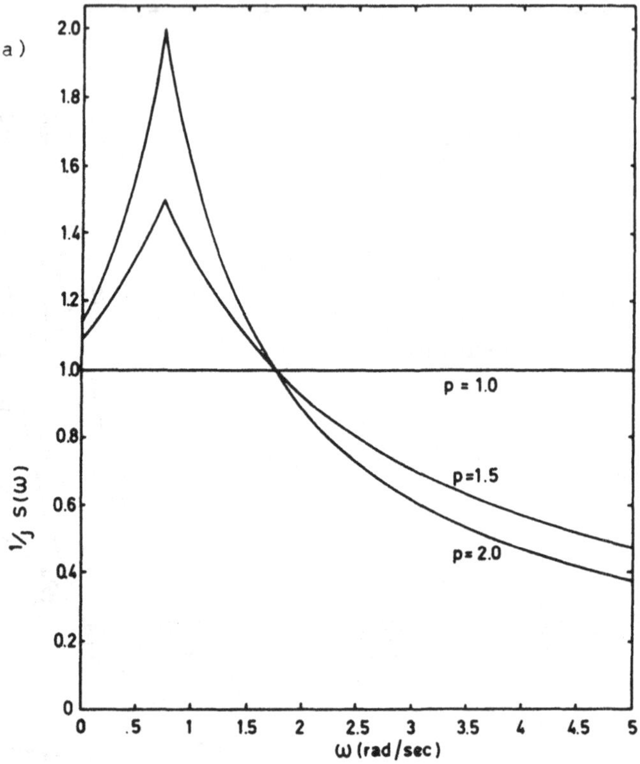

(b)

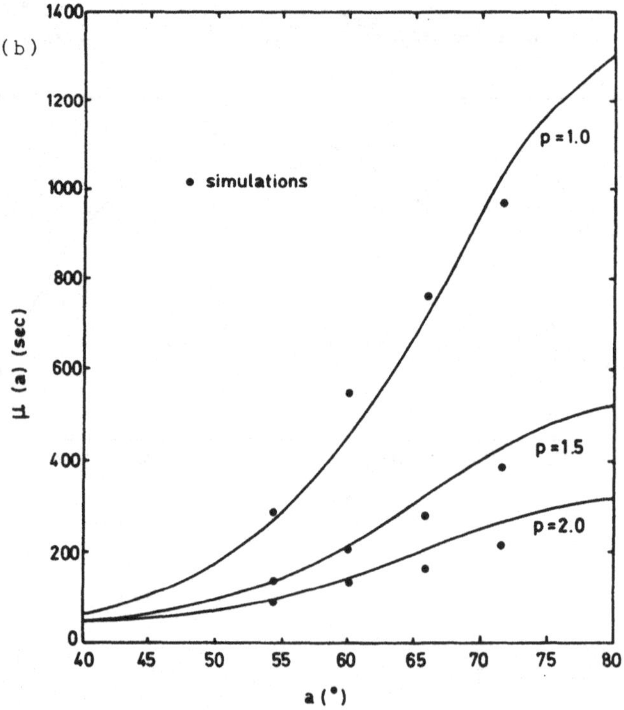

Fig.3 (a) Test spectrum, (b) Results

(a)

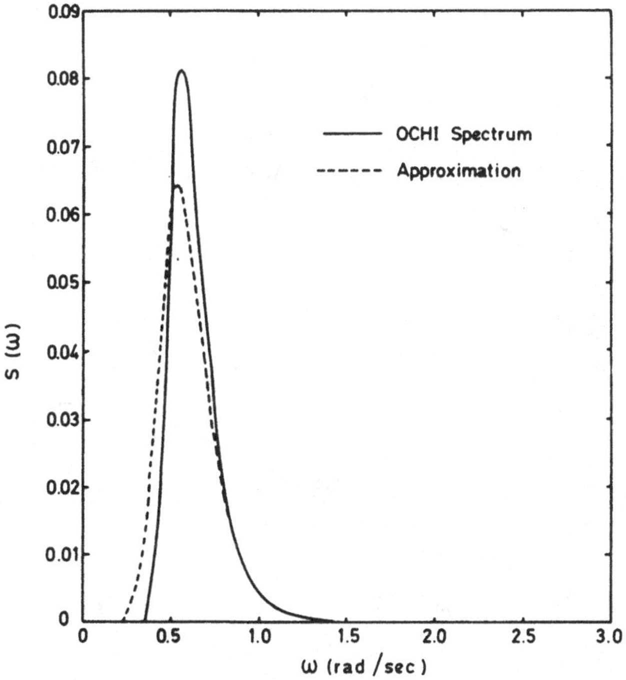

(b)

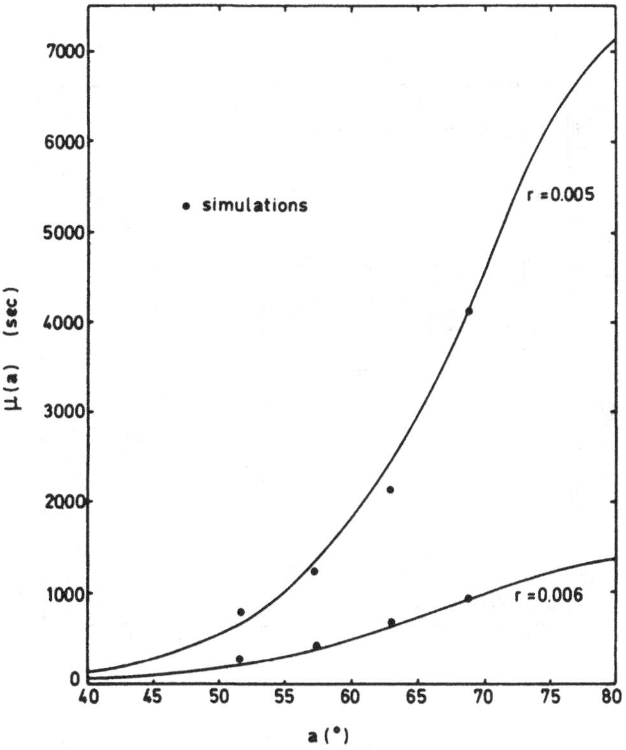

Fig.4 (a) Ochi spectrum, (b) Results

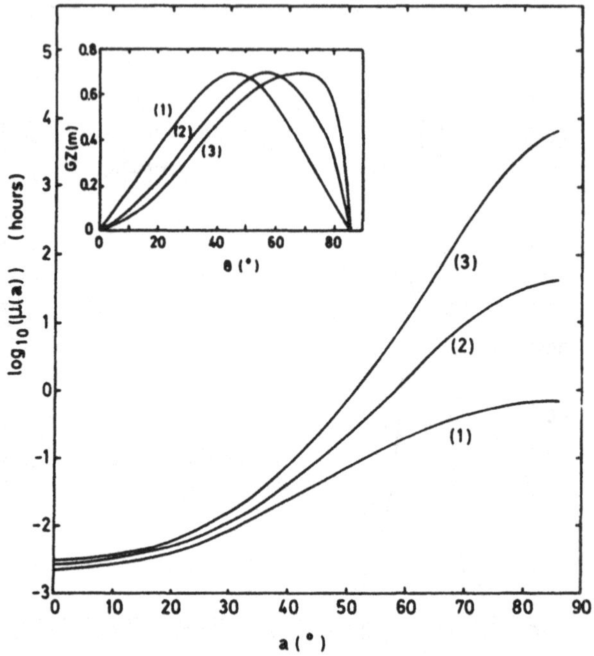

Fig.5 Sensitivity to restoring function

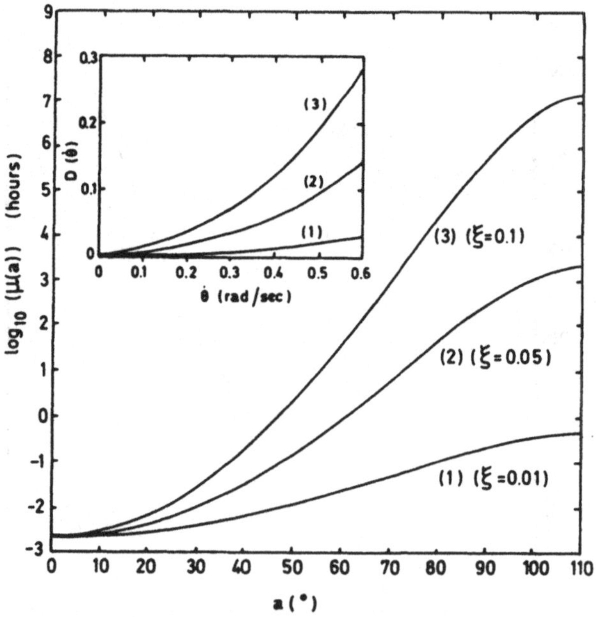

Fig.6 Sensitivity to damping function

ASYMPTOTIC APPROXIMATIONS FOR MULTINORMAL DOMAIN AND SURFACE INTEGRALS

K.Breitung
c/o Lehrstuhl für Massivbau
Technische Universität München
Arcisstr.21
D-8000 München 2, FRG

1. INTRODUCTION

In many reliability problems it is necessary to calculate the probability content of parts of a n-dimensional probability space.Especially,if the dimension of the space is large,the direct computation by numerical or Monte-Carlo integration is too time consuming.Therefore,approximation methods have been developed (/4/,/5/,/7/,/8/,/9/,/13/).The basic idea of these methods is to transform the random variables into a probability space,where they are mutually independent, standardized,normal distributed random variables and then to approximate the integral over the n-dimensional standard normal density by simplifying the boundary of the integration domain.For independent variables with continuous densities the transformation is simple /8/.A possible approach in the case of dependent variables is also described in /8/.

In the following it will be assumed,that such a transformation was performed.The random influences acting on the structure or system are then described by a n-dimensional random vector with independent,standard normal distributed components.The reliability depends on the behaviour of the random influences.If they stay within certain limits,the structure will remain intact,if not,it will fail.The safe domain (the values of the random influences,which do not cause a failure) is given by the limit state function $g(\underline{x})$ ($\underline{x}=(x_1,..,x_n)$) by the points with $g(\underline{x})>0$.The space is divided into the safe domain $S=\{\underline{x};g(\underline{x})>0\}$,the failure domain $F=\{\underline{x};g(\underline{x})<0\}$ and the limit state surface $G=\{\underline{x};g(\underline{x})=0\}$.In the time-invariant case the probability of failure P_f is given by the integral :

$$P_f = \int_{g(\underline{x})<0} p_n(\underline{x})d\underline{x} \tag{1}$$

with $p_n(\underline{x})=(2\pi)^{-\frac{n}{2}}\exp(-(\sum_{i=1}^{n} x_i^2)/2)$ the n-dimensional standard normal density. The approximations are now obtained in the following way.First,all the points $\underline{x}_1,..,\underline{x}_k$ on the limit surface G with minimal distance to the origin are computed.(A globally convergent algorithm to minimize functions on hypersurfaces which can be used to find these points,is given in /12/. The algorithm proposed in /13/ for this task is a simplification of the method described in /12/ and convergence problems may occur.)Approximations are now obtained by replacing $g(\underline{x})$ at the points $\underline{x}_1,..,\underline{x}_k$ by its Taylor expansion to the first or second order,taking the so found hyperplanes and quadratic surfaces as approximations for the surface G and computing the probability content of the domains bounded by these surfaces or planes as estimate of P_f.The usual method described in /8/,/9/ and /13/ ("first-order-reliability-method") is to take only the first derivatives of $g(\underline{x})$,i.e. replace the surface G by a hyperplane.Quadratic expansions have been proposed in /7/.In the cited papers no theoretical justification for the methods is given.

For a fixed failure domain the error of the approximation can be found only by comparison with the exact result obtained by numerical integration.But in the case,that the probability content of the failure domain is small as usual in reliability problems methods of asymptotic analysis (see the appendix) can be used to obtain approximations and error estimates.The approximations derived in the following are asymptotic approximations for $\beta\to\infty$ (β=distance of G to the origin),i.e. for $\beta\to\infty$ the relative approximation error approaches zero.The derived formulae show, that it is necessary to take into account the second derivatives to obtain an asymptotically negligible error.If the above described method of taking only the first derivatives is used,the relative error will be somewhere between zero and infinity (for a discussion see /5/).

2. DERIVATION OF THE ASYMPTOTIC APPROXIMATION FOR DOMAIN INTEGRALS

Let a limit state function $g(\underline{x})$ with $g(\underline{0})>0$ be given in the n-dimensional space.Let β_o be the distance of the limit state surface to the origin. This surface is now imbedded in a sequence of surfaces.These surfaces $G(\beta)$ are defined by :

$$G(\beta)=\{\underline{x};g(\underline{x};\beta)=0\} \text{ with } g(\underline{x};\beta)= g((\beta_o/\beta)\underline{x}) \tag{2}$$

The function $P(\beta)$ is defined by :

$$P(\beta) = \int_{g(\underline{x};\beta)<0} p_n(\underline{x})\, d\underline{x} \tag{3}$$

$P(\beta)$ is the failure probability for the limit state function $g(\underline{x};\beta)$. Clearly for β_o $g(\underline{x};\beta_o)=g(\underline{x})$.By a simple substitution the integrals,whose integration domains depend on β,can be transformed into a sequence of integrals over a fixed domain.The transformation is :

$$\underline{x} \to \quad \underline{y}= \beta^{-1}\underline{x} \quad \text{with } \det(\frac{\partial x_l}{\partial y_m}) = \beta^n \tag{4}$$

Then :

$$P(\beta) = \beta^n \int_{g(\beta_o \underline{y})<0} p_n(\beta \underline{y})d\underline{y} = \beta^n (2\pi)^{-n/2} \int_{g(\beta_o \underline{y})<0} \exp(-\beta^2 |\underline{y}|^2/2)d\underline{y} \quad (5)$$

According to the appendix,this is a Laplace type integral,if we identify

$$\lambda = \beta^2, \ f(\underline{y}) = -|\underline{y}|^2/2, \ D=\{\underline{y};\ g(\beta_o \underline{y})<0\}. \quad (6)$$

where λ, $f(\underline{y})$ and D are used in the appendix following the notation in /2/,chap.8.The asymptotic behaviour of this integral depends only on the behaviour of the functions $-|\underline{y}|^2/2$ and $g(\beta_o \underline{y})$ at the points at the boundary of $\{\underline{y}; g(\beta_o \underline{y})<0\}$,where $-|\underline{y}|^2/2$ achieves its maximum.This function is proportional to the negative of the distance of $\underline{y}$ to the origin; therfore,it is maximal,if the distance is minimal.Since the origin is not in the integration domain,these points $\underline{y}_1,..,\underline{y}_k$ $(k\geq 1)$ have to be located on the hypersurface defined by $g(\beta_o \underline{y})=0$.This corresponds to find on the hypersurface defined by $g(\underline{z})=0$ all points $\underline{z}_1,...,\underline{z}_k$ with minimal distance to the origin,which are given by $\underline{z}_i=\beta_o \underline{y}_i$.Assume now,that there are k points $\underline{z}_1,..,\underline{z}_k$ with minimal distance.Then,by using eq.(22) given in the appendix and the relation between the $\underline{z}_i$ and the $\underline{y}_i$,the following asymptotic approximation is obtained :

$$\int_{g(\beta_0 \underline{y})<0} p_n(\beta \underline{y})\ d\underline{y} \sim$$

$$\sim \exp(-\beta_o^2/2)\ \left(\sum_{i=1}^{k} |J_i|^{-1/2}\right)\ (2\pi/\beta)^{(n-1)/2} \beta^{-1} \quad (\beta\to\infty) \quad (7)$$

with :

$$J_i = \beta_o^{-2}\ \underline{z}_i^{\ T}\ \underline{\underline{C}}^i\ \underline{z}_i$$

$$\underline{\underline{C}}^i = (\mathrm{cof}[\ -\delta_{lm} - K^i g_{lm}])_{l,m=1,..,n}$$

$\mathrm{cof}[d_{lm}]$=the cofactor of the element d_{lm} in the matrix $(d_{lm})_{l,m=1,..,n}$

δ_{lm}= Delta-Kronecker-symbol

$$g_{lm} = \beta_o^2 \left.\frac{\partial^2 g(z)}{\partial z_l\ \partial z_m}\right|_{\underline{z}=\underline{z}_i}$$

$K^i = |\nabla g(\underline{z}_i)|^{-1} \beta_o^{-1}$, with $\nabla g(\underline{z})$ the gradient of $g(\underline{z})$

This gives for $P(\beta)$:

$$P(\beta) \sim (2\pi)^{-1/2} \exp(-\beta_o^2/2)\beta^{-1} \left(\sum_{i=1}^{k} |J_i|^{-1/2}\right) \quad (\beta\to\infty) \quad (8)$$

and for $P(\beta_o)$,using the relation $\Phi(-\beta) \sim (2\pi)^{-1/2}\exp(-\beta^2/2)\beta^{-1}$ $(\beta\to\infty)$, with $\Phi(x)$ denoting the standard normal integral :

$$P(\beta_o) \sim \Phi(-\beta_o)\left(\sum_{i=1}^{k} |J_i|^{-1/2}\right) \quad (9)$$

Although the factors J_i can be computed without difficulty, their meaning is best appreciated by the following considerations (for details see /4/ and /5/). By a suitable rotation the point $\underline{z}_i$ can be transformed into the point $\beta_o \underline{e}_n$ ($\underline{e}_n$ the unit vector in the direction of the x_n-axis) so, that the main curvature axes of the transformed hypersurface are in the directions of the new n-1 first coordinate axes. Now, in this coordinate system, the mixed second derivatives g_{lm} are zero, and since $\beta_o \underline{e}_n$ is a point with minimal distance to the origin with respect to the surface, only the first derivative in the direction of the x_n-axis is not zero due to the Lagrange multiplier theorem. This yields :

$$J_i = \prod_{j=1}^{n-1} (1 - \beta_o \kappa_{i,j}) \tag{10}$$

with the $\kappa_{i,j}$ (j=1,.,n-1) the main curvatures of the hypersurface $G(\beta_o)$ defined by $g(\underline{z})=0$, at $\underline{z}_i$. Therefore, the factor J_i depends on the curvatures of $g(\underline{z})$ at $\underline{z}_i$ multiplied with the distance of $\underline{z}_i$ to the origin.

3. THE TIME-VARIANT CASE

In the case of time-variant random influences these can be modeled in most cases by a random vector process. Until now almost only Gaussian vector processes are used, since only for them the mathematical theory is available. In the following only stationary Gaussian vector processes with zero mean values and continuously differentiable sample paths will be considered. Let the process be denoted by $\underline{x}(t)=(x_1(t),..,x_n(t))$ and its derivative process by $\underline{x}'(t)=(x_1'(t),..,x_n'(t))$. By a suitable coordinate change (see/15/) it can be achieved, that:

$$\begin{aligned} \mathrm{cov}(x_l(t)x_m(t)) &= \delta_{lm} \\ \mathrm{cov}(x_l'(t)x_m'(t)) &= \delta_{lm}\,\sigma_m^2 \end{aligned} \tag{11}$$

with the σ_m's being positive constants. As in /15/ it will be assumed, that $\underline{x}(t)$ and $\underline{x}'(t)$ are uncorrelated, i.e. $\mathrm{cov}(x_l(t)x_m'(t))=0$ for l,m=1,.n.

As before let be given a limit state function $g(\underline{x})$. The probability, that that the structure remains intact during the time interval [0,T], if it was intact at time 0, is given by $P(g(\underline{x}(t))>0$ for all $t\in[0,T])$. Let $P_f(T)$ denote the probability, that the vector process $\underline{x}(t)$ leaves the safe domain during the time interval, i.e. the structure fails, then $P_f(T)$ can be bounded by (see /3/ and /15/) :

$$P_f(T) \le P(g(\underline{x}(0))<0) + \nu_+ T \tag{12}$$

with ν_+ denoting the mean outcrossing rate of the vector process $\underline{x}(t)$ out of the safe domain.

This mean outcrossing rate ν_+ is given by the surface integral (/15/):

$$\nu_+ = (2\pi)^{-1/2} \int_{g(\underline{x})=0} \left(\sum_{j=1}^{n} n_j(\underline{x})\sigma_j^2\right)^{1/2} p_n(\underline{x})\, ds(\underline{x}) \tag{13}$$

with : $n(\underline{x})=(n_1(\underline{x}),\dots,n_n(\underline{x}))$ the surface normal vector of the surface $G=\{\underline{x};g(\underline{x})=0\}$ at $\underline{x}$ with direction towards $F=\{\underline{x};g(\underline{x})<0\}$,$ds(\underline{x})$ denoting surface integration over G.This formula was derived by Belyaev/1/(in a more general form including dependencies between $\underline{x}(t)$ and $\underline{x}'(t)$).Lindgren/10/ proved it under less restrictive conditions than Belyaev.For this surface integral an asypmtotic approximation can be derived.

4. DERIVATION OF THE ASYMPTOTIC APPROXIMATION FOR SURFACE INTEGRALS

For the surface integral in eq.(13) several heuristic approximations have been proposed,which are based on the idea of estimating the integral by a weighted average of the values of the integrand at several points (see /3/,p.391 and /15/).The following derivation gives a theoretical foundation for these methods and shows,how to choose the points and the weights to obtain an asymptotic correct result.

Let β_o be the distance of the surface G to the origin and let be on the surface k points $\underline{z}_1,\dots,\underline{z}_k$ with minimal distance β_o to the origin.If $\psi:[0,1]^{n-1}\to G,\underline{u}\to \underline{\psi}(\underline{u})=(\psi_1(\underline{u}),\dots,\psi_n(\underline{u}))$ is a global parametrization of the surface G,then the function $f(\underline{u})= -\frac{1}{2}\sum_{j=1}^{n}(\psi_j(\underline{u}))^2\beta_o^{-2}$ has k global maxima at the points $\underline{u}_1,\dots,\underline{u}_k$ defined by $\underline{\psi}(\underline{u}_i)= \underline{z}_i$ with $f(\underline{u}_i)=-\frac{1}{2}$.Writing the surface integral in the parametric form and $p_n(\underline{z})$ explicitly :

$$\nu_+ =(2\pi)^{-(n+1)/2}\int_{[0,1]^{n-1}} \left(\sum_{j=1}^{n} n_j^2(\psi(\underline{u}))\sigma_j^2\right)^{1/2}\exp(-\beta_o^2 f(\underline{u}))\,|t(\underline{u})|\, d\underline{u} \tag{14}$$

with $t(\underline{u})$ the transformation determinant (for its form see /14/). Using the same method as in the case of the domain integrals,i.e. imbedding G into a sequence of surfaces,which are obtained by taking the β in the exponent as a variable, eq.(20) in the appendix can be applied and the asymptotic approximation is found :

$$\nu_+ \sim (2\pi)^{-1/2} p_1(\beta_o)\left(\sum_{i=1}^{k} |t(\underline{u}_i)/\det(f_{lm}(\underline{u}_i))^{1/2}|\left(\sum_{j=1}^{n} n_j^2(\underline{z}_i)\sigma_j^2\right)^{1/2}\right) \tag{15}$$

with $\det(f_{lm}(\underline{u}_i))$ the determinant of the matrix of the second derivatives of the function $f(\underline{u})$ at $\underline{u}_i$.In /2/,p.340,eq.(8.3.64) it is shown,that :

$$|J_i|^{-1/2}= |t(\underline{u}_i)/\det(f_{lm}(\underline{u}_i))^{1/2}| \tag{16}$$

with J_i defined as in eq.(7).The final result is then :

$$\nu_+ \sim (2\pi)^{-1/2} p_1(\beta_o)\left(\sum_{i=1}^{k} |J_i|^{-1/2}\left(\sum_{j=1}^{n} n_j^2(\underline{z}_i)\sigma_j^2\right)^{1/2}\right) \tag{17}$$

5. SUMMARY AND CONCLUSIONS

A method has been outlined for obtaining simple asymptotic approximations for multinormal domain and surface integrals by expanding the function defining the boundary of the domain or the given surface at the points with minimal distance to the origin.The results show,that it is necessary to take in account the curvatures of the surface at these points,even when they are small.Therefore,the application of the so-called "first-order-reliability-methods" as described in /8/,/9/ and /13/, where the probability content of a domain in the space of n independent standard normal distributed variables is estimated by $\Phi(-\beta)$, where β is the distance of the domain to the origin, does not give correct results.

The formula for the crossing rates of Gaussian vector processes is valid only,if the value of the process $\underline{x}(t)$ and its derivative process $\underline{x}'(t)$ at each time t are independent.Recently,Ditlevsen/6/ has obtained results for the crossing rates of Gaussian vector processes with dependencies out of domains with linear boundaries.The general case,including dependencies and curved boundaries, can be treated also with the methods described here and will be published in a forthcoming paper.

APPENDIX

THE ASYMPTOTIC EVALUATION OF MULTIPLE INTEGRALS

In the following we consider the asymptotic behaviour of multiple integrals of the form

$$I(\lambda) = \int_D \exp(\lambda f(\underline{x}))\, f_0(\underline{x})\, d\underline{x} \tag{19}$$

with D a domain in the n-dimensional space.We shall restrict our considerations to the case of a real parameter λ for $\lambda \to \infty$ .These integrals are called Laplace type integrals.

It will be assumed,that the boundary ∂D of is given by $\partial D = \{\underline{x}; g(\underline{x}) = 0\}$ and that the functions $f(\underline{x})$, $f_0(\underline{x})$ and $g(\underline{x})$ are sufficiently differentiable functions to allow the following operations.

First it will be asumed,that there is only one point in $\overline{D}$,where the global maximum of $f(\underline{x})$ is achieved.Then there are three cases possible, excluding the case of a global maximum,where also the second derivatives vanish and the local behaviour of the function $f(\underline{x})$ near this maximum is determind by higher derivatives.Let $\underline{x}_0$ denote the point,where the global maximum occurs.Then :

1) $\underline{x}_o$ is an interior point of D.

$$I(\lambda) \sim \frac{\exp(\lambda f(\underline{x}_o)) f_o(\underline{x}_o)}{(|\det(f_{lm}(\underline{x}_0))|)^{1/2}} \left(\frac{2\pi}{\lambda}\right)^{n/2} \qquad (\lambda \to \infty) \tag{20}$$

with the $f_{lm}(\underline{x}_o)$ (l,m=1,..,n) the second derivatives of $f(\underline{x})$ at $\underline{x}_o$.

2) $\underline{x}_o$ is on the boundary ∂D and $\nabla f(\underline{x}_0)=0$.

$$I(\lambda) \sim \frac{1}{2} \frac{\exp(\lambda f(\underline{x}_0)) f_o(\underline{x}_o)}{(|\det(f_{lm}(\underline{x}_0))|)^{1/2}} \left(\frac{2\pi}{\lambda}\right)^{n/2} \qquad (\lambda \to \infty) \tag{21}$$

3) $\underline{x}_o$ is on the boundary ∂D and $\nabla f(\underline{x}_o) \neq 0$.

$$I(\lambda) \sim \frac{\exp(\lambda f(\underline{x}_o))}{|J|^{1/2}} \quad \frac{f_o(\underline{x}_o)}{\lambda} \left(\frac{2\pi}{\lambda}\right)^{(n-1)/2} \qquad (\lambda \to \infty) \tag{22}$$

with: $J = \sum_{l=1}^{n} \sum_{m=1}^{n} f_l(\underline{x}_0) f_m(\underline{x}_0) \operatorname{cof}[f_{lm}(\underline{x}_o) - K g_{lm}(\underline{x}_o)]$

K is defined by the equation $\nabla f(\underline{x}_o) = K \nabla g(\underline{x}_o)$

$\operatorname{cof}[d_{lm}]$ denotes the cofactor of the element d_{lm} in the matrix $(d_{lm})_{l,m=1,..,n}$.

A discussion of these formulae and the special cases not considered here,can be found in /2/ and the literature cited therein.In the case, that there are several points,where the global maximum is achieved,the approximation is obtained by adding the contributions from each point.

LITERATURE REFERENCES

/1/ Belyaev Yu.K.,On the number of exits across a boundary of a region by a vector stochastic process,Theory Probability Appl. 13,(1968) 320-324.

/2/ Bleistein,N. and Handelsman,R.A.,Asymptotic Expansions of Integrals (Holt,Rinehard and Winston,New York,1975).

/3/ Bolotin,V.V.,Wahrscheinlichkeitsmethoden zur Berechnung von Konstruktionen(VEB-Verlag für das Bauwesen,Berlin,GDR,1981).

/4/ Breitung,K.,An asymptotic formula for the failure probability, DIALOG 82-6,Dept. of Civil Engineering,Danmarks Ingeniørakademi, Lyngby,Denmark,1982,19-45.

/5/ Breitung,K.,Asymptotic approximations for multinormal integrals,to appear in the Journal of the Engineering Mechanics Div.,ASCE.

/6/ Ditlevsen O.,Gaussian outcrossings from safe convex polyhedrons,J. of the Eng. Mech. Div.,ASCE,109(1983),127-148.

/7/ Fiessler,B.,Neumann,H.-J. and Rackwitz,R.,Quadratic limit states in structural reliability,J. of the Eng. Mech. Div.,ASCE,105(1979),661-676.

/8/ Hohenbichler,M. and Rackwitz,R.,Non-normal dependent vectors in structural safety,J. of the Eng. Mech. Div.,ASCE(1981),1227-1241.
/9/ Hohenbichler,M. and Rackwitz,R.,First-Order concepts in system reliability,to appear in Structural Safety 1(1983).
/10/Lindgren,G.,Model processes in nonlinear prediction with applications to detection and alarm,Ann. Probability 8(1980),775-792.
/11/Lindgren,G.,Extreme values and crossings for the χ^2-process and other functions of multidimensional Gaussian processes with reliability applications,Adv. Appl. Prob. 12(1980),746-774.
/12/Psenicnyj,B.N.,Algorithms for general mathematical programming problems,Cybernetics 6(1970),120-125.
/13/Rackwitz,R. and Fiessler,B.,Structural reliability under combined random sequences,Computers and Structures 9(1978),489-494.
/14/Thorpe,J.A.,Elementary Topics in Differential Geometry (Springer, New York,1979).
/15/Veneziano,D.,Grigoriu,M. and Cornell,C.A.,Vector process models for system reliability,J. of the Eng. Mech. Div.,ASCE(1977),441-460.

MODEL UNCERTAINTY FOR BILINEAR HYSTERETIC SYSTEMS

J. D. Sørensen and P. Thoft-Christensen
Aalborg University Centre
Aalborg, Denmark

1. INTRODUCTION

In structural reliability analysis at least three types of uncertainty must be considered, namely physical uncertainty, statistical uncertainty, and model uncertainty (see e.g. Thoft-Christensen & Baker [1]). The physical uncertainty is usually modelled by a number of basic variables. The statistical uncertainty - due to lack of information - can e.g. be taken into account by describing the variables by predictive density functions, Veneziano [2].

In general, model uncertainty is the uncertainty connected with mathematical modelling of the physical reality.

When structural reliability analysis is related to the concept of a failure surface (or limit state surface) in the n-dimensional basic variable space then model uncertainty is at least due to the neglected variables, the modelling of the failure surface and the computational technique used. A more precise definition is given in section 2, where some different methods to treat model uncertainty are described. In section 3 a new method based on subjectively modelled conditional density functions is presented. It is shown that in some special cases this method is equivalent to existing more simple methods.

In the analysis of dynamically loaded structures it is often assumed that the loading and the response can be modelled by stationary stochastic processes. Further, it is assumed that the structures can be modelled by non-linear systems showing hysteresis. This non-linear behaviour is essential to the design procedure from an economic and reliability point of view. In section 4 it is shown how the probability of failure of a simple bilinear oscillator can be estimated and in section 5 it is demonstrated by numerical examples how model uncertainty can be included in the calculations.

2. MODEL UNCERTAINTY

Usually mathematical models describing the relations between the basic variables are deterministic models although there is a great deal of uncertainty associated with them. They may be based on a good understanding of the mechanical problem, but they will usually to some degree be imperical. In this paper model uncertainty is uncertainty in relation to the stochastic structure of the basic variables and the choice of failure surface. The last-mentioned uncertainty in relation to the failure surface is due to a number of neglected variables and also to the mathematical expressions chosen. The choice of density functions for the basic variables is of great importance due to tail sensitivity. Usually very little is known of the shape of the density functions in the significant intervals. Selection of density functions has been treated by Grigoriu, Veneziano & Cornell [11].

It has been suggested to evaluate the model uncertainty by comparing different mathematical models or using experimental data from existing structures or laboratory tests. However, this type of comparison will also be uncertain. Use of experience from other types of structure will also be uncertain due to the great variety of structures of interest.

It is clearly of importance to include model uncertainty in such a way that the estimation of the structural reliability is not getting too complicated. Further model uncertainty should be included in a form which is invariant to mathematical transformations of the equations in question. As emphasized by Ditlevsen [7] this is obtained if the model uncertainty can be related directly to the basic variables.

Model uncertainty can be included in level 1 methods simply by adding constants to the basic variables or multiplying the basic variables by constants. The magnitude of these constants must be determined by a subjective judgement of the model uncertainty. In level 2 methods (first order - second moment methods) the same technique may be used if the constants are substituted by stochastic variables with subjectively modelled second order moments (see Ang [3], Milford [4], NKB-recommendations [5]).

Let the idealized failure surface be given by

$$g(\underset{\sim}{x}) = 0 \tag{1}$$

where $\underset{\sim}{x}$ is a realization of the basic variables $\underset{\sim}{X} = (X_1, \ldots, X_n)$. Ditlevsen [7] has suggested that the model uncertainty can be modelled by assuming the failure surface to be stochastic in the basic variable space. According to the extended level 2 methods the basic variables are transformed into variables which are normally distributed with the expected values $\underset{\sim}{\mu}_X$ and the matrix of covariance $\underset{\sim}{C}_X$ (see Ditlevsen [8]). In this space the stochastic failure surface is modelled by a linear transformation of the normally distributed variables

$$\underset{\sim}{M}(\underset{\sim}{X}) = \underset{\sim}{A}\,\underset{\sim}{X} + \underset{\sim}{Z} \tag{2}$$

where $\underset{\sim}{A}$ is a matrix with constant elements and $\underset{\sim}{Z}$ a normally distributed stochastic vector with expected value $\underset{\sim}{\mu}_Z$ and covariance $\underset{\sim}{C}_Z(\underset{\sim}{x})$. By letting $\underset{\sim}{C}_Z$ be dependent on $\underset{\sim}{x}$ local variations in the model uncertainty can be included. The parameters in (2) are determined subjectively. The stochastic failure surface is defined by $g(\underset{\sim}{M}(\underset{\sim}{x})) = 0$. Assume $\underset{\sim}{X}$ and $\underset{\sim}{Z}$ to be independent, then the expected value and covariance of $M(\underset{\sim}{X})$ are $\underset{\sim}{A}\underset{\sim}{\mu}_X + \underset{\sim}{\mu}_Z$ and $\underset{\sim}{A}\underset{\sim}{C}_X\underset{\sim}{A}^T + \underset{\sim}{C}_Z$, respectively. Therefore, by this technique the model uncertainty is included with only a small increase in the calculation time. This formulation is in principle equal to the method suggested in the NKB-regulations [5].

Several techniques to include model uncertainty have been suggested in level 3 reliability methods. Consider a number of mathematical models with corresponding reliability estimates. Let the probability that such a model is correct be given by a formal probability. Then a weighted estimate of the reliability can be constructed by the total probability theorem. A second technique is based on modelling the uncertain variables by fuzzy sets (see Blockley & Ellison [9]). The model uncertainty is included in this technique by subjectively modelled conditional fuzzy sets characterized by membership functions.

3. A MODELLING METHOD BASED ON CONDITIONAL DENSITY FUNCTIONS

This section describes a method to include model uncertainty in level 3 reliability methods, where the model uncertainty is modelled by conditional density functions $f_{\underset{\sim}{X}'|\underset{\sim}{X}}(\underset{\sim}{x}'|\underset{\sim}{x})$, where $\underset{\sim}{X}' = (X_1', \ldots, X_n')$ can be considered as stochastic variables modelled in such a way that the physical uncertainty corresponding to $\underset{\sim}{X}$ and the model uncertainty are included in their characterization. The conditional density function is determined on the basis of subjective estimates and should be chosen in such a way that the calculation of the reliability is only slightly increased. Note that the model uncertainty is modelled in the basic variable space so that this formulation is invariant to transformations of the failure function (1). The failure probability is now given by

$$P_f = \int_{g(\underset{\sim}{x}')<0} \int_{R^n} f_{\underset{\sim}{X}'|\underset{\sim}{X}}(\underset{\sim}{x}'|\underset{\sim}{x})\, f_{\underset{\sim}{X}}(\underset{\sim}{x})\, d\underset{\sim}{x}\, d\underset{\sim}{x}' \tag{3}$$

If model uncertainty can be neglected then

$$f_{\underset{\sim}{X}'|\underset{\sim}{X}}(\underset{\sim}{x}'|\underset{\sim}{x}) = \delta(\underset{\sim}{x}' - \underset{\sim}{x}) \tag{4}$$

where $\delta(\cdot)$ is Dirac's delta-function, and the probability of failure is defined by the usual expression

$$P_f = \int_{g(\underset{\sim}{x})<0} f_{\underset{\sim}{X}}(\underset{\sim}{x})\, d\underset{\sim}{x} \tag{5}$$

Evaluation of the multi-integral in (3) will of course in general be very expensive. However, if the conditional density function is chosen in such a way that $f_{\underset{\sim}{X}'}$ can be calculated without using numerical integration then inclusion of model uncertainty will not increase the computer time. The probability of failure can then be calculated by a formula like (5). This procedure corresponds to using natural conjugated families of density functions in Bayesian statistics.

Let $\underset{\sim}{X}$ be normally distributed with expected values $\underset{\sim}{\mu}_X$ and covariance $\underset{\sim}{C}_X$

$$f_{\underset{\sim}{X}}(\underset{\sim}{x}) = N(\underset{\sim}{x}; \underset{\sim}{\mu}_X, \underset{\sim}{C}_X) \tag{6}$$

and let the conditional density function be given by

$$f_{\underset{\sim}{X}'|\underset{\sim}{X}}(\underset{\sim}{x}'|\underset{\sim}{x}) = N(\underset{\sim}{x}'; \underset{\sim}{\mu}_X^* + \underset{\sim}{A}\underset{\sim}{x}, \underset{\sim}{C}_X^*) \tag{7}$$

where $\underset{\sim}{A}$ is a quadratic matrix with constant elements. By the assumption (7) the model uncertainty is included through a linear transformation from $\underset{\sim}{X}$ to $\underset{\sim}{X}'$ and is modelled by $\underset{\sim}{\mu}^*_X$, $\underset{\sim}{A}$ and $\underset{\sim}{C}^*_X$. By inserting (6) and (7) in (3) one gets

$$P_f = \int_{g(\underset{\sim}{x}')\leq 0} N(\underset{\sim}{x}'; \underset{\sim}{\mu}^*_X + \underset{\sim}{A}\underset{\sim}{\mu}_X, \underset{\sim}{C}^*_X + \underset{\sim}{A}\underset{\sim}{C}_X\underset{\sim}{A}^T)d\underset{\sim}{x}' \tag{8}$$

This formulation is equivalent to the modelling (2) if $\underset{\sim}{\mu}^*_X = \underset{\sim}{\mu}_Z$ and $\underset{\sim}{C}^*_X = \underset{\sim}{C}_Z$. The failure probability (8) can be calculated either by numerical integration or by e.g. the extended first order second moment method (see [8]). Local variations in the model uncertainty can e.g. be included by letting $\underset{\sim}{\mu}^*_X$ and $\underset{\sim}{C}^*_X$ be dependent on $\underset{\sim}{x}$. P_f can also be written

$$P_f = \int_{R^n} f_{\underset{\sim}{X}}(\underset{\sim}{x}) \int_{g(\underset{\sim}{x}')\leq 0} f_{\underset{\sim}{X}'|\underset{\sim}{X}}(\underset{\sim}{x}'|\underset{\sim}{x})\, d\underset{\sim}{x}'d\underset{\sim}{x} = \int_{R^n} f_{\underset{\sim}{X}}(\underset{\sim}{x})P(\underset{\sim}{x})d\underset{\sim}{x} \tag{9}$$

where

$$P(\underset{\sim}{x}) = \int_{g(\underset{\sim}{x}')\leq 0} f_{\underset{\sim}{X}'|\underset{\sim}{X}}(\underset{\sim}{x}'|\underset{\sim}{x})d\underset{\sim}{x}' \tag{10}$$

is the failure probability given $\underset{\sim}{x}$. If the model uncertainty is neglected then

$$P(\underset{\sim}{x}) = \begin{cases} 1 & \text{if } g(\underset{\sim}{x}) \leq 0 \\ 0 & \text{if } g(\underset{\sim}{x}) > 0 \end{cases} \tag{11}$$

Example 1

Let $n = 2$ and X_1 and X_2 be independent and normally distributed $N(0, 1)$. Let $f_{\underset{\sim}{X}'|\underset{\sim}{X}}(\underset{\sim}{x}'|\underset{\sim}{x}) = N(x_1'; x_1, \sigma) \cdot N(x_2'; x_2, \sigma)$ and let the failure event be defined by $\{x_1, x_2 \mid x_1 + 3 \leq 0 \ \vee \ x_2 + 3 \leq 0\}$.

Then from (9)

$$P_f = \int_{-\infty}^{\infty}\int_{-\infty}^{\infty} N(x_1; 0, 1)N(x_2; 0, 1)P(x_1, x_2)dx_1dx_2 \tag{12}$$

where

$$P(x_1, x_2) = 1 - (1 - \Phi(\frac{-3 - x_1}{\sigma}))(1 - \Phi(\frac{-3 - x_2}{\sigma})) \tag{13}$$

The function P is shown in figure 1 for $\sigma = 0$, 0.5, and 1. For $\sigma = 0$ (corresponding to no model uncertainty) $P(x_1, x_2)$ is a step function. For increasing σ the function $P(x_1, x_2)$ is becoming still »smoother». Therefore, alternatively, the model uncertainty can be modelled by prescribing $P(\underset{\sim}{x})$. By this formulation $P(\underset{\sim}{x})$ is a function modelling the failure surface uncertainty. In the same figure the generalized reliability index β_s is shown. As expected β_s is decreasing with increasing σ (increasing model uncertainty).

4. ELASTO-PLASTIC HYSTERETIC SYSTEMS

This section describes a method to estimate the reliability of a structure which can be modelled as an elasto-plastic simple oscillator loaded by white-noise. Failure is defined as the event that the permanent deformation crosses a critical deterministic value b. The yield limit of the oscillator is called y.

The equation of motion for a linear time invariant system with one degree of freedom is

$$\ddot{x}_0 + 2\zeta_0\omega_0\dot{x}_0 + \omega_0^2 x_0 = f \tag{14}$$

where ω_0 is the undamped eigenfrequency, ζ_0 the damping ratio, $x_0(t)$ a realization of the response process $\{X_0(t), t \in [0, \infty[\}$ and $f(t)$ a realization of the load process $\{F(t), t \in [0, \infty[\}$. The load process is assumed to be stationary and Gaussian with the spectrum $S_F(\omega) = s_F$. Then the stationary variance of X_0 is

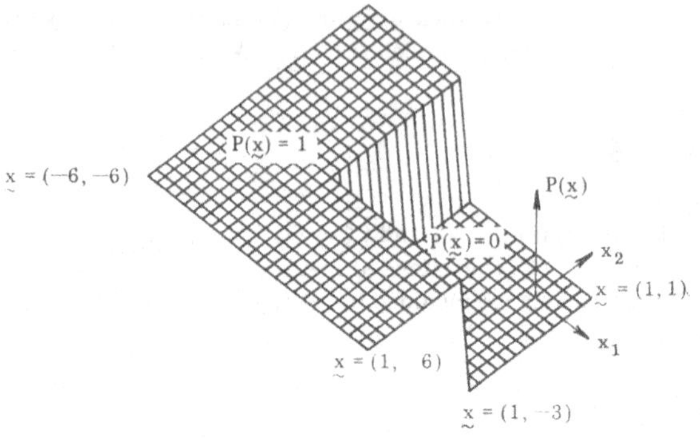

a) $\sigma = 0,\ \beta_s = -\Phi^{-1}(P_f) = 2.78$

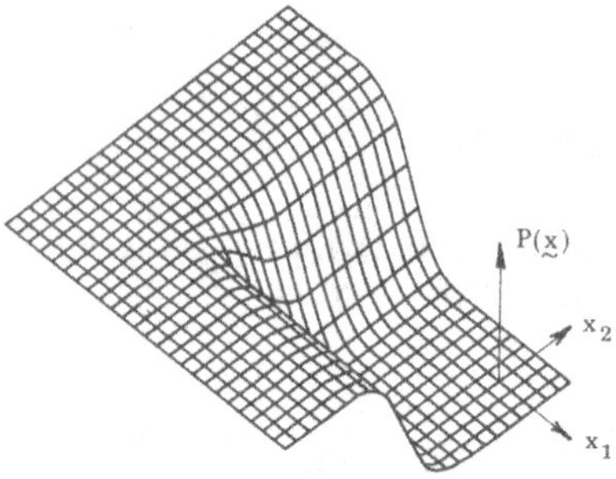

b) $\sigma = 0.5,\ \beta_s = -\Phi^{-1}(P_f) = 2.44$

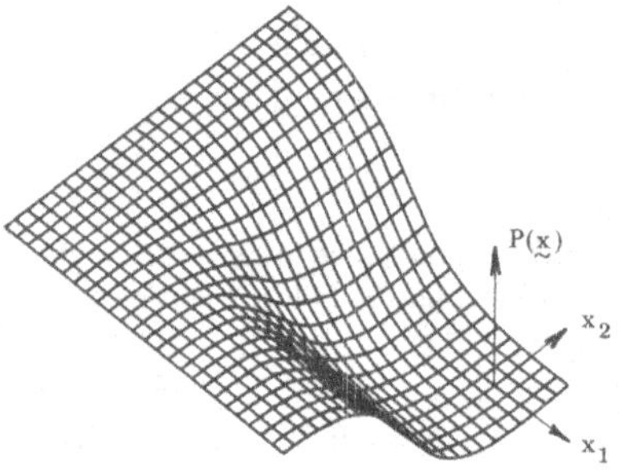

c) $\sigma = 1,\ \beta_s = -\Phi^{-1}(P_f) = 1.83$

Figure 1. The function $P(x_1, x_2)$.

$$\sigma^2_{X_0} = \frac{\pi S_F}{2\zeta_0\omega_0^3} \tag{15}$$

For slightly damped structures realizations of the response process show that peaks outside $]-y, y[$ tend to come in »clumps». Vanmarche [12] has suggested the following approximation for the expected number $E[N_y]$ of consecutive peaks outside $]-y, y[$

$$E[N_y] = [1 - \exp(-\sqrt{\frac{\pi}{2}}\ rq)]^{-1} \tag{16}$$

where $r = y/\sigma_X$ and $q \cong 2\sqrt{\zeta_0/\pi}$. The expected number of »clumps» per unit time can be determined by, Vanmarche [12],

$$\mu_y = \frac{\omega_0}{\pi}(1 - \exp(-\sqrt{\frac{\pi}{2}}\ rq))\exp(-\frac{r^2}{2}) \tag{17}$$

Next it is assumed that important statistical characteristics for an elasto-plastic system can be determined from a linear system with the equivalent system characteristics ζ_e and ω_e calculated by the method of Krylov and Boguliubov (see Caughey [13]). It is here assumed that the response is narrow-banded and the expected accumulated permanent deformation is 0. By using the Stratonovitch-Khasminskii limit theorem Roberts [14] has shown on the same assumptions that the simultaneous density function for the amplitude A and the phase Ψ for the narrow-banded response process of the elasto-plastic system is

$$f_\Psi(\psi)f_A(a) = \begin{cases} \frac{1}{2\pi} ca \exp(-\frac{a^2}{2\sigma_0^2}) & , \quad 0 \leqslant a \leqslant y\ ,\ \psi \in [0, 2\pi] \\ \frac{1}{2\pi} ca \exp(-\frac{a^2}{2\sigma_0^2} - \frac{a-y}{\delta})(\frac{a}{y})^{y/\delta} & , \quad a > y \quad ,\ \psi \in [0, 2\pi] \end{cases} \tag{18}$$

where

$$\sigma_0^2 = \sigma_{X_0}^2(\frac{\omega_0}{\omega_e})^2 \tag{19}$$

$$\delta = \frac{\pi\zeta_0\omega_0\sigma^2_{X_0}}{2\omega_e y} \tag{20}$$

By using (18) the distribution function F_V for the velocity of the response by crossing into the plastic area can be determined (see Stratonovich [15])

$$F_V(v) = \frac{F_A(\sqrt{y^2 + (v/\omega_e)^2}) - F_A(y)}{1 - F_A(y)} \tag{21}$$

Let the response cross into the plastic area at the time $t = t_0$ with the velocity V and leave it again at $t = t_0 + \Delta T$. If the damping and inertia terms in the energy equation are neglected then it can be proved that

$$F_{\Delta D_0}(d) \cong F_V(\omega_0\sqrt{2yd}) \quad , \quad d > 0 \tag{22}$$

where $F_{\Delta D_0}$ is the distribution function for the increment in plastic deformation ΔD_0. It is seen from (19) that f_A for $y/\sigma_{X_0} \to \infty$ is Rayleigh distributed. (22) then shows that ΔD_0 is exponentially distributed with the expected value $\sigma^2_{X_0}/y$.

Let ΔD be the permanent deformation from a single outcrossing of the elastic range. It is then reasonable to approximate $f_{\Delta D}$ by the Laplace distribution

$$f_{\Delta D}(d) = \frac{1}{\sqrt{2}\,\sigma_{\Delta D}} \exp(-\frac{|d|\sqrt{2}}{\sigma_{\Delta D}}) \tag{23}$$

where

$$\sigma^2_{\Delta D} = \int_0^\infty x^2 f_{\Delta D_0}(x)dx \tag{24}$$

The accumulated permanent deformation D(t), $t \in \,]0, T]$ can be written

$$D(T) = \sum_{i=1}^{N(T)} \Delta D_i^* \tag{25}$$

where $\{N(t), t \in \,]0, T]\}$ is a stochastic counting process, and ΔD_i^* is the increment from »clump» number i. When y is large compared with σ_{X_0} then it is reasonable to expect this counting process to be a Poisson process with the intensity ν and to assume the individual terms in (25) to be independent. D(t) is then modelled by a filtered Poisson process.

Let $f_{\Delta D^*}$ be given by (23). The characteristic function for ΔD^* is then

$$\varphi_{\Delta D^*} = \left(1 + \frac{1}{2}(\sigma_{\Delta D^*} u)^2\right)^{-1} \tag{26}$$

and the density function $f_{D_n} = f_{D(T)}$ for N(T) = n is

$$f_{D_n}(x) = \frac{1}{2\pi}\int_{-\infty}^{\infty} e^{-iux}\left(1 + \frac{1}{2}(\sigma_{\Delta D^*} u)^2\right)^{-n} du$$

$$= \frac{1}{\pi}\int_{0}^{\infty}\left(1 + \frac{1}{2}(\sigma_{\Delta D^*} u)^2\right)^{-n} \cos ux \, du = \frac{\sqrt{2}}{\sigma_{\Delta D^*}\sqrt{\pi}}\left(\frac{x}{\sqrt{2}\,\sigma_{\Delta D^*}}\right)^{n-\frac{1}{2}} K_{n-\frac{1}{2}}\left(\frac{\sqrt{2}x}{\sigma_{\Delta D^*}}\right) \tag{27}$$

where K_n is the modified Bessel function. For $\nu \to \infty$, $\sigma_{\Delta D^*} \to 0$ and $\sigma^2_{\Delta D^*}\nu$ constant it is seen that

$$\log \varphi_{D(T)}(u) \to -\frac{1}{2}\sigma^2_{\Delta D^*}\nu T \tag{28}$$

(28) shows that D(T) is normally distributed $N(\cdot\,; 0, \sigma_{\Delta D^*}\sqrt{\nu T})$ at time T. If it is assumed that D(0) = 0 and that D(t) has stationary independent increments, then $\{D(t)\}$ is approximating a Wiener process when the number of outcrossings approaches ∞ and the standard deviation for the increments approaches 0.

The intensity ν is put equal to μ_y (see (17)) and the standard deviation $\sigma_{\Delta D^*}$ of the increment in permanent deformation per »clump» is chosen approximately equal to $\sqrt{E[N_y]}\,\sigma_{\Delta D}$, where $E[N_y]$ is given by (16).

The elasto-plastic oscillator is assumed to fail if the accumulated permanent deformation leaves the safe interval $S = \,]-b, b[$. It is shown by Nielsen, Sørensen and Thoft-Christensen [16] that the first passage density $f_0(t)$ can be determined on the basis of the Markov property of the process

$$f_0(t) = \nu \sum_{n=0}^{\infty} \frac{(\nu t)^n \exp(-\nu t)}{n!}(1 - \lambda_n)\prod_{\ell=0}^{n-1}\lambda_\ell \tag{29}$$

where

$$\lambda_\ell = \frac{\int_{-b}^{b} f_{D_n}(x)\int_{-b}^{b} f_{\Delta D^*}(x - z)\,dz\,dx}{\int_{-b}^{b} f_{D_n}(x)\,dx} \tag{30}$$

The probability of failure in the time interval $]0, T[$ can be estimated by

$$P_f(t) = \int_0^T f_0(t)\,dt \tag{31}$$

if $D_0 \in S$.

Example 2

Let $\zeta_0 = 0.02$, $\omega_0 = 10\pi\,s^{-1}$, b = 0.02, $\sigma_{X_0} = 0.04$, $y/\sigma_{X_0} = 2.5$, and P(D(0) = 0) = 1. It is seen from (17) and (24) that $\nu = 0.173$ and $\sigma_{\Delta D^*} = 0.0105$. The estimate (31) is shown in figure 2 and compared with simulation estimates. The simulation estimates are shown as vertical bars corresponding to the 95% confidence intervals. The small horizontal bars are the point estimates. In this example the agreement is very good.

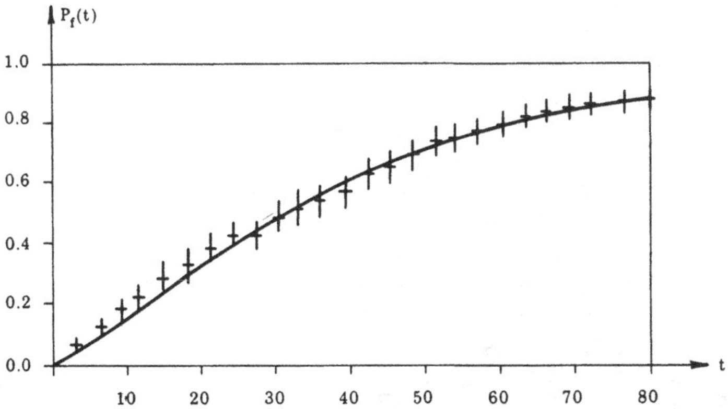

Figure 2. Failure probability $P_f(t)$ for an elasto-plastic hysteretic system.

5. APPLICATIONS OF MODELLING OF MODEL UNCERTAINTY

This section shows in two examples how the modelling method introduced in section 3 can be used for elasto-plastic hysteretic systems.

Example 3

The same system as in example 2 is considered but now model uncertainty is taken into account. Model uncertainty can in this case e.g. be due to

- the stress-strain relation is not perfectly elasto-plastic
- the response cannot be modelled sufficiently accurately by the first mode of vibration, i.e. the structure should be modelled by an n-dimensional system.

Let the permanent deformation ΔD^* at a single »clump» be a basic variable with a density function given by (23). The conditional density function in (3) is chosen as

$$f_{\Delta D'|\Delta D^*}(d'|d) = \frac{1}{2s}\exp(-\frac{1}{s}|d'-d|) \tag{32}$$

where the influence from the model uncertainty is eliminated for $s \to 0$. From (23) and (32) it is seen that

$$f_{\Delta D'}(d) = \int_{-\infty}^{\infty} f_{\Delta D'|\Delta D^*}(d|x) f_{\Delta D^*}(x)dx$$

$$= \begin{cases} \dfrac{\frac{1}{2s}\exp(-\frac{|d|}{s})}{1-(\frac{\sigma_{\Delta D^*}}{\sqrt{2}\,s})^2} + \dfrac{\frac{1}{\sqrt{2}\,\sigma_{\Delta D^*}}\exp(-\frac{\sqrt{2}\,|d|}{\sigma_{\Delta D^*}})}{1-(\frac{\sqrt{2}\,s}{\sigma_{\Delta D^*}})^2} & \text{for } s \neq \dfrac{\sigma_{\Delta D^*}}{\sqrt{2}} \\[2ex] \dfrac{1}{4s}(1+\dfrac{|d|}{s})\exp(-\dfrac{|d|}{s}) & \text{for } s = \dfrac{\sigma_{\Delta D^*}}{\sqrt{2}} \end{cases} \tag{33}$$

Note that in accordance with (3) and (5) one gets for $s \to 0$ that

$$f_{\Delta D'}(d) \to \frac{1}{\sqrt{2}\,\sigma_{\Delta D^*}}\exp(-\frac{\sqrt{2}\,|d|}{\sigma_{\Delta D^*}}) \tag{34}$$

The conditional density function $f_{\Delta D'|\Delta D^*}$ and the density function $f_{\Delta D'}$ are shown in figure 3 for different values of s ($\sigma_{\Delta D^*} = 0.0105$). The probability of failure $P_f(t)$ is shown in figure 4 for s = 0, 0.005 and 0.01. It is seen from figure 4 that the model uncertainty has most influence with small failure probabilities. The computer calculation is increased very little in this example, since (33) can be used directly instead of $f_{\Delta D^*}$ in (27) - (31).

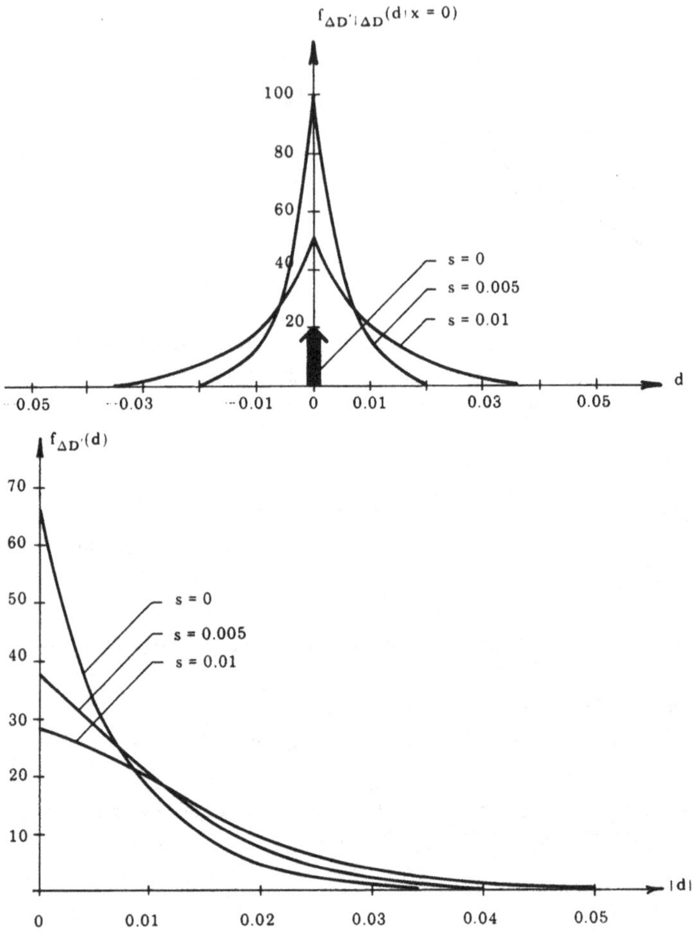

Figure 3. Density functions $f_{\Delta D'|\Delta D}$ and $f_{\Delta D'}$.

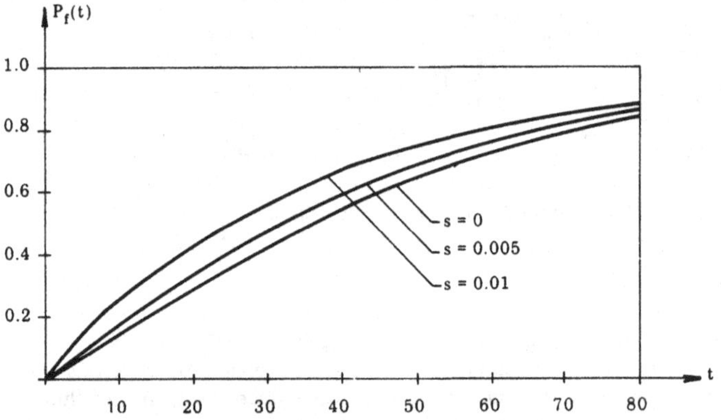

Figure 4. The failure probability for different levels of model uncertainty.

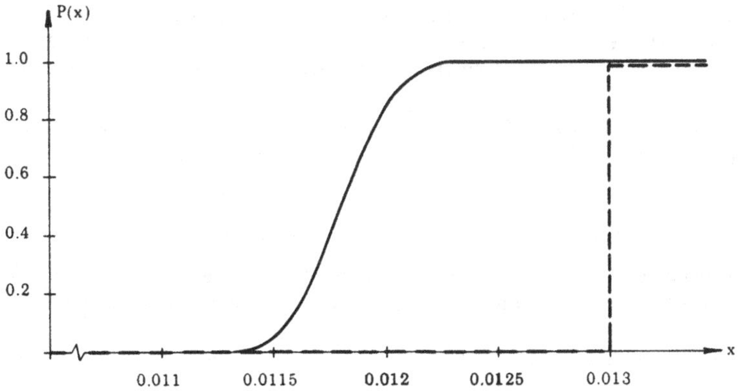

Figure 5. ——— with model uncertainty included
– – – without model uncertainty included.

Example 4
Consider the case where the probability of getting more than one upcrossing in the time interval $]0, T[$ is negligible. Let the failure function be

$$g(d) = d - d^* \quad , \quad d \geqslant 0 \tag{35}$$

where d^* is the maximum permissible permanent deformation by a single outcrossing of the elastic region and where d is a realization of the stochastic variable ΔD_0 (see (22)). Let the model uncertainty be modelled by the conditional density function

$$f_{\Delta D_0' | \Delta D_0}(d_0' | d_0) = N(d_0' \, ; 1.1\, d_0, 0.0002)/\Phi(5500\, d_0) \quad , \quad d' > 0 \, , d > 0 \tag{36}$$

Then it is seen from (9) that

$$P_f = \int_{d^*}^{\infty} \int_0^{\infty} f_{\Delta D_0' | \Delta D_0}(x' | x) f_{\Delta D}(x) dx dx' = \int_0^{\infty} f_{\Delta D_0}(x) P(x) dx \tag{37}$$

where

$$P(x) = 1 - \frac{\Phi\left(\dfrac{d^* - 1.1\, x}{0.0002}\right)}{\Phi(5500\, x)} \tag{38}$$

P(x) is shown in figure 5 for $d^* = 0.013$. P(x) corresponding to negligible model uncertainty is shown as a step function. The failure probability $P_f = 0.202$ when model uncertainty is included and $P_f = 0.162$ without model uncertainty included.

6. CONCLUSIONS

A new model to include model uncertainty in reliability analysis is presented and investigated in some details. The model uncertainty is modelled by a conditional density function. If this density function is chosen in an appropriate way corresponding to naturally conjugated density functions in Bayesian statistics then the computational work is only slightly increased. The model uncertainty is modelled in the physical basic space so that invariance is secured.

Further a method to estimate the reliability of structures modelled by a simple elasto-plastic oscillator is presented. The accumulated permanent deformations are modelled by a Markov process and failure defined by crossing of a critical limit. This method is compared with simulation estimates and good agreement is achieved for a structure with slight damping.

Finally, it is demonstrated by two examples how model uncertainty can be included in the estimate of reliability of elasto-plastic oscillators.

7. ACKNOWLEDGEMENT

Part of this investigation has been supported by the Danish Council for Scientific and Industrial Research. The method presented in section 4 is based on unpublished work by Nielsen, Sørensen & Thoft-Christensen [17].

8. REFERENCES

[1] Thoft-Christensen, P. & Baker, M. J.: *Structural Reliability Theory and Its Applications.* Springer-Verlag, Berlin - Heidelberg - New York, 1982.

[2] Veneziano, D.: *A Theory of Reliability which Includes Statistical Uncertainty.* Proc. ICASP-2, Appl. of Statist. and Prob. in Soil and Struct., Aachen, Germany, 1975, pp. 231-249.

[3] Ang, A. H.-S.: *Structural Risk Analysis and Reliability-Based Design.* ASCE, Journal of the Structural Division, Vol. 99, No. ST9, Sept. 1973, pp. 1891 - 1910.

[4] Milford, R. V.: *Structural Reliability and Crosswind Response of Tall Chimneys.* Eng. Struct., Vol. 14, Oct. 1982, pp. 263 - 270.

[5] The Nordic Committee on Building Regulations (NKB), The Loading and Safety Group: *Recommendations for Loading and Safety Regulations for Structural Design.* NKB-Report No. 36, Nov. 1978.

[6] Ditlevsen, O.: *A Collection of Notes Concerning Structural Reliability.* Note 6: Some Remarks on Simulating Model Uncertainty in Structural Reliability Analysis. DIALOG 2-76, Danish Engineering Academy, Denmark, 1976.

[7] Ditlevsen, O.: *Model Uncertainty in Structural Reliability.* Structural Safety, Vol. 1, No. 1, Sept. 1982, pp. 73-86.

[8] Ditlevsen, O.: *Basic Reliability Concepts.* Proceedings NATO Advanced Study Institute. P. Thoft-Christensen (ed.), Martinus Nijhoff Publishers, The Netherlands, 1983, pp. 1-55.

[9] Blockley, D. I. & Ellison, E. G.: *A New Technique for Estimating System Uncertainty in Design.* Proceedings of the Institution of Mechanical Engineers, Vol. 193, No. 5, 1979, pp. 159 - 168.

[10] Guttman, I.: *Statistical Tolerance Regions: Classical and Baysian.* Griffins Statistical Monographs & Courses, No. 26, Stuart Ed., Hafner Publ. Co., London, 1970.

[11] Grigoriu, M., Veneziano, D. & Cornell, C. A.: *Probabilistic Modelling as Decision Making.* ASCE, Journal of the Engineering Mechanics Division, Vol. 105, No. EM4, Aug. 1979, pp. 585-597.

[12] Vanmarcke, E. H.: *Properties of Spectral Moments with Applications to Random Vibration.* ASCE, Journal of the Engineering Mechanics Division, Vol. 98, No. EM2, 1972, pp. 425 - 446.

[13] Caughey, T. K.: *Random Excitation of a System with Bilinear Hysteresis.* ASME Journal of Applied Mechanics, Vol. 27, 1960, pp. 649-652.

[14] Roberts, J. B.: *The Yielding Behaviour of a Randomly Excited Elasto-Plastic Structure.* Journal of Sound and Vibration, Vol. 72, No. 1, 1980, pp. 71-85.

[15] Stratonovich, R. L.: *Topics in the Theory of Random Noise.* Vol. I and II, Gordon and Breach, 1963.

[16] Nielsen, S. R. K., Sørensen, J. D. & Thoft-Christensen, P.: *Lifetime Reliability Estimate and Extreme Permanent Deformations of Randomly Excited Elasto-Plastic Structures.* Reliability Engineering, Vol. 4, No. 2, 1983, pp. 85-103.

[17] Nielsen, S. R. K., Sørensen, J. D. & Thoft-Christensen, P.: *Plastic Deformation of Bilinear Oscillators due to Random Excitation.* Aalborg University Centre, 1982 (not published).

A STOCHASTIC ALGORITHM FOR THE OPTIMIZATION OF SIMULATION PARAMETERS

F. Archetti
Department of Mathematics
University of Milan
Milano, Italy

M. L. Nitti
CNR - IAMI
Via Cicognara, 7
Milano, Italy

Introduction

In this paper the authors are concerned with selecting tnat design x^* of a system which is optimal according to a performance criterion $f(x)$, whose values, for $x \in K$, the feasible domain of design parameters, cannot be computed by a deterministic procedure, but only approximated, in a statistical sense, by a Monte Carlo estimator $u(x,w)$ such that

$$f(x) = E_w[u(x,w)]$$

When the selection of x^* takes place among N alternative designs, i. e. $K=\{x_1,x_2,\ldots,x_N\}$, one could use a method suggested in Iglehart (1977) to choose x^* such that, with a given probability level,

$$f(x^*) = \min_{i=1,..,N} f(x_i)$$

or an iterative procedure developed in Rubinstein (1980) that selects the optimal design x^* with probability 1.

In this paper the authors consider the case in which a finite number of alternative feasible designs is not fixed a priori and K should therefore be considered a compact set in R^N.

The determination of the optimal design x^* can thus be modelled as the stochastic optimization problem

$$\text{P1)} \quad \min_{x \in K} f(x) = E_w[u(x,w)] = \int u(x,w)P(dw)$$

where P is some probability measure. Once the distribution $P(w)$ and the expression $u(.,w)$ are known any method of constrained optimization could be employed, at least in principle, to solve P1.

Still this approach is numerically unfeasible because standard optimization methods require the avalability of the values of $f(x)$ which in turn requires, after P1, the computation of multiple integrals whose dimension is the same as the number of "elementary random events" used in the process of obtaining $u(x,w)$.

Moreover $u(x,w)$ is obviously non differentiable in x, actually not even continuous, so that the computation of the gradient of $f(x)$ should be done in an approximate way, by finite differences of $f(x)$, leading to the computation of another multiple integral.

A more sensible approach calls for replacing P1 with the "proxy" problem:

$$\min_{x \in K} g(x) = \psi[u(x,w_1),\ldots,u(x,w_k)]$$

where g(x) is an unbiased estimator of f(x) derived from the sample $\{u(x,w_i)\}$, i=1,...,k, whose accuracy can be easily evaluated, in the form of a confidence interval , at a given probability level, under the normality assumption usually accepted in a simulation experiment.

Here we shall not consider the application of a standard procedure to the minimization of g(x): even if it can work, at least in some cases, it cannot be mathematically justified and moreover requires g(x) to be an unnecessarily close approximation to f(x), also far from x^*, resulting in prohibitively long simulation runs.

In this paper the authors take the view that in order to design a suitable minimization algorithm one must consider a model of g(x) which takes into account the stochastic nature of the simulation process.

A simple such model introduced in Archetti (1979) is considered in sect. 1: an iterative optimization algorithm is subsequently derived which allows an adaptive control of the length of the simulation run. The application of the algorithm to the optimization of the simulated performance of a queueing system considered in Betro' (1982) is reported in sect. 2.

Sect. 1 - The structure of the optimization algorithm

Let $g(x_1), g(x_2),\ldots, g(x_n)$ be the observed values of g(x) at $x_1, x_2,\ldots, x_n \in K$: our model of g(x) assumes that, for any $x \in K$, $x \neq x_i$, i=1,...,n g(x) is a normal variable with expected value

$$\mu_n(x) = \frac{\sum_{i=1}^{n} \frac{f(x_i)}{\| x-x_i \|}}{\sum_{i=1}^{n} \frac{1}{\| x-x_i \|}} \tag{1.1}$$

and variance

$$\sigma_n^2(x) = \sigma^2 \min_{i=1,..,n} \| x-x_i \| \tag{1.2}$$

where $\|.\|$ is the euclidean norm and σ^2 is a parameter of the model.

A sensible value of σ^2 can be given, evaluating g(x) at the points $y_j \in K$, j=1,....,m , by the expression

$$\bar{\sigma}^2 = \sum_{j=1}^{m} \frac{[g(y_j)-\mu_n(y_j)]^2}{\min_{i=1,..,n} \| y_j - x_i \|} \tag{1.3}$$

Now we shall discuss how the model given by (1.1), (1.2) and (1.3) can be exploited in the design of an optimization algorithm.

Let

$$g_n^* = \min_{i=1,.,n} g(x_i)$$

and x_n^* the point where the value g_n^* has been observed; moreover we assume that the simulation output includes a confidence interval (l_i,u_i), at a given probability level, of $g(x_i)$, $i=1,...,n$ and we denote by (l_n^*, u_n^*) the confidence interval of g_n^*.

A step of the optimization algorithm requires two decisions to be taken: first, whether the current value x_n^*, g_n^* should be accepted as the final approximation to the optimum, at a given confidence level and within a prefixed accuracy δ; second, in case x_n^*, g_n^* is not accepted, where the next observation of g(x) is to be taken, in order to maximize the probability of improving over g_n^*.

In order to perform the first decision, we compute, for every $x \epsilon K$, $x \neq x_i$, the probability for g(x) to be less than l_n^*.

After (1.1) and (1.2) this probability, denoted by $\psi_n(x, l_n^*)$ is given by:

$$\Phi\,(l_n^*, \mu_n(x), \sigma_n(x))$$

where $\Phi(x;a,b)$ is the normal distribution with mean a and variance b.

Let

$$E^n(l_n^*, \varepsilon_1) = \{x \epsilon K \;/\; \psi_n(x, l_n^*) > \varepsilon_1\}$$

Given a value $\varepsilon_2 > 0$, we want to decide whether

$$\mu(E^n(l_n^*, \varepsilon_1)) < \varepsilon_2$$

where $\mu(.)$ denotes the measure of the set.

The value $\mu(E^n(l_n^*, \varepsilon_1))$ cannot be evaluated analitically; an approximation to it can be obtained by a MonteCarlo method.

Let z be a random variable uniformly distributed in K, y a Bernoulli random variable such that its realizations are:

$$y_i = 1 \text{ when } \psi_n(z_i, l_n^*) > \varepsilon_1 \quad \text{and}$$
$$y_i = 0 \text{ when } \psi_n(z_i, l_n^*) <= \varepsilon_1$$

where z_i are independent realizations of z. The expected value E(y) is the probability for a point uniformly distributed in K of falling in $E^n(l_n^*, \varepsilon_1)$ i.e.

$$E(y) = \mu(E^n(l_n^*, \varepsilon_1))$$

What we are really interested in is not the approximation of the value E(y) but rather deciding whether $E(y) < \varepsilon_2$: this can be done transforming the problem into that of discriminating, by a Sequential Probability Ratio Test (SPRT), (Ghosh 1970) between the two hypotheses

$H_0 : p <= \varepsilon_2 = p_0$ $\qquad$ $H_1 : p >= \varepsilon_2 + \eta = p_1$

where $0 < \varepsilon_1 < \varepsilon_2 + \eta < 1$ and the interval $(\varepsilon_2, \varepsilon_2 + \eta)$ is called the "indifference region" for the test.

In order to perform SPRT one has to fix two values α and β, the "errors" of the test, which are upper bounds respectively for the probability of accepting H_0 when H_1 is true and the probability of rejecting H_0 when it is true and compute, for every t, the values

$$a(t) = h_r(\alpha, \beta, p_0, p_1) + st$$

and

$$b(t) = h_a(\alpha, \beta, p_0, p_1) + st$$

where

$$h_a = \frac{b^*}{\log \frac{p_1(1-p_0)}{p_0(1-p_1)}}, \quad h_r = \frac{a^*}{\log \frac{p_1(1-p_0)}{p_0(1-p_1)}}$$

$$b^* = \log \frac{\beta}{1-\alpha} \qquad a^* = \log \frac{1-\beta}{\alpha}$$

$$s = \frac{\log \frac{1-p_0}{1-p_1}}{\log \frac{(1-p_0)p_1}{p_0(1-p_1)}}$$

The test is performed comparing the random variable $\Sigma_t = \sum_{i=1}^{t} y_i$ according to the following scheme:

i) accept H_0 if $\Sigma_t <= b(t)$

ii) accept H_1 if $\Sigma_t >= a(t)$

iii) continue by observing y_{t+1} if $b(t) < \Sigma_t < a(t)$

If H_0 is accepted, the "error" $d = g_n^* - l_n^*$, at the assumed confidence level, is compared with the positive value δ. If $d <= \delta$, then the algorithm is terminated and x_n^*, g_n^* are accepted as final approximation to the optimum.

If $d > \delta$, we perform a new statistically independent simulation run in x_n^* obtaining a new estimate $g(x_n^*)$ and a new confidence interval. The SPRT is then performed again about $\mu(E^n(l_n^*, \varepsilon_1))$, after the values g_n^*, l_n^*, and possibly x_n^*, have been modified according to the latest simulation runs.

If H_1 is accepted, i. e. the decision is taken that there is enough room for further improvement, g(x) is to be observed in a new point x_{n+1} given by the condition:

$$\psi_n(x_{n+1}, l_n^*) = \max_{x \varepsilon K} \psi_n(x, l_n^*) \tag{1.4}$$

Since $\psi_n(x, l_n^*)$ will generally exibit several minima rather than trying a local search from different starting point, (1.4) is replaced by the approximate condition:

$$\psi_n(x_{n+1}, l_n^*) = \max_i \psi_n(z_i, l_n^*) \tag{1.5}$$

where the values $\psi_n(z_i, l_n^*)$ are already available from the previous stage.

Sect. 2 - The simulation model and the numerical results

We consider the case of a single isolated intersection of two one-way streets, a main street M and a cross street C.

On the main street a lane is reserved for buses. Cars on M and C and buses are assumed to arrive at the crossing independently as Poisson processes with rate, respectively, λ_M, λ_C and μ; when the green is turned on, cars leave the stop line equispaced of one time unit.

We assume a simple preemption strategy activated by bus arrivals.

Let B the length of the basic cycle (i. e. not preempted) of the traffic lights. We indicate by G, A and R the lengths of the green, amber and red phases within the basic cycle. As phases can be different on M and C, they will be distinguished by the subfix M and C. We assume that the amber phase has the same length on both streets. Thus

$$B = G_M + A + R_M = G_C + A + R_C$$

The detector location D will be assumed on M at a distance S_{2A} from the stop line, where S_{2A} is the space covered by a bus during a time period 2A (see fig.1). We will indicate by D' the point where buses arrive at A instants after being detected at D. We assume that buses have the same known behaviour past D, so that, on the basis of the observations at the detector, it is possible to predict the position of a bus between D and the stop line at any instant.

If a bus arrives at D' when the traffic light signal on M is red and the one on C is amber, then no modification of the phase is needful, as the bus will arrive at the stop line with the green. If the bus arrives at D' while the signal on M is red and the one on C is green, then, in order to let the bus not to stop, the amber on C must be turned immediately, so that the signal on M will turn green when the bus arrives at the stop line. If the bus arrives at D' when the signal on M is green, then, if the green will end after the bus arrives at the stop line, no phase modification is needful; otherwise, the green on M must be extended. As successive green phases must be spaced in time at least 2A (allowing the amber on M and the amber on C), green will be actually turned off only when there are no buses between D and the stop line after last bus arrival at the stop line. In other words, green lasts until an instant t ($>G_M$) if and only if there was an arrival at D at time t-2A and then no arrivals at D until t.

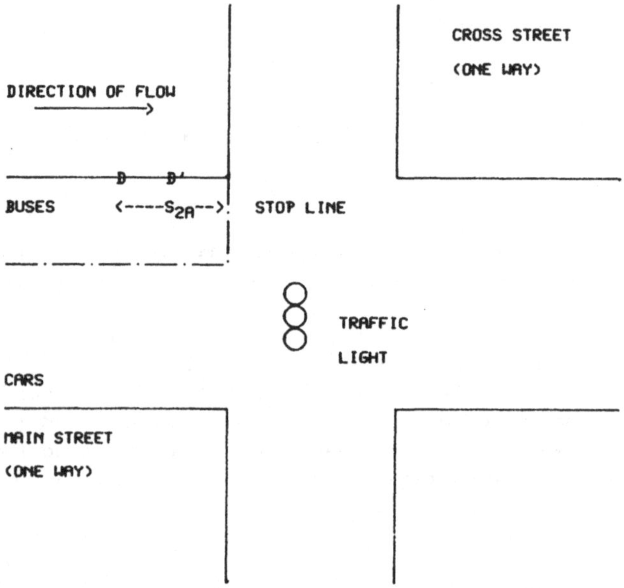

fig. 1

We remark that the length of a phase, both red and green, only depends on bus arrivals after the beginning of the phase itself. Therefore, for the properties of Poisson processes, phases legths are independent r. v. and their distribution does not depend on the time at which the phase begins.

The analysis of this model, through some mathematical intricacies, is carried in Betro' (1982) to the point of relating, in a complex expression, the average aggregate delay experienced by cars during a cycle to the parameters of the model.

The numerical computation of the delay, according to this expression, is a rather difficult task, so that MonteCarlo simulation looks like a natural tool for the analysis of this model.

The simulation has been performed on a SEL 32/27 computer using DESFOR (Discrete Event Simulation on FORtran), (Bruno 1983) a language for the simulation of discrete systems, created in the Departement of Informatic at Turin University.

DESFOR follows the method know as "process view of simulation", that relies on concurrent programming techniques. The simulation evolves with asynchronous timing. The main features of the package DESFOR are: i) quasi parallel programming: in fact, the simulation program is decomposed into quasi-parallel processes (or coroutines); ii) predefined constructs for modelling interactions among processes, such as competitions for acquiring limited resources, exchanges of items through limited buffers and so on; iii) random number generation and collection and printing of statistical data.

DESFOR has been further developed at IAMI allowing the user to perform a particular simulation run to establish, at some fixed probability level, the reset point for further runs and providing the statistical analysis of the simulation output, namely confidence interval, at a prefixed probability level, for the expected value and the variance of a choosen variable.

We simulated the behaviour of the "average aggregate delay", when the red phase R and the green phase G (with respect to the main street) vary in a specified range. So, the objective function to minimize was the following:

$$f(R,G) = 1/NCAR \sum_{i=1}^{NCAR} w(i)$$

where:
R = length of the red phase within the cycle
G = length of the green phase within the cycle
NCAR = number of departures
w(i) = waiting time of the i-th car before crossing

The algorithm has been used selecting at random 10 points for the computation of the model parameters.

The values of the parameters choosen for the optimization runs were:

probability levels(SPRT):	$p_0 = 0.05$ $(= \varepsilon_2)$
	$\eta = 0.01$, $p_1 = 0.06$
errors (SPRT)	: $\alpha = 0.05$, $\beta = 0.05$
delta value	: $\delta = 0.05$
eps1 value	: $\varepsilon_1 = 0.05$
confidence level	: 90 %

- average interarrival times
  - λ_C : 2.8
  - λ_M : 2.
  - μ : 25.

- simulation times

basic simulation time	: 150
replication time	: 50
reset time	: 100

- phases (clock simulation units)

length of amber phase	: 1
range for the green phase:	$4 <= G <= 8$
range for the red phase :	$4 <= R <= 8$

The optimization run gave these results:

number of iteration	x* R	x* G	f*	confidence interval (90%)
1	7.54	5.73	2.76	2.53 - 2.99
2	5.90	4.71	2.64	2.41 - 2.87
3	4.01	4.52	2.15	2.11 - 2.19

The final point was reached after 10 function evaluations.

Acknowledgements

This research was supported by the CNR project: Progetto Finalizzato Trasporti, Sottoprogetto II, Metodologie.

The authors wish to thank Dr. B. Betro'(CNR-IAMI, Milan) and Ing. G. Bruno (University of Turin) for their helpful and constructive remarks.

References

Archetti F., Betro' B. , A stopping criterion for global optimization algorithms, Quaderno N.61 del Dipartimento di Ricerca Operativa e Scienze Statistiche, Universita' di Pisa, 1979

Betro' B., Speranza M. G., A model for bus preemption at a semaphorized intersection , Quaderno IAMI 82.6, 1982

Bruno G., Canuto E., Simulation of production systems with DESFOR, in: "Efficiency of manufactory systems", pp.203-216, Ed. Wilson, Berg, French ,Plenum Press, NY 1983

Ghosh B. K., Sequential tests of statistical hypotheses, Addison-Wesley, Calif., 1970

Iglehart, D. L., Simulating stable stochastic system,VII Selecting best system, Algorithmic Methods in Probability, vol.7, North-Holland, Amsterdam, 1977, 37-50

Rubinstein, Y. R., Selecting the best stable stochastic system, Stochastic Processes Appl., 10, 1980, 75-85

APPROXIMATIONS AND BOUNDS IN DISCRETE STAGE MARKOV DECISION PROCESSES

Pierre L'Ecuyer
Département d'informatique,
Université Laval, Pavillon Pouliot,
Ste-Foy, Québec, Canada - G1K 7P4

1. INTRODUCTION AND SUMMARY

A great many real life problems can be modelized as Markov or Semi-Markov Decision Processes, for which the decisions are made at discrete points in time. Very often, the state or action space of the model is very large, sometimes infinite, and it is not possible to apply the dynamic programming algorithm without using some form of approximation.

In this paper, an approximate version of the value iteration dynamic programming algorithm is proposed. Each iteration of that method consists roughly in computing an approximation V_{n+1} of $T(V_n)$, where V_n is the current cost-to-go function and T is the usual dynamic programming operator. The method of approximation to be used is not fixed; it can be choosen among many available methods (spline interpolation or approximation, finite elements methods, etc.) and may vary from iteration to iteration.

The model presented here is a discrete stage markovian decision model with Borel state and action spaces, and with state dependent discounting. It generalizes common semi-markov or markov renewal decision models with discounting.

If bounds (or estimate bounds) are available at the current iteration for $T(V_n)-V_{n+1}$, then one can compute bounds (or estimate bounds) for the optimal cost-to-go function V_*. Also, at each iteration, one can use the current function V_n to obtain a policy μ and, under a simple condition on μ, one can compute bounds for the difference between V_* and the expected cost-to-go function V_μ corresponding to the policy μ. Sufficient conditions for the algorithm to yield an ε-optimal policy in a finite number of iterations are also given.

The Discrete Stage Markov Decision (DSMD) model is described in section 2. Results concerning approximations and bounds are given in section 3, and a numerical illustration is provided in section 4. Other illustrations can be found in [5].

2. THE MODEL

Let X and A be two Borel spaces, respectively called the <u>state</u> and <u>action</u> spaces. Let Γ, the set of admissible state-action couples, be an analytic subset of $X \times A$. For each state x in X, A(x) denotes the x-slice $\{a \in A \mid (x,a) \in \Gamma\}$ in Γ, assumed to be nonempty, which represents the set of actions that are <u>admissible</u> when we are in state x. The one stage <u>cost function</u> is a lower semi-analytic function $g:\Gamma \to \mathbb{R}$, and the transition kernel Q is a Borel measurable stochastic kernel on X given $X \times A$. For (x,a) in Γ, g(x,a) represents the (expected) cost incurred for the current stage, and $Q(\cdot|x,a)$ is the probability law according to which the next state is generated. The discounting function is a Borel measurable function β: $X \to (0,1]$. A cost incurred in state x is discounted to the "origin" by the factor $\beta(x)$.

At each of a sequence of stages, labeled 0, 1, 2, ..., the state x of the system is observed and an action a is choosen in A(x). A cost g(x,a) is incurred for the current stage, and the next state is generated randomly according to the probability measure $Q(\cdot|x,a)$. That new state x' is then observed, a new action a' is choosen in A(x'), and so on. All current value costs are discounted at a given point of reference, called the "origin", by the discount factor $\beta(x)$, which depends on the current state of the system. The actions are choosen dynamically by a decision maker, who is trying to minimize the expected total discounted cost over an infinite horizon.

This model has been studied in more detail in [5] and [6]. In this section, we present a summary of the results that have been proved in these two references.

A <u>policy</u> is a universally measurable function $\mu:X\to A$ such that $\mu(x)\in A(x)$ for all x in X. Here, we restrict to non random, markovian stage-stationary policies; as shown in [5, 6] for this model, it is of no benefit to allow for a larger class of policies.

It could be shown that for any policy μ and initial state x, there is a unique probability measure $P_{\mu,x}$ over the space of sequences $(x_0, a_0, x_1, a_1, \ldots)$ in $X\times A\times X\times A\times\ldots$, and a corresponding mathematical expectation $E_{\mu,x}$.

We define, when it exists, $V_\mu(x)$ as the expected present value cost-to-go from state x when we use policy μ, and $V_*(x)$ as the optimal expected present cost-to-go from state x.

$$V_\mu(x) \triangleq \frac{1}{\beta(x)} E_{\mu,x}\left[\sum_{n=0}^{\infty} \beta(x_n) g(x_n, a_n)\right] \tag{1}$$

$$V_*(x) \triangleq \inf_\mu V_\mu(x). \tag{2}$$

We say that a policy μ is optimal [ε-optimal] if $V_\mu(x) = V_*(x)$ [$\le V_*(x)+\varepsilon$] for all x.

In order to guarantee the existence of $V_\mu(x)$ and $V_*(x)$, we need additional assumptions. Each of the two following sets of assumptions, called C and LC, is a sufficient set to guarantee the existence of these quantities.

Let us define a function $\alpha:\Gamma\to[0,1]$ as

$$\alpha(x,a) \triangleq \frac{1}{\beta(x)}\int_X \beta(x')Q(dx'|x,a) , \tag{3}$$

which represents the expected discount factor from the next stage to the present stage, if we are in state x and choose action a.

<u>Assumption C</u>: There exists $\alpha_1 < 1$, $g_0 \le 0$, $g_1 \ge 0$ such that for all (x,a) in Γ,

$$0 \le \alpha(x,a) \le \alpha_1 \tag{4}$$

$$g_0 \le g(x,a) \le g_1 \tag{5}$$

<u>Assumption LC</u>: There exists $\delta_1 < 1$, $g_1 \ge 0$, K_1 and K_2 in $\mathbb{R}$, and a policy $\tilde{\mu}$ such that

$$\alpha(x,\tilde{\mu}(x)) \le \delta_1 \quad \text{for all x in X} \tag{6}$$

$$g(x,\tilde{\mu}(x)) \le g_1 \quad \text{for all x in X} \tag{7}$$

$$Q(\{x'|\beta(x') \le \beta(x)\}|x,a) = 1 \quad \text{for all (x,a) in } \Gamma \tag{8}$$

$$g(x,a) \ge K_1 + K_2\, \alpha(x,a) \quad \text{for all (x,a) in } \Gamma \tag{9}$$

$$K_1 + K_2 > 0. \tag{10}$$

A DSMD model satisfying assumption C is called <u>contracting</u>. This is the standard assumption permitting to use the contraction mapping properties, as in Denardo [2]. A DSMD model satisfying assumption LC is called <u>locally contracting</u>. This locally contracting model encompasses many concrete situations. The example considered in section 4 is an instance of a locally contracting model which is not contracting. Other examples are given in [5].

We now define B_2 as the set of lower semi-analytic functions $V:X\to\mathbb{R}$, bounded by $\underline{V}$ and $\overline{V}$, where

$$\underline{V} \triangleq \begin{cases} g_0/(1-\alpha_1) & \text{under assumption C} \\ K_1 + \min(0,K_2) & \text{under assumption LC} \end{cases} \tag{11}$$

$$\overline{V} \triangleq \begin{cases} g_1/(1-\alpha_1) & \text{under assumption C} \\ g_1/(1-\delta_1) & \text{under assumption LC} \end{cases} \tag{12}$$

For any functional V in B_2, we define

$$H(V)(x,a) \triangleq g(x,a) + \int_X \frac{\beta(x')}{\beta(x)} V(x')Q(dx'|x,a) \quad \text{for all (x,a) in } \Gamma \tag{13}$$

$$T_\mu(V)(x,a) \triangleq H(V)(x,\mu(x)) \quad \text{for any policy } \mu \text{ and x in X} \tag{14}$$

$$T(V)(x) \triangleq \inf_{a\in A(x)} H(V)(x,a) \quad \text{for all x in X.} \tag{15}$$

$T: B_2 \to B_2$ is the usual dynamic programming operator. The value iteration dynamic programming procedure consists in choosing an initial V in B_2 and applying T successively. Under assumption C or LC, we can prove the following theorems (see [6]).

THEOREM 1. For any V in B_2, we have:

(a) $T(V) = V$ iff $V = V_*$

(b) $\lim_{n\to\infty} \| T^n(V) - V_* \| = 0$, where $\|\cdot\|$ denotes the supremum norm and T^n denotes the composition of n times T.

(c) V_* is in B_2

(d) A policy μ is optimal iff $T_\mu(V_*) = V_*$, and iff $T(V_\mu) = V_\mu \in B_2$.

THEOREM 2.

(a) For any $\varepsilon > 0$, there exists an ε-optimal policy

(b) There exists an optimal policy iff the infimum in $\inf_{a\in A(x)} H(V_*)(x)$ is attained for all x in X.

These theorems form the basis of the dynamic programming value iteration algorithm. However, this algorithm is not always implementable in its pure form. First, one will obviously have to stop after a finite number of iterations. Second, if the state space is infinite, or finite but too large, which is often the case, one will not be able to compute T(V) for every state x.

3. APPROXIMATIONS AND BOUNDS

MacQueen [7], Denardo [2], and Porteus [8], [9] have obtained bounds for $\| T^n(V)-V_*\|$ and $\| V_\mu-V_* \|$ where μ is the policy retained after the n-th iteration, and T is applied exactly at each iteration. These bounds converge geometrically to 0 as n tends to infinity.

When the state space X is infinite, it is necessary to use an approximate computation of T(V).

A common approach makes use of a discretization procedure, which consists in selecting a finite partition of X and a representing state for each part. This yields a finite state model, called an approximating model. Fox [3], Bertsekas [1], Whitt [10], Hinderer [4], and several others, have obtained various properties of such an approximation scheme. Some of them also considered a similar discretization for the action space. Typically, they obtained bounds for the norms $\| V_*-\tilde{V}_* \|$ and $\| V_*-V_\mu\|$, where $\tilde{V}_*$ is an extension to the state space X of the optimal value function of an approximate model, constant on each subset of the partition, and μ is the optimal policy of the approximate model, also extented to X as a piecewise constant function.

Knowing that piecewise constant approximation is not always the best mean to approximate a function, one might consider more sophisticated schemes, like spline interpolation or approximation, finite element methods, etc. The idea is to compute T(V) at a finite number of points, then approximate T(V) over the entire state space by a function V_1, take V_1 as our new V, and repeat.

Many questions can be raised about such an approach. (1) When does such a procedure converge? (2) Can we obtain bounds for the difference between the current cost function V_1, obtained after a certain iteration, and the optimal cost function V_*? (3) Can we know how good is the policy retained at the end of the algorithm? That is, if μ is the policy retained at the last iteration, can we bound $V_\mu-V_*$?

The following algorithm provides answers to these questions.

ALGORITHM.

1. Initialization.

For model C, let $n_0 = 1$ and α_1 as in assumption C. For model LC, choose α_1 in (0,1) and

$$n_0 \triangleq \frac{g_1/(1-\delta_1) - K_1 - \min(0,K_2)}{(K_1+K_2)\alpha_1} . \quad (16)$$

Set $\underline{V}$ and $\overline{V}$ as in (11, 12). Take any function V in B_2 as an initial guess for V_*. Choose real values Δ_1 and Δ_2 for the stopping tests.

2. Compute T(V) at a finite number of points. Then choose V_1 in B_2, $\underline{V} \le V_1 \le \overline{V}$, as an approximation of T(V) on X.

3. Obtain δ^- and δ^+ such that

$$-\delta^- \le T(V)-V_1 \le \delta^+ . \quad (17)$$

Compute

$$\varepsilon^- = n_0\delta^- + (n_0-1)\,\|(V-V_1)^+\| \quad (18)$$

$$\varepsilon^+ = n_0\delta^+ + (n_0-1)\,\|(V_1-V)^+\| \quad (19)$$

$$\underline{V} := \max(\underline{V},\ V_1 - \varepsilon^- - \|(V-V_1+\varepsilon^-)^+\|\alpha_1/(1-\alpha_1)) \quad (20)$$

$$\overline{V} := \min(\overline{V},\ V_1 + \varepsilon^+ + \|(V_1+\varepsilon^+-V)^+\|\alpha_1/(1-\alpha_1)) . \quad (21)$$

4. Stopping test: stop if $\overline{V}-\underline{V} \le \Delta_1$.

5. Find $\varepsilon_0 \ge 0$ and a "best current policy" μ such that

$$T_\mu(V) \le V_1 + \varepsilon_0 . \quad (22)$$

Let α be the contracting factor of T_μ. Set $\varepsilon=\infty$ if $\alpha=1$, otherwise

$$\varepsilon = \varepsilon^- + \|(V-V_1+\varepsilon^-)^+\|\alpha_1/(1-\alpha_1) + \varepsilon_0 + \|(V_1+\varepsilon_0-V)^+\|\,\alpha/(1-\alpha) . \quad (23)$$

6. Stopping test: Stop if $\varepsilon \le \Delta_2$. Otherwise, set V := V_1 and return to step 2.

THEOREM 3

(a) After step 3, we have $\underline{V} \le V_* \le \overline{V}$. Hence, if we stop at step 4, then the retained V_1 satisfies $\| V_1 - \overline{V}_* \| \le \Delta_1$.

(b) A policy μ retained at step 5 is ε-optimal. Hence, if we stop at step 6, then the retained policy is Δ_2 -optimal.

(c) If the sequence of values of δ^- and δ^+ converge to 0, then $\| V-V_* \| \to 0$, that is the sequence of cost functions provided by the algorithm converge uniformly to V_*.

(d) Under assumption C, if the sequence of values of δ^-, δ^+ and ε_0 in the algorithm converge to 0, then for any $\varepsilon > 0$, an ε-optimal policy is obtained in a finite number of iterations.

This Theorem is proven in [5].

Notice that some steps of the algorithm can be implemented in different ways, which makes the algorithm very flexible. For instance, one can choose the method of approximation as he wishes, and may even change the method from one iteration to the other.

The main originality of this algorithm is that it provides bounds which takes into account the fact that only a finite number of iterations are done, and also the error of approximation at each iteration.

One of the most difficult implementation aspects is certainly the computation of δ^-, δ^+ and ε_0. In many cases, one will content himself with "reasonable estimations" of these values, obtaining thus "estimate bounds". For instance, one way to estimate δ^- and δ^+, when T(V) and V_1 are reasonably well behaved, is to recompute T(V) at a very large number of new points, and compute the real approximation error at these points. If these points are well distributed and numerous, then the smallest and largest of these errors could estimate δ^- and δ^+.

4. A NUMERICAL ILLUSTRATION

Consider a system comprised of 3 identical components, working independently. The system is observed at discrete times: (a) when a component fails, (b) when the repairman decides to perform preventive replacements. Each component has an age dependent failure rate function $\lambda(t) = .02\, t$. For $i = 1, 2, 3$, the state s_i of component i is $s_i = \infty$ if component i is failed, s_i = the age of component i otherwise. The state of the system is $x = (\tau, s)$, where τ is the current time and $s = (s_1, s_2, s_3)$. The state space is $X = [0,\infty) \times [0,\infty]^3$. At each observation time, the repairman chooses an action $a = (\ell, d)$, that is (a) a number ℓ of components to replace, which is at least one, and (b) a time interval d until the next planned intervention. The action space is $A = \{1, 2, 3\} \times [0,\infty]$.

The cost of an intervention comprises a fixed cost $c_i = 1$, and a replacement cost $c_r = 1$ for each component replaced. A failure cost $c_f = 2$ is also incurred each time a component fails, and all the costs are discounted at rate $\rho = 0.1$. The discounting function here is then

$$\beta(\tau, s) = e^{-\rho\tau}. \tag{24}$$

One can easily define Γ, g and Q using the above quantities, and obtain a locally contracting DSMD model, with $K_1 = 2$, $K_2 = 0$, $g_1 = 6$, $\delta_1 = .6201$, $\underline{V} = 2$, $\overline{V} = 15.88$, $\alpha_1 = .5$ and $n_0 = 14$.

Since the components are identical and the failure rate is non decreasing, then at each intervention time, we will certainly replace at least the oldest component (which is the failed one if there is one). Hence, it is enough to consider only the states (ages) of the two others, in decreasing order. Also, since this problem gives rise to a stationary model, the cost-to-go functional will not depend on the current time. We can then restrict ourselves to cost-to-go functionals V defined on the cone

$$S = \{(r_1, r_2) \in \mathbb{R}^2 \mid r_1 \ge r_2 \ge 0\}.$$

At each iteration of the algorithm, in step 2, we can choose $0 = p_1 < p_2 < \ldots < p_n$ and define

$$\Omega = \{(p_i, p_j) \mid 1 \le j \le i \le n\}$$

as the finite set where T(V) is to be evaluated (Fig. 1).

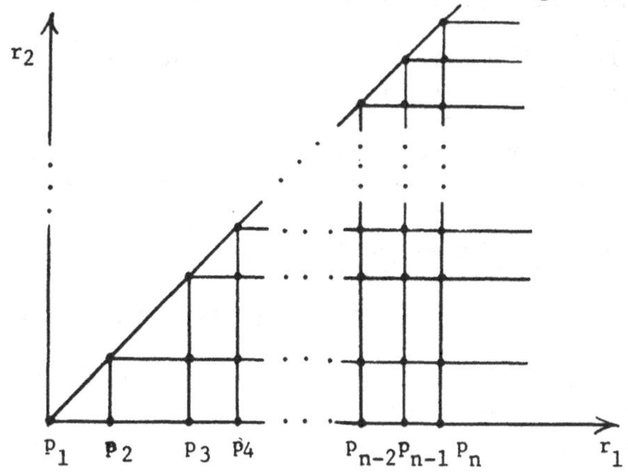

Figure 1. A partition of S.

These points determine a covering of S by rectangles, triangles and unbounded polyedra. We then choose the approximating function V_1 as the one that interpolates T(V) on Ω, is affine on each triangle, bilinear on each bounded rectangle, and affine constant in r_1 on each imbounded polyedra.

We can do 30 iterations with $n = 5$ and $p_i = 2.5\ (i - 1)$ for $i = 1, \ldots, 5$; then 15 iterations with $n = 22$ and $p_i = (i-1)/4$ for $i = 1, \ldots, 22$. At the last iteration, we obtain $\| (V_1-V)^+ \| = .00008$, $\| (V-V_1)^+ \| = 0.0$, and $V_1(0,0) = 11.148$.

In order to obtain values of δ^- and δ^+, we now reevaluate $T(V)-V_1$ on a very fine grid, and take the minimum and maximum values of $T(V)-V_1$ on that new grid. This procedure is reasonable, since V_1 and T(V) are smooth monotonous functions. Taking $n = 88$ and $p_i = (i-1)/16$, $i = 1, \ldots, 88$ to define the new grid, this yields $\delta^- = .00073$ and $\delta^+ = 0.0$. Using (20,21), one easily computes the "estimate" bounds: $-.021 \le V_* - V_1 \le .002$. It says that the relative error on V_* is at most 0.2%.

Also, using the last cost-to-go function V, it is easy to determine a "best current policy". For a given pair of ages (r_1, r_2), its says how many components to replace and the optimal delay until the next planned intervention (which can be) approximated in the same way as T(V)). Notice that this optimal delay need to be computed only over the surface in figure 1 where only one component has to be replaced.

We then compute $T_\mu(V)-V_1$ on the very fine grid, to estimate ε_0, and use (23) to compute ε. We obtain $\varepsilon = .021$, suggesting that the retained policy is at worst .021 -optimal.

ACKNOWLEDGEMENT
The author wish to thank professor Alain Haurie for his continuous support and interest. This research was also supported by NSERC-Canada Grant #A5463.

REFERENCES

[1] D.P. Bertsekas, "Convergence of Discretization Procedures in Dynamic Programming", IEEE Trans. Automat. Contr., vol. AC-20, pp. 415-419, 1975.

[2] E. V. Denardo, "Contraction Mappings in the Theory Underlying Dynamic Programming", SIAM Review, 9, pp. 165-177, 1967.

[3] B.L. Fox, "Discretizing Dynamic Programs", J. Optim. Theory Appl., 11, pp. 228-234, 1973.

[4] K. Hinderer, "On Approximate Solutions of Finite State Dynamic Programs", in Dynamic Programming and its Applications, M.L. Puterman ed., Academic Press, pp. 289-317, 1978.

[5] P. L'Ecuyer, "Processus de décision markoviens à étapes discrètes: application à des problèmes de remplacement d'équipement", Les cahiers du GERAD, no. G-83-06, Ecole des H.E.C., Montréal, 1983 (in French).

[6] P. L'Ecuyer and A. Haurie, "Discrete Event Dynamic Programming in Borel Spaces with State Dependent Discounting", Dept. Informatique, U. Laval, Québec, Report no. DIUL-RR-8309, 1983.

[7] J. MacQueen, "A Modified Dynamic Programming Method for Markovian Decision Problems", J. Math. Anal. Appl., 14, pp. 38-43, 1966.

[8] E. Porteus, "Some Bounds for Discounted Sequential Decision Processes", Man. Sci., 18, pp. 7-11, 1971.

[9] E. Porteus, "Bounds and Transformations for Discounted Finite Markov Decision Chains", Oper. Res., 23, pp. 761-784, 1975.

[10] W. Whitt, "Approximation of Dynamic Programs I and II", Math. of Oper. Res., vol. 3, pp. 231-243, 1978 and vol. 4, pp. 179-185, 1979.

OVERALL CONTROL OF AN ELECTRICITY SUPPLY AND DEMAND SYSTEM : A GLOBAL FEEDBACK FOR THE FRENCH SYSTEM

P. LEDERER, Ph. TORRION, JP.BOUTTES
Electricité de France
Paris, France

1. INTRODUCTION

a) The French System in the Future

- a continuous improvement of the daily load factor, which is already greater than 0.9, owing to a tariff policy which induces in particular the development of storage systems for heating uses ;
- increased seasonal variations of demand ;
- an increasing sensitivity of the load to temperature changes : night hours of a cold winter day are now at the same level as the peak hours of a mild winter day.

As a consequence the peak hours of the year will cover a large number of hours per day during the coldest days of the year, which occur of course at unforeseeable dates.

Thus, the marginal cost of generation can cover, for any winter day, the whole range of fuel costs, according to the realization of random perturbations and to the operation of regulation facilities.

b) Towards an Improved Expansion and Operation Planning of the Supply and Demand System

Electricité de France thus faces the problem of improving the control of the generating and consumption system seen as a whole.

In order to meet instantaneous variations in demand while minimizing the associated long range total cost for the national economy in the expansion process of the system, several actions are available on the supply as well as on the demand side :
- the addition of peaking generating facilities ;
- the addition of hydro storage (conventional or pumped hydro) ;
- peak-load pricing schemes (such as the new "peak-day withdrawal option" included in EDF's revised tariff system).

These regulation facilities have to be operated near to real time in order to take advantage of the latest available information on the realization of random phenomena -such as temperature- wich cannot be accurately forecasted several days in advance.

The detailed representation of random phenomena affecting the balance between supply and demand, together with the optimization of management strategies capable of responding to these perturbations, have become essential in view of the expansion problem described above.

The problem to be solved is thus a stochastic, large scale, non linear system.

c) Earlier Work on Related Problems

This problem of optimizing the long-term (yearly) operation of a multistorage power system has received considerable attention in recent research, but has not yet been completely solved.

The choice of methods for solving such a stochastic large system is indeed limited to dynamic programming which is usually not feasible for more than three state variables due to the exponential growth of the computing time and storage requirements.

Since there is no alternative, compromises must be made. Unfortunately, classical compromises such as :

- deterministic techniques,
- aggregation applied by Massé (1946) and his followers...,
- relaxation techniques applied by Pronovost and Boulva (1978) or Delebecque and Quadrat (1978),
- optimization in a parametrized class of feedbacks applied at E.D.F. by Colleter and Lederer (1981) and further developed by Quadrat (1982) ,

suffer major drawbacks when applied to our problem.

Determistic techniques are obviously unsuitable since they completely overlook the importance of random phenomena in electrical systems.

When aggregating all regulation facilities into a unique storage, all the information related to local constraints is lost. It is therefore clear that aggregation can be successfully applied only to a system composed of roughly similar reservoirs, i.e. reservoirs which have about the same time-scale -ratio between the energy in the full storage and the maximum capacity of the power plant-, and similar patterns of inflows. This is not the case in the French system which includes seasonal conventional hydro, seasonal pumped-hydro, weekly pumped-hydro, and peak-day withdrawal options.

Relaxation techniques can be used to determine the optimal local feedback operating policy for a system of parallel reservoirs, taking into account the stochastic inflows. However, in the present case where the most important random phenomena -such as temperature and thermal forced outages- are coupling the operation of the different storages, the only possible way to use relaxation techniques would consist of repeatedly solving the problem at the beginning of each period of time, thus implementing an open-closed loop solution. Such an approach could be suitable for operating purposes, but not for expansion planning purposes where a closed-loop operating policy is clearly needed.

Sometimes there is, a priori, an idea of what a "good" feedback would look like. This a priori information can be used to solve a simpler problem than the general one. The feedback is parametrized and the open-loop parameters are optimized. Unfortunately in this problem we are dealing with several time-scales, several types of reserves and the available a priori information is very poor.

Actually our system should be more precisely defined at this point as a multi type, multi time-scale, multi reservoir power system (Table 2).

Table 2

Type of regulation facility	Time scale
Conventional hydro (seasonal)	1 100 hours
Pumped hydro (seasonal)	160 hours
Pumped hydro (weekly)	35 hours
Pumped hydro (daily)	10 hours
Real-time tariff (peak day withdrawal option)	400 hours

Type and time-scale are the key-concepts that characterize a regulation facility. They are the guidelines to the approach we developed. This approach is made of two successive steps :

- System modelling with an initial reduction of the natural dimension of the original system.

Here we keep the best part of aggregation by aggregating together all facilities with the same type and time-scale.

In this way the natural dimension of the French system is reduced from about 60 to 4 or 5 (conventional hydro, seasonal pumped hydro, weekly pumped-hydro, optional real-time tariffs, ...) i.e. the number of the different types of regulation facilities in the supply and demand system.

Optimal operation of the system

In this paper we investigate a global sub-optimal operation feedback of the supply and demand system which may help the utility to better appreciate the interactions between the different regulation facilities in the system, i.e. to better estimate their profitability and better operate them.

2. PROBLEM FORMULATION

a) System Modelling

As usual in tightly interconnected systems, a single node model is assumed whith all generators, connected to a common node, meeting the total demand for electric power. The total demand is accordingly adjusted to take transmission losses into account.

The Cost Function

The classical piecewise linear approximation of the thermal power cost curve is adequate for the long term regulation problem. The outage cost function used by E.D.F. is a function of the amount of unsupplied energy, growing with the depth of the outage. It is represented in the model by fictitious generating facilities thus allowing for optimal management of failures.

The Hydropower System

According to the nature of the problem under consideration, the dimension of the system is first reduced by aggregating all the regulation facilities of identical type and time-scale :

- conventional hydroplants are thus represented, according to a classical approach used by E.D.F. (Meslier, 1978), by three fictitious hydroplants : a seasonal reservoir, a weekly reservoir, an a run of the river plant ;

- pumped hydroplants are represented by three different plants with seasonal, weekly and daily regulating cycles.

b) The problem

The system under consideration consists of N independent reservoirs that must compete with thermal units to meet the demand for electricity so as to minimize the yearly mathematical expectation of generation and unsupplied energy costs.

These reservoirs are defined by :

- their maximum energy content : X_i^{max} $i = 1,N$;
- their maximum generating capacity : U_i^{max} $i = 1,N$;
- their maximum pumping capacity if applicable : U_i^{min} $i = 1,N$;
- their natural inflows (random variables) : a_i^t $i = 1,N$ during period t.

Let us denote :

X_i^t : the energy in reservoir i at the beginning of period t ; $0 \leqslant x_i^t \leqslant X_i^{max}$;

U_i^t : the generation (or demand for pumping, if negative) of reservoir i at period t $U_i^{min} \leqslant U_j^t \leqslant U_i^{max}$;

d_i^t : the energy spillage from reservoir i, at period t ;

D^t : the demand for electrical energy at period t ;

C^t : the generation and outage cost function (which depends on thermal availability).

The original problem is, thus, a N-state variable problem. We are looking for a closed-loop operating policy so as to :

$$(P) \left\{ \text{Minimize} \sum_{t=0}^{t_f} E\, C^t \left(D^t - \sum_{i=1}^{N} U_i^t\right) \right.$$

under the following constraints :

$$(C) \left\{ \begin{array}{ll} X_i^{t+1} = x_i^t + a_i^t - U_i^t - d_i^t & \\ 0 \leqslant x_i^{t+1} \leqslant X_i^{max} & i = 1, \ldots, N \\ U_i^{min} \leqslant U_i^t \leqslant U_i^{max} & t = 0, \ldots, t_f \\ d_i^t \geqslant C & \end{array} \right.$$

Owing to the importance of random phenomena affecting temperature and thermal forced outages, the control law has to be a function of the demand for electricity and of thermal availability.

In order to obtain such a control law, problem (P) is to be solved by dynamic programming. This consists of solving the following functional equation recursively, going backwards in time from period t_f :

$$(1)\quad V^t(x_1^t, \ldots, x_N^t) = E\left(\underset{U_1^t,\ldots,U_n^t}{\text{minimum}} \left[C^t\left(D^t - \sum_{i=1}^{N} U_i^t\right) + V^{t+1}\left(x_1^{t+1}, \ldots, x_N^{t+1}\right)\right]\right.$$

with respect to set of constraints (C) at period t. As a result, we would have :

$$\forall i,\ i = 1, \ldots, N \qquad U_i^t = U_i^t\left(x_1^t, \ldots, x_N^t, D^t, T^t\right)$$
$$\forall t,\ t = 1, \ldots, t_f$$

where T^t is the vector of available thermal generating capacities.

Unfortunately, since N is greater than three in our case, computing time and storage requirements are far too large to make this method feasible. Although, in the French system, the knowledge of the total energy in all the reservoirs at period t allows a first rough estimate of the future expected production cost, aggregation cannot help determine a feasible operation of each reservoir.

Therefore, the control law U_i should at least be a function of :
- the energy in reservoir i,
- the total energy in all the reservoirs.

Hence, we developed a method, starting from an idea of Turgeon (1980), who suggested a way to obtain such a control law in a large multi-reservoir system.

3. OPTIMIZATION : PARTIAL AGGREGATIONS

This idea consists of breaking up the original system of N reservoirs into N independent systems of two reservoirs and then solving each of them by stochastic dynamic programming.

Each problem (P) is derived from the original problem (P) by aggregating all reservoirs except one, i.e. reservoir i.

Let us denote :

$$\bar{x}_i^t = \sum_{j \neq i} x_j^t, \quad \bar{X}_i^{max} = \sum_{j \neq i} X_j^{max}$$

$$\bar{U}_i^t = \sum_{j \neq i} U_j^t, \qquad \begin{cases} \bar{U}_i^{max} = \sum_{j \neq i} U_j^{max} \\ \bar{U}_i^{min} = \sum_{j \neq i} U_j^{min} \end{cases}$$

$$\bar{a}_i^t = \sum_{j \neq i} a_j^t$$

For problem (P_i), we thus obtain the following formulation :

$$(P_i) \qquad \underset{U_i^t,\, \bar{U}_i^t}{\text{Min}} \sum_{t=0}^{t_f} E\, C^t(D^t - U_i^t - \bar{U}_i^t)$$

With respect to constraints :

$$(C_i) \begin{cases} x_i^{t+1} = x_i^t + a_i^t - U_i^t - d_i^t \quad ; \quad \bar{x}_i^{t+1} = \bar{x}_i^t + \bar{a}_i^t - \bar{U}_i^t - \bar{d}_i^t \\ 0 \leqslant x_i^t \leqslant X_i^{max}, \quad 0 \leqslant \bar{x}_i^t < \bar{X}_i^{max} \quad ; \quad U_i^{min} \leqslant U_i^t \leqslant U_i^{max}, \ \bar{U}_i^{min} \leqslant \bar{U}_i^t \leqslant \bar{U}_i^{max} \\ d_i^t \geqslant 0 \quad ; \quad \bar{d}_i^t \geqslant 0 \qquad t = 0, \ldots, t_f \end{cases}$$

The dimension of the state vector in problem (P_i) is two.

We are, then, capable of completely solving it with the following recursive equation :

$$V_i^t\,(x_i^t, \bar{x}_i^t) = E\left(\underset{U_i^t,\bar{U}_i^t}{\text{Min}} \left[C^t(D^t - U_i^t - \bar{U}_i^t) + V_i^{t+1}\,(x_i^{t+1}, \bar{x}_i^{t+1})\right]\right)$$

where U_i^t, $\bar{U}_i^t$ must satisfy the set of constraints (C_i) at period t.

Since, for all i, the constraint set (C_i) of problem (P_i) is larger than the constraint set (C) of the original problem (P), it is obvious that :

$$(2) \qquad V_i^t (x_i^t, \bar{x}_i^t) \leqslant V^t (x_1^t, x_2^t, \ldots, x_N^t) \qquad \begin{array}{l} i = 1, \ldots, N \\ t = 0, \ldots, t_f \end{array}$$

Transition optimisation

The problem here is to obtain the control law as a function of demand and thermal availability, taking into account stochastic inflows, and is further complicated by pumped-storage operation.

Due to the effect of daily temperature on the load curve, time discretization must be as fine as possible that is several time-steps a day all over the year.

In order to perform the whole optimization within an acceptable computing time, we had to devise a very quick calculation method for the transition optimization (equation 1). For a detailed description see LEDERER, TORRION and BOUTTES (1983).

4. SIMULATION : THE CHOICE OF A GLOBAL FEEDBACK

The question, at this point, is : how to obtain the best possible closed-loop policy with the only available information i.e. the N functions V_i^t ?

As mentioned above, the calculation of the true marginal values of energy $-\frac{\partial V^{t+1}}{\partial x_i}$ i = 1, ..., N is impossible since the function V^{t+1} is unknown.

Now, let us consider the derivatives $-\frac{\partial V_i^{t+1}}{\partial x_i}$ i = 1, ..., N. It is clear that $\frac{\partial V_i^{t+1}}{\partial x_i}$ is the true marginal value of energy in reservoir i, but when others are aggregated into a single reservoir. Hence $\frac{\partial V_i^{t+1}}{\partial x_i}$ takes into account :

- local constraints on reservoir i which express its type and time-scale ;
- total energy left in all the reservoirs, which is, as known before, a very valuable state- information.

Therefore, we will choose the operating policy defined by :

$$(PS) \quad \underset{U_1^t, U_2^t, \ldots, U_N^t}{\text{Min}} \quad C^t(D^t - \sum_{i=1}^{N} U_i^t) - \sum_{i=1}^{N} \frac{\partial V_i^{t+1}}{\partial x_i} \quad U_i^t$$

A Comment On Turgeon's Paper

Turgeon suggest to choose the operating policy of the N-reservoir system so as to :

$$\underset{U_1^t, \ldots, U_N^t}{\text{Minimize}} \left[C^t(D^t - \sum_{i=1}^{N} U_i^t) + \frac{1}{N} \sum_{i=1}^{N} V_i^{t+1} (x_i^{t+1}, \bar{x}_i^{t+1}) \right]$$

with respect to the set of constraints (C).

In other words, Turgeon assumes that the unknown function V^t is best approximated by the average of the N fuctions V_i^t, that is :

$$V^t(x_1^t, x_2^t, \ldots, x_N^t) \simeq \frac{1}{N} \sum_{i=1}^{N} V_i^t\,(x_i^t, \bar{x}_i^t) = V_m^t\,(x_1^t, \ldots, x_N^t).$$

Actually, Turgeon does not seem to be able to produce any general reason for that assumption. On the contrary, we found drawbacks in that choice :

From inequality (2), it it quite clear that V_m^t will never be the best available estimate of V^t ;

From the definition of V_m^t as the Bellman function, the marginal value of energy in the reservoir i is :

$$m_i = -\frac{1}{N}\left[\frac{\partial V_i^t}{\partial x_i}(x_i^t, \bar{x}_i^t) + \sum_{j \neq i} \frac{\partial V_j^t}{\partial x_j}(x_j^t, \bar{x}_j^t)\right]$$

that is, m_i is taken as an average of :

. the true marginal value of energy in reservoir i, but when others are aggregated into a single reservoir :

$$\text{i.e. } \frac{\partial V_i^t}{\partial x_i}(x_i^t, \bar{x}_i^t) \;;$$

. (N-1) marginal values of energy stored in larger reservoirs (complementary to reservoir j, j ≠ i) of which the reservoir i is only a part :

$$\text{i.e. } \frac{\partial V_j^t}{\partial \bar{x}_j}(x_j^t, \bar{x}_j^t).$$

Now, it is obvious that the larger is N, the closer the characteristics of complementary reservoirs x_j are ; hence the closer the derivatives $\frac{\partial V_j^t}{\partial x_j}$ and the heavier their weight in m_i. As a result, the larger is N, the closer are the m_i's for i = 1, 2, ..., N.[1]

Therefore, such an operation choice can lead to near optimal policy when characteristics of the different reservoirs are similar (since, the true marginal values of energy will be actually close). As shown before, such conditions do not prevail in the French system.

At last, another cumbersome consequence of that choice is the instability of the m 's when adding a(N+1)th marginal reservoir, which obviously is not true for the real marginal values of energy in the reservoirs i, i = 1, ..., N.

5. NUMERICAL RESULTS

In this section, we present some numerical results based on the demand forecast of year 2000 with the associated optimal generating facilities.

Different numerical methods are compared on this test problem.

We simulated 400 years of realization of all random variables in order to estimate the annual operating costs. As expected the results rank as follows :

Cost of fuell aggregation < Cost of partial aggregation i < Cost of the real system

Thus providing a lower bound of the true unreachable optimum cost. The sub-optimality of the global feedback that we developed and of the associated feasible operating policy is thus qualified by this lower bound.

Results

	Total aggregation	Partial aggregation*	Real system
Operating Cost (Fuel + Outage) 10^6.FF	31,530	32,410	32,670
Outage Cost 10^6.FF	1,150	1,520	1,740
Unsupplied Energy GWh	285	350	400
Total Hydro Generation GWh	19,130	18,430	17,880
Total Hydro Pumping GWh	8,430	7,430	6,650

* We chose here the 2 state model which led to the highest $V^o(X^o, \bar{X}^o)$ in order to obtain through simulation the best possible lower bound.

We thus obtain the following differences between :

Total aggregation vs. real system : - 3.5 %
Partial aggregation vs. real system : - 0.8 %.

It should be emphasized that this 0.8 % tells us that the cost of our operating policy is at worst this far from the ideal optimum.

6. CONCLUSION

From the test problem solved above, it is clear that the method described in this paper brings, in the French case, significant advances :

- the electrical supply and demand system can be modelled while specifying, with one component of the state vector, each different type of regulation facility. Implementing the global feedback obtained yields a feasible operating policy ;

- this policy is very close to the optimum (which cannot be obtained in the needed closed-loop form due to the large scale of the system under study) ;

- a detailed representation of random phenomena can be performed as the model works on a daily basis thanks to the quick analytic optimization method developed ;

- finally -as Turgeon points out- the method is quite flexible as its computing time increases only linearly with the number of regulation facilities under study.

REFERENCES

COLLETER-LEDERER
Optimal Operation Feedbacks for the French Hydro-power System, CORS-TIMS-ORSA, Nat. Meeting, May 1981.

DELEBECQUE-QUADRAT
Contribution of stochastic Control Singular Perturbation Averaging and Team Theories to an Example of Large Scale System : Management of Hydropower Production, IEEE AC, April 1978.

ERNOULT-MESLIER
Analysis and Forecast of Electrical Energy Demand, R.G.E., April 1982.

KUSHNER, M.
Probabilistic Methods for Approximations in Stochastic Control, Academic Press, 1977.

MASSE, P.
Les Réserves et la Régulation de l'Avenir, Hermann, Paris, 1944.

MESLIER, F.
Simulation of the Optimal Control of a Power System : Latest Developments in the Greta Model, IPC, PSCC 6, 1978.

PRONOVOST, R. and J. BOULVA
Long Range Operation Planning of a Hydro-Thermal System Modelling and Optimization. Meeting of the Canadian Electrical Association, Toronto, Ontario, March 1978.

TURGEON, A.
Optimal Operation of Multireservoir Power Systems with Stochastic Inflows, Water Resources Research, April 1980.

KOHN D.
Calcul Analytique de l'Espérance d'un Coût de gestion et de Défaillance Internal EDF Report - December 1978.

JENKINS, and BAYLESS
More than you Really Need to Know About the Cumulant Method, WASP Conference, Colombus, March 1981.

LEDERER, TORRION, BOUTTES
A Global Feedback for the French System, Internal Report, EDF, 1983.

OPTIMAL MAINTENANCE POLICIES FOR MODULAR STANDBY SYSTEMS

S. G. Tzafestas
Control Systems Laboratory
Electrical Engineering Dept.
Patras University
Patras, Greece

C. A. Botsaris
Operational Research Unit
Hellenic Airforce Academy
Dekelia Airbase
Athens, Greece

ABSTRACT

The problem of determining optimal maintenance policies for a class of repairable standby systems with a hierarchical or modular design is considered. The reliability measures of this system were derived in a previous work by a direct use of the discrete-state continuous-time Markov model of the system behaviour. Here, standard optimal control theory, specifically a generalized version of Pontryagin's minimum principle, based on an integral Hamiltonian functional, is applied to this model for deriving both time-variable and fixed maintenance policies over the whole mission time. To this end an appropriate cost function, involving a maintanance cost term and a down-time cost term, is minimized with respect to the vector repair rate function u(t) subject to the practical constraint $0 \leq u(t) \leq U$, where U is a given upper repair-rate limit. A particular nontrivial numerical example is worked out.

1. INTRODUCTION

In this paper we derive optimal maintenance policies for a repairable standby system of items, called assemblies, with a hierarchical or modular design. By a hierarchically designed repairable assembly we mean an item that has components which are also repairable items. In a previous paper [1] the basic reliability measures, namely (i) the probability $P_{ffo}(t)$ of failure-free operation, the probability Q(t) of failure, the meantime to first failure: MTFF, the mean up time: MUT, the mean down time: MDT, the mean cycle time: MCT, and the system availability A have been determined. The analysis was carried out describing the system as a discrete-state continuous time Markov process. Also a preliminary study on the optimal configuration determined by maximizing the MTFF subject to the following budget constraint $k_\alpha n+k_c m \leq C$ (k_α is the cost of each assembly; k_c is the cost of each component ;n,m are integers, C is the maximum allowed cost) was made through a numerical procedure. Our purpose here is to apply optimal control theory to this model, specifically a generalized version of Pontryagin's minimum principle based on an integral Hamiltonian functional, for deriving both fixed and time-dependent maintenance policies over the whole operation time, that minimize a mixture of maintenance and down-time costs [2-4]. To illustrate the general results a specific nontrivial example is worked out. Computational results regarding this and other examples will be reported elsewhere.

2. THE SYSTEM MODEL

Consider a repairable standby system of items (assemblies) with a hierarchical (or modular) structure in which the items have components that are also repairable. The system consists of a stock of m independent components and n assemblies, with one assembly operating and n-1 assemblies in standby (ready to replace failed assemblies). Items in standby are assumed failure-free. Each assembly failure is assumed to occur by a failure of at most a single component. A failed assembly is returned to operation by removing the defective component and replacing it by a functional one.

The defective component is then repaired and put back to the standby stock of components. If no stock of operative components is available, the assembly repair cycle is lengthened until an operative component becomes available through the repair procedure.

Denote by S(k,i) the state where there are k failed components and i operative assemblies in the system. Then S(k,0), k=0,..., n+m are the states in which all n assemblies are failed. These failed states can be *absorbing* or *reflecting*. Now, under the assumption that at t=0 our system is in an operating state, the state propability equations of the system are

$$\frac{d}{dt}\begin{bmatrix} p_w(t) \\ \hline p_f(t) \end{bmatrix} = \left[\begin{array}{c|c} E & F \\ \hline G & H \end{array}\right] \begin{bmatrix} p_w(t) \\ \hline p_f(t) \end{bmatrix} \tag{1}$$

where $p_w(t)\{p_f(t)\}$ is the vector with components the probabilities that the system is in a working failed state, and

$$E = \begin{bmatrix} A_0 B_1 & & & & 0 \\ L_0 A_1 B_2 & & & & \\ & \ddots & & & \\ 0 & & L_{n+m-3} & A_{n+m-2} & B_{n+m-1} \\ & & & L_{n+m-2} & A_{n+m-1} \end{bmatrix}, \quad F = \begin{bmatrix} F_0 \\ F_1 \\ \vdots \\ F_{n+m-1} \end{bmatrix}$$

$$G = [G_0 G_1 \dots G_{n+m-1}]$$

$$H = \begin{bmatrix} -(n+m)\mu_c & & & & & 0 \\ (n+m)\mu_c & \ddots & & & & \\ & & -[k\mu_c+r(k)\mu_\alpha] & & & \\ & & k\mu_c & \ddots & & \\ & & & & -\mu_c+n\mu_\alpha & \\ 0 & & & & & \mu_c-n\mu_\alpha \end{bmatrix} \tag{2}$$

with $\lambda(k,i)$, $\mu_\alpha(k,i)$, $\mu_c(k,i)$ being the failure rate, assembly service rate, and component service rate in state $S(k,i)$, respectively. Note that given the number k of failed components, the number i of operative assemblies assumes only the values $i=0,\ldots,r(k)$, where $r(k)=\min\{n,n+m-k\}$ $k=0,1,\ldots,n+m$. The matrices A_k, L_k, B_k, F_k, and G_k have the following form

$$A_k = \begin{bmatrix} -(\lambda+k\mu_c) & \mu_\alpha & & & 0 \\ & -(\lambda+k\mu_c+\mu_\alpha) & 2\mu_\alpha & & \\ & & \ddots & & \\ 0 & & & -[\lambda+k\mu_c+(r(k)-2)\mu_\alpha] & [r(k)-1]\mu_\alpha \\ & & & & -[\lambda+k\mu_c+(r(k)-1)\mu_\alpha] \end{bmatrix}$$

$$L_k = \left[\begin{array}{c|c} 0 & 0 \\ \hline \Lambda_k & 0 \end{array}\right], \quad 0\le k<m, \quad L_k=[\Lambda_k|0], \quad m\le k<n+m-1 \tag{3}$$

$$\Lambda_k = \text{diag}\ (\lambda), \qquad \Lambda_k \in R^{(r(k)-1)x(r(k)-1)}$$

$$B_k = M_k, \quad 0<k\le m, \quad B_k = \left[\frac{0}{M_k}\right], \quad m<k\le n+m-1$$

$$M_k = \text{diag}\ (k\mu_c), \qquad M_k \in R^{r(k)xr(k)}$$

$$G_k = \left[\frac{D_k}{0}\right], \quad D_k = \begin{bmatrix} 0 & \ldots & 0 \\ \vdots & & \vdots \\ 0 & \ldots & \lambda \\ \vdots & & \vdots \\ 0 & & 0 \end{bmatrix} \leftarrow (k+1), \quad 0\le k\le n+m-1, \quad D_k \in R^{(n+m)xr(k)}$$

$$F_k = [0|V_k], \quad V_k = \begin{bmatrix} 0 & \ldots & 0 & \ldots & 0 \\ \vdots & & \vdots & & \vdots \\ 0 & . & r(k)\mu_\alpha & . & 0 \end{bmatrix}, \quad 0\le k\le n+m-1, \quad V_k \in R^{r(k)x(n+m)}$$

$$\uparrow (k+1)$$

Denoting by p_{ki} the probability that the system is in state $S(k,i)$ at time $t=0$, where by assumption

$$\sum_{k=0}^{n+m-1} \sum_{i=1}^{r(k)} p_{ki} = 1 , \quad p_{ki} = p(k,i,0) \tag{4}$$

the reliability measures of the system can be determined by solving (1) under the initial condition (4). One observes that E is a q×q block tridiagonal matrix $(q=n(m+1)+n(n-1)/2$ and that the column sums of the system state transition matrix $A = \begin{bmatrix} E & F \\ G & H \end{bmatrix}$ and equal to zero.

Here we assume unrestricted repair, i.e.

$$\lambda(k,i) = \lambda$$

$$\mu_\alpha(k,i) = [r(k)-i]\mu_\alpha \quad , \qquad r(k) = \min\{n,n+m-k\}$$

$$\mu_c(k,i) = k\mu_c$$

Let now Q(t) be the probability that the system is in a failed state at time $t \geq 0$. Clearly

$$Q(t) = \sum_{k=0}^{n+m} p(k,0;t) \tag{5a}$$

Denoting by T the random variable respresenting the time interval from t=0 to system failure, the probability density function f(t) of T is given by

$$f(t) = \dot{Q}(t) = \sum_{k=0}^{n+m} \dot{p}(k,0;t) \tag{5b}$$

As is proved in [1] the mean value E(T) of T is given by the sum

$$E(T) = \sum_{k=0}^{n+m-1} \sum_{i=1}^{r(k)} z(k,i) \tag{5c}$$

where the q-dimensional vector z with components z(k,i) is the solution of the linear system

$$Ez = -p_w(0) \tag{5e}$$

Moreover,

$$MUT = u_q^T \pi_w / u_{n+m+1}^T G \pi_w \tag{5f}$$

and

$$MDT = u_{n+m+1}^T \pi_f / u_q^T F \pi_f \tag{5g}$$

In the above π_w and π_f are the stationary values of $p_w(t)$ and $p_f(t)$, respectively.

In [1] a model was also developed for determining the optimal configuration of the system subject to some budget constraint. The model has the form

$$\begin{aligned} &\max \text{ MTFF} \\ &\text{subject to } k_\alpha n + k_c m \leq B \\ &n, m \text{ integers} \end{aligned} \tag{5h}$$

where n and m are the decision variables, k_α is the cost of each assembly, k_c is the cost of each component and B is the available budget. Given the values of n and m, the mean time to first failure is found from the relation

$$MTFF = \sum_{k=0}^{n+m-1} \sum_{i=1}^{r(k)} z(k,i) \tag{5i}$$

where the vector z with components z(k,i) is the solution of the linear system

$$Ez = -(1,\ldots,0)^T \tag{5j}$$

obtained using a numerical procedure which does not require a matrix to be stored.

3. FORMULATION OF THE OPTIMAL MAINTENANCE PROBLEM

Let $F \subset \Omega$ be the set of failed states (Ω is the system's finite state space), and $\tilde{F} \subset F$ be the set of reflecting failed states from which the system can be returned to a working state by a maintenance policy. Now if $P_\alpha(t)$ is the probability that the system is in state $S_\alpha \in \Omega$, and $u_\alpha(t)$ is the maintenance (repair) rate vector $[\mu_\alpha, \mu_c]^T$, then the model (1) can be written in the compact form

$$\frac{dP(t)}{dt} = A(\tilde{u}_F, t)P(t) \quad , \quad P(0) = P_0 \tag{6}$$

where $P=\{P_\alpha, S_\alpha \epsilon \Omega\}$, $\tilde{u}_F=\{u_\alpha, S_\alpha \epsilon \tilde{F}\}$, and

$$\sum_{\Omega} P_\alpha(t)=1, \qquad t \geq 0 \tag{7}$$

The maintenance action $u_\alpha(t)$, which here plays the role of the control variable, is allowed to be fixed throughout the operation (control) interval $[0,T]$ or variable depending on the failure rates of the individual components. The optimal maintenance control problem to be solved here is: For system (6) choose $u_\alpha(t)$ such that the following total cost is minimezed

$$J=\frac{1}{T}\int_0^T \sum_{S_\alpha \epsilon \tilde{F}} [q_\alpha^T(t)P_\alpha(t)+r_\alpha^T(t)u_\alpha(t)]dt \tag{8}$$

subject to the constraints

$$0 \leq u_\alpha \leq U_\alpha(t), \qquad t \geq 0, \qquad S_\alpha \epsilon \tilde{F} \tag{9}$$

Obviously, one can also treat integral constraints without special difficulty. The term involving $q_\alpha^T(t)P_\alpha(t)$ in (8) represents the average downtime cost, and the term involving $r_\alpha^T(t)u_\alpha(t)$ represents the average maintenance cost, over the desired operation period T. Care must be given in selecting the weighting cost functions $q_\alpha(t)$ and $r_\alpha(t)$.

4. SOLUTION OF PROBLEM

The solution will be found in the following two cases:

Case A: Maintenance action $u_\alpha(t)$, $t\epsilon[0,T]$ fixed.
Case B: Maintenance action $u_\alpha(t)$, $t\epsilon[0,T]$ variable.

To cover both the fixed and variable control cases we introduce the following generalized (integral) Hamiltonian functional (y is the costate vector):

$$\hat{H}=\frac{1}{T}\int_0^T H dt, \qquad \text{where} \qquad H=y^T AP-\sum_{\alpha \epsilon \tilde{F}} (q_\alpha^T P_\alpha + r_\alpha^T u_\alpha) \tag{10}$$

which is to be maximized (since the minus sign is used) over all admissible maintenance control policies $u_{\tilde{F}}$ in $[0,T]$. Here H is the standard Hamiltonian functional.

The state and costate equations (in terms of H) are:

$$\frac{dP(t)}{dt} = \frac{\partial H}{\partial y(t)} , \quad P(0) = P_0 \tag{11a}$$

$$\frac{dy(t)}{dt} = - \frac{\partial H}{\partial P(t)} \quad \text{(Final state determined from transversality conditions)} \tag{11b}$$

In the variable maintenance case without integral constraints it is sufficient to maximize the standard Hamiltonian H, i.e. the optimal variable maintenance policy $u^0_{\alpha v}$ is given by

$$u^0_{\alpha v} = \{u_\alpha : \text{ H is maximized}\} \tag{12a}$$

Correspondingly, the optimal fixed maintenance policy $u^0_{\alpha f}$ is given by

$$u^0_{\alpha f} = \{u_\alpha : \hat{H} \text{ is maximized}\} \tag{12b}$$

Clearly, the integral Hamiltonian $\hat{H}$ is more general than H, but leads to weaker conditions. The maximum principle based on $\hat{H}$ is known as extended (or generalized) maximum principle [5-6].

4. A PARTICULAR EXAMPLE

Consider the case where n=2 and m=1, i.e. a system with two assemblies (one operating and one in failure-free standby) each one consisting of a single component.

Then form (1)-(3) we get

$$\frac{d}{dt}\begin{bmatrix} p_w \\ p_f \end{bmatrix} = \begin{bmatrix} E & F \\ G & H \end{bmatrix} \begin{bmatrix} p_w \\ p_f \end{bmatrix} , \quad p_w = \begin{bmatrix} P_1 \\ P_2 \\ P_3 \\ P_4 \\ P_5 \end{bmatrix} , \quad p_f = \begin{bmatrix} P_6 \\ P_7 \\ P_8 \\ P_9 \end{bmatrix} \tag{13}$$

$$E = \left[\begin{array}{cc|cc|c} -\lambda & \mu_\alpha & \mu_c & 0 & 0 \\ 0 & -(\lambda+\mu_\alpha) & 0 & \mu_c & 0 \\ \hline 0 & 0 & -(\lambda+\mu_c) & \mu_\alpha & 0 \\ \lambda & 0 & 0 & -(\lambda+\mu_\alpha+\mu_c) & 2\mu_c \\ \hline 0 & 0 & \lambda & 0 & -(\lambda+2\mu_c) \end{array}\right] \qquad F = \left[\begin{array}{cc|cc} 0 & 0 & 0 & 0 \\ 0 & 0 & 0 & 2\mu_\alpha \\ \hline 0 & 0 & 0 & 0 \\ 0 & 0 & 2\mu_\alpha & 0 \\ \hline 0 & \mu_\alpha & 0 & 0 \end{array}\right]$$

$$
G = \left[\begin{array}{cc:cc:c} 0 & 0 & 0 & 0 & \lambda \\ 0 & 0 & 0 & \lambda & 0 \\ \hdashline 0 & \lambda & 0 & 0 & 0 \\ 0 & 0 & 0 & 0 & 0 \end{array}\right] , \quad H = \left[\begin{array}{cc:cc} -3\mu_c & 0 & 0 & 0 \\ 3\mu_c & -(2\mu_c+\mu_\alpha) & 0 & 0 \\ \hdashline 0 & -2\mu_c & -(\mu_c+2\mu_\alpha) & 0 \\ 0 & 0 & \mu_c & -2\mu_\alpha \end{array}\right]
$$

Here the set Ω of states is

$$\Omega = \{S_1, S_2, S_3, S_4, S_5, S_6, S_7, S_8, S_9\}$$

and the set F of failed states is

$$F = \{S_6, S_7, S_8, S_9\}$$

The state transition diagram of this system is shown in Fig. 1, from which one can easily see that the set of reflecting states from which the system can be returned to a working state with some maintenance action is

$$\tilde{F} = \{S_6, S_7, S_8, S_9\}$$

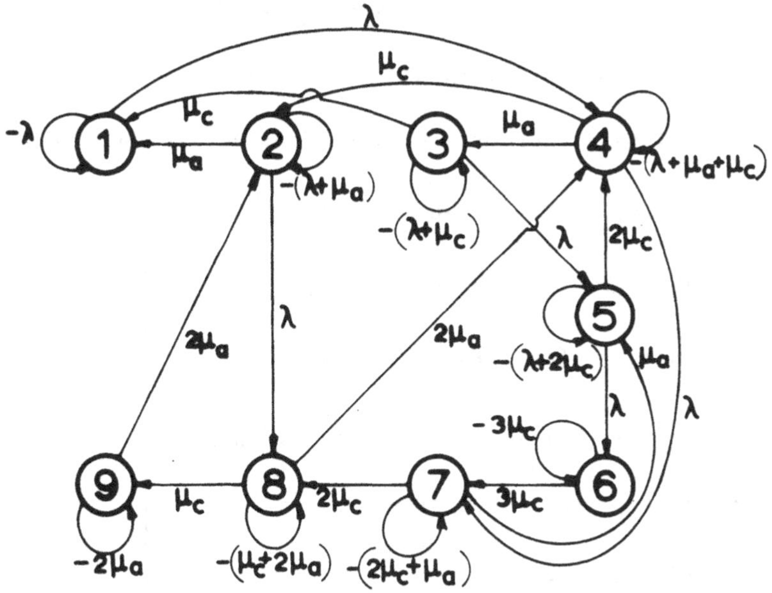

Fig. 1. System state transition diagram.

i.e. all failed states are reflecting states.

The cost function to be minimized here is

$$J(u) = \frac{1}{T} \int_0^T \{q_6(t)P_6(t)+q_7(t)P_7(t)+q_8(t)P_8(t)$$

$$+q_9(t)P_9(t)+r_\alpha(t)\mu_\alpha(t)+r_c(t)\mu_c(t)\}dt \tag{15}$$

where $u_\alpha = [\mu_\alpha, \mu_c]^T$.

The standard Hamiltonian function H is

$$H = y^T \begin{bmatrix} E & F \\ G & H \end{bmatrix} \begin{bmatrix} P_w \\ P_f \end{bmatrix} \{q_6P_6+q_7P_7+q_8P_8+q_9P_9+r_\alpha\mu_\alpha+r_c\mu_c\}$$

$$= (-\lambda P_1+\mu_\alpha P_2+\mu_c P_3)y_1+\{-(\lambda+\mu_\alpha)P_2+\mu_c P_4+2\mu_\alpha P_9\}y_2$$

$$+\{-(\lambda+\mu_c)P_3+\mu_\alpha P_4\}y_3+\{\lambda P_1-(\lambda+\mu_\alpha+\mu_c)P_4+2\mu_c P_5+2\mu_\alpha P_8\}y_4$$

$$+\{\lambda P_3-(\lambda+2\mu_c)P_5+\mu_\alpha P_7\}y_5+(\lambda P_5-3\mu_c P_6)y_6$$

$$+\{\lambda P_4+3\mu_c P_6-(2\mu_c+\mu_\alpha)P_7\}y_7$$

$$+\{\lambda P_2+2\mu_c P_7-(\mu_c+2\mu_\alpha)P_8\}y_8+(\mu_c P_8-2\mu_\alpha P_9)y_9$$

$$-(q_6P_6+q_7P_7+q_8P_8+q_9P_9+r_\alpha\mu_\alpha+r_c\mu_c) \tag{16}$$

The canonical (state and costate) equations are found to be:

$$\left.\begin{aligned}
dP_1/dt &= \partial H/\partial y_1 = -\lambda P_1+\mu_\alpha P_2+\mu_c P_3 \\
dP_2/dt &= \partial H/\partial y_2 = -(\lambda+\mu_\alpha)P_2+\mu_c P_4+2\mu_\alpha P_9 \\
dP_3/dt &= \partial H/\partial y_3 = -(\lambda+\mu_c)P_3+\mu_\alpha P_4 \\
dP_4/dt &= \partial H/\partial y_4 = \lambda P_1-(\lambda+\mu_\alpha+\mu_c)P_4+2\mu_c P_5+2\mu_\alpha P_8 \\
dP_5/dt &= \partial H/\partial y_5 = \lambda P_3-(\lambda+2\mu_c)P_5+\mu_\alpha P_7 \\
dP_6/dt &= \partial H/\partial y_6 = \lambda P_5-3\mu_c P_6 \\
dP_7/dt &= \partial H/\partial y_7 = \lambda P_4+3\mu_c P_6-(2\mu_c+\mu_\alpha)P_7 \\
dP_8/dt &= \partial H/\partial y_8 = \lambda P_2+2\mu_c P_7-(\mu_c+2\mu_\alpha)P_8 \\
dP_9/dt &= \partial H/\partial y_9 = \mu_c P_8-2\mu_\alpha P_9
\end{aligned}\right\} \tag{17}$$

$$\left.\begin{aligned}
dy_1/dt &= -\partial H/\partial P_1 = -(-\lambda y_1+\lambda y_4)\\
dy_2/dt &= -\partial H/\partial P_2 = -\{\mu_\alpha y_1-(\lambda+\mu_\alpha)y_2+\lambda y_8\}\\
dy_3/dt &= -\partial H/\partial P_3 = -\{\mu_c y_1-(\lambda+\mu_c)y_3+\lambda y_5\}\\
dy_4/dt &= -\partial H/\partial P_4 = -\{\mu_c y_2+\mu_\alpha y_3-(\lambda+\mu_\alpha+\mu_c)y_4+\lambda y_7\}\\
dy_5/dt &= -\partial H/\partial P_5 = -\{2\mu_c y_4-(\lambda+2\mu_c)y_5+\lambda y_6\}\\
dy_6/dt &= -\partial H/\partial P_6 = -\{-3\mu_c y_6+3\mu_c y_7-q_6\}\\
dy_7/dt &= -\partial H/\partial P_7 = -\{\mu_\alpha y_5-(2\mu_c+\mu_\alpha)y_7+2\mu_c y_8-q_7\}\\
dy_8/dt &= -\partial H/\partial P_8 = -\{2\mu_\alpha y_4-(\mu_c+\mu_\alpha)y_8+\mu_c y_9-q_8\}\\
dy_9/dt &= -\partial H/\partial P_9 = -\{2\mu_\alpha y_2-2\mu_\alpha y_9-q_9\}
\end{aligned}\right\} \qquad (18)$$

The Hamiltonian function can be written as

$$H = h_0+h_\alpha\mu_\alpha+h_c\mu_c$$

$$\triangleq h_0+h^T u_\alpha$$

where

$$h^T = [h_\alpha, h_c] \ , \qquad u_\alpha^T = [\mu_\alpha, \mu_c] \ . \qquad (18a)$$

$$h_0 = -(q_6P_6+q_7P_7+q_8P_8+q_9P_9)+\lambda P_1(y_4-y_1)$$

$$+\lambda P_2(y_8-y_2)+\lambda P_3(y_5-y_3)+\lambda P_4(y_7-y_4)+\lambda P_5(y_6-y_5) \qquad (18b)$$

$$h_\alpha = P_2(y_1-y_2)+P_4(y_3-y_4)+P_7(y_5-y_7)$$

$$+2P_8(y_4-y_8)+2P_9(y_2-y_9) - r_\alpha \qquad (18c)$$

$$h_c = P_3(y_1-y_3)+P_4(y_2-y_4)+2P_5(y_4-y_5)$$

$$+3P_6(y_7-y_6)+2P_7(y_8-y_7)+P_8(y_9-y_8)-r_c \qquad (18d)$$

The problem is to maximize H with respect to u_α(i.e. w.r.t. μ_α and μ_c) subject to the constraint $\{M^T = [M_\alpha, M_c]\}$

$$0 \le u_\alpha \le M \quad \text{or} \quad 0 \le \mu_\alpha \le M_\alpha \ , \quad 0 \le \mu_c \le M_c \tag{19}$$

This means that the linear function of u_α should be maximum. The optimum maintenance policy is given by

$$u_\alpha^0(t) = \begin{bmatrix} \mu_\alpha^0 \\ \mu_c^0 \end{bmatrix} \tag{20}$$

where

$$\mu_\alpha^0 = \frac{M_\alpha}{2}\{1+\text{sgn}(h_\alpha)\} \tag{21a}$$

$$\mu_c^0 = \frac{M_c}{2}\{1+\text{sgn}(h_c)\} \tag{21b}$$

and sgn(·) is the standard signum function defined as

$$\text{sgn}\omega(t) = \begin{cases} +1 , & \omega(t)>0 \\ \text{unspecified}, & \omega(t)=0 \\ -1 , & \omega(t)<0 \end{cases} \tag{21c}$$

In order to implement this optimum maintenance policy one has to solve the canonical equations (17) and (18) with μ_α and μ_c being replaced by the optimum ones μ_α^0 and μ_c^0. The only difficulty is that the canonical equations form a two-point-boundary value problem (TPBVP), i.e. P(t) is determined at the initial time, say

$$P_1(0) = 1 \ , \quad P_i(0) = 0 \ , \quad i = 2,\dots,q \tag{22a}$$

while y(t) is determined at the final time.

Such TPBVPs can be integrated by two main methods: (i) hill climbing on an initial guess $u^{(1)}(t)$, $t \in [0,T]$ of the optimal policy, and (ii) hill climbing for improving an initial guess of the initial value y(0) of the costate vector [7-10].

The final condition y(T) of y(t) is determined from the associated transversality condition.

Here, if P(T) is free the $P_i(T)$ must satisfy the condition

$$\sum_{S_i \in \Omega} P_i(T) = 1$$

and the transversality condition implies that the final value $\lambda(T)$ of the costate vector must be orthogonal to the tangent manifold M_t of the manifold $M = \{P(T) \in R^N : P_i(T) \geq 0, \sum_{\Omega} P_i(T) = 1\}$. Obviously here $P_1+P_2+\ldots+P_q=1$ represents a line in the q-dimensional Euclidean space, and so the tangent manifold M_t is identical to the manifold M itself. Thus any point of M_t is of the form $[z_1, z_2, \ldots, z_8, 1-z_1-z_2-\ldots-z_8]$ with $z_i \geq 0$, $i=1,2,\ldots,$ z_8 and $z_1+z_2+\ldots+z_8 \leq 1$. This implies that

$$y_1(T) = y_2(T) = \ldots = y_q(T) = 0 \tag{22b}$$

We have therefore seen that the canonical equations (17) and (18) must be integrated by using the initial-final conditions (22a,b).

5. CONCLUSIONS

Among the primary problems in the planning, design, and management of modern multicomponent systems is the problem of maximizing the overall system reliability and availability or minimizing the consumption resources subject to imposed reliability/availability constraints. The two typical ways for improving system reliability are: (i) using redundant components, and (ii) using repair maintenance policies.

Optimal control theory has been used to optimize system reliability by choosing appropriate maintenance policies or to optimize a certain objective function choosing simultaneously the maintenance policy and the life time (sale date) of the system at hand [2,3].

Here we have applied a generalized version of Pontryagin's minimum principle to a repairable standby system of items with a modular structure. The objective function considered is a linear mixture of maintenance and down-time costs. Work is in progress in two directions; first to generalize in various ways the repairable standby model in order to cover other practical cases, and second to find maintenance policies by minimizing objective functions of other types.

REFERENCES

1. C.A. Botsaris and S.G. Tzafestas: "Reliability Analysis and Optimization of a Repairable Standby System with Modularly Designed-Items", <u>7th Symp. über Operations Research</u>, Hochschule St. Gallen, Aug.1982.

2. S.G. Tzafestas: "Optimization of System Reliability: A Survey of Problems and Techniques", Int.J.Systems Sci., Vol. 11, No. 4, 445-486, 1980.
3. S.G. Tzafestas: "Optimal Control Policies in System Reliability and Maintenance", in "Optimization and Control of Dynamic O.R. Models" (S.G. Tzafestas, Ed.), Ch. 12, North-Holland, 1982.
4. N.U. Ahmed and K.F. Schenk: "Optimal Availability of Maintenance Systems", IEEE Trans. Reliab., Vol. R-27, No. 1, 41-45, 1978.
5. H.H. Yeh and J. Tou: "On the General Theory of Optimal Processes", Int. J. Control, Vol. 9, 443-451, 1969.
6. H.H. Yeh and R. Kuhler: "Additional Properties of an Extended Maximum Principle", Int. J. Control, Vol. 17, 1281-1286, 1973.
7. S.G. Tzafestas: "Optimal Distributed-Parameter Control Using Classical Variational Theory", Int.J.Control, Vol. 12, 593-608, 1970.
8. S.G. Tzafestas: "Final-Value Control of Nonlinear Composite Distributed-and Lumped-Parameter Systems, J.Franklin Inst., Vol. 290, 439-451, 1970.
9. S.G. Tzafestas and D. Efthymiatos: "A Newton-Like Optimization Algorithm for Digital Systems with Time Delay", Int.J.Systems Sci., Vol. 2, 389-394, 1972.
10. S.G. Tzafestas: Automatic Control Systems, Patras Univ. (Ch. 3 Sec. 5.5), 1981.

PROBABILISTIC ANALYSIS OF SOME TRAVELLING SALESMAN HEURISTICS

Hoon Liong Ong
Department of Industrial & Systems Engineering
National University of Singapore
Kent Ridge, Singapore 0511

In this paper we analyze the probabilistic performance of some heuristics for the travelling salesman problem (TSP) and the maximum weight travelling salesman problem (MWTSP). If the edge weights for a TSP graph are independent and identically distributed according to the uniform or exponential distribution then the expected tour length produced by the nearest neighbour heuristic is $O(\log n)$, where n is the number of cities to be visited. It is further shown that the probability that the tour length produced by this heuristic exceeds any fixed percentage of the expected tour length goes to zero when n tends to infinity. For the MWTSP, if the edge weights are drawn independently from a uniform distribution on [0,1] then the expected tour length produced by the best neighbour heuristic is $O(n-\ln(n))$ and the tour length produced by this heuristic converges to the expected tour length in probability.

1. INTRODUCTION

The travelling salesman problem (TSP) is one of the combinatorial optimization problems that has been extensively studied because of its ease of statement and great difficulty of solution as well as its many applications and generalizations to other combinatorial optimization problems. The problem can be formulated as follows: Given a complete graph G = (V,E) containing n vertices and a distance function defined on the edge set E, find a shortest tour passing through every vertex of G exactly once. There are several variants of the TSP. A variant is to formulate the problem as a maximization problem (MWTSP). That is, to find a maximum weight Hamiltonian circuit on G. A problem stated in this form can be transformed to the classical TSP through a simple linear transformation and hence, the two problems are mathematically equivalent.

It has been shown that the TSP is NP-hard [8, 13] meaning it cannot be solved by a polynomial algorithm unless P = NP. Thus, instead of solving the problem optimally, one looks for some fast heuristics to provide good approximate solutions. The performance of a heuristic can be evaluated by analyzing its worst-case and average-case approximate solutions.

Worst-case analysis for several TSP heuristics has been successfully performed by a number of researchers [2, 4, 5, 6, 7, 10, 11, 20]. The best known worst-case approximation result is due to Christofides [2], who gives an algorithm that runs in time $O(n^3)$ and produces a solution bounded above by 1.5 times the optimum.

In probabilistic analysis, one assumes a probability distribution on the problem data and tries to establish some probabilistic properties of a heuristic such as the expected run time, the expected solution value and the probability that the heuristic finds a solution within a pre-specified percentage of optimality.

The approach of probabilistic analysis has been quite popular since the notable papers of Karp on asymptotically optimal heuristics for the TSP [1, 9, 14, 15, 16, 18, 19, 21, 22]. Most of these results prove that for certain probability distribution assumptions on the input data the optimal solution value and the heuristic value are asymptotically close. When the probability of such events happening for all values of n converge sufficiently quickly to 1 the heuristic is called asymptotically optimal almost everywhere. The interest in these results stems from the fact that heuristic that fare badly according to worst-case analysis seem to perform satisfactorily according to probabilistic analysis.

Lucker [18, 19] and Weide [22] have analyzed the asymptotic behaviour of the optimal solution to the TSP with unit normal edge weights and showed that the optimal solution converges to $-n(2 \log n)^{\frac{1}{2}}$ almost everywhere. However, their analysis cannot be easily applied to establish a similar result for other type of edge weight distribution. It appears that tight bounds are particularly difficult to obtain in the case in which edge weights are chosen from a uniform distribution.

In this paper we analyze the behaviour of some TSP heuristics for the case in which edge weights are chosen independently from various distributions. In particular, the expected tour lengths produced by these heuristics for the uniform and exponential distributions on edge weights are derived and we further show that the tour length converges to the expected heuristic solution in probability. Therefore, given a TSP graph, one can estimate explicitly the expected travel cost incurred which provides a very useful information in the management of a distribution system.

In Section 2, we analyze the expected performance of the nearest neighbour heuristic for both the symmetric and asymmetric TSPs. Section 3 provides a general framework for analyzing the performance of the savings heuristic. Finally, in Section 4, we analyze the expected performance of the best neighbour heuristic for the maximum weight TSP.

2. EXPECTED PERFORMANCE OF NEAREST NEIGHBOUR HEURISTIC

In this section, we analyze the performance of a greedy heuristic for the TSP. The symmetric TSP will be considered first and then extended to the asymmetric case.

Let V be the set of n vertices and c a cost function defined in the edge set E of an undirected complete graph G = (V,E). Suppose the edge costs of G are independent and identically distributed with a common density function f(x) defined on the interval $[0,\infty)$. Then both the optimal tour length and the tour length produced by a heuristic are random variables. Let H(n) denote the tour length produced by a heuristic H and E(H(n)), VAR(H(n)) the expected value and variance of H(n) respectively.

Consider the nearest neighbour heuristic of the TSP which starts at an arbitrary vertex and chooses an edge of minimum cost adjacent to it. It then continues to choose edges of minimum cost adjacent to the vertex just reached subject to the requirement that the collection of selected edges be contained in a tour.

Let e_i, $1 \leq i \leq n$, be the edges selected by the nearest neighbour heuristic in that order. Since all the edges in E are independent random variables and when an edge is to be selected, the heuristic only examines those edges which are adjacent to the vertex just visited, it follows that the costs of the n edges e_i except the last edge e_n are independent random variables.

To obtain the distributions of $c(e_i)$, we need the following elementary lemma [12, 17].

<u>Lemma 1</u>. Let $X_1, X_2, \ldots, X_m$ be m independent random variables having the same distribution density function f(x), then the first order statistic Y_1 and the mth order statistic Y_m of the random variables X_i have the following distributions:

$$G_1(y) = 1 - (1-F(y))^m \ , \quad G_m(y) = (F(y))^m$$

where F(x) is the distribution function of X_i.

Consider the situation of the nearest neighbour heuristic after the first i edges in the tour have been chosen. The next edge will be one of the n - i edges between the current vertex and an as yet unvisited vertex. It is noted that as these n - i edges are not the candidates of the choices made for the first i edges, the weights of these n - i edges are still independent random variables. Thus, $c(e_i)$ is the first order statistic among these adjacent edges. It follows from Lemma 1 that $c(e_i)$, $1 \leq i \leq n-1$, have the following distributions:

$$G_i(y) = Pr(c(e_i) \leq y) = 1 - (1-F(y))^{n-i} \ . \tag{1}$$

The distribution of $c(e_n)$ is given as follows:

$$G_n(y) = \Pr(c(e_n) \le y | c(e_n) \ge c(e_1))$$
$$= 1 - \Pr(c(e_n) \ge y,\ c(e_n) \ge c(e_1))/\Pr(c(e_n) \ge c(e_1))$$

$$\Pr(c(e_n) \ge c(e_1)) = \int_0^\infty \int_0^y dG_1(z)\ dF(y) = 1 - 1/n$$

$$\Pr(c(e_n) \ge y,\ c(e_n) \ge c(e_1)) = \int_0^\infty \Pr(c(e_n) \ge \max(y,z),\ c(e_1)=z)\ dG_1(z)$$
$$= \int_0^y (1-F(y))\ dG_1(z) + \int_y^\infty (1-F(z))\ dG_1(z)$$
$$= 1-F(y) - (1-F(y))^n/n\ .$$

Thus we have

$$G_n(y) = (n\cdot F(y) + (1-F(y))^n - 1)/(n-1)\ . \tag{2}$$

Let NEARNEIGHBOUR(n) be the tour length produced by this heuristic. We have the following results:

Theorem 1. If the edge costs of a TSP are drawn independently from a uniform distribution on [0,1], then

$$\lim_{n\to\infty} E(\mathrm{NEARNEIGHBOUR}(n))/\ln(n) = 1 \quad \text{and}$$

$$\lim_{n\to\infty} \mathrm{VAR}(\mathrm{NEARNEIGHBOUR}(n)) = \frac{4}{3} - \sum_{i=1}^{\infty} i^{-2} \le \frac{5}{6}\ .$$

Theorem 2. If the edge costs for a TSP are drawn independently from a uniform distribution on [0,1], then for every $\varepsilon > 0$,

$$\lim_{n\to\infty} \Pr(|\mathrm{NEARNEIGHBOUR}(n))/\ln(n) - 1| < \varepsilon) = 1\ .$$

Proof of Theorem 1:

It follows from (1) and (2) that

$$G_i(y) = 1 - (1-y)^{n-i}\ , \quad \text{for} \quad 1 \le i \le n-1 \tag{3}$$

and $G_n(y) = (ny + (1-y)^n - 1)/(n-1)$. (4)

It follows that

$$E(c(e_i)) = \int_0^1 y\ dG_i(y) = 1 - \int_0^1 G_i(y)\ dy$$

$$= \begin{cases} \dfrac{1}{n-i+1} & \text{if} \quad 1 \le i \le n-1 \\[2mm] \dfrac{n+2}{2(n+1)} & \text{if} \quad i = n\ . \end{cases} \tag{5}$$

Therefore the expected tour length of the nearest neighbour is given by

$$E(\mathrm{NEARNEIGHBOUR}(n)) = \sum_{i=1}^{n} E(c(e_i)) = \frac{n+2}{2(n+1)} + \sum_{i=2}^{n} \left(\frac{1}{i}\right)$$

$$= \frac{1}{2(n+1)} + \sum_{i=1}^{n} (1/i) \to \ln(n) \text{ as } n \to \infty .$$

It follows from (3) and (4) that

$$E((c(e_i))^2) = \int_0^1 y^2 \, dG_i(y)$$

$$= \begin{cases} \frac{2}{(n-i+1)(n-i+2)} & \text{if } 1 \le i \le n-1 \\ \frac{1}{3} + \frac{n+4}{3(n+1)(n+2)} & \text{if } i = n \end{cases} \qquad (6)$$

It then follows from (5) and (6) that

$$VAR(c(e_i)) = E((c(e_i))^2) - (E(c(e_i)))^2$$

$$= \begin{cases} \frac{2}{n-i+1} - \frac{2}{n-i+2} - \frac{1}{(n-i+1)^2} & \text{if } 1 \le i \le n-1 \\ \frac{1}{12} + \frac{1}{2(n+1)} - \frac{2}{3(n+2)} - \frac{1}{4(n+1)^2} & \text{if } i = n \end{cases} \qquad (7)$$

Since all the edge costs $c(e_i)$, $c(e_j)$ are mutually independent except for the two edge costs $c(e_1)$ and $c(e_n)$, the covariances for these edge costs are given by

$$COVAR(c(e_i),c(e_j)) = E(c(e_i)\cdot c(e_j)) - E(c(e_i))\cdot E(c(e_j))$$

$$= \begin{cases} \frac{1}{8} - \frac{1}{n} + \frac{5}{n+1} - \frac{1}{n+2} + \frac{3}{8(n+3)} & \text{if } (i,j) = (1,n) \text{ or } (n,1) \\ 0 & \text{if } i \ne j \text{ and } (i,j) \notin \{(1,n),(n,1)\}. \end{cases}$$

Therefore, the variance of the nearest neighbour heuristic solution is given by

$$VAR(NEARNEIGHBOUR(n)) = 2\, COVAR(c(e_1),\, c(e_n)) + \sum_{i=1}^{n} VAR(c(e_i))$$

$$= \frac{4}{3} + O(1/n) - \sum_{i=2}^{n} i^{-2} \to \frac{4}{3} - \sum_{i=2}^{\infty} i^{-2} \le \frac{5}{6} .$$

The last inequality follows from $1/2 \le \sum_{i=2}^{\infty} i^{-2} \le 1$, completing the proof.

Proof of Theorem 2:

$$Pr(|NEARNEIGHBOUR(n)/\ln(n) - 1| \ge \varepsilon)$$
$$\le Pr(|NEARNEIGHBOUR(n)/\ln(n) - E(NEARNEIGHBOUR(n))/\ln(n)|$$
$$+ |E(NEARNEIGHBOUR(n))/\ln(n) - 1| \ge \varepsilon) .$$

Since $E(NEARNEIGHBOUR(n))/\ln(n)$ converges to 1, for sufficiently large n, we have

$$|E(NEARNEIGHBOUR(n))/\ln(n) - 1| \le \varepsilon/2 .$$

It then follows that

$$\begin{aligned} &Pr(|NEARNEIGHBOUR(n)/\ln(n) - 1| \geq \varepsilon) \\ &\leq Pr(|NEARNEIGHBOUR(n)/\ln(n) - E(NEARNEIGHBOUR(n))/\ln(n)| \geq \varepsilon/2) \\ &\leq VAR(NEARNEIGHBOUR(n)/\ln(n))/(\varepsilon/2)^2 \\ &\leq (10/3)/(\varepsilon\cdot\ln(n))^2 \to 0 \quad \text{as} \quad n \to \infty . \end{aligned}$$

The second last inequality follows from the Chebyshev inequality [12, 17] and the last inequality follows from the result of Theorem 1, completing the proof.

Consider the asymmetric TSP. If the edge costs are drawn independently from the uniform distribution on the interval [0,1], then the costs of the n edges selected by the nearest neighbour heuristic are all mutually independent. Furthermore, the costs of the first n - 1 selected edges have the same distributions as (1) while the cost of the edge e_n has the uniform distribution over the unit interval [0,1]. Therefore, it is easy to verify that the results of Theorems 1 and 2 are also true for the asymmetric TSP.

The performance of the nearest neighbour heuristic for other distibutions of the problem data can be derived in a similar way. For example, if the edge costs have the following exponential distribution:

$$F(x) = 1 - e^{-\alpha x} \quad \text{for} \quad 0 \leq x < \infty , \tag{8}$$

then the results which hold for both the symmetric and asymmetric TSPs are given in the following theorems.

Theorem 3. If the edge costs are drawn independently from the exponential distribution of (8) then

$$\lim_{n\to\infty} E(NEARNEIGHBOUR(n))/\ln(n) = 1/\alpha \quad \text{and}$$

$$\lim_{n\to\infty} VAR(NEARNEIGHBOUR(n)) = \alpha^{-2}(1 + \sum_{i=1}^{\infty} i^{-2}) < 3\cdot\alpha^{-2} .$$

Theorem 4. If the edge costs are drawn independently from the exponential distribution of (8) then for every $\varepsilon > 0$,

$$\lim_{n\to\infty} Pr(|NEARNEIGHBOUR(n)/\ln(n) - 1/\alpha| < \varepsilon) = 1 .$$

It is interesting to note that the variances of the nearest neighbour solutions for the uniform and exponential distributions both converge to some small constant independent of n.

3. EXPECTED PERFORMANCE OF SEQUENTIAL SAVINGS HEURISTIC

In this section, we outline in some detail the probabilistic analysis of the sequential savings heuristic. All the results given below hold for both symmetric and asymmetric TSPs.

The sequential savings heuristic of the TSP is similar to the nearest neighbour method [10]. The algorithm starts at an arbitrary vertex v_1 and choose an arbitrary vertex v_0 as the 'depot' vertex. The edges e_i are selected sequentially by the algorithm in the following way: The savings, $s(v_i,v) = c(v_i,v_0) + c(v_0,v) - c(v_i,v)$, for all the unvisited vertices v with respect to the depot vertex v_0 and the vertex v_i just visited are computed; the vertex v_{i+1} which yields the largest saving is selected as the next vertex to be visited. The process continues until all the vertices in $V - \{v_0\}$ are visited and finally the two edges (v_0,v_1), (v_{n-1}, v_0) are added to the selected edges to form a connected tour.

Notice that for the sequential savings method, the selection of a vertex v so that $s(v_i,v)$ is maximized is equivalent to the selection of a vertex v such that $d(v_i,v) = c(v_0,v) - c(v_i,v)$ is maximized.

Let $e_i(v_i,v_{i+1})$ for $1 \leq i \leq n-2$ and $e_{n-1} = (v_{n-1}, v_0)$, $e_n = (v_0,v_1)$ be the edges selected by the savings heuristic. Let $d(e_i)$, $1 \leq i \leq n-2$, denote the n - 2 values $d(v_i,v_{i+1})$. Then the total savings, SAVINGS(n), obtained by the heuristic is given by

$$\text{SAVINGS}(n) = \sum_{i=1}^{n-2} s(v_i,v_{i+1}) \sum_{i=1}^{n-2} c(v_i,v_0) + \sum_{i=1}^{n-2} d(e_i) \qquad (9)$$

Suppose the edge costs of G are drawn independently from a common distribution F(x) on the interval $[0, \infty)$. Then the costs $c(v_i,v_0)$ and the values $d(e_i)$ in (9) are all independent random variables. The costs $c(v_i,v_0)$ have the same distribution F(x) while the random variables $d(e_i)$ have the following distributions:

$$G_i(y) = (F_i(y))^{n-i-1} ,$$

where $F_i(y)$ is the distribution of the random variable $d(v_i,v)$ which can be computed as follows:

$$F_i(y) = \Pr(c(v_0,v) - c(v_i,v) \leq y)$$

$$= \begin{cases} \int_0^\infty F(x+y)\, dF(x) & \text{if } y \geq 0 \\ \int_{-y}^\infty F(x+y)\, dF(x) & \text{if } y < 0 . \end{cases} \qquad (10)$$

It then follows that

$$E(SAVINGS(n)) = (n-2)\mu + \sum_{i=1}^{n-2} E(d(e_i)) \quad \text{and}$$

$$VAR(SAVINGS(n)) = (n-2)\sigma^2 + \sum_{i=1}^{n-2} VAR(d(e_i))$$

where μ is the expected value and σ^2 the variance of the distribution $F(x)$. Hence, the expected savings and its variance can be calculated from (10).

We observe that the tour length, TOUR(n), produced by the savings heuristic can be expressed in terms of the total savings SAVINGS(n) as follows:

$$TOUR(n) = SAVINGS(n) - \sum_{i=1}^{n-1} c(v_0,v_i) - \sum_{i=1}^{n-1} c(v_i,v_0) .$$

It follows that the expected tour length produced by the heuristic is given by $E(TOUR(n)) = E(SAVINGS(n)) - 2(n-1)\mu$.

The expected performance for other TSP heuristics is difficult to analyze, since the dependecny between successive edge selection operations is tightened. The statistical approach may be an easier way to analyze the performance of these heuristics.

4. EXPECTED PERFORMANCE OF THE BEST NEIGHBOUR HEURISTIC

In this section, we extend the analysis given in Section 2 to the maximum weight travelling salesman problem (MWTSP).

Suppose the edge weights of G are drawn independently from a uniform distribution on the unit interval [0,1]. If the weight $w(e)$ for each of the edges in E is transformed to the cost $c(e) = 1 - w(e)$, then the solution of the MWTSP is equivalent to the solution of the transformed TSP. Furthermore, the transformed costs are also independent random variables having the same uniform distribution over the unit interval [0,1]. Consequently, we have the following results:

Theorem 5. If the costs of a TSP and the weights of a MWTSP are drawn independently from a uniform distribution on the interval [0,1], then

$$E(OPTIMAL(MWTSP,n)) = n - E(OPTIMAL(TSP,n)) ,$$

where OPTIMAL(MWTSP,n) and OPTIMAL(TSP,n) are the optimal solutions of a random MWTSP and a random TSP respectively.

Consider the best neighbour heuristic which starts at an arbitrary vertex and chooses an edge of maximum weight adjacent to it. It then continues to select an edge of maximum weight adjacent to the vertex just reached subject to the requirement that the collection be contained in a tour.

Let BESTNEIGHBOUR(n) denote the tour length produced by the best neighbour heuristic. Suppose the edge weights of a MWTSP are drawn independently from an identical distribution F(x) defined on the interval $[0, \infty)$, then the edges selected by the best neighbour heuristic except the last are all independent random variables. Their distributions are given as follows:

$$G_i(y) = (F(y))^{n-i} \quad \text{for all} \quad 1 \leq i \leq n-1 , \tag{11}$$

and
$$\begin{aligned} G_n(y) &= \Pr(w(e_n) \leq y \mid w(e_n) \leq w(e_1)) \\ &= \int_0^\infty \Pr(w(e_n) \leq y \mid w(e_1) = z)\, dG_1(z) \\ &= (n \cdot F(y) - (f(y))^n/(n-1) . \end{aligned} \tag{12}$$

The performance of the best neighbour heuristic for the uniform distribution is given in the following theorems. The results can either be derived from the distribution (11) and (12) in a way similar to that of the nearest neighbour heuristic or they can be obtained directly from the results of Theorems 1, 2 and 3 by transforming the MWTSP to a TSP as discussed at the beginning of this section.

<u>Theorem 6</u>. If the costs of a maximum weight TSP are drawn independently from a uniform distribution on the interval [0,1], then

$$\lim_{n\to\infty} E(\text{BESTNEIGHBOUR}(n))/(n-\ln(n)) = 1 \quad \text{and}$$

$$\lim_{n\to\infty} VAR(\text{BESTNEIGHBOUR}(n)) = \frac{4}{3} - \sum_{i=1}^{\infty} i^{-2} \leq \frac{5}{6} .$$

<u>Theorem 7</u>. If the edge costs for a TSP are drawn independently from a uniform distribution on [0,1], then for every $\varepsilon > 0$,

$$\lim_{n\to\infty} \Pr(|\text{BESTNEIGHBOUR}(n)/(n-\ln(n)) - 1| < \varepsilon) = 1 .$$

REFERENCES

[1] Beardwood, J., Halton, J.H. and Hammersely, J.D., "The Shortest Path Through Many Points", <u>Proc. Cambridge Philosophical Society</u>, 55 (1959), 299-328

[2] Christofides, N., "Worst-Case Analysis of a New Heuristic for the Travelling Salesman Problem", Working paper 62-75-76, February 1976, Graduate School of Industrial Administration, Carnegie-Mellon University, Pittsburgh, Pennsylvania

[3] Clarke, G. and Wright, J.W., "Scheduling of Vehicles from a Central Depot to a Number of Delivery Points", Ops. Res., 12 (1963), 568-581

[4] Cornuejols, G. and Nemhauser, G.L., "Tight Bounds for Christofides' Travelling Salesman Heuristic", Math. Programming, 14 (1978), 116-121

[5] Fisher, M.L., Nemhauser, G.L. and Wolsey, L.A., "An Analysis of Approximations for Finding a Maximum Weight Hamiltonian Circuit", Ops. Res., 27 (1979), 799-809

[6] Frederickson, G.N., Hecht, M.S. and Kim, C.E., "Approximation Algorithms for Some Routing Problems", SIAM J. on Computing, 7 (1978), 178-193

[7] Frieze, A.M., "Worst-Case Analysis of Algorithms for Travelling Salesman Problems", Operations Research Verfahren, 32 (1978), 94-112

[8] Garey, M.R. and Johnson, D.S., Computers and Intractability: A Guide to the Theory of NP-Completeness, Freeman, San Francisco, Calif., 1979

[9] Garfinkel, T.J. and Gilbert, K.C., "The Bottleneck Travelling Salesman Problem: Algorithms and Probabilistic Analysis", ACM, 25 (1978), 435-448

[10] Golden, B., "Evaluating a Sequential Vehicle Routing Algorithm", AIIE Transactions, 9 (1977), 204-208

[11] Golden, B., Bodin, L. and Stewart, W., "Approximate Travelling Salesman Algorithms", Management Sci., 28 (1980), 694-711

[12] Hogg, R.V. and Craig, A.T., Introduction to Mathematical Statistics, The Macmillan Company, New York, 1965

[13] Karp, R.M., "Reducibility Among Combinatorial Problems", in Complexity of Computer Computations, R.E. Miller and J.W. Thatcher, eds., Plenum Press, New York, 1972, 85-104

[14] Karp, R.M., "The Probabilistic Analysis of Some Combinatorial Search Algorithms", Proc. Sympos. on New Directions and Recent Results in Algorithms and Complexity, Academic Press, N.Y., 1976

[15] Karp, R.M., "Probabilistic Analysis of Partitioning Algorithms for the Travelling Salesman Problem in the Plane", Math. Operations Res., 2 (1977), 209-224

[16] Karp, R.M., "A Patching Algorithm for the Nonsymmetric Travelling Salesman Problem", SIAM J. Comput., 8 (1979), 561-573

[17] Loeve, M., Probability Theory, Van Nostrand, 1963

[18] Lucker, G.S., "Maximization Problems on Graphs with Edge Weights Chosen From a Normal Distribution", Proc. 10th Annual ACM Symp. on Theory of Computing, 1978, 13-18

[19] Lucker, G.S., "Optimization Problems on Graphs with Independent Random Edge Weights", SIAM J. Comput., 10 (1981), 338-351

[20] Rosenkrantz, D.J., Stearns, R.E. and Lewis, P.M., "An Analysis of Several Heuristics for the Travelling Salesman Problem", SIAM J. Comput., 6 (1977), 563-581, 1978, 13-18

[21] Stein, D.M., "An Asymptotic Probabilistic Analysis of a Routing Problem", Math. of Operations Res., 3 (1978), 89-101

[22] Weide, B.W., Statistical Methods in Algorithm Design and Analysis, Ph.D. Thesis, Dept of Computer Science, Carnegie-Mellon University, August 1978

AN OPTIMAL METHOD FOR THE MIXED POSTMAN PROBLEM

N.Christofides Imperial College of London.UK

E.Benavent, V.Campos, A.Corberán and E.Mota Universidad de Valencia.Spain

1. INTRODUCCION

The problem of traversing every arc and edge of a mixed graph (with nonnegative associated costs) by a tour of minimum total cost, is known as the Mixed Chinese Postman Problem (MCPP). It is a more general case than the nondirected (CPP) and totally directed (DCPP) versions of this problem, and which were already solved efficiently by Edmonds and Johnson [1] .

The particular case of the MCPP in which every vertex of the graph is incident with an even number of lines (arcs or edges), is the only one that can be solved in polynomial time ([1] , [5]). The general case of the MCPP is an NP-Complete problem, as shown by Papadimitriou [7] .

In this paper we present an exact algorithm for the MCPP based on a branch and bound algorithm using Lagrangean bounds and also report some computational results.

In Section 2 we present an integer formulation of the MCPP which is equivalent to the one presented in [4] , and in which some redundant constraints have been added. In Section 3 we describe the heuristic algorithm used for finding an initial feasible solution, which can be considered equivalent to the one proposed by Frederickson [2] (known as MIXED 2).

2. A FORMULATION OF THE MCPP

Let $G=(N,A,E)$ be a strongly connected mixed graph, where N is a set of vertices, A is a set of arcs and E is a set of edges.

We represent each arc $a \in A$, directed from vertex i to vertex j, by the ordered pair (i,j), and each edge $e \in E$, having terminal vertices i and j, by the nonordered pair $<i,j>$. In addition, we suppose that every line of graph G has a nonnegative associated cost c_{ij}.

Let $\Gamma_A(i)$ be the set of all final vertices of the arcs leaving vertex i, let $\Gamma_A^{-1}(i)$ be the set of all initial vertices of the arcs entering vertex i and let $\Gamma_E(i)$ be the set of all vertices linked with vertex i by an edge. The MCPP can be formulated as an integer problem as follows:

$$(P)\ \operatorname{Min}\{ \sum_A c_{ij}x_{ij} + \sum_E c_{ij}(y_{ij}+y_{ji}) + \sum_A c_{ij} \}$$

s.t.:

$$\sum_{j\in\Gamma_A(i)} (1+x_{ij}) + \sum_{j\in\Gamma_E(i)} y_{ij} = \sum_{j\in\Gamma_A^{-1}(i)} (1+x_{ji}) + \sum_{j\in\Gamma_E(i)} y_{ji} \qquad \forall i\in N \qquad (1)$$

$$y_{ij} + y_{ji} \geq 1 \qquad \forall e=<i,j>\varepsilon E \qquad (2)$$

$$x_{ij},\ y_{ij},\ y_{ji} \geq 0 \text{ and integer} \qquad \forall (i,j)\varepsilon A,\ \ \forall e=<i,j>\varepsilon E \qquad (3)$$

$$\sum_{j\varepsilon\Gamma_A(i)} x_{ij} + \sum_{j\varepsilon\Gamma_A^{-1}(i)} x_{ji} + \sum_{j\varepsilon\Gamma_E(i)} (y_{ij}+y_{ji}-1) - 2w_i = b_i \qquad \forall i\varepsilon N \qquad (4)$$

$$w_i \geq 0 \text{ and integer} \qquad \forall i\varepsilon N \qquad (5)$$

where the arc variables x_{ij} represent the number of extra times that arc (i,j) must be traversed by the solution circuit; edge variables y_{ij}, y_{ji} represent the number of times that edge<i,j> must be traversed (from i to j and j to i, respectively) by the solution circuit, and: b_i= 0 if vertex i is incident with an even number of lines in G, and b_i= 1 if vertex i is incident with an odd number of lines in G.
Constraints (1) state that, relative to the solution tour, every vertex is symmetric (i.e. the indegree is equal to the outdegree). Constraints (2) force every edge of graph G to be traversed by the tour at least once. Constraints (4) and (5) are redundant and express the requirement that every vertex must (relative to the tour) have an even number of lines incident with it. The importance of these redundant constraints will become clear later (see Section 4).
Every feasible solution to problem P produces a feasible solution to the MCPP on G and viceversa. It can also be shown ([4] , [6]) that, if in an optimal solution of the MCPP the two variables associated with a given edge <i,j> are both non-zero, then both of them must be equal to one.

2.1 Graph transformations

(a) Replacing edges with arcs

Under certain conditions it is possible to transform the original graph G by converting some edges into arcs thus reducing the number of constraints of type (2). The conditions that must be hold for such a transformation to be possible are given by the following Propositions.

Definition 1. An edge <i,j> is a "cut-edge in the direction i to j " in G iff replacing it by the arc (j,i), the resulting graph is not strongly connected.

Definition 2. An edge <i,j> is a "cut-edge" in G iff it is a cut-edge in its two possible directions.

Proposition 1. If an edge <i,j> of graph G is a cut-edge only in one direction, then it will not be traversed in the opposite direction by any optimal solution of the MCPP.

Proposition 2. If an edge <i,j> of graph G is a cut-edge, then it must be traversed exactly once in each direction by every optimal solution of the MCPP.

The purpose of the apriori transformation of edges into arcs -implied by Propositions 1 and 2- is to try to convert the mixed CPP into the purely directed CPP which can then be easily solved.

(b) Repeating existing arcs

Definition 3. An arc $(i,j)\varepsilon A$ is a "cut-arc" iff by deleting (i,j) the resulting graph is not strongly connected.

The above definition implies that for a cut-arc (i,j) there exists in the graph a cut $(S,\bar{S})$, $i\varepsilon S$, $j\varepsilon\bar{S}$ such that: (i) the arc (i,j) is the only one directed from S to $\bar{S}$, and (ii) there is no edge between S and $\bar{S}$. This, in turn, implies that if the number of arcs from $\bar{S}$ to S is k, the cut-arc (i,j) must be traversed at least k times in any feasible solution. Hence, the lower bound on the variables x_{ij} in constraints (3) can be increased from 0 to k-1.

Note that transformations of type (a) above involve the substitution of some y-variables with x-variables in the formulation of problem P. Henceforth, when we refer to the graph and the problem formulation it will be relative to the graph that is produced after the above apriori transformations are applied.

3. A HEURISTIC ALGORITHM

As already pointed out in Section 1, the MCPP can be efficiently solved in the case of an even graph. Minieka [5] presents an algorithm for even graphs based on the resolution of a minimum cost flow problem on a transformed graph. From the solution to the flow problem, a direction is assigned to every edge of the original graph, transforming it into a symmetric graph. The algorithm can be extended to the general case of the MCPP, in order to obtain feasible solutions, and is as follows:

Step 1: Assign a direction to every edge of G, thus transforming edges into arcs and obtaining the directed graph $G_d=(N,A_d)$. Compute $\xi(i)=|\Gamma^{-1}_{A_d}(i)| - |\Gamma_{A_d}(i)|$ $\forall i\varepsilon N$.

Step 2: Construction of the directed graph $G_F=(N,A\cup A_F)$.
Corresponding to every edge $<i,j>$ εE, put two arcs in parallel one in each direction in A_F, with costs equal to the cost of the edge. In addition, put an "artificial" arc of zero cost in the opposite direction to the one previously assigned to the edge in G_d. Arcs of A appear with their original costs.

Step 3: Solve in G_F the minimum cost flow problem, in which:

(a) every vertex $i\varepsilon N$ with $\xi(i)>0$ is a source with supply $\xi(i)$.

(b) every vertex $i\varepsilon N$ with $\xi(i)<0$ is a sink with demand $-\xi(i)$.

(c) the capacity of each artificial arc is 2 and that of the remaining arcs ∞.

Step 4: Construction of the symmetric mixed graph $G_S=(N,A_S,E_S)$.
Let $f(i,j)$ and $f^a(i,j)$ represent the number of flow units circulating in the optimal solution through non-artificial and artificial arcs, respectively. From the resolution of the flow problem defined in Step 3, construct graph G_S as follows:

(a) Referring to the flow circulating through the arcs representing edges:

  (1) if $f^a(i,j)=0$, put $1+f(j,i)$ arcs (j,i) in A_S.

  (2) if $f^a(i,j)=2$, put $1+f(i,j)$ arcs (i,j) in A_S.

  (3) if $f^a(i,j)=1$, put edge $<i,j>$ in E_S.

(b) Referring to the flow circulating through the remaining arcs of G_F, put $1+f(i,j)$ arcs (i,j) in A_S.

Step 5: Obtaining a feasible solution.

If $E_S=\phi$, G_S represents the optimal solution to the MCPP on G.

If $E_S\neq\phi$, solve a minimum cost perfect matching on the odd vertices of the graph induced by E_S on G_S, adding to E_S a copy of every edge duplicated by the matching.

The output is an even and symmetric graph and, therefore, a feasible solution to the MCPP on G.

4. AN EXACT ALGORITHM FOR THE MCPP

We present in this Section a branch and bound algorithm for solving the MCPP which is based in the computation of lower bounds from two possible Lagrangean relaxations of problem P.

4.1 Lower bounds

Two lower bounds can be obtained to the value of the optimal solution of the MCPP. The first one is obtained by relaxing, in a Lagrangean way, the symmetry constraints (1), using multipliers u_i and solving the minimum cost perfect matching. The second one is obtained via the Lagrangean relaxation of constraints (2), using multipliers λ_{ij} (eliminating (4) and (5)) and resolving a minimum cost flow problem, as described below.

4.1.1 Lagrangean relaxation of symmetry constraints

Consider the original problem P and problem $PR_1(x,y,u)$ obtained by relaxing, in a Lagrangean way, constraints (1) using multipliers u_i. The formulation of problem $PR_1(x,y,u)$ is then as follows:

$$PR_1(x,y,u) \quad \text{Min} \left\{ \sum_A (c_{ij}+u_j-u_i)x_{ij} + \sum_{E_d} (c_{ij}+u_j-u_i)y_{ij} + \sum_{i\varepsilon N} D(i)u_i + \sum_A c_{ij} \right\}$$

s.t.: (2),(3),(4),(5) and u_i non-restricted.

where $E_d=\{ (i,j), (j,i) \mid \forall <i,j>\varepsilon E \}$ and $D(i)=|\Gamma_A^{-1}(i)| - |\Gamma_A(i)|$.

A. Solving problem $PR_1(x,y,u)$

Obviously, problem $PR_1(x,y,u)$ is bounded only when the multipliers u produce nonnegative modified costs $c'_{ij}=c_{ij}+u_j-u_i$.

Since the edge-variables y_{ij}, y_{ji}, associated to every edge <i,j> of G always appear as $y_{ij}+y_{ji}$, in the constraints of problem $PR_1(x,y,u)$, at least one of these two variables must be nonzero, and that variable with the lowest corresponding cost will certainly take a nonzero value. Therefore, defining variables $h_{ij}=y_{ij}+y_{ji}-1$, for every edge e= <i,j>εE, with associated cost $\hat{c}_e=\min\{c'_{ij}, c'_{ji}\}$, we obtain the following minimization problem, which is equivalent to $PR_1(x,y,u)$:

$$\text{Min}\{\sum_{A} c'_{ij}x_{ij} + \sum_{E} \hat{c}_e(h_{ij}+1) + \sum_{i\varepsilon N} D(i)u_i + \sum_{A} c_{ij}\}$$

s.t.:

$$\sum_{j\varepsilon\Gamma_A(i)} x_{ij} + \sum_{j\varepsilon\Gamma_A^{-1}(i)} x_{ji} + \sum_{j\varepsilon\Gamma_E(i)} h_{ij} - 2w_i = b_i \quad \forall i\varepsilon N$$

$$x_{ij},\ h_{ij},\ w_i \geqq 0, \text{ and integer } \forall i,j.$$

which is easily recognized as a minimum cost perfect matching on the odd degree vertices of graph G, ignoring the direction of the arcs. The modified costs are c'_{ij}, $\hat{c}_e$ and there is a constant term in the objective function (for a given u), i.e. :

$$K(u) = \sum_{E} \hat{c}_e + \sum_{i\varepsilon N} D(i)u_i + \sum_{A} c_{ij}.$$

The optimal solution to problem $PR_1(x,y,u)$, given u, will be obtained by assigning to the arc variables x_{ij} the values 1 or 0, depending on whether arc (i,j) has been used, or not, by the perfect matching. The values of y_{ij} are now given by: $y_{ij}=2$ if $c'_{ij}=\hat{c}_e$ and $h_{ij}=1$; $y_{ij}=1$ if $c'_{ij}=\hat{c}_e$ and $h_{ij}=0$; $y_{ij}=0$ otherwise.
The cost of the solution can be expressed as: $v(PR_1(x,y,u))=c(M^*(u)) + K(u)$, where $M^*(u)$ represents the set of arcs and edges in the optimal solution to the minimum cost perfect matching and $c(M^*(u))$ is the cost of the arcs in this set.

B. Selection of multipliers u

Several difficulties appear when applying subgradient optimization,in order to obtain a good approximation to the problem of $\max_u v(PR_1(x,y,u))$ and hence a good bound. The main difficulty is to compute at each iteration all the shortest distances among the odd degree vertices in graph G, before solving a minimum cost perfect matching problem. In addition, not all the multipliers produced by the subgradients can be used unchanged, since the modified costs c'_{ij} should always remain non-negative.
An alternative easy way of selecting multipliers u is to consider only one term of the objective function and to solve the problem

$$\max_u \sum_{i\varepsilon N} D(i)u_i$$

$$\text{s.t.: } u_i - u_j \leqq c_{ij} \qquad \forall (i,j)\varepsilon A\cup E_d.$$

The constraints being the non-negativity restrictions on the arc/edge modified costs. The optimal solution to the above problem can easily obtained from the values of the variables of its dual problem, which corresponds to the minimum cost flow problem defined by the objective function of P and constraints (1).
However, this selection of multipliers u has also not performed very well and has been computationally surpassed by a simple greedy procedure which tries to optimize approximately problem Q: $\underset{u}{\text{Max}}\ K(u)$

$$\text{s.t.: } u_i - u_j \leqq c_{ij} \qquad \forall (i,j)\varepsilon A\cup E_d$$

This was the method finally adopted and the computational results reported later were derived using this procedure.

4.1.2 Lagrangean relaxation of constraints (2)

In problem P, ignoring the redundant constraints (4) and (5), we can relax, in a Lagrangean fashion, constraints (2), associating to each one a non-negative scalar λ_{ij}, to obtain the minimization problem:

$$PR_2(x,y,\lambda) \quad \text{Min} \sum_A c_{ij}x_{ij} + \sum_E (c_{ij}-\lambda_{ij})(y_{ij}+y_{ji}) + \sum_E \lambda_{ij} + \sum_A c_{ij}$$

s.t.: (1) and (3).

A. Solving problem $PR_2(x,y,\lambda)$

Given a vector of multipliers $\lambda \geq 0$, constraints (1) and the non-negativity of variables x and y, define a minimum cost flow problem. Therefore, the optimal solution of $PR_2(x,y,\lambda)$ consists of the following assignment of values: $x_{ij}=f(i,j)$ $\forall (i,j) \varepsilon A$ and $y_{ij}=f(i,j)$ $\forall (i,j) \varepsilon E_d$, where f(i,j) represents the number of flow units transported through the arc (i,j), or through the edge <i,j> from i to j, in the optimal solution. The above problem is solved without capacity requirements on the arcs and edges of graph G, and from its solution we can construct a symmetric graph by repeating arcs (i,j), f(i,j) times. Note, however, that not every edge of G was necessarily traversed by the optimal flow f(i,j).

On the other hand, since demands and supplies are integer-valued, the problem $PR_2(x,y,\lambda)$ has the integrality property and, therefore (see [3]), the lower bound that can be obtained from this relaxation, though easily computed, will be, at most, equal to the bound obtained from the linear relaxation of P, ignoring constraints (4) and (5).

B. Selection of multipliers λ

The trivial selection $\lambda=0$, may produce a feasible solution to problem P, in which case the subproblem is solved optimally and the node is fathomed. This occurs when every edge of graph G has been traversed by a nonzero number of flow units in the solution (x°,y°) to the problem $PR_2(x,y,0)$. Otherwise, a heuristic procedure is applied to obtain new multipliers $\bar{\lambda}_{ij}$ as follows:

Let $E_0=\{ \langle i,j\rangle \varepsilon E \mid y^\circ_{ij}+y^\circ_{ji}=0 \}$: Non traversed edges.

$E_1=\{ \langle i,j\rangle \varepsilon E \mid y^\circ_{ij}+y^\circ_{ji}=1 \}$: Edges traversed once.

$E_2=\{ \langle i,j\rangle \varepsilon E \mid y^\circ_{ij}+y^\circ_{ji}\geq 2 \}$: Edges traversed more than once.

We now penalize edges <i,j> according to the following choice of $\bar{\lambda}_{ij}$:

$\bar{\lambda}_{ij}=c_{ij}$ $\forall \langle i,j\rangle \varepsilon E_0$; $\bar{\lambda}_{ij}=c_{ij}/2$ $\forall \langle i,j\rangle \varepsilon E_1$; $\bar{\lambda}_{ij}=0$ $\forall \langle i,j\rangle \varepsilon E_2$.

These multipliers $\bar{\lambda}_{ij}$ have been used as the initial ones in the subgradient method, producing good computational results.

4.2 An upper bound

The upper bound used in the branch and bound method is obtained using the heuristic algorithm described in Section 3. The values of the upper bound shown in the computational results, obtained on a collection of test problems, are given in Section 5. In 11 of the total of 34 tested problems, this bound was optimal. On average, this bound is within 3.11% of the optimal value.

4.3 The tree search

The branching is made on the edge-variables. In this way, each node is separated in two successors, one of them obtained by adding constraint $y_{ij} \geqq 1$, whereas the other, by adding the alternative constraint $y_{ij}=0$.
Each constraint $y_{ij} \geqq 1$ allows us to bound (at this node and its successors) the "paired" variable y_{ji} by 1, thus improving the lower bounds produced by $PR_2(x,y,\lambda)$.
On the other hand, constraint $y_{ij}=0$ transforms the associated edge into the arc (j,i), allowing the appearance of new cut-arcs and cut-edges, at this node and its successors. At the level in which all edges of graph G have an assigned direction, the MCPP becomes the DCPP which is optimally solved by calculating the corresponding minimum cost flow problem.
We have used the depth-first branching strategy, due to its smaller memory requirements. The order in which the branching variables are chosen is made on the expectation of a similarity between the direction assigned to an edge by the upper bound and the one assigned to it by the optimal solution to the problem. The Table in Section 5 shows that, in most of the tested problems the optimal solution is found very quickly, with respect to the total number of nodes, justifying the adopted rule.

5. COMPUTATIONAL RESULTS

The algorithm described in the previous Section has been tested on a collection of 34 problems, randomly generated and costs chosen between 1 and 20. The algorithm was coded in FORTRAN and run on the UNIVAC 1100/60 computer. The times shown in the table are CPU seconds of the above computer and correspond to total execution time.
Problems up to 50 vertices, 66 arcs and 39 edges could be completely solved with a maximum time of 500 seconds.
The maximum number of subgradient iterations allowed to obtain the Lagrangean bounds was 20 at the root node, and only 10 at the internal ones. Finally, we present in the table the computational results, in which the three numbers in the first column represent, respectively, the total number of : vertices/arcs/edges in the original graph.

6. REFERENCES

[1] Edmonds, J. and Johnson, E. "Matching, Euler tours and the Chinese Postman", Math. Prog. 5, 1973, p. 88-124.
[2] Frederickson, G. N. "Approximation Algorithms for some Postman Problems", Journal of A.C.M., vol.26, 1979, p. 538-554.
[3] Geoffrion, A. "Lagrangean Relaxation for Integer Programming", Math. Prog. Study 2, 1974, p. 82-114.

[4] Kappauf, C. H. and Koehler, G. J. "The Mixed Postman Problem", Discrete App. Math. 1, 1979, p. 89-103.

[5] Minieka, E. "Optimization Algorithms for Networks and Graphs". Ed. Marcel Dekker, New York, 1978.

[6] Minieka, E. "The Chinese Postman Problem for Mixed Networks", Manag. Science, vol.25, 1979, p. 643-648.

[7] Papadimitriou, C. "On the Complexity of Edge Traversing", Journal of A.C.M., vol. 23, 1976, p. 544-554.

Problem	Upper bound	Opt. value	L.B. matching	L.B. flow	Total number nodes	Number of opt. node	Fathomed nodes by L.B.'s	CPU seconds
8/ 7/ 4	83	83	-	83	1	0	1	0.03
7/ 3/ 6	36	36	-	36	1	0	1	0.09
9/ 7/ 5	54	54	53	51	7	0	0	0.18
13/14/ 7	31	31	30	30	7	0	2	0.32
10/ 8/ 6	55	55	55	50	1	0	1	0.07
13/13/ 8	140	140	140	129	1	0	1	0.27
10/ 7/10	122	122	109	100	55	0	16	0.84
8/ 5/10	95	89	89	75	17	17	2	0.68
10/ 9/10	123	119	107	107	33	17	13	0.65
19/15/15	191	191	158	174	43	0	21	1.16
13/12/13	122	122	122	112	1	0	1	0.04
16/16/17	205	190	190	177	28	28	6	1.02
16/16/17	184	184	184	170	1	0	1	0.07
18/16/16	194	181	174	160	81	54	39	3.35
15/ 9/20	235	211	211	191	16	16	0	0.53
25/25/20	394	386	364	353	141	22	59	10.54
33/48/22	811	811	634	782	319	0	156	30.53
17/19/20	257	253	241	237	229	37	114	8.58
20/15/24	317	303	287	290	175	39	80	8.62
27/43/20	657	630	569	590	593	31	273	47.65
43/74/20	887	881	790	847	315	50	152	51.54
15/10/21	275	257	257	232	23	23	0	0.62
30/46/23	610	601	559	575	363	25	178	33.11
32/55/26	972	964	890	908	1631	23	808	194.06
21/12/26	338	308	308	260	102	102	31	5.22
22/19/25	1393	1190	1178	1073	507	118	247	30.82
28/47/25	793	791	677	748	2427	26	1192	232.31
41/52/29	1062	1040	898	1016	501	25	280	74.60
38/61/27	826	814	728	791	463	48	228	72.06
30/47/28	726	704	688	664	483	91	234	55.07
48/78/32	1170	1162	1056	1149	361	57	173	79.98
25/23/29	1438	1331	1331	1207	286	286	133	18.10
50/85/36	1598	1580	1449	1540	1917	42	949	507.37
50/66/39	1381	1339	1206	1309	1943	38	968	500.57

Table : Computational results

MODELING AND ANALYSIS OF COMPUTER AND COMMUNICATIONS SYSTEMS WITH QUEUEING NETWORKS: AN ANALYTICAL STUDY

Thien Vo-Dai
Département d'informatique,
Université Laval, Pavillon Pouliot,
Ste-Foy, Québec, Canada - G1K 7P4

1. INTRODUCTION

Closed queueing networks have been of interest to computer scientists since as early as 1966 when Smith [SMIT66], Moore [MOOR71], Buzen [BUZE71] and others recognized that closed queueing systems were suitable for modeling the behaviour of multiple ressource computer systems. In the last decade, dramatic results have been obtained in the algebraic solution of, in computational algorithms for, and in the application of closed queueing networks as computer system models. From mathematical and practical points of view, the most significant achievement has been, on one hand, the identification of a rather large class of closed queueing networks which admit a product form solution [CHAN72, BASK75, REIS75, CHAN77, NOET79 and CHAN83] on the other hand, the discovery of efficient computational algorithms for calculating performance measures in closed queueing network with product form solution [BUZE71, CHAN75, REIS75, REIS76, REIS78 and CHAN80]. Some notable techniques for obtaining approximate solution to more general networks include the application of difusion approximation [KOBA74a and KOBA74b] and the use of decomposition principle [COUR75 and COUR77].

Closed queueing networks with product form solution are important for modeling computer systems not only because they lend themselves to exact solution and feasible numerical computation but also because they have operational justification as computer system models. In fact, operational analysis as proposed by Buzen [BUZE76] and developed by Denning and Buzen [DENN78] may be viewed as an attempt to caracterize the operational aspect of computer systems by queueing networks with product form solution.

Thus, closed queueing networks with product form solution should be fully explored both analytically and computationally. By an <u>analytical treatment</u>, we mean a treatment, we mean a treatment which allows us to obtain closed-form analytical expressions for network performance measures of interest in contrast to a computational approach in which the main interest is to compute numerically network performance measures. In the development of queueing theory for application in other fields than computer science, such as telephony, the analytical approach has always preceeded the computational approach in the sense that analytical formulas for quantities of interest were derived first and then numerical analyses of the formulas were performed to seek efficient computational algorithms. This has not been the case for the development of queueing networks theory for application in computer science. In fact, researchers and practitionners working with queueing networks have hardly seen an analytical formula for the average marginal queue lenghts of a closed queueing network. Exceptions to this assertion may be applied to the works of Moore [MOOR72] and Kamoun and Kleinrock [KAMO80] in which the generating function technique was used to obtain closed form expressions for the network state probability normalization constant and average queue lengths of some closed queueing networks.

The main objective of this paper is to present an analytical treatment of closed queueing networks with state independent service rates. Two basic considerations led to the choice of this restricted class of closed queueing networks. Firstly, this class of closed queueing networks is perhaps the only one with a structure simple enough for a general analytical treatment. Secondly, closed queueing networks with state independent service rates are of great practical interest and often constitute very large subnetworks in queueing network models of computer systems or other types of systems. The motivation for an analytical treatment is manifold.

Firstly, analytical formulas allow us to gain insight into the behavior of the system. Secondly, efficient and accurate computational algorithm can be sought from analytical formulas. Thirdly, analytical treatment is likely to be called for in some types of investigation, example of which may be sensivity analysis, system optimation and asymptotic analysis. Lastly, it is possible for a performance analyst to demonstrate to himself or to his clients the behavior of the system under study by maniable mathematical formulas rather than by numbers given by a computer.

The organization of this paper is as follows. In Section 2, we introduce the concept of unnormalized probability generating function (P.G.F.) for a closed queueing network and obtain the $\mathcal{Z}$ - transform for the unnormalized P.G.F. In Section 3, we examine the problem of inverting the $\mathcal{Z}$ - transform of a function having the form of the unnormalized P.G.F. In Section 4, we present some new computational algorithms for the network state probability normalization constant. In Section 5, we treat marginal queue length distributions and average queue lengths. Significant results are summarized and discussed in Section 6.

2. PROBABILITY GENERATING FUNCTION (P.G.F.)

2.1 Network with fixed service rates and one job class

Consider a network consisting of M servers whose service-time distributions and service disciplines satisfy the conditions specified by Basket and al [BASK75]. Let n be the total number of jobs in the network and let the service rates at each server be independent of the network state.

The state of the network is described by a vector:

$$\underline{n} = (n_1, n_2, \ldots, n_M)$$

where n_i is the number of jobs at server i, and the vector $\underline{n}$ must satisfy the contraint:

$$n_1 + n_2 + \ldots + n_M = n \tag{2.1}$$

The equilibrium-state probability distribution for the network is given by [BASK75]:

$$P(\underline{n}) = \frac{1}{G(n)} \prod_{i=1}^{M} a_i^{n_i} \tag{2.2}$$

where a_i is the relative traffic intensity at server i. Employing the computer performance terminology, we will refer to the a_i as relative workloads.

Let $\underline{s}$ be a M-dimensional vector defined as:

$$\underline{s} = (s_1, s_2, \ldots, s_M)$$

and introduce the unnormalized probability generating function (P.G.F.) for the network state vector as:

$$G_M(n, \underline{s}) = \sum_{\underline{n}} \prod_{i=1}^{M} a_i^{n_i} s_i^{n_i}$$

where the sum extends to all vectors $\underline{n}$ satisfying the constraint (2.1).

The usefulness of the unnormalized P.G.F. as a performance predicting tool is the same as that of the usual normalized P.G.F. except that quantities obtained from the former must be devided by the normalization constant G(n) in order to be meaningful. The normalization constant is in turn directly obtainable from the unnormalized P.G.F. as:

$$G(n) = G_M(n, \underline{1}),$$

when $\underline{1} = (1, 1, \ldots, 1)$ is the unit M-dimensional vector. Thus, all meaningful performance measures are obtainable from the unnormalized P.G.F.

Putting $a_i s_i = w_i$ and defining the function $G_i(n, \underline{w})$, where $\underline{w} = (w_1, \ldots w_M)$, as:

$$G_i\ (n,\underline{w}) = \sum_{\underline{n}} \prod_{k=1}^{i} w_i^{\ n_i} ,$$

our problem is to determine an analytical expression for the function G_i.
Let H_i ($\underline{w}$, z) be the Ƶ - transform for the function G_i, i.e.:

$$H_i\ (\underline{w},\ z) = \sum_{n=0}^{\infty} G_i\ (n,\underline{w})\ z^n.$$

From the define of G_i and Eq. (2.1) we have:

$$H_i\ (n,\ \underline{w}) = \sum_{k=0}^{n} w_i^{\ k}\ H_{i-1}\ (n-k,\ \underline{w}). \tag{2.3}$$

Using the fact that

$$\sum_{k=0}^{\infty} w_i^{\ k}\ z^k = \frac{1}{1-w_i z}\ ,\quad w_i z < 1,$$

and applying the convolution theorem to Eq. (2.3) we obtain:

$$H_i(\underline{w},\ z) = H_{i-1}(\underline{w},\ z)\ (1 - w_i z)^{-1}\ .$$

Solving this difference equation with boundary condition:

$$H_0\ (\underline{w},\ z) = 1,$$

we obtain:

$$H_i\ (\underline{w},\ z) = \prod_{k=1}^{i} (1 - w_i z)^{-1}\ . \tag{2.4}$$

This equation is a generalization of the notion of independance to closed queueing networks having product-form solution. It is the Ƶ-transform of the P.G.F. that has the product form rather than the P.G.F. itself.

2.2 Networks with multiple job classes

For the sake of simplicity, we will consider networks with two job classes. Generalization to networks with more than two job classes is trivial. We will assume that all service rates are state independant and that jobs of one class are not allowed to switch to other job classes. Since any network with job class switching can be transformed to a network with no class switching [BALBO80], we will not discuss network with class switching.

Let n_j, j=1, 2, be the total number of class-j jobs. The state vector of the network is described by:

$$\underline{n} = (\underline{n}_1,\ \underline{n}_2,\ \ldots,\ \underline{n}_M)$$

where each $\underline{n}_i$ is a two-dimensional vector:

$$\underline{n}_i = (n_{i1},\ n_{i2})$$

where n_{ij} is the number of class-j jobs at server i. The state vector must satisfy the constraint:

$$\sum_{i=1}^{M} n_{ij} = n_j\ ,\ j=1,\ 2. \tag{2.5}$$

The network equilibrium-state probability distribution [BASK75] is:

$$P(\underline{n}) = \frac{1}{G(n_1,n_2)} \prod_{i=1}^{M} f_i(\underline{n}_i), \tag{2.6}$$

where:

$$f_i(\underline{n}_i) = \frac{(n_{i1} + n_{i2})\,!}{n_{i1}!\ \ n_{i2}!}\ a_{i1}^{\,n_{i1}}\ a_{i2}^{\,n_{i2}}$$

where a_{ij} is the relative workload (traffic intensity) of class-j jobs at server i.

Let $\underline{s} = (\underline{s}_1, \underline{s}_2)$ be a 2M-dimensional vector where $\underline{s}_i = (s_{i1}, s_{i2})$ is a two-dimensional vector. Similarly we define $\underline{a} = (\underline{a}_1, \underline{a}_2)$ and $\underline{a}_i = (a_{i1}, a_{i2})$.

Let us introduce the unnormalized P.G.F. as:

$$G_M(n_1, n_2, \underline{s}) = \sum_{\underline{n}} \prod_{i=1}^{M} f(\underline{n}_i)\ s_{i1}^{\,n_{i1}}\ s_{i2}^{\,n_{i2}}$$

where the sum extends to all possible vector $\underline{n}$ subject to the constraint (2.5).

Putting $a_{ij}\ s_{ij} = w_{ij}$, defining the function G_i as:

$$G_i(n_1, n_2, \underline{w}) = \sum_{\underline{n}} \prod_{k=1}^{i} \frac{(n_{i1} + n_{i2})\,!}{n_{i1}!\ \ n_{i2}!}\ w_{i1}^{\ n_{i1}}\ w_{i2}^{\ n_{i2}}$$

and letting $H_i(\underline{w}, \underline{z})$ be the Z-transform of $G_i(\underline{n}, \underline{w})$, i.e.:

$$H_i(\underline{w}, \underline{z}) = \sum_{n_1,\, n_2\ \ 0} G(n_1, n_2, \underline{w})\ z_1^{\,n_1}\ z_2^{\,n_2},$$

we can show that:

$$H_i(\underline{w}, \underline{z}) = \frac{H_{i-1}(\underline{w}, \underline{z})}{1 - w_{i1}z_1 - w_{i2}\,z_2}\ . \tag{2.7}$$

Solving this difference equation with boundary condition:

$$H_0(\underline{w}, \underline{z}) = 1,$$

we obtain:

$$H_i(\underline{w}, \underline{z}) = \prod_{k=1}^{i} (1 - w_{k1}\,z_1 - w_{k2}\,z_2)^{-1}\ . \tag{2.8}$$

This equation is a generalization of eq. (2.4) to networks with multiple service classes. An analytical expression for $H_i(\underline{w}, \underline{z})$ will be derived in the next section.

3. INVERTING THE Z-TRANSFORM

In this section, we will discuss methods for inverting the Z-transform of the unnormalized P.G.F.

3.1 Networks with a single job class

When all the w_i are distinct, a partial-fraction expansion of the righ hand side of Eq. (2.4) leads to:

$$G_M(n,\underline{w}) = \sum_{i=1}^{M} A_i\ w_i^{\,n}\ , \tag{3.1}$$

where

$$A_i = \prod_{\substack{j=1\\ j\neq i}}^{M} (1 - \frac{w_j}{w_i})^{-1}\ . \tag{3.2}$$

The unnormalized P.G.F. is obtained by substituting w_i for $a_i\ s_i$ in the above equations.

The case in which some w_i are equal is not excluded. In fact, when some of the relative workloads a_i are equal, the corresponding w_i will be equal at the point $s_1 = s_2 = \dots = 1$. Such a point is unfortunately encountered when we wish to calculate the normalization constant and marginal mean queue lenghts from the unnormalized P.G.F.

When some of the relative workloads are equal, we consider the reduced network consisting of only servers with distinct relative workloads.

Let M now be the number of such servers and $a_1, a_2, \dots, a_M$ be the distinct relative workloads. Using the theory of partial-fraction expansion we can write the unnormalized P.G.F. as:

$$G\ (n,\ \underline{w})\ =\ \sum_{i=1}^{M}\ A_i\ (n)\ w_i^{\ n} \tag{3.3}$$

where

$$A_i\ \ (n)\ =\ \sum_{j=1}^{m_i}\ \binom{n+j-1}{j-1}\ B_{ij}$$

where m_i is the number of servers having the same workload a_i, and:

$$B_{ij}\ =\ \frac{1}{(m_i-j)!\ (-w_i)^{m_i-j}}\ \ (\frac{d}{dz})^{\ m_i-j}\ \ H_M^{\ i}\ \ (\underline{w},\ \frac{1}{w_i})$$

where

$$H_M^{\ i}\ (\underline{w},\ z)\ =\ \prod_{\substack{j=1\\ j\neq i}}^{M}\ \ (1\ -\ w_j z)^{-m_j}.$$

Although B_{ij} does not have a general analytical expression, a simple iterative algorithm for calculating the B_{ij}, analytically or numerically, will be given in the next section. (See Eq. 4.6)

The functions $A_i(n)$ in Eq. (3.3) are polynomials of degree at most equal to m_i. The coefficients of the polynomials are completely determined by Eq. (3.3) by the values of $G(0,\ \underline{w})$, $G(1,\underline{w})$, ..., $G(M-1,\ \underline{w})$. Thus, performance properties of a closed network with M servers are completely determined in principle by those obtained from the same network for the cases where the total number of jobs is less than M. This observation constitutes a method of determining performance measures for networks of arbitrary number of jobs from performance measures of networks where the total number of jobs are less than M.

Another method of determining the unnormalized P.G.F. for a general network consists in differentiating the unnormalized P.G.F. for the "reduced" network which is given by $G_M(n,\ \underline{w})$ in Eq. (3.1). In fact, it can be shown easily that:

$$G(n,\ \underline{w})\ =\ \prod_{i=1}^{M}\ \frac{1}{m_i!}\ \ (\frac{\partial}{\partial w_i})^{m_i}\ \ G_M\ (n\ +\ m_1\ +\ \dots\ +\ m_i,\ \underline{w}) \tag{3.4}$$

Eqs. (3.3) and (3.4) are equivalent and can be used interchangedly depending on circumstances.

3.2 Networks with multiple job classes

For networks with multiple job classes, the unnormalized P.G.F. is given by Eq. (2.8) so that we must invert the multidimensional Z-transform. Unfortunately, the partial-fractional expansion method is not applicable to the multi-dimensional situation and we must find a different method.

By inspecting Eq. (2.7) we find that the function $G_i(n_1, n_2, \underline{w})$ must satisfy the following system of difference equation:

$$G_i(n_1, n_2, \underline{w}) - G_{i-1}(n_1, n_2, \underline{w}) = w_{i1}\, G_i(n_1-1, n_2, \underline{w}) + w_{i2}\, G_i(n_1, n_2-1, w) \quad (3.5)$$

with boundary condition:

$G_i(0, 0, \underline{w}) = 1$

Inversely, we can show that $H_i(\underline{w}, \underline{z})$ satisfies Eq. (2.7) if $G_i(n_1, n_2, \underline{w})$ satisfies Eq. (3.5). Our problem is thus to find a solution to Eq. (3.5) with the correspondings boundary condition. Such a solution is of course unique.

We try a solution of the form:

$$G_i(n_1, n_2, \underline{w}) = \sum_{k=1}^{i} A_k^{\,i} \frac{n!}{n_1!\, n_2!} w_{k1}^{\,n_1} w_{k2}^{\,n_2} \quad (3.6)$$

Substituting Eq. (3.6) into Eq. (3.5) we find the condition:

$$A_k^{\,i} = \frac{A_k^{\,i-1}}{1 - \frac{1}{n}\left(n_1 \frac{w_{i1}}{w_{k1}} + n_2 \frac{w_{i2}}{w_{k2}}\right)} \quad (3.7)$$

By induction on Eq. (3.7) and by noting that the boundary condition implies

$$A_k^{\,0} = 1 ,$$

we obtain:

$$A_k^{\,i} = \prod_{\substack{m=1 \\ m\neq k}}^{i} \left[1 - \frac{1}{n}\left(n_1 \frac{w_{m1}}{w_{k1}} + n_2 \frac{w_{m2}}{w_{k2}}\right)\right]^{-1} . \quad (3.8)$$

Thus, the unnormalized P.G.F. for networks with multiple job classes is given by Eqs. (3.6) and (3.8). In particular, the normalization constant is given by:

$$G(n_1, n_2) = \frac{n!}{n_1!\; n_2!} \sum_{i=1}^{M} A_i\, a_{i1}^{\,n_1} a_{i2}^{\,n_2} \quad (3.9)$$

when

$$A_i = \prod_{\substack{k=1 \\ k\neq i}}^{M} \left[1 - \frac{1}{n}\left(n_1 \frac{a_{k1}}{a_{i1}} + n_2 \frac{a_{k2}}{a_{i2}}\right)\right]^{-1} \quad (3.10)$$

It must be observed that Eqs. (3.6) - (3.10) are valid only for the case in which all relative workload vectors, $\underline{a}_i$, are distinct. The more general case in which some of the relative workload vectors are equal will be treated in section 5.3.

4. COMPUTATIONAL ALGORITHMS

In this section we will present some algorithms for evaluating the normalization constant based on its Z-transform.

4.1 Single job-class networks

(1) First, we observe that the normalization constant is given by:

$$G(n) = \frac{1}{n!} \left.\frac{d^n H(z)}{dz^n}\right|_{z=1} \quad (4.1)$$

where:

$$H(z) = \prod_{i=1}^{M} (1 - a_i z)^{-m_i} .$$

Since:

$$\frac{dH(z)}{dz} = h(z)\, H(z),$$

where:

$$h(z) = \sum_{i=1}^{M} \frac{m_i a_i}{1 - a_i z} ,$$

we have:

$$\frac{d^n H(z)}{dz^n} = \frac{d^{n-1}}{dz^{n-1}} h(z)\, H(z) = \sum_{k=0}^{n-1} \binom{n-1}{k} D^k h(z)\, D^{n-1-k} H(z), \; n \geq 1, \tag{4.2}$$

where we have defined $D^k \equiv d^k / dz^k$. On the other hand, we have:

$$D^k h(z) = k! \sum_{i=1}^{M} \frac{m_i a_i^{k+1}}{(1 - a_i a)^{k+1}}$$

Combining this equation with Eqs. (4.1) and (4.2) we obtain:

$$G(n) = \frac{1}{n} \sum_{k=1}^{n-1} g(k)\, G(n-1-k), \; n \geq 1 \tag{4.3}$$

where

$$g(k) = \sum_{i=1}^{M} m_i \left[\frac{a_i}{1-a_i}\right]^{k+1}$$

Eq. (4.3) together with the boundary condition:

$$G(0) = 1 \tag{4.4}$$

provide us with a recurrence relation for calculating the normalization constant. An algorithm based on this relation requires only one memory array of n elements, Mn multiplications and Mn additions. It is thus the most efficient algorithm known to date in terms of memory requirement.

(ii) Another algorithm can be derived directly from Eq. (2.4). It is easy to see that this equation implies that the normalization constant $G_i(n)$ must satisfy the difference equation:

$$G_i(n) = G_{i-1}(n) + a_i\, G_i(n-1), \tag{4.5}$$

with the initial condition (4.4). The algorithm based on Eq. (4.5) was first discovered by Buzen [BUZE71].

(iii) Another algorithm consists of using Eq. (3.3) with B_{ij} calculated recursively. In fact, we have

$$B_{i,m_i} = \prod_{\substack{j=1 \\ j \neq i}}^{M} \left(1 - \frac{a_k}{a_i}\right)^{-m_i} .$$

By using the same kind of argument which led us to Eq. (4.3), we can show that:

$$B_{i,m_i-j} = \frac{1}{m_i-j+1} \sum_{k=0}^{m_i-j} h(k)\, a_i^{k+1}\, B_{i,m_i-j+k+1} . \tag{4.6}$$

This equation together with the values B_{i,m_i} allow us to calculate recursively all the coefficient $B_{i,j}$ in the partial-fractional expansion. It also constitues an efficient algorithm for calculating the coefficients in any partial-fractional expansion. To our knowledge, this efficient algorithm for partial-fractional expansion has not been discussed elsewhere.

(iv) Lastly, a different algorithm results from the fact that the functions A_i (n) in Eq. (3.3) are polynomials of degree less than m_i. By using the m values G(0), G(1), ... G (n-1) obtained by any conceivable methods, the coefficients of polynomials $A_i(n)$ can be determined and so is the function G(n).

4.2 Multiple job-class networks

(i) The first algorithm can be obtained from Eq. (3.5):

$$G_i\ (n_1, n_2) = G_{i-1}\ (n_1, n_2) + a_{i1}\ G_i\ (n_1-1, n_2) + a_{i2}\ G_i\ (n_1, n_2-1). \tag{4.7}$$

This equations together with the initial condition:

$$G_i\ (0,0) = 1 \tag{4.8}$$

determine iteratively the normalization constant. This algorithm is a generalization of algorithm already proposed by Buzen discussed earlier for single job-class networks and has been discussed by Balbo [BALB80].

(ii) Another algorithm can be obtained by noting that from Eq. (2.4) we have:

$$G\ (n,m) = \frac{1}{n!\ m!}\ D_1^{\ n}\ D_2^{\ m}\ H(\underline{a},\ \underline{z})\ \Big|_{\underline{z}\ =\ \underline{1}}$$

where

$$D_i^{\ n} = (\frac{\partial}{\partial z_i})^n$$

and

$$H(\underline{a},\ \underline{z}) = \prod_{i=1}^{M} (1 - a_{i1}a_1 - a_{i2}z_2)^{-1} .$$

It is easy to derive the relation:

$$D_2^{\ m}\ D_1^{\ n}\ H(\underline{a},\ z) = \sum_{\ell=0}^{m} \sum_{k=0}^{n-1} \Big[\binom{m}{\ell}\binom{n-1}{k}\ D_2^{\ell}\ D_1^{\ k}\ h(\underline{a},\ \underline{z})\ D_2^{\ m-\ell}\ D_1^{\ n-1-k}\ H(\underline{a},\ \underline{z})\Big]$$

where:

$$h(\underline{a},\ \underline{z}) = \sum_{i=1}^{M} \frac{a_{i1}}{1 - a_{i1}z_1 - a_{i2}z_2} ,$$

so that:

$$D_2^{\ell}\ D_1^{\ k}\ h(\underline{a},\ \underline{z}) = \sum_{i=1}^{M} \frac{a_{i1}^{\ k+1}\ a_{i2}^{\ \ell}\ k!\ \ell!}{(1 - a_{i1}z_1)^{k+1}\ (1 - a_{i2}z_2)^{\ell}}$$

From the above equations, we have:

$$G(n,m) = \frac{1}{n} \sum_{\ell=0}^{m} \sum_{k=0}^{n-1} g(k,\ \ell)\ G(n-k-1,\ m-\ell),\ n,\ m \geq 1.$$

$$g(k,\ell) = \sum_{i=1}^{M} \frac{a_{i1}^{\ k+1}\ a_{i2}^{\ \ell}\ k!\ \ \ell!}{(1 - a_{i1})^{k+1}\ (1 - a_{i2})^{\ell}} \tag{4.9}$$

This equation together with the initial condition (4.8) offer a recurrence algorithm for calculating the normalization constant. A computer implementation of this algorithm requires only one memory array of n × m words.

5. MARGINAL DISTRIBUTIONS AND PERFORMANCE MEASURES

5.1 Multiple Job Class Networks With Distinct Relative Workloads

Let P_{ij} (n_1, n_2, i_1, i_2) be the unnormalized probability of having i_1 class -1 jobs and i_2 class -2 jobs at server i and let Q_{ij} $(n_1, n_2, s_{i1}, s_{i2})$ be the P.G.F. for the P_{ij}. The $\mathcal{Z}$ - transform for the Q_{ij} can be obtained from H_M $(\underline{w}, \underline{z})$ in Eq. (2.8) by letting:

$$s_{kj} = 1 \quad \text{if} \quad k \neq i .$$

Thus:

$$\mathcal{Z}\{Q_{ij}(n_1, n_2, s_{i1}, s_{i2})\} = \frac{1}{1 - a_{i1} s_{i1} z_1 - a_{i2} s_{i2} z_2} \prod_{\substack{k=1 \\ k\neq i}}^{M} \frac{1}{1 - a_{k1}\mathcal{Z}_1 - a_{k2}\mathcal{Z}_2} \tag{5.1}$$

Let G^i (n_1, n_2) be the inverse $\mathcal{Z}$ - transform of the factor $\Pi^M_{k=1,\ k\neq i}$ $(1 - a_{k1}\mathcal{Z}_1 - a_{k2}\mathcal{Z}_2)^{-1}$. Eq. (5.8) implies:

$$\mathcal{Z}\{Q_{ij}(n_1, n_2, s_{i1}, s_{i2})\} = \sum_{i_2=0}^{n_2} \sum_{i_1=0}^{n_1} \frac{(i_1 + i_2)!}{i_1!\ i_2!} (a_{i2}\ s_{i2})^{i_1} (a_{i2}\ s_{i2})^{i_2} . G^i(n_1 - i_1, n_2 - i_2)$$

From this equation, the unormalized queue length distribution at server i is obtained as:

$$P_{ij}(n_1, n_2, i_1, i_2) = \frac{(i_1 + i_2)!}{i_1!\ i_2!}\ a_{i1}^{i1}\ a_{i2}^{i2}\ G^i\ (n_1 - i_1, n_2 - i_2) \tag{5.2}$$

The function G^i can be obtained from the normalization constant G as (see Eq. 3.5):

$$G^i\ (n_1, n_2) = G(n_1, n_2) - a_{i1}\ G(n_1 - 1, n_2) - a_{i2}\ G\ (n_1, n_2 - 1)$$

This means that the function $G(n_1, n_2)$ determines completely the queue-length marginal distributions.

An analytical expression for the function G^i can be derived using results of section (3.2) as:

$$G^i\ (n_1, n_2) = \frac{(n_1 + n_2)!}{n_1!\ n_2!} \sum_{\substack{j=1 \\ j\neq i}}^{M} A_j^i\ a_{j1}^{n_1}\ a_{j2}^{n_2} , \tag{5.3}$$

where:

$$A_j^i = \prod_{\substack{k=1 \\ k\neq i \\ k\neq j}}^{M} [1 - \frac{1}{n_1 + n_2} (n_1 \frac{a_{k1}}{a_{j1}} + n_2 \frac{a_{k2}}{a_{j2}})]^{-1} . \tag{5.4}$$

From Eqs. (5.2) and (5.3), we obtain an expression for $L_i(n_1, n_2, i_1, i_2)$, the <u>normalized</u> marginal distribution at server i:

$$L_i(n_1,n_2,i_1,i_2) = \frac{(n_1 + n_2)!}{n_1!\ n_2!} \frac{1}{G(n_1,n_2)} . \sum_{\substack{j=1 \\ j\neq i}}^{M} A_j^i\ a_{i1}^{n_1}\ a_{i2}^{n_2} \left(\frac{a_{i1}}{a_{j1}}\right)^{i_1} \left(\frac{a_{i2}}{a_{j2}}\right)^{i_2} \tag{5.5}$$

where $G(n_1, n_2)$ is given by Eq. (3.9).

Let $\overline{N}_{ij}$ (n_1, n_2) be the <u>unormalized</u> average queue length of class -j job at server i. By differentiating Eq. (5.1) with respect to s_{ij} and let $s_{ij} = 1$ we have:

$$\mathcal{Z}\{\overline{N}_{i1}\ (n_1, n_2)\} = a_{ij} z_1 \frac{H^M(n_1, n_2, \underline{a}, \underline{z})}{1 - a_{i1}z_1 - a_{i2}z_2} \tag{5.6}$$

From this equation, we infer that:

$$\overline{N}_{i1}\ (n_1, n_2) = 0\ , \text{ if } n_1 = 0$$

$$\bar{N}_{i1}(n_1,n_2) = \sum_{k_1=1}^{n_i} \sum_{k_2=0}^{n_2} [G(n_1 - k_1 - n_2 - k_2) \frac{k_1}{k_1 + k_2} \frac{(k_1 + k_2)!}{k_1!\, k_2!} a_{i1}^{k_1} a_{i2}^{i_2}],\ n_1 > 0. \tag{5.7}$$

Using Eq. (3.9), we have:

$$\bar{N}_{i1}(n_1,n_2) = \sum_{j=1}^{M} \sum_{k_1=1}^{n_1} \sum_{k_2=0}^{n_2} A_j\, a_{j1}^{n_1}\, a_{j2}^{n_2} \frac{k_1}{k_1 + k_2} \frac{(n_1 - k_1 + n_2 - k_2)!}{(n_1 - k_1)!\,(n_2 - k_2)!}$$

$$\cdot \frac{(k_1 + k_2)!}{k_1!\, k_2!} \left(\frac{a_{i1}}{a_{j1}}\right)^{k_1} \left(\frac{a_{i2}}{a_{j2}}\right)^{k_2}, \tag{5.8}$$

where A_j is given by Eq. (3.10).

We close this subsection by noting that analytical formulas given by Eqs. (5.5) and (5.7) are valid only when the A_j^i and A_j are defined. A_j^i and A_j may not be defined if, for some k and h, we have (see Eqs. 3.10 and 5.4):

$$a_{kj} = a_{hj} \text{ for all } j.$$

This will happen if some servers have identical workload for each job class. If this happens, the method presented in the next subsection may be used.

5.2 Networks in which some relative workloads are equal

When the workload vectors at some servers are equal, Eq. (5.2) for marginal distribytion and Eq. (5.7) for average marginal queue lengths still hold provided that the function $G(n_1, n_2)$ is given. To obtain the function G for the general case, we can use the following procedure.

Define $G^S(n_1, n_2)$ be the normalization constant for any network consisting of S servers having workload vectors $\underline{a}_1, \underline{a}_2, \ldots \underline{a}_S$. Let N_i^S be the unnormalized total average queue length at server i. From Eq. (5.6) we have:

$$\mathcal{Z}\{N_i^S(n_1, n_2)\} = \frac{H^S(n_1, n_2, \underline{a}, z)}{1 - a_{i1}z_1 - a_{i2}z_2} - H^S(n_1, n_2, \underline{a}, z). \tag{5.9}$$

Let $G^{S+\{i\}}$ be the normalization constant of the network consisting of S servers defined previously plus a server having workload vector a_i. Then Eq. (5.9) implies:

$$N_i^S(n_1, n_2) = G^{S+\{i\}}(n_1, n_2) - G^S(n_1, n_2).$$

If we define n_i^S as the normalized average queue length at server i for the network consisting of s servers, the above equation implies:

$$G^{S+\{i\}}(n_1, n_2) = \bar{n}_i^S\, G^S(n_1, n_2) + 1\ . \tag{5.10}$$

Eq. (5.10) serves as basis to obtain marginal distributions and average queue lengths for any network from those of networks having only <u>distinct</u> workload vectors. The procedure is as follows:

<u>Step 1</u>: Consider the "reduced" network consisting only of servers with distinct workload vectors $a_1, a_2, \ldots, a_S$. Let m_i be the number of servers having $\underline{a}_i$ as workload vector. Use Eq. (3.9) to obtain G^S and Eq. (5.8) to obtain $\bar{n}_i^S$. Set i = 1.

<u>Step 2</u>: Use Eq. (5.10) to obtain the function $G^{S+\{1\}}$ and Eq. (5.7) to obtain the marginal queue $\bar{n}_i^S$ at server having workload vector $\underline{a}_i$. Repeat this step $m_i - 1$ times.

<u>Step 3</u>: Set $i \leftarrow i + 1$ and return to step 2 if $i < S$, otherwise determine marginal distributions and other performance measures. Stop.

To implement this procedure on a computer, we need only two $n_1 \times n_2$ dimensional arrays

arrays, one for $\bar{n}_i^S$ and one for G_i^S.

6. DISCUSSION AND CONCLUSION

We have introduced in section 2 the concept of unnormalized P.G.F. and found that the unnormalized P.G.F. turned out to have the form of the so called normalization constant used in previous works on closed queueing networks. The unnormalized P.G.F. has helped us to derive many results concerning marginal distributions and average queue lengths in a simple manner. Our effort to derive analytical expression for the P.G.F. in section 3 has met only with moderate success because we were able to obtain to obtain closed form expressions only for networks in which all servers must have distinct workload vectors. However, we have shown that the P.G.F. for any network can be obtained analytically or numerically by a procedure presented in section 5.3. Furthermore, an interesting observation has been made for networks with a single job class, namely, that the P.G.F. for any total number of jobs in the network depend only on those P.G.F. for networks in which the total number of jobs is less than the number of servers. We have not explored this observation to obtain practical results nor have we known how to generalize this observation to networks with multiple service classes. Our use of the Z - transform has allowed us to discover a number of efficient computational algorithms for the normalization constant in section 4. These algorithms are superior to known algorithms in term of computer memory requirement. In particular, the algorithm expressed in Eq. (4.6)for partial-fractional expansion is useful to other problems than the one considered in this paper. In section 5., we were able to obtain closed form expressions for marginal queue length distributions and average queue lengths only for networks consisting of servers with distinct workload vectors, but an efficient procedure has been offered to treat more general networks. Insight has been gained into the behavior of a single server in a closed network, namely, that marginal probability distributions are superpositions of geometric distributions in the single job class case and of multinomial distributions in the multiple job class case. Since geometric and multinomial distributions have well-known properties, this result is useful for calculating higher moments of marginal distributions or deriving waiting time distributions. We have obtained closed form expression for the Z - transforms of marginal probability distributions and average queue lengths. These formulas may be explored in works on sensitivity analysis, asymptotic analysis, cycle-time distribution, etc.

REFERENCES

[BASK75] F. Baskett, K. M. Chandy, R.R. Muntz and F. Palacios - Gomez, "Open Closed, and Mixed Network of Queues with Different Class of Customers", JACM 22, 2 (April 1975).

[BUZE71] J. P. Buzen, "Queueing Network Models of Multiprogramming", Ph.D. Thesis, Harvard University, Cambridge, Mass. (1971).

[CHAN77] K. M. Chandy, J. H. Howard and D. F. Towsley, "Product Form and Local Balance in Queueing Networks", JACM 24, 2, pp. 250-263 (April 1977).

[CHAN80] K. M. Chandy and C. H. Sauer, "Computational Algorithms for Product Form Solution Networks", CACM 23, 10 (October 1980), pp. 573-583.

[CHAN83] K. M. Chandy and A. J. Martin, "A Characterization of Product-Form Queueing Networks", JACM 30, 2, pp. 286-299.

[COURT77] P. J. Courtois, Decomposability: Queueing and Computer System Applications, New York: Academic Press (1977).

[GORD67] W. T. Gordon and G. F. Nowell, "Closed Queueing Systems with Exponential Servers", Operations Res., 15 (April 1967), pp. 252-265.

[JACK63] J. R. Jackson, "Jobshop-like Queueing Systems", Management Sci., 10 (Oct. 1963), pp. 131-142.

[KAMO80] F. Kamoun and L. Kleinrock, "Analysis of Shared Finite Storage in a Computer Network Node Environment Under General Traffic Conditions", IEEE Com-28, 7 (July 1980), pp. 992-1003.

[KOBA74a] H. Kobayashi, "Application of Diffusion Approximation to Queueing Networks, Part I". JACM 21 (1974), pp. 316-328.

[KOBA74b] H. Kobayashi, "Application of Diffusion Approximation to Queueing to Queueing Networks, Part II, JACM 21, 2 (1974), pp. 459-469.

[MOOR61] C. G. III Moore, "Network Models for Large-Scale Time-Sharing Systems", Technical Report No. 71-1, Department of Industrial Engineering, University of Michigan, Ann Arbor, Michigan (April 1961).

[MOOR72] F. R. Moore, "Computational Model of a Closed Queueing Networks with Exponential Servers", IBM Johr. Res. Dev. 16, 6 (1972), (June 1972), pp. 567-572.

[NOET79] A. S. Noetzel, "A Generalized Queueing Discipline for Product Network Solutions", JACM 26, 4 (October 1979), pp. 779-793.

[REIS75] M. Reiser and H. Kobayashi, "Queueing Network with Multiple Closed Chains: Theory and Computational Algorithms", IBM J. of Research and Development 15, 3 (May 1975).

[REIS80] M. Reiser and S. S. Lavenberg, "Mean Value Analysis of Closed Multichain Queueing Networks", JACM 27, 2 (April 1980), pp. 313-322.

[WILL7] A. C. Williams and R. A. Bhandiwad, "A Generating Function Approach to Queueing Network Analysis of Multiprogrammed Computers", Networks, 6 (1976), pp. 1-22.

A HIERARCHICAL ALGORITHM FOR LARGE-SCALE SYSTEM OPTIMIZATION PROBLEMS WITH DUALITY GAPS

P.Tatjewski
Institute of Automatic Control
Technical University of Warsaw
00-665 Warszawa, Poland

1. INTRODUCTION

Large-scale interconnected system optimization problems can be nonlinear and nonconvex, especially when they result from engineering applications. In such cases duality gaps can arise. To resolve dual gaps application of augmented Lagrangian is the main approach. Main trouble in the use of the augmented Lagrangian for hierarchical (multilevel) optimization is lack of separability of that Lagrangian. And hierarchical optimization algorithms are important for large-scale interconnected system optimization problems not only computationally, but also as a basis for development of control and management structures advantageous for several reasons (see e.g. Findeisen et al.[1]). To overcome the problem of nonseparability of the augmented Lagrangian some specific linearization of its nonseperable terms can be applied (Stephanopoulos and Westerberg [2], Stoilov [3], Findeisen et al.[1]). First algorithms based on that concept were, however, rather unefficient and without convergence analysis. Only specific algorithm given by Stoilov [3] was more satisfactory. In this paper new approach based on linearization of nonseparable terms of the augmented Lagrangian is presented, together with thorough analysis of its applicability conditions. Structure of the presented algorithm seems to be most naturall and efficient generalization of the structure of well known price method of two-level optimization, to the augmented Lagrangian case.

The paper is organized as follows. In section 2, first, large-scale interconnected system description and the concept of linearization of the augmented Lagrangian to get its separability are presented. Then the structure of the algorithm is described. In section 3 convergence conditions of the algorithm are analysed. In section 4 it is demonstrated by numerical examples how parameters of the algorithm influence its convergence and how they should be adjusted.

2. DESCRIPTION OF THE ALGORITHM

It is assumed that the interconnected system optimization problem has the form (see e.g. Findeisen et al.[1])

$$\begin{cases} \min \sum_{i=1}^{N} Q_i(c_i,u_i) \\ y_i=F_i(c_i,u_i),\ u_i=H_i y,\ (c_i,u_i)\in CU_i,\quad i=1,\dots,N, \end{cases} \tag{1}$$

where c_i,u_i and y_i are, respectively, i-th subsystem control, interaction input and interaction output vector variables from appropriate finite-dimensional spaces R^{n_i}, R^{r_i} and R^{s_i}. $F_i: R^{n_i} \times R^{r_i} \to R^{s_i}$ denote subsystem output mappings and H_i subsystem output matrices composed from zeros and ones, $i=1,\dots,N$. Local constraint sets of each subsystem are denoted by CU_i, $i=1,\dots,N$.

Problem (1) is allowed to have duality gap and augmented Lagrangian approach is applied. The following augmented Lagrangian is used (it does not include local constraints):

$$L_a(c,u,\lambda,\varsigma) \triangleq \sum_{i=1}^{N} Q_i(c_i,u_i) + \lambda^T(u-HF(c,u)) + \tfrac{1}{2}\varsigma\|u-HF(c,u)\|^2, \tag{2}$$

where $c^T \triangleq (c_1^T,\dots,c_N^T)$, $u^T \triangleq (u_1^T,\dots,u_N^T)$, $F^T \triangleq (F_1^T,\dots,F_N^T)$, $H^T \triangleq [H_1^T,\dots,H_N^T]$ and $\lambda^T \triangleq (\lambda_1^T,\dots,\lambda_N^T)$, $\lambda_i \in R^{r_i}$, are multipliers. To apply multilevel approach function (2) should be separable with respect to (c_i,u_i), $i=1,\dots,N$. But it is not the case, since by direct calculation we get

$$L_a(c,u,\lambda,\varsigma) = \sum_{i=1}^{N} [Q_i(c_i,u_i) + \lambda_i^T u_i - \sum_{j=1}^{N} \lambda_j H_{ji} F_i(c_i,u_i) +$$

$$+ \tfrac{1}{2}\varsigma(\|u_i\|^2 + \|F_i(c_i,u_i)\|^2)] - \varsigma u^T HF(c,u),$$

and the last term is not separable. To overcome this difficulty a specific linearization of that term (in some point (c^k,u^k)) can be used (Stephanopoulos and Westerberg [2])

$$u^T HF(c,u) \cong -u^{k^T} HF(c^k,u^k) + u^T HF(c^k,u^k) + u^{k^T} HF(c,u). \tag{3}$$

Using (3) augmented Lagrangian (2) has the following, approximate but separable form, denoted by Λ_a:

$$\Lambda_a(c,u,\lambda,\varsigma,c^k,u^k) \triangleq$$

$$\triangleq \sum_{i=1}^{N} [Q_i(c_i,u_i) + \lambda_i^T u_i - \sum_{j=1}^{N} \lambda_j H_{ji} F_i(c_i,u_i) + \tfrac{1}{2}\varsigma(\|u_i\|^2 + \|F_i(c_i,u_i)\|^2) +$$

$$-\varsigma(u_i^T H_i F(c^k,u^k) + \sum_{j=1}^{N} u_j^{k^T} H_{ji} F_i(c_i u_i))] + \varsigma u^{k^T} HF(c^k,u^k) \triangleq$$

$$\triangleq \sum_{i=1}^{N} \Lambda_{ai}(c_i,u_i,\lambda,\varsigma,c^k,u^k) + \varsigma u^{k^T} HF(c^k,u^k) \tag{4}$$

Hence, the following <u>local optimization problems</u> can be formulated

$$\min_{(c_i,u_i)\in CU_i} \Lambda_{ai}(c_i,u_i,\lambda,\varrho,c^k,u^k), \qquad i=1,\dots,N. \tag{5}$$

It is interesting to note that minimization of Λ_{ai} is equivalent to minimization of the function

$$L^k_{ai}(c_i,u_i,\lambda,\varrho,c^k,u^k) \triangleq$$

$$\triangleq L_a(c^k_1,\dots,c^k_{i-1},c_i,c^k_{i+1},\dots,c^k_N,u^k_1,\dots,u^k_{i-1},u_i,u^k_{i+1},\dots,u^k_N,\lambda,\varrho),$$

$i=1,\dots,N$ - what can be proved by direct calculation.

The following <u>structur of the proposed algorithm</u> can be now formulated (with penalty coefficient ϱ kept fixed):

1° state $\varepsilon>0,\alpha$, $n=0$, $k=0$, λ^0, c^0, u^0.

2° for given value of λ^n and (c^k,u^k) solve local optimization problems (5), separately and simultaneously when possible, with results $\hat{c}^{nk}_i,\hat{u}^{nk}_i$, $i=1,\dots,N$.

If $\|(\hat{c}^{nk},\hat{u}^{nk})-(c^k,u^k)\|\le\varepsilon$ (6)

then set $(\hat{c}^n,\hat{u}^n)=(\hat{c}^{nk},\hat{u}^{nk})$ and go to 3°.

Otherwise, set $(c^{k+1},u^{k+1})=(\hat{c}^{nk},\hat{u}^{nk})$, (7)

$k=k+1$, and repeat 2°.

3° check the overall stop criterion. If not satisfied, then modify coordination variables (multipliers) λ according to some formula f,

$$\lambda^{n+1}=f(\lambda^n,\hat{c}^n,\hat{u}^n,c^k,u^k), \tag{8}$$

set $n=n+1$, $k=0$, $(c^0,u^0)=(\hat{c}^n,\hat{u}^n)$, and go to 2°.

The above algorithm has the same basic structure as ordinary, well known and justified price method of hierarchical optimization (see e.g. Findeisen et al.[1]), hence it will be called <u>augmented price method</u> in the sequel. The only significant difference is in step 2° constituting <u>inner iteration loop</u>, instead of one local problems evaluation in standard price method. The point obtained as a result of inner iterations approximates, with accuracy determined by the value of ε, solution of the problem

$$\min_{(c,u)\in CU} L_a(c,u,\lambda,\varrho), \tag{9}$$

where $CU\triangleq CU_1\times\dots\times CU_N$. When ε is sufficiently small then the results of inner iterations can be treated as solutions of (9). Hence, adjusting formulae for λ derived in well known augmented Lagrangian theory for optimization problems with constraints can be applied, e.g., the Hestenes-Powell multiplier rule

$$\lambda^{n+1}=\lambda^n+\alpha\varrho(\hat{u}^n-HF(\hat{c}^n,\hat{u}^n)). \tag{10}$$

3. CONVERGENCE ANALYSIS

The key problem in analyzing the described class of algorithms are conditions assuring convergence of inner iteration loop. Before stating these conditions let us introduce some notation. Let us denote

$$Q(c,u) \triangleq \sum_{i=1}^{N} Q_i(c_i,u_i)$$

$$h(c,u) \triangleq HF(c,u)$$

$$CU \triangleq \{(c,u): g(c,u) \leq 0\}.$$

Let us denote optimal solution of (1) by $(\hat{c},\hat{u})$, with corresponding vector of multipliers $\hat{\lambda}$. Let us denote by g_A vector of active constraints at the optimal point, with corresponding vector of Kuhn-Tucker multipliers $\hat{\mu}_A$. Let us denote also

$$C \triangleq \nabla^T h(\hat{c},\hat{u})$$

$$D \triangleq \nabla^T g_A(\hat{c},\hat{u})$$

$$I_o \triangleq [0_{rxn} I_{rxr}], \text{where } r=\dim u,\ n=\dim c,$$

$$\mathcal{L}(c,u,\lambda,\mu) \triangleq Q(c,u) + \lambda^T(u-h(c,u)) + \mu^T g(c,u).$$

Theorem 1.

Assume that

(i) Q, h and g are of class C^2 in some neighbourhood of $(\hat{c},\hat{u})$,

(ii) gradients of all active constraints are linearly independent, i.e.

$$[(I_o-C)^T \; D^T]\alpha = 0 \Rightarrow \alpha = 0.$$

(iii) second order sufficient optimality conditions with strict complementarity are fulfilled at $(\hat{c},\hat{u})$, i.e.,

$$\mathcal{L}'_{(c,u)}(\hat{c},\hat{u},\hat{\lambda},\hat{\mu}) = 0,$$

$$\hat{u}-h(\hat{c},\hat{u}) = 0,$$

$$g(\hat{c},\hat{u}) \leq 0,\ \hat{\mu}_j g_j(\hat{c},\hat{u}) = 0,\ \hat{\mu}_j \geq 0,\ \mu_{A_j} > 0,$$

$$(c,u)^T \mathcal{L}''_{(c,u)(c,u)}(\hat{c},\hat{u},\hat{\lambda},\hat{\mu})(c,u) > 0 \text{ for } (c,u) \neq 0$$

and satisfying $(I_o-C)(c,u)=0$, $D(c,u)=0$,

(iv) $$(c,u)^T \mathcal{L}''_{(c,u)(c,u)}(\hat{c},\hat{u},\hat{\lambda},\hat{\mu})(c,u) > 0 \text{ for } (c,u) \neq 0$$

and satisfying $(I_o+C)(c,u)=0$, $D(c,u)=0$.

Then for sufficiently great values of ρ there is a neighbourhood of $\hat{\lambda}$ such that for every λ from this neighbourhood the inner iteration loop algorithm is locally well defined and convergent to solution $(\hat{c}(\lambda),\hat{u}(\lambda))$ of problem (9).

Outline of the proof is given in Appendix. ■

Assumptions (i), (ii) and (iii) are standard sufficient conditions assuring existence of saddle point of the augmented Lagrangian. Only assumption (iv) is a new one, connected with the necessity to assure convergence of inner iterations. It is a sufficient but important condition, what can be shown on the following simple example.

Example.

Let us take $Q(c,u)\triangleq c.u$, $h(c,u)\triangleq c+\frac{1}{2}u$, $CU=R\times R$, hence (1) has the form

$$\begin{cases} \min\ [c.u] \\ u-c-\frac{1}{2}.u=0. \end{cases}$$

It can be easily verified that the solution is $(\hat{c},\hat{u})=(0,0)$, together with multiplier $\hat{\lambda}=0$. Considering inner iteration loop (for $\lambda=\hat{\lambda}=0$) we get

$$\begin{bmatrix} u^{k+1} \\ c^{k+1} \end{bmatrix} = \begin{bmatrix} \frac{\varrho(\varrho-2)}{2(\varrho^2-\varrho-1)} & \frac{\varrho^2}{\varrho^2-\varrho-1} \\ \frac{3\varrho-4}{4(\varrho^2-\varrho-1)} & \frac{-\varrho-2}{2(\varrho^2-\varrho-1)} \end{bmatrix} \begin{bmatrix} u^k \\ c^k \end{bmatrix} = P(\varrho) \begin{bmatrix} u^k \\ c^k \end{bmatrix}$$

The eigenvalues s_1, s_2 of matrix $P(\varrho)$ are

$$s_{1,2}=\frac{\varrho}{\varrho^2-\varrho-1}\cdot(-1\pm\sqrt{\varrho^2-\varrho}),$$

hence, the algorithm does not converge for $\varrho>2$. It can be easily verified that the example problem satisfies all assumptions of Theorem 1, except of (iv).

The second important question are convergent coordination strategies, i.e., adjusting rules for multipliers. As mentioned earlier, if accuracy ε of inner iteration loop stop criterion (6) is chosen sufficiently small, then performing inner iteration loop can be treated as solving problem (9), i.e., evaluating dual function for given value of λ (and ϱ). In this convenient case known adjusting formulae can be used, see e.g. Bertsekas [5]. For testing our algorithm Hestenes-Powell multiplier rule will be chosen below, it seems to be suitable due to its simple form not involving first or second derivatives. Convergence results for optimization algorithms with this formula, with $\alpha=1$, are known, see reference given above. The following a bit more general result can be obtained using argument similar as in Theorem 1.

Result.

If the assumptions (i), (ii) and (iii) of Theorem 1 are fulfilled, then the coordination algorithm (10) is locally convergent to $\hat{\lambda}$ for $\alpha\in(0,2)$, provided inner iteration stop criterion (6) is accurate.

4. EXAMPLE NUMERICAL RESULTS

Some numerical results obtained for a very simple interconnected sys-

Table 1. Convergence for $\varrho=10$, $\alpha=1$ and various values of ε.

$\varepsilon = 0.1$			$\varepsilon = 0.05$			$\varepsilon = 0.01$		
TNE	Q	DN	TNE	Q	DN	TNE	Q	DN
17	-27.3970	1.4664	17	-27.3970	1.4664	19	-27.7808	1.5034
21	-6.0130	0.9561	21	-6.0130	0.9561	25	-5.5647	1.0416
27	-12.4759	0.3800	28	-12.6389	0.3809	34	-13.2260	0.4162
29	-8.6570	0.1533	32	-8.0848	0.2395	39	-7.8155	0.2789
31	-8.8225	0.0867	35	-9.4006	0.1168	44	-9.5907	0.1299
32	-8.6812	0.0532	37	-8.3159	0.0601	48	-8.1621	0.0755
33	-8.1587	0.0628	39	-8.4005	0.0363	52	-8.5184	0.0414
35	-8.3219	0.0238	40	-8.3098	0.0216	54	-8.1910	0.0162
36	-8.1780	0.0097	41	-8.1093	0.0232	56	-8.1740	0.0119
37	-8.0617	0.0118	42	-8.1705	0.0133	57	-8.1431	0.0070
38	-8.0995	0.0069	43	-8.0360	0.0214	59	-8.0675	0.0064
39	-8.0402	0.0103	44	-8.1450	0.0187	60	-8.0922	0.0024
40	-8.0990	0.0097	45	-8.0085	0.0224	61	-8.0686	0.0038
41	-8.0328	0.0110	46	-8.1415	0.0225	62	-8.0826	0.0028
42	-8.1005	0.0114	47	-7.9914	0.0250	63	-8.0571	0.0039
43	-8.0258	0.0124	49	-8.1521	0.0118	64	-8.0776	0.0036
44	-8.1040	0.0131	50	-8.0577	0.0011	65	-8.0518	0.0042
45	-8.0195	0.0141	51	-8.0535	0.0018	66	-8.0772	0.0043
⋮	⋮	⋮	⋮	⋮	⋮	⋮	⋮	⋮
56	-8.1531	0.0293	61	-8.0753	0.0042	76	-8.0608	0.0007
57	-7.9661	0.0314	62	-8.0586	0.0044	77	-8.0649	0.0007
58	-8.1660	0.0335	63	-8.0772	0.0047	78	-8.0603	0.0007
59	-7.9525	0.0358	64	-8.0467	0.0051	79	-8.0650	0.0008
60	-8.1807	0.0383	65	-8.0793	0.0051	80	-8.0598	0.0008

tem will be presented. The system consists of two subsystems connected as shown in Fig. 1. The performance functions Q_i, input-output mappings

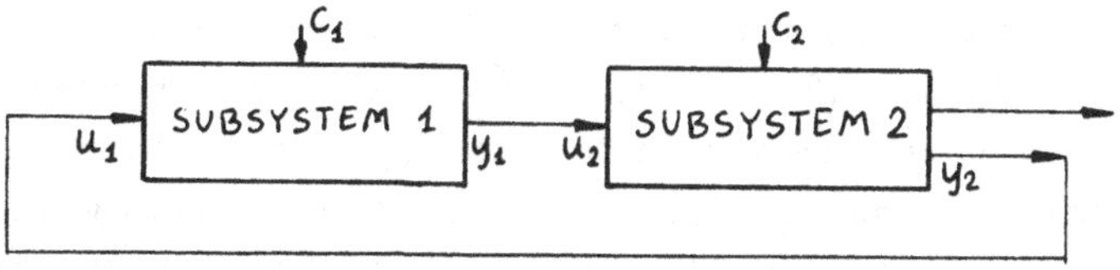

Fig. 1. The example system structure

F_i and local constraints sets CU_i are as follows, $i=1,2$:

<u>Subsystem 1</u>:

$$Q_1 = 32c_1^2 - 16c_1 + (2c_1 + u_1 - 1)^2$$

$$F_1:\ y_1 = 2c_1 + u_1$$

$$CU_1:\ 2c_1 + u_1 \leq 2.25$$

<u>Subsystem 2</u>:

$$Q_2 = 10c_2^2 + 4c_2u_2 - 0.5(2c_2 + 2u_2)^2$$

$$F_2:\ y_2 = 0.5c_2 + 0.5u_2$$

$$CU_2 = R^2$$

Table 2. Convergence for various values of ϱ and α.

$\alpha=1$ $\varepsilon=0.01$				$\varrho=10$ $\varepsilon=0.05$					
$\varrho=20$		$\varrho=30$		$\alpha=0.6$		$\alpha=0.8$		$\alpha=1.2$	
TNE	Q	TNE	Q	TNE	Q	TNE	Q	TNE	Q
34	-16.8664	49	-14.1057	17	-27.3970	17	-27.3970	17	-27.3970
40	-8.7886	55	-8.7858	21	-13.2368	21	-9.3741	22	-3.0399
46	-9.0019	61	-8.4511	24	-10.9910	25	-10.6149	34	-18.1978
49	-8.3963	65	-8.1923	27	-10.3176	28	-9.8112	38	-3.7257
52	-8.1902	67	-8.1113	29	-9.8447	30	-9.2523	48	-13.2922
54	-8.1328	68	-8.0648	31	-9.4961	32	-8.9402	52	-5.1402
56	-8.0986	70	-8.0469	33	-9.1864	33	-8.7490	59	-10.8649
57	-8.0739	71	-8.0652	34	-9.0028	35	-8.3169	63	-6.1991
58	-8.0610	72	-8.0636	35	-8.6790	36	-8.3488	69	-9.6838
59	-8.0633	73	-8.0666	36	-8.5534	37	-8.2215	72	-6.9493
60	-8.0591	74	-8.0607	37	-8.4007	38	-8.2165	75	-9.0536
61	-8.0645	75	-8.0661	38	-8.3386	39	-8.1294	78	-7.2935
62	-8.0593	76	-8.0585	39	-8.2618	40	-8.1393	80	-8.6390
63	-8.0656	78	-8.0536	40	-8.2247	41	-8.0878	81	-7.8343
65	-8.0514	79	-8.0673	.		.		.	
66	-8.0635	80	-8.0642	.		.		.	
67	-8.0667			.		.		.	
68	-8.0642			50	-8.0736	50	-8.0675	90	-8.1226
69	-8.0645			.		.		.	
70	-8.0621			.		.		.	
				.		.		.	
				60	-8.0632	60	-8.0633	101	-7.9278
				61	-8.0630	61	-8.0618	102	-8.1145
				62	-8.0629	62	-8.0630	103	-8.1393
				63	-8.0628	63	-8.0620	104	-7.9873
				64	-8.0627	64	-8.0629	106	-8.1815
				65	-8.0626	65	-8.0621	107	-8.0220

The problem is nonconvex and with duality gap. The optimal value of the overall performance function $Q = Q_1+Q_2$ is $\hat{Q} = -8.0625$, with optimal point $\hat{c}_1=0.5$, $\hat{c}_2=0.25$, $\hat{u}_1=1.25$, $\hat{u}_2=2.25$. Chosen results obtained using augmented price method with coordination formula (10) are presented in tables 1 and 2. In all cases the algorithm was started from values of all variables (multipliers, controls, inputs) equal to zero. Table 1 presents performance function value Q and discoordination norm DN,

$$DN \triangleq |u_1-y_2| + |u_2-y_1|,$$

versus total number of local problems (5) evaluations TNE. Subsequent lines in the table correspond to subsequent modifications (10) of multipliers, difference between two subsequent values of TNE gives number of steps in inner iteration loop. Results given in table 1 show that too great value of ε leads to some oscillations in a neighbourhood of the optimal point. Decreasing ε diminishes this oscillations to acceptable limits (performance function accuracy of about 0.1%, i.e.

$-8.0545 \leq Q \leq -8.0705$, results in quite satisfactory accurracy of optimal point evaluation) leading, on the other hand, to some increase in computational effort. Using values of α less then 1 results in more monotonous and faster convergence, even for a bit greater value of ε , as table 2 shows. First columns of that table show that the algorithm is rather not sensitive to the choice of ϱ ,greater ϱ leads only to some increase in computational effort. In table 2 behaviour of DN was not shown - it was similar for all cases except of that with $\alpha = 1.2$ where values of DN, like those of Q, were not convergent satisfactorily.

5. CONCLUSIONS

General structure of the augmented price method was presented in the paper. It is the most natural generalization of well known price method of multilevel optimization, to the cases of problems with duality gaps. The key problem of the presented algorithm structure, convergence of inner iterations, was investigated. It was shown on an example that standard conditions used in augmented Lagrangians techniques are not sufficient. A suitable additional simple condition was formulated and convergence of inner iterations was proved. The presented augmented price method constituts in fact some class of algorithms, differing in various coordination strategies. For a chosen coordination strategy (Hestenes-Powell multiplier rule (10)) the resulting algorithm was tested numerically showing its rather good properties.

Another algorithms resulting from using another coordination strategies are also possible. Analysing explicitely finite accuracy of inner iterations seems also to be interesting - algorithm given by Stoilov [3] can be considered as somewhat related to this. It has, however, different structure with both multipliers and approximation points as coordination variables.

The whole analysis was done in the paper for interconnected system optimization problem. However, similar results seem to be true for another important classes of problem formulations, e.g., optimization problem with subsystems coupled by resource constraint.

The presented method occured to be important not only for pure optimization purposes. It is a basis for a very efficient on-line optimizing control algorithm with feedback, for large scale nonlinear interconnected systems working in steady-state, see (Tatjewski [6]).

6. APPENDIX

Outline of the proof of Theorem 1:

The proof of existence of unique solutions of optimization problems

(5) of inner iteration loop is rather standard and will be omitted. It is known, see (Bertsekas [5]) for list of references, that under assumptions (i),(ii) and (iii) optimization problem (9) has locally unique solution $(\hat{c}(\lambda),\hat{u}(\lambda),\hat{\mu}_A(\lambda))$ for every λ from some neighbourhood of $\hat{\lambda}$, where $\hat{\mu}_A(\lambda)$ is Kuhn-Tucker vector corresponding to active constraints g_A. Moreover, mappings $\hat{c}(.),\hat{u}(.)$ and $\hat{\mu}_A(.)$ are locally continuous.

Let us describe one step of inner iteration loop algorithm (part 2° of the algorithm structure given in section 2) implicitely as

$$G(c^{k+1},u^{k+1},\mu_A^{k+1},c^k,u^k,\mu_A^k)=0,$$

where μ_A^{k+1} is Kuhn-Tucker vector for constraints g_A in $(c^{k+1},u^{k+1})\triangleq(\hat{c}^k,\hat{u}^k)$. The operator G is described by the set of equations

$$\begin{cases} \nabla Q(c^{k+1},u^{k+1})+(I_o-\nabla h(c^{k+1},u^{k+1}))^T(\lambda+\varrho(u^{k+1}-h(c^{k+1},u^{k+1})))+\\ +\varrho(\nabla h(c^{k+1},u^{k+1})(u^{k+1}-u^k)+I_o^T(h(c^{k+1},u^{k+1})-h(c^k,u^k)))+\\ +g_A(c^{k+1},u^{k+1})\mu_A^{k+1}=0\\ g_A(c^{k+1},u^{k+1})=0 \end{cases}$$

Using now theorem 10.3.4 from (Ortega and Rheinboldt [7]),continuity of $\hat{c}(.),\hat{u}(.),\hat{\mu}_A(.)$ and the fact that determinant and eigenvalues of a matrix are continuous functions of its elements it can be easily shown that for every λ from some neighbourhood of $\hat{\lambda}$ the inner iteration algorithm is locally convergent to the solution of problem (9) if $\det M(\varrho)>0$ and $\mathrm{sr}\ P(\varrho)<1$, where sr(.) denotes spectral radius, and

$$M(\varrho)\triangleq\begin{bmatrix} A+\varrho E & D^T\\ D & 0\end{bmatrix},\qquad P(\varrho)\triangleq M(\varrho)^{-1}\begin{bmatrix}\varrho B & 0\\ 0 & 0\end{bmatrix},$$

$$A\triangleq Q''(\hat{c},\hat{u})+h''(\hat{c},\hat{u})\hat{\lambda}+g_A''(\hat{c},\hat{u})\hat{\mu}_A,$$
$$E\triangleq(I_o-C)^T(I_o-C)+B,$$
$$B\triangleq C^TI_o+I_o^TC.$$

Matrix $M(\varrho)$ is nonsingular iff the set of equations

$$(A+\varrho E)x+D^Ty=0$$
$$Dx=0$$

has not nonzero solution. It can be shown to be true for sufficiently great ϱ, using assumptions (i)-(iii).

It will be proved now that for sufficiently great ϱ $\mathrm{sr}\ P(\varrho)<1$. Let δ be nonzero eigenvalue of $P(\varrho)$ with corresponding eigenvector (r,t). Then we have

$$\delta(A+\varrho E)r+\delta D^Tt=\varrho Br$$
$$\delta Dr=0.$$

Due to (ii) r must be nonzero, hence the above set of equations can be transformed to

$$\delta r^T(A+\varrho E)r=\varrho r^T Br \qquad (11)$$

If δ is a complex number, $\delta=\alpha+j\beta$, then (11) gives

$$\alpha r^T(A+\varrho E)r=\varrho r^T Br$$

$$\beta r^T(A+\varrho E)r=0$$

and thus δ is real since $r^T(A+\varrho E)r>0$ for every $r\neq 0$ satisfying $Dr=0$, if ϱ is sufficiently great.

Let $\delta=1+\Delta\delta$, then from (11)

$$r^T(A+\varrho(E-B)r=-\Delta\delta r^T(A+\varrho E)r,$$

thus $\Delta\delta<0$ since $r^T(A+\varrho(E-B))r>0$ for every $r\neq 0$ satisfying $Dr=0$, if ϱ is sufficiently great. Let $\delta=-1+\Delta\delta$, then from (11)

$$r^T(A+\varrho(E+B)r)=\Delta\delta r^T(A+\varrho E)r.$$

Using now additionally assumption (iv) it can be easily shown that $r^T(A+\varrho(E+B))r>0$ for every $r\neq 0$ satisfying $Dr=0$, for sufficiently great ϱ. Hence $\delta>-1$, what completes the proof.

REFERENCES

[1] Findeisen,W., F.F.Bailey,M.Brdyś,K.Malinowski,P.Tatjewski and A.Woźniak: Control and Coordination in Hierarchical Systems. John Wiley, London, 1980.

[2] Stephanopoulos,G., and A.Westerberg: The use of Hestenes method of multipliers to resolve dual gaps in engineering system optimization. JOTA, 15 (1975), pp.285-309.

[3] Stoilov,E.: Augmented lagrangian method for two-level static optimization. Arch.Automatyki i Telemechaniki, 22(1977),pp.210-237 (in Polish).

[4] Tatjewski,P.: Multilevel optimization techniques. In: Second Workshop on Hierarchical Control, W.Findeisen, ed. Technical University of Warsaw, 1979, pp.241-266.

[5] Bertsekas,D.P.: Multiplier methods: a survey.Automatica,12(1976), pp. 133-145.

[6] Tatjewski,P.: On-line steady-state control of large-scale nonlinear interconnected systems using augmented interaction balance method with feedback. Proceedings of 3rd IFAC/IFORS Symposium on Large Scale Systems, Warsaw, July 1983. Pergamon Press.

[7] Ortega,J.M., and W.C.Rheinboldt: Iterative Solution of Nonlinear Equations im Several Variables. Academic Press, New York and London, 1970.

AGGREGATION BOUNDS IN STOCHASTIC PRODUCTION PROBLEMS

John R. Birge
The University of Michigan
Ann Arbor, Michigan, U.S.A.

1. INTRODUCTION

The optimization of multi-stage production-inventory control systems with uncertain demands and capacitated production represents a challenging problem that most production planning methods have not considered. This problem is an example of a multi-stage stochastic program for which few computational procedures exist. When only two periods are present, the methods of El Agizy [6], Everett and Ziemba [7] and Wets [9] may be applied. For three or four period examples, Birge's [3] general stochastic programming code may be applied for random variables with discrete realizations. The only approaches to specifically consider the production problem are Beale, Forrest, and Taylor's [2] and Ashford's [1] approximations for normally distributed demands. Large general problems are still, however, not readily solved.

To simplify these models to some solvable form, the general approach of aggregating variables and constraints (Zipken [10, 11]) may be applied. When the weights of these aggregations coincide with the distributions of the random variables, it has been shown (Birge [4]) that the resulting aggregate problem is the stochastic problem with expected values replacing random variables. This expected value problem has been analyzed in Bitran and Yanasse [5] under different assumptions about the distributions of the random variables.

Other types of aggregation are also possible. In this paper, we show that bounds on the value of the full stochastic problem can be found from solutions of aggregate problems that combine both random variables and time periods. The assumptions usually necessary for these bounds in general aggregation are shown to be true by virtue of the problem structure in Section 2. In Section 3, a specific aggregation for combining random variables and time periods is given and the bounds resulting from this aggregate problem are presented. Section 4 presents an example and other potential aggregations.

2. PROBLEM DEFINITION AND VARIABLE BOUNDS

The formulation we consider is similar to that in Beale, Forrest and Taylor [2] and Bitran and Yanasse [5]. We write the single product multistage stochastic production problem as

$$\max \quad z = [E(\sum_{t=1}^{T} \rho^{t-1}(p_t\, x_t - q_t\, o_t - h_t\, i_t^+)] \tag{1}$$

$$\text{subject to} \quad x_t - o_t \le k_t, \tag{1.1}$$

$$y_{t-1} + x_t - y_t = 0, \tag{1.2}$$

$$y_t \ge b_t, \tag{1.3}$$

$$+i_{t-1}^+ - i_{t-1}^- + x_t - i_t^+ + i_t^- = d_t, \tag{1.4}$$

$$x_t, o_t, y_t, i_t^+, i_t^- \ge 0, \; t = 1,\dots,T;$$

where the decision variables are x_t, production in time period t, o_t, overtime used in period t, i_t^+, inventory after period t, i_t^-, back orders after period t, and y_t, total production through period t. p_t is net production revenue, q_t is overtime cost and h_t is inventory cost. These costs, the capacity k_t and minimum total production b_t are assumed known. The demand d_t is a random variable defined on an interval (d_t^{min}, d_t^{max}) with distribution function $F_t(d_t)$. The random variables $d_1,\dots d_t$ have a joint distribution function $F(d_1,\dots,d_t)$. E [ ] signifies mathematical expectation with respect to these random variables. The decision variables depend on past outcomes so x_t, for example, is really $x_t(d_1,\dots d_t)$. The expected value can then be written as

$$E\,[\sum_{t=1}^{T} \rho^{t-1}(p_t x_t - q_t o_t - h_t\, i_t^+)] \tag{1}$$

$$= \sum_{t=1}^{T} \rho^{t-1} \{\int_{d_t^{min}}^{d_t^{max}} \cdots \int_{d_1^{min}}^{d_1^{max}} (p_t x_t(d_1,\dots,d_t) - q_t o_t(d_1,\dots,d_t) - h_t\, i_t^+(d_1,\dots,d_t)\; dF(d_1,\dots,d_t).$$

The constraint (1.3) has been added to (1) as in Bitran and Yanassee [5] as an alternative formulation of a constraint for demand satisfaction with some confidence. Constraint (1.2) is used to keep the staircase structure of the problem so that period t is only linked directly to periods t - 1 and t + 1 through the constraints.

In order to obtain bounds on the optimal value,

$$z^* = E\,[\sum_{t=1}^{T} \rho^{t-1}(p_t x_t^* - q_t o_t - h_t\, i_t^{+*})] \;,$$

bounds on optimal primal variable levels, $x_t^*, o_t^*, y_t^*, i_t^{+,*}, i_t^{-,*}$, must be found. The general conditions in Zipken [11] and Birge [4] may be met by assuming the variables are bounded. The structure of problem 1, however, provides bounds on the variables without extra assumptions on the variable values. For dual variables, $(\pi_t, \nu_t, \sigma_t, \mu_t)$ associated with (1.1), (1.2), and (1.4) respectively, optimal dual variable levels $\pi_t^*, \nu_t^*, \sigma_t^*$, and μ_t^* can also be found.

First, note that total production will never exceed total demand over the planning horizon for 0 to T. Hence,

$$\sum_{t=1}^{T} x_t^* \leq \sum_{t=1}^{T} d_t^{max}. \tag{2}$$

From (2), we also obtain

$$y_t \leq \sum_{t=1}^{T} d_t^{max}; \; t = 1,\ldots,T, \tag{3}$$

$$\sum_{t=1}^{T} o_t^* \leq \sum_{t=1}^{T} d_t^{max} - \min_{1 \leq t \leq} k_t, \tag{4}$$

where (4) follows because x_t^* must be nonzero in at least one period so that at least one period's capacity (without overtime) was used in producing the total production.

Constraint (1.3) forces back orders to be bounded above by

$$i_t^- \leq \sum_{\tau=1}^{t} d_\tau^{max} - b_t. \tag{5}$$

Inventory is also constrained in the last period by

$$i_T^+ \leq d_T^{max} - d_T^{min}, \tag{6}$$

since having any inventory in the event of d_t^{min} would be sub-optimal. For periods $t < T$, the inventory is at most the remaining demand,

$$i_t^+ \leq (d_t^{max} - d_t^{min}) + \sum_{\tau=t+1}^{T} d_\tau^{max}. \tag{7}$$

The dual variables can be bounded in a similar manner. First, observe that $- \rho^{t-1} q_t \leq - \pi_t$, hence

$$0 \leq \pi_t \leq \rho^{t-1} q_t, \tag{8}$$

for all t. Dual feasibility in the inventory variables in period T implies

$$\mu_T \leq \rho^{t-1} h_t, \tag{9}$$

$$\mu_T \geq 0. \tag{10}$$

Iterating backwards, we obtain

$$\mu_f \leq \sum_{\tau=t}^{T} \rho^{\tau-1} h_\tau, \tag{11}$$

$$\mu_t \geq 0. \tag{12}$$

Constraints on ν_t and σ_t are obtained through dual feasibility corresponding to the x and y variables. Note that

$$\sigma_t \leq 0, \tag{13}$$

$$\nu_t \leq \sigma_T \leq 0, \tag{14}$$

$$\nu_{t-1} - \nu_t \leq \sigma_{t-1}, \; t = 2,\ldots,T. \tag{15}$$

(13), (14), and (15) imply

$$\nu_t \leq 0, \; t = 1,\ldots,T. \tag{16}$$

For the dual constraint corresponding to x_t,

$$\nu_t \geq \rho^{t-1} p_t - \rho^{t-1} q_t - \sum_{\tau=t}^{T} \rho^{\tau-1} h_\tau. \tag{17}$$

From (14), (15), and (17),

$$\sigma_t \geq \rho^{t-1} p_t - \rho^{t-1} q_t - \sum_{\tau=t}^{T} \rho^{\tau-1} h_\tau. \tag{18}$$

Equation (8) - (18) represent upper and lower bounds on the dual variables that will be used in obtaining bounds on z*.

3. STOCHASTIC PRODUCTION AGGREGATION

Problem (1) will be simplified by aggregating both random variables and time periods. The resulting problem will be a single period deterministic approximation of the original multi-stage stochastic problem. Our procedure is similar to those in Zipken [10, 11] and Birge [4]. This will represent the most extreme aggregation possible, although less extreme aggregations involving conditional means and the aggregation of a subset of the periods may be possible.

For constraints (1.1), (1.3) and (1.4), our aggregation procedure is similar to the method in Birge [4]. We define

R_i = {Rows corresponding to constraint (1.i) in periods 1,...,T for i = 1, 2, 3, 4}.

The rows in R_i are summed together using a weighting function $f^\beta(i_t)$ where i_t is some constraint (1.i) in period t and

$$f^\beta(i_t) = \rho^{t-1} dF(d_2,\ldots,d_t), \tag{19}$$

for i = 1, 3, 4.

Columns are grouped together as

S_i = {Columns corresponding to variables i = 1,...,5 in periods t = 1,...,T},

where i = 1, 2, 3, 4, 5 corresponds to x_t, o_t, y_t, i_t^+, and i_t^- respectively. These columns are summed directly with a weight $g^\alpha \equiv 1$ for each column.

The rows corresponding to constraints (1.2) are treated differently because of the problem structure. The y variables represent total production and, therefore, should increase each period. If a given x is used throughout the horizon then the y variables increase exactly in multiples of x. This observation leads to a definition of weights α(t) in (1.2) where

$$f^\beta(i_t) = \alpha(t)dF(d_2,\ldots,d_t) \qquad \text{for } i = 2.$$

To see this, we first observe that constraint (1.3) in the aggregate problem becomes

$$\left(\sum_{t=1}^{T}\rho^{t-1}\right) y \geq \sum_{t=1}^{T} b_t, \tag{20}$$

where y is the aggregate variable. The left hand side considering disaggregated variables is

$$\left(\sum_{t=1}^{T} \rho^{t-1}\right) y = \sum_{t=1}^{T}\rho^{t-1} y_t. \tag{21}$$

Now, in constraint (1.2) in the aggregate problem, we have

$$\left(\sum_{t=1}^{T} \alpha(t)\right) X-Y = 0. \tag{22}$$

Now, we would like $X = y_1$ and $y_t = t\,X$. From (21), this would lead to

$$Y = \left(\frac{\sum_{t=1}^{T} \rho^{t-1}\, t}{\sum_{t=1}^{T} \rho^{t-1}}\right) X. \tag{23}$$

By defining,

$$\alpha(t) = \left(\frac{\sum_{\tau=t}^{T} \rho^{t-1}}{\sum_{t=1}^{T} \rho^{t-1}}\right), \tag{24}$$

(23) is obtained and (20) and (22) are, therefore, consistent with the problem structure.

Given these weighting functions, the aggregate problem is:

$$\max \quad z = (\sum_{t=1}^{T} \rho^{t-1} p_t) \;X - (\sum_{t=1}^{T} \rho^{t-1} q_t) \;O - (\sum_{t=1}^{T} \rho^{t-1} h_t) \;I^+$$

subject to

$$(\sum_{t=1}^{T} \rho^{t-1}) \;X - (\sum_{t=1}^{T} \rho^{t-1}) \;O \quad \le \sum_{t=1}^{T} \rho^{t-1} k_t, \qquad (24.1)$$

$$(\sum_{t=1}^{T} \alpha(t))X \quad - \quad Y \quad = 0, \qquad (24.2)$$

$$(\sum_{t=1}^{T} \rho^{t-1})Y \quad \ge \sum_{t=1}^{T} \rho^{t-1} b_t, \qquad (24.3)$$

$$(\sum_{t=1}^{T} \rho^{t-1})X \quad -I^+ + I^- = \sum_{t=1}^{T} \rho^{t-1} \bar{d}_t, \qquad (24.4)$$

$$X,\ O,\ Y,\ I^+,\ I^-,\ \ge 0,$$

where $\bar{d}_t = E[d_t]$. Optimal primal variables in (24) are X^*, O^*, Y^*, $I^{+,*}$ $I^{-,*}$ and the optimal dual variables are Π^*, N^*, Σ^*, M^*. The disaggregated solution obtained from (24) is $(\hat{x}_t, \hat{o}_t, \hat{i}_t^+, \hat{i}_t^-)=(X^*, O^*, I^{+,*}, I^{-,*})$ and $\hat{y}_t = tX^*$ for all $d_1,\ldots,d_t$. For dual variable disaggregation, we have

$$(\hat{\pi}_t,\hat{\gamma}_t,\hat{\sigma}_t,\hat{\mu}_t) = (\rho^{t-1} \pi^*, \alpha(t)N^*, \rho^{t-1}\Sigma^*, \rho^{t-1} M^*)dF(d_2,\ldots,d_t).$$

Note that disaggregating is not performed exactly as in aggregation due to particular problem structure of (1).

We let $\hat{z} = \sum_{t=1}^{T} \rho^{t-1} (c_t \hat{x}_t - q_t \hat{o}_t - h_t \hat{i}_t^+)$ and wish to find ε^+ and ε^- such that

$$\hat{z} - \varepsilon^- \le z^* \le \hat{z} + \varepsilon^+. \qquad (26)$$

To simplify the exposition of these bounds, we assume first that the problem data is stationary, except for demand. That is, $p_t = p$, $q_t=q$, $h_t=h$, $k_t=k$, $b_t=tb$ for all t. Without this assumption, bounds are still attainable, but they will involve more complicated formulas for ε^- and $\varepsilon+$.

In obtaining values for ε^+ and ε^-, we first check primal and dual feasibility. By stationarity and the definition of $\hat{x}_t$ and $\hat{o}_t$, (1.1) is always satisfied. Also by the definition of $\hat{x}_t$ and $\hat{y}_t$, (1.2) is satisfied. For constraint (1.3), note that (24.2) and the definition of $\alpha(t)$ imply

$$Y^* = \frac{\sum_{t=1}^{T} \rho^{t-1} t\, x^*}{\sum_{t=1}^{T} \rho^{t-1}},$$

so (24.3) can be re-written as $(\sum_{t=1}^{T} \rho^{t-1}t)\; X^* \ge (\sum_{t=1}^{T} \rho^{t-1}t)\; b.$

Hence, $y_t = tx^* \geq tb$, and constraint (1.3) is satisfied. The only remaining infeasibilities may occur in (1.4) as demand varies.

For dual feasibility, stationarity implies that

$$\hat{\pi}_t = \rho^{t-1} \Pi^* \leq \rho^{t-1} q, \text{ and } \hat{\pi}_t \geq 0. \tag{27}$$

For constraints associated with the variables y_t, for feasibility, we need

$$0 \leq - \hat{\nu}_t+\hat{\sigma}_t+\hat{\nu}_{t+1}, \text{ or } \rho^t N^*(1-\rho) \leq \rho^t \Sigma^*, \text{ or } N^*(1-\rho) \leq \Sigma^*. \tag{28}$$

However, dual feasibility in (24) implies $N^* \left(\frac{1}{\sum_{t=1}^{T} \rho^{t-1}} \right) \leq \Sigma^*$

and $1/ \sum_{t=1}^{T} \rho^{t-1} \leq (1-\rho)$ implies (28). We also have $\hat{\sigma}_t = \rho^{t-1} \Sigma^* \leq 0$.
A similar argument applies for the i_t^+ variables where we want

$$-\rho^t h \leq - \hat{\mu}_t + \hat{\mu}_{t+1}, \quad \text{or} \quad M^*(1-\rho) \leq h. \tag{29}$$

From (24), we have $M^* \leq \sum_{t=1}^{T} \rho^{t-1} h$, which implies (29).

For i_t^-, we require $0 \leq \hat{\mu}_t - \hat{\mu}_{t+1}$ or $0 \leq M^*(1-\rho)$, (30)

which, for $\rho < 1$ and dual feasibility in (24), is true.

The only dual infeasibilities may then incur in the constraint associated with the x_t variables. The possibility for this infeasibility must be considered in calculating bounds on z^*.

We are now able to state bounds on the optimal value of the solution in (1). We first assume that no duality gap exists in (1) by assuming that $d_t^{max} < +\infty$ for all t. In this case, for any realization of $d_1,\ldots,d_t$, the function

$$\sum_{t=1}^{T} \rho^{t-1}(p\, x_t^*(d_1,\ldots,d_t) - q o_t^* (d_1,\ldots d_t) - h\, i_t^*(d_1,\ldots,d_t))$$

is bounded. This implies that no dualtiy gap exists in (1) (Rockafellar and Wets [8]) and that

$$z^* = \sum_{t=1}^{T} (\int_{d_t^{min}}^{d_t^{max}} \ldots \int_{d_1^{min}}^{d_1^{max}} (\pi_t^*(d_1,\ldots d_t)k + \sigma_t^* (d_1,\ldots,d_t)\, b_t$$

$$+ \mu_t^*(d_1,\ldots d_t)d_t)\},$$

for $(\pi_t^*, \nu_t^*, \sigma_t^*, \mu_t^*)$ optimal in the dual of (1).

Proposition: The optimal value z^* of (1) is bounded by

$$\hat{z} - \varepsilon^- \leq z^* \leq \hat{z} + \varepsilon^+,$$

where

$$\varepsilon^- = \sum_{t=1}^{T} \int_{d_t} \max\{-d_t + X^* - I^{+,*} + I^{-,*}, 0\} \left(\sum_{\tau=t}^{T} \rho^{\tau-1} h_\tau\right) dF_t(d_t) \tag{30}$$

and

$$\varepsilon^+ = \max_{1 \leq t \leq T} \{\rho^{t-1}(p - \Pi^* - N^* - M^*), 0\} \sum_{t=1}^{T} d_t^{max}. \tag{31}$$

Proof: To show (30), we first note that

$$* \geq \sum_{t=1}^{T} \Big[\int_{d_t^{min}}^{d_t^{max}} \cdots \int_{d_1^{min}}^{d_1^{max}} (\pi_t^*(d_1,\ldots,d_t)k + \sigma_t^*(d_1,\ldots,d_t)b_t + \mu_t^*(d_1,\ldots,d_t)d_t)$$

$$+ \sum_{t=1}^{T} \Big\{ \int_{d_t^{min}}^{d_t^{max}} \cdots \int_{d_1^{min}}^{d_1^{max}} (\rho^{t-1}p - \pi_t^*(d_1,\ldots,d_t) - \nu_t^*(d_1,\ldots,d_t) - \mu_t^*(d_1,\ldots,d_t)) \hat{x}_t(d_1,\ldots,d_t)$$

$$+ \sum \Big\{ \int_{d_t^{min}}^{d_t^{max}} \cdots \int_{d_1^{min}}^{d_1^{max}} (-\rho^{t-1}q + \pi_t^*(d_1,\ldots d_t)) \hat{o}_t(d_1,\ldots,d_t) \Big\}$$

$$+ \sum_{t=1}^{T} \Big\{ \int_{d_t^{min}}^{d_t^{max}} \cdots \int_{d_1^{min}}^{d_1^{max}} (-\rho^{t-1} h + \nu_t^*(d_1,\ldots,d_t) - \nu_{t+1}^*(d_1,\ldots,d_t)) \hat{i}_t^+(d_1,\ldots,d_t) \Big\}$$

$$+ \sum_{t=1}^{T} \Big\{ \int_{d_t^{min}}^{d_t^{max}} \cdots \int_{d_1^{min}}^{d_1^{max}} (+\nu_t^*(d_1,\ldots,d_t) - \sigma_t^*(d_1,\ldots,d_t) - \nu_{t+1}^*(d_1,\ldots,d_t)) \hat{y}_t(d_1,\ldots,d_t) \Big\}$$

$$+ \sum_{t=1}^{T} \Big\{ \int_{d_t^{min}}^{d_t^{max}} \cdots \int_{d_1^{min}}^{d_1^{max}} (-\mu_t^*(d_1,\ldots,d_t) + \mu_{t+1}^*(d_1,\ldots,d_t)) \hat{i}_t^-(d_1,\ldots,d_t) \Big\}$$

$$= \hat{z} + \sum_{t=1}^{T} \int_{d_t^{min}}^{d_t^{max}} \cdots \int_{d_1^{min}}^{d_1^{max}} (\pi_t^*(d_1,\ldots,d_t)(k + \hat{o}_t(d_1,\ldots,d_t) - \hat{x}_t(d_1,\ldots,d_t))$$

$$+ \sum_{t=1}^{T} \int_{d_t^{min}}^{d_t^{max}} \cdots \int_{d_1^{min}}^{d_1^{max}} \nu_t^*(d_1,\ldots,d_t)(\hat{y}_t(d_1,\ldots,d_t) - \hat{x}_t(d_1,\ldots,d_t) - \hat{y}_{t-1}(d_1,\ldots d_t))$$

$$+ \sum_{t=1}^{T} \int_{d_t^{min}}^{d_t^{max}} \cdots \int_{d_1^{min}}^{d_1^{max}} \sigma_t^*(d_1,\ldots,d_t)(b_t - \hat{y}_t(d_1,\ldots d_t))$$

$$+ \sum_{t=1}^{T} \int_{d_t^{min}}^{d_t^{max}} \cdots \int_{d_1^{min}}^{d_1^{max}} \mu_t^*(d_1,\ldots,d_t)(d_t - \hat{i}_t^-(d_1,\ldots,d_t) + \hat{i}_t(d_1,\ldots,d_t)$$

$$- \hat{x}_t(d_1,\ldots,d_t) + \hat{i}_{t-1}(d_1,\ldots,d_t)$$

$$- \hat{i}_{t-1}(d_1,\ldots,d_t)) \qquad (32)$$

$$\geq \hat{z} - \varepsilon^-,$$

since only the last term in (32) can be negative and μ_t^* is bounded as in (11).

For the upper bound, we follow a similar development.

$$z^* \leq \sum_{t=1}^{T} \Big[\int_{d_t^{min}}^{d_t^{max}} \cdots \int_{d_1^{min}}^{d_1^{max}} \rho^{t-1}(p\, x_t^*(d_1,\ldots,d_t) - q\, o_t^*(d_1,\ldots,d_t)$$

$$- h\, i_t^*(d_1,\ldots,d_t))dF(d_1,\ldots,d_t)\Big]$$

$$+ \sum_{t=1}^{T} \Big(\int_{d_t^{min}}^{d_t^{max}} \cdots \int_{d_1^{min}}^{d_1^{max}} \hat{\pi}_t(d_1,\ldots,d_t)(k + o_t^*(d_1,\ldots,d_t) - x_t^*(d_1,\ldots d_t)$$

$$+ \int_{d_t^{min}}^{d_t^{max}} \cdots \int_{d_1^{min}}^{d_1^{max}} \hat{\nu}_t(d_1,\ldots,d_t)(y_t^*(d_1,\ldots,d_t) - x_t^*(d_1,\ldots,d_t)$$

$$- y_{t-1}^*(d_1,\ldots d_t))$$

$$+ \int_{d_t^{min}}^{d_t^{max}} \cdots \int_{d_1^{min}}^{d_1^{max}} \hat{\sigma}_t(d_1,\ldots d_t)(b_t - y_t^*\, (d_1,\ldots,d_t))$$

$$+ \int_{d_1^{min}}^{d_1^{max}} \cdots \int_{d_1^{min}}^{d_1^{max}} \hat{\mu}_t(d_1,\ldots,d_t)(d_t - i_t^{-,*}(d_1,\ldots,d_t) + i_t^{+,*}(d_1,\ldots d_t$$

$$- x_t^*(d_1,\ldots d_t) + i_{t-1}^{-,*}(d_1,\ldots,d_t) - i_t^{+,*}(d_1,\ldots,d_t))$$

$$\leq \hat{z} + \varepsilon^+,$$

by rearranging terms and noting dual feasibility in all but the constraints corresponding to x variables. ■

The bounds given for the aggregate problem represent the simplest proble attainable from (1) with the same basic structure. Other possible aggregations are presented in Birge [4]. One of these amounts to the expected value approach in Bitran and Yanasse [5] . Another possibility is to allow the random variables to remain but to aggregate time periods The result from this aggregation is a simple recourse problem that can b solved by the methods in Everett and Ziemba [7] or Wets [9] . The problem has the basic form:

$$\max z = \left(\sum_{t=1}^{T} \rho^{t-1} p\right) X - \left(\sum_{t=1}^{T} \rho^{t-1} q\right) 0 + E\left[-\left(\sum_{t=1}^{T} \rho^{t-1} h\right) I^{+}(D)\right]$$

subject to

$$\begin{aligned}
&\left(\sum_{t=1}^{T} \rho^{t-1}\right) X - \left(\sum_{t=1}^{T} \rho^{t-1}\right) 0 && \le \sum_{t=1}^{T} \rho^{t-1} k, \\
&\left(\sum_{t=1}^{T} \alpha(t)\right) X \qquad -Y && = 0, \\
&\sum_{t=1}^{T} \rho^{t-1} Y && \ge \sum_{t=1}^{T} \rho^{t-1} b_t, \\
&\left(\sum_{t=1}^{T} \rho^{t-1}\right) X \qquad - I^{+}(D) + I^{-}(D) && = D, \\
&X,\ 0,\ Y,\ I^{+}(D),\ I^{-}(D) \ge 0,
\end{aligned} \tag{33}$$

where D is random variable equal to $\sum_{t=1}^{T} \rho^{t-1} d_t$. The value of (33) can then be used as in the development provided above to give a tighter bound on the optimal value z*.

4. EXAMPLE AND EXTENSIONS

In this section, we present an example to demonstrate how the bounds may apply. The example is similar to those in Bitran and Yanasse [5] where they have given bounds from the expected value problem for various distribution assumptions. We assume in this example that p represents cost of production. The parameters are

$p = -19$/unit produced
$h = .4$/unit/time period
$o = 1.9$/unit of overtime
$k = 20000$ units/month
$b = 9500$ units,

and demand is uniformly distributed on [8000, 10000].

The aggregate problem as in (24) is

$$\max z = -51.5X - 5.150 - 1.08 I^{+} \tag{34}$$

subject to

$$\begin{aligned}
&2.71X - 2.710 && \le && 54{,}200, \\
&1.93X \qquad -Y && = && 0, \\
&\qquad 2.71Y && \ge && 49{,}685, \\
&2.71X \qquad - I^{+} + I^{-} && = && 24{,}390, \\
&X,\ 0,\ U,\ I^{+},\ I^{-} \ge 0.
\end{aligned}$$

The value obtained from (34) is $\hat{z} = -490684$. From Proposition, we obtain

$\varepsilon^- = 11$ and $\varepsilon^+ = 243600$, hence

$$-490695 \leq z^* \leq -247084.$$

In this example, ε^+ is very large because of the loose bound in (2) on x_t^*. If some other bound (such as $x_t^* \leq d_t^{max}$) is available then this error could be reduced significantly. This shows that additional information may help bound the problem. In many problems, less extreme aggregations, such as keeping a larger number of periods or possible values for the random variables, may be used.

The general approach in bounding z* may also be used for multi-product multi-stage stochastic production problems. The basic difference in these problems would be in identifying constraints such as (1.2) which require special consideration in aggregations. Otherwise, the general procedures of Birge [4] may be used.

5. CONCLUSIONS

A method for simplifying multi-stage capacitated stochastic production problems has been presented. The method employs the principle of aggregation and combines both random variables and time period. The solution of the aggregated problem provides bounds on the optimal value of the original problem. These bounds may be improved by using the same principles on less extremely aggregated problems that may allow for more than one period and for some randomness in the demands.

References

[1] Ashford, P. W., "A stochastic programming algorithm for production planning", Scicon report, 1982.

[2] Beale, E. M. L., J. J. H. Forrest and C. Taylor, "Multi-time period stochastic programming", in Stochastic Programming, M. A. H. Dempster (ed.), Academic Press, New York, 1980.

[3] Birge, J. R., "Decomposition and partitioning methods for multi-stage stochastic linear programs", Department of Industrial and Operations Engineering, The University of Michigan, Technical Report 82-6, 1982.

[4] Birge, J. R., "Aggregation bounds in stochastic linear programming", Department of Industrial and Operations Engineering, The University of Michigan, Technical Report 83-1, 1983.

[5] Bitran, G. R. and H. H. Yanasse, "Deterministic approximations to stochastic production problems", Massachusetts Institute of Technology, 1982.

[6] ElAgizy, M., "Two-stage programming under uncertainty with discrete distribution function", Operations Research 15, 55-70, 1967.

[7] Everitt, R., and W. T. Ziemba, "Two period stochastic programs with simple recourse", Operations Research 27, 485-502, 1979.

[8] Rockafellar, R. T. and R. J.-B. Wets, "Nonanticipativity and L^1 Martingales in stochastic optimization problems", Mathematical Programming Study 6, 170-187, 1976.

[9] Wets, R. J. -B., "Solving stochastic programs with simple recourse", Stochastics, 1983.

[10] Zipkin, P., "Bounds on the effect of aggregating variables in linear programming", Operations Research 28, 403-418, 1980.

[11] Zipkin, P., "Bounds for row-aggregation in linear programming", Operations Research 28, 903-916, 1980.

AN ALLOCATION PROBLEM IN THE DESIGN OF A CLASS OF LARGE-SCALE SYSTEMS: MODEL AND ALGORITHM

Z. Strezova
Scientific Center for MIS
Sofia, Bulgaria

INTRODUCTION

Most complex systems as, e.g., energy systems, economic and management systems, air and urban traffic control systems, computer networks, etc., represent large-scale systems with decentralized structure (DLSS). The design and the functioning of such systems very often concern resource allocation problems whose solving is of great importance in today's world of limited and dwindling resources.

A class of DLSS which has been of interest the last decade comprises the so-called Management Information Systems (MIS). The purpose of a MIS is information support for management activities and functions of a managed system, e.g. economic, industrial, organizational or social system. When a MIS is intended for a large company, an industrial branch or a regional economy its structure includes information and computer resources, distributed geographically, decentralized decision-making, distributed computations.

Because of organizational change and unexpected events, today's management problems are very often essentially unstructured. Therefore the "classical" consideration of MIS as a large-scale, fully integrated system for coordinating all aspects of organizational activity in the managed system could lead to failures. A MIS, in particular a DMIS, must be considered and used more as Decision Support System.

In this paper an approximate approach to the design of DMIS is generally described. A stage of the design process of such systems, according to this approach, is studied in detail. It concerns the solution of a class of optimization problems with special structure by a heuristic algorithm. The structure of the algorithm is outlined and its efficiency is discussed.

AN APPROXIMATE APPROACH TO THE DESIGN OF DMIS

In Fig. 1 is shown a detail of the structure of a DMIS. The system consists of two subsystems: information subsystem (ISS) and decision subsystem (DSS). The latter can be described as an entity of decision-making units (DMU), a

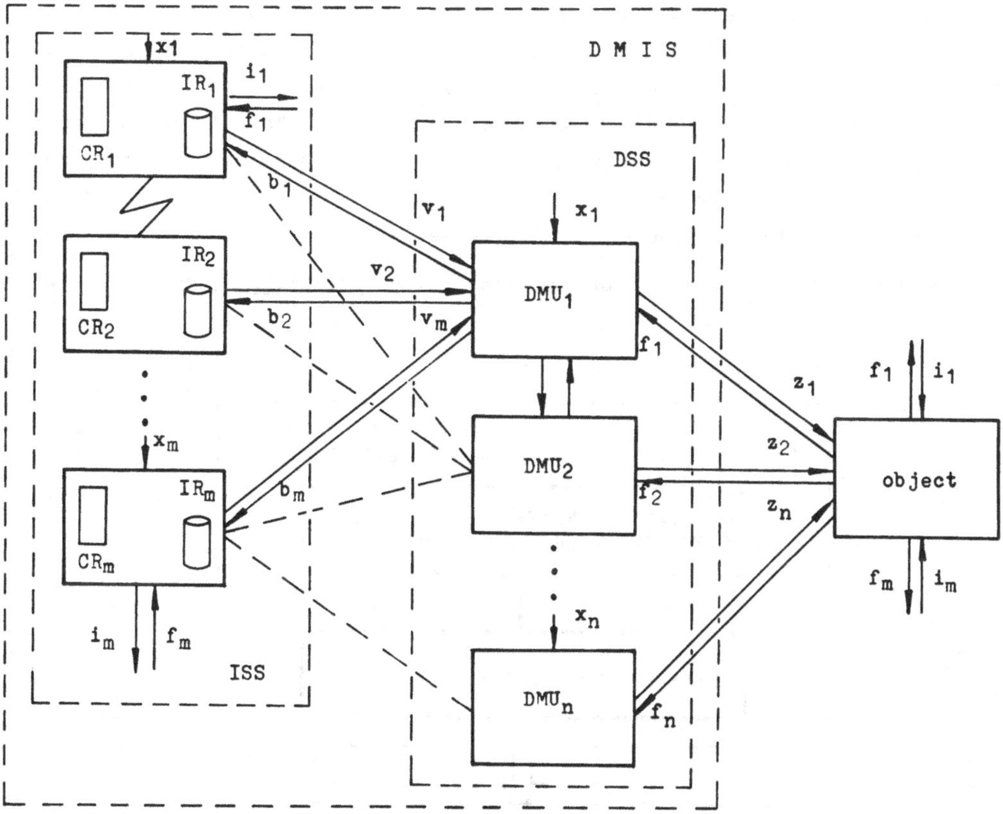

Fig. 1. Structure of a decentralized MIS

set of decision tasks (DT), a set of input variables x (desirable state variables), a set of input information variables v, a set of decision variables z, and a set of feedback variables f (object output).

The ISS is described as an entity of information resources (IR), an entity of computer resources (CR), a set of input variables f, a set of feedback variables b (decision state variables), a set of output information variables v (concerning decisions in DSS) and a set of output variables i (direct information to the object).

Considering the general functioning of a DMIS from a control viewpoint (Strezova /1/) a management problem (by analogy with a control problem) can be stated. Its solution is identified with solving the DMIS structure design problem. Generally stated, this problem consists in defining such structures of the DSS and ISS and their elements so that a set of criteria of DMIS functioning efficiency be satisfied.

As has appeared, many difficulties exist when systems of the kind described are handled by the known methods and techniques, i.e. by exact methods, explicit models and algorithms, well-defined criteria. These are not applicable, since the DLSS in general, and particularly the DMIS, are "soft" systems, with ill-defined structure and functioning. Such systems require new tools: approximate methods for system state description, approximate criteria, implicit models and algorithms instead of explicit ones, etc.

Following these considerations an approach, called approximate, to the design

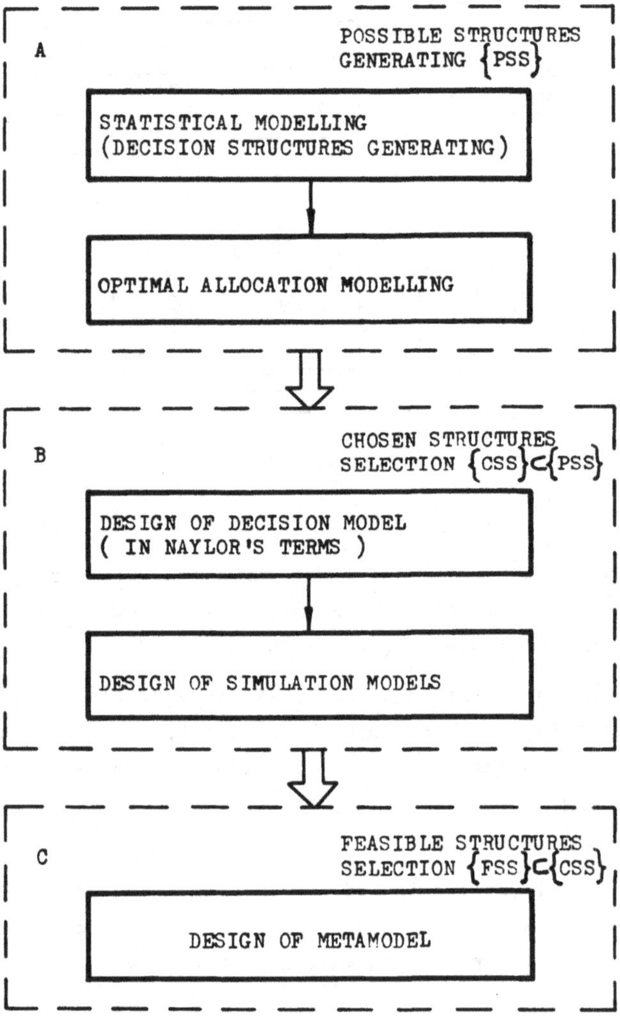

Fig. 2. Multistage procedure for DMIS structures modelling and evaluation

of large MIS with decentralized structure has been suggested earlier (Strezova /2/, /3/). This approach is based on several concepts and policies: the concept of suboptimality, the interval programming concept, and the policy, called "design by alternatives". The approach in point assumes that the problem of a DMIS structure design cannot be stated as an optimization problem, and the system functioning cannot be modelled explicitly. Instead, an architecture of models is to be designed, related to the structure and the functioning of the subsystems and the system as a whole.

The approach developed suggests a multistage procedure for generating and evaluation of alternative system structures. The general stages of this procedure are shown in Fig. 2. The stage A includes two phases. Through the first of them using statistical modelling (discriminant analysis) a set of decision structures (DS) can be modelled. A DS is an entity of DMU and relevant DT.

The second phase of the stage A consists in allocation modelling. After describing a set of computer structures (CS), i.e. configurations of computer resources, communications, etc., an optimal allocation problem can be stated. It concerns the distribution of DT (elements of the DS) among a set of CR. Using the full combination of the sets of DS and CS, a set of possible (alternative) system structures $\{PSS\}$ are modelled.

The second stage (stage B) represents an evaluation of the consequences of each alternative and a selection of a smaller set of DMIS structures. The first phase of this stage consists in finding a decision model (in the terms of Naylor's methodology). The decision model describes the inputs and outputs of the modelled in the previous stage structures. It also supposes the measures for evaluating system structures.

The next phase of the second stage represents design of simulation models by which all of the alternatives described in the terms of the decision model will be simulated. Using the measures supposed in the decision model, an analysis of all simulated alternatives is to be performed. As a result a set of chosen system structures $\{CSS\}$, $\{CSS\} \subset \{PSS\}$, will be selected.

The third stage (stage C) of the DMIS design consists in further assessment and selection of the chosen alternative structures. To the purpose a metamodel performing sensitivity analysis on the simulated structures is used. In a result, a final set of feasible DMIS structures $\{FSS\}$, $\{FSS\} \subset \{CSS\}$, is selected. Each one of these structures is a "candidate" to be designed and implemented. The final choice of the "best feasible" system structure is a result of human decision-making.

As it is clear,all stages of the procedure described include modelling activities. In fact, this is a procedure for system models design, as it supposes activities previous to the very system design.

More details concerning the approach described above and its implementation can be found, e.g., in Strezova /2/, /3/, /7/, Sage & Thissen /4/.

AN ALLOCATION PROBLEM IN MODELLING A SET OF ALTERNATIVE DMIS STRUCTURES

In this section is considered in some detail the second phase of stage A of the process of DMIS structures modelling and analysis, described briefly above.

Statement of the problem

As it was mentioned, the structure of a DMIS comprises entities of DMU, DT, CR, IR. After describing the sets of decision and computer structures, next problem consists in modelling a set of alternative system structures. It is assumed a set I of DT which can be processed by means of a set J of CR. Each DT could be solved according to one of the algorithms of the set k_i, by the jth CR, with an efficiency a_{ikj}. A resource m_{ik} is needed for solving the ith DT according to the k_i algorithm; each jth CR is restricted to m_j; the expected expense for the ith DT, solved using the jth CR, according to the k_i^{th} algorithm, is e_{ikj}; the total expenses made for the jth CR are R_j. The expenses for solving all DT cannot exceed the value E, given a priori. Under these assumptions the problem consists in allocation of the set of DT among the set of CR so that the total efficiency of the DMIS functioning be maximal on given constraints.

The model

$$\left.\begin{array}{lll}
\max \sum\limits_{ikj} \left[a_{ikj}(\omega)\, x_{ikj} - R_j\, x_{ikj} \right] & i=1,..,I & (1) \\
\text{s.t.} & & \\
\sum\limits_{j} x_{ikj} = 1 & i=1,..,g \quad g \in I & (2) \\
1 < \sum\limits_{j} x_{ikj} \leqslant J & i=g+1,..,I \quad g \in I & (3) \\
\sum\limits_{ik} m_{ik}\, x_{ikj} \leqslant m_j & i=1,..,I & (4) \\
\sum\limits_{ikj} e_{ikj}\, x_{ikj} \leqslant E & i=1,..,I & (5)
\end{array}\right\} \begin{array}{l} k=1,..,k_i \\ j=1,..,J \end{array}$$

$$0 \leqslant x_{ikj} \leqslant 1 \qquad (6)$$

x - integer for all i, k, j.

Here the constraint (2) ensures that a part of DT is assigned to one CR only;

(3) allows another part of DT to be solved by more CR; (4) defines the restrictions on the CR load; (5) ensures that the expenses for the solution of all DT do not exceed a given value E; (6) is an integrality condition.

Thus, the problem (1) ÷ (6) is stated as an Integer Linear Programming problem, in particular 0 - 1 LP problem.

Features of the problem

a) in real-world situations the problem (1)÷(6) is one of stochastic nature, i.e. the coefficients of the objective function are uncertaint. In (1) ω is a random parameter or a set of random parameters (e.g. the frequency of DT solving, expenses for DT solving, etc.);
b) the model includes equality and inequality constraints of the kind (2) and (3). The classical assignment model requires equality constraints of the kind (2) for all i, i=1,..,I;
c) the model (1)÷(6) presents a three-index problem.

Because of the aforementioned features one can consider the problem described as a specially structured problem. For comparison the canonical statement of the 0-1 LP problem is given below:

$$\max \quad ax \tag{7}$$

s.t.

$$cx \leqslant b \tag{8}$$

$$0 \leqslant x \leqslant 1, \quad x \text{ - integer} \tag{9}$$

where: a is a (n x 1) vector
c is a (m x n) matrix
b is a (m x 1) vector
x is a (n x 1) vector

Problems of the kind (1)÷(6), especially these of medium and large size, cannot be solved with "a reasonable amount of effort" (Geoffrion & Marsten /5/), i.e. within a reasonable computation time and expenses, using the methods and algorithms known for ILP problems. Obviously, new tools are needed for treating a number of problems with special structure.

A HEURISTIC ALGORITHM

For solving ILP problems of the type (1)÷(6) a heuristic algorithm has been developed and applied to a number of test and real-world cases. The procedures used in the algorithm are as follows:

A. Modification of the model

Such a modification requires two approximations to be performed:
1. Approximation concerning the o.f. coefficients

This procedure leads to a deterministic description of the objective function. To the purpose an assumption and some heuristic rules are needed. The assumption is as follows: The o.f. coefficients are independent, identically distributed random variables with uniform distribution, known in certain bounds d_1 and d_2 ($d_2 > d_1 \geqslant 0$). In real-world problems of the type in point this assumption holds. The heuristic rules are of following kind:

a) if the relative difference δ^* between the upper and lower bounds d_1^{ikj} and d_2^{ikj} of the distribution function of each variable a_{ikj} is in the limits of 10%, then the respective mathematical expectations are substituted for the coefficients a_{ikj}; if that difference is larger than 10%, the use of mathematical expectations is not recommended.

Such a substitution is often considered as unfavourable in all cases (see e.g. Zimmermann & Pollatschek /6/). It should be noted, however, that the efficiency of this approximation depends on the nature of the parameters bringing about stochastic o.f. coefficients. So, for a long-time period, e.g. five years or an year, the values of these parameters cannot be determined with a certainty, but for short-time periods (a quarter, a month) they are known. From this suggestion follows the second heuristic rule;

b) a good enough approximation consists in a choice of determinate values of the coefficients based on the values known for the nearest time-period. The change of the coefficients values through the successive solutions of the problem can be made with a given step, as follows: (i) if the first known value of a given coefficient is lower than the mathematical expectation value (assuming known distribution function of the coefficients for further time-periods) the next coefficients values are obtained by increasing that first known value; (ii) if it is larger than the m.e., then the step is negative, i.e. one should go to the lower bound d_1 of the distribution function.

The procedures above would be more efficient if the information available from earlier versions of the problem solving is exploited entirely. Some considerations on this point can be found in Dillon /8/.

Using the heuristic rules mentioned above, a set of discrete values of the o.f. coefficients will be chosen. Thus, the problem (1)+(6) can be considered as a deterministic problem.

2. Reduction of the problem data

This is a measure of decreasing the size of the problem. A number of transformations has been proposed and used (concerning the problem in point):

* relaxation of some constraints by decreasing the number of indices, i.e. transformation of three-index constraints in two-index ones
* prediction of null variables
* choice of null efficiencies (o.f. coefficients); thus the density of

the matrix $\|a_{ik_j}\|$ will be decreased.

The heuristics used for reduction of the problem data are usually based on a prior information on the statement and the solution of a comparable problem.

B. Modification of a known enumerative algorithm

In the case considered this modification consists in a choice of heuristics, added to a version of the Land & Doig algorithm. This algorithm is taken as a basis, because it is the only generally successful approach known presently. The following heuristics are used:

1. k_2 - assumption (Zimmermann & Pollatschek /6/). This heuristic consists in prediction of the number of ones in the optimal problem solution. If k_1 and k_2 are the lower and the upper bounds of the number of ones ($0 \leqslant k_1 \leqslant k_2 \leqslant n$, where n is the number of variables), the value of k_2 can be predetermined exactly or within close bounds. For most real-world scheduling problems this assumption holds.

2. Upper bounding the ILP problem solution. To the purpose, information from a relaxed solution of the problem is used. This is the solution of the LP version of the problem considered, obtained by relaxation of the integer constraints. The following relationships between the objective functions (in a case of maximization) hold:

$$F(X'') \geqslant F(X') \geqslant F(Y') \qquad (10)$$

where X'' is the optimal solution of the LP problem, X' is the optimal solution of the ILP problem and Y' is an approximate ILP problem solution.

Thus, on the basis of X'' an approximate upper bound of the o.f. of the ILP problem can be defined. This heuristic is an effective measure of limitation of the solution search and it is included in a lot of modern ILP packages.

3. Stopping the optimal solution search

In the cases when the duration of the computing time of the problem solving seems unreasonable, the iteration process can be stopped using some heuristic rules. In the case considered two conditions should be fulfiled simultaneously:

$$\frac{N^k(It) - N^{k-1}(It)}{N^k(It)} 100 = \triangle(It) \geqslant 50\% \qquad (11)$$

$$\frac{F^k(Y) - F^{k-1}(Y)}{F^k(Y)} 100 = \triangle(F) \leqslant 1\% \qquad (12)$$

where $N^k(It)$, $N^{k-1}(It)$ are the numbers of iterations of two successive feasible solutions, F^k (Y), F^{k-1} (Y) are the values of the o.f. of these same feasible solutions.

The heuristic algorithm, described above, leads in most cases to obtaining the following problem solutions:
LP solution - 1st feasible ILP solution ... (k-1)th ILP feasible solution - kth ILP feasible solution ... reasonable ILP feasible solution ... last ILP feasible solution - optimal ILP solution.

The good sense in using such a heuristic algorithm requires a reasonable solution to be chosen in both cases: (i) when no optimal solution is known, and (ii) when the optimal solution is obtained with an unreasonable amount of effort.

Estimation of the heuristic algorithm

Three approaches are available for measuring the properties of a heuristic algorithm: empirical testing, probabilistic analysis, worst-case analysis. All these approaches have advantages and deficiencies (see e.g. Geoffrion & Nauss /9/, Garey, Graham & Johnson /10/, Fisher /11/, Silver, Vidal & Werra /12/). As it seems, more preferable is the worst-case analysis which provides a guaranteed percent deviation from the optimal solution for any problem instance within the class studied.

In the case considered in this paper, the estimation of the developed heuristic algorithm has been performed, firstly, by an empirical testing. A considerable number of computational experiments have been run, using simultaneously all of the heuristics described or a part of them. Test and real-world problems of medium size (nearly 150 variables and 40-50 constraints) have been studied.The results prove a good applicability of the heuristics, and a good efficiency of the algorithm as a whole. The computational experiments are discussed in some detail in Strezova /13/.

Worst-case analysis has been also applied in studying the performance of the algorithm. However, the results obtained are specific for some problem instances, and so, not allowing a generalization for the present.

CONCLUDING REMARKS

In this paper some real-world problems are discussed: design of a particular class of LSS, and solving a specially structured optimization problem. The tools suggested and applied for treating both problems can be qualified as implicit ones: an approximate approach to the modelling and analysis of DMIS, and a heuristic algorithm for solving the stated allocation problem. The development and the use of such tools are a result of the new trends in LSS area, from the one hand, and in the mathematical programming, from the other hand, when dealing with real-world problems.

As concerns the heuristic algorithm described, it is based on a passive approach to obtaining feasible solutions,since the problem solution accepted is a side product of a non-terminate process of the optimum search. A number of approximations and heuristics used in this algorithm reduces the computational difficulty of analysing the model at the expense of introducing a certain error. Defining the value of this error, caused by one or another heuristic, will improve essentially the quality of the algorithm. Of particular interest is the influence of the modification of the model structure, especially the reduction of the problem data, on the solution.

When real-world problems in the LSS area are treated, a compromise between the statement of a given problem and the choice of a method and algorithm for solving this problem is necessary. From the one hand, the model, describing such a problem, should be appropriate from mathematical and computational viewpoint, even if, as a result, the real situation is modified to some extent. On the other hand, the methods and algorithms, used for solving such problems, have often some rigorous mathematical basis because of the heuristics included. These methods and algorithms, however, provide feasible solutions of problems intractable by exact and more elegant tools.

Some challenging research topics in the field discussed can be outlined, e.g. development of effective methods and algorithms for special applications, design of particular models, tractable by heuristics, development of more explicit guidelines for designing and testing heuristics, etc.

REFERENCES

1. Strezova, Z. Control problems in management systems. Proceedings of the 7th IFAC Congress, vol. 2. Pergamon Press, 1979.

2. Strezova, Z. An approximate approach to the design of decentralized management systems structures. Proceedings of the 8th IFAC Congress, vol. 7. Pergamon Press, 1982.

3. Strezova, Z. Complex systems modelling and analysis: a methodology and a software system. In A. Javor (Ed.), Discrete Simulation and Related Fields. North-Holland Publ., 1982.

4. Sage, A.P. and Thissen, W.A.H. Methodologies for system simulation. In Oren, Shub and Roth (Eds.), Simulation with Discrete Models: A State-of-the-Art View. University of Ottawa, Ottawa, 1980.

5. Geoffrion, A.M. and Marsten, R.E. Integer programming algorithms: a framework and state-of-the-art survey. Management Science, vol. 18, No. 9, May 1972.

6. Zimmermann, H.-J. and Pollatschek, M.A. The probability distribution function of the optimum of a 0-1 LP with randomly distributed coefficients of the objective function and the right-hand side. Operations Research, vol. 23, 137-142.

7. Strezova, Z. Modelling in management information systems design. Preprints of the 3rd IFAC/IFORS Symposium LSSTA, Warsaw, July, 1983.

8. Dillon, M. Heuristic selection of advanced bases for a class of LP models. Operations Research, vol. 21, 90-100.

9. Geoffrion, A.M. and Nauss, R. Parametric and postoptimality analysis in integer linear programming. Management Science, vol. 23, No. 5, January 1977.

0. Garey, M.R., Graham, R.L. and Johnson, D.S. Performance garantees for scheduling algorithms. Operations Research, vol.26, 3-21.

1. Fisher, M.L. Worst-case analysis of heuristic algorithms. Management Science, vol. 26, No. 1, January 1980.

2. Silver, E.A., Vidal, R.V.V. and de Werra, D. A tutorial on heuristic methods. EJOR, vol.5, No. 3, September 1980.

3. Strezova, Z. A heuristic branch-and-bound algorithm for solving assignment problems. Intern. Conference on Numerical Methods in Operational Research, Sofia, November 1980.

AN IMMUNE LYMPHOCYTE CIRCULATION MODEL[1]

R.R. Mohler[2], Z. Farooqi and T. Heilig
Department of Electrical and Computer Engineering
Oregon State University
Corvallis, OR 97331

1. INTRODUCTION

Mathematical models are developed here to help explain the transient behavior of circulating lymphocytes which are basic to the immune response. It is intended that the development of such a model will be used for predictive purposes in experimental planning. Eventually such modelling, analysis and experimentation may have an impact on tumor control, cancer and immunology in general. In the long run, such research could relate system control theory to effective immunotherapy and chemotherapy. Radiation and chemotherapy adversely affects the immune response in a manner which may be similar to pollutant effects. Systemic immune research could help explain such effects and provide a base for improved treatment.

An excellent introduction to immunology is given by Roitt [1], and a treatment of more detailed aspects of its theory is edited by Bell, et al [2]. An overview of mathematical system theory in immunology and in disease control is given by Mohler, Bruni and Gandolfi [3] and by Marchuk [4], respectively.

One interest of lymphoblast migration is created by the migratory behavior of a transferable lymphoblastic leukemia. Experimentally, this was developed in an HO rat following the injection of a β-emitting isotope into the spleen and lymph nodes by Roser and Ford [5]. When transferred by intravenous injection, the spleen and the lymph nodes swelled as the leukemia developed, but the coeliac lymph nodes grew way out of proportion. Experimental evidence suggests that this excessive enlargement is caused by a preferential migration of lymphoma cells from the blood. Such data is available for analysis from experiments conducted by Smith and Ford [6] at the University of Manchester and studied in the present paper. The lymphoblasts seem to be transient visitors to the lymph node, arriving at afferent lymphatics from the liver and departing by efferent lymphatics available after a few hours. Several apparent conclusions may be made directly from the tracer data and the mathematical analysis given here.

It has been speculated that the high frequency of coeliac node involvement in

---

[1]Research sponsored by National Science Foundation Grant No. ECS-8215724.

[2]NAVELEX Professor of Electrical Engineering, Naval Postgraduate School, Monterey, CA 93943, 1983-1984.

Hodgkin's disease may be explained by the migration from the primary lymph-node (LN) site via blood, the liver and hepatic lymph rather than by retrograde lymphatic spread [7, 8].

Apparently, there is a need to better understand migratory patterns of immune mechanisms. A preliminary, three-compartment, humoral model is studied by Mohler, et al [3, 7, 8]. The present analysis studies only lymphocyte migration which is basic to the humoral process but includes mostly thymus-derived T cells. Also included here are twelve compartments and extensive experimental data.

2. EXPERIMENTAL SUMMARY

Details of the experiment from which the data is derived are given by Smith and Ford [6]. Briefly, the data were taken from a uniform strain of rats as near to the natural physiological state as possible. Lymphocytes were taken from the thoracic duct of a donor, radioactively labelled in vitro, passaged from blood to lymph in an intermediate rat and finally injected into a series of recipients for examination at thirteen time points from one minute to one day. Thirteen tissues were examined from sacrificed rats at each time.

To better understand the system dimensions, it is interesting to note that 180 rats (AO female) were used in the experiment with blood sampled and cells injected on the venous side of the right heart. The total pool of recirculating lymphocytes number about $1.2(10)^9$ with about $40(10)^6$/hour circulating through the thoracic duct and other efferent lymphatics each (see Fig. 1). The coeliac LN weigh only about 8 mg out of a total LN weight of 700 to 800 mg.

At 1, 2 and 5 minutes after injection most of the labelled cells are in blood, lungs and liver, [6]. Concentrations in these compartments subsided during the ensuing 25 minutes as more cells entered the spleen, lymph nodes and Peyer's patches where they peaked between 1 and 18 hours. The migratory pattern of lymphocytes is summarized in Fig. 1 with experimental data compared to model-simulated data in Figs. 2 to 4 for spleen, bone marrow and lungs which exemplify the other organs.

3. COMPARTMENTAL MODEL

The models studied here consist of separate compartments and states for blood, bone marrow, lungs, liver, spleen, lymph nodes, Peyer's patches, gut and miscellaneous tissues. The lymph nodes were further broken down into mesenteric, coeliac, subcutaneous, right and left popliteal, and deep and superficial cervical lymph nodes for certain data collection.

The most common approach to compartmental models assumes lumped time-invariant linearity with diffusions proportional to the concentration differences between the compartments. A slight generalization of this leads to the following time-variant linear model:

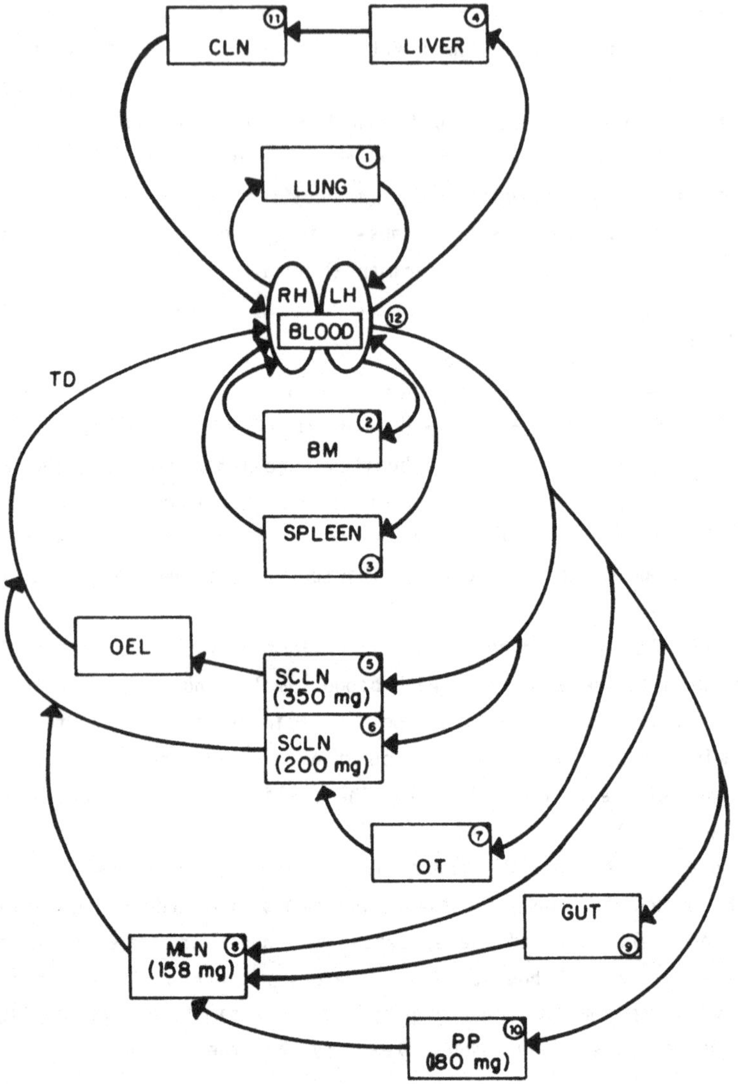

TD: Thoracicduct
CLN: Coeliac lymph nodes
RH,LH: Right, left heart
BM: Bone marrow
OEL: Other efferent lymphatics
SCLM: Subcutaneous lymph nodes
OT: Other tissue
MLN: Mesenteric lymph nodes
PP: Peyer's patches

Fig. 1. Lymphocyte Circulation Paths

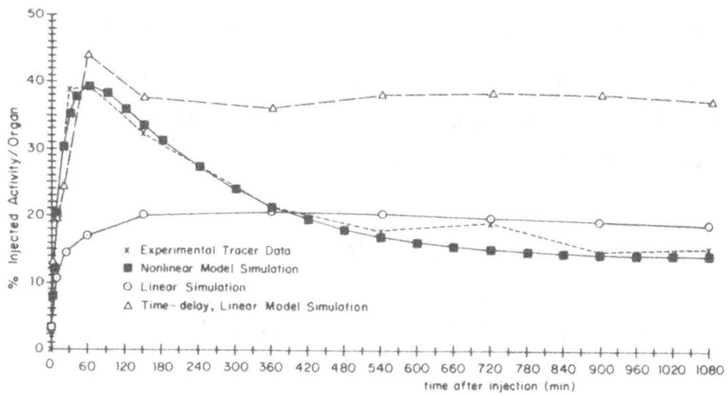

Fig. 2. Rat spleen lymphocyte response

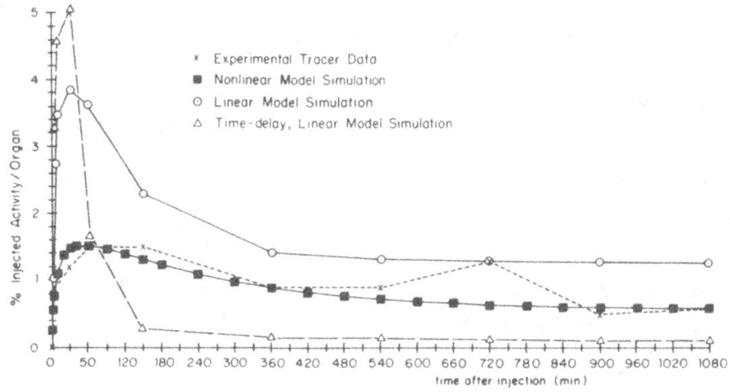

Fig. 3. Rat Bone-Marrow, Lymphocyte Response

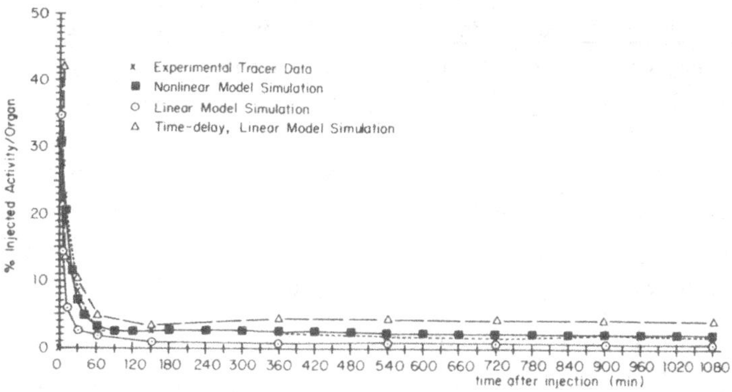

Fig 4. Rat Lungs, Lymphocyte Response

$$\frac{dx_i}{dt} = \alpha_i x_{12} - \beta_i g_i(t) x_i \tag{1}$$

$i = 1, 2, 3, 4, 5, 7, 9, 10$. Roughly, the α_i, β_i represent directional permeabilities and may include a lumped inverse capacitance which is proportional to flow rate divided by volume.

$$\frac{dx_6}{dt} = \alpha_6 x_{12} - \beta_6 g_6(t) x_6 + \beta_7 g_7(t) x_7, \tag{2}$$

$$\frac{dx_8}{dt} = \alpha_8 x_{12} - \beta_8 g_8(t) x_8 + \beta_9 g_9(t) x_9 + \beta_{10} g_{10}(t) x_{10}, \tag{3}$$

$$\frac{dx_{11}}{dt} = \beta_4 g_4(t) x_4 - \beta_{11} g_{11}(t) x_{11}, \tag{4}$$

$$\begin{aligned}\frac{dx_{12}}{dt} = -[\dot{x}_1 + \dot{x}_2 + \dot{x}_3 + \dot{x}_5 + (\alpha_4 + \alpha_6 + \alpha_7 + \alpha_8 + \alpha_9 + \alpha_{10}) x_{12} - \\ \beta_6 g_6(t) x_6 - \beta_8 g_8(t) x_8 - \beta_{11} g_{11}(t) x_{11}].\end{aligned} \tag{5}$$

Here, the designation is

1. lungs
2. bone marrow
3. spleen
4. liver
5. subcutaneous lymph nodes (SCLN) with other lymphatics
6. SCLN with other tissue
7. miscellaneous tissue
8. mesenteric lymphnodes (MLN)
9. gut
10. Peyer's patches
11. coeliac LN
12. blood

For the <u>time-invariant linear model</u> (TIL Model),

$$g_i(t) = 1 \tag{6}$$

$i = 1, \ldots, 11$.

The <u>time-variant linear model</u> (TVL Model) is used to approximate time delays, time-variant capacitance and the presence of subcopartmental modes such as analyzed by Hammond for the spleen [9]. Consequently, the selected time-variant coefficients are given by

$$g_i(t) = 1 - e^{-t/\tau_i}, \tag{7}$$

for $i = 2, 4, 5, 6, \ldots, 11$, (6) for $i = 1$, and

$$g_3(t) = \beta_3 \left[1 + \frac{\gamma_3}{\beta_3} (1 - e^{-t/\tau_3})\right]. \tag{8}$$

By "cut and try" along with "immunologically reasonable" parameter values, the selected TIL Model and TVL Model parameters are given in Tables I and II.

Table I. TIL Model Parameters

i	1	2	3	4	5	6	7	8	9	10	11
α_i	1.8	0.02	0.055	0.25	0.0045	0.002	0.15	0.003	0.015	0.004	
β_i	2.0	0.016	0.03	0.07	0.0002	0.03	0.004	0.001	0.005	0.0006	1.00

Table II. TVL Model Parameters

i	1	2	3	4	5	6	7	8	9	10	11
α_i	1.8	0.02	0.08	0.1	0.004	0.002	0.09	0.005	0.015	0.003	
β_i	1.7	1.00	0.0015	2.02	0.0005	0.05	0.01	0.005	0.05	0.003	1.00
γ_i			0.008								
$\tau_{i,min}$		540	60	400	500	200	200	200	360	540	20

The time-responses of circulating lymphocytes for these two models are compared with the experimental data in Figs. 2 to 4 for spleen, bone marrow and lungs. Since the fit is not very accurate, a nonlinear model is derived which considers saturation. While conventionally, modeled by a multiplicative parametric term of the form of $(1 - y)^{-1}x$ approximation terms involving products of x and y are used to formulate the following nonlinear (NL) model.

Lungs: As (1) and (6) above with i = 1.

Bone marrow:

$$\frac{dx_2}{dt} = \alpha_2(1 - \gamma_2 x_2)x_{12} - \beta_2(1 - e^{-t/\tau_2})x_2. \tag{9}$$

Spleen: As (1) and (8) above with i = 3. (10)

Liver:

$$\frac{dx_4}{dt} = \alpha_4(1 - \delta_4 x_4)x_{12} - \beta_4 x_4 + [\beta_3 + \gamma_3(1 - e^{-t/\tau_3})]x_4 \tag{11}$$

Lymph nodes: i = 5 combines i = 5, 6, 8, 11 above with

$$\frac{dx_5}{dt} = \alpha_5 x_{12} + \gamma_5(1 - \delta_5 x_5)x_9 - \beta_5(1 - e^{-t/\tau_5})x_5. \tag{12}$$

Miscellaneous tissues: As (1) and (6) above with i = 7.

Gut:

$$\frac{dx_9}{dt} = \alpha_9 x_{12} + \gamma_9(1 - \delta_9 x_9)x_7 x_9 - \beta_9(1 - e^{-t/\tau_9})x_9. \tag{13}$$

Peyer's patches:

$$\frac{dx_{10}}{dt} = \alpha_{10} x_{12} + \gamma_{10}(1 - \delta_{10} x_{10})x_{12} x_{10} - \beta_{10}(1 - e^{-t/\tau_{10}})x_{10}. \tag{14}$$

Blood:

$$\frac{dx_{12}}{dt} = -(\dot{x}_1 + \dot{x}_2 + \dot{x}_4 + \dot{x}_5 + \dot{x}_9 + \dot{x}_{10}). \tag{15}$$

Parameter values for this NL model are given in Table III. Again, they were arrived at by educated guesses, trial and error.

Table III. NL Model Parameters

i	1	2	3	4	5	7	9	10
α_i	1.5	0.004	0.055	0.5	0.009	0.025	0.0015	0.0011
β_i	2.0	0.016	0.0015	0.42	0.001	0.002	0.003	0.0006
γ_i		0.55	0.0075		0.0035		0.01	0.008
δ_i				0.055	0.047		1.0	0.16
$\tau_{i,min}$		360	150		720		360	540

It is obvious from Figs. 2 to 4 that the NL model provides a more accurate description of lymphocyte circulation than the linear models according to the experimental data given here. Similar conclusions are made for the other organs although not presented here. It should be realized, however, that this is only a preliminary study. Future efforts should include a thorough statistical analysis of the experimental data along with a more detailed derivation of a compartmental lymphocyte circulation model based more completely on immunological principles. Admittedly, certain liberties were taken in the parameter selection here which cannot be immunologically verified. In particular, the liver presents some problem as discussed below in the conclusion. The δ_4 and τ_3 terms to approximate saturation and subcompartment capacitance or time delay are questionable. Similarly, the values of τ_2 for bone and τ_{10} for Peyer's patches may be too large, and the saturation by δ_5, δ_9, δ_{10} for LN, gut and Peyer's patches may be too constraining. These aspects need to be more throroughly investigated.

4. CONCLUSION

The nonlinear model seems to mimick lymphocyte circulation quite accurately in showing the following experimental results. After a rapid lymphocyte exchange with the lungs during the first couple minutes after injection, the level of blood lymphocytes decreases for the next hour with a half life of approximately 16 minutes. As lymphocytes return to blood (particularly from spleen), the exponential decay ceases between 1 and 2.5 hours after injection followed by a slow rise to near equilibrium at 6 hours onward. Localization of lymphocytes in the liver is somewhat similar to blood in its time response, but has a slower terminal decay of approximately a 24-minute half life. Approximately 40 percent of the injected lymphocytes are found in the spleen at the 30 minute mark. This is followed by a decay mode of approximately 300-minute half life. LN lymphocytes gradually build up to almost 60 percent in about 18 hours. MLN and SCLN responses have similar shape. Peyer's patch level builds up to about 7 percent in about 1.5 hours. Modeling of the liver is particularly complicated due apparently to three independent phenomena involving the lymphocyte migration. First, there is intravascular pooling similar to lungs which results in rapid initial response. Then there is genuine recirculation blood to liver to coeliac LN, to thoracic duct and back to blood again. Finally, there is an accumulation of dying cells in liver which involves only about 1% to 2% of the total population per day. Still, the latter can be a substantial part of the liver response itself. This may very well account for the long term error which was found between the model simulation and certain other compartments such as liver.

ACKNOWLEDGEMENT

The authors are grateful to W. L. Ford for his detailed explanation of circurlatory immunology and the corresponding experiments.

REFERENCES

1. I. Roitt, Essential Immunology, Blackwell Scientific, Oxford and London, 1974.
2. G.I. Bell, A.S. Perelson and G.H. Pimbley, Jr. (eds.), Theoretical Immunology, Marcel Dekker, New York, 1978.
3. R.R. Mohler, C. Bruni and A. Gandolfi, "A Systems Approach to Immunology," IEEE Proc. 68, 964-990, 1980.
4. G.I. Marchuk, Mathematical Models in Immunology, Science Press, Moscow, 1980 (In Russian; English version to appear, Springer-Verlag, 1984).
5. B.J. Roser and W.L. Ford, "Prolonged Lymphocytopenia in the Rat. I.," Aust. J. Exp. Biol. Med. Ser. 50, 165-184, 1972.
6. M.E. Smith and W.L. Ford, "The Recirculating Lymphocyte Pool of the Rat: A systematic description of the migratory behaviour of recirculating lymphocytes," Immunology 49, 1983 (to appear).

7. M.E. Smith, A.F. Martin and W.L. Ford, "The Migration of Lymphoblasts in the Rat," in Essays on the Anatomy and Physiology of Lymphiod Tissue, S. Karger AG Basel. (to appear).
8. H.S. Kaplan, R.F. Dorman, T.S. Nelsen and S.A. Rosenberg, "Staging Laparatomy and Splenectomy in Hodgkins' Disease: Analysis of indications and patterns of involvement in 285 consecutive, unselected patients," Nat. Canc. Inst. Mono. 36, 291, 1973.
9. R.R. Mohler and C.F. Barton, "Compartmental Control Model of the Immune Process," Proc. 8th IFIP Optimiz. Conf., Springer-Verlag, New York, 1978.
10. R.R. Mohler and C.S. Hsu, "Systems Compartmentation in Immunological Modeling," Proc. Rome Conf. Sys. Theory in Immunology, Springer-Verlag, New York, 1978.
11. B.J. Hammond, "A Compartmental Analysis of Circulatory Lymphocytes in the Spleen," Cell Tissue Kinet. 8, 153-169, 1975.

MATHEMATICAL MODELING OF INFECTIOUS DISEASES: PRESENT STATE, PROBLEMS AND PROSPECTS

G.I. Marchuk, L.N. Belykh, S.M. Zuev
Department of Numerical Mathematics
Gorky Str. 11, Moscow 103905, U.S.S.R.

1. INTRODUCTION

In this paper we represent a new mathematical model of antiviral immune response and discuss its properties and possible expansions. We also discuss the problem of stability, numerical analysis and statistical estimation of parameters for models of this kind. The survey of our previous results of a disease modeling is given in [1,2].

2. MATHEMATICAL MODEL OF ANTIVIRAL IMMUNE RESPONSE

Immune response to stimulants of viral infections in the organism, such as influenza, measles, poliomyelitis, viral hepatitis and others, includes two types: "humoral" response, when the system of B-lymphocytes produces antibodies; and "cellular" response, when cytotoxic T-lymphocyte-effectors accumulate in the organism. It is cellular response that secures the defense of the organism. Antibodies neutralize viral particles circulating in the blood but they are not capable of freeing the organism from infection since virions multiply inside the cells sensitive to a given virus. Antibodies do not penetrate inside the cells. Cytotoxic lymphocyte-effectors which have accumulated after the immune response, detect cells affected by the virus and kill them, in their role as killers of cells of the host organism. Thus an antiviral immune response of the cellular type seems to be of autoimmune nature. This, however, is not to be confused with the real autoimmune reaction. The latter involves pathological reactions of the immune system against normal (unchanged) cells or normal cellular antigenic substances. By their antiviral immunity lymphocyte-killers destroy the cells of the host organism affected by the virus. Apparently, this is the only way to clean the organism from viruses.

Based on this facts and modern immunological knowledge of viral infection dynamics we distinguish the following main model variables:

$V_f(t)$ - concentration of "free" viruses (viral particles freely circulating in the body) which are capable of multiplying in the cells of the organ sensitive to a given type of viruses;

$M_v(t)$ - concentration of stimulated macrophages which have interacted with free viruses);

$H_E(t)$ - concentration of T-lymphocyte-helpers , participating in a cellular type of immune response;

E (t) - concentration of T-cell-effectors (killers);

B(t) - concentration of immunocompetent B-lymphocytes capable of adopting the stimulation signal from stimulated macrophages M_v and helpers H_B(T-lymphocytes taking part in a humoral response);

P(t) - concentration of plasma cells (antibody producents);

F(t) - concentration of antibodies;

$C_v(t)$ - concentration of organ's cells infected with viruses;

m(t) - non-functioning part of the organ damaged by viruses.

Now we introduce the following assumptions:

1. The quantities of "virgin" macrophages in the body M and of organ's cells C are considered constant and sufficiently large for the increase in stimulated macrophages M_v and in infected cells C_v to be proportional to the quantity of free viruses V_f.
2. The adoption of a stimulation signal by lymphocytes leads after a certain period of time necessary for their division and proliferation to the formation of the terminal cells ' clone. To stimulate helpers a single signal from M_v is necessary and the double one (from M_v and a corresponding helper) for the stimulation of E and B cells.
3. Part of the formed clone of terminal cells can be stimulated to form a new clone under a corresponding signal. The remaining part executes other immune functions such as help at stimulation, killers' effect and antibody production.
4. The living cycle of the lymphocyte-helpers H_E and H_B is over after the interaction with lymphocytes E and B respectively (after helping E and B lymphocytes).
5. During a certain period of time infected cells C_v execute their normal functions. Their death is due either to the development of irreversible viral infection or to their elimination by effectors E. The damaged mass of the organ is therefore the value of cells killed by viruses plus the value of cells killed by lymphocyte-effectors.

According to these assumptions we constructed [2] the mathematical model of antiviral immune response which has the form:

$$
\begin{aligned}
&\dot V_f = n b_E C_V E + \rho b_m C_V - \gamma_m M V_f - \gamma_f V_f F - \kappa \sigma C V_f , \\
&\dot M_V = \gamma_M M V_f - \alpha_m M_V , \; P_1(t) = M_V H_E , \; P_2(t) = M_V H_E E , \; P_3(t) = M_V H_B , \; P_4(t) = M_V H_B B , \\
&\dot H_E = b_H [\xi(m) P_1(t-\tau_H) - P_1(t)] - b_p P_2(t) + \alpha_H (H_E^* - H_E), \\
&\dot H_B = b_H^{(B)} [\xi(m) P_3(t-\tau_H^{(B)}) - P_3(t)] - b_p^{(B)} P_4(t) + \alpha_H^{(B)} (H_B^* - H_B), \\
&\dot E = b_p [\xi(m) \rho_E P_2(t-\tau_E) - P_2(t)] - b_E C_V E + \alpha_E (E^* - E), \\
&\dot B = b_p^{(B)} [\xi(m) \rho_B P_4(t-\tau_B) - P_4(t)] + \alpha_B (B^* - B), \qquad (1) \\
&\dot P = b_p^{(P)} \xi(m) \rho_B P_4(t-\tau_B) + \alpha_P (P^* - P), \\
&\dot F = \rho_f P - \eta_f \gamma_f V_f F - \alpha_f F , \\
&\dot C_V = \sigma C V_f - b_E C_V E - b_m C_V , \\
&\dot m = \mu b_E C_V E + \eta b_m C_V - \lambda m .
\end{aligned}
$$

It can be easily seen that the stationary model solution corresponding to the healthy body state is the following:

$$
\begin{aligned}
&V_f = M_V = C_V = m = 0 , \; H_E = H_E^* , \; H_B = H_B^* , \qquad (2) \\
&E = E^* , \; P = P^* , \; B = B^* , \; F = \rho_f P^* / \alpha_f = F^* .
\end{aligned}
$$

As before [2] we are interested in simulating the entirely natural situation - the infection of a healthy body with a small dose of free viruses V_f° at time $t=t^\circ=0$. This means that the system is in a stationary state (2) before infection, i.e. at $t<t^\circ$, but at $t=t^\circ$ the infection with a small dose $V_f(t^\circ) = V_f^\circ$ takes place. Other components at $t=t^\circ$ reserve their stationary values.

For system (1) we have proved the theorems of global (i.e. for all $t \geq t^\circ = 0$) existence of the unique solution and its non-negativity and have derived the stability condition for solution (2). A simple numerical analysis showed that this model reproduces the main forms of a disease, i.e. subclinical, acute with the recovery, chronic and lethal forms. We will use this model for simulating the disease course under the immunodeficiencies of different types. While the analysis of this model is not finished we are going to expand it introducing a specific organ (liver, lungs) rather than the abstract one as we have now and the local places (lymph nodes) where the immune response is developed. Thus our model will acquire the immuno-physiological meaning.

3. ON MATHEMATICAL PROBLEMS OF MODELING

All our models including the system (1) have the form

$$\dot{x} = F(x(t), x(t-\tau))$$
$$x(t) = \varphi(t) \qquad t \in [t^0-\tau, t^0] \qquad (3)$$

and satisfy the following main features of adequacy to a real biological process:

1. Global (at $t \geq t^0$) existence and uniqueness of solution.
2. Non-negativity of solution at $t \geq t^0$.
3. Existence of stability conditions for the stationary solution interpreted as a healthy body state.

The global existence guarantees that none of the process components achievesinfinity during the finite period of time. Uniqueness promotes the reproduction of model experiments. Non-negativity emphasizes the biological meaning of model variables. The third feature gives the existence of an attraction zone for the healthy body state which is called the immunological barrier V^*. This means that the disease will not develop if immunological barrier V^* is not passed ($V^0 < V^*$) and for it to develop it is necessary either to pass this barrier $V^0 > V^*$ or to violate the stability conditions. If the immunological barrier does not exist then according to the models all people must be sick. But the real situation is quite different.

Usually we prove global existence and uniqueness basing on the fact that our model (3) is dominated by a linear system,non-negativity is proved using continuity of solution [2]. As far as stability is concerned we have proved the following theorem which is formulated here (for the sake of simplicity) for the system of two ordinary differential equations. Its expansion on the system (3) is apparent.

Theorem 1. Let the system of equations

$$\dot{v} = f(v,u), \quad \dot{u} = \varphi(v,u) \qquad (4)$$

have the stationary solution $x^* = \{v^*, u^*\} = \{0, u^*\}$ and functions f and φ provide the existence of the unique solution of this system. If this stationary solution x^* is exponentially stable in the system

$$\dot{v} = f(v, u^*) = f_1(v), \quad \dot{u} = \varphi(0,u) = \varphi_1(u) \qquad (5)$$

then it is asymptotically stable in the system (4).

This theorem enabled us to essentially simplify the stability analysis of complex models such as model (1). It should be noted that our models are very naturally decomposed on two subsystems of type (4) which have "biological sense" - "disease system" and "defence system". For example, for model (1) $v = \{V_f, M_V, C_V, m\}$ - "disease vector" and $u=\{H_E, H_B, E, B, P, F\}$ - "defense vector". It would be quite important to prove such decomposition theorems for existence, uniqueness and non-negativity of solutions. The three features mentioned above are very important but too general. To know more about specific properties of the solution we must use the numerical analysis of the models. Remembering the general form of our models (3) we note that the solution of systems of this kind can have at some points $\{t_k\}$ where $t_k=t^o+(K-1)\tau$ and K=1,2,3 the finite discontinuances of its derivatives $\Delta^i_k= x^i(t_k+0)-x^i(t_k-0)$ where $x^i(t)=d^ix/dt^i$ and i=1,2,3,... Thus, we cannot use the classical computational methods for ordinary differential equations based on the classical Taylor's series. To avoid this we proved the following theorem [3] which enabled us to modify classical methods and apply them for systems of type (3).

Theorem 2. For any given in (a,b) piece-wise smooth function $f(t)$ having l derivatives ($l>1$) on the differentiation intervals and finite discontinuances of its derivatives $\Delta f^i_k (i=\overline{1,l}; k=\overline{1,N})$ at points $\{t_k\}$ from (a,b) there exists the unique function g(t) computed according to formula

$$g(t)=\sum_{K=1}^{N}\sum_{i=1}^{l-1}\Delta f^i_K \frac{(t-t_K)^i}{i!}\theta(t-t_K)$$

such that the difference $u(t)=f(t)-g(t)$ is ($l-1$) times continuously differentiable function in (a,b).

The last problem we want to discuss is connected with the statistical estimation of our model's coefficients. Since our model's solutions are deterministic the real observable trajectories does not belong to the set of the model's solution. We,therefore, need a stochastic model describing real trajectories in a certain meaning. Assuming that our deterministic model describes the dynamics of a real stochastic system in average we can construct a stochastic model which corresponds to the deterministic one linear on the coefficients vector[2,4]. Investigation of the stochastic model properties enabled us to apply a maximum likelihood method for estimating model coefficients on experimental data.

REFERENCES

1. Marchuk G.I. Mathematical Modeling in Immunology. Publications Di-

vision of Optimization Software Inc., New York, 1983, 250 p.
2. Belykh,L.N., Zuev, S.M., Marchuk,G.I. On the Mathematical Modeling of Disease. Proc. of the IFIP TC-7 Conf. on Recent Advantages in System Modeling and Optimization, Hanoi, Vietnam,1983,Springer Verlag,1983.
3. Belykh,L.N. On the Computational Methods in Disease Models. In: Mathematical Modeling in Immunology & Medicine (Eds. Marchuk,G.I and Belykh,L.N.), North Holland, Amsterdam,1983,pp. 79-84.
4. Zuev,S.M. Statistical Estimation of Immune Response Mathematical Models Coefficients. Ibid.,pp. 255-264.

HYPERTHERMIA CANCER THERAPY: MODELLING, PARAMETER ESTIMATION AND CONTROL OF TEMPERATURE DISTRIBUTION IN HUMAN TISSUE.

Morten Knudsen and Leif Heinzl
Institute of Electronic Systems
Aalborg University
Aalborg, Denmark

1. INTRODUCTION

The heating of tumor tissue to a temperature of 41-45°C over a period of ½ to 1 hour is known to have a destructive effect on cancer. The effect increases dramatically with increasing temperature, so the temperature in malignant tissue should be as high as possible, while the temperature in healthy tissue should stay within certain bounds.

With the hyperthermia system, described in section 2 of this paper, heat can be applied and controlled selectively to obtain a desired temperature distribution in human tissue. The system is being developed at the Institute of Electronic Systems, Aalborg University in cooperation with the Institute of Cancer Research in Aarhus.

With an early version of the system, not including the microcomputer, a total of 42 heat treatments has been given to 10 patients with superficial tumors. [1]

In the new system the microcomputer controls the temperature distribution in the tissue by manipulating heating and cooling effects. The control has been tested clinically with the aid of a pig.

To design the controller a dynamic model of the thermal process has been developed. The model parameters are determined from simple experiments or by letting the computer perform the system identification.

2. HYPERTHERMIA

Cancer therapy with hyperthermia has been known for several thousand years, but only in the last few years has there been substantial progress, due to global medical and biological research and new technology in electronics.

The destructive effect of heating living cells - malignant as well as normal - is a function of time (t) and temperature (T). For a constant temperature the relative number of destroyed cells per time unit is constant [2]. The number of surviving cells (N_s) is then

$$N_s(t) = N_o \cdot e^{-\frac{t}{\alpha}} \qquad (1)$$

where $\alpha=\alpha(T)$ is a temperature dependant time "constant", indicating how long it takes to reduce the number of surviving cells by a factor e (e=2.72). Figure 1 shows

α as a function of temperature for asynchronous L_1A_2 cells. The data is taken from [2].

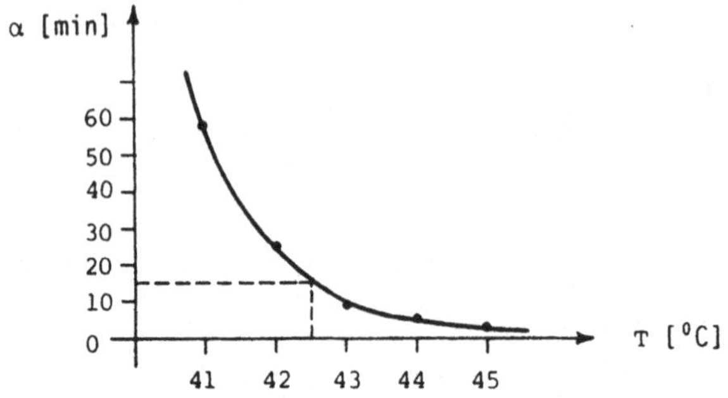

Fig. 1. Time "constant" α as of function of temperature.

It appears that α decreases almost exponentially with increasing temperature - a factor 2 pr. degree. If f.ex. a 86% cell destruction in 30 min. is specified, $\frac{N_S}{N_0} = 0{,}14 = e^{-\frac{30}{\alpha}} \Rightarrow \alpha = 15$ min. and, according to fig. 1, the temperature T=42,5°C. Treatment for 60 min. at 42,5° gives 98% cell destruction.

Cancer cells are often more heat sensitive than normal cells, because of the environment in poorly vascularized part of solid tumors. Moreover, the temperature in these parts will tend to be higher than in healthy tissue during hyperthermia treatment, due to less blood cooling.

Hyperthermia and radiotherapy is a very advantageous combination. Firstly because the most radioresistant cells are also the most heat sensitive, and secondly because hyperthermia has a radiosensitizing effect.

3. THE HYPERTHERMIA SYSTEM

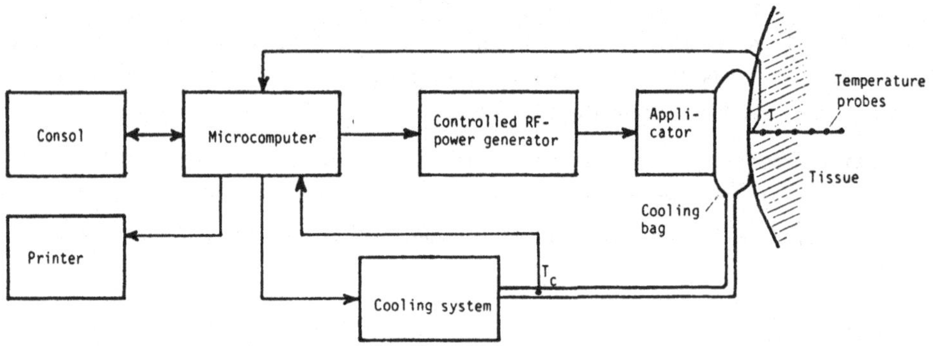

Fig. 2. The hyperthermia system

The following four functions are performed by the system:

- heating: a noninvasive inductive applicator converts RF-power to heat in the tissue. The frequency range is 20 to 200 MHz, and at present the maximum RF-power is 150 W [1].
- surface cooling: a plastic bag with recirculating destilled water. The temperature T_c is controlled by the computer via a heat exchanger.
- temperature measurement: invasive thermistor or thermocouple multiprobe. The RF-heating is disconnected for a short period while the temperatures are sampled [1].
- operator communication, data processing and control is carried out by the computer.

4. PROBLEM DEFINITION

Along an axis through the middle of the tumor and the applicator, perpendicular to applicator and skin surface, the effect of heating and of cooling will decrease with distance. The combined effect can cause a temperature maximum in a fairly superficial point (maximum depth 2-3 cm).

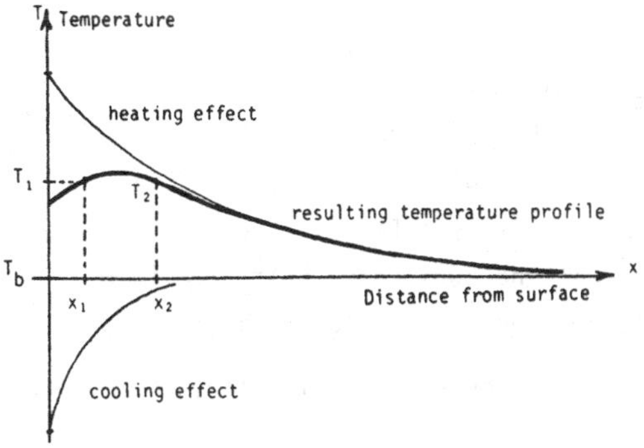

Fig. 3. Temperature distribution

If the tumor is located between x_1 and x_2, fig. 3, the desired values of the temperatures T_1 and T_2 at these depths are specified as TR_1 and TR_2. According to section 2 a reasonable choice would be

$$TR_1 = TR_2 = TR = 42{,}5^0$$

Although the temperature distribution is considered and measured in one dimension only, it is assumed that this is tantamount to the desired situation:

Temperature in tumor > TR

Temperature in healthy tissue < TR

Obtaining this situation is now a traditional two-variable control problem, as illustrated by fig. 4

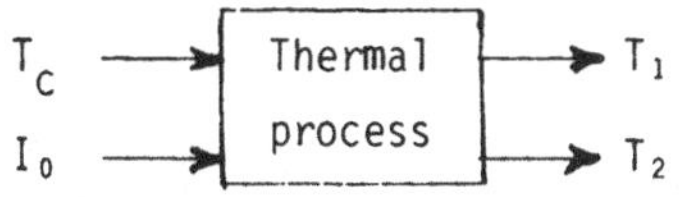

Fig. 4 Control plant

T_c is cooling water temperature

I_0 is transmitted RF-power

T_1 and T_2 are temperatures in two specified depths

To obtain a high therapeutic effect in a short treatment period, the specified temperatures must be reached quickly and maintained within a narrow range, in spite of disturbances, e.g. from variation in the blood cooling. This indicates the necessity of good control, which again necessitates a reliable dynamic model of the thermal process.

5. MATHEMATICAL MODEL AND COMPUTER SIMULATION

The one-dimensional heat conduction equation for a semi-infinite homogeneous volume, the so-called bioheat equation, is a widely used linear thermal model for human tissue. In this case it takes the form:

$$\frac{\partial T}{\partial t} = \frac{\gamma}{\rho_t c_t} \frac{\partial^2 T}{\partial x^2} - \frac{\rho_b c_b}{c_t} m (T - T_b) + \frac{I_0}{\rho_t c_t L} e^{-x/L} \qquad (2)$$

where

- x is distance from surface [m]
- t is time [s]
- $T = T(x,t)$ is tissue temperature [^{0}C]
- T_b is temperature of blood entering the region [^{0}C]
- γ is thermal conductivity of tissue [W/(m · ^{0}C)]
- ρ_t and ρ_b are densities of tissue and blood [kg/m^3]
- c_t and c_b are specific heat of tissue and blood [W · s/(kg · ^{0}C)]
- m is flow rate of blood pr unit mass of tissue [m^3/(s · kg)]
- $I_0 = I_0(t)$ is transmitted RF-power at the surface pr unit area [W/m^2]
- L is the depth where the power is reduced by a factor e [m].

In (2) the first term on the right side represents heat diffusion, the second blood cooling and the third RF-heating

The surface cooling gives a boundary condition:

$$\left.\frac{\partial T}{\partial x}\right|_{x=0} = -\frac{k}{\gamma} (T_c - T(o,t)) \qquad (3)$$

where

- k is heat transfer coefficient between cooling water and tissue [W/(m^2 · ^{0}C)]
- T_c is cooling water temperature [^{0}C]

The steady state solution to (2) and (3), corresponding to the resulting temperature profile in fig. 3, is easily found to be:

$$T(x,\infty) = I_0 \frac{1}{\gamma L(b-1/L^2)} \left[e^{-x/L} - \frac{\gamma/L+k}{\gamma\sqrt{b}+k} e^{-\sqrt{b}x} \right] + (T_c - T_b) \frac{k}{\gamma\sqrt{b}+k} e^{-\sqrt{b}x} + T_c \qquad (4)$$

where $b = \rho_b c_b \rho_t m/\gamma$

The transfer function from heating takes the form:

$$\frac{T(x,s)}{I_0(s)} = \frac{K_1(x)}{1 + s\,\tau_1(x)} \qquad (5)$$

while the transfer function from cooling $\frac{T(x,s)}{T_c(s)}$ is an irrational function of the Laplace operator s. For that reason a program for digital simulation of the distributed parameter model has been constructed. It is based on a lumped parameter and discrete time approximation, where section width is chosen to Δx = 2mm, and time step i determined from the stability criteria $\Delta t < const \cdot \Delta x^2$ [3]. Inhomogeneous tissue can be simulated as well as homogeneous.

The simulation model is used for:

i) developing a simple model approximation for controller design. The transfer function from cooling is approximated by a time constant and a time delay:

$$\frac{T(x,s)}{T_c(s)} = \frac{K_c(x)}{1 + s\,\tau_c(x)} e^{-s\tau_0(x)} \qquad (6)$$

ii) testing control, and

iii) testing identification

The models based on the one dimensional bioheat equation gives a very simplified description of the thermal system, as this is threedimensional, very inhomogeneous and nonlinear. As the usability of a model decreases with increasing complexity, it is our contention, however, that these models serve the purpose. In particular it should be noted, that in the controller design only the structure of the theoretical model is used, while the model parameters are determined experimentally (cf. section 7).

6. CONTROL

For the simple model a digital controller has been designed.

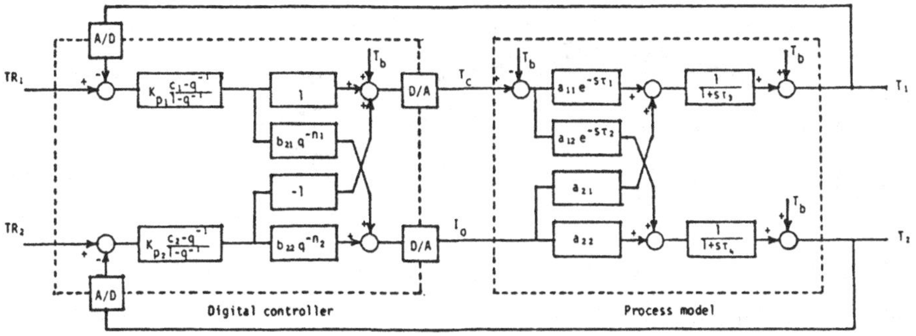

Fig. 5. Control system (q^{-1} is the backward shift operator, s is the Laplace operator)

By dynamic decoupling the cross coupling in the process is eliminated, so the temperature reference TR_1 effects the output temperature T_1 only, and TR_2 affects T_2 only. Besides the controller has proportional-integral (PI) action.

The controller is tested clinically and with the simulation model. Fig. 6 shows responses to set point steps TR(t) = 36⁰ for t < 0 and TR(t) = 43,5⁰ for t ≥ 0.

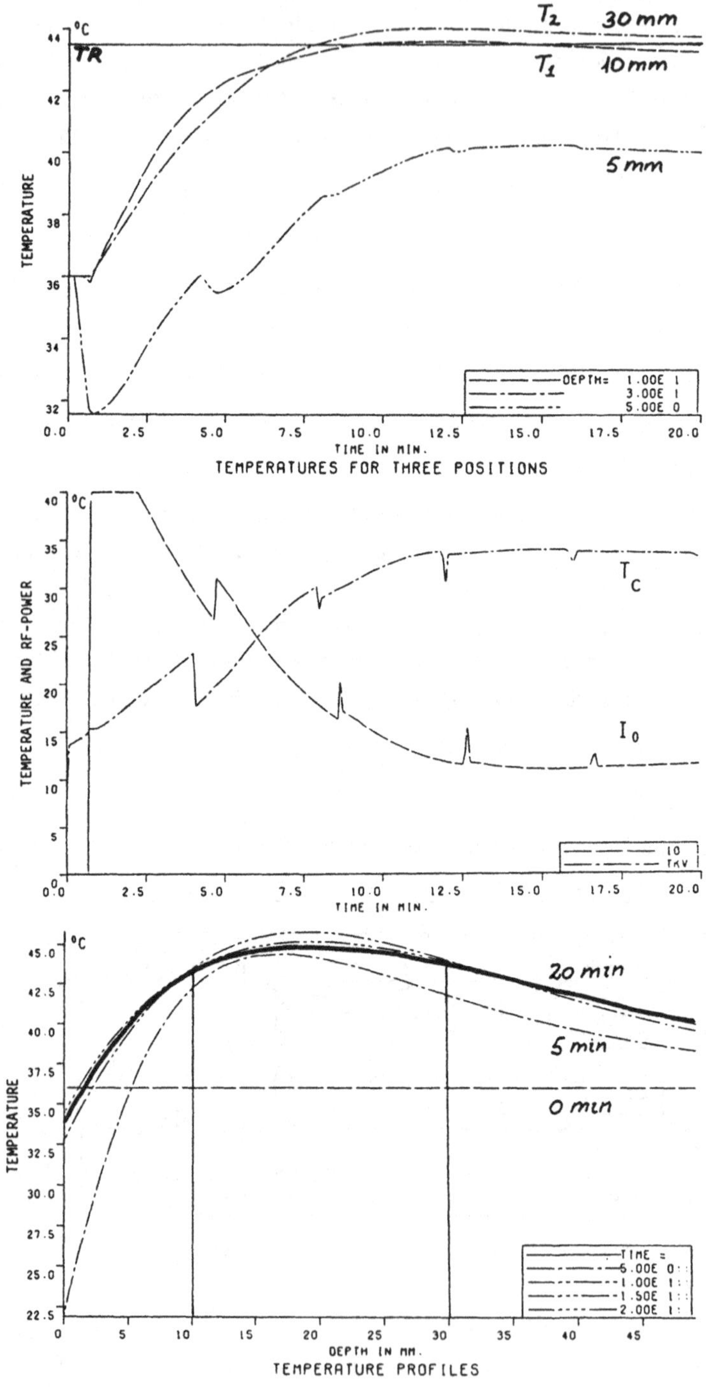

Fig. 6. Control simulation. Closed loop step responses: from a steady state situation with T=TR=36⁰, TR is changed to 43,5⁰.

7. PARAMETER ESTIMATION

In the clinical application of the control, the process model parameters are determined from step response experiments using a graphical technique. This method and calculation of corresponding control-parameters are, however, very time consuming, so it is advantageous to let the computer perform the system indentification.

For this purpose a stocastic discrete-time model of the process is used [4]

$$T(n+1) = -A(q^{-1})T(n) + B(q^{-1})U(n) + C(q^{-1})E(n)$$

where $T(n) = \begin{bmatrix} T_1(n) \\ T_2(n) \end{bmatrix}$ and $U(n) = \begin{bmatrix} T_c(n) \\ T_0(n) \end{bmatrix}$ are the output and input vectors,

and $\{E(n)\}$ is a sequence of independent random vectors, representing system noise. A, B and C are matrix transfer operators. The parameters in A and B of the simulation model have been estimated with the "prediction error method", combined with a minimum search to estimate the delays. In most cases C was asumed to be the unit matrix, as this gives the particulary simple Least Squares method.

Two different types of input signals have been used. The input signals giving time-optimal control are highly correlated, resulting in large parameter variance. Pseudo Random Binary Sequences (PRBS) gave more distinct minima and, for reasonable amplitude, smaller parameter variance.

Fig. 7 shows simulation results of identification and control. During identification T_C is a step function and I_0 is a PRBS. The controller is a Smith predictor, which appears to be superior to the PI-controller when the model is fairly accurate.

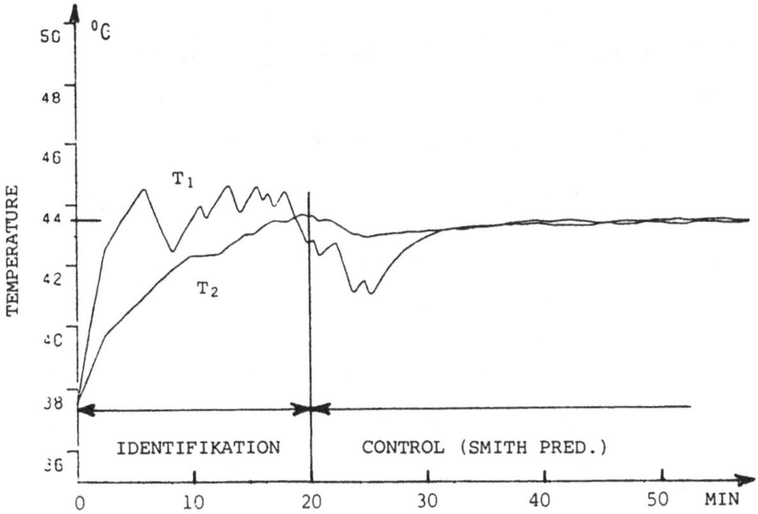

Fig. 7. Output temperatures during identification and control

8. CONCLUSION

The described control strategy appears to be adequate for controlling the temperature profile in normal tissue. The presence of tumor tissue can be expected to cause a further improvement of the profile, as the lower blood cooling will give a more distinct temperature maximum in the tumor.

Traditional determination of the model parameters from step response experiments are too time consuming. Computer estimation of the model parameters based on input-output measurements, i.e. systems identification, is, however, a promising solution to that problem.

9. ACKNOWLEDGEMENT

Important parts of the work described have been done in student projects. Anders Hornemann has programmed the simulation model and carried out some initial work on controller design. Claus E. Jensen and Bjarne G. Pedersen have constructed the microcomputer programs and designed the cooling system.

Dr. Jens Overgaard at The Cancer Research Institute is responsible for the animal tests.

10. REFERENCES

[1] J. Bach Andersen, L. Heinzl et al, "A hyperthermia system using a new type of inductive applicator". IEEE Transactions on Biomedical Engineering vol. BME-31, 1, Jan. 1984.

[2] O.S. Nielsen and J. Overgaard, "Effect of Extracellular pH on Thermotolerance and Recovery of Hyperthermic Damage in Vitro", Cancer Research 39, 2772-2778, July, 1979.

[3] H.S. Carslaw and J.C. Jæger, "Conduction of Heat in Solids", London: Oxford 1959.

[4] G.C. Goodwin and R.L. Payne, "Dynamic System Identification", Academic Press, 1977.

OPTIMAL CONTROL OF THE HEEL-OFF TO LIFT-OFF PHASE OF TWO MAXIMUM HEIGHT JUMPS

W.S. Levine*, F.E. Zajac+, Y.M. Cho* and M.R. Zomlefer+

*Dept. of EE, Univ. of Maryland, College Park, MD 20742, USA

+Mech. Eng. Dept.-Design, Stanford University, Stanford, CA 94305, USA & R.R.&D. Center (153), VA Medical Center, Palo Alto, CA 94304, USA

ABSTRACT

A Mathematical model is first developed for the following two tasks. Jump as high as possible starting from a deep crouch with your feet flat on the floor. Jump as high as possible starting from a deep crouch with your weight balanced on your toes (heels off the ground throughout). Then, the mathematical model, which takes the form of an optimal control problem, is solved for the optimal muscle "activation" and trajectories. The analytical results are compared with the experimental results.

1. INTRODUCTION

We have been studying maximum height jumps for several years for reasons that are explained in detail in (Zajac and Levine, 1979). Briefly, the major motivation for this work has been twofold. Firstly we believe that it is important to study the entire limb, if not the entire body, in order to understand mammalian motor control. In other words, motor control is fundamentally a multi-input and multi-output nonlinear control problem. Secondly, it is relatively convenient to analyze multivariate nonlinear control problems when there is a well defined control task such as maximizing some performance measure. Incidentally, the maximal height jump is also experimentally convenient. It is comparatively easy to train both humans and cats to perform the task. Accurate measurements of ground reaction forces can be obtained. EMG's and kinematic data are also obtainable.

Other people have recognized the advantages of using optimal control to study the multivariable control aspects of mammalian movement. Hatze (Hatze, 1980) has a quite complete and rather critical survey of the previous results. The main thrust of Hatze's criticism is that most of the previous work utilized models of mammalian movement that were far too oversimplified. Hatze himself has tried to develop very realistic models of the human motor control system (Hatze, 1976). However, we believe the opposite criticism is relevant to his work. The mathematical model is so complex that it is difficult to obtain insight from the model.

Acknowledgement: This research was supported, in part, by NIH under grant NS 11762.

We have tried to steer a course between these two extremes. Our hope is to ultimately arrive at the simplest mathematical model that replicates the experimental data to within experimental accuracy. Thus, we have been gradually adding complexity to our model and both accuracy and detail to our experiments.

The main emphasis in this paper is on the modeling and analysis associated with the heel-off to lift-off phase of two human jumps. Our subjects were told to jump as high as possible. Thus, the performance measure is unambiguous. Our subjects were asked to begin their jump in a deep crouch (a squat). By insisting on these initial conditions we avoid the difficult job of modeling actively lengthening muscle. Our subjects were also asked to keep their hands above their heads and motionless. The main purpose of this instruction was to eliminate the need to consider arms and torso separately. Furthermore, our subjects were asked to begin their jumps from two different initial positions (a) flat feet on the ground (flat foot jump) or (b) balanced on their toes (toe jump). We did not allow our "toe jumpers" to touch the ground with their heels prior to lift-off.

This paper is organized as follows. The second section summarizes our earlier work and the rationale for the research reported here. This is followed by a description of the mathematical model and its solution. The fourth section contains a comparison of experimental and analytical results. The paper concludes with some indications of the next steps in our research.

2. REVIEW OF PREVIOUS RESULTS

We began by studying the problem of propelling a baton to a maximum height by applying a bounded torque and letting one end of the baton rest (push) against the ground (Roberts et al, 1979; Levine et al, 1983a). The main purpose of these studies was to provide insight into the analytical problems associated with more realistic models of jumping animals. We were able to obtain a complete analytical solution to this problem.

With the baton results in mind we asked human subjects to jump as high as possible using only their ankle muscles for propulsion while keeping the rest of their body rigid (Levine et al, 1983b; Zajac et al, 1983a). The main rationale for this experiment, as for the baton problem, was that it would be easier to analyze than a more normal jump. However, we also felt that the experiment was a useful test of some of our modeling assumptions.

We were again able to obtain a complete solution to the mathematical optimal control problem although, in this case, the solution was partly computational. The optimal control divided into three phases:

(1) Phase 1 begins with the subject standing still in some arbitrary position. He then moves by any of a large number of trajectories (many controls are optimal) to a state in which his center of mass is stationary over his toes.

(2) During phase 2 there is a unique optimal trajectory which corresponds to moving slowly while keeping the center of pressure under the tips of the toes. If the control were torque, the optimal control would be unique but not bang-bang. When we include a dynamical model for muscle and regard muscle activation as the control, the uniqueness of the optimal control depends on muscle time constants. In any case, this phase terminates at a unique state in which the heel is on the ground.

Our subjects appear to follow the mathematically optimal strategy although the experimental resolution is not adequate to verify the phase 2 result (Levine et al, 1983b). Based on these results, a natural hypothesis is that all maximum height jumps can be divided into three phases that are very similar to those described above. However, there are some technical difficulties associated with the computation of the optimal control in Phases 1 and 2. In contrast, the maximum principle and Hamilton-Jacobi-Bellman results apply during phase 3. Thus, computation of the optimal control in phase 3 is relatively straightforward. Hence, we were led to the experiment and analysis described in this paper.

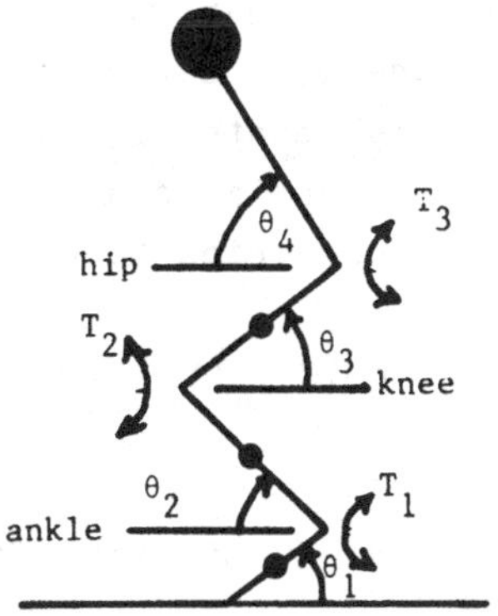

Figure 1 - Mechanical Model of Human

3. Mathematical Analysis

The mathematical model of these experiments can be divided into three parts, (a) the basic dynamics, (b) the performance criterion and (c) the mathematical model for muscle dynamics and activation. The basic dynamics are derived from the stick figure representation of the jumper shown in Figure 1. We believe that lossless pin joints are adequate as a model for the basic dynamics (see e.g. Hatze, 1980).

Thus, the basic dynamics are (see Figure 1 for definitions of $\underline{\theta}(t)$, $\underline{u}(t)$, (Cho, 1983).

$$\dot{\underline{\theta}}(t) = \underline{\omega}(t) \tag{1}$$

$$\dot{\underline{\omega}}(t) = \underline{A}^{-1}(\underline{\theta}(t))[\underline{B}(\underline{\theta}(t))\underline{\omega}^2(t) + \underline{C}(\underline{\theta}(t)) + \underline{D}\ \underline{T}(t)] \qquad (2)$$

where

$\underline{\theta}(t) = [\theta_1(t)\ \theta_2(t)\ \theta_3(t)\ \theta_4(t)]'$

$\underline{\omega}^2(t) = [\omega_1^2(t)\ \omega_2^2(t)\ \omega_3^2(t)\ \omega_4^2(t)]'$

$\underline{T}(t) = [T_1(t)\ T_2(t)\ T_3(t)]'$

$T_1(t)$ = ankle torque, $T_2(t)$ = knee torque, $T_3(t)$ = hip torque.

See (Cho, 1983) for the exact expressions for $\underline{A}(\theta(t))$, $\underline{B}(\theta(t))$, $\underline{C}(\theta(t))$, $\underline{D}$ as well as the numerical values we used.

The performance criterion is also straightforward. We (Levine et al, 1983a) showed that there is very little difference between the optimal control which maximizes the peak height reached by the center of mass of a baton and that which maximizes the peak height reached by the upper tip. Since more segments should result in less overall angular rotation about the center of mass we believe this similarity carries over to multi-segment models of maximum height jumping as well. Thus, the performance criterion becomes

$$J(\underline{\theta}, \underline{\omega}, \underline{T}, t_f) = y_c(t_f) + \frac{1}{2g}\dot{y}_c^2(t_f) \qquad (3)$$

where

t_f = the instant of lift-off

y_c = the vertical position of the center of mass

g = gravitational acceleration

It is possible to express $y_c(t)$ in terms of $\underline{\theta}(t)$. See (Cho, 1983) for details.

We remark that lift-off coincides with the instant at which the vertical ground reaction force, F_v, is less than zero.

$$F_v(\underline{\theta}, \underline{\omega}, \underline{u}) = mg + \sum_{i=1}^{4} d_i(\dot{\omega}_i\cos\theta_i - \omega_i^2\sin\theta_i) \qquad (4)$$

where

m = total mass and d_i are constants.

The final component of the mathematical model is a mathematical model for muscle dynamics. There exist many such models. We felt that our muscle models for the ankle, knee and hip should be simple, yet complex enough to capture the following important muscle properties

1. isometric, steady-state force during maximum activation depends on muscle length;
2. isometric force does not develop instantaneously upon muscle activation;
3. active shortening muscle develops less force than isometric muscle; and
4. isometric muscle force depends on the level of muscular activation.

To account for these basic muscle properties, a model for torque generation was used with dynamics

$$\dot{T} = -\alpha T + \alpha[1 - f(\tilde{\omega})]T_0(\tilde{\theta})u(t) \quad (5)$$

where T = joint torque

$\tilde{\theta}$ = joint angle (i.e., $\tilde{\theta}_1 = \theta_1+\theta_2; \tilde{\theta}_2=\theta_2+\theta_3; \tilde{\theta}_3 = \theta_3+\theta_4$)

$\tilde{\omega}$ = joint angular velocity (i.e., $\underline{\tilde{\omega}} = \underline{\dot{\tilde{\theta}}}$)

$u(t)$ = muscle activation

$\frac{1}{\alpha} \triangleq \tau$ = time constant for development of isometric torque.

Notice that $T_0(\tilde{\theta})$ corresponds to the isometric, steady-state torque-joint angle curve, since we let $f(0) = 0$. $T_0(\tilde{\theta})$ for the knee and hip were estimated from records obtained when our jumping subjects exerted maximum voluntary isometric contractions while strapped in a Cybex II instrument. For the ankle, $T_0(\tilde{\theta})$ was estimated from data reported by others (Fugl-Meyer et al, 1979, 1980, 1981; Sale et al, 1982). We estimated $f(\tilde{\omega})$ for the knee and hip from data obtained from these same subjects during maximum voluntary isokinetic (isovelocity) contractions. By measuring T and $\tilde{\theta}$ when peak torque is reached (i.e., when $\dot{T} = 0$) and by assuming $u(t)=1$ for maximum voluntary contractions, we used eq. (5) to calculate $f(\tilde{\omega})$ for each isovelocity studied. For the ankle, $f(\tilde{\omega})$ was similarly estimated from data reported by others (Fugl-Meyer et al, 1979, 1980, 1981). For all three joints, $f(\tilde{\omega})$ is monotone increasing from 0 at $\tilde{\omega} = 0$ to 1 at $\tilde{\omega}$ high speeds. (See Cho, 1983 for exact values of f and T_0).

Thus, the optimal control problem is to maximize the performance measure (Eq. (3)) subject to the dynamics (Eqs. (1), (2) and (5)), and terminal condition (Eq. (4)).

Note that the problem formulation above is incomplete in that we have not specified the initial conditions, that is, the state at heel-off which is the initial state for phase 3. This initial state was estimated from the experimental data. Since it is difficult to detect either the precise time at which heel-off occurs or the precise heel-off state these estimates include uncertainty. Our estimates are

Heel-Off State toe jump	Heel-Off State flat-foot jump
$\underline{\theta}' = (57 \quad 40 \quad 40 \quad 63)$ deg	$\underline{\theta}' = (34 \quad 56 \quad 39 \quad 53)$ deg
$\underline{\omega}' = (0 \quad 72 \quad 214 \quad 72)$ deg/sec	$\underline{\omega}' = (0 \quad 5 \quad 228 \quad 110)$ deg/sec
$\underline{T}' = (135 \quad 180 \quad 90)$ N-m	$\underline{T}' = (172 \quad 125 \quad 201)$ N-m

Once the initial conditions are specified the problem is completely formulated and, in principle, solveable. We found that the Differential Dynamic Programming algorithm (Jacobson and Mayne, 1970; Mayne & Polak, 1975; Polak and Mayne, 1975, 1977) solved this problem efficiently and reliably. See (Cho, 1983) for the details of our implementation. Our results are described in the following section.

4. Results

A natural, although somewhat superficial comparison between experimental and analytical results is to compare the stick figures of the jumper as is done in Figure 2 for the toe jump and in Figure 3 for the flat foot jump. Note that the upper sequence is the experimental result and the lower sequence is the analytical result based on our model. Notice that, although one can detect differences, the sequences are substantially the same. In part, this is because joint angles are not very sensitive to changes in the control.

Another area in which experiment and analysis agree quite closely is in the jump heights as can be seen in Table 1 below. Note that the experimental jumper out performs the model slightly.

Table 1

	Flat Foot Jump	Toe Jump
experimental total jump height	1.6±.05 m	1.5 ±.05 m
analytical total jump height	1.49 m	1.46 m
experimental jump height above ground	.26 m	.16 m
analytical jump height above ground	.28 m	.24 m
experimental horizontal distance jumped	0 m	0 m
analytical horizontal distance jumped	.26 m	.16 m

The fact that the model still needs improvement is apparent from Figures 4 and 5. The vertical force exerted on the ground is the best experimental measurement. Note that this also is the most sensitive comparison since the vertical force involves the highest mechanical derivative, acceleration.

The major cause of the discrepancies between experimental and analytical vertical force is the mathematical model for muscle. Although this is not very obvious from the results presented here Cho's (Cho, 1983) much more detailed results make it quite obvious. The fundamental difficulty is that the torque produced by our model muscles does not decrease nearly as rapidly with increasing joint angle and angular velocity as the experimental joint torques. There are several possible explanations for this. One such explanation is that various joint protection mechanisms become active at large joint angles and angular velocities. These have not been included in our model. Another possibility is that our model does not really capture some aspect of muscle dynamics. Deciding between these alternatives requires internal measurements of forces on tendons which are impossible with human subjects. Finally, the theoretical optimal controls are:

toe jump

$$u_1^*(t) = \begin{cases} 0 & 0 < t < 14 \text{ msecs} \\ 1 & 14 < t \text{ msecs} \end{cases}$$

$$u_2^*(t) = u_3^*(t) = 1 \qquad 0 < t$$

flat foot jump

$$u_1^*(t) = 1 \qquad 0 < t \text{ msecs}$$

$$u_2^*(t) = \begin{cases} 1 & 0 < t < 2 \text{ msecs} \\ 0 & 2 < t < 39 \text{ msecs} \\ 1 & t < 39 \text{ msecs} \end{cases} \qquad u_3^*(t) = \begin{cases} 1 & 0 < t < 43 \text{ msecs} \\ 0 & 43 < t74 \text{ msecs} \\ 1 & 74 < t \text{ msecs} \end{cases}$$

There are two differences between these results and our experimental data. Experimental muscle activations do not appear to switch as much as those for the analytical flat foot jump. Some extensor muscles appear to turn off just prior to lift-off in the experiments. However, small changes in the initial conditions for both jumps cause the analytically optimal controls to be fully on ($\underline{u}^*(t) = 1$) throughout the heel-off to lift-off phase. The experimentally observed switching off of some extensors just prior to lift-off is believed to be part of the joint protection mechanisms mentioned above. Thus, we believe that the experimentally observed muscle activation and the analytically optimal muscle activation agree fairly well.

5. SUGGESTIONS FOR FURTHER RESEARCH

At this time we believe that at least two improvements need to be made in the model. The most obvious requirement is to improve the mathematical model for muscles. We believe the best way to do this is to directly model the force production dynamics of each muscle group acting to extend the lower limb and torso. Then, the force to torque characteristics of each of these muscle groups is modeled from knowledge of the anatomy of muscle connections. We have begun to do this. A preliminary report can be found in Zajac et al (Zajac et al, 1983b). It should be noted that, with the improved muscle models, our new model is somewhat closer to Hatze's than the model presented in this paper.

It would also be very desireable to model and analyze less constrained jumps. If a human is simply told to jump as high as possible the jump is usually begun with a countermovement. That is, the subject starts by standing fairly erect and then rapidly contracts his body into a crouch, the countermovement. Then, without stopping he rapidly extends his body and leaves the ground. During the countermovement some muscles are both active and lengthening. There are many interesting questions associated with both the countermovement and actively lengthening muscle. We plan to improve our muscle model so as to include the active lengthening characteristics of muscle. We have already begun to study countermovement jumps. We also believe that there is an important unanswered question associated with the analysis. We would very much like to be able to analyze maximum height jumps from their static initial condition to lift-off rather than from heel-off to lift-off. The difficulty is that prior to heel-off the dynamics normally include a hard-limit. That is, the foot angle cannot decrease (because the foot rests on the ground) no matter what the control. We have shown (Levine et al, 1983a) that these hard limits cause serious analytical difficulties. For example, the optimal cost to go is discontinuous along an optimal trajectory and there is a region where the optimal control is not bang-bang even though the optimal control problem is monotone in the controls. Note that this is not a singular control situation because the costate is not well defined for these trajectories. We are currently working on techniques to handle hard-limits analytically and computationally.

REFERENCES

Cho, Y.M. (1983). The optimal control of multi-segment inverted pendula, Ph.D. Thesis, Dept. of EE, University of Maryland.

Fugl-Meyer, A.R., Sjostrom, M. annd Wahley, L (1979). Human plantar flexion strength and structure. Acta Physiol. Scand. 107:47-56.

Fugl-Meyer, A.R., Gustafsson, L. and Burstedt, Y. (1980). Isokinetic and static plantar flexion characteristics Eur. J. Appl. Physiol. 45: 221-234.

Fugl-Meyer, A.R. (1981). Maximum isokinetic ankle plantar and dorsal flexion torques in trained subjects. Eur. J. Appl. Physiol. 47:393-404.

Hatze, H. (1980). Neuromusculoskeletal control systems modelling -- a critical survey of recent developments, IEEE Trans. on AC, Vol. AC-25, No. 3, 375-384.

Hatze, H. (1976). A myocybernetic control model of skeletal muscle, CSIR Special Report WISK 220, National Research Institute for Mathematical Sciences, South Africa.

Jacobson, D.H. and D.W. Mayne (1970). "Differential Dynamic Programming," American Elsevier Pub. Comp., New York.

Levine, W.S., M. Christodoulou and F.E. Zajac, (1983a). On propelling a rod to a maximum vertical or horizontal distance. Automatica, 19, 321.

Levine, W.S., F.E. Zajac, M.R. Belzer and M.R. Zomlefer (1983b). Ankle controls that produce a maximal vertical jump when other joints are locked. IEEE Trans. Aut. Control, AC-28, (to appear in November).

Mayne, D.Q. and E. Polak (1975). "First-order Strong Variation Algorithms for Optimal Control," J. of Opt. Theory and Applications, Vol. 16.

Polak, E. and D.Q. Mayne (1975). "First-order Strong Variation Algorithms for Optimal Control Problems with Terminal Inequality Constraints," J. of Opt. Theory and Applications, Vol. 16.

Polak, E. and D.Q. Mayne (1977). "A Feasible Directions Algorithm for Optimal Control Problems with Control and Terminal Inequality Constraints," IEEE Trans. Auto. Cont., Vol. AC-22, No. 5.

Roberts, W.M., W.S. Levine and F.E. Zajac, III (1979). Propelling a torque controlled baton to a maximum height. IEEE Trans. Aut. Control, AC-24, 779.

Sale, D., J. Quinlan, E. Marsh, A.J. McComos and E.Y. Belanger (1982). Influence of joint position on ankle plantar flexion in humans. J. Appl. Physical: Reprint. Environ. Exercise Physiol. 52:1636-1642.

Zajac, F.E. and W.S. Levine (1979). Novel experimental and theoretical approaches to study the neural control of locomotion and jumping. In R.E. Talbott and D.R. Humphrey (Eds.), Posture and Movement. Raven Press, New York, pp. 259-279.

Zajac, F.E., R. Wicke and W.S. Levine (submitted 1983a). Dependence of jumping performance on muscle properties when humans use only calf muscles for propulsion. Biomech.

Zajac, F.E., Levine, W.S., Chapelier, J.P., and Zomlefer, M.R. (1983b). Neuromuscular and musculoskeletal control models for the human leg. Proc. Am. Cont. Conf. 1:229-234.

Zajac, F.E., Levine, W.S., Chapelier, J.P., and Zomlefer, M.R. (1983b). Neuromuscular and musculoskeletal control models for the human leg. Proc. Am. Cont. Conf. 1:229-234.

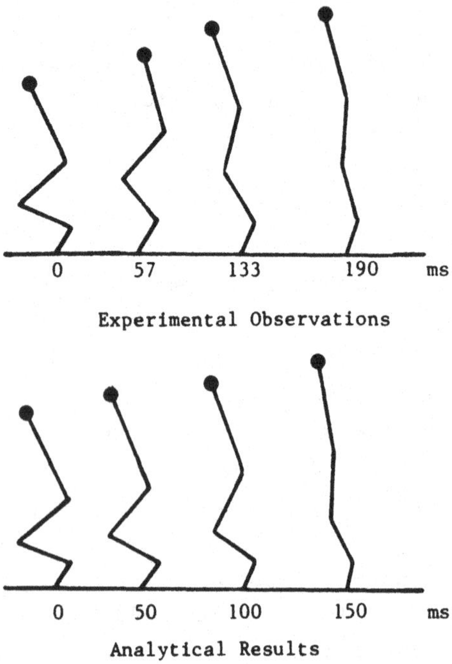

Figure 2 Comparisons for Toe Jump

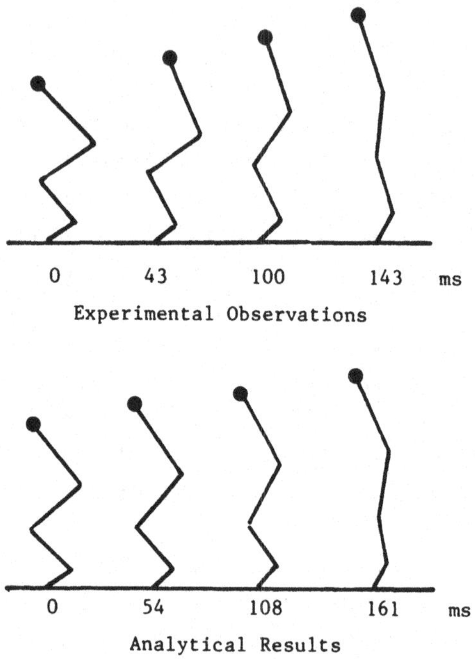

Figure 3 Comparisons for Flat Foot Jump

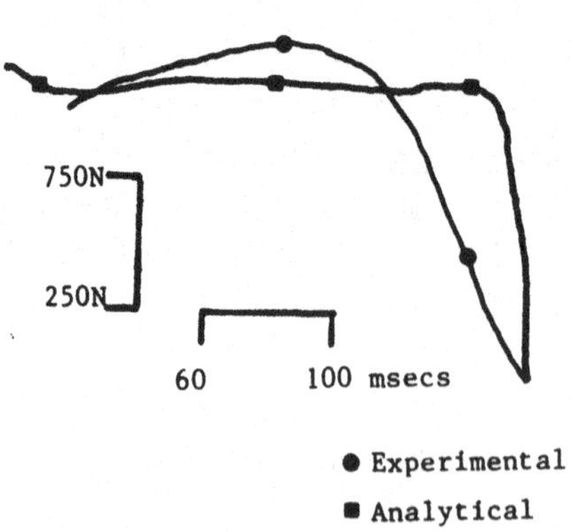

Figure 4 Trajectories of $F_v(t)$ Toe Jump

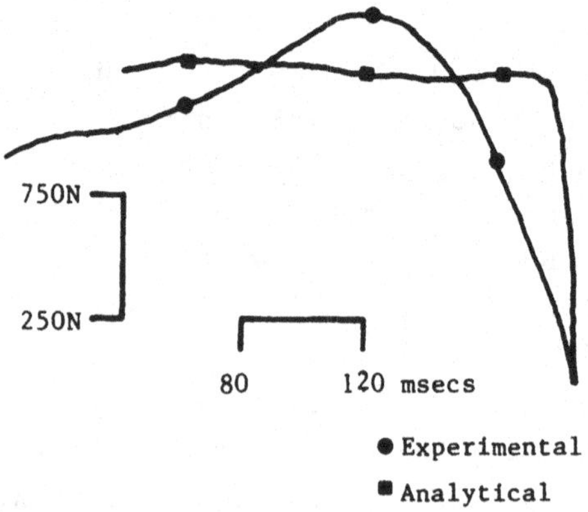

Figure 5 Trajectories of $F_v(t)$ Flat Foot Jump

THEORETICAL ANALYSIS OF THE SLIDING FILAMENT MODEL FOR THE EVALUATION OF MUSCLE MACROSCOPIC PERFORMANCE

G. Avanzolini
Dipartimento di Sistemi e Informatica
University of Florence
50139 Firenze, ITALY

A. Cappello
Dipartimento di Elettronica, Informatica e Sistemistica
University of Bologna
40136 Bologna, ITALY

1. INTRODUCTION.

The sliding filament theory has been recognized to be of great significance for the purpose of interpreting macromolecular contraction mechanisms. The starting point of this approach was the A.F.Huxley's mathematical formulation (1957,[1]) that unified the then existing knowledge about muscle structure, biochemistry, mechanics and energetics. Afterwards, a series of increasingly sophisticated kinetic models have appeared which attempt to predict the mechanical behaviour of the muscle on the basis of chemical actin-myosin interactions [2,3].

With respect to the classic two-element model introduced by A.V.Hill (1938,[4]), these models have the advantage of being based directly on current physiological knowledge about mechanisms of muscle contraction, and they are able to reproduce various aspects of muscle behaviour realistically. Unfortunately, these models are expressed by partial differential equations on the cross-bridge distribution functions and consequently they require a large computational effort to simulate even simple experiments. Such elaborate computations would be prohibitive for studies of ventricular mechanics, limb dynamics and related areas where length-dependent behaviour, time-varying stimulation, as well as passive viscoelastic properties, play a role. The efficient analysis of these more complex situations needs an approximated (and lumped-parameter) model which retains some of the realistic behaviour of the kinetic models and their intimate connection with the underlying microscopic phenomena while eliminating much of the mathematical complexity.

In this direction, a mathematical approach, termed the distribution-moment approximation (Zahalak 1981,[5]), has been suggested, where the exact cross-bridge distribution functions are approximated by "a priori" postulated functional forms. This approach, leading to a set of first-order ordinary differential equations on the low-order moments of the approximating distributions, gives results in good agreement with the original Huxley's model if the stimulation is kept at a constant level for a sufficiently long time.

The different approach followed in this paper allows to overcome some of the limits concerning the above-mentioned method; in particular, the need of a subjective evaluation of the approximate distribution functions is reduced substantially. In fact, the proposed procedure starts from the analytical form of the cross-bridge distribution function as obtained by the characteristics method.

Afterwards, a local linearization of the characteristic lines leads to a simple expression for the distribution function, governed by low-order ordinary differential equations. Taking advantage of the knowledge acquired about the form of the cross-bridge distribution function, it is possible to obtain a formulation approximating the two-state model suitably not only in the constant stimulation case but also when the sti-

mulation is time-varying. In the first case, the contractile element is described by a model of the first order as the two-element Hill's representation is, but exhibits a more realistic behaviour[5]. Furthermore, if the contractile velocity too is imposed at a constant level, the approximated model and the exact one assume the same formulation since the characteristics are rectilinear.
This procedure of approximation, differently from Zahalak's one, is not yet directly extensible to sophisticated multi-state kinetic models[6] since it requires that one knows the analytical solution of the partial differential equations and further elaborations are necessary to achieve this aim. Nevertheless, these models require too much informations which are, for the most part, not directly measurable by experiments. Consequently, for the moment, they do not seem to present a great interest from a macroscopic point of view.
In this paper, the approximation procedure is developed in detail for the shortening case only, but it may be extended to study the lengthening phase in a straightforward manner. The resulting approximated model is simple enough to be helpful in studies of ventricular dynamics, even if further elaborations are required to account for length dependence of the activation and rate parameters, and it retains an intimate connection with the underlying microscopic contraction mechanisms.

2. MODEL OF THE SARCOMERE

2.1 Model formulation

The sarcomeres, the subcellular contractile units of muscle, form a repeating linear array along the fibre.
Each sarcomere is in turn composed of an interdigitating array of thick (myosin) and thin (actin) filaments (Fig.1). The thick filament projects a sort of protrusion (called cross-bridge) at regular intervals. Activation of muscle makes the cross-bridge attachment to a site on the thin filament possible. The attached cross-bridge produces a longitudinal elastic force which may be towards or against contraction (Fig.2). The total force given by all the attached cross-bridges tends to produce a sliding motion of one set of filaments relative to the other. According to this mechanism the sarcomere, and so the muscle fibre, exerts a force and/or varies its length in response to activation.

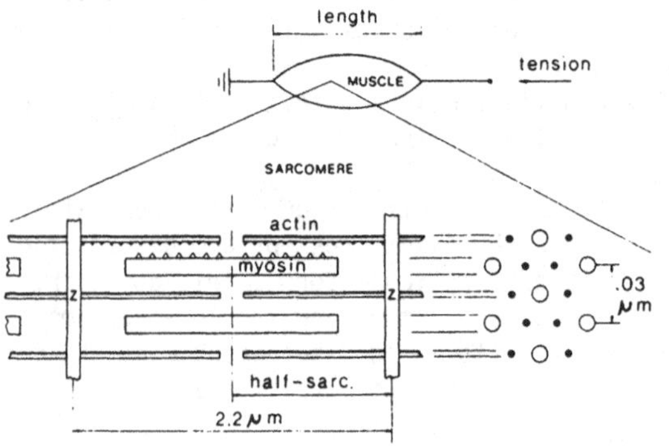

Fig.1. Structure of the sarcomere.

Some definitions are introduced usually in order to represent the actin-myosin interaction in terms of a model:

x = distance (with sign) of a generical site A on the actin filament from a generical site M (cross-bridge) on the myosin filament.

v(t) = dx/dt = sliding velocity or, equivalently, half-sarcomere lengthening velocity.

n(x,t) = probability distribution function that a cross-bridge M is attached at time t to a site A with distance between x and x+dx.

f(x,t) = probability per unit time for cross-bridge attachment.

g(x) = probability per unit time for cross-bridge detachment.

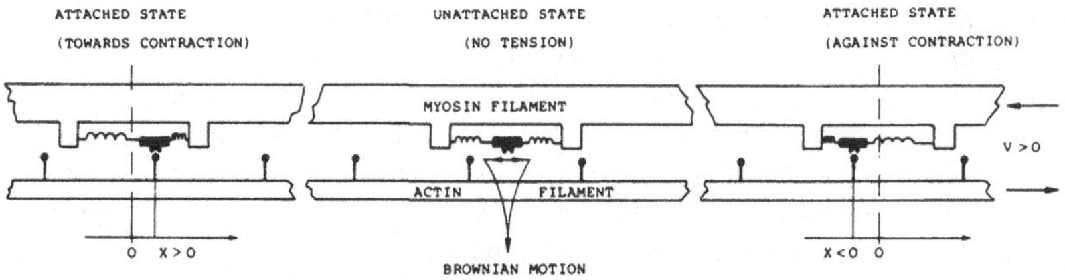

Fig.2. Actin-myosin interaction mechanism.

Extension of the original Huxley's formulation to the study of non-steady state muscle behaviour has been obtained here by introducing explicit time dependence in the attachment probability[2].

The kinetics of cross-bridge attachment and detachment is schematically represented by a two-state diagram (Fig.3).

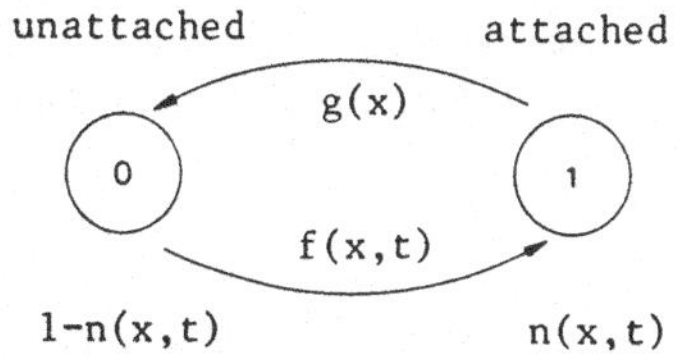

Fig.3. Two-state diagram.

Under classical assumptions[7], the actin-myosin bonding reaction is described by the partial differential equation:

$$\frac{D}{Dt}\, n(x,t)=[1-n(x,t)]\, f(x,t) - n(x,t)g(x), \quad (1)$$

where $D/Dt = \partial/\partial t + v(t)\partial/\partial x$ is the time derivative moving with the sliding filaments at velocity $v(t)$.

2.2 Model solution

Let us briefly present now an outline of the analytical procedure followed for the integration of (1) from an initial cross-bridge distribution function $n(x,0)$.

The hyperbolic differential equation transforms into an ordinary differential equation along this family of characteristic curves:

$$C:\ x(t,x_o) = x_o+\int_0^t v(\tau)d\tau = x_o+ \ell_c(t) \quad (2)$$

where the parameter x_o is the initial value of x and ℓ_c is the half-sarcomere lengthening or, equivalently, the total relative displacement of filaments.

Integration of (1) between a generical time $t_i(x_o)$ and t gives, after easy manipulations:

$$\begin{aligned} n(x,t) &= \Phi(x_o,t,t_i)n(x_o+\ell_c(t_i),t_i)+\int_{t_i}^{t} f(x_o+\ell_c(\tau),\tau)\Phi(x_o,t,\tau)d\tau = \\ &= \Phi(x_o,t,t_i)n(x_o+\ell_c(t_i),t_i)+\int_{t_i}^{t} a(x_o+\ell_c(\tau),\tau)(d\Phi/d\tau)d\tau \quad , \end{aligned} \quad (3)$$

where the transition function

$$\Phi(x_o,t,t_i)=\exp\left\{-\int_{t_i}^{t} [f(x_o+\ell_c(\xi),\xi)+g(x_o+\ell_c(\xi))]\ d\xi\right\}$$

is the solution of the homogeneous equation relative to (1) with the condition $\Phi(x_o,t_i,t_i)=1$, and where $a=f/(f+g)$, $x_o=x-\ell_c(t)$.
In order to study the properties of the solution found, it is convenient a choice of the rate functions which satisfies both criteria of reasonableness and mathematical simplicity. To this purpose we assume for $f(x,t)$ and $g(x)$ the particular forms suggested by Julian[2] and shown in Fig.4a. In this figure, x has been normalized to the maximum distance h for cross-bridge attachment. So, in the following, all the microscopic lengths and velocities will be assumed as h-normalized.
The analysis is limited here to the case of muscle shortening ($v(t)\leq 0$, $\forall t\geq 0$ and then $\ell_c(t)\leq 0$), but the study of the lengthening phase is quite similar.

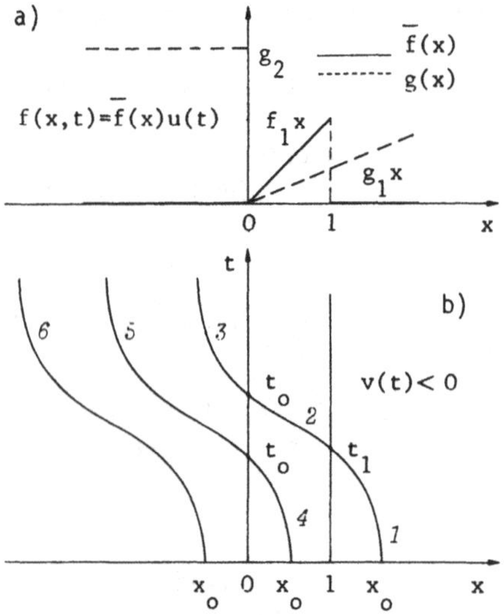

Fig.4. a) Rate functions
b) Family of characteristics

Let us denote (Fig.4b) by $t_o(x_o)$ and $t_1(x_o)$ the times at which the generic characteristic line (2) crosses the axes x=0 and x=1 respectively. The two points of discontinuity of the rate functions lead to subdivide the family of characteristics into six distinguished subfamilies (Fig.4b).
Starting from a steady-state initial condition $n(x,0)=n_o$ for $0\leq x\leq 1$ ($n(x,0)=0$ elsewhere) and observing that

$$a(x_o+\ell_c(\tau),\tau)=a(\tau)=f_1u(\tau)/\left[f_1u(\tau)+g_1\right] \quad ,$$

$$\Phi(x_o,t,\tau)=-\int_\tau^t [g_1+f_1u(\xi)]\,[x_o+\ell_c(\xi)]\,d\xi \quad ,$$

$$t_o(x_o)=\ell_c^{-1}(-x_o) \quad ,$$

$$t_1(x_o)=\ell_c^{-1}(1-x_o) \quad ,$$

$$x_o=x-\ell_c(t) \quad ,$$

the expressions of $n(x,t)$ are given by

$$
\begin{aligned}
&1)\ n(x,t)=0 && 1\leq x\\
&2)\ n(x,t)=\int_{t_1}^{t} a(\tau)d\,e^{\Phi(x_o,t,\tau)} && \max\left[0,1+\ell_c(t)\right]\leq x\leq 1\\
&3)\ n(x,t)=e^{-g_2(t-t_o)}\int_{t_1}^{t_o} a(\tau)d\,e^{\Phi(x_o,t_o,\tau)} && 1+\ell_c(t)\leq x\leq 0\\
&4)\ n(x,t)=n_o e^{(x_o,t,0)}+\int_0^t a(\tau)d\,e^{\Phi(x_o,t,\tau)} && 0\leq x\leq 1+\ell_c(t)\\
&5)\ n(x,t)=e^{-g_2(t-t_o)}\left[n_o e^{\Phi(x_o,t,0)}+\int_0^{t_o} a(\tau)d\,e^{\Phi(x_o,t_o,\tau)}\right] && \ell_c(t)\leq x\leq \min[0,\ 1+\ell_c(t)]\\
&6)\ n(x,t)=0 && x\leq \ell_c(t)
\end{aligned}
\qquad (4)
$$

The expressions (4) correspond to an infinite order model.

2.3 Model approximation

The main purpose of this section is to obtain, starting from eqs.(4), two different low-order approximated models characterized by significative parameters. To this end, application of the mean value theorem to every integral term gives:

$$\int_{t_{inf}}^{t_{sup}} a(\tau) d\, e^{\Phi(x_o,t_{sup},\tau)} = a(\eta)\left[1- e^{\Phi(x_o,t_{sup},t_{inf})}\right] \qquad \text{and}$$

$$\Phi(x_o,t_{sup},t_{inf}) = -\int_{t_{inf}}^{t_{sup}} [g_1+f_1u(\xi)]\,[x_o+\ell_c(\xi)]\,d\xi =$$

$$= -\,[g_1+f_1u(\psi)]\,[x-\ell_c(t)+\bar{\ell}_c(t_{sup},t_{inf})]\;(t_{sup}-t_{inf})$$

where $\bar{\ell}_c(t_{sup},t_{inf}) = \frac{1}{t_{sup}-t_{inf}}\int_{t_{inf}}^{t_{sup}} \ell_c(\tau)\, d\tau$.

η e ψ depend on x and t generally and satisfy the inequalities $t_{inf} \le \eta \le t_{sup}$, $t_{inf} \le \psi \le t_{sup}$.

A first low-order approximated model descends immediately from the following three assumptions:

i) $$\bar{\ell}_c(t_{sup},t_{inf}) \cong \frac{\ell_c(t_{sup})+\ell_c(t_{inf})}{2} \quad . \tag{5}$$

ii) Local linearization of characteristic lines. From (2) :

$$\ell_c(\xi) \cong \ell_c(t)+v(t)(\xi-t) \quad . \tag{6}$$

Consequently, from definition of $t_o(x_o)$ and $t_1(x_o)$:

$$t_o \cong t-\frac{x}{v} \quad ; \qquad t_1 \cong t+\frac{1-x}{v} \quad . \tag{7}$$

iii) $$a(\eta) \cong \bar{a}(t_{sup},t_{inf}) = \frac{1}{t_{sup}-t_{inf}}\int_{t_{inf}}^{t_{sup}} a(\tau)\, d\tau \quad ,$$

$$u(\psi) \cong \bar{u}(t_{sup},t_{inf}) = \frac{1}{t_{sup}-t_{inf}}\int_{t_{inf}}^{t_{sup}} u(\tau)\, d\tau \quad , \tag{8}$$

where t_{sup},t_{inf} are evaluated from a zero-order approximation of t_o, t_1 with respect to x:

$$t_o \cong t_1 \cong t \quad .$$

Obviously, when the contractile velocity v is constant,the relationships (5),(6),(7) are satisfied exactly. The same consideration applies, when the activation u is con-

stant, to the relationship (8).
In the general case, the resulting approximated expressions of the cross-bridge distribution function are given by:

$$
\begin{array}{lll}
1) & n(x,t)=0 & 1\le x \\
2) & n(x,t)=a[1-e^{(g_1+f_1u)(1-x^2)/2v}] & \max(0,1+\ell_c)\le x<1 \\
3) & n(x,t)=a[1-e^{(g_1+f_1u)/2v}]e^{-g_2x/v} & 1+\ell_c\le x<0 \\
4) & n(x,t)=\bar{a}+(n_o-\bar{a})\,e^{(g_1+f_1\bar{u})(\ell_c-2x)t/2} & 0\le x<1+\ell_c \\
5) & n(x,t)=[\bar{a}+(n_o-\bar{a})\,e^{(g_1+f_1\bar{u})(x-\ell_c)(x-vt)/2v}]\,e^{-g_2x/v} & \ell_c\le x<\min(0,1+\ell_c) \\
6) & n(x,t)=0 & x<\ell_c
\end{array}
\tag{9}
$$

where

$$
a(t)=f_1u(t)/[g_1+f_1u(t)]\ ;\ \bar{u}(t)=\frac{1}{t}\int_0^t u(\tau)\,d\tau;\ \bar{a}(t)=\frac{1}{t}\int_0^t a(\tau)\,d\tau\ ;\ \ell_c(t)=\int_0^t v(\tau)\,d\tau\ .
$$

They correspond to a third-order, non-linear, non-stationary model which reduces to th rigorous solution if both the contractile velocity and the activation are constant.
In the following we will refer to expressions (9) as Model 1.

A second roughly approximated model may be obtained from Model 1 introducing the further hypothesis

$$
\ell_c \to -\infty \quad \text{(in practice } \ell_c \ll -1)\ .
$$

Observing that, in this case, the domains of the subfamilies 4), 5), 6) degenerate, Model 2 is given by:

$$
\begin{array}{lll}
1) & n(x,t)=0 & 1\le x \\
2) & n(x,t)=a\,[1-e^{(g_1+f_1u)(1-x^2)/2v}] & 0\le x\le 1 \\
3) & n(x,t)=a\,[1-e^{(g_1+f_1u)/2v}]\,e^{-g_2x/v} & x\le 0
\end{array}
\tag{10}
$$

These expressions may be found directly from the original differential equation assuming $\partial n(x,t)/\partial t=0$ and correspond to a zero-order model.
Obviously this approximation is satisfactory for large times only. We can observe also that the initial condition plays no role on the solution. The resulting effect depends as we will see after, both on the activation and the loading condition.

3. MACROSCOPIC MODEL OF THE MUSCLE FIBRE

This model relates the sarcomere contraction processes to muscle macroscopic performance.
Assuming that the force-displacement relationship of a cross-bridge is linear with a spring constant k, the developed tension, stiffness and total rate of energy liberatio are defined [1] by low-order moments of n(x,t):

$$\text{Tension} = T(t) = M\int_{-\infty}^{+\infty} kx\; n(x,t)dx \quad ,$$
$$\text{Stiffness} = S(t) = M\int_{-\infty}^{+\infty} n(x,t)dx \quad , \qquad (11)$$
$$\text{Total rate of energy liberation} = E(t) = \varepsilon\int_{-\infty}^{+\infty} g(x)\; n(x,t)dx \quad ,$$

where M and ε are appropriate scale factors.

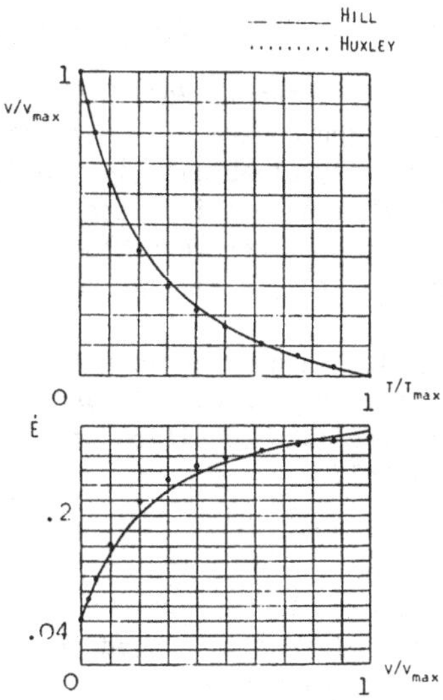

Fig.5. Steady-state relations.

It is interesting to evaluate the steady-state values of these moments when activation and velocity are constant. In this situation, the rigorous Huxley model, Model 1 and Model 2 give the same results, which are consistent with the macroscopic data obtained by A.V.Hill previously [1,4] (Fig.5).

Another equation

$$\varphi(T,\ell_c,v) = 0 \qquad (12)$$

is needed before the model formulation is mathematically complete. This relation represents the mechanical constraints prescribed by the loading condition. In the next section we shall analyse two meaningful loading conditions.

4. RESULTS

4.1 Isovelocity shortening

In this experiment, the muscle is fully activated (u=1) while its length is mantained constant (isometric contraction). When steady-state is reached (T=Mka/2), muscle is allowed to shorten at constant velocity $v=v_0 \leq 0$. The constant velocity case is very attractive from a mathematical point of view since the characteristic curves become straight lines. In this case ($u=1, v=v_0$), the rigorous model and the approximated Model 1 are equivalent. Then, introducing

$$u=1;\; \bar{a} = a = n_0 = f_1/(f_1+g_1);\; v=v_0;\; \ell_c=v_0 t \;,$$

eqs. (9) transform into

$$\begin{array}{lll}
1) & n(x,t)= 0 & 1\leq x \\
2) & n(x,t)= a\,[1- e^{-f_1(x^2-1)/2av_0}] & \max(0,1+v_0 t)\leq x\leq 1 \\
3) & n(x,t)= a\,[1- e^{f_1/2av_0}]\, e^{-g_2 x/v_0} & 1+v_0 t\leq x\leq 0 \\
4) & n(x,t)= a & 0\leq x\leq 1+v_0 t \\
5) & n(x,t)= a\, e^{-g_2 x/v_0} & v_0 t\leq x\leq \min(0,1+v_0 t) \\
6) & n(x,t)= 0 & x\leq v_0 t
\end{array} \qquad (13)$$

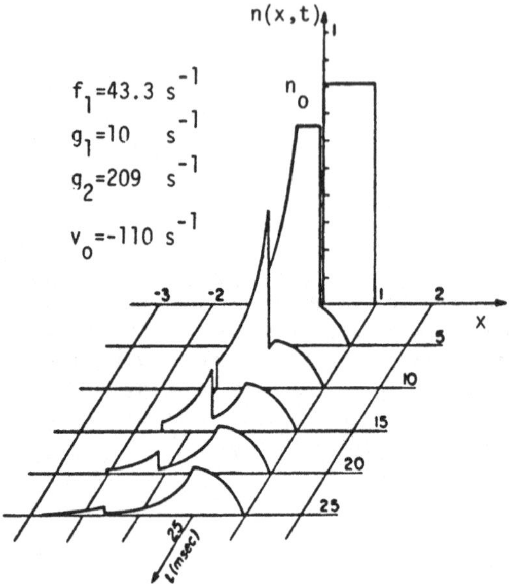

Fig.6. Time-evolution of n(x,t) [5].

In this case, the solution depends on time through the subfamily definition domains only.

The time evolution of n(x,t) is shown in Fig.6 [5], where the propagation of the initial data discontinuity is due to the hyperbolic character of the equation (1). Eqs. (13) are easily integrated to calculate the contractile tension for different values of the shortening velocity. The normalized contractile tension is plotted in Fig.7 and four shortening velocity values are considered (solid line).

Dashed lines correspond to Model 2. In the last model, the contractile tension depends on velocity algebraically. Then, it is obvious that a velocity step causes a tension step. Therefore Model 2 is not suited to study a similar condition.

These results are in good agreement with the curves obtained by Wood and Mann [8] who simulated the same condition on a multi-state model.

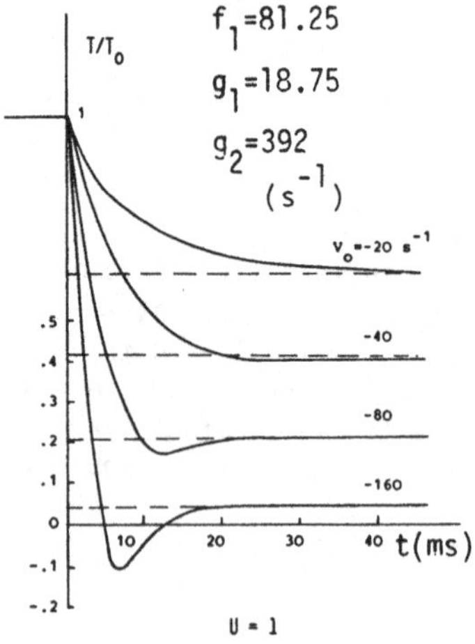

Fig.7. Contractile tension during isovelocity shortening (solid line:exact-Model 1, dashed line:Model 2)

4.2 Isometric contraction

In this case, the contractile element is placed in series to a viscoelastic element which simulates the effect of both distributed elasticity along the fibre and concentrated at the terminal ends.

The total length of the fibre is kept constant and so the loading condition (12) may be written as:

$$T = \alpha_s \left(e^{-\beta_s \ell_c} - 1 \right) - \mu_s v \tag{14}$$

assuming an exponential tension-length relationship and a linear damping for the series element.

The normalized activation function u is assumed to be a biexponential curve [9]:

$$u(t) = K_1 \left[e^{-K_2 t^2} - e^{-K_3 t^2} \right] \tag{15}$$

representing the course of the Ca^{++} concentration in the sarcomere.

This is a typical physiologic condition for the cardiac muscle during the isovolumic contraction.

The time courses of the contractile tension and velocity obtained integrating eqs.(1), (14) by a finite difference method from zero initial conditions are shown in Fig.8 (arbitrary units).

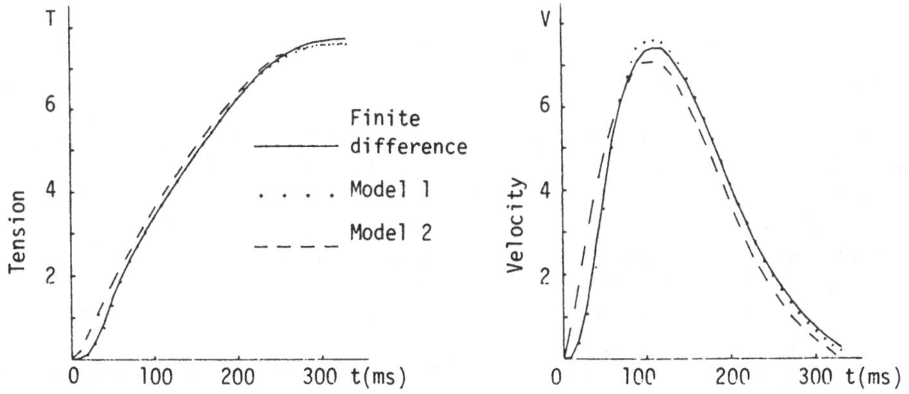

Fig.8. Time-courses of the contractile tension and velocity in isometric conditions (arbitrary units).

On the same figure, and for comparison, the results of the two approximated models are presented.
The agreement of Model 1 to the numerical solution is quite satisfactory during the whole contraction phase.
Model 2, as it would been expected, gives good results only after a starting phase in which both tension and velocity are overstimated. Nevertheless, the behaviour of Model 2 is acceptable, expecially if compared with the previous analysed condition.

5. CONCLUSIONS

Two approximated models with significative parameters have been obtained from the exact solution of Huxley's equation.
Model 1, of the third-order, provides a satisfactory agreement with the rigorous solution even with time-dependent activation. Moreover, it reduces to the exact solution if both the contractile velocity and the activation are constant.
Model 2, of the zero-order, is coincident with Model 1 and the exact solution only in stationary conditions. Nevertheless, it may be preferred, for its simplicity, when activation and loading conditions present only gradual variations, as in the case of cardiac muscle.
These low-order models can be used with advantage in studies of ventricular mechanics, limb dynamics and related areas where Huxley's original formulation appears too complex since many other complicating factors are present.
Additional applications of these approximated models include the analysis of the influence of the muscle parameters on the dynamics of contraction, a very important step before dealing with parametric identification problems.

ACKNOWLEDGEMENTS

This work was supported by the Italian Ministry of Education (M.P.I.).

REFERENCES

[1] A.F.Huxley, Muscle structure and theories of contraction, *Prog. Biophys. Biophys. Chem.* 7, 255-318 (1957).

[2] F.J.Julian, Activation in a skeletal muscle contraction model with a modification for insect fibrillar muscle, *Biophys.J.* 9, 547-570 (1969).

[3] T.L.Hill, E.Eisenberg, Y-D Chen and R.J.Podolsky, Some self-consistent two-state sliding filament models of muscle contraction, *Biophys.J.* 15, 335-372 (1975).

[4] A.V.Hill, The heat of shortening and the dynamic constants of muscle, *Proc. Roy. Soc. Ser. B.* 126, 136-195 (1938).

[5] G.I.Zahalak, A distribution-moment approximation for kinetic theories of muscular contraction, *Math. Biosc.* 55, 89-114 (1981).

[6] E.Eisenberg, T.L.Hill and Y-D Chen, Cross-bridge model of muscle contraction, *Biophys.J.* 20, 195-227 (1980).

[7] T.S.Feit, Numerical study of the behavior of an activation parameter in a cardiac muscle model, *Math. Biosc.* 34, 251-265 (1977).

[8] J.E.Wood and R.W.Mann, A sliding-filament cross-bridge ensemble model of muscle contraction for mechanical transients, *Math. Biosc.* 57, 211-263 (1981).

[9] A.Y.K.Wong, Application of Huxley's sliding-filament theory to the mechanics of normal and hypertrophied cardiac muscle, in Cardiac Mechanics, I.Mirsky, D.N.Ghista and H.Sandler Editors, J.Wiley, New York (1974).

OR IS WHAT OR DOES

D A Conway
Centre for OR Studies
The Hatfield Polytechnic
Hatfield, England

1. INTRODUCTION

Virtually ever since the inception of Operational Research there has been an intermittent but continuing discussion about the question "What is OR?." This has involved such well-known names as Blackett [1], Waddington [2] and Ackoff [3, 4] as well as many lesser-known contributors. This has continued to be very evident in the United Kingdom OR community, involving at times major questions as to the current status of the subject (e.g. Dando and Bennett [5]). In particular, there has often seemed to be a major difference of opinion in Britain between practitioners and academics (e.g. Hirschheim [6]). This paper considers the nature of this difference and casts some new light on the problem based on a survey, carried out by the author, of a large sample of OR groups in Britain.

In seeking to answer the question, "What is OR?", several sources may be considered. The most obvious source is any formally adopted definition. Such a definition is printed every month in the Journal of the Operational Research Society. In essence it states "Operational Research is the application of ... science to ... problems arising in industry, business, government and defence. The distinctive approach is ... the scientific model ... " In this compressed form this definition is very similar to that adopted by the Operations Research Society of America. Two essential concepts emerge from these definitions,

a) OR is scientific, i.e. uses the methods of science,

b) OR investigations normally involve the modelling of the system under review.

This concept of OR will be called the "problem-solving" concept.

An alternative source for a definition of OR is to look at the principal journals and textbooks in the subject. Although this paper does not represent a thorough review of such sources, it is widely accepted that any such review would result in

OR being described as a collection of mathematical techniques, such as Mathematical Programming, Inventory Control Theory, Queueing Theory, etc., which may be applied to certain types of problems. It is also widely accepted that this is the traditional "academic" picture of OR obtained by most students from courses in higher education. In this picture, the concept of science has been narrowed down to Mathematics, and the idea of the model has been replaced (via the taxonomy developed by Ackoff [7]) by the concept of a set of techniques which is dominant in current textbooks. This concept of OR will be called the "academic" concept.

2. THE SURVEY

This problem of defining a discipline or subject is by no means unusual. One way out of the problem that is often adopted in such debates is to say, mutatis mutandis, "Physics is what physicists do." In order to adopt this approach for OR it is necessary to answer the question, "What does OR do in practice?" It was partly to answer this question that a survey was carried out of practising, in-house, OR groups in the United Kingdom in 1980. The results of the survey cover 27 groups and (with the exception of one very large group) all the projects that were current at the time of the visit or had been completed in the previous six months - 785 projects in all. The survey was designed to give a good representation of British OR and the distribution of OR groups interviewed by type of organisation, by industry and by size of group can be seen in Tables 1 - 3 below.

Public Sector			Private Sector	
Local Government	Central Government	Nationalised Industries	Large Companies	Medium Companies
3	4	9	9	2
Administration		Industry and Commerce		

Table 1. Coverage by Type of Organisation

Central Govt.	Local Govt.	Steel	Manu-facturing	Food	Energy	Finance	Transport
4	3	2	4	3	5	5	1

Table 2. Coverage by Industry

1-5	6-10	11-15	16-20	21-25	26-30	31-35	>35
6	7	6	1	2	-	2	3

Table 3. Coverage by Size of Group

Another factor to be considered is the organisational maturity of the Group within its parent organisation. Using this concept, developed by Pettigrew [8] and extended by Conway [9], the distribution of groups is shown in Table 4 below.

Pioneering	Maladaptive Response	Adaptive Response	Specialisation	Demise by Default	Planned Absorption	Consoli-dation and Renewal
1	1	6	1	2	2	4

Table 4. Coverage by Organisational Maturity of Group

It can therefore be seen that the survey covers a wide range of in-house OR groups of varying types and in a wide range of industries and types of organisations.

3. THE RESULTS

Whilst a wide range of information was obtained for each Group and project, some information is more directly relevant to the question "What is OR?" than others. One question asked about the techniques used in each project. If OR in practice is based on the set of techniques contained within the "academic" concept then the survey should show standard techniques such as Linear Programming, Optimisation, Queueing, etc. as the dominant ones. The survey results on the usage of techniques are shown in Table 5 below.

It can be seen that the traditional set of techniques of the "academic" concept account for only 15.2%. Even if the usage of the other traditional disciplines of Mathematics and Statistics is added to that figure, it still only rises to 32.8%. On the other hand, the newer, and strongly problem oriented techniques account for 53.1%. This strongly implies that, in practice, "in-house" OR in the United Kingdom is strongly problem oriented and conforms closely to the "problem solving" concept outlined above.

		Number	%
Disciplines:			
Mathematics		33	3.7
Statistics		123	13.9
	Sub Total	156	17.6
Traditional Techniques:			
Linear Programming and Optimisation		25	2.8
Queueing		7	0.8
Inventory Control		31	3.5
Replacement		7	0.8
Game Theory		3	0.3
Forecasting		43	4.8
Scheduling & Project Network Analysis		20	2.2
	Sub Total	136	15.2
Newer Techniques:			
Decision Support Systems & Computing		123	13.8
Financial Planning		16	1.8
Simulation		76	8.5
Special Models and Heuristics		259	29.0
	Sub Total	474	53.1
Miscellaneous:			
None		46	5.2
Others		78	8.5
	TOTAL	890	100.0

Table 5. Usage of Techniques

This is reinforced by a closer examination of the figures, which throws up several interesting facts. Firstly, by far the largest individual group is Special Models and Heuristics, most of which have been separately developed in relation to the individual problem facing the organisation. Secondly, no techniques were used in more projects than were any of the traditional set of techniques. Thirdly, within the traditional set of techniques, the two which are most used are those where the technique is still very closely problem oriented, viz. Inventory Control and Forecasting.

One factor which might affect the results is the proportion of failed projects. This, however, can have produced very little bias as only 10.2% of the projects were not successfully completed or currently active. Further, as can be seen from Table 6 below 5.7% were temporarily stopped, usually because of decisions about the priorities to be applied to the use of scarce OR resources. This leaves only 4.5% which were unsuccessful.

		Number	%
Temporarily Inactive:			
Negotiating terms of Reference		11	1.4
Stopped due to OR priorities		27	3.4
Stopped due to management priorities		5	0.6
Others		2	0.3
	Totals	45	5.7
Abandoned as Unsuccessful:			
Due to			
Technical Failure		6	0.7
Loss of Management Interest		18	2.3
Rejection of Recommendations		7	0.9
External Factors		2	0.3
Other Reasons		2	0.3
	Totals	35	4.5

Table 6. Non-active, Non-successful Projects

4. CONCLUSIONS

Two alternative definitions for OR have been considered; a set of techniques and a problem solving, modelling activity. These have been set against the results of a survey of "in-house" OR Groups in Great Britain, which have shown that the set of techniques concept - the "academic" concept - is not a valid description of in-house OR in Britain. Instead the traditional concept that OR is a problem-solving, model-building, activity has been shown to fit the survey results much more accurately.

Furthermore, the survey has shown that the work undertaken by these Groups is successful in the great majority of cases considered. Undoubtedly the extensive use of purpose built, non-standard, models is a major factor contributing to this success.

References

[1] Blackett, P M S (1943) A Note on Certain Aspects of the Methodology of Operational Research, in P M S Blackett (1948), Operational Research, The Advancement of Science, April 1948, 26-38.

[2] Waddington, C H (1951) Operational Research, Animal Breeding Abstracts, 19, 1 (Dec), 409-415.

[3] Ackoff, R L (1951) Principles of Operations Research, in Proceedings of First Seminar in Operations Research, Case Institute of Technology, Cleveland, Ohio, 8-16.

[4] Ackoff, R L (1979) The Future of Operational Research is Past, J.Opl.Res.Soc. 30, 2, 93-104, and Resurrecting the Future of Operational Research, J.Opl.Res. Soc., 30, 3, 189-200.

[5] Dando, M R and Bennett, P G (1981) A Kuhnian Crisis in Managment Science? J.Opl.Res.Soc., 32, 2(Feb), 91-103.

[6] Hirschheim, R A (1983) Systems in OR: Reflections and Analysis, J.Opl.Res.Soc. 34, 8(Aug), 813-818.

[7] Ackoff, R L (1956) The Development of Operations Research as a Science, Op.Res., 4, 3, 265-295.

[8] Pettigrew, A M (1975) Strategic Aspects of the Management of Specialist Activity, Personnel Review, 4, 1, 5-13.

[9] Conway, D A (1983) Planning the Project Mix: A Practical Use for Methodology, to Sixth European Congress on Operational Research, Vienna, July 1983.

ON THE DEVELOPMENT OF LARGE-SCALE PERSONNEL PLANNING MODELS

Saul I. Gass
College of Business and Management
University of Maryland
College Park, MD 20742/USA

1. INTRODUCTION

The U.S. Army is developing an integrated personnel planning model for the management of its enlisted, officer and civilian personnel. The system is called FORECAST, which is an abbreviated term for the U.S. Army Strength and Personnel Management Actions Forecasting System. The basic use of the FORECAST system will be as an analysis procedure to support the planning, programming and budgeting system of the U.S. Department of Defense. With respect to Army personnel requirements, FORECAST will serve the following uses:

1. To develop, analyze and evaluate policy alternatives.
2. To evaluate and analyze the impact of changes that effect the size, composition and readiness of the force to meet Army requirements.
3. To aid in the management of the force by grade, skill, years-of-service, unit and other variables that influence the Army's ability to meet its requirements.

The principle computational functions are projecting Army personnel strength over a seven year time horizon (by grade, skill and years-of-service) and forecasting personnel management data such as attrition rates and accessions (gains). In this paper we shall review some model development procedures associated with the officer portion of FORECAST.

2. THE OFFICER FORECAST SYSTEM

The management of officer personnel is an example of a hierarchical, decentralized control system. The U.S. Congress, Department of Defense and Army Headquarters set specific personnel force targets over a 5-7 year planning horizon. The Army Office of Deputy Chief of Staff for Personnel is responsible for overall personnel policy and the setting of specific personnel policies and related personnel actions, with such activities being taken so as to best meet the force targets throughout the planning horizon. In turn, the day-to-day management and decisions are delegated to Army functional offices and managers. These groups are concerned with setting promotion rates and making promotions; recruiting new personnel (gains); determining skills required; training to meet skill requirements; assigning officers to duty stations (distribution); determining causes of attrition and attrition rates to manage losses; accounting for individuals in transit, hospitals, etc. (individual accounts); and the analysis of Congressional and Department of Defense directives and budget authorizations to determine the force targets over the planning horizon (target analysis). Each functional manager has local decision-making authority, but decisions are constrained by the overall system requirements and must be consistent with the decisions of the other managers. The coordination of such activities, especially ensuring that common data and parameters are used by all, is the task of the strength manager. All these functions will be integrated into the computer-based officer FORECAST system. Conceptually, this is shown in Fig. 1.

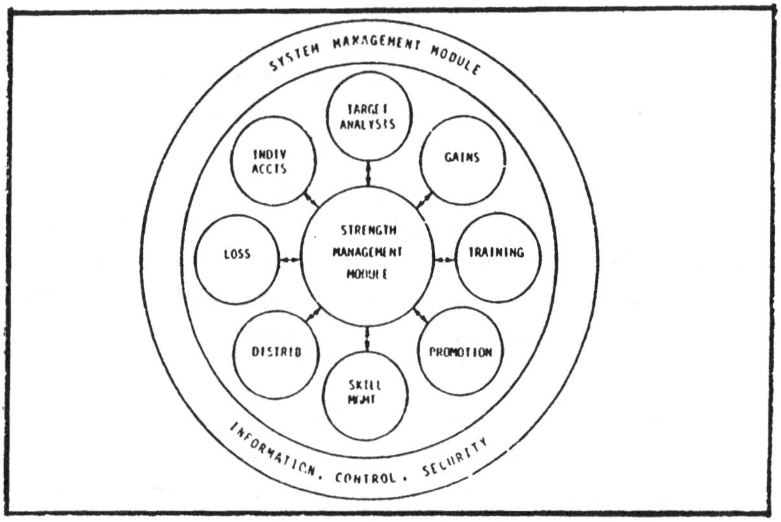

Figure 1. Forecast System Structure

In the fully automated FORECAST system, each functional manager will have a full array of data-processing, statistical, and prescriptive decision models and algorithms that have been integrated into a consistent management structure. Two of the main modeling activities are (1) the strength management model that will be used to study the impact of policy decisions on the Army's ability to recruit, promote and train officers to meet skill and personnel targets, and (2) the officer distribution model that will be used to determine the flow of personnel between Army commands to meet grade, skill and assignment policies. Both models can be viewed as goal-programming network-flow models, with complicating features. As the number of decision variables and constraints are quite large, and available network computational procedures are extremely efficient, it is a decided advantage to have the resultant formulation be the type that can be solved by network techniques. Of course, we need to be careful not to distort the problem statement so that the model is an inaccurate representation of the decision situation. The purpose of this paper is to illustrate how we evolved the network model of the officer distribution plan (ODP) problem. It is difficult to impart the complete rationale for why certain assumptions and structures were made, changed, corrected and finally adopted. We trust that the reader will be able to discern much of what happened by the following discussion and by the study of the figures that follow.

From the start, it was clear that most of the problem conditions and policy-evaluation decision-making needs could be captured by a modified minimum-cost network flow structure. The modifications were basically those that accounted for the fact that the initial personnel inventory (by grade and skill) and its distribution (by grade, skill and command) could not be adjusted in the short-term so as to meet the many thousand personnel targets that represent the Army's desired force structure over the planning horizon. The function of the strength management model is to develop a time-sequenced plan that best meets the force requirements from year one to year seven; while the distribution model assigns the resulting new inventories to positions. For both models, the exact meeting of the proposed force and distribution targets would be a rare event, i.e., from a linear-programming point-of-view, the problem is infeasible. Thus, the challenge is to "best" meet the targets, and this is defined and accomplished by using a goal-programming network formulation.

3. THE ODP PROBLEM STATEMENT

The primary purpose of the ODP is to provide for effective planning and control of the distribution process for commissioned and warrant officers. In this discussion, we are concerned with the distribution problem, i.e. the determination of the flow of personnel among commands as a function of grade and skill. The resultant flow information is then used to make personnel assignments to fill specific requisitions.

At the beginning of a planning cycle, we assume each command knows the following information for every grade/skill combination:

1. its current actual strength;
2. the number of personnel available for transfer out of the command;
3. the number of personnel that must be transferred out of the command;
4. the minimum number of personnel required--the minimum fill goal; and
5. the percent of operating strength that the command is allowed--maximum authorized goal.

Items 1) through 3) represent data that will be available from the proposed Officer FORECAST system. Items 4) and 5) are policy parameters set by the Army.

We need to differentiate between U.S. - based (CONUS) and overseas (OS) commands. OS commands have special Army personnel policies that must be observed. They are the following: (a) for warrant officers each grade/skill combination has an upper bound that is applied to the total number of overseas personnel (total from all commands) with that grade/skill combination; (b) for each grade/skill combination, an OS command must rotate a determined number of personnel to some CONUS command. Besides individual grade/skill authorized inventory goals, both OS and CONUS commands have total personnel authorized inventory levels.

Under the above conditions (and others to be introduced), an ODP determines the flow of personnel among commands for each grade/skill combination. The starting grade/skill inventories of the planning cycle represent the numbers and types of personnel at a command after attrition, promotion, and other personnel actions that are expected to occur during the planning cycle have been accounted for. New personnel are brought into the system (new gains by grade/skill) during a planning cycle and distributed to CONUS and OS commands based on new gain totals and Army movement restrictions for new gain personnel.

The ODP transfers of personnel must be accomplished so that at the end of the planning cycle the new grade/skill inventories meet as "best possible" the minimum fill goals and authorized strength goals. These goals and bounds often conflict with one another and, in many instances, cannot all be satisfied simultaneously due to shortages in some grade/skill categories and surpluses in others. Thus, it should be recognized that there is no best or optimum solution to this problem. There is a set of acceptable solutions to this problem, with each solution deviating from the goals and bounds based on the importance of weights (priorities) established by Army policy. In fact, one use of the ODP mathematical model is to measure the impact of policy changes (as reflected by changing the weights or having a policy-free structure) on a planning cycle's distribution of personnel.

For discussion purposes we assume 6 grades, and 20 commands, and up to 300 skills. In what follows we describe a network optimization model that is based on standard linear-programming concepts, augmented by goal-programming deviation and weighting procedures. The resultant model will be able to handle simultaneously all grade/skill combinations and all commands. We next introduce the necessary notation, network concepts, and then illustrate these ideas with a simple network. We then introduce a more general minimum-cost network model that encompasses all the ODP requirements except substitution of one grade for another.

4. NETWORK CONCEPTS

The analytical structure of the movement of personnel in an organization from position and/or location over a time horizon can, in most instances, be represented by flows in a network; or more generally, by a linear-programming model. The network yields a picture of the organizational structure in terms of personnel movements and is a natural one to use if the problem assumptions allow it. Also from a computational point of view, when there are a large number of personnel categories and many organizational units to represent, and/or an extended time horizon to contend with, a network model can be solved much more rapidly than the comparable linear-programming model. Even so, the ability to formulate a network-flow model with the complexity of the Army ODP that is computationally tractable depends on how the flows are aggregated and distributed, with each variation yielding a network of different size and resolution. For example, if we wanted to know the exact number of personnel by grade and skill that move from one command to any other command, it would require a network with 75,000 nodes (equations) and 125,000 arcs (variables). This richness of detail is not necessary, and the data-handling and solution evaluation procedures would be quite burdensome. Although the network model we are proposing is still rather large (19,000 nodes and 75,000 arcs), we feel that it can be solved efficiently using existing network codes. We illustrate some of the above concerns by next discussing basic network concepts, model structures, and the proposed ODP network-flow model.

The mathematical structure of the ODP is that of a minimum-cost network-flow model. A network consists of arcs and nodes, with arcs representing flows of personnel (decision variables) and nodes representing constraints (equations that force personnel balances to be maintained). Some nodes are classified as sources at which initial personnel inventories are assumed (e.g., an overseas command); some nodes are sinks at which final inventories are established (e.g., a CONUS command); with the rest of the nodes being intermediate or transfer nodes established to maintain personnel balances and to force personnel inventories to meet the various grade/skill and command inventory goals. A simple network with two sources, seven intermediate nodes, and two sinks is shown in Figure 2:

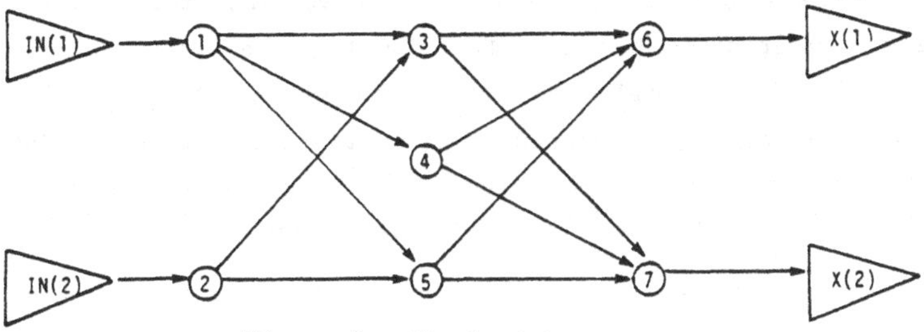

Figure 2. Simple ODP Network

Here IN(1) and IN(2) represent initial inventories and X(1) and X(2) represent final inventories. Flows (movement of personnel) take place only across the indicated arcs.

It is convenient to assume that each arc (variable) of the network has lower and upper bounds for its flow, and a cost or weight per unit of flow. For example, arc (1,3), i.e., variable x_{13}, in the above figure could have a lower bound of 10, upper bound of 20, and a unit cost of 2. This would be represented by the number triple (10, 20, 2) and is shown on the arc as follows:

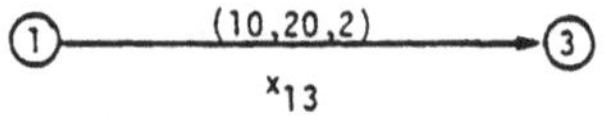

This means that $10 \leq x_{13} \leq 20$ and the contribution to the cost function is $2x_{13}$. Note that these are linear conditions. In the ODP, the costs will be weights or priorities whose values will be set to force the flows to achieve various goals, if possible.

A basic assumption for our class of minimum-cost network models is that of conservation of flow through a node. This means that the value of the flow (personnel) that enters a network at its sources is the same as the value that leaves the network at its sinks. There are no losses (e.g., attrition) or gains (e.g., input from a school) as the flow moves from the sources through the nodes (transfer points) to the sinks. For example, at a node n we might have three flows in and two flows out as shown in Figure 3. The flows in, the x_{in} from nodes 1,2,3, and the flows out, the x_{nj} to nodes 4,5, must balance as follows: $x_{1n} + x_{2n} + x_{3n} = x_{n4} + x_{n5}$.

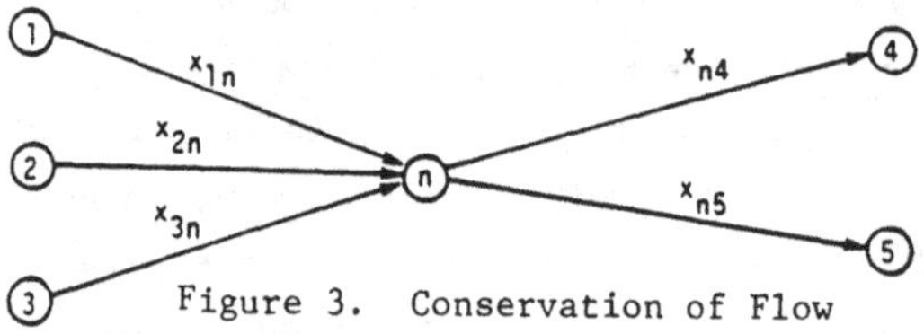

Figure 3. Conservation of Flow

Note that this is a simple linear balance (accounting) equation that states the conservation of flow assumption at node n. Each variable would also be constrained by its corresponding bounds. For the ODP, the conservation of flow assumption means that the total number of personnel in CONUS, overseas, and those about to enter the system as new gains must be redeployed to CONUS and overseas commands. Any attrition, other losses, or other gains are factored into the initial inventories before redeployment occurs.

Another network assumption is that the flow through a network is a homogeneous product, e.g., personnel of the same type. For our purposes, like the conservation of flow, this is not a restrictive assumption. However, depending on how detailed we need to identify a personnel movement, maintaining homogeneity will impact the network size. For example, it is simpler to identify a personnel movement by stating that it is a shipment coming from a specific CONUS command and going to a specific overseas command. In contrast, as the ODP must be done by grade and skill combinations, it is more complex to keep each combination separate so as to be able to address the grade/skill goals by commands. However, the grade/skill command identification can be simplified to only a command identification when the total strength by command goals are being addressed.

The identification of the flows as they move through the network is accomplished at each node. For example, consider two overseas commands 1 and 2 that are shipping personnel with grade g and skill s to CONUS command j. We also have CONUS command 3 shipping to command j. These shipments are denoted by y(g,s,1,j), y(g,s,2,j), and y(g,s,3,j), respectively. The resultant flow (inventory) at the CONUS command is denoted by x(g,s,i). The situation is shown in Figure 4:

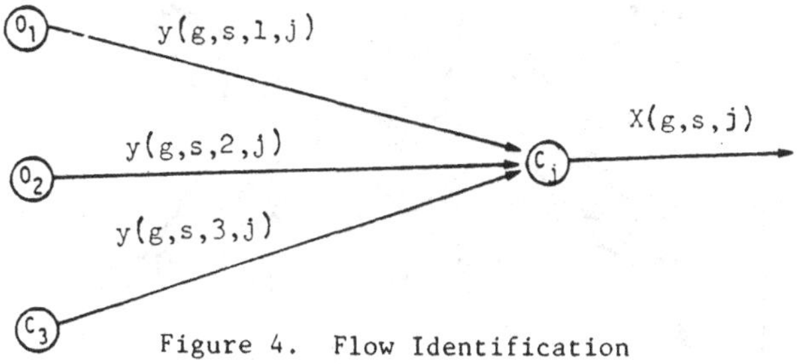

Figure 4. Flow Identification

Note that the flow out of node j has lost the overseas 1 and 2 and CONUS command 3 identifications; we now have personnel that can be identified only by grade g, skill s, and command j. The identifications of the flows out of a node is limited to the common identifications of the flows into the node. Once a level of identification is lost, it cannot be regained.

5. ODP NETWORK MODELS

A first, illustrative approach to the development of the proposed minimum-cost network problem is shown in Figure 5. We use this network to discuss the model formulation and some of the basic conditions of the ODP.

The network in Figure 5 is for a single grade/skill combination (g,s) and four commands (1,2,3,4). Each command has a beginning inventory of IN(g,s,i), i=1,2,3,4. Here we do not distinguish between CONUS and overseas commands, and we assume that each command can ship to every other command. For discussion purposes, the nodes are divided into five node types. Each command has corresponding node types 1 to 4 associated with it. Node type 1 is a source node from which the initial inventories are split between personnel

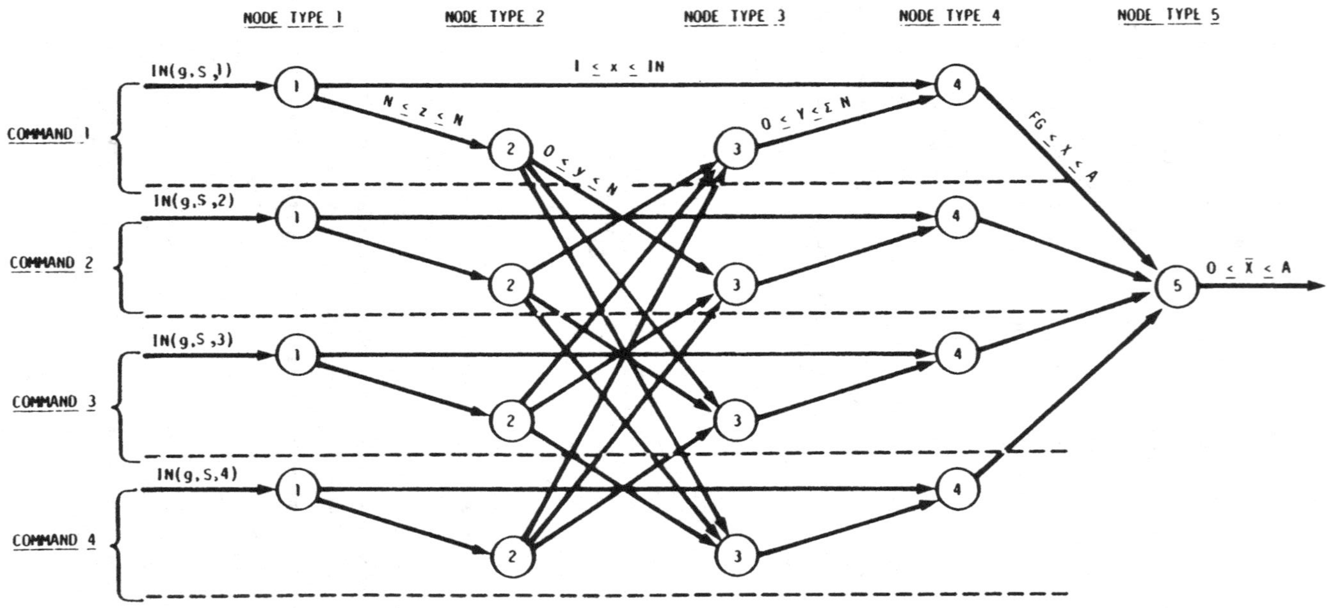

Figure 5. Illustrative Officer Distribution Network

(flows) that stay at the command and those that are to be shipped to the other commands. The first flow (x) is from node 1 to node 4, and the second flow (Z) is from node 1 to node 2. At a node type 2, the deployment of personnel from a command to the other commands takes place. These flows (y) go from node type 2 to node type 3. Note that a flow from 2 to 3 maintains the originating command identity. Each node type 3 sums all the personnel shipments going to the corresponding command. The flow out of node type 3 (Y) has lost the originating command identification and just represents the number of new personnel with grade g and skill s going to the corresponding command. At node type 4, the flow is the number of personnel that stayed at the command (x) plus the new shipment (Y); these numbers are summed to $X = x + Y$. Note that X is constrained between the minimum fill goal FG and the strength maximum bound A for that command. Node type 5 illustrates how the totals for each command can also be summed so that they can be constrained by some total strength upper bound. Note that the individual shipments x, Z, y, Y are also bounded by lower and upper bounds, as appropriate.

The illustrative network of Figure 5 does not indicate how the flows are constrained to meet the minimum fill goals FG, authorized inventory goals by grade and skill, and total force goals. These are accomplished by goal-programming relationships translated into network structures. Figure 6 describes a minimum fill goal situation, and Figure 7 describes a total force goal situation. These structures are modified later to take into account any overseas force that cannot be redeployed and which account for some of the minimum fill goal.

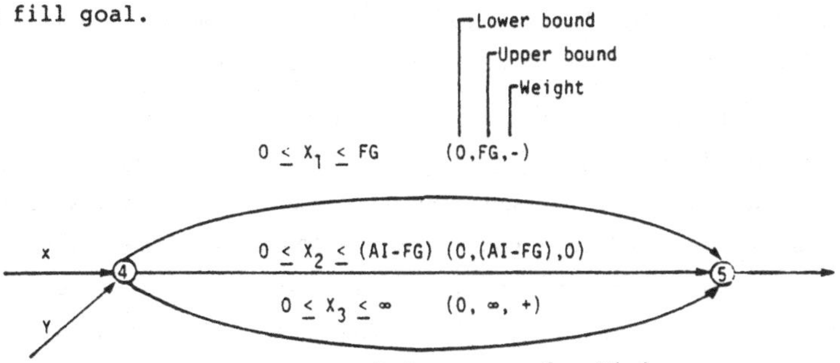

Figure 6. Network Structure for Minimum Fill Goal

In Figure 6, the top arc with negative weight allows (x + Y) to reach the minimum fill FG, if possible. X_1 represents the flow up to FG. If $X_1 < FG$, we undershoot the minimum fill goal. The middle arc with zero cost allows the flow to reach the inventory goal AI. Note: $X_2 = 0$ as long as $X_1 \leq FG$. The bottom arc with positive cost represents any overage above the inventory goal AI. Note: $X_3 = 0$ as long as $X_2 \leq (AI-FG)$. Here AI is the authorized inventory goal for each command.

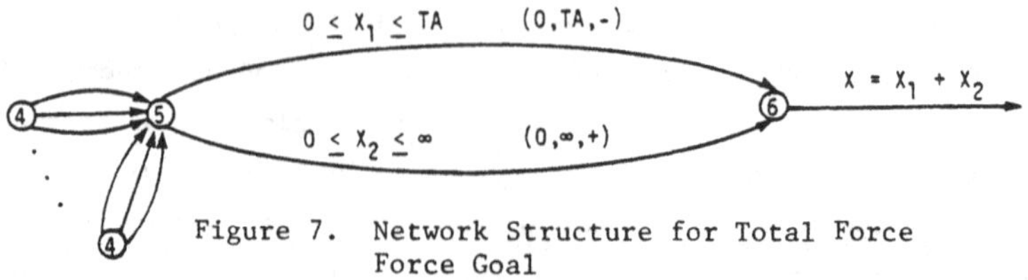

Figure 7. Network Structure for Total Force Force Goal

In Figure 7, TA is total force authorization goal for a group of commands, e.g., overseas, and the flows into node type 5 are for different grade and skill combinations.

The above network structure, with the goal-programming modifications, can encompass most of the conditions and constraints of the ODP. The major missing item is the handling of the overseas upper bound limitation for each grade and skill combination when dealing with Warrant Officers. To do this, all overseas shipments for a grade and skill (g,s) combination must be summed and restricted to be less than or equal to the upper bound AS(g,s). We show how this is done in Figure 8. This figure illustrates a two overseas command, two CONUS commands, and new gains structure for two grade/skill combinations. It is based on the previous approach in which we kept track of the individual shipments (y) between all commands. Here overseas collection and distribution nodes (0) are introduced. For each (g,s) combination, the collection nodes sum all overseas shipments ($\overline{X}$) and force $\overline{X} \leq$ AS. At the distribution nodes, the total is split among the overseas nodes to meet the individual (g,s) minimum fill goals. Finally, the total overseas personnel are summed by command and restricted by the total force goal TA for the command. Similar conditions apply to CONUS commands, exept no upper bound applies.

The ODP network model of Figure 8 is an adequate representation of the total ODP problem, without substitution. The difficulty with this formulation is due to its size. To encompass all commands and all grade/skill combinations simultaneously, we would require 75,000 nodes and 125,000 arcs. The solution time for such a large network would be too long. Thus, in Figure 9, we offer a modification that still enables us to handle simultaneously all commands and grade/skill combinations. The modification causes us to lose some of the detail information, especially knowledge of the individual shipments between commands. Figure 9 is a network structure for two overseas commands, two CONUS commands, and new gains for two grade/skill combinations. It is readily generalized to all commands and all grade/skill combinations. The Figure 9 structure is the current version of our proposed ODP network model. Note that here all overseas and CONUS commands are collapsed into collective personnel pool (source) nodes for each grade and skill. This reduces the size of the problem to 19,000 nodes and 75,000 arcs.

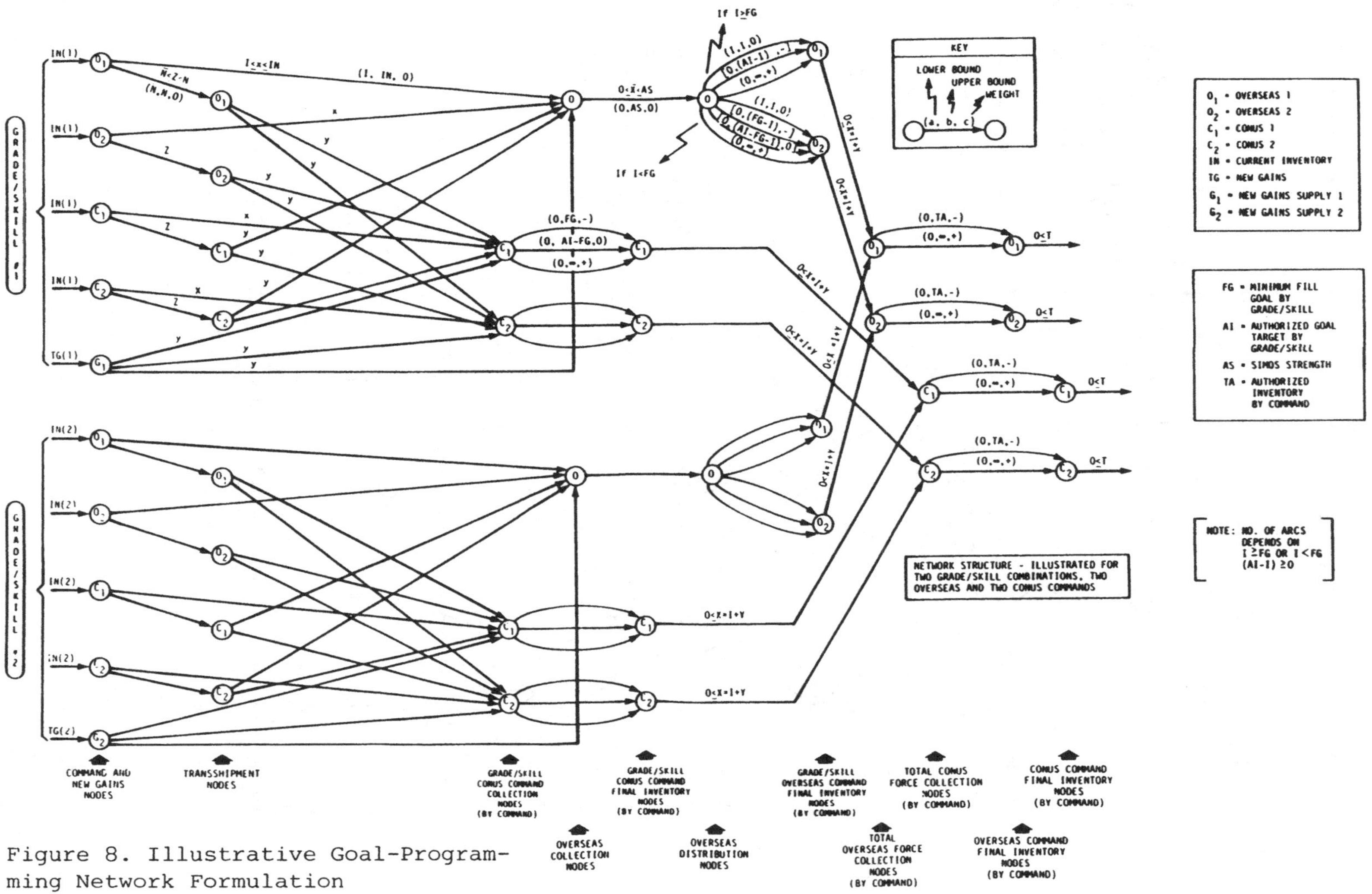

Figure 8. Illustrative Goal-Programming Network Formulation

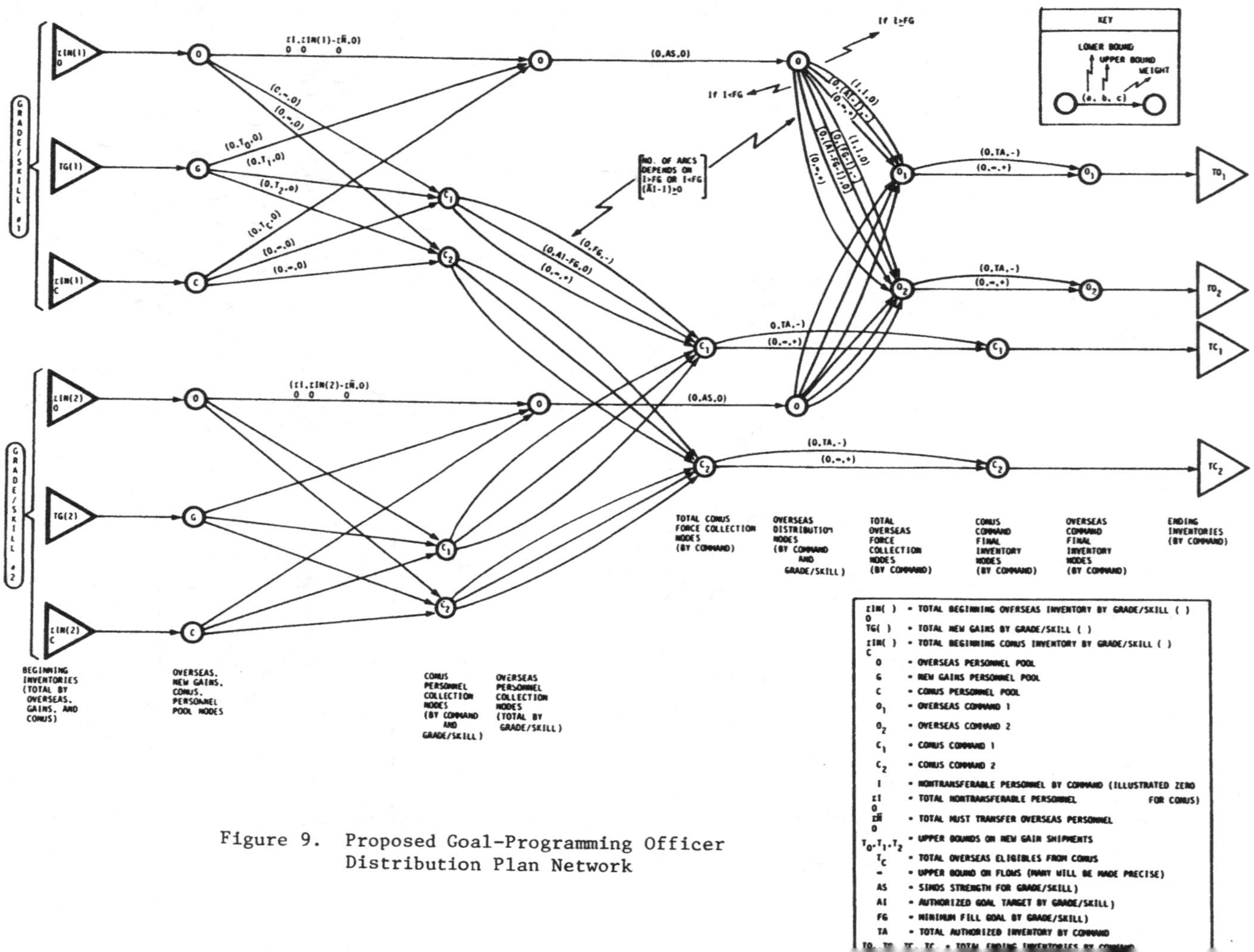

Figure 9. Proposed Goal-Programming Officer Distribution Plan Network

Modeling Dynamic Systems of Variable Structure

Alexander Umnov,
International Institute for
Applied Systems Analysis,
A-2361, Laxenburg, Austria

Introduction

This paper is devoted to the mathematical modelling of discrete dynamic systems, *i.e.* systems, states of which are defined for specific fixed points in time.

The method generally used to solve this problem consists in building an integrated model, which describes both the states for all points in time under consideration and the links between these states. This approach has several advantages. The main one is the comparatively high efficiency of usage of computer resources. The field of application of the method is sometimes restricted, for example, when the number of the time points considered is unknown or variable, as in minimal-time control problems.

The static approach to the problem of modelling a discrete dynamic system requires that the structure of the model be changed whenever the number of time points is changed. This, in its turn, is a source of significant computer difficulties. This paper describes another approach to the problem, which has been shown to be more appropriate in practice.

The main idea of this method is to build a set of consistent submodels, each of them describing the state of the modeled system for a fixed point in time. This requires the solution of a system of problems concerned with finding states of the modeled object with the required properties . The solution procedures for these problems are informatically linked and synchronized and may be solved in an independent way. Using this method, the effect of changing the number of time points considered is reduced to changes in the number of problems included in the system and, therefore, the number of independently-run computer processes. A more attractive idea is to use not only logical but physical disaggregation of these calculation processes, or, in other words, to use parallel working processors or computers.

The main question in the new approach described is how to organize the information exchange between the submodels, which will direct the whole set to the required states. The following trick is suggested here. Let us consider that all the processes of solving subproblems are imbedded in a common information 'environment', the state of which is described by a vector. Let the components of this vector be exogenous parameters of the linked submodels. An appropriate change in the values of these parameters is expected to bring all submodels to the required states.

In this situation each of the submodels interacts only with the 'environment', receives exogenous information from it, and initiates changes in the state of the 'environment'. Information links between different submodels do not explicitly appear in this scheme: their function is performed implicitly via the information 'environment'. This makes the changing the number of submodels much easier.

Next, we need to define a mathematical form for the interaction. The most natural way to do this is to include the components of the vector of the 'environmental' state in the description of the submodels to be linked. This makes it possible for the 'environment' to influence the states of the submodels. On the other hand, the level of consistency of the submodels and the state of the 'environment' will be interdependent, and the interdependence may be used for iterative changes of this state to achieve better consistency. A special method, called 'compact modelling', will be applied here to construct a computer procedure to handle the iterative change.

Detailed descriptions of all these ideas are given in the paper.

1. Statement of the Initial Problem

We will consider a mathematical model of a discrete dynamic system for the following fixed points in time

$$t = 1,2,3,\ldots,T.$$

Let the state of the model be described by a vector x^t, which has n components defined for all t, and by a vector u^t, which has l components defined for $t = [1,T-1]$. The formal difference between vectors x^t and u^t is that the first depends on x^{t-1} (*i.e.* pre-history) in an explicit way, but the second does not.

This means that there exists a system of equations

$$x^{t+1} = f^t(x^t,u^t) \qquad (1)$$

for all $t = [1,T-1]$ describing the dynamic links of the object under investigation.

Moreover, the components of x^t and u^t must satisfy the following set of constraints

$$y_s^t(x^t,u^t) \geq 0,\ s = [1,m] \qquad (2)$$

In particular, these constraints may include boundary conditions, if these exist.

Finally, it is assumed a functional exists, which defines the quality of the synthesized sequence of the vectors x^t and u^t

$$\Phi(x^1,x^2,\ldots,x^T,u^1,u^2,\ldots,u^{T-1}) \to \min \qquad (3)$$

The initial problem (1)-(3) consists in finding sequences $\{ x^t, t = [1,T] \}$ and $\{ u^t, t = [1,T-1] \}$, which satisfy the constraints (1) and (2) and define extremes for the functional (3) .

The conditions (1)-(3) describe the statements of a sufficiently wide class of problems of practical value. Moreover, in some cases it is possible to reduce the statements of mathematical models investigated to the form of (1)-(3) by means of various artificial transformations.

2. The Information 'Environment' and Parallel Processes

The main idea in the approach considered consists in changing an integrated model of the discrete dynamic system into a set of linked submodels, the number of which can be easily varied.

The original statement (1)-(3), however, is not very suitable for this purpose. Therefore we reformulate this statement to an equivalent one as follows. Let us introduce an auxiliary vector V, the components of which may be considered as exogenous parameters of the linked submodels

$$V^{t+1} = f^t(x^t, u^t), \ t = [1, T-1] \tag{4}$$

$$V^t = x^t, \ t = [1, T] \tag{5}$$

$$y_s^t(x^t, u^t) \geq 0, \ s = [1, m] \tag{6}$$

for all $t = [1, T]$. If the system of constraints (4)-(6) is feasible for a fixed V, then it is also possible to evaluate the functional of the model as a dependance of V.

Relations (4)-(6) may be considered as an implicit definition of the dependances $x^t(V^t, V^{t+1})$ and $u^t(V^t, V^{t+1})$, and we can try to reduce the process of solving the initial problem (1)-(3) to a two-level scheme; on the first level, components of V will be changed, and x^t and u^t will be calculated on the second one. It may thus be possible to bring all the submodels to the desired states by appropriate variations of the components of V. In other words, we have constructed an information 'environment', which is single-valued as described by the vector V and in which all submodels are supposedly imbedded.

As the subproblems can be solved independently subject to V being fixed, the next main problem is to find an optimal state of the 'environment'. This state must be such that all subproblems have solutions with the required properties. In the present case, these solutions have to be considered for different points in time and the functional has to have an extremal value.

In practice, it seems there is only one way to find this optimal state of the information 'environment'. This consists in a step-by-step movement towards optimality. It is clear that in such an approach it is necessary to formulate rules on how to find the local direction of the improvement and how to determine the value of the step along this direction. Technical aspects of this problem will be discussed in the next sections.

Now let us demonstrate why this scheme makes it possible to change the number of considered points in time comparatively easily.

We naturally utilize the possibility of solving subproblems (4)-(6) independently, when V is fixed. We need to initiate a set of parallel computer processes, each of them solving the subproblem for a given t. We also need to have a process that finds the optimal state of the information 'environment'. All these processes have to be linked informationally and must be synchronized. This means that the main process should be be able to initiate all required processes at any time when they become necessary and to wait for the results.

In this scheme, changing the number of subproblems T is reduced to changing the number of these parallel processes. The number of components of vector V and lengths of auxiliary arrays for the main process need to be also changed. All these changes may be easily made, of course, by means of the standard tools of programming languages and operating systems.

Synchronization of the processes and information exchange between them can also be implemented with the help of commands from the operating system used. In practice, this will only require the opening and closing of files.

More detailed discussion of all these questions is presented in the following sections.

3. The Method of Compact Modelling

Now we consider the problem of finding the optimal state of the information 'environment'.

The most obvious approach is to try to use an iterative scheme, which converges to the desired state. First of all, however, we need to have a mathematical statement of this searching problem in terms of components of the vector V. The simplest way to formulate this is to exclude x^t and u^t from the statement of (4)-(6), but this is only possible in the case of some relatively trivial problems.

We therefore employ the more complex, but more versatile method known as *compact modelling*. This method is designed for the analysis of systems described by complex mathematical models.

The method is based on the assumption that, even in the case of a very complex model, the user is generally interested in the interdependance of only a comparatively small number of input and output data. In other words, the user implicitly wishes to reformulate the model in terms of these data alone.

Now we will consider a more formal description of the compact modelling approach.

Let us assume that there is a high-dimensional vector X, describing a state of a model M. Futher, let also vector X^* be a solution of a problem solved by means of the model M and having the required properties. Then the process of solution may be formulated as the relation

$$M(X^\circ) \to X^*$$

where X° is an initial state of the model.

Let us assume that the low-dimensional vectors V and W describe the input and output data respectively. The user is actually interested in the dependence between W and V. We also postulate two conversion processes. The first permits the generation of the initial state of the model, using the initial input data

$$G(V) \to X^\circ$$

and the second converts the final state X^* into the output data

$$S(X^*) \to W.$$

Finally, the dependance between the input and the output data can be written as

$$W = S(M(G(V))).$$

The operator $\hat{M} = S(M(G(\cdot)))$ will be referred to as *the compact image of the model M*

It is unnecessary to prove that the explicit form of this compact image cannot be built for the majority of mathematical models of practical value. Hence, it is necessary to find a form of the approach that would be acceptable in practice.

A useful idea in this connection is that this image might be built locally for the immediate vicinity of a current V, rather than globally for the whole set of V under consideration. Here the following analogy is relevant. The use of the local compact image may be compared with using a part of a power series as a local approximation of a function. This part has a convenient and simple description in comparison with the description of the original function, but this simple description is different for different points. Simply stated, the coefficients of the power series are different.

It could be said, in fact, that the compact modelling approach is a method for the modelling of models. The approximating model must be low-dimensional and as simple as possible due to the local nature of the approximation.

At the same time, this analogy has only a restricted validity. In fact the dependance $W = \hat{M}(V)$ is not correctly described by, for example, Taylor approximations, even if the model $X^* = M(X)$ is described by smooth enough functions.

There are three main reasons for this inconvenient property of the dependance $W = \hat{M}(V)$

- there may not exist a feasible state X of the model M for a given vector V of the input data

- the solution x^* of the model M may be non-unique

- the dependance of X^* on V may be nondifferentiable, even if the functions in (4)-(5)-(6) are themselves differentiable.

These properties are present even in the case of such simple models as linear programming problems or systems of linear inequalities.

In order to make use of the standard Taylor approximations, it is necessary to transform the original model M into a new one $\bar{M}$ possessing the following properties:

- the $\bar{M}$ has a solution $\bar{X}$ for any vector V

- this solution $\bar{X}$ must be close (in the sense of a metric) to the original solution X^* everywhere, wherever the latter exists

- the dependance of W on V must be smooth enough for the use of standard Taylor approximations.

Local approximations of this new form of the model $\bar{M}$ are these very compact images of the original mathematical model, which make it possible to investigate the interdependence of input and output data in a desirable form.

It should be emphasized here that the *practical* effectiveness of the compact modelling approach depends mainly on the specific method used to transform the original model M into $\bar{M}$. Here this problem is considered only for finite-dimensional optimization mathematical models, to which class the model (4)-(6) belongs.

4. The Method of Smooth Penalty Functions

Now we shall consider how the compact modelling approach may be used in practice to find consistent solutions of the submodels described by relations (4)-(6). To simplify the description we will assume that the functional of the whole model is not defined at all.

The compact modelling approach appears suitable for use here because it enables us to link together *the compact images* of the submodels, even if the submodels in their original form cannot be linked. In the specific case under consideration the input data are described by the exogenous parameters V. The output data measure the degree of inconsistency of the solutions found for different submodels.

The description of the submodel for a point in time t is

$$\begin{cases} V^{t+1} = f^t(x^t,u^t) \\ V^t = x^t \\ y_s^t(x^t,u^t) \geq 0,\ s = [1,T] \end{cases} \tag{7}$$

If $X^t = \begin{bmatrix} x^t \\ u^t \end{bmatrix}$ is the solution of system (7), then the relation $\begin{bmatrix} V^t \\ V^{t+1} \end{bmatrix} \to \begin{bmatrix} x^t \\ u^t \end{bmatrix}$ describes the operator $M(G(\bullet))$ for the submodel (7).

A norm of vector $V^t - x^t$ may be taken as a measure of the inconsistency of the optimal states of the linked submodels. This norm may be calculated by, for example, a Euclidean or some other metric. Therefore, the vector W of the output data will have components $W_t = || V^t - x^t ||, t = [1,T]$.

It is not very difficult to demonstrate that the dependance between X and V possesses all the undesirable properties mentioned in the previous section. For example, it is possible to choose input vector V, such that the system (7) will be infeasible. In other words, this dependance is not defined for all V.

On the other hand, the dependance may also be non-unique, because the system (7) may have a non-unique solution for a given V.

Finally, even if this dependance exists and is unique-valued, the function $X^*(V)$ may be nondifferentiable, because of the fact that system (7) has inequality-type constraints.

This means that all of these submodels must be transformed into the new form $\overline{M}$.

As discussed earlier, the model $\overline{M}$ has to reproduce exactly all the properties of the original model M as well as having an input-output dependance possessing "good" properties,*i.e.* permiting the use of Taylor approximations.

We propose here that an auxiliary function, created for system (7) according to the rules of the smooth version of the Penalty Functions Method, be taken as the transformed model $\overline{M}$ [A.Fiacco, G.McCormick, 1968].

This auxiliary function is

$$E^t = P_=(\tau, || V^{t+1} - f^t(x^t,u^t) ||) + P_=(\tau, || V^t - x^t ||) + \sum_{s=1}^{m} P_\geq (\tau, y_s), \tag{8}$$

where the so-called *penalty function* $P_\geq(\tau,\alpha)$ is defined, is smooth enough for all α and $\tau > 0$, and satisfies the following relation

$$\lim_{\tau \to +0} P_{\geq}(\tau, \alpha) = \begin{cases} 0, & \alpha \geq 0 \\ +\infty, & \alpha < 0 \end{cases}$$

The function $P_{=}$ is usually defined as

$$P_{=}(\tau,\alpha) = \frac{1}{2}[\, P_{\geq}(\tau,\alpha) + P_{\geq}(\tau,-\alpha)].$$

Now we will demonstrate that the auxiliary function (8) may be used as the transformed submodel $\overline{M}$.

Firstly, the value of E^t may be used as a measure of the degree of the inconsistency of the submodels, because of the equivalence of $E^t = 0$ and $W_t = 0$.

Secondly,

$$\begin{bmatrix} \bar{x}^t \\ \bar{u}^t \end{bmatrix} = \underset{\begin{bmatrix} x^t \\ u^t \end{bmatrix}}{argmin}\, E^t(\tau, x^t, u^t)$$

exists for any vector of the input data as a minimum point for a continuous function, which is bounded-below.

Thirdly, $\overline{E}^t = E^t(\tau, \bar{x}^t, \bar{u}^t)$ exists and is unique-valued for any V.

Finally, $\overline{E}^t$ will be a differentiable function of V. This stems from the fact that the *implicit functions theorem* is applicable to the condition of the stationarity of auxiliary function (8)

$$\nabla_{\begin{bmatrix} x^t \\ u^t \end{bmatrix}} \overline{E}^t(\tau, x^t, u^t) = 0, \tag{9}$$

which defines the implicit functions $\bar{x}^t(V)$ and $\bar{u}^t(V)$, subject to all functions P, f^t, and y_s being smooth enough.

In other words, the vector $\begin{bmatrix} \bar{x}^t \\ \bar{u}^t \end{bmatrix}$ can be considered as a smoothed and predetermined image of the vector $X^{t*}(V)$. Hence, we may use local Taylor approximations of $\overline{E}^t(V)$ to analyse the properties of the submodels (7).

5. Searching for the Optimal State of the Information 'Environment'

Now we may formulate a mathematical statement of the problem of finding the optimal state of the information 'environment'. In order that the values of the components of V guarantee consistency between the linked submodels, it is sufficient to have a situation where the minimum of

$$\varepsilon = \sum_{t=1}^{T} \overline{E}^t(V) \tag{10}$$

equals zero. This value will be essentially different from zero only if there is no consistency between the submodels, or if their internal constraints are contradictory. Therefore, the problem of searching for the optimal state of the 'environment' is equivalent to the minimization of (10).

This minimization procedure may be carried out according to any numerical scheme. The method of compact modelling does not restrict the choice to any particular scheme. To illustrate this we will consider the use of the Newton method, where quadratic approximations of the function to be minimized are used. This requires the calculation of all partial derivatives up to and including the second order.

We will use the standard method to calculate the gradient and Hessian of the function ε. We have

$$\nabla_V \varepsilon = \frac{\partial \varepsilon}{\partial V} + H_{xV}\nabla_x \varepsilon + H_{uV}\nabla_u \varepsilon,$$

where the following notation is used

$$\frac{\partial \varepsilon}{\partial V} = \left\| \frac{\partial \varepsilon}{\partial V_i^t} \right\|; \quad H_{xV} = \left\| \frac{\partial \bar{x}_j^t}{\partial V_i^t} \right\|; \quad H_{uV} = \left\| \frac{\partial \bar{u}_j^t}{\partial V_i^t} \right\|$$

for all t, i, j.

Substituting here (10) and taking into account that $\nabla_{x^t}\bar{E}^t = 0$ and $\nabla_{u^t}\bar{E}^t = 0$ from (9), we find

$$\nabla_V \varepsilon = \frac{\partial \varepsilon}{\partial V} + \sum_{t=1}^{T}(H_{xV}^t \nabla_{x^t}\bar{E}^t + H_{uV}^t \nabla_{u^t}\bar{E}^t) = \frac{\partial \varepsilon}{\partial V}.$$

This result is of great practical importance. It means, that, firstly, it is not necessary in calculating the gradient to know the elements of the sensitivity matrices H_{xV} and H_{uV} and, secondly, formulae for the gradient may be written in an explicit form as a function of vectors $V, \bar{x}^t$, and $\bar{u}^t$

$$\nabla_V \varepsilon = \sum_{t=1}^{T} \frac{\partial \bar{E}^t}{\partial V} \tag{11}$$

In an analogous way let us find the second partial derivatives of the minimized function ε. For the Hessian of E we have

$$\nabla_V^2 \varepsilon = \frac{\partial^2 \varepsilon}{\partial V^2} + H_{xV}\frac{\partial^2 \varepsilon}{\partial V \partial x} + H_{uV}\frac{\partial^2 \varepsilon}{\partial V \partial u} +$$
$$H_{xV}(\nabla_x \varepsilon)'_V + H_{uV}(\nabla_u \varepsilon)'_V +$$
$$(H_{xV})'_V \nabla_x \varepsilon + (H_{uV})'_V \nabla_u \varepsilon$$

It is easy to see that, by virtue of (9), only the first three terms will be nonzero here. Hence, we have finally

$$\nabla_V^2 \varepsilon = \frac{\partial^2 \varepsilon}{\partial V^2} + H_{xV}\frac{\partial^2 \varepsilon}{\partial V \partial x} + H_{uV}\frac{\partial^2 \varepsilon}{\partial V \partial u} \tag{12}$$
$$= \sum_{t=1}^{T}\left(\frac{\partial^2 \bar{E}^t}{\partial V^2} + H_{xV}^t \frac{\partial^2 \bar{E}^t}{\partial V \partial x} + H_{uV}^t \frac{\partial^2 \bar{E}^t}{\partial V \partial u}\right),$$

where matrices $\frac{\partial^2 \varepsilon}{\partial V^2}$, $\frac{\partial^2 \varepsilon}{\partial V \partial x}$ and $\frac{\partial^2 \varepsilon}{\partial V \partial u}$ have the components $\frac{\partial^2 \varepsilon}{\partial V_j^t \partial V_i^q}$, $\frac{\partial^2 \varepsilon}{\partial V_i^t \partial x_j^q}$ and $\frac{\partial^2 \varepsilon}{\partial V_i^t \partial u_j^q}$ for all feasible indexes.

It is known that the quadratic approximation

$$\bar{E}^t(V_0) + (V - V_0)\,\nabla_V \bar{E}^t + \frac{1}{2}(V - V_0)\nabla_V^2 \bar{E}^t (V - V_0)$$

may be constructed if all components of $\bar{E}^t(V)$, $\nabla_V \bar{E}^t$ and $\nabla_V^2 \bar{E}^t$ have been calculated for a given point V_0. This approximation is the *compact local image* of the submodels (7) for the vicinity of the point V_0:

Theoretically it is possible to imagine a case where an explicit form for the function $E(V)$ can be found. But in practice, anyway, we will use a numerical algorithm for the minimization. Hence, it will be sufficient to have a local approximation, which is required in this method.

The whole procedure of searching for the optimal state of the information 'environment' is an iterative process, at each step of which the following elements are calculated:

- vectors $\bar{x}^t, \bar{u}^t$ and , if necessary, sensitivity matrices H^t_{xV} and H^t_{uV},
- a direction ω, along which there is a trend to a decrease in the auxiliary function E,
- ρ - the length of a step along this direction ω, which guarantees the convergence of the whole process.

Finally, vector V^{k+1} describes a new 'improved' state of the information 'environment'

$$V^{k+1} = V^k + \rho\omega,$$

after the kth iteration

Related Work

In this paper the compact modelling approach has been applied to the analysis of discrete dynamic models. It has also been succesfully applied to various other problems where the user is interested in the interdependence of a relatively small number of inputs and outputs of a complex model. Examples of these problems include linkage procedures for optimization mathemetical models [Umnov A., Albegov M., 1981] and optimization of the share of the Pareto set in multicriteria mathematical models [Umnov A., 1982] . The most detailed description of practical aspects for using the compact modelling approach in the case of large-scale models can be found in Umnov [1983].

Acknowledgments

The author would like to thank Ing. Miloslav Lenko, for all his work on coding the algorithm and carrying out numerical experiments.

References

Fiacco, A.V. and McCormick, G.P. 1968. Nonlinear Programming: Sequential Unconstrained Minimization Techniques. New York. Wiley.

Umnov, A.E. and Albegov, M.M. 1981. An Approach to Distributed Modelling. Behavioral Science, Vol. 26, pp. 354-365.

Umnov, A.E. 1982. Software for Regional Studies: Analysis of Parametrical Multicriteria Models, WP-82-66, International Institute for Applied Systems Analysis, Laxenburg, Austria.

Umnov, A.E. 1983. Impacts of Price Variations on the Balance of World Trade,PP-83-03,International Institute for Applied Systems Analysis, Laxenburg, Austria.

OPTIMAL STRUCTURAL DESIGN FOR MAXIMUM DISTANCE BETWEEN ADJACENT EIGENFREQUENCIES

N. Olhoff
The Technical University of Denmark
DK-2800 Lyngby, Denmark

R. Parbery
The University of Newcastle
NSW 2308, Australia

1. INTRODUCTION

Most problems of optimal design against structural vibration and instability dealt with in the literature consist in maximizing a fundamental eigenvalue for given structural volume, or in minimizing the volume for prescribed fundamental eigenvalue. The reader is referred to review papers [1-6]. The practical significance of optimizing with respect to a fundamental eigenvalue, i.e., the fundamental natural vibration frequency of a bar, beam or plate, or the first critical whirling speed of a rotating shaft, is that one obtains a design of minimum weight (or cost) against resonance or whirling instability, respectively, subject to all external vibration frequencies or service speeds within a large range from zero and up to the particular fundamental eigenvalue.

However, for problems of resonance due to external excitation frequencies, or whirling instability at service speeds, where the external frequencies or the service speeds are confined within a given range of *finite* upper and lower limits, much more competitive designs may be obtained by maximizing the distance between two adjacent natural frequencies or critical speeds; this being done in such a way that the excitation frequency range lies between the two natural frequencies in question.

The present paper is based on work reported in Ref. [7]. We consider problems of determining the distribution of structural material of transversely vibrating beams or of rotating circular shafts, such that maximum difference $\omega_n - \omega_{n-1}$ is obtained between two adjacent natural frequencies or critical whirling speeds ω_n and ω_{n-1} of given orders n and $n-1$. The volume, length, and boundary conditions are assumed to be given for the beam or shaft, which may be equipped with a given set of non-structural masses or disks. A geometric constraint, namely a minimum allowable value for the cross-sectional area, is used in our formulation for optimal design. The two different types of physical problem under consideration are governed by the same set of non-linear integro-differential equations.

If the cross-sectional area function is unconstrained (except for the given volume), then the solutions to our optimization problem are the same as the solutions to the different problem of maximizing a single, higher order natural frequency ω_n of given order n, for specified volume and length of our beam or shaft. The problem of maximizing a single, higher order frequency ω_n is treated in Ref. [8], where a number of optimal designs are available. The reason why the two different optimization problems have identical solutions, is the following: when a single, higher order natural frequency ω_n is maximized without specification of a geometric minimum constraint, *all* the natural frequencies of orders lower than n become associated with rigid body motions and attain zero value, see [8]. Obviously, maximum ω_n implies maximum difference $\omega_n - \omega_{n-1}$ under such conditions. In that problem, singularities of zero cross-sectional area occur, giving rise to inner separations and hinges.

In Ref. [9], single, higher order natural frequencies ω_n are maximized while a geometric minimum constraint is taken into account, and the lower order natural frequencies then all become finite. It was conjectured in Ref. [9] that the frequency differences found for this somewhat simpler problem, are close to the maximum obtainable. This conjecture is confirmed by way of examples in the present study.

2. FREE TRANSVERSE VIBRATION OF BEAMS AND WHIRLING OF SHAFTS

We consider a Bernoulli-Euler beam or shaft of length L and structural volume V, made of a linearly elastic material with Young's modulus E and the mass density ρ. The structure has variable, but geometrically similar cross-sections (in the case of a shaft, we assume circular cross-sections) with the relationship $I = cA^2$ between the moment of inertia $I(X)$ and the area $A(X)$, where the positive constant c is given by the cross-sectional geometry. A number, M, of given non-structural masses Q_i, $i=1,..,M$, (circular disks in the case of a shaft) are assumed to be attached to the structure at prescribed points $X = X_i$, $i=1,..,M$, along the coordinate axis, X.

Denoting by Ω_j the j-th natural angular frequency of transverse vibrations or the j-th critical angular whirling speed if the structure is a beam or a rotating shaft, respectively, we now introduce the following dimensionless quantities: coordinate x by $x = X/L$ $(0 \le x \le 1)$, cross-sectional area $\alpha(x)$ by $\alpha = AL/V$, non-structural masses q_i by $q_i = Q_i/\rho V$ $(x = x_i$, $i=1,..,M)$, and natural frequencies (or critical speeds) ω_j by

$$\omega_j^2 = \frac{\Omega_j^2 \rho L^5}{c E V} . \tag{1}$$

The eigenfunction (i.e., vibration or whirling mode) y_j associated with the j-th natural frequency or critical speed ω_j of the transversely vibrating beam or rotating shaft, is governed by the dimensionless differential equation

$$(\alpha^2 y_j'')'' = \omega_j^2 \alpha y_j , \tag{2}$$

in which rotational inertia in the case of a beam and gyroscopic effects for a shaft are neglected. In addition to (2), appropriate boundary conditions must be specified. We assume, for simplicity, that the structure has clamped, simply supported or free ends, i.e., that no flexible or intermediate supports are present. If the cross-sectional area is everywhere larger than zero, the eigenfunction y_j, its first derivative y_j' and the bending moment $\alpha^2 y_j''$ are continuous throughout, but the shear force $(\alpha^2 y_j'')'$ has discontinuities (jumps) given by

$$(\alpha^2 y_j'')'\Big|_{x=x_i^+} - (\alpha^2 y_j'')'\Big|_{x=x_i^-} = \omega_j^2 q_i y_j(x_i) , \quad i=1,..,M , \tag{3}$$

at the points $x = x_i$, $i=1,..,M$, where the masses q_i are attached to the structure.

Multiplying (2) by y_j, integrating twice by parts over the interval $0 \le x \le 1$ and taking (3) and the boundary conditions into account, we obtain the following Rayleigh expression for the eigenvalue ω_j^2,

$$\omega_j^2 = \frac{\int_0^1 \alpha^2 y_j''^2 \, dx}{\int_0^1 \alpha y_j^2 \, dx + \sum_{i=1}^{M} q_i y_j^2(x_i)} . \tag{4}$$

If ω_j^2 is a higher order eigenvalue $(j > 1)$, then Rayleigh's principle states that the right hand side of (4), i.e., the Rayleigh quotient, is stationary at the eigenvalue ω_j^2 corresponding to the actual eigenfunction y_j among all kinematically admissible deflection functions that are orthogonal to all the lower eigenfunctions y_k, $k=1,..,j-1$. Together with a normalization condition for y_j, which makes the denominator of the Rayleigh quotient in (4) equal to unity, these orthogonality conditions can be written as

$$\int_0^1 \alpha y_j y_k \, dx + \sum_{i=1}^{M} q_i y_j(x_i) y_k(x_i) = \delta_{jk} , \quad k=1,..,j , \tag{5}$$

where δ_{jk} denotes Kronecker's delta.

We consider the problem of finding the distribution of cross-sectional area, $\alpha(x)$, of a beam or rotating shaft of given length and volume, for which the difference between two adjacent higher order natural frequencies ω_n and ω_{n-1} is to be maximized for given $n > 1$. We shall assume that both frequencies are single modal and that $\omega_{n-1} > 0$. Noting that in nondimensional form the length of the beam or shaft is unity, we may express the volume constraint after nondimensionalization by

$$\int_0^1 \alpha\, dx = 1 \ . \tag{6}$$

In addition, we consider a geometric minimum constraint, $\alpha(x) \geq \bar{\alpha}$, to be specified for the cross-sectional area function $\alpha(x)$, where the minimum allowable value $\bar{\alpha}$, $0 < \bar{\alpha} < 1$, is assumed to be given. By introduction of the real slack variable $g(x)$, the constraint $\alpha(x) \geq \bar{\alpha}$ can be expressed via the equality constraint

$$g^2(x) = \alpha(x) - \bar{\alpha} \ . \tag{7}$$

3. SOLUTION OF THE OPTIMIZATION PROBLEM

A variational formulation of the optimization problem stated above is employed, using the functional

$$\begin{aligned} F = {} & \left[\int_0^1 \alpha^2 y_n''^2\, dx\right]^{\frac{1}{2}} - \left[\int_0^1 \alpha^2 y_{n-1}''^2\, dx\right]^{\frac{1}{2}} \\ & - \sum_{k=1}^{n} \Lambda_k \left[\int_0^1 \alpha\, y_n y_k\, dx + \sum_{i=1}^{M} q_i\, y_n(x_i)\, y_k(x_i) - \delta_{nk}\right] \\ & - \sum_{k=1}^{n-1} \lambda_k \left[\int_0^1 \alpha\, y_{n-1} y_k\, dx + \sum_{i=1}^{M} q_i\, y_{n-1}(x_i)\, y_k(x_i) - \delta_{(n-1)k}\right] \\ & - \sum_{k=1}^{n-2} \int_0^1 \mu_k(x) \left[(\alpha^2 y_k'')'' - \omega_k^2\, \alpha\, y_k\right] dx \\ & - \beta\left[\int_0^1 \alpha\, dx - 1\right] - \int_0^1 \kappa(x)\left[g^2(x) - \alpha(x) + \bar{\alpha}\right] dx \ , \end{aligned} \tag{8}$$

where the frequency difference $\omega_n - \omega_{n-1}$ is expressed as a functional via (4) with unit value of the denominator, and is augmented by the orthonormality conditions (5) for the mode y_n, similar conditions for the mode y_{n-1}, the differential constraints (2) for the lower order modes y_k, $k = 1,..,n-2$, the volume constraint (6), and the geometric minimum constraint (7). These side conditions are introduced by means of the Lagrangian multipliers Λ_k $(k = 1,..,n)$, λ_k $(k = 1,..,n-1)$, $\mu_k(x)$ $(k = 1,..,n-2)$, β, and $\kappa(x)$, respectively. In view of Rayleigh's principle for higher order eigenvalues and our introduction of the Lagrangian multipliers, we may take the variations of F with respect to y_k $(k = 1,..,n)$, ω_k^2 $(k = 1,..,n-2)$, α and g independently in the following.

The derivation of the Euler-Lagrange equations expressing the stationarity of the functional F in (8) with respect to the variables can be found in Ref. [7]. For reasons of brevity only the results will be given here.

We find that the modes y_n and y_{n-1} together with the frequencies ω_n and ω_{n-1}, respectively, must satisfy the same differential equation (2), jump condition (3) and boundary conditions as the lower order modes y_k and associated frequencies ω_k, $k = 1,..,n-2$. The Lagrangian multipliers are determined as $\Lambda_n = \omega_n/2$, $\Lambda_k = 0$ $(k = 1,..,n-1)$; $\lambda_{n-1} = -\omega_{n-1}/2$, $\lambda_k = 0$ $(k = 1,..,n-2)$; and $\mu_k(x)$ all vanish in the interval $0 \leq x \leq 1$, i.e., $\mu_k(x) = 0$ $(k = 1,..,n-2)$. After elimination of the slack variable $g(x)$ and the Lagrangian multiplier $\kappa(x)$, we find [7] the following equations for the optimal cross-sectional area function $\alpha(x)$,

$$\alpha^{-3}\left[\omega_n^{-1} m_n^2 - \omega_{n-1}^{-1} m_{n-1}^2\right] - \frac{1}{2}\left[\omega_n y_n^2 - \omega_{n-1} y_{n-1}^2\right] = \beta \quad (\text{if} \quad \alpha > \bar{\alpha}) \ , \ x \in x_u \ ;$$
$$\alpha = \bar{\alpha} \quad , \quad x \in x_c \ . \tag{9}$$

Here, x_u and x_c denote the union(s) of sub-intervals of $0 \leq x \leq 1$ where the cross-sectional area function $\alpha(x)$ is unconstrained and constrained, respectively, and m_n and m_{n-1} are the dimensionless bending moment functions defined by

$$m_n(x) = \alpha^2 y_n'' \quad , \quad m_{n-1}(x) = \alpha^2 y_{n-1}'' \ . \tag{10}$$

Our problem is now as follows: for given mode order n, minimum allowable cross-sectional area $\bar{\alpha}$ and set of attached masses q_i, $i = 1,..,M$, we are required to find the eigenfunctions y_j and eigenfrequencies ω_j, $j = 1,..,n$, the Lagrangian multiplier β, and optimal cross-sectional area function $\alpha(x)$ together with the constrained and unconstrained sub-intervals x_c and x_u.

The solution must satisfy the optimality condition (9), the volume constraint (6), the differential equation (2) for all modes, the "jump" conditions (3), the Rayleigh expression (4) and the orthonormality conditions (5), as well as suitable boundary conditions. The examples reported later deal with a cantilever beam or shaft for which the deflection and slope are zero at the fixed end and the moment and shear force are zero at the free end, i.e.,

$$y_j(0) = y_j'(0) = \alpha^2 y_j''(1) = (\alpha^2 y_j'')'_{x=1} = 0 \quad , \quad j = 1,..,n \ . \tag{11}$$

In order to put the optimality condition in the form used in the computational algorithm we consider (9) in the region $x \in x_u$. Dividing each side by β, raising each side to a power r and multiplying by α, we obtain

$$\alpha_{i+1}(x) = \begin{cases} \alpha_i\left[\beta^{-1}\left\{\alpha_i^{-3}[(\omega_n)_i^{-1}(m_n)_i^2 - (\omega_{n-1})_i^{-1}(m_{n-1})_i^2] - \frac{1}{2}[(\omega_n)_i(y_n)_i^2 - (\omega_{n-1})_i(y_{n-1})_i^2]\right\}\right]^r & (\text{if} > \bar{\alpha}) \ , \ x \in (x_u)_{i+1} \ ; \\ \bar{\alpha} \ , & x \in (x_c)_{i+1} \ . \end{cases} \tag{12}$$

Here, the subscript i refers to the i-th iteration.

The solution is based on a discrete representation of the variables at a large number of equally spaced points in the interval $0 \leq x \leq 1$. The integrations and other computations are performed numerically. The computation scheme is as follows:

(i) Assume starting values for $\alpha(x)$.

(ii) Determine the set of orthonormal eigenfunctions y_j, their derivatives y_j', y_j'', and associated eigenvalues ω_j, $j = 1,..,n$, from the equations (2), (3), (4), (5) and (11).

(iii) Calculate new values of $\alpha(x)$, and the sub-intervals x_u and x_c, from (12), adjusting the value of β so that (6) is satisfied.

(iv) Repeat from (ii) until the iterates become stationary.

The procedure is started by assuming $\alpha(x)$, and in many of the examples calculated α was simply set equal to unity throughout the interval $0 \leq x \leq 1$. However, when a series of examples was to be calculated, with the same value of n and the same set of non-structural masses, it was usually economical in terms of computer time to use one of solutions already found as a starting point.

A common procedure for computing eigenfunctions and eigenvalues associated with a given cross-sectional area function is used in step (ii). The orthonormalization is performed by the well-known Gram process. In step (iii) the value of β is found by a simple step search. The power r in equation (12) must be chosen to give a balance between stability and speed of convergence. The most suitable value appears to change with changes in $\bar{\alpha}$, the attached masses q_i and the number of intervals used in the calculation. Values of r between 0.05 and 0.2 were used in the examples reported here.

The convergence was rather slow and in most cases it was necessary to apply a very sharp convergence criterion to ensure that the process was not stopped before the solution was reached. In some cases the iteration converged to a stationary solution which was not the global optimum. This could be avoided to a large degree by changing the value of r and/or using a different starting approximation.

4. RESULTS

We present the results for two sets of examples, namely, $n = 3$ with a single non-structural mass at the free end $x = 1$, and $n = 5$ with two equal masses at $x = \frac{1}{2}$ and at $x = 1$. In each case solutions were found for values of mass q from 0 to 100, and for values of $\bar{\alpha}$ from 0.1 to 1.

The limiting case $\bar{\alpha} = 1$ corresponds to fully constrained, i.e., uniform design with $\alpha(x) = 1$. On the other hand, the solution for the limiting case with $\bar{\alpha} = 0$ is geometrically unconstrained and corresponds to the solution of the problem discussed in reference [8], as explained in the Introduction.

In Figs. 1 and 2 the non-dimensional eigenfrequencies are plotted against the value of the minimum allowable cross-sectional area $\bar{\alpha}$. In Figs. 3 and 4 we present the optimal shape of optimal beams obtained for $\bar{\alpha} = 0.1$. It should be noted that the linear dimensions of the cross-sections are proportional to $\sqrt{\alpha}$.

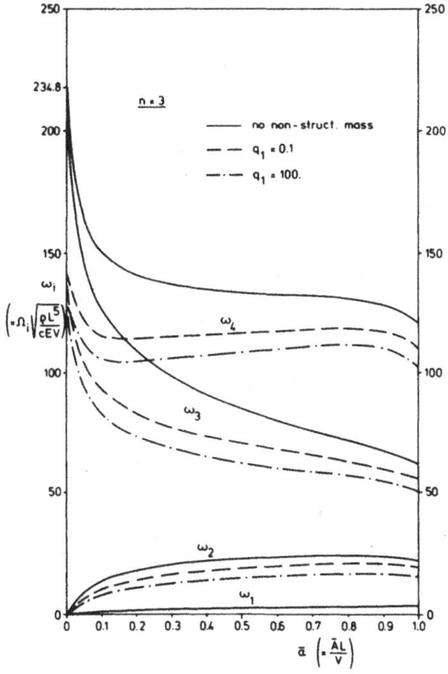

Fig. 1. Eigenfrequencies of optimal $\omega_3 - \omega_2$ cantilevers vs. minimum cross-sectional area constraint $\bar{\alpha}$.

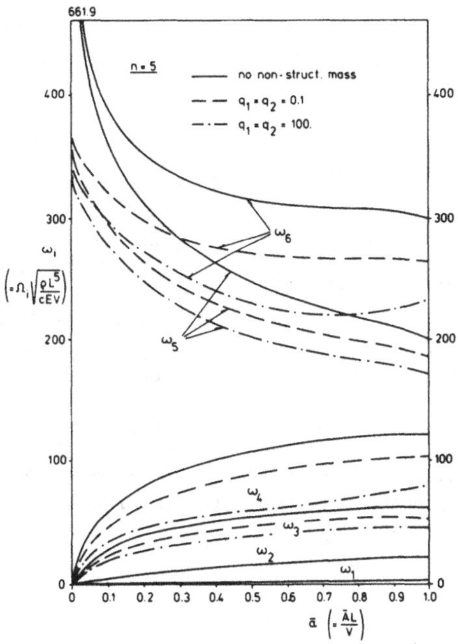

Fig. 2. Eigenfrequencies of optimal $\omega_5 - \omega_4$ cantilevers vs. minimum cross-sectional area constraint $\bar{\alpha}$.

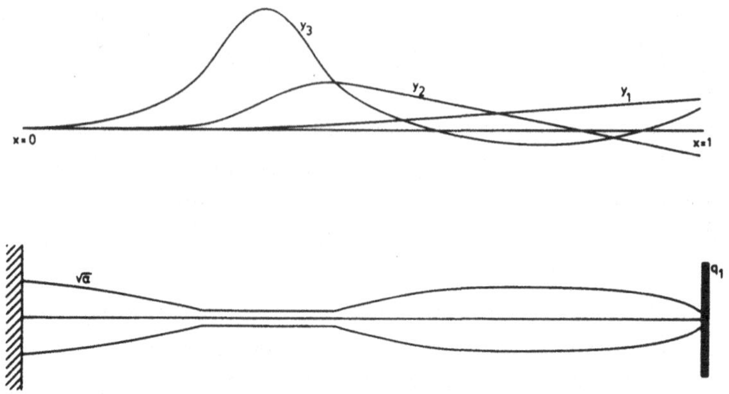

Fig. 3. Optimal $\omega_3 - \omega_2$ cantilever with tip mass $q_1 = Q_1/\rho V = 0.1$ and minimum area constraint $\bar{\alpha} = \bar{A}L/V = 0.1$. The solution has $\omega_3 = \Omega_3(\rho L^5/cEV)^{\frac{1}{2}} = 92.4$, $\omega_2 = 10.4$, and $(\omega_3 - \omega_2)/(\omega_3^u - \omega_2^u) = \Delta\omega/\Delta\omega^u = 2.27$.

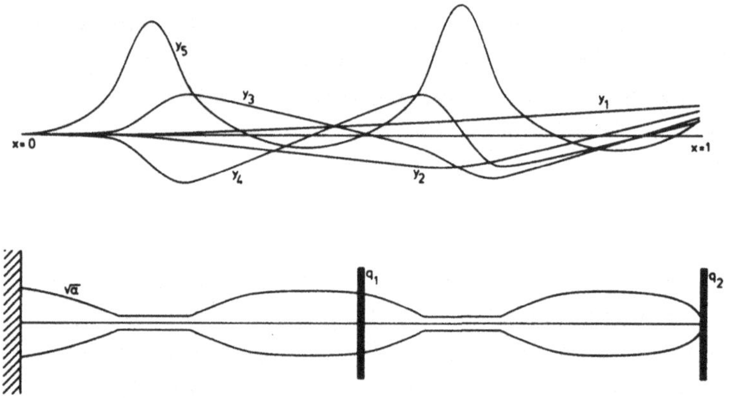

Fig. 4. Optimal $\omega_5 - \omega_4$ cantilever with $\bar{\alpha} = 0.1$ and tip mass $q_1 = q_2 = 0.1$ at $x = \frac{1}{2}$ and $x = 1$. The solution has $\omega_5 = 296$, $\omega_4 = 48$ and $\Delta\omega/\Delta\omega^u = (\omega_5 - \omega_4)/(\omega_5^u - \omega_4^u) = 3.03$.

In order to assess the possible improvement brought about by the optimization, it is interesting to compare the optimum results with those of other designs. This is done in Figs. 5 and 6, and Tables I and II. Figs. 5 and 6 show, for $n = 3, 5$ respectively, the frequency separation of the optimal beam compared with that of a uniform reference cantilever of the same volume, length, and non-structural mass. The results are plotted against $\bar{\alpha}$. Tables I and II show the optimal frequency difference obtained here compared to that obtained in the simpler problem of maximizing ω_n .

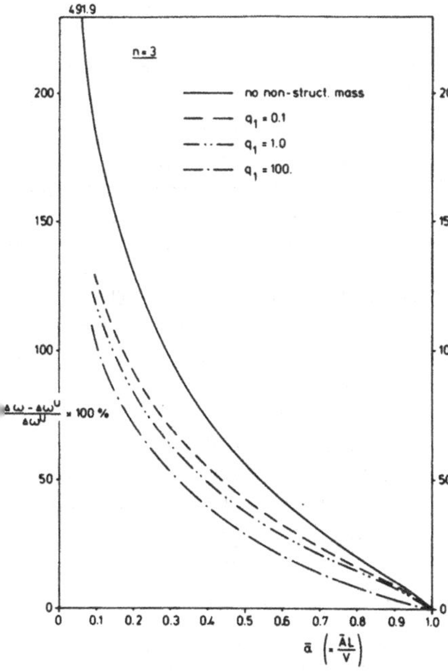

Fig. 5. Percentage improvement in frequency difference for $n = 3$. The solid line corresponds to optimal beams without non-structural mass and the other curves for beams with a single mass $q_1 = Q_1/\rho V$ at the tip.

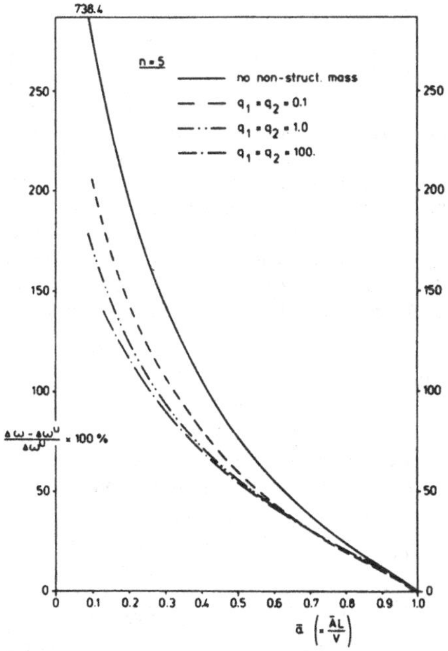

Fig. 6. Percentage improvement in frequency difference for $n = 5$. The solid line corresponds to optimal beams without non-structural mass and the other curves for beams with added masses $q_1 = q_2 = 0.1$ at $x = \frac{1}{2}$ and $x = 1$, respectively.

Table I. Percentage improvement in the case $n = 3$ of Frequency difference of Optimal Beam compared with that of Beam with maximum value of ω_n.

	q			
$\bar{\alpha}$	0.	0.1	1.	100.
0	0%	0%	0%	0%
0.2	1.3	0.94	0.95	1.0
0.4	1.6	1.5	1.5	1.6
0.6	1.4	1.3	1.3	1.5
0.8	0.7	0.5	0.5	0.6
1.0	0	0	0	0

Table II. Percentage improvement in the case $n = 5$ of Frequency difference of Optimal Beam compared with that of Beam with maximum value of ω_n.

	q_1, q_2			
$\bar{\alpha}$	0.	0.1	1.	100.
0	0%	0%	0%	0%
0.2	4.0	3.2	3.2	3.1
0.4	5.4	4.5	5.6	5.4
0.6	6.7	5.0	6.9	7.3
0.8	5.5	4.0	6.2	6.3
1.0	0	0	0	0

5. CONCLUSION

This paper demonstrates the maximization of the difference between adjacent higher natural frequencies of given order for transversely vibrating beams and rotating shafts. In comparison with a uniform beam or shaft, the frequency difference is substantially increased, particularly for small minimum cross-sectional area constraint values, $\bar{\alpha}$. The comparison with the beam or shaft having maximum value of ω_n (see Refs. 8 and 9) shows small, but consistent increases in frequency difference for the optimal design. However, it may be seen that neither the optimal shapes nor the natural frequencies differ greatly from those of the beam or shaft having maximum ω_n . Generally the value of ω_n for the present problem is marginally less than that obtained when ω_n alone is maximized, while the value of ω_{n-1} is slightly depressed compared to the simpler problem.

REFERENCES

1. F.I. Niordson and P. Pedersen, A review of optimal structural design. In *Applied Mechanics*, Proc. XIIIth Int. Cong. Th. Appl. Mech., Moscow, 1972 (Edited by E. Becker and G.K. Mikhailov), pp. 264-278. Springer-Verlag, Berlin (1973).
2. G.I.N. Rozvany and Z. Mróz, Analytical methods in structural optimization. *Appl. Mech. Rev.* 30, 1461-1470 (1977).
3. V.B. Venkayya, Structural optimization: A review and some recommendations. *Int. J. Num. Meth. Engrg.* 13, 203-228 (1978).
4. E.J. Haug, A review of distributed parameter structural optimization literature. In *Optimization of Distributed Parameter Structures*, Proc. Nato ASI Meeting, Iowa City, 1980, (Edited by E.J. Haug and J. Cea), pp. 3-74. Sijthoff & Noordhoff, The Netherlands (1981).
5. B.L. Pierson, A survey of optimal structural design under dynamic constraints. *Int. J. Num. Meth. Engrg.* 4, 491-499 (1972).
6. N. Olhoff, Optimal design with respect to structural eigenvalues. In *Theoretical and Applied Mechanics*, Proc. XVth Int. Cong. Th. Appl. Mech., Toronto, 1980 (Edited by F.P.J. Rimrott and B. Tabarrok), p. 133-149. North-Holland, Amsterdam (1980).
7. N. Olhoff and R. Parbery, Designing vibrating beams and rotating shafts for maximum difference between adjacent natural frequencies. *Int. J. Solids Struct.* (in press).
8. N. Olhoff, Optimization of vibrating beams with respect to higher order natural frequencies. *J. Struct. Mech.* 4, 87-122 (1976).
9. N. Olhoff, Maximizing higher order eigenfrequencies of beams with constraints on the design geometry. *J. Struct. Mech.* 5, 107-134 (1977).

EXISTENCE PROOFS FOR A CLASS OF PLATE OPTIMIZATION PROBLEMS

Martin Philip Bendsøe
Mathematical Institute
The Technical University of Denmark
DK-2800 Lyngby, Denmark

1. INTRODUCTION

In this paper we shall discuss optimization of thin, solid elastic plates of given domain and elastic properties. The design variable is the variable thickness of the plate and the problem is thus a distributed parameter optimization problem.

Numerical experiments ([9]) have shown that one cannot guarantee existence of solutions to for example the minimum compliance (maximum stiffness) problem if the admissable designs considered correspond to thickness functions in a bounded, closed and convex set of the bounded, measurable functions on the plate domain. This means that this type of optimization problem has to be regularized and in this paper we suggest doing this by imposing an extra (physically reasonable) constraint on the admissable designs in the form of a bound on the average of the gradient of the thickness function, i.e. we consider thickness functions in a bounded subset of the Sobolev space H^1. For this type of admissable designs we obtain the existence for not only the traditional static optimization problems such as the minimum compliance problem, but also for dynamic problems.

2. THE MINIMUM COMPLIANCE PROBLEM

The minimum compliance plate optimization problem can in general terms be formulated as: For a given plate domain $\Omega \subseteq R^2$, load $p \in L^2(\Omega)$, boundary conditions and admissable set of rigidity tensors (the admissable designs) $U_{ad} \subseteq (L^\infty(\Omega))^{16}$ find the solution $D^*_{\alpha\beta\kappa\gamma} \in U_{ad}$ to

$$\text{minimize} \quad \Pi(w) = \int_\Omega p\, w\, d\Omega \ , \quad \text{where}$$

$$w \in \left\{ w \in V \subseteq H^2(\Omega) \;\middle|\; a_D(w,v) = b(v) \quad \text{for all} \quad v \in V ,\, D \in U_{ad} \right\} , \tag{1}$$

with the Kirchhoff plate equation written in its variational form, i.e.

$$\begin{aligned} a_D(w,v) &= \int_\Omega D_{\alpha\beta\kappa\gamma} \frac{\partial^2 w}{\partial x_\alpha \partial x_\beta} \frac{\partial^2 v}{\partial x_\kappa \partial x_\gamma} d\Omega \ ^{*)} \\ b(v) &= \int_\Omega p\, v\, d\Omega \end{aligned} \tag{2}$$

for $w, v \in H^2(\Omega)$, and the set $V \subseteq H^2(\Omega)$ is the set of kinematically admissable dis-

*) usual summation convention on $\{1,2\}$.

placements defined in accordance with the imposed (homogeneous) boundary conditions. We note that the elements D of U_{ad} usually arise as functions D(u) of some control u , and in the traditional formulation of the problem of minimum compliance for given volume the admissable set is:

$$U_1 = \left\{ h^3 \tilde{D}_{\alpha\beta\kappa\gamma} \;\middle|\; h \in L^\infty(\Omega) \; , \; \int_\Omega h \, d\Omega = \gamma \; , \; 0 < h_{min} \le h \le h_{max} \right\} \tag{3}$$

where the non-zero elements of the constant tensor $\tilde{D}_{\alpha\beta\kappa\gamma}$ are given by

$$\begin{aligned} \tilde{D}_{1111} &= \tilde{D}_{2222} = E/12(1-\nu^2) = k \; , \\ \tilde{D}_{1212} &= \tilde{D}_{1221} = \tilde{D}_{2112} = \tilde{D}_{2121} = (1-\nu)\,\frac{k}{2} \; , \\ \tilde{D}_{1122} &= \tilde{D}_{2211} = \nu k \end{aligned} \tag{4}$$

in which E and ν are Youngs modulus and Poisson's ratio, respectively, for an isotropic, linearly elastic material. In the formulation above U_1 consists of homogeneous plate designs made up of a given material and the thickness function (height) h of the plate is the design variable. Here, h identifies the distance between the upper and lower plate surface, which are assumed to be disposed symmetrically with respect to the plate midplane.

The numerical experiments of Ref. [9] shows that with $U_{ad} = U_1$ in problem (1) , solutions obtained numerically using a finite difference (or finite element) approximatio of the state equation do not, for large values of h_{max}/h_{min} , converge with respect to the mesh size of the discretization; for decreasing mesh-size integral stiffeners o height h_{max} are formated and the objective decreases. The limit of such plate geometries cannot be described by a rigidity tensor in U_1 and it follows that problem (1) with the set U_1 of admissable designs does not always have a solution.

In contrast to the situation described above, we note that with a set of admissable rigidity tensors of the form

$$\tilde{U}_1 = \left\{ D^o_{\alpha\beta\kappa\gamma} + u^i D^i_{\alpha\beta\kappa\gamma} \;\middle|\; u^i \in L^\infty(\Omega) , \, a^i \le u^i \le \beta^i \; , \; \int u^i \, d\Omega = \gamma^i \; , \; i = 1, \ldots , k \right\} \tag{5}$$

existence of solutions to (1) are indeed ensured. This is based on the fact that the load linear form b(•) is equal to the functional Π ([8]) and it is thus not a property that generalizes to most other types of optimization problems (various applications of this idea can be found in Refs. [2], [11] and [13]). It does, however, show that the minimum compliance plate optimization problem has a solution for sandwic design and design using Young's modulus or the shear and dilation moduli as the design variable. For problems where only small design variations from a given design are allo ed (remodelling) linearization of (1) gives a problem for which existence can be prove and a linearization also simplifies the analysis ([6], [7]).

One method of regularizing the full problem (1) with thickness as the design variable is to extend the set U_1 of admissable designs by including in the extended set the limits of (minimizing) sequences of designs in U_1. The functional Π is a (weakly) continuous function of the deflection w and this implies that G-limits is the relevant limiting process to be considered, i.e. a sequence of designs G-converges if the corresponding sequence of deflections converges in $H^2(\Omega)$-weak ([24]). For plates with stiffeners in one direction G-limits can be found by homogenization ([3]) of the plate equation ([12]) and the numerical work of Ref. [10] indicates that including these limits in the set of admissable designs regularizes the problem in the case of axisymmetry. For the general problem the G-closure of U_1 has yet to be characterized, but for some other special plate design problems the G-closure has been found ([18]). Exchanging U_1 with the G-closure of U_1 is in fact sufficient in order to regularize problem (1): The condition $h \geq h_{min} > 0$ implies that the set W_1 of deflections corresponding to designs in U_1 is bounded and thus weakly precompact in $H^2(\Omega)$ (the set $\{a_D(\cdot,\cdot) \mid D \in U_1\}$ is a family of uniformly strongly elliptic operators) and taking G-closure extends W_1 to its weak closure in $H^2(\Omega)$; existence then follows from the fact that Π is weakly continuous.

A regularization of the problem (1) can also be achieved by restricting the design space and this restricted design space has the property that existence of solutions is also obtained for other functionals than compliance. Specifically we impose constraints on the gradient of the thickness functions in the design space so that the resulting set of admissable rigidity tensors is given by:

$$U_2 = \left\{ h^3 \tilde{D}_{\alpha\beta\kappa\gamma} \;\middle|\; h \in H \right\} \quad , \tag{6}$$

where

$$H = \left\{ h \in H^1(\Omega) \;\middle|\; \|h\|_{H^1} \leq M, 0 < h_{min} \leq h \leq h_{max}, \int_\Omega h \, d\Omega = \gamma \right\}$$

and the following properties hold ([2]):

A. H is convex, closed and bounded in $H^1(\Omega)$, and thus weakly compact in $H^1(\Omega)$.

B. U_2 is strongly G-closed ([24]), i.e. if $h_n \to h$ in H, then for the corresponding deflection:

$$w_n \to w \text{ , weakly in } H^2(\Omega) \text{ ,} \tag{7}$$

$$h_n^3 \tilde{D}_{\alpha\beta\kappa\gamma} \frac{\partial^2 w_n}{\partial x_\alpha \partial x_\beta} \longrightarrow h^3 \tilde{D}_{\alpha\beta\kappa\gamma} \frac{\partial^2 w}{\partial x_\alpha \partial x_\beta} \quad \text{in } L^1_{loc}(\Omega)\text{-weak} \quad . \tag{8}$$

Property B implies that the set W_2 of deflections corresponding to designs in U_2 is weakly compact in $H^2(\Omega)$, and thus (1) has a solution with $U_{ad} = U_2$.

Imposing a constraint on the gradient of the thickness function h is reasonable from a technical point of view and has likewise been proposed in Ref. [20] as well as in Refs. [15] - [18].

3. OTHER STATIC PROBLEMS

We note that the proof of existence of solution to problem (1) on the space U_2 holds for any functional Π that is a weakly continuous function of the deflections w. This means that problem (1) with Π expressing maximum deflection at a point, average deflection or average curvature has a solution on U_2 and the convergence (8) ensures that also the problem of minimizing the average momentum over the plate has a solution on U_2.

The condition $\int_\Omega h\,d\Omega = \gamma$ in the definition of the set H means that we have considered min-problems for given volume; property $\underline{A}$ for the set H gives that the dual problems of minimizing volume for given upper value of the mechanical properties, which in (1) were minimized, also have solutions for thickness functions with constrained H^1-norm. Thus for example the problem

$$\begin{aligned} &\underset{h \in H^1}{\text{minimize}} \quad \int_\Omega h\,d\Omega \;, \\ &\text{subject to:} \quad \|h\|_{H^1} \le M \;, \\ &\qquad\qquad 0 < h_{min} \le h \le h_{max} \;, \\ &\qquad\qquad a_D(w,v) = b(v) \quad \text{for all} \quad v \in V \;, \\ &\qquad\qquad \int_\Omega p\,w\,d\Omega \le \Pi_o \end{aligned} \tag{9}$$

has a solution.

Finally we note, that certain min-max problems also have a solution for $U_{ad} = U_2$, e.g. the problem of minimizing the maximum deflection

$$\begin{aligned} &\min_{w \in W_2} \max_{x \in \Omega} |w| \;, \quad \text{where} \\ &W_2 = \left\{ w \in V \subseteq H^2(\Omega) \;\middle|\; a_D(w,v) = b(v) \quad \text{for all} \quad v \in V \,,\; D \in U_2 \right\} \end{aligned} \tag{10}$$

has a solution. This is seen from property $\underline{A}$ and $\underline{B}$ and by rewriting (10) as ([23])

$$\begin{aligned} &\min \beta \;, \quad \text{where} \\ &\int_\Omega (w - \beta)^{+2}\,d\Omega = 0 \;, \\ &\int_\Omega (-w - \beta)^{+2}\,d\Omega = 0 \;, \\ &w \in W_2 \;. \end{aligned} \tag{11}$$

Here f^+ denotes the positive part of a function f.

4. EIGENVALUE PROBLEMS

The property that the set U_2 of rigidity tensors is strongly G-closed implies convergence of eigenvalues and corresponding eigenspaces for eigenvalue problems involving the plate operator with rigidity tensors in U_2, with the limit values being given by the operator corresponding to the limit of the rigidity tensors ([4], [14], [24]). From this it follows that we have existence of solutions to the problem

$$\max_{D \in U_2} \min_{w \in V} \frac{\int_\Omega D_{\alpha\beta\kappa\gamma} \frac{\partial^2 w}{\partial x_\alpha \partial x_\beta} \frac{\partial^2 w}{\partial x_\kappa \partial x_\gamma} d\Omega}{\int_\Omega h w^2 d\Omega} \tag{12}$$

of maximizing the lowest natural frequency of free vibrations of the plate, as well as to the problem

$$\max_{D \in U_2} \min_{w \in V} \frac{\int_\Omega D_{\alpha\beta\kappa\gamma} \frac{\partial^2 w}{\partial x_\alpha \partial x_\beta} \frac{\partial^2 w}{\partial x_\kappa \partial x_\gamma} d\Omega}{\int_\Omega N_{\alpha\beta} \frac{\partial w}{\partial x_\alpha} \frac{\partial w}{\partial x_\beta} d\Omega} \tag{13}$$

of maximizing the buckling load of the plate ([2]). In (13), $N_{\alpha\beta}$ denotes the prebuckling stress field which depends on D and which via the Airy stress function is given by a fourth order equation of the same form as the plate equation. (The formulation in (12) and (13) is based on the Rayleigh expression for eigenvalues).

5. AN OBSTACLE PROBLEM

Consider the problem of minimizing the compliance of a plate that is deflected over an obstacle φ. This can be stated as

$$\min_{w \in W_3} \int_\Omega p\, w\, d\Omega \quad , \text{ where}$$

$$W_3 = \left\{ w \in K \subseteq H^2(\Omega) \;\middle|\; a_D(w, v-w) \geq b(v-w) \quad \text{for all} \quad v \in K,\ D \in U_{ad} \right\} \tag{14}$$

$$K = \left\{ v \in V \;\middle|\; v(x,y) \geq \varphi(x,y) \quad \text{a.e. in } \Omega \right\}$$

where the state equation has been written as a variational inequality. From this formulation and the property of G-convergence of (minimizing) sequences in U_2 gives that also (14) has a solution for $U_{ad} = U_2$ ([4]); this can also be shown directly as in Ref. [5].

6. ALTERNATIVE METHODS OF REGULARIZATION

The plate optimization problems described above can alternatively be regularized by pertubation of the functionals Π. The work of Ref. [1] shows the existence of an element $g \in L^\infty$, such that for an $\varepsilon > 0$, the functional

$$\Pi + \varepsilon \|h - g\|_{L^\infty}$$

has an extremum on U_1 , but the proof of existence of the element g is not constructive and this method is thus not applicable for problems where a solution is to be computed. A method which in practice is similar to the method described above is to employ a functional ([21])

$$\Pi + \varepsilon \|h\|_{H^1}$$

on $U_1 \cap H^1(\Omega)$, and this has been used in Refs. [19], [22] for min-max problems and eigenvalue problems.

Acknowledgement - Work leading to the developments reported in this paper was supported by the Danish Council for Scientific and Industrial Research, grant. No. 16-3019. M-745.

REFERENCES

[1] J. Baranger: Existence de solutions pour des problèmes d'optimisation non convexe. J. Math. pures et appl., Vol. 52 (1973), pp. 377-405.

[2] M.P. Bendsøe: On Obtaining a Solution to Optimization Problems for Solid, Elastic Plates by Restriction of the Design Space. J. Struct. Mech. Vol. 11, no. 4. (to appear).

[3] A. Bensousson, J.-L. Lions, G. Papanicolaou: Asymptotic Analysis for Periodic Structures. North-Holland, Amsterdam, 1978.

[4] L. Boccardo, P. Marcellini: Sulla convergenza delle soluzioni di disequazioni variazionali. Annali di Math. Pura ed Appl., Vol. 110 (1976), pp. 137-159.

[5] I. Bock, J. Lovišek: An Optimal Control Problem for an Elliptic Variational Inequality. Math. Slovaca, Vol. 33 (1983), pp. 23-28.

[6] A.S. Bratus: Asymptotic Solutions in Problems of the Optimal Control of the Coefficients of Elliptic Operators. Soviet Math. Dokl., Vol. 24 (1981), pp. 145-148.

[7] A.S. Bratus, V.M. Kartevelishvili: Approximate Analytic Solutions in Problems of Optimizing the Stability and Vibrational Frequencies of Thin-Walled Elastic Structures. Mechanics of Solids (MTT), Vol. 16 (1981), No. 6, pp. 102-118.

[8] J. Cea, K. Malanowski: An Example of a Max-Min Problem in Partial Differential Equations. SIAM J. Control, Vol. 8 (1970), pp. 305-316.

[9] K.T. Cheng, N. Olhoff: An Investigation Concerning Optimal Design of Solid Elastic Plates. Int. Solids Struct., Vol. 17 (1981), pp. 305-323.

[10] K.T. Cheng, N. Olhoff: Regularized Formulation for Optimal Design of Axisymmetric Plates. Int. J. Solids Struct., Vol. 18 (1982), pp. 153-169.

[11] N.I. Didenko: Optimal Distribution of Bending Stiffness on an Elastic Freely

Supported Plate. Mech. Solids (MTT), Vol. 16, No. 1. (1981), pp. 133-143.

[12] G. Duvaut, A.-M. Metellus: Homogénéisation d'une plaque mince en flexion de structure périodique et symétrique. C.R. Acad. Sc. Paris, t. 283 (1976), Serie A, pp. 947-950.

[13] M. Goebel: Optimal Control of Coefficients in Linear Elliptic Equations I. Existence of Optimal Control. Math. Operationsforsh. Statist. Ser. Optim., Vol. 12 (1981), pp. 525-533.

[14] S. Kesavan: Homogenization of Elliptic Eigenvalue Problems. Appl. Math. Optim., Vol. 5 (1979), pp. 153-167 (part I) and pp. 197-216 (part II).

[15] V.G. Litvinov: Optimal Control of the Natural Frequency of a Plate of Variable Thickness. U.S.S.R. Comput. Maths. Math. Phys., Vol. 19 (1980), pp. 70-86.

[16] V.G. Litvinov, A.D. Panteleev: Optimization Problem for Plates of Variable Thickness. Mechanics of Solids (MTT), Vol. 15, No. 2. (1980), pp. 140-146.

[17] V.G. Litvinov: Optimal Problems on Eigenvalues. Ukrainian Math. J., Vol. 33 (1981), pp. 465-468.

[18] K.A. Lurie, A.V. Cherkaev, A.V. Fedorov: Regularization of Optimal Design Problems for Bars and Plates. JOTA, Vol. 37 (1982), pp. 499-522 (Part I), and pp. 523-543 (Part II), and preprint (Part III).

[19] A. Myslinski: Nonsmooth Optimal Design of Vibrating Plates. Preprint, Osaka University. Faculty of Engineering, Japan 1983.

[20] F.I. Niordson: Optimal Design of Plates with a Constraint on the Slope of the Thickness Function. Int. J. Solids Struct., Vol. 19 (1983), pp. 141-151.

[21] J. Sokołowski: Optimal Control in Coefficients for Weak Variational Problems in Hilbert Space. Appl. Math. Optim., Vol. 7 (1981), pp. 283-293.

[22] J. Sokolowski, A. Myślinśki: Non-differentiable Optimization Problems for Elliptic System. Preprint, Polish Academy of Sciences, 1983.

[23] J.E. Taylor, M.P. Bendsøe: An Interpretation for Min-Max Structural Design Problems Including a Method for Relaxing Constraints. Int. J. Solids Struct. (to appear).

[24] V.V. Žikov, S.M. Kozlov, O.A. Oleĭnik, Ha T'en Ngoan: Averaging and G-convergence of Differential Operators. Russian Math. Surveys, Vol. 34 (1979), No. 5., pp. 69-148.

SHORT TERM PRODUCTION SCHEDULING OF THE PULP MILL
- A DECENTRALIZED OPTIMIZATION APPROACH

Kauko Leiviskä
University of Oulu
Oulu, Finland

1. INTRODUCTION

In pulp and paper industry the term 'production control' usually refers to operative control activities carried out by the staff members from shift foremen to production managers. In pulp mills, production targets are based on customer orders or on the paper mill requirements. Production control includes coordination and scheduling tasks that aim to maintain the prescheduled delivery times of the end-products keeping simultaneously the production costs moderate. Coordination refers to actions that are carried out in order to reduce the effects of shutdowns, production rate changes and other disturbances occurring in one process upon other processes, thus minimizing the costs of these disturbances. Scheduling, on the other hand, is a task that reduces customer orders or the long term production plan to the short term production schedules of the separate pulp mill processes. Coordination is improved by increasing the efficiency of the mill information system and scheduling, respectively, by applying computerized scheduling aids.

Production control operations take place continuously. The time span is relatively short: for coordination (e.g. status reporting) from minutes to hours, for scheduling up to one week and for history reports up to one year.

From the production control viewpoint the pulp mill is usually described as a series of processes and storage tanks between them. Storage tanks are used in compensating for the unbalanced operation of the processes. Their compensating capacity is, however, limited. Also process capacities are limited depending on the built capacity and disturbances. The energy production system can also be a limiting factor. Because of the above mentioned facts a proper storage tank control is of vital importance, because the disturbances in one part of the mill tend to propagate and influence other parts of the mill. This leads to unnecessary production rate changes that further result in high production losses, quality impairment, and increasing environmental load.

Computerized scheduling aids can be divided into three levels of increasing complexity, namely:

(1) Utilization of the computerized mill information system providing the staff with status and trend information on the storage tanks and production rates.

(2) Utilization of simulation models to produce additional information and to test different alternate schedules.

(3) Utilization of optimization methods connected with simulation to produce the optimal production schedules.

In most practical cases the methods at levels (1) and (2) are sufficient. Optimization methods considered in this paper are especially important in cases where, for instance, several production lines are to be dealt with. It must be, however, pointed out that both the simulation and the optimization methods use basically the same mathematical formulation. This paper is mainly based on Ref. /1/. Because of space limitations some details have been omitted. An example concerning scheduling of parallel fibre lines has been worked out for this paper.

2. MATHEMATICAL FORMULATION

2.1 Pulp mill model

For production scheduling purposes the pulp mill is usually described as a system of processes and storage tanks between them /1/. The model flows are of pulp and liquors. Concentrations of chemicals are assumed to be constant. The dynamics of individual processes are neglected. The ratio between flows around each process is assumed to be constant, i.e. the processes use raw materials and prepare products always in the same ratio, when the same quality is produced.

For the modelling common state space notations are used. The state of the system is described by the amount of the material in each storage tank i.e. by the state vector $\bar{x}(t)$. The production rates of the processes are chosen as control variables; ie. as the contol vector $\bar{u}(t)$. The given pulp production is taken as a deterministic disturbance to the system. It is denoted by the disturbance vector $\bar{v}(t)$.

The relationship between state variables $\bar{x}(t)$, control variables $\bar{u}(t)$ and disturbance variables $\bar{v}(t)$ is described by the vector-matrix differential equation

$$\frac{d\bar{x}(t)}{dt} = B\bar{u}(t) + C\bar{v}(t). \quad (1)$$

B and C are coefficient matrices describing the relationships between the model flows. Since most storage tanks have only one input flow and one output flow, most elements in B and C matrices equal zero.

The variables of the model are constrained by the capacity limits of processes and storage tanks. The restrictions can be written as follows

$$\bar{x}^{Min} \leq \bar{x}(t) \leq \bar{x}^{Max} \text{ and} \quad (2)$$
$$\bar{u}^{Min} \leq \bar{u}(t) \leq \bar{u}^{Max}. \quad (3)$$

Example 1:

As an example of the modelling approach let us consider a simple case with one process and one storage tank. It could describe the bleach plant and the bleached pulp storage of the pulp mill.

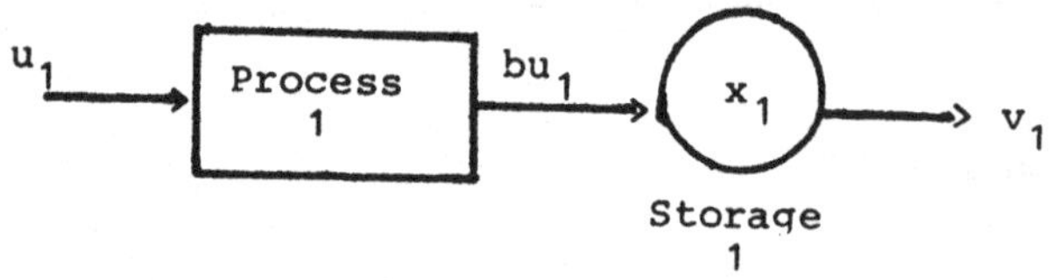

Figure 1. A simple process-storage system.

Now a system equation can be written either in continuous form

$$\frac{dx_1(t)}{dt} = b_1u_1(t) - v_1(t) \quad (4)$$

or in discrete form

$$x_1(k+1) = x_1(k) + bu_1(k) - v_1(k). \quad (5)$$

In these equations b denotes the transfer ratio of the process and it can be determined by writing the steady-state material balance of the process /1/.

2.2 The objective function

In this study a linear quadratic objective function is used

$$J = \frac{1}{2}\{||\bar{x}(K) - \bar{x}^o(K)||^2_{Q(K)} + \sum_{k=o}^{K-1} [||\bar{x}(k) - \bar{x}^o(k)||^2_{Q(k)} + ||\bar{u}(k) - \bar{u}^o||^2_{R(k)}]\} \quad (6)$$

where $||\bar{x}||^2_Q = \bar{x}^T Q\bar{x}$ and $||\bar{u}||^2_R = \bar{u}^T R\bar{u}$. The values of $\bar{x}^o(k)$ are determined a priori values so that they are statistically the most advantageous as for unplanned shut-downs.

The reference values, $\bar{u}^o$, are determined so that the required pulp production over the scheduling period is assured

$$u^o_1 = [\gamma_1 \sum_{k=o}^{K-1} v_1(k)]/K \quad (7)$$

$$u^o_i = \gamma_i u^o_{i-1}; \qquad i=2,3,... \quad (8)$$

Constants γ are used to convert the pulp production to the corresponding production of the process in question.

The optimization problem is now

$$\min_{\bar{x},\bar{u}} J \text{ , when} \quad (9)$$

$\bar{x}(o)=\bar{x}_o$ and the model equations given before are satisfied.

In the objective function, Equation 6, the term $||\bar{u}(k)- \bar{u}^o||^2_{R(k)}$ represents the costs of production rate changes. It is difficult to express these costs in monetary terms. Preferably, the relative values of the weighting factors are approximated, based on a priori knowledge of the costs. Usually, this means that the production rate changes are first allowed in processes, where they can be carried out most easily. The term $||\bar{x}(k) - \bar{x}^o(k)||^2_{Q(k)}$ in the objective function represents the risk of filling or emptying the storage tanks, which further results in production losses, shut-downs of the processes or in production rate changes.

3. PRODUCTION SCHEDULING METHOD

Pioneering work in pulp mill production scheduling was done in Sweden at the end of 60's /2-6/. In these studies both simulation and optimization methods were considered. The heuristic simulation approach presented in /7/ is, in a way, an extension of these simulation studies. In Finland at the beginning of 70s production scheduling methodology was studied from a systems theoretical viewpoint /8, 9/. An application of a network flow algorithm has been considered in /10-13/. A method using maximum principle with multiple constraints on state and control variables to pulp mill production scheduling was presented in /14/.

After the literature survey and some preliminary tests /15/, Tamura's time delay algorithm was selected for further development. This algorithm solves discrete dynamical problems with pure time delays for state and control variables using a Goal Coordination structure. Usually these kinds of problems are solved by augmentation of the state space thus introducing additional variables for the delays. This, however, increases the dimensionality of the system. By using Goal Coordination the optimization can be performed without increasing the dimensionality of the system. Mathematical formulation of this algorithm is given elsewhere /16, 17/. This study has also been reported in /18-21/.

4. PROBLEM FORMULATION

Several factors must be taken into account in solving the production scheduling problems and in the following they are summarized as a list of general production scheduling goals. Details are given in Ref. /1/, but generally each calculated schedule must fulfil following:

1. The planned pulp production is produced in due time.
2. Storage capacity between the processes is totally used before production rate changes are allowed.
3. Planned shut-downs are included in production scheduling and so their effect is compensated for.
4. Unplanned disturbances, i.e. disturbances that leave only a short time or no time at all to prepare for compensa-

tion of their effect are taken into account by utilizing an appropriate storage tank control strategy.

5. The capacity constraints of processes and storage tanks are included in scheduling.

6. It is possible to change the target level of the storage tanks during scheduling, e.g. for preparing for a planned shut-down.

7. Possible delays after production rate changes are included in scheduling.

8. The possibility to store steam indirectly in black liquor or in pulp is included.

Example 2:

Figure 2 shows a pulp mill with two fibre lines and one chemical recovery. In this case the production rate of the bleached pulp is to be maximized using the method shown in Ref. /1/. Therefore an auxiliary control variable is needed, i.e.

$$u_8(k) = v_1(k).$$

The pulp flows v_2 and v_3 are mixed together so that the mixed product includes 15 % of bleached pulp, or

$$v_3 = 0.15(v_2+v_3) \Rightarrow v_3 = 0.177\ v_2$$

Now the model equation in this case is as follows

$$\frac{d\bar{x}(t)}{dt} = \begin{bmatrix} 0.99 & 0 & 0 & 0 & 0 & 0 & 0 & -1 \\ -1 & 1 & 0 & 0 & 0 & 0 & 0 & 0 \\ 0 & 1.2758 & -1 & 0 & 0 & 0 & 1.0403 & 0 \\ 0 & 0 & 0.2154 & -1 & 0 & 0 & 0 & 0 \\ 0 & 0 & 0 & 2.00 & -1 & 0 & 0 & 0 \\ 0 & -0.4633 & 0 & 0 & 0.8450 & 0 & -0.3785 & 0 \\ 0 & 0 & 0 & 0 & 0 & 1 & 0 & 0 \\ 0 & 0 & 0 & 0 & 0 & -1 & 1 & 0 \end{bmatrix} \bar{u}(t) + \begin{bmatrix} -0.177 \\ 0 \\ 0 \\ 0 \\ 0 \\ 0 \\ 0 \\ -1 \\ 0 \end{bmatrix} v_2(t)$$

A scheduling period of 2 days divided into 6 intervals 8 hours each is considered. Now, consider an 8 hours shut-down in the production (v_2+v_3) during the third interval. The corresponding production rates of other processes together with storage tank levels are shown in Figure 3.

5. CONCLUSION

In this paper a method based on a well-known hierarchical optimization algorithm for pulp mill production scheduling has been introduced. This algorithm has the following advantages compared with other hierarchical algorithms /1/.

1. It is written in discrete time. Pulp mill production scheduling occurs at discrete time events and therefore the discrete time presentation is the most natural way to describe the production scheduling problems.
2. The process delays can be taken into account without increasing the dimensionality of the system.
3. The constrained nature of the state and control variables can also be included in scheduling.
4. Because a linear-quadratic objective function is used, algebraic equations are solved on the lower level and a simple gradient updating is needed on the upper level. This guarantees a simple solution approach, which is easy to understand for the user and results in reasonable computing times for complex systems.
5. In Tamura's algorithm the decomposition is made only with respect to time. Therefore no additional interconnection variables are required as in standard Goal Coordination methods and all the variables have a clear physical meaning.
6. The solution is optimal.

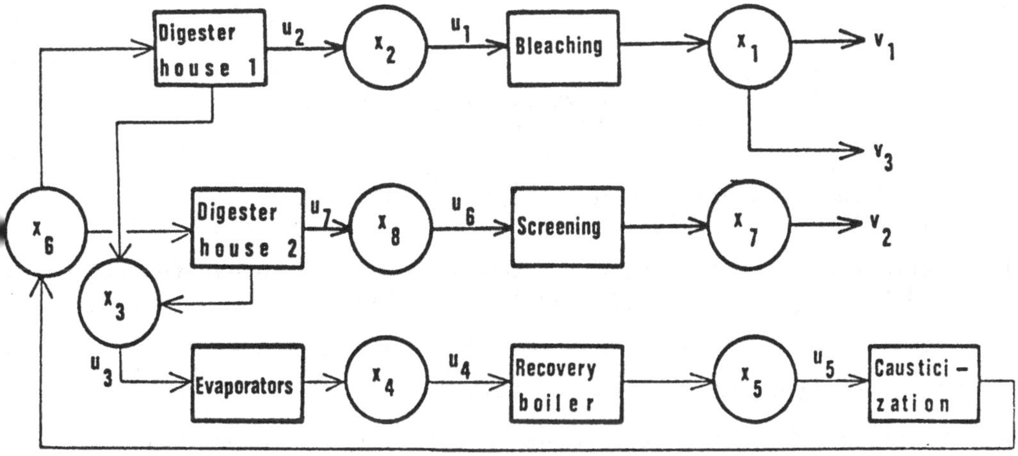

Figure 2. The flow diagram of the pulp mill with two fibre lines and one chemical recovery.

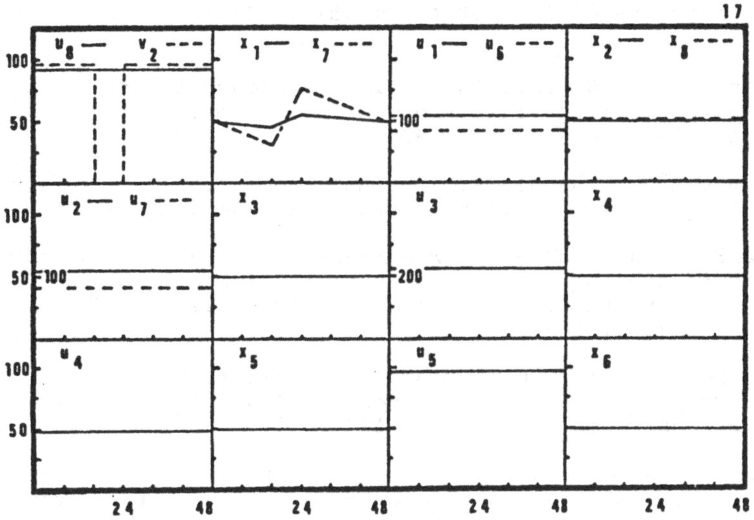

Figure 3. Optimal production schedules and storage tank levels in the case, where the production of bleached pulp is to be maximized and a shut-down of the drying plant 2 occurs. The production rate of the drying plant 1 is denoted by u_8.

REFERENCES

1. Leiviskä K., Short term production scheduling of the pulp mill. Acta Polytechnica Scandinavica, Mathematics and Computer Science Series No. 36, Helsinki 1982, 80 pp.

2. Alsholm O., Pettersson B., Integration of pulp mill control with the production planning of the paper mill. Trans. of the BP&BMA Symposium, 1969, 209-231.

3. Pettersson B., Production control of a complex integrated pulp and paper mill. Tappi 52(1969)11, 2155-2159.

4. Pettersson B., Mathematical methods of a pulp and paper mill scheduling problem. Report 7001, Division of Automatic Control, Lund Institute of Technology, 1970.

5. Pettersson B., Production control of a pulp and paper mill. Report 7007. Lund Institute of Technology, Division of Automatic Control. Lund, 1970.

6. Pettersson B., Optimal production schemes coordinating subprocesses in a complex integrated pulp and paper mill. Pulp and Paper Magazine of Canada 71(1970)5, 59-63.

7. Leiviskä K., Komokallio H., Aurasmaa H., Uronen P., Heuristic algorithm for production control of an integrated pulp and paper mill. Preprints of 2nd LSSTA Symposium, Toulouse, France, 1980, 551-561.

8. Golemanov L.A., Systems Theoretical Approach in the Projecting and Control of Industrial Production Systems. Doctoral Thesis, EKONO-publication No. 113, 1972.

9. Golemanov L.A., Koivula E., Optimal production control and mathematical programming. Paperi ja Puu 55(1973)2, 53-67.

10. Edlund S.G., Johansson R., Experiences with a multiple unit production control system for the optimization of mill operations. Preprints, International Symposium on Process Control, CPPA, Vancouver, B.C., 1977, 36-39.

11. Edlund S.G., Kallmén C., Production control in pulp and paper mills. Preprints of British Paper and Board Industry Federation International Symposium, Kent, 1974, 11 p.

12. Edlund S.G., Rigerl K.H., A computer-based production control system for the coordination of operations in pulp and paper mills. Preprints of 7th Triennial IFAC World Congress, Helsinki, 1978, 221-227.

13. Ranki J., Experiences with computerized production control and supervision in a pulp and board mill. Tappi 66(1981)6, 115-118.

14. Chalaye G., Foulard C., Hierarchical control of a pulp mill. Proc. of 3rd International IFAC PRP Conference. Brussels, 1976, 109-119.

15. Leiviskä K., Comparison of optimal and suboptimal methods for pulp mill production control. 9th IFIP Conference on Optimization Techniques, Warzawa, 1979, 9 p.

16. Tamura H., A discrete dynamic model with distributed transport delays and its hierarchical optimization for preserving stream quality. IEEE Trans. Systems, Man and Cybernetics SMC-4(1974)5, 424-431.

17. Tamura H., Decentralized optimization for distributed-lag models of discrete systems. Automatica 11(1975)6, 593-602.

18. Leiviskä K., Uronen P., Dynamic optimization of a sulphate mill pulp line. Preprints of IFAC/IFORS Symposium, Toulouse, France, 1979, 25-32.

19. Leiviskä K., Uronen P., Hierarchical control of an integrated pulp and paper mill. Report No. 113. PLAIC, Purdue University, West Lafayette, Indiana, 1979.

20. Leiviskä K., Uronen P., Dynamic optimization of a sulphate pulp mill. Preprints of the 2nd IFAC/IFORS Symposium on Optimization Methods - Applied Aspects. Varna, Bulgaria, 1979, 291-300.

21. Leiviskä K., Uronen P., Different approaches for the production control of a pulp mill. Preprints of the 4th PRP Conference, Ghent, 1980, 151-160.

SHAPE OPTIMIZATION FOR CONTACT PROBLEMS

B. Benedict
University of Iowa
Iowa City, IA 52242
USA

J. Sokolowski
Systems Research
Institut of Polish
Academy of Sciences
Warszawa, POLAND

J.P. Zolesio
Univ. de Nice,
Parc Valrose
06034 Nice Cedex
FRANCE

I. INTRODUCTION

Shape optimization for linear elastic bodies in frictionless contact is studied. The nonpenetration condition leads to a unilateral constraint on the displacement fields. The governing system is then described by a variational inequality. We will consider two geometric cases:

Case A

The unilateral constraint is applied over the domaine. Examples are problems involving strings, membranes, beams and plates. Consider a membrane in possible contact with a rigid obstacle (Fig. 1). The system is governed by a variational inequality with the constraint on admissible displacements

$$u \leq g \quad \text{a.e. in } \Omega$$

where u is the normal displacement of the membrane and g describes the shape of the obstacle. For any solution of the analysis problem we may define the contact region Z as in the subdomaine of Ω where $u = g$. Since Σ, the boundary of this subdomaine, is unknown before the contact problem is solved, it is called a free boundary. For the problems considered here to set $\Gamma \cup \Sigma$ is assumed to be empty.

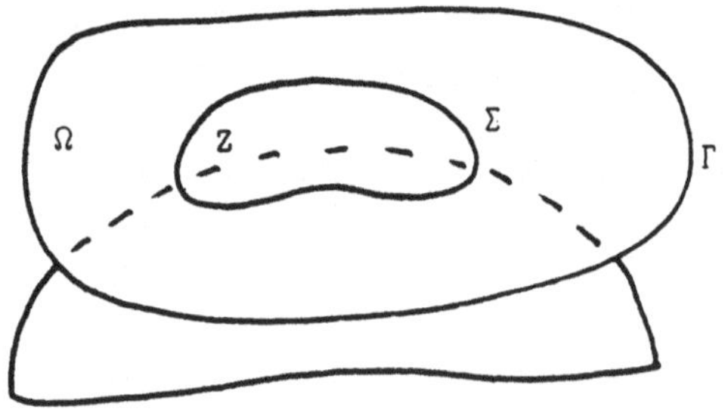

Figure 1

Case B

The unilateral constraint is applied on the boundary of the domaine. An example is one of planar elasticity (Fig. 2). The boundary may be divided into three regions:

$$\Gamma = \Gamma^0 \cup \Gamma' \cup \Gamma^c$$

with boundary conditions:

bilateral displacement conditions on Γ^0

stress conditions on Γ^1

unilateral displacement conditions on Γ^c

The contact region Z will Be a subset of Γ^c and the free boundary reduces to the end points of Z.

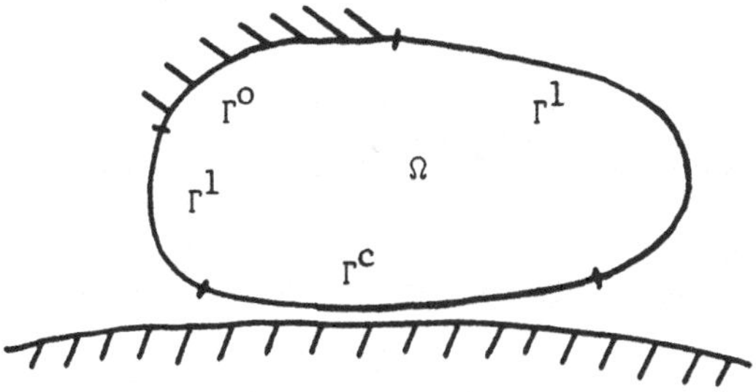

Figure 2

In each case the design problem is to minimize on the shape some cost funcitonal J which may depend on the shape and the state within some design constraints on the shape and the state. An engineering approach to these problems using the necessary conditions was presented by Benedict [1] and a discrete space programming approach by Haug and Kwak [2] and Conry [3].

II. Shape Optimization Approach

The "speed" method of shape sensitivity analysis [4] is used here. In this approach the deformation of the domaine Ω is related to a "velocity" field V(t,n) over R^n. The position of a point in the deformed domaine x is given by the solution of the ordinary differential equation

$$\frac{d}{dt} x(t) = V(t,x(t)) \quad , \qquad x(0) = X$$

(see Fig. 3). $T_t(V)$ is the transformation which depends on V such that

$$X \mapsto x = x(t,X)$$

$$\Omega_t = T_t(V)(\Omega)$$

$$\Gamma_t = T_t(V)(\Gamma) \quad .$$

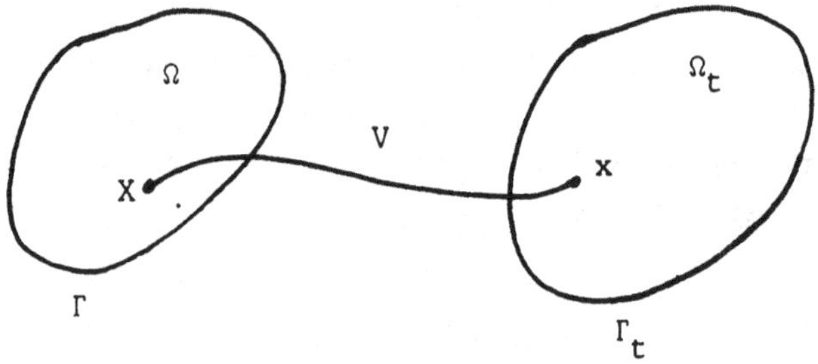

Figure 3

For

$$u \in H^M(\Omega_t)$$

consider

$$u^T = u_t \circ T_t(V)$$

where u^t is also an element of H^m. Using the idea of material derivative from fluid mechanics, in the direction V

$$\dot{u} = \lim_{t \to 0} \frac{u^t - u}{t}$$

where the limit is taken in the H^m norm.

The shape derivative is an element of the $H^{m-1}(\Omega)$ Sobolev space

$$u' = \dot{u} - \nabla u \cdot V(0)$$

We will need several results from previous work (Zolesio [5]).

Theorem 1. In "general" u' depends on only the normal component of V(0) on the boundary Γ ($v = V(0) \cdot n$).

The Eulerian semi-derivative of the cost J is

$$dJ(\Omega;V) = \lim_{t\to 0} \frac{J(\Omega_t) - J(\Omega)}{t} \quad .$$

By definition J is differentiable if the mapping

$$V \mapsto dJ(\Omega;V)$$

is linear.

Theorem 2: If J is differentiable

$$dJ(\Omega;V) = \int_\Gamma gv \, d\Gamma$$

where g is the density gradient.

III. DIFFERENTIABILITY OF CASE A

Explicit results are presented for a model problem. Consider a domaine Ω in which lies a membrane that may come in contact with a rigid obstacle. We call the PACKAGING problem the design problem of minimizing the area of Ω within the constraint that the contact range contain some specified subdomaine (Fig. 4). That is:

$$\begin{array}{ll} \min |\Omega| & \\ \text{subject to} & Z \supset \Omega_0 \quad . \end{array}$$

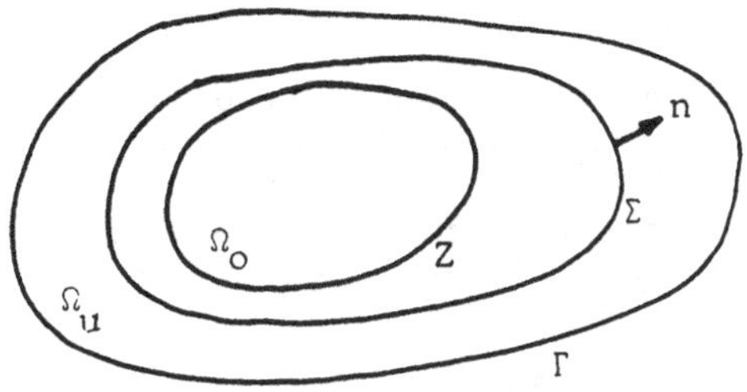

Figure 4

To enforce the constraint we introduce a penalty functional. For $\alpha > 0$ the cost functional becomes

$$J_\alpha(\Omega) = |\Omega| + \alpha \int_{\Omega_0} (g - u) d\Omega \qquad (1)$$

In addition to its role as a penalty multiplier α may be considered as a parameter which controls a tradeoff between area covered and the amount of material used.

The displacement u is the solution of the variational inequality

$$u \in H^1_0(\Omega), \quad \int_\Omega \nabla u \nabla(\phi - u)d\Omega \geq \int_\Omega f(\phi - u)d\Omega \qquad \phi \in H^1_0(\Omega)$$

For smooth data this implies

$$-\Delta u = f \quad \text{in} \quad \Omega u \qquad (\Omega_u = \Omega \setminus z)$$

$$u = 0 \quad \text{on } \Gamma$$

$$u = g \quad \text{on } \Sigma$$

$$\frac{\partial u}{\partial n} = \frac{\partial g}{\partial n} \qquad \text{on } \Sigma$$

The shape derivative u' has been characterized by Sokolowski and Zolesio [6] as the solution of the variational inequality:

$$u' \in S_v(\Omega) = \{\phi \in H^1(\Omega) \mid \phi = -\frac{\partial u}{\partial n} v \quad \text{on } \Gamma \quad ,$$

$$\int_Z f\phi \; d\Omega = 0 \quad , \qquad \phi \geq 0 \quad \text{q.e. on } Z\}$$

$$u' = P_{S_v(\Omega)}(0)$$

That is: the projection of zero on the closed convex subset S_V of $H^1(\Omega)$.

A stronger result for strictly positive loads is from Zolesio [7]. In this case

$$S_v(\Omega) = \{\phi \in H^1\Omega \mid \phi = \frac{\partial u}{\partial n} v \quad \text{on } \Gamma, \quad \phi = 0 \quad \text{on } z\}$$

then the shape derivative u' is the solution of the Dirichlet problem

$$-\Delta u' = 0 \quad \text{in } \Omega_u$$

$$u' = 0 \quad \text{on } \Sigma$$

$$u' = -\frac{\partial u}{\partial n} V \quad \text{on } \Gamma$$

Recall that the field $V = V(t,X)$ builds the new boundary of the body Γ_t. The free boundary Σ_t is built by the field

$$W(t,X) = \frac{\partial}{\partial n_t} u_t'(x) \frac{1}{f(x) + \Delta g(x)} n_t(x)$$

where n_t is the outer normal to Σ_t (Fig. 4).

Remark: The noncontact domain

$$\Omega_{u_t} = \{u | u_t(n) > g\}$$

is built by any field of the form

$$Y = r_t V + (1 + r_t) W$$

$$\text{with } r_t(x) = \begin{cases} 1 & \text{on } \Gamma_t \\ 0 & \text{on } \Sigma_t \end{cases}$$

Consider the solution on the noncontact domaine as the solution of a usual bilateral Dirichlet problem:

$$U_t = t_t \Big|_{\Omega_{u_t}} .$$

Recall that the shape derivative U' in the direction of the field Y is the solution of

$$\Delta U' = 0 \quad \text{in } \Omega_u$$

$$U' = \frac{\partial}{\partial n} (g - U) \quad , \qquad Y \cdot n = 0 \quad \text{on } \Sigma$$

$$U' = - \frac{\partial U}{\partial n} v \quad \text{on } \Gamma$$

Then the shape derivative may be expressed as

$$U_Y' = (u_V') \Big|_{\Omega_u} .$$

We may express the objective functional (1) as

$$J(\Omega_t) = \int_{\Omega_t} d\Omega + \alpha \int_{\Omega_{u_t}} R_o (g - V_t) d\Omega \quad , \qquad R_o = \begin{cases} 1 & \text{on } \Omega_o \\ 0 & \text{out of } \Omega_o \end{cases}$$

Then the classical Eulerian semi-derivative computation leads to

$$dJ(\Omega;V) = \int_\Gamma v\, d\Gamma - \alpha \int_{\Omega_u} R_o U'\, d\Omega + \alpha \int_{\partial\Omega_u} R_o(g - U)Y\cdot n\, ds. \tag{2}$$

The last term may be decomposed as

$$\alpha \int_\Gamma R_o(g - V)v\, d\Gamma + \alpha \int_\Sigma R_o(g - U)W(0)\cdot n\, d\Sigma$$

where the first term vanishes since R_o is the zero on Γ and the second term vanishes since $g - U$ is zero on Σ.

Introducing the adjoint equation

$$-\Delta p = R_o \quad \text{in} \quad \Omega_u$$

$$p = 0 \quad \text{on} \quad \Sigma \cup \Gamma$$

we may express (2) as

$$dJ(\Omega;V) = \int_\Gamma \left(1 + \alpha \frac{\partial p}{\partial n}\frac{\partial u}{\partial n}\right) v\, d\Gamma \quad .$$

This expression is linear in v and the density gradient can be explicitly computed and used in a descent technique to minimize J. Note that for this objective functional the adjoint relation is a variational equality.

IV. NONDIFFERENTIABILITY OF CASE B

In geometric Case B the unilateral constraint is applied on the boundary of the domaine of problem definition. Consider for a planar elasticity problem the spaces

$$H(\Omega_t) = \{\phi_t \in H^1(\Omega_t)^2 \mid \phi_t|_{\Gamma_t^0} = 0\}$$

$$K(\Omega_t) = \{\phi_t \in H^1(\Omega_t) \mid n_t\cdot\phi_t \leq 0 \quad \text{a.e. on } \Gamma_t^c\}$$

the bilinear form

$$a^0(\phi_t,\psi_t) = \int_{\Omega_t} \varepsilon(\phi_t)\cdot\cdot C\cdot\cdot\varepsilon(\phi_t)d\Omega$$

where

$$C = C_{ijk\ell} \quad , \qquad u\cdot\cdot C\cdot\cdot v = u_{ij}\, C_{ijk\ell}\, v_{k\ell}$$

$$\varepsilon(\phi) = 1/2(\phi_{i,j} + \phi_{j,i})$$

and the load potential

$$\langle F_t, \phi_f \rangle_t = \int_{\Omega_t} f\cdot\phi_t \, d\Omega + \int_{\Gamma_t^1} P\cdot\phi_t \, d\Gamma$$

The equilibrium displacement field y_t $K(\Omega_t)$ is the solution of the variational inequality

$$a_t(y_t, \psi_t - y_t) \geq \langle F_t, \psi_t - y_t \rangle_t \qquad \forall \psi \in K(\Omega_t) \quad .$$

We may consider some y_t to be the projection onto $K(\Omega_t)$ of some of $q_t \in H(\Omega_t)$:

$$y_t = P_{k(\Omega_t)}(q_t) \quad .$$

Shape Derivative of y

For a velocity field V, the domaine is transformed as

$$\Omega_t = T_t(v)\Omega \quad , \qquad \Gamma_t^i = T_t(v)(\Gamma^i) \quad , \qquad (i = 1,2,c)$$

and the new solution is, of course, a member of the transformed space

$$y_t \in K(\Omega_t) \quad .$$

Consider

$$y^t \in K(\Omega)$$

defined as

$$y^t = (DT_t)^{-1}\cdot y_t\cdot T_t \quad .$$

Theorem 3 [6]

y^t is right-side differentiable and $\partial y = \frac{d}{dt} y^t \big|_{t=0}$ is the solution of the variational inequality

$$\partial y \in S(\Omega) = \{\phi \in H(\Omega) \,|\, n\cdot\phi \leq 0 \;\; \text{a.e. on } Z(y)\subset\Gamma^c \text{ and } a(y,\phi) = \langle F \;, \phi_0 \rangle\}$$

$$a(\partial y, \phi - \partial y) \geq \langle F' - B\cdot y, \; \phi - \partial y \rangle \qquad \phi \quad S(\Omega)$$

$S(\Omega)$ is a fixed (relative to V) convex subset of $H(\Omega)$. F' and B linearly depend on V(0) and have explicit expressions given in [6]. We may also consider ∂y as a projection:

$$\partial y = P_{S(\Omega)}(F' - B\cdot y) \quad .$$

Cost Functionals

One of the most interesting cost functionals in this case is the maximum of the normal stress on the contact domaine [8]:

$$\sigma_n(t_{n\cdot n})\cdot\cdot C\cdot\cdot\varepsilon(y) = C_{ijk\ell}\, n_i n_j\, \varepsilon_{kl}$$

that is

$$J_\infty(\Omega_t) = \max_{x\ \Gamma_t^c} \sigma_n(x)$$

For this functional the Eulerian semiderivative is [4]

$$\partial J_\infty(\Omega,V) = \max_{x\ \gamma} \dot{\sigma}_n(x)$$

where γ is the subset of Γ^c where σ_n achieves its maximum and $\dot{\sigma}_n$ is the material derivative of σ_n. After a long computation we find

$$\dot{\sigma}_n = L(V) + (t_{n\cdot n})\cdot\cdot C\cdot\cdot\varepsilon(\partial y))$$

where L(V) depends linearly on the vector field V.

To more clearly reveal the nature of the sensitivity calculation for this geometric case, consider a simpler objective functional

$$J_p(\Omega) = \int_{\Gamma^c} (\sigma_n(x))^p \, d\Gamma$$

where $1 < p < \infty$. As p becomes large J_p approaches J_∞.

For the choice p = 1

$$J_1 = \int_{\Gamma^c} \sigma_n \, dx \qquad \text{and}$$

$$dJ_1(\Omega;V) = \int_{\Gamma^c} L_2(V)\, d\Gamma + \int_{\Gamma^c}(t_{n\cdot n})\cdot\cdot C\cdot\cdot\varepsilon(\partial y))\, d\Gamma \quad ,$$

where $L_2(V)$ depends linearly on V. However,

$$\partial y = P_{S(\Omega)}(F' - B \cdot y)$$

does not depend linearly on V. Although F' and B depend linearly on V their projecton does not. Then even this simple objective functional does not possess a differential that is linear on the field V.

Other nondifferentiable situations occur in structures optimization problems. The most studied are those due to repeated eigenvalues. In that case the differential is concave on the velocity field V [7] and the sensitivity may be computed. For the geometric Case B considered here, there is neither a convex nor concave property underlying the dependence on V. We have no immediate characterization of the generalized gradient.

References

1. Benedict, R.L., "Stiffness Optimization for Elastic Bodies in Contact," Journal of Mechanical Design, Vol. 104, No. 4, 1982.

2. Haug, E.J. and Kwak, B.M., "Contact Stress Minimization by Contour Design," Numerical Methods in Engineering, Vol. 12, 1978.

3. Conry, T.F., The use of mathematical programming in design for uniform load in nonlinear elastic systems, University of Wisconsin, Ph.D. Thesis, 1970.

4. Zolesio, J.P., "The Material Derivative (or Speed) Method for Shape Optimization," Optimization of Distributed Parameter Structures, Sijthoff & Noordhoff, 1981.

5. Zolesio, J.P., Identification de deomain per deformation, University of Nice, Thesis, 1979.

6. Sokolowski, J. and Zolesio, J.P., "Shape Sensitivity Analysis for Variational Inequalities," Proceedings of Tenth IFIP Conference on Optimization, Springer Verlag, 1981.

7. Zolesio, J.P., "Semi Derivatives of Repeated Eigenvalues," Optimization of Distributed Parameter Structures, Sijthoff & Noordhoff, 1981.

8. Benedict, R.L. and Taylor, J.E., "Optimal Design for Elastic Bodies in Contact," Optimization of Distributed Parameter Structures, Sijthoff & Noordhoff, 1981.

VARIATIONAL APPROACH TO OPTIMAL DESIGN AND SENSITIVITY ANALYSIS OF ELASTIC STRUCTURES

K. Dems
Łódź Technical University
ul. Żwirki 36
Łódź, Poland

1. INTRODUCTION

There are numerous problems of structural mechanics where we need to asses the variation of any local or global structural response characteristic due to variation of structural stiffness or structure shape parameters. For instance, in the redesign or optimization procedures, the variation of material stiffness or boundary shape is introduced and the respective variation of displacement or stress fields and an objective functional are evaluated at each redesign step. Similarly, in the identification procedure, the scalar distance functional between theoretical and experimental fields is minimized and its variation is to be determined for consecutive iterative steps.

Our analysis will be restricted to the linear elastic structures for which the design variables or parameters can be classified into the following two groups:

i/ material or cross-sectional variables specifying material or cross-sectional properties,

ii/ shape variables specifying shape of external boundaries or interfaces of a structure.

Thus, in the first case the problem is formulated within specified domain of a structure with variable material properties, whereas in the second case the external or internal boundaries of a structure with specified material properties are allowed to vary.

Generally, if G denotes any scalar functional and b_k are the design parameters of a structure, the sensitivity analysis is aimed at expressing the variation of G explicitly in terms of polynomial expansion

$$\Delta G = S_i \delta b_i + \frac{1}{2} R_{ij} \delta b_i \delta b_j + \ldots = \delta G + \frac{1}{2} \delta^2 G + \ldots \quad , \qquad (1)$$

where S_i and R_{ij} are respectively the first-order sensitivity vector

and the second-order sensitivity matrix.
To derive the first and second variations of any integral functional, we shall use the concept of an adjoint structure, that is the structure of the same form and material properties as primary structure, subjected to initial strain or stress fields within its domain and surface tractions or displacements on its boundary. The respective variations of G will be expressed using the solutions of boundary-value problems for primary and adjoint structures and will be dependent explicitly on variations of design variables or parameters.
In general, functional G can be assumed in the two following forms

$$G_1(\underline{\sigma},\underline{u},\underline{T},\underline{\varphi}) = \int\psi(\underline{\sigma},\underline{\varphi})\,dV + \int h(\underline{u},\underline{\varphi})\,dV + \int g(\underline{u})\,dS_T + \int f(\underline{T})\,dS_u \quad , \tag{2}$$

and

$$G_2(\underline{\varepsilon},\underline{u},\underline{T},\underline{\varphi}) = \int\phi(\underline{\varepsilon},\underline{\varphi})\,dV + \int h(\underline{u},\underline{\varphi})\,dV + \int g(\underline{u})\,dS_T + \int f(\underline{T})\,dS_u \quad , \tag{3}$$

where $\underline{\sigma}$, $\underline{\varepsilon}$, $\underline{u}$, $\underline{T}$ denote stress, strain, displacement and surface traction fields and $\underline{\varphi}$ is a set of design variables which can be dependent on a set of design parameters b_m, so that

$$\varphi_k = \varphi_k(\underline{x}, b_m) \quad , \qquad \delta\varphi_k = \frac{\partial\varphi_k}{\partial b_m}\delta b_m \quad . \tag{4}$$

2. SENSITIVITY ANALYSIS FOR VARYING MATERIAL PROPERTIES

Consider first the case of a fixed domain of a structure, Fig. 1a, for which the stress and displacement conditions are specified, $\underline{\sigma}\cdot\underline{n} = \underline{T}^o$ on S_T, $\underline{u} = \underline{u}^o$ on S_u, and the stiffness and compliance matrices depend on a set of design variables $\underline{\varphi}$, so that the Hooke's law and its invertion take the following form

$$\underline{\sigma} = \underline{D}(\varphi_k)\cdot\underline{\varepsilon} \quad , \qquad \underline{\varepsilon} = \underline{E}(\varphi_k)\cdot\underline{\sigma} \quad , \qquad \underline{D}\cdot\underline{E} = \underline{I} \quad . \tag{5}$$

The first variations of functionals G_1 and G_2 can be generally expressed as follows

$$\delta G_1 = \frac{\partial G_1}{\partial\underline{\sigma}}\cdot\delta\underline{\sigma} + \frac{\partial G_1}{\partial\underline{u}}\cdot\delta\underline{u} + \frac{\partial G_1}{\partial\underline{T}}\cdot\delta\underline{T} + \frac{\partial G_1}{\partial\underline{\varphi}}\cdot\delta\underline{\varphi} \quad , \tag{6}$$

and

$$\delta G_2 = \frac{\partial G_2}{\partial \underline{\varepsilon}} \cdot \delta \underline{\varepsilon} + \frac{\partial G_2}{\partial \underline{u}} \cdot \delta \underline{u} + \frac{\partial G_2}{\partial \underline{T}} \cdot \delta \underline{T} + \frac{\partial G_2}{\partial \underline{\varphi}} \cdot \delta \underline{\varphi} \quad , \tag{7}$$

and are dependent on variations of both state and design variables.

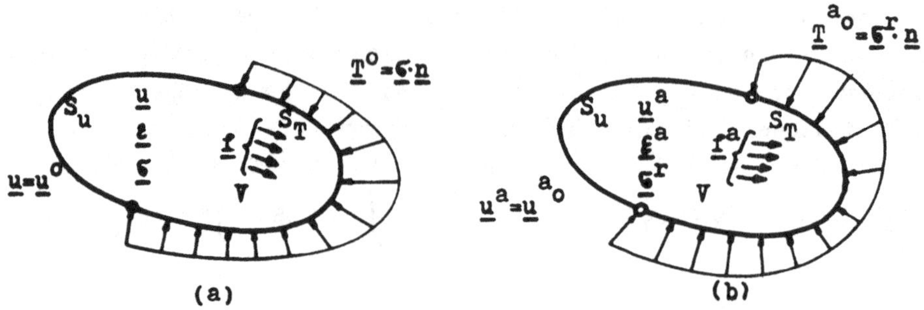

Fig. 1.(a) Primary structure subjected to variation of material parameters; (b) Adjoint structure.

To eliminate the terms involving the variations of state variables, let us introduce the adjoint structure, Fig. 1b, with the following boundary conditions

$$\underline{T}^{a_o} = \frac{\partial g}{\partial \underline{u}} \quad \text{on} \quad S_T \quad , \quad \underline{u}^{a_o} = -\frac{\partial f}{\partial \underline{T}} \quad \text{on} \quad S_u \quad , \quad \underline{f}^a = \frac{\partial h}{\partial \underline{u}} \quad \text{within} \quad V \quad , \tag{8}$$

and with the imposed initial strain or stress field within V, defined by the rule

$$\underline{\varepsilon}^i = \frac{\partial \psi}{\partial \underline{\sigma}} \qquad \text{or} \qquad \underline{\sigma}^i = \frac{\partial \phi}{\partial \underline{\varepsilon}} \qquad \text{within} \quad V \quad . \tag{9}$$

Since $\underline{\varepsilon}^i$ does not satisfy the compatibility conditions, it will induce the residual stress field $\underline{\sigma}^r$ and total strain field $\underline{\varepsilon}^a$, so that

$$\begin{aligned} \underline{\varepsilon}^a &= \underline{\varepsilon}^i + \underline{\varepsilon}^r \quad , \quad & \underline{\sigma}^r &= \underline{D} \cdot \underline{\varepsilon}^r \quad , \\ \underline{\sigma}^a &= \underline{\sigma}^i + \underline{\sigma}^r \quad , \quad & \underline{\sigma}^a &= \underline{D} \cdot \underline{\varepsilon}^a \quad , \end{aligned} \tag{10}$$

and

$$\operatorname{div} \underline{\sigma}^r + \underline{f}^a = 0 \quad \text{within} \quad V \quad , \qquad \underline{\sigma}^r \cdot \underline{n} = \underline{T}^{a_o} \quad \text{on} \quad S_T \quad . \tag{11}$$

These relations are shown on Fig. 2 for one-dimensional case.

Using now the state fields of the primary and adjoint structures, as well as the proper virtual work equations, we can eliminate the terms involving the variations of state variables and the variations of G_1

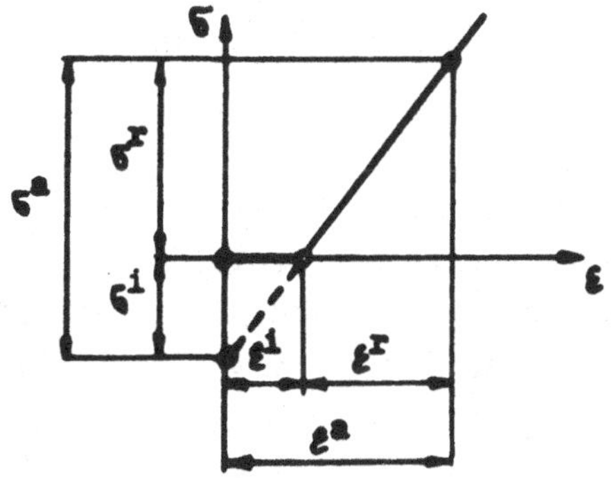

Fig. 2. Stress and strain in the adjoint structure.

and G_2 can be finally expressed as follows [1]

$$\delta G_1 = \int \left(\frac{\partial \psi}{\partial \varphi_k} + \frac{\partial h}{\partial \varphi_k} + \frac{\partial \underline{f}}{\partial \varphi_k} \underline{u}^a + \underline{\sigma}^r : \frac{\partial \underline{E}}{\partial \varphi_k} \cdot \underline{\sigma} \right) \delta \varphi_k \, dV \quad , \tag{12}$$

and

$$\delta G_2 = \int \left(\frac{\partial \phi}{\partial \varphi_k} + \frac{\partial h}{\partial \varphi_k} + \frac{\partial \underline{f}}{\partial \varphi_k} \underline{u}^a - \underline{\varepsilon}^a : \frac{\partial \underline{D}}{\partial \varphi_k} \cdot \underline{\varepsilon} \right) \delta \varphi_k \, dV \quad . \tag{13}$$

It is seen that the variation of G is expressed in terms of stress fields $\underline{\sigma}$, $\underline{\sigma}^r$ or strain fields $\underline{\varepsilon}$, $\underline{\varepsilon}^a$ of the primary and adjoint structures and is explicitly related to the variation of the design variables. Thus, the first-order sensitivity vector of functional G is obtained as the result of solutions of only two problems for primary and adjoint structures.

Derivation of second variation of G_1 or G_2 follows the similar steps. Consider, for instance, the functional G_1 and assume that design variables $\underline{\varphi}$ depend on a set of design parameters b_m, $m = 1, 2, \ldots, p$. Since $\delta^2 G_1$ depend on variation of state variables for both of primary and adjoint structures as well as variations of design parameters, we must introduce now two sets of p adjoint structures, in order to eliminate the terms involving the variations of state variables.

To eliminate the terms involving $\delta \underline{\sigma}^r$ and $\delta \underline{u}^a$ let us introduce the set of p initial adjoint structures with following boundary conditions

$$\overset{a_o}{\underline{T}}_{1m} = 0 \quad \text{on} \quad S_T \;, \quad \overset{a_o}{\underline{u}}_{1m} = 0 \quad \text{on} \quad S_u \;, \quad \underline{f}^a_{1m} = \frac{\partial \underline{f}}{\partial b_m} \quad \text{within} \quad V \;, \tag{14}$$

and with the imposed fields of initial strains

$$\underline{\varepsilon}^i_{1m} = \frac{\partial \underline{E}}{\partial b_m} \cdot \underline{\sigma} \qquad \text{within} \quad V \quad . \tag{15}$$

Furthermore, to eliminate the terms involving the variations of state variables of primary structure, the additional set of p terminal adjoint structures is introduced, for which the following boundary conditions

$$\overset{a_o}{\underline{T}}_{2m} = \frac{\partial^2 g}{\partial \underline{u} \partial \underline{u}} \cdot \underline{u}^a_{1m} \quad \text{on} \quad S_T \;, \quad \overset{a_o}{\underline{u}}_{2m} = -\underline{T}^a_{1m} \cdot \frac{\partial^2 f}{\partial \underline{T} \partial \underline{T}} \quad \text{on} \quad S_u \;,$$

$$\underline{f}^a_{2m} = \frac{\partial^2 h}{\partial b_m \partial \underline{u}} + \frac{\partial^2 h}{\partial \underline{u} \partial \underline{u}} \cdot \underline{u}^a_{1m} \quad \text{within} \quad V \;, \tag{16}$$

and the imposed fields of initial strains

$$\underline{\varepsilon}^i_{2m} = \frac{\partial^2 \psi}{\partial b_m \partial \underline{\sigma}} + \underline{\sigma}^r_{1m} \cdot \frac{\partial^2 \psi}{\partial \underline{\sigma} \partial \underline{\sigma}} + \underline{\sigma}^r \cdot \frac{\partial \underline{E}}{\partial b_m} \quad \text{within} \quad V \;, \tag{17}$$

are introduced. Using now the static and kinematic fields of primary, adjoint, initial adjoint and terminal adjoint structures, the second variation of G_1 can be explicitly expressed in terms of variations of material compliance parameters, and can be written as follows [2]

$$\delta^2 G_1 = \Big[\int \Big(\frac{\partial^2 \psi}{\partial b_m \partial b_n} + \underline{\sigma}^r_{1m} \cdot \frac{\partial^2 \psi}{\partial \underline{\sigma} \partial b_n} + \frac{\partial^2 h}{\partial b_m \partial b_n} + \frac{\partial^2 h}{\partial \underline{u} \partial b_n} \cdot \underline{u}^a_{1m} + \frac{\partial^2 \underline{f}}{\partial b_m \partial b_n} \cdot \underline{u}^a + \frac{\partial \underline{f}}{\partial b_n} \cdot \underline{u}^a_{1m} +$$

$$+ \underline{\sigma}^r \cdot \frac{\partial^2 \underline{E}}{\partial b_m \partial b_n} - + \underline{\sigma}^r_{1m} \cdot \frac{\partial \underline{E}}{\partial b_n} \cdot \underline{\sigma}^r + \underline{\sigma}^r_{2m} \cdot \frac{\partial \underline{E}}{\partial b_n} \cdot \underline{\sigma} \Big) dV \Big] \delta b_m \delta b_n \quad . \tag{18}$$

Thus, the second-order sensitivity matrix of functional G_1 is obtained as the result of solutions of $(2p+2)$ problems. The second variation of strain functional G_2 can be evaluated in a similar way.

The derived formulae for the second variations of an arbitrary functional may appear to be unnecessary complex. The complexity results from a general formulation of the problem. However, in most applications, many of the terms that appear will vanish and many others will have a simple form. Even in the most general case, all terms occuring are computable and the first and second variations may be calculated analytically or numerically.

3. SENSITIVITY ANALYSIS FOR VARYING STRUCTURAL SHAPE

Consider now the case when the material properties and boundary conditions of a structure are specified, whereas the actual body shape is defined by the transformation field $\underline{\varphi}(P)$ measured from the assumed basic reference configuration, so that typical point P at the reference configuration passes into the point P^* after transformation according to the rule, cf. Fig. 3

$$P \rightarrow P^*: \qquad \underline{x}^*(P^*) = \underline{x}(P) + \underline{\varphi}(P) \quad , \tag{19}$$

where $\underline{x}$ and $\underline{x}^*$ are the position vectors of P and P^*. The components φ_k /k=1,2,3/ of the vector $\underline{\varphi}$ can be regarded as design variables, for which the variation of any functional G is to be determined.
Consider now the functional G_1 and assume that transformation field modifies only the loaded boundary portion S_T, and the other boundary portions are fixed. The general form of first variation of G_1 takes now the following form

$$\delta G_1 = \int\left(\frac{\partial \psi}{\partial \underline{\sigma}}\cdot\delta\underline{\sigma} + \frac{\partial h}{\partial \underline{u}}\cdot\delta\underline{u}\right)dV + \int(\psi + h)\,\delta(dV) + \int\frac{\partial g}{\partial \underline{u}}\cdot\delta\underline{u}\,dS_T + \int g\,\delta(dS_T) + \int\frac{\partial f}{\partial \underline{T}}\cdot\delta\underline{T}\,dS_u \quad . \tag{20}$$

Introducing, as previously, the adjoint structure with boundary conditions (8) and imposed initial strain field (9), the first variation of any functional G can be expressed solely in terms of variations of components of transformation field $\underline{\varphi}$ and is dependent on states of primary and adjoint structures. The first variation of G_1 is thus expressed as follows [3]

$$\delta G_1 = \int \delta\underline{T}^o\cdot\underline{u}^a\,dS_T + \int\left\{\left[\psi + h + g_{,n} + (\underline{T}^o\cdot\underline{u}^a)_{,n} - 2\left(g + \underline{T}^o\cdot\underline{u}^a\right)H - \underline{\sigma}\cdot\underline{\varepsilon}^a + \underline{f}\cdot\underline{u}^a\right]n_k - \underline{T}^o_{,k}\cdot\underline{u}^a\right\}\delta\varphi_k\,dS_T \quad , \tag{21}$$

where $\delta \underline{T}^{o}$ is the total variation of surface tractions due to shape variation of a boundary S_T, and H denotes the mean curvature of S_T. The variation of functional G_2 is expressed by the formula similar to (21), where the function ψ is replaced by ϕ.
The similar expressions can be also obtained for the case of shape variation of supported boundary S_u or interface S_c between materials of different stiffness moduli within a structure.
Thus, similarly as for fixed domain of a structure, the first-order sensitivity vector of functionals G_1 or G_2 can be obtained, basing on so-

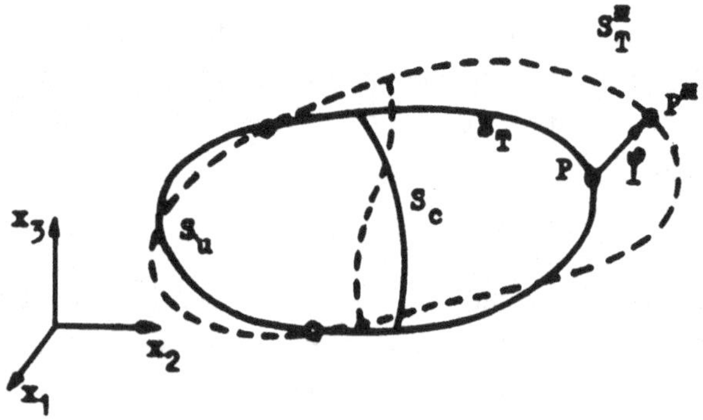

Fig. 3. Primary structure of varying shape.

lutions of only two boundary-value problems for primary and adjoint structures. The second variation of G and the second-order sensitivity matrix can be obtained using the similar approach as in the case of material stiffness variation. This problem is solved at present and the results will be presented in future.

4. STATIONARITY CONDITIONS IN OPTIMAL SHAPE DESIGN

In order to illustrate the applicability of derived expressions for variations of functionals G_1 and G_2, let us discuss the optimal shape design problem. Assume the cost of the structure proportional to the material volume, thus

$$C = \int c(\underline{\varphi})\, dV \quad , \quad \delta C = \int \frac{\partial c}{\partial \underline{\varphi}} \cdot \delta \underline{\varphi}\, dV \quad , \tag{22}$$

in the case of varying material parameters, and

$$C = c\int dV \quad , \quad \delta C = c\int n_k \delta\varphi_k \, dS \quad , \tag{23}$$

in the case of varying structural shape. Here c denotes the material cost per unit volume of a structure. The problem is then reduced to minimizing or maximizing the objective functional G with a specified upper bound on the structure cost, thus

$$\text{min. or max. of } G \quad \text{subject to } C \leqslant C_o \quad , \tag{24}$$

where C_o is a given quantity. Introducing the functional

$$G' = G + \lambda(C - C_o) \quad , \tag{25}$$

where λ is the Lagrange multiplier, the condition of stationarity of G' is expressed as follows

$$\delta G' = \delta G + \lambda \delta C + \delta\lambda(C - C_o) = 0 \quad , \tag{26}$$

and

$$\delta G = -\lambda \delta C \quad , \quad \delta\lambda(C - C_o) = 0 \quad . \tag{27}$$

The second equality requires either $C = C_o$ or $\delta\lambda = 0$. The first condition can be expressed explicitly by using the respective expression for variation of G given in the form of (12), (13) or (21) and variation of structural cost defined by (22) or (23), respectively.

In the next Section, a simple example of application of the derived optimality conditions is presented. Further examples can be found in previous works [1 - 4].

5. EXAMPLE

Consider a prismatic bar of elliptic cross-section, subjected to combined torsion and bending by the moments M_t and M_b, Fig. 4. We shall look for an optimal cross sectional shape within the class of elliptic shapes of specified cross-section area A_o and for the stress constraint

$$\sigma_e = \left[\sigma_{33}^2 + 3(\sigma_{13}^2 + \sigma_{23}^2)\right]^{1/2} \leqslant \sigma_o \quad , \tag{28}$$

where σ_{13}, σ_{23} and σ_{33} are the non-vanishing shear and normal stress components within the bar, refered to the coordinate system (x_1, x_2, x_3). Instead of the condition (28), we shall minimize the functional

$$G_1 = \int \left(\frac{\sigma_e}{\sigma_o} \right)^m dA = \int \frac{\left[\sigma_{33}^2 + 3(\sigma_{13}^2 + \sigma_{23}^2) \right]^{\frac{m}{2}}}{\sigma_o^m} dA \longrightarrow \text{min.} \quad , \tag{29}$$

subject to the constraint

$$\pi a_1 a_2 = A_o = \text{const.} \tag{30}$$

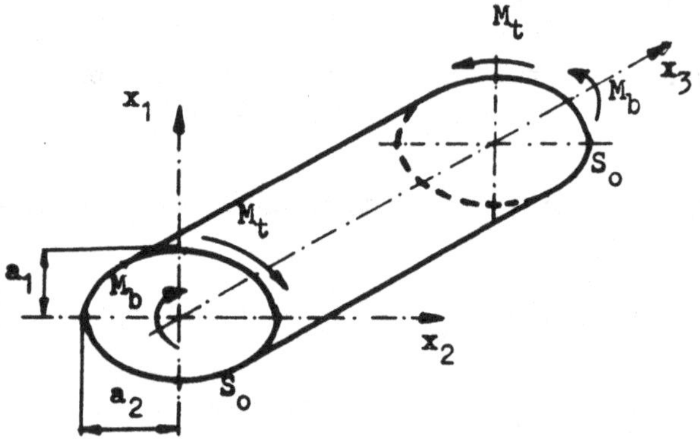

Fig. 4. Prismatic bar under torsion and bending.

Let us note that for $m \to \infty$ the functional G_1 represents the local effective stress defined by (28). The shape variation now depends on two parameters a_1 and a_2, and the stationarity condition (27), in view of (21) and (23) now takes the form

$$\int (\psi - \underline{\sigma} \cdot \underline{\varepsilon}^a) \delta\varphi_n \, dS_o = -\lambda c \int \delta \varphi_n \, dS_o \quad , \tag{31}$$

where

$$\psi = \left(\frac{\sigma_e}{\sigma_o} \right)^m \quad . \tag{32}$$

Using the solution for adjoint bar, the optimality condition (31) takes the form

$$(m-1) \int \left(\frac{\sigma_e}{\sigma_o} \right)^m \frac{x_i}{a_i} n_i \, dS_o = -\lambda c \int \frac{x_i}{a_i} n_i \, dS_o \quad /i=1,2/ \ , \quad \pi a_1 a_2 = A_o \ . \tag{33}$$

The optimal values of a_1 and a_2 are calculated from (33)

$$a_2 = \sqrt{\frac{A_o}{\pi}}\ \sqrt[4]{\frac{4}{3}\left(\frac{M_b}{M_t}\right)^2 + 1}\quad , \qquad a_1 = \frac{A_o}{\pi a_2} \quad . \tag{34}$$

It can be shown that the effective stress σ_e is constant for the shape specified by (34).

5. CONCLUDING REMARKS

The present work summerizes briefly the essential results of a systematic variational approach to first and second-order sensitivity analysis for arbitrary volume or surface integrals. The concept of adjoint structures, with clear engineering interpretation, can easily be applied in order to derive the expressions for first and second variations of these integrals. This variational approach can be applied in many problems of structural mechanics, as for instance in structural optimization, optimal identification of structural parameters, problems of phase transformation in solids or growth of biological structures, for which the transformation field $\underline{\varphi}(\underline{x})$ is governed by a constitutive rate equation.

REFERENCES

1. K. Dems and Z. Mróz, Variational approach by means of adjoint systems to structural optimization and sensitivity analysis - I, Int. J. Solids Struct., 19(8), 677 - 692, 1983
2. K. Dems and Z. Mróz, Variational approach to first- and second-order sensitivity analysis of elastic structures, Int. J. Num. Meth. Eng. /submitted for publication/
3. K. Dems, Optimal shape design of loaded boundaries, Arch. Mech. Stos., 33, 343 - 260, 1981
4. K. Dems and Z. Mróz, Optimal shape design of multicomposite structures, J. Struct. Mech., 8, 309 - 329, 1980

SHAPE OPTIMAL DESIGN OF A RADIATING FIN[1,2]

M.C.Delfour
Centre de recherche de mathématiques appliquées
Université de Montréal
C.P. 6128, Succ.A
Montréal, Québec H3C 3J7

G. Payre
Département de Génie Chimique
Université de Sherbrooke
Sherbrooke, Québec
Canada J1K 2R1

J.-P. Zolésio
Département de Mathématiques
Université de Nice
06034 - Nice - Cédex
France

1. INTRODUCTION. Current trends indicate that future communications satellites and spacecrafts will grow ever larger, consume ever more electrical power, and dissipate larger amounts of thermal energy. Various techniques and devices can be employed to condition the thermal environment for payload boxes within a spacecraft, but it is desirable to employ those which offer good performance for low cost, low weight and high reliability.

A thermal radiator (or radiating fin) which accepts a given thermal power flux (TPF) from a payload box and radiates it directly to space, can offer good performance and high reliability at low cost. However, without careful design, such a radiator can be unnecessarily bulky and heavy. It is the mass-optimized design of the thermal radiator which is the problem at hand. We may assume that the payload box presents a uniform TPF (typ. 0.1 to 1.0 W/cm^2) into the radiator at the box/radiator interface. The radiating surface is a second surface mirror which consists of a sheet of glass whose inner surface has silver coating. We may assume that the TPF out of the radiator/space interface is governed by the T^4 radiation law, although we must account also for a constant TPF (typ. 0.01 W/cm^2) into this interface from the sun. Any other surfaces of the radiator may be treated as adiabatic. Two constraints restrict freedom in the design of the thermal radiator:

(i) The maximum temperature at the box/radiator interface is not to exceed some constant (typ. 50°C), and

(ii) no part of the radiator is to be thinner than some constant (typ. 1mm).

A similar problem with a convective rather than a radiative condition has been considered by G.E.Schneider, M.M.Yovanovich and R.L.D.Cane [1] in connection with the cooling of banks of electronic circuitry where extensive use is made of integrated circuit devices. This work is also connected with the design of solar collectors and collector plates.

Shape optimal design techniques (cf. J.Céa [1,2,3]) have been successfully used in the design of mass-optimized thermal diffusers (cf. M.Delfour, F.Payre and J.P.Zolésio [1], and Ph.Destuynder [1]).

---

1. This problem has been suggested by A.S.Jones (Spar Aerospace Ltd., Toronto, Canada.)The above detailed statement has been provided by Dr.V.A.Wehrle (Communications Research Centre, Department of Communications, Ottawa, Canada).

2. This research has been supported by the Natural Sciences and Engineering Research Council of Canada, Strategic Grant G-0654 in Communications.

In this paper we present an efficient iterative method of solving the finite element approximation to the non-linear boundary value problem describing the temperature distribution in the radiating fin. This basic tool will be used many times to find the solution of the optimal design problem.

For the optimal design problem, the theory and gradient computations are obtained for the finite element model. Shapes are restricted to a family of polygonal domains characterized by a finite number of shape parameters. Gradient computations involve partial derivatives of the position of the nodes with respect to the shape parameters and partial material derivatives of the state variable with respect to the position of each node. Both ingredients are combined to obtain the gradient of the penalized cost function with respect to the shape parameters. A few steps of some very preliminary computations are presented. However, at this stage, the program has not yet been carefully checked and is not completely operational.

Notation. Let $\mathbb{R}$ be the field of all real numbers. For an integer $K \geq 1$, $\mathbb{R}^K$ will be the K-dimensional Euclidean space. Given an open domain $\Omega \subset \mathbb{R}^K$, $L^p(\Omega)$ will be the space of p-integrable functions from Ω into $\mathbb{R}$, $1 \leq p \leq \infty$, and $L^\infty(\Omega)$ the space of essentially bounded functions from Ω into $\mathbb{R}$. $W^{1,p}(\Omega)$, $1 \leq p \leq \infty$, will denote the Sobolev space of $L^p(\Omega)$ functions with distributional derivatives of the first order in $L^p(\Omega)$. Define $H^1(\Omega) = W^{1,2}(\Omega)$. $C^0(\Omega)$ will denote the space of bounded continuous functions from Ω into $\mathbb{R}$.

2. STATEMENT OF THE PROBLEM. We assume that the radiator is a volume Ω symmetrical about the z-axis (cf. Figure 1) whose boundary surface Σ is made up of three regular pieces: the contact surface Σ_1 (a disk perpendicular to the z-axis with center at the point $(r,z) = (0,0)$, the lateral adiabatic surface Σ_2 and the radiating surface Σ_3 (a disk perpendicular to the z-axis with center at $(r,z) = (0,L)$). More precisely

$$\Sigma_1 = \{(x,y,z) \mid z=0 \text{ and } x^2+y^2 \leq R_0^2\}, \quad \Sigma_2 = \{(x,y,z) \mid x^2+y^2=R(z)^2, \ 0 \leq z \leq L\}$$

(1)
$$\Sigma_3 = \{(x,y,z) \mid z=L, \ x^2+y^2 \leq R(L)^2\}$$

where the radius $R_0 > 0$ (typ. 10cm) the length $L > 0$ and the function

(2)
$$R:[0,L] \to \mathbb{R}, \quad R(0) = R_0, \quad R(z) > 0, \quad 0 \leq z \leq L$$

are given. ($\mathbb{R}$, the field of real numbers).

The temperature distribution (in Kelvin degrees) over this volume Ω is the solution of the stationary heat equation

(3)
$$\Delta T = 0 \quad \text{(the Laplacian of T)}$$

with the following boundary conditions on the surface $\Sigma = \Sigma_1 \cup \Sigma_2 \cup \Sigma_3$ (the boundary of Ω):

(4)
$$k\frac{\partial T}{\partial n} = q_{in} \text{ on } \Sigma_1, \quad k\frac{\partial T}{\partial n} = 0 \text{ on } \Sigma_2, \quad k\frac{\partial T}{\partial n} + \sigma\varepsilon T^4 = q_s \text{ on } \Sigma_3,$$

where n always denotes the outward normal to the boundary surface Σ and $\partial T/\partial n$ is the normal derivative on the boundary surface Σ. The parameters appearing in (1) to (6) are k = thermal conductivity (1.8 W/cm×°C), q_{in} = uniform inward thermal power flux at the contact surface (typ. 0.1 to 1.0 W/cm^2), σ = Boltzmann's constant (5.67×10^{-8} W/m^{2}K^4), ε = surface emissivity (typ. 0.8), q_s = solar inward thermal power flux (0.01 W/cm^2).

The optimal design problem consists in minimizing the volume

(5) $$J(R,L) = \pi\int_0^L R(z)^2 dz$$

over all lengths $L > 0$ and shape functions R subject to the constraint

(6) $$T(x,y,z) \le T_f \quad (\text{typ. } 50°\text{C}), \quad \forall\ (x,y,z) \in \Omega.$$

In the present analysis we drop the requirement (ii) in section 1.

3. SCALING AND VARIATIONAL FORMULATION OF THE NON-LINEAR BOUNDARY VALUE PROBLEM. It is convenient to introduce the following changes of variables (cf. Figure 2)

(7) $$\zeta = z/L, \quad \xi_1 = x/R_0, \quad \xi_2 = y/R_0, \quad \lambda = L/R_0$$

(8) $$y(\xi_1,\xi_2,\zeta) = \lambda^{1/3}(\sigma\varepsilon R_0/k)^{1/3}\, T(R_0\xi_1, R_0\xi_2, L\zeta).$$

This defines the *new shape function*

(9) $$\rho(\zeta) = R(L\zeta)/R_0, \quad \rho: [0,1] \to R_+, \quad \rho(0) = 1, \quad \rho(\zeta) > 0 \text{ in } [0,1],$$

and the *dimensionless parameter* $\lambda > 0$. The new volume $\tilde{\Omega}$ and its boundary $\tilde{\Sigma}$ are given by

(10) $$\tilde{\Omega} = \{(\xi_1,\xi_2,\zeta)\,|\,0 < \zeta < 1,\ \xi_1^2+\xi_2^2 < \rho(\zeta)^2\}, \quad \tilde{\Sigma} = \tilde{\Sigma}_1 \cup \tilde{\Sigma}_2 \cup \tilde{\Sigma}_3$$

(11) $$\tilde{\Sigma}_1 = \{(\xi_1,\xi_2,0)\,|\,\xi_1^2+\xi_2^2 \le 1\}, \quad \tilde{\Sigma}_2 = \{(\xi_1,\xi_2,\zeta)\,|\,0\le\zeta\le 1,\ \xi_1^2+\xi_2^2=\rho(\zeta)^2\}$$
$$\tilde{\Sigma}_3 = \{(\xi_1,\xi_2,1)\,|\,\xi_1^2+\xi_2^2 \le \rho(1)^2\}.$$

Equations (3) and (4) become

(12) $$A(y) = -\left[\lambda^2\left(\frac{\partial^2 y}{\partial\xi_1^2} + \frac{\partial^2 y}{\partial\xi_2^2}\right) + \frac{\partial^2 y}{\partial\zeta^2}\right] = 0 \text{ in } \tilde{\Omega},$$

(13) $$\frac{\partial y}{\partial\nu_A} = \tilde{q}_{in} \text{ on } \tilde{\Sigma}_1, \quad \frac{\partial y}{\partial\nu_A} = 0 \text{ on } \tilde{\Sigma}_2, \quad \frac{\partial y}{\partial\nu_A} + y^4 = \tilde{q}_s \text{ on } \tilde{\Sigma}_3$$

where

(14) $$\beta = (\sigma\varepsilon R_0^4/k^4)^{1/3}, \quad \tilde{q}_{in} = \beta q_{in}\lambda^{4/3}, \quad \tilde{q}_s = \beta q_s\lambda^{4/3}.$$

The solution y only depends on λ, βq_{in}, βq_s and the shape function ρ. Once βq_{in} and βq_s have been fixed, the *optimal design problem* consists in finding the parameter λ and the shape function ρ which minimizes the volume

(15) $$J(\lambda,\rho) = \pi\lambda\int_0^1 \rho(\zeta)^2 d\zeta$$

subject to the constraint

(16) $$\sup\{y(\sigma)\,|\,\sigma \in \tilde{\Sigma}_1\} \le \tilde{y}_1, \quad \tilde{y}_1 = T_f(\sigma\varepsilon R_0/k)^{1/3}\lambda^{1/3}.$$

It was shown in Delfour, Payre and Zolesio [2], that the solution

(17) $$y(r,\zeta) = y(\xi_1,\xi_2,\zeta), \quad r = \sqrt{\xi_1^2+\xi_2^2}$$

(in cylindrical coordinates) of the above boundary-value problem coincides with the minimizing element of the functional j

(18) $$j(\varphi) = \frac{1}{2}a_\lambda(\varphi,\varphi)+\int_0^{\rho(1)} 2\pi r dr[\frac{1}{5}|\varphi|^5-\tilde{q}_s\varphi]-\int_0^1 2\pi r dr\tilde{q}_{in}\varphi$$

(19) $$a_\lambda(\varphi,\psi) = \int_0^1\int_0^{\rho(\zeta)}[\lambda^2 \frac{\partial\varphi}{\partial r}\frac{\partial\psi}{\partial r} + \frac{\partial\varphi}{\partial\zeta}\frac{\partial\psi}{\partial\zeta}]2\pi r dr d\zeta,$$

over the function space

(20) $$W(A) = \{\varphi|\sqrt{r}\,\varphi,\ \sqrt{r}\frac{\partial\varphi}{\partial r},\ \sqrt{r}\frac{\partial\varphi}{\partial\zeta} \in L^2(A),\ \sqrt[5]{r}\,\varphi|_{S_3} \in L^5(S_3)\}.$$

The functional j is Gateau differentiable and its derivative at φ in the ψ direction is given by

(21) $$dj(\varphi;\psi) = a_\lambda(\varphi,\psi)+\int_0^{\rho(1)} 2\pi r dr[|\varphi|^3\varphi-\tilde{q}_s]\psi-\int_C^1 2\pi r dr\tilde{q}_{in}\psi.$$

The minimizing element y is completely characterized by the variational equation

(22) $$dj(y;\psi) = 0, \quad \forall\psi \in W(A).$$

Moreover the function y is positive over the closure $\bar{A}$ of A. The Hessian can also be explicitly computed:

(23) $$d^2j(\varphi;\psi,\eta) = a_\lambda(\eta,\psi)+\int_0^{\rho(1)} 2\pi r dr\ 4|\varphi|^2\varphi\eta\psi.$$

4. FINITE ELEMENT APPROXIMATION OF THE NLBVP. The function space W(A) is approximated in two steps. First an approximation ρ_h of the shape function ρ as a continuous piecewise linear function. The function ρ_h generates an approximation A_h

(24) $$A_h = \{(r,\zeta)|0 < \zeta < 1, \quad 0 < r < \rho_h(\zeta)\}$$

to the cross-section A; by revolution about the ζ-axis, A_h generates the volume $\tilde{\Omega}_h$.

Then, define a triangulation τ_h on the cross-section A_h and associate with it the finite element approximation

(25) $$W_h = \{\varphi_h|\varphi_h \in C^0(\bar{A}_h),\ \varphi_h \text{ linear on each triangle in } \tau_h\}.$$

The finite element approximation y_h of the minimizing function y is given by the solution of the variational equation

(26) $$dj_h(y_h;\psi_h) = 0, \quad \forall\psi_h \in W_h,$$

where j_h is defined by (18) with ρ_h and A_h instead of ρ and A. Various non-linear programming algorithms have been developed in Delfour-Payre-Zolésio [2] to compute y_h. In this paper we use the following new iterative scheme. Given the real function

(27) $$f(x) = (|x|^3x-c^4)/(x-c), \quad c^4 = \tilde{q}_s,$$

we consider the following sequence of minimization problems: at step $n \geq 1$, y_n is known and $y_{n+1} \in W_h$ is constructed as follows:

(28) $$j_n(y_{n+1}) = \text{Inf}\{j_n(\varphi_h)|\varphi_h \in W_h\},$$

where

$$j_n(\varphi) = \frac{1}{2}a_\lambda(\varphi,\varphi) + \int_0^{\rho_h(1)} f(y_n)\frac{1}{2}|\varphi - c|^2 2\pi r dr - \int_0^1 \tilde{q}_{in}\varphi 2\pi r dr. \tag{29}$$

5. DISCRETE OPTIMAL DESIGN PROBLEM. Given the function ρ_h, the cross-section A_h and the triangulation τ_h, the temperature distribution y_h in W_h is the solution of (26). The "discrete optimal design problem" is the same as the one in section 3 except for the fact that ρ is replaced by ρ_h in (15) and (16). This constrained problem is solved by penalization of (16):

$$J_\varepsilon(\lambda,\tau_h) = J(\lambda,\rho_h) + \frac{1}{\varepsilon}\int_0^1 \{[y_h - \tilde{y}_1]^+\}^2 2\pi r dr, \quad \varepsilon > 0 \tag{30}$$

where $[u]^+ = \max\{u,0\}$ and τ_h indicates that J_ε depends not only on ρ_h but also on the triangulation through the state y_h.

6. CONSTRUCTION OF THE TRIANGULATION τ AND INTRODUCTION OF THE SHAPE PARAMETERS. For simplicity we drop the subscript "h" associated with the "discrete problem". It has been noticed that the state y does not only depend on the surface A but also on the chosen triangulation τ and, a fortiori, on the set of nodes $\vec{M} = \{M_i | 1 \le i \le m\}$ (m, an integer) which defines τ. It is not computationally and physically desirable to leave all the nodes completely free. So we introduce a fixed set of shape parameters, $\vec{\ell} = \{\ell_k | 1 \le k \le p+q\}$ (p and q, integers) to "control" $\vec{M}$.

The domain A is divided into two parts by introducing a focus $F = (F_r, 0)$, $F_r > 1$, and two sets of nodes $\vec{E}_p = \{E_i | 0 \le i \le p\}$ $(E_0 = (0,0),\ E_p = (F_r, 1))$ and $\vec{E}_q = \{E_i | p < i \le p+q\}$ $(E_i = (E_{ir}, 1))$ distributed along the ζ-axis and the boundary S_3 (cf. Figure 3). In the first part, rays are drawn from F to each node E_i, $0 \le i \le p$; in the second part, lines are drawn parallel to the ζ-axis through each point of $\vec{E}_q$. The boundary S_2 and, a fortiori, the function ρ are defined by the set of positive parameters $\vec{\ell}$ in the following way. The boundary points are

$$B_i = \begin{cases} E_i + (F - E_i)\ell_i / |F - E_i|, & 1 \le i \le p, \\ E_i - (0, \ell_i), & p < i \le p+q \,. \end{cases} \tag{31}$$

By joining the points $B_0 = (1,0)$, B_1, $B_2, \ldots, B_{p+q}$ and E_{p+q} by straight lines, the function ρ_h and the boundary S_2 are generated. In the actual computations the first component of the node E_{p+q} was also considered as a shape parameter controlling the nodes E_i, $p < i < p+q$. However, for lack of space, we do not describe this construction here.

7. ELEMENTS IN THE COMPUTATION OF THE GRADIENT OF THE COST FUNCTION WITH RESPECT TO THE SHAPE PARAMETERS. Though the surface A is completely determined by the boundary nodes, the finite element state y depends on the whole triangulation τ and thence on the set of nodes $\vec{M}$. In general we can write

$$y = y(\vec{M}), \quad A = A(\vec{M}) \text{ and } J_\varepsilon = J_\varepsilon(A(\vec{M})). \tag{32}$$

By construction the nodes $\vec{M}$ are completely determined by the fixed nodes $\vec{E}_p$, $\vec{E}_q$ and

F and the shape parameters $\vec{\ell}$. Therefore we write

(33) $$\vec{M} = \vec{M}(\vec{\ell}),$$

and the cost function becomes a function L of $\vec{\ell}$

(34) $$L(\vec{\ell}) = J_\varepsilon(A(\vec{M}(\vec{\ell}))).$$

As a result, using the chain rule,

(35) $$\frac{\partial L}{\partial \ell_k} = \sum_{i=1}^{m} \left[\frac{\partial J_\varepsilon}{\partial r_i}\frac{\partial r_i}{\partial \ell_k} + \frac{\partial J_\varepsilon}{\partial \zeta_i}\frac{\partial \zeta_i}{\partial \ell_k}\right], \quad M_i = (r_i,\zeta_i), \quad 1 \le r \le m,$$

where the pairs $\{(\partial J_\varepsilon/\partial r_i, \partial J_\varepsilon/\partial \zeta_i) \mid 1 \le i \le m\}$ are the partial material derivatives of J_ε with respect to the node M_i and the partial derivatives of r_i and ζ_i with respect to ℓ_k. The computation of (35) involves the *partial material derivatives* $(\dot{y}_{r_i}, \dot{y}_{\zeta_i})$ of the state y with respect to the position of each node M_i and the introduction of an adjoint state. This will require the construction of appropriate deformation fields.

8. PARTIAL MATERIAL DERIVATIVES.

We briefly recall the speed method for boundary value problems over smooth domains Ω. Given a smooth deformation vector field V defined in a neighbourhood of Ω, each point X in Ω at time t=0 is transported into a point x(t) at time t > 0 through the differential equation

(36) $$dx(t)/dt = V(t,x(t)), \quad x(0) = X.$$

This induces a smooth transformation $T_t(V)X = x(t)$ which maps Ω onto $\Omega_t = T_t(V)\Omega$. The Eulerian derivative of the cost function J at Ω for the field V is defined as (cf. J.P.Zolésio [1,2])

(37) $$dJ(\Omega;V) = (d/dt)J(\Omega_t)|_{t=0}.$$

In the discrete case, the state y depends on the nodes $\vec{M}$ through the triangulation τ. Given a node $M_i = (r_i,\zeta_i)$ and a small $t > 0$ we perturb the nodes $\vec{M}$ into

(38) $$\vec{M}^t_{r_i} = \{M_j + t(\delta_{ij},0) \mid 1 \le j \le m\} \quad \text{or} \quad \vec{M}^t_{\zeta_i} = \{M_j + t(0,\delta_{ij}) \mid 1 \le j \le m\}$$

where δ_{ij} is the Kronecker index function. Now we want to construct in each case vector fields which will transport triangles in τ onto a new set of triangles τ^t and the shape functions $\vec{e} = \{e_j \mid 1 \le j \le m\}$ in W_h for τ onto shape functions $\vec{e}^t = \{e^t_j \mid 1 \le j \le m\}$ in W^t_h for τ^t:

(39) $$e_j(M_i) \quad (\text{resp. } e^t_j(M^t_i)) = \delta_{ij}, \quad 1 \le i,j \le m.$$

J.P.Zolésio [4] has shown that an appropriate choice is

(40) $$V_{r_i}(r,\zeta) = (e_i(r,\zeta),0) \quad (\text{resp. } V_{\zeta_i}(r,\zeta) = (0,e_i(r,\zeta))).$$

Such a field maps each triangle of τ onto a triangle of τ^t and each basis element e_j onto a basis element e^t_j. Moreover if y_t is a solution of the boundary value problem in W^t_h, then

(41) $$y^t_{r_i} = y_t \circ T_t(V_{r_i}) \quad (\text{resp. } y^t_{\zeta_i} = y_t \circ T_t(V_{\zeta_i}))$$

belongs to W_h. Thus the *partial material derivative* of y is an element of W_h

(42) $$\dot{y}_{r_i} = dy^t_{r_i}/dt|_{t=0} \quad (\text{resp. } \dot{y}_{\zeta_i} = dy^t_{\zeta_i}/dt|_{t=0})$$

and the *partial Eulerian derivative* of J_ε is given by

(43) $$\partial J_\varepsilon/\partial r_i = dJ_\varepsilon(A;V_{r_i}) \quad (\text{resp. } \partial J_\varepsilon/\partial\zeta_i = dJ_\varepsilon(A;V_{\zeta_i})).$$

It can be shown that for vector fields $V = (V^r, V^\zeta)$ of this type (that is, V_{r_i} or V_{ζ_i}), the Eulerian derivative of J_ε is given by (cf. J.P.Zolésio [1,4])

(44) $$dJ_\varepsilon(A;V) = \int_A 2\pi\,[V^r + r(\partial_r V^r + \partial_\zeta V^\zeta)]\,dr\,d\zeta + \int_A \langle \mathcal{A}'\nabla y, \nabla p\rangle 2\pi\,dr\,d\zeta + \int_{S_3}(y^4 - \tilde{q}_s)\,p\,(V^r + r\partial_r V^r)\,dr,$$

where ∂_r and ∂_ζ denote partial derivatives and $\mathcal{A}'$ is the 2×2 matrix

(45) $$\mathcal{A}' = (\operatorname{div}V)B - (DV)B - B(DV)^* + V^r E, \quad B = rE, \quad E = \begin{bmatrix} \lambda^2 & 0 \\ 0 & 1 \end{bmatrix}$$

DV is the Jacobian matrix of V, $(DV)^*$ its transpose and p is the solution in W_h of the variational equation

(46) $$d^2 j(y;p,\psi) + \frac{2}{\varepsilon}\int_0^1 [y - \tilde{y}_1]^+ \psi\, 2\pi r\,dr = 0, \quad \forall\psi \in W_h.$$

The substitution of $V = V_{r_i}$ (resp. $V = V_{\zeta_i}$) will yield an explicit expression for (43) in terms of e_i, y and p. Combining this with the chain rule (35), the optimal design problem can be solved by any gradient method.

9. NUMERICAL RESULTS. The results presented here are very preliminary. They only show a few iterations with a computer program which has not yet been carefully checked and which is not completely operational. The triangulation is made up of 167 nodes and 275 elements (linear on each triangle).

The tests have been divided into two steps. Firstly the geometry of the boundary S_2 has been chosen as two straight lines (cf. Figure 4): one from (1,0) to (R,Z), 0 < Z < 1 (R is the r-component of the point E_{p+q} = (R,1)) and one from (R,Z) to E_{p+q} = (R,1). After a series of evaluations of the volume and the constraint, the parameters λ, R and Z were chosen in the following way: λR_0 = L = 1cm, ZL = 0,9 cm, RR_0 = 18,75 cm, R_0 = 5 cm. During the iterations, λ, R and Z were fixed and the algorithm began to dig into the surface S_2. Five such iterations are shown in Figure 5. The evolution of the volume, the constraint and the cost as a function of the iteration number are shown in Figure 6. The temperature profiles at iteration 5 are shown in Figure 7. The hottest profile is 49,75° on the side of the boundary $\tilde{\Sigma}_3$ near the ζ-axis. The shape obtained at iteration 5 is not necessarily optimal since iterations were not continued. Nevertheless the algorithm seems to converge towards a "physical" solution.

REFERENCES

J. Céa [1], Une méthode numérique pour la recherche d'un domaine optimal, Publication IMAN, Université de Nice, 1976.

[2], Optimization, theory and algorithm, Tata Institute, Springer Verlag, 1978.

[3], Problem of Shape Optimal Design, pp. 1005-1048; Numerical Methods of Shape Optimal Design, pp. 1049-1088, in "Optimization of distributed parameter structures", Vol.II, E.Haug, and J.Céa, eds., Sijthoff and Noordhoff, Alphen aan den Rijn, The Netherlands, 1981.

M.C. Delfour, G.Payre and J.P. Zolésio [1], Optimal design of a minimum weight
    thermal diffuser with constraint on the output thermal power flux, J.Applied
    Math. and Optimization 9 (1983), 225-262.
    [2], Optimal design of a radiating fin for communications satellites, Proc.
    IFAC Workshop on Appl.Nonlinear Programming to Optimization and Control,
    H.Rauch,ed., San Francisco, U.S.A., June 1983 (to appear).
Ph.Destuynder [1], Etude théorique et numérique d'un algorithme d'optimisation de
    structures. Thèse de docteur ingénieur, Université de Paris VI, 1976.
G.E.Schnéider, M.M. Yovanovich and R.L.D. Cane [1], Thermal resistance of a con-
    vectively cooled plate with nonuniform applied flux, J.Spacecraft and Rockets
    17, 372-376, 1980.
J.P.Zolésio [1], The material derivative (or speed) method for shape optimization.
    In "Optimization of distributed parameter structures" Vol.II, pp. 1089-1151,
    E.J.Haug and J.Céa,eds., Sijthoff and Noordhoff, Alphen aan den Rijn, The
    Netherlands, 1981.
    [2], Identification de domaines par déformation. Thèse de doctorat d'état,
    Université de Nice, 1979.
    [3] In Proceedings of a conference held at Murat le Quaire, A.Auslender,ed.,
    pp. 200-230, Lecture Notes in Economical and Mathematical Systems no.14,
    Springer Verlag, New York, 1977.
    [4], Les dérivées par rapport aux noeuds des triangularisations et leur uti-
    lisations en identification de domaine, Ann.Sc.Mat.Québec, to appear.

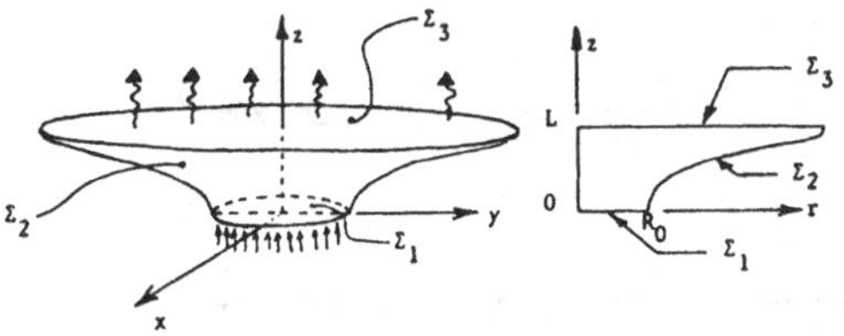

Figure 1. Domain Ω and its
          generating surface.

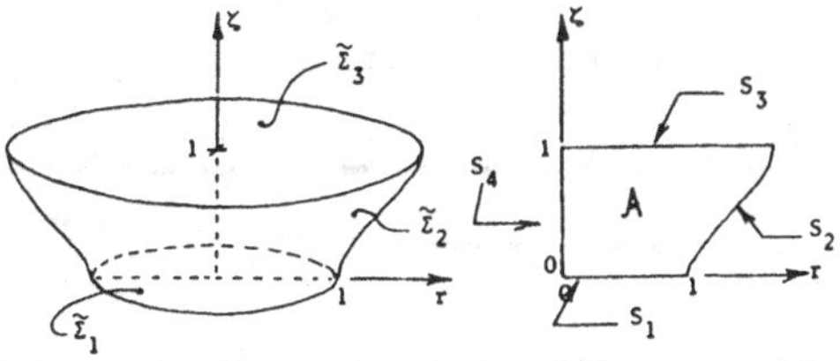

Figure 2. Domain $\widetilde{\Omega}$ and its
          generating surface A.

$$M_i = (r_i, \widetilde{z}_i) \quad \left(\frac{\partial r_i}{\partial \ell_5}, \frac{\partial \widetilde{z}_i}{\partial \ell_5} \right) = \tfrac{2}{3}(\cos\theta_5, -\sin\theta_5)$$

$$M_j = (r_j, \widetilde{z}_j) \quad \left(\frac{\partial r_j}{\partial \ell_{p+2}}, \frac{\partial \widetilde{z}_j}{\partial \ell_{p+2}} \right) = \left(0, -\tfrac{2}{3}\right)$$

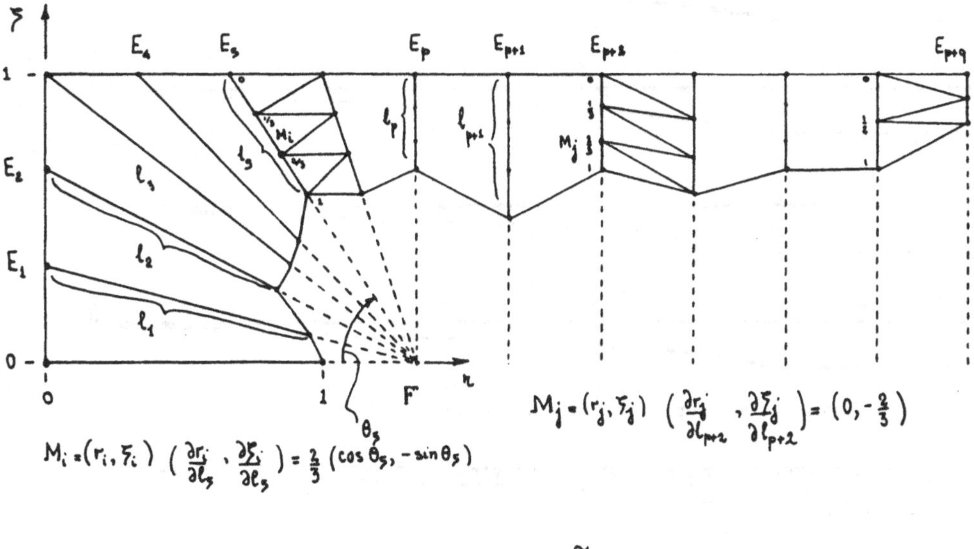

Figure 3. Triangulation τ and shape parameter $\widetilde{\ell}$.

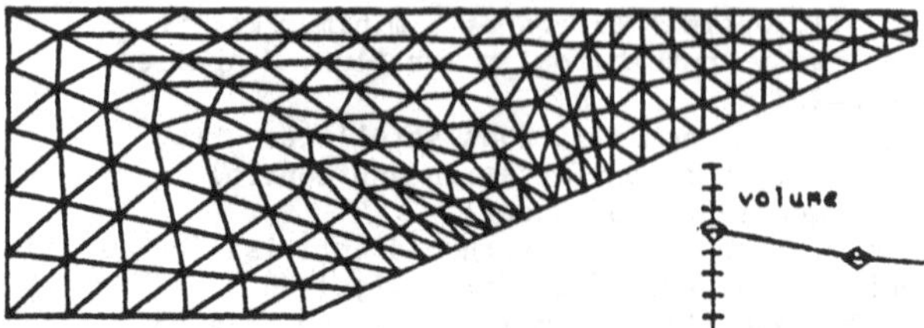

Figure 4. Triangulation.

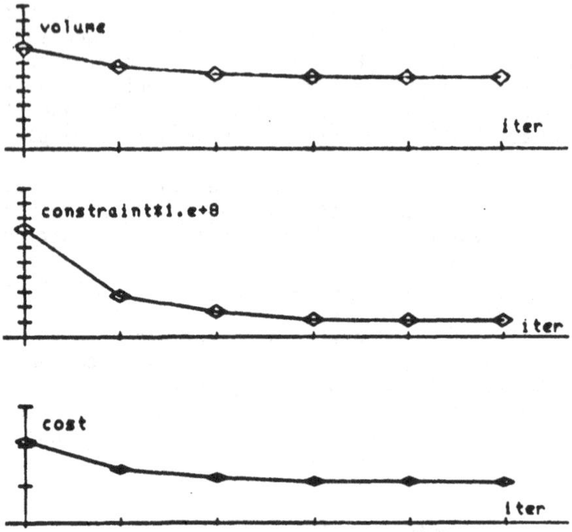

Figure 6. Volume, constraint and penalized cost versus iteration number.

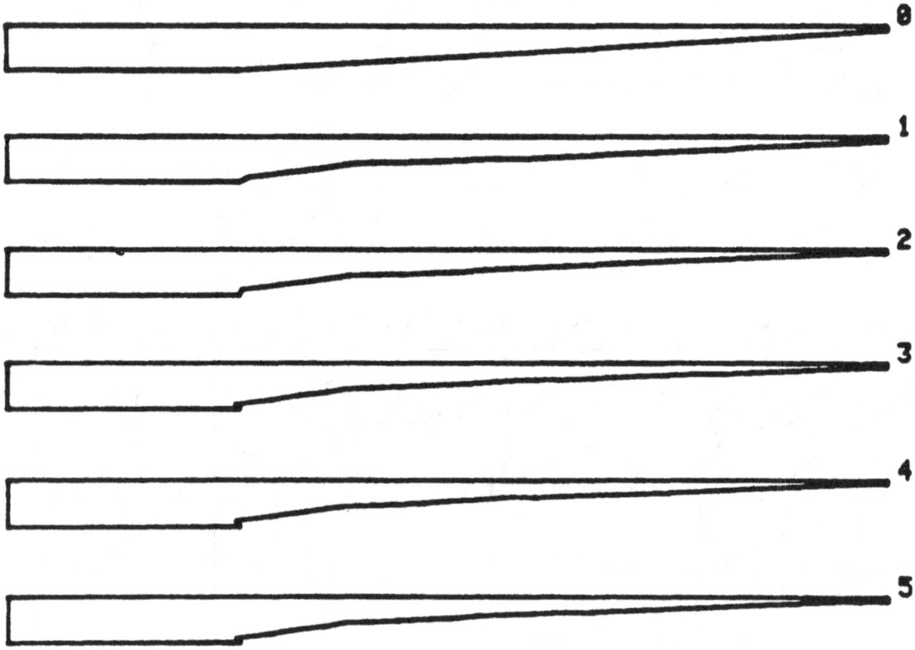

Figure 5. Successive shapes for the second set of five iterations.

Figure 7. Temperature profiles at iteration 5.

APPLICATION OF OPTIMIZATION PROCEDURES ON THE DESIGN OF VARIOUS SHELL STRUCTURES

H.A. Eschenauer and G. Kneppe
University of Siegen
FR Germany

1. INTRODUCTION

Computational as well as experimental investigations on structural optimization of machine and supporting structure components are carried out at the Institute of Mechanics and Control Engineering of the University of Siegen, sponsored by the Deutsche Forschungsgemeinschaft (German Research Community) within the scope of a research project.

For the development of complex supporting structures with extremely high specification demands, vector optimization presents a valuable method for the finding of optimal designs [1], [2], [3]. Especially within the scope of variant search it yields effective aid when deciding between different alternative proposals in the phases of feasibility, project definition, and development.

In this paper, optimization procedures for various shell structures influenced by deterministic variables are demonstrated. For that,

1) the behaviour of these supporting structures under various loads has to be determined as exactly as possible;

2) a realistic preference function has to be formulated;

3) an effective algorithm for the solution of the nonlinear optimization problem has to be available.

2. MODELLING OF OPTIMIZATION PROBLEMS WITH MULTIPLE OBJECTIVES

2.1 FORMULATION OF MULTIOBJECTIVE OR VECTOR-OPTIMIZATION PROBLEMS (VOP)

In practice, decision problems are often characterized by the fact that the decision maker is not easily able or willing to commit himself to one objective. Such optimization problems with several objective functions lead to an objective conflict, i.e., alternatives being especially favourable with regard to one objective yield relatively bad values for other objectives.

With

$\mathbb{R}$	the set of real numbers,
$\underline{f}$	the vector of k objective functions,
$\underline{x} \in \mathbb{R}^n$	the vector of n design variables,
$\underline{g}$	the vector of p inequality constraints,
$\underline{h}$	the vector of q equality constraints (e.g., the system equations for determining stresses and deformations),

$\mathbb{R}^n_z := \left\{\underline{x} \in \mathbb{R}^n : \underline{h}(\underline{x}) = 0,\ \underline{g}(\underline{x}) \leqq 0\right\}$ the "feasible" domain,

the classical vector optimization problem is defined

$$\text{"minimize"} \quad \left\{\underline{f}(\underline{x}) : \underline{h}(\underline{x}) = 0,\ \underline{g}(\underline{x}) \leqq \underline{0}\right\} \quad . \tag{1}$$

The aim is to find a design vector $\underline{x}$ (cross sections, geometry, material) making the components of the objective vector $\underline{f}$ as small as possible, and fulfilling all equality and inequality constraints (stresses, deformations, dimensions, etc.).

The objective functions $\underline{f}(\underline{x})$ are nonlinear, continuous functions of the design variable vector $\underline{x}$ in $\mathbb{R}$ with continuous derivations of at least second order. A condition for finding optimal solutions is that the existence of local minima and the characteristics of convexity are available.

The solution of the vector optimization problem (1) is the set of the functional-efficient points $\underline{x}^* \in \mathbb{R}^n_z$ or the complete solution set X^* [4], [5]. For all those vectors that are not functional-efficient, the value of at least one objective function f_ℓ can be reduced without increasing the function values of the remaining components at the same time.

By determining the complex solution set X^*, the number of all feasible solutions is indeed considerably limited by the elimination of not functional-efficient solutions; it does not lead, however, to an unique solution of the problem. Rather, one needs additional assumptions to be able to determine an objective vector as "optimal compromise solution".
By transformation into a substitute problem, the vector optimization problem is reduced to a scalar optimization problem. Thus a feasible way is given to get a compromise solution by applying approved optimization algorithms.

2.2 PREFERENCE FUNCTIONS

With the help of preference functions, the reduction of the vector optimization problem VOP (1) into a scalar optimization problem SOP is achieved. It thus becomes possible to determine a compromise solution $\underline{\tilde{x}}$ from the complete solution set X^*.

The problem

$$\text{"minimize"} \quad \left\{ p[\underline{f}(\underline{x})] \; : \; \underline{h}(\underline{x}) = \underline{0}, \; \underline{g}(\underline{x}) \leqq \underline{0} \right\} \tag{2}$$

is called a substitute problem for the vector optimization problem exactly then if an $\underline{\tilde{x}}$ of the following quality exists

$$p[\underline{f}(\underline{\tilde{x}})] = \min \left\{ p[\underline{f}(\underline{x})] \; : \; h(\underline{x}) = \underline{0}, \; \underline{g}(\underline{x}) \leqq \underline{0} \right\} \tag{3a}$$

with

$$\underline{\tilde{x}} \in X^*. \tag{3b}$$

$p[\underline{f}(\underline{x})]$ is called preference function.

In practice, this method is often used especially for economic problems, although it is less based on the search for a scientifically founded compromise than on the simple reduction of the vector optimization problem onto a problem with a single objective function.

Depending on the kind of substitute formulation describing the preference behaviour, different ways of proceeding are distinguished.
In previous application, objective weighting has had the greatest importance in most cases. The substitute objective function here consists of the sum of the weighted single objective functions. A further possibility of reduction is the model of distance functions based on a given demand level vector and on the definition of a special metric. The

transformation into a SOP can also be achieved by constraint oriented transformations. Herewith only one objective function is minimized and the rest are upwardly confined by constraints [5].

In our following investigations we used objective weighting and the constraint oriented transformation:

a) Objective Weighting

The single objective functions are related to the values of the initial design. The preference conceptions of the decision maker are expressed by means of weighting factors w_i and the specification values f_{io}:

$$p[\underline{f}(\underline{x})] := \sum_{i=1}^{k} w_i \frac{f_i(\underline{x})}{f_{io}} , \quad w_i \geqq 0 \text{ and } \sum_{i=1}^{k} w_i = 1. \tag{4}$$

b) Constraint Oriented Transformation

A constraint vector optimization problem is transformed into a scalar substitute problem if only one objective function is minimized and the other objectives are limited by upper bounds:

$$\begin{array}{ll} \text{"minimize"} & f_i(\underline{x}) \quad , \\ \text{with respect to} & f_j(\underline{x}) \leqq \overline{f}_j, j=1,2,\ldots,k, \ j \neq i. \end{array} \tag{5}$$

The objective function f_i is called main objective, and the f_j are called secondary objectives. The constraint levels $\overline{f}_j$ contain the preference behaviour of the decision maker.

In the present case - that means for multiobjective optimization of shells of revolution - the following single objective functions $f_i(\underline{x})$ are the most important ones:

- weight of the shell,
- maximum stiffness resp. minimum deformations.

3. OPTIMIZATION ALGORITHMS

For optimization computations of complex shell structures it is necessary to provide more effective algorithms. Previous research has shown that methods of 0th order for structural problems increase a lot in computing time at a higher number of constraints and design variables. The inclusion of constraints into the VOP, and the transfer to an SOP, already bring forth further improvements in computing time; but nevertheless it is necessary to improve the existing algorithms as well as to develop new optimization procedures.

On the basis of numerical tests and comparisons the multiplier method LPNLP and the sequential linearization procedure SEQLI were chosen for the present study.

3.1 AUGMENTED LAGRANGIAN METHOD (LPNLP)

The optimization procedure LPNLP by PIERRE and LOWE [6] is based on an augmented Lagrangian formulation. It serves to minimize multivariable, nonlinear problems with equality and inequality constraints.

The nonlinear problem is:

$$\text{"minimize"} \quad \left\{ p[\underline{f}(\underline{x})] \; : \; \underline{h}(\underline{x}) = \underline{0}, \; \underline{g}(\underline{x}) \leqq \underline{0} \right\} . \tag{6}$$

Sufficient optimality conditions for this NLP are obtained by way of an augmented Lagrangian formulation:

$$L_a(\underline{x},\underline{\alpha},\underline{\beta},\underline{\omega}) := L(\underline{x},\underline{\alpha},\underline{\beta}) + \omega_1 P_1 + \omega_2 P_2 + \omega_3 P_3 \qquad (7)$$

with the Lagrangian function

$$L(\underline{x},\underline{\alpha},\underline{\beta}) := p[f(\underline{x})] + \Sigma\, \beta_j g_j(\underline{x}) + \sum_k \alpha_k h_k(\underline{x})$$

and the penalty like terms

$$P_1 := \sum_k h_k(\underline{x})^2 \quad ,$$

$$P_2 := \sum_{j \in C_a} g_j(\underline{x})^2 \text{ with } C_a = \left\{ j \mid \beta_j > 0 \right\} \quad ,$$

$$P_3 := \sum_{j \in C_b} g_j(\underline{x})^2 \text{ with } C_b = \left\{ j \mid \beta_j = 0 \text{ and } g_j \geqq 0 \right\} \quad .$$

For large, but finite penalty-factors ω_i, L_a satisfies second order sufficient conditions. $\underline{\alpha}$ and $\underline{\beta}$ are the Lagrange multipliers.

The algorithm LPNLP minimizes the extended Lagrange function L_a (7) by way of an iterative search strategy. Usually, the multipliers relevant in the optimum $(\underline{\alpha}^*,\underline{\beta}^*)$ are not available from the start. Therefore, besides the penalty factors ω_i also $(\underline{\alpha},\underline{\beta})$ are systematically varied in the course of the optimization in order to get consistent values for x^* and $(\underline{\alpha}^*,\underline{\beta}^*)$ fulfilling the optimality conditions. The algorithm stands out because of high reliability.

3.2 SEQUENTIAL LINEARIZATION TECHNIQUE (SEQLI)

Especially for the shape optimization of complex plate and shell structures, a sequential linearization procedure is combined with a finite-element programme modified and extended for this purpose [7]. A special advantage of this system is the treatment of structures with relatively large-scale structure analyses that are commonly not convenient for optimization procedures because of too high computing times. Also, interactive operations allow a systematic influencing of the optimization process.

The nonlinear problem (6) is solved by way of a series of linear substitute problems. The nonlinear functions p, g_j, h_k are developed in Taylor's series in the neighbourhood of a reference point $\underline{x} = (\underline{x}^{(k)} + \delta\underline{x})$. Only the linear terms are considered:

$$\text{"minimize" } z = \underline{c}^T \underline{y} \quad , \quad \begin{aligned} \underline{a}_j^T \underline{y} &\leqq b_j \quad , \\ \underline{a}_k^T \underline{y} &= b_k \quad , \\ \underline{a}_m^T \underline{y} &\geqq b_m \quad , \\ y_i &\geqq 0 \quad , \end{aligned} \qquad (8)$$

with $\underline{y} := \delta\underline{x} + (\underline{x}^{(k)} - \underline{x}_\ell)$,

$\underline{c} := \nabla p\ [\underline{f}(x^{(k)})]$,

$\underline{a}_j := \nabla g_j\ (\underline{x}^{(k)})$, $b_j := -g_j(\underline{x}^{(k)}) + \underline{a}_j^T(\underline{x}^{(k)} - \underline{x}_\ell)$,

$\underline{a}_k := \nabla h_k\ (\underline{x}^{(k)})$, $b_k := -h_k(\underline{x}^{(k)}) + \underline{a}_k^T(\underline{x}^{(k)} - \underline{x}_\ell)$,

$a_{im} := \begin{cases} -1 & \text{for } i=m \\ 0 & \text{for } i \neq m \end{cases}$, $b_m := -x_{mu} + x_{m\ell}$,

(Index k: kth iteration step).

The way of procedure is demonstrated in fig. 1. A problem of the application of sequential linearization is the choice of appropriate move limits $\underline{x}_\ell$ and $\underline{x}_u$. In the optimization programme SEQLI, an adaptive correction rule is applied [8]. Also the convergence of the method is improved by additional cubic interpolation steps. For practical applications it is advantageous that the algorithm is able to start with feasible as well as nonfeasible initial designs.

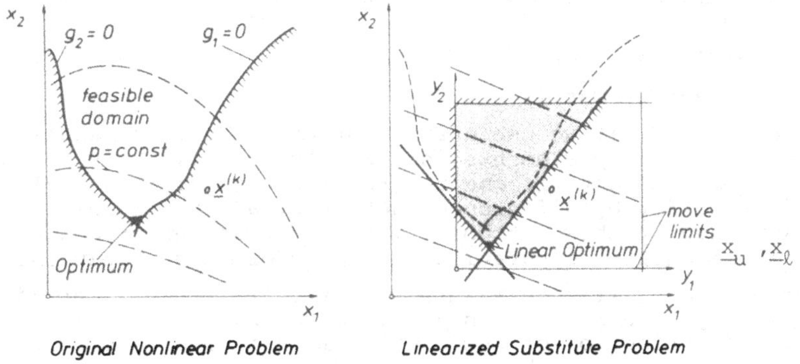

Figure 1: Sequential Linear Programming

4. APPLICATION ON SPHERICAL SHELLS

For the following studies we presuppose a spherical shell under axisymmetric loads clamped at the inner edge and free at the outer edge (see fig. 2). The purpose of the optimization is to find a shape of the shell in such a way that the competing objectives "minimal deformation under dead weight $w(\underline{x})$" and "minimal weight (equivalent to minimal volume $V(\underline{x})$)" are satisfied in the best possible way.

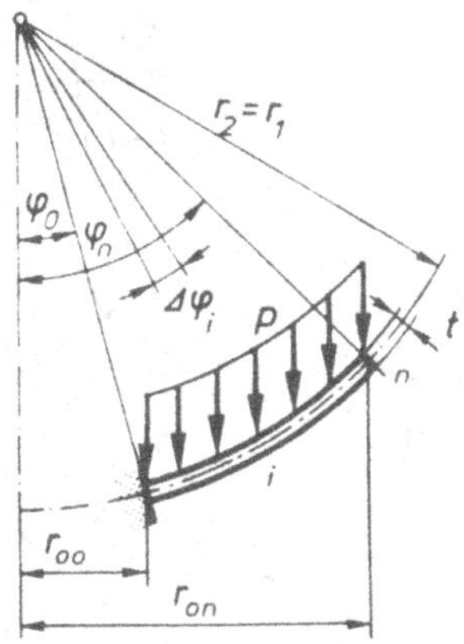

Figure 2: Spherical shell under deadweight

4.1 STRUCTURAL ANALYSIS

The deformations of the spherical shells are analyzed by way of a transfer method and an FE-method. Both methods will be briefly discussed below.

Starting points for the transfer method are the equilibrium conditions, strain-displacement equations, and the elastic law of a shell of revolution [3] constituting a system of differential equations of 1st order.

With the state vector

$$\underline{z}^T = (N_{\varphi\varphi}, Q_{\varphi}, M_{\varphi\varphi}, u, w, \chi) \in \mathbb{R}^6 \quad , \tag{9}$$

it can be written in an abbreviated form:

$$\frac{d\underline{z}}{d\varphi} = \underline{\underline{A}}\ \underline{z} + \underline{\underline{B}}\ \frac{d\underline{z}}{d\varphi} + \underline{b} \quad , \tag{10}$$

where $\underline{\underline{A}}$ and $\underline{\underline{B}}$ are (6x6)-matrices with variable coefficients and $\underline{b}$ represents the vector of external loads. The form of the matrices $\underline{\underline{A}}$ and $\underline{\underline{B}}$ as well as the inclusion of the load vectors into these matrices lead to a simplification of the equations (10):

$$\frac{d\underline{z}}{d\varphi} = \underline{\underline{C}}^* \ \underline{z}^* \tag{11}$$

with

$$\underline{z}^* = \begin{bmatrix} \underline{z} \\ \hline 1 \end{bmatrix} , \quad \underline{\underline{A}}^* = \left[\begin{array}{c|c} \underline{\underline{A}} & \underline{b} \\ \hline \underline{0} & 1 \end{array}\right] , \quad \underline{\underline{B}}^* = \left[\begin{array}{c|c} \underline{\underline{B}} & \underline{0} \\ \hline \underline{0}^T & 0 \end{array}\right] ,$$

$$\underline{\underline{C}}^* = (\underline{\underline{I}} + \underline{\underline{B}}^*)\ \underline{\underline{A}}^*$$

and the (7x7)-identity matrix $\underline{\underline{I}}$.

For the solution of (11) we use a transfer procedure. The shell is divided into single step elements, so that the coefficients of the matrix $\underline{\underline{C}}^*$ will be constant. If one now replaces the differential coefficients of 1st order by difference quotients, all variables at point i can be expressed by those at point i-1. The transfer procedure allows a sufficiently accurate computation of the deformation behaviour [3].

As a second method of structural analysis the FE-method is applied. A quarter of the spherical shell is approximated by finite elements (fig. 3). The distribution of the element mesh in φ-direction is selected more precisely close by the clamped boundary in order to determine the large gradients of the bending moments with sufficient accuracy. In ϑ-direction an equidistant distribution is sufficient. The computations are carried out with the FE-programme SAPIV [9].

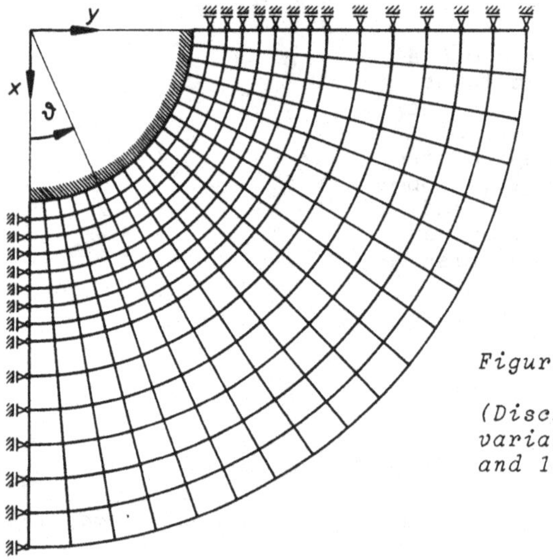

Figure 3: Finite Element Model for spherical shell (Discretization for 10 design variables, 285 nodes, 252 elements and 1568 degrees of freedom)

In meridional direction stiffened spherical shells are analyzed by means of the described model and additional beam elements. The cross-sectional areas, the moments of inertia, and the elasticity modulus of the beam elements here represent the stiffness qualities of the stiffeners.

4.2 OPTIMIZATION RESULTS

For unstiffened spherical shells, the wall thicknesses of the shell are selected as design variables. The vector optimization problem - minimal deformation and minimal weight - is formulated according to (4). Constraints are represented in form of restrictions of the design variables. The choice of 5 variables and a variation of the weighting factors, i.e., different preference conceptions, lead to the functional-efficient boundary represented in fig. 4. Depending on the choice of weighting factors, the initial design $f(\underline{x}_o)$ results in different optimal designs $f(\underline{x})$.

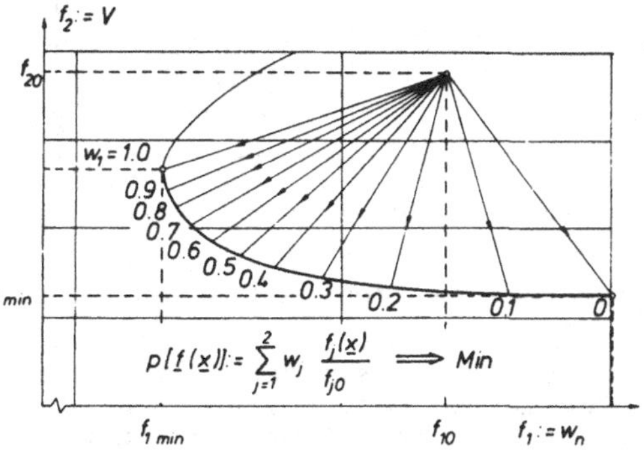

Figure 4: Functional-efficient boundary for various weighting factors

By means of a constraint-oriented transformation, the VOP can also be reduced to an SOP. Here, a limitation of the volume is defined as an additional constraint, and the deformation is defined as objective function. Optimization results for this problem formulation are listed in fig. 5. The starting point is always an initial design with equal wall thicknesses of 150 mm for all step elements. Using 5 design variables, the wall thicknesses of the optimal design vary between 178 mm at the inner region and 50 mm at the outer region. We see that the maximal deformation normal to the middle surface of the shell is reduced by about 38 % and the weight by even 50 %. Choosing a greater number of design variables, the deformation behaviour can be improved. Interesting is the "contradiction" in the inner region which can be recognized already by n=10 variables and is clearly identified by n=15 variables. At further low weight this optimal shape leads to more favourable deformation values.

The comparison of the two methods of structural analysis, namely transfer matrices (TM) and Finite Elements (FE), is interesting. The optimal values for w_{max} and V are sufficiently congruent, as well as the number of the necessary function analyses n_F and gradient analyses n_G. There are, however, substantial differences in computing time. The computing effort of the transfer matrices is distinctly lower (factor ≈ 200). The smaller expense for the structural analyses by means of the transfer matrices results in substantially more economic computing times. In general, many problems can only be approached by using simple computation concepts. The selection of an appropriate

	Initial Design	Optimal Designs					
		5 Variables		10 Variables		15 Variables	
		TM	FE	TM	FE	TM	FE
w_{max} [mm]	0.3220	0.2209	0.2133	0.2050	0.1991	0.2024	0.1975
V [m^3]	32.51	24.52	24.51	24.51	24.48	24.52	24.48
n_F	-	11	10	11	13	12	14
n_G	-	8	10	11	10	12	10
computing time [s] (Cyber 76)	-	12	2283	24	3996	41	9545
principal sketch	x_i = 150mm	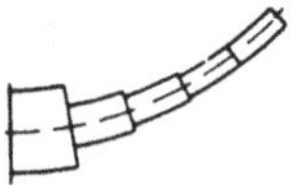		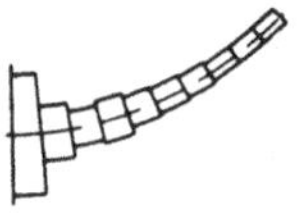		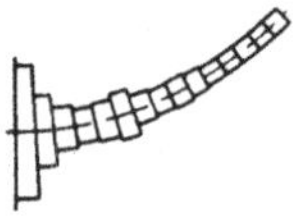	

Figure 5: Optimization results of an unstiffened spherical shell under deadweight

method for structural analysis is therefore as important as an efficient optimization algorithm.

Comparing the results of the two structural analysis methods, it has to be added, however, that the transfer method can only be used for spherical shells with axisymmetric loads [3]. The FE-method, on the other hand, permits the analysis of arbitary loads and boundary conditions without additional calculation steps. Also the shape can be varied, e.g. conical and cylindrical shells. Aim of future research is therefore to apply the transfer method on not axisymmetric cases, too.

As a second example, a spherical shell stiffened in meridional direction is discussed. For the structural analysis the transfer method and the FE-method are applied. Design variables are the heights of the ribs. The number of the stiffeners (n=36) and the width (b=639 mm) remain constant. Objective functions are the minimal deformation and the minimal weight. As constraints lower bounds of the design variables have to be considered.

The optimization results are compared to the respective results of the unstiffened shell in fig. 6. For the same volume, the stiffened shell produces substantially better deformation values. Therefore, except for some possible fabrication problems, the stiffened version is preferable to the unstiffened one, due to better material efficiency.

Also, a comparison of the deformation curves in fig. 6 shows that the point of maximum deformation is shifted to the free boundary when using a stiffened shell instead of an unstiffened one. Distinctive for the optimal designs of the spherical shells is the occurence of "increased peaks" in the maximum ranges of deformation, see fig. 7.

Graphically, these ranges can be interpreted as rigid clamping, as here

$$\left.\frac{dw}{d\varphi}\right|_{\varphi = \varphi(w_{max})} = 0 .$$

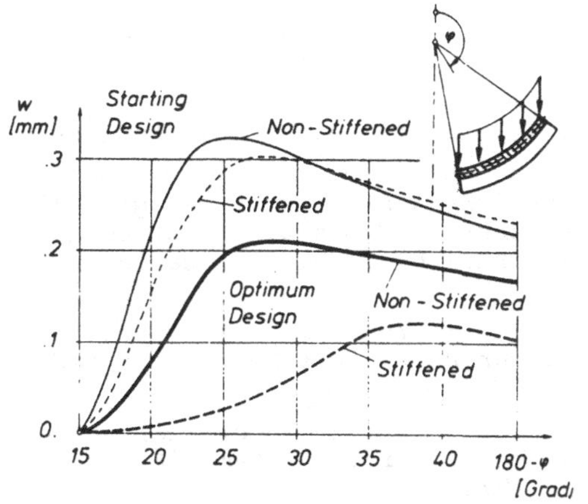

		Non-Stiffened	Stiffened
Starting Design	$w_{0\,max}$ [mm]	.3220	.3027
	V_0 [m³]	32.51	18.83
Optimum Design	w_{max} [mm]	.2133	.1219
	V [m³]	24.51	24.52

Figure 6: Optimization results for an unstiffened and a stiffened spherical shell

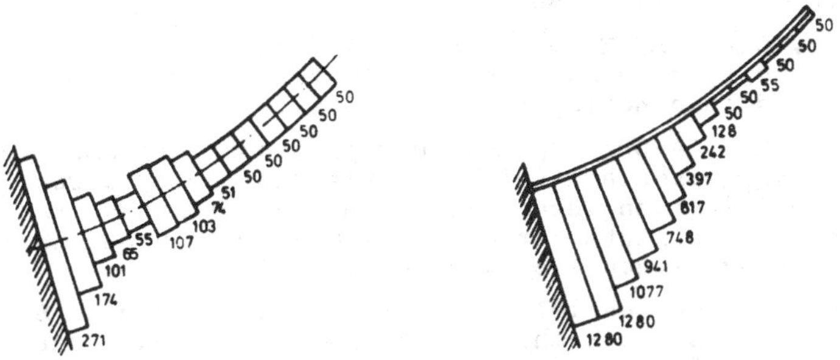

Figure 7: Optimal designs for an unstiffened and a stiffened spherical shell

5. CONCLUSION

This paper shows that an optimization of shell structures demands effective optimization algorithms as well as efficient structure analysis programmes. An unstiffened spherical shell and a spherical shell stiffened in meridional direction exemplify the shape optimization of the meridional type for the both objectives "minimal volume" and "minimal deformation". As preference functions the Objective Weighting as well as the Constraint Oriented Transformation are applied. Both ways can be very effectively integrated into the optimization algorithms LPNLP (Augmented Lagrangian Method by PIERRE and LOWE [6]) and SEQLI (Sequential Linear Programming [8]).

Besides the FE-programme SAPIV, transfer matrices are applied for the structural analysis computations. The last-named method was developed on the basis of the fundamental equations of the theory of the shells of revolution. For the examples discussed here, the inclusion of the transfer matrices in the optimization procedure leads to an essential reduction in computing time.

REFERENCES

[1] ESCHENAUER, H.: Anwendung der Vektoroptimierung bei räumlichen Tragstrukturen. Der Stahlbau 4 (1981) 110-115

[2] ESCHENAUER, H.: Vector-Optimization in Structural Design and its Application on Antenna Structures. Proceedings Euromech-Colloquium 164. Mannheim, Wien, Zürich: BI-Wissenschaftsverlag (1983) 146-155

[3] ESCHENAUER, H.; SATTLER, H.J.: Optimierung von Schalenbauteilen für mehrere Ziele. Zeitschrift für angewandte Mathematik und Mechanik, 63 (1983) T404-406

[4] FANDEL, G.: Optimale Entscheidungen bei mehrfacher Zielsetzung. Lecture Notes in Economics and Mathematical Systems 76. Berlin, Heidelberg, New York: Springer 1972

[5] SATTLER, H.J.: Ersatzprobleme für Vektoroptimierungsaufgaben und ihre Anwendung in der Strukturmechanik. Universität-GH-Siegen: Dr.-Thesis 1982

[6] PIERRE, D.A.; LOWE, M.J.: Mathematical Programming via Augmented Lagrangians. London, Amsterdam, Ontario, Sydney, Tokio: Addison Wesley P.C., Inc. 1975

[7] KNEPPE, G.: SAPOP - Optimierungsprogramm für komplexe Gestaltsoptimierungsprobleme auf der Basis des FE-Verfahrens SAPIV und einer erweiterten sequentiellen Linearisierungsstrategie. Bericht des Instituts für Mechanik und Regelungstechnik der Universität-GH Siegen 1983 (unpublished)

[8] KNEPPE, G.: Sequentielle Linearisierungsstrategie SEQLI. Bericht des Instituts für Mechanik und Regelungstechnik der Universität-GH Siegen 1983 (unpublished)

[9] BATHE, K.J.; PETERSON, F.E. and WILSON, E.L.: SAPIV, A Structural Analysis Program for Static and Dynamic Response of Linear Systems. Earthquake Engineering Research Center, University of California, Berkeley 1974

On Nondifferentiable Plate Optimal Design Problems

A. Myśliński */
Osaka University
Faculty of Engineering
Science, Department of
Control Engineering
Toyonaka, Osaka 560,
Japan

J. Sokołowski
Systems Research Institute
Polish Academy of Sciences
01-447 Warszawa, ul. Newelska 6
Poland

J.P. Zolesio
Universite de Nice
Departement de Mathematiques
Parc Valrose, 06034-Nice CEDEX
France

1. Introduction

We consider the problem (P1) of optimal thickness distribution of a solid Kirchoff plate which for given plate maximum and minimum thickness and for specified volume of the plate provides the minimal value of the maximal deflection of the plate corresponding to the prescribed load distribution. Another optimum design problem (P2) considered in the paper consists in maximizing the fundamental frequency for free, transverse vibrations of the plate.

We present several results for these optimum design problems concerning in particular, the existence of generalized optimal solutions, necessary optimality conditions, shape sensitivity analysis of cost functionals as well as numerical results for two examples. Problems of determining optimal thickness distribution for thin, solid elastic plates have been investigated in many papers, e.g. |1|-|7|, |9|-|11|, |14|-|18|.

In order to investigate the existence of generalized solutions to such optimization problems the so- called G-convergence of elliptic operator is used |8|,|10|,|16|.

---

*/ on leave from: Systems Research Institute, Polish Academy of Sciences, 01-447 Warszawa, ul. Newelska 6, Poland.

The directional differentiability of multiple eigenvalues with respect to coefficients of a fourth order elliptic operator is proved in |21|, |22|, see also |5| for some related results. The method of Zolesio is exploited in |16| in order to obtain necessary optimality conditions for nondifferentiable optimization problems for fourth order elliptic system. We present such necessary optimality conditions for the optimum design problems under consideration. Our numerical results confirm that numerical methods of nonsmooth optimization |12| can be used effectively for solving nonsmooth optimum design problems.
The outline of the paper is the following. Section 2 describes theoretical results obtained for the problem (P1). Section 3 containes results obtained for the problem (P2). In section 4 the results of computations for two examples are presented.
We introduce the necessary notation.

Notation

Let $\Omega \subset R^2$ be a domain with the boundary $\Gamma = \partial\Omega$. Sobolev spaces $H^2(\Omega)$, $H^2_o(\Omega)$ are defined as follows:

$$H^2(\Omega)=\{\phi \in L^2(\Omega) \mid \phi_{,i}, \phi_{,ij} \in L^2(\Omega),\ i,j=1,2\}$$

$$H^2_o(\Omega)=\{\phi \in H^2(\Omega) \mid \phi = \frac{\partial\phi}{\partial n} = 0 \quad \text{on } \Gamma\}$$

where

$$\phi_{,i} = \frac{\partial\phi}{\partial x_i}$$

$$\phi_{,ij} = \frac{\partial^2\phi}{\partial x_i \partial x_j} \qquad i,j=1,2$$

and $\frac{\partial\phi}{\partial n}$ denote the normal derivative of ϕ on Γ, n denotes the outward unit normal vector on Γ. Let $U_{ad} \subset L^\infty(\Omega)$ be a nonempty, closed and convex set of the form:

$$(1) \quad U_{ad} = \{h \in L^\infty(\Omega) \mid 0 < h_{min} \le h(x) \le h_{max} \quad \text{a.e. in } \Omega, \quad \int_\Omega h(x)dx = c\}$$

where h_{min}, h_{max}, c are given constants.

Denote by $a_\Omega(h;.,.)$: $H^2_o(\Omega) x H^2_o(\Omega) \longrightarrow R$ the bilinear form:

$$(2)\ a_\Omega(h;y,z)=\int_\Omega h^3\left[y_{,11}z_{,11}+y_{,22}z_{,22}+\nu y_{,11}z_{,22}+\nu y_{,22}z_{,11}+2(1-\nu)y_{,12}z_{,12}\right]dx$$

$\forall h \in U_{ad}, \forall y,z \in H_o^2(\Omega)$

Let $f \in H^{-2}(\Omega)=(H_o^2(\Omega))'$ be given element, consider elliptic equation

$$(3) \quad \begin{cases} a_\Omega(h;w_\Omega(h),z)=\int_\Omega f(x)z(x)dx \\ w_\Omega(h;.) \in H_o^2(\Omega) \qquad \forall z \in H_o^2(\Omega) \end{cases}$$

It can be shown that there exists a unique solution to Eq.(3). Equation (3) describes deflection of a thin, solid plate loaded by the perpendicular force $f(x)$, $x \in \Omega$.

2. Minimization of the maximal deflection

In this section we present results concerning the existance of generalized solutions and necessary optimality conditions for the optimum design problem (P1) as well as a result concerning local shape sensitivity analysis of the cost functional (4).

Denote by $J_\Omega(h)$ the maximal deflection of the plate loaded by a fixed perpendicular force $f(.) \in H^{-2}(\Omega)$.

$$(4) \quad J_\Omega(h)=\sup\{\,|w_\Omega(h;x)|\;|\;x \in \Omega\}, \; \forall h \in U_{ad}$$

where $w_\Omega(h;.)$ satisfies Eq.(3).

In this section the problem (P1) of minimizing the maximal deflection $J_\Omega(h)$ with respect to $h \in U_{ad}$ is considered:

$$(P1) \quad \inf\{J_\Omega(h) \;|\; h \in U_{ad}\}$$

Theorem 1

Let $\{h_m\} \subset U_{ad}$, $m=1,2,...$ be a minimizing sequence for the problem (P1). There exists a subsequence, still denoted $\{h_m\}$, $m=1,2,...$ and elements $\bar{w}_\Omega \in H_o^2(\Omega)$, $q_{ijkl} \in L^\infty(\Omega)$, $i,j,k,l=1,2$ such that

$$(5) \quad \lim_{m\to\infty} J_\Omega(h_m)=\inf\{J_\Omega(h) \;|\; h \in U_{ad}\}=\sup\{\bar{w}_\Omega(x) \;|\; x \in \Omega\}$$

$$(6) \quad w_\Omega(h_m) \longrightarrow \bar{w}_\Omega \quad \text{weakly in } H_o^2(\Omega)$$

(7) $$h_m^3(w_\Omega(h_m),_{11}+\nu w_\Omega(h_m),_{22}) \longrightarrow \sum_{i,j=1}^{2} q_{11ij}\, \bar{w}_{\Omega,ij} \quad \text{weakly in } L^2(\Omega)$$

(8) $$h_m^3(w_\Omega(h_m),_{22}+\nu w_\Omega(h_m),_{11}) \longrightarrow \sum_{i,j=1}^{2} q_{22ij}\, \bar{w}_{\Omega,ij} \quad \text{weakly in } L^2(\Omega)$$

(9) $$(1-\nu)h_m^3 w_{\Omega,12} \longrightarrow \sum_{i,j=1}^{2} q_{12ij}\, \bar{w}_{\Omega,ij} \quad \text{weakly in } L^2(\Omega)$$

Remark 1

The tensor function $\{q_{ijkl}\} \subset [L^\infty(\Omega)]^{16}$ depends only on the subsequence $\{h_m\}$ and it does not depend on the element $f \in H^{-2}(\Omega)$. It can be shown |16| that the following condition is satisfied:

(10) $$q_{ijkl}(x)=q_{jikl}(x)=q_{klij}(x) \quad \text{a.e. in } \Omega$$

It can be verified that the element $\bar{w}_\Omega \in H_0^2(\Omega)$ is a unique solution to the elliptic equation:

(11) $$\sum_{i,j,k,l=1}^{2} (q_{ijkl}(x)\bar{w}_{\Omega,ij}(x)),_{kl}=f(x) \quad \text{in } \Omega$$

In particular in |16| some variational estimates for tensor function $\{q_{ijkl}\} \subset [L^\infty(\Omega)]^{16}$ are provided.

Theorem 2

If there exists an optimal solution $\hat{h} \in U_{ad}$ to problem (P1) then the element $\hat{h}$ satisfies the optimality system:

(12) $$a_\Omega(\hat{h};w(\hat{h}),z) = \int_\Omega fz\,dx, \quad \forall z \in H_0^2(\Omega)$$

(13) $$a_\Omega(\hat{h};p(s),z) = \operatorname{sign}(w(\hat{h};s))z(s) \quad \forall z \in H_0^2(\Omega)$$

(14) $$\sup_{s \in \Omega^*(\hat{h})} \int_\Omega \hat{h}^2(x)(h(x)-\hat{h}(x))\big[w(\hat{h}),_{11}p(s),_{11}+w(\hat{h}),_{22}p(s),_{22}+ +\nu w(\hat{h}),_{11}p(s),_{22}+\nu w(\hat{h}),_{22}p(s),_{11}+2(1-\nu)w(\hat{h}),_{12}p(s),_{12}\big]dx \geq 0$$

$$\forall h \in U_{ad}$$

where $w(\hat{h}) = w_\Omega(\hat{h})$ and

(15) $$\Omega^*(h) = \{s \in \Omega \mid J_\Omega(h) = w_\Omega(h;s)\}$$

The proof is given in |16|.

Consider the shape sensitivity analysis of the maximal deflection $J_\Omega(h)$
To this end, we recall the so - called speed method |21| in the shape sensitivity analysis.
Let Q be an open neighbourhood of the set $\bar{\Omega}$. Given vector field

(16) $\underset{\sim}{V}(.,.) : [0,\bar{\delta}) \times Q \longrightarrow R^2$

where $\bar{\delta} > 0$. Denote by $\Omega_\varepsilon \subset R^2$, $\varepsilon \in [0,\delta)$, $\delta > 0$, $0 < \delta < \bar{\delta}$, a family of domains of the form

(17) $\Omega_\varepsilon = T(\underset{\sim}{V})(\Omega) =$

$= \{ x \in R^2 \mid \exists X \in \Omega$ such that $x = x(\epsilon),\ x(0) = X \}$

where

$$(18)\quad \begin{cases} \frac{dx}{dt}(t) = \underset{\sim}{V}(t,x(t)) , & t \in (0,\delta) \\ x(0) = X \end{cases}$$

Assume that there are given elements $f \in L^2(Q)$ and $h \in C^1(Q)$ such that $h|_\Omega \in U_{ad}$, $h_{min} \leq h(x) \leq h_{max}$ a.e. in Q. Denote $h_\varepsilon = h|_{\Omega_\varepsilon}$, $h = h|_\Omega$ and let

(19) $j(\varepsilon) = J_{\Omega_\varepsilon}(h_\varepsilon) = \sup\{ w_{\Omega_\varepsilon}(h_\varepsilon;x) \mid x \in \Omega_\varepsilon \}$

<u>Theorem 3</u>

For $\varepsilon > 0$, ε small enough:

(20) $j(\varepsilon) = j(0) + \varepsilon \sup\{ \mathrm{sign}(w_\Omega(h;x)) w'_\Omega(x) \mid x \in \Omega(h) \} + o(\varepsilon)$

where $o(\varepsilon)/\varepsilon \longrightarrow 0$ with $\varepsilon \downarrow 0$.

The element $w'_\Omega \in H^2(\Omega)$ denotes so-called domain derivative of the solution to (3).
The domain derivative w'_Ω satisfies the following elliptic boundary-value problem:

$$(21)\quad \begin{cases} a_\Omega(h;w'_\Omega,z) = 0 , & \forall z \in H_0^2(\Omega) \\ w'_\Omega = 0 \quad \text{on } \Gamma \\ \frac{\partial w'_\Omega}{\partial n} = \Delta w_\Omega(h) < \underset{\sim}{V}(0), n >_{R^2} \quad \text{on } \Gamma \end{cases}$$

where $< \underset{\sim}{V}(0), n >_{R^2}$ denotes the normal component of $\underset{\sim}{V}(0,.)$ on Γ.

For related results see also |6|, |19|.

3. Maximization of the smallest eigenvalue

In this section we present the results on the existence of generalized solutions and necessary optimality conditions for problem (P2). Shape sensitivity analysis of the smallest eigenvalue $\lambda_\Omega(h)$ is investigated. Let us consider the following optimization problem:

$$(P2) \qquad \sup \{ \lambda_\Omega(h) \mid h \in U_{ad} \}$$

where $\lambda_\Omega(h)$ denotes the smallest eigenvalue to the eigenvalue problem:

$$(22) \qquad a_\Omega(h;\eta,\phi) = \lambda_\Omega(h)\int_\Omega h(x)\eta(x)\phi(x)dx \qquad \forall\phi \in H_0^2(\Omega)$$

Denote by $M_\Omega(h) \subset H_0^2(\Omega)$ the set of eigenfunctions corresponding to the eingenvalue $\lambda_\Omega(h)$, normalized by the condition:

$$\int_\Omega h(x)(\eta(x))^2 dx = 1 , \qquad \forall\eta \in M_\Omega(h)$$

<u>Theorem 4</u>

Let $\{h_m\} \subset U_{ad}$, $m=1,2,\ldots$ be a maximizing sequence for the problem (P2) and let $\eta_m \in M_\Omega(h_m)$ be a sequence of eigenfunctions.

There exists a subsequence, still denoted $\{h_m\}$, elements $\lambda^* \in R$, $h_o \in U_{ad}$, $\eta_o \in H_0^2(\Omega)$ and a tensor function $\{q_{ijkl}\} \subset [L^\infty(\Omega)]^{16}$ such that the following convergences take place:

$$(23) \qquad \lim_{m\to\infty} \lambda_\Omega(h_m) = \sup\{\lambda_\Omega(h) \mid h \in U_{ad}\} = \lambda^*$$

$$(24) \qquad h_m \longrightarrow h_o \qquad \text{weakly - } (*) \text{ in } L^\infty(\Omega)$$

$$(25) \qquad \eta_m \longrightarrow \eta_o \qquad \text{weakly in } H_0^2(\Omega)$$

$$\left.\begin{aligned} (26) \qquad & h_m^3(\eta_{m,11}+\nu\eta_{m,22}) \rightarrow \sum_{i,j=1}^{2} q_{11ij}\,\eta_{o,ij} \\ (27) \qquad & h_m^3(\eta_{m,22}+\nu\eta_{m,11}) \rightarrow \sum_{i,j=1}^{2} q_{22ij}\,\eta_{o,ij} \\ (28) \qquad & (1-\nu)h_m^3\eta_{m,12} \rightarrow \sum_{i,j=1}^{2} q_{12ij}\,\eta_{o,ij} \end{aligned}\right\} \text{ weakly in } L^2(\Omega)$$

The proof is given in |16|

Remark

It can be shown that the tensor function $\{q_{ijkl}\} \in [L^\infty(\Omega)]^{16}$ satisfies condition (10). The elements $\lambda^* \in R$ and $\eta_o \in H_o^2(\Omega)$ are, respectively, the smallest eigenvalue and the corresponding eigenfunction to the eigenvalue problem:

$$(29) \qquad \sum_{i,j=1}^{2} \sum_{k,l=1}^{2} \int_\Omega q_{ijkl}(x)\eta_{o,ij}(x)\phi_{,kl}(x)dx = \lambda^* \int_\Omega h_o(x)\eta_o(x)\phi(x)dx,$$

$$\forall \phi \in H_o^2(\Omega)$$

Theorem 5

If there exists an optimal solution $h^* \in U_{ad}$ to problem (P2) then the following optimality condition is satisfied:

$$(30) \qquad \inf_{\eta \in M_\Omega(h^*)} \int_\Omega (h^*(x))^2(h(x)-h^*(x))\Big[(\eta_{,11})^2 + (\eta_{,22})^2 + 2\nu\eta_{,11}\eta_{,22} + 2(1-\nu)(\eta_{,12})^2 - \lambda_\Omega(h^*)(\eta)^2\Big]\,dx \geq 0$$

$$\forall h \in U_{ad}$$

The proof follows from the results given in |22|, |16|.

We recall the shape sensitivity analysis |19|, |20|, |21| of the smallest eigenvalue $\lambda_\Omega(h)$. We use the notation of the preceding section.

Theorem 6

For $\varepsilon > 0$, ε small enough:

$$(31) \qquad \lambda_{\Omega_\varepsilon}(h_\varepsilon) = \lambda_\Omega(h) + \varepsilon \inf_{\phi \in M_\Omega(h)} \Big\{ -\int_\Gamma \langle \underset{\sim}{V}(0), n \rangle_{R^2}\big[(\phi_{,11})^2 + (\phi_{,22})^2 + 2\nu\phi_{,11}\phi_{,22} + 2(1-\nu)(\phi_{,12})^2\big]d\Gamma\Big\} + o(\varepsilon)$$

where $o(\varepsilon)/\varepsilon \to 0$ with $\varepsilon \downarrow 0$.

Numerical results.

Problems (P1) and (P2) were solved numerically using the finite element method combined with Lemarechal's ε-subgradient method of nondifferentiable optimization and the shifted penalty function method. Our numerical results confirm that problems (P1) and (P2) are nondifferentiable from the numerical viewpoint, hence the utilization of Lemarechal's method instead of a gradient type method for these problems is fully justified. For problem (P1) the computations were performed for a rectangular plate divided into 56 elements. The plate is loaded by a constant perpendicular force $f(x)$, $x \in \Omega=(0,1)x(0,1)$. The constants $h_{min}=0.8$, $h_{max}=1,2$ and $c=1$ determine the set (1) in our examples. One quarter of the optimal shape plate for problem (P1) is shown in Fig.1.

The maximal deflection of the optimal shape plate is attained in more then one point. The gain of optimization computed relatively to the plate of uniformly constant thickness is equal to 63%. The material concentrates in the middle of the boundaries and around the middle of the plate.

One quarter of the optimal shape plate for problem (P2) is shown in Fig.2. The algorithm QZ |13| was used for solving the eigenvalue problem. The gain of optimization computed relatively to the plate with uniformly constant thickness is 102%. The material concentrates in the middle of the boundaries and around the middle of the plate. The smallest eigenvalue of the optimal shape plate has the multiplicity of two. For further numerical results we refer the reader to |14|, |15|.

References

|1| N.V.Banitchuk, Optimization of Shape of Elastic Bodies, Nauka, Moscow, 1980 /in Russian/

|2| M.P.Bendsøe, On Obtaining a Solution to Optimization Problems for Solid Elastic Plates by Restriction of the Design Space, to appear

|3| K.T.Cheng and N.Olhoff, An Investigation Concerning Optimal Design of Solid Elastic Plates, Int. J.Solids Structures vol.17 (1981), pp.305-323

|4| K.T.Cheng and N.Olhoff, Regularized Formulation for Optimal Design of Axisymmetric Plates, Int. J. Solids Structures vol.18 (1982), No.2, pp.153-169.

|5| E.J.Haug and B.Rousselet, Design Sensitivity Analysis in Structural Mechanics, Part 2: Eigenvalue Variations, J. Struct. Mech. 8 (1980) pp.161-186.

|6| E.J.Haug and J.Cea eds., Optimization of Distributed Parameter Structures, Sijthoff & Noordhoff, Alphen aan den Rijn, Netherlands (1981).

|7| E.J.Haug and K.K.Choi, Systematic Occurence of Repeated Eigenvalus in Structural Optimization, JOTA, vol.38 (1982) No.2

|8| R.V.Kohn and M.Vogelius, A New Model for Thin Plates with Rapidly Varying Thickness, to appear

|9| W.Kozlowski and Z.Mroz, Optimal Design of Solid Plates, Int. J. Solids Structures 5(1969), pp.781-794.

|10| K.A.Lurie, A.V.Cherkaev and A.V.Fedorov, Regularization of Optimal Design Problems for Bars and Plates, JOTA, vol.37 (1982) No.4 pp.499-543.

|11| V.G.Litwinov, Optimal Control of the Natural Frequency of a Plate of Variable Thickness, Comput. Math. Math. Phys., vol.19 (1980) pp.133-143 /in Russsian/.

|12| C.Lemarechal, An Extension of Davidon Method to Nondifferentiable Problems, Mathematical Programming Study 3: Non-differentiable Optimization, North - Holland (1975), pp.95-109.

|13| C.B.Moller and G.W.Stewart, An Algorithm for Generalized Matrix Eigenvalue Problems, SIAM J. on Num. Anal. 10 (1973), pp.241-256.

|14| A.Myśliński, A Nonsmooth Optimal Design Problem of Bending Thin Plates Control and Cybernetics, vol.11 (1982), No.3-4, pp.121-134.

|15| A.Myśliński, Nonsmooth Optimal Design of Vibrating Plates, to appear.

|16| A.Myśliński and J.Sokołowski, Nondifferentiable Optimization Problems for Elliptic Systems, SIAM Journal on Control and Optimization, to appear.

|17| N.Olhoff, K.A.Lurie, A.V.Cherkaev and A.V.Fedorov, Sliding Regimes and Anisotropy in Optimal Design of Vibrating Axisymmetric Plates, Int. J.Solids Structures vol.17 (1981), No.10, pp.931-948.

|18| N.Olhoff, Optimal Design of Vibrating Rectangular Plates, Int. J. Solids Structures, 10 (1974), pp.93-109.

|19| B.Rousselet, Quelques resultats en optimisation de domaines, These d'Etat, Universite de Nice, (1982).

|20| J.Sokołowski and J.P.Zolesio, Shape Sensitivity Analysis of Elastic Structures, to appear.

|21| J.P.Zolesio, Identification de Domaines par Deformation, These d'Etat, Universite de Nice, (1979).

|22| J.P.Zolesio, Semi Derivatives of Repleated Eigenvalues, in |6|, pp.1457-1473.

Acknowledgment

Lemarechal's implementation of his method was used for computations.

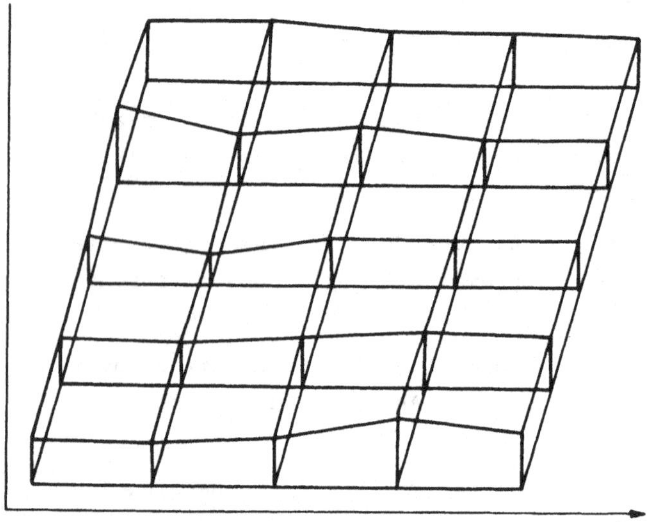

Figure 1. Problem (P1), one quarter of optimal shape plate

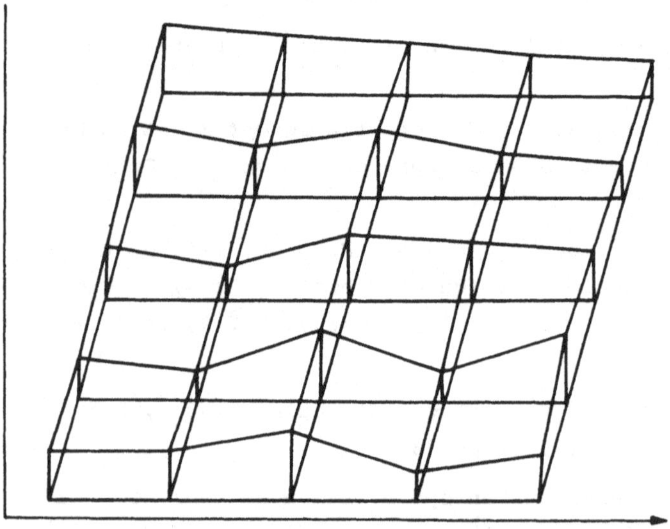

Figure 2. Problem(P2) one quarter of optimal shape plate.

OPTIMUM GEOMETRY MODELING FOR MINIMIZING WEIGHT OF PLATE BENDING STRUCTURE WITH SUBSTRUCTURES

D.T. Nguyen
Department of Civil Engineering
Northeastern University
Boston, MA 02115

J.S. Arora
Division of Materials Engineering
The University of Iowa
Iowa City, IA 52242

I. INTRODUCTION

A triangular plate bending element with shear deformation is used as part of the structural analysis - synthesis procedure to obtain the mimimum weight design of plate bending structures. Such an element has been successfully employed in Ref. [1]. In this paper, the design variables include not only thickness of the plate as most commonly discussed but also include nodal coordinates with respect to a global reference frame. Thus the geometry of the structure is allowed to change at the same time with thickness of the plate during the optimization process. Constraints can be imposed on stresses, displacements, frequency and lower and upper bounds on design variables. The material of this paper is focussed on the presentation of the analytical derivatives of the cost and constraint functions. Such analytical derivatives are desirable since derivatives obtained by finite difference approach have been proven to be inefficient [2]. Once the cost and constraint functions and their gradients have been evaluated, any gradient based algorithm can be used to optimize the structure. Definitions of the design problem and sensitivity analysis is briefly described in Section II. More details can be found in Ref. [3]. Element stiffness and stress matrices are summarized in Section III and finally, analytical derivatives for design sensitivity analysis are derived in Section IV.

II. DESIGN PROBLEM AND SENSITIVITY ANALYSIS

2.1 Design Problem

The most practical method of analyzing the behavior of a complex structure is the finite element approach. The governing equilibrium equation for a finite element model of a structure subjected to quasi-static loads is $K(b)z = R(b)$. Or, in the partitioned form, it is

$$\begin{bmatrix} K_{BB} & K_{BI} \\ K_{IB} & K_{II} \end{bmatrix} \begin{bmatrix} Z_B \\ Z_I \end{bmatrix} = \begin{bmatrix} R_B \\ R_I \end{bmatrix} \tag{2.1}$$

where subscripts B and I represent boundary and interior quantities, respectively, and

$K(b)$ = an nxn symmetric nonsingular structural stiffness matrix
$R(b)$ = an n-vector representing equivalent nodal loads for the finite element model
$Z(b)$ = an n-vector of nodal displacements
b = a k-vector of design variables for the problem

Another equilibrium equation that governs the free vibration response of a structure or its buckling behavior is the eigenvalue problem:

$$K(b)y = \xi M(b)y \tag{2.2}$$

where:

$M(b)$ = an nxn structural mass matrix for the vibration problem or the geometric stiffness matrix for buckling problem.

ξ = an eigenvalue related to the natural frequency or the buckling load for the problem.
y = an eigenvector

A mathematical model for optimal design of linearly elastic structural systems is defined as follows:

Find a design variable vector b that minimize a cost function

$$\psi_o(b,Z_B,Z_I,\xi) \tag{2.3}$$

satisfies Eqs. (2.1) and (2.2) and the following constraints on stresses, displacements, eigenvalues and design variables

$$\psi_i(b,Z_B,Z_I,\xi) \leq 0 \ , \ i = 1,2,. \ . \ . \ m \tag{2.4}$$

2.2 Design Sensitivity Analysis

An efficient method for design sensitivity analysis with substructures has been presented in Refs. [4,5]. Therefore the general approach for sensitivity analysis is only summarized in this section.

Let $\psi(b, Z_B, Z_I, \xi)$ be a general function that may represent the cost or any constraint function. When the design is changed by a small δb, the displacements Z_B and Z_I, and the eigenvalue ξ will also change by small amounts δZ_B, δZ_I and $\delta\xi$ due to the well-posed nature of the structural analysis problem. Let $\delta\psi$ represent a first order change in the function ψ. Taking b, Z_B, Z_I and ξ as independent variables, $\delta\psi$ is given as:

$$\delta\psi = \frac{\partial\psi}{\partial b}\delta b + \frac{\partial\psi}{\partial Z_B}\delta Z_B + \frac{\partial\psi}{\partial Z_I}\delta Z_I + \frac{\partial\psi}{\partial\xi}\delta\xi \tag{2.5}$$

The term δZ_B, δZ_I and $\delta\xi$ appearing in the above equation can be expressed in term of δb by taking the first variation with respect to b of the state equation (2.1) and the eigenvalue equation (2.2). Equation (2.5) finally can be put in the form [3,4]

$$\delta\psi = G^T\delta b \tag{2.6}$$

where:

$$G = \frac{\partial\psi^T}{\partial b} + C_2\lambda_I + C^T\lambda_B + \Lambda^e \quad ; \quad \Lambda^e = \frac{\frac{\partial}{\partial b}[Ky - \xi My]^T y\frac{\partial\psi}{\partial\xi}}{y^T My} \tag{2.7}$$

$$C_1 = \frac{\partial R_B}{\partial b} - \frac{\partial}{\partial b}[K_{BB}\,Z_B] - \frac{\partial}{\partial b}[K_{BI}\,Z_I] \quad ; \quad C_2 = \frac{\partial R_I}{\partial b} - \frac{\partial}{\partial b}[K_{IB}\,Z_B] - \frac{\partial}{\partial b}[K_{II}\,Z_I] \tag{2.8}$$

$$C = C_1 + Q^T C_2 \quad ; \quad Q = -K_{II}^{-1}\,K_{IB} \tag{2.9}$$

and λ_I, λ_B are solutions of the adjoint equations:

$$K_{II}\,\lambda_I = \frac{\partial\psi^T}{\partial Z_I} \quad ; \quad K_B\,\lambda_B = \frac{\partial\psi^T}{\partial Z_B} - K_{BI}\,\lambda_I \tag{2.10}$$

where: $$K_B = K_{BB} + K_{BI}Q \tag{2.11}$$

Equation (2.6) is the desired relationship between the design change and the change in a member force, a nodal displacement, the cost function and/or eigenvalue. The vector G of Eq. (2.7) is the required sensitivity vector. In subsequent sections, Eq. (2.7) will be used for different type of constraint ψ.

III. STIFFNESS AND STRESS MATRICES OF THE TRIANGULAR PLATE BENDING ELEMENT

Detail of element properties of the triangular plate bending with shear deformation has been presented in Ref. [1]. Thus, in this section, the derivations of element stiffness and stress matrices is summarized and only those formulas which are required for the design sensitivity analysis in the next section are presented.

The local coordinate system of a triangular plate bending element is shown in Figure 1. The force-displacement relationship for the element can be written as:

$$f_e = K_{ee}\, u_e \tag{3.1}$$

where K_{ee} is the element stiffness matrix, u_e is the element nodal displacement and f_e is element nodal force. It should be noted that among the 9DOF (Degree Of Freedom) for the triangular bending element, 3 describe rigid body motion. Equation (3.1) can therefore be partitioned as follows:

$$\begin{bmatrix} f_i \\ f_a \end{bmatrix} = \begin{bmatrix} K_{ii} & K_{ia} \\ K_{ai} & K_{aa} \end{bmatrix} \begin{bmatrix} u_i \\ u_a \end{bmatrix} \tag{3.2}$$

where the flexible body nodal coordinates u_i and rigid body nodal coordinates u_a are defined as:

$$u_i = [w_b, \alpha_b, \beta_b, w_c, \alpha_c, \beta_c]^T \quad ; \quad u_a = [w_a, \alpha_a, \beta_a]^T \tag{3.3}$$

Figure 1 defines the translation w along the z direction and the rotations α and β about the x and y axes, respectively. The subscripts a, b and c refer to the nodes of the triangular finite element.

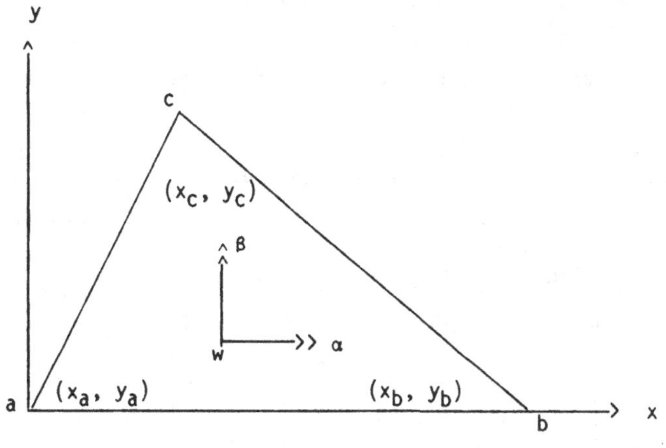

Figure 1 Local Coordinate System of Triangular Bending Plate Element abc

The nodal coordinates u_i and u_a are related by the following transformation:

$$u_i = S\, u_a \tag{3.4}$$

where the matrix S can be calculated from kinematics and is given as:

$$S = \begin{bmatrix} 1 & 0 & 0 & 1 & 0 & 0 \\ 0 & 1 & 0 & y_c & 1 & 0 \\ -x_b & 0 & 1 & -x_c & 0 & 1 \end{bmatrix}^T \tag{3.5}$$

where x_b is the local x-coordinate of node b, x_c and y_c are the x and y local coordinate of node c, respectively (Refer to Figure 1).

Equation (3.2) can be arranged to the following form:

$$\begin{bmatrix} u_i \\ f_a \end{bmatrix} = \begin{bmatrix} K_{ii}^{-1} & -K_{ii}^{-1} K_{ia} \\ K_{ai} K_{ii}^{-1} & K_{aa} - K_{ai} K_{ii}^{-1} K_{ia} \end{bmatrix} \begin{bmatrix} f_i \\ u_a \end{bmatrix} \tag{3.6}$$

When there are no forces on the u_i coodinates, i.e. $f_i = 0$, by comparing Eq. (3.4) with the top half of Eq. (3.6), one obtains:

$$K_{ia} = -K_{ii} S \tag{3.7}$$

From the bottom half of Eq. (3.6) and realizing that there is no strain energy associated with rigid body motion, one has:

$$K_{aa} = K_{ai} K_{ii}^{-1} K_{ia} \tag{3.8}$$

Substituting Eq. (3.7) into Eq. (3.8), one obtains: $K_{aa} = S^T K_{ii} S$ (3.9)

Thus the main effort in the following subsection is to calculate the submatrix K_{ii} because once K_{ii} is known, Eqs. (3.7) and (3.9) can be used to calculate K_{ia} and K_{aa}, respectively.

The displacement w along the z-direction is assumed to be given by the following polynomial:

$$w = \gamma_x x + \gamma_y y + q_1 x^2 + q_2 xy + q_3 y^2 + q_4 x^3 + q_5 xy^2 + q_6 y^3 \tag{3.10}$$

The coefficients γ_x and γ_y are related to the transverse shear strains and $q_r = [q_1, q_2 \dots q_6]^T$ are the generalized coordinates.

The rotations α and β (about the x and y axis, respectively) are given as:

$$\alpha = \frac{\partial w}{\partial y} - \gamma_y \quad ; \quad \beta = \gamma_x - \frac{\partial w}{\partial x} \tag{3.11}$$

Nodal coordinates u_i are related to the relative displacements u_r by:

$$u_i = S\, u_a + u_r \tag{3.12}$$

When there is no rigid body displacement, $u_a = 0$. Thus $u_i = u_r$.

Substituting the coordinates of nodes b and c into Eqs. (3.10) and (3.11), one has:

$$u_r = u_i = \begin{bmatrix} w_b \\ \alpha_b \\ \beta_b \\ w_c \\ \alpha_c \\ \beta_c \end{bmatrix} = \begin{bmatrix} x_b & 0 \\ 0 & 0 \\ 0 & 0 \\ x_c & y_c \\ 0 & 0 \\ 0 & 0 \end{bmatrix} \begin{bmatrix} \gamma_x \\ \gamma_y \end{bmatrix} + \begin{bmatrix} x_b^2 & 0 & 0 & x_b^3 & 0 & 0 \\ 0 & x_b & 0 & 0 & 0 & 0 \\ -2x_b & 0 & 0 & -3x_b^2 & 0 & 0 \\ x_c^2 & x_c y_c & y_c^2 & x_c^3 & x_c y_c^2 & y_c^3 \\ 0 & x_c & 2y_c & 0 & 2x_c y_c & 3y_c^2 \\ -2x_c & -y_c & 0 & -3x_c^2 & -y_c^2 & 0 \end{bmatrix} \begin{bmatrix} q_1 \\ q_2 \\ q_3 \\ q_4 \\ q_5 \\ q_6 \end{bmatrix} \tag{3.13}$$

or: $u_r = H_{u\gamma}\, \gamma + \bar{H}\, q_r$ (3.14)

where vector γ and matrices $H_{u\gamma}$ and $\bar{H}$ can be easily identified from Eq. (3.13).

Transverse shear strain vector γ is related to the shear forces. Shear forces are related to the moments through the equilibrium equations. These moments are then related to the curvatures and finally, the curvatures can be expressed in terms of the generalized coordinates q_r [see Eq. (3.28)]. Thus:

$$\gamma = H_{\gamma q}\, q_r \tag{3.15}$$

where:

$$H_{\gamma q} = -\begin{bmatrix} 0 & 0 & 0 & 6(J_{11}D_{11} + J_{12}D_{13}) & J_{11}(2D_{12} + 4D_{33}) + 6J_{12}D_{23} & 6(J_{11}D_{23} + J_{12}D_{22}) \\ 0 & 0 & 0 & 6(J_{12}D_{11} + J_{22}D_{13}) & J_{12}(2D_{12} + 4D_{33}) + 6J_{22}D_{23} & 6(J_{12}D_{23} + J_{22}D_{22}) \end{bmatrix} \tag{3.16}$$

$$D_{ij} = \begin{bmatrix} \frac{E}{1-\nu^2} & \frac{\nu E}{1-\nu^2} & 0 \\ \frac{\nu E}{1-\nu^2} & \frac{E}{1-\nu^2} & 0 \\ 0 & 0 & G \end{bmatrix} * \frac{t^3}{12} \equiv [\ \overline{Q}\ ] * \frac{t^3}{12} \tag{3.17}$$

$$J_{ij} = \begin{bmatrix} 1 & 0 \\ 0 & 1 \end{bmatrix} * \frac{1}{Gt} \equiv [G^*]^{-1} \tag{3.18}$$

Substituting Eq. (3.15) into (3.14), one obtains: $u_r = H\ q_r$ (3.19)

where: $H = H_{u\gamma}\ H_{\gamma q} + \overline{H}$ (3.20)

and

$$\overline{H} = \begin{bmatrix} x_b^2 & 0 & 0 & x_b^3 & 0 & 0 \\ 0 & x_b & 0 & 0 & 0 & 0 \\ -2x_b & 0 & 0 & -3x_b^2 & 0 & 0 \\ x_c^2 & x_c y_c & y_c^2 & x_c^3 & x_c y_c^2 & y_c^3 \\ 0 & x_c & 2y_c & 0 & 2x_c y_c & 3y_c^2 \\ -2x_c & -y_c & 0 & -3x_c^2 & -y_c^2 & 0 \end{bmatrix} \tag{3.21}$$

The stored elastic energy is given as:

$$V_e = \frac{1}{2} \iint q_r^T\ k^q\ q_r\ dA \tag{3.22}$$

where k^q is element stiffness matrix per unit area (with respect to the generalized coordinates q_r). Element stiffness sub-matrix corresponding to <u>flexible</u> motion is therefore

$$K^q = \iint k^q\ dA \tag{3.23}$$

Using Eqs. (3.19) and (3.22), one obtains:

$$K_{ii} = H^{-1^T}\ K^q\ H^{-1} \tag{3.24}$$

where

$$K^q = K^\gamma + K^x \tag{3.25}$$

$$K^\gamma = A\ H_{\gamma q}^T\ G^*\ H_{\gamma q} \tag{3.26}$$

and $H_{\gamma q}$ and G^* have been defined in Eqs. (3.16) and (3.18), respectively.

$$K^x = 4A \begin{bmatrix} D_{11} & D_{13} & D_{12} & 3\overline{x}D_{11} & \overline{x}D_{12} + 2\overline{y}D_{13} & 3\overline{y}D_{12} \\ & D_{33} & D_{23} & 3\overline{x}D_{13} & \overline{x}D_{23} + 2\overline{y}D_{33} & 3\overline{y}D_{23} \\ & & D_{22} & 3\overline{x}D_{12} & \overline{x}D_{22} + 2\overline{y}D_{23} & 3\overline{y}D_{22} \\ & & & 9\rho_x^2 D_{11} & 3\rho_x^2 D_{12} + 6\rho_{xy}^2 D_{13} & 9\rho_{xy}^2 D_{12} \\ & \text{SYM.} & & & \rho_x^2 D_{22} + 4\rho_{xy}^2 D_{23} + 4\rho_y^2 D_{33} & 3\rho_{xy}^2 D_{22} + 6\rho_y^2 D_{23} \\ & & & & & 9\rho_y^2 D_{22} \end{bmatrix} \tag{3.27}$$

The curvatures may be related to the generalized coordinates as following:

$$X = \begin{bmatrix} X_x \\ X_x \\ X_{xy} \end{bmatrix} = \begin{bmatrix} -\frac{\partial\beta}{\partial x} \\ \frac{\partial\alpha}{\partial y} \\ \frac{\partial\alpha}{\partial x} - \frac{\partial\beta}{\partial y} \end{bmatrix} = \begin{bmatrix} 2 & 0 & 0 & 6x & 0 & 0 \\ 0 & 0 & 2 & 0 & 2x & 6y \\ 0 & 2 & 0 & 0 & 4y & 0 \end{bmatrix} \begin{bmatrix} q_1 \\ q_2 \\ \cdot \\ \cdot \\ q_6 \end{bmatrix} \quad (3.28)$$

or: $X = H_{xq}\, q_r$ (3.29)

where $H_{xq} \equiv f$ and can be identified from Eq. (3.28). The quantities appearing in Eq. (3.27) are defined as:

$$A \;= \text{area of triangle abc} = \frac{1}{2}x_b y_c \quad ; \quad \overline{x} = \frac{1}{3}(x_b + x_c) \; ; \; \overline{y} = \frac{1}{3}y_c \; ; \; \rho_y^2 = \frac{y_c^2}{6}$$

$$\rho_{xy} = \frac{y_3}{12}(x_b + 2x_c) \quad ; \quad \rho_x^2 = \frac{1}{6}(x_b^2 + x_b x_c + x_c^2) \quad (3.30)$$

IV. DESIGN SENSITIVITY ANALYSIS

Since sensitivity analysis of stress constraint is more complex than other type of constraints, only analytical differentiation of stress constraint with respect to design variable vector b is discussed in this section. In addition, much of the information presented in this section can be also used for other type of constraints and the cost function. Coordinates of the ith node may be expressed in global coordinates $R_i = [R_{iX}, R_{iY}, R_{iZ}]^T$ or in local coordinates $r_i = [r_{ix}, r_{iy}, r_{iz}]^T$. These coordinate systems are related by the following transformation.

$$R_i = R_1 + Lr_i \quad (4.1)$$

where: $L = [\hat{x},\hat{y},\hat{z}]$ (4.2)

and $\hat{x},\hat{y},\hat{z}$ are the unit vectors of the local x,y,z axes expressed in global coordinates. It is noted that $L^{-1} = L^T$, so from Eq. (4.1), one has

$$r_i = L^T(R_i - R_1) \quad (4.3)$$

Differentiating r_i with respect to c_j:

$$\frac{\partial r_i}{\partial c_j} = \frac{\partial L^T}{\partial c_j} * (R_i - R_1) + L^T\frac{\partial}{\partial c_j}(R_i - R_1) \quad (4.4)$$

where: $c_j = [R_{1X}, R_{1Y}, R_{1Z}, R_{2X}, R_{2Y}, R_{2Z}, R_{3X}, R_{3Y}, R_{3Z}]^T$

Differentiating L with respect to c_j: $\frac{\partial L}{\partial c_j} = \left[\frac{\partial\hat{x}}{\partial c_j}, \frac{\partial\hat{y}}{\partial c_j}, \frac{\partial\hat{z}}{\partial c_j}\right]$ (4.5)

A typical term of Eq. (4.5) is given as:

$$\frac{\partial\hat{x}}{\partial c_j} = \frac{1}{\|\vec{R}_2 - \vec{R}_1\|} * \frac{\partial}{\partial c_j}(\vec{R}_2 - \vec{R}1) + (\vec{R}_2 - \vec{R}_1) * \frac{(-1)}{(\|\vec{R}_2 - \vec{R}_1\|)^2} * \frac{\partial}{\partial c_j}\|\vec{R}_2 - \vec{R}_1\| \quad (4.6)$$

Derivatives in Eq. (4.6) and in the 2nd term of Eq. (4.4) can be evaluated easily. For examples:

$$\frac{\partial}{\partial c_j}(\vec{R}_2 - \vec{R}_1) = \begin{bmatrix} -1 & 0 & 0 & 1 & 0 & 0 & 0 & 0 & 0 \\ 0 & -1 & 0 & 0 & 1 & 0 & 0 & 0 & 0 \\ 0 & 0 & -1 & 0 & 0 & 1 & 0 & 0 & 0 \end{bmatrix}$$

$$\frac{\partial}{\partial c_j}\left\|\vec{R}_2 - \vec{R}_1\right\| = \frac{1}{\left\|\vec{R}_2 - \vec{R}_1\right\|} *$$

$$\left[(R_{2X} - R_{1X})\frac{\partial}{\partial c_j}(R_{2X} - R_{1X}) + (R_{2Y} - R_{1Y})\frac{\partial}{\partial c_j}(R_{2Y} - R_{1Y}) + (R_{2Z} - R_{1Z})\frac{\partial}{\partial c_j}(R_{2Z} - R_{1Z})\right]$$

Using strain-displacement and stress-strain relationship, one obtains:

$$\sigma_{x-y} = z\,\bar{Q}\,f\,H^{-1}\,(\,\bar{\lambda}_i\,\bar{u}_i - S\bar{\lambda}_a\,\bar{u}_a) \tag{4.7}$$

where z is the distance from the plate neutral surface to the point of interest, $\bar{u}_i$ and $\bar{u}_a$ are nodal displacements in global coordinates. $\bar{\lambda}_i$ and $\bar{\lambda}_a$ are local to global transformation correspond to $\bar{u}_i$ and $\bar{u}_a$, respectively, and $\sigma_{x-y} = [\sigma_x, \sigma_y, T_{xy}]^T$.

As can be seen from Eqs. (2.7) to (2.11), it is required to calculate the following quantities for sensitivity analysis:

$$\frac{\partial\Psi}{\partial b},\ \frac{\partial\Psi}{\partial u},\ \frac{\partial R}{\partial b}\ \text{and}\ \frac{\partial K}{\partial b} \tag{4.8}$$

where:

$$b \equiv \left[\frac{t}{c_j}\right] = \left[\frac{\text{thickness of the plate}}{\text{nodal coordinates}}\right]$$

$$u \equiv \left[\frac{\bar{u}_i}{\bar{u}_a}\right]\ ;\ R = \left[\frac{R_B}{R_I}\right]$$

For many cases, $\frac{\partial R}{\partial b} = 0$. However, calculation of the remaining terms in Eq. (4.8) are quite lengthy. The following subsections will only summarize the calculations of $\frac{\partial\Psi}{\partial u}$, $\frac{\partial\Psi}{\partial b}$ and $\frac{\partial K}{\partial b}$. More details can be found in Ref. [6].

In Eq. (4.7), only the scalar z and matrix H^{-1} are function of the thickness t. Define:

$$B2 = \bar{Q}f \quad ; \quad \lambda = B2\ H^{-1} \quad ; \quad B1 = \bar{\lambda}_i\bar{u}_i - S\bar{\lambda}_a\bar{u}_a \tag{4.9}$$

Thus Eq. (4.7) becomes: $\sigma_{x-y} = \lambda B1 z$ (4.10)

In Eq. (4.7), only the matrices f, H^{-1}, S, $\bar{\lambda}_i$ and $\bar{\lambda}_a$ are functions of nodal coordinates c_j. Define:

$$P1 = \lambda\bar{\lambda}_i\bar{u}_i \quad ; \quad P2 = \lambda S\bar{\lambda}_a\bar{u}_a \tag{4.11}$$

Hence, Eq. (4.7) can still be put in another form: $\sigma_{x-y} = z(P1 - P2)$ (4.12)

From Eq. (4.7), one has:

$$\frac{\partial\Psi}{\partial b} = \frac{\partial\Psi}{\partial\sigma_x}\frac{\partial\sigma_x}{\partial b} + \frac{\partial\Psi}{\partial\sigma_y}\frac{\partial\sigma_y}{\partial b} + \frac{\partial\Psi}{\partial T_{xy}}\frac{\partial T_{xy}}{\partial b} \tag{4.13}$$

$$\frac{\partial\Psi}{\partial u} = \frac{\partial\Psi}{\partial\sigma_x}\frac{\partial\sigma_x}{\partial u} + \frac{\partial\Psi}{\partial\sigma_y}\frac{\partial\sigma_y}{\partial u} + \frac{\partial\Psi}{\partial T_{xy}}\frac{\partial T_{xy}}{\partial u} \tag{4.14}$$

In Eqs.(4.13) and (4.14), $\frac{\partial\Psi}{\partial\sigma_x}$, $\frac{\partial\Psi}{\partial\sigma_y}$ and $\frac{\partial\Psi}{\partial T_{xy}}$ are known explicitly once a particular form of stress constraint has been selected. Thus, the problem remained to be solved is to evaluate the derivatives of stress quantities with respect to design variable vector b and nodal displacement u.

(i) Calculation of $\frac{\partial \Psi}{\partial u}$

Derivatives of stress quantities with respect to nodal displacement u are simple since stresses are given explicitly in term of nodal displacements according to Eq. (4.7).

(ii) Calculation of $\frac{\partial \Psi}{\partial b}$

Evaluation of $\frac{\partial \Psi}{\partial b}$ requires "chain" calculation and is summarized as follows:

Step (1) To calculate $\frac{\partial \Psi}{\partial b}$, it is required by Eq. (4.13) to evaluate $\frac{\partial \sigma_x}{\partial b}$, $\frac{\partial \sigma_y}{\partial b}$ and $\frac{\partial \tau_{xy}}{\partial b}$

Step (2) Considering the case where b is the thickness of the plate, hence b = t, then

Step (3) To calculate derivatives of stress quantities with respect to the thickness t (as shown in step 1), it is required by Eq. (4.10) to evaluate $\frac{\partial \lambda}{\partial t}$ and $\frac{\partial z}{\partial t}$. It should be noted that $\frac{\partial z}{\partial t}$ can be easily obtained since z is explicitly a function of thickness t.

Step (4) From Eqs. (4.9), to calculate $\frac{\partial \lambda}{\partial t}$ (as shown in step 3) it is required to have $\frac{\partial H}{\partial t}$

Step (5) From Eq. (3.20), to calculate $\frac{\partial H}{\partial t}$, it is required to have $\frac{\partial H_{\gamma q}}{\partial t}$

Step (6) From Eq. (3.16), to calculate $\frac{\partial H_{\gamma q}}{\partial t}$, it is required to have $\frac{\partial J}{\partial t}$ and $\frac{\partial D}{\partial t}$

Step (7) Finally, since the material matrices D and J are given explicitly in terms of thickness t as shown in Eqs. (3.17) and (3.18), thus $\frac{\partial J}{\partial t}$ and $\frac{\partial D}{\partial t}$ can be easily obtained.

Step (8) Considering the case b is the global nodal coordinates, hence b = c_j where c_j has been defined in Eq. (4.4)

Step (9) From Eq. (4.12), the derivatives of stress quantities with respect to nodal coordinates c_j require the calculation of $\frac{\partial P1}{\partial c_j}$ and $\frac{\partial P2}{\partial c_j}$

Step (10A) From Eqs. (4.11) and (4.2), calculation of $\frac{\partial P1}{\partial c_j}$ and $\frac{\partial P2}{\partial c_j}$ require to have $\frac{\partial \lambda}{\partial c_j}$, $\frac{\partial L}{\partial c_j}$ where $\frac{\partial L}{\partial c_j}$ has been obtained from Eq. (4.5), and

Step (10B) calculate $\frac{\partial S}{\partial c_j}$

Step (11) From Eqs. (4.9), to calculate $\frac{\partial \lambda}{\partial c_j}$ it is required to have $\frac{\partial B2}{\partial c_j}$ and $\frac{\partial H}{\partial c_j}$

Step (12) From Eqs. (4.9) and (3.20), it is required to have $\frac{\partial f}{\partial c_j}$ and $\frac{\partial H_{u\gamma}}{\partial c_j}$, $\frac{\partial \bar{H}}{\partial c_j}$

Step (13) Finally, quantity appearing in step (10B) can be expressed in term of $\frac{\partial r_i}{\partial c_j}$ by refering to Eqs. (3.5) and (4.3) where $\frac{\partial r_i}{\partial c_j}$ has been given explicitly in Eq. (4.4).
Also, quantities appearing in Step (12) can be expressed in term of $\frac{\partial r_i}{\partial c_j}$ by refering to Eqs. (3.29), (3.13), (3.21) and (4.3)

(iii) Calculation of $\frac{\partial K_{ee}}{\partial b}$

Step (1) As can seen from Eq. (3.2), it is required to calculate $\frac{\partial K_{ii}}{\partial b}$, $\frac{\partial K_{ia}}{\partial b}$ and $\frac{\partial K_{aa}}{\partial b}$

Step (2A) To calculate $\frac{\partial K_{ii}}{\partial b}$

Step (2B) From Eq. (3.24), it is required to evaluate $\frac{\partial K^q}{\partial b}$ and $\frac{\partial H}{\partial b}$ where $\frac{\partial H}{\partial b}$ has been calculated in subsection (ii).

Step (2C) From Eq. (3.25), it is required to evaluate $\frac{\partial K^\gamma}{\partial b}$ and $\frac{\partial K^x}{\partial b}$

Step (2D) To evaluate $\frac{\partial K^\gamma}{\partial b}$, it is required by Eq. (3.26) to calculate $\frac{\partial H_{\gamma q}}{\partial t}$ [already known by refering to subsection (ii)], $\frac{\partial G^*}{\partial t}$ [easily obtained by using Eq. (3.18)] and $\frac{\partial A}{\partial c_j}$ [easily obtained by using Eqs. (3.30), (4.3) and (4.4)].

Step (2E) To evaluate $\frac{\partial K^x}{\partial b}$, it is required by Eq. (3.27) to calculate $\frac{\partial D}{\partial t}$ [using Eq. (3.17)], $\frac{\partial A}{\partial c_j}$ [see step (2D)], $\frac{\partial \bar{x}}{\partial c_j}$, $\frac{\partial \bar{y}}{\partial c_j}$, $\frac{\partial \rho^2_{xy}}{\partial c_j}$, $\frac{\partial \rho^2_x}{\partial c_j}$ and $\frac{\partial \rho^2_y}{\partial c_j}$ [using Eqs. (3.30), (4.3) and (4.4)]

Step (3A) To calculate $\frac{\partial K_{ia}}{\partial b}$

Step (3B) It is required by Eq. (3.7) to evaluate $\frac{\partial K_{ii}}{\partial b}$ [refer to step (2A) through step (2E)] and $\frac{\partial S}{\partial b}$ [$\frac{\partial S}{\partial c_j}$ has already been discussed in subsection (ii), $\frac{\partial S}{\partial c_j}$ can be easily calculated by using Eqs. (3.5), (4.3) and (4.4)]

Step (4A) To calculate $\frac{\partial K_{aa}}{\partial b}$

Step (4B) It is required by Eq. (3.9) to evaluate $\frac{\partial K_{ii}}{\partial b}$ [refer to step (2A) through step (2E)] and $\frac{\partial S}{\partial b}$ [refer to step (3B)]

In summary, the derivatives of stress constraint with respect to plate thickness t and with respect to nodal coordinates c_j has been discussed with details in the above subsections. It is noted that Eq. (4.4) has been used extensively. Also, it should be mentioned that only <u>simple</u> derivatives of material matrices J, D and derivatives of Area of the plate element need be given <u>explicitly</u> in the final stage of the procedure.

REFERENCE

1. Macneal, R. H., The NASTRAN Theoretical Manual, COSMIC, University of Georgia, Athens, GA.

2. Giles, G.L. and Rogers, J.L., Jr., "Implementation of Structural Response Sensitivity Calculations in a Large-Scale Finite-Element Analysis System." NASA Langley Research Center, Hampton, VA, March 1982.

3. Nguyen, D.T. and Arora, J.S., "Fail-Safe Optimal Design of Complex Structures with Substructures." ASME Journal of Mechanical Design, Vol. 104, No. 4, October 1982.

4. Haug, E.J. and Arora, J.S., Applied Optimal Design: Mechanical and Structural System, John Wiley and Sons, 1979.

5. Nguyen, D.T., Belegundu, A.D. and Arora, J.S., "Design Optimization Codes for Structures: DOCS Computer Program." Presented at the Second JTCG/AS Workshop on Survivability and Computer Aided Design. Wright-Patterson AFB, Ohio, May 18-20, 1982.

6. Nguyen, D.T. and Arora, J.S., "Optimum Geometry Modeling for Minimizing Weight of Plate Bending Structures." Technical note No. 1, Civil Engineering Department, Northeastern Univeristy, Boston, MA (1982).

OPTIMAL MANAGEMENT OF AN ALMOST PURELY HYDRO SYSTEM : THE IVORY COAST CASE

Patrick COLLETER
Electricité de France
Paris, France

1. INTRODUCTION :

Hydroelectric power is for many countries a major supply for demand and more precisely for developping countries in Afrika, or South America, where great hydroelectric possibilities make the construction of big dams possible ; but how to plan the optimal development of the production facilities, in such systems ?

If this problem can be solved for thermal production facilities thanks to the standardization of the plants, for hydroelectric power plants several difficulties arise ; what choices should be made, concerning : power and storage, commissionning dates, system operating, marginal cost pricing... ?

The purpose of this paper is first to give a framework to solve these questions, and then to focus on the optimal operation of an almost purely hydro system including a pluriannual dam.

2. THE EXPANSION PLANNING PROBLEM FOR A PLURIANNUAL HYDRO POWER SYSTEM

2.1 Discontinuities in power and energy

The purpose of expansion planning consists in minimizing the sum of operating and investment costs through a given horizon (usually several decades). In an hydraulic system, the projects are generally non marginal in the system and the stepwise growth of the production set can be sketched as below.

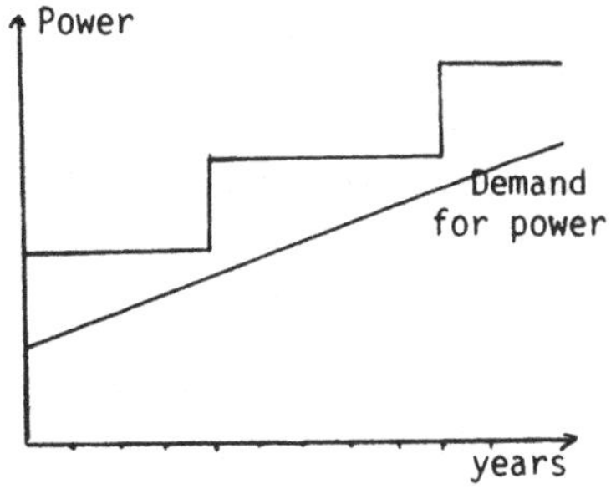

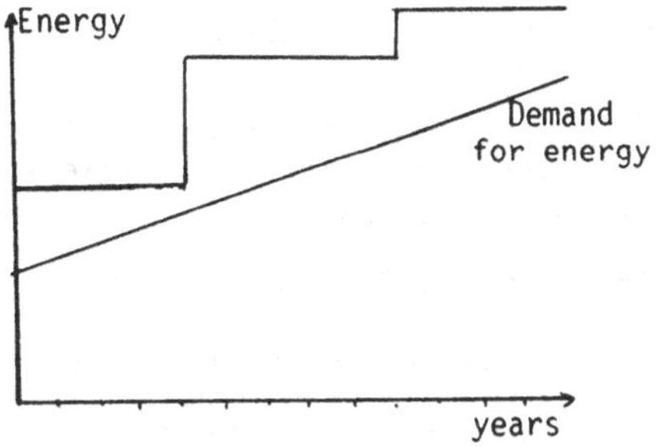

The discontinuties in the supply demand balance make the system swing from under equipped to overequipped situations. The only way to avoid this is an interannual storage : a good management of a pluriannual reservoir will tend to levelize the discontinuities in yearly energy availability. Expansion and system operation will be strongly connected, and the first step towards a global solution is to take this fact into account. Our purpose is here to make a model which will perform the optimal operation of the system for a given expansion planning. Thanks to this model, it will be possible to compare various expansion projects, and to define marginal cost even when there is no thermal plant running.

3. OPTIMAL OPERATION PROBLEM IN THE IVORIAN SYSTEM

3.1 The Ivorian system
3 11 The hydro plants

The Ivorian System consists of four existing reservoirs, located on three rivers according to fig 1. :

BUYO	KOSSOU	AYAME
SOUBRE (1986)	TAABO	
SASSANDRA RIVER	BANDAMA RIVER	BIA RIVER

Figure 1

A fifth reservoir, SOUBRE, is planned for the coming years on the SASSANDRA RIVER. Table 1 gives the main features of each reservoir. Dynamic analysis and time constants allow us to consider them as following :
- KOSSOU is a pluriannual dam,
- BUYO and AYAME are of annual type,
- TAABO and SOUBRE are of weekly type.

Table 1

	Volume 10^6 m^3	Power MW	Energy GWh	Evaporation (m)	TMAX/V	$\bar{A}$ / V
KOSSOU	27 680	179	500	1,5	1 %	0,2
TAABO	631	210	1 000	1,5	74 %	9,1
BUYO	8 300	165	850	1,45	5 %	1,6
SOUBRE	1 030	328	1 500	1,45	58 %	5,2
AYAME	1 015	50	230	1,05	7 %	2,3

TMAX = maximum weekly release.
A = average inflows (excluding releases from upstream reservoirs).
V = maximum capacity.

3.12 Thermal plants and shortage

The four thermal plants located at VRIDI (214 MW, 15 F/kWh) and a 12 MW Westinghouse (32 F/kWh), are the only flame - powered plants of the system. Three groups of shortage situations are considered :
- from zero up to 25 MW (45 F/kWh),
- from 25 MW up to 75 MW (90 F/kWh),
- more than 75 MW (200 F/kWh).

3.131 Demand

Table 2 gives the projected annual demand for the coming years. In the model, the demand will be described by a 3 blocks load duration curve.

Table 2

Year / Demand	1982	1983	1984	1985	1986	1987	1988
Energy (GWh)	1995	2233	2496	2795	3083	3427	3764
Peak (MW)	317	359	406	461	514	578	643

3.3 The optimal operation problem

3.31 Criterion and economic horizon

Classically defined as meeting demand at minimum cost (fuel + shortage cost), the optimal operation of the system on a given horizon is the solution of a dynamic stochastic problem.
As a matter of fact, hydro power plants are most of time able to meet demand without any thermal help (except during dry years or just before the commissioning of SOUBRE).Satisfying the demand, a given week or year, will thus be often possible at zero cost, and this for different uses of each dam. To make a difference between such solutions, it will be necessary to take into account the future of the system, and this on a several year horizon. A second reason for choosing this economic horizon is the non-stationnarity resulting from a discontinuous development of the hydro facilities (e.g. SOUBRE). After testing several cases, we choosed a 14 years horizon.

3.32 System modelling : slow and fast modes

3.321 Weekly dams

As previously described, we have a five dimensional dynamic stochastic problem, which is too big to be treated by global dynamic programming. Nevertheless, the particular structure of the problem will enable us to decrease this dimension down to 3. Namely, recent developments in control theory indicate that parametrizing the operation of fast modes (weekly dams) of a system as function of the slow modes (seasonal dams) can yield good solutions [5][6]. Moreover, as weekly dams are incapable of seasonal storage, their average level should be almost constant (except during the wet season), and thus their releases depend only on the upstream discharge and on the intermediate inflows. The operation rule for each weekly dam will thus be given by a target level, computed by dynamic programming to maximize the produced energy.

3.33 Seasonal and interannual dams

The system can now be described as 3 independant dams, fed by stochastic inflows. Evaporation and strong variation of the water conversion coefficient with the elevation makes it very difficult to find a good aggregation of the system. Stochastic relaxation yielding local feedbacks could have been carried out 3 , although Turgeon's works [4] seem to point out some potential difficulties, and price decomposition as used in [1] does not apply either. The best solution is of course a three dimensional dynamic programming, provided the CPU time is not to great ; the two main advantages of this method being :
- quality of the operating strategy : each reservoir will be operated according to its level, and according to the levels of the others. This is naturally crucial in

a system where the inflows are highly stochastic ;
- easy-to-handle model, because the Bellman's values derived from optimization do not need any updating, unless data such as thermal availability or demand are subject to unforeseen and significative variations.

4. SOLVING THE OPTIMAL CONTROL PROBLEM : THE MOGLI MODEL

In this section, we only give general indications. For more details, see annex 1.

$S_i(t)$	volume stored,	for dam i
$A_i(t)$	cumulated inflows till instant t,	
$U_i(t)$	turbined volume,	
$e\ (S_i,t)$	evaporation,	
$P_i(U_i,S_i)$	power, as function of U and S.	

The inflows are given by a stochastic process with independant increments :

$$dA_i = a_i(t)dt + \sigma_i(t)dW$$

where W is a normalized Brownian motion :

$$E\ (dW) = 0, \qquad E(dW^2\ (t)) = dt \qquad E\left[dW(t+dt)\ dW(t)\right] = 0$$

dynamic equation :

$$dS_i = dA_i - U_i dt - e(S_i)\ dt \qquad S_i \min \leq S_i \leq S_i \max$$

$$U_i \min \leq U_i \leq U_i \max$$

Criterion to be minimized

$$J = E \int_0^T C\left[D(t) - \sum_i P_i(U_i, S_i)\right] dt$$

where D(t) is the demand for power and C the thermal cost, E stands for the expectation with respect to inflows. T will be equal to 14 years. The time step is the week;
Let us denote by $V(S_1, S_2, S_3, t)$ the Bellman's value of the problem, then V is solution of the following equation [8] :

$$(1) \quad \frac{\partial V}{\partial t} + \frac{1}{2} \sum \frac{\partial^2 V}{\partial S_i^2} \sigma_i^2 + \min_{(u_i)} \left[C(D - \sum_i P_i(U_i, S_i)) + \sum_i \frac{\partial V}{\partial S_i} (a_i - e_i - U_i)\right] = 0.$$

This equation is then solved with the discretization technique described by Kushner in [7] and already used in [2] (more insight on this point is given in the mathematical annex). Let us notice that spatial correlations between inflows at the dams can be taken into account with a slight modification of equation (1).

4.2 Economic interpretation of the optimum

4.21 Shadow price of hydro energy

Optimizing the control at time t for given levels S_i consists in minimizing the Hamiltonian H :

$$H = C\ (D - \sum P_i\ (U_i, S_i)) + \sum \frac{\partial V}{\partial S_i} (a_i - u_i - e_i)$$

let us assume for clarity that $P_i = k_i U_i$, and denote by : T_j the power delivered by unit j, at proportional cost C_j. Denoting $\rho_i = \frac{-1}{k_i} \frac{\partial V}{\partial S_i}$, the Hamiltonian reads :

$$H = \sum_j C_j T_j + \sum_i \rho_i P_i - \sum_i \rho_i\ (a_i - e_i) \text{ and } \sum_j T_j + \sum_i P_i = D.$$

ρ_i appears as the shadow price of energy in reservoir i at time t and levels (S_1, S_2, S_3) and allows us to solve the problem as an all thermal one by piling up the different means of production, according to increasing proportional costs (for thermal plants) and shadow prices (for hydro plants). This accounts for a large part in the small CPU time needed by the model (1.5 mn/year on IBM 3081), although the real optimization is slightly more complex (See annex 1).

4.23 Definition of marginal cost from Bellman's value

The concept of marginal cost is generally defined as the proportional cost of the last running unit. When the demand is satisfied by hydro plants only, the shadow price of hydro energy enables us to define the marginal cost.

Economically speaking, ρ_i is the expected discounted gain related to the last kWh in dam i. When dam i is marginal in the load diagram, the marginal cost at this instant is ρ_i : producing one more kWh will cost in the future ρ_i, even if at the moment this kWh does not imply a thermal cost.

5. MOGLI'S ROLE IN DECISION MAKING AND RESULTS

Using the Bellman's values given by optimization, the simulation of the optimal controlled system through the observed time series of inflows is computed and yields the following results :
- optimal law of each storage (expectation, and standard deviation) ;
- optimal laws of the delivered energy, of marginal costs ;
- global cost for each time series and expectation.

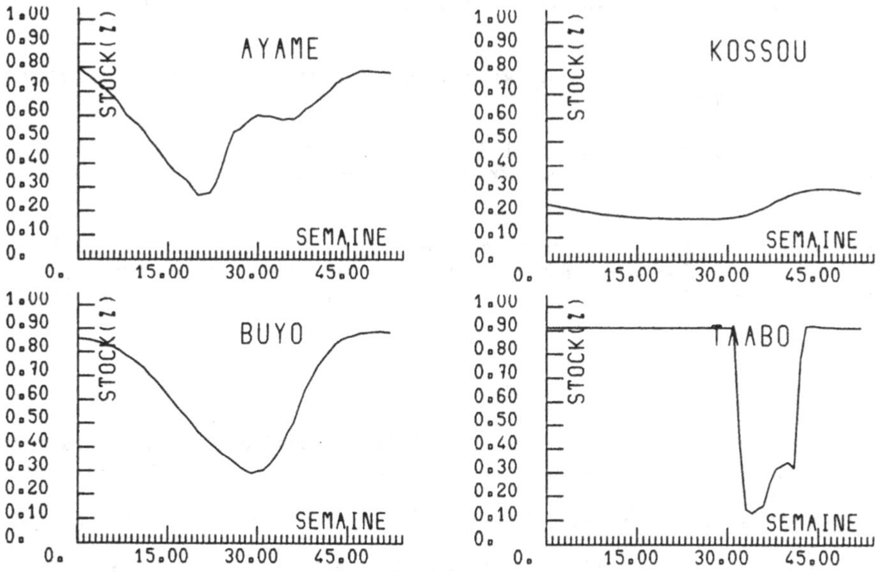

5.1 KOSSOU's operation : saving investment with the optimal operating strategy

The main problem in an almost purely hydro system including a pluriannual reservoir lies in the strong interaction between operation and expansion planning, when the planned units are not marginal in the existing system. The MOGLI Model computes the optimal use of the pluriannual reservoir. Comparisons of different expansion prog rams are thus possible.As an example the following graph shows us how KOSSOU should be operated if SOUBRE's coming is planned at the beginning of 1986 (figure 3) :

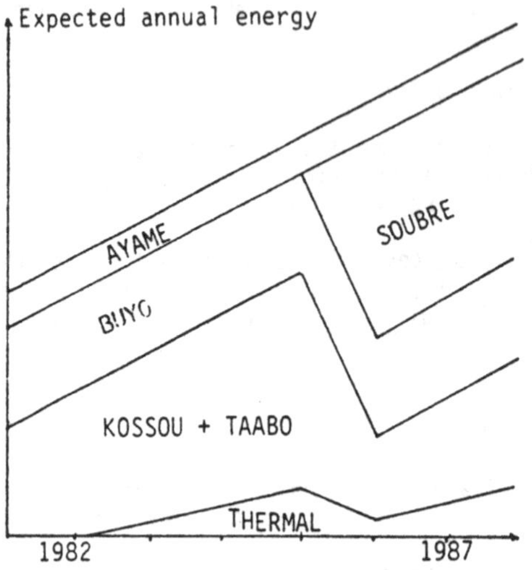

figure 3

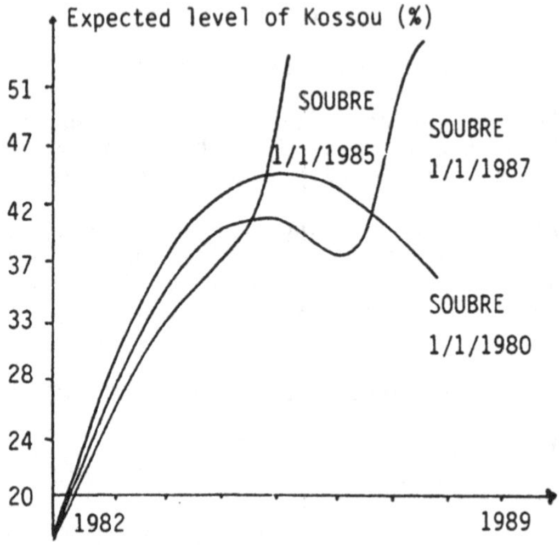

figure 4

KOSSOU's role appears clearly : storing energy up to 1984, it transfers it to 1985 just before SOUBRE's coming.

In the same way, fig. 4 shows the sensitivity of the operation of KOSSOU according to the commissioning of SOUBRE.

These two sets of results show us that KOSSOU must be used to damp the energy discontinuities of the system, either from production structure or hydrological events (dry and wet years).

Concerning the weekly operation, we are also able to give the following indications :
- energy and power dam by dam, for each hour block of the load duration curve ;
- values (shadow prices) of these energies, giving an order of merit for the overall system.

CONCLUSION

The MOGLI Model enables the operating people to make the good decisions in reservoir operation, and this is important because the system is almost purely hydraulic with an interannual storage.
Moreover, performing simulations including SOUBRE is a great help to get acquainted with the future production system.
And last but not least, this model can help decision makers to choose among future investment hypothesis, by giving them the optimal operating cost for each considered program.

The MOGLI Model is thus the first step towards a global model for optimal planning.

ACKNOWLEDGMENT

This study has been carried out with the efficient help of Michel MONTEIL from E.D.F-International.
We also want to thank MM. NDA and TANO from E.E.C.I. (Electricité de Côte d'Ivoire), for their valuable advices and for the success of the implementation of the model on E.E.C.I.'s computer.

REFERENCES

1 Optimal Operation Feedbacks for the french Hydropower System
P. COLLETER, P. LEDERER
Tims Orsa Joint Meeting, Toronto 1981.

2 Application of Stochastic Control Methods to the Managagement of EnergyProduction in New Caledonia.
P. COLLETER, F. DELEBECQUE, JP. QUADRAT, F. FALGARONE
In Applied Stochastic Control in Economics and Management Science, North Holland 1980.

3 Contribution of Stochastic Control, Singular Perturbations Averaging and Team Theories to an Example of Large Scale System : Management of Hydropower Production.
F. DELEBECQUE, JP. QUADRAT
IEEE Automatic Control AC 28, n° 2, p.209-222, april 1980.

4 Optimal Operation of Multireservoir Power System with stochastic Inflows
A. TURGEON
Water Ressources Research, vol. 16 , n°2, April 1980.

5 A. Decomposition of Near Optimum Regulators for Systems with Slow and Fast Modes.
J.H. CHOW, P.V. KOKOTOVIC
IEEE Automatic Control, October 1976.

6 Optimal Control of Markov Chains with Strong and Weak Interactions.
F. DELEBECQUE, J.P. QUADRAT.
Automatica, March 1981, Vol. 17 n°2, p. 281-296.

7 Probabilistic Methods for Approximation in Stochastic Control
H. KUSHNER
AcademicPress 1977.

Annexe 1

DISCRETIZATION OF BELLMAN'S EQUATION

1. THE CONTINUOUS PROBLEM

According to our notations the continuous problem reads :

$$dS_i = dA_i - u_i dt - e_i(S_i)dt \qquad dA_i = a_i dt + \sigma_i dw$$

$$j = \int_0^T c(D - \sum_i P_i(u_i, S_i))\, dt$$

denoting by $V(S_1, S_2, S_3, t)$ the Bellman's value, we have :

$$\frac{\partial V}{\partial t} + \frac{1}{2} \sum_i \frac{\partial^2 V}{\partial S_i^2} \sigma_i^2 + \min_{u_i} \left[C(D - P_i(U_i)) + \sum_i \frac{\partial V}{\partial S_i} (a_i - u_i - e_i) \right] = 0.$$

Following Kushner, we solve this equation by approximating the derivatives with the derivatives with finite differences. Let Δt and ΔS_i be the steps in time and volume

$$\frac{\partial V}{\partial S} (a - u - e) = \left[\left(\frac{\Delta V}{\Delta S}\right)^+ (a - u - e)^+ + \left(\frac{\Delta V}{\Delta S}\right)^- (a - u - e)^- \right].$$

With :

$$(a - u - e)^+ = a - u - e \quad \text{if} \quad u < a - e, \quad 0 \text{ if not.}$$
$$(a - u - e)^- = a - u - e \quad \text{if} \quad u > a - e, \quad 0 \text{ if not.}$$

$$\left(\frac{\Delta V}{\Delta S}\right)^+ = \frac{V(is + 1, it + 1) - V(is, it + 1)}{\Delta S}$$

$$\left(\frac{\Delta V}{\Delta S}\right)^- = \frac{V(is, it + 1) - V(is - 1, it + 1)}{\Delta S}$$

$\frac{\partial^2 V}{\partial S_i^2}$ is approximated in the same way, and interpreting the coefficients of V as probabilities we come to the optimal control of a Markov chain, approximating the continuous problem [9], [4].

2. TRANSITION OPTIMIZATION

For a given point (volume and time), we have to minimize H :

$$H = C(D - \sum P_i(U_i))\, \Delta t + \sum \left(\frac{\Delta V}{\Delta S_i}\right)^+ (a_i - u_i - e_i)^+ \Delta t + \sum \left(\frac{\Delta V}{\Delta S_i}\right)^- (a_i - u_i - e_i)^- \Delta t.$$

Let us denote :

x_i the conversion coefficient at level S_i, for dam i,

x_t, x_s the conversion coefficient for TAABO and SOUBRE,

U_t the maximum release at TAABO,

U_s the maximum release at SOUBRE.

$$P_1 = (x_1 + x_t)\, u_1 \qquad \text{if} \quad u_1 \leq u_t$$

$$P_1 = x_1\,(u_1 - u_t) + (x_1 + x_t)\, u_t \qquad \text{if} \quad u_1 \geq u_t$$

$$P_2 = (x_2 + x_s)\, u_2 \qquad \text{if} \quad u_2 < u_s$$

$$= x_2(u_2 - u_s) + (x_2 + x_s)\, u_s \qquad \text{if} \quad u_2 > u_s$$

$$P_3 = x_3\, u_3.$$

Denoting $\psi_i^+ = -\frac{\Delta V^+}{\Delta S_i}$, we can write after dropping the constant terms :

$$H = \sum_j c_j T_j + \sum_i \psi_i^+ u_i^+ + \sum_i \psi_i^- u_i^-$$

with :

$$u_i^+ < a_i - e_i \text{ and } u_i^+ + u_i^- = u_i, \quad \text{and : } \sum_j T_j + \sum_i P_i(u_i) = D.$$

We now need a description of the solution in every possible case. We only describe it when $u_t \leq a_1 - e_1$. Then :

$$\psi_i^+ u_i^+ + \psi_i^- u_i^- = \psi_i^+ \frac{p_1^+}{x_1 + x_t} + \psi_i^t \frac{P_1^o}{x_1} + \psi_i^- \frac{P_1^-}{x_1},$$

Where :

P_i^+ = power provided by KOSSOU and TAABO, together and below the limit of inflows,

P_i^o = power provided by KOSSOU alone, below the limit of inflows,

P_i^- = power provided by KOSSOU beyond the limit of inflows,

This is sketched on the following graph :

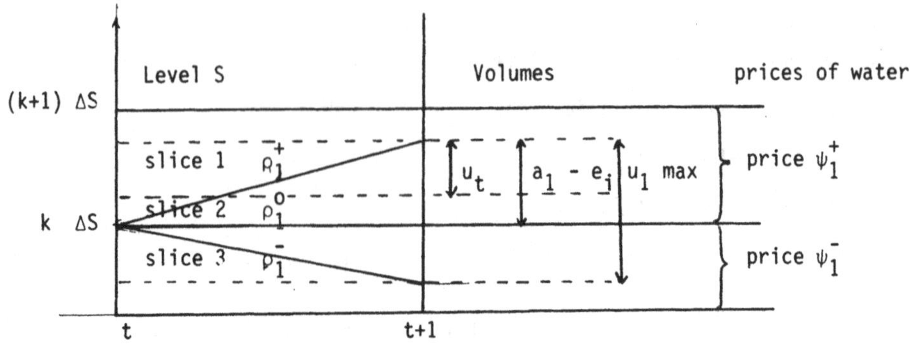

Reservoir 1 is thus locally discretized in 3 energy slices, the price of each slice is given by :

$$\rho_i^+ = \frac{\psi_1^+}{x_1 + x_t} \qquad \rho_i^o = \frac{\psi_1^+}{x_1} \qquad \rho_i^- = \frac{\psi_1^-}{x_1}$$

The expression to be minimized becomes :

$$\Sigma\, C_j\, T_j + \Sigma\, \rho_i^+\, P_i^+ + \sum_i \rho_i^o\, P_i^o + \sum_i \rho_i^-\, P_i^- .$$

With :
$$\sum_j T_j + \sum_i P_i^+ + \sum_i P_i^o + \sum_i P_i^- = D .$$

The optimum is clearly obtained by piling up the different slices according to increasing price. The prices are either c_i (thermal cost) or ρ_i (shadow price of water divided by the associated conversion coefficient).

From P_i's we are then able to derive the values of u_i^*'s, and we have to compute ;

$$V(S_1, S_2, S_3, t) = c(D - P_i(u_i^*))\, \Delta t + E(V(S_1, S_2, S_3, t+1) \,|\, u_i^*)$$

where the expectation is computed by the Markov transition matrix approximating the continuous problem.

3. REMARKS

- Let us point out that this discretization allows very fast computations because the optimization is made with the expected value of inflows ; the standard deviation is outside the minimization and adds a corrective term on V $(\frac{1}{2}\, \sigma^2 \frac{\partial^2 V}{\partial S^2})$.

- Note that Δt and ΔS should be sized such that the coefficients of V after discretization are bounded by 0 and 1. Nevertheless, in some cases we can avoid the use of a too small Δt (increasing the CPU time), by discretizing the equation around a different point. (Example : if $\Delta S < a\, \Delta t < 2\, \Delta S$, we can discretize around $S + \Delta S$ instead of S).

REAL-TIME OPTIMAL ENERGY MANAGEMENT BY MATHEMATICAL PROGRAMMING IN INDUSTRIAL PLANTS

K. Yamashita
Industrial and Public Facilities Control System Department
Fuchu Works, TOSHIBA Corporation
Toshiba-cho, Fuchu, Tokyo, JAPAN 183.

T. Watanabe and T. Katoh
Heavy Apparatus Engineering Laboratory
Fuchu Works, TOSHIBA Corporation
Toshiba-cho, Fuchu, Tokyo, JAPAN 183.

1. INTRODUCTION

Continuous escalation of energy costs since 1973 necessitates changes in energy utilization policies among energy intensive industries. The recent trend in Japan is to introduce computer-based energy management system with optimizing functions for in-plant power generating facility operation. The optimization problem in this case is dynamic by nature since the process to be controlled is dynamic and the energy demand variation is also dynamic. However, we formulated the problem in static framework and applied nonlinear programming on real-time by using mini-computer. In fact, the static formulation seems the only feasible solution because of the process nonlinearity and available computer capacity.

In this paper, we present a computer-based energy management system. The paper focuses on the problems faced in applying nonlinear programming to dynamic process optimization and how we coped with them. Section 2 explains how the optimization problem is formulated. Software configuration of the system is described in Section 3, where pre-processing of measured data is also discussed. Section 4 is the main part of the paper. Difficulties faced are first explained and the measures taken are presented with field data demonstrating the effect. The system evaluation is discussed in Section 5.

2. PROBLEM FORMULATION

In this section, we show problem formulation to apply mathematical programming. Fig. 1 is a conceptual diagram for optimizing energy supply. Here the energy consuming processes are pulp mill, paper mill, petrochemical plant, steel plant and cement mill. These production plants require electric power and steam at different presures. A typical energy bill is one hundred million dollars per year. The purpose of optimization is to minimize energy supply cost consisting of power bill and fuel consumed by boilers. The minimum cost is achieved by optimizing power load dispatching to generators and purchased power, load allocation to boilers and turbines, and fuel allocation to multi-fuel fired boilers. The optimization should be carried out under numerous ristrictions due to demand limit to the purchased power, machine capacity, environmental regulations and so on. It is noted that the cost of purchased power varies during the day and week, requiring a choice between buying and generating.

Now, we formulate the problem. First, it is assumed that the energy demand is constant for a certain period, usually 5 - 30 minutes. Secondly, we take a stance that our purpose is the steady state optimization and not the transient. This assumption is not impractical as process transients or disturbances can be handled by local and dedicated controllers. In Fig. 2, an example of powerhouse configuration is shown. There are four boilers and six turbine-generators. In this case, steam is supplied at two different pressures. In the following, let us use this powerhouse as an example to demonstrate the problem formulation procedure.

Variables : x First of all, variables have to be specified. This is shown in Fig. 2. There are altogether fourteen variables x_1, x_2, ..., x_{14}. If boilers are multi-fuel fired, new variables have to be brought in for boiler fuels.

Objective function : $f(x)$ The object of optimization is to minimize energy supply cost. Therefore, it is necessary to express the energy cost by using the variables, i.e., x_1, x_2, x_5, x_6 and x_{14}. Since boiler fuels are not given explicitly in this example, it is necessary to express them by the variables x_1, x_2, x_5 and x_6, where boiler models are needed. (Models will be discussed later.)

Inequality constraints : $g_i(x) \leq 0$, $i = 1,2,\ldots,m$. Most of the inequality constraints are upper and lower limits to each variable, i.e.,

$$x_{min} \leq x \leq x_{max} \tag{1}$$

or equivalently,

$$(x_{min} - x)(x_{max} - x) \leq 0 \tag{2}$$

These constraints are due to machine capacity and operational limits. The other constraints are expressed by functions of several variables. For instance, the constraint to the 3TG power output is expressed by the variables x_5 and x_7, where a turbine-generator model is needed. Environmental restrictions such as SOx density also constitute inequality constraints expressed as functions of several variables.

Equality constraints : $h_j(x) = 0$, $j = 1,2,\ldots,k$, In the case of Fig. 2, there are at least four equality constraints. These are steam mass balances at three pressure headers and electric power balance. This is where the energy demand comes in. For instance, one of the steam balances is given by

$$(x_3 - c_1) + (x_4 - c_2) + x_6 + x_7 + x_{12} -x_8 -x_9 - x_{10} - x_{13} - S_1 = 0 \tag{3}$$

where c_1 : steam leak from 1T
c_2 : steam leak from 2T
S_1 : steam demand at the pressure concerned

Boiler model A typical model used is basically an efficiency model given by

$$E_{ff} = a_0 + a_1 S_B + a_2 S_B^2 = \frac{S_B(i_s - i_w) - Q}{KF} \tag{4}$$

where E_{ff} : boiler efficiency, S_B : boiler steam
i_s : steam enthalpy, i_w : feed water enthalpy
Q : calorie brought in by boiler auxiliary steam
K : fuel calorie, F : fuel flow

The coefficients a_0, a_1, and a_2 vary depending on the mixing rate of fuels for multi-fuel fired boilers. Eq. (4) is a nonlinear model relating F and S_B. In practice, the model and its parameters are modified, on real-time, by other factors affecting efficiency.

Turbine-generator model Fig. 3 shows a characteristic curve for a turbine-generator without steam extraction, where S_T is the main steam flow and P the power output. S_T is adjusted by three control valves, which are opened one by one to increase the steam intake. The power loss due to valve opening is zero when the valve is fully open as is shown in Fig. 3. The model gives polynomial expression for P with respect to S_T, i.e.,

$$P = a_0 + a_1 S_T + \ldots + a_n S_T^n \tag{5}$$

For the turbine-generator in Fig. 3, several sets of parameters a_0, a_1, ..., a_n are needed. These parameters are, if necessary, modified with respect to steam pressure variation.

3. SYSTEM CONFIGURATION

In Fig. 4, software configuration of the system is shown. Some of the functions are explained in the following.

Data pre-processing Fig. 5 is a diagram showing the concept of data pre-processing to reduce measurement error. Averaging is necessary since we treat the optimization problem in static framework. Either moving average or first order low-pass filtering is used. In stead of using averaged values directly to identify the current process status, we introduce estimation process as a part of data pre-processing. The basic idea is to use interrelations of the variables and weighting based on sensor accuracy. Use of the former means the use of measurement redundancy due to interdependency of variables. Some of the examples of interrelations are steam mass balances of turbines, steam flow - power output relations of turbine-generators and steam flow - control valve opening relations. The information on sensor accuracy is used to determine weighting factors to evaluate individual measurements in the error function for estimation. The estimation algorithm used is the constrained weighted least squares method. Nonlinear least squares method is also applied to treat nonlinear relations.

Model management This function enables engineers or operators to tune-up and update model parameters from the video display terminals. Parameter estimating function is also provided.

Energy demand forecasting Long term forecast is based on production schedule and short range forecast, say 5 - 30 minutes, is based on current energy demand. Current energy demand can be obtained by using pre-processed data. This is done by using steam balance equations at common pressure headers and power balance equations, in stead of directly gathering energy consumption data scattered around the production process. Adjustment to the current demand to obtain forecast can be easily done from video display terminals. Our experience shows that the data pre-processing described earlier is very effective in obtaining reliable energy demand forecast.

Optimization More than one level of optimization can be considered. The top level optimization is to coordinate energy supply and production scheduling. The second level is the energy load dispatching to various energy supply sources such as power company, thermal powerhouses and hydraulic power stations. The third level is for single energy supply source operation, and this is where real-time optimization is required. It is often the case that the second level and the third level are combined, if there are no hydraulic power stations. For the top level optimization, we used both linear programming and dynamic programming. In this paper, we focus on real-time optimization to which nonlinear programming is applied. The algorithm chosen is the multiplier method by Powell [1] and Hestenes [3] (see Fig. 6). For the optimization of augmented objective function given in Eq. (5), the conjugate gradient method [2] is used.

$$L(x) = f(x) + \sum_{j=1}^{m} \mu_j h_j(x) + \frac{1}{2}\sum_{j=1}^{m} r_j h_j^2(x)$$

$$+ \frac{1}{2}\sum_{i=1}^{k} \frac{1}{t_i}\left[[\max(0, \lambda_i + t_i g_i(x))]^2 - \lambda_i^2\right] \qquad (5)$$

where μ_j, λ_i, r_j and t_i are multipliers and penalty parameters. The result of optimization at this level is used to determine set points for the dedicated controllers.

4. REAL-TIME OPTIMIZATION

In this section, we discuss the problems faced and the countermeasures taken in applying nonlinear programming to optimization of powerhouse operation on real-time. The constraints for the real-time optimization are

1) limited computer capacity due to the use of minicomputer
2) limited computation time
3) need for moderate control

The first constraint will become less restrictive with the development of minicomputers in general. The second constraint is unavoidable and the time for computation is usually limited to 1 - 5 minutes depending on the energy consuming process. The third constraint is important as the large and frequent load shifting for turbines and boilers must be avoided. The large load shifting imposes too much thermal stress on the machines and boiler efficiency decreases during the transient operation.

The problems faced are as follows.

1) selection of initial search point
2) multi-modal objective function
3) large load shifting for boilers and turbines
4) computing time and optimization accuracy

These problems are closely related each other. As the computation time is limited, an effective initial search point must be chosen. The objective function turned out to be multi-modal, having local minimums. One of the obvious causes is the turbine-generator characteristics shown in Fig. 3. The most serious problem was the large perturbation by optimization result. During the field test, we observed that the load shifting being too large and optimization based control being almost bang-bang.

To cope with the above problems, we reached the following measures after painful field tuning.

1) set the initial search point to the current plant operating point
2) obtain sub-optimum point, which could also be the global optimal point, in the neighbourhood of the current plant operating point
3) introduce dynamic specification of feasible domain based on the current operating point
4) modify the multiplier method algorithm to fully make use of the available computing time

By taking the initial search point at the current plant operating point, one can expect the following advantages. The solution obtained is not too far from the current operating point and the initial search point is within the feasible domain, resulting in less computing time. In engineering applications, it is not always necessary to obtain global optimum. What is expected is an improvement by applying non-linear programming. Furthermore, we have to take into account of other factors such as computation time and moderate control.

The dynamic specification of feasible domain is done by the following. In stead of using inequality constraint given by Eq. (1), we use

$$\text{Max}\ (x_{min},\ x_c - c_1) \leq x \leq \text{Min}\ (x_{max},\ x_c + c_2) \tag{6}$$

where x_c is the current status, and c_1 and c_2 are positive constants. x_{min} and x_{max} are mainly due to machine capacity and operational limits. Eq. (6) ensures that the solution obtained, x^*, will be within $(x_c - c_1) \leq x^* \leq (x_c + c_2)$, i.e., the set point change will be atmost by c_1 or c_2. We named it dynamic specification because the lower and upper limits given by Eq. (6) vary depending on the current plant status on real-time.

The modification to the multiplier method algorithm is done by using time limit stopper in stead of constraint tolerance stopper and storing the best solution on the way that satisfies the constraints within the specified tolerance. Fig. 6 gives the concept of the modification.

Fig. 7 shows the effect of the measures described by field data. In this case, two turbines have identical specification. Therefore, almost equal load allocation is applied in conventional operation. On the other hand, optimal control brings one of the turbines to valve full opening position and the other to take the load variation. The load shifting to each turbine is done gradually, achieving moderate control. In this case, c_1 and c_2 in Eq. (6) were set to 10 T/H for the main steam flow variables of two turbines.

5. SYSTEM EVALUATION

To evaluate the optimization function, we carried out system evaluation test in two different sites, one in a paper mill and the other in a petrochemical plant. Two methods are used. In one of the methods, the test was carried out while the system was on-line, performing optimal control. The conventional load allocation was simulated on computer and the energy supply costs were compared. In the other method, the powerhouse operators were told that the computer system was down so that they had to operate the plant manually. However, we kept the system and the calculation for optimization running. In this case, too, the energy costs were compared. The results showed that in both sites, the energy costs were reduced by 1 - 2%. For a plant consuming $100 million/year, this reduction corresponds to $1 - 2 million/year.

6. CONCLUSIONS

We have presented a real-time optimal energy management system and shown that nonlinear programming can be applied to dynamic process optimization on real-time by using minicomputer. We believe that the combination of static problem formulation and measures described made our energetic attempt successful. The system developed and installed in industrial plants have been in commercial run for more than a year and half, reducing energy cost by 1 - 2%. Furthermore, the system reduces the personnel, provides more plant information and makes operation easier.

REFERENCES

[1] Powell, M.J.D., "A Method for Nonlinear Constraints in Minimization Problems", in Optimization, R. Fletcher (ed.), Academic Press, 1969

[2] Fletcher, R., and Reeves, C.M., "Function Minimization by Conjugate Gradients", Computer J., Vol. 7, pp. 149-154, 1964

[3] Hestenes, M.R., "Multiplier and Gradient Methods", J. Optimization Theory (2)

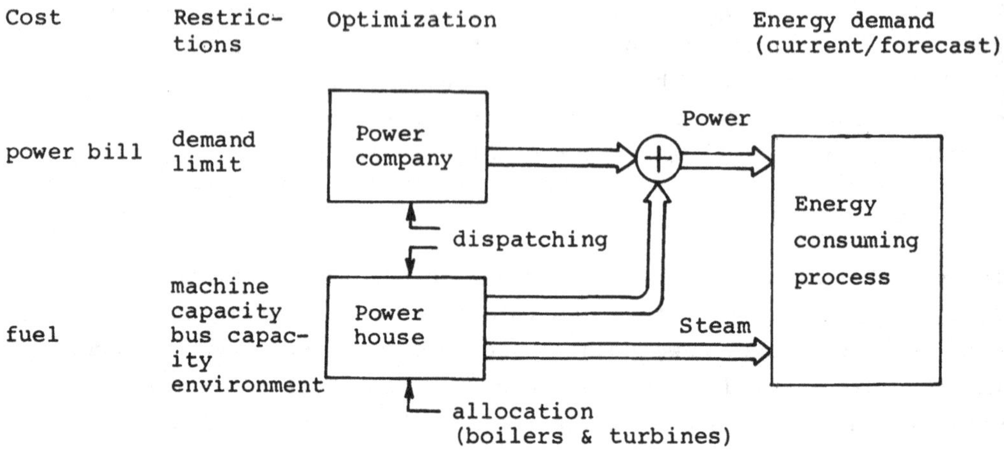

Fig. 1 Conceptual diagram for optimizing energy supply

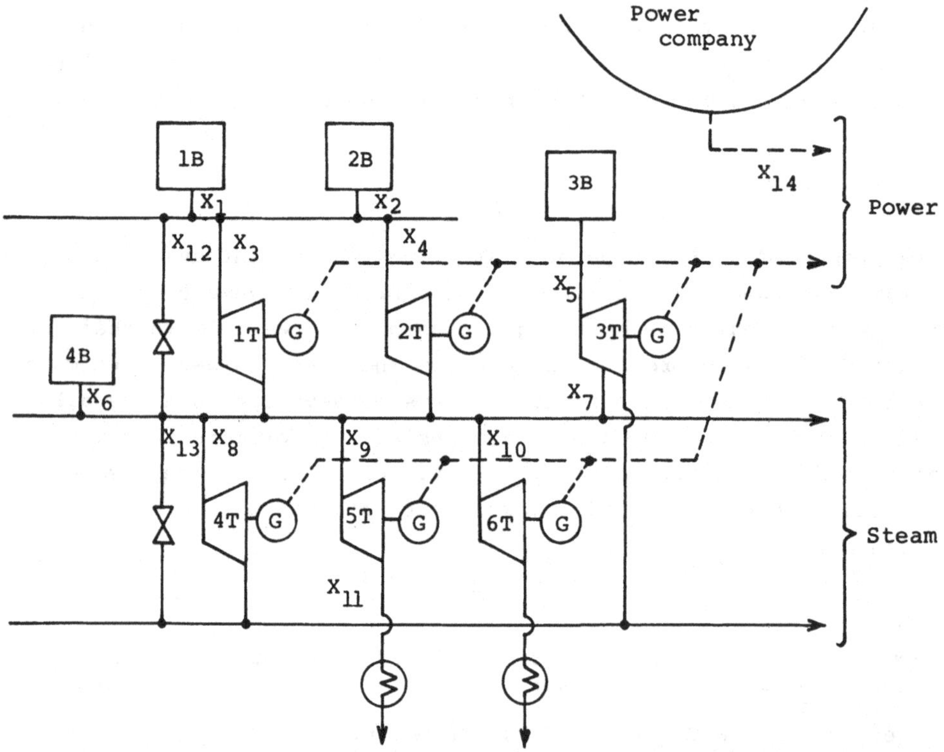

Fig. 2 An example of powerhouse configulation

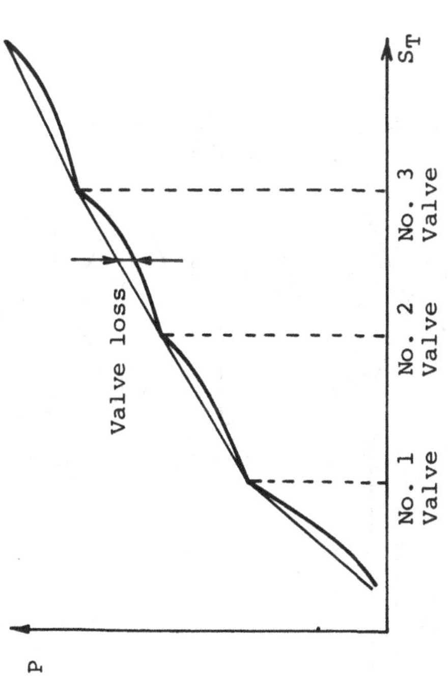

Fig. 3 Turbine-generator characteristic curve

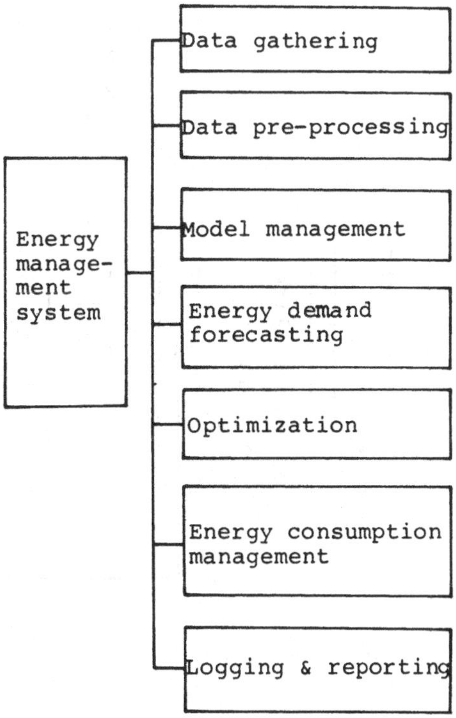

Fig. 4 Software configuration for energy management system

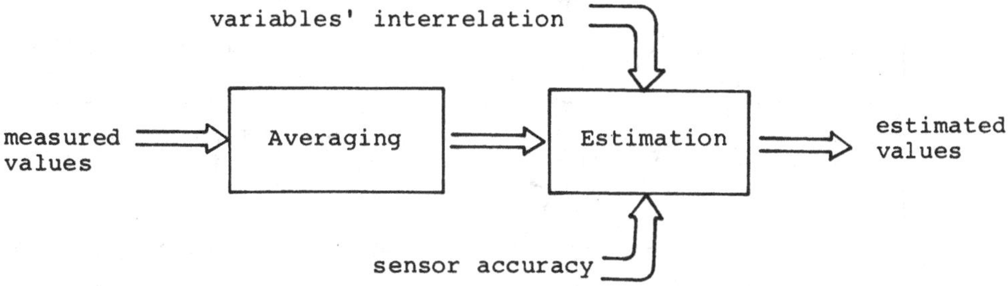

Fig. 5 Data pre-processing to reduce measurement error

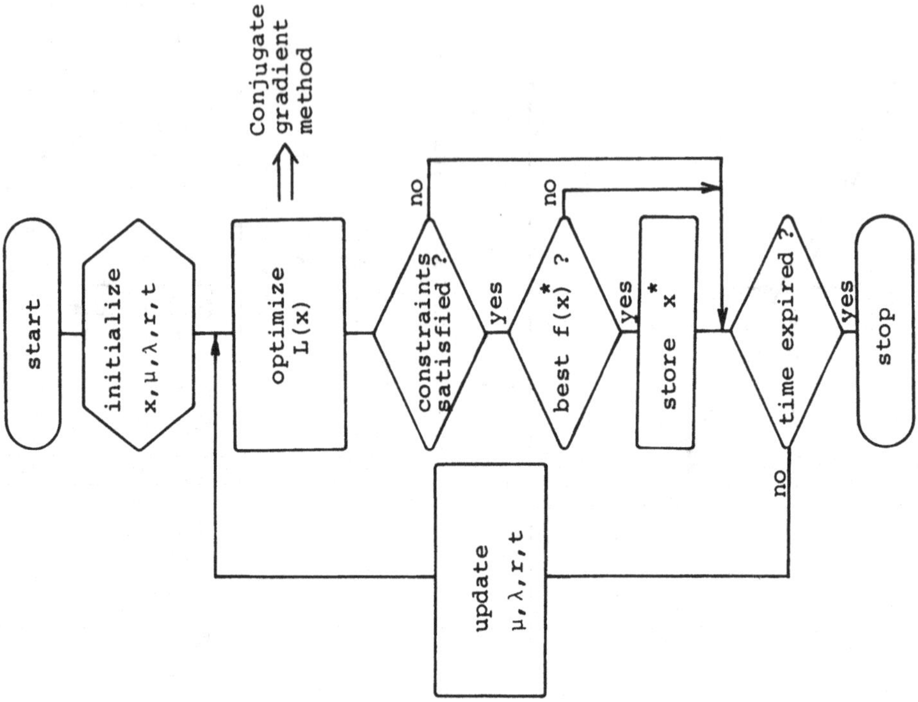

Fig. 6 Multiplier Method

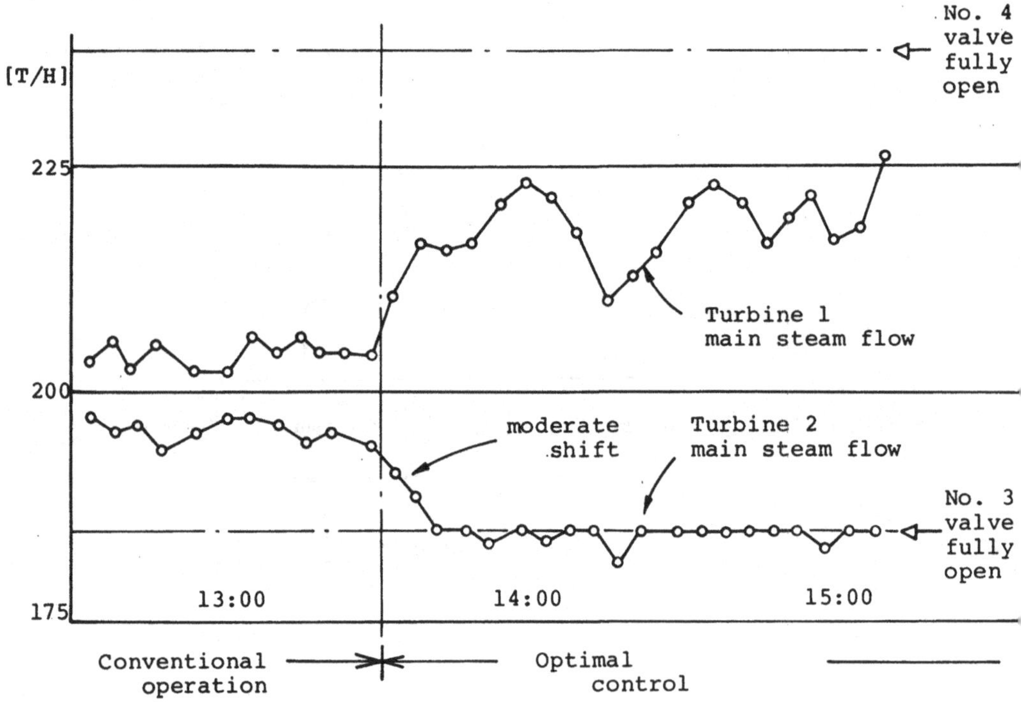

Fig. 7 Moderate control - field data

A COMPUTERIZED-OPTIMIZED STUDY ON FILM COOLING TECHNIQUE (PART III)

Y. Sidrak*
Laboratory for Applied Industrial Control
A. A. Potter Engineering Center
Purdue University
West Lafayette, IN 47907, USA

N. Sh. Matta
Mechanical Engineering Department
Faculty of Engineering and Technology
El-Minia University
El-Minia, Egypt

ABSTRACT

In Part I of this study, the authors considered the industrial applications of the film cooling technique and obtained the optimal velocity and temperature distributions in the boundary layer region. In Part II, the authors could optimize the conditions of experimental fluid injection through holes.

The present part is extended to study the modified technique of film cooling in which additional slots, either discontinuous or continuous, are made behind the original holes. In this case, the optimal values of several engineering design parameters for both holes and slots, in addition to blowing ratio and velocity of injected fluid, are determined using a search technique. The aim of the search technique used in this paper is to find the maximum of a multivariable, non-linear function subject to non-linear inequality constraints:

Maximize $F(Z_1,Z_2,\ldots,Z_N)$

Subject to $G_k \leq Z_k \leq H_k$, $k = 1,2,\ldots,M$

The implicit variables $Z_{N+1},\ldots,Z_M$ are dependent functions of the explicit independent variables $Z_1,Z_2,\ldots,Z_N$. The upper and lower constraints H_k and G_k are either constants or functions of the independent variables.

*Visiting Associate Professor on Sabbatical Leave from El-Minia University, El-Minia, Egypt.

The optimization technique utilizes the empirical formula for film cooling effectiveness and the formula for mass flow rate of coolant fluid as objective functions and the recommended experimental values as implicit and explicit constraints.

Some experimental results which were obtained from the study of three cases of experimental injection in the research laboratory at El-Minia University, are also included.

INTRODUCTION

Liquid film cooling provides an attractive means of protecting the internal surfaces of chemical and nuclear rocket motors [1-3]. It protects a given surface by interposing a thin continuous liquid film between the surface and the hot gaseous stream. The coolant is injected through slots (discontinuous or continuous) or holes or a combination of them onto the surface to be cooled and is swept downstream by the action of the main gas stream until it is consumed by evaporation as in Figure 1. The same technique is also very important in cooling certain structural elements, combustion chamber walls, nozzles, or gas-turbine blades, in the development of aircraft power plants against the influence of hot gases.

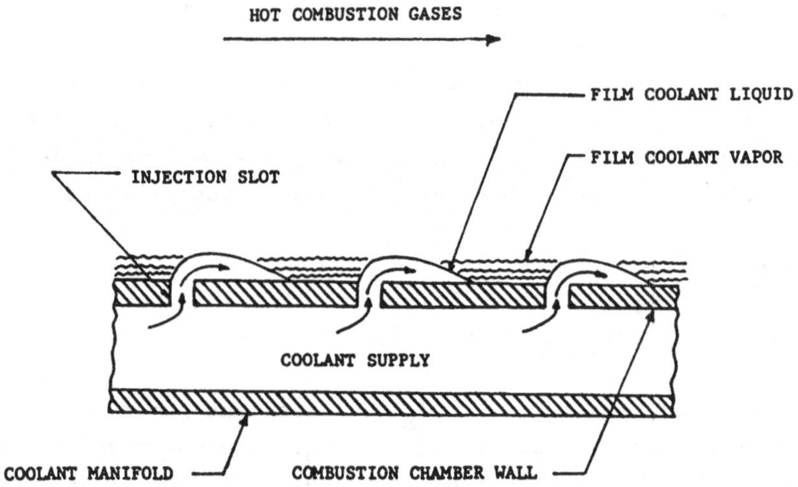

FIGURE 1
FILM-COOLED COMBUSTION CHAMBER

Figure 2 shows sketches of cooling methods. The upper left-hand sketch indicates the standard convection cooling method. The upper right-hand arrangement is the film-cooling process in which a stream of coolant is blown through a series of slots in a direction that is tangential to the surface. The coolant film is gradually destroyed by mixing with the hot gases so that its effectiveness decreases in the downstream direction. This disadvantage can be avoided by the process indicated in the lower left-hand sketch and called transpiration cooling. In this method the wall is manufactured from a porous material and the coolant is blown through the pores. The coolant film on the hot-gas side is, therefore, continuously renewed and the cooling effectiveness can be made to stay constant along the surface. In the discussion of film and transpiration cooling it was assumed that the coolant as well as the hot medium are either both gases or both liquids. When the hot medium is a gas, the cooling effectiveness obtained in both methods can be very much increased by using a liquid as coolant. In this case a liquid film is created on the hot-gas side, the liquid is evaporated on its surface, and the heat absorbed by the evaporation process substantially increases the effectiveness of this method. This method is called evaporative film cooling or evaporative transpiration cooling, depending on whether the coolant is discharged through slots or through a porous wall. The evaporative film cooling technique, with the coolant liquid blown through holes and slots, is the one which is going to be analyzed in this paper.

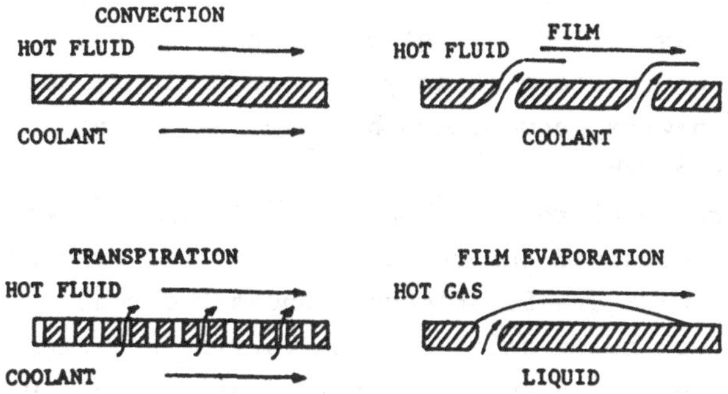

FIGURE 2
DIFFERENT METHODS OF COOLING

It can be shown that in theory the ideal method of film cooling would be by transpiration. With gas cooling, however, this system is scarcely practicable because it is difficult, if not impossible, to provide sufficient flow area whil still retaining some strength in the wall.

As the jet of the fluid leaves the hole it retards the main stream along the upstream side of the jet causing an increased pressure there, whereas rarefaction occurs at the downstream side. This pressure difference provides the force which deflects and deforms the jet. Circulatory motions are caused by the intensive intermixing of the two flows. The deformation of the jet is strongly affected by the blowing rate parameter. In general, one has to expect that the injected fluid penetrates farther into the stream when the blowing parameter is higher.

There have been many studies on heat transfer with film cooling, a good survey on this is reported [4-10].

EXPERIMENTAL

Figure 3 represents a sketch of injection geometry for the 3 cases of coolant liquid injection studied experimentally by the authors. Table I shows some of the values of film cooling effectiveness, η, at different values of blowing ratio, M, for the following 3 cases:

Case I - Injection is performed through holes only.

Case II - Injection is performed through holes and discontinuous slots.

Case III - Injection is performed through holes and continuous slots.

Although Case III always gives the highest value of film cooling effectiveness under the same injection conditions, however, continuous slots are not always recommended because they cause the wall to lose some of its strength.

The experimental results together with the corresponding best-fitting curves obtained through a curve fitting procedure, are computer-plotted in Figure 5. Figure 4 shows the results of a case of injection through holes only but with varying P/D parameter. It can be clearly shown from Figure 4 that decreasing the P/D parameter always improves the cooling effectiveness, but there is a minimum value of this parameter which should be kept in order to retain some strength in the wall.

It can now be clearly shown that the following facts should be kept in mind during any optimization procedure:

1. Film cooling effectiveness, η, is to be maximized.

2. Coolant liquid flow rate should be kept at the value greater than which deformation of the jet can take place.

3. The P/D parameter should give the highest possible flow area through the holes but not on the account of the strength of the walls.

4. The flow area through the slots, which is expressed as L.t.n, where L is the length of one slot, t is the width and n is the number of slots in one row, should be optimized to give the highest possible film cooling effectiveness but not to decrease the strength of the walls.

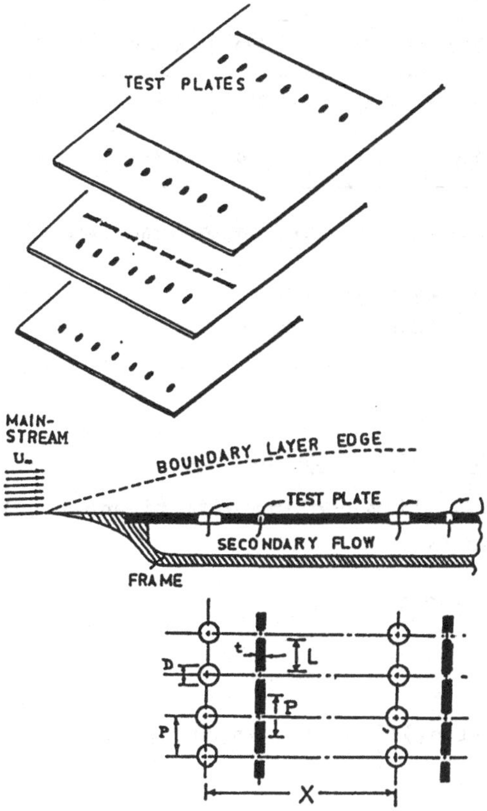

FIGURE 3
A SCHEMATIC SKETCH OF INJECTION GEOMETRY

TABLE I
VALUES OF FILM COOLING EFFECTIVENESS AT DIFFERENT BLOWING RATIOS

M \ η	X/D = 15 P/D = 5.3			X/D = 5 P/D = 5.3		
	CASE I	CASE II	CASE III	CASE I	CASE II	CASE III
0.250	0.116	0.120	0.130	0.325	0.450	0.462
0.375	0.125	0.135	0.150	0.350	0.480	0.530
0.500	0.130	0.157	0.170	0.360	0.500	0.600
0.625	0.132	0.170	0.175	0.340	0.490	0.570
0.750	0.130	0.170	0.175	0.320	0.470	0.550
0.875	0.125	0.160	0.170	0.275	0.430	0.510
1.000	0.110	0.150	0.160	0.230	0.400	0.475

X = Longitudinal Distance in X - Direction From the Leading Edge.

D = Diameter of Injection Holes in a Certain Row.

P = Pitch in a Row of Holes of Diameter D.

COMPUTATIONS PRINCIPLES

According to Best [10], film cooling effectiveness (η), for one injection stage, is defined as:

$$\eta = G_4 + e^{-G5\sqrt{X}})/(1+G_4) \quad (1)$$

and for n stages,

$$\eta_{n+1} = \eta_1+\eta_2(1-\eta_1)+\eta_3(1-\eta_1)(1-\eta_2)+\ldots+\eta_n(1-\eta_1)(1-\eta_2)\ldots(1-\eta_{n-1}) \quad (2)$$

$$G_4 = \dot{m}_{inj}C_{p_{inj}}/\dot{m}_\infty C_{p_\infty} \quad (3)$$

$$G_5 = 4.e^{-3M_{inj}}+3.0 \quad (4)$$

where:

X = distance in streamwise direction,

$C_{p_{inj}}$ = specific heat of injected coolant liquid,

C_{p_∞} = specific heat of main stream,

$\dot{m}_\infty$ = mass flow rate of main stream,

$\dot{m}_{inj}$ = mass flow rate of injected coolant liquid, and

M_{inj} = blowing ratio

Mass flow rate of main stream is defined as:

$$\dot{m}_\infty = (\pi d^2/4) \cdot u_\infty \cdot \rho_\infty \tag{5}$$

Mass flow rate of injected liquid is defined as:

$$\dot{m}_{inj} = \rho_{inj} \cdot \left[(\pi D^2/4) + L.t\right] \cdot n \cdot u_{inj} \tag{6}$$

The blowing ratio is defined as:

$$M_{inj} = \rho_{inj} \cdot u_{inj}/\rho_\infty \cdot u_\infty \tag{7}$$

where:

d = diameter of the chimney,

u_∞ = linear velocity of the main stream,

ρ_∞ = density of the main stream,

ρ_{inj} = density of the injected coolant liquid,

D = diameter of one hole in a band of holes,

L = length of one slot in a band of slots,

t = width of one slot in a band of slots,

n = number of holes in a band of holes,

= number of slots in a band of slots, and

u_{inj} = linear velocity of injected coolant liquid

Central distance between holes in a band of holes, or central distance between slots in a band of slots, P, is defined as:

$$P = \pi d/n$$

Equations 1-7 are the governing equations for the coolant injection through holes and discontinuous slots.

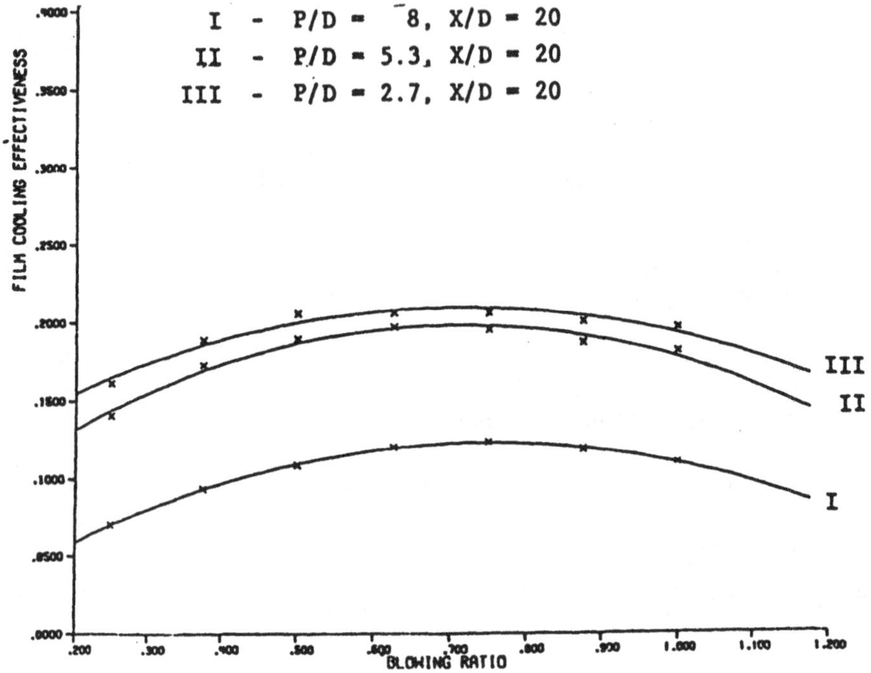

FIGURE 4
RELATIONSHIP OF FILM COOLING EFFECTIVENESS TO BLOWING RATIO

The aim of the optimization technique [11] used in this paper is to find the maximum of a multivariable, nonlinear function subject to nonlinear inequality constraints:

Maximize $F(Z_1, Z_2, \ldots, Z_N)$

Subject to $G_k \leq Z_k \leq H_k \qquad k = 1, 2, \ldots, M$

The implicit variables $Z_{N+1}, \ldots, Z_M$ are dependent functions of the explicit independent variables $Z_1, Z_2, \ldots, Z_N$. The upper and lower constraints H_k and G_k are either constants or functions of the independent variables.

Equation (6) has been modified to be in the form:

$$F = 1 - [\,0.5234(D)(D.n) + 0.6664(L.n)(t)\,]M_{inj}$$

(assuming that the main stream is hot air at 900°F) where F is a function of $\dot{m}_{inj}$. Maximizing F enables the optimal mass flow rate of the coolant liquid to be calculated. The constraints are:

$$0.0 < D \leq 0.08d$$

$$0.3d < D.n \leq 0.9d$$

$$0.3d < L.n \leq 0.9d$$

$$0.0 < t \leq 0.03d$$

$$0.0 < M_{inj} \leq 2.5$$

The optimal values of $D,n,L,t,P,M_{inj},u_{inj}$, and $\dot{m}_{inj}$ could then be calculated. Values of G_4 and G_5, defined by equations (3) and (4), respectively, could also be calculated.

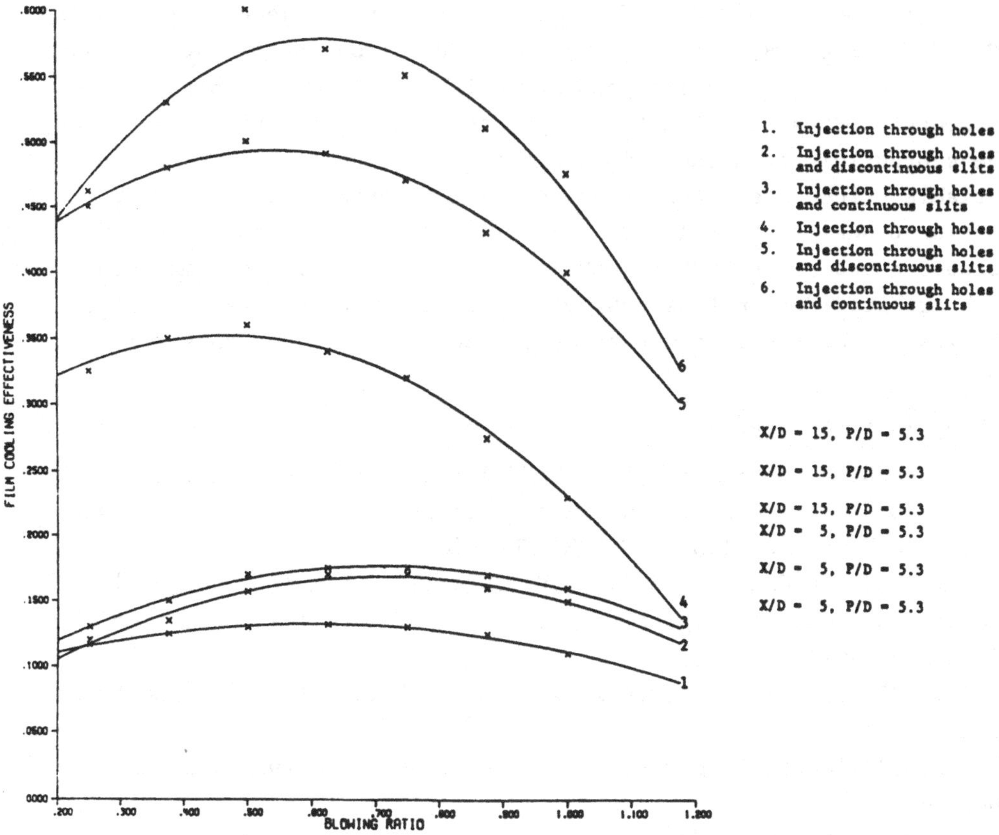

FIGURE 5
RELATIONSHIP OF FILM COOLING EFFECTIVENESS TO BLOWING RATIO

Film cooling effectiveness, defined by equation (1), is calculated assuming the presence of only the first band of holes and slots, and at the desired distance in the streamwise direction with the constraint that $5 < (X/D) \leq 40$. Taking into consideration that the overall film cooling effectiveness, defined by equation (2), should not be smaller than a predetermined value (say 0.75), η_2 can be calculated and, applying the same optimization technique [11] using equation (1), the location of the second band of holes and slots along the chimney can be determined in addition to the parameters D,L,t,n, and P at that band.

PROGRAM DESCRIPTION

Usage

The program consists of a main program, three general subroutines (CONSX,CHECK,CENTR) and two user supplied subroutines (FUNC,CONST). Initial guesses of the independent variables, random numbers, solution parameters, dimension limits and printer code designation are passed to the subroutines from the main program. Final function and independent variable values are transferred to the main program for printout. Subroutine CONSX is the primary subroutine and coordinates the special purpose subroutines (CHECK, CENTR, FUNC and CONST). Intermediate printouts are provided in this subroutine, if the user desires. Format changes may be required depending on the problem under consideration.

Subroutines Required

SUBROUTINE CONSX (N,M,ITMAX,ALPHA,BETA,GAMMA,DELTA,X,R,F,IT,IEV2, NO,G,H,XC,IPRINT) called from main program and coordinates all special purpose subroutines (CHECK,CENTR,FUNC,CONST).

SUBROUTINE CHECK (N,M,K,X,G,H,I,KODE,XC,DELTA,K1) checks all points against explicit and implicit constraints and applies correction if violations are found.

SUBROUTINE CENTR (N,M,K,IEV1,I,XC,X,K1) calculates the centroid of points.

SUBROUTINE FUNC (N,M,K,X,F,I) specifies objective function user-supplied).

SUBROUTINE CONST (N,M,K,X,G,H,I) specifies explicit and implicit constraint limits (user-supplied), order explicit constraints first.

Description of Parameters

N	Number of explicit independent variables-Define in main program
M	Number of sets of constraints-Define in main program
K	Number of points in the complex-Define in main program
ITMAX	Maximum number of iterations-Define in main program
IC	Number of implicit variables-Define in main program
ALPHA	Reflection factor-Define in main program
BETA	Convergence parameter-Define in main program
GAMMA	Convergence parameter-Define in main program
DELTA	Explicit constraint violation correction-Define in main program
IPRINT	Code to control printing of intermediate iterations. IPRINT=1 causes intermediate values to print on each iteration. IPRINT=0 suppresses printing until final solution is obtained-Define in main program
X	Independent variables-Define initial values in main program
R	Random numbers between 0 and 1-Define in main program
F	Objective function-Define in subroutine FUNC
IT	Iteration index-Defined in subroutine CONSX
IEV2	Index of point with maximum function value-Defined in subroutine CONSX
IEV1	Index of point with minimum function value-Defined in subroutines CONSX and CHECK
NI	Card reader unit number-Define in main program
NO	Printer unit number-Define in main program
G	Lower constraint-Define in subroutine CONST
H	Upper constraint-Define in subroutine CONST
XC	Centroid-Defined in subroutine CENTR
I	Point index-Defined in subroutine CONSX
KODE	Key used to determine if implicit constraints are provided-Defined in subroutines CONSX and CHECK
K1	Do loop limit-Defined in subroutine CONSX

Summary of User Requirements

(a) Determine values for N,M,K,ITMAX,IC,ALPHA,BETA,GAMMA,DELTA, and IPRINT. Guidelines for specifying the parameters are as follows:

K = N+1

ALPHA = 1.3

BETA = Some small number, say magnitude of function times 10^{-4}

GAMMA = 5

DELTA = Some small number, say magnitude of X vector times 10^{-4}

(b) Determine initial estimates for optimum values of independent variables: enter as (X(1,J), J=1,N). Initial point must satisfy explicit and implicit constraints.
(c) Obtain random numbers between 0 and 1: enter as ((R(II,JJ), JJ=1,N; II=2,K).
(d) Adjust DIMENSION and FORMAT statements as necessary.
(e) Specify objective function by writing SUBROUTINE FUNC.
(f) Define H (upper constraints) and G (lower constraints) in SUBROUTINE CONST. Explicit constraints must precede implicit constraints.
(g) Specify NI and NO.

PROGRAM FLOW CHART AND LISTING

These are available on request from the authors.

REFERENCES

1. Elverum, G. W., and Standhammer, P., Progress Report 30-4, Jet Propulsion Laboratory, August 1959.

2. Forde, J. M., Molder, S., and Szpiro, E. J., J. Spacecr. Rockets, Vol. 3, No. 8, pp. 1172-1176, 1966.

3. Geery, E. L., and Margetts, M. J., J. Spacecr. Rockets, Vol. 6, No. 1, pp. 79-81, 1969.

4. Metzger, D. E., Am. Soc. Mech. Eng., 67-WA/GT-1, 1967.

5. Goldstein, R. J., Adv. Heat Transfer, Vol. 7, pp. 269-321, 1971.

6. Eckert, E. R. G., J. Heat Transfer, Vol. 99, pp. 620-627, 1977.

7. Leontyev, A. I., Heat Transfer-Sov. Res., Vol. 10, No. 5, 1978.

8. Crawford, M. E., Kays, W. M., and Moffat, R. J., Am. Soc. Mech. Eng., 80-GT-43, 1980.

9. Sidrak, Y., and Matta, N. Sh., Proceedings of the 14th European Symposium on Computerized Control and Operation of Chemical Plants, Verein Österreichischer Chemiker, Vienna, pp. 251-255, 1981.

10. Best, R., Wärme-und Stoffubertragung, Vol. 15, pp. 79-91, 1981.

11. Kuester, J. L., and Mize, J. H., Optimization Techniques With FORTRAN, McGraw-Hill, Inc., New York, p. 368, 1978.

OPTIMIZATION OF RESOURCE ALLOCATION FOR LARGE SCALE PROJECTS

K. Ishido and T. Yoshida
Engineering System Center, Technical Headquarters,
Mitsubishi Heavy Industries, Ltd.
Wadasaki, Hyogo-Ku, Kobe, 652 Japan

1. Motivation and Problem Statements

The number of large-scale projects has increased year by year especially in the field of plant construction. In order to accommodate ourselves to this trend, we must develop effective and efficient planning methods to cope with difficulties resulting from large-scale complexity. Since the starting and finishing times of the project and the quantity of work are usually given as constraints at the early stages of the planning process, resource allocation is the most important factor of optimizing the project plan. In fact, about half of the total expenditure depends on resource allocation.

We consider the problem called PMP which is to minimize the variable expenditure. Since the two kinds of variables ((1) numbers of workers and (2) daily capacities of machines + facilities) can be treated in the same manner, here only the PMP for the first will be discussed. Therefore the variable expenditure will here be considered a function of the peak numbers of workers with various skills.

A PMP is defined as follows :

(1) a project consists of J jobs;

(2) job-sequence can be described in the form of a PERT-type network, and the jobs are expressed by arrows in the network;

(3) Q_j^k : man-days (Q) of appointed skill (k) for job (j) is given.

(4) Additional constraints on the term of execution time of job (j) may be given, such as : $LT_j \leq t_{fj} - t_{sj} \leq UT_j$ (1 - 1)

where LT_j : lower limit of execution time of job (j)

UT_j : upper limit of execution time of job (j)

t_{fj} : finishing time of job (j)

t_{sj} : starting time of job (j)

(5) Additional constraints on the relations between jobs may be given, such as : $LT' \leq t_j - t_{j'} \leq UT'$ (1 - 2)

where LT' : lower limit of time

UT' : upper limit of time

t_j : starting or finishing time of job (j)

$t_{j'}$: starting or finishing time of job (j')

(6) Additional constraints on numbers of workers may be given, such as : $LN \leq m_j^k \leq UN$, $LN' \leq \sum_k m_j^k \leq UN'$ (1 - 3)

where m_j^k : number of workers with skill (k) for job (j)

LN, LN' : lower limits of number of workers

UN, UN' : upper limits of number of workers

Solving the PMP, we can get the optimal resource allocation which shows the following:

(1) the starting time of every job;

(2) the finishing time of every job;

(3) the numbers of workers with the required skills for every job.

2. Network Models

Before we formulate a mathematical programming problem, we need to define some network models and analyze them.

2.1 Original Network (ON)

A project is supposed to consist of J jobs, and the job-sequence can be described in the form of a PERT-type network called an ON (Original Network) and made up of arrows (representing jobs) and nodes (which are the starting/ending points of arrows). An ON may include some dummy arrows and nodes. Fig. 1 is an example, and the following notation is used

j_i : job (i)

k_l : skill (l)

t_n : time when node (n) is realized

Q_j^k : man-days of appointed skill (k) to be required by job (j)

(n1,n2) : arrow or job from node (n1) to node (n2)

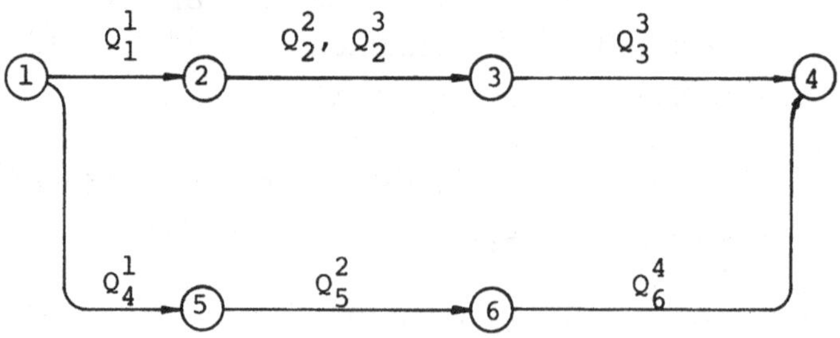

Fig. 1 Example of an ON (Original Network)

2.2 Condensed Network (CN) and Augmented Network (AN)

In order to lower the peak number of workers, we try to assign many jobs to the same workers. It is possible only when these jobs can be done in an appropriate order. For example, if t3 $\leq$ t5 in Fig. 1, the workers assigned to Q_2^2 can do Q_5^2. This is expressed by the additional arrow (3,5) in Fig. 2.

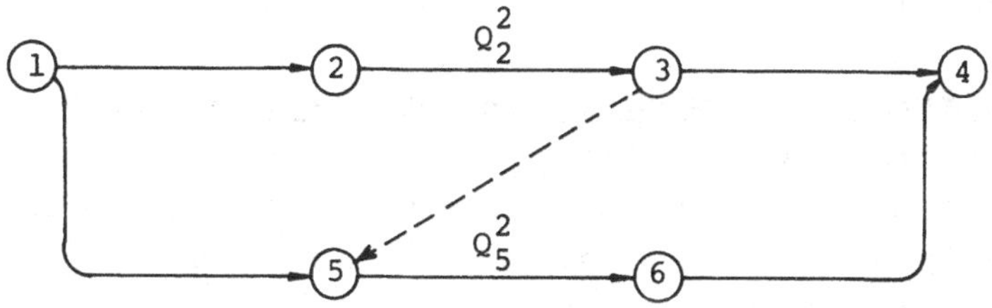

Fig. 2. Example of an additional arrow.

There are many other possible additional arrows with respect to Fig. 1 and all of these are shown in Fig. 3. The addition of these arrows changes an ON to an AN (Augmented Network).

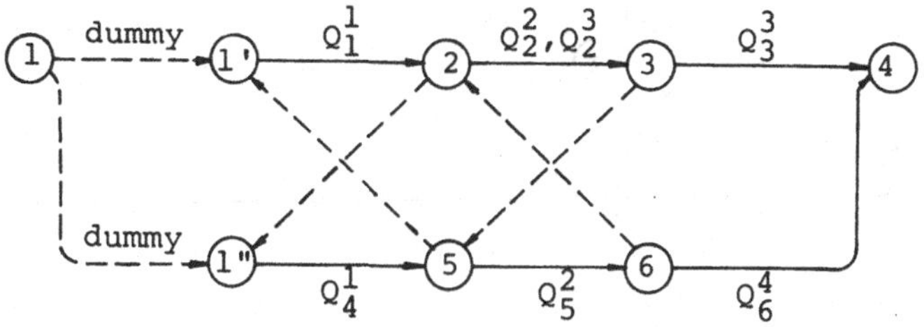

Fig. 3 Example of an AN (Augmented Network)

A CN (Condensed Network) is necessary in order to make an AN based on

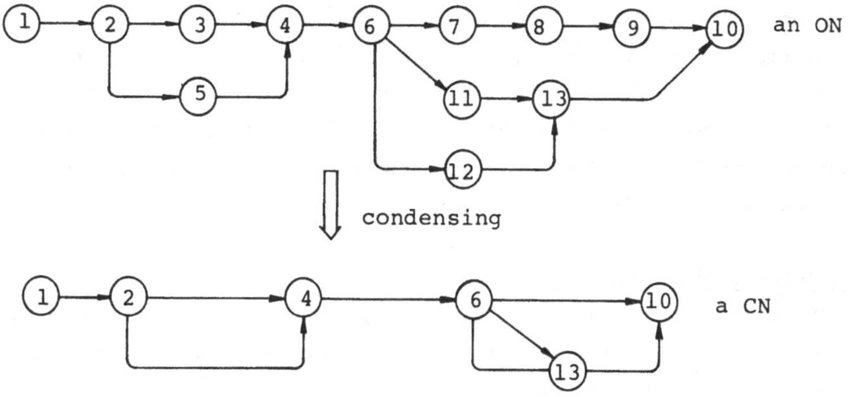

Fig. 4. Example of condensing

a complicated ON. Condensing means to eliminate every node that has onl one incoming arrow and only one outgoing arrow. Fig. 4 shows an example of condensing. After condensing we identify all combinations of paralle arrows on the CN that have the same starting node or the same ending nod in common. Additional arrows on an AN can exist only between the nodes in the parallel sequences which correspond to a combination of parallel arrows on the CN.

2.3 Locally Maximal and Feasible Augmented Network (LMFAN)

An AN includes all possible additional arrows, but not all of them can coexist in the real project. For example, if the additional arrows (3,5) and (6,2) in Fig. 3 coexist, a cycle : n2→n3→n5→n6→n2 comes out, a execution of the project is impossible. Only some combinations of additional arrows are feasible. When the number of additional arrows in a network is maximal and they are feasible, we call it a Locally Maximal a Feasible Augmented Network (LMFAN).
The LMFAN in Fig. 5 can be produced from Fig. 3.

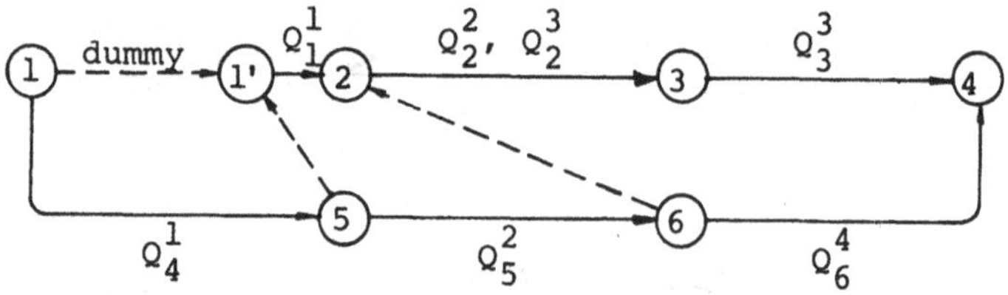

Fig. 5. An example of LMFAN

2.4 Workers as flows on an LMFAN

Let us suppose that the workers with various skills are the various flows from the starting node to the finishing node of a project. Then the summation of flows from the starting node gives the summation of the peak numbers of workers. However, the well known min-flow max-cut theor [3] cannot be used to solve the problem of minimizing the summation of peak values multiplied by coefficients of cost, because our problem has many constraints of different kinds.

2.5 Reduced LMFAN (RLMFAN)

Let us suppose that there are two paths from n1 to n2 and that one the paths doesn't include any real work, then elimination of this path makes no change in the value of the minimized summation of the peak valu of numbers of workers. This elimination is called reduction and contrib utes to the simplification of an LMFAN for optimization. Fig. 6 shows an example.

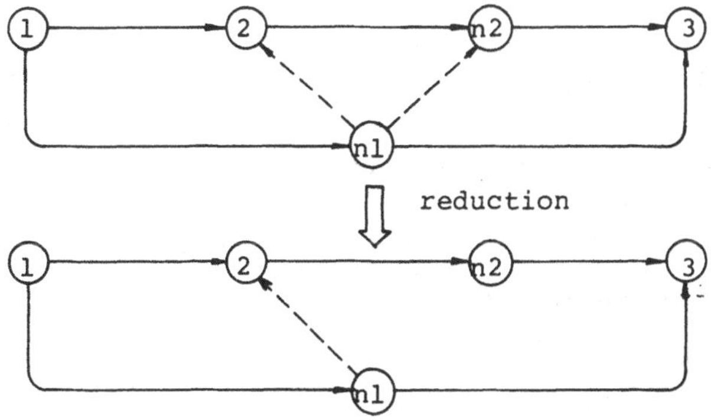

Fig. 6. Example of reduction of LMFAN

3. Mathematical Formulation

3.1 Objective and Constraints

We consider the problem of minimizing the summation of the peak numbers of workers multiplied by coefficients of cost on every RLMFAN. We call this problem an LPMP (Local Problem of Minimizing a summation of Peak values multiplied by coefficients of cost). The starting time, the finishing time and the numbers of workers with various skills for every job are the only planning variables of this LPMP. And the solution of this LPMP gives the local optimum of the ON.

The new notation in this chapter is as follows :

W_k : coefficient of cost for skill (k)

n_s : the starting node of a project

n_f : the finishing node of a project

$\{n^+\}$: the set of all nodes that immediately succeed node (n)

$\{n^-\}$: the set of all nodes that immediately precede node (n)

$\{1,2,\cdots,N\}$: the set of all nodes of a project

$\{1,2,\cdots,J\}$: the set of all jobs of a project

$\{1,2,\cdots,A\}$: the set of all additional arrows of an RLMFAN

$\{1,2,\cdots,K\}$: the set of all skills

C_i : given constants

(1) Objective function

$$\text{minimize } f = \sum_{k=1}^{K} W_k \cdot \sum_{n \in \{n_s{}^+\}} m^k_{(n_s,\, n)} \qquad (3-1)$$

(2) Constraints

(i) Constraints on execution of work

for $\forall j = (n_1, n_2) \in \{1,2,\cdots,J\}$, $\forall k \in \{1,2,\cdots,K\}$;

$$m_j^k \cdot (t_{n2} - t_{n1}) \geq Q_j^k \qquad (3 - 2)$$

(ii) Constraints on time period of project

$$t_{ns} = C_1,\ t_{nf} = C_2 \qquad (3 - 3)$$

(iii) Constraints on job-sequence

for $\forall j = (n_1, n_2) \in \{1,2,\cdots,J\}$;

$$t_{n1} \leq t_{n2} \qquad (3 - 4)$$

(iv) Constraints on additional arrows

for $\forall a = (n_1, n_2) \in \{1,2,\cdots,A\}$;

$$t_{n1} \leq t_{n2} \qquad (3 - 5)$$

(v) Constraints on flow balance

for $\forall n \in \{1,2,\cdots,N\}$, $\forall k \in \{1,2,\cdots,K\}$;

$$\sum_{n_1 \in \{n^-\}} m^k_{(n_1,\, n)} = \sum_{n_2 \in \{n^+\}} m^k_{(n,\, n_2)} \qquad (3 - 6)$$

where $\{n_2^-\} = n_f,\ \{n_f^+\} = n_s$

(vi) Additional constraints on the starting time or the finishing time of jobs

for $\exists n_1, \exists n_2, \exists n_3, \exists n_4 \in \{1,2,\cdots,N\}$;

$$t_{n1} \leq C_3,\ C_4 \leq t_{n2},\ C_5 \leq t_{n3} - t_{n4} \leq C_6 \qquad (3 - 7)$$

(vii) Additional constraints on the peak numbers of workers with a particular skill

for $\exists k \in \{1,2,\cdots,K\}$; $\displaystyle\sum_{n \in \{n_s^+\}} m^k_{(n_s,\, n)} \leq C_7 \qquad (3 - 8)$

(viii) Additional constraints on the number of workers to be assigned to a particular job

for $\exists j_1, \exists j_2 \in \{1,2,\cdots,J\}$, $\exists k \in \{1,2,\cdots,K\}$;

$$C_8 \leq m_{j1}^k \leq C_9,\ C_{10} \leq \sum_{k=1}^{K} m_{j2}^k \leq C_{11} \qquad (3 - 9)$$

3.2 Linearization

All constraints of the LPMP are linear except the constraints on execution of work (3 - 2). The feasible region can be expressed by the following constraints :

for $\forall r \in \{1,2,\cdots,R\}$;

$$t_{n2} - t_{n1} \geq Q_j^k \cdot a_r \cdot m_j^k + Q_j^k \cdot b_r \tag{3 - 10}$$

where a_r, b_r : coefficients of piece-wise linearization
R : number of linear constraints for approximation

The error produced by approximation always tends to increase the number of workers, thus insuring a margin of safety with respect to the execution of the project. And the error can be decreased to the required level by an increase in R. Coefficients a_r and b_r are given as the optimal solution of the nonlinear programming problem as follows:

minimize E (3 - 11)

subject to :

for $\forall r \in \{1,2,\cdots,R\}$;

$$E \geq \frac{(V_{r-1} + V_r)^2}{4 \cdot V_{r-1} \cdot V_r} - 1 \tag{3 - 12}$$

$$a_r = -V_{r-1} \cdot V_r \tag{3 - 13}$$

$$b_r = V_{r-1} + V_r \tag{3 - 14}$$

$$V_{r-1} \geq V_r \tag{3 - 15}$$

where E : maximal relative error of approximation

$$V_0 = \frac{1}{\text{lower bound of m}} \tag{3 - 16}$$

$$V_R = \frac{1}{\text{upper bound of m}} \tag{3 - 17}$$

3.3 Solving the linear programming problems

After linearization, every LPMP can be solved easily by the usual linear programming algorithms. Because the LPMPs of a project are solved in series and the difference between two serial LPMPs is not usually very large, various techniques, such as the dual simplex method, can be used to decrease the amount of calculation.

4. Procedure of Optimization

Fig. 7 shows the outline of the procedure which is used in practice.

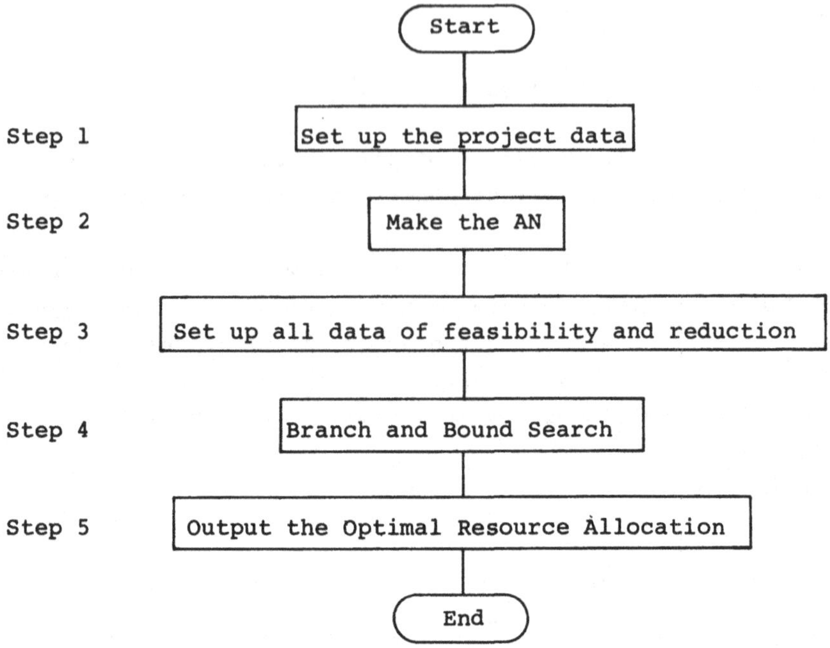

Fig. 7 Outline of the procedure

The concrete steps are as follows:

(1) Step 1 :
Assemble the project data such as the starting time, the finishing time of the project, job-sequence, man-days for every job, and additional constraints.

(2) Step 2 :
(i) Make the CN from the ON set up in the Step 1.
(ii) Identify all combinations of parallel arrows on the CN.
(iii) Find all additional arrows with respect to every combination identified in the preceding substep.

(3) Step 3 : Set up all data about feasibility and reduction
(i) Identify all combinations of additional arrows that make a cyc except disjunctive pairs (Two additional arrows directed oppositely between two nodes make a disjunctive pair. [1]).
(ii) Set up all data about reduction.
(iii) Sort all additional arrows into decreasing order of the value of the sum of
(a) the number of cycle combinations in which the additional arrow is included, and

(b) the number of additional arrows which can be reduced by the additional arrow.

(4) Step 4 : Branch and Bound Search

(i) Set the ON as a waiting node at the root of a searching tree.

(ii) Choose a waiting node which has the smallest upper limit in the searching tree.

(iii) Pick up an additional arrow lexicographically as an branching variable, make two FAN (Feasible Augmented Network)s, and execute reduction.

(iv) Set up two LPMPs basd on them.

(v) Linearize and solve the LPMPs. In this substep post optimization technique is sometimes used.

(vi) Set two waiting nodes with the upper and lower limits.

(vii) Execute bounding.

(viii) If the current best solution is not optimal, return to Substep (ii).

(5) Step 5 :

Output the optimal resource allocation through the various interface devices.

5. Discussion of Results

5.1 Discussion of the method

By introduction of new concepts, such as AN, LMFAN, and RLMFAN, it has become possible to formulate an LPMP whose optimal solution gives the locally minimal value of the summation of peak values multiplied by coefficients of cost. Nonlinear constraints of the LPMP are linearized by using a nonlinear programming algorthm. A branch and bound method is contributory to lessen the amount of calculation. This new method has proved to be valid for optimizing resource allocation for a project which can be described in the form of a PERT-type network.

5.2 Discussion of application to actual projects

It has been proved that application of this method to actual projects brings us very successful results. Table 1 shows one example of its effectiveness on an actual project. The virtue of this method is that the optimum given by this method includes allocation of machines and facilities in addition to workers.

Table 2. Example of effectiveness of optimization

	value of objective function
manual allocation by skilled planners	1
optimal allocation using this method	0.586

5.3 Future topics

(1) Extension of the objective function

Although at this stage the objective function is a summation of pea values multiplied by coefficients of cost, it can be extended to a non-linear separable function of peak values.

(2) Other applications of this method

This method can be used to solve many problems in other fields, suc as line balancing, factory planning, and so on.

Acknowledgements

The authors are indebted to their colleagues for their cooperation. Special thanks are due to prof. Hisasi Mine of Kyoto University for his encouragement and suggestions.

References

[1] Balas, E. : Machine Sequencing via Disjunctive Graphs : An Implicit Enumeration Algorithm. Operations Research, 1969, vol. 17, pp. 941 957.

[2] Busacker, R.G., and T.L. Saaty : Finite Graphs and Networks : An Introduction and Applications. McGrow-Hill, 1965.

[3] Ford, L.R., Jr., and D.R. Fulkerson : Flows in Networks. Princeton University Press, Princeton, N.J., 1962.

[4] Gill, P.E., and W. Murray (edited) : Numerical Methods for Constrained Optimization. Academic Press, 1974.

[5] Ishido, K. and T. Yoshida : A Practical Optimization on Project Planning (in Japanese). Proc. 2nd Mathematical Programming Symposium-Japan, 1981, pp. 149-158.

EXPERIENCE RUNNING OPTIMISATION ALGORITHMS ON PARALLEL PROCESSING SYSTEMS

L C W Dixon, K D Patel and P G Ducksbury
The Numerical Optimisation Centre
The Hatfield Polytechnic
P.O. Box 109
Hatfield, Herts. AL10 9AB
United Kingdom

ABSTRACT

The availability of parallel processing machines makes it possible to envisage solving four types of numerical optimisation problems that still present difficulties on fast sequential machines. These four types of problems are described, then two parallel processing machines, ICL DAP and NEPTUNE are briefly described, and ways of mapping the problems onto ICL DAP and NEPTUNE are discussed. Finally, our experience in implementing parallel algorithms on ICL DAP and NEPTUNE are detailed.

1. INTRODUCTION

We are concerned with solving the optimisation problem

$$\min f(\underline{x}) \; , \qquad \underline{x} \in R^n$$

subject to simple upper and lower bounds, i.e. $l_i \leq x_i \leq u_i$.

There exists a number of sequential algorithms with superlinear convergence, based on the iterative scheme,

$$\underline{x}^{(k+1)} = \underline{x}^{(k)} + \alpha \underline{p}^{(k)}$$

where $\underline{p}^{(k)}$ is the search direction and α the step size. These algorithms can solve most optimisation problems. Given this situation the question arises as to why then should we be interested in introducing the parallel processing concept in numerical optimisation. The main reasons are:

a) The processing time for solving some of the industrial optimisation problems is too long,
b) There are some problems that cannot be tackled because their size leads to storage difficulties on sequential systems,
c) People only tend to pose optimisation problems that they feel might be soluble.

By introducing the parallel processing concept to numerical optimisation we hope to reduce some of the difficulties encountered in the sequential approach.

The four different situations where we feel improvements are most likely in the solution of optimisation problems with the currently available parallel processing machines are:

a) Small dimensional problems

Small dimensional problems, $n < 100$, where the cost of computing the objective function/gradient vector/Hessian matrix greatly exceeds the overheads in the optimisation part of the algorithm.

b) Large dimensional problems

Large dimensional problems, $n > 2000$, where the combined processing time and storage requirements causes difficulties.

c) On-line optimisation

On-line optimisation problems, like the optimisation of car fuel consumption, cannot always be solved using the existing sequential algorithm because of the time factor and in these problems a saving of the order of four or five could be sufficient.

d) Multiextremal global optimisation problems

Problems where the objective function has many local minima. For these problems the sequential algorithms for locating the global minimum are still relatively unsatisfactory and expensive in computer time.

Later in this paper we will consider each classification in more detail; describe parallel algorithms and present numerical results. Before we consider the interaction of the four classifications mentioned above and the parallel processing computers, we will briefly described the parallel processing computers available to us.

2. CLASSIFICATION OF PARALLEL COMPUTERS

We classify parallel computers as either SIMD (Single Instruction Multiple Data) or MIMD (Multiple Instruction Multiple Data).

2.1 SIMD systems

We subdivide the SIMD machines as vector and array computers. The difference is based primarily on the way data is communicated to elements of the system.

2.1.1 Vector processors

Vector processing is the application of arithmetic and logical operators simultaneously to components of vectors. This leads to a straightforward reduction in time at the arithmetic and logical operations stage whenever a vector processing system is used. Examples of vector processors are CRAY-1/S and CYBER 205 systems. Both contain pipelined vector arithmetic processors capable of performing operations on arrays at very high speeds. In one strict sense, pipelined computers are not parallel computers as the parallelism occurs within the fundamental arithmetic operations. However, their efficient use depends heavily on the redesign of algorithms to make efficient use of the parallel arithmetic option.

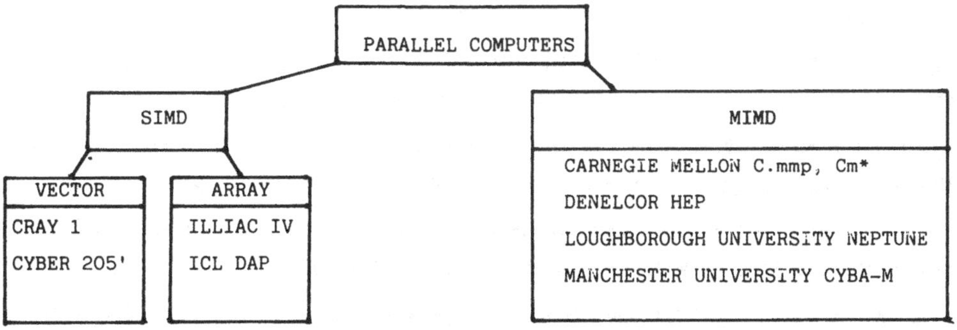

Figure 1. Classification of parallel computers

Most of the research in optimisation using vector processors is in the area of solving very large optimisation problems. The underlying optimisation approach is the Conjugate Gradient algorithm. The principle is to 'vectorise' the arithmetic within conjugate gradient algorithms. To make an efficient use of a vector processor, an algorithm must be arranged to do nearly all its arithmetic and logical operators on long vectors. Various researchers, Rodrigue and Greenbaum [1], Rodrigue et al [2], Schreiber [3], have reported their experience in implementing conjugate gradient algorithms on 'vector' machines.

2.1.2 Array processors

An SIMD array machine is defined as a computer with a single master control unit and multiple directly connected processing elements (see Figure 2). Each processing element is independent, i.e. has its own registers and storage, but only operates on command from the master control unit. Data access is from its own local memory and that of its nearest neighbour. Each processor carries out the same instruction as all the others but on its own specific data set. Because of the simplicity of each processing element, the number of processors that can be connected together in an SIMD array sense can be very large.

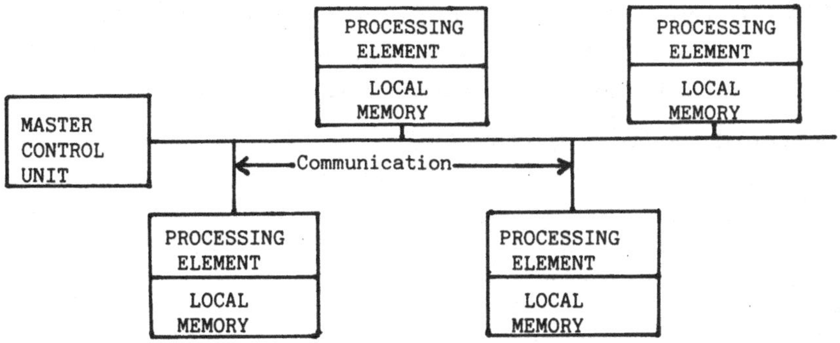

Figure 2. Array processor

Some of the application areas for an array processor are:

a) large problems such as weather data processing,

b) problems that have a large number of independent data sets.

The earliest SIMD array machine was the 64 processor ILLIAC IV. The most recent one in the UK is the 4096 processor - ICL Distributed Array Processor (DAP).

As one of the parallel systems we have used for the implementation of parallel algorithms is the ICL DAP, we will briefly describe the ICL DAP.

The ICL DAP at Queen Mary College, London, comprises 4096 elements each originally with a local store of 4K. (This has now been increased to 16K). All the processors obey a single instruction stream broadcast by a Master Control Unit. Processors are arranged in a 64 x 64 matrix form and each processor has access to the four neighbouring elements and to their stores.

A high level, special purpose language, called DAP-FORTRAN, is available on the DAP; this is an extension of FORTRAN with additional features that enable parallel algorithms to be expressed naturally and efficiently. DAP-FORTRAN has about 50 built-in functions for manipulation of vector and matrices. The implication of the hardware structure is that 64 x 64 matrices, 4096 long element vectors and 64 element vectors can be processed as single entities.

For example the DAP-FORTRAN declaration

```
REAL A(,), B(,), G()
```

is equivalent to the standard FORTRAN declaration

```
REAL A(64,64), B(64,64), G(64)
```

and the DAP-FORTRAN statement:

```
A = A + B
```

is equivalent to the standard FORTRAN loop:

```
DO   10   I = 1,64
DO   10   J = 1,64
10 A(I,J) = A(I,J) + B(I,J).
```

The statement A = A + B represents one operation performed simultaneously in all 4096 processing elements. The example below illustrates how some of the DAP-FORTRAN built-in aggregate functions can be utilised in numerical optimisation.

Consider a 64-dimensional Rosenbrock function

$$f(\underline{x}) = \sum_{i=1}^{32} 100(x_{2i-1}^2 - x_{2i})^2 + (1 - x_{2i-1})^2$$

The first four components of the gradient vector are:

$$g_1(\underline{x}) = 400x_1(x_1^2 - x_2) - 2(1-x_1)$$

$$g_2(\underline{x}) = -200(x_1^2 - x_2)$$

$$g_3(\underline{x}) = 400x_3(x_3^2 - x_4) - 2(1 - x_3)$$

$$g_4(\underline{x}) = -200(x_3^2 - x_4).$$

The DAP-FORTRAN code for the analytic gradient vector is

```
G = MERGE(400.0*X*(X**2 - SHLP(X,1)) - 2.0*(1.0 - X), - 200.0*(SHRP(X**2,1) - X),
     .NOT.ALT(1)
```

where G and X are declared as

```
REAL G(), X().
```

The components of G, the gradient vector, are computed in parallel.

We briefly summarise the effect of the DAP-FORTRAN built-in aggregate function used in the gradient evaluation.

SHLP - This function returns a vector value that is equal to the first argument shifted a number of places to the left, using plane geometry. The number of places by which the vector is to be shifted is given by the second argument.

SHRP - as for SHLP except that the vector is shifted to the right.

MERGE - returns a vector (or matrix) result, components of which are selected from corresponding components of the first or second argument, depending on whether the corresponding component of the third argument is .TRUE. or .FALSE. respectively.

ALT - function takes a single integer scalar argument and returns a logical vector value. If the value of the argument modulo 64 is i, the logical vector will have its first i components .TRUE, and so on in alternation until all the components of the vector have a value.

e.g. let $Z = (z_1, z_2, z_3, z_4)$

$Y = (y_1, y_2, y_3, y_4)$

$SHLP(Z,1) = (z_2, z_3, z_4, 0)$

$SHRP(Y,1) = (0, y_1, y_2, y_3)$

$MERGE\ (Z,Y, ALT(1)) = (y_1, z_2, y_3, z_4).$

Details of all the DAP-FORTRAN built-in aggregate functions can be found in the ICL DAP manuals [21].

2.2 MIMD systems

An MIMD system is defined as a series of processors (minis/micros) working independently in parallel. The number of processors cannot reach the same order as an Array processor because of the complexity in communications between processors and the difficulties introduced by data access from a global data set. Early examples of MIMD systems were the C.mmp and Cm* designed at Carnegie-Mellon University, Pittsburgh A more recent example is the Heterogeneous Element Processor (HEP) of Denelcor HEP which typically consists of from one to sixteen processors.

Another parallel system we have used for the implementation of parallel algorithms is the LUT NEPTUNE, which is briefly described below. The system contains 4 processors (Texas Instruments 990/110 minicomputers) and the current configuration is shown in Figure 3. Each processor has access to 96Kb of memory (local memory)

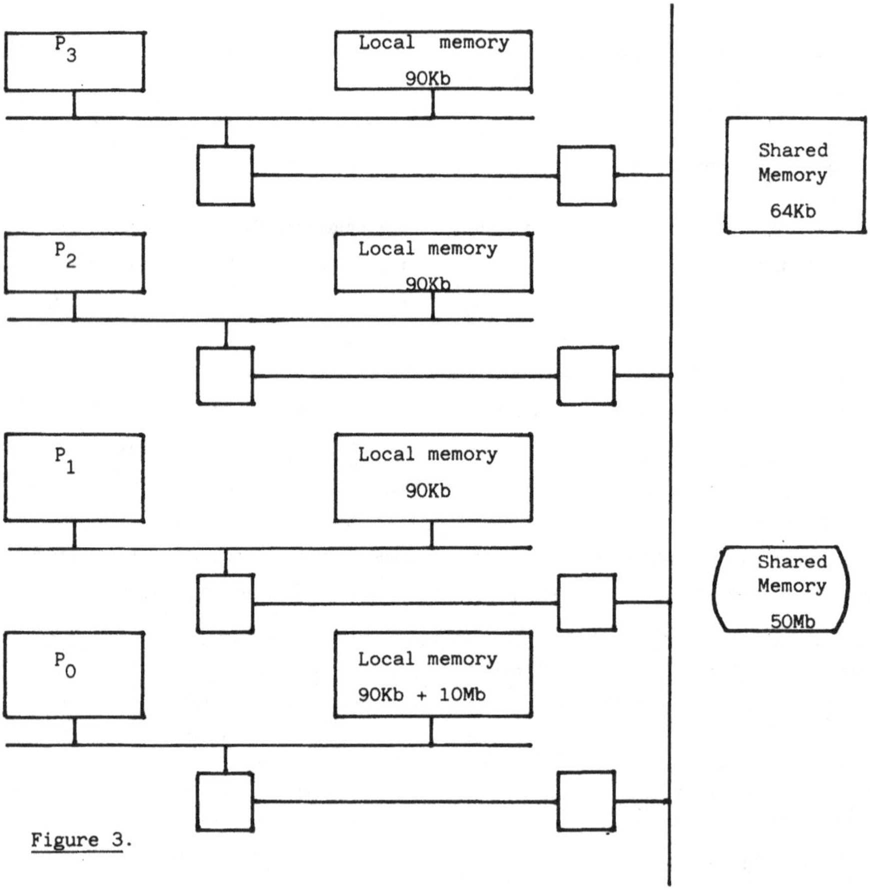

Figure 3.

and one processor (numbered 0) also has access to 10Mb of memory on a disc drive. There is also 64Kb of shared memory and 50Mbdisc drive. Hence each processor has a nett memory of 160Kb. The 50Mb drive can be accessed by each processor with disc interrupts being sent to them all.

The software used for executing a program on the NEPTUNE multiprocessor system is standard FORTRAN and some pseudo-FORTRAN syntax constructs.

The creation and termination of paths with identical code is defined using

```
    $DOPAR   10 I = 1, N1, N2, N3
    'code'.
    10 $PAREND.
```

The creation of paths with different code is defined using

```
    $FORK 10, 20, 30, 40
 10 'code 1'
    GOTO 40
 20 'code 2'
    GOTO 40
 30 'code'
 40 $JOIN
```

Creation of paths with the same code and with each processor forced to execute the code once and once only is defined using

```
    $DOALL 10
    'code'.
 10 $PAREND.
```

Shared data is defined using

```
   $SHARED variable list.
```

All other data, including program code, is held in local memory.

For critical sections of the program the user has available up to 8 'resources' which can only be owned by one of the processors at any one time. Resources used are declared using

```
    $REGION list of names
```

and claimed or released using

```
    $ENTER name or $EXIT name.
```

A fuller description of the NEPTUNE multiprocessor system can be found in Evans, et al [20].

New MIMD architectures, called data flow computers, which depart dramatically from the classical Von Neumann architecture are in the design stage. The data flow approach to computing involves evaluating statements in a program as and when their input operands become available. There has been a lot of theoretical work on the data flow architecture. Texas Instruments designed, built and benchmarked a four laboratory model of a data flow computer to assess the benefits that might arise, Sauber [5]. The main conclusions from their experience in using a data flow computer were:

a) given sufficient program parallelism, the processing power grows linearly with the number of processors,

b) given sufficient parallelism, all execution overhead is overlappable with useful work, and

c) a computer can find the necessary parallelism in ordinary programs, expose it, and then partition it automatically.

It is too early to say how numerical optimisation algorithms can benefit from data flow computers and no data flow system was available for our use.

The design of parallel algorithms for SIMD and MIMD requires two quite distinctive approaches. The main difference is, for a SIMD system we require that tasks should be identical, for a MIMD system we can dispense with that requirement. For a MIMD system we can also dispense with the requirement that tasks should be synchronous, at the cost of memory congestion and convergence problems.

Other distinctions will become apparent when we describe the parallel algorithms.

3. PERFORMANCE MEASUREMENT

Measuring the performance of a parallel algorithm is dependent on the type of parallel system used.

3.1 MIMD system

Two concepts available for measuring the performance of theoretical parallel algorithms are 'SPEED-UP' and 'EFFICIENCY'. They are defined as follows:

Let $\tau(P)$ be the processing time using P identical processors. Then the 'speed-up' factor over a single identical processors is

$$S_p = \frac{\tau(1)}{\tau(P)}$$

and the 'efficiency' is

$$E_p = \frac{S_p}{P} \leq 1$$

Ideally we might expect the 'speed-up' ratio, S_p, to be p and hence the efficiency E_p to be unity. In general, however, some degradation must be expected. The main factors that contribute to this degradation are:

a) at the system level:

i) the actual processing speed of the processors differ,

ii) input/output interrupts,

iii) memory contention,

iv) bottleneck in the data transfer.

b) at algorithmic level:

i) synchronisation losses, if, say, p tasks are to be performed, all must wait for the slowest,

ii) critical section losses, if many processors are trying to access the global data set at a particular instant, then only one processor can get such access at any instant, so while that processor is accessing the global data set most of the other processors are idle. This implies a time delay. We can minimise the waiting time by ensuring that access to the global data set is kept to a minimum.

3.2 SIMD system

The 'speed-up' and 'efficiency' ratios defined above are meaningless concepts for a SIMD type architecture. the performance of a parallel algorithm, for the SIMD type architecture is usually measured in relation to a sequential system. The equivalent sequential algorithm is on a sequential system. The performance ratio, τ , is given by:

$$\tau = \frac{\text{processing time on a sequential system}}{\text{processing time on a parallel (SIMD) system}}$$

This does, of course, depend on the sequential system chosen as the basis - in the results quoted in this paper the DEC system 1091 at The Hatfield Polytechnic was chosen. An additional complication in any such comparison arises because the optimum parallel algorithm for an SIMD system often differs considerably from the optimum sequential algorithm; and the choice of sequential code can, therefore, affect the result. In particular, should the best sequential code be used or one similar in principle to that implemented on the MIMD system.

4. INTERACTION OF PROBLEM CLASSIFICATION AND PARALLEL PROCESSING

Now let us consider the interaction of three of the four classifications, mentioned in Section 2, with the type of parallel processing computers we have considered. In particular, we will concentrate on the SIMD ICL DAP and MIMD NEPTUNE systems.

4.1 : Small expensive problems

Calculating function in parallel

Consider the situation of an extreme class of industrial problem in which, when attempting a solution by a sequential algorithm, the time spent in the function/constraint evaluation routine is, say, over 1000 times as long as that spent in the optimisation part of the algorithm.

A particular example of this is the case of optimisation of aircraft fuel engine performance subject to noise constraints and performance specifications. Details can be found in A H O Brown [6] of Rolls Royce Ltd. To parallelise such a problem we look for parallel features in the function evaluation routine. there is an obvious feature in the performance specification which applies to eleven different in-flight conditions and they can be computed in parallel. As each parallel task is independent, this approach is more applicable to an MIMD system than to an SIMD system.

When the function evaluation can be separated into a number of identical tasks as in a data fitting problem

$$\min \quad F(\underline{x}) = \sum_{k=1}^{M} S_k^2(\underline{x})$$

where each $S_k(x)$ is an identical calculation using different data, then we have an obvious calculation where (if $M \leq P$, the number of processors) an SIMD machine can be used.

Sargan, Chong and Smith [4] have solved large scale nonlinear econometric models on SIMD DAP by separating the objective function into identical tasks.

The efficient sequential approaches for solving small dimensional unconstrained optimisation problems are:

Modified Newton techniques: $2 \leq n \leq 5$
Variable Metric techniques: $5 \leq n \leq 120$
Conjugate Gradient techniques: $n \geq 120$
Conjugate Direction techniques: $n \leq 30$.

Nearly all efficient methods for solving unconstrained optimisation require the gradient vector **g** to be evaluated at $\underline{x}^{(k)}$. for some very expensive industrial problems, which may include simulation, the calculation of **g** can be very time consuming and is much longer than the overheads associated with calculating the search direction $\underline{p}$ given **g**. The calculation of **g** is therefore the part of the algorithm where most benefit from parallelism can be expected to occur on such problems. The components of **g** are not identical, even for a simple function. If we can obtain an analytic expression for **g** the components can be computed in parallel on MIMD machines but not on SIMD machines. For an MIMD machine the parallel tasks are the calculation of the components of **g**.

For most industrial optimisation problems, obtaining the analytic expression for **g** can be a very tedious process. We can use estimated values of g, obtained by some finite differences scheme, and this would be ideal for SIMD machines. There are efficient sequential variable metric and conjugate gradient codes which use a central difference scheme to approximate **g**, so this may be expected to be acceptable on a parallel machine.

Let us consider the implementation of a modified Newton algorithm on the ICL DAP. On a sequential system the modified Newton techniques for solving unconstrained optimisation problems is one of the most efficient available for small dimensional problems where $2 < n \leq 5$. By introducing parallelism into Modified Newton technique, suitable for implementation on the ICL DAP, we would expect to increase the range for which Modified Newton technique is most efficient.

The Modified Newton approach is a second derivative method, i.e. not only do we need to calculate g, but we also need to calculate the Hessian G. The elements of G are given by

$$G_{ij} = \frac{\partial^2 f}{\partial x_i \partial x_j} .$$

We could approximate $\underline{g}$ and G by some finite differences scheme and this would be an ideal parallel calculation on the DAP.

The overheads in solving

$$G\underline{p} = - \underline{g}$$

to determine the search direction $\underline{p}$ will also be reduced on the DAP.

The final stage in the modified Newton algorithm is to select the step size α. This is often done by a search along a preselected curve. For the DAP implementation we consider three possible searches.

a) '1 - D' search (line search)
b) '2 - D' search)
c) '4 - D' search) appropriate choice of planes.

The '4 - D' search seems to be the most attractive proposition. For problems with $4 < n \leq 64$, the parallel algorithms based on this idea will be considerably more efficient and also reduce the number of Newton iterations. In the next section we describe the parallel Modified Newton algorithm which we implemented on the DAP.

Parallel (SIMD) Modified Newton algorithm

The parallel Modified Newton algorithm consists of the following steps:

Step 1: Initial guess $\underline{x}^{(o)}$; h, the step length and ε, the tolerance.

Step 2(a): Calculate $f(\underline{x} \pm ha_i \pm ha_j)$ all $j > i$, where a_i is a unit vector along the ith axis. This function evaluation is a parallel calculation.

Step 2(b): Calculate the gradient vector $\underline{g}$ and the Hessian matrix G by the following finite differences scheme:

$$g_i = [f(\underline{x} + ha_i) - f(\underline{x} - ha_i)]/2h$$

$$G_{ij} = [f(\underline{x} + ha_i + ha_j) - f(\underline{x} + ha_i) - f(\underline{x} + ha_j) + f(\underline{x})]/h^2 \qquad i \neq j$$

$$G_{ii} = [f(\underline{x} + ha_i) - 2f(\underline{x}) + f(\underline{x} - ha_i)]/h^2$$

Step 3: Stop if $\max |g_i| < \varepsilon$

Step 4: Solve a set of linear simultaneous equations:

$G\underline{p} = - \underline{g}$, using DAP library subroutine FØ4GJNLE64.

Step 5: We considered three possible searches

(i) 4-dimensional grid search

The iterative step given by

$$x^{(k+1)} = \underline{x}^{(k)} + (A1)\alpha_1 \underline{p}^{(k)} + (A2)\alpha_2 \underline{g}^{(k)} + (A3)\alpha_3 d_3^{(k)} + (A4)\alpha_4 d_4^{(k)}$$

generates 4096 points

where $\alpha_1 = \alpha_1^{initial}$

$$\alpha_2 = \alpha_1 * \frac{g^T g}{g^T G g}$$

$$\alpha_3 = - \alpha_1 * \frac{d_3^T g}{d_3^T G d_3} \qquad (4.1)$$

$$\alpha_4 = - \alpha_1 * \frac{d_4^T g}{d_4^T G d_4}$$

$$\underline{d}_3 = \underline{x}^{(k)} - \underline{x}^{(k-1)}$$

$$\underline{d}_4 = \underline{x}^{(k-1)} - \underline{x}^{(k-2)}$$

A1, A2, A3 and A4 are (64 x 64) matrices given by

$A1_{ith\ row}$ = [-2,-2,-2,-2,-2,-2,-2,-2,⋮-1,-1,-1,-1,-1,-1,-1,-1⋮
0, ...0⋮1, ... 1⋮2, ... 2⋮3, ... 3⋮4, ... 4⋮
5, ... 5] (4.2)

$A2_{ith\ row}$ = [-2,-1,0,1,2,3,4,5⋮-2,-1,0,1,2,3,4,5⋮ ....
-2,-1,0,1,2,3,4,5] (4.3)

$A3_{ith\ col}$ = [-3,-3,-3,-3,-3,-3,-3,-3⋮-2,-2,-2,-2,-2,-2,-2,-2⋮
1, ... 1⋮0, ... 0⋮1, ...1⋮2, ... 2⋮3, ... 2⋮
3, ... 3⋮4, ... 4] (4.4)

$A4_{ith\ col}$ = [-3,-2,-1,0,1,2,3,4⋮-3,-2,-1,0,1,2,3,4⋮ ...... ⋮
-3,-2,-1,0,1,2,3,4] (4.5)

The 4096 points generated are ${}^{i}x$, i=1.2. ... 4096

if $\min_{i=1,...4096} f({}^{i}X^{(k+1)}) = f(\underline{x}^{(k)})$

then set $\alpha_1 = \frac{\alpha_1}{10}$ and recompute search step.

As soon as

$$\min_{i=1,...4096} f({}^{i}X^{(k+1)}) < f(\underline{x}^{(k)})$$

then reset $\alpha_1 = \alpha_1^{initial}$

(ii) 2-dimensional grid search

The iterative step is

$$X^{(k+1)} = \underline{x}^{(k)} + (A1)\alpha_1\underline{p}^{(k)} + (A2)\alpha_2\underline{g}^{(k)}$$

where matrices A1 and A2 are given by:

(a) (4.2) and (4.3) respectively. (We call this VERSION A).

(b) $A1_{ith\ col}$ = [-16,-15.5,-15,-14.5,-14, ... 14.5,15,15.5]

$A2_{ith\ row}$ = [-16,-15.5,-15,-14.5,-14, ... 14.5,15,15.5]

(We call this VERSION B)

The iterative step will generate 4096 points; in fact, only 64 points are distinct for VERSION A, the rest are identical. The α's are given by (4.1).

For the 2-dimensional grid search, we considered the plane defined by $\underline{p}^{(k)}$ and $\underline{g}^{(k)}$, as this plane contains the directions usually chosen in sequential codes to make progress when G is nearly singular.

(iii) 1-dimensional search

The iterative step is

$$X^{(k+1)} = \underline{x}^{(k)} + (A1)\alpha_1\underline{p}^{(k)}$$

where the matrix A1 is given by

(a) (4.2) (VERSION A)

(b) A1 has the values -102(0.05)102.75 in long vector order (VERSION B).

The iterative step will generate 4096 points in Version B but only 8 distinct points for VERSION A.

Step 6: $\underline{x}^{(k+1)} = \text{Arg} \min_{i=1,\ldots 4096} f({}^{i}X^{(k+1)})$

- minimising over a preselected grid.

Step 7: Return to Step 2 with k = k+1.

The strong feature of the parallel algorithm is the parallel function evaluation in Steps 2 to 6. The DAP can perform 4096 function evaluations in parallel. It would be naive to assume that since the DAP performs 4096 function evaluations in parallel, the processing time would be of the order 4096 faster than the sequential calculation. In fact the arithmetic operations in the processing elements of the DAP are slower than the fast sequential machines (CDC 7600) due to the bit serial nature of each of the 4096 processing elements in the DAP. Let the ratios of the speed be τ. To get a rough idea of the value of τ, we obtained processing times for 4096 function evaluations of five 64-dimensional test problems (specified in Appendix A) on ICL DAP, ICL 2980 and DEC 1091. The processing times are displayed in Table 4.1.

Functions	Parallel evaluations	Sequential evaluations	
	DAP	DEC 1091	ICL 2980
Quadratic	0.045832	2.601	1.00928
Rosenbrock	0.109032	3.963	1.315296
Powell	0.067200	4.453	1.563352
Box (M)	11.62174	189.779	93.995920
Trignometric	0.351760	51.267	15.554040
Average ratio	1	20.7	9.3

<u>Table 4.1</u>. The processing times for 4096 function evaluations (time in seconds).

Experimental Results

Performance measurement.

The criteria we have used for measuring the performance of the parallel algorithm is to compare the processing times obtained on the DAP in relation to a sequential system. The sequential system we used was the DEC 1091 at The Hatfield Polytechnic. The codes used for the sequential system were:

(a) The standard Newton-Raphson method from the NAG Library, routine EØ4EBF [Ref. 12]. This is the nearest equivalent sequential algorithm.

(b) Variable Metric algorithm, the usually recommended approach on a sequential system for a 64 dimensional problem. We selected the Numerical Optimisation Centre's OPTIMA Library program OPVM [Ref. 13], an implementation of the Broyden-Fletcher-Shanno variable metric algorithm.

(c) Noticing that the structure of some of the test functions was symmetric and noting that this symmetry would favour a conjugate gradient approach, the Harwell Library routine VA14A [Ref. 14] was also used.

Numerical Results

The five test problems (specified in Appendix A) were run on the

i) DAP, using the parallel algorithm

ii) DEC 1091, using

  (a) Modified Newton-Raphson, NAG routine, EØ4EBF

  (b) Variable Metric, NOC OPVM

  (c) Conjugate Gradient, Harwell VA14A.

We use approximate values of $\underline{g}$, the gradient vector and G, the Hessian, obtained using finite difference schemes described earlier in the section. For each test problem we used two sets of starting points, a symmetric and a nonsymmetric set. The numerical results are displayed in Tables A1 to A8.

Table A1 displays the processing times for the sequential codes.

Table A2 displays the DAP processing times for '1-D' search (VERSION A).

Table A3 displays the DAP processing times for '2-D' search (VERSION B).

Table A4 displays the DAP processing times for '2-D' search (VERSION A).

Table A5 displays the DAP processing times for '2-D' search (VERSION B).

Table A6 displays the DAP processing times for '4-D' search.

The performance ratio is displayed in Tables A7 and A8.

It will be seen from the numerical results that the parallel algorithm consistently outperformed the Newton-Raphson and Variable Metric sequential algorithms. In fact the DAP performed extremely well compared with the sequential Newton-Raphson algorithm.

Note of caution: Although the sequential Newton-Raphson is the nearest equivalent sequential algorithm to the parallel one, it is rarely used to solve a 64-dimensional problem. It is therefore fairer to use the Variable Metric method in the comparison.

For problems with special symmetry in the objective function, the parallel algorithm only just performs better than the Conjugate Gradient sequential algorithm (in terms of performance ratio); but for the nonsymmetric test function (Trignometric functtion) the parallel algorithm performed considerably better than the sequential Conjugate Gradient algorithm. Consider the '1-D' search algorithm: VERSION A performed better than VERSION B. For the '2-D' search algorithm, VERSION B performed better than VERSION A. Looking at the performance ratio and the number of iterations, there is not much to choose between the '2-D' search and the '4-D' search; for some problems the '2-D' search performed better than the '4-D' search and for other problems the '4-D' search performed better than the '2-D' search. This indicates that considerable further research is needed into the choice of the extra directions if the possible benefits of the 4-D search are to be obtained.

4.2 : Large dimensional problems

For large dimensional problems, say $n > 120$, the parallel Modified Newton approach would not be practicable, due to shortage of store on the DAP.

A different approach has been implemented on the ICL DAP for solving optimisation problems formulated by solving partial differential equations by finite elements as it was felt these would typify this class. The solution method follows the finite element approach and each processor of the DAP handles its own finite element. The solution is then completed by implementing linear or nonlinear versions of the conjugate gradient method as appropriate.

Consider the solution of partial differential equations

$$(a) \quad \frac{\partial}{\partial x} K_x \frac{\partial T}{\partial x} + \frac{\partial}{\partial y} K_y \frac{\partial T}{\partial y} = - Q \tag{5.1}$$

$$(b) \quad \frac{\partial^2 y}{\partial x^2} + \frac{\partial^2 u}{\partial y^2} - Ru \frac{\partial u}{\partial x} = 0. \tag{5.2}$$

It is wellknown that both the equations can be solved by minimising a functional

$$I = \iint F.dv.$$

The problem

We will briefly describe a solution method for solving the 2-D heat conduction equation (5.1). The solution to equation (5.1) will occur at the minimum of

$$I = \iint K_x(\frac{\partial T}{\partial x})^2 + K_y(\frac{\partial T}{\partial y})^2 - 2QT \qquad dv \tag{5.3}$$

The objective function becomes one of solving

$$\min \quad I = \int F(T(x)) \quad dV \tag{5.4}$$

Subject to the appropriate boundary conditions, Newman type boundary conditions were imposed. Now when using finite elements the domain V is covered with a set of grid points x_k and then divided into a set of elements V_i (of rectangular shape) with grid points at intersections. For each element there are shape functions ϕ_{ki} such that

$$\phi_{ki}(x) = 1 \quad \text{at} \quad x_k$$

$$\phi_{ki}(x) = 0 \quad \text{at} \quad x_j \text{ for } j \neq k \text{ and at all points } x \notin V_i$$

i.e. $\phi_{ki}(x) = (1 - \frac{x_k \pm x_i}{h})(1 - \frac{y_k \pm y_i}{h})/4.$

It is now possible to approximate T(x) by a linear combination of the shape functions.

$$\sum_k T_k \phi_{ki}(x) \tag{5.5}$$

so that equation (5.4) becomes approximately

$$\min I = \int F(\sum_k T_k \phi_{ki}(x)) \, dV. \tag{5.6}$$

On substituting (5.5) into equation (5.3) we get

$$F(T) = \iint K_x\{\sum_i \frac{\partial\phi_i}{\partial x} T_i\}^2 + K_y\{\sum_i \frac{\partial\phi_i}{\partial y} T_i\}^2 - 2Q\sum_i \phi_i T_i \quad dV \tag{5.7}$$

and on differentiating this with respect to T_i to get ∇F_i

$$\nabla_i F(T) = \iint 2K_x \frac{\partial\phi_i}{\partial x} \{\sum_j \frac{\partial\phi_j}{\partial x} T_j\} + 2K_y \frac{\partial\phi_i}{\partial y} \{\sum_j \frac{\partial\phi_j}{\partial y} T_j\} - 2Q\sum_i \phi_i \quad dV. \tag{5.8}$$

Differentiating once more gives $\nabla^2 F_{ij}$

$$\nabla^2_{ij} F(T) = \iint 2K_x \frac{\partial\phi_i}{\partial x} \frac{\partial\phi_j}{\partial x} + 2K_y \frac{\partial\phi_i}{\partial y} \frac{\partial\phi_j}{\partial y} \quad dV.$$

When the Q_i's are point sources at the nodes they can be moved outside the integration and then equation (5.7) becomes

$$F(T) = \sum_{el} \iint_{el} K_x\{\sum_i \frac{\partial\phi_i}{\partial x} T_i\}^2 + K_y\{\sum_i \frac{\partial\phi_i}{\partial y} T_i\}^2 \quad dV - 2Q_i T_i \tag{5.7a}$$

equation (5.8) becomes

$$\nabla_i F(T)_{el} = \iint_{el} 2K_x \frac{\partial\phi_i}{\partial x} \{\sum_j \frac{\partial\phi_j}{\partial x} T_j\} + 2Ky \frac{\partial\phi_i}{\partial y} \{\sum_j \frac{\partial\phi_i}{\partial y} \{\sum_j \frac{\partial\phi_j}{\partial y} T_j\} \, dV - 2Q_i \tag{5.8a}$$

and equation (5.9) becomes

$$\nabla^2_{ij} F(T)_{el} = \iint_{el} 2K_x \frac{\partial\phi_i}{\partial x} \frac{\partial\phi_j}{\partial x} + 2K_y \frac{\partial\phi_i}{\partial x} \frac{\partial\phi_j}{\partial y} \, dV. \tag{5.9a}$$

The solution of problem (5.6) then becomes equivalent to the solution of the set of equations (5.7). If the partial differential equations are linear this is a set of linear simultaneous equations which could be written Au = f.

Solution of the set of equations

When solving equations of the form

$A\underline{u} = \underline{f}$

where A is a real N x N matrix

$\underline{u}$ an N unknown vector

$\underline{f}$ an N known vector

if N is large there may be problems with computer storage. In the finite element approach the matrix A will be held as

$\sum_{e=1}^{E} A^e$ (namely, the sum of all the elements matrices).

All the values in A^e willl be zero except for those occurring in rows/columns corresponding to variables in the eth element. It is, therefore, possible to solve the above set of equations without assembling the original matrix A. We have used the element matrices individually to evaluate $A\underline{p}$ element by element, by taking each element matrix in turn and multiplying it with the correct elements of the vector $\underline{p}$, the answers being stored in the appropriate processors. This makes the use of the conjugate gradient method advantageous.

Conjugate Gradient method

The conjugate gradient method is an iterative method that converges to the true solution in a finite number of iterations assuming no rounding errors. The idea was initially presented by Hestenes and Stiefel [9] and subsequently modified for optimisation purposes by Fletcher and Reeves [8].

We have successfully implemented parallel conjugate gradient algorithms, linear and nonlinear cases, on the ICL DAP parallel processing computer. We will briefly describe sequential linear and nonlinear conjugate gradient algorithms and then show how the solution of the partial differential equation (5.1) has been mapped onto the ICL DAP. For a brief description of the ICL DAP and the language for implementation refer to Section 2.

The linear conjugate gradient algorithm

The basic linear algorithm used for solving the set of equation $A\underline{u} = \underline{f}$ is described below.

1. Evaluate $\nabla^2 F$ (A).
2. Initialise the right hand side vector $\underline{f}^{(o)}$ (it is set equal to the source/sink points and zero elsewhere).
3. Set $\underline{u}^{(1)} = \underline{f}^{(o)}$ and $\underline{p}^{(o)} = \underline{f}^{(o)}$ ($\underline{u}$ being the unknown and $\underline{p}$ the search direction).
4. Evaluate $\underline{w}^{(o)} = A\, \underline{p}^{(o)}$. (We can perform this operation without having to assemble the matrix A).
5. Evaluate $\underline{f}^{(1)} = \underline{f}^{(o)} - \underline{w}^{(o)}$.
6. If $\underline{f}^{(1)T}.\underline{f}^{(1)} < \varepsilon$ then stop.
7. Set $\underline{p}^{(1)} = \underline{f}^{(1)}$ and k = 1.

8. Evaluate $\underline{w}^{(k)} = A.\underline{p}^{(k)}$.
9. If $\underline{p}^{(k)T}.\underline{w}^{(k)} < \epsilon$ then stop.
10. Set $\alpha^{(k)} = \dfrac{\underline{f}^{(k)T}.\underline{f}^{(k)}}{\underline{p}^{(k)T}.A.\underline{p}^{(k)}}$.
11. Update the unknowns $\underline{u}^{(k+1)} = \underline{u}^{(k)} + \alpha^{(k)}\underline{p}^{(k)}$

    and $\underline{f}^{(k+1)} = \underline{f}^{(k)} - \alpha^{(k)}\underline{w}^{(k)}$.
12. Set $\beta^{(k)} = \dfrac{\underline{f}^{(k)T}.\underline{f}^{(k)}}{\underline{f}^{(k-1)T}.\underline{f}^{(k-1)}}$.
13. Update the search direction $\underline{p}^{(k+1)} = \underline{f}^{(k+1)} + \beta^{(k)}.\underline{p}^{(k)}$.
14. If $\underline{p}^{(k+1)T}\ \underline{p}^{(k+1)} < \epsilon$ then stop.
15. Set k = k+1, go to Step 8.

Note than an upper bound on the number of iterations is also imposed as a termination criteria.

On the parallel processor we need to evaluate the following products:

i) $\underline{f}^T\underline{f}$
ii) $\underline{p}^T.A.\underline{p}$
iii) $A.\underline{p}$
iv) $\underline{p}^T\underline{p}$.

Each of these can be subdivided into contributions from separate elements performed in parallel on separate processors.

The nonlinear conjugate gradient algorithm

The basic nonlinear conjugate gradient due to Fletcher and Reeves [8]is described below. Initially we guess values for $u_i^{(1)}$, i=1,2, ... N and then evaluate ∇^2F at $\underline{u}^{(1)}$.

1. Set k = 1.
2. Evaluate $\nabla F^{(k)}$.
3. If $\nabla F^{(k)T}.\nabla F^{(k)} < \epsilon$ then stop.
4. If k=1

   then set $\underline{p}^{(k)} = -\nabla F^{(k)}$

   else set $\beta^{(k)} = \dfrac{\nabla F^{(k)T}.\nabla F^{(k)}}{\nabla F^{(k-1)T}.\nabla F^{(k-1)}}$

   and $\underline{p}^{(k)} = -\nabla F^{(k)} + \beta^{(k)}\underline{p}^{(k-1)}$.
5. Evaluate α as $\alpha^{(k)} = \arg\min\ (F + \alpha\underline{p})$.
6. Update the unknown $\underline{u}^{(k+1)} = \underline{u}^{(k)} + \alpha^{(k)}\underline{p}^{(k)}$.
7. Set k = k+1.
8. If $k \leq k\ max$ then go to Step 2

   else stop.

Again in this case we need to calculate the following:

i) ∇F)

ii) $\underline{p}$)

iii) $\nabla F^T.\underline{p}$)

iv) $\nabla F^T.\nabla F$)

and these can be computed in parallel in a similar manner to that used for the linear case.

Test problems and results

The main problems come from Stone [11] who ran an n x n problem for n = 11,21,31.

He considered equation (5.1); in this equation Q is a point source; at one point on the boundary T is fixed and the corresponding component of g set to zero; at the remaining points on the boundary the solution should set the normal component of ∇T to zero. The distribution of K_x and K_y will be described later.

Point sources

The problems all had three point sources and two point sinks and these were located as below:

For 11 x 11 at (2,2), (2,10), (8,2), (6,6), (10,10)

For 21 x 21 at (2,2), (2,20), (14,3), (10,11), (20,20)

For 31 x 31 at (3,3), (3,27), (23,4), (14,15), (27,27)

For 41 x 41 at (6,6), (6,36), (28,6), (20,20), (36,36) and

For 64 x 64 at (9,9), (9,60), (54,9), (34,34), (60,60) with values of

1.0 0.5 0.6 -1.83 -0.27 respectively.

The boundary conditions were imposed by fixing the temperature at one point on the boundary to 1.0 and the corresponding value of the gradient to 0.0.

Note that in our nonlinear algorithm the values of Q were divided by 4 since each Q comes into 4 sets of equations, its overall effect must be Q and not 4Q.

Conductivities

Four different conductivity distributions were use in the tests. They were specified independently in the x and y directions and located at the centre of an element.

PROBLEM 1. The model problem with K_x and K_y equal to unity over the entire region.

PROBLEM 2. The generalised model problem with K_x and K_y both constant, but with K_x 100 times greater than K_y.

PROBLEM 3. The heterogeneous test problem, here the region is divided into a number of smaller regions.

PROBLEM 4. This was the same as for 3, except that in one of the regions, where K_x and K_y had been set to 1.0, the conductivities came via random numbers.

PROBLEM 5. For the nonlinear p.d.e. equation (5.2) the least squares formulation was adopted. The variables were augmented by introducing

$$a = \frac{\partial u}{\partial x}, \quad b = \frac{\partial u}{\partial y}$$

and the objective function I minimised over the values of u, a and b at the nodes, where I is given by

$$I = \iint (a - \frac{\partial u}{\partial x})^2 + (b - \frac{\partial u}{\partial u})^2 + (\frac{\partial a}{\partial x} + \frac{\partial b}{\partial y} - Rua)^2 \, dV.$$

Numerical Results

The problems were run on the DAP using the two codes, linear and nonlinear, and for comparison purposes the corresponding sequential versions were run on the Hatfield Polytechnic DEC 1091. The linear results are shown graphically in Figure B1 where the grid size is on the x-axis and the LOG of time on the y-axis. (Note that the LOG of time had to be used because of the large range of times obtained - particularly for the sequential cases). In all but a few cases (which have been indicated) the tolerance for termination was 1E-6. A suitable upperbound to the interations was set at 2* no. of points.

It is apparent from the graph that the parallel implementations are considerably faster than the sequential ones, Problem 1 was taken as far as a 64 x 64 problem for the parallel case just to prove that it was possible within a reasonable time but this could not be loaded on our sequential computer. The curves/lines for the parallel case increase only slightly in comparison with the sequential runs whose times soon become large as the number of unknowns is increased or as the ill-conditioning gets worse.

The results of the nonlinear problem are shown in Figure B2 which emphasise even more the benefits in time that can be obtained by solving the problem cn the DAP rather than the sequential machine. It will be noted that the sequential runs were so expensive that the comparison tests had to be discontinued at a dimension where the problem was small enough only to require an eighth of the DAP.

Conclusions

The two parallel implementations have exhibited the fact that the solution of p.d.e.'s using this method of approach is very suitable to the SIMD class of machines and in particular the DAP.

Almost certainly one of the main advantages of the DAP in this case was the fact that the processors are connected to their nearest four neighbours, via row and column data highways, and the powerful shift indexing facilities make good use of this. Thus we have a simple but an extremely effective means of communicating with other nodes and neighbouring elements.

This facility can be compared to the two sequential programs which need to keep, for each element, a separate record of the four neighbouring (local) nodes, which must be set up and then indexed correctly. This is an overhead for the sequential case, not to mention the fact that it is a complication in the writing and checking of the code which is not present in the algorithms. From the DAP's point of view it makes no significant difference whether it is calculating just 4 element matrices (a 3 x 3 problem) or 3,969 element matrices (a 64 x 64 problem) as either most or none of the processors will be switched off (masked out).

The algorithms that were employed for these solutions are basic with no sophisticated improvements though many are known. In theory we should terminate with the correct solution in at most N iterations (where N is the number of unknowns) but this assumes we use exact arithmetic with no rounding errors, which in practice is not the case. Improvements to the basic algorithms can be made, see for instance Powell [10].

Other major improvements are wellknown such as preconditioning, the multigrid approach, etc., where the system of equations

$$A\underline{x} = \underline{y}$$

is transformed into a new system

$$\hat{A}\hat{\underline{x}} = \hat{\underline{y}}$$

which will have a much smaller condition number in order to speed up the convergence. Parallel implementation of such a method will be the basis of additional work.

4.3 : Multiextremal global optimisation

In this section we begin with a brief introduction in which we define the global optimisation problem, give a classification of methods and briefly describe the nature of one class of methods.

Introduction

The global optimisation problem is posed as follows:

Consider the problem

$$\min f(\underline{x}), \quad S \subset R^n$$

where $S = \{\underline{x}: \; U_i \leq x_i \leq l_i, \; i = 1,2, \ldots n\}$.

We shall assume that the objective function $f(\underline{x})$ is nonconvex and possesses more than one local minimum. The problem is to locate that local minimum point $\underline{x}^*$ with the least function value. For some problems we have more than one local minimum that is global.

The methods for solving global optimisation problems are classified as deterministic or probabilistic. It is generally accepted that deterministic methods can only be applied to a restricted subset of objective functions $f(\underline{x})$, whereas more general statements can sometimes be made for a probabilistic method. Hence for a general problem probabilistic methods are often preferred. In this paper we will only consider methods belonging to that classification. Probabilistic methods rely on the following result:

Assume that a finite number of points chosen at random are distributed uniformly over a finite region S. If A is a subset of S with a measure m such that

$$\frac{m(A)}{m(S)} \geq \alpha > 0$$

and p(A,N) is the probability that at least one point of a sequence of N random points lies in A then

$$\lim_{N \to \infty} p(A,N) = 1.$$

A description of some of the probabilistic methods can be found in Dixon & Szegö [16].

Although many alternative probabilistic methods have been suggested none could claim to be the 'best' as no agreed measure of performance is available. In practice very few real multiextremal nonconvex global optimisation problems have been solved. The reason for this is the high 'computational costs' involved when implementing the probabilistic algorithm on a sequential system. One of the main contributing factors for this is the time of repeated function evaluations, the more complicated the nature of the objective function the more these costs increase. The clustering type of probabilistic algorithms are among the most successful and these generally consist of three phases:

a) selection of N trial points chosen at random from the region of interest and the computation of the associated function values,
b) local search phase. A few iterations of a local minimisation routine are run from each of the trial points. This is done to push the points near to a local minimum,
c) clustering phase. Find clusters amongst the resulting points using clustering analysis techniques and to return to phase b with a reduced number of points.

The choice of N, the number of trial points, is quite critical. If it is too small, then the exploration of the region of interest will be insufficient and possible optimum locations may be overlooked. So the greater the value of N the more likelihood there is of the algorithm locating a global minimum. Using a parallel system, say, MIMD or SIMD in nature, we could choose large N (and perform function evaluations in parallel), thus increasing the probability of the algorithm locating a global minimum.

A number of sequential probabilistic algorithms have been suggested, e.g. Price [17], Torn [18] and Boender [19]. We will consider the Controlled Random search algorithm of Price, intended for a mini-computer, and describe how we implemented the parallel versions of it on an SIMD array system and an MIMD system. The exploitation of parallelism that we will be describing will be at the level of designing parallel algorithms. When designing a parallel algorithm we shall bear in mind the target architecture. First we will summarise the sequential algorithm and present the numerical results for ten test problems.

The sequential Price's CRS algorithm consists of the following steps:

Step 1: A sample of N points from the domain of an objective function is taken. The points are uniformly distributed over the feasible region, S. We have chosen $N = 10(n + 1)$. Evaluate the function at each point.

Step 2: Choose the point M which has the greatest function value, fM (worst point). Choose the point L which has the least function value, fL (best pqint).

Step 3: Choose n distinct points, $R_2, R_3, \ldots R_{n+1}$, at random from the set of N points. Let $R_1 = L$. Determine G, the centroid of the first n points $R_1, R_2, \ldots R_n$. The next trial point is defined to be

$$P = 2G - R_{n+1}.$$

External reflection of R_{n+1} about G (pictorial illustration below for n = 2)

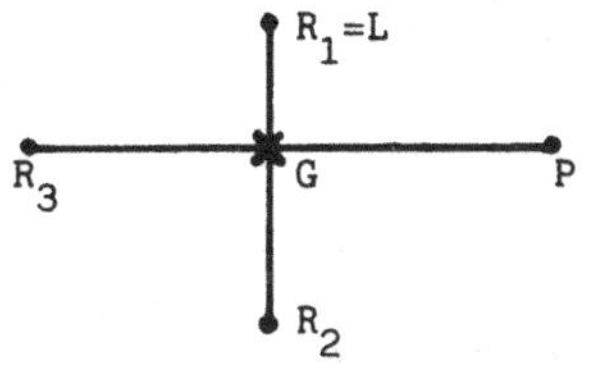

Step 4: If the trial point P is within the feasible region S then proceed to Step 5; otherwise repeat Step 3.

Step 5: Determine f_p, the function value at the point P. If $f_p < fM$, accept point P and proceed to Step 6; otherwise repeat Step 3.

Step 6: Replace the worst point M in the set of N points, by point P.

Step 7: If termination criterion is satsfied, stop; otherwise repeat from Step 2. The termination criteria we have used is:
at the ith iteration
if $|f_i - f_{i-1}| < \varepsilon$ and $|f_{i-1} - f_{i-2}| < \varepsilon$ then stop,
where ε is some tolerance.

Numerical Results

The sequential algorithm was coded in FORTRAN and run on the Hatfield Polytechnic DEC 1091. The test problems chosen are described in Appendix C. Computational results of tests on the standard test functions are collected in Table C1. The CRS procedure has the advantage of being simpler and requires less computer storage than the more sophisticated clustering algorithm.

Parallel (SIMD) Controlled Random search algorithm

Consider the interaction of the CRS algorithm with the SIMD ICL DAP. A parallel version of the CRS algorithm suggested by Dixon and Patel [15] and modified by Ducksbury is shown in Figure 4.3.

Numerical Results

The results of this parallel algorithm are displayed in Table C2 (sequential against parallel). It is observed that the parallel algorithm performed better than the sequential algorithm, in terms of function evaluation and performance ratio. The performance ratio ranges from 3.2 to 68. The lower ratio is for the test problem

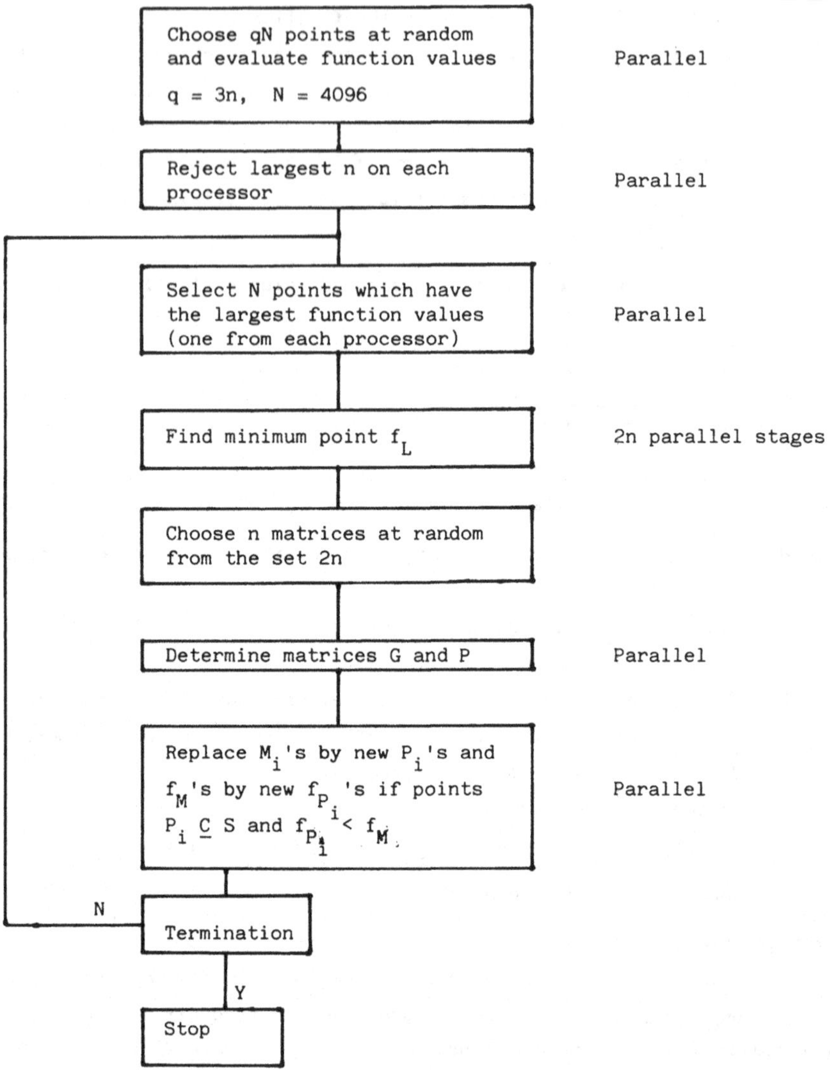

Figure 4.3

where there is only one global minima and the larger ratio is for the problems which possess multiple global minima. The algorithm is only suitable for small values of n, in the range $2 \leq n \leq 5$. This restriction was due to shortage of store on the DAP at the time these tests were undertaken. The store on the DAP has subsequently been increased.

Parallel (MIMD) Controlled Random Search algorithm

A parallel version of the Controlled Random Search algorithm for inplementation on the multiprocessor NEPTUNE system consists of the following steps:

Step 1: Choose N points at random in the region S to be searched, where $N = 10(n+1)$.

Evaluate the function at each point. This is a sequential step, but it can be parallelised.A set of N points and their associated function values are stored in the global memory, i.e. accessible to all the processors. Processor $i(i=0,1,2,3)$ then executes the following:

Step 2: Processor i chooses $(i+1)^{st}$ worst point, M_i, with function value f_{M_i}.

$f_{M_o} = \max (f_j, j=1,2, \ldots N)$

$f_{M_k} = \max (f_j, j=1,2, \ldots N, j \neq M_t, t = 0,1, \ldots (k-1))$.

Choose the point L which has the least function value, f_L (best point)

$f_L = \min (f_j, j=1,2, \ldots N)$.

Steps 3 and 4: Generate a new point, P, and check for its feasibility, as in the sequential algorithm.

Step 5: Determine f_p, the function value at the point P. If $f_p < f_{M_i}$ accept point P and proceed to Step 6; otherwise repeat Step 3.

Step 6: Replace the $(i+1)^{st}$ worst point M_i in the set of N points stored in the global memory by the new point P.

Step 7: Terminate or repeat from Step 2, as in the sequential algorithm.

The content of the global memory is continually being modified by each of the processors and there is no communication between the processors. Hence the process is an asynchronous process.

Numerical Results

The parallel algorithm was run on the four processor NEPTUNE system at Loughborough University of Technology. The test problems chosen are described in Appendix C.

For each test problem, the parallel algorithm was first run on one processor (processor 0), then on two processors (different combinations of the processors, i.e. 0,1; 0,2; 0,3), then on three processors (0,1,2; 0,1,3; 0,2,3) and finally on all four processors. A number of runs for each test problem were carried out. The numerical results are displayed in Table C3. The performance of the parallel algorithm, in terms of speed-up and number of function evaluations, improves as we increase the number of processors. We observe that the algorithm converges faster (i.e. less function evaluations) as we increase the number of processors.

5. CONCLUSIONS

In this paper we have presented results indicating that in 3 very separate circumstances the use of parallel processing systems can greatly affect the computer time involved in solving optimisation problems. This subject appears a very fruitful area for future research.

6. ACKNOWLEDGEMENTS

The work in this report was carried out under two research grants from the SERC, nos.

GR/B/1794/5 and GR/B/4665/5. The DAP Support Unit at Queen Mary College, University of London are gratefully acknowledged for their permission to use the DAP and Loughborough University for their permission to use the NEPTUNE system.

REFERENCES

1] Rodrigue, G and Greenbaum, A, The incomplete Cholesky Conjugate Gradient method for the STAR, Lawrence Livermore Lab. Report, UCID-17574, 1977.

2] Rodrigue, G, Dubois, P F and Greenbaum, A, Approx. the inverse of a matrix for use in iterative algorithms on a vector processor, Computing 22, pp 257-268, 1979.

3] Schreiber, R S, The implementation of the conjugate gradient method on a vector computer, Stanford University, CA 94305, 1981.

4] Sargan, J D, Chong, Y Y and Smith, K A, Nonlinear Econometric Modelling on a Parallel Processor, DAP Support Unit, Queen Mary College, London, October 1981.

5] Sauber, W, A data flow architecture implementation, Technical Report, Texas Instruments Incorporated, Austin, Texas, 1980.

6] Brown, A H O, The development of computer optimisation procedures for use in aero engine design, in 'Optimisation in Action,' ed. L C W Dixon, Academic Press, 1976.

7] Dixon, L C W, Global optima without convexity, in 'Design and Implementation of Optimisation Software,' 1978.

8] Fletcher, R and Reeves, C M, Function minimisation by conjugate gradients, Computer Journal, Vol.7, 1964.

9] Hestenes, M and Stiefel, E, Methods of conjugate gradients for solving linear systems, Journal Res.Nat.Bu.Standards, No.49, 1952.

10] Powell, M J D, Restart procedures for the conjugate gradient method, Math. Programming, Vol.12, pp 221-154, 1975.

11] Stone, H L, Iterative solution of implicit approximations of multi-dimensional PDE's, SIAM Num.Anal., Vol.5, No.3, September 1968.

12] NAG Manuals (Mark 8).

13] OPTIMA Manual, Numerical Optimisation Centre, The Hatfield Polytechnic, Issue No.4, 1982.

14] Harwell Subroutine Library, 1981.

15] Dixon, L C W and Patel, K D, The place of parallel computation in numerical optimisation II: The multiextremal global optimisation problem, TR117, Numerical Optimisation Centre, Hatfield Polytechnic, 1981.

16] Dixon, L C W and Szegö, G P, The global optimisation problem: An introduction, pp 1-15, in 'Towards Global Optimisation 2," eds. Dixon and Szegö, North Holland Publishing Company, 1978.

17] Price, W L, A new version of the controlled random search procedure for global optimisation, TR, Enginerring Dept., University of Leicester, 1981.

[18] Törn, A A, A search clustering approach to global optimisation, in 'Towards Global Optimisation 2,' eds. Dixon and Szegö, North Holland, 1978.

[19] Boender, C G E, et al, A stochastic method for global optimisation, Math. Programming, 1983.

[20] Evans, D J, Barlow, R H, Newman I A and Woodward, M C, A guide to using the NEPTUNE parallel processing system, Dept. of Computer Studies, Loughborough University of Technology, 1982.

[21] DAP Manuals, Technical Publications, TP 6755, TP 6918, TP 6920, ICL, Putney, London SW15.

APPENDIX A

The following five 64-dimensional test problems were used:

a) Quadratic function

$$f(\underline{x}) = \sum_{i=1}^{32} 100x_{2i}^2 + (1-x_{2i-1})^2 .$$

b) Rosenbrock function

$$f(\underline{x}) = \sum_{i=1}^{32} 100(x_{2i-1}^2 - x_{2i})^2 + (1-x_{2i-1})^2 .$$

c) Powell function

$$f(\underline{x}) = \sum_{i=1}^{16} [(x_{4i-3} + 10x_{4i-2})^2 + 5(x_{4i-2} - x_{4i})^2 + (x_{4i-2} - 2x_{4i-1})^4$$

$$+ 10(x_{4i-3} - x_{4i})^4].$$

d) Box (M) function

$$f(\underline{x}) = \sum_{j=1}^{32} \sum_{i=1}^{10} [\exp(-x_{2j-1}\, t_i) - 5\exp(-x_{2j}\, t_i) - \exp(-t_i) + 5\exp(-10t_i)]^2 .$$

$$t_i = \frac{i}{10},\ i=1,2,\ \ldots\ 10.$$

e) Trigonometric function

$$f(\underline{x}) = \sum_{i=1}^{64} [64 + i - \sum_{j=1}^{64} (A_{ij}\, \mathrm{Sin}(x_j) + B_{ij}\, \mathrm{Cos}(x_j))]^2 .$$

$$A_{ij} = \delta_{ij},\quad B_{ij} = i\ \delta_{ij} + 1.$$

$$\delta_{ij} = \begin{cases} 0 & \text{if } i \neq j \\ 1 & \text{if } i = j. \end{cases}$$

The first four problems have symmetric objective functions.

Table A1

Processing times for the sequential codes (time in seconds)

Function Starting Point $\underline{x}^{(0)}$	Newton-Raphson CPU Time	Variable Metric			Conjugate Gradient		
		CPU Time	No. of Funct. Calls	No. of Grad. Calls	CPU Time	Iter-ation	No. of Funct. & Grad. Calls
Quadratic							
$(0,1,....)^T$	15.00	1.80	8	5	1.23	3	9
non-symmetric	16.60	2.10	12	7	1.18	4	9
Rosenbrock							
$(0,1,....)^T$	154.23	72.11	414	210	5.67	19	48
$(-1.2,1,....)^T$	145.97	103.35	683	292	8.36	27	72
non-symmetric	140.98	80.78	496	249	11.36	49	103
Powell							
$(3,-1,0,3....)^T$	112.47	53.16	298	142	5.91	22	49
non-symmetric	134.97	41.28	218	114	11.06	35	87
Trigonometric							
$(1/64,1/64,...)^T$	F	66.18	74	43	37.56	14	27
non-symmetric	2604.31	286.65	455	191	78.68	24	52
Box(M)							
$(1,2,1,2,...)^T$	4181.86	1161.27	429	227	137.81	6	16
non-symmetric	7196.09	3194.39	1263	621	354.02	19	40

Table A2

DAP processing times for '1-D' search [VERSION A]

Function	Starting point $\underline{x}^{(0)}$	initial α_1	h	$\in$	Iter.	DAP time (secs.)
Quadratic	$(0,1,....)^T$	1.0	0.1	0.0001	2	0.8919624
	non-symmetric	1.0	0.1	0.0001	2	0.803285
Rosenbrock	$(0,1,0,1,...)^T$	0.5	0.001	0.001	14	5.532199
	$(-1.2,1.0,...)^T$	0.5	0.001	0.001	20	8.258047
	non-symmetric	1.0	0.001	0.001	26	9.842461
Powell	$(3,-1,0,3,....)^T$	1.0	0.01	0.0001	5	1.746016
	non-symmetric	1.0	0.01	0.0001	3	1.248776
Trigonometric	$(1/64,....)^T$	1.0	0.001	0.001	12	9.647102
	non-symmetric	0.5	0.001	0.001	14	13.33931
Box (M)	$(1,2,1,2,....)^T$	0.5	0.01	0.0001	18	452.6223
	non-symmetric	0.125	0.01	0.0001	21	508.0827

Table A3

DAP Processing times for '1-D' search [VERSION B]

Function	Starting point $\underline{x}^{(0)}$	initial α_1	h	$\in$	Iter.	DAP time (secs.)
Quadratic	$(0,1,....)^T$	0.5	0.1	0.001	2	0.79532
	non-symmetric	0.5	0.1	0.001	2	0.774928
Rosenbrock	$(0,1,0,1,....)^T$	0.5	0.001	0.001	8	3.60468
	$(-1.2,1.0,...)^T$	0.5	0.001	0.001	25	9.791973
	non-symmetric	0.125	0.001	0.001	34	14.340800
Powell	$(3,-1,0,3,....)^T$	1.0	0.01	0.001	16	5.338949
	non-symmetric	1.0	0.01	0.001	10	3.542872
Trigonometric	$(1/64,....)^T$	1.0	0.001	0.001	13	11.07125
	non-symmetric	0.5	0.001	0.001	14	13.379770
*Box (M)	$(1,2,1,2,....)^T$	-	0.001	0.001	-	F
	non-symmetric	-	0.001	0.001	-	F

* overflow in function evaluation

Table A4

DAP processing times for '2-D' search [VERSION A]

Function	Starting point $\underline{x}^{(0)}$	initial α_1	h	$\in$	Iter.	DAP time (secs.)
Quadratic	$(0,1,\ldots\ldots)^T$	1.0	0.1	0.0001	2	0.985168
	non-symmetric	1.0	0.1	0.0001	2	0.966224
Rosenbrock	$(0,1,\ldots\ldots)^T$	0.5	0.001	0.001	12	6.486191
	$(-1.2,1.0,\ldots)^T$	0.5	0.001	0.001	11	5.86484
	non-symmetric	0.5	0.001	0.001	71	36.626720
Powell	$(3,-1,0,3,\ldots)^T$	1.0	0.01	0.0001	4	1.929656
	non-symmetric	1.0	0.01	0.0001	3	1.500664
Trigonometric	$(1/64,\ldots\ldots)^T$	0.5	0.001	0.001	11	11.10786
	non-symmetric	0.5	0.001	0.001	17	17.57672
Box (M)	$(1,2,1,2,\ldots)^T$	0.25	0.01	0.001	5	130.9869
	non-symmetric	0.25	0.01	0.001	8	202.3719

Table A5

DAP processing times for '2-D' search [VERSION B]

Function	Starting point $\underline{x}^{(0)}$	initial α_1	h	$\in$	Iter.	DAP time (secs.)
Quadratic	$(0,1,\ldots\ldots)^T$	1.0	0.1	0.001	2	0.968968
	non-symmetric	1.0	0.1	0.001	2	0.949536
Rosenbrock	$(0,1,\ldots\ldots)^T$	1.0	0.001	0.001	6	3.282704
	$(-1.2,1.0,\ldots)^T$	1.0	0.001	0.001	4	2.290592
	non-symmetric	0.5	0.001	0.001	25	12.886690
Powell	$(3,-1,0,3,\ldots)^T$	1.0	0.001	0.001	7	3.154320
	non-symmetric	1.0	0.001	0.001	7	3.164520
Trigonometric	$(1/64,\ldots\ldots)^T$	0.5	0.001	0.001	10	10.048290
	non-symmetric	0.5	0.001	0.001	15	15.523480
Box (M)	$(1,2,1,2,\ldots)^T$	0.25	0.01	0.001	4	107.820100
	non-symmetric	0.0625	0.01	0.001	10	251.929100

Table A6

DAP processing times for '4-D' search

Function	Starting point $\underline{x}^{(0)}$	initial α_1	h	$\in$	Iter.	DAP time (secs)
Quadratic	$(0,1,\ldots.)^T$	1.0	0.1	0.0001	2	1.237488
	non-symmetric	1.0	0.1	0.0001	2	1.210280
Rosenbrock	$(0,1,\ldots.)^T$	1.0	0.001	0.001	6	4.279285
	$(-1.2,1.0,\ldots)^T$	1.0	0.001	0.001	7	4.949902
	non-symmetric	0.5	0.001	0.001	41	28.211770
Powell	$(3,-1,0,3,\ldots.)^T$	1.0	0.01	0.0001	4	2.52372
	non-symmetric	1.0	0.01	0.0001	3	1.920984
Trigonometric	$(1/64,\ldots.)^T$	1.0	0.001	0.001	12	13.998930
	non-symmetric	0.5	0.001	0.001	7	8.657191
Box (M)	$(1,2,\ldots.)^T$	0.5	0.01	0.001	4	107.9466
	non-symmetric	0.125	0.01	0.001	13	323.9529

Table A7

Performance measurement [VERSION A]

$$\text{Define 'speed-up' ratio,} = \frac{\text{processing time using a sequential system}}{\text{processing time using the DAP}}$$

Function	Starting point $\underline{x}^{(0)}$	Newton-Raphson DAP			Variable-Metric DAP			Conjugate-Gradient DAP		
		1-D	2-D	4-D	1-D	2-D	4-D	1-D	2-D	4-D
Quadratic	$(0,1,....)^T$	16.8	15.2	12.1	2.0	1.8	1.4	1.4	1.2	0.99
	non-symmetric	20.7	17.2	13.7	2.6	2.2	1.7	1.5	1.2	0.97
Rosenbrock	$(0,1,...)^T$	27.9	23.8	36.0	13.0	11.1	16.8	1.0	0.9	1.3
	$(-1.2,1.0,...)^T$	17.7	24.9	29.5	12.5	17.6	20.9	1.0	1.4	1.7
	non-symmetric	14.3	3.6	5.0	8.2	2.2	2.9	1.1	0.3	0.4
Powell	$(3,-1,0,3,...)^T$	64.4	58.3	44.6	30.4	27.6	21.1	3.4	3.1	2.3
	non-symmetric	108.1	89.9	70.3	33.1	27.5	21.5	8.9	7.4	5.8
Trigonometric	$(1/64,....)^T$	-	-	-	6.9	5.9	4.7	3.9	3.4	2.7
	non-symmetric	195.2	148.2	300.8	21.5	16.3	33.1	5.9	4.5	9.1
Box (M)	$(1,2,....)^T$	9.8	31.9	38.7	2.7	8.9	10.8	0.3	1.1	1.3
	non-symmetric	14.2	35.6	22.2	6.3	15.9	9.9	0.7	1.7	1.1

Table A8

Performance measurement [VERSION B]

Define 'speed-up' ratio, $= \dfrac{\text{processing time using a sequential system}}{\text{processing time using the DAP}}$

Function	Starting point $\underline{x}^{(0)}$	Newton-Raphson DAP		Variable Metric DAP		Conjugate gradient DAP	
		1-D	2-D	1-D	2-D	1-D	2-D
Quadratic	$(0,1,....)^T$	18.9	15.5	2.3	1.9	1.5	1.3
	non-symmetric	21.4	17.5	2.7	2.2	1.5	1.2
Rosenbrock	$(0,1,...)^T$	42.8	47.0	20.0	22.0	1.6	1.7
	$(-1.2,1.0,...)^T$	14.9	63.7	10.6	45.1	0.9	3.6
	non-symmetric	9.8	10.9	5.6	6.3	0.8	0.9
Powell	$(3,-1,0,3,...)^T$	21.1	35.7	10.0	16.9	1.1	1.9
	non-symmetric	38.1	42.7	11.7	13.0	3.1	3.5
Trigonometric	$(1/64,....)^T$	-	-	6.0	6.6	3.4	3.7
	non-symmetric	194.6	167.8	21.4	18.5	5.9	5.1
Box (M)	$(1,2,....)^T$	-	38.8	-	10.8	-	1.3
	non-symmetric	-	28.6	-	12.7	-	1.4

APPENDIX B

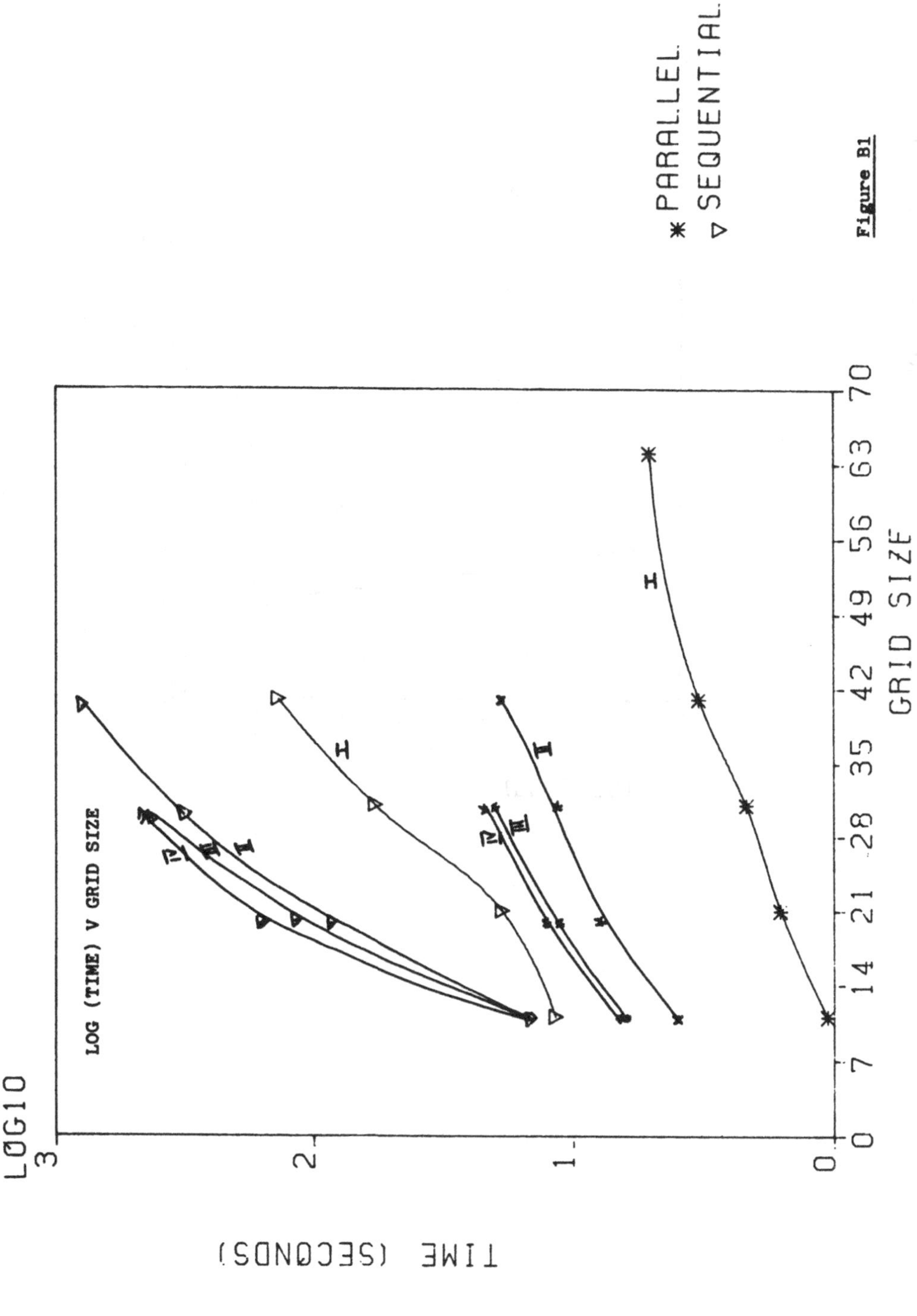

Figure B1

UNKNOWNS V CPU TIME

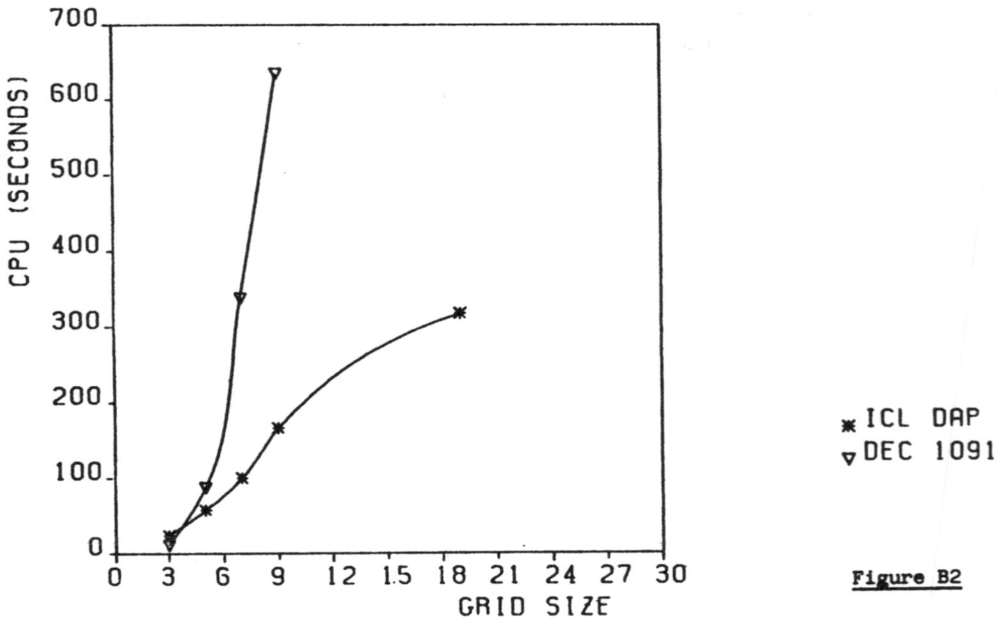

Figure B2

APPENDIX C

1] Three Hump Camel-back function

$f(\underline{x}) = ax_1^2 + bx_1^4 + cx_1^6 - x_1x_2 + dx_2^2 + ex_2^4$

where a = 2, b = -1.05, c = $^1/6$, d = 1, e = 0.

2] Six Hump Camel-back function

$f(\underline{x}) = ax_1^2 + bx_1^4 + cx_1^6 - x_1x_2 + dx_2^2 + ex_2^4$

where a = 4, b = -2.1, c = $^1/3$, d = -4, e = 4.

3 Treccani function

$f(x) = x_1^4 + 4x_1^3 + 4x_1^2 + x_2^2$.

The rest of the test functions, namely, Goldstein and Price, Branin, Hartman's family and Shekel's family are described in Towards Global Optimisation 2 (Dixon and Szegö [16]).

Table C1

Numerical results for the sequential CRS algorithm

Number of global points = 10(n+1)

Tolerance, ε = 0.000001

function	No. of function evaluations	x*	f(x*)
Goldstein and Price	792	(0, -1.0)	3.0
Branin	744	(-3.14163, 12.27467)	0.39789
	410	(3.13630, 2.26562)	0.39820
		(9.46559, 2.45409)	0.40897
Hartman's family			
m=4, n=3	1276	(0.11514, 0.55568, 0.85253)	-3.86278
m=9, n=6	3804	(0.40474, 0.88258, 0.84349, 0.13724, 0.03849)	-3.20316
m=4, n=6	6416 ε=0.0000001	(0.20168, 0.15004, 0.47699, 0.27515, 0.31168, 0.65732)	-3.32237
Shekel's family			
n=4, m=5	3996	(4.0, 4.0, 4.0, 4.0)	-10.15320
n=4, m=7	2872	(4.00069, 4.00080, 3.99939, 3.99957)	-10.40294
n=4, m=10	3284	(4.00066, 4.00058, 3.99965, 3.99954)	-10.53641
3 Hump Camel-back	706	(0.00000, 0.00044)	0.00000
6 Hump Camel-back	676	(-0.09002, -0.71249)	-1.03163
		(0.08999, 0.71341)	-1.03162
Treccani	412	(-0.00043, -0.00009)	0.00000

Table C2

Numerical results for parallel (SIMD) CRS algorithm

Function	Sequential		Parallel (SIMD)		Performance ratio
	Time(s)	fn.evals.	Time(s)	fn.evals.	
Six Hump Camel-back (2 global minima)	0.36 0.81 1.17	288 798 1086	0.093	7 Note (1)	12.5
Branin (3 global minima)	0.47 0.97 1.73 3.17	360 654 1024 2038	0.118	8 Note (1)	26.8
Goldstein and Price (1 global minimum)	0.45	481	0.13	8	3.5
Hartman's n=3 (1 global minimum)	0.88	503	0.276	11	3.2
Shekel Note (3)					
m=5	3.89	2732	0.453	14	8.6
m=7	3.82	2423	0.55	14	6.9
m=10	4.31	2567	0.693	14	6.2
(1 global minimum)					
Shubert 2-D (18 global minima)	average 1.58/min total 15.8 Note (2)	average 995	0.233 Note (2)	8	68

Notes:

(1) Multiple global minima are identified in one run.

(2) The parallel algorithm picked up 11 minima in one run hence the same 11 minima from the sequential version were used for the comparison (these 11 being picked up in 10 runs).

(3) Located the approximate global minimum.

Table C3

Numerical results for parallel (MIMD) CRS algorithm

function	No. of Processors	Speed-up range	fn.evals. for the best run Proc. 0	Proc. 1	Proc. 2	Proc. 3
3 Hump Camel-back	1		432			
	2	1.21 - 2.45	181	154		
	3	1.53 - 2.64	180	149	158	
	4	2.01 - 2.65	185	152	154	159
6 Hump Camel-back	1		434			
	2	1.29 - 1.86	232	212		
	3	1.80 - 1.99	234	193	212	
	4	2.05 - 3.21	170	128	128	129
Treccani	1		458			
	2	1.46 - 1.86	264	262		
	3	2.11 - 2.27	233	215	196	
	4	2.48 - 2.78	148	138	133	138
Goldstein and Price	1		547			
	2	1.38 - 2.11	303	271		
	3	1.96 - 2.75	212	182	185	
	4	2.11 - 3.01	224	189	186	210
Branin	1		475			
	2	1.27 - 1.69	257	221		
	3	1.81 - 2.39	205	170	180	
	4	2.37 - 3.74	139	105	107	120
Shekel's family						
n=4, m=5	1		3746			
	2	1.43	2388	2355		
	3	1.91	1818	1771	1751	
	4	2.76	1073	1014	1065	1101
n=9, m=7	1		2932			
	2	1.51	1962	1969		
	3	1.83	1547	1546	1542	
	4	2.23	1290	1259	1234	1286
n=4, m=10	1		2836			
	2	1.45	1941	1901		
	3	1.82	1602	1611	1609	
	4	2.41	1170	1146	1191	1183
Hartman's family						
m=4, n=3	1		1022			
	2	1.29 - 1.83	557	516		
	3	2.15 - 2.49	389	368	371	
	4	2.98 - 3.80	271	238	253	244
m=4, n=6	1		3348			
	2	-	-			
	3	2.17 - 2.37	1370	1325	1369	
	4	2.39 - 2.91	1154	1095	1116	1141

INDEX OF AUTHORS

Lecture Notes in Control and Information Sciences

Edited by A. V. Balakrishnan and M. Thoma

Lecture Notes in Control and Information Sciences

Edited by A. V. Balakrishnan and M. Thoma

Vol. 43: Stochastic Differential Systems
Proceedings of the 2nd Bad Honnef Conference
of the SFB 72 of the DFG at the University of Bonn
June 28 – July 2, 1982
Edited by M. Kohlmann and N. Christopeit
XII, 377 pages. 1982.

Vol. 44: Analysis and Optimization of Systems
Proceedings of the Fifth International
Conference on Analysis and Optimization of Systems
Versailles. December 14–17, 1982
Edited by A. Bensoussan and J. L. Lions
XV, 987 pages, 1982

Vol. 45: M. Arató
Linear Stochastic Systems
with Constant Coefficients
A Statistical Approach
IX, 309 pages. 1982

Vol. 46: Time-Scale Modeling of Dynamic Networks
with Applications to Power Systems
Edited by J. H. Chow
X, 218 pages. 1982

Vol. 47: P. A. Ioannou, P. V. Kokotovic
Adaptive Systems with Reduced Models
V, 162 pages. 1983

Vol. 48: Yaakov Yavin
Feedback Strategies for Partially
Observable Stochastic Systems
VI, 233 pages, 1983

Vol. 49: Theory and Application of Random Fields
Proceedings of the IFIP-WG 7/1
Working Conference
held under the joint auspices of the
Indian Statistical Institute
Bangalore, India, January 1982
Edited by G. Kallianpur
VI. 290 pages. 1983

Vol. 50: M. Papageorgiou
Applications of Automatic Control Concepts
to Traffic Flow Modeling and Control
IX, 186 pages. 1983

Vol. 51: Z. Nahorski, H.F. Ravn, R.V.V. Vidal
Optimization of Discrete Time Systems
The Upper Boundary Approach
V, 137 pages 1983

Vol. 52: A. L. Dontchev
Perturbations, Approximations and Sensitivity Analysis
of Optimal Control Systems
IV, 158 pages. 1983

Vol. 53: Liu Chen Hui
General Decoupling Theory of Multivariable
Process Control Systems
XI, 474 pages. 1983

Vol. 54: Control Theory for Distributed
Parameter Systems and Applications
Edited by F. Kappel, K. Kunisch,
W. Schappacher
VII, 245 pages. 1983.

Vol. 55: Ganti Prasada Rao
Piecewise Constant Orthogonal Functions
and Their Application to Systems and Control
VII, 254 pages. 1983.

Vol. 56: Dines Chandra Saha, Ganti Prasada Rao
Identification of Continuous
Dynamical Systems
The Poisson Moment Functional
(PMF) Approach
IX, 158 pages. 1983.

Vol. 57: T. Söderström, P. G. Stoica
Instrumental Variable Methods
for System Identification
VII, 243 pages. 1983.

Vol. 58: Mathematical Theory of
Networks and Systems
Proceedings of the MTNS-83 International
Symposium
Beer Sheva, Israel, June 20–24, 1983
Edited by P. A. Fuhrmann
X, 906 pages. 1984

Vol. 59: System Modelling and Optimization
Proceedings of the 11th IFIP Conference
Copenhagen, Denmark, July 25-29, 1983
Edited by P. Thoft-Christensen
IX, 892 pages. 1984

: United States
ısters